Conversion Factors

Mass

1 g $= 10^{-3}$ kg $= 6.85 \times 10^{-5}$ slug

1 kg $= 10^3$ g $= 6.85 \times 10^{-2}$ slug

1 slug $= 1.46 \times 10^4$ g $= 14.6$ kg

1 u $= 1.66 \times 10^{-24}$ g $= 1.66 \times 10^{-27}$ kg

1 metric ton $= 1000$ kg

Length

1 cm $= 10^{-2}$ m $= 0.394$ in.

1 m $= 10^{-3}$ km $= 3.28$ ft $= 39.4$ in.

1 km $= 10^3$ m $= 0.62$ mi

1 in. $= 2.54$ cm $= 2.54 \times 10^{-2}$ m

1 ft $= 12$ in. $= 30.48$ cm $= 0.3048$ m

1 mi $= 5280$ ft $= 1609$ m $= 1.609$ km

1 Å $= 10^{-10}$ m $= 10^{-8}$ cm

Area

1 cm^2 $= 10^{-4}$ m^2 $= 0.1550$ in.^{2} $= 1.08 \times 10^{-3}$ ft^2

1 m^2 $= 10^4$ cm^2 $= 10.76$ ft^2 $= 1550$ in.^{2}

1 in.^{2} $= 6.94 \times 10^{-3}$ ft^2 $= 6.45$ cm^2 $= 6.45 \times 10^{-4}$ m^2

1 ft^2 $= 144$ in.^{2} $= 9.29 \times 10^{-2}$ m^2 $= 929$ cm^2

Volume

1 cm^3 $= 10^{-6}$ m^3 $= 3.53 \times 10^{-5}$ ft^3 $= 6.10 \times 10^{-2}$ in.^{3}

1 m^3 $= 10^6$ cm^3 $= 10^3$ L $= 35.3$ ft^3 $= 6.10 \times 10^4$ in.^{3} $= 264$ gal

1 liter $= 10^3$ cm^3 $= 10^{-3}$ m^3 $= 1.056$ qt $= 0.264$ gal

1 in.^{3} $= 5.79 \times 10^{-4}$ ft^3 $= 16.4$ cm^3 $= 1.64 \times 10^{-5}$ m^3

1 ft^3 $= 1728$ in.^{3} $= 7.48$ gal $= 0.0283$ m^3 $= 28.3$ L

1 qt $= 2$ pt $= 946.5$ cm^3 $= 0.947$ L

1 gal $= 4$ qt $= 231$ in.^{3} $= 3.785$ L

Time

1 h $= 60$ min $= 3600$ s

1 day $= 24$ h $= 1440$ min $= 8.64 \times 10^4$ s

1 year $= 365$ days $= 8.76 \times 10^3$ h $= 5.26 \times 10^5$ min $= 3.16 \times 10^7$ s

Angle

1 rad $= 57.3°$

$1° = 0.0175$ rad	$60° = \pi/3$ rad
$15° = \pi/12$ rad	$90° = \pi/2$ rad
$30° = \pi/6$ rad	$180° = \pi$ rad
$45° = \pi/4$ rad	$360° = 2\pi$ rad

1 rev/min $= \pi/30$ rad/s $= 0.1047$ rad/s

Speed

1 m/s $= 3.60$ km/h $= 3.28$ ft/s $= 2.24$ mi/h

1 km/h $= 0.278$ m/s $= 0.621$ mi/h $= 0.911$ ft/s

1 ft/s $= 0.682$ mi/h $= 0.305$ m/s $= 1.10$ km/h

1 mi/h $= 1.467$ ft/s $= 1.609$ km/h $= 0.447$ m/s

60 mi/h $= 88$ ft/s

Force

1 N $= 10^5$ dynes $= 0.225$ lb

1 dyne $= 10^{-5}$ N $= 2.25 \times 10^{-6}$ lb

1 lb $= 4.45 \times 10^5$ dynes $= 4.45$ N

Equivalent weight of 1-kg mass $= 2.2$ lb $= 9.8$ N

Pressure

1 Pa (N/m^2) $= 1.45 \times 10^{-4}$ lb/in.^{2} $= 7.5 \times 10^{-3}$ torr (mm Hg) $= 10$ dynes/cm^2

1 torr (mm Hg) $= 133$ Pa (N/m^2) $= 0.02$ lb/in.^{2} $= 1333$ dynes/cm^2

1 atm $= 14.7$ lb/in.^{2} $= 1.013 \times 10^5$ N/m^2 $= 1.013 \times 10^6$ dynes/cm^2 $= 30$ in. Hg $= 76$ cm Hg

1 bar $= 10^6$ dynes/cm^2 $= 10^5$ Pa

1 millibar $= 10^3$ dynes/cm^2 $= 10^2$ Pa

Energy

1 J $= 10^7$ ergs $= 0.738$ ft-lb $= 0.239$ cal $= 9.48 \times 10^{-4}$ Btu $= 6.24 \times 10^{18}$ eV

1 kcal $= 4186$ J $= 4.186 \times 10^{10}$ ergs $= 3.968$ Btu

1 Btu $= 1055$ J $= 1.055 \times 10^{10}$ ergs $= 778$ ft-lb $= 0.252$ kcal

1 cal $= 4.186$ J $= 3.97 \times 10^{-3}$ Btu $= 3.09$ ft-lb

1 ft-lb $= 1.36$ J $= 1.36 \times 10^7$ ergs $= 1.29 \times 10^{-3}$ Btu

1 eV $= 1.60 \times 10^{-19}$ J $= 1.60 \times 10^{-12}$ ergs

1 kWh $= 3.6 \times 10^6$ J

Power

1 W $= 0.738$ ft-lb/s $= 1.34 \times 10^{-3}$ hp $= 3.41$ Btu/h

1 ft-lb/s $= 1.36$ W $= 1.82 \times 10^{-3}$ hp

1 hp $= 550$ ft-lb/s $= 745.7$ W $= 2545$ Btu/h

Mass-Energy Equivalents (at rest)

1 u $= 1.66 \times 10^{-27}$ kg $\leftrightarrow 931.5$ MeV

1 electron mass $= 9.11 \times 10^{-31}$ kg $= 5.49 \times 10^4$ u $\leftrightarrow 0.511$ MeV

1 proton mass $= 1.672 \times 10^{-27}$ kg $= 1.007276$ u $\leftrightarrow 938.28$ MeV

1 neutron mass $= 1.674 \times 10^{-27}$ kg $= 1.008665$ u $\leftrightarrow 939.57$ MeV

Temperature

$T_F = \dfrac{9}{5}T_C + 32$

$T_C = \dfrac{5}{9}(T_F - 32)$

$T_K = T_C + 273.16$

College Physics

Second Edition

Jerry D. Wilson

Lander University

Prentice Hall, Englewood Cliffs, New Jersey 07632

Acquisition Editor: Ray Henderson
Editor in Chief: Tim Bozik
Development Editor: Dan Schiller
Production Project Manager: Barbara DeVries
Marketing Manager: Leslie Cavaliere
Assistant Editors: Mary Hornby/Wendy Rivers
Product Manager: Trudy Pisciotti
Design Director: Florence Dara Silverman
Text and Cover Designer: Lee Goldstein
Page Layout: Lee Goldstein
Cover Photographs: background: J. D. Cuban, Allsport USA;
    inserts: Michael Ponzini, Focus on Sports/Robert Tringati, Jr., Sportschrom
Photo Editor: Lorinda Morris-Nantz
Photo Researcher: Tobi Zausner
Editorial Assistant: Pamela Holland-Moritz
Copy Editor: Cheryl Smith
Art Studio: Network Graphics
Text Composition: Monotype Composition Company, Inc.

Printed in the United States of America
10 9 8 7 6 5 4 3 2 1

ISBN 0-13-146168-0

Prentice-Hall International (UK) Limited, *London*
Prentice-Hall of Australia Pty. Limited, *Sydney*
Prentice-Hall Canada Inc., *Toronto*
Prentice-Hall Hispanoamericana, S.A., *Mexico*
Prentice-Hall of India Private Limited, *New Delhi*
Prentice-Hall of Japan, Inc., *Tokyo*
Simon & Schuster Asia Pte. Ltd., *Singapore*
Editora Prentice-Hall do Brasil, Ltda., *Rio de Janeiro*

Brief Contents

Contents

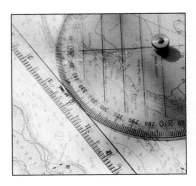
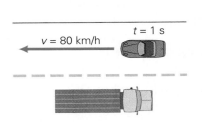

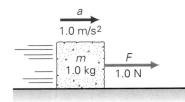

5 Work and Energy 138

6 Momentum and Collisions 172

7 Circular Motion and Gravitation 207

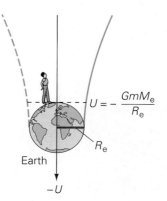

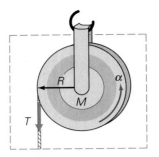

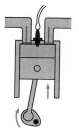

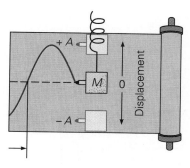

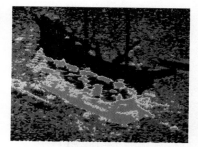

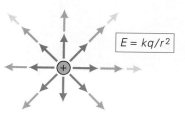

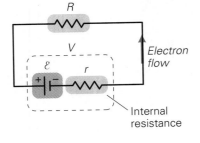

Electron flow

Internal resistance

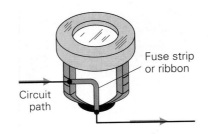

Fuse strip or ribbon

Circuit path

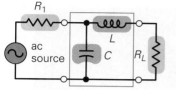

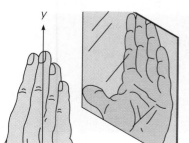

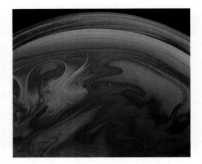

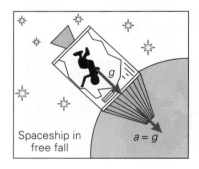

Spaceship in free fall $a = g$

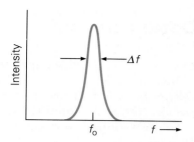

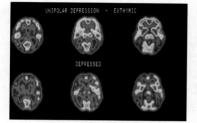

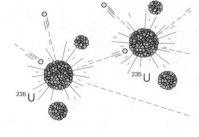

Demonstrations

Conceptual Examples

Applications

Insights in **boldface**

Topic

Preface

The Issue

There are two things that any introductory physics course must accomplish, regardless of its approach, its emphasis, or what else it may try to do:

- Impart an understanding of basic physical principles
- Enable students to solve a variety of reasonable problems in the areas covered by the text.

These aims are linked. An understanding of physical principles is of limited use if it does not enable students to solve problems. Physics is a problem-solving science—and students will be evaluated on their ability to produce correct answers, on final exams or on the MCAT. Yet few people would consider that learning to solve problems by rote is the same thing as learning physics. Knowing and doing, insight and skill, must go hand in hand.

Ironically, it is precisely in meeting these two goals that, by their own admission, many students and instructors alike feel they are failing—and that, all too often, their texts are failing them as well. Textbooks, which should be part of the solution, are too often part of the problem.

Any deficiency in meeting the first goal is likely to be obvious. Test scores quickly get the attention both of those who take the tests and those who grade them. Low grades demoralize instructors while discouraging students, who quite understandably conclude that physics is "too hard" for any but the phenomenally gifted. Yet nearly all instructors agree that the physics taught in a course such as this one is *not intrinsically difficult.* Why this paradox?

Deficiencies in meeting the second goal tend to be subtler. Recent research in physics education has shown that a surprising number of the students who *do* learn to solve typical problems well enough to pass examinations do so without ever arriving at a real understanding of the most elementary physical concepts. Such students often get high marks on exams, yet when asked to answer simple, qualitative questions designed to test their grasp of basic principles and their ability to reason from such principles, they betray a surprising lack of insight. Simply put, they are getting the right answers, but they do not understand how they are getting them or why they are right.

What We Have Done

The approach taken in this text is dictated by two simple propositions:

- An effective course in physics must strike a *balance* between fostering an understanding of principles and developing problem-solving skills.
- While placing more emphasis on problem solving per se will not necessarily improve students' understanding of principles, giving students a more secure grasp of principles will almost invariably enhance their problem-solving abilities. In other words, *conceptual insight must drive the development of practical skills.*

To maintain a sharp focus on essential concepts, a book should be concise, with a minimum of superfluous material. Along with balance, we have therefore set ourselves two additional goals: *brevity* and a strong emphasis on *basics*. Topics of marginal interest have been avoided, along with ones that present formal or mathematical difficulties for most students. Similarly, we have in many instances not wasted space deriving relationships when the derivations shed no additional light on the principles involved. It is usually more important for students in a course such as this one to understand what a relationship means and how it can be used than to see what mathematical or analytical techniques were employed to arrive at it.

For the same reason, we have (somewhat reluctantly) limited the amount of historical background provided except where it illuminates important principles. (Thus we have retained a discussion of the difference between the Aristotelian and Galilean conceptions of motion, and of the conceptual problems that led up to the Michelson-Morley experiment.) Interesting as historical material can be, for many students it is simply a distraction. Nor have we attempted to include interviews with or guest essays by living scientists. Digressions have been avoided. The pages saved in this way have been devoted to more generous explanation of basic principles and of the examples used to illustrate them.

Features of this Book

Most of the specific features of this text can best be understood in the light of the aims stated above. Central to these aims is a new approach to the development of problem-solving skills that stresses understanding of basic concepts as the essential foundation, rather than mechanical imitation of examples or the rote use of formulas from the text ("plug-and-chug"). This approach, which informs the entire text, is specifically embodied in several key features:

Suggested Problem-Solving Procedure

An extensive section in Chapter 1 (Sec. 1.7) provides a framework for thinking about problem solving. This section includes

- an overview of problem-solving strategy;
- a 7-step procedure that is general enough to be applicable to most problems in physics but concrete enought to be easily used;
- three Examples that illustrate the solution process, showing how the general procedure is applied in practice.

Problem-Solving Hints and Strategies

This initial treatment of problem solving is followed up throughout the text with an abundance of suggestions, tips, cautions, shortcuts, and useful techniques for solving specific kinds of problems. These help the student to apply general principles to specific contexts, as well as to avoid common pitfalls and misunderstandings. For example, there is an illustrated, step-by-step procedure for drawing free-body diagrams (see p. 111).

Conceptual Examples

Nearly ten percent of the text's many worked-out Examples are conceptual rather than computational. These examples ask the student to reason about a physical situation and choose the correct prediction based on an understanding of relevant principles. The discussion that follows (Reasoning and Answer) explains clearly how the correct answer can be identified as well as why the others are wrong. Each such Example is accompanied by a Follow-Up Exercise that requires similar reasoning and conceptual insight. (Reasoning and answers for these Exercises are given at the back of the text.)

More Explanation in Examples

Too many example solutions in texts rely on formulas such as "From Equation 6.7 we have. . . ." We have tried to make the solutions to in-text Examples as clear, patient, and detailed as possible. The aim is not merely to show the student what formulas to use but to explain the strategy being employed and the role of each step in the overall plan. Students are encouraged to learn the *why* of each step along with the *how*. This approach will make it easier for a student to apply the techniques being demonstrated *to other problems that are not identical in structure*. (See, for instance, Example 5.3 on pp. 141–142.)

Interactive Examples and Exercises

Over 120 of the in-text Examples and end-of-chapter Exercises can be dynamically simulated using the *Interactive Physics™ II Player*. This unique software package provides students with special insights into problem solving that would be difficult to obtain in any other way. The program not only simulates physical situations but allows students to perform their own experiments by varying key parameters. In this way the student can explore beyond specific solutions to grasp the general form of a solution.

Integration of Questions and Problems

In order to help break down the artificial and ultimately counterproductive barrier between "Questions" (qualitative or conceptual in nature) and "Problems" (calculations), we have eliminated these categories in the end-of-chapter Exercises. Instead, each section begins with a series of multiple-choice and short answer questions that provide content review, test conceptual understanding, and ask students to reason from principles. The quantitative problems follow without any break. The aim is to show students that the same kind of conceptual insight is required regardless of whether the desired answer involves words, equations, or numbers.

Insights

Applications are both intrinsically interesting and pedagogically useful; they satisfy the student's curiosity about the role of physics in the real world while reinforcing material presented in the text. This book includes 41 boxed Insight features, dealing with such diverse topics as the greenhouse effect and global warming (p. 381), automobile air bags (p. 186), space colonies and artificial gravity (p. 240), the compact disc (p. 820), magnetic stud sensors (p. 554), and observing black holes (p. 891)

Demonstrations

Photographs have an immediacy that gives them great pedagogical value. They can help students understand that the equations on the page describe phenomena in the real word—sometimes very dramatic ones. The 19 Demonstrations in the text make physical principles come to life in striking ways.

Pedagogical Apparatus

The end-of-chapter material was designed to support the dual aims of enhancing conceptual understanding and developing problem-solving skills.

Important Concepts: A listing of the key concepts introduced in the chapter that students should be able to define and explain.

Relationships for Review: A listing of the major laws and mathematical relationships introduced, cross-referenced to the chapter. Specific applicability and limiting conditions are clearly stated for each expression. *These relationships are presented as more than a convenient list of equations to be used in solving problems.* Students are urged to make sure that they understand the symbols used and can explain the relationship that each expresses.

End-of-Chapter Exercises

- Exercises are organized by chapter section
- Each chapter includes a supplemental section of Additional Exercises drawn from all sections of the chapter
- Each section integrates conceptual and calculational exercises, as described above
- Problems are graded by level of difficulty
- A computer icon identifies problems simulated by Interactive Physics™ II Player
- Each section includes at least one set of paired problems that deal with similar situations. The first is solved in the Study Guide, the second allows students to practice the same method (answer at back of book)
- Additional problems solved in the Study Guide are identified with an icon
- Answers appear with all exercises in the AIE; answers to odd-numbered exercises are included at the back of the book, along with answers to Conceptual Example Follow-Up Exercises from the text.

Supplements for the Instructor

Annotated Instructor's Edition (0-13-146168-0)

One way to help students learn is to help instructors teach. The margins of the Annotated Instructor's Edition contain an abundance of suggestions for classroom demonstrations and activities, along with teaching tips, points to emphasize, discussion suggestions, and common misunderstandings to avoid. In addition, the AIE contains:

- Marginal icons that identify each figure reproduced as a transparency in the Transparency Pack.
- Answers to end-of-chapter Exercises (following each Exercise).

Instructor's Solutions Manual (0-13-146176-1)

This supplement provides answers to all end-of-chapter questions and complete worked-out solutions to all end-of-chapter problems.

Physics You Can See Video Demonstrations (0-205-12393-7)

Physics You Can See demonstrates 11 classical physics experiments, each 2–5 minutes in length. The experiments include "Coin & Feather" (acceleration due to gravity), "Monkey & Gun" (rate of vertical free fall), "Swivel Hips" (conservation of momentum), and "Collapse a Can" (atmospheric pressure).

Transparencies (0-13-146366-7)

The transparency pack provides 150 full-color acetates of figures from the text.

Test Item File (0-13-146184-2)

The test item file provides a menu of over 1400 multiple-choice test items.

Prentice Hall *Test Manager 2.0*
IBM 3.5 (0-13-146192-3) or MAC (0-13-146200-8)

The latest release of Prentice Hall's exclusive computerized testing package includes full control over printing including print preview, complete mouse support, on-screen VGA graphics with import capabilities for .TIFF and .PCX file formats, ability to export files to WordPerfect, Word, and ASCII, and context-sensitive help and toll-free technical support.

Prentice Hall's Telephone Testing

This unique service allows instructors to create tests with one toll-free phone call. Users simply pick the questions they want from the hard-copy test item file and call 1-800-842-2958. Our testing staff will take the order, create the test (multiple versions if you like) and send it within 48 hours.

Supplements for the Student

Interactive Physics™ II Player

INTERACTIVE PHYSICS™ II PLAYER (IPP) is a text-specific, run-time version of Knowledge Revolution's INTERACTIVE PHYSICS™ II that simulates and animates over 120 exercises and worked examples from the pages of Wilson's *College Physics, Second Edition*. Its accessibility allows users to easily manipulate values to create "what if" scenarios, helping build physical intuition. IPP brings unparalleled flexibility to problem-solving and the physics classroom as:

- a home study aid for students' personal computers
- the ideal working demonstration for classroom lectures
- a hands-on experiment tool in the lab
- Macintosh Software (0-13-143314-8)
- Macintosh Demo (0-13-146150-8)
- Windows Software (0-13-146218-0)
- Windows Demo (0-13-146143-5)

Student Study Guide/Solutions Manual (0-13-147612-2)

Wilson's study guide/solutions manual presents chapter-by-chapter reviews—including chapter summaries—key terms, additional worked problems, and solutions to selected exercises (including the first of each paired set).

Arco's Physics for the MCAT for Wilson's *College Physics*, Second Edition (0-13-145716-0)

This supplement gives students the in-depth guidance needed for success on the Medical College Admission Test's physics section, using material *directly* from Wilson's *College Physics*, Second Edition.

Prentice Hall/*The New York Times*

THE NEW YORK TIMES and PRENTICE HALL are sponsoring Themes of the Times, a program designed to enhance student access to current information of relevance in the classroom.

Through this program, the core subject matter provided in the text is supplemented by a collection of time-sensitive articles from one of the world's most distinguished newspapers, THE NEW YORK TIMES. These articles demonstrate the vital, ongoing connection between what is learned in the classroom and what is happening in the world around us.

To enjoy the wealth of information of THE NEW YORK TIMES daily, a reduced subscription rate is available in deliverable areas. For information, call toll-free: 1-800-631-1222.

PRENTICE HALL and THE NEW YORK TIMES are proud to co-sponsor Themes of The Times. We hope it will make the reading of both textbooks and newspapers a more dynamic, involving process.

Acknowledgments

I would like to acknowledge the generous assistance that I received from many people during the preparation of this edition. First, my sincere thanks to Professor Al Hilgendorf of the University of Wisconsin for his vital contributions to the *Annotated Instructor's Edition*, and to Professor Bo Lou of Ferris State University for his meticulous and conscientious help with the galleys. The work of Professors Mehdi Habibi (Suffolk County Community College), Rex Joyner (Indiana Institute of Technology), Bo Lou, and Stan Shepherd (Pennsylvania State University) in checking the solutions to the end-of-chapter exercises is also greatly appreciated. A special word of thanks is due John Kinard, who coordinated these independent checks to make the solutions as error-free as possible. John's ever-diligent work on the *Annotated Instructor's Edition, Instructor's Solutions Manual*, and *Student Study Guide/Solutions Manual* was invaluable. I am also greatly indebted to the colleagues listed below, who read and commented on various portions of the second edition manuscript. Their suggestions were thoughtful and constructive, and this book benefited greatly from their work.

At Prentice Hall, the editorial staff was particularly helpful. I am deeply indebted to Barbara DeVries, production project manager, whose cheerful competence and experienced hand on the tiller kept this complex endeavor under control and moving forward at a brisk pace. Special recognition goes to Dan Schiller, Senior Development Editor, whose queries, insight, and creativity helped this second edition to "bloom." The indispensable support and guidance of Ray Henderson, Senior Editor, and Tim Bozik, Editor-in-Chief, have been greatly appreciated. I would also like to express my gratitude to designer Lee Goldstein and photo researcher Tobi Zausner for their outstanding work in helping to make this book as attractive and pedagogically useful as possible.

Finally, I would like to urge anyone using this book—student or instructor—to pass on to me any suggestions that you may have for its improvement. I welcome your comments.

Reviewers of the Second Edition

CHARLES BACON
Ferris State College

WILLIAM BERRES
Wayne State University

BENNET BRABSON
Indiana University

MICHAEL BROWNE
University of Idaho

PHILIP A. CHUTE
University of Wisconsin—Eau Claire

LATTIE F. COLLINS
East Tennessee State University

DAVID M. CORDES
Belleville Area Community College

JAMES R. CRAWFORD
Southwest Texas State University

J.P. DAVIDSON
University of Kansas

RICHARD DELANEY
College of Aeronautics

ROBERT J. FOLEY
University of Wisconsin—Stout

DONALD R. FRANCESCHETTI
Memphis State University

SIMON GEORGE
California State—Long Beach

BARRY GILBERT
Rhode Island College

RICHARD GRAHAM
Ricks College

ROBERT GRAHAM
University of Nebraska

TOM J. GRAY
Kansas State University

DOUGLAS A. HARRINGTON
Northeastern State University

AL HILGENDORF
University of Wisconsin—Stout

JOSEPH M. HOFFMAN
Frostburg State University

OMAR AHMAD KARIM
University of North Carolina—Wilmington

S.D. KAVIANI
El Camino College

RUBIN LAUDAN
Oregon State University

BRUCE A. LAYTON
Mississippi Gulf Coast Community College

ROBERT MARCH
University of Wisconsin—Madison

JOHN D. McCULLEN
University of Arizona

J. RONALD MOWREY
Harrisburg Area Community College

R. DARYL PEDIGO
Austin Community College

T.A.K. PILLAI
University of Wisconsin—La Crosse

DARDEN POWERS
Baylor University

DONALD S. PRESEL
University of Massachusetts—Dartmouth

E.W. PROHOFSKY
Purdue University

DAN R. QUISENBERRY
Mercer University

GEORGE RAINEY
California State Polytechnic University

WILLIAM RILEY
Ohio State University

R.S. RUBINS
The University of Texas at Arlington

SID RUDOLPH
University of Utah

RAY SEARS
University of North Texas

FREDERICK J. THOMAS
Sinclair Community College

PIETER B. VISSCHER
University of Alabama

JOHN C. WELLS
Tennessee Technical University

DEAN ZOLLMAN
Kansas State University

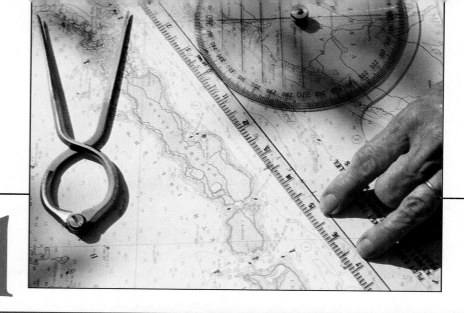

Units and Problem Solving

Not everyone has had the experience of laying out a course on a marine chart, as in the photo above. But most people have done something quite similar with a road map—plotting the shortest distance between two cities, measuring it, and estimating how long the trip will take. You might be planning a trip of 500 miles; traveling at an average speed of 50 mi/h, the trip would then take about 10 hours. Knowing your car's fuel economy, you might also estimate the amount of gas you expect to need and what it is likely to cost. (At 25 mpg and $1.25 per gallon, how much should you be prepared to spend?)

Measurement and problem solving are not confined to science—they are part of our lives. But they play a particularly central role in our attempts to describe and understand the physical world, as we shall see in this chapter.

1.1 ■ Why and How We Measure

> *Weights and measures may be ranked among the necessities of life to every individual of human society. . . . The knowledge of them, as established in use, is among the first elements of education, . . .*
>
> John Quincy Adams, *Report to Congress*, 1821

> *Measurement began our might. . . .*
>
> W. B. Yeats, "Under Ben Bulben," 1938

Chapter 1 introduces
measurement, systems of
measurement, unit analysis, and
problem-solving techniques.

Imagine that someone is giving you directions to her house. Would you find it helpful to be told, "Drive along Elm Street for a little while and turn right at one of the lights. Then keep going for quite a long way." Or suppose that you were baking a cake. How would you like to use a recipe that said, "Beat up some eggs, add a little sugar and some butter and a *whole lot* of flour, and bake for a while in a pretty hot oven." Similarly, would you want to deal with a bank that sent you a statement at the end of the month saying, "You still have some money left in your account. Not a great deal, though."

Measurement is important to all of us. It is one of the concrete ways in which we deal with our world. This is particularly true in physics. *Physics is concerned with the description and understanding of nature,* and measurement is one of its most important tools.

In everyday life you might use a yardstick to measure the length of a room, step onto a scale to measure your weight, or look at your watch to measure how many hours have passed since lunchtime. Measurements in physics are often more precise than those you would typically make in daily life, and they sometimes involve quantities much larger or smaller than those we ordinarily encounter. As a physicist, you might find yourself measuring the diameter of an atom, the mass of an entire galaxy—or the elapsed time of an event, to the nearest thousandth of a second! But the process of measurement is essentially no different.

There are certainly ways of describing the physical world that do not involve measurement. For instance, we might talk about the color of a flower or a dress. But color perception is subjective and may vary from one person to another. Indeed, many people suffer from color blindness and cannot tell certain colors apart. Objectively, light can be described in terms of wavelengths and frequencies. Different wavelengths are associated with different colors. We perceive color because of the physiological response of our eyes to light waves. But unlike the sensations or perceptions of color, the wavelengths can be measured. They are the same for everyone. In other words, they are *objective*. *Physics attempts to describe nature in an objective way through measurements.*

Measurements are expressed in unit values, or units. As you are probably aware, a large variety of units are used to express measured values. Some of the earliest units of measurement were referenced to parts of the human body; for instance, the hand is still used as a unit to measure the heights of horses. If a unit becomes officially accepted, it is called a **standard unit.** Traditionally, a government or international body establishes standard units.

Standard unit

System of units

A group of standard units and their combinations is called a **system of units.** Two major systems of units are in use today—the metric system and the British system. The latter is still widely used in the United States but will probably be phased out in favor of the metric system, which is used predominantly throughout the world.

Ask students to bring a straight
edge calibrated in centimeters
with them to class. Clear plastic
with both cm and inches is best.
A protractor will be needed later.

Different units in the same system or units from different systems can be used to describe the same thing. For example, your height can be expressed in inches, feet, centimeters, or meters. It is always possible to convert from one unit to another, however, and this is sometimes necessary.

Measured quantities are frequently used in calculations. If you know the length and width of a solar panel, you can calculate its area. If you can also measure the intensity of sunlight falling on the panel and you know the efficiency with which it converts light into electricity, you can calculate the amount of power that the unit generates, the amount of work that it will enable you

One of the first measurements of the circumference of the Earth was done sometime before 200 B.C. by the Greek scientist Eratosthenes (276–196 B.C.). His method, although quite simple, shows a great deal of insight and ingenuity. The basis for Eratosthenes' technique lies in the understanding that because the distance between the Earth and the Sun is so great compared to our planet's size, the Sun's rays arriving at the Earth are virtually parallel. Of course, the Sun would have to be at an infinite distance for the rays to be exactly parallel, but the divergence is so small that it is insignificant for practical purposes.

Eratosthenes lived in Alexandria, Egypt. He had learned that on the first day of summer at the town of Syene (now Aswan), some 5000 stadia to the south of Alexandria, the noon Sun was directly overhead. (A stadium, plural stadia, was a Greek unit of length believed to be equal to about ⅛ km.) The Sun's position overhead was apparent because deep wells at Syene were lighted all the way to the bottom, and a vertical stick would cast no shadow at noon on that day.

However, at Alexandria, some 800 km to the north, Eratosthenes observed that a vertical stick *did* cast a noon-day shadow with a length ⅛ that of the stick. This corresponds to an angle of 7.1° between the Sun's rays and the vertical (see Fig. 1 insert). Since lines extending through the well at Syene and through the vertical stick at Alexandria intersect at the Earth's center, the angular distance between these locations is also 7.1° (Fig. 1).

Then Eratosthenes reasoned that the 800 km arc length was to the 7.1° angle as the Earth's total circumference was to 360°, the number of degrees in a complete circle. That is, in ratio form,

$$\frac{\text{circumference}}{360°} = \frac{800 \text{ km}}{7.1°}$$

and

$$\text{circumference} = \frac{360°}{7.1°} \times 800 \text{ km} = 40,600 \text{ km}$$

The Earth's radius (R) can easily be calculated from the

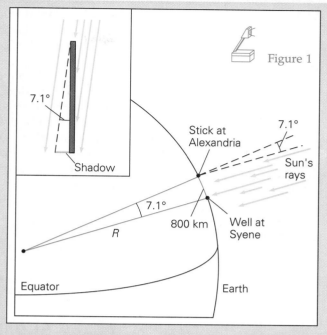

FIGURE 1 A 2000-year-old calculation Observations on the incidence of sunlight allowed the Greek scientist Eratosthenes to calculate the circumference and radius of the Earth sometime before 200 B.C. (See Insight text for description.)

circumference (c), since $c = 2\pi r$:

$$R = \frac{c}{2\pi} = \frac{40,600 \text{ km}}{2\pi} = 6,460 \text{ km}$$

This is quite close to the modern value of 6,378 km for the Earth's equatorial radius (Appendix I). We cannot be certain of the accuracy of Eratosthenes' result because several different Greek stadia of various lengths were in use at that time. (You can see why it is important to have *standard* units.) Even so, his method was sound, and the circumference of the Earth was measured over 2000 years ago.

to perform in a given period of time, and the money you would have to spend to purchase an equivalent amount of electricity from your local power company.

You may be surprised at how easy it is to determine very basic facts about our world through simple measurements and calculations that almost anyone can perform. Some 2200 years ago, a Greek scientist, using only some basic facts of geometry, was able to calculate the size of the Earth with considerable accuracy (see the Insight feature).

Modern science may utilize more sophisticated mathematics and technology, but the approach is still the same. Much of our knowledge rests on a foundation of ingenious measurement and simple calculation. Consider another solar example. If you know the intensity of solar radiation at the top of the Earth's atmosphere (which can be measured by orbiting satellites), along with the distance of the Earth from the Sun, you can calculate the total power output of the Sun. Then, another simple calculation will tell you how much material (mass) is being converted into energy each day by nuclear processes going on in the Sun's interior (Chapter 29). This turns out to be millions of tons per day! Scientists perform such calculations in their attempts to better understand aspects of the physical world, such as the forces that control the Earth's climate, the processes that cause the stars to shine, or the age and ultimate fate of the Sun.

In a sense, mathematics is the language of physics. Principles can be expressed in words, but are stated most clearly—and usually most concisely—in mathematical terms. One equation is worth a thousand words, so to speak. But, in order to use equations to do calculations, you must have the proper background and understanding. Toward this end, you will begin your study of physics with a look at some basic quantities and units and an approach to problem solving. A firm understanding of these will benefit you throughout this entire course of study.

1.2 ■ SI Units for Length, Mass, and Time

Length, mass, and time are fundamental physical quantities that describe a great many objects and phenomena. In fact, the topics of mechanics (the study of motion and force) covered in the first part of this book require *only* these physical quantities. The system of units used by scientists to represent these and other quantities is based on the metric system.

Historically, the metric system was the outgrowth of proposals for a more uniform system of weights and measures in France during the seventeenth and eighteenth centuries. The modernized version of the metric system is called the **International System of Units,** officially abbreviated as **SI** (from the French, *Système International des Unités*).

The SI includes *base quantities* and *derived quantities,* which are described by base units and derived units, respectively. Base units such as the meter and kilogram are easily represented by standards. Other quantities that may be expressed in terms of combinations of base units are called derived units. (As a familiar analogy, think of how we commonly measure the length of a trip in miles and the time in hours. We therefore express how fast we travel in the derived unit of miles/hour, which represents distance per unit of time, or distance/time.)

One of the refinements of the SI was the adoption of new standard references for some base units, including those for length, mass, and time.

The SI

Length

Length is the base quantity used to measure distances or dimensions in space. We commonly say that length is the distance between two points. But the

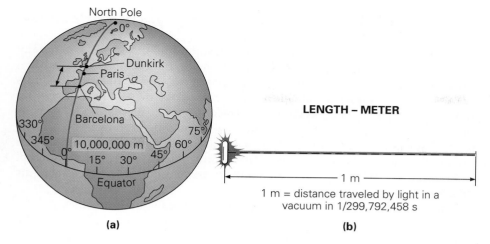

LENGTH – METER

1 m

1 m = distance traveled by light in a vacuum in 1/299,792,458 s

(a)   (b)

distance between any two points depends on how space is traversed, which may be in a straight or a curved line.

The SI unit of length is the **meter (m).** The meter was originally defined as 1/10,000,000 of the distance from the North Pole to the Equator along a meridian running through Paris (Fig. 1.1a). A portion of this meridian between Dunkirk, France, and Barcelona, Spain, was surveyed to establish the standard length, which was assigned the name *metre* (from the Greek word *metron*, meaning "a measure"; the American spelling is *meter*). A meter is 39.37 inches—slightly longer than a yard.

The length of the meter was initially preserved in the form of a material standard: the distance between two marks on a metal bar (made of a platinum-iridium alloy) that was stored under controlled conditions. However, it is not desirable to have a reference standard which changes with external conditions, such as temperature. Research provided more accurate standards. In 1983, the meter was redefined in terms of an unvarying property of light: the length of the path traveled by light in a vacuum during an interval of 1/299,792,458 of a second (Fig. 1.1b). Another way of saying this is that light travels 299,792,458 meters in a second, or that the speed of light in a vacuum is defined to be 299,792,458 meters per second. Note that the length standard is referenced to time, which is one of the most accurate measurements.

Mass

Mass is the base quantity used to describe amounts of matter. The more massive an object, the more matter it contains. (We will encounter more precise definitions of mass in Chapters 4 and 7.)

The SI unit of mass is the **kilogram (kg).** The kilogram was originally defined in terms of a specific volume of water, but is now referenced to a specific material standard: the mass of a prototype platinum-iridium cylinder kept at the International Bureau of Weights and Measures in Sèvres, France.

MASS - KILOGRAM

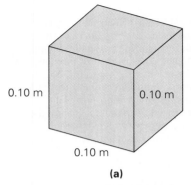

(a)

(b)

●**FIGURE 1.2 The SI mass standard: the kilogram** **(a)** The kilogram was originally defined in terms of a specific volume of water, that of a cube 0.10 m on a side, thereby associating the mass standard with the length standard. The standard kilogram is now defined by a metal cylinder. **(b)** Prototype Kilogram No. 20 is the national standard for mass for the United States. It is kept under a double bell jar in a vault at the National Institute of Standards and Technology (NIST, formerly the National Bureau of Standards) in Washington, D.C.

The United States has a duplicate of the prototype cylinder, which serves as the standard mass for this country (● Fig. 1.2).

You may have noticed that the phrase "weights and measures" is generally used instead of "masses and measures." In the SI, mass is the base quantity, but in the familiar British system, weight is used instead to describe amounts of mass—for example, weight in pounds instead of mass in kilograms. By definition, weight is the gravitational attraction that a celestial body, such as Earth, exerts on an object. For example, when you weigh yourself on a scale, your weight is a measure of the downward gravitational force exerted on you by the Earth. Since the gravitational force on any object is directly proportional to the object's mass, we can determine the mass of an object by measuring its weight.

But treating weight as the base quantity creates some problems. A base quantity is naturally most useful if its value is the same everywhere. This is the case with mass—an object has the same mass, or amount of matter, regardless of its location. *But this is not true of its weight.* Weight (w) and mass (m) are related by the formula $w = mg$, where g represents the acceleration of gravity. (As we shall see in Chapter 2, this is the acceleration produced by gravity acting on a body falling freely in a vacuum.) To find an object's mass from its weight measurement, we compute $m = w/g$.

We commonly take g to be a constant on the Earth's surface. However, its actual value varies slightly at different locations, because the Earth's mass distribution is not perfectly uniform and because not all places are precisely the same distance from the Earth's center. This means that weight will also vary.

The problem becomes more severe when we think about the weight of an object on the moon, or on another planet. The value of g (and therefore weight) is directly related to the mass of the celestial object that supplies the gravitational attraction. Since the moon is less massive than the Earth, the value of g is less on the moon than on Earth—only about one-sixth as great. Thus an object with a given amount of mass has a particular weight on Earth, but on the moon, the same amount of mass will weigh only about one-sixth as much. Clearly, therefore, *weight is closely related to mass, but they are not the same.* It is much more useful to take mass as the base quantity, as the SI does. Base quantities remain the same regardless of where they are measured, under normal or stationary conditions.*

The distinction between mass and weight will be more fully explained in a later chapter. Our discussion until then will be chiefly concerned with mass.

Time

Time is a difficult concept to define. (Try it.) A common definition is that time is the forward flow of events. This is not so much a definition as an observation that time has never been known to run backwards (as it might appear to do when you view a film run backwards in a projector). Time is sometimes said to be a fourth dimension, accompanying the three dimensions of space. Thus, if something exists in space, it also exists in time. In any case, events can be

---

*We will see in Chapter 25 that when objects move at high speeds, approaching the speed of light, their masses increase, as the theory of relativity predicts.

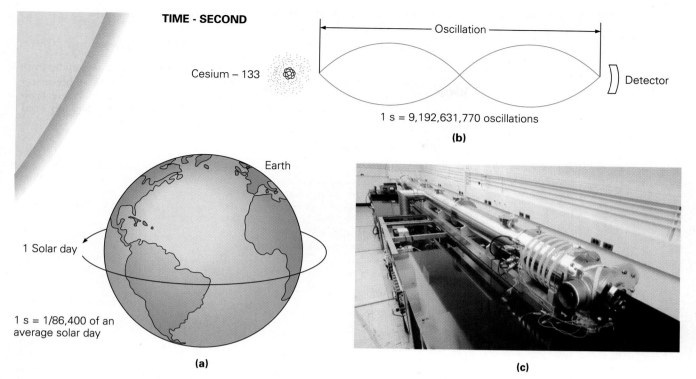

TIME - SECOND

Cesium – 133

Oscillation

Detector

1 s = 9,192,631,770 oscillations

(b)

Earth

1 Solar day

1 s = 1/86,400 of an
average solar day

(a)

(c)

●**FIGURE 1.3 The SI time standard: the second** **(a)** The second was originally
defined in terms of the average solar day. **(b)** It is currently defined in terms of the
frequency of the radiation associated with an atomic transition. **(c)** This atomic "clock" is
the primary frequency standard at the National Institute of Standards and Technology.
The device keeps time with an accuracy of about three millionths of a second per year.

used to mark time measurements. The events are analogous to the marks on
a meterstick used for length measurements.

The SI unit of time is the **second(s).** The solar "clock" was originally used
to define the second. A solar day is the interval of time that elapses between
two successive crossings of the same meridian by the Sun, at its highest point
in the sky at that meridian (●Fig. 1.3a). A second was fixed as 1/86,400 of this
apparent solar day (1 day = 24 h = 1440 min = 86,400 s). However, the
elliptical path of the Earth's motion around the Sun causes apparent solar days
to vary in length.

As a more precise standard, an average, or mean, solar day was computed
from the lengths of the apparent solar days during a solar year. In 1956, the
second was referenced to this mean solar day. But the mean solar day is not
exactly the same for each yearly period because of minor variations in the
Earth's motions and a steady slowing of its rate of rotation due to tidal friction.
So scientists kept looking for something better.

In 1967, an atomic standard was adopted as a more precise reference.
Currently, the second is defined as the duration of 9,192,631,770 cycles (periods)
of the radiation associated with a particular transition of the cesium-133 atom.
The atomic "clock" used to maintain this standard doesn't look much like an
ordinary clock (●Fig. 1.3b,c).

> Many students now have digital
> watches that can be used as stop
> clocks. This is a good way to get
> student participation in simple
> demonstrations involving time
> measurements.

TABLE 1.1 ■ The Seven Base Units of the Modernized Metric System (SI)

Name of Unit (abbreviation)	Property Measured	Definition	Additional Information
meter (m)	length	The meter is defined as the distance traveled by light in a vacuum during 1/299,792,458 of a second. Thus time, our most accurate measurement, is used to define length. In effect, this definition establishes the speed of light as 299,792,458 meters per second.	The SI unit of area is the **square meter** (m^2). The SI unit of volume is the **cubic meter** (m^3). The liter (0.001 cubic meter), although not an SI unit, is commonly used to measure fluid volume.
kilogram (kg)	mass	The standard for the unit of mass, the kilogram, is a cylinder of platinum-iridium alloy kept by the International Bureau of Weights and Measures at Paris. A duplicate in the custody of the National Institute of Standards and Technology in Washington, DC serves as the mass standard for the United States. (The kilogram is the only base unit still defined by an artifact.)	The SI unit of force is the **newton** (N). One newton is the force which, when applied to a 1-kg mass, will give the mass an acceleration of 1 (meter per second) per second: $$1\ N = 1\ \frac{kg\text{-}m}{s^2}$$ The SI unit for pressure is the **pascal** (Pa): $$1\ Pa = 1\ N/m^2$$ The SI unit for work and energy of any kind is the **joule** (J): $$1\ J = 1\ N\text{-}m$$ The SI unit for power of any kind is the **watt** (W): $$1\ W = 1\ J/s$$
second(s)	time	The second is defined as the duration of 9,192,631,770 cycles of the radiation associated with a specified transition of the cesium-133 atom.	The number of periods or cycles per second is called *frequency*. The SI unit for frequency is the **hertz** (Hz). One hertz equals one cycle per second.
ampere (A)	electric current	The ampere is defined as that current which, if maintained in each of two long parallel wires separated by one meter in free space, would produce a force between the two wires (due to their magnetic fields) of 2×10^{-7} newton for each meter of length.	
kelvin (K)	temperature	The kelvin is defined as the fraction 1/273.15 of the thermodynamic temperature of the triple point of water. The temperature 0 K is called "absolute zero."	On the commonly used Celsius temperature scale, water freezes at about 0°C and boils at about 100°C. The °C is defined as an interval of 1 K, and the Celsius temperature 0°C is defined as 273.15 K. 1.8 Fahrenheit degrees are equal to 1.0C° or 1.0 K; the Fahrenheit scale uses 32°F as a temperature corresponding to 0°C.
mole (mol)	amount of substance	The mole is the amount of substance of a system that contains as many elementary entities as there are atoms in 0.012 kg of carbon-12 (6.02×10^{23}).	When the mole is used, the elementary entities must be specified and may be atoms, molecules, ions, electrons, other particles, or specified groups of such particles.
candela (cd)	luminous intensity	The candela is defined as the luminous intensity of 1/600,000 of a square meter of a blackbody at a temperature of 2045 K.	The SI unit of light flux is the **lumen** (lm). A source having an intensity of one candela in all directions radiates a light flux of 4π lumens. (A 100-watt light bulb emits about 1700 lumens.)

SI Base Units

The complete SI has the seven base quantities and units listed in Table 1.1. This is thought to be the smallest number of base quantities needed for a full description of everything observed or measured in nature. There are also two supplemental units for angular measure: the two-dimensional plane angle and the three-dimensional solid angle. There has not been general agreement about whether these geometric units are base or derived.

SI base units

1.3 ▪ More about the Metric System

The metric system involving the standard units of length, mass, and time now incorporated in the SI was once called the **mks system** (for *meter-kilogram-second*). Another metric system that has been used in dealing with relatively small quantities is the **cgs system** (for *centimeter-gram-second*). In the United States, the system still generally in use is the British (or English) engineering system, in which the standard units of length, mass, and time are foot, slug, and second, respectively. You may not have heard of the slug, because gravitational force, or weight, is commonly used instead of mass (pounds instead of slugs) to describe quantities of matter. As a result, the British system is sometimes called the **fps system** (*foot-pound-second*).

The metric system is predominant throughout the world and is coming into increasing use in the United States. Primarily because of its mathematical simplicity, it is the preferred system of units for science and technology. SI units are used throughout most of this book. All quantities can be expressed in SI units. However, some units from other systems are accepted for limited use as a matter of practicality—for example, the time unit of hour and the temperature unit of degree Celsius. British units will sometimes be used in the early chapters for comparison purposes, since these units are still employed in everyday activities and for many practical applications.

The increasing worldwide use of the metric system means that you should be familiar with it. One of the greatest advantages of the metric system is that it is a decimal, or base-10, system. This means that larger or smaller units are obtained by multiplying or dividing a base unit by powers of 10. A list of the various multiples and corresponding prefixes is given in Table 1.2. In decimal measurements, the prefixes milli-, centi-, and kilo- are the ones most commonly used. The decimal characteristic makes it convenient to change measurements from one size metric unit to another. With the familiar British system, different factors must be used, such as 16 for converting pounds to ounces and 12 for converting feet to inches.

You are already familiar with one base-10 system—U.S. currency. Just as a meter can be divided into 10 decimeters, 100 centimeters, or 1000 millimeters, the "base unit" of the dollar can be broken down into 10 "decidollars" (dimes), 100 "centidollars" (cents), or 1000 "millidollars" (tenths of a cent, or mils, used in figuring property taxes and bond levies). Since the metric prefixes are all powers of 10, there are no metric analogs for quarters or nickels.

The official metric prefixes can help eliminate confusion. In the United States, a billion is a thousand million (10^9); in Great Britain, a billion is a million

In general, SI units will be used throughout the text. British units will be used in a few problems.

A straight edge that has both centimeters and inches will show the difference between a base ten system and the English system of quarters, eighths, etc. A clear plastic straight edge can be displayed on the overhead projector.

TABLE 1.2 ■ Multiples and Prefixes for Metric Units*

Multiple†	Prefix (and Abbreviation)	Pronunciation	Multiple†	Prefix (and Abbreviation)	Pronunciation
10^{24}	yotta- (Y)	yot′ta (*a* as in *a*bout)	10^{-1}	deci- (d)	des′i (as in *deci*mal)
10^{21}	zetta- (Z)	zet′ta (*a* as in *a*bout)	10^{-2}	centi- (c)	sen′ti (as in *senti*mental)
10^{18}	exa- (E)	ex′a (*a* as in *a*bout)	10^{-3}	milli- (m)	mil′li (as in *mili*tary)
10^{15}	peta- (P)	pet′a (as in *peta*l)	10^{-6}	micro- (μ)	mi′kro (as in *micro*phone)
10^{12}	tera- (T)	ter′a (as in *terra*ce)	10^{-9}	nano- (n)	nan′oh (*an* as in *an*nual)
10^{9}	giga- (G)	ji′ga (*ji* as in *ji*ggle, *a* as in *a*bout)	10^{-12}	pico- (p)	pe′ko (*peek-oh*)
10^{6}	mega- (M)	meg′a (as in *mega*phone)	10^{-15}	femto- (f)	fem′toe (*fem* as in *fem*inine)
10^{3}	kilo- (k)	kil′o (as in *kilo*watt)	10^{-18}	atto- (a)	at′toe (as in an*ato*my)
10^{2}	hecto- (h)	hek′to (*heck-toe*)	10^{-21}	zepto- (z)	zep′toe (as in *zep*pelin)
10	deka- (da)	dek′a (*deck* plus *a* as in *a*bout)	10^{-24}	yocto- (y)	yock′toe (as in s*ock*)

*For example, 1 gram (g) multiplied by 1000 (10^3) is 1 kilogram (kg), or multiplied by 1/1000 (10^{-3}) is 1 milligram (mg).
†The most commonly used prefixes are printed in color.

 Table 1.2

million (10^{12}). The use of metric prefixes eliminates any confusion, since giga- indicates 10^9 and tera- stands for 10^{12}.

In the SI, the standard unit of volume is the cubic meter (m^3). Since this is a rather large quantity, it is often more convenient to use the nonstandard unit of volume (or capacity) of a cube 1 dm (decimeter), or 10 cm (centimeters) on a side (●Fig. 1.4). This volume unit was given the name litre, which is spelled liter in the United States. The volume of a **liter (L)** is 1000 cm^3 (10 cm × 10 cm × 10 cm), and since 1 L = 1000 mL (milliliters), it follows that 1 mL = 1 cm^3 (cubic centimeter is sometimes abbreviated cc, particularly in chemistry

The liter

Liter is sometimes abbreviated with a lowercase "ell" (l), but a capital "ell" (L) is preferred in the United States so that the abbreviation is less likely to be confused with the numeral one. (Isn't 1 L clearer than 1 l?)

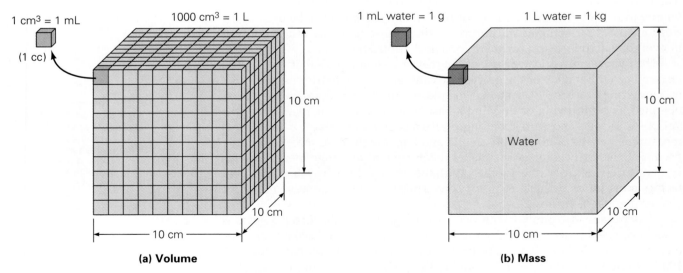

1 cm3 = 1 mL
(1 cc)
1000 cm3 = 1 L
10 cm
10 cm
10 cm

1 mL water = 1 g
1 L water = 1 kg
Water
10 cm
10 cm
10 cm

(a) Volume   **(b) Mass**

 Figure 1.4

●**FIGURE 1.4 The liter and the kilogram**   Other metric units are derived from the meter. **(a)** A unit of volume (capacity) was taken to be the volume of a cube 1.0 dm (10 cm or 0.10 m) on a side and was given the name liter. **(b)** The mass of a liter of water (at its maximum density) was defined to be 1 kg. Note that the decimeter cube contains 1000 cm^3, or 1000 mL. Thus, 1 cm^3, or 1 mL, of water has a mass of 1 gram.

and biology). The use of the liter is becoming quite common in the United States, as ● Fig. 1.5 indicates.

The standard unit of mass, the kilogram, was originally defined to be the mass of a cubic volume of water 0.10 m or 10 cm on a side, or 1 *liter* of water, at its maximum density (at ≅4°C). Since 1 kg = 1000 g and 1 L = 1000 cm³, then *1 cm³ (or 1 mL) of water has a mass of 1 g.* Large quantities of mass are sometimes expressed in metric tons. A metric ton (or *tonne*) is 1000 kg.

Because the metric system is coming into increasing use in this country, you may find it helpful to have an idea of how metric and British units compare. The relative sizes of some units are illustrated in ● Fig. 1.6. The mathematical conversion from one unit to another will be discussed shortly.

● **FIGURE 1.6 Comparison of some SI and British units** The bars illustrate the relative magnitudes of each pair of units. (The comparison scales are different in each case.)

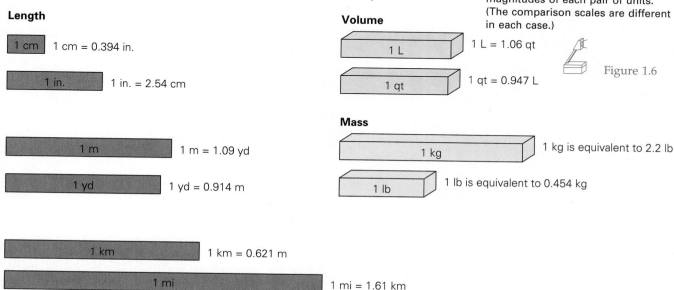

Length

1 cm	1 cm = 0.394 in.
1 in.	1 in. = 2.54 cm
1 m	1 m = 1.09 yd
1 yd	1 yd = 0.914 m
1 km	1 km = 0.621 m
1 mi	1 mi = 1.61 km

Volume

1 L	1 L = 1.06 qt
1 qt	1 qt = 0.947 L

Mass

1 kg	1 kg is equivalent to 2.2 lb
1 lb	1 lb is equivalent to 0.454 kg

Figure 1.6

1.4 ■ Dimensional Analysis

The fundamental, or base, quantities used in physical descriptions are called *dimensions*. For example, length, mass, and time are dimensions. You could measure the distance between two points and express it in units of meters, centimeters, or feet. In any case, the quantity would have the dimension of length.

It is common to express dimensional quantities by bracketed symbols, such as [L], [M], and [T] for length, mass, and time, respectively. Derived quantities are combinations of dimensions; for example, velocity (v) has the dimensions [L]/[T] (think of miles/hour or meters/second), and volume (V) has the dimensions [L] × [L] × [L], or [L³]. Addition and subtraction can only be done with quantities that have the same dimensions, for example, 10 s + 20 s = 30 s, or [T] + [T] = [T].

Dimensional analysis is a procedure by which the dimensional consistency of any equation may be checked. You have used equations and know that an equation is a mathematical equality. Since physical quantities used in equations have dimensions, *the two sides of an equation must be equal not only in numerical magnitude, but also in dimensions. And dimensions can be treated as algebraic quantities.* That is, they can be multiplied or divided. For example, suppose the quantities in an equation have the dimension of length: 3.0 m × 4.0 m = 12 m². This is [L] × [L] = [L²] dimensionally; thus both sides of the equation are *numerically* and *dimensionally* equal.

One of the major advantages of dimensional analysis is its usefulness for checking whether an equation that has been derived or is being used in solving a problem has the correct form. For example, suppose that you think you can solve a problem about the distance an object travels using the equation $x = at$, where x represents distance, or length, and has the dimension [L]; a represents acceleration, which, as we will see in Chapter 2, has the dimensions [L]/[T²]; and t represents time and has the dimension [T]. First, before trying to use numbers with the equation, you can check to see if it is dimensionally correct:

$$x = at$$

is dimensionally

$$[L] = \frac{[L]}{[T^2]} \times [T] \qquad \text{or} \qquad [L] = \frac{[L]}{[T]}$$

which is obviously not true. Thus, $x = at$ *cannot* be a correct equation.

Dimensional analysis will tell you if an equation is incorrect, but a dimensionally consistent equation may *not* correctly express the real relationship of quantities. For example,

$$x = at^2$$

is dimensionally

$$[L] = \frac{[L]}{[T^2]} \times [T^2] \qquad \text{or} \qquad [L] = [L]$$

This equation is dimensionally correct. But, as you will see in Chapter 2, it is not physically correct. The correct form of the equation—both dimensionally and physically—is $x = \frac{1}{2}at^2$. The fraction ½ has no dimensions; it is a dimen-

Dimensional analysis

Dimensional analysis is a very helpful tool in problem solving.

sionless constant, like the constant π. [For a circle, the circumference (c) is related to the diameter (d) by $c = \pi d$, or $\pi = c/d$. Do you see why π is dimensionless?]

Using symbols for dimensions is fine, but in actual practice it is often more convenient to use units, which are often written into an equation along with the numerical magnitudes of the quantities. Units can also be treated as algebraic quantities in a form of dimensional analysis that can be called **unit analysis.** The following example demonstrates unit analysis.

Unit analysis

EXAMPLE 1.1 ■ Checking with Units

A professor puts two equations on the board: (a) $v = v_0 + at$ and (b) $x = v/2a$, where x is distance in meters (m), v and v_0 are velocities or speeds in meters/second (m/s), a is acceleration in (meters/second)/second or meters/second^2 (m/s^2), and t is time in seconds (s). Are the equations dimensionally correct? Use unit analysis to find out.

Solution.
(a) The equation is

$$v = v_0 + at$$

Inserting units for the physical quantities gives

$$\frac{m}{s} = \frac{m}{s} + \left(\frac{m}{s^2} \times s\right) \quad \text{or} \quad \frac{m}{s} = \frac{m}{s} + \left(\frac{m}{s \times s} \times s\right)$$

Notice that units cancel like numbers in a fraction. Then, we have

$$\frac{m}{s} = \frac{m}{s} + \frac{m}{s} \quad \begin{bmatrix} \text{Dimensionally} \\ \text{correct} \end{bmatrix}$$

The equation is dimensionally correct, since the units on each side are meters per second. If an incorrect unit had been used in the equation, the unit analysis would have shown the error.

(b) The equation

$$x = \frac{v}{2a}$$

with unit analysis is

$$m = \frac{\left(\dfrac{m}{s}\right)}{\left(\dfrac{m}{s^2}\right)} = \frac{m}{s} \times \frac{s^2}{m} \quad \text{or} \quad m = s \quad \begin{bmatrix} \text{Not dimensionally} \\ \text{correct} \end{bmatrix}$$

Meters (m) cannot equal seconds (s), so in this case, the equation is dimensionally incorrect. A dimensionally correct form of the equation is $x = v^2/2a$. (Do you agree?)

Mixed Units

Unit analysis also allows you to check for mixed units. In general, in working problems, you should always use the same unit for a given dimension throughout a problem. For example, suppose that you wanted to find the floor area of a rectangular room in order to buy a new carpet to fit it. To compute the area, you would measure the lengths of two adjacent sides of the room and multiply them. Would you measure the sides in different units such as 10 ft $\times$ 3.0 m and express the area as 30 ft-m? Probably not. Such mixed units are not usually very useful.

Let's look at mixed units in an equation. Suppose that you used centimeters as the unit for x in this equation:

$$v^2 = v_o^2 + 2ax$$

and the units for the other quantities as in Example 1.1. Dimensionally, this would give

$$\left(\frac{m}{s}\right)^2 = \left(\frac{m}{s}\right)^2 + \left(\frac{m}{s^2} \times cm\right)$$

or

$$\frac{m^2}{s^2} = \frac{m^2}{s^2} + \frac{m \times cm}{s^2}$$

which is dimensionally correct. That is, it is equivalent to

$$\frac{[L^2]}{[T^2]} = \frac{[L^2]}{[T^2]} + \frac{[L^2]}{[T^2]}$$

But the units are mixed (m and cm). The terms on the right-hand side could not be added together without first changing one of the units.

Determining the Units of Quantities

Another aspect of unit analysis that is very important in physics is the determination of the units of quantities from defining equations. For example, **density** (ρ) is defined by the equation

$$\rho = \frac{m}{V} \tag{1.1}$$

where m is mass and V is volume. (Density is the mass per unit volume and is a measure of the compactness of the mass of an object or substance.) What are the units of density? Since mass is measured in kilograms and volume in cubic meters in SI units, the defining equation

$$\rho = \frac{m}{V} \left(\frac{kg}{m^3}\right)$$

gives the derived SI unit for density as kilograms per cubic meter (kg/m^3).

What are the units of π? Since the relationship of the circumference and the diameter of a circle is given by the equation $c = \pi d$, then $\pi = c/d$, and if length is measured in meters,

$$\pi = \frac{c}{d} \left(= \frac{m}{m} \right)$$

Thus, the constant π has no units because they cancel out. It is unitless.

1.5 ■ Unit Conversions

Because different units in the same system or in different systems can express the same quantity, it is sometimes necessary to convert the units of a quantity from one unit to another, for example, from feet to yards or from inches to centimeters. You already know how to do many unit conversions. For example, if a room is 12 feet long, what is its length in yards? Your immediate answer is 4 yards.

What you do in this operation is to use a **conversion factor.** You know that 1 yd = 3 ft,* and this can be written as a ratio: 1 yd/3 ft or 3 ft/1 yd. (The one is often omitted in the denominator of such ratios, for example, 3 ft/yd.) Note that dividing the expression 1 yd = 3 ft by 3 ft on both sides gives

Conversion factors

$$\frac{1 \text{ yd}}{3 \text{ ft}} = \frac{3 \text{ ft}}{3 \text{ ft}} = 1$$

This illustrates that a conversion factor (ratio) always has a magnitude of 1. You can multiply any quantity by 1 without changing its magnitude, or size. Thus, a *conversion factor simply lets you express a quantity in terms of other units without changing its magnitude.*

What is done in converting 12 feet to yards may be expressed mathematically as follows:

$$12 \text{ ft} \times \frac{1 \text{ yd}}{3 \text{ ft}} = 4 \text{ yd} \quad \text{(cancel feet)}$$

The feet cancel, giving yd = yd.

Suppose you are asked to convert 12.0 inches to centimeters. You may not know the conversion factor in this case, but you can get it from a table (such as the one that appears inside the front cover of this book): 1 in. = 2.54 cm or 1 cm = 0.394 in. It makes no difference which form of the conversion factor is used. The question is whether to divide or multiply the given quantity to make the conversion. *In doing unit conversions, you should take advantage of unit analysis.*

Note that you may use either 1 in./2.54 cm or 2.54 cm/in. for the conversion factor in ratio form. The unit analysis tells you that in this case the second form is the appropriate one:

$$12.0 \text{ in.} \times \frac{2.54 \text{ cm}}{\text{in.}} = 30.5 \text{ cm}$$

Alternatively, you can express the conversion factor as follows:

$$12.0 \text{ in.} \times \frac{1 \text{ cm}}{0.394 \text{ in.}} = 30.5 \text{ cm}$$

*Although we commonly express relationships such as 1 yd = 3 ft in equation form, this is not really an equation but rather an *equivalence statement*. That is, we mean that 1 yd is equivalent to, or equal in length to, 3 ft.

Students often ask "Do we have to show all of the work?" Insist that they show the work, and encourage them to analyze the unit cancellations. Also insist that students perform the unit conversions using a horizontal division line as shown in the text—not a slanted division sign, which obscures the numerator and denominator positions.

What would be the result if you had used the other form of the first conversion factor, 1 in./2.54 cm? Let's see:

$$12.0 \text{ in.} \times \frac{1 \text{ in.}}{2.54 \text{ cm}} = 4.72 \frac{\text{in}^2}{\text{cm}}$$

This is dimensionally correct (why?), but it is not a conversion to centimeters. The units in²/cm for length are nonstandard and confusing. To use this ratio properly, you must *divide* a quantity in inches by the conversion factor:

$$\frac{12.0 \text{ in.}}{\left(\dfrac{1 \text{ in.}}{2.54 \text{ cm}} \right)} = 30.5 \text{ cm}$$

(Compare this procedure with the first conversion of 12.0 in. above.)

A few commonly used conversion factors are not dimensionally correct; for example, 1 kg = 2.2 lb is used for conversions, but the kilogram is a unit of mass and the pound is a unit of weight. What the conversion factor really means is that 1 kilogram is equivalent to 2.2 pounds, or a 1-kilogram *mass* has a *weight* of 2.2 pounds. In any case, 1 kg/2.2 lb is a convenient conversion factor for converting kilograms of mass to pounds of weight, and vice versa. (As we saw earlier, mass and weight are related by g, the acceleration of gravity, a factor that is taken to be constant at the Earth's surface. This is taken into account in the ratio given as the conversion factor.)

1 kg has an equivalent weight of 2.2 lb

EXAMPLE 1.2 ■ Converting Units

Do the following unit conversions: (a) 15 meters to feet, (b) 30 days to seconds, and (c) 50 miles/hour to meters/second.

Solution.
(a) From the conversion table, 1 m = 3.28 ft, so

$$15 \text{ m} \times \frac{3.28 \text{ ft}}{\text{m}} = 49 \text{ ft}$$

(b) The conversion factor for days and seconds is available from the table (1 day = 86,400 s), but you may not always have a table handy, so instead use several better-known conversion factors to get the result:

$$30 \text{ days} \times \frac{24 \text{ h}}{\text{day}} \times \frac{60 \text{ min}}{\text{h}} \times \frac{60 \text{ s}}{\text{min}} = 2.6 \times 10^6 \text{ s}$$

Note how unit analysis checks the conversion factors for you. The rest is simple arithmetic.
(c) In this case, from the conversion table, 1 mi = 1609 m and 1 h = 3600 s. (The latter is easily computed.) These ratios are used to cancel the units that are to be changed, leaving behind the ones that are wanted:

$$\frac{50 \text{ mi}}{\text{h}} \times \frac{1609 \text{ m}}{\text{mi}} \times \frac{1 \text{ h}}{3600 \text{ s}} = 22 \text{ m/s}$$

Notice that a direct conversion factor is available from the table, 1 mi/h = 0.447 m/s, and could have been used.

EXAMPLE 1.3 ■ Converting Units of Area

A hall bulletin board has an area of 2.5 m². What is this area in square centimeters (cm²)?

Solution. A common error in such conversions is the use of incorrect conversion factors. Because 1 m = 100 cm, it is sometimes assumed that 1 m² = 100 cm², which is *wrong*. The correct area conversion factor may be obtained directly from the correct linear conversion factor, 100 cm/1m or 10^2 cm/1 m, by squaring it:

$$\left(\frac{10^2 \text{ cm}}{1 \text{ m}}\right)^2 = \frac{10^4 \text{ cm}^2}{1 \text{ m}^2}$$

Hence, 1 m² = 10^4 cm² (= 10,000 cm²). We can therefore write:

$$2.5 \text{ m}^2 \times \left(\frac{10^2 \text{ cm}}{1 \text{ m}}\right)^2 = 2.5 \text{ m}^2 \times \frac{10^4 \text{ cm}^2}{\text{m}^2} = 2.5 \times 10^4 \text{ cm}^2$$

(Similarly, 1 m³ is equivalent to 10^6 cm³. Prove this to yourself.)

EXAMPLE 1.4 ■ Selecting Units

You want to express the fuel economy of a car so it will sound the most impressive. Which set of units would give the greatest numerical value?
(a) mi/gal, (b) km/gal, (c) mi/L, (d) km/L. (Note: 1 mi = 1.61 km.) *Clearly establish the reasoning used in determining your answer before checking it below. That is, **why** did you select your answer?*

Reasoning and Answer. Since 1 mi is equivalent to 1.61 km, (b) would have a larger numerical value than (a), and (d) would have a larger value than (c). That is, 1 mi/gal = 1.61 km/gal and 1 mi/L = 1.61 km/L. Hence, the length unit would be km, and we must decide which volume unit would give the larger fuel economy value—km/gal or km/L. Since a liter is about a quart, a gallon is much larger than a liter (about 4 times larger). A gallon of gas would therefore run a car many more kilometers than a liter of gas. Thus the answer is (b).

Follow-up Exercise. (a) Which set of units would give the *smallest* numerical value? (b) Given that 1 L = 0.264 gal, justify your answer mathematically. (*Reasoning and answers may be found in the Answers to Exercises section at the back of the book.*)

1.6 ▪ Significant Figures

> *"What is the meaning of it all, Mr. Holmes?" "Ah. I have no data. I cannot tell," he said.*
>
> Arthur Conan Doyle,
> *The Adventure of the Copper Beeches*, 1892

Pass out 3 × 5 cards to the students and ask them to use a calibrated straight edge to measure their width and length. Have them record those values on the card. Then collect some (or all) of the cards and record the values on the board for discussion of significant digits. Some students will round; others will truncate; and others will read the measurement correctly.

Unlike Sherlock Holmes, you will be given numerical data when asked to solve a problem. In general, such data are either exact numbers or measured numbers (quantities). **Exact numbers** are those without any uncertainty or error. These include numbers such as the 100 as used to calculate a percentage and the 2 in the equation $r = d/2$ relating the radius and diameter of a circle. **Measured numbers** are those obtained from measurement processes and so generally have some degree of uncertainty or error.

When calculations are done with measured numbers, the error of measurement is propagated, or carried along, by the mathematical operations. A question of how to report a result arises. For example, suppose that you are asked to find time (t) from the formula $x = vt$, and you are given that $x = 5.3$ m and $v = 1.67$ m/s. Then

$$t = \frac{x}{v} = \frac{5.3 \text{ m}}{1.67 \text{ m/s}} = ?$$

A calculator may have a floating decimal point and list up to ten figures or digits. The number of digits displayed on some calculators can be preset.

Significant figures

Doing the division operation on a hand calculator yields a result such as 3.173652695 (see ● Fig. 1.7). How many figures, or digits, should you report in the answer?

The error or uncertainty of the result of a mathematical operation may be computed by statistical methods. A simpler, widely used procedure for making an estimate of this uncertainty involves the use of **significant figures (sf)**, sometimes called significant digits. The degree of accuracy of a measured quantity depends on how finely divided the measuring scale of the instrument is. For example, you might measure the length of an object as 2.5 cm with one instrument and 2.54 cm with another; the second instrument provides more significant figures and a greater degree of accuracy. In general, *the number of significant figures of a numerical quantity is the number of reliably known digits it contains*. For a measured quantity, this is usually defined as all of the digits that can be read directly from the instrument used in making the measurement plus one uncertain digit that is obtained by estimating the fraction of the smallest division of the instrument's scale.

General-purpose electric meters often have more than one scale for different sensitivities. Display one of these (an overhead projection meter if available) and discuss the difference in significant figures. Also consider car speedometers that read in mi/h and km/h.

The quantities 2.5 cm and 2.54 cm have two and three significant figures, respectively. This is rather evident. However, some confusion may arise when a quantity contains one or more zeroes. For example, how many significant figures does the quantity 0.0254 m have? What about 104.6 m? 2705.0 m? In such cases, we will use the following rules:

Rules for determining the number of significant figures in quantities with zeroes

1. Zeroes at the beginning of a number are not significant. They merely locate the decimal point.

   0.0254 m   three significant figures   (2, 5, 4)

2. Zeroes within a number are significant.

   104.6 m   four significant figures   (1, 0, 4, 6)

3. Zeroes at the end of a number after the decimal point are significant.

2705.0 m   five significant figures   (2, 7, 0, 5, 0)

4. In whole numbers without a decimal point that end in one or more zeroes (trailing zeroes)—for example, 500 kg—the zeroes may or may not be significant. In such cases, it is not clear which zeroes serve only to locate the decimial point and which are actually part of the measurement. That is, if the first zero from the left (5̲00 kg) was the estimated digit in the measurement, then only two digits are reliably known and there are only two significant figures. Similarly, if the last zero was the estimated digit (50̲0 kg), then there are three significant figures. This ambiguity may be removed by using scientific (powers-of-ten) notation:

5.0×10^2 kg   two significant figures

5.00×10^2 kg   three significant figures

This is helpful in expressing the results of calculations with the proper numbers of significant figures, as we shall see shortly.

Note: To avoid confusion with numbers having trailing zeroes used as given quantities in text examples and problems, we will consider the trailing zeroes to be significant. For example, a time of 20 s should be assumed to have two significant figures, even if it is not written out as 2.0×10^1 s.

• FIGURE 1.7 Significant figures and insignificant figures For the division operation of 5.3/1.67, a calculator with a floating decimal point gives many figures. A calculated quantity can be no more accurate than the least accurate quantity involved in the calculation, so this result should be rounded off to two significant figures, that is, 3.2.

It is important to report the results of mathematical operations with the proper number of significant figures. The following general rule tells how to determine the number of significant figures in the result of a multiplication or division:

The final result of a multiplication or division operation should have the same number of significant figures as the quantity with the least number of significant figures that was used in the calculation.

Multiplication and division with significant figures

What this means is that the result of a calculation can be no more accurate than the least accurate quantity used; that is, you cannot gain accuracy in performing mathematical operations. Thus, the result that should be reported for the division operation discussed at the beginning of this section is

$$\frac{5.3 \text{ m}}{1.67 \text{ m/s}} = 3.2 \text{ s}$$

The result is rounded off to two significant figures (see • Fig. 1.7).

For multiple operations, rounding off to the proper number of significant figures should not be done at each step since rounding errors may accumulate. It is usually suggested that one or two insignificant figures be carried along, or if a calculator is being used, rounding off may be done on the final result of the multiple calculations.

The rules for rounding off numbers are as follows:

Rounding off of results of intermediate steps of multiple operations will be done in some examples for instructional purposes

Rounding rules

1. If the next digit after the last significant figure is 5 or greater, the last significant figure is increased by 1.

2. If the next digit after the last significant figure is less than 5, the last significant figure is left unchanged.

EXAMPLE 1.5 ■ Using Significant Figures in Multiplication and Division

The following operations are performed and the results rounded off to the proper number of significant figures (sf).

Multiplication

$$2.4 \text{ m} \times 3.65 \text{ m} \ (= 8.76 \text{ m}^2) = 8.8 \text{ m}^2 \quad \text{(rounded to 2 sf)}$$
(2 sf)   (3 sf)

Division

(4 sf)

$$\frac{725.0 \text{ m}}{0.125 \text{ s}} \ (= 5800 \text{ m/s}) = 5.80 \times 10^3 \text{ m/s} \quad \text{(represented with 3 sf)}$$
(3 sf)

There is also a general rule for determining the number of significant figures for the results of addition and subtraction.

Addition and subtraction with significant figures

The final result of the addition or subtraction of numbers should have the same number of decimal places as the quantity with the least number of decimal places that was used in the calculation.

This rule is applied by rounding off numbers to the least number of decimal places before adding or subtracting. The rounding off ensures that a result will have no more reliability than the quantity read from the measuring instrument with the least fine scale. Another way of looking at this is that the rounding off is determined by the quantity with the first doubtful or estimated figure from the left. (Why?)

EXAMPLE 1.6 ■ Using Significant Figures in Addition and Subtraction

The following operations are performed by finding the entry that has the first doubtful figure (counting from the left) and rounding the other numbers to this column. All the digits in this rightmost column are then considered least significant. (Units have been omitted for convenience.)

Addition
In the numbers to be added, note that 23.1 has the first doubtful figure from the left, the 1.

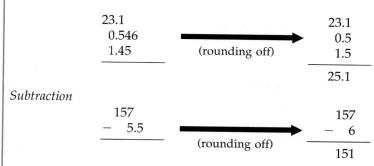

$$
\begin{array}{c}
23.1 \\
0.546 \\
\underline{1.45}
\end{array}
\quad \xrightarrow{\text{(rounding off)}} \quad
\begin{array}{c}
23.1 \\
0.5 \\
\underline{1.5} \\
25.1
\end{array}
$$

Subtraction

$$
\begin{array}{r}
157 \\
- \quad 5.5 \\
\hline
\end{array}
\quad \xrightarrow{\text{(rounding off)}} \quad
\begin{array}{r}
157 \\
- \quad 6 \\
\hline
151
\end{array}
$$

In this subtraction example, the first doubtful figure from the left is the 7 in the 157. The decimal point is understood in aligning the numbers.

Basically, the number of digits reported in a result depends on the number of digits in the given data. The rules for rounding off will generally be observed in the solutions given for the examples in this book. Rounding off results is a good practice to follow when you work problems. In general, the examples and problems will involve quantities with two or three significant figures.

Students should practice this method in the end-of-chapter exercises.

1.7 ■ Problem Solving

An important aspect of physics is problem solving. In general, this involves the application of physical principles and equations to data from a particular situation in order to find some unknown or wanted quantity. There is no universal method for approaching a problem that will automatically produce a solution. A few general points, though, are worth keeping in mind:

- *Make sure you understand the problem.* The foundation of successful problem solving is having a thorough grasp of the problem. All too often a student will start working on a problem without really understanding the situation described or knowing what is to be found. (Basically, if you don't know exactly what you want, it is very difficult to find it!)

- To get from the given data to the ultimate solution, you may have to devise a strategy or plan. Many problems cannot be solved merely by finding one equation and "plugging in" the given quantities to get the solution. Often you will need to perform intermediate steps, each of which will bring you closer to the final answer. The process is a lot like crossing a stream by stepping from stone to stone—you need to keep in mind where you are and where you're going, and as you plan each step you must think ahead to the next one.

- Remember that equations are expressions of physical principles. Solving problems involves translating principles into the "language" of equations, but you should also be able to look at an equation and say what physical relationship it embodies. If you use an equation without knowing what it means, you *may* be lucky and get the right answer, but you will not be learning much physics—and you probably won't be able to repeat your success when confronted with a slightly different problem.

- A common cause of failure in problem solving is the *misapplication* of physical principles or equations. Principles and equations are generally subject to certain conditions or are limited in their applicability to physical situations.

- Many problems can be solved by more than one method. As with many things in life, however, there is often a hard way and an easy way. If you understand the problem and the relevant principles well, you can probably figure out how to solve it in the fastest and easiest way.

Although there is no magic formula for problem solving, there are some sound practices that can be very useful. The steps in the following procedure are intended to provide you with a framework that can be applied to solving most of the problems that you will encounter in this text. We will generally use these steps in dealing with the Example problems throughout this text. Additional helpful problem-solving hints will be given where appropriate in subsequent chapters.

Suggested Problem-Solving Procedure

Say it in words

1. *Read the problem carefully and analyze it. Write down the given data and what it is you are to find.* Some data may not be given explicitly in numerical form. For example, if a car "starts from rest," its initial speed is zero ($v_o = 0$). In some instances, you may be expected to know certain quantities or to look them up in tables.

Say it in pictures

2. *Draw a diagram as an aid in visualizing and analyzing the physical situation of the problem where appropriate.* (This may not be necessary in every case.)

Say it in equations

3. *Determine what principle(s) and equation(s) are applicable to this situation, and how they can be used to get from the information given to what is to be found.* Keep in mind that many problems cannot be solved simply by plugging all of the given data into one equation; you may have to devise a strategy involving several steps.

Simplify the equations

4. *Simplify mathematical expressions as much as possible through algebraic manipulation before inserting actual values.* Trigonometric relationships (summarized in Appendix I) can also be used to simplify equations sometimes. The less calculation you do, the less likely you are to make a mistake—so *don't put the numbers in until you have to.*

Check the units

5. *Check units before doing calculations.* Make unit conversions if necessary so that all units are in the same system and quantities with the same dimensions have the same units (preferably standard units). This avoids mixed units and is helpful in unit analysis. (Unit checking and conversions are often done in Step 1 when writing the data.)

Insert numbers and calculate; check significant figures

6. *Substitute given quantities into equation(s) and perform calculations. Report the result with proper units and number of significant figures.*

7. *Consider whether the result is reasonable.* That is, does the answer have an appropriate magnitude? (This means is it "in the right ball park.") For example, if a person's calculated mass turns out to be 2.30×10^2 kg, the result should be questioned, since 230 kg corresponds to a weight of 506 lbs.

Figure 1.8 recapitulates these steps in the form of a flow chart. The following examples illustrate the procedure. The steps are stacked to help you follow along.

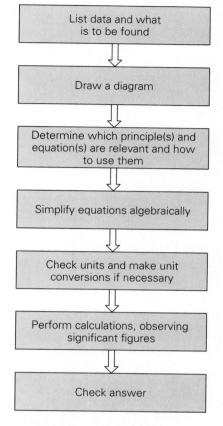

FIGURE 1.8 A flow chart for the suggested problem-solving procedure.

EXAMPLE 1.7 ■ Finding the Area of a Rectangle

Two students measure the lengths of adjacent sides of their dorm room. One reports 15 ft, 8 in., and the other reports 4.25 m. What is the area of the room in square meters?

Solution.

1. Adjacent sides of a room give its length and width, so we may write
   Given: Length = l = 15 ft, 8 in.   *Find:* Area (in m²)
   Width = w = 4.25 m

2. Sketch a diagram to help visualize the situation (Fig. 1.9).

3 and 4. For this simple situation, the required equation is well known. The area (A) of a rectangle is $A = l \times w$, both of which are given.

5. It is obvious that a unit change is necessary. Let's first convert the length measurement to inches and then inches to meters:

$$15 \text{ ft} + 8 \text{ in.} = \left(15 \text{ ft} \times \frac{12 \text{ in.}}{\text{ft}} \right) + 8 \text{ in.} = 188 \text{ in.}$$

Figure 1.8

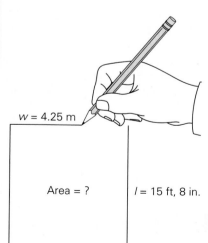

w = 4.25 m

Area = ?   *l* = 15 ft, 8 in.

FIGURE 1.9 A helpful step in problem solving   Drawing a diagram helps you to visualize and better understand the situation. See Example 1.7.

and

$$188 \text{ in.} \times \frac{2.54 \text{ cm}}{\text{in.}} = 478 \text{ cm} = 4.78 \text{ m}$$

Notice how easy it is to convert units in the decimal metric system (cm to m). You should perform the conversion explicitly if necessary, using the conversion factor (1 m/100 cm).

6. Now perform the calculation.

$$A = l \times w = 4.78 \text{ m} \times 4.25 \text{ m}$$
$$= 20.315 \text{ m}^2 = 20.3 \text{ m}^2 \quad \text{(computed value rounded to 3 sf)}$$

7. The answer appears reasonable. Since 1 m ≈ 3 ft, the dorm room would be about 13 ft by 14 ft, which is about right (but, as always, too small). Suppose you had inadvertently punched 47.8 instead of 4.78 on your calculator. The result would be $A = 47.8 \text{ m} \times 4.25 \text{ m} = 203 \text{ m}^2$. A room with an area of about 200 m² would have dimensions of about 10 m by 20 m, which is roughly 30 ft by 60 ft. Since this is not the size of a typical dorm room, the magnitude of the result should make you suspect there may be an error.

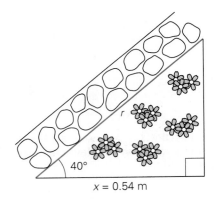

● FIGURE 1.10 The length of a triangle's side See Example 1.8.

This problem gives the instructor a great opportunity to review some trigonometry. Also, doing it graphically provides another opportunity to discuss significant figures.

EXAMPLE 1.8 ■ Finding the Length of One Side of a Triangle

A flower bed is laid out in the form of a triangle as shown in ● Fig. 1.10. What is the length of the side of the bed that runs along the walkway?

Solution.

1 and 2. In some problems we are given diagrams with data. Here we have
   Given: $x = 0.54$ m   *Find:* r (long side of triangle)
   $\theta = 40°$

3 and 4. Noting in the figure that the flower bed is a right triangle (as indicated by the right angle symbol at the large angle), we can use a trigonometric function to find the hypotenuse r. (For reference and review, important trigonometric relationships are summarized in Appendix I.) Recalling that x and r are associated with the cosine for a standard right triangle, we may write

$$\cos \theta = \frac{x}{r} \quad \text{or} \quad r = \frac{x}{\cos \theta}$$

5 and 6. The units are OK. (Since x is in meters, r will come out in meters.) So put in the data:

$$r = \frac{x}{\cos \theta} = \frac{0.54 \text{ m}}{\cos 40°} = \frac{0.54 \text{ m}}{0.766} = 0.70 \text{ m}$$

7. The magnitude of r seems reasonable compared to the value of x.

It is understood that you do not need to explicitly write out the steps of problem-solving procedure each time. However, it is a good practice to run through them mentally when working a problem. In the examples in the following chapters, the problem-solving steps will not always be listed as was done here for illustration. Even so, you should be able to see the general pattern outlined above. In some instances, especially when introducing the application of a new principle or concept, problem-solving steps or hints will be listed to promote better understanding.

The main point of this section is that in solving problems you should have some systematic procedure for analyzing the situation and extracting the information wanted. The suggested problem-solving procedure given here is one option. Let's review it again.

EXAMPLE 1.9 ■ Finding the Capacity of a Cylinder

A cylindrical container with a height of 28.5 cm and an inside diameter of 10.4 cm is filled with water. What is the mass of the water in kilograms?

Solution.

1. *Read the problem carefully and analyze it.*
   From the problem, we have
   Given: $h = 28.5$ cm (height)   **Find:** m (mass in kg)
          $d = 10.4$ cm (diameter)

2. *Draw a diagram as an aid in visualizing and analyzing the situation.*
   Make a sketch of the cylinder showing its dimensions, and adding any general information you might know about a cylinder. For example, the circular base has an area of $A = \pi r^2$, where r is the radius and $r = d/2$. (See ● Fig. 1.11.)

3. *Determine what principle(s) and equation(s) are applicable and how to use them, and* 4. *Simplify expressions algebraically if possible.*
   We are given the height and diameter of the cylinder, so we can find its volume, which is the volume of the water it contains. But how is volume related to mass, which is what we want? The link is density (ρ), which we saw earlier (Eq. 1.1) is mass per unit volume:

$$\rho = \frac{m}{V} \quad \text{or} \quad m = \rho V$$

The density of water is not given. However, this can be looked up in a table, or you may remember from the definition of the kilogram that $\rho_{H_2O} = 1.00$ g/cm³ (at maximum density, which will be assumed here). So, knowing the volume and the density of water, we can find the mass.

Should you not remember the formula for the volume of a cylinder, your sketch can be helpful. The area of the circular end of the cylinder is $A = \pi r^2 = \pi d^2/4$ (since $r = d/2$), and the volume is this area times the height of the cylinder, or $V = Ah$. (Formulas for areas and volumes of common shapes can be found in Table 5 of Appendix I.)

● **FIGURE 1.11 The capacity of a cylinder** Drawing a diagram helps in problem solving. See Example 1.9.

4. *We can therefore write an equation for the mass entirely in terms of known quantities:*

$$m = \rho V = \rho A h = \rho \left(\frac{\pi d^2}{4} \right) h$$

5. *Check units before doing calculations.*
   The cylinder dimensions are in cm and the density in g/cm^3. Hence, the mass will come out in grams, which we can convert to kg as requested in the problem. (Another option would be to convert to meter and kilogram units first, but this would require more steps.) *Note: If the given data have mixed units, it is a good practice to convert to consistent units when writing down what is given. This avoids mistakes that might be made later if this step were overlooked.*

6. *Substitute the given quantities into the equation and perform calculations, observing significant figures.*
   Using the equation derived above and inserting the known quantities, we have

$$m = \rho V = \rho A h = \rho \left(\frac{\pi d^2}{4} \right) h$$

$$= \left(\frac{1.00 \text{ g}}{cm^3} \right) \left[\frac{\pi (10.4 \text{ cm})^2}{4} \right] (28.5 \text{ cm}) = 2.42 \times 10^3 \text{ g}$$

$$= 2.42 \text{ kg}$$

7. *Consider whether the result is reasonable.*
   It may be difficult sometimes to determine whether or not a result is reasonable because of unfamiliarity with units. Here, one might calculate that about 2.5 kg of water has a weight of about 5.0 lb (since 1 kg = 2.2 lb or $\simeq$ 2.0 lb). This seems to be a reasonable weight for this volume of water. (The dimensions are approximately those of a 2-L soda bottle.)

Important Concepts

You should be able to define and explain these chapter concepts clearly.

standard unit	second (s)	unit analysis
system of units	mks system	density (ρ)
International System of Units (SI)	cgs system	conversion factor
SI base units	fps system	exact number
SI derived units	liter (L)	measured number
meter (m)	dimensional analysis	significant figures (sf)
kilogram (kg)		

Important Relationships for Review

These relationships are mathematical statements of the concepts and principles presented in the chapter. You should be able to identify the symbols and to explain the relationships before proceeding to the Exercises. In-text equation reference numbers are given for convenience.

$$\text{Density: } \rho = \frac{m}{V} \quad \left(\frac{\text{mass}}{\text{volume}}\right) \tag{1.1}$$

Steps in Suggested Problem-Solving Procedure

1. Read the problem and analyze it carefully. Write down the given data and what it is you are to find.

2. Draw a diagram as an aid in visualizing and analyzing the physical situation of the problem (where appropriate).

3. Decide what principle(s) and equation(s) are applicable to this situation, and how they can be used to arrive at the solution.

4. Simplify mathematical expressions using algebra and trigonometric relationships.

5. Check units before making calculations. Make unit conversions if necessary.

6. Substitute given quantities into equation(s) and perform calculations, reporting the result with the proper number of significant figures.

7. Consider whether the result is reasonable, that is, whether it has an appropriate magnitude.

Exercises

Throughout the text, many exercise sections will include "paired" exercises. These exercise pairs, identified with blue numbers, are intended to assist you in problem solving and learning. In such a pair, the first exercise is worked out in the Study Guide so that you can consult it should you need assistance in its solution. The second exercise is similar in nature and its answer is given at the back of the book. Other exercises marked with a blue dot are also solved in the Study Guide.

1.2 ▪ Units

1 The SI has how many base units: (a) 3, (b) 5, (c) 7, (d) 9? (c)

2 Which of the following is *not* an SI base unit: (a) length, (b) weight, (c) time, (d) mass? (b)

3 What makes a quantity a fundamental or base quantity? see ISM

4 In the British system, 16 oz = 1 pt and 16 oz = 1 lb. Is there something wrong here? Explain. ounce used for both volume and weight measurements

5 The metric system is a decimal (base-10) system and the British system is in part a duodecimal (base-12) system. Discuss the ramifications if our monetary system had a duodecimal base. What would be the possible values of our coins if this were the case? see ISM

1.4 ▪ Dimensional Analysis*

● **6** Both sides of an equation are equal in (a) magnitude, (b) units, (c) dimensions, (d) all of the preceding. (d)

7 If you apply unit analysis to an equation to learn the units of a particular quantity and the result turns out to be m^2/m^2, then the quantity would (a) have units of m^2, (b) have units of length, (c) have units of m^4, (d) be unitless. (d)

8 Can dimensional analysis tell you whether you have used the correct equation in solving a problem? Explain. no; only tells if the equation is dimensionally correct

9 Must an equation be correct both by dimensional analysis and unit analysis? Explain. yes (see ISM)

10 Is the unit equation kg-cm-s/g-min² = m/s dimensionally correct? If so, explain how this can be with so many different units. yes; dimensions cancel, but not correct by unit analysis

11 ▪ Show that the equation $x = x_0 + vt$, where v is velocity and x and x_0 are lengths, is dimensionally correct. [L] = [L] + [L]

● **12** ▪ Use SI unit analysis to show that the equation $A = 4\pi r^2$, where A is the area and r is the radius of a sphere, is dimensionally correct. $[L]^2 = [L]^2$

*Dimensions and/or units of velocity and acceleration are given in the chapter.

13 ■ Is the equation $V = \pi d^2/4$, where V is the volume and d is the diameter of a sphere, dimensionally correct? (Use SI unit analysis to find out.) If not, how could it be made so, assuming that the left side is correct? no; $[L]^3 \neq [L]^2$; $V \propto d^3$

14 ■ Is the equation $t = 2x/v$, where t is time, x is length, and v is velocity, dimensionally correct? Show by using unit analysis. yes; s = s

15 ■■ Show that the equation $x = x_0 + v_0 t + \frac{1}{2}at^2$ is dimensionally correct (a is acceleration, v_0 is velocity, and x and x_0 are lengths). $[L] = [L] + [L] + [L]$

●**16** ■■ If $x = gt^2/2$, where x is length and t is time, is dimensionally correct, what are the SI units of the constant g? Is g a particular derived quantity? If so, identify it. (m/s²); acceleration due to gravity

17 ■■ Is the equation $v = v_0 \sin\theta - gt$ dimensionally correct? Show by using SI unit analysis. (v and v_0 are velocities, θ is an angle, t is time, and g is the same as in Exercise 16.) yes; m/s = m/s − m/s

18 ■■ Using unit analysis, determine whether the following equations are dimensionally correct (t is time, x is length, a is acceleration, and v is velocity): (a) $t = \sqrt{2x/a}$, and (b) $v = \sqrt{2ax}$. (See Section 1.4 for units.) (a) yes; s = s (b) yes; m/s = m/s

19 ■■ Is the equation for the area of a trapezoid, $A = \frac{1}{2}a(b_1 + b_2)$, where a is the altitude and b_1 and b_2 are the bases, dimensionally correct? (See ●Fig. 1.12.) If not, how should it be changed to correct it? yes; $[L]^2 = [L]^2 + [L]^2$

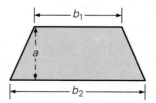

●**FIGURE 1.12 The area of a trapezoid** See Exercise 19.

20 ■■ Using unit analysis, determine if the following equation is dimensionally correct: $v = \sqrt{2ax} - v_0 t$, where a is acceleration, v and v_0 are velocities, and x is length. no; see ISM

21 ■■ The formulas for a straight line, a parabola, and a hyperbola are commonly given as (a) $y = mx$, (b) $y = bx^2$, and (c) $y = c/x$, respectively, where m, b, and c are constants. What are the SI units of each of the constants if x and y are lengths? (a) dimensionless (b) 1/m (c) m²

●**22** ■■ In the equation $mg = kx$, the g is acceleration and k is a constant called the spring constant. What are the SI units of the spring constant? kg/s²

23 ■■ The equation for the distance traveled by an object is given by $x = v_0 t + bt^2$, where v_0 is velocity, t is time, and b is a constant. What are the SI units of b? m/s²

24 ■■ (a) Show that the equation $A = 2(\pi r^2) + 2\pi rL$, where r is a radius and L is a length, is dimensionally correct. (b) Prove that this formula gives the total area of a solid circular cylinder. (a) m² = m² + m² (b) see ISM

25 ■■■ Newton's second law of motion is expressed by the equation $F = ma$, where F represents force, m is mass, and a is acceleration. (a) The SI unit of force is, appropriately, the newton (N). What are the units of the newton in terms of base quantities? (b) Using the result of part (a), show that the equation $Ft = mv$, where v is velocity and t is time, is dimensionally correct by using unit analysis. (a) kg-m/s² (b) see ISM

26 ■■■ Newton's law of gravitation is expressed by the equation $F = Gm_1m_2/r^2$, where F is force, G is a constant (called the universal gravitational constant), m_1 and m_2 are masses, and r is a distance. (a) What are the SI units of the constant G when the force is in newtons? (b) As given in Exercise 25, $F = ma$, where m is mass and a is acceleration. What are the units of G in SI base quantities? (a) N-m²/kg² (b) m³/kg-s²

1.5 ■ Unit Conversions*

27 A conversion factor written as 1 in. = 2.54 cm means that (a) 1 in. is equivalent to 2.54 cm, (b) this is a true equation, (c) 1 cm = 2.54 in. (d) none of the preceding. (a)

28 A good way to ensure proper unit conversion is to (a) use another measurement instrument, (b) always work in one system of units, (c) use unit cancellation, (d) use dimensional analysis. (c)

29 The width of the nail of your little finger is about the size of a common metric unit. Which one? cm

30 If you wanted to express your height as a larger number, which unit in each of the following pairs would you use: (a) meter or yard; (b) decimeter or foot; (c) centimeter or inch? (a) yd (b) dm (c) cm

31 ■ How many meters are there in (a) 30 ft and (b) 5280 ft? (a) 9.1 m (b) 1.61 × 10³ m

32 ■ Do the following conversions: (a) 25 m to feet, (b) 12 in. to centimeters, and (c) 14 days to seconds. (a) 82 ft (b) 30 cm (c) 1.2 × 10⁶ s

33 ■ (a) What is the mass in kilograms of a 154-lb person? (b) What is your mass in kilograms? (a) 70 kg (b) answers will vary

●**34** ■ Which is longer and by how many centimeters, a 100-m dash or a 100-yd dash? 100 m is longer by 856 cm

35 ■■ In converting to the metric system, a road sign has been only partially changed. One town is indicated to be 60 km away, and another, 50 mi away. Which town is farther, and by how many meters? 50 mi by 2.1 × 10⁴ m

*Conversion factors are listed inside the front cover and in Table 2 of Appendix I.

36 ■■ An automobile speedometer is shown in ● Fig. 1.13. (a) What would the equivalent scale readings be in km/h? (b) What would a 55 mi/h speed limit be in km/h?
(a) 16 km/h for each 10 mi/h   (b) 89 km/h

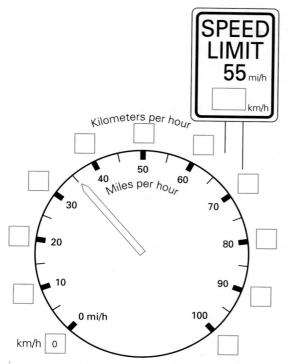

FIGURE 1.13 Speedometer readings   See Exercise 36.

37 ■■ Suppose that a 16-oz soda and a 500-mL soda sell for the same price. Which would you choose to get the most for your money and how much more (in milliliters) would you get? [*Hint:* 1 pint = 16 oz.]   27 mL more in 500 mL

38 ■■ A professor regularly buys 12 gal of gas, but the gas station has installed new pumps that deliver liters. How many liters of gas (rounded off to a whole number) should she ask for?   45 L

● **39** ■■ An automobile owner checks his car's gas consumption and finds that 30.0 gal of gasoline were used to travel 750 mi. How many miles per gallon (mpg) does the car average? What is this in km/L? In m/mL?   25.0 mi/gal; 10.6 km/L;   10.6 m/mL

40 ■■ Some common product labels are shown in ● Fig. 1.14. From the units given on the labels, form a set of conversion factors.   (a) 1 fl. oz. = 29.7 mL   (b) 1 oz = 28.3 g

41 ■■ A particular student was 20 in. long when she was born. She is now 5 ft 4 in. and 18 years old. How many centimeters a year did she grow on the average?   6.2 cm/y

42 ■■ A basketball team from the United States has a center who is 6 ft 9 in. tall and weighs 200 lb. If the team plays

● **FIGURE 1.14 Conversion factors**   See Exercise 40.

exhibition games in Europe, what values will be listed for the center's height and mass on programs for fans there?   2.06 m; 91 kg

43 ■ A European contractor builds a house from a blueprint with a scale of 0.50 cm : 1.00 m ($\frac{1}{2}$ cm to the meter). The floor plan of a room on the blueprint is 3.3 cm × 4.1 cm. What is the actual area of the floor in (a) square meters and (b) square yards?   (a) 54 m^2 (b) 65 yd^2

● **44** ■■ (a) A football field is 300 ft long and 160 ft wide. What are the field's dimensions in meters and its area in square centimeters? (b) A football is 11 to 11$\frac{1}{4}$ in. long. What is its length in centimeters?   (a) 91.4 m by 48.8 m; 4.46 × 10^7 cm^2 (b) 28.0 cm to 28.6 cm

45 ■■ Suppose that when the United States goes completely metric, the dimensions of a football field are established as 100 m by 54 m. What would be the difference between the area of the metric football field and that of a current football field?   9.4 × 10^2 m^2

46 ■■ The width and length of a room are 3.2 yd and 4.0 yd. If the height of the room is 8.0 ft, what is the volume of the room in (a) cubic meters and (b) cubic feet?
(a) 2.6 × 10^1 m^3 (b) 9.2 × 10^2 ft^3

47 ■■ In the Bible, Noah is instructed to build an ark 300 cubits long, 50.0 cubits wide, and 30.0 cubits high (see ● Fig. 1.15). A cubit was a unit of length based on the length of the forearm and equal to half of a yard. (a) What would the dimensions of the ark be in meters? (b) What would its volume be in cubic meters? Assume that the ark was rectangular.   (137 m)(22.9 m)(13.7 m) = 4.30 × 10^4 m^3

48 ■■■ The metal mercury has a density of 13.6 g/cm^3. How much mercury would be required to fill a 0.355-L container? Compare this to the mass of the same volume of a soft drink. (Use the density of water.)   4.83 × 10^3 g; mass of mercury is 13.6 times greater than that of water

49 ■■■ Engineers often express the density of a substance as a weight density, for example, in lb/ft^3. (a) What is the weight density of water? (b) What is the weight of a gallon of water?   (a) 62 lb/ft^3 (b) 8.3 lb

FIGURE 1.15 Noah and his ark See Exercise 47.

1.6 ■ Significant Figures

50 Significant figures are used in (a) making an estimate of the uncertainty involved in measurements, (b) expressing exact numbers, (c) expressing completely reliable figures, (d) identifying computation errors. (a)

51 Which of the following has the greatest number of significant figures: (a) 0.254 cm, (b) 0.00254 × 10^2 cm, (c) 254 × 10^{-3} cm, (d) all are the same? (d)

52 If a measured length is reported as 25.483 cm, could this length have been measured with an ordinary meterstick whose smallest division is millimeters? Discuss in terms of significant figures. no; only one doubtful digit; best 25.48 cm

● **53** ■ Determine the number of significant figures in the following measured numbers: (a) 1.007 m, (b) 8.03 cm, (c) 16.272 kg, (d) 0.015 μs (microseconds). (a) 4 (b) 3 (c) 5 (d) 2

54 ■ Express each of the numbers in Exercise 53 with two significant figures. (a) 1.0 m (b) 8.0 cm (c) 16 kg (d) 1.5 × 10^{-2} μs

55 Which of the following quantities has three significant figures: (a) 305.0 cm, (b) 0.0500 mm, (c) 1.00081 kg, (d) 8.06 × 10^4 m^2? (b) and (d); (a) has four and (c) has six

56 ■ Express each of the following numbers with three significant figures: (a) 10.072 m, (b) 775.4 km, (c) 0.002549 kg, (d) 93,000,000 mi. (a) 10.1 m (b) 775 km (c) 2.55 × 10^{-3} kg (d) 9.30 × 10^7 mi

57 ■■ Express the length 50,570 μm (micrometers) in centimeters, decimeters, and meters to three significant figures. 5.06 cm; 5.06 × 10^{-1} dm; 5.06 × 10^{-2} m

● **58** ■■ A circular flower bed has a radius of 4.25 m. Compute its (a) circumference and (b) area. (a) 26.7 m (b) 56.7 m^2

59 ■■ A triangle has a base with a length of 10.5 cm and an altitude of 8.7 cm (Fig. 1.16). What is the area of the triangle? 46 cm^2

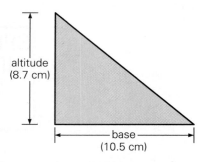

FIGURE 1.16 The area of a triangle See Exercise 59.

60 ■■ The parallel bases of a trapezoid are 7.75 cm and 4.6 cm, respectively, and the altitude is 3.25 cm. What is the area of the trapezoid? [*Hint:* see Exercise 19.] 20 cm^2

61 ■■ A spherical communications satellite has a diameter of 1.35 m. How much space does the satellite occupy? 1.29 m^3

62 ■■ The outside dimensions of a cylindrical soda can are reported as 12.559 cm for the diameter and 5.62 cm for the height. What is the total outside area of the can? 470 cm^2

63 ■■ Assume that the radius of the Earth is 4.00 × 10^3 mi and that it is a perfect sphere. How many cubic meters of space does the Earth occupy? 1.12 × 10^{21} m^3

64. ■■■ Work the following exercise by two procedures as directed and comment on and explain any difference in the answers. Use your calculator for the calculations. Compute $p = mv$, where $v = x/t$. (a) First compute v, and then p. (b) Compute $p = mx/t$ without an intermediate step. Given: $x = 8.5$ m, $t = 2.7$ s, and $m = 0.66$ kg. (a) 2.0 kg-m/s (b) 2.1 kg-m/s, rounding difference

1.7 ■ Problem Solving

65 The first suggested step in problem solving is (a) write down what is given, (b) write down what is to be found, (c) read and analyze the problem, (d) decide on a consistent set of units to use. (c)

66 An important step in problem solving before actually mathematically solving an equation is (a) checking units, (b) checking significant figures, (c) consulting with a friend, (d) checking to see if the result is reasonable. (a)

67 Explain how you might determine whether a calculated result is reasonable. compare to typical, familiar value

● **68** ■ The mass of the Earth is 6.0 × 10^{24} kg, and it has a volume of 1.1 × 10^{21} m^3. What is the Earth's average density? 5.5 × 10^3 kg/m^3

69 ■ The metal mercury has a density of 13.6 g/cm^3. What is the mass of a liter of mercury? 1.36 × 10^4 g

70 ■ Determine whether each of the following reported measurements has an appropriate magnitude. (a) A subcompact car has a mass of 2050 kg. (b) A person is 183 cm tall. (c) A runner runs a mile in 3.00×10^8 μs. (d) A normal highway driving speed is 25 m/s. (e) A dachshund has a tail 280 mm long and stands 5.0 dm off the ground. (f) A professor has a mass of 98 kg and a 50-cm waist. (a) no; 4.5×10^3 lb (b) yes; 6.0 ft (c) yes; 5.0 min (d) yes; 90 km/h (e) no; 11 in. and 20 in. (f) no; 216 lb and 20 in.

71 ■■ A car travels with a constant speed of 65 km/h. How many kilometers will it travel in 4 h, 45 min? 3.1×10^2 km

72 ■■ The thickness of a textbook, not including the covers, is measured as 3.75 cm. If the last page of the book is numbered 859, what is the average thickness of a page? 8.72×10^{-3} cm/page

73 ■■ A light year is a unit of distance corresponding to the distance light can travel in a vacuum in 1 year. If the speed of light is 3.0×10^8 m/s, what is the length of a light year in kilometers? 9.5×10^{12} km

● **74** ■■ Three pounds of butter cost $6.50. Some European visitors want to know the price per kilogram. What would this be? $4.78/kg

75 ■■ Joe's Pizza Palace sells a 9.0-in. (diameter) pizza for $7.95 and a 12-in. pizza for $13.50. Which is the better buy? larger is the better buy

76 ■■■ A flat porch roof measuring 5.0 m by 9.0 m collapsed when snow had accumulated on it to a depth of 44 cm. If the average density of the snow was 15 kg/m³, what was the weight in pounds of the snow on the roof when it collapsed? 6.6×10^2 lb

77 ■■■ In 1986, the experimental aircraft *Voyager* flew around the world without refueling. It used approximately 1.2×10^3 gallons of fuel to travel 2.5×10^4 miles in 9.1 days. (a) What was *Voyager's* average mileage rating in mi/gal and km/L? (b) What was its average fuel consumption in gal/h and L/h? (c) What was its average speed in mi/h and km/h? (a) 21 mi/gal and 8.9 km/L (b) 5.5 gal/h and 21 L/h (c) 1.1×10^2 mi/h and 1.8×10^2 km/h

■ **Additional Exercises**

78 There seem to be some inconsistencies in the labeling of the products shown in ● Fig. 1.17. What is the problem and which are correct? [*Hint:* 1 lb = 0.4535924 kg, and 1 L = 1.057618 qt.] 16 oz = 454 g; 12 fl. oz = 355 mL

79 The general equation for a parabola is $y = ax^2 + bx + c$, where a, b, and c are constants. What are the units of each constant if y and x are in meters? (a) 1/m (b) dimensionless (c) m

80 A rectangular block has the dimensions 4.85 cm, 6.5 cm, and 15.51 cm. What is the volume of the block in cubic centimeters? 4.9×10^2 cm³

81 Which is greater and by how much, a metric ton or a U.S. ton (2000 lb)? metric ton by 200 lb or 91 kg

82 Using unit analysis, show that the equation $t =$

● **FIGURE 1.17 An extra gram and milliliter?** See Exercise 78.

$x/(v_o + at/2)$ (where v_o is velocity, a is acceleration, x is length, and t is time) is dimensionally correct. s = s

● **83** A solid sphere has a radius of 12 cm. What is its surface area in (a) square centimeters and (b) square meters? (c) If it has a mass of 4.0 kg, what is its density in kg/m³? (a) 1.8×10^3 cm² (b) 0.18 m² (c) 5.6×10^2 kg/m³

84 A fathom is a nautical measure of length or depth and is equal to 6 ft. A furlong is a unit of distance used in horse-racing and is equal to $\frac{1}{8}$ mi. What are the values of each of these distances expressed in meters using a metric prefix that allows the numerical part of the result to be as close as possible to 1? 1 fathom = 0.1829 dam; 1 furlong = 0.201 km

85 The top of a rectangular table measures 1.245 m by 0.760 m. (a) What is the smallest division on the scale of the measurement instrument? (b) What is the area of the tabletop? (a) cm (b) 0.946 m²

86 A cylindrical drinking glass has an inside diameter of 8.0 cm and a depth of 12 cm. If a person drinks a completely full glass of water, how much (in liters) will be consumed? 0.60 L

87 When computing the average speed of a cross-country runner, a student gets 25 m/s. Is this a reasonable result? Justify your answer. not reasonable (56 mi/h)

88 Using the conversion factor 1 in. = 2.54 cm and other necessary factors, convert 1.05×10^2 mi/h to m/s. 46.9 m/s

89 The average density of the moon is 3.3 g/cm³, and it has a diameter of 2160 mi. What is the total mass of the moon? 7.3×10^{22} kg

90 Is the equation $v^2 = (4x^2/t^2) - v_o^2$, where v and v_o are velocity, x is length, and t is time, dimensionally correct? yes; $[L/T]^2 = [L/T]^2$

91 A thick-walled spherical metal shell has an inner diameter of 18.5 cm and an outer diameter of 24.6 cm. What is the volume occupied by the shell itself? 4.48×10^3 cm³

92 Which holds more soda and how much more, a half-gallon bottle or a two-liter bottle? 2 L by 0.11 L

2

INSIGHT

- Galileo Galilei and the Leaning Tower of Pisa

Kinematics: The Description of Motion

The study of motion is most important to the new physics student. Students generally encounter their first major difficulties with problem solving in this chapter.

The description of motion involves the representation of a restless world. The camera can seem to "freeze" an instant in time, but we know that in reality nothing is ever perfectly still. The cyclist in the photo is obviously in motion—but so, less obviously, are you, reading this book. You sit, apparently at rest, but your blood flows, and air moves into and out of your lungs. The air is composed of gas molecules moving at different speeds and in different directions. And, while you experience stillness, you, your chair, the building, and the air you breathe are all revolving through space with the Earth, part of a solar system in a spiraling galaxy in an expanding universe.

*The branch of physics concerned with the study of motion and what produces and affects it is called **mechanics**. The roots of mechanics and of human interest in motion go back to early civilizations. The study of the motions of heavenly bodies, or celestial mechanics, grew out of the need to measure time and location. Several early Greek scientists, notably Aristotle, put forth theories of motion that were useful descriptions but were later proved to be inaccurate. Currently accepted concepts of motion were formulated in large part by Galileo (1564–1642) and Isaac Newton (1642–1727).*

*Mechanics is usually divided into two parts: kinematics and dynamics. **Kinematics** deals with the description of the motion of objects without consideration of what causes the motion. **Dynamics** analyzes the causes of motion. This chapter covers kinematics and reduces the description of motion to its simplest terms by considering the simple case of motion in a straight line, or linear motion, which is motion in one dimension (of space). Chapter 3 focuses on motion in two dimensions (which can easily be extended to three dimensions). Chapter 4 investigates dynamics to show what causes motion.*

2.1 ■ A Change of Position

What is motion? This seems a simple question, but you might have some difficulty giving an immediate answer. After a little thought, you should conclude that **motion** is the changing of position. Motion may be described in part by specifying *how far* something travels in changing position—that is, the distance it travels. **Distance** is simply the *total path length* traversed in moving from one location or point to another. For example, you may drive to school from your hometown and express the distance traveled in miles or kilometers. In general, the distance between two points depends on the path traveled (● Fig. 2.1).

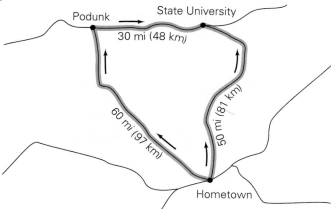

Vectors and Scalars

Distance is a *scalar* quantity. It has a magnitude, or numerical value, only (plus the units in which it is expressed—e.g., 25 km). Physics also uses *vector* quantities. These have magnitude *plus* direction.

The concept of vector quantities may seem strange at first. We generally use scalar quantities in our everyday expressions and calculations. But the distinction between scalars and vectors is quite simple, and vectors are not as unusual as they may appear.

A **scalar quantity** is one with magnitude (numerical value) only, along with units. That is, a "scalar" (short for scalar quantity) can be described by a number (along with units) that gives its magnitude or size. Measurements or quantities such as 25 km, 3.0 kg, and 20°C are examples of scalars. A **vector quantity** is one with both magnitude and *direction*. For example, when we express a distance traveled, say in an airplane, as 25 km north (magnitude and direction), we are giving a vector description. Other vector quantities include force, and as we shall shortly discover, velocity and acceleration.

Graphically, vectors are represented as arrows. Quite logically, the direction of the arrow gives the direction of the vector (● Fig. 2.2). The length of the arrow may be drawn to scale and made proportional to the magnitude or numerical value of the vector. Note in the figure how one vector is drawn to be twice as long as the other.

Because of the property of direction, vectors are added and subtracted differently from scalars. Vector addition and subtraction will be considered in more detail in Chapter 3. Until then, we will generally limit our discussion of vectors to the description of motion in one dimension—that is, in a straight line.

● **FIGURE 2.1 Distance—total path length** In driving to school, one student may take the shortest route and travel a distance of 50 mi. Another student takes a longer route in order to visit a friend in Podunk before returning to school. The trip is in two segments; the distance traveled is the total path length, 60 mi + 30 mi = 90 mi.

Emphasize through this and future chapters which quantities are scalars and which quantities are vectors.

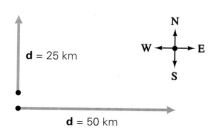

● **FIGURE 2.2 Vector representation** A vector quantity is represented graphically by an arrow. The vector direction is indicated by the arrow direction, and the vector magnitude is proportional to the length of the arrow.

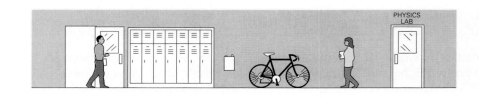

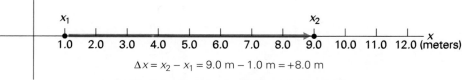

(a) Distance (magnitude or numerical value)

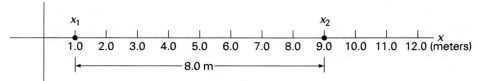

$\Delta x = x_2 - x_1 = 9.0\ \text{m} - 1.0\ \text{m} = +8.0\ \text{m}$

(b) Displacement (magnitude and direction)

• **FIGURE 2.3 Distance (scalar) and displacement (vector)**
(a) The distance (straight-line path length) between x_1 and x_2 is 8.0 m and is a scalar quantity. (b) The displacement of the person going toward the physics lab between these two points is +8.0 m (in the positive x direction) and is a vector quantity. (c) The displacement of the person going away from the physics lab between these two points is −8.0 m (in the negative x direction) and is a vector. Note how the sign gives the direction of the motion.

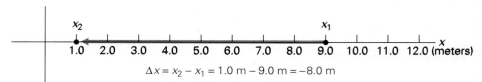

$\Delta x = x_2 - x_1 = 1.0\ \text{m} - 9.0\ \text{m} = -8.0\ \text{m}$

(c) Displacement (magnitude and direction)

Displacement

For straight-line, or linear, motion, it is convenient to specify position using the familiar Cartesian coordinate system with x and y axes at right angles. The straight-line path may be in any direction, but for simplicity we usually choose to orient the axes so that the motion is along one of them. We define **displacement** as the straight-line distance between two points, together with the direction. Hence, displacement is a vector quantity with both magnitude and direction. A linear displacement, say along the x axis, is given by

$$\Delta x = x_2 - x_1 \tag{2.1}$$

where x_1 and x_2 are the initial and final positions, respectively. The symbol Δ (delta) means difference, or change in. Here Δx indicates a change in position along the x axis. In this form, the direction of the displacement is given by the sign (+ or −) associated with Δx.

Suppose a person moves in a straight line from x_1 ro x_2, toward the physics lab, as shown in ● Fig. 2.3. With $x_1 = 1.0$ m and $x_2 = 9.0$ m, the displacement is

$$\Delta x = x_2 - x_1$$

$$= 9.0\ \text{m} - 1.0\ \text{m} = +8.0\ \text{m}$$

The person's displacement (magnitude and direction) is then 8.0 m in the

Displacement: How far, and in what direction

positive x direction, as indicated by the positive result. (The plus sign is often omitted, being taken as understood.) This displacement is illustrated with a vector arrow in Fig. 2.3b.

Suppose the other person in the figure moves in the opposite direction, from $x_1 = 9.0$ m to $x_2 = 1.0$ m. In this case, the displacement is

$$\Delta x = x_2 - x_1$$
$$= 1.0 \text{ m} - 9.0 \text{ m} = -8.0 \text{ m}$$

The minus sign indicates that the direction of the displacement is in the negative x direction, and a representative vector arrow would point this way (Fig. 2.3c).

Notice how vectors give more information than scalars. Both persons walked the same distance (8.0 m), but the displacements, with directional signs, tell you the direction of the motion of each, and hence that they walked in opposite directions.

A common designation for a vector quantity is a boldface symbol, e.g., $\Delta \mathbf{x}$, or for a general vector, $\mathbf{A}$. The magnitude *and* direction of the vector must be expressed explicitly—for example, with numerical magnitudes and directional signs, as we have done for linear displacements. In our discussion of straight-line motion we will continue to use this magnitude-sign notation to describe vector quantities wherever appropriate.

2.2 ■ Speed and Velocity

When something is in motion, its position changes with time. That is, it moves a certain distance in a given time. Both length and time are therefore important considerations in describing motion. For example, imagine a car and a pedestrian moving down a street and traveling a distance (length) of one block. In general, the car travels faster, or covers the distance in a shorter time, than the person does. This can be expressed by using length and time to give the *time rate of change in position* or **speed** for each.

Average speed is the distance traveled divided by the total time elapsed in traveling that distance.

Speed: How fast

Average speed

$$\text{average speed} = \frac{\text{distance traveled}}{\text{total time to travel that distance}}$$

$$\text{or} \qquad \bar{s} = \frac{\Delta d}{\Delta t} \qquad\qquad (2.2)$$

where the symbol d is used for distance, the actual path length. (A bar over a symbol, as $\bar{s}$ here, is a common way of indicating an average value.) The SI standard unit for speed is m/s (length/time), although km/h is used in many everyday applications. The British standard unit is ft/s, but we often use mi/h.

Since distance is a scalar quantity (as is time), speed is also a scalar. The distance does not have to be in a straight line (see Fig. 2.1). For example, you probably have computed the average speed for an automobile trip using the distance obtained from the starting and ending odometer readings. Suppose these readings were 17,455 km and 17,775 km, respectively, for a 4-hour trip.

Instantaneous speed: How fast right now

Velocity: How fast and in what direction

Average velocity

●**FIGURE 2.4 Instantaneous speed** The speedometer of a car gives the speed over a very short interval of time, so its reading approaches the instantaneous speed. Add the direction of the car, and you have a good approximation of the instantaneous velocity.

(We'll assume that you have a foreign car with odometer readings in kilometers.) Subtracting the readings gives a traveled distance of 320 km, so the average speed for the trip is $\bar{s} = \Delta d/\Delta t = 320 \text{ km}/4.0 \text{ h} = 80 \text{ km/h}$ (or 50 mi/h).

Average speed gives a general description of motion over a time interval Δt. In the case of the auto trip with an average speed of 80 km/h, the car's speed wasn't *always* 80 km/h. With various stops and starts on the trip, the car must have been moving more slowly than the average speed part of the time. It therefore had to be moving more rapidly than the average speed part of the time. With an average speed you really don't know how fast the car was moving at any particular time during the trip. Similarly, the average test score for a class doesn't tell you the score of any particular student.

If the time interval (Δt) considered becomes smaller and smaller and approaches zero, the speed calculation is an **instantaneous speed**. This is how fast something is moving at a particular instant of time. The speedometer of a car gives an approximate instantaneous speed. For example, the speedometer shown in ●Fig. 2.4 indicates a speed of about 44 mi/h or 70 km/h. If the car travels with a constant speed (so that the speedometer reading does not change), then the average and instantaneous speeds will be equal. (Do you agree? Think of the average test score analogy.)

As we have seen, speed, like the distance it incorporates, is a scalar quantity—it has magnitude only. Another quantity used to describe motion is velocity. Speed and velocity are often used synonomously, but the terms have different meanings in physics. Basically, speed is a scalar and **velocity** is a vector—it has both magnitude and direction. Velocity tells you how fast *and* in what direction. And just as we can speak of average and instantaneous speeds, we have average and instantaneous velocities, incorporating vector displacements. The **average velocity** is the displacement divided by the total travel time.

$$\text{average velocity} = \frac{\text{displacement}}{\text{total travel time}}$$

$$\bar{v} = \frac{\Delta x}{\Delta t} = \frac{x - x_0}{t - t_0}$$

(2.3)

The vector difference, $\Delta x = x - x_0$, is simply the displacement between the initial and final positions. In Fig. 2.3, x_2 and x_1 denote the positions, but x and x_0 are used in Eq. 2.3. These are the more general symbols for position and are more convenient because they involve the use of fewer subscripts. The subscripts on x_0 and t_0 stand for *original* and indicate that these quantities refer to original, or initial, position and time. The symbols x and t represent position and time at some arbitrary later time and are sometimes called the final position and time.

As discussed in Section 2.1, for motion in one dimension, it is convenient to use plus and minus signs (+ and −) to indicate the directions of displacements and velocities along the positive and negative axes— for example, $+x$ and $+\bar{v}$ (or simply x and $\bar{v}$) and $-x$ and $-\bar{v}$. Also, it is common to take $x_0 = 0$ and $t_0 = 0$, so Eq. 2.3 becomes

$$\bar{v} = \frac{x}{t} \quad \text{or} \quad x = \bar{v}t$$

(2.4)

As can be readily seen from this equation, the standard units for velocity are the same as those for speed: m/s or ft/s.

You might be wondering whether there is a relationship between average speed and average velocity. A quick look at Fig. 2.1 will show you that *if* the motion is in one direction, the distance is equal to the magnitude of the displacement, and the average speed is the magnitude of the average velocity. However, be careful. This is not true if there is motion in both directions, as the following example shows.

EXAMPLE 2.1 ■ **Computing Average Speed and Velocity**

A jogger jogs from one end to the other of a straight 300-m track (from point A to point B in ● Fig. 2.5) in 2.50 min and then turns around and jogs 100 m back toward the starting point (to point C) in another 1.00 min. What are the jogger's average speeds and velocities in going (a) from A to B and (b) from A to C?

Solution. From the problem we have

Given: $x_B = +300$ m (from A to B)   *Find:* Average speeds and
$x_C = -100$ m (from B to C)   average velocities
$t_B = (2.50$ min$)(60$ s/min$) = 150$ s $\rbrace$ (conversion to
$t_C = (1.00$ min$)(60$ s/min$) = 60.0$ s $\rbrace$ standard units)

(a) The jogger's average speed in going from A to B is easily computed.

$$\bar{s}_B = \frac{\Delta d}{\Delta t} = \frac{\Delta x}{\Delta t}$$

$$= \frac{300 \text{ m}}{150 \text{ s}} = 2.00 \text{ m/s} \quad \text{(a scalar)}$$

The average velocity in going from A to B with $x_o = 0$ is also easily found, but direction must be indicated. Using Eq. 2.4,

$$\bar{v}_B = \frac{x_B}{t_B}$$

$$= \frac{+300 \text{ m}}{150 \text{ s}} = +2.00 \text{ m/s} \quad \text{(a vector)}$$

The positive direction is to the right in Fig. 2.5. Note that the average speed is equal to the magnitude of the average velocity in this instance.

(b) The average speed in going from A to C involves the *total* distance traveled, so

$$\bar{s}_C = \frac{\Delta x}{\Delta t}$$

$$= \frac{300 \text{ m} + 100 \text{ m}}{150 \text{ s} + 60.0 \text{ s}} = 1.90 \text{ m/s}$$

where there are no directional signs. (Why?)

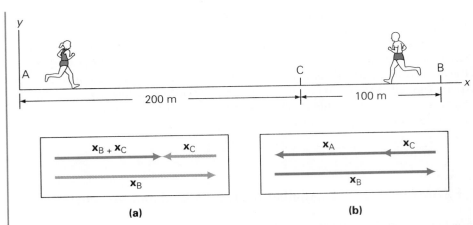

(a)   (b)

FIGURE 2.5 Average speed and velocity   (a) The jogger starts at A, jogs to B, then back to C. What are the average speeds and velocities for the segments of the lap? See Example 2.1. (b) If the jogger returns to the point from which he started (A), his total displacement, $x_C + x_B + x_A$, is zero.

The average velocity, on the other hand, involves the sum of the vector displacements:

$$\bar{v}_C = \frac{x_B + x_C}{t_B + t_C}$$

$$= \frac{+300 \text{ m} - 100 \text{ m}}{150 \text{ s} + 60.0 \text{ s}} = +0.952 \text{ m/s}$$

Note that direction makes a difference; the average speed is *not* equal to the magnitude of the average velocity in this case. This is because the displacements add vectorially. Notice in ● Fig. 2.5a, which illustrates this simple case of vector addition, that adding x_B and x_C gives an effective or resultant displacement from the initial starting point (A) to the stopping point (C), or +200 m.

Suppose that next the jogger sprints from point C back to the starting point (A) in 0.500 min. What are the average speed and average velocity for the whole lap? The additional data are $x_A = -200$ m (Fig 2.5) and $t_A = 30.0$ s. Then, for the round trip, the average speed is

$$\bar{s}_A = \frac{\Delta x}{\Delta t}$$

$$= \frac{300 \text{ m} + 100 \text{ m} + 200 \text{ m}}{150 \text{ s} + 60.0 \text{ s} + 30.0 \text{ s}} = 2.50 \text{ m/s}$$

The average velocity is

$$\bar{v}_A = \frac{x_B + x_C + x_A}{t_B + t_C + t_A}$$

$$= \frac{300 \text{ m} - 100 \text{ m} - 200 \text{ m}}{240 \text{ s}} = 0 \text{ m/s}$$

The average velocity in this case is zero!

The total displacement is measured from the initial starting point to the final stopping point. When the jogger comes back to the starting point, the

FIGURE 2.6 Back home again!   Despite having covered nearly 110 m on the base paths (and regardless of whether he is safe or out), at the moment he slides through the batter's box (his original position), into home the runner's displacement is zero—at least if he is a right-handed hitter. Therefore, no matter how fast he ran the bases, his average velocity for the round trip is also zero. (What would you estimate his displacement to be if he bats lefty? Could you also estimate his average velocity in this case?)

displacement is zero and so, therefore, is the average velocity. This is true for any round trip (● Fig. 2.6). Notice in Fig. 2.5b that the (vector) addition of $x_C + x_A$ gives a combined or resultant vector that is equal and opposite (in direction) to x_B. The sum of all three vectors is zero.

As Example 2.1 shows, average velocity provides only a very general description of motion. One way to obtain a closer look at motion is to take smaller time segments, that is, to let the time of observation (Δt) become smaller. As with speed, when Δt approaches zero, we obtain the **instantaneous velocity**, which describes how fast something is moving and in what direction at a particular instant of time. When using instantaneous velocity, we may write Eq. 2.4 as $x = vt$, since the average bar over the v is not needed. The instantaneous velocity is a vector with the instantaneous speed as its magnitude. The direction of the vector is the direction of the motion at that particular instant.

In some instances, the motion of an object may be uniform, which means that the velocity or speed (or both) is constant. For example, the car in ● Fig. 2.7 has a uniform velocity (as well as a uniform speed). It travels the same distance in equal time intervals (50 km each hour), and the direction of its motion does not change.

Instantaneous velocity: How fast and in what direction *right now*

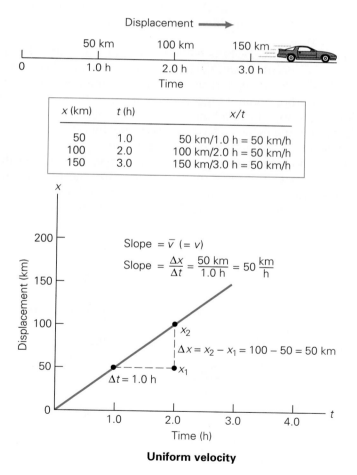

x (km)	t (h)	x/t
50	1.0	50 km/1.0 h = 50 km/h
100	2.0	100 km/2.0 h = 50 km/h
150	3.0	150 km/3.0 h = 50 km/h

Slope $= \bar{v}\,(= v)$

Slope $= \dfrac{\Delta x}{\Delta t} = \dfrac{50\ \text{km}}{1.0\ \text{h}} = 50\ \dfrac{\text{km}}{\text{h}}$

$\Delta x = x_2 - x_1 = 100 - 50 = 50$ km

$\Delta t = 1.0$ h

Uniform velocity

Figure 2.7

● **FIGURE 2.7 Uniform linear motion—constant velocity** In uniform linear motion, an object travels with a constant velocity, covering the same distance in equal time intervals. Here, a car travels 50 km each hour. An x versus t plot is a straight line, since equal displacements are covered in equal times. The numerical value of the slope of the line is equal to the magnitude of the velocity and the sign of the slope gives its direction. (Average velocity equals instantaneous velocity in this case.)

Graphical analysis is often helpful in understanding motion and its related quantities. In this case, the car's motion may be represented on a plot of displacement versus time, or x versus t. As can be seen from Fig. 2.7, a straight line is obtained for a uniform, or constant, velocity on such a graph (and for uniform speed on a plot of distance versus time).

Recall from Cartesian graphs of y versus x that the slope of a straight line is given by $\Delta y / \Delta x$. Here, with a plot of x versus t, the slope of the line, $\Delta x / \Delta t$, is equal to the average velocity ($\bar{v} = \Delta x / \Delta t$). For uniform motion ($v = x/t$), this is equal to the instantaneous velocity. That is, $\bar{v} = v$. (Why?) The numerical value of the slope is the magnitude of the velocity and the sign of the slope gives direction. A positive slope indicates that x increases with time, so the motion is in the positive x direction.

Suppose that a plot of displacement versus time for a car's motion was a straight line with a negative slope, as in ●Fig. 2.8. What does this indicate? As can be seen in the figure, the position (x) values get smaller with time, indicating that the car was traveling in uniform motion in the negative x direction.

In most instances the motion of an object is nonuniform, meaning that different distances are covered in equal intervals of time. An x versus t plot for such motion in one dimension is a curved line, as illustrated in Fig. 2.9. The average velocity for a particular interval of time is the slope of a straight line between the two points on the curve corresponding to the starting and ending times of the interval. In the figure, the average velocity for the total trip is the slope of the straight line joining the beginning and ending points of the curve (t_1 and t_2).

The instantaneous velocity is equal to the slope of a straight line tangent to the curve at a specific point. Several tangent lines are shown in ●Fig. 2.9. At (1), the slope is positive, and the motion is in the positive x direction. At (2), the slope of a horizontal tangent line is zero, so there is no motion. That is, the object has stopped instantaneously ($v = 0$). At (3), the slope is negative, so the object is moving in the negative x direction. Thus, it stopped and changed direction at point (2). What happens at point (4)?

Drawing various tangent lines along the curve, we see that their slopes vary, indicating that the instantaneous velocity (or speed) is changing with time. An object in nonuniform motion speeds up and/or slows down. How motion with a change in velocity is described is the topic of the next section.

Figure 2.8

●**FIGURE 2.8 Displacement versus time graph for an object in uniform motion in the** $-x$ **direction** A straight line on an x versus t plot with a negative slope indicates uniform motion in the $-x$ direction. Note that the displacement decreases uniformly with time. At $t = 4.0$ h, the object is at the origin ($x = 0$). How does the graph look if the motion continues for $t > 4.0$ h?

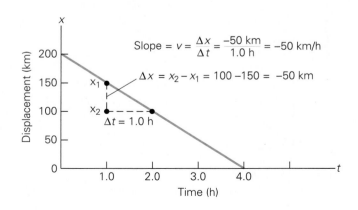

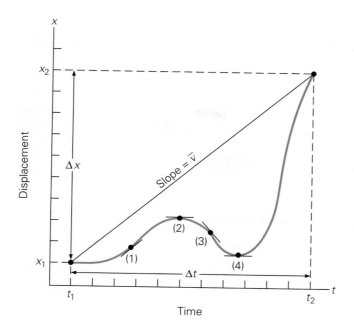

Figure 2.9

FIGURE 2.9 Displacement versus time graph for an object in nonuniform linear motion For a nonuniform velocity, an *x* versus *t* plot is a curved line. The slope of the line between two positions is the average velocity between those positions, and the instantaneous velocity is the slope of a line tangent to the curve at any point. Four such lines are shown in the figure. How does the motion vary through this sequence?

2.3 ■ Acceleration

The basic description of motion involves the time rate of change of position, which may be expressed by velocity. Going one step further involves considering how the *rate of change* changes. Suppose that something is moving at a constant velocity and then the velocity changes; this is an acceleration. The gas pedal on an automobile is commonly called the accelerator. When you push down on the accelerator, the car speeds up; and when you let up on the accelerator, the car slows down. That is, there is a change in velocity with time, or an acceleration. Specifically, **acceleration** is the time rate of change of velocity.

> *An inclined air track or ramp and cart with a strobe light can be used to demonstrate accelerated motion.*

Acceleration: A change in velocity

Average acceleration

Analogous to average velocity is the **average acceleration**, or the change in velocity divided by the time taken to make the change.

$$\text{average acceleration} = \frac{\text{change in velocity}}{\text{time to make the change}}$$

(2.5)

or

$$\bar{\mathbf{a}} = \frac{\Delta \mathbf{v}}{\Delta t} = \frac{\mathbf{v} - \mathbf{v_o}}{t - t_o}$$

where $\mathbf{v}$ and $\mathbf{v_o}$ are instantaneous velocities—the velocities at times t and t_o. Here we use the boldface vector notation for, in general, the velocities may be in different (nonlinear) directions. Since velocity is a vector quantity, so is acceleration. Analogous to instantaneous velocity is **instantaneous acceleration**, which is the acceleration at a particular instant of time.

Instantaneous acceleration

The dimensions of acceleration are (length/time)/time (as is obvious from $\Delta v / \Delta t$). The SI units for acceleration are therefore (m/s)/s, or m/s-s, commonly written m/s^2 (and read as "meters per second squared"). In the British system, the units are ft/s^2.

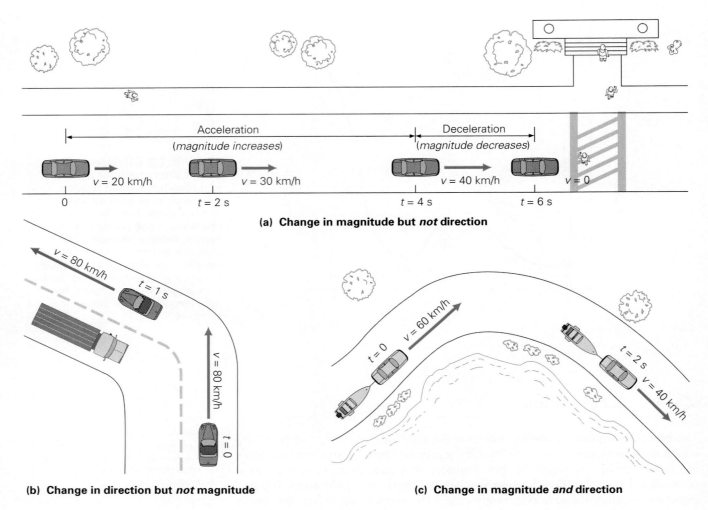

(a) Change in magnitude but *not* direction

v = 80 km/h   t = 1 s

v = 80 km/h   t = 0

(b) Change in direction but *not* magnitude

t = 0   v = 60 km/h

t = 2 s   v = 40 km/h

(c) Change in magnitude *and* direction

Figure 2.10

FIGURE 2.10 Acceleration—the time rate of change of velocity Since velocity is a vector quantity with magnitude and direction, an acceleration can occur when there is **(a)** a change in magnitude but not direction, **(b)** a change in direction but not magnitude, or **(c)** a change in both magnitude and direction.

Since velocity is a vector quantity, having both magnitude and direction, a change in velocity may involve either or both of these factors. An acceleration, therefore, may result from a change in *speed* (magnitude), a change in *direction*, or a change in *both*, as illustrated in ● Fig. 2.10.

For the special case of straight-line or linear motion, plus and minus signs will be used to indicate velocity directions, as was done for linear displacements. Then, Eq. 2.5 can be written as

$$\bar{a} = \frac{v - v_o}{t} \tag{2.6}$$

where t_o is taken to be zero (v_o may not be zero, so it cannot generally be omitted).

EXAMPLE 2.2 ■ Finding Average Acceleration

An automobile traveling on a straight road at 90 km/h slows down to 40 km/h in 5.0 s. What is its average acceleration?

Solution. From the problem, we have the following data. [With the motion in a straight line, the instantaneous velocities are assumed to be in the positive direction, and conversions to standard units (km/s to m/s) are made right away since it is noted that the time is given in seconds. In general, one always works with acceleration in standard units.]

Given: $v_0 = (90 \text{ km/h}) \left(\dfrac{0.278 \text{ m/s}}{1 \text{ km/h}} \right)$ *Find:* $\bar{a}$ (average acceleration)

$\qquad = 25 \text{ m/s}$

$\qquad v = (40 \text{ km/h}) \left(\dfrac{0.278 \text{ m/s}}{1 \text{ km/h}} \right)$

$\qquad = 11 \text{ m/s}$

$\qquad t = 5.0 \text{ s}$

Given the initial and final velocities and the time interval, the average acceleration may be found using Eq. 2.6.

$$\bar{a} = \frac{v - v_0}{t}$$

$$= \frac{11 \text{ m/s} - 25 \text{ m/s}}{5.0 \text{ s}} = -2.8 \text{ m/s}^2$$

The minus sign indicates the direction of the (vector) acceleration. In this case, the acceleration is opposite to the direction of the initial motion ($+v_0$), and it slows the car. Such an acceleration sometimes is called a *deceleration*.

A negative acceleration does not necessarily mean that a moving object is slowing down or its velocity decreasing. The $+$ and $-$ signs indicate the vector directions with respect to the reference axis. If the velocity and acceleration are in *opposite* directions, a moving object will slow down. However, suppose that a car traveling in the negative x direction ($-v_0$) experiences an acceleration in the positive x direction ($+a$). The car would slow down. So a positive acceleration ($+a$) can produce a deceleration. Similarly, if the velocity and acceleration are both in the *same* direction, the car will speed up. For example, if the car is initially traveling in the negative x direction ($-v_0$), a negative acceleration ($-a$) will speed it up in that direction.

Although acceleration can vary with time, our study of motion will be restricted to constant accelerations for simplicity. (An important constant acceleration is the acceleration due to gravity near the Earth's surface, which will be considered in the next section.) Since for a constant acceleration the average is equal to the constant value ($\bar{a} = a$), the bar over the acceleration in Eq. 2.6 may be omitted.

Thus, for a constant acceleration, the equation relating velocity, acceleration, and time is commonly written

$$\boxed{v = v_0 + at} \quad \text{(constant acceleration only)} \quad (2.7)$$

EXAMPLE 2.3 ■ Motion with Constant Acceleration

A drag racer starting from rest accelerates in a straight line at a constant rate of 5.5 m/s² for 6.0 s. (a) What is the racer's velocity at the end of this time? (b) If a parachute deployed at this time causes the racer to slow down uniformly at a rate of 2.4 m/s², how long will it take the racer to come to a stop?

Solution. Notice that the racer first speeds up and then slows down, so we must pay close attention to the directional signs of the vector quantities. Taking the initial motion to be in the positive direction, we have,

Given: (a) $v_0 = 0$ (at rest)   *Find:* v (final velocity)
$a = 5.5$ m/s²
$t = 6.0$ s

(b) $v_0 = v$ (from part a)   *Find:* t (time)
$v = 0$ (comes to stop)
$a = -2.4$ m/s² (opposite direction of v_0)

The data have been listed in two parts. This helps avoid confusion with symbols. Note that the initial velocity for (b) is the final velocity in (a), which is what is to be found there.

(a) To find v, we use Eq. 2.7 directly:

$$v = v_0 + at = 0 + (5.5 \text{ m/s}^2)(6.0 \text{ s}) = 33 \text{ m/s}$$

(b) Here we want time, so solving Eq. 2.6 for t and using $v_0 = 33$ m/s from part (a), we have

$$t = \frac{v - v_0}{a} = \frac{0 - 33 \text{ m/s}}{-2.4 \text{ m/s}^2} = 14 \text{ s}$$

Note that the time comes out positive, as it should. We start implicitly at zero time ($t_0 = 0$) and time goes forward or in a "positive direction."

Motions with constant accelerations are easy to represent graphically. A v versus t plot is a straight line whose slope is equal to the acceleration, as illustrated in ● Fig. 2.11. Note that Eq. 2.7 can be written $v = at + v_0$, which as you may recognize, has the form of an equation for a straight line, $y = mx + b$. In Fig. 2.10 (a), the motion is in the positive direction, and the accleration adds to the velocity for a time t, as illustrated by the vertical arrows at the right of the figure. In (b), the negative slope indicates a negative acceleration that produces a slowing down or deceleration. However, (c) illustrates how a negative acceleration can speed things up (for motion in the negative direction). Suppose that an object was set into motion from rest. How would this be indicated on a v versus t graph?

When an object moves with a constant acceleration, its velocity changes by the same amount in each time unit. For example, if the acceleration is 10 m/s², the object's velocity increases by 10 m/s in each second. Suppose that the object has an initial velocity (v_0) of 20 m/s at $t_0 = 0$. Then, for $t = 0, 1.0, 2.0, 3.0$, and 4.0 s, the velocities are 20, 30, 40, 50, and 60 m/s, respectively.

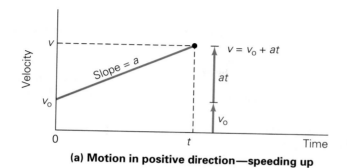

(a) Motion in positive direction—speeding up

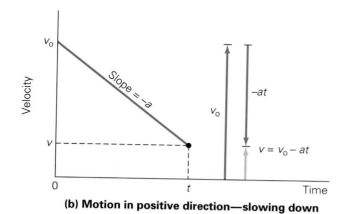

(b) Motion in positive direction—slowing down

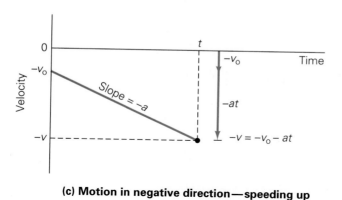

(c) Motion in negative direction—speeding up

Figure 2.11

● **FIGURE 2.11 Velocity vs. time graphs for motions with constant accelerations** The slope of a v versus t plot is the acceleration. **(a)** A positive slope indicates an increase in the velocity in the positive direction. The vertical arrows to the right indicate how the acceleration adds to the initial velocity v_0. **(b)** A negative slope indicates a decrease in the initial velocity v_0, or a deceleration. **(c)** Here a negative slope indicates a negative acceleration, but the initial velocity is in the negative direction, $-v_0$, so the velocity of the object *increases* in that direction.

The average velocity over the 4-s interval is $\bar{v} = 40$ m/s. This average may be computed in the regular manner, or you may immediately recognize that the uniformly increasing series of numbers 20, 30, 40, 50, and 60 has an average value of 40 (the midway value of the series). Note that the average of the extreme (initial and final) values also gives the average of the series, that is, $(20 + 60)/2 = 40$. This is true in general: When the velocity changes at a uniform rate because of a constant acceleration, $\bar{v}$ will be the average of the initial and final velocities.

$$\bar{v} = \frac{v + v_0}{2} \quad \text{(constant acceleration only)} \quad (2.8)$$

EXAMPLE 2.4 ■ Finding the Distance Traveled by a Constantly Accelerating Motorboat

A motorboat starting from rest on a lake accelerates in a straight line at a constant rate of 3.0 m/s² for 8.0 s. How far does the boat travel during this time?

Solution. Reading the problem and summarizing the given data and what is to be found, we have,

Given: $v_o = 0$ Find: x (distance)
$a = 3.0$ m/s²
$t = 8.0$ s

(It is noted that all of the units are standard.)

In analyzing the problem, the reasoning might go something like this: To find x, we will need to use Eq. 2.4, $x = \bar{v}t$. This is the only equation we have for finding distance. (The average velocity $\bar{v}$ must be used because the velocity is changing and not constant.) With t given, the solution of the problem then involves finding $\bar{v}$. By Eq. 2.8, $\bar{v} = (v + v_o)/2$, so with $v_o = 0$, we need only find the final velocity v and the problem is solved. Eq. 2.7, $v = v_o + at$, provides this from the given data. So, computing the required quantities in reverse order to the reasoning in solving the problem, we have:

The velocity of the boat at the end of 8.0 s is

$$v = v_o + at = 0 + (3.0 \text{ m/s}^2)(8.0 \text{ s}) = 24 \text{ m/s}$$

The average velocity over that time interval is

$$\bar{v} = \frac{v + v_o}{2} = \frac{24 \text{ m/s} + 0}{2} = 12 \text{ m/s}$$

Finally, the magnitude of the displacement or the distance traveled is

$$x = \bar{v}t$$
$$= (12 \text{ m/s})(8.0 \text{ s}) = 96 \text{ m}$$

2.4 ■ Kinematic Equations

The description of motion in one dimension with constant acceleration requires only three basic equations. From previous sections, these are

$$x = \bar{v}t \tag{2.4}$$

$$\bar{v} = \frac{v + v_o}{2} \quad \text{(constant acceleration only)} \tag{2.8}$$

$$v = v_o + at \quad \text{(constant acceleration only)} \tag{2.7}$$

Point out which variables are scalars and which are vectors.

(Keep in mind that the first equation is general and is not limited to situations where there is constant acceleration as the latter two are.)

However, as Example 2.4 showed, the description of motion in some instances requires multiple applications of these equations, which may not be obvious at first. It would be helpful if there were a way to reduce the number of operations in solving kinematic problems, and there is—combining equations algebraically.

For instance, combining the above equations involves first substituting for $\bar{v}$ from Eq. 2.8 into 2.4:

$$x = \bar{v}t = \left(\frac{v + v_0}{2}\right)t$$

Then substituting for v from Eq. 2.7 gives

$$x = \left(\frac{v + v_0}{2}\right)t = \left[\frac{(v_0 + at) + v_0}{2}\right]t$$

Simplifying gives

$$x = v_0 t + \tfrac{1}{2}at^2 \qquad (2.9)$$

Essentially, this series of steps was done in Example 2.4. This combined equation allows the distance traveled by the motorboat in that example to be computed directly.

$$x = v_0 t + \tfrac{1}{2}at^2 = 0 + \tfrac{1}{2}(3.0 \text{ m/s}^2)(8.0 \text{ s})^2 = 96 \text{ m}$$

Another possibility is to use Eq. 2.7 to eliminate time (t), rather than the final velocity (v), by writing that equation in the form $t = (v - v_0)/a$. Then, as before, substituting for $\bar{v}$ in Eq. 2.4 from 2.8 gives

$$x = \bar{v}t = \left(\frac{v + v_0}{2}\right)t$$

But then substituting for t gives

$$x = \left(\frac{v + v_0}{2}\right)t = \left(\frac{v + v_0}{2}\right)\left(\frac{v - v_0}{a}\right)$$

Simplifying gives

$$v^2 = v_0^2 + 2ax \qquad (2.10)$$

Notice how the units reduce to a unit of distance.

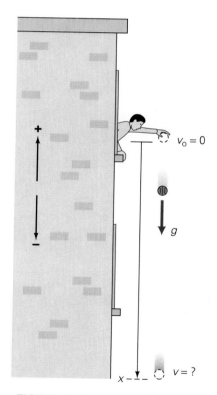

● **FIGURE 2.12 A dropped ball** A sketch to help visualize Example 2.5.

EXAMPLE 2.5 ■ Finding the Velocity of a Dropped Ball

A ball is dropped out of a window near the top of a building. If it accelerates toward the ground at a rate of 9.80 m/s², what is its velocity when it has fallen 4.00 m?

Solution. Here, we list the given data and make a sketch to help visualize the situation (● Fig. 2.12).

Given: $v_0 = 0$ *Find:* v (velocity)
       $a = -9.80 \text{ m/s}^2$
       $x = -4.00 \text{ m}$

TABLE 2.1 ■ Equations for Linear Motion with Constant Acceleration*

$x = \bar{v}t$	(2.4)†
$\bar{v} = \dfrac{v + v_0}{2}$	(2.8)
$v = v_0 + at$	(2.7)
$x = v_0 t + \frac{1}{2}at^2$	(2.9)
$v^2 = v_0^2 + 2ax$	(2.10)

*It is assumed that $x_0 = 0$ and the velocity is v_0 at $t_0 = 0$. The initial position and time, x_0 and t_0, may be included for general cases, for example, $x - x_0 = \bar{v}(t - t_0)$.
†Note that Eq. 2.4 is not limited to constant acceleration but applies generally.

As is customary, upward is taken as the positive direction and downward as the negative direction. The magnitudes of a and x are then written with minus signs to indicate their directions. No directional sign is explicitly assigned to the velocity we wish to find. Its direction will be given in the result of the mathematical operations. Of course, in this case you should realize that the final velocity will be negative or in the downward direction.

The velocity may be obtained directly from Eq. 2.10:

$$v^2 = v_0^2 + 2ax$$
$$= 0 + 2(-9.80 \text{ m/s}^2)(-4.00 \text{ m}) = 78.4 \text{ m}^2/\text{s}^2$$

and

$$v = \pm \sqrt{78.4 \text{ m}^2/\text{s}^2} = -8.85 \text{ m/s}$$

where the negative root is taken because the motion is downward.

Using the combined forms of the equations can often save you a lot of steps and calculations.

The equations of motion for linear motion with constant acceleration are summarized in Table 2.1 for your convenience. Note that only the first three are basic equations; the last two are convenient combinations.

EXAMPLE 2.6 ■ Squared Quantities

During some time trials, race car A, starting from rest, accelerates uniformly along a straight, level track for a particular time interval. Similarly, car B accelerates at the same rate but for twice the time. Then, at the ends of their respective acceleration periods, which of these statements is true: (a) car A has traveled a greater distance, (b) car B has traveled twice as far as car A, (c) car B has traveled four times as far as car A, (d) both cars have traveled the same distance? *Clearly establish the reasoning for your selected answer before checking it below. That is, **why** did you select your answer?*

Reasoning and Answer. It is given that $v_0 = 0$ and the acceleration is the same for both cars. To find the distances traveled in a time t, you would use Eq. 2.9, which, with $v_0 = 0$, becomes $x = \frac{1}{2}at^2$. The important thing to note here is that the distance increases at t^2. That is, if you double the time, the distance increases by a factor of four (quadruples).

In this example, car B accelerates twice as long as car A, or $t_B = 2t_A$, so car B travels four times as far as car A and the answer is (c). Expressed mathematically,

$$x_A = [\tfrac{1}{2}at_A^2] \quad \text{and} \quad x_B = \tfrac{1}{2}at_B^2 = \tfrac{1}{2}a(2t_A)^2 = \tfrac{1}{2}a(4t_A^2) = 4[\tfrac{1}{2}at_A^2] = 4x_A.$$

What if $t_B = 3t_A$?

Follow-up Exercise. How would the speeds of the cars compare at the ends of the acceleration periods? (*Reasoning and answer may be found in the Answers to Exercises section at the back of the book.*)

EXAMPLE 2.7 ■ Vehicle Stopping Distance

The stopping distance for a vehicle after the brakes have been applied is an important factor in highway safety. This distance depends on the initial velocity (v_0) and the braking capacity, or deceleration, $-a$, which is assumed to be constant. (Recall that the minus sign indicates that the acceleration is in the negative direction, which is opposite that of the velocity. Thus the car slows to a stop.) Express the stopping distance x in terms of these quantities.

Solution. Here, we are working with variables, so we can represent quantities only in symbolic form.

Given: v_0
     $-a$ (opposite direction of v_0)
     $v = 0$ (comes to stop)

Find: x (in terms of the given variables)

Again, it is helpful to make a sketch of the situation, particularly when directional vector quantities are involved (see ● Fig. 2.13). Since Eq. 2.10 has the variables we want, it should allow us to find the stopping distance. Expressing the negative acceleration explicitly gives

$$v^2 = v_0^2 + 2\,(-a)x$$

or

$$v^2 = v_0^2 - 2ax$$

Since the vehicle comes to a stop ($v = 0$), we can solve for x:

$$x = \frac{v_0^2}{2a}$$

This gives us x expressed in terms of the vehicle's initial velocity and stopping acceleration.

Notice that the stopping distance x is proportional to the square of the initial velocity. Doubling the initial velocity therefore increases the stopping distance by a factor of 4 (for the same deceleration). That is, if the stopping

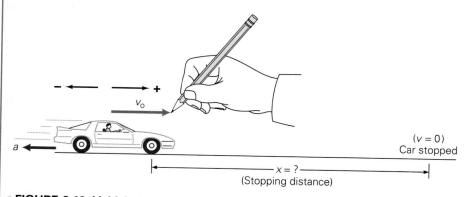

● **FIGURE 2.13 Vehicle stopping distance** A sketch to help visualize Example 2.7.

distance is x_1 for an initial velocity of v_1, then for a two-fold initial velocity, $v_2 = 2v_1$, the stopping distance would be four-fold as shown by the proportions

$$x_1 \propto v_1^2 \quad \text{and} \quad x_2 \propto v_2^2 = (2v_1)^2 = 4v_1^2$$

so

$$x_2 = 4x_1$$

Or, directly by ratio,

$$\frac{x_2}{x_1} = \frac{v_2^2}{v_1^2} = \left(\frac{v_2}{v_1}\right)^2 = \left(\frac{2v_0}{v_0}\right)^2 = 4.$$

Do you think this is an important consideration in setting speed limits, for example, in school zones? (The driver's reaction time should also be considered. A method to approximate a person's reaction time is given in the next section.)

2.5 ■ Free Fall

One of the more familiar cases of constant acceleration is that due to gravity near the Earth's surface. When an object is dropped, its initial velocity is zero (at the instant it is released), but at a later time during falling, it has a non-zero velocity. There has been a change in velocity and, by definition, an acceleration. This **acceleration due to gravity** (g) has an approximate value (magnitude) of

$$g = 9.80 \text{ m/s}^2 \quad \text{acceleration due to gravity}$$

or 980 cm/s² and is directed downward (toward the center of the Earth). In British units, the value of g is about 32 ft/s².

The values given here for g are only approximate because the acceleration due to gravity varies slightly at different locations as a result of differences in elevation and regional average mass density of the Earth. These small variations will be ignored in this book unless otherwise noted. (Gravitation is studied in more detail in Chapter 7.) Air resistance is another factor that affects the acceleration of a falling object. But for relatively dense objects and over the short distances of fall commonly encountered, air resistance produces only a small effect, which will also be ignored here for simplicity. (The frictional effect of air resistance will be considered in Chapter 4.)

Objects in motion solely under the influence of gravity are said to be in free fall. The acceleration due to gravity g is the *constant* acceleration for all free-falling objects, regardless of their mass or weight. It was once thought that heavier bodies fell faster than lighter bodies. This was part of Aristotle's theory of motion. You can easily observe that a coin falls faster than a sheet of paper when dropped simultaneously from the same height. But in this case air resistance plays a noticeable role. If the paper is crumpled into a compact ball, it gives the coin a better race. Similarly, a feather "floats" down much more slowly than a coin falls. However, in a near-vacuum, where there is negligible

Stand on top of the lecture desk and drop objects of different masses at the same time. Air resistance has little effect on such things as coins, marbles, stones, etc., so they hit at the same time.

Also, drop a book and a piece of paper to show the effect of air friction. Place a sheet of paper on top of the book and another sheet on the bottom of the book and drop them flat side down. Explain.

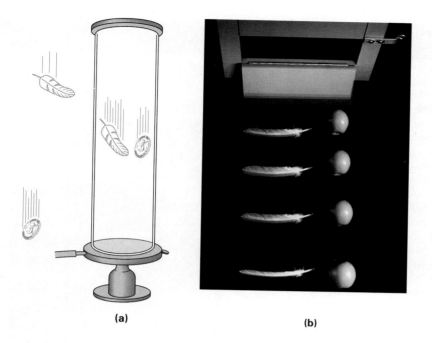

(a)   (b)

● FIGURE 2.14 Free fall and air
resistance   (a) When dropped
simultaneously from the same
height, a feather falls more slowly
than a coin because of air resistance.
But when both objects are dropped
in an evacuated container with a
good partial vacuum, where air
resistance is negligible, the feather
and the coin fall together with a
constant acceleration. (b) An actual
demonstration with multiflash
photography. An apple and a feather
are released simultaneously through
a trap door into a large vacuum
chamber, and they fall together—
almost. Although the chamber has a
partial vacuum, there is still some air
resistance. How can you tell?

air resistance, the feather and the coin fall with the same acceleration—the
acceleration due to gravity (● Fig. 2.14).

Video demonstration 1.

Astronaut David Scott performed a similar experiment on the moon in
1971 by simultaneously dropping a feather and a hammer from the same height.
Of course, he did not need a vacuum pump since the moon has no atmosphere
and no air resistance. The hammer and the feather reached the lunar surface
together, but fell at a slower rate than on Earth. The acceleration due to gravity
near the moon's surface is approximately one-sixth of that near the Earth's
surface ($g_m \cong \frac{1}{6} g$).

Currently accepted ideas about the motion of falling bodies are due in
large part to Galileo. He challenged Aristotle's theory and experimentally investigated the motion of objects. Legend has it that he studied the accelerations
of falling bodies by dropping objects of different weights from the top of the
Leaning Tower of Pisa (see the Insight feature).

The words "free fall" bring to mind dropped objects that are moving
downward under the influence of gravity ($g = 9.80$ m/s² in the absence of air
resistance). However, the term can be applied in general to any vertical motion
under the influence of gravity. An object with an initial velocity, directed either
up or down, may be thought of as being projected in one dimension and having
an acceleration equal to g. (Even when an object projected upward is traveling
upward, *it is accelerating downward.*) Thus, the set of equations for motion in
one dimension (in Table 2.1) can be used to describe generalized free fall.

A projected object while traveling
upward (+ v) is accelerating
downward, or has an acceleration in
the downward direction (− g).

It is customary to use y to represent the vertical direction and to take
upward as positive (as with the vertical y axis of Cartesian coordinates). Since
the acceleration due to gravity is always downward, it is in the negative direction. This negative acceleration, $a = -g = -9.80$ m/s², may be substituted
each time into the equations of motion. However, the relationship $a = -g$
may be expressed explicitly in the equations for linear motion (see Table 2.1):

Galileo Galilei was born in Pisa, Italy, in 1564 during the Renaissance. Today, he is known throughout the world by his first name and often referred to as the father of modern science or the father of modern mechanics and experimental physics, which attests to the magnitude of his scientific contributions (Fig. 1).

One of Galileo's greatest contributions to science was the establishment of the scientific method, that is, investigation through experiment. In contrast, Aristotle's approach was based on logical deduction. By the scientific method, for a theory to be valid it must predict or agree with experimental results. If it doesn't, it is invalid or requires modification. Galileo said, ''I think that in the discussion of natural problems we ought not to begin at the authority of places of Scripture, but at sensible experiments and necessary demonstrations.''*

Probably the most popular and well-known legend about Galileo is that he performed experiments with falling bodies by dropping objects from the Leaning Tower of Pisa (see Fig. 2). There is some doubt as to whether Galileo actually did this, but there is little doubt that he questioned Aristotle's view on the motion of falling objects. In 1638, Galileo wrote:

> Aristotle says that an iron ball of one hundred pounds falling from a height of one hundred cubits reaches the ground before a one-pound ball has fallen a single cubit. I say that they arrive at the same time. You find, on making the experiment, that the larger outstrips the smaller by two finger-breadths, that is, when the larger has reached the ground, the other is short of it by two finger-breadths; now you would not hide behind these two fingers the ninety-nine cubits of Aristotle.†

This and other writings show that Galileo was aware of the effect of air resistance.

The experiments at the Tower of Pisa were supposed to have taken place around 1590. In his writings of about that time, Galileo mentions dropping objects from a high tower, but never specifically names the Tower of Pisa. A letter written to Galileo from another scientist in 1641 describes the dropping of a cannon ball and a musket ball from the Tower of Pisa. The first account of Galileo doing a similar experiment was written a dozen years after his death by Vincenzo Viviani, his last pupil and first biographer. Viviani related that the falling bodies ''all moved at the same speed'' and that Galileo demonstrated ''this with repeated experiments from the height of the Campanile [Tower] of Pisa in the presence of the other teachers and philosophers, and the whole assembly of

FIGURE 1 Galileo
Galileo is alleged to have performed free-fall experiments by dropping objects off the Leaning Tower of Pisa.

FIGURE 2 The Leaning Tower of Pisa

students.'' However, there is no other record of this event, which seems odd given that a crowd of people supposedly witnessed it.

It is not known whether Galileo told this story to Viviani in his declining years or Viviani created this picture of his former teacher.

The important point is that Galileo recognized (and probably experimentally showed) that free-falling objects fall at the same rate regardless of their mass or weight (see Fig. 2.14). Galileo gave no reason why all objects in free fall have the same acceleration, but Newton did, as you will learn in a later chapter.

* From *Growth of Biological Thought: Diversity, Evolution & Inheritance*, by F. Meyr (Cambridge, MA: Harvard University Press, 1982).

† This and the next quotation are both from *Aristotle, Galileo, and the Tower of Pisa*, by L. Cooper (Ithaca, NY: Cornell University Press, 1935).

$$y = \bar{v}t \qquad\qquad (2.4')$$

$$\bar{v} = \frac{v + v_o}{2} \qquad\qquad (2.8')$$

Free-fall equations

$$v = v_o - gt \qquad \text{with } -g \text{ expressed} \qquad (2.7')$$

$$y = v_o t - \tfrac{1}{2}gt^2 \qquad \text{explicitly} \qquad (2.9')$$

$$v^2 = v_o^2 - 2gy \qquad\qquad (2.10')$$

The origin ($y = 0$) of the reference frame is usually taken to be at the initial position of the object. Since upward is generally taken to be in the positive direction ($+y$ axis on a graph), writing $-g$ explicitly in the equations reminds you of directional differences and avoids bracketed minus signs in calculations [i.e., (-9.80 m/s^2)]. However, the choice is arbitrary. The equations can be written with $a = g$, for example, $v = v_o + gt$, with the directional minus sign associated directly with g (that is, $g = -9.80$ m/s^2). In this case, a value of -9.80 m/s^2 must be substituted for g each time.

Note that you have to be explicit about the directions of vector quantities when using these equations. The displacement y and the velocities v and v_o may be positive or negative, depending on the direction of motion. The use of these equations and the sign convention is illustrated in the following examples.

EXAMPLE 2.8 ■ An Object Thrown Downward

A boy on a bridge throws a stone vertically downward toward the river below with an initial velocity of 14.7 m/s. If the stone hits the water 2.00 s later, what is the height of the bridge above the water?

Solution. As usual, we first read the problem and write down what is given and what is to be found.

Given: $v_o = -14.7$ m/s (downward taken as   *Find:* y (height)
the negative direction)

$t = 2.00$ s

$g\ (= 9.80$ m/s^2)

Notice that g is now just a number, since the directional minus sign has already been put into the previous equations of motion. After a while, you will probably just write down the symbol g, since you will be familiar with its numerical value. This time, draw a sketch of your own to help analyze the situation.

Considering which equation(s) will provide the solution using the given data, it should become evident that the distance the stone travels in a time t is given directly by Eq. 2.9':

$$y = v_o t - \tfrac{1}{2}gt^2 = (-14.7 \text{ m/s})(2.00 \text{ s}) - \tfrac{1}{2}(9.80 \text{ m/s}^2)(2.00 \text{ s})^2$$

$$= -29.4 \text{ m} - 19.6 \text{ m} = -49.0 \text{ m}$$

The minus sign indicates that the displacement is downward, which agrees with what you know from the statement of the problem. (Could you find how long it would take the stone to reach the river if the boy had dropped it rather than thrown it?)

● **FIGURE 2.15 Reaction time** A person's reaction time can be measured by having her (or him) grasp a dropped ruler. Example 2.9.

Explain how units can be used in verifying a solution. Notice how the units in this example reduce to (s), which is a time as expected.

Offer to give a dollar bill to a student who can catch it in free fall. Hold the bill so that the bottom edge is between the student's thumb and finger. Release the bill quickly. Only a very fast student can catch the bill.

EXAMPLE 2.9 ■ Measuring Reaction Time

Reaction time is how long it takes a person to notice, think, and act in response to a situation—for example, the time between first observing and then responding to something happening on the road ahead while driving an automobile. Reaction time varies with the complexity of the situation (and the individual). In general, the largest part of a person's reaction time is spent thinking, but practice in dealing with a given situation can reduce this time.

A person's reaction time for a simple situation may be measured by having another person drop a ruler (without warning) through the thumb and forefinger as shown in ● Fig. 2.15. The falling ruler is grasped by the first person as quickly as possible, and the length of the ruler below the top of the finger is noted. If the ruler descends 18 cm on the average before it is grasped, what is the person's average reaction time?

Solution. Notice that only the drop distance is given. However, we know a couple of other things, such as v_0 and g, so

Given: $y = -18 \text{ cm} = -0.18 \text{ m}$ *Find:* t (reaction time)
$v_0 = 0$
$g \, (= 9.80 \text{ m/s}^2)$

(Note that the distance y has been converted directly to meters.) We can see that Eq. 2.9′ applies here:

$$y = v_0 t - \tfrac{1}{2} g t^2$$

or $y = -\tfrac{1}{2} g t^2$ (with $v_0 = 0$)

Solving for t gives

$$t = \sqrt{\frac{2y}{-g}}$$
$$= \sqrt{\frac{2(-0.18 \text{ m})}{-9.80 \text{ m/s}^2}} = 0.19 \text{ s}$$

Try this with a fellow student and measure your reaction time. Why do you think another person should drop the ruler rather than yourself? Substitute a dollar bill for the ruler. Are you fast enough to catch it?

EXAMPLE 2.10 ■ Free Fall Up and Down

A worker on a scaffold on a billboard throws a ball straight up. It has an initial velocity of 11.2 m/s when it leaves his hand at the top of the billboard (● Fig. 2.16). (a) What is the maximum height the ball reaches relative to the top of the billboard? (b) How long does it take to reach this height? (c) What is the position of the ball at $t = 2.00$ s?

Solution. It might appear that all that is given in the general problem is the initial velocity v_0. However, a couple of other things are implicitly "given"

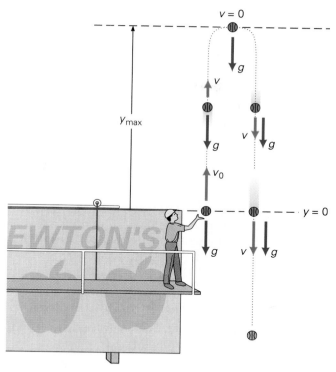

● **FIGURE 2.16 Free fall up and down** (The ball is horizontally displaced for illustration.) Note the lengths of the velocity and acceleration vectors at different times.

because they are understood. One is the acceleration g, and the other is the velocity at the maximum height where the ball stops. Here, in changing direction, the velocity of the ball is momentarily zero, so we have

> **Given:** $v_o = 11.2$ m/s   **Find:** (a) y_{max} (maximum height)
> $g\ (= 9.80$ m/s^2)   (b) t_u (time upward)
> $v = 0$ (for part a)   (c) y (at $t = 2.00$ s)
> $t = 2.00$ s (for part c)

(a) Notice that we reference the height ($y = 0$) to the top of the billboard. For this part of the problem, we need only be concerned with the upward motion—a ball is thrown upward and it stops at its maximum height y_{max}. With $v = 0$ at this height, y_{max} may be found directly from Eq. 2.10′:

$$v^2 = 0 = v_o^2 - 2\,g y_{max}$$

and

$$y_{max} = \frac{v_o^2}{2\,g}$$

$$= \frac{(11.2 \text{ m/s})^2}{2(9.80 \text{ m/s}^2)} = 6.40 \text{ m}$$

relative to the top of billboard ($y = 0$, see figure).

(b) The time the ball travels upward is designated t_u. This is the time it takes to reach y_{max} where $v = 0$. Then, knowing v_o and v, the time t_u may be found directly from Eq. 2.7':

$$v = 0 = v_o - gt_u$$

and

$$t_u = \frac{v_o}{g}$$

$$= \frac{11.2 \text{ m/s}}{9.80 \text{ m/s}^2} = 1.14 \text{ s}$$

(c) The height of the ball at $t = 2.00$ s is given directly by Eq. 2.9':

$$y = v_o t - \tfrac{1}{2}gt^2$$
$$= (11.2 \text{ m/s})(2.00 \text{ s}) - \tfrac{1}{2}(9.80 \text{ m/s}^2)(2.00 \text{ s})^2 = 22.4 \text{ m} - 19.6 \text{ m} = 2.8 \text{ m}$$

Note that this is 2.8 m above, or measured upward from, the reference point ($y = 0$). The ball has reached its maximum height and is on its way back down.

Considered from another reference point, this situation is like dropping a ball from a height of y_{max} above the top of the billboard with $v_o = 0$ and asking how far it falls in a time $t = 2.00 \text{ s} - t_u = 2.00 \text{ s} - 1.14 \text{ s} = 0.86 \text{ s}$. The answer is

$$y = v_o t - \tfrac{1}{2}gt^2$$
$$= 0 - \tfrac{1}{2}(9.80 \text{ m/s}^2)(0.86 \text{ s})^2 = -3.6 \text{ m}$$

This is the same as the position found above but is measured with respect to the maximum height as the reference point; that is,

$$y_{max} - 3.6 \text{ m} = 6.4 \text{ m} - 3.6 \text{ m} = 2.8 \text{ m}$$

■ Problem-Solving Hint

When working vertical projection problems involving motions up and down, it is often convenient to divide the problem into two parts and consider each separately. As seen in the previous example, for the upward part of the motion, the velocity is zero at the maximum height. A zero quantity simplifies the calculations. Similarly, the downward part of the motion is analogous to that of an object dropped from a height, where the initial velocity is zero.

However, as the previous example shows, the appropriate equations may be used directly at any position or time of the motion. For instance, note in part (c) of Example 2.10 that the height was found directly for a time *after* the ball had reached the maximum height. The velocity of the ball at that time could also have been found directly from Eq. 2.7', $v = v_o - gt$.

EXAMPLE 2.11 ■ Interesting Facts about Vertical Free-Fall Projections

Referring to the thrown ball in Example 2.10 and Fig. 2.16, (a) compare the travel time upward (t_u) with the travel time downward (t_d) required for the ball to return to its starting point. (b) Compare the initial velocity of the ball with the velocity it has when back at the starting point.

Solution. The given data in this case will include some of the results of Example 2.10. There, t_u and y_{max} were computed.

Given: $v_o = 11.2$ m/s
$t_u = 1.14$ s
$y_{max} = 6.40$ m

Find: (a) t_d, and compare with t_u
(b) v (returning velocity), and compare with v_o

(a) The time to reach the maximum height is $t_u = 1.14$ s. The time for the return trip downward, t_d, is the time it takes the ball to fall 6.40 m (the maximum height) from rest ($v = 0$ at maximum height). (Refer to Fig. 2.16.) This situation is exactly the same as if the ball were dropped and we were asked to find the time it would take to fall 6.40 m. With $v_o = 0$ for this downward part of the motion, Eq. 2.9′ applies:

$$y = v_o t_d - \tfrac{1}{2} g t_d^2 = 0 - \tfrac{1}{2} g t_d^2 \quad \text{or} \quad y = -\tfrac{1}{2} g t_d^2$$

and

$$t_d = \sqrt{\frac{2y}{-g}}$$
$$= \sqrt{\frac{2(-6.40 \text{ m})}{-9.80 \text{ m/s}^2}} = 1.14 \text{ s}$$

Hence, $t_u = t_d$, or *the time of flight upward and the time to return to the starting point are the same.*

Notice that the total time for the ball to return to its starting point is given by

$$t = t_u + t_d = 1.14 \text{ s} + 1.14 \text{ s} = 2.28 \text{ s}$$

If we had been asked to compute this total time, it would not have been necessary to do this in two steps (time up and time down). Another way is to note that when the ball returns to its original starting point, it is again at $y = 0$, and then use Eq. 2.9′ to find the total time:

$$y = v_o t - \tfrac{1}{2} g t^2 = (v_o - \tfrac{1}{2} g t) t = 0$$

where v_o is the initial upward velocity. This equation is satisfied when $t = 0$, which is when the ball is initially at $y = 0$, and when $(v_o - \tfrac{1}{2} g t) = 0$, which is when the ball is again at $y = 0$ on the return trip. From the latter,

$$t = \frac{2v_o}{g}$$
$$= \frac{2(11.2 \text{ m/s})}{9.80 \text{ m/s}^2} = 2.29 \text{ s}$$

Note that the answer is slightly different. This is an example of a rounding difference, which may occur when a problem is solved by different methods.

(b) The velocity of the ball when it returns to its starting point ($y = 0$) may be found using Eq. 2.10′:

$$v^2 = v_o^2 - 2gy = v_o^2 - 0 \quad \text{or} \quad v^2 = v_o^2$$

Taking the square root of both sides, we have

$$v = -v_o = -11.2 \text{ m/s}$$

where the negative root was taken to indicate the proper direction. (The positive root, $v = +v_o$, gives the initial velocity, since this is also a velocity of the ball at $y = 0$.) Hence, the velocities are equal and opposite, and the ball returns to the starting point *with the same speed it had initially.*

Notice that the returning velocity could also be found by considering the velocity of the ball at time t_d, or 1.14 s after falling from its maximum height. As for a dropped ball, $v_o = 0$, and Eq. 2.7' gives

$$v = v_o - gt$$
$$= 0 - (9.80 \text{ m/s}^2)(1.14 \text{ s}) = -11.2 \text{ m/s}$$

Keep in mind that, as the examples in this section show, there is often more than one approach to solving a problem.

Follow-up Exercise. For the situation in Fig. 2.16, the ball continues falling after $t = 2.28$ s. How would you find its position at some later time before it hits the ground? (*Reasoning and answer may be found in the Answers to Exercises section at the back of the book.*)

Important Concepts

You should be able to define and explain these chapter concepts clearly.

mechanics	scalar (quantity)	instantaneous speed	average acceleration
kinematics	vector (quantity)	velocity	instantaneous acceleration
dynamics	displacement	average velocity	acceleration due to gravity
motion	speed	instantaneous velocity	free fall
distance	average speed	acceleration	

Important Relationships for Review

These relationships are mathematical statements of the concepts and principles presented in the chapter. You should be able to identify the symbols and to explain the relationships before proceeding to the Exercises. In-text equation reference numbers are given for convenience.

Average speed:

$$\bar{s} = \frac{\Delta d}{\Delta t} \tag{2.2}$$

Kinematic Equations for Linear Motion with Constant Acceleration:

$$x = \bar{v}t \tag{2.4}$$

(general, not limited to constant acceleration)

$$\left. \begin{array}{l} \bar{v} = \dfrac{v + v_o}{2} \\[2ex] v = v_o + at \\[2ex] x = v_o t + \frac{1}{2}at^2 \\[2ex] v^2 = v_o^2 + 2ax \end{array} \right\} \begin{array}{l} \\ \text{constant} \\ \text{acceleration} \\ \text{only} \\ \end{array}$$

$$\bar{v} = \frac{v + v_o}{2} \tag{2.8}$$

$$v = v_o + at \tag{2.7}$$

$$x = v_o t + \frac{1}{2}at^2 \tag{2.9}$$

$$v^2 = v_o^2 + 2ax \tag{2.10}$$

Acceleration Due to Gravity:

$$g = 9.80 \text{ m/s}^2 = 980 \text{ cm/s}^2 \approx 32 \text{ ft/s}^2$$

Kinematic Equations Applied to Free Fall:
(with downward taken as the negative direction and g expressed explicitly):

$$y = \bar{v}t \tag{2.4'}$$

$$\bar{v} = \frac{v + v_o}{2} \tag{2.8'}$$

$$v = v_o - gt \tag{2.7'}$$

$$y = v_o t - \frac{1}{2}gt^2 \tag{2.9'}$$

$$v^2 = v_o^2 - 2gy \tag{2.10'}$$

Exercises

2.1 ▪ A Change of Position

1 Distance is always (a) equal to the magnitude of the corresponding displacement, (b) less than or equal to the magnitude of the corresponding displacement, (c) greater than or equal to the magnitude of the corresponding displacement. (c)

2 Two people choose different reference points to specify an object's position. Does this affect their coordinate descriptions of the object? Explain. different coordinates for same location

3 Can a displacement from one point to another be zero, yet the distance involved in moving between these points be nonzero? Can the distance between two points be zero, yet the displacement from one point to the other be nonzero? Explain. yes, for round trip; no, distance always greater than or equal to magnitude of displacement

4 You are told that a person walks 500 m. What can you safely say about the person's final position relative to the starting point? no final position can be given; may be 0 to 500 m

2.2 ▪ Speed and Velocity

5 A vector quantity has (a) magnitude, (b) direction, (c) units, (d) all of these. (d)

● **6** For a constant velocity, the speed is (a) continually changing, (b) less than the magnitude of the velocity, (c) greater than the magnitude of the velocity, (d) equal to the magnitude of the velocity vector. (d)

7 When is speed equal to the magnitude of velocity? Give two instances. zero velocity and speed; constant velocity

8 When is the average velocity equal to the instantaneous velocity? constant velocity

9 ▪ A student walks 0.30 km to class in 5.0 min. What is the student's average speed in m/s? 1.0 m/s

10 ▪ A small airplane flies in a straight line at a speed of 140 km/h. How long does it take the plane to fly 400 km? 2.86 h

11 ▪ A bus travels on an interstate highway at an average speed of 90 km/h. How far does the bus travel in 15 min on the average? Would this be the actual distance? Explain. 23 km (see ISM)

● **12** ▪ A motorist drives 125 km from one city to another in 2.0 h, but makes the return trip in only 1.5 h. What are the average speeds for (a) each half of the round trip and (b) the total trip? (a) 63 km/h, 83 km/h (b) 71 km/h

13 ▪▪ Is the average speed of the jogger in Example 2.1 going from A to C equal to the average of the average speeds in going from A to B and from B to C? Justify your answer. s_{AC} = 1.90 m/s; s_{av} = 1.84 m/s

14 ▪▪ A runner makes three complete laps around a track 1.00 km in length in 13.3 min. What are her (a) average speed and (b) average velocity? (a) 3.77 m/s (b) zero since the displacement is zero

15 ▪▪ An automobile travels along a straight, level highway uniformly covering a distance of 88 ft each second. (a) What is the car's speed in mi/h? (b) What are the car's average speed and average velocity in m/s? (c) What can you say about the car's instantaneous speed and velocity? (a) 60 mi/h (b) 27 m/s

● **16** ▪▪ A distance of 1.50 km is covered in making one lap around an oval dirt bike track. If a rider going at a constant speed makes one lap in 1.25 min, what is the speed of the bike and rider in m/s? Is the velocity of the bike also constant? Explain. 20.0 m/s; velocity not constant since direction changes

17 ▪▪ A circular track has a diameter of 0.50 km. A go-cart with a constant speed of 7.0 m/s makes two complete laps around the track. How long does it take the cart to complete the two laps? 4.4 × 10² s

18 ▪▪ A plot of displacement versus time is shown in ● Fig. 2.17 for an object in linear motion. (a) What are the average velocities for the segments AB, BC, CD, DE, EF, FG, and BG? (b) State whether the motion is uniform or nonuniform in each case. (c) What is the instantaneous velocity at point D? (a) v_{AB} = 0; v_{BC} = 3.0 m/s; v_{CD} = 1.3 m/s; v_{DE} = −1.3 m/s; v_{EF} = −1.7 m/s v_{FG} = 0; v_{BG} = 0.10 m/s (b) CD, DE nonuniform (c) zero

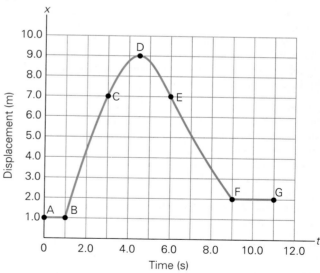

● **FIGURE 2.17 Displacement vs. time** See Exercise 18.

19 ▪▪ In demonstrating a dance step, a person moves in one dimension as shown in ● Fig. 2.18. What are (a) the average speed and (b) the average velocity for each phase of the motion? (c) What are the instantaneous velocities at t = 1.0 s, 2.5 s, 4.5 s, and 6.0 s? (d) What is the average velocity for the interval between t = 4.5 s and t = 9.0 s? [*Hint:*

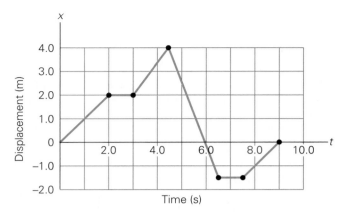

FIGURE 2.18 Displacement vs. time See Exercise 19.

Recall that the effective displacement is the displacement between the starting point and the stopping point.] see ISM

20 ■■ On a walk in the country, a couple goes 1.80 km east along a straight road in 20.0 min and then 2.40 km directly north in 35.0 min. Answer the following using units of km/h. (a) What is the average velocity for each segment of their walk? (b) What is the average speed for the total distance walked? (c) If they walk along a straight-line path back to their original starting place in 25.0 min, what is the average speed and average velocity for the total trip? (a) $v_1 = 5.40$ km/h east; $v_2 = 4.11$ km/h north (b) 4.58 km/h (c) 5.40 km/h

21 ■■ The Indianapolis 500, a 500-mi auto race, was first run in 1911 in a time of 6 h, 42 min, and 8 s. In 1992, the race was run in a time of 3 h, 7 min, and 12 s. (a) What are the average speeds for the Indy 500 for these years in mi/h? (Ignore significant figures.) (b) What is the percentage change in the average speed from 1911 to 1992? (a) $v = 74.6$ mi/h; $v = 134.4$ mi/h (b) +80%

22 ■■ According to the theory of continental drift and plate tectonics, the continents were once parts of a single giant supercontinent, called Pangaea, which broke up. Recent measurements of sea-floor spreading along mid-oceanic ridges show the continental drift to be on the order of 4 cm per year. Assume that this rate has been constant throughout the past, and take the current distance between South America and Africa to be 5000 mi. Approximately how long ago did Pangaea break up? 2.0×10^8 y

23 ■■ On a still lake, a boat travels with a constant speed of 55.5 km/h at full throttle. The boat is then taken to a river which has a flow rate (current) of 2.5 km/h. Assuming the same boat speed in the water, what would be the boat's speed as observed by a person on the river bank when it is traveling directly (a) downstream and (b) upstream? [Hint: Think of vector addition.] (a) 58.0 km/h (b) 53.0 km/h

24 ■■ A student driving home for the holidays starts at 8:00 a.m. to make the 675-km trip, practically all of which is on nonurban interstate highway. If she wants to arrive home no later than 3:00 p.m., what must her average speed be, at minimum? Will she have to exceed the 65 mi/h speed limit? 60.0 mi/h; no

● 25 ■■■ Two runners approaching each other on a straight track have constant velocities of +4.50 m/s and −3.50 m/s, respectively, when they are 100 m apart. How long will it take for the runners to meet and at what position will this occur? 12.5 s; 56.3 m (relative to first runner)

26 ■■■ Two motorcyclists race against the clock on a 40-km cross-country route. The first cyclist travels the route with an average speed of 55 km/h. The second cyclist starts 3.5 min after the first but crosses the finish line at the same time. What is the average speed of the second cyclist? 60 km/h

2.3 ■ Acceleration

27 The gas pedal of an automobile is commonly referred to as the accelerator. Which of the following might also be called an accelerator: (a) the brakes, (b) the steering wheel, (c) both, (d) neither? (c)

28 For a constant linear acceleration, which of the following changes uniformly with time: (a) distance, (b) displacement, (c) velocity, (d) all of these? (c)

29 An object traveling with a constant velocity v_o experiences a constant acceleration in the same direction for a time t. Then, an acceleration of equal magnitude is experienced in the opposite direction for the same time t. What is the velocity after this? v_o

● 30 What would be the general form of a v versus t plot for an object in linear motion with (a) a nonconstant acceleration in the direction of the velocity, and (b) a nonconstant acceleration in the direction opposite the velocity? (c) Describe the motions of the two objects that have the v versus t plots shown in ● Fig. 2.19. (a) concave up (b) concave down (c) see ISM

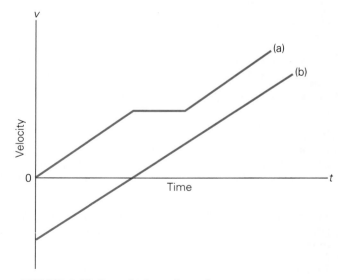

FIGURE 2.19 Description of motion See Exercise 30.

31 ▪ An automobile traveling at 25.0 km/h along a straight road accelerates to 50.0 km/h in 5.00 s. What is the average acceleration? 1.39 m/s²

32 ▪ A car moving with a velocity of 24 m/s on a one-way street must be brought to a stop in 4.0 s. What is the required average acceleration? -6.0 m/s²

33 ▪▪ If an automobile moving with a velocity of 40 km/h along a straight road is braked uniformly to rest in 5.0 s, by how much must the velocity change each second? -2.2 m/s each s

34 ▪▪ A drag racer starting from rest reaches a speed of 180 km/h in 6.75 s along a straight track. (a) What is the average acceleration of the racer? (b) Assuming that the acceleration is constant, what is the average velocity of the racer? (a) 7.41 m/s² (b) 90.0 km/h

35 ▪▪ What is the acceleration for each graph segment in Fig. 2.20? Describe the motion of the object over the total time interval. $a_{0-4} = 2.0$ m/s²; $a_{4.0-10} = 0$; $a_{10-17} = -1.1$ m/s²

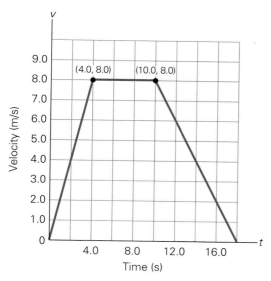

FIGURE 2.20 Velocity vs. time See Exercise 35.

36 ▪▪ An object moves in the negative x direction with a speed of 2.5 m/s. If an acceleration of 0.50 m/s² is experienced in the positive x direction, after how long a time will the object have a positive velocity? 5.0 s

37 ▪▪ A uniform acceleration of -4.0 m/s² is applied to a moving object, bringing it to rest. Its average velocity during deceleration is 10 m/s. Is the deceleration adequate to bring the object to rest in 4.0 s? (Justify your answer.) If it is not adequate, what is the required deceleration? -5.0 m/s²

38 ▪▪ ● Figure 2.21 shows a plot of velocity versus time for an object in linear motion. (a) Compute the acceleration for each phase of motion. (b) Describe how the object moves during the last time segment. (a) $a_{0-1} = 0$; $a_{1-3} = 4.0$ m/s²; $a_{3-8} = -4.0$ m/s²; $a_{8-9} = 8.0$ m/s²; $a_{9-12} = 0$ (b) constant velocity of -4.0 m/s

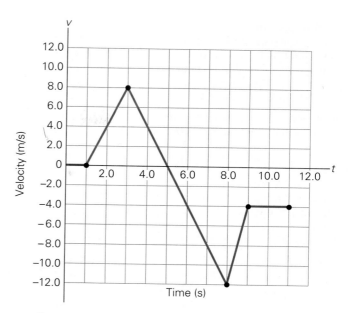

FIGURE 2.21 Velocity vs. time See Exercise 38.

2.4 ▪ Kinematic Equations

39 An object accelerates uniformly from rest for t seconds. The average speed for this time interval is (a) $\frac{1}{2}at$, (b) $\frac{1}{2}at^2$, (c) $2at$, (d) none of the preceding. (a)

40 The distance traveled by an object moving with a constant acceleration varies as (a) t, (b) t^2, (c) v, (d) none of the preceding. (b)

41 A classmate states that a negative acceleration always means that a moving object is decelerating. Is this true? Explain. no; would accelerate on an object having $-v$ or $v_0 = 0$

42 ▪ A motorboat traveling on a straight course slows down uniformly from 70 km/h to 35 km/h in a distance of 50 m. What is the acceleration? -2.7 m/s²

43 ▪ A car initially at rest rolls down a hill with a uniform acceleration of 5.0 m/s². How long will it take the car to travel 150 m, the distance to the bottom of the hill? 7.7 s

44 ▪ A sprinter accelerates from rest at a uniform rate of 1.75 m/s² for 4.00 s to reach his maximum speed. What is that speed in km/h and mi/h? 25.2 km/h; 15.7 mi/h

45 ▪ A car accelerates uniformly from rest at a rate of 5.25 m/s². (a) How far does the car travel in 7.00 s? (b) What is the speed of the car at that time? (a) 129 m (b) 36.8 m/s

46 ▪▪ The driver of a pickup truck going 100 km/h applies the brakes, giving the truck a uniform deceleration of 6.50 m/s² while it travels 20.0 m. (a) What is the velocity of the truck in km/h at the end of this distance? (b) How much time has elapsed? (a) 81.4 km/h (b) 0.800 s

47 ▪▪ Large, ocean-going oil tankers may weigh as much as 640,000 tons. Such a ship requires 3.0 miles to stop from a top speed of 15 knots. Assuming that the ship slows down at a constant rate in a straight line, what would be the ship's acceleration in m/s² in such a case? (A knot is a nautical unit of speed, nautical mile/hour, and 1 nautical mile = 6067 ft = 1.15 mi. Our common mile (5280 ft) is properly called a statute mile.) -6.2×10^{-3} m/s²

● 48 ■■ A bullet traveling horizontally with a speed of 25 m/s strikes a tree and penetrates to a depth of 6.0 cm before coming to rest. Assuming a constant acceleration, how long after hitting the tree does the bullet come to a stop?
4.8×10^{-3} s

49 ■■ A bullet traveling horizontally with a speed of 35.0 m/s hits a board perpendicular to the surface, passes through it, and emerges on the other side with a speed of 21.0 m/s. If the board is 4.00 cm thick, how long does the bullet take to pass through it? 1.43×10^{-3} s

🏷50 ■■ A motorboat is traveling due north in calm water at a speed of 6.0 m/s. The operator puts the motor in reverse and the boat receives an acceleration of −1.5 m/s². (a) How long after the boat is put into reverse does it momentarily come to a halt? (b) What is the position of the boat when it momentarily comes to a halt relative to the point where the motor was put into reverse? (a) 4.0 s (b) 12 m north

51 ■■ An object moves in the positive x direction with a speed of 40 m/s. As it passes through the origin, it starts to experience a constant acceleration of 3.5 m/s² in the negative x direction. How much time elapses before the object returns to the origin? 23 s

● 52 ■■ The speed limit in a school zone is 40 km/h (about 25 mi/h). A driver traveling at this speed sees a child run into the road 17 m ahead of his car. He applies the brakes, and the car decelerates at a uniform rate of 8.0 m/s². If the driver's reaction time is 0.25 s, will the car stop before hitting the child? yes; $x = 10.4$ m

53 ■■ Assuming a driver's reaction time of 0.25 s and construct a table of the stopping distances for an automobile traveling at initial velocities of 36 km/h, 72 km/h, and 90 km/h with uniform decelerations of (a) 8.0 m/s² (on a dry pavement) and (b) 4.0 m/s² (on a wet pavement). Also list the initial velocities in mi/h and the stopping distances in ft in parentheses. see ISM

54 ■■ A sports car is advertised to be able to accelerate uniformly from rest to 100 km/h in 8.20 s. (a) What is the acceleration, and (b) how far would the car travel in this time? (a) 3.39 m/s² (b) 114 m

55 ■■ An automobile is braked to a stop with a uniform deceleration in a time of t_s. Show that the distance traveled during this time is given by $x = (v_0^2/a) - \frac{1}{2}at_s^2$, where the positive direction is taken to be in the direction of the initial motion. see ISM

● 56 ■■ Show that the area under the curve of a v versus t plot for a constant acceleration is equal to the displacement. Do this for the cases where (a) $a = 0$, (b) $v_0 = 0$, and (c)

$v_0 \neq 0$. [Hint: The area of a triangle is $\frac{1}{2}ab$, or one-half the altitude times the base.] (a) A $= vt$ (b) A $= \frac{1}{2}at^2$ (c) A $= v_0t + \frac{1}{2}at^2$

57 ■■ Compute the distance traveled for the motion represented by Fig. 2.20. 96 m

58 ■■ ● Fig. 2.21 shows velocity versus time for an object in linear motion. (a) What are the instantaneous velocities at $t = 8.0$ s and $t = 11.0$ s? (b) Compute the final displacement of the object. (c) Compute the total distance the object travels. (a) $v_8 = -12.0$ m/s; $v_{11} = -4.0$ m/s (b) -18 m (c) 50 m

59 ■■ A train on a straight, level track has an initial speed of 45.0 km/h. A uniform acceleration of 1.50 m/s² is applied while the train travels 200 m. (a) What is the speed of the train at the end of this distance? (b) How long did it take for the train to travel the 200 m? (a) 27.5 m/s (b) 10.0 s

● 60 ■■■ An object moves in the positive x direction with a constant acceleration. At $x = 5.0$ m, its speed is 10 m/s. In 2.5 s, the object is at $x = 65$ m. What is its acceleration? 11 m/s²

🏷61 ■■■ A car and a motorcycle start from rest at the same time on a straight track, but the motorcycle is 25.0 m behind the car (● Fig. 2.22). The car accelerates at a uniform rate of 3.70 m/s², and the motorcycle at a uniform rate of 4.40 m/s². (a) How much time elapses until the motorcycle overtakes the car? (b) How far will each have traveled during that time? (c) How far ahead of the car will the motorcycle be 2.00 s later? (Both vehicles are still accelerating.)
(a) 8.45 s (b) $x_m = 157$ m; $x_c = 132$ m (c) 13 m

62 ■■■ A pitcher throws a fastball toward home plate 60.5 ft away with an initial velocity of 90.0 mi/h in the horizontal direction. (a) If the batter's combined reaction and swing times total 0.350 s, how long can the batter watch the ball after it has left the pitcher's hand before making a decision to swing? (A major reason why some players illegally use hollowed out bats with cork or superball filling is that such bats are lighter, which decreases the swing time, allowing more time to follow the ball and make a decision to swing.) (b) How far does the ball drop from its original horizontal line in traveling to the plate? [Hint: Consider the motions in the x and y directions independently.] (a) 0.108 s (b) 3.36 ft

2.5 ■ Free Fall

63 A dropped object in free fall (a) falls 9.8 m each second,

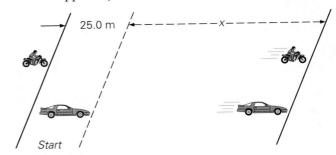

● **FIGURE 2.22 A tie race** (drawing not to scale.) See Exercise 61.

(b) falls 9.8 m during the first second, (c) has an increase in speed of 9.8 m/s each second, (d) has an increase in acceleration of 9.8 m/s each second.　(c)

64 When an object is thrown vertically upward, neglecting air resistance, its (a) velocity changes nonuniformly, (b) maximum height is independent of the initial velocity, (c) travel time upward is slightly greater than its travel time downward, (d) speed on returning to its starting point is the same as its initial speed.　(d)

65 Given the data for a vertically projected object, you are asked to find the time the object is at height y. The equation $y = v_0t - \frac{1}{2}gt^2$ is solved using the quadratic formula and two roots are obtained. Does this mean there are two answers? Explain. [*Hint:* Do the roots of a quadratic equation necessarily have to be both positive and negative?]　yes; y up and down

66 For the motion of a dropped object in free fall, sketch the general forms of the graphs of (a) v versus t, and (b) y versus t.　(a) linear, slope g　(b) parabola

67 Sketch the general forms of the graphs of (a) v versus t, and (b) y versus t for an object projected vertically upward.　(a) inverted V　(b) parabola (see ISM)

68 ■ If a dropped object falls 19.6 m in 2.00 s, how far will it fall in 4.00 s?　78.4 m

69 ■ How long would it take a dropped object to reach a speed of 60 mi/h? (Make a guess before you do the calculations.)　2.8 s

● **70** ■ An object dropped from the top of a cliff takes 1.80 s to hit the water in the lake below. What is the height of the cliff above the water?　−15.9 m

71 ■ A stone is dropped from a height of 20.0 m. (a) How long does it take for the stone to hit the ground? (b) What velocity does it have just before hitting the ground?　(a) 2.02 s　(b) 19.8 m/s downward

72 ■ With what speed must an object be projected vertically upward for it to reach a maximum height of 12.0 m above its starting point?　15.3 m/s

73 ■■ When checking reaction times as described in Example 2.8, some students decide to calibrate the ruler in time units so they can obtain a quick approximation without having to do a calculation each time the test is made. They want to calibrate the ruler in time intervals of 20 ms. Show on a sketch of a ruler where the time gradations would be for the first 20 cm.　$y_1 = 0.20$ cm; $y_2 = 0.78$ cm; $y_3 = 1.8$ cm; $y_4 = 3.1$ cm

74 ■■ For an object dropped from rest, what is the distance it will fall *during* the second second (that is, between $t = 1.00$ s and $t = 2.00$ s)? Does this distance double during the fourth second of fall? Explain.　no, $\Delta y_{12} = 14.7$ m; $\Delta y_{34} = 34.3$ m; more than doubles

75 ■■ The ceiling of a classroom is 3.75 m above the floor. A student tosses an apple vertically upward, releasing it

0.50 m above the floor. What is the maximum magnitude of the initial velocity that can be given to the apple if it is not to touch the ceiling?　slightly less than 8.0 m/s

● **76** ■■ A stone is thrown vertically downward with an initial speed of 12.4 m/s from a height of 65.0 m above the ground. (a) How far does the stone travel in 2.00 s? (b) What is its velocity when it hits the ground?　(a) −44.4 m　(b) 37.8 m/s downward

77 ■■ A ball is projected vertically downward with a speed of 3.00 m/s. (a) How far does the ball travel in 1.80 s? (b) What is the velocity of the ball at that time?　(a) −21.3 m　(b) 20.6 m/s downward

78 ■■ At what rate would a car have to accelerate from rest on a straight, level road to have the same speed in the same amount of time that an object dropped from a height of 20 m would have just before hitting the ground?　same; 9.8 m/s²

79 ■■ A ball is thrown upward with an initial speed of 6.0 m/s by someone on the top of a building 34 m tall who is leaning over the edge so that the ball will not strike the building on the return trip. (a) How far above the ground will the ball be at the end of 1.0 s? (b) What is the ball's velocity at that time? (c) When and with what speed will the ball strike the ground?　(a) 36 m　(b) −3.8 m/s　(c) $v = -27$ m/s; $t = 3.4$ s

● **80** ■■ A baseball thrown vertically upward is caught at the same height 3.20 s later. What are (a) the initial velocity of the ball and (b) its maximum height above its starting point?　(a) 15.7 m/s　(b) 12.6 m

81 ■■ A certain person can jump a vertical distance of 0.85 m. (a) What is the total time the person is off the ground? (b) With what velocity does the person hit the ground?　(a) 0.84 s　(b) 4.1 m/s

82 ■■ Suppose that the boy on the bridge in Example 2.8 throws the stone vertically upward instead of downward. How long will the stone be in flight before it hits the river?　5.00 s

83 ■■ Referring to the situation in Fig. 2.16, (a) what would be the position of the ball 3.00 s after it was thrown upward, and (b) if the top of the billboard is 20.0 m above the ground, what is the ball's total time of flight, and what is its velocity just before hitting the ground?　(a) 10.5 m below the initial point　(b) 3.46 s; −22.7 m/s

84 ■■ Draw plots of y versus t and v versus t for the following: (a) an object dropped from rest from a height of 30 m above the ground, and (b) an object projected vertically upward with an initial velocity of 34.3 m/s and its return to the same point. (c) Draw plots of v and a versus t for these motions.　graphs (see ISM)

85 ■■ In throwing an object vertically upward with a velocity of 7.25 m/s from the top of a tall building the thrower leans over the edge so that the object will not strike the building on the return trip. (a) What is the velocity of the object when it has traveled a total distance of 25.0 m? (b) How long does it take to travel this distance?　(a) −20.9 m/s　(b) 2.87 s

86 ▪▪ An aluminum ball with a mass of 4.0 kg and an iron ball of the same size with a mass of 11.6 kg are dropped simultaneously from a height of 49 m. (a) Neglecting air resistance, how long does it take the aluminum ball to fall to the ground? (b) How much later does the heavier iron ball strike the ground? (a) 3.2 s (b) 3.2 s

●**87** ▪▪ A photographer in a helicopter ascending vertically at a constant rate of 1.75 m/s accidentally drops a camera when the helicopter is 50.0 m above the ground. (a) How long will it take the camera to reach the ground? (b) What will its speed be when it hits? (a) 3.38 s (b) −31.4 m/s

88 ▪▪ The acceleration due to gravity on the moon is one-sixth of that on Earth. (a) If an object were dropped from the same height on the moon and on the Earth, how much longer (by what factor) would it take it to hit the surface of the moon? (b) For a projectile with an initial velocity of 18.0 m/s upward, what would be the maximum height and the total time of flight on the moon and on the Earth?
(a) $t_M = (2.4) t_E$ (b) moon: $y = 99.2$ m; $t = 22.0$ s Earth: $y = 16.5$ m; $t = 3.67$ s

89 ▪▪▪ A student at a window on the second floor of a dorm sees his math professor coming along the walkway beside the building. He drops a water balloon from 18.0 m above the ground when the prof is 1.00 m from the point directly beneath the window. If the prof is 170 cm tall and walks at a rate of 0.450 m/s, does the balloon hit his head? Does it hit him at all? hits 14 cm in front

90 ▪▪▪ A dropped object passes a window that is 1.35 m tall in 0.210 s. From what height above the top of the window was the object released? y (above top) = 0.14 m

91 ▪▪▪ A student drops a stone from the top of a building 26.0 m tall. Another student simultaneously throws a second stone downward from the same height, and the thrown stone hits the ground 0.30 s before the dropped stone does. What was the initial velocity of the thrown stone? −3.2 m/s

▪ Additional Exercises

92 After landing, a jet plane comes uniformly to rest along a straight runway with an average velocity of −35.0 km/h. If this takes 7.00 s, what is the plane's acceleration? 2.7 m/s²

93 A vertically moving projectile reaches a maximum height of 17.5 m above its starting position. (a) What was the projectile's initial velocity? (b) What is its height above the starting point at $t = 2.45$ s? (a) 18.5 m/s (b) 15.9 m

94 Given that the speed of sound is 340 m/s, how much time will elapse between seeing a lightning flash and hearing the resulting thunder if the lightning strikes 3.50 km away? (The speed of light is 3.00×10^8 m/s, or about 186,000 mi/s, so the lightning flash is seen instantaneously.)
$t_L = 1.16 \times 10^{-5}$ s; $t_s = 10.3$ s

95 A drag racer traveling on a straight track with a speed of 200 km/h ejects a parachute and slows uniformly to a speed of 20 km/h in 12 s. (a) What is the racer's acceleration? (b) How far does it travel in the 12-s interval? (a) −4.2 m/s² (b) 365 m

96 On a cross-country trip, a couple drives 500 mi in 10 h on the first day, 380 mi in 8.0 h on the second day, and 600 mi in 15 h on the third day. What was their average speed?
45 mi/h

97 A projectile is given an initial velocity of 8.0 m/s directly upward. Use the quadratic formula (with $g = 10$ m/s² to simplify the calculations) to determine (a) the time when the projectile is 1.4 m above its starting point and (b) the time when it is 3.5 m above its starting point. (c) Suppose that the projectile is instead given an initial velocity of 8.0 m/s directly downward. At what time is the projectile a distance of 1.8 m below its starting point? (Comment on both roots of the quadratic equation in each case.) (a) 0.20 s and 1.4 s (b) v_o is not enough to reach the height (c) 0.20 s; the negative root has no physical significance

98 A spring-loaded gun shoots a 0.0050-kg bullet vertically upward with an initial velocity of 21 m/s. (a) What is the height of the bullet 4.0 s after firing? (b) At what times is the bullet 12 m above the muzzle of the gun? (a) 5.6 m (b) $t_{up} = 0.71$ s; $t_{down} = 3.6$ s

99 A car going 85 km/h on a straight road is brought uniformly to a stop in 10 s. How far does the car travel during that time? 1.2×10^2 m

100 An object initially at rest experiences an acceleration of 1.5 m/s² for 6.0 s and then travels at that constant velocity for another 8.0 s. What is the object's average velocity over the 14-s interval? 7.1 m/s

101 (This is an old one.) In order to find the depth of the water surface in a well, a person drops a stone from the top of the well and simultaneously starts a stopwatch. The watch is stopped when the splash is heard, giving a reading of 3.65 s. The speed of sound is 340 m/s. Find the depth of the water surface below the top of the well. Take the person's reaction time for stopping the watch to be 0.250 s. 52.4 m

102 An arrow shot from a bow vertically upward has an initial velocity of 50.0 m/s. (a) What is the maximum height of the arrow above its launch point? (b) How long does it take for the arrow to return to its launch point? (Neglect air resistance and assume that the arrow travels in a straight line.) (a) 128 m (b) 10.2 s

103 A rocket car travels with a constant velocity of 250 km/h on a salt flat. The driver gives the car a reverse thrust, and it experiences a continuous and constant deceleration of 8.25 m/s². How much time elapses until the car is 175 m from the point where the reverse thrust is applied? 3.09 s and 13.8s

104 An object starting from rest and traveling in a straight line has velocities of 5.0 m/s, 10 m/s, 15 m/s, 20 m/s, and 25 m/s at the ends of the first, second, third, fourth, and fifth seconds, respectively. (a) What is the acceleration of the object? (b) What is its average velocity for the 5-s interval? (c) Sketch a graph of v versus t. (d) Sketch a graph of x versus t. (a) 5.0 m/s² (b) 13 m/s (c) and (d) see ISM

3 Motion in Two Dimensions

Could you describe the motion of the ball in the photograph above? Obviously, this is not a case of straight-line, or linear, motion in one dimension like those we considered in the last chapter. Here, we have motion in a plane—that is, in two dimensions. In this chapter we will extend our study of motion to two-dimensional cases, such as the soccer ball in the photo or a car rounding a curve on a level road. These are examples of what is called curvilinear motion. The analysis and understanding of curved-path motions will eventually lead you to the study of more dramatic things, such as planetary and atomic orbits, in later chapters.

Curvilinear motion is quite easy to analyze using rectangular components of motion. Essentially, you break down, or resolve, the curved motion into rectangular (x and y) components and look at the straight-line motion in each dimension. To those components you can apply the kinematic equations introduced in Chapter 2. To find the position of an object moving in a curved path, for example, you simply find x and y coordinates at any time; then the object's position is given as the point (x, y).

Components of motion can be conveniently represented in vector notation: A displacement vector for example, has, or is made up of, the position components x and y. Similarly, a velocity vector has the velocity components v_x and v_y; an acceleration vector has the acceleration components a_x and a_y. In some instances, aspects of motion can be analyzed by adding vectors directly. Because every vector has both magnitude and direction, the scalar addition of numbers you learned in grade school does not apply. In this chapter, you'll learn how to add and subtract vectors, operations that take direction into account.

Chapter 3 deals with the very important topics of vectors and projectile motion.

3.1 ■ Components of Motion

An object moving in a straight line was considered in Chapter 2 to be moving along one of the Cartesian axes (x or y). But what if the motion is not along an axis? For example, consider the situation illustrated in ● Fig. 3.1. Here, the balls are moving uniformly across a tabletop. The ball rolling in a straight line along the side of the table, designated as the x direction, is moving in one dimension. That is, its motion can be described with a single coordinate, x, as was done for motions in Chapter 2. Similarly, the motion of the ball rolling in the y direction can be described by a single y coordinate. However, both x and y coordinates are needed to describe the motion of the ball rolling diagonally across the table. We call this motion in two dimensions.

You might observe that if the diagonally-moving ball were the only one, the x axis could be chosen in that direction and the motion reduced to one dimension. This is true, but once the coordinate axes are fixed, motions not along the axes must be described with two coordinates (x, y) or in two dimen-

Review Section 2.1

Figure 3.1

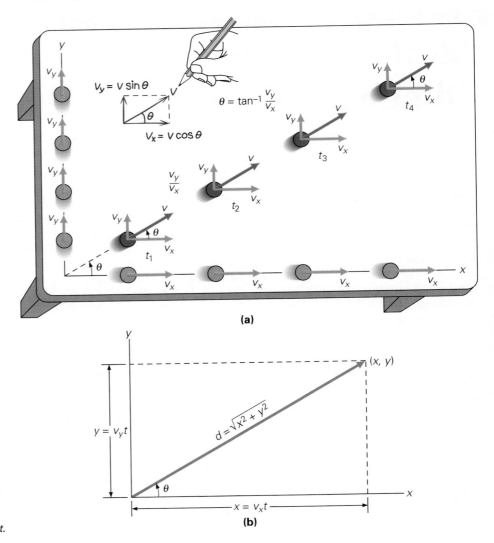

(a)

(b)

● **FIGURE 3.1 Components of motion** **(a)** The velocity (and displacement) for uniform straight-line motion, may have x and y components because of the chosen orientation of the coordinate axes. Note that the velocity and displacement of the ball in the x direction are exactly the same as those that a ball rolling along the x axis with a uniform velocity of v_x would have. A comparable relationship holds true for the ball's motion in the y direction. Since the motion is uniform, the ratio v_y/v_x (and therefore θ) is constant. **(b)** The coordinates (x, y) of the ball's position and the distance d it has traveled may be found for any time t.

sions. Also, keep in mind that all motions in a plane (two dimensions) are not in straight lines. Think about the path of a ball you toss to another person.

In considering the motion of the ball moving diagonally across the table in Fig. 3.1a, one can think of the ball moving in the x and y directions simultaneously. That is, it has a velocity in the x direction (v_x) *and* a velocity in the y direction (v_y) at the same time. The combined velocities describe the actual motion of the ball. If the ball has a constant velocity (**v**) in a direction at an angle θ relative to the x axis, then the velocities in the x and y directions are obtained by resolving, or breaking down, the velocity vector into components in these directions. As may be seen from the pencil drawing in Fig. 3.1a, these components have magnitudes of

$$v_x = v \cos \theta \qquad (3.1a)$$

$$v_y = v \sin \theta \qquad (3.1b)$$

Velocity components

(Notice that $v = \sqrt{v_x^2 + v_y^2}$, so v is a combination of the x and y velocities.)

You have no doubt used similar two-dimensional components before, for example, length or displacement components. The position of the ball, (x, y), or the distance traveled in each of the component directions at time t, is given by

$$\boxed{\begin{aligned} x &= v_x t \\ y &= v_y t \end{aligned}} \quad \text{(zero acceleration only)} \quad \begin{aligned} (3.2a) \\ (3.2b) \end{aligned}$$

Distances or displacement components

(with $x_o = y_o = 0$ at $t = 0$). The ball's straight-line distance from the origin is then $d = \sqrt{x^2 + y^2}$ (Fig. 3.1b).

EXAMPLE 3.1 ■ Using Components of Motion

If the diagonally-moving ball in Fig. 3.1 has a velocity of 0.50 m/s at an angle of 37° relative to the x axis, find how far it travels in 3.0 s using x and y components.

Solution. Organizing the data, we have

Given: $v = 0.50$ m/s   *Find:* d (distance)

$\theta = 37°$

$t = 3.0$ s

The distance in terms of the x and y components is given by $d = \sqrt{x^2 + y^2}$. So to find x and y as given by Eq. 3.2, we first compute the velocity components v_x and v_y (Eq. 3.1).

$$v_x = v \cos 37° = (0.50 \text{ m/s})(0.80) = 0.40 \text{ m/s}$$

$$v_y = v \sin 37° = (0.50 \text{ m/s})(0.60) = 0.30 \text{ m/s}$$

Then, the component distances are

$$x = v_x t = (0.40 \text{ m/s})(3.0 \text{ s}) = 1.2 \text{ m}$$

$$y = v_y t = (0.30 \text{ m/s})(3.0 \text{ s}) = 0.90 \text{ m}$$

and the actual path distance is

$$d = \sqrt{x^2 + y^2} = \sqrt{(1.2 \text{ m})^2 + (0.90 \text{ m})^2} = 1.5 \text{ m}$$

(Note that for this simple case, the distance can also be obtained directly from $d = vt = (0.50 \text{ m/s})(3.0 \text{ s}) = 1.5 \text{ m}$. However, we have solved this Example in a more general way to illustrate the use of components of motion.)

Kinematic Equations for Components of Motion

The preceding example involved two-dimensional motion in a plane. With a constant velocity (constant components v_x and v_y), the motion is in a straight line. The motion may also be accelerated. For motion in a plane *with a constant acceleration* having components a_x and a_y, the displacement and velocity components are given by the kinematic equations of Chapter 2 for the x and y directions:

Kinematic equations for displacement and velocity components

$$x = v_{x_0} t + \tfrac{1}{2} a_x t^2 \qquad (3.3a)$$

$$y = v_{y_0} t + \tfrac{1}{2} a_y t^2 \qquad (3.3b)$$

(constant acceleration only)

$$v_x = v_{x_0} + a_x t \qquad (3.3c)$$

$$v_y = v_{y_0} + a_y t \qquad (3.3d)$$

If an object moving with a constant velocity experiences an acceleration (a vector) in the direction of the velocity (0°) or opposite to it (180°), it would continue in a straight-line path, either speeding up or slowing down, respectively.

But motion can also be along a curved path. For the motion of an object to be curvilinear, that is, to vary from a straight-line path, an acceleration is required. In this case, however, the acceleration vector must be at some angle to the velocity vector other than 0° or 180°. Such a deflecting acceleration produces a curved path, for which the ratio of the velocity components varies with time. That is, the direction of the motion, $\theta = \tan^{-1}(v_y/v_x)$, varies with time, so the motion is not in a straight line.

Consider a ball initially moving along the x axis, as illustrated in ● Fig. 3.2. Assume that starting at a time $t_0 = 0$ it receives a constant acceleration a_y in the y direction. (You'll learn what produces an acceleration in the next chapter.) The magnitude of the x component of the ball's displacement is given by $x = v_x t$, since there is no acceleration in the x direction. Prior to t_0, the motion is in a straight line along the x axis. But at any time after t_0, there is a y displacement with a magnitude of $y = \tfrac{1}{2} a_y t^2$ (with $v_{y_0} = 0$). The result is a curved path for the ball.

Note that the length (magnitude) of the velocity component v_y changes with time. The total velocity vector (**v**) at any time is said to be tangent to the curved path of the ball. It is at an angle θ relative to the x axis, given by $\theta = \tan^{-1}(v_y/v_x)$.

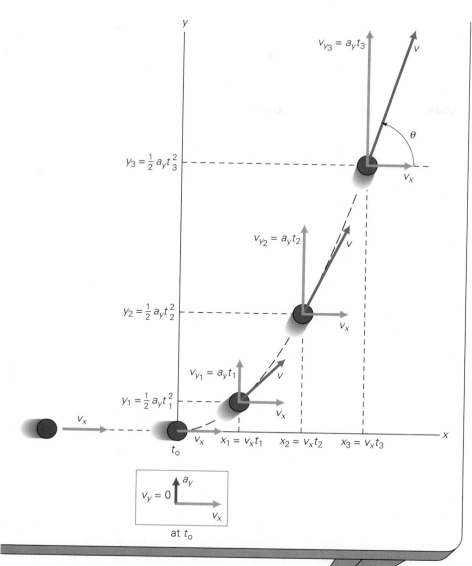

$$v_{y_3} = a_y t_3$$

$$y_3 = \frac{1}{2} a_y t_3^2$$

$$v_{y_2} = a_y t_2$$

$$y_2 = \frac{1}{2} a_y t_2^2$$

$$v_{y_1} = a_y t_1$$

$$y_1 = \frac{1}{2} a_y t_1^2$$

$$x_1 = v_x t_1 \qquad x_2 = v_x t_2 \qquad x_3 = v_x t_3$$

$v_y = 0$

at t_o

Figure 3.2

• FIGURE 3.2 Curvilinear motion An acceleration not parallel to the instantaneous velocity produces a curved path. Here an acceleration ($\mathbf{a}_y$) is applied at t_o to a ball moving with a constant velocity ($\mathbf{v}_x$). The result is a curved path with the velocity components as shown.

EXAMPLE 3.2 ■ Vector Components of Curvilinear Motion

Suppose that the ball in Fig. 3.2 has an initial velocity of 1.50 m/s along the x axis and starting at t_o receives an acceleration of 2.80 m/s² in the y direction. (a) What is the position of the ball 3.00 s after t_o? (b) What is the velocity of the ball at that time?

Solution. Referring to Fig. 3.2, we have the following:

Given: $v_{x_0} = v_x = 1.50$ m/s
$\qquad v_{y_0} = 0$
$\qquad a_x = 0$
$\qquad a_y = 2.80$ m/s²
$\qquad t = 3.00$ s

Find: (a) (x, y)
$\qquad$ (position
$\qquad\qquad$ coordinates)
$\qquad$ (b) $\mathbf{v}$
$\qquad\qquad$ (velocity)

(a) At 3.00 s after $t_o = 0$, the ball has traveled the following distances from the origin in the x and y directions:

$$x = v_{x_o}t + \tfrac{1}{2}a_x t^2 = (1.50 \text{ m/s})(3.00 \text{ s}) + 0 = 4.50 \text{ m}$$

$$y = v_{y_o}t + \tfrac{1}{2}a_y t^2 = 0 + \tfrac{1}{2}(2.80 \text{ m/s}^2)(3.00 \text{ s})^2 = 12.6 \text{ m}$$

Thus, its position is $(x, y) = (4.50 \text{ m}, 12.6 \text{ m})$. If you computed the distance, $d = \sqrt{x^2 + y^2}$, what would it be? (Note that this is not the actual distance the ball has traveled in 3.00 s, but the magnitude of the displacement from the origin at $t = 3.00$ s.)

(b) The x component of the velocity is given by

$$v_x = v_{x_o} + a_x t = 1.50 \text{ m/s} + 0 = 1.50 \text{ m/s}$$

(This is constant since there is no acceleration in the x direction.) The y component of the velocity is

$$v_y = v_{y_o} + a_y t = 0 + (2.80 \text{ m/s}^2)(3.00 \text{ s}) = 8.40 \text{ m/s}$$

The velocity therefore has a magnitude of

$$v = \sqrt{v_x^2 + v_y^2} = \sqrt{(1.50 \text{ m/s})^2 + (8.40 \text{ m/s})^2} = 8.53 \text{ m/s}$$

and its direction relative to the x axis is

$$\theta = \tan^{-1}\left(\frac{v_y}{v_x}\right) = \tan^{-1}\left(\frac{8.40 \text{ m/s}}{1.50 \text{ m/s}}\right) = 79.9°$$

> Don't confuse the direction of the velocity with the direction of the displacement from the origin. The direction of the velocity must always be tangent to the path.

3.2 ■ Vector Addition and Subtraction

Many physical quantities are vectors. You have worked with a few related to motion (displacement, velocity, and acceleration) and will encounter more during this course of study. A very important technique in the analysis of many physical situations is the addition (and subtraction) of vector quantities. By adding or combining such quantities (**vector addition**), you can obtain the overall, or net, effect that occurs—the resultant, as we call the vector sum.

See Fig. 2.5

You have already been adding vectors. In Chapter 2, displacements were added to get the net displacement. In this chapter, vector components of motion were added to give net effects. Notice in the preceding Example 3.2 the velocity components v_x and v_y were added to give the resultant velocity.

In this section, we will look at vector addition and subtraction in general, along with common vector notation. As you will learn, these operations are not the same as scalar or numerical addition and subtraction, with which you are already familiar. Vectors have magnitudes *and* directions, so different rules apply.

In general, there are geometric (graphic) methods and analytical (calculation) methods of vector addition. The geometric methods are useful in helping to visualize the concepts of vector addition. Analytical methods are more commonly used, however, because they are faster and more precise. One thing that scalar addition and vector addition have in common is that the quantities being added must have the same units. You cannot meaningfully add a displacement vector to a velocity vector, just as you cannot add the scalars of distance and speed and obtain a result with any physical meaning.

Vector Addition: Geometric Methods

Triangle Method. To add two vectors, for example, to add **B** to **A** (that is, to find **A** + **B**) by the **triangle method**, you first draw **A** on a sheet of graph paper to some scale. For example, if **A** is a displacement in meters, a convenient scale is 1 cm : 1 m, or 1 cm of vector length on the graph corresponds to 1 meter of displacement (●Fig. 3.3a). As shown in Fig. 3.3b, the direction of the **A** vector is specified as being at an angle (θ) relative to a coordinate axis, usually the x axis.

Next draw **B** with its tail starting at the tip of **A**. (Thus, this method is also called the *tip-to-tail method*.) The vector from the tail of **A** to the tip of **B** is then the vector sum, or the resultant of the two vectors: **R** = **A** + **B**.

If drawn to scale, the magnitude of **R** can be found by measuring its length and using the scale conversion. In such a graphical approach, the direction angle (θ_R) is measured with a protractor. Knowing the magnitudes and directions (θ's) of **A** and **B**, the magnitude and direction of **R** can also be found analytically using trigonometric methods. For the non-right triangle in Fig. 3.3b, the laws

Figure 3.3

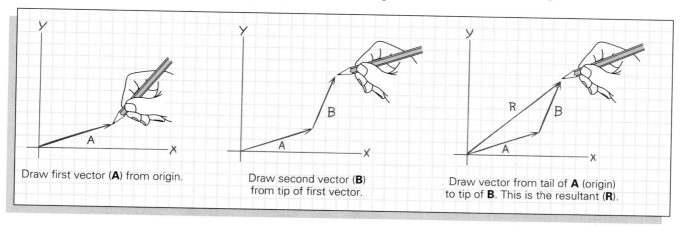

Draw first vector (**A**) from origin.

Draw second vector (**B**) from tip of first vector.

Draw vector from tail of **A** (origin) to tip of **B**. This is the resultant (**R**).

(a)

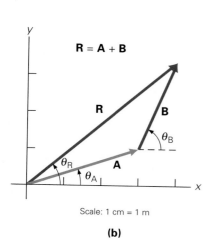

Scale: 1 cm = 1 m

(b)

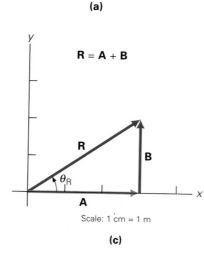

Scale: 1 cm = 1 m

(c)

●**FIGURE 3.3 Triangle method of vector addition** (a) The vectors **A** and **B** are placed tip to tail. The vector from the tail of **A** to the tip of **B**, forming the third side of the triangle, is the resultant, **R** = **A** + **B**. (b) When drawn to scale, the magnitude of **R** can be found by measuring its length and using the scale conversion, and the direction angle θ_R may be measured with a protractor. Analytical methods may also be used. For a non-right triangle as in (b), the laws of sines and cosines would be used. If the vector triangle is a right triangle as in (c), **R** is easily obtained with the Pythagorean theorem and the direction angle is given by an inverse trigonometric function.

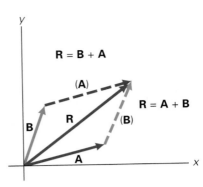

FIGURE 3.4 Parallelogram method of vector addition The diagonal of the parallelogram formed after **A** and **B** are drawn with their tails at the origin is the resultant, **R** = **A** + **B**. When drawn to scale, the magnitude and direction of **R** can be measured directly from the diagram as in the triangle method. Notice how similar the parallelogram method is to the triangle method. Shifting **B** to the right forms the same **A** + **B** vector triangle shown in Fig. 3.3b. Shifting **A** up has the same effect where **R** = **B** + **A**.

of sines and cosines would be used (see Appendix I). The resultant of the vector right triangle in Fig. 3.3c would be much easier to find using the Pythagorean theorem for the magnitude and an inverse trig function to find the direction angle. [Notice how **R** in this case is made up of x and y components (**A** and **B**).]

Parallelogram Method. Another graphical method of vector addition similar to the triangle method is the **parallelogram method**. In Fig. 3.4, **A** and **B** are drawn tail to tail, and a parallelogram is formed as shown. The resultant **R** lies along the diagonal of the parallelogram. Drawing the diagram to scale with proper orientations, the magnitude and direction of **R** can be measured directly from the diagram as in the triangle method.

Notice that **B** could be moved over to the other side of the parallelogram to form the **A** + **B** triangle. In general, a vector (arrow) can be moved around in vector addition methods. As long as you don't change its length (magnitude) or direction, you don't change the vector. In Fig. 3.4, this shifting of vector arrows shows that **A** + **B** = **B** + **A**—that is, the vectors can be added in either order.

Polygon Method. The tip-to-tail method can be extended to include the addition of any number of vectors. The method is then called the **polygon method** because the resulting graphical figure is a polygon. This is illustrated for four vectors in Fig. 3.5, where **R** = **A** + **B** + **C** + **D**. Note that this addition is essentially three applications of the triangle method. The length and direction of the resultant could be found analytically by successive applications of the laws of sines and cosines, but an easier analytical method, the component method, is described on page 73. Also, as in the parallelogram method, the four vectors (or any number of vectors) can be added in any order.

Vector Subtraction. Vector subtraction is a special case of vector addition.

$$\mathbf{A} - \mathbf{B} = \mathbf{A} + (-\mathbf{B})$$

That is, to subtract **B** from **A**, a *negative* **B** is added to **A**. In Chapter 2, you

A common error some students make when using the polygon method is to put the arrow head on the wrong end of the resultant vector. Make sure all vectors are well constructed with an arrow showing their direction. Clearly differentiate the resultant (e.g., by color).

FIGURE 3.5 Polygon method of vector addition The vectors to be added are placed tip to tail. The resultant **R** is the vector from the tail of the first vector (**A**) to the tip of the last vector (**D**), which completes the polygon. This method is essentially multiple applications of the triangle method: (**A** + **B**) + **C**, and so on.

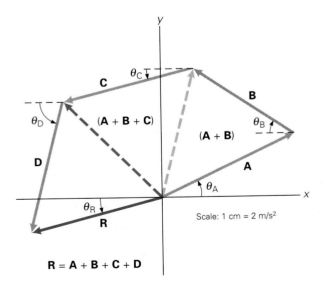

learned that a minus sign simply means that the direction of a vector is opposite to that of one with a plus sign, e.g., $+x$ and $-x$. The same is true for a general vector. The vector $-\mathbf{B}$ has the same magnitude as the vector $\mathbf{B}$, but is in the opposite direction (see ● Fig. 3.6). The vector diagram in the figure provides a graphical representation of $\mathbf{A} - \mathbf{B}$.

See Section 2.1.

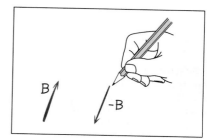

EXAMPLE 3.3 ■ Vector Subtraction

Consider yourself in a car (A) traveling along a straight, level highway with a speed $v_A = 75$ km/h. Another car (B) travels at a different speed $v_B = 90$ km/h. Find the differences in the velocities, $\mathbf{v}_{BA} = \mathbf{v}_B - \mathbf{v}_A$ when (a) the other car travels in the same direction in front of you, and (b) the other car is approaching you traveling in the opposite direction.

Solution. Taking the direction in which you are traveling as positive, we have

Given: $v_A = (+)\,75$ km/h   *Find:* $\mathbf{v}_{BA} = \mathbf{v}_B - \mathbf{v}_A$
     (a) $v_B = (+)\,90$ km/h (same direction)
     (b) $v_B = -90$ km/h (opposite direction)

(a) With both cars traveling in the same $(+)$ direction,

$$\mathbf{v}_{BA} = \mathbf{v}_B - \mathbf{v}_A = 90\ \text{km/h} - (+75\ \text{km/h}) = 15\ \text{km/h}$$

(b) When traveling in opposite directions,

$$\mathbf{v}_{BA} = \mathbf{v}_B - \mathbf{v}_A = 90\ \text{km/h} - (-75\ \text{km/h}) = 165\ \text{km/h}$$

Here, $\mathbf{v}_{BA}$ is what is called a *relative* velocity—the velocity of B relative to A. In part (a), relative to you in car A, car B appears to travel at a speed of 15 km/h. However, when approaching you, car B appears to travel at 165 km/h. It's as though you consider yourself or your car's reference frame to be at rest. What are the given velocities (the 90 km/h and 75 km/h) referenced to? (We will explore the subject of relative velocity further in Section 3.3.)

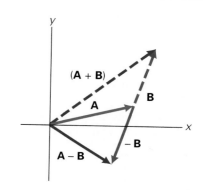

● **FIGURE 3.6 Vector subtraction** Vector subtraction is a special case of vector addition; that is, $\mathbf{A} - \mathbf{B} = \mathbf{A} + (-\mathbf{B})$, where $-\mathbf{B}$ has the same magnitude as $\mathbf{B}$ but is in the opposite direction (see sketch). Note that $\mathbf{A} + \mathbf{B}$ is *not* the same as $\mathbf{A} - \mathbf{B}$.

Vector Components and the Analytical Component Method

Probably the most widely used analytical method for adding multiple vectors is the **component method**. It will be used again and again throughout the course of our study, so a basic understanding of the method is *essential*. Learn this section well.

Adding Rectangular Vector Components. By rectangular components, we mean those at right (90°) angles to each other and usually taken in the rectangular coordinate x and y directions. You have already had an introduction to the addition of such components in the discussion on the velocity components of motion. For the general case, suppose that $\mathbf{A}$ and $\mathbf{B}$, two vectors at right angles, are added, as illustrated in ● Fig. 3.7a. The right angle makes things easy. The magnitude of $\mathbf{C}$ is given by the Pythagorean theorem:

$$C = \sqrt{A^2 + B^2} \tag{3.4a}$$

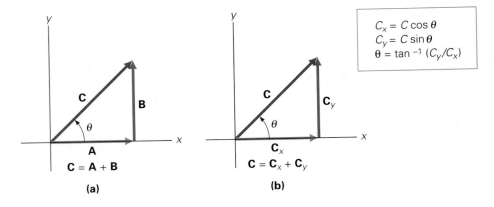

$$C_x = C \cos \theta$$
$$C_y = C \sin \theta$$
$$\theta = \tan^{-1}(C_y/C_x)$$

(a) $\mathbf{C} = \mathbf{A} + \mathbf{B}$

(b) $\mathbf{C} = \mathbf{C}_x + \mathbf{C}_y$

• FIGURE 3.7 Vector components (a) The vectors **A** and **B** along the x and y axes, respectively, add to give **C**. (b) A vector **C** may be resolved into rectangular components $\mathbf{C}_x$ and $\mathbf{C}_y$.

The orientation of **C** to the x axis is given by the angle

$$\theta = \tan^{-1}\left(\frac{B}{A}\right) \tag{3.4b}$$

This is how a resultant is expressed in **magnitude-angle form**.

Magnitude-angle form of vector representation

Resolving a Vector into Rectangular Components; Unit Vectors. Resolving a vector into rectangular components is essentially the reverse of adding the components. Given a vector **C**, Fig. 3.7b illustrates how it may be resolved into x and y components, $\mathbf{C}_x$ and $\mathbf{C}_y$. You simply complete the vector triangle with x and y components. From the diagram, it can be seen that the magnitudes or vector lengths of these components are given by

> Emphasize correct trigonometry notation. In most calculators the angle is entered first and then the function button is pressed. Thus some students tend to write them in that order on their paper, which is incorrect and misleading.

$$C_x = C \cos \theta \tag{3.5a}$$
$$C_y = C \sin \theta \tag{3.5b}$$

magnitudes of components

(similar to $v_x = v \cos \theta$ and $v_y = v \sin \theta$ in Example 3.1). The angle of direction for **C** can also be expressed in terms of the components since $\tan \theta = C_y/C_x$, or

Vector components

$$\theta = \tan^{-1}\left(\frac{C_y}{C_x}\right) \qquad \begin{array}{l}\text{direction of vector} \\ \text{from components magnitudes}\end{array} \tag{3.6}$$

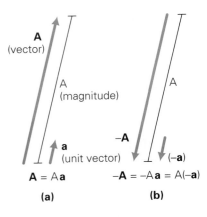

$\mathbf{A} = A\mathbf{a}$

$-\mathbf{A} = -A\mathbf{a} = A(-\mathbf{a})$

(a)   **(b)**

• FIGURE 3.8 Unit vectors (a) A unit vector **a** has a magnitude of unity, or one, and so simply indicates a vector's direction. Written with the magnitude A, it represents the vector **A**, that is, $\mathbf{A} = A\mathbf{a}$. (b) For the vector $-\mathbf{A}$, the unit vector is $-\mathbf{a}$, and $-\mathbf{A} = -A\mathbf{a} = A(-\mathbf{a})$. The magnitude A is a positive number, i.e., $|A|$.

A general notation for expressing the magnitude and direction of a vector involves the use of unit vectors. For example, as illustrated in •Fig. 3.8, a vector **A** may be written as $\mathbf{A} = A\mathbf{a}$. The numerical magnitude is represented by A, and **a** is called a **unit vector**. That is, it has a magnitude of unity, or one, and so simply indicates the vector's direction. For example, a velocity along the x axis might be written $\mathbf{v} = 4.0$ m/s **x** (that is 4.0 m/s magnitude in the +x direction).

Note in the figure how $-\mathbf{A}$ would be represented in this notation. Although the minus sign is sometimes put in front of the numerical magnitude, this is an absolute number; the minus actually goes with the unit vector; $-\mathbf{A} = -A\mathbf{a} = A(-\mathbf{a})$.* That is, the unit vector is in the $-\mathbf{a}$ direction (opposite **a**). A velocity

*The notation is sometimes written $\mathbf{A} = |A|\,\mathbf{a}$ or $-\mathbf{A} = -|A|\,\mathbf{a}$ to express the absolute value of the magnitude.

of $\mathbf{v} = -4.0$ m/s $\mathbf{x}$ has a magnitude of 4.0 m/s in the $-x$ direction, $\mathbf{v} = 4.0$ m/s $(-\mathbf{x})$.

This notation can be used to express explicitly the rectangular components of a vector. For example, the ball's displacement from the origin in Example 3.2 could be written $\mathbf{d} = (4.50$ m$)$ $\mathbf{x} + (12.6$ m$)$ $\mathbf{y}$, where $\mathbf{x}$ and $\mathbf{y}$ are unit vectors in these directions. In some instances it may be more convenient to express a general vector in this **component form:**

$$\mathbf{C} = C_x\mathbf{x} + C_y\mathbf{y} \qquad (3.7)$$

Component form of vector representation

Vector Addition Using Components. The **analytical component method** of vector addition involves resolving the vectors into rectangular components and adding the components for each axis independently. This is illustrated graphically in ● Fig. 3.9 for two vectors, $\mathbf{F}_1$ and $\mathbf{F}_2$. The sums of the x and y components of the vectors being added are then equal to the corresponding components of the resultant vector.

Component method of vector addition

Suppose that you are given three (or more) vectors to add. You could find the resultant by applying the graphical tip-to-tail method as illustrated in ● Fig. 3.10a. However, this involves drawing the vectors to scale and using a protractor to measure angles, which is time-consuming. Note that the magnitude of the directional angle θ is not obvious in the figure and would have to be measured (or computed trigonometrically).

However, if you use the component method, you do not have to draw the vectors tip to tail. In fact, it is usually more convenient to put all of the tails

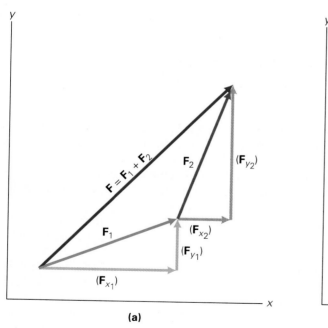

(a)

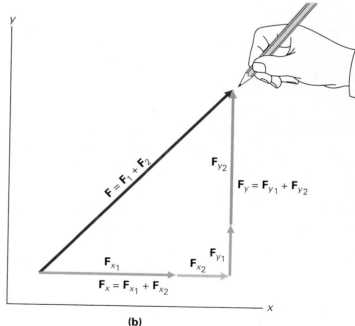

(b)

● **FIGURE 3.9 Component addition** (a) In adding vectors by the component method, each vector is first resolved into its x and y components. (b) The sums of the x and y components of $\mathbf{F}_1$ and $\mathbf{F}_2$ are the components of the resultant $\mathbf{F}$; that is, $\mathbf{F}_x = \mathbf{F}_{x_1} + \mathbf{F}_{x_2}$ and $\mathbf{F}_y = \mathbf{F}_{y_1} + \mathbf{F}_{y_2}$.

Figure 3.9

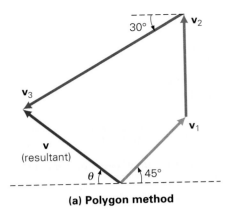

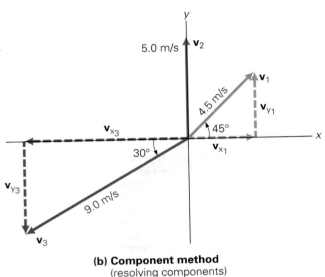

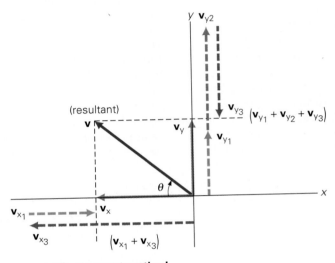

FIGURE 3.10 Component method of vector addition (a) Several vectors may be added graphically to find the resultant **v**, but this is time-consuming. In the analytical component method **(b)** the vectors to be added are all first placed with their tails at the origin so that they may be easily resolved into rectangular components. **(c)** The respective summations of all the *x* components and all the *y* components are then added to give the resultant **v**. Notice how the negative components subtract. See Example 3.4.

(a) Polygon method

(b) Component method
(resolving components)

(c) Component method
(adding *x* and *y* components, shown as offset dashed arrows, and finding resultant)

   Figure 3.10

together at the origin, as shown in Fig. 3.10b. Also, the vectors do not have to be drawn to scale, since the sketch is just a visual aid in applying the analytical method.

In the component method, you simply resolve the vectors to be added into their *x* and *y* components, add the respective components, and recombine to find the resultant. This procedure is illustrated in Fig. 3.10c. Notice that it is just linear vector addition in the component directions, and that components may add or subtract. A vector in the *x* or *y* direction has a "component" only in that direction (e.g., v_2 in the figure).

Notice also in Fig. 3.10c that the directional angle θ of the resultant is referred to the *x* axis, as are the individual vectors in Fig. 3.10b. In adding vectors by the component method, we will reference all vectors to the *nearest* *x* axis, i.e., $+x$ or $-x$. As will be seen, this eliminates angles greater than 90° (as occurs when customarily measuring angles counterclockwise from the $+x$ axis) and the use of double angle formulas, e.g., $\cos(\theta + 90°)$. The recommended procedures for adding vectors analytically by the component method can be summarized as follows:

Procedures for Adding Vectors by the Component Method

1. Resolve the vectors to be added into their x and y components. Use the acute angles (those less than 90°) between the vectors and the x axis, and indicate the components' directions by plus and minus signs. (See Fig. 3.11.)

2. Add all of the x components together and all of the y components together vectorially to obtain the x and y components of the resultant, or vector sum. (This is done algebraically using plus and minus signs.)

3. Express the resultant vector using:
   (a) the component form—e.g., $\mathbf{C} = C_x\,\mathbf{x} + C_y\,\mathbf{y}$—or
   (b) the magnitude-angle form.

For the latter, find the magnitude of the resultant using the summed x and y components and the Pythagorean formula.

$$C = \sqrt{C_x^2 + C_y^2}$$

Find the angle of direction (relative to the x axis) by taking the arctangent of the *absolute value* of the ratio of the y and x components.

$$\theta = \tan^{-1}\left|\frac{C_y}{C_x}\right|$$

Designate the quadrant in which the resultant lies. This is obtained from the signs of the summed components or a sketch of their addition using the triangle (or rectangle) method (see Fig. 3.11). The angle θ is the angle between the resultant and the x axis in that quadrant.

Figure 3.11

Taking the absolute value means to ignore any minus sign (e.g., $|-3| = 3$). This is done to avoid negative values and angles greater than 90°.

FIGURE 3.11 Vector addition by the analytical component method (a) Resolve the vectors into their x and y components. (b) Add all of the x components and all of the y components together vectorially to obtain the x and y components (C_x and C_y) of the resultant. Express the resultant in either component form or magnitude-angle form. See text for more detailed description.

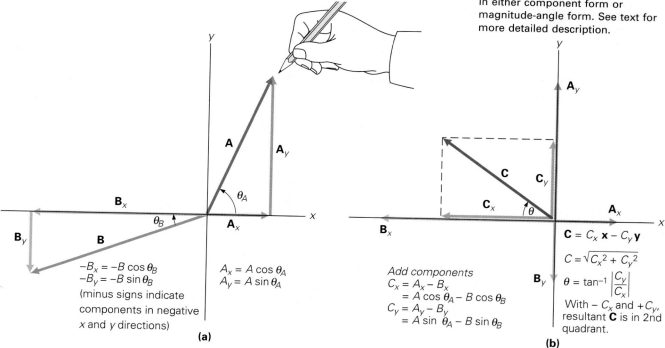

(a)

$-B_x = -B \cos \theta_B$
$-B_y = -B \sin \theta_B$
(minus signs indicate components in negative x and y directions)

$A_x = A \cos \theta_A$
$A_y = A \sin \theta_A$

Add components
$C_x = A_x - B_x$
$\quad = A \cos \theta_A - B \cos \theta_B$
$C_y = A_y - B_y$
$\quad = A \sin \theta_A - B \sin \theta_B$

(b)

$\mathbf{C} = C_x\,\mathbf{x} - C_y\,\mathbf{y}$

$C = \sqrt{C_x^2 + C_y^2}$

$\theta = \tan^{-1}\left|\frac{C_y}{C_x}\right|$

With $-C_x$ and $+C_y$, resultant $\mathbf{C}$ is in 2nd quadrant.

EXAMPLE 3.4 ■ Applying the Analytical Component Method

Let's apply the procedural steps of the component method to the addition of the vectors in Fig. 3.10b (reproduced here for convenience). The vectors with units of m/s represent velocities.

Solution.

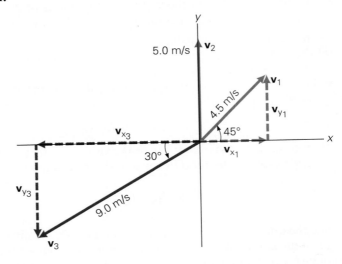

1. The rectangular components of the vectors are shown in the figure.
2. Summing these components gives

$$\mathbf{v} = v_x\mathbf{x} + v_y\mathbf{y} = (v_{x_1} + v_{x_2} + v_{x_3})\,\mathbf{x} + (v_{y_1} + v_{y_2} + v_{y_3})\,\mathbf{y}$$

where

$$v_x = v_{x_1} + v_{x_2} + v_{x_3} = v_1\cos 45° + v_2\cos 90° - v_3\cos 30°$$
$$= (4.5\ \text{m/s})(0.707) + (5.0\ \text{m/s})(0) - (9.0\ \text{m/s})(0.866) = -4.6\ \text{m/s}$$

$$v_y = v_{y_1} + v_{y_2} + v_{y_3} = v_1\sin 45° + v_2\sin 90° - v_3\sin 30°$$
$$= (4.5\ \text{m/s})(0.707) + (5.0\ \text{m/s})(1) - (9.0\ \text{m/s})(0.50) = 3.7\ \text{m/s}$$

In tabular form, the components are:

	x components		y components	
v_{x_1}	$v_1\cos 45° = +3.2$ m/s	v_{y_1}	$v_1\sin 45° = +3.2$ m/s
v_{x_2}	$v_2\cos 90° = 0$	v_{y_2}	$v_2\sin 0° = +5.0$ m/s
v_{x_3}	$v_3\cos 30° = -7.8$ m/s	v_{y_3}	$v_3\sin 30° = -4.5$ m/s
Sums: v_x	$= -4.6$ m/s	v_y	$= +3.7$ m/s

The directions of the components are indicated by signs (the + sign usually omitted as being understood) and that v_2 has no x component. Note that in general for the analytic component method, the x components are cosine functions and the y components are sine functions.

3. In component form, the resultant vector is

$$\mathbf{v} = (-4.6 \text{ m/s})\mathbf{x} + (3.7 \text{ m/s})\mathbf{y}$$

In magnitude-angle form, the resultant velocity has a magnitude of

$$v = \sqrt{(v_x^2) + v_y^2} = \sqrt{(-4.6 \text{ m/s})^2 + (3.7 \text{ m/s})^2} = 5.9 \text{ m/s}$$

Since the x component is negative and the y component is positive, the resultant lies in the second quadrant at an angle of

$$\theta = \tan^{-1}\left|\frac{v_y}{v_x}\right| = \tan^{-1}\left(\frac{3.7 \text{ m/s}}{4.6 \text{ m/s}}\right) = 39°$$

relative to the negative x axis (see Fig. 3.10c).

Although this discussion is limited to motion in two dimensions (in a plane), the component method is easily extended to three dimensions. For a velocity in three dimensions, the vector has x, y, and z components; that is, $\mathbf{v} = v_x\mathbf{x} + v_y\mathbf{y} + v_z\mathbf{z}$.

3.3 ■ Relative Velocity

Measurements must be made with respect to some reference. This is usually taken to be the origin of a coordinate system. The point you designate as the origin of a set of coordinate axes is arbitrary and entirely a matter of choice. For example, you may "attach" the coordinate system to the road or the ground and then measure the displacement or velocity of a car relative to these axes.

A situation may be analyzed from any frame of reference. For example, the origin of the coordinate axes may be attached to a car moving along a highway. In analyzing motion from another reference frame, you do not change the physical situation or what is taking place, only the point of view from which you choose to describe it. Hence, we say that motion is *relative* (to some reference frame), and we refer to **relative velocity.** Since velocity is a vector, vector addition and subtraction are helpful in determining relative velocities.

Relative Velocities in One Dimension

When the velocities are linear (along a straight line in the same or opposite directions) and all have the same reference (such as the Earth), relative velocities can be found simply by vector subtraction. As an illustration, consider cars moving with constant velocities along a straight, level highway, as in ● Fig. 3.12. The velocities shown in the figure are *relative to the Earth or ground*, as indicated by the reference set of coordinate axes in Fig. 3.12a. They are also relative to the stationary observers standing by the highway and sitting in the parked car A. That is, these observers see the cars moving with the velocities indicated. The relative velocity of two objects is given by the velocity (vector) difference. For example, the velocity of car B *relative to car A* is given by

$$\mathbf{v}_{BA} = \mathbf{v}_B - \mathbf{v}_A = +90 \text{ km/h} - 0 \text{ km/h} = +90 \text{ km/h}$$

Thus, a person sitting in car A would see car B move away at a speed of 90 km/h (in the positive *x* direction). For this linear case, the directions of the velocities are indicated by plus and minus signs (in addition to the minus sign in the formula).

Similarly, the velocity of car C relative to an observer in car A is

$$\mathbf{v}_{CA} = \mathbf{v}_C - \mathbf{v}_A = -60 \text{ km/h} - 0 \text{ km/h} = -60 \text{ km/h}.$$

The person in car A would see car C approaching with a speed of 60 km/h (in the negative *x* direction).

But suppose that you want to know the velocities of the other cars *relative to car B*, that is, from the point of view of an observer in car B, or relative to a set of coordinate axes with the origin fixed to car B (Fig. 3.12b). Relative to those axes, car B is not moving and acts as the fixed reference point. The other cars are moving relative to car B. The velocity of car C relative to car B is

$$\mathbf{v}_{CB} = \mathbf{v}_C - \mathbf{v}_B = -60 \text{ km/h} - (+90 \text{ km/h}) = -150 \text{ km/h}$$

Similarly, car A has a velocity relative to car B of

$$\mathbf{v}_{AB} = \mathbf{v}_A - \mathbf{v}_B = 0 \text{ km/h} - (+90 \text{ km/h}) = -90 \text{ km/h}$$

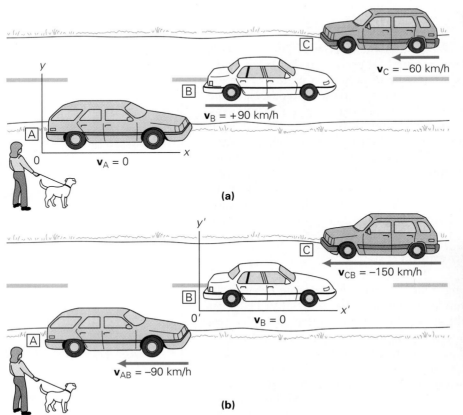

(a)

(b)

(c)

FIGURE 3.12 Relative velocity The observed velocity of a car depends on, or is relative to, the frame of reference. The velocities shown in **(a)** are relative to the ground or to the parked car A. In **(b)** the frame of reference is with respect to car B, and the velocities are those that the driver of car B would observe. **(c)** These aircraft, performing air-to-air refueling, are normally described as traveling at hundreds of km/h. To what frame of reference do these velocities refer? What is their velocity relative to each other?

Notice that relative to B, the other cars are both moving in the negative x direction. That is, C is approaching B with a velocity of 150 km/h in the $-x$ direction and A appears to be receding from B with a velocity of 90 km/h in the $-x$ direction. (Imagine yourself in car B, and take that position as stationary. Car C would appear to be coming toward you at a high rate of speed, and car A would be getting farther and farther away, as though it were moving backward relative to you.) Note that in general,

$$\mathbf{v}_{AB} = -\mathbf{v}_{BA}$$

What about the velocities relative to car C? From the point of view (or reference point) of car C, cars A and B would both appear to be moving in the positive x direction. For B relative to C,

$$\mathbf{v}_{BC} = \mathbf{v}_B - \mathbf{v}_C = 90 \text{ km/h} - (-60 \text{ km/h}) = +150 \text{ km/h}$$

Can you show that $\mathbf{v}_{AC} = +60$ km/h?

In some instances, we may need to work with velocities that do not all have the same reference. In such cases, relative velocities can be found by means of vector addition. To solve problems of this kind, *it is essential to identify the velocity references with care.*

Multiple reference frames

Let's look first at a one-dimensional (linear) example. Suppose that a straight moving walkway in a major airport moves with a velocity of $\mathbf{v}_{wg} = +1.0$ m/s, where the subscripts indicate the velocity of the walkway (w) with respect to or relative to the ground (g). A passenger (p) on the walkway trying to make a flight connection walks with a velocity of $\mathbf{v}_{pw} = +2.0$ m/s relative to the walkway. What is the passenger's velocity relative to an observer standing by the walkway (that is, relative to the ground)?

Use subscripts carefully! v_{AB} = velocity *of A relative to B*

The velocity we are seeking, $\mathbf{v}_{pq}$ is given by

$$\mathbf{v}_{pg} = \mathbf{v}_{pw} + \mathbf{v}_{wg} = +2.0 \text{ m/s} + 1.0 \text{ m/s} = +3.0 \text{ m/s}$$

Thus the stationary observer sees the passenger traveling with speed of 3.0 m/s down the walkway. (Make a sketch and show how the vectors add.)

Problem-Solving Hint

Notice the pattern of the subscripts in this example. On the right side of the equation, the two inner subscripts are the same (w). The outer subscripts (p and g) are sequentially the same as those for the relative velocity on the left side of the equation. When adding relative velocities, always check to make sure that the subscripts have this relationship—it indicates that you have set up the equation correctly.

What if a passenger got on the walkway going in the opposite direction and walked with the same speed as that of the walkway? Now it is essential to indicate the direction in which the passenger is walking by means of a minus sign: $\mathbf{v}_{pw} = -1.0$ m/s. In this case, relative to the stationary observer,

$$\mathbf{v}_{pg} = \mathbf{v}_{pw} + \mathbf{v}_{wg} = -1.0 \text{ m/s} + 1.0 \text{ m/s} = 0$$

so the passenger is stationary with respect to the ground and the walkway acts as a treadmill. (Excellent exercise!)

Relative Velocities in Two Dimensions

Of course, velocities are not always in the same or opposite directions. However, knowing how to use rectangular components to add or subtract vectors, we can solve problems involving relative velocities in two dimensions, as the following examples show.

EXAMPLE 3.5 ■ Relative Velocity and Components of Motion: Across and Down the River

The current of a 500-m wide straight river has a flow rate of 2.55 km/h. A motorboat that travels with a constant speed of 8.00 km/h in still water crosses the river (●Fig. 3.13). (a) If the boat's bow points directly across the river toward the opposite shore, what is the velocity of the boat relative to the stationary observer at the bridge? (b) How far downstream will the boat's landing point be from the point directly opposite its starting point? (c) What is the distance traveled by the boat in crossing the river? (Assume that the boat comes instantaneously to rest on a sandy shore.)

Solution. Here again, identifying the data is important. As indicated in the figure, we take the river's flow velocity ($\mathbf{v}_{rs}$) to be in the x direction and the boat's velocity ($\mathbf{v}_{br}$) to be in the y direction. Note however, that the river's

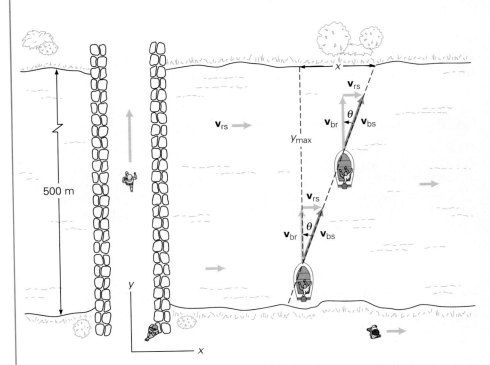

●**FIGURE 3.13 Relative velocity and components of motion** As the boat moves across the river, it is carried downstream by the current. Example 3.5.

Figure 3.13

flow velocity is *relative to the shore,* and that the boat's velocity is *relative to the river* or water. This is indicated by the subscripts. Listing the data, we have

Given: y_{max} = 500 m (river width)
v_{rs} = 2.55 km/h **x**
= 0.709 m/s **x**
(velocity of river *relative to shore*)
v_{br} = 8.00 km/h **y**
= 2.22 m/s **y**
(velocity of boat *relative to river*)

Find: (a) v_{bs} (velocity of boat relative to shore)
(b) x (distance downstream)
(c) d (distance traveled by boat)

(The conversion factor from km/h to m/s has been omitted for convenience.)

Notice that as the boat moves toward the opposite shore, it is also carried downstream by the current. These velocity components would be clearly apparent relative to the jogger crossing the bridge and to the person sauntering downstream in Fig. 3.13. If both stay even with the boat, the velocity of each will match one of the components of the boat's velocity. Since the velocity components are constant, the boat travels in a straight line diagonally across the river (much like the ball rolling across the table in Example 3.1).

(a) The velocity of the boat relative to the shore (v_{bs}) is given by vector addition. In this case, we have

$$\mathbf{v}_{bs} = \mathbf{v}_{br} + \mathbf{v}_{rs}$$

Since the velocities are not along one axis as in the previous example of the cars, they cannot be added directly. Notice in Fig. 3.13 that the vectors form a right triangle, so we can apply the Pythagorean theorem to find the magnitude of $\mathbf{v}_{bs}$:

$$v_{bs} = \sqrt{(v_{br})^2 + (v_{rs})^2} = \sqrt{(2.22 \text{ m/s})^2 + (0.709 \text{ m/s})^2}$$

$$= 2.33 \text{ m/s}$$

This is in the direction defined by

$$\theta = \tan^{-1}\left(\frac{v_{rs}}{v_{br}}\right) = \tan^{-1}\left(\frac{0.709 \text{ m/s}}{2.22 \text{ m/s}}\right) = 17.7°$$

(b) To find the distance x that the current carries the boat downstream, we use components. Note that in the y direction, $y_{max} = v_{br}t$, and

$$t = \frac{y_{max}}{v_{br}} = \frac{500 \text{ m}}{2.22 \text{ m/s}} = 225 \text{ s}$$

which is the time it takes the boat to cross the river.

During this time, the boat is carried downstream by the current a distance of

$$x = v_{rs}t = (0.709 \text{ m/s})(225 \text{ s}) = 160 \text{ m}$$

(c) We could find the distance d traveled by the boat using x and y_{max} with the Pythagorean theorem again, but let's use the magnitude of the relative velocity and time:

$$d = v_{bs}t = (2.33 \text{ m/s})(225 \text{ s}) = 524 \text{ m}$$

(Check this answer using x and y_{max}. It may be slightly different. Why?)

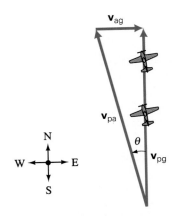

FIGURE 3.14 Flying into the wind To fly directly north, the plane's heading must be west of north. Example 3.6.

An air-speed indicator in a light plane gives its speed relative to the air. A fully instrumented plane will also have a ground-speed indicator.

EXAMPLE 3.6 ■ Relative Velocity: Flying Into the Wind

An airplane with an air speed of 200 km/h (its speed in still air) flies in a direction such that with a 50.0-km/h west wind blowing, it travels in a straight line northward. (Wind direction is specified by the direction *from* which the wind blows, so a west wind blows from west to east.) To maintain its course due north, the plane must fly at an angle into the wind as illustrated in ● Fig. 3.14. (a) What is the speed of the plane along its straight-line path? (b) What is the plane's heading (direction) in order to fly this route?

Solution. As always, it is important to identify what the given velocities are relative to.

Given: $\mathbf{v}_{pa}$ = 200 km/h at angle θ (velocity of plane with respect to still air, i.e., air speed)

$\mathbf{v}_{ag}$ = 50 km/h east (velocity of air with respect to Earth or ground, i.e., wind speed)

plane flies due north with velocity $\mathbf{v}_{pg}$

Find: (a) v_{pg} (ground speed of plane)

(b) θ (heading)

(a) The speed of the plane with respect to the Earth or ground, v_{pg}, is called its ground speed, whereas v_{pa} is its air speed. Vectorially, the respective velocities are related by

$$\mathbf{v}_{pg} = \mathbf{v}_{pa} + \mathbf{v}_{ag}$$

As can be seen, if there were no wind blowing ($v_{ag} = 0$), the ground speed and air speed would be equal. However, a head wind (one blowing directly toward the plane) would cause a slower ground speed; and a tail wind, a faster ground speed. The situation is analogous to that of a boat going upstream and downstream.

The vector relationship applies to our situation. Here $\mathbf{v}_{pg}$ is the resultant of the other two vectors, which can be added by the triangle method. We use the Pythagorean theorem to find v_{pg}, noting that v_{pa} is the hypotenuse of the triangle.

$$v_{pg} = \sqrt{v_{pa}^2 - v_{ag}^2} = \sqrt{(200 \text{ km/h})^2 - (50.0 \text{ km/h})^2} = 194 \text{ km/h}$$

(Note that it was convenient to use the units of km/h, since the calculation did not involve any other standard units.)

(b) To fly in a straight line, the plane must head obliquely into the wind at an angle. This head is given by

$$\theta = \tan^{-1}\left(\frac{v_{ag}}{v_{pg}}\right) = \tan^{-1}\left(\frac{50.0 \text{ km/h}}{194 \text{ km/h}}\right) = 14.5°$$

That is, 14.5° west of north.

3.4 ■ Projectile Motion

A familiar example of two-dimensional, curvilinear motion is the motion of objects that are thrown or projected by some means. The motion of a stone thrown across a stream or a golf ball driven off a tee is **projectile motion.** A special case of projectile motion in one dimension occurs when an object is projected vertically upward. This case was treated in Chapter 2 in terms of free fall with air resistance neglected. We will generally neglect air resistance here also so that the only acceleration acting on a projectile is that due to gravity.

Review Section 2.5

Projectile motion is easily analyzed using vector components. You simply break up the motion and look at its individual one-dimensional components.

Horizontal Projections

It is worthwhile to first analyze the special case of the motion of an object projected horizontally, or parallel to a level surface. Suppose that you throw an object horizontally with an initial velocity v_{x_0} (Fig. 3.15a). Projectile motion

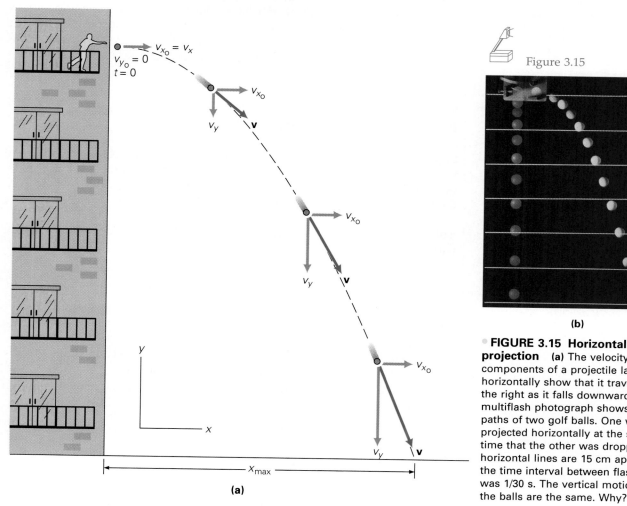

(a)

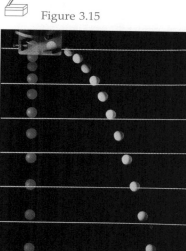

Figure 3.15

(b)

● **FIGURE 3.15 Horizontal projection** **(a)** The velocity components of a projectile launched horizontally show that it travels to the right as it falls downward. **(b)** A multiflash photograph shows the paths of two golf balls. One was projected horizontally at the same time that the other was dropped. The horizontal lines are 15 cm apart, and the time interval between flashes was 1/30 s. The vertical motions of the balls are the same. Why?

is analyzed beginning at the instant of release ($t = 0$). Once the object is released, the horizontal acceleration is zero ($a_x = 0$), so the horizontal velocity is constant: $v_x = v_{x_0}$.

According to the equation $x = v_x t$ (Eq. 3.2a), the projected object would continue to travel in the horizontal direction indefinitely. However, you know that this is not what really happens. As soon as the object is projected, it is in free fall in the vertical direction, with $v_{y_0} = 0$ (as though it were dropped) and $a_y = -g$. The result is a curved path as illustrated in Fig. 3.15. (Compare the motions in Fig. 3.15 and Fig. 3.2. Do you see any similarities?)

Essentially, the projected object travels with a uniform velocity in the horizontal direction while *at the same time* undergoing acceleration in the downward direction under the influence of gravity. The result is a curved path. If there were no horizontal motion, the object would simply drop to the ground in a straight line, and in fact, the time of flight of the projected object is exactly the same as if it were falling vertically.

Note the components of the velocity vector in Fig. 3.15a. The length of the horizontal component of the velocity vector remains the same, but the length of the vertical component increases with time. What is the instantaneous velocity at any point along the path? (Think in terms of vector addition, covered in the last section.) The photo in Fig. 3.15b shows the actual motions of a horizontally projected golf ball and one that is simultaneously dropped from rest. The horizontal reference lines show that the balls fall vertically at the same rate. The horizontally projected ball also travels to the right as it falls.

> Demonstration/Activity: The constant horizontal speed in Fig. 3.15 b can be shown by placing a blank transparency over the photograph and marking the position of the balls. Then place the transparency on the overhead projector and construct straight vertical lines through the balls and observe their equal spacing.

> Demonstration/Activity: While standing on a desk, throw an object such as a marble horizontally outward while dropping an identical object at the same time. Listen for the sound of the objects striking the floor. They will tend to do so at the same time.

EXAMPLE 3.7 ■ Horizontal Projection

Suppose that the ball in Fig. 3.15a is projected from a height of 25.0 m above the ground and is thrown with an initial horizontal velocity of 8.25 m/s. (a) How long is the ball in flight before striking the ground? (b) How far from the building does the ball strike the ground?

Solution. Writing the data with downward taken as the negative direction, we have

Given: $y = -25.0$ m   *Find:* (a) t (time of flight)
$v_{x_0} = 8.25$ m/s   (b) x (horizontal distance)
$a_x = 0$
$v_{y_0} = 0$
$a_y = -g$

(a) The time of flight is the same as the time it takes for the ball to fall vertically to the ground. This may be found using the equation $y = v_{y_0}t - \frac{1}{2}gt^2$, in which the negative direction of g is expressed explicitly as was done in Chapter 2. With $v_{y_0} = 0$,

$$y = -\tfrac{1}{2}gt^2$$

and

> Notice how the expression for time is first obtained algebraically from the fundamental expression prior to substituting in the values and calculating. Encourage students to practice this analytical technique.

$$t = \sqrt{\frac{2y}{-g}} = \sqrt{\frac{2(-25.0 \text{ m})}{-9.80 \text{ m/s}^2}} = 2.26 \text{ s}$$

(b) The ball travels in the x direction for the same amount of time it travels in the y direction (that is, 2.26 s); so, with $a_x = 0$,

$$x = v_{x_o} t = (8.25\ \text{m/s})(2.26\ \text{s}) = 18.6\ \text{m}$$

Projections at Arbitrary Angles

The general case of projectile motion involves an object projected at an arbitrary angle θ relative to the horizontal; for example, a golf ball struck by a club (Fig. 3.16). This motion is also analyzed using its components. As before, upward is taken as the positive direction and downward as the negative direction.

First, the initial velocity v_o is resolved into rectangular components:

$$v_{x_o} = v_o \cos\theta \quad (3.8a)$$
$$v_{y_o} = v_o \sin\theta \quad (3.8b)$$
Initial velocity components ($t = 0$)

Remind students to treat the x and y components of motion independently.

Since there is no horizontal acceleration and gravity acts in the negative y direction, the x component of the velocity is constant and the y component varies with time (see Eq. 3.3):

$$v_x = v_{x_o} = v_o \cos\theta \quad (3.9a)$$
$$v_y = v_{y_o} - gt = (v_o \sin\theta) - gt \quad (3.9b)$$
velocity components ($t > 0$)

The components of the instantaneous velocity at various times are illustrated in Fig. 3.16. The instantaneous velocity is the sum of these components and is tangent to the curved path of the ball at any point.

Similarly, the displacement components are given by

$$x = v_{x_o} t = (v_o \cos\theta)t \quad (3.10a)$$
$$y = v_{y_o} t - \tfrac{1}{2} g t^2 = (v_o \sin\theta)t - \tfrac{1}{2} g t^2 \quad (3.10b)$$
displacement components

Figure 3.16

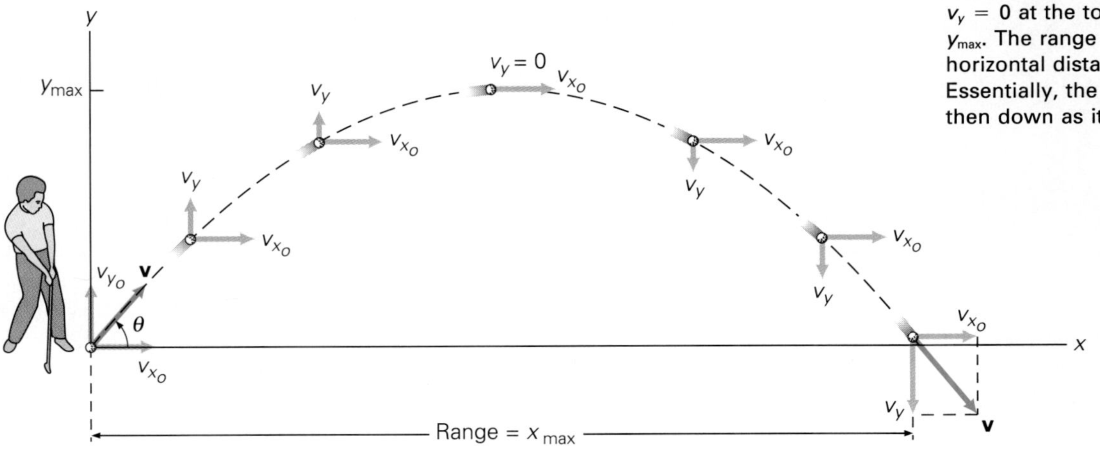

FIGURE 3.16 Projection at an angle The velocity components are shown for various times. Note that $v_y = 0$ at the top of the arc, or at y_{max}. The range (R) is the maximum horizontal distance, or x_{max}. Essentially, the ball travels up and then down as it travels to the right.

The curve produced by these equations, or the path of motion of the projectile, is called a **parabola**. The path of projectile motion is often referred to as a parabolic arc (●Fig. 3.17).

During projectile motion, the object travels up and down while traveling horizontally with a constant velocity. As in the case of horizontal projection, *time is the common feature shared by the components of motion.* Aspects of projectile motion that may be of interest in various situations include the time of flight, the maximum height reached, and the **range**, which is the maximum horizontal distance traveled.

(a)

(b)

●**FIGURE 3.17 Parabolic arcs**
(a) Fireworks and **(b)** fountain streams all move in parabolic arcs, though air resistance may distort the trajectories to some extent.

EXAMPLE 3.8 ■ Projection at an Angle

The golf ball in Fig. 3.16 leaves the tee with an initial velocity of 30.0 m/s at an angle of 37° to the horizontal. (a) What is the maximum height reached by the ball? (b) What is its range?

Solution.

Given: $v_o = 30.0 \text{ m/s}$ *Find:* (a) y_{max}
$\quad\quad\quad \theta = 37°$ (b) $R = x_{max}$
$\quad\quad\quad a_y = -g$

Computing v_{x_o} and v_{y_o} explicitly eliminates the sine and cosine terms in the kinematic equations:

$$v_{x_o} = v_o \cos 37° = (30.0 \text{ m/s})(0.799) = 24.0 \text{ m/s}$$

$$v_{y_o} = v_o \sin 37° = (30.0 \text{ m/s})(0.602) = 18.1 \text{ m/s}$$

(a) Just as for an object thrown vertically upward, $v_y = 0$ at the maximum height (y_{max}). Thus, the time to reach the maximum height (t_m) is given by Eq. 3.9b:

$$v_y = 0 = v_{y_o} - gt_m \quad \text{and} \quad t_m = \frac{v_{y_o}}{g} = \frac{18.1 \text{ m/s}}{9.80 \text{ m/s}^2} = 1.85 \text{ s}$$

As in the case of vertical projection, the time in going up is equal to the time in coming down, so the total time of flight is $t = 2t_m$ (to return to the elevation from which the object was projected).

The maximum height (y_{max}) is then obtained by substituting t_m into Eq. 3.10b:

$$y_{max} = v_{y_o}t_m - \tfrac{1}{2}gt_m^2 = (18.1 \text{ m/s})(1.85 \text{ s}) - \tfrac{1}{2}(9.80 \text{ m/s}^2)(1.85 \text{ s})^2 = 16.7 \text{ m}$$

The maximum height could also be obtained directly from $v_y^2 = v_{y_o}^2 - 2gy$ with $y = y_{max}$ and $v_y = 0$. However, the method of solution used here illustrates how the time of flight is obtained.

(b) The range (R) is equal to the horizontal distance traveled (x_{max}), which is easily found by substituting the total time of flight $t = 2t_m = 2(1.85 \text{ s}) = 3.70 \text{ s}$ into Eq. 3.10a:

$$R = x_{max} = v_x t = v_{x_o}(2t_m) = (24.0 \text{ m/s})(3.70 \text{ s}) = 88.8 \text{ m}$$

EXAMPLE 3.9 ■ Projectile Coordinates

What is the location of the golf ball in the preceding example at $t = 3.00$ s?

Solution. At this time, the ball is still in flight and its location at that instant is given by its Cartesian coordinates (x, y). Using data from Example 3.8, we have:

$$x = v_{x_0}t = (24.0 \text{ m/s})(3.00 \text{ s}) = 72.0 \text{ m}$$

$$y = v_{y_0}t - \tfrac{1}{2}gt^2 = (18.1 \text{ m/s})(3.00 \text{ s}) - \tfrac{1}{2}(9.80 \text{ m/s}^2)(3.00 \text{ s})^2$$
$$= 10.2 \text{ m}$$

So the ball is at $(x, y) = (72.0 \text{ m}, 10.2 \text{ m})$ on its way to its maximum horizontal distance or range.

The range of a projectile is an important consideration in various applications. This is particularly true in those sports where a maximum range is desired, such as when driving a golf ball or throwing a javelin.

At what angle should an object be projected so as to attain its maximum range? To answer this requires looking more closely at the equation used in Example 3.8 to calculate the range, $R = v_x t$. First let's look at the expressions for v_x and t. We know that

$$v_x = v_{x_0} = v_0 \cos \theta$$

and $t = 2t_m$, where

$$t_m = \frac{v_{y_0}}{g} = \frac{v_0 \sin \theta}{g}$$

Then,

$$R = v_x t = (v_0 \cos \theta)(2t_m) = (v_0 \cos \theta)\left(\frac{2v_0 \sin \theta}{g}\right) = \frac{2v_0^2 \sin \theta \cos \theta}{g}$$

Using the trigonometric identity $\sin 2\theta = 2 \cos \theta \sin \theta$ (Appendix I) we have

$$\boxed{R = \frac{v_0^2 \sin 2\theta}{g}} \qquad \text{Projectile range } (x_{max}) \text{ for } y_{initial} = y_{final} \quad (3.11)$$

Note that the range depends on the magnitude of the initial velocity (or speed, v_0) and the angle of projection, θ (g is assumed to be constant). Keep in mind that this equation applies only to the special, but common, case of $y_{initial} = y_{final}$—that is, the landing point is at the same height as the launch point.

For a particular v_0, the range is a maximum (R_{max}) when $\sin 2\theta = 1$, since this is the maximum value of the sine function (varies from 0 to 1), and

$$R_{max} = \frac{v_0^2}{g} \qquad (3.12)$$

Demonstration/Activity: A water hose directing a steady stream of water into the air can be used to demonstate the parabolic path of a projectile. Using an aquarium or sink to catch the water and a chalkboard protractor, show how the range can vary with the angle. Try illuminating it in a dark room with a strobe light!

This equation should be used only when the net y displacement is zero. Students often try to use this equation even when there is a net y displacement.

Special equations such as the range and maximum height are fun to derive and use. But discourage students from memorizing them. Understanding how to derive them is much more important than recognizing them.

Since this occurs for sin $2\theta = 1$, and sin $90° = 1$, we have

$$2\theta = 90°$$

or

$$\theta = 45°$$

for the maximum range for a given initial speed when the projectile returns to the elevation from which it was projected. At a greater or smaller angle, for a projectile with the same initial speed, the range will be less, as illustrated in Fig. 3.18. Also, the range is the same for angles equally above and below 45°, such as 30° and 60°. (You can use Eq. 3.11 to show that this is the case.)

Thus, to get the maximum range, a projectile should *ideally* be projected at an angle of 45°. However, air resistance has been neglected. In actual situa-

Video demonstration 2, showing the monkey gun, relates to this material. The monkey gun is also the subject of end-of-chapter Exercise 78.

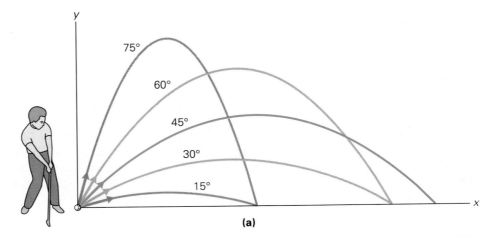

(a)

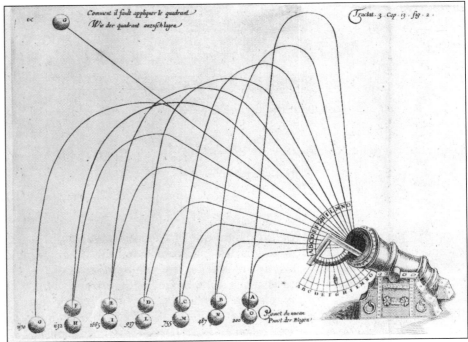

(b)

FIGURE 3.18 Range **(a)** For a projectile with a given initial speed, the maximum range is ideally attained with a projection angle of 45°. For projection angles above and below 45°, the range is shorter, and equal for angles equally different from 45° (for example, 30° and 60°) **(b)** The latter fact has been known for some time, as illustrated in this 1621 book on artillery. The general shapes of the (parabolic) paths, however, were not so well known.

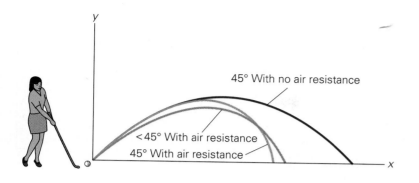

45° With no air resistance

<45° With air resistance

45° With air resistance

tions, such as when a ball or object is thrown or hit hard, this factor may have a significant effect. Air resistance reduces the speed of the projectile, thereby reducing the range. As a result, the angle of projection for maximum range is less than 45° when air resistance is a factor (Fig. 3.19). Other factors, such as spin and wind, may affect the range of a projectile. For example, backspin on a driven golf ball provides lift, and the projection angle for the maximum range may be considerably less than 45°.

EXAMPLE 3.10 ■ Extended Projectile Range

A stone thrown from a bridge 20 m above a river has an initial velocity of 12 m/s at an angle of 45° to the horizontal (Fig. 3.20). What is the range of the stone?

Solution.

> *Given:* $y = -20$ m (river distance)   *Find:* $R = x_{max}$ (range)
> $v_o = 12$ m/s
> $\theta = 45°$
> $a_y = -g$

Computing the velocity components:

$$v_{x_0} = v_o \cos 45° = (12 \text{ m/s})(0.707) = 8.5 \text{ m/s}$$

$$v_{y_0} = v_o \sin 45° = (12 \text{ m/s})(0.707) = 8.5 \text{ m/s}$$

(It is often convenient to compute these initially as given data.)

There is more than one way to work this problem. Let's divide the range into two parts: (a) x_1, the symmetrical part of the path, which ends where the projectile returns to the same elevation as the starting point, and (b) x_2, the projectile's path below the launch point, such that $R = x_1 + x_2 = x_{max}$ (see figure).

(a) The range of the first part of the motion, $x_1 = R_1$ may be found using Eq. 3.11. (This equation does not apply to the total range. Why?)

$$x_1 = R_1 = \frac{v_o^2 \sin 2\theta}{g} = \frac{(12 \text{ m/s})^2 \sin (2 \times 45°)}{9.8 \text{ m/s}^2} = 15 \text{ m}$$

(b) As with the initial and returning velocities for vertical projections (see

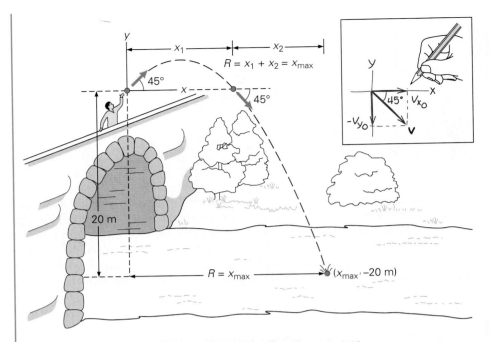

FIGURE 3.20 Extended range of a projectile Example 3.10.

Example 2.11), the magnitude of the velocity when the projectile returns to the elevation from which it started is the same as that of the initial velocity v_0. Also, the magnitude of the directional angle is the same, but below the x axis (Fig. 3.20). Looking only at the x_2 part of the motion, therefore, the situation is the same as for an object with a downward initial projection at an angle of 45° below the horizontal. The "*Given*" data for this part of the problem is then

Given: $y = -20$ m   *Find:* x_2
$v_{x_0} = 8.5$ m/s
$v_{y_0} = -8.5$ m/s (downward)
$a_y = -g$

This part of the problem, too, can be solved in a couple of ways. We will do it in two steps. Time is the connecting feature, so let's first find the y component of the velocity with which the stone hits the water.

$$v_y^2 = v_{y_0}^2 - 2gy$$

and

$$v_y = \sqrt{(-8.5 \text{ m/s})^2 - 2(9.8 \text{ m/s}^2)(-20 \text{ m})}$$

$$= -22 \text{ m/s}$$

where the negative root is the relevant one because we know that the y component of the velocity is in the downward direction.

Then, the time of flight (or fall) is found by solving $v_y = v_{y_0} - gt$ for t:

$$t = \frac{v_{y_0} - v_y}{g} = \frac{-8.5 \text{ m/s} - (-22 \text{ m/s})}{9.8 \text{ m/s}^2} = 1.4 \text{ s}$$

The horizontal distance for this part of the motion is then

$$x_2 = v_{x_0}t = (8.5 \text{ m/s})(1.4 \text{ s}) = 12 \text{ m}$$

Hence, the total range is

$$R = x_{max} = x_1 + x_2 = 15 \text{ m} + 12 \text{ m} = 27 \text{ m}$$

Notice in part (b) that the time for the motion could have been found from $y = v_{y_0}t - \frac{1}{2}gt^2$ or $-20 \text{ m} = (-8.5 \text{ m/s})t - \frac{1}{2}(9.8 \text{ m/s}^2)t^2$. This is a quadratic equation ($at^2 + bt + c = 0$). You can use the quadratic formula (Appendix I), to find the roots, and in this case, the positive root gives the time of fall. (Why?) Also, the total time for the projectile motion could have been found using a quadratic equation, as demonstrated in the next example.

Example 3.10 shows that there are several approaches to solving problems involving projectile motion. There we used a two-step procedure to avoid dealing with a quadratic equation. Now let's look at the most direct approach to this situation.

EXAMPLE 3.11 ■ **Example 3.10 Revisited**

Find the range of the thrown stone in Example 3.10 by the least-step method.

Solution.

Given: (from Example 3.10) *Find:* $R = x_{max}$
$$y = -20 \text{ m}$$
$$v_{y_0} = 8.5 \text{ m/s} \quad (t = 0)$$
$$v_{x_0} = 8.5 \text{ m/s} \quad (t = 0)$$
$$a_y = -g$$

To find the range, we need to know the total time of flight in the horizontal direction. This is the same time as the time for the combined vertical motions (upward and downward). This *total* time of flight may be found from $y = v_{y_0}t - \frac{1}{2}gt^2$:

$$y = v_{y_0}t - \frac{1}{2}gt^2$$

or $$-20 \text{ m} = (8.5 \text{ m/s})t - \frac{1}{2}(9.8 \text{ m/s}^2)t^2$$

Writing this in general quadratic equation form ($at^2 + bt + c = 0$), with the units omitted for convenience,

$$4.9\,t^2 - 8.5\,t - 20 = 0$$

Then, using the quadratic formula (Appendix I),

$$t = \frac{-b \pm \sqrt{b^2 - 4ac}}{2a}$$

$$= \frac{-(-8.5) \pm \sqrt{(-8.5)^2 - 4(4.9)(-20)}}{2(4.9)}$$

$$= \frac{8.5 \pm 21.5}{9.8} = 3.1 \text{ s}$$

where the positive root is taken.

Therefore, $R = x_{max} = v_{x_0}t = (8.5 \text{ m/s})(3.1 \text{ s}) = 26 \text{ m}$

Recall that the answer in Example 3.10 was $R = 27$ m. Is something wrong? Not really. This is an example of how rounding significant figures in problem steps can affect the final answer. If three significant figures were assumed for the data in Example 3.10, the result would be $R = 26.0$ m. (Try it and see.)

Significant figures and rounding are discussed in Section 1.6,

It should be noted that a projection angle of 45° does not give the maximum range for the situation in Fig. 3.20. For an object projected from a height above the ground (or above the landing elevation), the angle of projection for maximum range is less than 45°. (See Exercise 77.)

EXAMPLE 3.12 ■ Angle of Impact

Once again referring to the situation in Fig. 3.20, at what angle does the stone strike the water?

Solution. The data needed to find the angle of impact are available from Example 3.10. We know that $v_x = v_{x_0} = 8.5$ m/s for the total flight, and it was found that the y component of the velocity on impact was $v_y = -22$ m/s. Hence, the velocity on impact is

$$\mathbf{v} = v_x\mathbf{x} + v_y\mathbf{y} = 8.5 \text{ m/s } \mathbf{x} - 22 \text{ m/s } \mathbf{y}$$

Relative to the river surface (x-axis), this vector is at an angle of

$$\theta = \tan^{-1}(v_y/v_x) = \tan^{-1}(-22/8.5) = -69°$$

where the minus sign indicates $\mathbf{v}$ is directed below the horizontal. Note that this angle is quite different from the angle of projection.

Important Concepts

You should be able to define and explain these chapter concepts clearly.

components of motion
vector addition (subtraction)
triangle method
parallelogram method

polygon method
component method
magnitude-angle form
   (of vector)

unit vector
component form (of vector)
analytical component method
relative velocity

projectile motion
parabola
range

Important Relationships for Review

These relationships are mathematical statements of the concepts and principles presented in the chapter. You should be able to identify the symbols and to explain the relationships before proceeding to the Exercises. In-text equation reference numbers are given for convenience.

Components of Initial Velocity:

$$v_{x_0} = v_0 \cos\theta \tag{3.1a}$$

$$v_{y_0} = v_0 \sin\theta \tag{3.1b}$$

Components of Displacement
(constant acceleration only):

$$x = v_{x_0}t + \tfrac{1}{2}a_x t^2 \qquad (3.3a)$$

$$y = v_{y_0}t + \tfrac{1}{2}a_y t^2 \qquad (3.3b)$$

Components of Velocity:

$$v_x = v_{x_0} + a_x t \qquad (3.3c)$$

$$v_y = v_{y_0} + a_y t \qquad (3.3d)$$

Component Method of Vector Addition:

$$\mathbf{v} = \mathbf{v}_1 + \mathbf{v}_2 + \mathbf{v}_3 + \cdots$$

$$v_x = \Sigma v_{x_i} \quad \text{and} \quad v_y = \Sigma v_{y_i}$$

$$\mathbf{v} = v_x\,\mathbf{x} + v_y\,\mathbf{y} \qquad \text{component form} \qquad (3.7)$$

$$v = \sqrt{v_x^2 + v_y^2} \;\Bigg\} \qquad (3.4a)$$

$$\theta = \tan^{-1}\left|\frac{v_y}{v_x}\right| \;\Bigg\} \quad \text{magnitude-angle form} \qquad (3.4b)$$

Projectile Range ($y_{\text{initial}} = y_{\text{final}}$ only):

$$R = \frac{v_0^2 \sin 2\theta}{g} \qquad (3.11)$$

Exercises

3.1 ■ Components of Motion

1 On Cartesian axes, the x component of a vector is generally associated with a (a) cosine, (b) sine, (c) tangent, (d) none of these. (a)

2 Nonlinear motion in two dimensions results from (a) constant unequal velocity components, (b) constant and equal acceleration components starting from rest, (c) constant and equal acceleration components with equal initial velocity components, (d) unequal acceleration components. (d)

3 Describe the motion of an object that is initially traveling with a constant velocity and then receives an acceleration (a) in a direction parallel to the initial velocity, (b) in a direction perpendicular to the initial velocity, and (c) that is always perpendicular to the instantaneous velocity or direction of motion. (a) linear velocity increases or decreases (b) moves in a parabolic path (c) moves in a circle

●4 When you are sitting in a stationary car and it is raining (with no wind), the rain falls straight down relative to the car and the ground. But when you're driving, the rain appears to hit the windshield at an angle, which increases as the velocity of the car increases. Explain this effect in terms of velocity components. [*Hint:* Consider the velocity of the rain relative to you.] see ISM

5 ■ A student walks two blocks west and one block south. If the length of a block is 60 m, what displacement will bring the student back to the starting point? 1.3×10^2 m, 27° N of E

6 ■ An object moves with a velocity of 6.0 m/s at an angle of 37° relative to the +x axis. What is the magnitude of the x component of the velocity? 4.8 m/s

7 ■ A motor boat travels with a speed of 40 km/h in a straight path on a still lake. Suddenly, a strong, steady wind pushes the boat perpendicularly to its straight-line path with a speed of 15 km/h for 5.0 s. Relative to its position just when the wind started to blow, where is the boat located at the end of this time? 60 m; 21°

8 ■ An object moves with a velocity of 7.5 m/s at an angle of 7.5° to the x axis. What are the x and y components of the velocity? $v_x = 7.4$ m/s; $v_y = 0.98$ m/s

9 ■■ The y component of a velocity that has an angle of 27° to the x axis has a magnitude of 3.8 m/s. (a) What is the magnitude of the velocity? (b) What is the magnitude of the x component of the velocity? (a) 8.4 m/s (b) 7.5 m/s

●10 ■■ The displacement vector of a moving object initially at the origin has a magnitude of 12.5 cm and is at an angle of 210° relative to the +x axis at a particular instant. What are the coordinates of the object at that instant? $x = -10.8$ cm; $y = -6.25$ cm

11 ■■ A ball rolls with a constant velocity of 2.40 m/s at an angle of 37° relative to the x axis in the 4th quadrant. Taking the ball to be at the origin at $t = 0$, what are its (x, y) coordinates 1.75 s later? $x = 3.34$ m; $y = -2.52$ m

12 ■■ A ball rolling on a table has velocity with rectangular components $v_x = 0.60$ m/s and $v_y = 0.80$ m/s. What is the displacement of the ball in an interval of 2.5 s? $x = 1.5$ m; $y = 2.0$ m; $d = 1.5$ m $\mathbf{x}$ + 2.0 m $\mathbf{y}$)

13 ■■ A small plane takes off with a constant speed of 130 km/h at an angle of 37°. In 3.00 s, (a) how high is the plane above the ground, and (b) what horizontal distance has it traveled from the lift-off point? (a) 65.0 m (b) 86.7 m

14 ■■ A ball moves diagonally across a table top from corner to corner with a constant speed of 0.75 m/s. The table top is 3.0 m wide and 4.0 m long. Another ball, starting at the same time as the first one, moves along the longer edge of the table. What constant speed must the second ball have in order to reach the corner at the same time as the ball traveling diagonally? 0.60 m/s

15 ■■ After being hit from a corner of the rink, a hockey puck has a constant velocity of 25.0 m/s at an angle of 40° to the side wall along the length of the rink. (Neglect friction.) (a) Using the wall as an axis, what are the components of the puck's velocity? (b) What are the displacement compo-

nents for the puck 2.5 s after it is hit? (Let the direction along the wall be the x axis.)   (a) v_x = 19 m/s; v_y = 16 m/s
(b) x = 48 m; y = 40 m

● 16 ■■■ A ball moving along the x axis with a speed of 1.5 m/s experiences an acceleration of 0.25 m/s^2 at an angle of 37° to the x axis, starting when the ball is at the origin (t = 0) and continuing without interruption. What are the coordinates of the ball at t = 3.0 s? x = 5.4 m; y = 0.68 m

3.2 ■ Vector Addition and Subtraction

17 Vector subtraction (a) must be done by the triangle method, (b) is a special case of vector addition, (c) invariably results in a negative vector, (d) can only be done analytically.   (b)

18 A vector may be resolved into (a) only two components, (b) only three components, (c) any number of components, (d) a number of components that depends on the direction of the vector.   (c)

19 In the procedure given for the component method the angle for the resultant is found by taking the arctangent of the absolute value of the ratio of the component sums. What might happen if the absolute value were not used? Suppose all angles were measured counterclockwise from the positive x axis. Would this complicate matters? Explain.   could obtain negative angles and require use of trigonometric double angle formulas

20 ■ Find the rectangular components of a vector **v** that has a magnitude of 12.0 m/s and is oriented at an angle of 60° relative to the positive x axis. v_x = 6.00 m/s; v_y = 10.4 m/s

21 ■ A vector **B** has an x component of −2.5 m and a y component of 4.2 m. What is **B** expressed in magnitude-angle form?   4.9 m, 59° below the +x axis

● 22 ■■ In Chapter 2, the displacement between two points on the x axis was written $\Delta x = x_2 - x_1$. Show what this means graphically in terms of vector subtraction.   see ISM

23 ■■ For the two vectors x_1 = 20 m **x** and x_2 = 15 m **x**, compute and show graphically (a) $x_1 + x_2$, (b) $x_1 - x_2$, and (c) $x_2 - x_1$.   (a) 35 m **x**   (b) 5 m **x**   (c) − 5 m **x**

24 ■■ For each of the following vectors, give a vector that when added to it yields a null vector (a vector with a magnitude of zero): (a) **A** = 4.5 cm **x**, (b) **B** = −10 cm **y**, (c) **C** = 8.0 cm at an angle of 60° relative to the negative x axis in the second quadrant.   (a) −4.5 cm **x**   (b) 10 cm **y**
(c) 8.0 cm, 60° (4th quadrant)

25 ■■ Find the resultant (or sum) of the three vectors **A** = 4.0 m **x** + 2.0 m **y**, **B** = −6.0 m **x** + 3.5 m **y**, and **C** = −5.5 m **y**.   −2.0 m **x**

● 26 ■■ (a) Find the resultant (or sum) of the vectors in Fig. 3.21. (b) If **F**₁ in the figure were at an angle of 27° with the x axis, instead of 37°, what would be the resultant (or sum) **A** and **B**?   (a) 14.4 N **y**   (b) 1.1 N **x** + 12.7 N

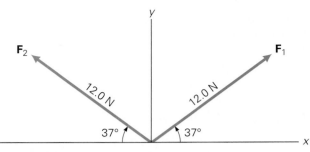

● FIGURE 3.21 Vector addition   See Exercises 26 and 27.

27 ■■ For the vectors in Fig. 3.21, determine (a) **F** = **F**₁ − **F**₂, and (b) **F** = **F**₂ − **F**₁. Do you think that the resultants will be the same? That is, does **F**₁ − **F**₂ = **F**₂ − **F**₁? (c) Is the following a true relationship? (**F**₁ − **F**₂) $\overset{?}{=}$ − (**F**₂ − **F**₁)
(a) 19 N **x**   (b) − 19 N **x**   (c) yes

28 ■■ An airplane takes off and flies directly north a distance of 50 km. Severe thunderstorms are reported in the flight path, so the pilot turns to a course of 40° east of north and flies 25 km in that direction, then takes a course of 20° west of north for 35 km. What is the plane's displacement from its starting point?   4 km **x** + 102 km **y**

29 ■■ Five vectors with equal magnitude of 5.0 m have the following directional angles with respect to the x axis: ±37° (1st and 2nd quadrants), ±45° (3rd and 4th quadrants), and −**y**. What is the resultant of the vectors?   −6.0 m **y**

30 ■■ A student works four different problems involving the addition of different vectors **F**₁ and **F**₂. He states that the magnitudes of the four resultants are given by (a) $F_1 + F_2$, (b) $F_1 - F_2$, (c) $\sqrt{F_1^2 + F_2^2}$, and (d) $F_1 + F_2 = F_1 - F_2$. Is this possible? If so, describe the vectors in each case.   (a) in the same direction   (b) in opposite directions   (c) at right angles   (d) only if F_2 = 0

31 ■■ For three arbitrary vectors, **F**₁, **F**₂, and **F**₃, show graphically that the resultant **F** is the same regardless of the order in which the vectors are added. (How many possibilities are there?)   6 possibilities; see ISM

● 32 ■■ A block rests on an inclined plane. Its weight is a force directed vertically downward, as illustrated in ● Fig. 3.22. Find the components of the force along the surface of the plane and perpendicular to it. $F_\parallel$ = w/2; $F_\perp$ = (0.87)w

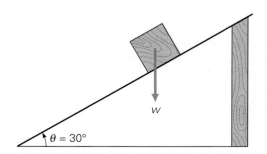

● FIGURE 3.22 Block on an inclined plane   See Exercise 32.

33 ■■ An amusement park roller coaster starts out initially on a level track 100 m long and then goes up a 75 m incline at an angle of 37° to the horizontal. It then goes down a 25 m ramp with an incline of 60° to the horizontal. What is the displacement of the roller coaster from its starting point when at the bottom of the ramp? **d** = 170m **x** + 23m **y**

34 ■■ A person walks from point A to point B as shown in ● Fig. 3.23. What is the person's displacement relative to A? **d** = −4.3m **x** − 26.7m **y** or 27 m, 81° (3rd quadrant)

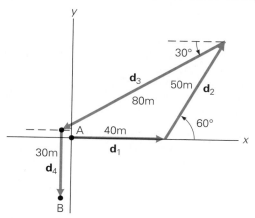

● **FIGURE 3.23 Walking along** See Exercise 34.

● **35** ■■ Find the resultant of the displacement vectors **d**$_1$ = 3.0 m **x** + 3.0 m **y**, **d**$_2$ = 2.5 m **x** − 6.0 m **y**, and **d**$_3$ = − 2.0 m **x** + 1.5 m **y**; that is, find **d**$_1$ + **d**$_2$ + **d**$_3$. (Express in both component notation and magnitude-angle form.)
d = 3.5 m **x** − 1.5 m **y** m or 3.8 m, 23° (4th quadrant)

36 ■■■ Find the resultant (or sum) of the four vectors in ● Fig. 3.24. **d** = 16.5 m **x** + 22.8 m **y**

37 ■■■ For the vectors in Fig. 3.24 find (**d**$_2$ − **d**$_4$) + (−**d**$_1$ + **d**$_3$). **d** = (−9.4 m **x** + 23.2 m **y**) m

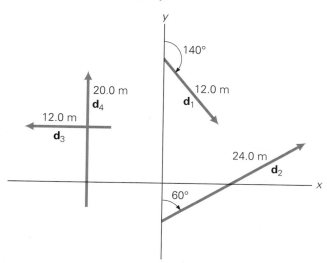

● **FIGURE 3.24 Vector addition** See Exercises 36 and 37.

3.3 ■ Relative Velocity

38 ■ While you are traveling in a car on a straight, level interstate highway at 100 km/h, another car passes; its speedometer reads 120 km/h. (a) What is your velocity relative to the other driver? (b) What are the velocities of both cars relative to a stationary observer standing by the road?
(a) −20 km/h   (b) v_1 = 100 km/h; v_2 = 120 km/h

39 ■■ A person riding in the back of a pickup truck traveling at 60 km/h on a straight, level road throws a ball with a speed of 20 km/h relative to the truck in the direction opposite its motion. (a) What is the velocity of the ball relative to a stationary observer by the side of the road? (b) What is the velocity of the ball relative to the driver of a car moving in the same direction as the truck at a speed of 90 km/h?
(a) 40 km/h   (b) −50 km/h

40 ■■ Two trains traveling with constant speeds of 90 km/h approach each other on straight, parallel tracks. If the trains are 5.0 km apart, how long will it take for them to meet? 1.7 min

41 ■■ For the situation described in Example 3.5, (a) in what direction should the boat be headed in order to land at a point directly across the river from its starting point? (b) How long would such a crossing take? (a) 18.7° upstream   (b) 238 s = 3.97 min

● **42** ■■ A pontoon boat that can move at 3.50 m/s in still water crosses a straight river 200 m wide and arrives at a point 25.0 m upriver from the point directly across from its starting point. If the boat is directed upriver at an angle of 40.0° (relative to a line directly across the river), what is the speed of the current? 1.91 m/s

43 ■■ If the flow rate of the current in a straight river is greater than the speed of a boat in the water, the boat cannot make a trip directly across the river. Prove this statement.
see ISM

44 ■■ A boat that travels at a speed of 6.75 m/s in still water goes directly across a river and back (● Fig. 3.25). The

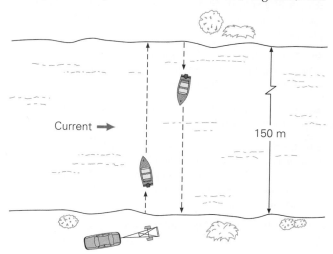

● **FIGURE 3.25 A boat race** See Exercise 44.

current flows at 0.50 m/s. (a) At what angle(s) must the boat be steered? (b) How long does it take to make a round trip? (Assume that the boat's speed is constant at all times and neglect turnaround time.)   (a) 4.25°   (b) 44.6 s

● **45** ■■ The current of a 200-m wide straight river has a flow rate of 2.50 km/h. A motorboat with a speed of 30 km/h in still water crosses the river. (a) If the boat is pointed directly across the river toward the opposite shore, how far will it travel across *and* down the river in 9.00 s? (b) How far downstream will its landing point be from the point directly opposite its starting point? (Assume that the boat comes instantaneously to rest on a sandy shore.)
(a) $y = 50.0$ m, $x = 6.25$ m   (b) 25.0 m

46 ■■ In Exercise 45, how long would it take the boat to travel *directly* across the river?   36.3 s

47 ■■ In Example 3.6, if the airplane heads directly north, what would be its position after a half hour of flight?
$x = 25.0$ km, W; $y = 100$ km, N

● **48** ■■ A moving walkway in an airport is 75 m long and moves with a speed of 0.30 m/s. A late passenger, after traveling 25 m standing on the walkway starts to walk with a speed of 0.50 m/s. How long does it take for the passenger to travel the total walkway distance?   146 s = 2.43 min

49 ■■ Two cars approach a street corner at right angles to each other. Car A travels northward with a speed of 50 km/h and car B travels eastward with a speed of 30 km/h. (a) What is the relative velocity of car A relative to car B? (b) What is the relative velocity of car B relative to car A?   (a) 58.3 km/h 59° S of E   (b) 58.3 km/h 59° W of N

3.4 ■ Projectile Motion*

50 If air resistance is neglected, the motion of an object projected at an angle consists of a uniform downward acceleration combined with (a) an equal horizontal acceleration, (b) a uniform horizontal velocity, (c) a constant upward velocity, (d) an acceleration always perpendicular to the path of motion.   (b)

51 For projectile motion, the displacement component(s) (a) both vary as t, (b) both vary as t^2, (c) of x varies as t^2 and y as t, (d) of x varies as t and y as t^2.   (d)

52 Is it possible for a projectile to travel in a circular arc rather than a parabolic one? Explain.   No

53 Using a sketch of components, verify that the velocity of a projectile at the same elevation as the initial projection ($y_{final} = y_{initial}$) has the same magnitude as the initial velocity and the same size directional angle, but directed below the horizontal.   see Fig. 3.16

● **54** ■ A ball with a horizontal speed of 1.5 m/s rolls off a bench with a height of 1.5 m. (a) How long will it take the

*Neglect air resistance unless stated otherwise.

ball to reach the floor? (b) How far from a point on the floor directly below the edge of the bench will it land?
(a) 0.55 s   (b) 0.83 m

55 ■ A spring-loaded gun can project a marble with an initial speed of 3.6 m/s. With the gun lying horizontally on a level table 1.0 m above the floor, what is the range of a marble fired from it?   1.6 m

56 An electron leaves an electron gun horizontally in a TV set with a speed of 1.5×10^6 m/s. If the viewing screen is 35 cm away, how far will the electron fall or travel in the vertical direction before hitting the screen? 2.6×10^{-13} m

✏ **57** ■■ One ball rolls on a table 1.0 m tall at a constant speed of 0.25 m/s, and another ball rolls on the floor directly under the first ball and with the same speed and direction. Will the balls collide after the first ball rolls off the table? If so, how far from the point directly below the edge of the table will they be when they strike each other?   yes, 0.11 m

● **58** ■■ While warming up, a pitcher throws a fast ball horizontally with a speed of 23.5 m/s. How far will the ball have dropped from its straight-line path in a distance of 18.5 m (the distance from the pitcher's mound to home plate)? Is the answer realistic? Explain.   3.03 m; not realistic

59 ■■ A student throws a softball horizontally from a dorm window 15.0 m above the ground. Another student standing 10.0 m away from the dorm catches the ball at a height of 1.50 m above the ground. What is the initial velocity of the ball?   6.02 m/s

60 ■■ In a movie, a monster climbs to the top of a building 30 m above the ground and hurls a boulder downward with a velocity of 20 m/s at an angle of 37° below the horizontal. How far from the building does the boulder land?   24 m

61 ■■ A football resting on the ground is kicked at an angle of 35.0° with an initial speed of 20.0 m/s. (a) What is the maximum height reached by the ball? (b) What is its range?   (a) 6.75 m   (b) 38.5 m

✏ **62** ■■ A soccer player kicks the ball, giving it a velocity of 18.0 m/s at an angle of 45.0° to the horizontal. (a) What is the maximum height reached by the ball? (b) What is its range? (c) How could the range be increased?   (a) 8.23 m
(b) 32.9 m   (c) kick the ball harder to increase v_o

63 ■■ An artillery shell with a muzzle velocity of 125 m/s is fired at an angle of 35.0° to the horizontal. If the shell explodes 10.0 s after being projected, where does the blast occur? $(x, y) = (1.02 \times 10^3$ m, 227 m)

● **64** ■■ Show that the path of a projectile is a parabolic arc. The general form of the equation for a parabola is $y = ax - bx^2$, where a and b are constants. [*Hint:* Use the appropriate forms of Eq. 3.3a or 3.3b for displacements.]   see ISM

65 ■■ For projectile motion in general at an angle θ relative to a level surface, sketch graphs of (a) y versus t, (b) x versus t, (c) v_x versus t, (d) v_y versus t, and (e) v versus t.   see ISM

66 ■■ Determine the initial velocity of the horizontally projected golf ball in Fig. 3.15(b). 2.2 m/s

67 ■■ A javelin is thrown at angles of 35° and 60° to the horizontal from the same height with the same speed in each case. For which throw does the javelin go farther and how many times farther? (Assume that the landing place is at the same height as the launching.) $R_{35°} = (1.1) R_{60°}$

68 ■■ An astronaut on the moon fires a projectile from a launcher sitting on a level surface so as to get the maximum range. If the launcher gives the projectile a muzzle velocity of 36 m/s, what is the range of the projectile? [*Hint:* Don't forget where the launcher is and gravitational effects.]
7.9×10^2 m

69 ■■ A diver cleaves the water at an angle of 45°, and the water is 10.0 m deep. Assume that she maintains a constant underwater velocity with a magnitude of 0.85 m/s. Will she reach the bottom of the pool in 4.0 s? No; 6.0 s

70 ■■ A firefighter holding the nozzle of a hose a horizontal distance of 50.0 m from a flaming building is just able to reach with maximum range a third-story window with a stream at a height of 11.0 m above the ground. If the firefighter is holding the nozzle 0.50 m above the ground, what is the speed of the water coming out of it? 20.1 m/s

71 ■■ A ditch 2.5 m wide crosses a trailbike path. An upward incline of 15° has been built up on the approach so that the top of the incline is level with the top of the ditch. What is the minimum speed at which trailbike must be moving to clear the ditch? (Add 1.4 m to the range for the back of the bike to clear the ditch safely.) 8.7 m/s

72 ■■ A quarterback passes a football with a velocity of 50 ft/s at an angle of 40° to the horizontal toward an intended receiver 30 yd downfield. The pass is released 5.0 ft above the ground. Assume that the receiver is stationary and that he will catch the ball if it comes to him. Will the pass be completed? the pass is short

73 ■■ A golf ball is hit and receives an initial velocity of 40.0 m/s at an angle of 40.0° to the horizontal. (a) In the absence of air resistance, what would be the ball's range? (b) If a light wind gave the ball a retarding horizontal acceleration of 0.0560 m/s², what would be the range?
(a) 161 m (b) 159 m

74 ■■■ Three identical balls are thrown or projected from a cliff at a particular height (*h*) above a horizontal plain. One ball is projected at an angle of 45° to the horizontal, the second ball is thrown horizontally, and the third is thrown at an angle of 30° below the horizontal (that is, downward). If all of the balls have the same initial speed, which one will have the greatest speed when it hits the ground? (Justify your answer mathematically.) all have the same speed v

75 ■■■ A pickup truck travels on a straight, level highway with a speed of 90.0 km/h. If a person riding in the back of the truck throws an object toward the side of the highway

with a speed of 10.5 m/s at an angle of 37.0° above the horizontal, where does the object land? Assume that the object is released 2.20 m above ground level and neglect air resistance. $(x,z) = $ (13.2 m, 39.3 m)

76 ■■■ A diver goes off a 4.0-m high diving board with a velocity of 8.0 m/s at an angle of 30° above the horizontal. (a) What is the maximum height of the diver above the water? (b) How far from a point directly below the edge of the board will the diver hit the water? (a) 4.8 m (b) 9.7 m

77 ■■■ Show that for an object projected from a height above the ground (or landing elevation), the angle of projection for maximum range is less than 45°. [*Hint:* Time of flight is obtained from $-y = v_{y_0}t - \frac{1}{2}gt^2$.] see ISM

■ **Additional Exercises**

78 The apparatus for a popular lecture demonstration is shown in ● Fig. 3.26. The gun is aimed directly at the can, and when the gun is fired, the can is simultaneously released. This so-called monkey gun won't miss as long as the initial velocity of the bullet is sufficient to reach the falling target before it hits the floor. Verify this, using the figure. [*Hint:* Note that $y_0 = x \tan \theta$.] see ISM

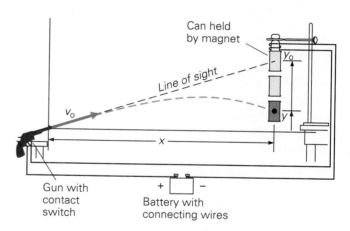

● **FIGURE 3.26 A sure shot** See Exercise 78.

79 Using the triangle method, show graphically that (a) **A** + **B** = **B** + **A**, and (b) if **A** − **B** = **C**, then **A** = **B** + **C**. see ISM

80 An object is projected with an initial speed of 24.0 m/s at an angle less than 45°. Its range at this angle is 51.0 m. At what angle greater than 45° could the object be projected at the same initial speed so that the range would be the same? 59.9°

81 A swimmer maintains a speed of 0.15 m/s relative to the water when swimming directly toward the opposite shore of a straight river with a current that flows at 0.75 m/s.

(a) How far downstream is the swimmer carried in 1.5 min?
(b) What is the velocity of the swimmer relative to an observer
on the shore?   (a) 68 m   (b) 79°

82 The shells fired from an artillery piece have a muzzle
speed of 1.80×10^2 m/s, and the target is at a horizontal
distance of 3.00 km. (a) At what angle relative to the horizontal should the gun be aimed? (b) Could the gun hit a
target 3.50 km away?   (a) 33°   (b) the target could not be hit

83 A ball moving along the y axis with a speed of 2.5 m/s
experiences an acceleration of 0.45 m/s^2 in the x direction
when it is at $(x, y) = (0, 1.0$ m). What are (a) the velocity
components and (b) the position of the ball 4.0 s after the
acceleration is applied?   (a) $\mathbf{v} = 1.8$ m/s $\mathbf{x}$ + 2.5 m/s $\mathbf{y}$
(b) $x = 3.6$ m; $y = 11$ m

84 How are the vectors $\mathbf{A} - \mathbf{B}$ and $\mathbf{B} - \mathbf{A}$ related? Write
an equation and also show graphically. $\mathbf{A} - \mathbf{B} = -(\mathbf{B} - \mathbf{A})$

85 A ball is thrown horizontally from the top of a building
at a height of 32.5 m above the ground and hits the level
ground 56.0 m from the base of the building. (a) What is the
initial velocity of the ball? (b) What is the velocity of the ball
just before it hits the ground?   (a) 21.7 m/s
(b) 33.3 m/s, $-49.2°$

86 A field goal is attempted with the football resting at the
center of the field 40 yd from the goalposts. If the kicker
gives the ball a velocity of 70.0 ft/s toward the goalposts at
an angle of 45° to the horizontal, will the kick be good? (The
crossbar of the goalposts is 10 ft above the ground, and the
ball must be higher than this when it reaches the goalposts
for the field goal to be good.)   the goal is good

87 An object is projected from a level surface with an initial
velocity of 15.0 m/s at an angle of 37.0° relative to the surface.
(a) What is the maximum height of the projectile? (b) What
is its range?   (a) 4.1 m   (b) 22 m

88 Find the resultant (or sum) of the three vectors in ● Fig.
3.27. (Express in both magnitude-angle form and component
notation.)   4.9 m/s 3.5° (3rd quadrant);
$v = -4.9$ m/s $\mathbf{x}$, -0.30 m/s $\mathbf{y}$

89 For the vectors in Fig. 3.27, find the magnitude and
orientation of $\mathbf{v}_1 - \mathbf{v}_2 - \mathbf{v}_3$.   10.9 m/s 1.6° (1st quadrant)

90 Is vector addition associative, that is, does $(\mathbf{A} + \mathbf{B}) +$
$\mathbf{C} = \mathbf{A} + (\mathbf{B} + \mathbf{C})$? Justify your answer.   yes; see ISM

91 The resultant of two rectangular component vectors has
a magnitude of 4.5 m and an angle of 37° relative to the
positive x axis in the fourth quadrant. What are the component vectors?   3.6 m $\mathbf{x}$, -2.7 m $\mathbf{y}$

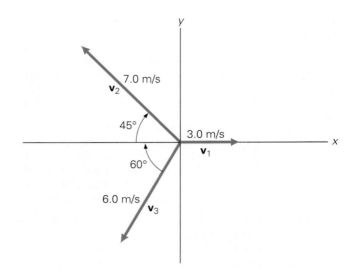

● **FIGURE 3.27 Vector addition**   See Exercises 88 and 89.

92 A package is to be dropped from an airplane to hit the
ground at a designated spot near some campers. The airplane
approaches the spot at an altitude of 0.500 km above level
ground, moving horizontally with a constant velocity of 140
km/h. Having the designated point in sight, the pilot prepares to drop the package. (a) What should the angle between
the horizontal and the pilot's line of sight be when the package is released? (b) What is the location of the plane when
the package hits the ground?   (a) 52°   (b) directly over
impact point

93 (a) An object moving in a straight line with a speed of
10 m/s has a velocity with an x component of 6.0 m/s. In
what direction is the object moving? (b) What is the value
of the y component of the velocity?   (a) 53°   (b) 8.0 m/s

94 At a track and field meet, the best long jump is measured
as 8.20 m. The jumper took off at an angle of 37° to the
horizontal. (a) What was the jumper's initial speed? (b) If
there were another meet on the moon and the same jumper
could attain only half of the initial earthly speed, what would
be the maximum jump there? (Air resistance does not have
to be neglected in part (b). Why?)   (a) 9.14 m/s   (b) 12.3 m

95 An airplane with a speed of 150 km/h heads directly
north while a west wind blows with a constant speed of 30
km/h. How far does the plane travel during the two hours
of flight?   306 km

96 For the vectors in Fig. 3.27, find (a) $\mathbf{v}_1 + \mathbf{v}_2$,
(b) $\mathbf{v}_1 + \mathbf{v}_3$, and (c) $\mathbf{v}_2 - \mathbf{v}_1$.   (a) -1.9 m/s $\mathbf{x}$ + 4.9 m/s $\mathbf{y}$
(b) -5.2 m/s $\mathbf{y}$ (c) -7.9 m/s $\mathbf{x}$ + 4.9 m/s $\mathbf{y}$

4 Force and Motion

An ace! But think for a moment about what actually happens when someone serves a tennis ball. The ball's slow motion in a vertical direction (the toss) is transformed into a very fast motion in a nearly horizontal direction. What brings about this change? You do not have to know much about either tennis or physics to answer: it is the force of the racket on the ball.

In Chapters 2 and 3 we learned how to analyze motion in terms of kinematics. Now we need to know more about the dynamics of motion—that is, what causes motion and changes in motion. This inquiry leads us to the concept of force.

If an object is at rest, a push or a shove may set it into motion, as when we push a sled or a stalled automobile. That is, a force is applied to the object. Similarly, an object in motion may be speeded up or slowed down by applying a force, as when we push a grocery cart or bring it to a stop. We can also change the direction of an object's motion by means of a force, as we saw in the example of the tennis serve. In fact, we shall find that all changes in motion result from the action of a force or forces.

There are several types of forces. For example, there is the gravitational force that attracts people and objects to the Earth, giving rise to the force we call weight. But this force varies a great deal from object to object, as we know from everyday experience. (Think of picking up a purse or gym bag versus a suitcase packed for a vacation trip.) There is also the common force of friction, which tends to oppose motion. For simplicity, frictional effects are sometimes ignored (as they were in Chapter 2 when air resistance was neglected in analyzing free fall). However, in many instances friction cannot be

> Chapter 4 deals with the study of Newton's laws of motion and force, solving problems with and without friction.

ignored and must be included in the analysis of motion. In general, it is important to recognize a force acting on an object and to be able to calculate the effects of the force.

The study of force and motion occupied many early scientists. It was Isaac Newton (1642–1727), the English scientist (● Fig. 4.1), who summarized the various relationships and principles into three statements, or laws, which not surprisingly are known as Newton's laws of motion. These laws sum up the concepts of dynamics. In this chapter, you'll learn what Newton had to say about force and motion.

4.1 ■ The Concept of Force and Net Force

Let's first take a closer look at the meaning of force. It is easy to give examples of forces, but how would you generally define this concept? An operational definition of force is based on observed effects. That is, a force is described in terms of what it does. From your own experience, you know that *forces can produce changes in motion*. A force can set a stationary object into motion. It can also speed up or slow down a moving object, or change the direction of its motion. In other words, a force can produce a change in velocity (speed and/or direction)—that is, acceleration. Therefore, an observed change in motion, including motion starting from rest, is evidence of a force. This leads to a common definition of **force**:

A force is something capable of changing an object's state of motion.

The word "capable" is very significant here. It takes into account the fact that a force may be acting on an object, but its capability to produce a change in motion may be balanced or canceled by another force or forces so that the net effect is zero. Thus, a force does not *necessarily* produce a change in motion.

Since a force may produce an acceleration—a vector quantity—it is itself a vector quantity, with both magnitude and direction. When several forces act on an object, you will often be interested in their combined effect, or the net force. The **net force** is the vector sum, $\Sigma_i \mathbf{F}_i$, or resultant, of all the forces acting on an object or system. As illustrated in ● Fig. 4.2a, the net force is zero when forces of equal magnitude act in opposite directions. Such forces are said to be *balanced forces*. A nonzero net force is referred to as an *unbalanced force*. In this case, *the situation can be analyzed as though only this single force were acting*. An unbalanced or net force produces an acceleration. In some instances, an applied unbalanced force may also deform an object, that is, change its size or shape (see Chapter 9). A deformation involves a change in motion for some part of an object, hence there is an acceleration.

Forces are sometimes divided into two types or classes. The more familiar of these consists of *contact forces*. Such forces arise because of physical contact between objects. For example, when you push on a door to open it or throw or kick a ball, you exert a contact force on the door or ball.

The other class of forces is called *action-at-a-distance forces*. Examples of these forces are gravity, the electrical force between two charges, and the magnetic force between two magnets. The moon is attracted to the Earth and maintained in orbit by a gravitational force, but there seems to be nothing physically transmitting that force. Later in the text (Chapter 29), you will learn the modern view of how action-at-a-distance forces are transmitted.

Now, with our better understanding of the concept of force, let's see how force and motion are related through Newton's laws.

●**FIGURE 4.1 Isaac Newton**
A modern portrait based on a contemporary original. Newton (1642–1727), one of the greatest scientific minds of all time, made fundamental contributions to mathematics, astronomy, and several branches of physics, including optics and mechanics. He formulated the laws of motion and universal gravitation (Chapter 7), and was one of the inventors of calculus. Some of his most profound work was done when he was in his middle twenties.

Net or unbalanced force

In the notation, $\Sigma_i \mathbf{F}_i$, the Greek letter sigma means the "sum of" the individual forces, as indicated by the *i* subscripts, i.e., $\Sigma_i \mathbf{F}_i = \mathbf{F}_1 + \mathbf{F}_2 + \mathbf{F}_3 + \ldots$. The *i* subscripts are sometimes omitted ($\Sigma \mathbf{F}$) as being understood.

Have students name different types of forces. Always try to identify what is exerting the force and what the force is acting on. Some students will tend to view a force as something abstract that merely exists without getting a feel for a real push or pull.

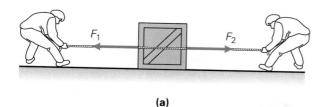

(a)

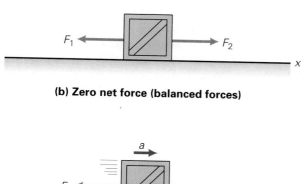

(b) Zero net force (balanced forces)

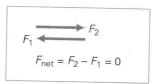

$F_{net} = F_2 - F_1 = 0$

Figure 4.2

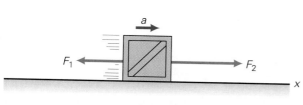

(c) Nonzero net force (unbalanced forces)

$F_{net} = F_2 - F_1 \neq 0$

FIGURE 4.2 Net force
(a) Opposite forces are applied to a block. **(b)** If the forces are equal, the vector resultant, or the net force acting on the block in the *x* direction, is zero. The forces acting on the block are said to be balanced. **(c)** If the forces are not equal, the resultant is not zero. A nonzero net force, or an unbalanced force, then acts on the block. This sets it into motion (a change in velocity, and hence an acceleration).

4.2 ■ Newton's First Law of Motion

The groundwork for Newton's first law of motion was laid by Galileo. In his experimental investigations, Galileo dropped objects to observe motion under the influence of gravity. However, the relatively large acceleration due to gravity causes dropped objects to move quite fast and quite far in a short time. From the kinematic equations in Chapter 2, you can see that 3 s after being dropped, an object in free fall has a speed of about 29 m/s (64 mi/h) and has fallen a distance of 44 m (about 48 yd, or almost half the length of a football field). Thus, experimental measurements of free-fall distance versus time were particularly difficult to make with the instrumentation available in Galileo's time.

To slow things down so that he could study motion, Galileo used balls rolling on inclined planes. He allowed a ball to roll down one inclined plane and then up another with a different degree of incline (● Fig. 4.3). Galileo noted that the ball rolled to approximately the same height in every case, but it rolled

FIGURE 4.3 Galileo's experiment A ball rolls farther along the upward incline as the angle of incline is decreased. On a smooth, horizontal surface, the ball rolls a great distance before coming to rest. How far would the ball travel on an ideal, perfectly smooth surface?

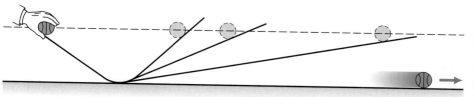

FIGURE 4.4 A difference in inertia The larger punching bag has more mass, and hence more inertia or resistance to a change in motion.

Demonstration/Activity: Place a demonstration cart that has some friction on the lecture desk and give it a push. Observe that it will coast to a stop. Does this verify Aristotle's theory of motion?

Mass and inertia

Newton's first law—the law of inertia

farther in the horizontal direction when the angle of incline was smaller. When allowed to roll onto a horizontal surface, the ball traveled a considerable distance and went even farther when the surface was made smoother. Galileo wondered how far the ball would travel if the horizontal surface could be made perfectly smooth (frictionless). Since this situation was impossible to attain experimentally, Galileo reasoned that in this ideal case with an infinitely long surface, the ball would continue to travel indefinitely with straight-line, uniform motion, since there would be nothing (no force) to cause its motion to change.

According to Aristotle's theory of motion, which had been accepted for about 1500 years prior to Galileo's time, the normal state of a body was to be at rest (with the exception of celestial bodies, which were naturally in motion). Aristotle probably observed that objects moving on a surface tend to slow down and come to rest, so this conclusion would have seemed logical to him. However, from his experiments, Galileo concluded that bodies in motion exhibit the behavior of maintaining that motion, and that if an object is initially at rest, it will remain so, unless something causes it to move.

Galileo called this tendency of an object to maintain its initial state of motion **inertia.** That is,

Inertia is the natural tendency of an object to maintain a state of rest or to remain in uniform motion in a straight line (constant velocity).

For example, if you've ever tried to stop a slowly rolling automobile by pushing on it, you felt its resistance to a change in motion, to slowing down. Physicists describe the property of inertia in terms of observed behavior, as they do for all physical phenomena. A comparative example of inertia is illustrated in Fig. 4.4. You may be quick to note that the large punching bag is heavier than the smaller one and to correlate weight with inertia. You're on the right track, but remember that mass is the fundamental property.

Newton related the concept of inertia to mass. Originally, he called mass a quantity of matter, but later redefined it as follows:

Mass is a measure of inertia.

That is, a massive object has more inertia, or resistance to a change in motion, than a less massive object does. For example, a car has more inertia than a bicycle.

Newton's first law of motion, sometimes called the law of inertia, summarizes these observations:

In the absence of an unbalanced applied force, a body at rest remains at rest, and a body already in motion remains in motion with a constant velocity.*

(See Demonstration 1.)

---

*What Newton actually said (in Latin) was this: "Every body preserves its state of rest, or of uniform motion in a right [straight] line unless it is compelled to change that state by forces impressed thereon." From *A Source Book in Physics,* by W. F. Magie (Cambridge, MA: Harvard University Press, 1963).

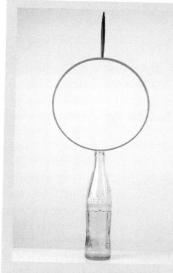

(a) A pen is at rest on an embroidery hoop on top of a bottle.

(b) The hoop is struck sharply and accelerates horizontally. Because the friction between the pen and hoop is small *and* acts only for a short time, the pen does not move appreciably in the horizontal direction. However, there is now an unbalanced force acting vertically upon it—gravity.

(c) The pen falls into the bottle.

According to Newton's first law, an object remains at rest or in motion with a constant velocity unless acted upon by an unbalanced force.

4.3 ■ Newton's Second Law of Motion

Since a change in motion, or an acceleration, is evidence of a force, it seems logical that the acceleration is directly proportional to the applied net force, that is,

$$\mathbf{a} \propto \mathbf{F}_{net}$$

where the boldface symbols indicate vector quantities. For example, if you hit a ball twice as hard (applied twice as much force), you would expect the acceleration of the ball to be twice as great.

However, as Newton recognized, the inertia or mass of the object also plays a role. For a given force, the more massive the object, the less its acceleration will be; that is, the acceleration and mass (m) of an object are inversely proportional:

$$a \propto \frac{1}{m}$$

> Use an air track to demonstrate Newton's first and second laws of motion.
>
> For demonstrating accelerated motion on the air track, use a small weight attached to a string hung over a pulley for a constant pulling force. Adding weight to the glider will show how the mass and acceleration are related. Attach a small air-bubble level to the glider for an interesting accelerometer.

For example, if you hit two different balls with the same force, the less massive ball would acquire a greater acceleration.

Then combining these relationships, we have,

$$\mathbf{a} \propto \frac{\mathbf{F}_{net}}{m}$$

or in words,

The acceleration of an object is directly proportional to the *net* force acting on it and inversely proportional to its mass. The direction of the acceleration is in the direction of the applied net force.

Figure 4.5 presents some illustrations of this principle. We can rewrite the preceding equation as

$$\mathbf{F}_{net} = \Sigma_i \mathbf{F}_i \propto m\mathbf{a}$$

The force has been expressed as $\mathbf{F}_{net} = \Sigma_i \mathbf{F}_i$ to emphasize that it is a net or unbalanced force—the vector sum ($\Sigma_i \mathbf{F}_i$) of the forces acting on the object—that produces an acceleration (**a**). It should be apparent that if the forces acting on an object are balanced, $\mathbf{F}_{net} = \Sigma_i \mathbf{F}_i = 0$, then there is no acceleration (**a** = 0). The net force is often written as just **F** for convenience, with the understanding that this is the vector sum ($\mathbf{F} = \Sigma_i \mathbf{F}_i$). This simpler notation will be generally used hereafter; however, you may wish to write $\Sigma_i \mathbf{F}_i$ explicitly as a reminder.

A proportion may be expressed as an equation through the use of an

Newton's second law—cause and effect

> Students find it difficult to distinguish between the net force and the individual forces. Writing Newton's second law as:
> sum of forces = ma
> net force = ma
> may help. Use of different colored chalk when constructing vectors may also help.

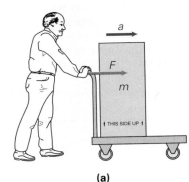

(a)
A net force accelerates the crate : $a \propto F/m$

Figure 4.5

FIGURE 4.5 Newton's second law The relationships among force, acceleration, and mass shown here are expressed by Newton's second law of motion.

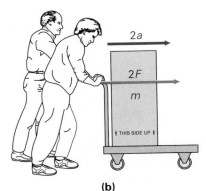

(b)
If the net force is doubled, the acceleration is doubled.

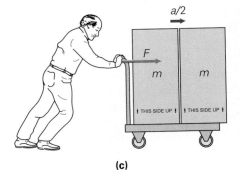

(c)
If the mass is doubled, the acceleration is halved.

appropriate constant (see marginal note). **Newton's second law of motion** is commonly expressed:

To change a proportion into an equation a constant of proportionality (*k*) is used. Thus, $F \propto ma$ is expressed as $F = kma$. As an analogy, the amount of deposit money ($) received for returnable bottles depends on the number (*n*) returned, $ \propto n$. But if you know that you get a nickel per bottle, you can write $ = kn$, where $k =$ $0.05/bottle. The *k* in $F = kma$ has been assigned a value of one by the definition of the force unit of newton (see text). That is, $k = 1$ N/kg-m/s², and therefore is omitted.

$$\boxed{F = ma} \qquad \text{Newton's second law} \qquad (4.1)$$

or in words, the force is equal to the mass times the acceleration.

$$\text{force} = \text{mass} \times \text{acceleration}$$

Thus, if the net force acting on an object is zero, its acceleration is zero, and it remains at rest or in uniform motion, which is consistent with the first law. For a nonzero net force (an unbalanced force), the resulting acceleration is in the same direction as the force.

The SI unit of force is, appropriately, the **newton (N).** As Eq. 4.1 shows, the base units of the newton are kilograms times meters per second squared:

The newton (N), unit of force

$$F = ma$$

$$N \equiv (kg)(m/s^2) = kg\text{-}m/s^2$$

That is, a force of 1 N gives a mass of 1 kg an acceleration of 1 m/s² (●Fig. 4.6). The British unit of force is the pound (lb), and 1 lb is equivalent to about 4.5 N (actually 4.448 N). An average apple weighs about 1 N.

Equation 4.1 shows how mass is related to weight. Recall that weight is the gravitational force of attraction that a celestial body exerts on an object. For us, this is the gravitational attraction of the Earth. Its effects are easily demonstrated: When you drop an object, it falls (accelerates) toward the Earth. Using **weight** (*w*) for *F* and the acceleration due to gravity (*g*) for *a* in Eq. 4.1, we can write in magnitude form

$$\boxed{w = mg} \qquad (4.2)$$

$$(F = ma)$$

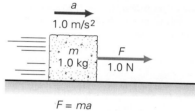

$$F = ma$$
$$1.0 \text{ N} = (1.0 \text{ kg})(1.0 \text{ m/s}^2)$$

●**FIGURE 4.6 The newton (N)**
A net force of 1.0 N acting on a mass of 1.0 kg produces an acceleration of 1.0 m/s².

The weight of 1.0 kg of mass is then $w = mg = (1.0 \text{ kg})(9.8 \text{ m/s}^2) = 9.8$ N.

Thus, 1.0 kg of mass has an equivalent weight of approximately 9.8 N, or 2.2 lb. But, although weight and mass are simply related through Eq. 4.2, keep in mind that *mass is the fundamental property.* Mass doesn't depend on the value of *g*, but weight does. As pointed out previously, the acceleration due to gravity on the moon is about $\frac{1}{6}$ that on Earth. The weight of an object on the moon would thus be different from its weight on Earth; but its mass, which reflects the quantity of matter it contains, would be the same in both places.

Bodies on the moon weigh about $\frac{1}{6}$ as much as they weigh on the Earth because the gravitational acceleration on the moon is about $\frac{1}{6}$ as much as it is on the Earth. But the mass of a body on the moon is no different from its mass on the Earth. Do we go on a diet to lose weight, or mass?

Newton's second law also shows why all objects in free fall have the same acceleration. Consider, for example, two falling objects, one with twice the mass of the other. The object with twice as much mass would have twice as much weight, or two times as much gravitational force acting on it. But the more massive object also has twice the inertia, so twice as much force is needed to give it the same acceleration. Expressing this relationship mathematically, for the smaller mass (*m*) we can write $F/m = g$, and for the larger mass (2*m*), we have the same acceleration: $2F/2m = g$ (●see Fig. 4.7).

Newton's second law allows you to analyze dynamic situations. In using this equation, you should keep in mind that *F* is the *magnitude of the net force*

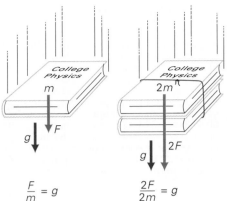

$$\frac{F}{m} = g \qquad \frac{2F}{2m} = g$$

FIGURE 4.7 Newton's second law and free fall In free fall, all objects fall with the same constant acceleration g. An object with twice the mass of another has twice as much gravitational force acting on it. But with twice the mass, it also has twice as much inertia, so twice as much force is needed to give it the same acceleration.

and m is the *total mass of the system*. The system is composed of all the objects or masses involved in the given situation. The boundaries defining a system may be real or imaginary. For example, a system might consist of all the gas molecules in a particular sealed vessel. But you might also define a system to be all the gas molecules in an arbitrary cubic meter of air. In studying dynamics we often have occasion to work with systems made up of one or more discrete masses: the Earth and moon, for instance, or a series of blocks on a table top, or a tractor and wagon.

EXAMPLE 4.1 ■ Newton's Second Law—Finding Acceleration

A garden tractor pulls a loaded wagon with a force of 15 N (● Fig. 4.8). If the total mass of the wagon and its contents is 60 kg, what is the wagon's acceleration? (Ignore any frictional forces.)

Solution. Listing the data from the example, we have

> *Given:* $F = 15$ N   *Find:* a (acceleration)
> $m = 60$ kg

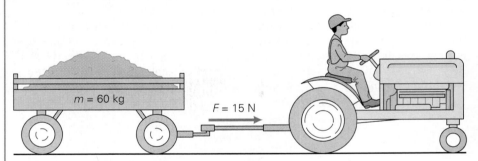

FIGURE 4.8 Force and acceleration Example 4.1.

It is noted that F is the net force in this case. (If an opposing frictional force were a consideration, the force exerted by the tractor and the frictional force on the wagon would have to be added to find the net force.)

The acceleration is given by Eq. 4.1, $F = ma$. Solving for a,

$$a = \frac{F}{m} = \frac{15\,\text{N}}{60\,\text{kg}} = 0.25\,\text{m/s}^2$$

in the direction that the tractor is pulling.

Note that m was the *total* mass of the wagon and its contents. If the masses of the wagon and the contents had been given separately, say $m_1 = 45$ kg and $m_2 = 15$ kg, respectively, they would be added together in Newton's law, $F = ma = (m_1 + m_2)a$.

With a constant force, the acceleration is also constant, so the kinematic equations of Chapter 2 can be applied. Suppose the wagon started from rest ($v_0 = 0$). Could you find how far it traveled in 4.0 s? Easily.

$$x = v_0 t + \tfrac{1}{2}at^2 = 0 + \tfrac{1}{2}(0.25\,\text{m/s}^2)(4.0\,\text{s})^2 = 2.0\,\text{m}$$

EXAMPLE 4.2 ■ Newton's Second Law—Finding Mass

A student weighs 588 N. What is her mass?

Solution.

   Given: $F = w = 588$ N   *Find:* m (mass)

Recall that weight is a (gravitational) force and Newton's second law, $F = ma$, can be written in the form $w = mg$ (Eq. 4.2), where g is the acceleration due to gravity (9.80 m/s²). Rearranging the equation, we have

$$m = \frac{w}{g} = \frac{588 \text{ N}}{9.80 \text{ m/s}^2} = 60.0 \text{ kg}$$

This is equivalent to 60.0 kg (2.2 lb/kg) = 132 lb. In metric countries, a person's "weight" is expressed as mass in kilograms. The student would weigh 60.0 "kilos."

   A dynamic system may consist of more than one discrete mass. In applications of Newton's second law, it is often advantageous, and sometimes necessary, to isolate a given mass within a system. This is possible because any part of a system can be treated as a discrete system to which the second law can be applied, as the following example shows.

EXAMPLE 4.3 ■ Newton's Second Law—All or Part of the System

Two masses, $m_1 = 2.5$ kg and $m_2 = 3.5$ kg, rest on a frictionless surface and are connected by a light string (Fig. 4.9). A horizontal force of 12.0 N is applied to m_1 as shown in the figure. (a) What is the magnitude of the acceleration of the masses (the system)? (b) What is the magnitude of the tension (T) in the string? (The tension is the force transmitted by the string.)

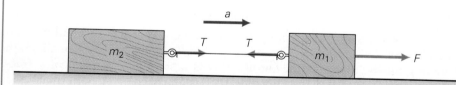

FIGURE 4.9 An accelerated system   Example 4.3.

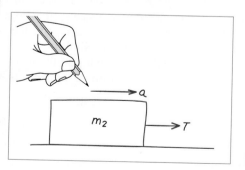

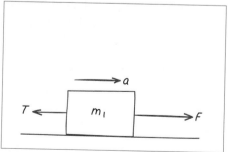

Isolating the masses

Free-body diagram for m_1

Solution. Carefully listing the data and what we want to find, we have

Given: $m_1 = 2.5$ kg   *Find:* (a) a (acceleration)
          $m_2 = 3.5$ kg   (b) T (tension force)
          $F = 12.0$ N

Given an applied force that produces motion, the acceleration of the masses can be found from Newton's second law. In using the second law, it is important to keep in mind that it applies to the total system *or to any part of it—* that is, to the total mass ($m_1 + m_2$), or to m_1 individually, or to m_2 individually. However, *we must be sure to identify correctly the appropriate force or forces in each case.* The net force acting on the combined masses, for example, is not the same as the net force acting on m_2 considered separately.

(a) Drawing an imaginary line around m_1 and m_2, we see that the force acting on this system is F. Representing the total mass simply as m, we can thus write

$$a = \frac{F}{m} = \frac{F}{m_1 + m_2} = \frac{12.0 \text{ N}}{2.5 \text{ kg} + 3.5 \text{ kg}} = 2.0 \text{ m/s}^2$$

The acceleration is in the direction of the applied force, as the figure indicates. Note that m is the *total* mass of the system, or all the mass that is accelerated. (The mass of the string is small enough to be ignored.)

(b) There is a force of magnitude T on each of the masses because of tension in the connecting string. The T forces on the masses are equal and opposite. (We will learn more about such equal and opposite forces in the next section.) Note that the tension forces did not appear in part (a), where the total system of both masses was considered. In this case, the *internal* equal and opposite T forces cancel each other.

However, each mass may also be considered a separate system, to which Newton's second law applies. In these systems, the tension force comes into play explicitly. Looking at the sketch of the isolated m_2 in Fig. 4.9, we see that the only force acting to accelerate this mass is T. Knowing the values of m_2 and a, the magnitude of this force is given directly by

$$T = m_2 a = (3.5 \text{ kg})(2.0 \text{ m/s}^2) = 7.0 \text{ N}$$

An isolated sketch of m_1 is also shown in Fig. 4.9, and the second law can equally well be applied to this mass to find T. Note the adjacent diagram showing all the forces acting on m_1. This is a simplified version of a **free-body diagram**, which represents an object as a "particle" or point mass and shows all the forces acting on it. Free-body diagrams, discussed more fully in the following Problem-Solving Hint, are very helpful in analyzing situations involving several bodies and multiple forces. (In this simple case, the free-body diagram for m_2 would have only one vector and is not shown.)

From the free-body diagram we can easily see that we must add the forces vectorially to get the net force that produces the acceleration of m_1. That is,

$$F_{\text{net}} = F - T = m_1 a \qquad \text{(direction of } F \text{ taken as positive)}$$

Then, solving for T,

$$T = F - m_1 a$$
$$= 12.0 \text{ N} - (2.5 \text{ kg})(2.0 \text{ m/s}^2) = 12.0 \text{ N} - 5.0 \text{ N} = 7.0 \text{ N}$$

When an object is described as being "light," you can ignore its mass in analyzing the problem situation. That is, the mass is negligible relative to the other masses.

Note that we are assuming the tension force to be transmitted *undiminished* through the string. That is, the magnitude of **T** acting on m_2 is the same as that acting on m_1. This is actually true only if the string has zero mass. If the mass of the string were taken into account, the magnitude of **T** would be different for m_1 and m_2; the difference would be the net force required to accelerate the mass of the string.

Problem-Solving Strategies: Free-body Diagrams

When working with problems in which two or more forces or components of force act on a body, it is convenient and instructive to draw a free-body diagram of the forces, as was done in the previous example. In such a diagram, we show all the forces acting on the body or object. If several bodies are involved, we may make a diagram for each body *separately*, showing all of the forces acting on each individual body.

In the figures that illustrate the physical situations, sometimes called *space diagrams*, force vectors may be drawn at different locations to indicate their points of application. However, since we are concerned only with linear motions, vectors in free-body diagrams may be shown emanating from a common point, which is chosen as the origin of x–y axes. One of the axes is generally chosen along the direction of the net force acting on a body, since that is the direction in which the body will move. Also, it is often important to resolve force vectors into components, and properly chosen x–y axes simplify this.

In a sketch of a free-body diagram, the vector arrows do not have to be exactly to scale. However, it should be made apparent if there is a net force, and whether forces balance each other in a particular direction. When the forces aren't balanced, we know from Newton's second law that there must be an acceleration.

In summary, the general steps in constructing and using free-body diagrams are as follows. (Refer to Fig. 4.10 as you read.)

1. Sketch a space diagram (if one is not already available) and identify the forces acting on each body of the system.

2. Isolate the body for which the free-body diagram is to be constructed. Draw a set of Cartesian axes with the origin at a point through which the forces act and one of the axes along the line of the body's motion. (This will be in the direction of the net force if there is one.)

3. Draw properly oriented force vectors on the diagram emanating from the origin of the axes. If there is an unbalanced force, indicate the direction of motion with an acceleration vector. (This direction may be arbitrarily selected if the direction of the net force is not evident.)

4. Resolve any forces that are not directed along the x or y axes into x or y components. Use the free-body diagram to analyze the forces in terms of Newton's second law of motion. (Note: If the acceleration is in the direction opposite that selected, this will be indicated by an acceleration with an opposite sign in the solution. For example, if the motion is taken to be in the positive direction and it is actually in the opposite direction, the acceleration will be negative.)

Free-body diagrams are a particularly useful way of following one of the Suggested Problem-Solving Procedures in Chapter 1: Draw a diagram as an aid in visualizing and analyzing the physical situation of the problem. *Make it a practice to draw free-body diagrams for force problems, as is done in the following examples.*

Figure 4.10

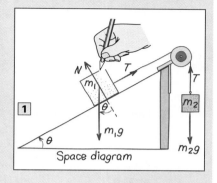

1 Space diagram

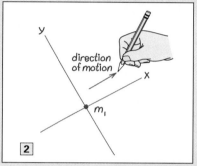

2

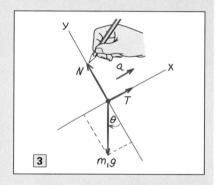

3

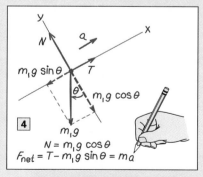

4

$N = m_1 g \cos\theta$

$F_{net} = T - m_1 g \sin\theta = ma$

FIGURE 4.10 Drawing a free-body diagram

The Second Law in Component Form

Not only does Newton's second law hold for any part of a system, it also applies to the components of motion. A force may be expressed in component notation in two dimensions as follows:

$$\mathbf{F} = m\mathbf{a}$$

and
$$F_x\,\mathbf{x} + F_y\,\mathbf{y} = m(a_x\,\mathbf{x} + a_y\,\mathbf{y}) \qquad (4.3)$$
$$= ma_x\,\mathbf{x} + ma_y\,\mathbf{y}$$

so

Component form of the second law

$$\boxed{\begin{aligned} F_x &= ma_x \\ F_y &= ma_y \end{aligned}} \qquad (4.3b)$$

and Newton's second law applies to each component of motion. (Similarly, $F_z = ma_z$ in three dimensions.) As pointed out in the preceding Problem-Solving Hint, free-body diagrams assist in resolving forces into rectangular components. An example of how the second law is applied to components follows.

EXAMPLE 4.4 ■ **Newton's Second Law and Components of Force**

A force of 15.0 N is applied at an angle of 30° to the horizontal on a 0.750-kg block at rest on a frictionless surface (● Fig. 4.11).

(a) What is the magnitude of the resulting acceleration of the block?
(b) If the force is applied for only 1.50 s, what happens after this?

Solution. First we write down the given data and what is to be found.

Given: $F = 15.0\text{ N}$ *Find:* (a) a (acceleration)
$\quad\quad m = 0.750\text{ kg}$ (b) Describe the motion when
$\quad\quad \theta = 30°$ the force is no longer ap-
$\quad\quad v_o = 0$ plied
$\quad\quad t = 1.50\text{ s}$

Then draw a free-body diagram as in Fig. 4.11. Note that the forces act in the same directions as in the space diagram, for example, the force N is upward.

(a) The acceleration of the block is given by Newton's second law. We choose our axes so that **a** is in the $+x$ direction; this is the direction in which the block will move along the surface. We can see that only a component (F_x) of the applied force F acts in this direction. From the figure, we see that the component of F in the direction of motion is $F_x = F\cos\theta$. Applying Newton's law for this component,

$$F_x = F\cos 30° = ma_x$$

so
$$a_x = \frac{F\cos 30°}{m} = \frac{(15.0\text{ N})(0.866)}{0.750\text{ kg}} = 17.3\text{ m/s}^2$$

This is the total acceleration of the block, since it does not move in the y direction. The sum of the forces in the y direction must then be zero. That is, the downward component of F acting on the block, F_y, and its downward weight force w must be balanced by the upward force N that the surface

112   Chapter 4   Force and Motion

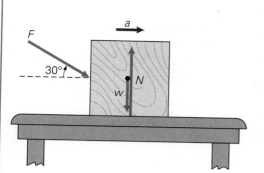

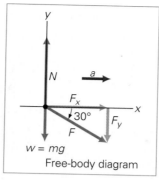

$w = mg$

Free-body diagram

exerts on the block. (N is commonly called the *normal force* because it is normal, or perpendicular to the surface.) If this were not the case, then there would be a net force and an acceleration in the y direction. (The nature of the normal force will become clearer with further explanation in Sections 4.5 and 4.6.)

Summing the forces in the y direction with upward as positive,

$$N - F_y - w = 0$$

or

$$N - F \sin 30° - mg = 0$$

and

$$N = F \sin 30° + mg$$
$$= (15.0 \text{ N})(0.500) + (0.750 \text{ kg})(9.80 \text{ m/s}^2) = 14.9 \text{ N}$$

The surface then exerts a force of 14.9 N upward on the block, which balances the downward forces acting on it.

(b) When the force stops acting at the end of 1.50 s, the block will have a velocity in the x direction with a magnitude of

$$v_x = v_{x_0} + a_x t = 0 + (17.3 \text{ m/s}^2)(1.50 \text{ s}) = 26.0 \text{ m/s}$$

By Newton's first law, the block will continue to travel with this constant velocity until acted on by another force.

In checking the block's speed, you might notice that 26.0 m/s is about 94 km/h (or 58 mi/h), which is a bit unrealistic for a block on a surface. However, such ideal (frictionless) examples are used to make things simple for illustration purposes. (On a real surface, the frictional force would have to be taken into account in computing the net force on the block; and after the applied force was removed, it would act to reduce the velocity, eventually decelerating the block to a halt. The reality of friction will be added later in the chapter.)

■ Problem-Solving Strategies: Newton's Second Law

There is no fixed way to go about solving a problem. However, there are general strategies or procedures that are helpful in solving problems involving Newton's second law. Using our Suggested Problem-Solving Procedures introduced in Chapter 1, you might include the following for this case:

• Draw a free-body diagram for each individual body, showing all of the forces acting on that body.

- Depending on what is to be found, Newton's second law may be applied to the system as a whole (in which case internal forces cancel) or to a part of the system. Basically, *you want to obtain an equation containing the quantity for which you want to solve.* Review Example 4.3. (If there are two unknown quantities, application of Newton's law to two parts of the system may give you two equations and two unknowns. See Example 4.6 in next section.)

- Keep in mind that Newton's second law may be applied to components of motion, and forces may be resolved into components to do this. Review Example 4.4.

4.4 ■ Applications of Newton's Second Law

The simple relationship expressed by Newton's second law, $F = ma$, allows the quantitative analysis of force and motion. This relationship can be thought of as a cause-and-effect one, with force being the cause and acceleration being the motional effect.

In general, you will be concerned with applications involving constant forces. Constant forces result in constant accelerations for moving objects and allow the use of the kinematic equations from Chapter 2 in analyzing the motion. When there is a variable force, Newton's second law holds for the *instantaneous* force and acceleration, but the acceleration will vary with time. This course of study will generally be limited to average and constant accelerations and forces, as expressed by $F = ma$.

This section presents several examples of applications of Newton's second law so that you may become familiar with its use. This small but powerful equation will be used again and again throughout this course of study.

Newton's second law gives the acceleration resulting from an applied force, but it can also be used to find the force from motional effects, as the following example shows.

> **EXAMPLE 4.5 ■ Finding a Force from Motional Effects**
>
> A car traveling 72.0 km/h along a straight level road is brought uniformly to a stop in a distance of 40.0 m. If the car weighs 8.80×10^3 N, what is the braking force?
>
> **Solution.** In bringing the car to a stop, the braking force caused an acceleration (actually a deceleration), as illustrated in ● Fig. 4.12. Listing what is given and wanted to find, we have
>
> > *Given:* $v_o = 72.0$ km/h $= 20.0$ m/s *Find:* F (braking force)
> > $v = 0$
> > $x = 40.0$ m
> > $w = 8.80 \times 10^3$ N
>
> We know that $F = ma$, so we can easily calculate F if we can find m and a.

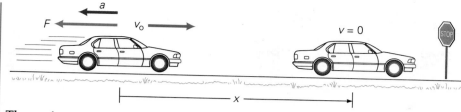

The car's mass m can be obtained from the given weight. The other given quantities should remind you of a kinematic equation from Chapter 2 from which the acceleration a may be found. Since the car is brought uniformly to a stop, the acceleration is constant and we may use Eq. 2.9, $v^2 = v_o^2 + 2ax$ to find a:

$$a = \frac{v^2 - v_o^2}{2x} = \frac{0 - (20.0 \text{ m})^2}{2(40.0 \text{ m})} = -5.00 \text{ m/s}^2$$

The minus sign indicates that the acceleration is opposite to v_o, as expected for a braking force, which slows the car.

The mass of the car is obtained from the weight: $w = mg$ or $m = w/g$. Using this expression along with the acceleration obtained previously gives a braking force of

$$F = ma = \left(\frac{w}{g}\right)a$$

$$= \left(\frac{8.80 \times 10^3 \text{ N}}{9.80 \text{ m/s}^2}\right)(-5.00 \text{ m/s}^2) = -4.49 \times 10^3 \text{ N}$$

EXAMPLE 4.6 ■ The Atwood Machine

The Atwood machine consists of two masses suspended from a fixed pulley, as shown in ●Fig. 4.13a. If $m_1 = 0.55$ kg and $m_2 = 0.80$ kg, what is the acceleration of the system? (Consider the pulley to be frictionless and the masses of the string and the pulley to be negligible.)

Demonstration/Activity: Connect a 1-kg mass to the end of a demonstration-type spring scale and observe the readings when the mass is stationary, moving up or down with constant velocity, and accelerated upward and downward.

Solution. Listing the two given quantities, we have
    Given: $m_1 = 0.55$ kg   *Find:* a (acceleration)
             $m_2 = 0.80$ kg

Since m_2 is greater than m_1, it is obvious that m_2 will fall and m_1 will rise with accelerations of the same magnitude. Looking at the two masses as separate systems (see the free-body diagrams), we are free to assign directional plus and minus signs arbitrarily. Here, the positive direction is taken to be the direction of motion for both masses. That is, for m_2, downward is taken to be the positive direction. The pulley is simply a direction changer; thus, the horizontal analog shown in Fig. 4.13b is equivalent to that in Fig. 4.13a, except for the changed directions of the accelerations. Then, applying Newton's second law to the system as a whole gives

$$F_2 - F_1 = m_2g - m_1g$$

$$= (m_1 + m_2)a$$

net force = total mass × acceleration

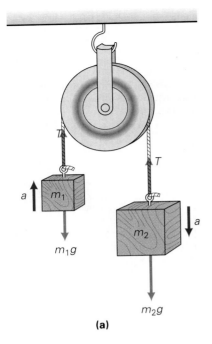

(a)

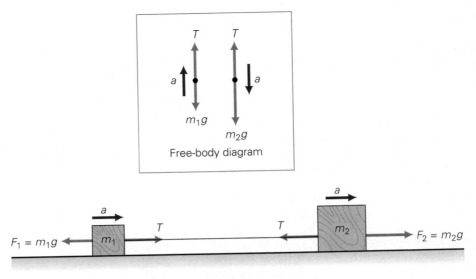

Free-body diagram

(b) Horizontal analogue

● **FIGURE 4.13 The Atwood machine** Example 4.6. **(a)** A single, fixed pulley is simply a direction changer. **(b)** An equivalent horizontal analog. (The surface is assumed to be frictionless.)

Solving for a gives

$$a = \frac{(m_2 - m_1)g}{m_1 + m_2}$$

$$= \frac{(0.80 \text{ kg} - 0.55 \text{ kg})(9.8 \text{ m/s}^2)}{0.55 \text{ kg} + 0.80 \text{ kg}} = 1.8 \text{ m/s}^2$$

Note that the tension forces in the string cancel out and need not be considered in finding the solution. However, the problem may also be worked by applying Newton's second law to each isolated mass on which the tension acts. (See the free-body diagrams in Fig. 4.13b.) This gives two equations (with two unknowns, a and T):

$$T - m_1g = m_1a$$
$$m_2g - T = m_2a$$

where the direction of the acceleration for each mass (upward for m_1 and downward for m_2) is taken as positive so as to avoid a minus sign for a. The magnitudes of the accelerations of the masses are equal. Eliminating T from the two equations gives the equation derived above:

$$m_2g - m_1g = m_2a + m_1a = (m_1 + m_2)a$$

Solving the problem this way gives you an equation containing T (actually, two of them) in case you need to find its value. For example, you might want to see whether it exceeds the tensile strength of the string (the force that would cause the string to break). Once the acceleration is found, you simply use it in either of the above two equations, solved for T. Using the first equation gives

$$T = m_1a + m_1g = m_1(a + g)$$
$$= (0.55 \text{ kg})(1.8 \text{ m/s}^2 + 9.8 \text{ m/s}^2) = 6.4 \text{ N}$$

Incidentally, the Atwood machine is named after George Atwood (1746–1807), who used the arrangement to study motion and measure the value of g. Obviously, the masses can be chosen to minimize the acceleration, making it easier to measure the time of fall. The next example concerns a variation of Atwood's machine, where one of the masses is on an inclined plane.

EXAMPLE 4.7 ■ Motion on a Frictionless Inclined Plane

Two masses are connected by a light string running over a light, frictionless pulley as illustrated in ● Fig. 4.14. One mass ($m_1 = 5.0$ kg) is on a frictionless 20° inclined plane, and the other ($m_2 = 1.5$ kg) is freely suspended. What is the acceleration of the masses?

Solution. Following our usual procedure, we write

> *Given:* $m_1 = 5.0$ kg   *Find:* a (acceleration)
> $m_2 = 1.5$ kg
> $\theta = 20°$

Figure 4.14

●**FIGURE 4.14 Application of Newton's second law** Example 4.7. (Drawing not to scale.)

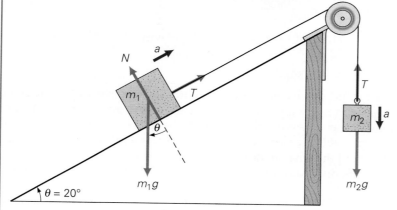

Space diagram

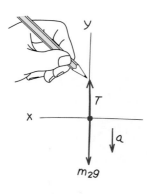

Draw free-body diagram for m_2

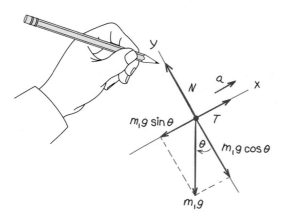

Draw free-body diagram for m_1

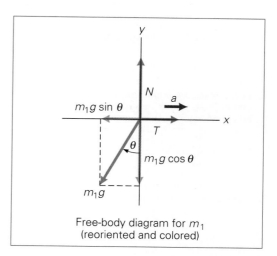

Free-body diagram for m_1
(reoriented and colored)

Then, to better visualize the forces involved, we isolate m_1 and m_2 and draw free-body diagrams for each mass. For mass m_1, there are three concurrent forces (forces acting through a common point). These are T, m_1g, and N, where T is the tension force in the string and N is the normal force of the table on the block. The forces are shown emanating from their common point of action. (Recall that a vector arrow can be moved as long as its direction is not changed.)

We will start by assuming that m_1, moves up the plane, which is taken to be in the x direction. (It makes no difference whether it is assumed that m_1 moves up or down the plane, as we shall see shortly.) Notice that m_1g is broken down into components. The x component is in the assumed direction of motion, and the y component acts perpendicularly to the plane and is balanced by the normal force N. (There is no motion in the y direction, so there is no net force in this direction.)

Applying the second law to the system as a whole (so that the T forces cancel), and neglecting the mass of the string and pulley, we have

$$F_{net} = m_2g - m_1g \sin 20° = (m_1 + m_2)a$$

(net force = $total$ mass × acceleration)

Solving for a,

$$a = \frac{m_2g - m_1g \sin 20°}{m_1 + m_2}$$

$$= \frac{(1.5\ \text{kg})(9.8\ \text{m/s}^2) - (5.0\ \text{kg})(9.8\ \text{m/s}^2)(0.342)}{5.0\ \text{kg} + 1.5\ \text{kg}}$$

$$= -0.32\ \text{m/s}^2$$

The minus sign indicates that the acceleration is opposite to the assumed direction. That is, m_1 actually moves down the plane and m_2 rises. As this example shows, if you assume the motion to be in the wrong direction, the sign on the result will give you the correct direction anyway.

Could you find the tension force T in the string if you were asked to do so? How this could be done should be quite evident from the free-body diagram.

Forces may act on an object without producing an acceleration. In such a case, with $a = 0$, we know from Newton's first law that

Translational equilibrium condition

$$\boxed{\Sigma \mathbf{F} = 0} \qquad (4.4)$$

That is, the vector sum of the forces, or the net force, is zero, so the object either remains at rest or moves with a constant velocity. In such cases, objects are said to be in **translational equilibrium**. When at rest, an object is said to be in *static translational equilibrium*.

In translational equilibrium, an object is at rest or moves with a constant velocity

It follows that the sums of the rectangular components of the forces for an object in equilibrium are also zero. (Why?)

Component condition for translational equilibrium

$$\boxed{\begin{aligned} \Sigma \mathbf{F}_x &= 0 \\ \Sigma \mathbf{F}_y &= 0 \end{aligned}} \qquad (4.5)$$

For three-dimensional problems, we would add $\Sigma \mathbf{F}_z = 0$. However, we will restrict our discussion of forces to two dimensions.

These equations give what is often referred to as the **condition for translational equilibrium**. (Another condition for rotational considerations will be given in Chapter 8.) Let's apply this translational condition to a static equilibrium case.

Video demonstration 5

EXAMPLE 4.8 ■ Static Translational Equilibrium

A 3.0-kg sign hangs in a hall in the Physics Department as shown in Fig. 4.15a. What is the minimum tensile strength necessary for the cord that is used to hang the sign?

Solution.

Given: $m = 3.0 \, \text{kg}$
$\theta_1 = \theta_2 = 45°$

Find: T_1 and T_2
(tensions in the cord)

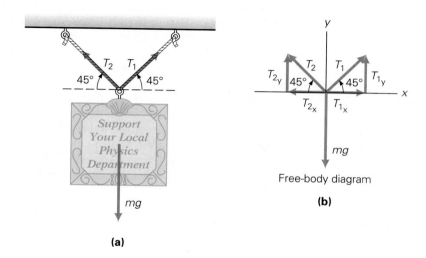

Free-body diagram

(b)

(a)

FIGURE 4.15 Static translational equilibrium Example 4.8.

The minimum tensile strength of the cord is just the amount of tension the cord must be able to support without breaking. With a cord of sufficient strength, the sign will hang in static equilibrium. We want to find the values of T_1 and T_2 that will *just* support the weight mg of the sign.

The rectangular components of the tensions are shown in the free-body diagram (Fig. 4.15b). Then, applying the component conditions for static translational equilibrium (Eq. 4.5), we have

$$\Sigma \mathbf{F}_x = T_{x_1} - T_{x_2} = 0$$

or

$$T_{x_1} = T_{x_2}$$

Hence, the magnitudes of the x components of the tensions are equal, as you might have expected. (They are the only forces in the x direction and they balance each other.) However, this gives no information about the magnitude of the tensions.

Since the angles θ_1 and θ_2 are equal, we also know that $T_{y_1} = T_{y_2}$. (Why?)

Applying the other component condition,

$$\Sigma \mathbf{F}_y = T_{y_1} + T_{y_2} - mg = 2T_{y_1} - mg$$
$$= 2(T_1 \sin 45°) - mg = 0$$

Solving for T_1,

$$T_1 = \frac{mg}{2 \sin 45°} = \frac{(3.0 \text{ kg})(9.8 \text{ m/s}^2)}{2(0.707)} = 21 \text{ N}$$

With the x and y tension components equal, $T_1 = T_2 = 21$ N, so the sign should be hung using a cord with a tensile strength of at least 21 N.

4.5 ■ Newton's Third Law of Motion

Newton formulated a third law that is as far-reaching in its physical significance as the first two. For a simple introduction to the third law, consider the forces involved in seat belt safety. When the brakes are suddenly applied in a moving car, you continue to move forward (the frictional force on the seat of your pants is not enough to stop you). In doing so, you exert forces on the seat belt and shoulder strap. The belt and strap exert corresponding reaction forces on you, causing you to slow down with the car. If you haven't buckled up, you may keep on going (Newton's first law) until another applied force, such as that applied by the dashboard or windshield, slows you down.

We commonly think of single forces. However, Newton recognized that it is impossible to have a single force. He observed that in any application of force, there is always a mutual interaction, and forces always occur in pairs. An example given by Newton was this: If you press on a stone with a finger, then the finger is also pressed by, or receives a force from, the stone.

Newton termed the paired forces action and reaction, and **Newton's third law of motion** is

> For every force (action) there is an equal and opposite force (reaction).

In symbol notation,

$$\mathbf{F}_{12} = -\mathbf{F}_{21}$$

F_{12} is the force exerted *on* object 1 *by* object 2, and $-F_{21}$ is the equal and opposite force exerted *on* object 2 *by* object 1. (The minus sign indicates the opposite direction.) *Which force is considered the action or the reaction is arbitrary.* F_{21} may be the reaction to F_{12} or vice versa.

The third law may seem to contradict the second law. If there are always equal and opposite forces, how can there be a nonzero net force? *An important thing to remember about a force pair of the third law is that the opposing forces do not act on the same object.* The second law is concerned with force(s) acting on a particular object (or system). The opposing forces of the third law act on different objects. Examples of this principle were encountered previously in our applications of Newton's second law. In both Examples 4.4 and 4.7, the normal force exerted by the surface *on the block* was the reaction force to the weight of the block *on the surface*.

In most instances when you are applying Newton's second law, you need

Newton's third law—action and reaction

Students have difficulty with contact forces between bodies. Slightly separating the objects on diagrams and constructing the two vectors with different colors may help. In discussions, have students state what is exerting the force, *and* which object the force is acting on.

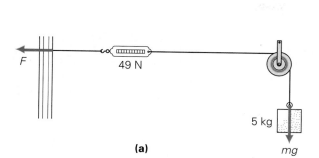
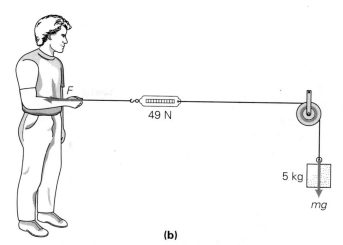

(a) 49 N

F

5 kg

mg

(b) 49 N

F

5 kg

mg

● **FIGURE 4.16 Third law forces** **(a)** The suspended mass exerts a force on the wall, via the string, that is equal to its weight (neglecting friction and the mass of the string, scale, and pulley). The wall exerts an equal and opposite force, which may not be obvious. **(b)** However, replace the wall with yourself and the equal and opposite force becomes apparent.

to consider only those forces acting on a single object. But the reaction forces of the third law are always there. For example, for the arrangement in ● Fig. 4.16a, you immediately perceive that the weight is pulling on the wall. But the wall also pulls on the weight (via the string), as you would realize if you were substituted for the wall (Fig. 4.16b). For an interesting look at a similar situation involving Newton's third law see Demonstration 2.

DEMONSTRATION 2 ■ Tension in a String

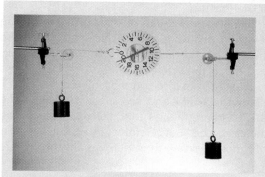

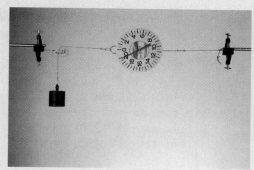

(a) Two 2-kg masses are suspended on each side of a scale (calibrated in newtons). The total suspended weight is $w = mg = (4.0 \text{ kg})(9.8 \text{ m/s}^2) = 39.2 \text{ N}$, yet, the scale reads about 20 N. Is something wrong with the scale?

(b) No, think of it in this manner. The weight of one mass supplies the reaction force that keeps the scale stationary, while the other mass stretches the scale spring giving a reading of 20 N [or $w = mg = (2.0 \text{ kg})(9.8 \text{ m/s}^2) = 19.6 \text{ N}$]. The reaction force can be equivalently supplied by a fixed support.

(c) A fixed pulley merely changes the direction of the force—the fixed support or reaction force can be vertical with the same effect. In all cases, the tension in the string is 19.6 N.

A demonstration showing action and reaction forces.

Figure 4.17

●**FIGURE 4.17 Force pairs of
Newton's third law** (a) When the
person holds the briefcase, there are
two force pairs: a contact pair and an
action-at-a-distance pair (gravity).
The net force acting on the briefcase
is zero: The upward contact force
balances the downward weight force.
Note, however, that these are *not* a
third-law pair. (b) When the briefcase
is falling, there is an unbalanced
force acting on the case (its weight
force), and it accelerates downward
(at *g* for free fall).

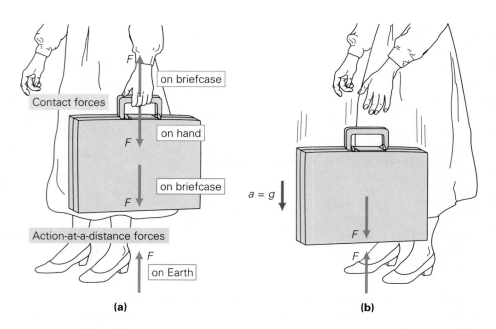

(a)   (b)

*A third-law force pair is as
follows: The Earth gravitationally
attracts you, while you exert an
attractive force of equal
magnitude on the Earth. Notice
that these two gravitational forces
act on* Point out *different objects.*

*Don't confuse the above force
pair with the following: The Earth
exerts a downward gravitational
force on you, while the surface of
the Earth exerts an upward
normal force on you. Notice that
the two forces act on the* Same
object.

*Video demonstration 6 relates to
this material.*

As another example, consider the situation in ● Fig. 4.17a. Two third-law
force pairs are acting when the person is holding the briefcase. There is a pair
of contact forces: The person's hand exerts an upward force on the handle,
and the handle exerts an equal downward force on the hand. That is,
$F_{\text{hand}} = -F_{\text{handle}}$. This is an action/reaction pair, with the forces acting on different objects. The other third-law force pair consists of action-at-a-distance forces
associated with gravitational attraction: The Earth attracts the briefcase (its
weight), and the briefcase attracts the Earth.

Concentrating on the briefcase in isolation, we can see that only two of the
four forces in Fig. 4.17a act on it—the upward force on the handle and the
downward force of its weight. These are *not* a third-law force pair, however,
because they act on the *same* object. Since the briefcase does not move, these
forces must be equal and opposite. Thus the net force on the isolated stationary
briefcase is zero, as required by Newton's first and second laws. When the
person drops the briefcase (Fig. 4.17b), there is then a nonzero net force (the
unbalanced force of gravity) on the case, and it accelerates. But there is still a
pair of third law forces *acting on different objects*—the briefcase and the Earth.

EXAMPLE 4.9 ■ Another Apple

An apple falls toward the ground. Which of the following statements is
correct? (a) Because the mass of the apple is infinitesimal compared with that
of the Earth, the force it exerts on the Earth while falling is much smaller
than that exerted on it by the Earth. (b) Because the third law forces are equal
and opposite, there is no net force on either object while falling. (c) The apple
and the Earth fall toward each other, but the Earth's motion is negligible
because of its extremely large mass. (d) The third law force pair act only
while the apple is falling and not after it strikes the ground. *Clearly establish*

the reasoning used in determining your answer before checking it below. That is, why did you select your answer?

Reasoning and Answer. While it is falling (and accelerating), there must be a net force on the apple. This is its weight, or the gravitational force of the Earth acting on the apple—one of the forces of the gravitational third law force pair between the apple and Earth. The other *equal* and opposite force acts on the Earth. These forces continue to act even after the apple has reached the ground. Hence, (a), (b), and (d) are incorrect. Because of their mutual gravitational attraction, the apple accelerates toward the Earth and the Earth (theoretically) accelerates toward the apple—that is, they "fall" toward each other. But the Earth's mass is so much greater than the apple's that its motion is unmeasurable. Thus the answer is (c). The weight of the apple and the contact force on the apple are not a third law force pair. Why?

Follow-up Exercise. What are the forces on the apple when it is hanging on the tree? (*Reasoning and answer may be found in the Answers to Exercises section at the back of the book.*)

4.6 ■ Friction

Friction refers to the ever-present resistance to motion that occurs whenever two materials, or media, are in contact with each other. This resistance occurs for all types of media—solids, liquids, and gases—and is characterized as the **force of friction**. Until this point, we have generally ignored all kinds of friction (including air resistance) in examples and problems for simplicity. Now that you know how to describe motion, we are ready to consider situations that are more realistic, in that the effects of friction are included.

In some real situations, we want to increase friction, for example by putting sand on an icy road or sidewalk to improve traction. This might seem contradictory, since an increase in friction presumably would increase the resistance to motion. However, consider the forces involved in walking as illustrated in ● Fig. 4.18. Without friction, the foot would slip backwards. (Think about walking on a slippery surface.) The force of friction prevents this, and sometimes needs to be increased on slippery surfaces (● Fig. 4.19a). In other situations, we try to reduce friction (Fig. 4.19b). For instance, we lubricate moving machine parts to allow them to move more freely, lessen wear, and reduce expenditure of energy. Automobiles would not run without friction-reducing oils and greases.

This section is concerned chiefly with friction between solid surfaces. All surfaces are microscopically rough, no matter how smooth they appear or feel. It was originally thought that friction was primarily due to the mechanical interlocking of surface irregularities, or asperities (high spots). However, research has shown that the friction between the contacting surfaces of ordinary solids (metals in particular) is mostly due to local adhesion. When surfaces are pressed together, local welding or bonding occurs in a few small patches where the largest asperities make contact. To overcome this local adhesion, a force great enough to pull apart the bonded regions must be applied. Once contacting surfaces are in relative motion, another form of friction may result when the asperities of a harder material dig into a softer material, with a "plowing" effect.

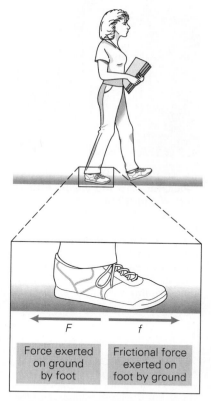

| Force exerted on ground by foot | Frictional force exerted on foot by ground |

● **FIGURE 4.18 Friction and walking** Note that the force of friction is shown in the direction of the walking motion. This may seem wrong at first glance, but it's not. The force of friction prevents the foot from slipping backwards while the other foot is brought forward.

(a)

(b)

FIGURE 4.19 Increasing and decreasing friction (a) The tread on automobile and truck tires is designed to increase friction between the tire and the road. This provides traction and prevents skidding during stops and turns. (b) Water serves as a good lubricant to reduce friction in rides like this one.

Frictional forces that obey the general equation $f = \mu N$ are said to represent Coulomb friction, and the equation is sometimes called Coulomb's law of friction. The French scientist Charles A. de Coulomb studied friction during the latter half of the eighteenth century. Although Coulomb gets the credit, the relationship was actually formulated earlier by Leonardo da Vinci.

Friction between solids is generally classified into three types: static, sliding (kinetic), and rolling. **Static friction** includes all cases in which the frictional force is sufficient to prevent relative motion between surfaces. **Sliding friction**, or **kinetic friction**, occurs when there is relative (sliding) motion at the interface of the surfaces in contact. **Rolling friction** occurs when one surface rotates and does not slip or slide at the point or area of contact with another surface. Rolling friction, such as occurs between a train wheel and a rail, is attributed to local deformations in the contact region. This type of friction is somewhat difficult to analyze.

Types of Frictional Forces; Coefficients of Friction

Here, we will consider the forces of friction on stationary and sliding objects. These are called the force of static friction and the force of kinetic (or sliding) friction. Experimentally, it is found that the force of friction depends on both the nature of the two surfaces and the load, or the force with which the surfaces are pressed together. For an object on a horizontal surface, this force is equal to the object's weight. However, as shown in Fig. 4.14, on an inclined plane only a component of the weight force contributes to the load. Thus, to avoid confusion, you should remember that the force of friction is proportional to the normal force ($f \propto N$). Keep in mind that the normal force is the force perpendicular to the surface acting on an object (the force exerted *by* the surface *on* the object). In the absence of other perpendicular forces, the normal force is equal in magnitude to the component of the weight force acting perpendicular to the surface by the third law.

The force of static friction (f_s) between parallel surfaces in contact is in the direction that opposes relative motion between the surfaces. The magnitude has different values such that

$$f_s \leq \mu_s N \qquad (4.6)$$

where μ_s is the **coefficient of static friction.** (Note that it is a dimensionless constant. Why?) You might wonder how the force of static friction can have more than one value, or different magnitudes. In Fig. 4.20a, one person pushes on a file cabinet and it doesn't move. With no motion, the net force on the cabinet is zero, and $F - f_s = 0$, or $F = f_s$. Suppose that a second person also pushes and the file cabinet still doesn't budge. Then f_s must increase, since the applied force has been increased. Finally, if the applied force is made large enough to overcome the static friction, motion occurs. The greatest, or maximum, force of static friction is exerted just before the cabinet starts to slide (Fig. 4.20b), and for this case Eq. 4.6 can be written with an equals sign:

$$f_{s\,max} = \mu_s N \qquad (4.7)$$

Once an object is in motion, or sliding, there is a force of kinetic friction (f_k) in the direction opposite to the motion and having a magnitude of

$$f_k = \mu_k N \qquad (4.8)$$

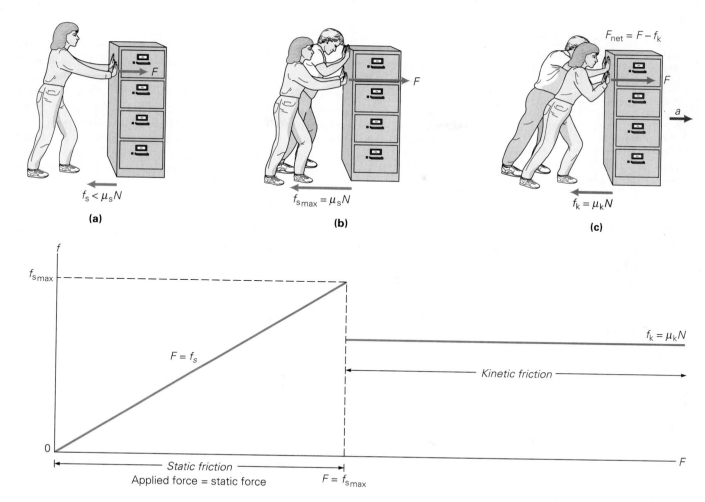

FIGURE 4.20 Force of friction versus applied force (a) In the static region of the graph, as the applied force F increases, so does f_s: that is $f_s = F$ and $f_s < \mu_s N$. (b) When the applied force F exceeds $f_{s_{max}} = \mu_s N$, the file cabinet is set into motion. (c) Once the cabinet is moving, the frictional force is decreased, since kinetic friction is less than static friction ($f_k < f_{s_{max}}$). Thus, if the applied force is maintained at $F = f_{s_{max}}$ there is a net force, and the cabinet is accelerated. For the cabinet to move with constant velocity, the applied force must be reduced to $F = f_k$.

Figure 4.20

where μ_k is the **coefficient of kinetic friction** (sometimes called the coefficient of sliding friction). Generally, the coefficient of kinetic friction is less than the coefficient of static friction ($\mu_k < \mu_s$) for two surfaces, which means that the force of kinetic friction is less than $f_{s_{max}}$ as illustrated in ● Fig. 4.20. The coefficients of friction between some common materials are listed in Table 4.1 on p. 126.

Note that the force of static friction (f_s) exists in response to an applied force. The magnitude of f_s and its direction depend on the magnitude and direction of the applied force. Up to its maximum value, the force of static friction is equal and opposite to the applied force (F), since there is no motion ($F - f_s = 0 = ma$). Thus, if the person in Fig. 4.20a pushed on the cabinet in the opposite direction, f_s would also be in the opposite direction. If there were no applied force F, then f_s would not exist. When F exceeds $f_{s_{max}}$, the block slides and kinetic friction comes into effect, with $f_k = \mu_k N$. If F is equal to f_k, the cabinet will slide with a constant velocity; and if F is greater than f_k, it will accelerate.

Also, it has been experimentally determined that the coefficients of friction (and therefore the forces of friction) are nearly independent of the size of the contact area between metal surfaces. This means that the force of friction between a brick-shaped metal block and a metal surface is the same regardless of whether the block is lying on a larger side or a smaller side. The observation

Friction between Materials	μ_s	μ_k
aluminum on aluminum	1.05	1.40
glass on glass	0.94	0.35
rubber on concrete		
dry	1.20	0.85
wet	0.80	0.60
steel on aluminum	0.61	0.47
steel on steel		
dry	0.75	0.48
lubricated	0.12	0.07
Teflon on steel	0.04	0.04
Teflon on Teflon	0.04	0.04
waxed wood on snow	0.05	0.03
wood on wood	0.58	0.40

is not generally valid for other surfaces, such as wood, and does not apply to plastic or polymer surfaces. The lack of dependence of friction on contact area is related to pressure, which is force per unit of area ($p = F/A$). If the smaller side of a metal block has an area that is half as large as the area of a larger side, it will have, on the average, only half as many asperities for local welding as the larger side. However, the pressure producing the welding will be twice as great on the smaller side (the same weight acting over half the area).

Finally, you should keep in mind that although the equation $f = \mu N$ holds in general for frictional forces, friction may not be linear over a wide range. That is, μ is not always constant. For example, the coefficient of kinetic friction varies somewhat with the relative speed of the surfaces. However, for speeds up to several meters per second the coefficients are relatively constant. Thus, this discussion will neglect any variations due to speed, and the forces of static and kinetic friction will depend only on the load and the nature of the materials (as expressed in the given coefficients of friction).

EXAMPLE 4.10 ■ Static and Kinetic Forces of Friction

(a) If the coefficient of static friction between the 40.0-kg crate in ● Fig. 4.21 and the floor is 0.650, with what horizontal force must the worker pull to move the crate? (b) If the worker maintains that force once the crate starts to move and the coefficient of kinetic friction between the surfaces is 0.500, what is the magnitude of the acceleration of the crate?

Solution. Listing the given data and what we want to find, we have

> *Given:* $m = 40.0\ \text{kg}$ *Find:* (a) F (force necessary to move crate)
> $\mu_s = 0.650$ (b) a (acceleration)
> $\mu_k = 0.500$

(a) The crate will not move until the applied force F slightly exceeds the

● **FIGURE 4.21 Forces of static and kinetic friction** Example 4.10.

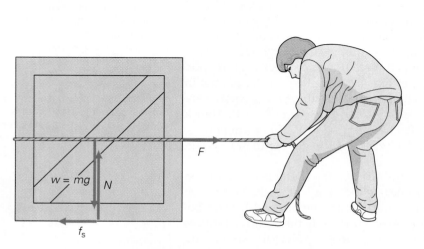

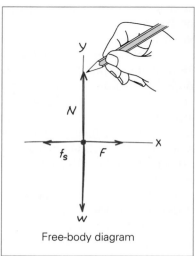

Free-body diagram

maximum static frictional force $f_{s\,max}$. So we must find $f_{s\,max}$ to see what force the worker must apply. The weight of the crate and the normal force are equal in magnitude in this case (see the free-body diagram in the figure), so the maximum force of static friction is

$$f_{s\,max} = \mu_s N = \mu_s(mg)$$
$$= (0.650)(40.0 \text{ kg})(9.80 \text{ m/s}^2)$$
$$= 255 \text{ N} \quad (\text{about 57 lb})$$

The crate moves if the applied force exceeds this maximum force.

(b) Now the crate is in motion and the worker maintains a constant applied force $F = f_{s\,max} = 255$ N. The force of kinetic friction f_k acts on the crate, but this is smaller than F since $\mu_k < \mu_s$. Hence, there is a net force, and the acceleration of the crate may be found by using Newton's second law:

$$F - f_k = F - \mu_k N = ma$$

or

$$a = \frac{F - \mu_k N}{m} = \frac{F - \mu_k(mg)}{m}$$

$$= \frac{255 \text{ N} - (0.500)(40.0 \text{ kg})(9.80 \text{ m/s}^2)}{40.0 \text{ kg}}$$

$$= 1.48 \text{ m/s}^2$$

Let's look at the worker and the crate again, but this time assume that he applies the force at an angle (Fig. 4.22).

EXAMPLE 4.11 ■ A Close Look at the Normal Force

A worker applies a force to a crate at an angle of 30° to the horizontal as shown in ● Fig. 4.22. How large a force must he apply to move the crate? (Before looking at the solution, would you expect that the force needed in this case would be greater or smaller than in the preceding example?)

●**FIGURE 4.22 Normal force** Example 4.11.

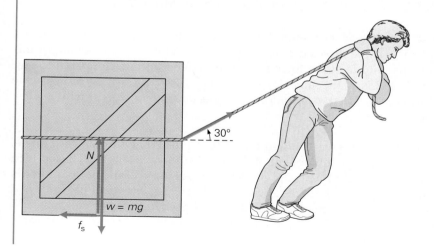

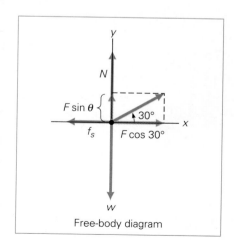

Free-body diagram

Solution. The data are the same as in Example 4.10, except that the force is applied at an angle.

> *Given:* $\theta = 30°$ *Find:* F (force necessary to move crate)

In this case, the crate will move when the *horizontal component* of the applied force, $F \cos 30°$, slightly exceeds the maximum static friction force. So, we may write for the maximum moment:

$$F \cos 30° = f_{s\,max} = \mu_s N$$

However, the magnitude of the normal force is *not* equal to that of the weight of the crate here because of the upward component of the applied force (see the free-body diagram in the figure). By the second law,

$$N + F \sin 30° - mg = 0$$

or

$$N = mg - F \sin 30°$$

In effect, the applied force partially supports the weight of the crate. Substituting this expression for N into the preceding equation gives

$$F \cos 30° = \mu_s(mg - F \sin 30°)$$

Solving for F gives

$$
\begin{aligned}
F &= \frac{mg}{(\cos 30°/\mu_s) + \sin 30°} \\[1em]
&= \frac{(40.0\ \text{kg})(9.80\ \text{m}/\text{s}^2)}{(0.866/0.650) + 0.500} \\[1em]
&= 214\ \text{N} \quad (\text{about 48 lb})
\end{aligned}
$$

Thus, less applied force is needed in this case since the frictional force is less because of the reduced load, or reduced normal force. As the angle between the applied force and the horizontal increases, the normal force gets smaller and so does $f_{s\,max}$.

EXAMPLE 4.12 ■ The Normal Force One More Time

Suppose that instead of pulling the crate as in Fig. 4.22, the worker pushes it, applying a force directed at an angle of 30° below the horizontal. In this case, the magnitude of the force needed to move the crate compared to that in Example 4.11 is (a) the same, (b) greater, (c) less. *Clearly establish the reasoning and physical principle(s) used in determining your answer. That is, why did you select your answer?*

Reasoning and Answer. When the crate is pushed at a downward angle, there is a downward component of the applied force, $F \sin 30°$. This adds to the load that the floor must support, so the magnitude of the normal force is increased: $N = mg + F \sin 30°$. Because N is greater, the frictional force $f_{s\,max}$ is also greater. Thus a greater applied force is needed to move the crate, and the correct answer is (b). (Note that the correct answer would be the same even in the absence of friction. Why?)

Follow-up Exercise. Suppose the crate is moved by two workers, one pulling as in Example 4.11 and one pushing as in Example 4.12. What would be the magnitude of the normal force in this case? (*Reasoning and answer may be found in the Answers to Exercises section at the back of the book.*)

Now let's look at a way to determine coefficients of friction.

EXAMPLE 4.13 ■ Experimental Determination of the Coefficient of Kinetic Friction

A block slides with a constant velocity down a plane inclined at 37° to the horizontal (●Fig. 4.23). What is the coefficient of kinetic friction between the block and the plane?

> *Students sometimes draw extraneous forces for masses moving down inclined planes.*

Solution.

Given: $a = 0$ (because v is constant)   **Find:** μ_k (coefficient of kinetic
$\qquad\theta = 37°$ friction)

Since the acceleration is zero, there is no net force on the block. So, noting the forces in the free-body diagram and using Newton's second law,

$$F_x = 0 = mg \sin \theta - f_k$$
$$F_y = 0 = N - mg \cos \theta$$

Rearranging the equations, we have

$$f_k = mg \sin \theta$$
$$N = mg \cos \theta$$

Then, since $f_k = \mu_k N$,

$$\mu_k = \frac{f_k}{N} = \frac{mg \sin \theta}{mg \cos \theta} = \tan \theta = \tan 37° = 0.75$$

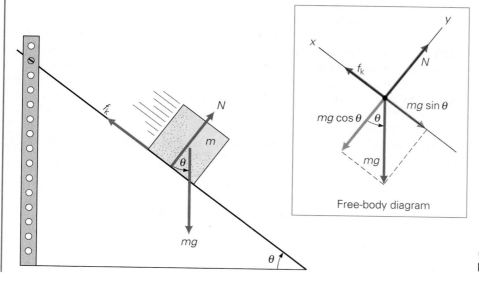

Free-body diagram

Figure 4.23

● **FIGURE 4.23 Coefficient of kinetic friction** Example 4.13.

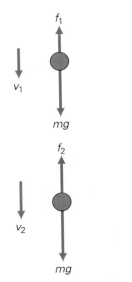

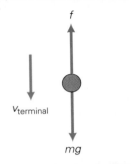

(a) As **v** increases, so does **f**.

(b) When **f = mg**, the object falls with a constant (terminal) velocity.

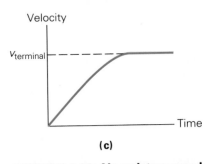

(c)

FIGURE 4.24 Air resistance and terminal velocity **(a)** As the speed of a falling object increases, so does the frictional force of air resistance. **(b)** When this force of friction equals the weight of the object, the net force is zero and the object falls with a constant (terminal) velocity. **(c)** A plot of velocity versus time showing these relationships.

Thus, adjusting the angle of incline until the velocity of the block sliding down the plane is constant allows μ_k to be determined experimentally from the angle of incline. From the preceding general result, we can write

$$\mu_k = \tan \theta$$

Suppose that μ_s between the plane and the block is 0.90. Can you determine the angle of incline at which the block will start to move down the plane? It will be an angle just slightly greater than that at which the component of the weight force down the plane equals the maximum force of static friction. That is,

$$mg \sin \theta = f_{s\,max} = \mu_s N = \mu_s (mg \cos \theta)$$

and

$$\frac{\sin \theta}{\cos \theta} = \tan \theta = \mu_s$$

Thus,

$$\theta = \tan^{-1} \mu_s = \tan^{-1}(0.90) = 42°$$

Therefore, the block will move if the angle of incline exceeds 42°. Adjusting the angle of incline until the block just starts to slide down the plane is an experimental way of approximating μ_s. (This critical angle is called the *angle of repose*.)

Notice how the difference in angles in this example points out the difference between static and kinetic friction. You must tilt the incline to 42° to start the block moving, but reduce it to 37° to keep the block's velocity constant.

Air Resistance

In analyses of free fall, you can generally ignore the effect of air resistance and still get valid approximations for objects falling relatively short distances. However, for longer distances, air resistance cannot be ignored.

Air resistance is the effect produced when a moving object collides with air molecules. Therefore, it depends on the object's shape and size (which determine the area exposed to collisions) as well as its speed. The larger the object and the faster it falls, the more collisions there will be with air molecules. (Air density is also a factor, but this can be assumed to be constant near the Earth's surface.) Since air resistance depends on velocity, as a falling object accelerates under the influence of gravity, the retarding force of air resistance increases (Fig. 4.24a). Eventually, the magnitude of the retarding force equals that of the object's weight force (Fig. 4.24b), so that the net force on the object is zero. It then falls with a maximum constant velocity, which is called the **terminal velocity**.

This can be easily seen from Newton's second law. For the falling object, we have

$$F_{net} = ma$$
$$mg - f = ma$$

where downward has been taken as positive for convenience. Solving for a,

$$a = g - \frac{f}{m}$$

where a is the *instantaneous* acceleration.

Notice that the acceleration for a falling object with air resistance is less than g, that is, $a < g$ or $a < 9.8$ m/s^2. As the object continues to fall, the force of air resistance f increases until $a = 0$ with $f = mg$. The object then falls at its constant terminal velocity.

For a skydiver with an unopened parachute, the terminal velocity is about 200 km/h (about 125 mi/h). To reduce the terminal velocity, so that it can be reached sooner and the time of fall be extended, a skydiver will try to increase exposed body area to a maximum by assuming a spread-eagle position (●Fig. 4.25). Doing this takes advantage of the shape and size dependence of air resistance. Once the parachute is open (giving a larger exposed area and a shape that catches the air), the additional air resistance slows the diver down to about 40 km/h (or 25 mi/h), which is preferable for landing.

● **FIGURE 4.25 Terminal velocity** Sky divers assume a spread-eagle position in order to maximize air resistance. This causes them to reach terminal velocity more quickly and prolongs the time of fall.

EXAMPLE 4.14 ■ Slowing It Down

From a high altitude, a balloonist simultaneously drops two balls of identical size but appreciably different weight. Assuming that both balls reach terminal velocity during the fall, (a) the heavier ball reaches terminal velocity first, (b) the balls reach terminal velocity at the same time, (c) the heavier ball hits the ground first, (d) the balls hit the ground at the same time. *Clearly establish the reasoning and physical principle(s) used in determining your answer before checking it below. That is,* **why** *did you select your answer?*

Reasoning and Answer. Terminal velocity is reached when the weight of a ball is balanced by the frictional air resistance. Both balls start to fall with the same acceleration, g, and their speeds and the retarding forces of air resistance increase at the same rate. The weight of the lighter ball will be balanced first, so (a) and (b) are incorrect. The lighter ball reaches terminal velocity ($a = 0$), but the heavier ball continues to accelerate and pulls ahead of the lighter ball. Hence, the heavier ball hits the ground first and the answer is (c).

Follow-up Exercise. Suppose the heavier ball were much larger than the lighter ball. How might this affect the outcome? (*Reasoning and answer may be found in the Answers to Exercises section at the back of the book.*)

Have the students sketch graphs of acceleration, velocity, and distance versus time for an object falling in air and reaching a terminal velocity. Compare these graphs with ones for free fall without friction.

Discuss how denser air at lower altitudes changes these graphs.

Important Concepts

You should be able to define and explain these chapter concepts clearly.

force	weight	static friction
net (unbalanced) force	free-body diagram	kinetic friction
inertia	translational equilibrium	rolling friction
Newton's first law of motion (law of inertia)	condition for translational equilibrium	coefficient of static friction
Newton's second law of motion	Newton's third law of motion	coefficient of kinetic friction
newton unit	force of friction	air resistance
		terminal velocity

Important Relationships for Review

These relationships are mathematical statements of the concepts and principles presented in the chapter. You should be able to identify the symbols and to explain the relationships before proceeding to the Exercises. In-text equation reference numbers are given for convenience.

Newton's Second Law:

$$\mathbf{F} = m\mathbf{a} \tag{4.1}$$

Weight:

$$w = mg \tag{4.2}$$

Component Form of Newton's Second Law:

$$\mathbf{F}_x\mathbf{x} + \mathbf{F}_y\mathbf{y} = ma_x\mathbf{x} + ma_y\mathbf{y} \tag{4.3}$$

Condition for Translational Equilibrium:

$$\Sigma\mathbf{F} = 0 \tag{4.4}$$

or

$$\Sigma\mathbf{F}_x = 0 \quad \text{and} \quad \Sigma\mathbf{F}_y = 0 \tag{4.5}$$

Force of Static Friction:

$$f_s \leq \mu_s N \tag{4.6}$$

$$f_{s\,\text{max}} = \mu_s N \tag{4.7}$$

Exercises

4.1 ▪ The Concept of Force and Net Force
4.2 ▪ Newton's First Law of Motion

1 Newton's first law is sometimes referred to as the law of inertia. A measure of an object's inertia is given by its (a) size, (b) speed, (c) shape, (d) mass.   (d)

● **2** In the absence of a net force, an object will always (a) be at rest, (b) be in motion with a constant velocity, (c) be accelerated, (d) none of the these.   (d)

3 An object weighs 300 N on Earth and 50 N on the moon. Does the object also have less inertia on the moon?
no, same mass, same inertia

4 It is sometimes said that Newton's first law of motion is a special case of Newton's second law. Explain. $F = 0, a = 0$

5 Consider an air-bubble level sitting on a horizontal surface ● (Fig. 4.26). (a) If you pushed the level with a horizontal force so as to accelerate it, would the bubble move forward as shown here, or is this a trick? Which way would the bubble move when the force is removed and the level comes to rest? (b) Such a level is sometimes used as an "accelerometer" to indicate the *direction* of the acceleration of an applied force. Explain the principle involved. [*Hint:* Think about pushing a pan of water.]   (a) forward, then backwards
(b) liquid inertia

6 As a follow-up to Exercise 5, consider the situation of a child holding a helium balloon in a closed car at rest. What would be observed when the car (a) accelerates from rest and (b) brakes to a stop. (The balloon does not touch the roof of the car.)   (a) forward   (b) backwards

7 ▪ A large car weighs 4850 lb and a compact car weighs 1900 lb. Which car has the greater inertia, and how many times greater? $w_{\text{large}} = (2.553)\,w_{\text{small}}$

8 ▪▪ Which has more inertia, 10 cm³ of gold or 20 cm³ of iron, and how many times more? (See Table 9.2.)
$m_{\text{gold}} = (1.2)\,m_{\text{iron}}$

9 ▪▪ Three forces act on an object that is moving in a straight line with a constant speed. If two of the forces are $F_1 = 4.5\,\text{N}\,\mathbf{x} - 1.5\,\text{N}\,\mathbf{y}$ and $F_2 = -3.5\,\text{N}\,\mathbf{x} - 1.0\,\text{N}\,\mathbf{y}$, what is the third force? $-1.0\,\text{N}\,\mathbf{x} + 2.5\,\text{N}\,\mathbf{y}$

● **10** ▪▪ A particle of mass 2.5×10^{-6} kg moves in the $+x$ direction with constant velocity. If applied forces of $F_1 = 5.0$ N at an angle of 30° to the $+x$ axis (first quadrant) and $F_2 = 3.0$ N at 37° to the $-x$ axis (third quadrant) act on the particle, what other force, applied simultaneously, will allow it to maintain a constant velocity? $F_3 = -0.19\,\text{N}\,\mathbf{x} - 0.69\,\text{N}\,\mathbf{y}$

11 ▪▪▪ A 5.0-kg block at rest on a frictionless surface is acted on by forces $F_1 = 5.5$ N and $F_2 = 3.5$ N, as illustrated in ● Fig. 4.27. What additional force will keep the block at rest? $F_3 = -7.4\,\text{N}\,\mathbf{x}$

● **FIGURE 4.26 An accelerometer**   See Exercise 5.

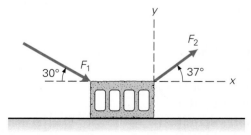

● **FIGURE 4.27 Two applied forces**   See Exercise 11.

12 ■■■ A 1.5-kg object moves up the y axis with a constant speed. When it reaches the origin, the forces $F_1 = 5.0$ N at 37° relative to the positive x axis (first quadrant), $F_2 = (2.5$ N$) x$, $F_3 = 3.5$ N at 45° relative to the negative x axis (third quadrant), and $F_4 = (-1.5$ N$) y$ are applied to it. (a) Will the object maintain a path along the y axis? (b) If not, what simultaneously applied force will keep it moving along the y axis with a constant speed? (a) no (b) $F_5 = -4.0$ N $x + 1.0$ N y, or 4.1 N at 14° (third quadrant)

4.3 ■ Newton's Second Law of Motion

4.4 ■ Applications of Newton's Second Law

Note: In Exercises with strings and pulleys, "ideal conditions" means that the masses of the string(s) and pulley(s) should be neglected, as well as the friction of the pulley.

13 Newton's second law of motion relates the acceleration of an object acted upon by a net force as being (a) inversely proportional to its mass, (b) zero, (c) inversely proportional to the net force, (d) independent of the mass. (a)

14 The newton unit of force is equivalent to (a) kg-m/s, (b) kg-m/s², (c) kg-m²/s, (d) none of these. (b)

15 If an Atwood machine is used to measure the value of g experimentally, how is the acceleration of the masses determined? see ISM

16 In general, this chapter considered forces that were applied to objects of constant mass. What would be the situation if mass were added or lost from a system while a force was being applied? Give examples of situations in which this might happen. see ISM

17 ■ Determine the net force required to give a 4.50-kg object an acceleration of 1.50 m/s². 6.75 N

18 ■ A worker pushes on a crate, and it experiences a net force of 100 N. If the crate moves with an acceleration of 0.750 m/s², what is its weight? 1.30×10^3 N

19 ■ (a) What are the mass in kilograms and the weight in newtons of a 155-lb person? (b) What are your mass and weight in these units? (a) $w = 690$ N; $m = 70.4$ kg

20 ■ A stalled 1800-kg automobile is towed by another car with a horizontal rope, and the cars accelerate at a rate of 0.55 m/s². Find the tension force in the rope. 9.9×10^2 N

21 ■ A net force of 40 N acts on a container containing 2.0 L of water. Determine the resulting acceleration. (Neglect the mass of the container.) 20 m/s² in direction of the force

22 ■ A force acts on a mass. If the force is tripled and the mass is halved, how is the acceleration affected? (Give a factor of change, e.g., 2 times as great or $\frac{1}{2}$ as great.) 6 times as great

23 ■■ At a party, 18 students lift a sports car. While holding the car off the ground, each student exerts an upward force

of 415 N. (a) What is the mass of the car in kilograms? (b) What is its weight in pounds? (a) 762 kg (b) 1.68×10^3 lb

24 ■■ What is the force acting on a 0.45-kg object in free fall? 4.4 N

25 ■■ A horizontal force of 12.0 N acts on an object resting on a level, frictionless surface on the moon, where the object has a weight of 98.0 N. (a) What is the acceleration of the object? (b) What would be the acceleration of the same object in a similar situation on Earth? (a) 0.200 m/s² (b) 0.200 m/s²

26 ■■ When a horizontal force of 300 N is applied to a 7.50-kg crate, it slides on a level floor opposed by a force of kinetic friction of 65.0 N. What is the acceleration of the crate? 31.3 m/s²

27 ■■ A professor's car, which has a weight of 1.38×10^4 N, moves with a constant speed of 55.0 km/h. Determine the unbalanced force acting on the car. $F_{net} = 0$

28 ■■ A motorboat on a lake traveling in a straight line with an initial speed of 50 km/h is slowed nonuniformly to a speed of 15 km/h in 3.0 s. If the motorboat has a mass of 65 kg, what is the net average force acting on it? 2.1×10^2 N

29 ■■ A large tanker weighs 600,000 metric tons. Coming to rest from a speed of 28 km/h in a straight line requires a distance of 5.0 km. What is the average net force acting on the tanker during this time? 3.7×10^6 N

30 ■■ Two blocks of ice, weighing 80 N and 50 N, sit side by side in contact with each other on a horizontal surface. (a) If a constant horizontal force of 40 N is applied to one of the blocks in the direction of the other block, what is the resulting acceleration? (Neglect friction.) (b) If the force is applied to the block at an angle of 25° below the horizontal, what would be the acceleration? (a) 3.1 m/s² (b) 2.8 m/s²

31 ■■ A 1600-kg car traveling at 90 km/h on a straight level road is brought uniformly to rest. What are the magnitude and the direction of the braking force if this is done in (a) a time of 5.0 s or (b) a distance of 50 m? (a) 8.0×10^3 N opposite direction of v_0 (b) 1.0×10^4 N opposite direction of v_0

32 ■■ A loaded jet plane with a weight of 2.75×10^6 N is ready for take-off. If its engines supply 6.35×10^6 N of net thrust, how long a runway will the plane need to reach its minimum take-off speed of 285 km/h? 139 m

33 ■■ A jet catapult on an aircraft carrier accelerates a plane weighing one metric ton uniformly from rest to a launch speed of 320 km/h in 2.00 s. What is the magnitude of the net force on the plane? 4.45×10^4 N

34 ■■ In catching a baseball traveling horizontally with a speed of 15.0 m/s, a player moves the glove straight backward 20.0 cm from the time of contact to the time the ball comes to rest. If the ball has a mass of 0.075 kg, what is the average force on the ball during that interval? 42.2 N

35 ■■ In serving, a tennis player accelerates a 75-g tennis ball horizontally from rest to a speed of 45 m/s. Assuming that the acceleration is uniform when the raquet is applied over a distance of 0.85 m, what is the magnitude of the force exerted on the ball by the raquet?   90 N

36 ■■ What is the net force required to uniformly stop an 1800-kg automobile traveling at 90.0 km/h in a distance of 60.0 m? 9.38×10^3 N

37 ■■ A 150-kg block slides down a 37.0° frictionless inclined plane. What is the block's acceleration?   5.90 m/s²

38 ■■ An object with a mass of 2.0 kg travels with a constant velocity of 4.8 m/s northward. It is then acted upon by a force of 6.5 N in the direction of the motion and a force of 8.5 N to the south, which continue even after the mass come momentarily to rest. (a) How far will the object travel before coming to rest? (b) What will be its position 1.5 s after coming momentarily to rest?   (a) 12 m north (b) 1.1 m south

39 ■■ A 75-kg gymnast hangs vertically from a pair of parallel rings. (a) If the ropes supporting the rings are attached to the ceiling directly above, what is the tension in the ropes? (b) If the ropes are supported so that they make an angle of 45° with the ceiling, what is the tension in the ropes? (c) Analyze the tension as the angle becomes smaller and smaller. (Why are telephone and electric lines allowed to sag between the poles? Wouldn't it take less wire and be more economical if they were strung tightly?)
(a) 3.7×10^2 N   (b) $T = 5.3 \times 10^2$ N   (c) $T \propto 1/\sin \theta$, T becomes large as θ becomes small

40 ■■ Two boats pull a 75.0-kg water skier as illustrated in ● Fig. 4.28. (a) If each boat pulls with a force of 600 N and the skier travels with a constant velocity, what is the magnitude of the retarding force between the water and the skis? (b) Assuming the retarding force remains constant, if the boats each pull with a force of 700 N, what is the acceleration of the skier?   (a) 849 N   (b) 1.88 m/s²

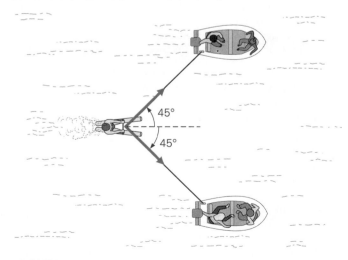

● **FIGURE 4.28 Double tow**   See Exercise 40.

41 ■■ An Atwood machine (see Fig. 4.13) has suspended masses of 0.20 kg and 0.15 kg. What will be the acceleration of the smaller mass with ideal conditions?   1.4 m/s²

42 ■■ For the arrangement in Fig. 4.14, with ideal conditions, what will be the acceleration of the system if $m_1 = 2.25$ kg, $m_2 = 1.65$ kg, and $\theta = 36.9°$?   0.754 m/s² up the plane

43 ■■ Suppose that $\theta = 30°$ and $m_2 = 2.5$ kg for the arrangement in Fig. 4.14. What should m_1 be if it is to move with an acceleration of (a) 0.25 m/s² down the plane, or (b) 0.10 m/s² down the plane?   (a) 5.4 kg   (b) 5.2 kg

44 ■■ Three blocks are pulled along a frictionless surface by a horizontal force, as shown in ● Fig. 4.29. (a) What is the acceleration of the system? (b) What are the tension forces in the light strings? (*Hint:* can T_1 equal T_2? Investigate by isolating m_2.)   (a) 3.0 m/s²   (b) $T_1 = 3.0$ N, $T_2 = 9.0$ N

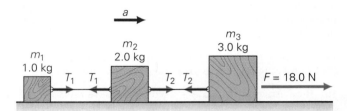

● **FIGURE 4.29 Three block system**   See Exercises 44 and 83.

45 ■■ Assume ideal conditions for the arrangement illustrated in ● Fig. 4.30. What is the acceleration of the system if (a) $m_1 = 0.25$ kg, $m_2 = 0.50$ kg, and $m_3 = 0.25$ kg, and (b) $m_1 = 0.35$ kg, $m_2 = 0.15$ kg, and $m_3 = 0.50$ kg?
(a) 2.5 m/s² right   (b) 2.0 m/s² left

46 ■■ A hoist is designed to give a maximum acceleration of 0.45 m/s² to a maximum load of 9.5×10^2 kg. What is the tension in the support cable when this load travels (a) upward and (b) downward?   (a) 9.7×10^3 N   (b) 8.9×10^3 N

47 ■■ A horizontal net force of 60 N acting on a block on a frictionless level surface produces an acceleration of 2.5 m/s². A second block with a mass of 4.0 kg is dropped

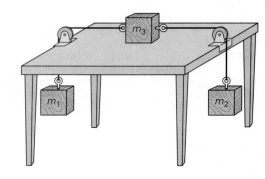

● **FIGURE 4.30 Which way will they go?**   See Exercises 45, 85, and 86.

onto the first. What is the acceleration of the combination if the same force continues to act? (Assume that the second block does not slide on the first block.)   2.1 m/s²

48   One mass, m_1 = 215 g, of an ideal Atwood machine rests on the floor 1.10 m below the other mass, m_2 = 255 g. (a) If the masses are released from rest, how long does it take m_2 to reach the floor? (b) How high will mass m_1 ascend from the floor? [*Hint:* Think about projectile motion.]
(a) 1.62 s   (b) 1.19 m

49   ■■■ A double Atwood machine is illustrated in Fig. 4.31. Assuming ideal conditions and letting m_3 = 4.0 kg and m_1 = m_2 = 3.0 kg, (a) what is the acceleration of m_3? (b) What are the magnitudes of the tensions in the strings?   (a) 2.0 m/s²   (b) 23 N

50   ■■■ What is the acceleration of the system if the masses for a double Atwood machine similar to that in Fig. 4.31 are (a) m_3 = 3.0 kg, m_2 = 4.0 kg, and m_1 = 3.0 kg, or (b) m_3 = 2.5 kg, m_2 = 3.5 kg, and m_1 = 4.0 kg? (Assume ideal conditions.)   (a) 3.9 m/s²   (b) 4.9 m/s²

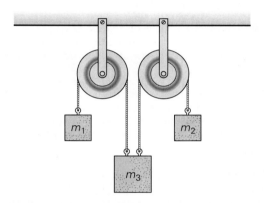

● **FIGURE 4.31 A double Atwood machine**   See Exercises 49, 50, 58, 59, and 89.

51   ■■■ A 70.0-kg person stands on a scale in an elevator. What weight in newtons does the scale read if the elevator is (a) at rest, (b) moving with an upward acceleration of 0.525 m/s², (c) moving with a downward acceleration of 0.525 m/s²?   (a) 686 N   (b) 723 N   (c) 649 N

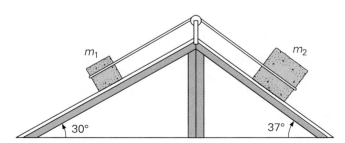

● **FIGURE 4.32 Double inclined plane**   See Exercises 52, 53, and 87.

● **52**   ■■■ Two blocks are on a frictionless double inclined plane as illustrated in (● Fig. 4.32). If one block (m_1) has a mass of 15 kg and the other (m_2) has a mass of 25 kg, what is the acceleration of the system? (Assume ideal conditions.)   1.8 m/s²

53   ■■■ If the blocks on the double frictionless incline in Fig. 4.32 are at rest or move with a constant speed, (a) which block is more massive? (b) How many times more massive?   (a) m_1 (b) m_1 = (1.2)m_2

4.5 ■ Newton's Third Law of Motion

54   The force pair of Newton's third law (a) consists of forces that are always opposite, but sometimes not equal, (b) always cancel each other when the second law is applied to a body, (c) always act on the same object, (d) consists of forces that are always identical in both magnitude and direction but act on different objects.   (d)

55   One force of the action-reaction force pair (a) never produces an acceleration, (b) is always greater than the other, (c) may or may not produce a change in velocity, (d) none of these.   (c)

56   Identify all of the forces in Fig. 4.9 in terms of Newton's third law. What forces are not explicitly illustrated in the figure?   see ISM

57   A canary sits on a perch in a cage with an open wire floor that is suspended from a delicate scale, on which the weight is noted. The canary then leaves the perch and flies around the cage. How, if at all, is the scale reading affected?   see ISM

● **58**   Analyze the forces acting in the system of the double Atwood machine in Fig. 4.31 in terms of Newton's third law. Take the system to be in static equilibrium.   see ISM

59   Analyze the forces acting in the system of the double Atwood machine in Fig. 4.31 in terms of Newton's third law with the masses accelerating.   see ISM

60   ■ Three blocks with masses of 1.0 kg, 2.0 kg, and 3.0 kg are stacked on a table with the smallest on top and the largest on the bottom. Make a sketch and analyze this system in terms of the force pairs of Newton's third law.   see ISM

61   ■ A rifle weighs 50.0 N, and its barrel is 0.750 m long. It shoots a 25.0-g bullet, which leaves the barrel with a speed of 300 m/s (muzzle velocity) after being uniformly accelerated. What is the magnitude of the reaction force on the rifle?   1.50 × 10³ N

4.6 ■ Friction

Note: Neglect air resistance unless otherwise stated.

62   In general, the frictional force (a) is greater for smooth

FIGURE 4.33 A downslope run See Exercises 65 and 66.

surfaces, (b) depends on slow sliding speeds, (c) is proportional to the load, (d) depends on the surface area. (c)

63 The coefficient of kinetic friction (a) is usually greater than μ_s, (b) usually equals μ_s, (c) equals the applied force that exceeds the maximum static force, (d) is unitless. (d)

64 In general, μ_k is less than μ_s. Speculate why this is the case. see ISM

65 Suppose the slope conditions for the skier shown in Fig. 4.33 are such that he travels with a constant velocity. Could you find the coefficient of kinetic friction between the snowy surface and the skis from the photo? If so, describe how this would be done. $\tan \theta$

66 (a) Explain why the skier in Fig. 4.33 is in a crouched position with arms tucked in. (b) Automobiles are streamlined so that they will get better gas mileage. Explain this in terms of air resistance. smaller exposed surface for air resistance

67 Consider the following frictional aspects. (a) We commonly say that friction opposes motion. Yet, when we walk, the frictional force is *in the direction* of our motion (see Fig. 4.18). Is there an inconsistency here in terms of Newton's second law? Explain. (b) What effects would wind have on air resistance? [*Hint:* The wind can blow in different directions.] (a) opposes slipping (b) can increase or decrease

68 ■ In moving a 35.0-kg desk from one side of a classroom to the other, a professor finds that a horizontal force of 275 N is necessary to set the desk in motion and a force of 195 N is necessary to keep it in motion with a constant speed. What are the coefficients of (a) static and (b) kinetic friction between the desk and the floor? (a) 0.802 (b) 0.569

69 ■ A skier on waxed skis coasts down a small slope with a constant velocity. What is the angle of the slope? [*Hint:* see Table 4.1.] 1.7°

70 ■ A 50-kg crate is at rest on a level surface. If the coefficient of static friction between the crate and the surface is 0.79, what horizontal force is required to move the crate? 3.9×10^2 N

71 ■ A wooden block is placed on an adjustable wooden inclined plane. (a) What is the angle of incline above which the block will start to slide down the plane? (b) Once in motion, at what angle of incline will the block slide down the plane at a constant speed? (a) 30° (b) 22°

72 ■■ A packing crate is placed on a 20° incline plane. If the coefficient of static friction between the crate and the plane is 0.65, will the crate slide down the plane? (Justify your answer.) $\theta = \tan^{-1}(0.65) = 33°$, will not move

73 ■■ A 30.0-kg block is pushed up an incline plane ($\theta = 25.0°$) with a constant velocity of 4.00 m/s by a force of 280 N parallel to the plane. What is the coefficient of kinetic friction between the block and the plane? 0.586

74 ■■ A glass paperweight with a weight of 3.5 N is placed on a pane of glass measuring 0.75 m × 0.75 m and elevated at one end to act as an inclined plane. What is the elevation of the higher end of the pane when the paperweight slides down with a constant velocity once it is in motion? 0.24 m

75 ■■ The coefficients of static and kinetic friction between a floor and a 25-kg crate are 0.68 and 0.34, respectively. If forces of (a) 150 N and (b) 175 N are applied horizontally, what is the net force on the crate in each case? (a) $F_{net} = 0$ if mass initially at rest (b) $F_{net} = 92$ N

76 ■■ A horizontal force of 80.0 N is required to set in motion a 5.25-kg wooden box sitting on a rough concrete floor. What is the coefficient of static friction between the box and the floor? 1.55

77 ■■ While being unloaded from an airplane, a 20.0-kg suitcase is placed on a flat ramp inclined at 37°. When released from rest, the suitcase accelerates down the ramp at 0.250 m/s². What is the coefficient of kinetic friction between the suitcase and the ramp? 0.720

78 ■■ Consider a system as in Fig. 4.14. (a) If the coefficient of static friction between $m_1 = 10$ kg and the plane surface is 0.80, what suspended mass (m_2) will just set m_1 into motion up in the plane? (b) What suspended mass is required to keep m_1 moving up the plane at a constant velocity if $\mu_k = 0.60$? (c) What suspended mass is required to keep m_1 moving down the plane with a constant velocity? (Assume ideal conditions for the string and pulley.) (a) 10.9 kg (b) 9.0 kg (c) friction too large for constant velocity

79 ■■ A crate containing machine parts sits unrestrained on the back of a flatbed truck traveling along a straight road at a speed of 80 km/h. The driver applies a constant braking force and comes to a stop in a distance of 22 m. What is the minimum coefficient of friction between the crate and the truck bed if the crate is not to slide forward? 1.0

80 ■■ A 1500-kg automobile travels at a speed of 90 km/h along a straight concrete highway. Faced with an emergency situation, the driver jams on the brakes and skids to a stop. What will the stopping distance be for (a) dry pavement and (b) wet pavement? (a) 26 m (b) 40 m

81 ■■ A school bus pulls into an intersection as a car approaches on an icy street at a speed of 40 km/h. Seeing the bus from 25 m away, the driver of the car locks the brakes causing the car to slide toward the intersection. If the coefficient of kinetic friction between the car's tires and the icy road is 0.25, does the car hit the bus? (You will learn in Chapter 8 why shorter stopping distances result from pumping the brakes.) no

82 ■■ For the situation shown in Fig. 4.27, what is the minimum coefficient of static friction between the block and the surface that will keep the block from moving? ($F_1 = 5.0$ N, $F_2 = 4.0$ N, and $m = 5.0$ kg.) 0.15

83 ■■ For the system illustrated in Fig. 4.29, if $\mu_s = 0.45$ and $\mu_k = 0.35$ between the blocks and the surface, what applied forces will (a) set the blocks in motion and (b) move the blocks with a constant velocity? (a) 26 N (b) 21 N

84 ■■ A ramp 135 m long is to be built for a ski jump. If a skier starting from rest at the top is to have a speed of at least 24 m/s at the bottom, should the angle of the incline be greater than 20°? Justify your answer. no

85 ■■ For the arrangement in Fig. 4.30, what is the minimum value of the coefficient of static friction between the block (m_3) and the table that would keep the system at rest if $m_1 = 0.25$ kg, $m_2 = 0.50$ kg, and $m_3 = 0.75$ kg? (Assume ideal conditions for the string and pulleys.) 0.34

● **86** ■■■ If the coefficient of kinetic friction between the block and the table in Fig. 4.30 is 0.560, and $m_1 = 0.150$ kg and $m_2 = 0.250$ kg, (a) what should m_3 be if the system is to move with a constant speed? (b) If $m_3 = 0.100$ kg, what is the acceleration of the system? (Assume ideal conditions for the string and pulleys.) (a) 0.179 kg (b) 0.80 m/s²

87 ■■■ For the double inclined plane illustrated in Fig. 4.32, if $m_1 = 1.5$ kg and $m_2 = 2.7$ kg, what is the minimum value of the coefficient of static friction between the blocks and the planes that will keep the system at rest? (Assume the same coefficient for both surfaces and ideal conditions for the string and pulleys.) 0.25

■ **Additional Exercises**

88 A person pushes on a block of wood that has been placed against a wall. Make a sketch, and analyze this situation in terms of Newton's third law. see ISM

89 If the masses for a double Atwood machine similar to the one shown in Fig. 4.31 are $m_1 = 2.0$ kg, $m_2 = 3.0$ kg, and $m_3 = 8.0$ kg, what is the acceleration of the system? (Assume ideal conditions.) 2.3 m/s²

90 Does it require more force to start a block of steel moving on a horizontal steel surface or on a horizontal aluminum surface? How many times as much? $f_{s\text{-}s} = (1.2)f_{s\text{-}a}$

91 The maximum load that can safely be supported by a rope in an overhead hoist is 300 N. What is the maximum acceleration that can safely be given to a 15-kg object being hoisted vertically upward? 10 m/s²

92 The coefficient of static friction between a 9.0-kg object and a horizontal surface is 0.45. (a) Would a force of 35 N applied horizontally cause the object to move from rest? (b) If not, could this force applied in another manner cause the object to move? Justify your answer. (a) no (b) yes, upward angle > 19°

93 In the operation of a machine, a 5.0-kg steel part moves on a horizontal steel surface. The force applied to the part is downward at an angle of 30° from the horizontal. (a) What is the magnitude of the applied force required to set the part in motion if the surface is dry? (b) If the surface is lubricated, by what factor is the needed force reduced? (a) 31 N (b) 0.16

94 A crate weighing 9.80×10^3 N is pulled up a 37° incline by a pulley arrangement. If the coefficient of kinetic friction between the crate and the surface of the plane is 0.750, what is the magnitude of the applied force (parallel to the plane) required to move the crate with a constant velocity? (Assume ideal conditions for the pulley arrangement.) 1.18×10^4 N

95 For an Atwood machine with suspended masses of 0.30 kg and 0.40 kg, the acceleration of the masses is measured as 0.95 m/s². What is the effective force of friction for the system? 0.31 N

96 A 0.45-kg shuffleboard puck is given an initial velocity of 4.5 m/s down the playing surface. If the coefficient of sliding friction between the puck and the surface is 0.20, how far will the puck slide before coming to rest? 5.1 m

97 An object is acted on by $F_1 = 4.0$ N at an angle of $+37°$ relative to the positive x axis and $F_2 = 6.0$ N along the negative x axis. (a) What single additional force will ensure that the particle has zero acceleration? (b) Describe the particle's motion in that case. (a) $\mathbf{F} = 2.9$ N $\mathbf{x} - 2.4$ N $\mathbf{y}$ (b) constant velocity if in motion; at rest if initially at rest

98 A hockey player hits a puck with his stick, giving it an initial velocity of 5.0 m/s. If the puck slows uniformly and comes to rest in a distance of 25 m, what is the coefficient of kinetic friction between the ice and the puck? 0.051

99 In a cheerleading act, two cheerleaders with masses of 55 kg and 63 kg, respectively, are thrown from the same height vertically upward with the same initial speed of 3.6 m/s. What is the maximum height above the point of release reached by each? $y = 0.66$ m for both (acceleration independent of mass)

100 Two blocks initially at rest on a horizontal frictionless surface are each acted on by a constant force of 20 N. One block has a mass of 4.0 kg, and the other a mass of 5.0 kg. If the identical forces act on the blocks for 3.0 s, which block will have traveled the greater distance at the end of this time, and how far ahead of the other block will it be? 4-kg mass 5 m ahead

101 A 4.0-kg block initially at rest at a height of 1.8 m on a frictionless inclined plane slides 2.5 m to the bottom of the plane and out on to a level surface. If the block experiences a constant frictional force of 2.5 N on the level surface, how far from the bottom of the inclined plane does the block travel before coming to rest? 28 m

5

Work and Energy

This chapter centers on two concepts that are important in both science and everyday life—work and energy. We commonly think of work as being associated with doing or accomplishing something. Because work makes us physically (and sometimes mentally) tired, we have invented machines and use them to decrease the amount of effort we expend personally. Energy, on the other hand, brings to mind the cost of fuel for transportation and heating, or of electricity. Food is the fuel that supplies the energy our bodies need to carry out life processes and to do work. We say that we are full of energy when ready (and willing) to do something.

Although, these notions do not really define work and energy, they point us in the right direction. As you might have guessed, and will learn in the chapter, work and energy are closely related. You will find that, in physics as in everyday life, when something possesses energy, it usually has the ability to do work. For example, the hammer in the photo above has energy of motion, and this energy allows it to do the work of driving a nail. Conversely, no work can be performed without energy.

Energy exists in various forms: There is mechanical energy, chemical energy, electrical energy, heat energy, nuclear energy, and so on. A transformation from one form to another may take place, but the total *amount of energy is conserved, or always remains the same. This is the point that makes the concept of energy so useful. When a physically measurable quantity is conserved, it not only gives us an insight that leads to a better understanding of nature, but also usually provides another approach to practical problems. (You will be introduced to other conserved quantities during the course of our study of physics.)*

This chapter is devoted to the concept of energy and the closely related concept of work. We will study the quantitative relationship between work and energy, as well as one of the cornerstones of physics, the law of conservation of energy. We will also see how these concepts provide a powerful tool for analyzing the motion of objects. To begin, let us look at the technical definition of work as done by a constant force.

Chapter 5 introduces the topics of work by constant and variable forces, kinetic and potential energies, and the law of conservation of energy. It includes a study of power and efficiency.

5.1 ▪ Work Done by a Constant Force

The word work is commonly used in a variety of ways: We go to work; We work on projects; We work at our desks or on computers; We work problems. In physics, however, work has a very specific meaning. Mechanically, **work** involves force and displacement, and we use this work to quantitatively describe what is accomplished when a force moves an object through a distance. In the simplest case of a *constant* force:

> The work done by a constant force in moving an object is equal to the product of the magnitudes of the displacement and the component of the force parallel to the displacement.

Work then involves moving an object through a distance. A force may be applied, as in (●Fig. 5.1a), but if there is no motion (no displacement), then no work is done. For a constant force **F** acting in the same direction as the displacement **d** (Fig. 5.1b), the work (W) is simply

$$W = Fd \tag{5.1}$$

In general, work is done when a force moves through a distance and some *component* of the force is along the line of motion (Fig. 5.1c). That is, if the force is at an angle θ to the displacement, then $F_\parallel = F \cos \theta$ is the component of force parallel to the displacement, and we have the more general equation

$$\boxed{W = F_\parallel d = (F \cos \theta)d = Fd \cos \theta} \tag{5.2}$$

Note that if $\theta = 0°$, Eq. 5.2 reduces to Eq. 5.1. The perpendicular component of the force, $F_\perp = F \sin \theta$, does no work since there is no displacement in this direction.

Notice that θ is the angle *between* the force and displacement vectors. If the force is in the opposite direction to the displacement (for example, a braking force that tends to slow down, or decelerate, an object) $\theta = 180°$ and $\cos 180° = -1$. Then the work is negative: $W = Fd \cos 180° = -Fd$.

Work is a scalar quantity. Since $W = Fd$, force is given in newtons, and displacement is given in meters, work has the SI unit of newton-meter (N-m). This unit is given the special name of **joule**:

$$Fd = W$$
$$\text{N-m} \equiv \text{joule (J)}$$

Work—force and displacement

> Point out that both force and displacement are vectors, but their vector product is a scalar quantity.
>
> Discuss whether carrying an object across the room involves work according to the definition given here.

●**FIGURE 5.1 Work—the product of the magnitudes of the parallel component of force and the displacement** (a) If there is no displacement, no work is done: $W = 0$. (b) For a constant force in the same direction as the displacement, $W = Fd$. (c) For a constant force at an angle to the displacement, $W = (F \cos \theta)d = Fd \cos \theta$.

$d = 0$

(a)

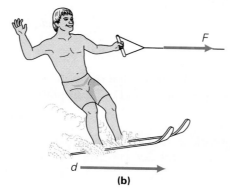

d

(b)

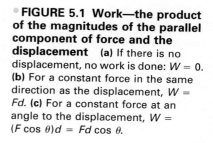

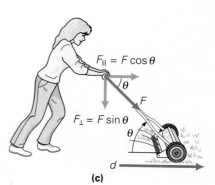

$F_\parallel = F \cos \theta$

θ

F

$F_\perp = F \sin \theta$

θ

d

(c)

The name joule (J) (pronounced "jool") was given in honor of James Prescott Joule (1818–1889), a British scientist who investigated work and energy.

For example, the work done by a force of 25 N moving an object through a distance of 2.0 m is $W = Fd = (25 \text{ N})(2.0 \text{ m}) = 50$ N-m or 50 J.

From the equation, we also see that in the British system work would have the unit pound-foot. However, this is commonly written in reverse. The British standard unit for work is the **foot-pound (ft-lb)**.

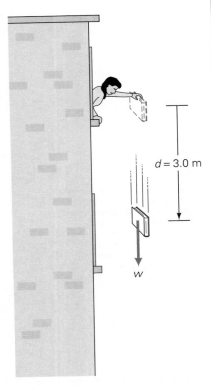

● **FIGURE 5.2 Mechanical work requires motion** Example 5.1.

$d = 3.0$ m

w

EXAMPLE 5.1 ■ Mechanical Work Requires Motion

A student holds her psychology textbook, which has a mass of 1.5 kg, out of a second-story dormitory window until her arm is tired, and then releases it (● Fig. 5.2). (a) How much work is done on the book by the student in simply holding it out the window? (b) How much work will have been done by the force of gravity during the time in which the book falls 3.0 m?

Solution. Listing the data, we have

Given: $v_0 = 0$ (initially at rest)   *Find:* (a) W (work holding)
$m = 1.5$ kg   (b) W (work)
$d = 3.0$ m

(a) Even though the student gets tired (because work is performed within the body to maintain muscles in a state of tension), she does *no* mechanical work in merely holding it stationary. She exerts an upward force on the book (equal in magnitude to its weight), but the displacement is zero in this case ($d = 0$). Thus, $W = Fd = F \times 0 = 0$.
(b) While the book is falling, the net force acting on it is the force of gravity, which is equal in magnitude to the weight of the book, $F = w = mg$ (neglecting air resistance). The displacement is in the same direction as the force and has a magnitude of $d = 3.0$ m, so

$$W = Fd = (mg)d = (1.5 \text{ kg})(9.8 \text{ m/s}^2)(3.0 \text{ m}) = 44 \text{ J}$$

EXAMPLE 5.2 ■ Parallel Component of a Force

If the person in Fig. 5.1(c) pushes on the lawn mower with a constant force of 90.0 N at an angle of 40° to the horizontal, how much work does he do in pushing it a horizontal distance of 7.50 m?

Solution

Given: $F = 90.0$ N   *Find:* W (work)
$\theta = 40°$
$d = 7.50$ m

Here, the horizontal component of the applied force, $F\cos\theta$, is parallel to the displacement, so Eq. 5.2 applies.

$$W = Fd \cos 40° = (90.0 \text{ N})(7.50 \text{ m})(0.766) = 517 \text{ J}$$

Work is done by the horizontal component of the force, but the vertical component does no work. Why?

We commonly specify what is doing work *on* what. For example, the force of gravity does work on a falling object, such as the book in Example 5.2. Also, when you lift an object, you do work *on* the object. We sometimes describe this as doing work *against* gravity because the force of gravity is in the direction opposite that of the applied lift force, and opposes it.

In both Examples 5.1 and 5.2, work was done by a single constant force. If more than one force acts on an object, the work done by each may be calculated separately. The total, or net, work is then the scalar sum of those quantities of work. Alternatively, you may find the vector sum of the forces and use this net force to compute the total, or net, work.

EXAMPLE 5.3 ■ Total or Net Work

A 0.75-kg block slides down a 20° inclined plane with a uniform velocity (Fig. 5.3). (a) How much work is done by the force of friction on the block as it slides the total length of the plane? (b) What is the net work done on the block? (c) Discuss the work done if the angle of incline is adjusted so that the block accelerates down the plane.

Solution. We list what is given, but equally importantly we want to know specifically what is to be found.

Given: $m = 0.75$ kg   Find: (a) W_f (work done by friction)
$\quad\quad\theta = 20°$ $\quad\quad\quad\quad\quad\quad$ (b) W_{net} (net work)
$\quad\quad L = 1.2$ m $\quad\quad\quad\quad\quad\quad$ (c) W (discuss work with block
$\quad\quad$ (from figure) $\quad\quad\quad\quad\quad\quad\quad$ accelerating)

(a) Note from Fig. 5.3 that only two forces do work because there are only two forces parallel to the motion: f_k, the force of kinetic friction; and $mg \sin \theta$, the component of the block's weight acting down the plane. We first find the work done by the frictional force,

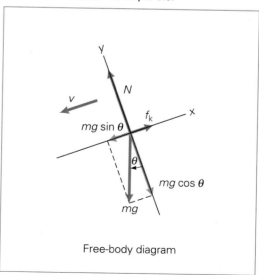

Figure 5.3

FIGURE 5.3 Total or net work Example 5.3.

$$W_f = f_k d \cos 180° = -f_k d = -\mu_k N d$$

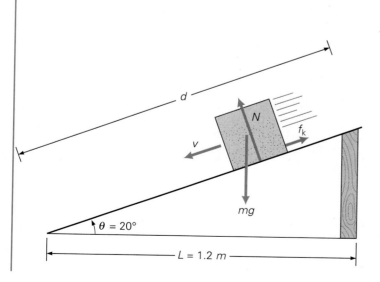

Recall the discussion of static friction in Section 4.6

where $f_k = \mu_k N$. The angle 180° indicates that the force and displacement are in opposite directions. (It is common in such cases to write $W_f = -f_k d$ directly, since friction typically opposes motion.) The magnitude of the normal force N is equal to that of the equal and opposite weight component:

$$N = mg \cos \theta$$

The distance d the block slides down the plane may be found using trigonometry. Note that $\cos \theta = L/d$, so

$$d = \frac{L}{\cos \theta}$$

But what is μ_k? It doesn't seem to be given (nor are the types of surfaces specified). This is an example in which a key piece of data is given to you indirectly in a problem. Note that the block slides down the plane with a uniform (constant) velocity. In Chapter 4, you learned that for uniform motion down a plane, $\mu_k = \tan \theta$. Thus,

This result was derived in Example 4.12

$$W_f = -\mu_k N d = -(\tan \theta)(mg \cos \theta)\left(\frac{L}{\cos \theta}\right) = -mgL \tan 20°$$

$$= -(0.75 \text{ kg})(9.8 \text{ m/s}^2)(1.2 \text{ m})(0.364) = -3.2 \text{ J}$$

(b) The net work may be directly seen to be zero. Since the block moves with a constant velocity, the net force acting on it must be zero (by Newton's first or second law), and so the net work, $W_{net} = F_{net} d = 0$. Let's show this explicitly. We just computed the work done by the frictional force. The only other work is that done by the component of the gravitational force down the block, W_g.

$$W_g = Fd = (mg \sin \theta)\left(\frac{L}{\cos \theta}\right) = mgL \tan 20° = 3.2 \text{ J}$$

Then,

$$W_{net} = W_g + W_f = 3.2 \text{ J} - 3.2 \text{ J} = 0$$

Remember that work is a scalar quantity, so scalar addition is used to find net work.

(c) If the block accelerates down the plane, then from Newton's second law, $mg \sin \theta - f_k = ma$. The component of the gravitational force is greater than the opposing frictional force, so there is net work done on the block. You might be wondering what the effect of nonzero net work is. As you will learn shortly, it causes a change in the amount of energy an object has.

Point out that friction does negative work on the object, while the downward component of the weight does positive work on the object. Therefore the net work is zero only when the forces are equal and opposite.

■ Problem-Solving Hint

Note how in part (a) the equation for W_f was easily simplified by using the algebraic expressions for N and d instead of computing these quantities initially. It is a good rule of thumb not to put numbers into an equation until you have to. Simplifying an equation through cancellation is easier with symbols, and saves computation time.

5.2 ■ Work Done by a Variable Force ◇

The discussion in the preceding section was limited to work done by constant forces. In general, however, forces are variable; that is, they change with time and/or position. For example, a force applied to an object to overcome the force of static friction may be increased, until it exceeds $f_{s_{max}}$. However, the force of static friction does no work because there is no motion or displacement.

An example of variable force doing work in stretching a spring is illustrated in ● Fig. 5.4. As a spring is stretched (or compressed) farther and farther, the restoring force of the spring gets greater and an increasing applied force is required. It is found that the applied force F is directly proportional to the displacement or the change in length of the spring. In equation form, this is expressed

$$F = k\Delta x = k(x - x_o)$$

or, with $x_o = 0$,

$$\boxed{F = kx} \tag{5.3}$$

As can be seen, the force varies with x. We describe this by saying that the *force is a function of position.*

The k in this equation is a constant of proportionality and is commonly called the **spring, or force, constant.** The greater the value of k, the stiffer or stronger is the spring. As you should be able to prove to yourself, the units of k are newtons per meter (N/m).

The relationship expressed by Eq. 5.3 holds only for ideal springs. Real springs approximate this linear relationship between force and displacement within certain limits. For example, if a spring is stretched beyond a certain

k, the spring constant

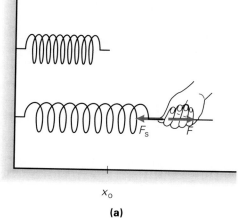

x_o

(a)

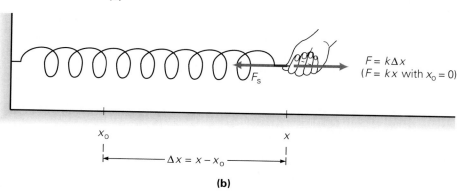

$F = k\Delta x$
($F = kx$ with $x_o = 0$)

x_o

x

$\Delta x = x - x_o$

(b)

● **FIGURE 5.4 Spring force**
(a) An applied force F stretches the spring, and the spring exerts an equal and opposite force F_s on the hand. **(b)** The magnitude of the force depends on the change in the spring's length. This is often referenced to the position of a mass on the end of the spring.

Figure 5.5

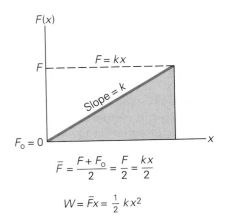

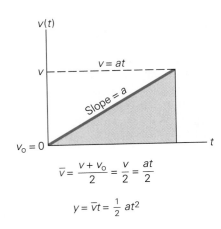

$$\overline{F} = \frac{F + F_0}{2} = \frac{F}{2} = \frac{kx}{2} \qquad \overline{v} = \frac{v + v_0}{2} = \frac{v}{2} = \frac{at}{2}$$

$$W = \overline{F}x = \frac{1}{2}kx^2 \qquad y = \overline{v}t = \frac{1}{2}at^2$$

FIGURE 5.5 Work done by a uniformly variable force The work done by a uniformly varying force of the form $F = kx$ is $W = \frac{1}{2}kx^2$. A plot of this special case of a variable force is graphically analogous to a plot of v versus t for a uniformly varying velocity: $v = at$. The work (W) and the distance (y) are equal to the areas under the respective lines.

Compare Eq. 2.7 and Fig. 2.10

Point out why constant acceleration expressions are not applicable when the force varies. Explain why it is not possible to derive an expression for the work versus distance for a bow, as can be done for the spring.

point, called its elastic limit, it will be pulled out of shape and $F = kx$ will no longer apply.

Note that a spring exerts an equal and opposite force, $F_s = -k\Delta x$, when its linear length is changed by an amount Δx. The minus sign indicates that this spring force is in the direction opposite to the displacement whether the spring is stretched or compressed. This equation is a form of what is known as *Hooke's law*, after Robert Hooke, a contemporary of Newton.

Our study of work will generally be limited to situations involving constant or average forces. As you learned in studying kinematics (Chapter 2), an average value may not be very useful except in special cases. One of these cases is the average velocity for a constant acceleration: $\overline{v} = (v + v_0)/2$. A special case of a variable force that can be analyzed without calculus is the spring force, where the spring constant is k. A plot of F versus x is shown in Fig. 5.5 (which also shows the analogous case of the plot of v versus t for a constant acceleration). The slope of the line is equal to k, and F increases uniformly with x. The average force is then

$$\overline{F} = \frac{F + F_0}{2}$$

or, if $F_0 = 0$,

$$\overline{F} = \frac{F}{2}$$

Thus, the work done in stretching or compressing the spring is

$$W = \overline{F}x = \frac{Fx}{2}$$

Since $F = kx$, the work done is

Work done in stretching (or compressing) a spring

$$\boxed{W = \tfrac{1}{2}kx^2} \qquad \begin{array}{l}\text{work done in stretching} \\ \text{(or compressing) a spring}\end{array} \qquad (5.4)$$

Note that the work is the area under the curve in Fig. 5.5.

EXAMPLE 5.4 ■ Determining the Spring Constant and the Work Done in Stretching a Spring

A 0.15-kg mass is suspended from a vertical spring and descends a distance of 4.6 cm, after which it hangs at rest (Fig. 5.6). An additional 0.50-kg mass is then suspended from the first. (a) What is the total extension of the spring? (b) How much work is done in stretching the spring?

Solution. The data given are as follows:

Given: $m_1 = 0.15$ kg   *Find:* (a) x (total stretch length)
   $x_1 = 4.6$ cm $= 0.046$ m   (b) W (work)
   $m_2 = 0.50$ kg

(a) The total stretch distance is given by $F = kx$, where F is the applied force or the weight of the mass suspended on the spring. However, the spring constant k, is not given. This may be found from the data pertaining to the suspension of m_1 and resulting displacment x_1. (This is a common method of determining spring constants.) As seen in Fig. 5.6a, the magnitudes of the weight force and the restoring spring force are equal, so we may write

and
$$F_s = m_1 g = kx_1$$

$$k = \frac{m_1 g}{x_1} = (0.15 \text{ kg})(9.8 \text{ m/s}^2)/0.046 \text{ m} = 32 \text{ N/m}$$

Then, knowing k, the total extension of the spring is found from the balanced force situation shown in Fig. 5.6b:

$$F = (m_1 + m_2)g = kx$$

Thus,

$$x = \frac{(m_1 + m_2)g}{k} = \frac{(0.15 \text{ kg} + 0.50 \text{ kg})(9.8 \text{ m/s}^2)}{32 \text{ N/m}}$$

$$= 0.20 \text{ m} \quad (\text{or } 20 \text{ cm})$$

(b) The work done in stretching the spring this distance is given by Eq. 5.4.

$$W = \tfrac{1}{2}kx^2 = \tfrac{1}{2}(32 \text{ N/m})(0.20 \text{ m})^2 = 0.64 \text{ J}$$

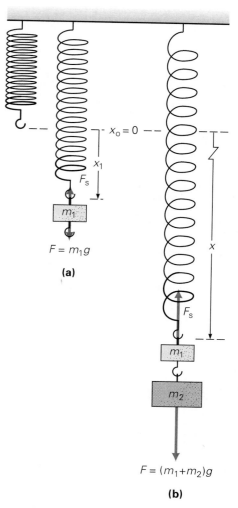

FIGURE 5.6 Determining the spring constant and the work done in stretching a spring Example 5.4.

■ Problem-Solving Hint

The reference position x_o for the change in length of a spring is arbitrary and is usually chosen for convenience. *The important quantity is the displacement difference Δx, or the net change in the length of the spring.* As shown in Fig. 5.7 for a mass suspended on a spring, x_o can be referenced to the unloaded length of the spring or to the loaded position, which may be taken as zero for convenience. In the preceding example, x_o was referenced to the end of the unloaded spring. Also, with the displacement only in one direction, downward was designated as the $+x$ direction to avoid minus signs.

When the net force on the suspended mass is zero, the mass is said to be at its equilibrium position (Fig. 5.7b). This position may also be taken as a zero reference ($x = x_o = 0$). The equilibrium position is a convenient reference point for a case in which the mass oscillates up and down on the spring. (We will describe this motion in a later chapter.) Notice that in general there are both positive and negative directions. Also, since the displacement is in the vertical direction, the x's are often replaced by y's.

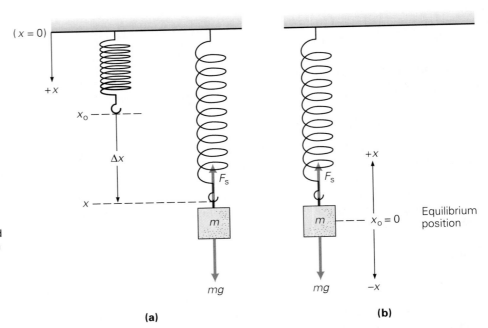

Figure 5.7

● FIGURE 5.7 Displacement reference The reference position x_0 is arbitrary and is usually chosen for convenience. It may be at the end of the spring at its unloaded position (a) or at the equilibrium position when a mass is suspended on the spring (b). The latter is particularly convenient in cases in which the mass oscillates up and down on the spring.

5.3 ■ The Work-Energy Theorem: Kinetic Energy

Now that we have an operational definition of work, we are ready to look at how work is related to energy. Energy is one of the most important concepts in science. We describe it as a quantity possessed by objects or systems. Basically, work is something that is *done* on objects, whereas energy is something that objects *have*.

One form of energy that is closely associated with work is kinetic energy. (Another form of energy, potential energy, will be described in the next section.) Consider a mass at rest on a frictionless surface. Let a horizontal force act on the mass and set it in motion. Work is done *on* the mass, but where does the work "go," so to speak? It goes into setting the mass into motion or changing its *kinetic* conditions. Because of its motion, we say the mass has energy—kinetic energy.

Kinetic energy is often called the energy of motion. It is defined mathematically as one-half of the product of the mass and the square of the (instantaneous) velocity of a moving object.

Kinetic energy—the energy of motion

$$\boxed{K = \tfrac{1}{2}mv^2} \quad \text{(kinetic energy)} \qquad (5.5)$$

For a constant net force doing work on a moving object, as illustrated in (● Fig. 5.8), the force does an amount of work, $W = Fx$. But what are the kinematic effects? The force causes the object to accelerate, and from the equation $v^2 = v_0^2 + 2ax$,

Eq. 2.9, Section 2.4

$$a = \frac{v^2 - v_0^2}{2x}$$

where v_0 may or may not be zero. Writing the force in its Newton's second law form ($F = ma$) and then substituting for a from the equation gives

$$W = Fx = max = m\left(\frac{v^2 - v_o^2}{2x}\right)x$$
$$= \tfrac{1}{2}mv^2 - \tfrac{1}{2}mv_o^2$$

In terms of kinetic energy, then,

$$W = \tfrac{1}{2}mv^2 - \tfrac{1}{2}mv_o^2 = K - K_o = \Delta K$$

Work done causes a change in kinetic energy. Discuss the process of pitching and catching a baseball. Catching a baseball involves negative work, which decreases kinetic energy. Toss a ball to one of the students in class to initiate the discussion.

or

$$\boxed{W_{net} = \Delta K} \qquad (5.6)$$

Work-energy theorem

This equation is called the **work-energy theorem** and relates the net work done on an object to the change in its kinetic energy. That is, *the net work done on a body by an external force is equal to the change in kinetic energy of the body.*

Keep in mind that the work-energy theorem is true in general and not just for the special case considered in deriving Eq. 5.6. As a concrete example, recall that in Example 5.1 the force of gravity did 44 J of work on a book that fell from rest through a distance of $y = 3.0$ m. At this position and instant, the falling book has 44 J of kinetic energy. This can be easily shown. Since $v_o = 0$ in this case, $v^2 = 2gy$, and

$$W = Fd = mgy = \frac{mv^2}{2} = K = 44\text{ J}$$

where $K_o = 0$.

As you can see from the work energy theorem, kinetic energy has the same units as work. You should also note that, like work, kinetic energy is a scalar quantity. (The squaring of the velocity (vectors), $\mathbf{v} \cdot \mathbf{v} = v^2$, is a special type of multiplication that gives a scalar.)

Kinetic energy is always a positive scalar quantity. Some students may incorrectly associate a positive and negative direction with kinetic energy.

What the work-energy theorem tells us is that when work is done, there is change or a transfer of energy. In general then, we might say that *work is a measure of the transfer of energy*. For example, a force doing work on an object that causes it to speed up gives rise to an increase in the object's kinetic energy. On the other hand, work done by the force of kinetic friction may cause a moving object to slow down and decrease its kinetic energy. So, for a change in or transfer of kinetic energy, there must be a net force and net work, as Eq. 5.6 tells us.

Work and energy

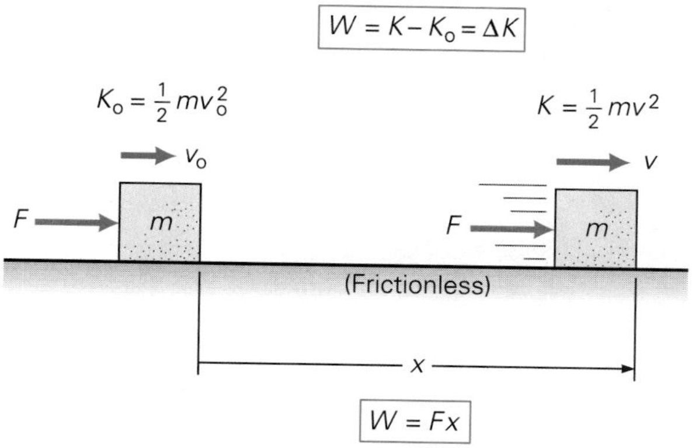

FIGURE 5.8 The relationship of work and kinetic energy The work done on a block in moving along a horizontal frictionless surface is equal to the change in its kinetic energy, $W = \Delta K$.

When an object is in motion, it possesses kinetic energy and has the capability to do work. For example, a moving automobile has kinetic energy and can do work in crumpling a fender in a fender-bender—not *useful* work in that case, but still work (● Fig. 5.9).

(a)

(b)

● **FIGURE 5.9 Kinetic energy and work** (a) A moving object, such as a wrecking ball, possesses kinetic energy and so can do work. (b) A great deal of kinetic energy was converted to work when a large meteorite made this crater in Arizona, perhaps 10,000 years ago. The crater is 1200 m in diameter and 180 m deep.

EXAMPLE 5.5 ■ Work and Kinetic Energy

A shuffleboard player (● Fig. 5.10) pushes a 0.25-kg puck in a way that causes a horizontal constant force of 6.0 N to act on it through a distance of 0.50 m.

(a) What are the kinetic energy and the speed of the puck when the force is removed? (b) How much work would be required to bring the puck to rest?

Solution. Listing the given data as usual, we have

Given: $m = 0.25$ kg   *Find:* (a) K (kinetic energy)
$F = 6.0$ N v (speed)
$d = 0.50$ m   (b) W (work done in
$v_o = 0$ stopping puck)

(a) Since the speed or velocity is not known, we cannot compute the kinetic energy ($K = \frac{1}{2}mv^2$) directly. However, kinetic energy is related to *work* by the work-energy theorem. The work done on the puck is

$$W = Fd = (6.0 \text{ N})(0.50 \text{ m}) = 3.0 \text{ J}$$

Then, by the work-energy theorem,

$$W = \Delta K = K - K_o = 3.0 \text{ J}$$

But, $K_o = \frac{1}{2}mv_o^2 = 0$, because $v_o = 0$, so

$$K = 3.0 \text{ J}$$

The speed may be found from the kinetic energy. Since $K = \frac{1}{2}mv^2$, we have

$$v = \sqrt{\frac{2K}{m}} = \sqrt{\frac{2(3.0 \text{ J})}{0.25 \text{ kg}}} = 4.9 \text{ m/s}$$

(Note how work-energy considerations were used to find speed. This could be done in another way. First, the acceleration could be found from $a =$

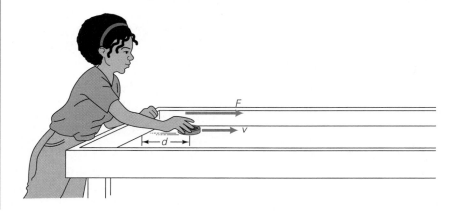

● **FIGURE 5.10 Work and kinetic energy** Example 5.5.

F/m, and then the kinematic equation $v^2 = v_0^2 + 2ax$ could be used to find v (where $x = d$ in the equation). The point is that problems often can be solved in different ways, and finding the fastest and most efficient way is often the key to success. As our discussion of energy progresses, you will see how useful and powerful the notions of work and energy are, both as theoretical concepts and as practical tools for solving many kinds of problems.)

(b) As you might guess, the work required to bring the puck to rest is equal to its kinetic energy (the amount of energy that must be "removed" from the puck to bring it to rest). In this case, $v_0 = 4.9$ m/s and $v = 0$, so

$$W = K - K_0 = 0 - K_0 = -\tfrac{1}{2}mv_0^2$$
$$= -\tfrac{1}{2}(0.25 \text{ kg})(4.9 \text{ m/s})^2 = -3.0 \text{ J}$$

The minus sign indicates that the puck loses energy as it slows down. The work is done *against* the motion of the puck; that is, the opposing force is opposite to the direction of motion. The force that does the work is friction, of course.

EXAMPLE 5.6 ■ Kinetic Energy's v^2 Dependence

In a football game, a 140-kg guard runs with a speed of 4.0 m/s and a 70-kg free safety moves at 8.0 m/s. Then, (a) both players have the same kinetic energy, (b) the safety has twice as much kinetic energy as the guard, (c) the guard has twice as much kinetic energy as the safety, (d) the safety has four times as much kinetic energy as the guard. *Clearly establish the reasoning and physical principle(s) used in determining your answer before checking it below. That is why did you select your answer?*

Reasoning and Answer. The kinetic energy of a body depends on both its mass and its speed or velocity. You might think that, with half the mass but twice the speed, the safety would have the same kinetic energy as the guard, but this is not the case. Kinetic energy, $K = \tfrac{1}{2}mv^2$, is directly proportional to the mass, but proportional to the *square* of the speed. Thus, halving the mass decreases the kinetic energy by a factor of two; so if the two athletes had equal speeds, the safety would have half as much kinetic energy as the guard.

However, doubling the speed increases the kinetic energy, not by a factor of 2, but by a factor of 2^2 or 4. Thus, the safety, with half the mass but twice the speed, would have $\tfrac{1}{2} \times 4 = 2$ times as much kinetic energy as the guard, and the answer is (b).

Note that to answer this question it was not necessary actually to calculate the kinetic energy of each player. We can do so, however, to check our conclusions:

$$K_s = \tfrac{1}{2}m_s v_s^2 = \tfrac{1}{2}(70 \text{ kg})(8.0 \text{ m/s})^2 = 2.2 \times 10^3 \text{ J}$$
$$K_g = \tfrac{1}{2}m_g v_g^2 = \tfrac{1}{2}(140 \text{ kg})(4.0 \text{ m/s})^2 = 1.1 \times 10^3 \text{ J}$$

Thus we see explicitly that our answer was correct.

Follow-up Exercise. Suppose that the safety's speed were only 50 percent greater than the guard's, or 6.0 m/s. Which athlete would then have the greater kinetic energy and how many times greater? (*Reasoning and answer may be found in the Answers to Exercises section at the back of the book.*)

5.4 ■ Potential Energy

Potential energy—the energy of position

An object in motion has kinetic energy. However, whether an object is in motion or not, it may have another form of energy—potential energy. As the name implies, an object having potential energy has the *potential* to do work. You can probably think of many examples: a compressed spring, a drawn bow, water held back by a dam, a wrecking ball poised to drop. In all such cases, the potential to do work derives from the position or configuration of bodies. The spring has energy because it is compressed, the bow because it is drawn, the water and the ball because they have been lifted above the surface of the earth (● Fig. 5.11). Consequently, **potential energy** (U) is often called the energy of position.

In a sense, potential energy can be thought of as stored work. You have already seen an example of potential energy in Section 5.2 when work was done in compressing a spring. Recall that the work done in such a case is $W = \tfrac{1}{2}kx^2$ (with $x_o = 0$). Note that the amount of work done depends on position (x). Because work is done, there is a *change* of position and a *change* in potential energy (ΔU), which is equal to the work done *by* the applied force in compressing the spring:

$$W = \Delta U = U - U_o = \tfrac{1}{2}kx^2 - \tfrac{1}{2}kx_o^2$$

Potential energy is a scalar quantity. However, unlike kinetic energy, it can be either positive or negative, as will be shown.

Thus, with $x_o = 0$, as it is commonly taken to be, the *potential energy of a spring* is

$$\boxed{U = \tfrac{1}{2}kx^2} \tag{5.7}$$

(a)

(b)

(c)

[Since the potential energy has x^2 as one term, the previous problem-solving hint applies when $x_\text{o} \neq 0$; that is, $x^2 - x_\text{o}^2 \neq (x - x_\text{o})^2$.]

Perhaps the most common type of potential energy is **gravitational potential energy**. In this case, position refers to the height of an object above some reference point such as the floor or the ground. Suppose that an object of mass m is lifted a distance Δh (Fig. 5.12). Work is done against the force of gravity, and an applied force at least equal to the object's weight is necessary to lift it, $F = w = mg$. The work done in lifting is then equal to the change in potential energy. Expressed in equation form, we have

$$\text{work} = \text{change in potential energy}$$

or

$$W = F\Delta h = \Delta U = U - U_\text{o} = mgh - mgh_\text{o}$$

FIGURE 5.11 Potential energy Potential energy has many forms. **(a)** Work must be done to bend the bow, giving it potential energy. That energy is converted into kinetic energy when the arrow is released. **(b)** Gravitational potential energy is converted into kinetic energy when an object falls. (Where did the gravitational potential energy of the water and the diver come from?) **(c)** A pole-vault involves a series of energy conversions. Several different kinds of potential energy play a part.

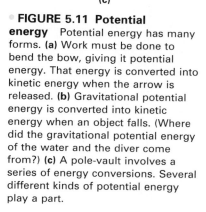

$$W = \Delta U = U - U_\text{o} = mgh - mgh_\text{o}$$

$U = mgh$

mg

h

$\Delta h = h - h_\text{o}$

m

$U_\text{o} = mgh_\text{o}$

h_o

FIGURE 5.12 Gravitational potential energy The work done in lifting an object is equal to the change in gravitational potential energy, $W = F\Delta h = mg(h - h_\text{o})$.

And, with $h_o = 0$, the gravitational potential energy is

$$\boxed{U = mgh} \quad \text{(gravitational potential energy)} \quad (5.8)$$

Because we commonly take upward to be the y direction, Eq. 5.8 is often written $U = mgy$.

EXAMPLE 5.7 ■ Kinetic and Potential Energies

A 0.50-kg ball is thrown vertically upward with an initial velocity of 10 m/s (see ● Fig. 5.13). (a) What is the change in the ball's kinetic energy between the starting point and its maximum height? (b) What is the ball's potential energy at its maximum height? (c) What are the total changes in the ball's kinetic energy (ΔK_T) and potential energy (ΔU_T) when it returns to its starting point? (Neglect air resistance.)

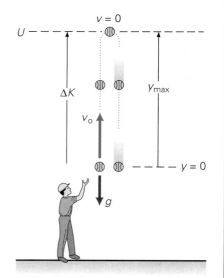

● FIGURE 5.13 Kinetic and potential energies
Example 5.7.

Solution. Studying Fig. 5.13 and listing the data, we have

Given: $m = 0.50$ kg
$v_o = 10$ m/s
$a = g$
$v = 0$, at y_{max}

Find: (a) ΔK (change in kinetic energy)
(b) U (potential energy) at y_{max}
(c) ΔK_T and ΔU_T
(total changes in KE and PE for round trip)

(a) To find the *change* in kinetic energy, we first compute the kinetic energy at each point. We know the initial velocity v_o, and at the maximum height, $v = 0$, so $K = 0$. Thus,

$$\Delta K = K - K_o = 0 - K_o = -\tfrac{1}{2}mv_o^2 = -\tfrac{1}{2}(0.50 \text{ kg})(10 \text{ m/s})^2 = -25 \text{ J}$$

That is, the ball loses 25 J of kinetic energy as negative work is done on it by the force of gravity (force and displacement are in opposite directions).
(b) To find the potential energy at the ball's maximum height, we need to know the ball's height above its starting point when $v = 0$. Using the equation $v^2 = v_o^2 - 2gy$ to find y_{max},

$$y_{max} = \frac{v_o^2}{2g} = \frac{(10 \text{ m/s})^2}{2(9.8 \text{ m/s}^2)} = 5.1 \text{ m}$$

Then, with $y_o = 0$, $\Delta U = U - U_o = U - 0$, and

$$\Delta U = U = mgy_{max} = (0.50 \text{ kg})(9.8 \text{ m/s}^2)(5.1 \text{ m}) = 25 \text{ J}$$

The potential energy increases by 25 J, as might be expected. Notice that this is the change in potential energy with respect to the release point, which was taken as the zero reference point ($y_o = 0$).
(c) The ball returns to its starting point with the same speed as it had initially, or $v = v_o$. Therefore, $K = K_o$ and $\Delta K_T = K - K_o = 0$ (that is, the ball gains 25 J of kinetic energy on the return trip—positive work is done on the ball by the force of gravity). It should be evident that the ball loses 25 J of potential energy on the return trip and $\Delta U_T = 0$ for the round trip (because $mg\Delta y = 0$ since $\Delta y = 0$).

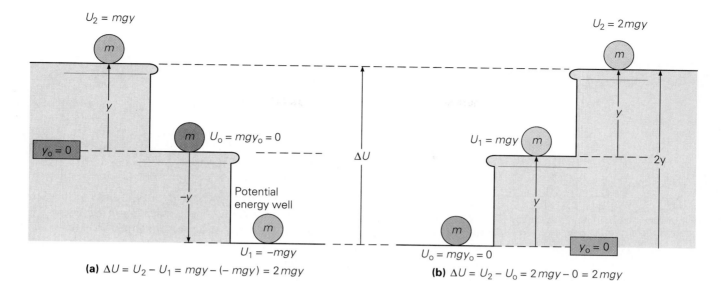

(a) $\Delta U = U_2 - U_1 = mgy - (-mgy) = 2mgy$

(b) $\Delta U = U_2 - U_o = 2mgy - 0 = 2mgy$

An important point is illustrated in the preceding example, namely, the choice of a zero reference point. Potential energy is the energy of *position*, and the potential energy at a particular position (U) is referenced to the potential energy at some other position (U_o). The reference position or point is arbitrary, as is the origin of a set of coordinate axes for analyzing a system. It is usually chosen with convenience in mind, for example, $y_o = 0$ or $h_o = 0$. The value of the potential energy at a particular position depends on the reference point used. However, the *difference or change in potential energy associated with two positions is the same regardless of the reference position.*

In Example 5.7 if gound level had been taken as the zero reference point, then U_o at the release point would not have been zero. However, U at the maximum height would have been greater and $\Delta U = U - U_o$ would have been the same. This concept is illustrated in Fig. 5.14. Note that the potential energy can be negative. When an object has a negative potential energy, it is said to be in a potential energy well, which is something like being in an actual well. Work is needed to raise the object to a higher position in the well or to get it out of the well.

Also, for gravitational potential energy, the path by which an object is raised (or lowered) makes no difference (see Fig. 5.15). That is, *the change in*

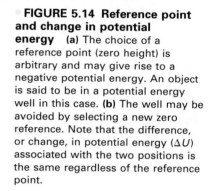

FIGURE 5.14 Reference point and change in potential energy (a) The choice of a reference point (zero height) is arbitrary and may give rise to a negative potential energy. An object is said to be in a potential energy well in this case. (b) The well may be avoided by selecting a new zero reference. Note that the difference, or change, in potential energy (ΔU) associated with the two positions is the same regardless of the reference point.

Figure 5.14

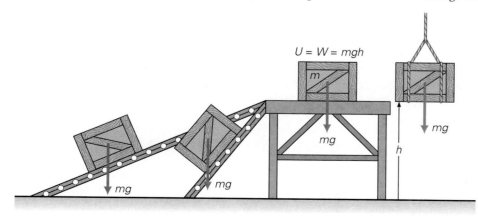

FIGURE 5.15 Path-independence of the change in gravitational potential energy When the crate is resting on the table, it has the same potential energy relative to the floor regardless of how it got there. Only the vertical component of the force that moves the crate does work against gravity.

gravitational potential energy is independent of path. As illustrated in the figure, an object raised to a height h has a change in potential energy of $\Delta U = mgh$ whether it is lifted vertically or moved along an inclined plane (or any other path). This is because the force of gravity always acts downward and only vertical displacement (or vertical component) is involved in doing work against gravity and changing the potential energy.

5.5 ■ The Conservation of Energy

Recall that a system is a physical situation with real or imaginary boundaries. A classroom might be considered a system and so might an arbitrary cubic meter of air.

Conservation laws are the cornerstones of physics, both theoretically and practically. Most scientists would probably name the conservation of energy as the most profound and far-reaching of these important laws. When we say something is *conserved,* we mean that it is constant, or has a constant value. Because so many things continually change in physical processes, conserved quantities are extremely helpful in our attempts to understand and describe the universe. Of course, quantities are generally conserved under special conditions. But even so, such quantities are given special status by being made the subject of conservation laws.

One of the most important conservation laws is that concerning conservation of energy. (You may have seen this coming in Example 5.7.) A familiar statement is that the total energy of the universe is conserved. This is true if the whole universe is taken to be a system. In effect, the universe is the largest possible isolated, or closed, system; that is, one whose particles may interact with each other but have absolutely no interaction with anything outside. In general then, the amount of energy remains constant when no mechanical work is done on or by the system, and no energy is transmitted to or from the system (including thermal energy and radiation).

Thus, the **law of conservation of total energy** may be stated:

Conservation of total energy

The total energy of an isolated system is always conserved.

Within such a system, energy may be converted from one form to another, but the total amount of energy is constant or unchanged.

Conservative and Nonconservative Forces

Friction was discussed in Section 4.6.

A general distinction among systems may be made by considering two categories of forces that may act within them: conservative and nonconservative forces. You have already been introduced to a couple of conservative forces: the force due to gravity and the spring force. A classic nonconservative force, friction, was considered in Chapter 4. Here's the distinction:

Conservative force—work done in round trip is zero

A force is said to be conservative if the work done by or against it in moving an object is independent of the object's path.

What this means is that the work done by a **conservative force** (or a transfer of energy), depends only on the initial and final positions of an object.

The concept of conservative and nonconservative forces is sometimes difficult to comprehend at first. Because this concept is so important in the conservation of energy, let's consider some illustrative examples to increase our understanding.

First, what does *independent of path* mean? An example was given in Fig. 5.15, where work was done against the conservative force of gravity. The figure illustrates that the work done in moving a crate to a table does not depend on the path, but only on the initial and final positions. The work done is equal to the change in potential energy, and in fact, potential energy is associated only with conservative forces. A change in potential energy may be defined as the work done by a conservative force.

On the other hand, work done against a **nonconservative force** such as friction does depend on path. If you moved the crate in Fig. 5.15 from one location to another on the tabletop, the work would depend on its path; the longer the path, the more work done. It should therefore be evident that the work done is not equal to a change in potential energy. In this case, the energy associated with the work would be lost as heat. Hence, in a sense, a conservative force allows you to conserve or store all of the energy as potential energy, whereas a nonconservative force does not.

Another approach to help you understand the distinction between conservative and nonconservative forces is through an equivalent statement of the previous definition:

> A force is conservative if the work done by or against it in moving an object through a round trip is zero.

Conservative force

Consider a book resting on a table. It has gravitational potential energy $U = mgh$, relative to $h_o = 0$. You could drop it on the floor, pick it up, and place it back at its original position; or you could pick it up from the table and carry it around with you all day, then place it back at its original position. Both are round trips, and the potential energy of the book is the same when it is returned to its original position. Thus, the change in its potential energy or the work done by the conservative force of gravity is $\Delta U = W = 0$. However, if you pushed the book around on the tabletop and eventually back to its original position, the work done against the nonconservative force of friction would depend on the path (the longer the path, the more work done), and energy would be lost.

Notice that for the conservative gravitational force, the force and displacement are sometimes in the same direction (in which case positive work is done) and sometimes in opposite directions (in which case negative work is done against the force) during a round trip. Think of the simple case of the book falling to the floor and being placed back on the table. With positive and negative work, the total can be zero. However, for a nonconservative force of kinetic friction, which always opposes the motion or is in the opposite direction to the displacement, the total work done in a round trip can never be zero.

Conservation of Total Mechanical Energy

The idea of a conservative force allows us to extend the conservation of energy to the special case of mechanical energy, which greatly helps us in many physical situations. The sum of the kinetic and potential energies is called the

Demonstration/Activity: To demonstrate the independence of path, use a demo cart of 2 to 5 kg on good wheels. First pull the cart up a ramp using a spring scale and record the distance and force. Then pull the cart horizontally (slowly, so that the force is minimal) the length of the ramp and lift it vertically to the same position with the scale showing its full weight. Compare the work done and final position of the cart in each case.

• **FIGURE 5.16 Conservative and nonconservative systems**

(a) Conservative system

(b) Nonconservative system

(a) The work done by a conservative force exchanges energy between kinetic and potential forms: that is, $\Delta K = -\Delta U$ or $(K - K_o = -(U - U_o))$. The total mechanical energy is conserved ($E = E_o$). (b) Part of the work done by a nonconservative force does not go into mechanical energy exchange, but is lost to friction or some other cause. The mechanical work is not conserved in this case.

Figure 5.16

Mechanical energy—kinetic plus potential

Conservation of mechanical energy

total mechanical energy:

$$E = K + U$$

$$\text{total mechanical energy} = \text{kinetic energy} + \text{potential energy}$$

For a **conservative system** (one in which only conservative forces do work), Eq. 5.9 says that the mechanical energy is constant, or conserved; that is,

$$E = E_o$$
$$K + U = K_o + U_o$$

or

$$\tfrac{1}{2}mv^2 + U = \tfrac{1}{2}mv_o^2 + U_o \qquad (5.10)$$

This is a mathematical statement of the **law of the conservation of mechanical energy**:

In a conservative system, the sum of the kinetic and potential energies is constant and equals the total mechanical energy of the system.

The kinetic and potential energies in a conservative system may change, but their sum is always constant. This is illustrated by the diagram in • Fig. 5.16.

Notice for a conservative system, when work is done and energy is transferred, we have

$$K_o + U_o = K + U$$
$$\text{initial energy} \qquad \text{final energy}$$

This equation can be rewritten as

$$K - K_o = -(U - U_o) \qquad (5.11)$$

or

$$\Delta K = -\Delta U$$

What this expression tells us is that these quantities are related in see-saw fashion: If there is a decrease in potential energy ($-\Delta U$), then the kinetic energy increases ($+\Delta K$) by an equal amount, and vice versa ($-\Delta K = \Delta U$). However, in a nonconservative system, some of the mechanical energy is lost (usually to the heat of friction). The following examples illustrate the conservation of mechanical energy for conservative systems.

Demonstration/Activity: A hot wheels track or any similar loop-the-loop apparatus can be used to demonstrate the conservation of energy. What occurs when the speed of the mass at the top of the circle is less than the critical speed to go around the loop? What is the critical speed? Also consider swinging a mass on a string in a vertical circle.

EXAMPLE 5.8 ■ Conservation of Mechanical Energy

A painter on a scaffold drops a 1.50-kg can of paint from a height of 6.00 m. (a) What is the kinetic energy of the can when it is at a height of 4.00 m? (b) With what speed will the can hit the ground? (Neglect air resistance.)

Graphing Exercise: A 10-kg mass is dropped from rest from a height of 80 m. Graph the potential, kinetic, and total energies as functions of the displacement and the time.

For a more challenging exercise, give the mass an initial horizontal or vertical velocity.

Solution. Listing what is given and what we are to find, we have

Given: $m = 1.50$ kg   Find: (a) K (kinetic energy)
$\quad\quad\quad h_o = 6.00$ m $\quad\quad\quad\quad$ (b) v (speed hitting the ground)
$\quad\quad\quad h = 4.00$ m
$\quad\quad\quad (v_o = 0)$

(a) First it is convenient to find the can's total mechanical energy, since this quantity is conserved while it is falling. Initially, with $v_o = 0$, the can's total energy is equal to its potential energy,

$$E = K + U = 0 + mgh_o = (1.50 \text{ kg})(9.80 \text{ m/s}^2)(6.00 \text{ m}) = 88.2 \text{ J}$$

The relation $E = K + U$ continues to hold while the can is falling, but now we know what the constant sum of K and U is. Rearranging the equation, we have $K = E - U$, and finding U at $h = 4.00$ m:

$$K = E - U = E - mgh = 88.2 \text{ J} - (1.50 \text{ kg})(9.80 \text{ m/s}^2)(4.00 \text{ m})$$
$$= 29.4 \text{ J}$$

Alternately, we could have computed the change in (in this case, the loss of) potential energy, $-\Delta U$. This is equal to the change in (gain of) kinetic energy, ΔK (Eq. 5.11). Then,

$$\Delta K = -\Delta U$$
$$K - K_o = -(U - U_o) = -(mgh - mgh_o)$$

and with $K_o = 0$ (since $v_o = 0$),

$$K = mg(h_o - h) = (1.50 \text{ kg})(9.8 \text{ m/s}^2)(6.00 \text{ m} - 4.00 \text{ m})$$
$$= 29.4 \text{ J}$$

(b) Just before the can strikes the ground ($h = 0$, $U = 0$), the total mechanical energy is all kinetic, or

$$E = K = \tfrac{1}{2}mv^2$$

Thus,

$$v = \sqrt{\frac{2E}{m}} = \sqrt{\frac{2(88.2 \text{ J})}{1.50 \text{ kg}}} = 10.8 \text{ m/s}$$

Basically, all of the potential energy of a free-falling object released from some height h is converted into kinetic energy just before it hits the ground, so

$$K = U$$
$$\tfrac{1}{2}mv^2 = mgh$$

or

$$v = \sqrt{2gh}$$

This result is also obtained from the kinematic equation $v^2 = v_o^2 + 2gy$ (Eq. 2.9′), with $v_o = 0$ and $y = h$.

EXAMPLE 5.9 ■ Speed and Energy

Three balls of equal mass m are projected with the same speed in different directions as shown in ● Fig. 5.17. Which ball strikes the ground with the greatest speed? (Neglect air resistance.) (a) ball 1, (b) ball 2, (c) ball 3, (d) all strike with the same speed. *Clearly establish the reasoning and physical principle(s) used in determining your answer before checking it below. That is, **why** did you select your answer?*

Reasoning and Answer. All of the balls have the same initial kinetic energy, $K_o = \frac{1}{2}mv_o^2$. (Recall that energy is a scalar quantity, and the different directions of projections do not produce any difference in the kinetic energies.) Regardless of their trajectories, all of the balls ultimately descend a distance h relative to their common starting point, so they all lose the same amount of potential energy. (U is energy of *position*, and thus *independent* of path—see Fig. 5.15.)

By the law of conservation of energy, the amount of potential energy each ball loses is equal to the amount of kinetic energy it gains. Thus all three balls must have equal kinetic energies just before striking the ground. This means that their speeds must be equal, so the answer is (d). Note that although balls 1 and 2 are projected at 45° angles, this is not a factor. Since the change in potential energy is independent of path, it is independent of the projection angle. The vertical distance between the starting point and the ground is the same (h) for projectiles at any angle.

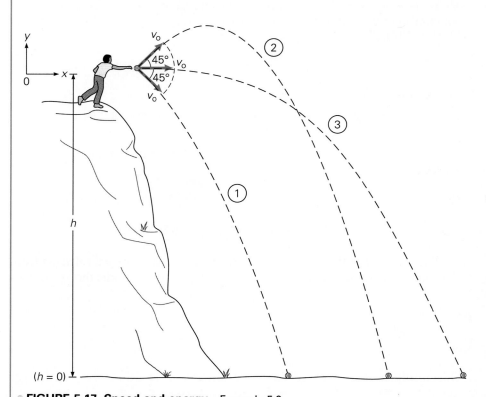

 Figure 5.17

● **FIGURE 5.17 Speed and energy** Example 5.9.

EXAMPLE 5.10 ■ Conservative Force and Mechanical Energy of a Spring

A 0.30-kg mass sliding on a horizontal frictionless surface with a speed of 2.5 m/s, as depicted in Fig. 5.18, strikes a spring, which has a spring constant of 3.0×10^3 N/m. (a) What is the total mechanical energy of the system? (b) What is the kinetic energy (K_1) of the mass when the spring is compressed a distance $x_1 = 1.0$ cm? (c) How far will the spring be compressed when the mass comes to a stop? (Assume that no energy is lost in the course of the collision.)

Students sometimes have trouble with problems involving springs. Consider drawing a series of diagrams: the mass immediately prior to hitting the spring, the mass and spring compressed a short distance, the mass and spring compressed the maximum distance, etc.

Solution.

Given: $m = 0.30$ kg
$\quad\quad\quad v_0 = 2.5$ m/s
$\quad\quad\quad k = 3.0 \times 10^3$ N/m
$\quad\quad\quad x_1 = 1.0$ cm $= 0.010$ m

Find: (a) E (total mechanical energy)
$\quad\quad\quad$ (b) K_1 (kinetic energy)
$\quad\quad\quad$ (c) x_{max} (distance)

(a) Before the mass makes contact with the spring, the total mechanical energy of the system is all kinetic energy, and

$$E = K_0 = \tfrac{1}{2}mv_0^2 = \tfrac{1}{2}(0.30 \text{ kg})(2.5 \text{ m/s})^2 = 0.94 \text{ J}$$

Since the system is conservative, this is the total energy at any time.
(b) When the spring is compressed a distance x_1, it has potential energy $U_1 = \tfrac{1}{2}kx_1^2$, and

$$E = K_1 + U_1 = K_1 + \tfrac{1}{2}kx_1^2$$

Solving for K_1, we have

$$K_1 = E - \tfrac{1}{2}kx_1^2$$
$$= 0.94 \text{ J} - \tfrac{1}{2}(3.0 \times 10^3 \text{N/m})(0.010 \text{ m})^2 = 0.94 \text{ J} - 0.15 \text{ J} = 0.79 \text{ J}$$

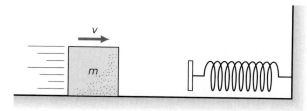

FIGURE 5.18 Conservative force and the mechanical energy of a spring Example 5.10.

(c) When the mass stops and the spring is compressed a distance x_{max}, all of the mechanical energy is stored in the spring as potential energy, and

$$E = U = \tfrac{1}{2}kx_{max}^2 = 0.94 \text{ J}$$

Then, solving for x_{max}

$$x_{max} = \sqrt{\frac{2E}{k}} = \sqrt{\frac{2(0.94 \text{ J})}{3.0 \times 10^3 \text{ N/m}}} = 0.025 \text{ m}$$

Total Energy and Nonconservative Forces

In the preceding examples, the force of friction, which is probably the most common nonconservative force, was ignored. In general, both conservative and nonconservative forces do work on objects. However, as you know, when nonconservative forces do work, the mechanical energy is not conserved. Mechanical energy is "lost" through the work done by the nonconservative force.

It might be thought that we can no longer use an energy approach to analyze problems, since energy is being lost—usually dissipated as heat (•Fig. 5.19). However, in some instances we can use the conservation of total energy to find out how much energy was lost to the work done by a nonconservative force. Suppose an object initially has mechanical energy and that nonconservative forces do work on it, causing it to lose an amount (Q) of its energy. Then, by the conservation of *total* energy,

$$K_o + U_o = K + U + Q \qquad (5.12)$$

initial mechanical energy = final mechanical energy + mechanical energy lost

•**FIGURE 5.19 Nonconservative force and energy loss** Friction is a nonconservative force—when it is present, mechanical energy is not conserved. Can you tell from the photo what is happening to the kinetic energy being supplied by the motor of this grinding wheel?

For example, when a ball is thrown vertically upward, air resistance retards its motion in both directions, and it returns to the starting position with less mechanical energy than it had initially. In such a case, the amount of work done by the nonconservative force is equal to the amount of mechanical energy lost.

Thus, if E_o and E are the initial and final total mechanical energies, respectively,

$$E_o = E + Q$$

or

$$Q = E_o - E \qquad (5.13)$$

The following example illustrates this concept.

EXAMPLE 5.11 ■ Work Done by a Nonconservative Force

A skier with a mass of 80 kg starts from rest and skis down a slope from an elevation of 110 m (•Fig. 5.20). The speed of the skier at the bottom of the slope is 20 m/s. (a) Show that the system is nonconservative. (b) How much work is done by the nonconservative force?

Solution.

Given: $m = 80$ kg   *Find:* (a) Show that E is not constant
$\quad\quad\quad v_o = 0$ (b) W (work done)
$\quad\quad\quad v = 20$ m/s
$\quad\quad\quad h = 110$ m

$v = 20$ m/s

110 m

FIGURE 5.20 Work done by a nonconservative force
Example 5.11.

(a) If the system is conservative, the total mechanical energy is constant. Initially,

$$E_o = U_o = mgh = (80 \text{ kg})(9.8 \text{ m/s}^2)(110 \text{ m}) = 8.6 \times 10^4 \text{ J}$$

At the bottom of the slope,

$$E = K = \tfrac{1}{2}mv^2 = \tfrac{1}{2}(80 \text{ kg})(20 \text{ m/s})^2 = 1.6 \times 10^4 \text{ J}$$

Therefore, this system is not conservative (as you might expect).
(b) The amount of work done by the nonconservative force of friction is equal in to the change in the mechanical energy, or to the amount of energy lost (Eq. 5.13):

$$W = Q = E_o - E = (8.6 \times 10^4 \text{ J}) - (1.6 \times 10^4 \text{ J}) = 7.0 \times 10^4 \text{ J}$$

This is over 80% of the initial energy. (Where did this energy actually go?)

Note that in a nonconservative system the total energy is conserved, but it is not all available for mechanical work. For a conservative system, you get back what you put in, so to speak. That is, if you do work on the system, the transferred energy is generally available to do work. Of course, conservative systems are idealizations, because all real systems are nonconservative to some degree. However, working with ideal conservative systems gives an understanding of the conservation of energy.

During this course of study, you will be concerned with other forms of energy, such as thermal, electrical, nuclear, and chemical energy. In general, on the microscopic and submicroscopic levels, these forms of energy can be described in terms of kinetic energy and potential energy. Also, you will learn that mass is a form of energy and that the law of the conservation of energy must take this into account to be applied to the analysis of nuclear reactions.

Demonstration/Activity: Elevate one end of an air track slightly and allow a glider to accelerate downward to hit a spring and rebound back up. This will show that even an air track has some friction. Or is energy absorbed by the spring?

To keep an object swinging around on a string, one must continually move the hand to 'pump' energy into the system.

5.6 ■ Power

A particular task may require a certain amount of work, but it might be done over different lengths of time or at different rates. For example, suppose that you have to mow a lawn. This takes a certain amount of work, but you might do the job in a half hour, or you might take an hour or two. There's a practical distinction to be made here. There is usually not only an interest in the amount of work done, but also an interest in how fast it is done, that is, the rate at which it is done. **Power** is *the rate of doing work.*

The average power is the work done divided by the time it takes to do the work, or work per unit of time:

$$\overline{P} = \frac{\Delta W}{\Delta t}$$

If W_o and t_o are taken to be zero,

$$\overline{P} = \frac{W}{t} \qquad (5.14)$$

If the work is done by a constant force,

$$\boxed{\overline{P} = \frac{W}{t} = \frac{Fd}{t} = F\overline{v}} \qquad (5.15)$$

where it is assumed that F is in the direction of the displacement. (If not, it is the component of the force in that direction, and $\overline{P} = (Fd \cos \theta)/t$.) Here $\overline{v}$ is the magnitude of the average velocity. If the velocity is constant, then $\overline{P} = P = Fv$.

As you can see from Eq. 5.16, the SI units of power are joules per second (J/s), but this is given another name:

$$J/s \equiv watt \ (W)$$

The SI power unit is called a **watt (W)** in honor of James Watt (1736–1819), a Scottish engineer who developed one of the first practical steam engines. A common unit of electrical power is the kilowatt (kW).

The British units are foot-pounds per second (ft-lb/s). However, a larger unit, the **horsepower (hp)** is more commonly used:

$$1 \ hp \equiv 550 \ ft\text{-}lb/s = 746 \ W$$

Power tells you how fast work is being done *or* how fast energy is transferred. For example, motors have power ratings (commonly given in horsepower). A 1-hp motor can do a given amount of work in half the time that a $\frac{1}{2}$-hp motor would take, or twice the work in the same amount of time; that is, a 1-hp motor is twice as powerful as a $\frac{1}{2}$-hp motor (see Example 5.13).

EXAMPLE 5.12 ■ Power delivery

A hoist crane like those shown in ● Fig. 5.21 lifts a load of 1.0 metric ton a vertical distance of 25 m in 9.0 s at a constant velocity. (a) How much work is done by the hoist? (b) What is the minimum horsepower delivered by the hoist motor?

The watt (W), unit of power

This unit was originated by James Watt. At the time, steam engines were replacing horses for work in mines and mills. To characterize the performance of his new engine, which was more efficient than other existing ones, Watt used the average rate at which a horse could do work as a unit—a horsepower.

Note that 1 metric ton (or tonne) is 1000 kg. This is equivalent to 2200 lb, which is the weight of a British long ton (a short ton is 2000 lb).

Solution.

Given: $m = 1.0$ metric ton   *Find:* (a) W (work)
        $= 1.0 \times 10^3$ kg   (b) P (power in hp)
        $h = 25$ m
        $t = 9.0$ s

(a) Since the load moves with a constant velocity, $\overline{P} = P$. (Why?) The work is done against gravity, so $F = mg$, and

$$P = \frac{W}{t} = \frac{Fd}{t} = \frac{mgh}{t}$$
$$= \frac{(1.0 \times 10^3 \text{ kg})(9.8 \text{ m/s})^2(25 \text{ m})}{9.0 \text{ s}}$$
$$= 2.7 \times 10^4 \text{ W} \quad \text{(or 27 kW)}$$

Note that the velocity has a magnitude of $v = d/t = 25$ m/9.0 s $= 2.8$ m/s, and the power is $P = Fv$.
(b) The minimum power refers to useful work output (neglecting frictional and other losses), and this was computed in part (a). In horsepower, it is

$$P = (2.7 \times 10^4 \text{ W})(1 \text{ hp}/746 \text{ W}) = 36 \text{ hp}$$

● **FIGURE 5.21 Power delivery** Example 5.12.

EXAMPLE 5.13 ■ Work and Time

Two motors have net power outputs of 1.00 hp and 0.500 hp. (a) How much work in joules can each motor do in 3.00 min? (b) How long does it take for each motor to do 56.0 kJ of work?

Solution.

Given: $P_1 = 1.00$ hp $= 746$ W   *Find:* (a) W (work)
        $P_2 = 0.500$ hp $= 373$ W   (b) t (time)
        $t = 3.0$ min $= 180$ s
        $W = 56.0$ kJ $= 56.0 \times 10^3$ J

(a) Since $P = W/t$,

$$W_1 = P_1 t = (746 \text{ W})(180 \text{ s}) = 1.34 \times 10^5 \text{ J}$$
$$W_2 = P_2 t = (373 \text{ W})(180 \text{ s}) = 0.67 \times 10^5 \text{ J}$$

Note that the smaller motor does half the work the larger one does in the same time, as you would expect.
(b) The times are given by $t = W/P$, and for the same amount of work,

$$t_1 = \frac{W}{P_1} = \frac{56.0 \times 10^3 \text{ J}}{746 \text{ W}} = 75.0 \text{ s}$$
$$t_2 = \frac{W}{P_2} = \frac{56.0 \times 10^3 \text{ J}}{373 \text{ W}} = 150 \text{ s}$$

Note that the smaller motor takes twice as long as the larger one to do the same amount of work.

Efficiency

Machines and motors are quite common, and we often talk about their efficiency. Efficiency involves work, energy, and/or power. Simple and complex machines that do work have mechanical parts that move, so some energy is always lost because of friction or some other cause (perhaps in the form of sound). Thus, not all of the input energy goes into doing useful work.

Mechanical efficiency is essentially a measure of what you get out for what you put in, that is, the useful work output for the energy input. **Efficiency (ε)** is given as a fraction (or percentage):

Efficiency—what you get out for what you put in

$$\boxed{\begin{aligned} \varepsilon &= \frac{\text{work output}}{\text{energy input}} \ (\times \ 100\%) \\ &= \frac{W_{out}}{E_{in}} \ (\times \ 100\%) \end{aligned}}$$

(5.16)

For example, if a machine has a 100-J (energy) input and a 50-J (work) output, then its efficiency is

$$\varepsilon = \frac{W_{out}}{E_{in}} = \frac{50 \ \text{J}}{100 \ \text{J}} = 0.50 \ (\times \ 100\%) = 50\%$$

An efficiency of 0.50 or 50% means that half of the energy input is lost because of friction or some other cause and doesn't serve its intended purpose.

You can also write efficiency in terms of power (P):

$$\boxed{\varepsilon = \frac{P_{out}}{P_{in}} \ (\times \ 100\%)}$$

(5.17)

Have students explain why the efficiency cannot be greater than 100%. Have them explain what condition(s) must exist if the efficiency is 100%. Many students already know that the efficiency of typical internal combustion engines is less than 50%.

Note that if both terms of the ratio in Eq. 5.16 are divided by time (t), $W_{out}/t = P_{out}$ and $E_{in}/t = P_{in}$.

EXAMPLE 5.14 ■ Mechanical Efficiency

A machine with an efficiency of 40% has a power input of 600 W. How much work is done by the machine in a time of 30 s?

Solution.

Given: $\varepsilon = 40\% = 0.40$ *Find:* W_{out} (work output)
$P_{in} = 600 \ \text{W}$
$t = 30 \ \text{s}$

Given the efficiency and power input, we can readily find the power output (P_{out}), and this, of course, is related to the work output ($P_{out} = W_{out}/t$). Then, rearranging Eq. 5.17,

$$P_{out} = \varepsilon P_{in} = (0.40)(600 \ \text{W}) = 2.4 \times 10^2 \ \text{W}$$

and

$$W_{out} = P_{out} t = (2.4 \times 10^2 \ \text{W})(30 \ \text{s}) = 7.2 \times 10^3 \ \text{J}$$

Important Concepts

You should be able to define and explain these chapter concepts clearly.

work	gravitational potential energy	law of the conservation of
joule (unit)	law of conservation of total	mechanical energy
foot-pound (unit)	energy	power
spring (force) constant	conservative force	watt (unit)
kinetic energy	nonconservative force	horsepower (unit)
work-energy theorem	total mechanical energy	efficiency
potential energy	conservative system	

Important Relationships for Review

These relationships are mathematical statements of the concepts and principles presented in the chapter. You should be able to identify the symbols and to explain the relationships before proceeding to the Exercises. In-text equation reference numbers are given for convenience.

Work:

$$W = Fd \cos \theta \qquad (5.2)$$

◇ *Hooke's Law (spring force):*

$$F = -k(x - x_o) \qquad (5.3)$$
$$F = -kx \qquad \text{(with } x_o = 0\text{)}$$

◇ *Work Done (or Potential Energy Change) in Stretching or Compressing a Spring:*

$$W = \tfrac{1}{2}k(x^2 - x_o^2) \qquad (5.4)$$
$$W = \tfrac{1}{2}kx^2 \qquad \text{(with } x_o = 0\text{)}$$

Kinetic Energy:

$$K = \tfrac{1}{2}mv^2 \qquad (5.5)$$

Work-energy Theorem:

$$W_{net} = K - K_o = \Delta K \qquad (5.6)$$

Elastic (Spring) Potential Energy:

$$U = \tfrac{1}{2}kx^2 \qquad \text{(with } x_o = 0\text{)} \qquad (5.7)$$

Gravitational Potential Energy:

$$U = mgh \qquad \text{(with } h_o = 0\text{)} \qquad (5.8)$$

Conservation of Mechanical Energy:

$$\tfrac{1}{2}mv^2 + U = \tfrac{1}{2}mv_o^2 + U_o \qquad (5.10)$$

Conservation of Energy (with a Nonconservative Force):

or
$$K_o + U_o = K + U + Q \qquad (5.12)$$
$$E_o = E + Q \qquad (5.13)$$

Power:

$$\overline{P} = \frac{W}{t} = \frac{Fd}{t} = F\overline{v} \qquad (5.15)$$

(constant force in direction of d and v; otherwise $F \cos \theta$)

Efficiency (Percent):

$$\varepsilon = \frac{W_{out}}{E_{in}} \, (\times 100\%) \qquad (5.16)$$

$$= \frac{P_{out}}{P_{in}} \, (\times 100\%) \qquad (5.17)$$

Exercises

5.1 ■ Work Done by a Constant Force

1 Two identical boxes are moved equal horizontal, straight-line distances. Box A experiences a constant net force F applied at an angle θ to the horizontal and box B experiences a constant force of $2F$ applied at an angle of 2θ to the horizontal. Then, (a) more work is done on A than on B, (b) more work is done on B than on A, (c) the same amount of work is done on each box, (d) it can't be determined on which box more work is done from the data given. (d)

2 In expressing work, you may use the units of (a) N-m, (b) J, (c) ft/lb, (d) both (a) and (b).   (d)

●3 Can work be done if there is no motion? Explain.
no motion, no displacement, therefore no work
4 How would *negative* work be represented on a graph of force versus distance?   area under curve in 2nd and 4th quadrants
5 ■ A crane lifts a 500-kg load a vertical distance of 20.0 m. If the speed of the load is constant, how much work is done in lifting it? 9.80×10^4 J

6 ■ A tractor exerts a constant force of 5.0×10^3 N on a horizontal chain while moving a load a horizontal distance of 25 cm. How much work is done by the tractor?
1.3×10^3 J
7 ■ Two constant horizontal forces have magnitudes of 50 N and 75 N. The 75-N force does 400 J of work. Through what horizontal distance would the 50-N force have to act to do the same work?   8.0 m

●8 ■■ A person pushing a lawnmower on a level lawn applies a constant force of 250 N at an angle of 30° to the horizontal. How far does the person push the mower in doing 1.44×10^3 J of work?   6.65 m

9 ■■ A 3.0-kg block slides down a frictionless plane inclined 20° to the horizontal. If the length of the plane's surface is 1.5 m, how much work is done and by what force?   15 J

10 ■■ The formula for work is sometimes written $W = F_\parallel d$, where $F_\parallel$ is the component of the force parallel to the displacement. Show that work is also given by $Fd_\parallel$, where $d_\parallel$ is the component of the displacement parallel to the force.   see ISM

11 ■■ A hot-air balloon with a mass of 500 kg ascends at a constant rate of 1.50 m/s for 20.0 s. How much work is done against gravity? (Neglect air resistance.) 1.47×10^5 J

12 ■■ A 1500-kg helicopter accelerates vertically upward at a constant rate of 2.0 m/s² for a distance of 30 m. (a) What is the net work done on the helicopter? (b) How much work is done by the helicopter motor and rotor?
(a) 9.0×10^4 J   (b) $W = 5.3 \times 10^5$ J
13 ■■ A 120-kg sleigh is pulled by one horse at a constant velocity for a distance of 0.75 km on a level snowy surface. The coefficient of kinetic friction between the sleigh runners and the snow is 0.25. (a) Calculate the work done by the horse. (b) Calculate the work done by friction.
(a) 2.2×10^5 J   (b) -2.2×10^5 J
●14 ■■ A father pulls his young daughter on a sled with a constant velocity on a level surface through a distance of 10 m, as illustrated in ●Fig. 5.22a. If the total mass of the sled and the girl is 35 kg and the coefficient of kinetic friction between the sled runners and the snow is 0.25, how much work does the father do? 7.5×10^2 J

15 ■■ A father pushes horizontally on his daughter's sled to move it up a snowy incline, as illustrated in Fig. 5.22b. If

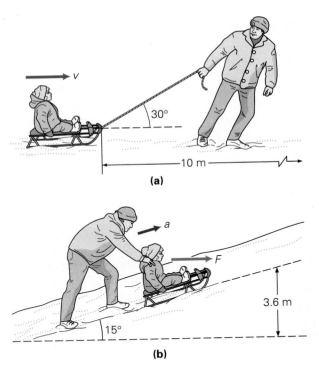

(a)

(b)

●FIGURE 5.22 Fun and work   See Exercises 14 and 15.

the sled goes up the hill with a constant acceleration of 0.25 m/s², how much work is done by the father in getting it from the bottom to the top of the hill? (Some necessary data are given in Exercise 14.) 2.5×10^3 J

16 ■■ A 50-kg crate slides down a 5.0 m loading ramp that is inclined at an angle of 25° to the horizontal. A worker pushes on the crate parallel to the ramp surface so that it slides down with a constant velocity. If the coefficient of kinetic friction between the crate and the ramp is 0.33, how much work is done by (a) the worker, (b) the force of friction, and (c) the force of gravity? (d) What is the net work?
(a) -3.0×10^2 J   (b) -7.4×10^2 J   (c) 1.04×10^3 J   (d) 0
17 ■■ A bicyclist exerts an average downward force of 75 N on alternate pedals during the downward half of each cycle. If a circle with a diameter of 40 cm is traced out each cycle, what is the total average work done by the rider in 5 cycles? 3.0×10^2 J

5.2 ■ Work Done by a Variable Force ◇

18 The work done by a variable force of the form $F = kx$ is equal to (a) kx^2 (b) kx, (c) the area under the F versus x curve, (d) none of these.   (c)

19 If a spring is compressed 2.0 cm from its equilibrium position and then compressed an additional 4.0 cm, how much more work is done in the second compression than in the first? (a) the same amount, (b) twice as much, (c) four times as much, (d) eight times as much.   (d)

20 In Example 5.4, a spring is stretched a distance x_1 by a mass m_1, and then an additional mass m_2 stretches the spring an additional distance x_2, such that the total stretching distance $x = x_1 + x_2$. Show that the total work W done in stretching the spring is *not* equal to $W_1 + W_2$, that is $W \neq W_1 + W_2$ or $\frac{1}{2}kx^2 \neq \frac{1}{2}kx_1^2 + \frac{1}{2}kx_2^2$. How would you calculate W_2 explicitly? see ISM

21 ■ A spring has a spring constant of 25 N/m. How much work is required to stretch the spring 3.0 cm? 1.1×10^{-2} J

22 ■■ A student suspends a 0.25-kg mass on a spring and it stretches a distance of 5.0 cm. The student then pulls down the suspended mass, doing 6.0 J of work in further stretching the spring. How far is the spring stretched as a result of the work done by the student? (Neglect gravitational potential energy in student stretching.) 0.50 m

23 ■■ When a 75-g mass is suspended from a vertical spring, the spring is stretched from a length of 4.0 cm to a length of 7.0 cm. If the mass is pulled downward an additional 10 cm, what is the *total* work done against the spring force? 0.21 J

24 ■■ A particular force is described by the equation $\mathbf{F} = (60\ \text{N})\mathbf{x}$. How much work is done when this force acts through the distance between (a) $x_0 = 0$ and $x = 0.25$ m, and (b) $x_0 = 0.15$ m and $= 0.35$ m? (a) 1.9 J (b) 3.0 J

25 ■■ A force $\mathbf{F} = (50\ \text{N})\mathbf{x}$ stretches a spring continuously from $x_0 = 0$ to $x_2 = 20$ cm. Is more work required to stretch the spring from $x_1 = 10$ cm to $x_2 = 20$ cm than from x_0 to x_1? Justify your answer mathematically. 0.50 J

26 ■■ Compute the work done by the variable force in the F versus x graph in Fig. 5.23. [*Hint:* The area of a triangle is $A = \frac{1}{2}ab$ ($\frac{1}{2}$ altitude × base).] −3.0 J more

27 ■■ A F versus x plot for a spring is a straight line that

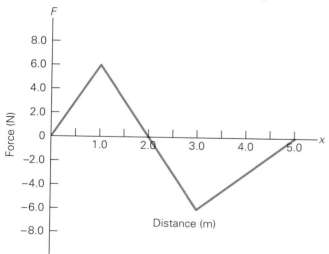

FIGURE 5.23 How much work is done? See Exercises 26 and 80.

makes an angle of 1.5° relative to the x axis when drawn to a scale with 1 N = 1 m. If a mass of 1.0 g is hung on the spring, how far will it be stretched? 0.38 m

28 ■■■ A spring force (in newtons) is given by $F = -\pi(x - 0.10)$. (a) How much work is done in stretching the spring from the position $x = 0$ to $x = 0.25$ m? (b) How much work is done if the position of the end of the spring is changed from $x_1 = 0.05$ m to $x_2 = 0.15$ m? (a) 0.098 J (b) 0.031 J

29 ■■■ A force (F) versus distance (x) graph has a half-circle curve with ends at (0 N, 0 m) and (0 N, 0.5 m) and a maximum force of 60 N. (a) Compute the approximate work done by using a single symmetrical triangle within the half-circle. (b) Compute the approximate work using straight line segments between points (0, 0) to (50 N, 0.10 m) to (50 N, 0.40 m) to (0,0.50 m). (c) Explain how a better approximation can be obtained. (a) 16 J (b) 20 J (c) use more segments

5.3 ■ The Work-Energy Theorem: Kinetic Energy

30 Work is (a) always negative, (b) a measure of transfer of energy, (c) always equal to the change in kinetic energy, (d) all of the preceding. (b)

31 Which of the following objects has the greatest kinetic energy: (a) an object of mass $4m$ and velocity v; (b) an object of mass $3m$ and velocity $2v$; (c) an object of mass $2m$ and velocity $3v$; (d) an object with mass m and velocity $4v$? (c)

32 ■ An automobile with a mass of 1.2×10^3 kg travels at a speed of 90 km/h. (a) What is its kinetic energy? (b) What is the minimum work that would be required to bring it to rest? (a) 3.8×10^5 J (b) -3.8×10^5 J

33 ■ A constant net force of 75 N acts on an object initially at rest and acts through a parallel distance of 0.60 m. (a) What is the final kinetic energy of the object? (b) If the object has a mass of 0.20 kg, what is its final speed? (a) 45 J (b) 21 m/s

34 ■■ A 3.0-g bullet traveling at 350 m/s hits a tree and penetrates a distance of 12 cm. What was the average force exerted on the bullet in bringing it to rest? -1.5×10^3 N

35 ■■ An electron ($m = 9.11 \times 10^{-31}$ kg) has 8.00×10^{-17} J of kinetic energy. What is the speed of the electron? 1.3×10^7 m/s

36 ■■ The stopping distance of a vehicle is an important safety factor. Assuming a constant braking force, show that a vehicle's stopping distance is proportional to the square of its instantaneous speed or velocity. see ISM

37 ■■ An automobile traveling at 45 km/h is brought to a stop in 60 m. Assume that the same conditions (same braking force, driver reaction time, and so on) hold for all cases. (a) What would be the stopping distance for an initial speed of 90 km/h? (b) What would be the initial speed for a stopping distance of 100 m? (a) 240 m (b) 58 km/h

38 ■■ A 0.50-kg object with an initial velocity of 4.5 m/s in the positive x direction is acted on by a force in the direction of the motion. The force does 2.0 J of work. What is the final velocity of the object?   5.3 m/s

39 ■■ A 1.5-kg block with an initial speed of 3.0 m/s slides to a stop in a straight line on a horizontal surface. If the coefficient of kinetic friction between the block and the surface is 0.42, how much work is done by friction? Compute this two ways using work and energy considerations.   −6.8 J, see ISM

40 ■■■ A 150-g block sliding on a frictionless surface moves directly toward a fixed, horizontal spring with a speed of 4.75 m/s. The spring has a spring constant of 200 N/m. (a) After the block makes contact with the spring, by how much will the spring be compressed when the block comes to rest? (b) What will the speed of the block be when the spring has been compressed 6.00 cm? (c) What will the kinetic energy and the speed of the block be when it recoils free of the spring? (Assume that no energy is lost in the collision.)   (a) 0.130 m   (b) 4.21 m/s   (c) $K = 1.69$ J; $v = 4.75$ m/s

5.4 ■ Potential Energy

41 The potential energy of a spring is proportional to (a) k^2, (b) $(x − x_0)^2$, (c) only x_0^2, (d) $x^2 − x_0^2$.   (d)

42 A change in gravitational potential energy (a) is always positive, (b) depends on the reference point, (c) depends on path, (d) depends only on the initial and final heights.   (d)

43 ■ To store exactly 1 J of potential energy in a spring with $k = 45$ N/m, how much would the spring have to be stretched beyond its equilibrium length?   0.21 m

44 ■ Sketch a plot of U versus x for a mass oscillating on a spring between the limits of $−x_{max}$ and $+x_{max}$.   see ISM

45 ■ How much more gravitational potential energy does a 1.0-kg hammer have when it is on a shelf 1.5 m high than when it is on a shelf 1.2 m high?   2.9 J

46 ■■ The floor of the basement of a house is 3.0 m below ground level and the floor of the attic is 4.5 m above ground level. (a) What are the potential energies of 1.5-kg objects on each floor, relative to ground level? (b) If the object in the attic were brought to ground level and the object in the basement were taken to the attic, what would be the change in potential energy for each?   (a) $U_b = −44$ J; $U_a = 66$ J (b) $\Delta U_{b\,to\,a} = 110$ J; $\Delta U_{a\,to\,g} = −66$ J

47 ■■ A fossilized dinosaur bone with a mass of 7.5 kg is dug out from a depth of 2.5 m below the ground surface. The bone is taken to the second story of a nearby building where a lab has been set up; it is placed on a table at a height of 9.0 m above ground level. (a) When the bone is in the dig hole, what is its potential energy relative to ground level? (b) What is the change in the bone's potential energy when it is taken from the dig hole to the lab table? (c) When the bone is on the table, what is its potential energy relative to

the bottom of the hole? [*Hint:* Make a sketch of the situation.]   (a) −1.8 × 10² J   (b) 8.5 × 10² J   (c) 8.5 × 10² J

48 ■■ A 0.20-kg stone is projected vertically upward with an initial velocity of 7.5 m/s, from a starting point 1.2 m above the ground. (a) What is the potential energy of the stone at its maximum height relative to the ground? (b) What is the change in the potential energy of the stone?   (a) 8.0 J   (b) 5.6 J

49 ■■ A 154-lb student climbs 5.0 m up a vertically hanging rope at a homecoming event. What is the least amount of work the student must do to climb this height? Explain why more work is probably done.   3.4 × 10³ J; swings horizontally

50 ■■ A 0.50-kg projectile is given an initial speed of 24 m/s at an angle of 37° to a horizontal surface and lands a certain distance (range) from its launch point. How much work is done on the projectile on landing? (Neglect air resistance.)   −1.4 × 10² J

51 ■■ A student has six textbooks, each with a thickness of 4.0 cm and a weight of 30 N. What is the minimum work the student would have to do to stack the books on top of one another?   21 J

52 ■■■ A 0.400-kg block attached to a fixed, horizontal spring that has a force constant of 7.00 N/m can slide back and forth on a tabletop. There is a coefficient of kinetic friction of 0.340 between the surfaces. (a) If the block is displaced and the spring is stretched 10.0 cm from equilibrium, how far will the spring be compressed when the block is released and travels to the other side of its equilibrium position? (b) How far will the spring be stretched on the return trip? (c) Is the same amount of energy lost to friction during each round trip, or cycle? Explain.   (a) 6.11   (b) 2.24 cm   (c) no, since the distance is decreasing

5.5 ■ The Conservation of Energy

53 If a nonconservative force acts on an object, (a) its kinetic energy is always greater than its potential energy, (b) the mechanical energy is conserved, (c) its kinetic energy is always conserved, (d) the total energy is conserved.   (d)

54 The speed of a pendulum is greatest (a) before it reaches the bottom of its swing, (b) when its acceleration is the greatest, (c) when its potential energy is the least, (d) when its kinetic energy is a minimum.   (c)

55 ■ A person standing on a bridge at a height of 115 m above a river drops a 0.250-kg rock. (a) What is the rock's mechanical energy at the time of release relative to the river? (b) What are the rock's kinetic and potential energies after it has fallen 75.0 m? (c) What are the rock's velocity and total mechanical energy just before hitting the water? (d) Answer parts (a)–(c) taking the reference point ($h = 0$) at the elevation where the rock is released.   (a) 282 J   (b) $K = 184$ J; $U = 98$ J   (c) $E = 282$ J; $v = 47.5$ m/s   (d) 0; 184 J; 47.5 m/s, 0

56 ■ A 0.50-kg ball thrown vertically upward has an initial kinetic energy of 80 J. (a) What are its kinetic and potential energies when it has traveled $\frac{3}{4}$ of the distance to its maximum height? (b) What is its velocity at this point? (c) What is its potential energy at the maximum height?   (a) $K = 20$ J; $U = 60$ J   (b) 8.9 m/s   (c) 80 J

57 ■■ A 0.20-kg projectile is given an initial velocity of 12 m/s at an angle of 37° to the horizontal. (a) What are the projectile's potential energy and kinetic energy at $t = 0.50$ s? (b) What is its kinetic energy at $t = 1.1$ s?   (a) $K = 10$ J; $U = 4.2$ J   (b) $K = 10$

58 ■■ A girl swings back and forth on a swing whose ropes are 4.00 m long. The maximum height she reaches is 2.50 m above the ground. At the lowest point of the swing, she is 0.50 m above the ground. What is the girl's maximum speed, and where does she attain it?   6.26 m/s

59 ■■ A roller coaster starts from rest at a point 45 m above the bottom of a dip. Neglecting friction, what will be the speed of the roller coaster at the top of the next slope, which is 30 m above the bottom of the dip?   17 m/s

● 60 ■■ A 28-kg child slides down a playground slide from a height of 3.0 m above the bottom of the slide. If her speed at the bottom is 2.5 m/s, how much energy is lost to friction? 7.3×10^2 J

61 ■■ An automobile is traveling at 90 km/h on level ground and then coasts up a hill. (a) Neglecting friction, how high above the level ground will the car go before stopping? (b) Assuming that 30% of the energy is lost to friction, how high will the car go?   (a) 32 m   (b) 22 m

62 ■■ In the high jump and pole vault events, athletes run and then convert kinetic energy into potential energy. Estimate the running speeds for a high jump and a pole vault of 2.3 m and 5.5 m, respectively.   high jump: $v = 6.7$ m/s; pole vault: $v = 10$ m/s

● 63 ■■ A simple pendulum has a length of 0.75 m and a bob whose mass is 0.15 kg. The bob is released from an angle of 25° relative to a vertical reference line (●Fig. 5.24). (a)

Show that the vertical height of the bob when released is $h = L(1 - \cos 25°)$. (b) What is the kinetic energy of the bob when the string is at an angle of 9.0°? (c) What is the speed of the bob at the bottom of the swing? (Neglect friction and the mass of the string.)   (a) see ISM   (b) 9.0×10^{-2} J   (c) 1.4 m/s

64 ■■ The grade of a hill is given by the rise (Δy) over the run (Δx), which is the slope of the incline, or the tangent of the angle of incline ($\tan \theta$), usually expressed as a percentage. If a car starts up a hill with a 3.0% grade traveling at 90 km/h in neutral gear, what vertical distance up the hill will it coast before coming to a stop? Assume a frictional energy loss of 40%.   19 m

65 ■■ On a snowy day when school has been canceled, a student on a sled starts from rest at a vertical height of 25 m above the horizontal base of the hill and slides down. If the sled and the student have a speed of 20 m/s at the bottom of the hill, is mechanical energy conserved? Justify your answer.   no since $K_{bottom} < U_{top}$

66 ■■ Suppose that at the base of the hill the student on the sled in Exercise 65 slides onto a rough horizontal surface with a coefficient of kinetic friction of $\mu_k = 1.5$. How far will the sled travel from the base of the hill before coming to rest?   14 m

67 ■■■ A ball with a mass of 0.360 kg is dropped from a height of 1.20 m above the top of a fixed vertical spring, whose force constant is 350 N/m. (a) What is the maximum distance the spring is compressed by the ball? (Neglect energy loss due to the collision.) (b) What is the speed of the ball when the spring has been compressed 5.00 cm?   (a) 0.166 m   (b) 4.82 m/s

68 ■■■ A hiker plans to swing across a ravine in the mountains on a rope, as illustrated in ●Fig. 5.25, and to drop when

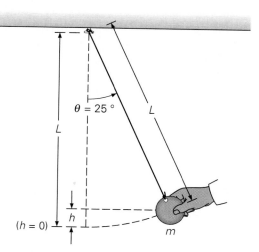

FIGURE 5.24 A pendulum swings   See Exercise 63.

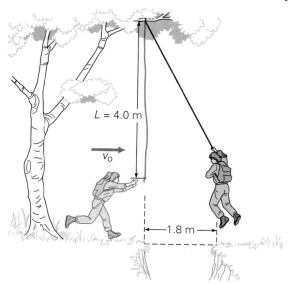

FIGURE 5.25 Can he make it?   See Exercise 68.

he is just above the far edge. (a) With what horizontal speed should he be moving when starting to swing? (b) Will he be in any danger of hitting the edge and falling into the ravine? Explain.   (a) 2.6 m/s   (b) yes; start with v_o > 2.6 m/s

5.6 ▪ Power

69 Which of the following is *not* a unit of power? (a) J/s, (b) W-s, (c) ft-lb/s, (d) hp   (b)

70 If machine A has an efficiency of 30% and machine B has an efficiency of 50%, you know that (a) B can do work faster than A, (b) B's power input is greater than A's, (c) B's power output is greater than A's, (d) none of the preceding.   (d)

71 Two students start at the same ground floor location at the same time to go to a classroom on the third floor by different routes. Who does more work against gravity? If they arrive at different times, which student will have expended more power? Explain.   see ISM

72 A factory worker may be paid either an hourly rate or by piece work. Are there any power considerations incorporated into these methods of paying for work? Discuss.
yes; piece work based on work/time
● **73** (a) Does efficiency describe how fast work is done? Explain. (b) What would the following conditions indicate physically: an efficiency of 100%, and an efficiency greater than 100%?   see ISM

74 ▪ What is the horsepower rating of a 100-W light bulb?   0.134 hp

75 ▪ (a) How many joules of electrical energy are used by a 1650-W hair dryer in 10 min? (b) If the dryer is operated a total of 4.5 h during one month, how much will this contribute to the electrical bill if the billing rate is 9¢/kWh?
(a) 9.90×10^5 J   (b) $0.67
76 ▪▪ A 1.0-kg stone is dropped and free falls a distance of 4.5 m. What is the average power developed by gravity?   49 W

77 ▪▪ A tractor pulls a wagon with a constant force of 700 N, giving the wagon a constant velocity of 20.0 km/h. (a) How much work is done by the tractor in 3.50 min? (b) What is the tractor's power output?   (a) 8.19×10^5 J
(b) 3.90×10^5 W
● **78** ▪▪ A self-propelled lawn mower with a power delivery of 3.5 hp travels at a constant rate of 1.5 ft/s. (a) What is the magnitude of the propelling force? (b) What is the magnitude of the frictional force? (c) How much work is done by each in 6.0 s?   (a) 1.3×10^3 lb   (b) -1.3×10^3 lb   (c) 7.8×10^3 ft-lb

79 ▪▪ A 1500-kg elevator accelerates upward at a constant rate of 0.50 m/s². How much power is developed on the average during the time the elevator's speed goes from 0.25 m/s to 0.75 m/s? 7.7×10^4 W

80 ▪▪ What is the average power developed or expended by the force in the graph in Fig. 5.23 if the force acts for 8.0 s? -0.38 W

81 ▪▪ An electric motor with a 2.0-hp output drives a machine with an efficiency of 45%. What is the energy output of the machine per second? 6.7×10^2 J/s

82 ▪▪ A constant horizontal force of 30 N moves a box along a rough surface with a constant speed. If the force does work at a rate of 50 W, (a) what is the box's speed? (b) How much work is done by the force in 2.5 s?   (a) 1.7 m/s
(b) 1.3×10^2 W
83 ▪▪ A constant applied force of 25 N acts downward at an angle of 30° at the horizontal on a 12-kg box on a rough surface, and the box moves with a constant speed of 0.15 m/s. (a) What is the power input of the applied force? (b) How much work is done by the frictional force in 4.0 s? (c) What is the coefficient of kinetic friction between the box and the surface?   (a) 3.2 W   (b) -13 J   (c) 0.21

● **84** ▪▪▪ A 3.25×10^3-kg aircraft takes 12.5 min to achieve its cruising altitude of 10.0 km and cruising speed of 850 km/h. If the plane's engines deliver, on the average, 1500 hp of power during this time, what is the efficiency of the aircraft?   48.9%

85 ▪▪▪ A horse pulls a sled whose total mass is 120 kg up a hill with a 15° incline, as illustrated in (● Fig. 5.26). (a) If the overall retarding frictional force is 950 N and the sled moves up the hill with a constant velocity of 5.0 km/h, what is the power output of the horse? (Express in horsepower, of course. Note the magnitude of your answer and explain.) (b) Suppose that in a spurt of energy the horse accelerates the sled uniformly from 5.0 km/h to 20 km/h in 5.0 s. What is the horse's maximum instantaneous power output? Assume the same force of friction.   (a) 2.4 hp   (b) 10 hp

● **FIGURE 5.26 A one-horse open sleigh**   See Exercise 85.

▪ Additional Exercises

86 Discuss all the different energy conversions that are involved in a dive from a spring board, as shown in ● Fig. 5.27. (Include the diver's running start.)   see ISM

● **FIGURE 5.27 Energy conversion(s)** See Exercise 86.

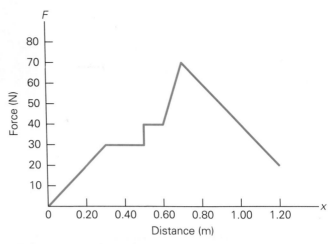

● **FIGURE 5.28 Work done by a variable force** See Exercise 91.

87 A large electric motor with an efficiency of 75% has a power output of 1.5 hp. If the motor is run steadily for 2.0 h and the cost of electricity is 12¢/kWh, how much does it cost to run the motor for this period? $0.36

88 When a certain rubber ball is dropped from a height of 1.25 m onto a hard surface, it loses 18.0% of its mechanical energy on each bounce. (a) How high will the ball bounce on the first bounce? (b) How high will it bounce on the second bounce? (c) With what speed would the ball have to be thrown downward to make it reach its original height on the first bounce? (a) 1.03 m (b) 0.84 m (c) 2.10 m/s

89 In planing a piece of wood 35 cm long, a carpenter applies a force of 40 N to a jack plane at an downward angle of 25° to the horizontal. How much work is done by the carpenter? 13 J

90 A sports car weighs a third as much as a large luxury car. (a) If the sports car is traveling at a speed of 90 km/h, at what speed would the larger car have to travel to have the same kinetic energy? (b) Suppose that the large car is traveling at a speed that gives it half the kinetic energy of the sports car. What is the speed of the large car? (a) 52 km/h = 14 m/s (b) 36 km/h = 10 m/s

91 Compute the work done by the force in the force versus distance graph in ● Fig. 5.28. 42.5 J

92 A water slide has a height of 4.0 m, and the people coming down the slide shoot out horizontally at the bottom, which is a distance of 1.5 m above the water in the swimming pool. If a person starts down the slide from rest, and there are no frictional losses, how far from a point directly below the bottom of the slide does the person land? Does it make any difference whether the person is a small child or an adult? 4.9 m; independent of mass

93 How much work does a 1000-W motor do in 5.0 min if it has an efficiency of 90%? 2.7 × 10⁵ J

94 A block of mass m may be put on a table of height h by three routes: by lifting it vertically and by sliding it up 37° and 60° frictionless inclined planes. Show that the work done

against gravity is the same in all cases. What does this tell you about the gravitational force? see ISM

95 A 10-g bullet is fired vertically upward with an initial speed of 200 m/s. If the bullet reaches a maximum height of 1.2 km, what percentage of mechanical energy is lost to air resistance? 41%

96 A spring compressed from $x_1 = 4.5$ cm to $x_2 = 9.5$ cm has a change in potential energy of 0.63 J. What is the spring constant? 1.8 × 10² N/m

97 A 0.75-kg block slides 1.2 m down a 20° inclined plane with a constant acceleration of 0.25 m/s². (a) How much work is done by the force of gravity? (b) What is the coefficient of kinetic friction between the block and plane? (a) 3.0 J (b) 0.33

98 A motorboat traveling on a lake with a constant speed of 2.50 m/s has a loss of 12.0 hp due to friction between the hull and the water. What is the average force of friction? −3.58 × 10³ N

99 The pedal arm on a five-speed bicycle is 7.0 in. long. When moving at a constant velocity, a cyclist exerts an average downward force of 5.0 lb on a pedal during each downward stroke (a half cycle). What is the average work done by the cyclist on the pedal during each cycle in (a) foot-pounds and (b) joules? (a) 5.8 ft-lb (b) 7.6 J

100 A 70-kg student who is late for a class runs up two flights of stairs whose combined vertical height is 7.0 m in 10 s (and then silently enters the classroom while the professor is writing on the blackboard). Compute the student's power output in doing work against gravity in (a) watts and (b) horsepower. (a) 4.8 × 10² W (b) 0.64 hp

101 A constant force of 6.0 N acts on a 1.0-kg object initiall at rest on a horizontal surface and gives it 50 J of kine' energy. How much power was expended in doing this? 29 W

102 A constant force of 50 N moves an object 4.0 m 5.0 s. (a) How much energy is expended? (b) What is average power developed? (a) 2.0 × 10² J (b) 4.0 × 1

Exercises

6 Momentum and Collisions

*duces the second
in mechanics—
vation of linear
se, the change
the center of
sed.*

If the bowling ball in the photo above bounced off the pins and rolled back toward you, you would probably be very surprised. But why? What leads us to expect that the ball will send the pins flying and continue on its way, rather than rebounding? You might say that the momentum of the ball carries it onward even after the collision (and you would be right)—but what does that really mean?

In this chapter, you will study the concept of momentum *and learn how it is particularly useful in describing and analyzing motion. Like energy, the momentum of a body or system is a* conserved *quantity under certain conditions, and this fact allows us to analyze a wide range of situations and solve many problems easily. The conservation of momentum is one of the most important principles in physics. In particular, it is used to analyze the collision of objects ranging from subatomic particles to automobiles in traffic accidents.*

The term momentum may bring to mind a football player running down the field, knocking down players who are trying to stop him. Or you might have heard someone say that a team lost its momentum (and so lost the game). Such everyday usages give some insight into the meaning of momentum. They suggest the idea of mass in motion, and therefore of inertia. We tend to think of heavy or massive objects in motion as having a great deal of momentum, even if they move very slowly. However, as you will learn from the technical definition of momentum given below, a light object can have just as much momentum as a heavier one, and sometimes more.

6.1 ■ Linear Momentum

Newton referred to what modern physicists term **linear momentum** as "the quantity of motion . . . arising from velocity and the quantity of matter conjointly." In other words,

The linear momentum of an object is the product of its mass and velocity.

$$\mathbf{p} = m\mathbf{v} \qquad (6.1)$$

It is common to refer to linear momentum as simply momentum. From Eq. 6.1, you can see that the SI units for momentum are kg-m/s. Momentum is a vector quantity having the same direction as the velocity. Like velocity, it can be resolved into rectangular components: $\mathbf{p}_x = m\mathbf{v}_x$ and $\mathbf{p}_y = m\mathbf{v}_y$.

Eq. 6.1 expresses the momentum of a single object or particle. For a system of more than one particle, the **total linear momentum** of the system is the vector sum of the momenta (plural of momentum) of the individual particles.

$$\mathbf{P} = \mathbf{p}_1 + \mathbf{p}_2 + \mathbf{p}_3 + \cdots \qquad (6.2)$$

Stress that linear momentum is a vector *quantity*.

Total linear momentum—a vector sum

EXAMPLE 6.1 ■ Comparison of Momenta

A 100-kg football player runs straight down the field with a velocity of 4.0 m/s. A 1.0-kg artillery shell leaves the barrel of a gun with a muzzle velocity of 500 m/s. Which has the greater momentum?

Solution. As usual, we first list the given data and what we are to find, using the subscripts p and s to refer to the player and shell, respectively.

Given: $m_p = 100$ kg *Find:* p_p and p_s
$v_p = 4.0$ m/s   (magnitudes of the momenta)
$m_s = 1.0$ kg
$v_s = 500$ m/s

The magnitude of the momentum of the player is

$$p_p = m_p v_p = (100 \text{ kg})(4.0 \text{ m/s}) = 400 \text{ kg-m/s}$$

and that of the shell is

$$p_s = m_s v_s = (1.0 \text{ kg})(500 \text{ m/s}) = 500 \text{ kg-m/s}$$

Thus, the much lighter, or less massive, shell has the greater momentum. Keep in mind that the magnitude of momentum depends on *both* the mass and the magnitude of the velocity.

EXAMPLE 6.2 ■ Momentum and Kinetic Energy: Some Ballpark Comparisons

(a) Consider the three objects shown in ● Fig. 6.1—a .22-caliber bullet, a cruise ship, and a glacier. Assuming each to be moving at its normal speed, which do you think has the greatest linear momentum? The least? Which has the greatest kinetic energy? The least? (List your choices before going on.)

(b) Using the estimates given below as your data, calculate the magnitudes of the momenta and the kinetic energies of the three objects. How accurate were your original answers—any surprises?

Bullet: A typical .22-caliber bullet would have a weight of about 30 grains and a muzzle velocity of about 1300 ft/s. (A grain is an old British unit used for pharmaceuticals, such as 5-grain aspirin tablets, and bullets; 1 lb = 7000 grains.)

Ship: A ship like the one shown would have a weight of about 70,000 tons and a speed of about 20 knots. (A knot is another old unit, still commonly used in nautical contexts; 1 knot = 1.15 mi/h.)

Glacier: The glacier might be 1 km wide, 10 km long, and 250 m deep, and might move at a rate of a meter per day. (Not surprisingly, there is much variation among glaciers. It is therefore obvious that these figures must involve more assumptions and rougher estimates than those for the bullet or ship. For example, we are assuming a uniform, rectangular cross-sectional area for the glacier. The depth is particularly difficult to estimate from a photograph; a minimum value is given by the fact that glaciers must be at least 50–60 m thick before they can "flow." Observed speeds range from a few centimeters to as much as 40 meters a day for valley glaciers like the one in the photograph. The value we have chosen is considered a typical one.)

Solution. Converting our data to metric units, we have:

Bullet: $m_b = 30\ gr \left(\dfrac{1\ lb}{7000\ gr}\right)\left(\dfrac{1\ kg}{2.2\ lb}\right) = 0.0019\ kg$

$$v_b = 1300\ ft/s \left(\dfrac{0.305\ m/s}{ft/s}\right) = 400\ m/s$$

Ship: $m_s = 7.0 \times 10^4\ ton \left(\dfrac{2.0 \times 10^3\ lb}{ton}\right)\left(\dfrac{1\ kg}{2.2\ lb}\right) = 6.4 \times 10^7\ kg$

$$v_s = 20\ knots \left(\dfrac{1.15\ mi/h}{knot}\right)\left(\dfrac{0.447\ m/s}{mi/h}\right) = 10\ m/s$$

Glacier: width $= 10^3$ m, length $= 10^4$ m, depth $= 2.5 \times 10^2$ m;

$$v_g = 1\ m/day \left(\dfrac{1\ day}{86,400\ s}\right) = 1.2 \times 10^{-5}\ m/s$$

We know the formulas for kinetic energy, $K = \frac{1}{2}mv^2$ (Eq. 5.5) and momentum, $p = mv$ (Eq. 6.1). We have all the speeds and masses except for m_g, the mass of the glacier. To calculate this we need to know the density of ice, since $m = \rho V$ (Eq. 1.1). The density of ice is less than that of water, but the two are not very different, so we will use the density of water, 1.0×10^3 kg/m³, to simplify the calculations. (Since we have said that the data are approximations, this shortcut should produce results that are good enough for our purposes.)

$$\begin{aligned} m_g &= \rho V = \rho(l \times w \times d) \\ &\approx (1.0 \times 10^3\ kg/m^3)(10^3\ m)(10^4\ m)(2.5 \times 10^2\ m) \\ &= 2.5 \times 10^{12}\ kg \end{aligned}$$

(a)

(b)

FIGURE 6.1 Three moving objects: a comparison of momenta and kinetic energies (a) A .22-caliber bullet shattering a ball point pen. (b) Cruise ship and glacier, Glacier Bay, Alaska. Example 6.2.

Then, calculating the magnitudes of the momenta of the objects, we have

Bullet: $p_b = m_b v_b = (0.0019 \text{ kg})(400 \text{ m/s}) = 0.76 \text{ kg-m/s}$

Ship: $p_s = m_s v_s = (6.4 \times 10^7 \text{ kg})(10 \text{ m/s}) = 6.4 \times 10^8 \text{ kg-m/s}$

Glacier: $p_g = m_g v_g = (2.5 \times 10^{12} \text{ kg})(1.2 \times 10^{-5} \text{ m/s}) = 3.0 \times 10^7 \text{ kg-m/s}$

These results are probably what you expected: the enormous differences in the masses of these objects outweigh the differences in their velocities. When we turn to the kinetic energies, however, the dependence on the *square* of the velocity, which we have already called attention to in Example 5.6, produces results that might surprise you:

Bullet: $K_b = \frac{1}{2}m_b v_b^2 = \frac{1}{2}(0.0019 \text{ kg})(400 \text{ m/s})^2 = 1.5 \times 10^2 \text{ J}$

Ship: $K_s = \frac{1}{2}m_s v_s^2 = \frac{1}{2}(6.4 \times 10^7 \text{ kg})(10 \text{ m/s})^2 = 3.2 \times 10^9 \text{ J}$

Glacier: $K_g = \frac{1}{2}m_g v_g^2 = \frac{1}{2}(2.5 \times 10^{12} \text{ kg})(1.2 \times 10^{-5} \text{ m/s})^2 = 1.8 \times 10^2 \text{ J}$

that is, $K_b \approx K_g$ and $K_g \approx 10^7 K_{sb} \approx 10^7 K_g$. Thus the bullet actually has about the same kinetic energy as the glacier—but the ship has more than 10 million times as much kinetic energy as either of them!

■ Problem-Solving Strategies: "Ballpark" and Order-of-Magnitude Calculations

Calculations such as we have performed above are sometimes known as "ballpark" or "back of the envelope" calculations. In performing such calculations we know that, since the data are only approximate (because they are not known precisely, or because they are being rounded off to simple whole numbers for the purpose of rapid calculation), the answer can't be taken as very accurate (see Section 1.6). However, it is often precise enough for our needs. In the example given above, some of the data were unavoidable approximations (the glacier obviously isn't a simple rectangular object), while others, such as the density of ice, were deliberately approximated to simplify calculation. Nevertheless, this method gave us a good idea of the relative momenta and kinetic energies of the three objects.

Physicists often find it convenient to use a formalized version of the "ballpark" approach, commonly known as *order-of-magnitude calculation*. In this method, all values are rounded off to the nearest "order of magnitude," or power of 10: 10^2, 10^3, 10^4, etc. Since it is very easy to multiply and divide powers of 10, this shortcut makes calculation very fast. However, although sometimes just the powers of 10 can be used in a calculation, it is usually a good idea to retain one significant figure of the prefix. This is particularly true when quantities are squared or raised to higher powers, as in the kinetic energy calculations. For instance, $v_s = 4 \times 10^2$ m/s or $\approx 10^2$ m/s. Then $v_s^2 = (4 \times 10^2 \text{ m/s})^2 = 16 \times 10^4 \text{ m}^2/\text{s}^2 \approx 10^5 \text{ m}^2/\text{s}^2$. However, if we had just used 10^2 m/s, we would have obtained $10^4 \text{ m}^2/\text{s}^2$.

In order-of-magnitude calculations, we can generally expect the answer to be correct to the nearest order of magnitude. In other words, if we get an answer of 1000, the true value might really be 750 or 1200, but we can be pretty confident that it is closer to 1000 than to 100 or to 10,000.

EXAMPLE 6.3 ■ Finding Total Momentum

What is the total momentum for each of the systems of particles illustrated in ● Fig. 6.2?

Solution.

Given: Magnitudes and directions of momenta from Fig. 6.2

Find: (a) Total momentum (**P**)
(b) Total momentum (**P**)

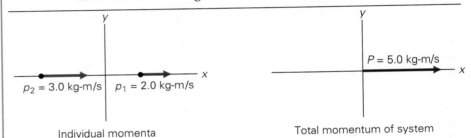

(a) **P = p₁ + p₂**

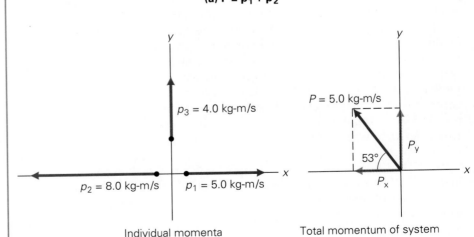

(b) **P = p₁ + p₂ + p₃**

● **FIGURE 6.2 Total momentum**
The total momentum of a system of particles is the vector sum of the individual momenta. Example 6.3.

(a) The total momentum of a system is the vector sum of the momenta of the individual particles, so

$$\mathbf{P} = \mathbf{p}_1 + \mathbf{p}_2 = (2.0 \text{ kg-m/s}) \mathbf{x} + (3.0 \text{ kg-m/s}) \mathbf{x} = (5.0 \text{ kg-m/s}) \mathbf{x}$$
$$(+x \text{ direction})$$

(b) Computing the total momenta in the x and y directions gives

$$\mathbf{P}_x = \mathbf{p}_1 + \mathbf{p}_2 = (5.0 \text{ kg-m/s}) \mathbf{x} - (8.0 \text{ kg-m/s}) \mathbf{x}$$
$$= -(3.0 \text{ kg-m/s}) \mathbf{x} \quad (-x \text{ direction})$$
$$\mathbf{P}_y = \mathbf{p}_3 = (4.0 \text{ kg-m/s}) \mathbf{y} \quad (+y \text{ direction})$$

Then

$$\mathbf{P} = \mathbf{P}_x + \mathbf{P}_y = (-3.0 \text{ kg-m/s}) \mathbf{x} + (4.0 \text{ kg-m/s}) \mathbf{y}$$

Expressed in magnitude-angle form, the total momentum is

$$P = \sqrt{P_x^2 + P_y^2} = \sqrt{(3.0 \text{ kg-m/s})^2 + (4.0 \text{ kg-m/s})^2}$$

$$= \sqrt{25 \text{ kg}^2\text{-m}^2/\text{s}^2} = 5.0 \text{ kg-m/s}$$

and

$$\theta = \tan^{-1} \left| \frac{P_y}{P_x} \right| = \tan^{-1} \left| \frac{4.0}{3.0} \right| = 53°$$

relative to the negative x axis in the second quadrant (see Fig. 6.2b).

In this example, the momenta were along the coordinate axes and are easily added. If the motion of one (or more) of the particles is not along an axis, its momentum may be broken up, or resolved, into rectangular components and added to the component sums—just as you learned to do with force components in Chapter 4. For example, suppose you were given a momentum vector with a magnitude of 5.0 kg-m/s at an angle of 53° relative to the negative x axis (like the total momentum vector in Fig. 6.2b). Then, its rectangular components are

Review Example 4.4 and Fig. 4.11.

$$p_x = -p \cos 53° = -(5.0 \text{ kg-m/s})(0.60) = -3.0 \text{ kg-m/s}$$
$$p_y = p \sin 53° = (5.0 \text{ kg-m/s})(0.80) = 4.0 \text{ kg-m/s}$$

where the signs give the directions. Each momentum vector would be broken down thus and the rectangular components added to find the vector sum.

Since momentum is a vector, a change in momentum can result from a change in magnitude and/or direction. Examples of changes in the momenta of particles because of changes of direction on collision are illustrated in Fig. 6.3. In the figure, the magnitude of a particle's momentum is taken to be the same before and after collision (as indicated by the arrows of equal length). Fig. 6.3a illustrates a direct rebound—a 180° change in direction. Note that the change in momentum ($\Delta \mathbf{p}$) is the vector difference, and directional signs for the vectors are important. Fig. 6.3b shows a glancing collision, for which the change in momentum is given by the component differences.

As you know, a change in velocity (an acceleration) requires a force. Similarly, since momentum is directly related to velocity (as well as mass), a change in momentum also requires a force. In fact, Newton originally expressed his second law of motion in terms of momentum rather than acceleration. The force momentum relationship may be seen by starting with $\mathbf{F} = m\mathbf{a}$ and using $\mathbf{a} = (\mathbf{v} - \mathbf{v}_0)/\Delta t$:

$$\mathbf{F} = m\mathbf{a} = \frac{m(\mathbf{v} - \mathbf{v}_0)}{\Delta t} = \frac{m\mathbf{v} - m\mathbf{v}_0}{\Delta t} = \frac{\mathbf{p} - \mathbf{p}_0}{\Delta t} = \frac{\Delta \mathbf{p}}{\Delta t}$$

Define Newton's second law of motion in terms of momentum.

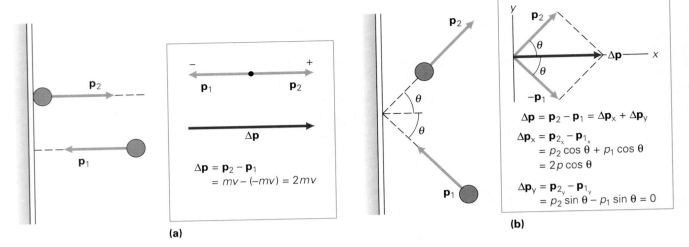

(a)

(b)

Figure 6.3

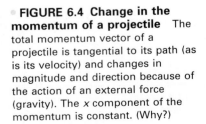

● FIGURE 6.3 **Change in momentum** The change in momentum is given by the *difference* in the momentum vectors. **(a)** Here the vector sum is zero, but the vector *difference*, or change in momentum, is not. (The particles are displaced for convenience.) **(b)** The change in momentum is found by using the change in the components.

or

$$\mathbf{F} = \frac{\Delta \mathbf{p}}{\Delta t} \qquad (6.3)$$

where **F** is the average force if the acceleration is not constant or the instantaneous force if Δt goes to zero.

Expressed in this form, Newton's second law is that *the net external force acting on an object is equal to the time rate of change of the object's momentum.* It is easily seen from the previous development that the equations $\mathbf{F} = m\mathbf{a}$ and $\mathbf{F} = \Delta \mathbf{p}/\Delta t$ are equivalent if the mass is constant. In some situations the mass may vary. This will not be a consideration here in our discussion of particle collisions, but a special case will be given later in the chapter.

Just as the equation $\mathbf{F} = m\mathbf{a}$ indicates that an acceleration is evidence of a force, the equation $\mathbf{F} = \Delta \mathbf{p}/\Delta t$ indicates that *a change in momentum is evidence of a force.* For example, as illustrated in ● Fig. 6.4, the momentum of a projectile is tangential to the projectile's parabolic path and changes in both magnitude and direction. The change in momentum indicates that there is a force acting on the projectile, which you know is the force of gravity. Changes in momentum were illustrated in Fig. 6.3. Can you identify the forces in these cases? Think in terms of Newton's third law.

● FIGURE 6.4 **Change in the momentum of a projectile** The total momentum vector of a projectile is tangential to its path (as is its velocity) and changes in magnitude and direction because of the action of an external force (gravity). The *x* component of the momentum is constant. (Why?)

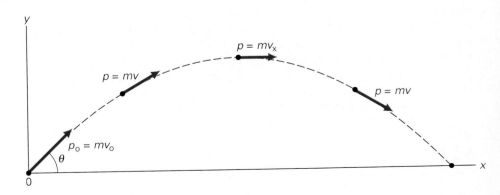

6.2 ■ The Conservation of Linear Momentum

For the linear momentum of a particle or an object to be conserved (to remain constant with time), a condition must hold. This condition is apparent from the momentum form of Newton's second law (Eq. 6.2). If the force acting on a particle is zero, that is,

$$\mathbf{F} = \frac{\Delta \mathbf{p}}{\Delta t} = 0$$

then

$$\Delta \mathbf{p} = 0 = \mathbf{p} - \mathbf{p}_o$$

where $\mathbf{p}_o$ is the initial momentum and $\mathbf{p}$ is the momentum at some later time. Since they are equal, the momentum is conserved.

$$\mathbf{p} = \mathbf{p}_o \tag{6.4}$$

or

$$m\mathbf{v} = m\mathbf{v}_o$$

Note that this is consistent with Newton's first law: an object remains at rest ($\mathbf{p} = 0$) or in motion with a *uniform* velocity (constant $\mathbf{p}$), unless acted on by a net external force.

The conservation of momentum may be extended easily to a system of particles by writing Newton's second law in terms of the sums (resultants) of the forces acting on the system and of the momenta of the particles: $\mathbf{F} = \Sigma \mathbf{F}_i$ and $\mathbf{P} = \Sigma \mathbf{p}_i = \Sigma m \mathbf{v}_i$. As stated in the development of Eq. 6.4, if there is no net force acting on the system, then $\mathbf{P} = \mathbf{P}_o$. This generalized condition is referred to as the law of **conservation of linear momentum**:

> **Conservation of momentum—no net external force**

The total linear momentum of a system is conserved if the net external force acting on the system is zero.

There are other ways to express this. For example, recall that an isolated system is one on which no work is done. Therefore there is no net external force acting on the system, so the total linear momentum of an isolated system is conserved.

> **This is the second law of conservation in mechanics.**

Within a system, internal forces may be acting, for example, when particles collide. These are force pairs of Newton's third law and there is a good reason why such forces are not explicitly referred to in the condition for the conservation of momentum. By Newton's third law, these internal forces are equal and opposite and vectorially cancel each other. Thus, the net internal force of a system is zero.

> **Third-law force pairs were discussed in Section 4.5.**

An important point to understand is that the momenta of individual particles or objects within a system may change. But in the absence of an external force, the vector sum of the momenta remains the same. If the objects are initially at rest (total momentum is zero) and then set in motion as the result of internal forces, the total momentum must still add up to zero. An example of this is illustrated in ● Fig. 6.5 and analyzed in Example 6.4. Objects in an isolated system may initially be in motion, and the motion may be changed by internal forces, but the total momentum after the changes must add up to the initial value (see Demonstration 3).

> **Video demonstration 6 relates to this material.**

The conservation of momentum is often a powerful and convenient tool for analyzing situations involving motion. Its application is illustrated in the following examples.

(a) Getting ready to push off. Note that the forces are internal to the system.

(b) Play ball! Here a ball is used to exchange momentum between parts of the system.

In both cases, the initial total momentum is zero. What would happen after the actions (and reactions) begin? Actually there is an external force (what is it?), so the momentum is not conserved and the motion is limited.

Other demonstrations of conservation of linear momentum:

Suspend a loosely stoppered test tube containing a small amount of water from strings. Heat it to produce steam pressure and a small explosion. Have students explain this in terms of momentum.

Suspend a block of wood from strings. Attach a glob of clay to one side. Then using a BB gun or other spring gun, fire a projectile at the clay and observe the motion of the block. Fire at the other, hard side so that the projectile rebounds and discuss the observed differences. Students are often surprised.

The air track can be used to provide quantitative data for analysis.

EXAMPLE 6.4 ■ Zero Total Momentum

Two masses, $m_1 = 1.0$ kg and $m_2 = 3.0$ kg, are held on either side of a light compressed spring by a light string joining them, as shown in Fig. 6.5. The string is burned (negligible external force), and the masses move apart on the frictionless surface, with m_1 having a velocity of 1.8 m/s to the left. What is the velocity of m_2?

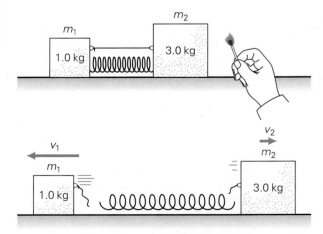

●**FIGURE 6.5 An internal force and the conservation of momentum** The spring force is an internal force, so the momentum of the system is conserved. Example 6.4.

Solution. Listing the given masses and speeds given, we have

> *Given:* $m_1 = 1.0$ kg $\qquad$ *Find:* v_2 (velocity—speed, and direction)
> $\qquad\quad m_2 = 3.0$ kg
> $\qquad\quad v_1 = 1.8$ m/s

(The term *light* indicates that the masses of the spring and string can be ignored.)

Here the system consists of the two masses and the spring. Since the spring force is internal to the system, the momentum of the system is conserved. It should be apparent that the initial total momentum of the system ($\mathbf{P_o}$) is zero, as is the final momentum. Thus, we may write

$$\mathbf{P_o} = \mathbf{P} = 0 \qquad \text{and} \qquad \mathbf{P} = \mathbf{p}_1 + \mathbf{p}_2 = 0$$

(The spring does not come into the equations since its mass is negligible.)
Thus,
$$\mathbf{p}_2 = -\mathbf{p}_1$$

which means that the momenta of m_1 and m_2 are equal and opposite. Using directional signs (with $+x$ to the right in figure) gives

$$p_2 - p_1 = m_2 v_2 - m_1 v_1 = 0$$

or
$$m_2 v_2 = m_1 v_1$$

and
$$v_2 = \left(\frac{m_1}{m_2}\right) v_1 = \left(\frac{1.0 \text{ kg}}{3.0 \text{ kg}}\right)(1.8 \text{ m/s}) = 0.60 \text{ m/s}$$

Thus, the velocity of m_2 is 0.60 m/s in the positive x direction, or to the right in the figure.

Note that the speeds are related by the ratio of the masses in this case; that is, $v_2/v_1 = m_1/m_2$. Suppose that one object were twice as massive as the other. What would the ratio of the velocities be then?

EXAMPLE 6.5 ■ Glancing Collision—Components of Momentum

A moving shuffleboard puck has a glancing collision with a stationary one of the same mass, as shown in ● Fig. 6.6. Considering friction to be negligible, what are the velocities of the pucks after the collision?

Solution.

> *Given:* $v_{1_o} = 0.95$ m/s $\qquad$ *Find:* v_1 and v_2 (velocities)
> $\qquad\quad \theta_1 = 50°$
> $\qquad\quad \theta_2 = 40°$
> $\qquad\quad m_1 = m_2$

The collision forces of the two-puck system are internal forces, so the total momentum is conserved (the external force of friction is negligible). This means that the rectangular components of momentum are also conserved. That is, $\mathbf{P} = \mathbf{P}_x + \mathbf{P}_y$ is a constant, so $\mathbf{P}_x$ and $\mathbf{P}_y$ must be constant. For the x component of the momentum with signs indicating directions (positive axes

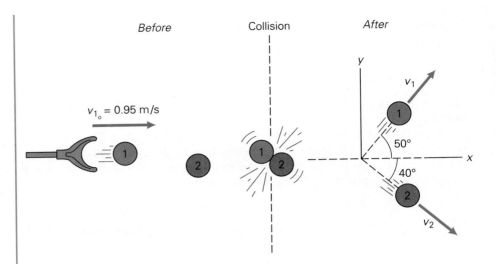

FIGURE 6.6 A glancing collision Momentum is conserved in an isolated system. The motion in two dimensions may be analyzed in terms of the components of momentum which are also conserved. Example 6.5.

shown in figure),

$$p_{x_0} = p_{x_1} + p_{x_2}$$

or

$$m_1 v_{1_0} = m_1 v_1 \cos 50° + m_2 v_2 \cos 40° \quad (1)$$

For the y component,

$$0 = p_{y_1} - p_{y_2}$$

or

$$0 = m_1 v_1 \sin 50° - m_2 v_2 \sin 40° \quad (2)$$

where the minus sign indicates that the component is in the negative y direction. With two equations and two unknowns, the rest is simple algebra. The masses cancel in Eq. 2 since they are equal ($m_1 = m_2$), and

$$v_2 = \left(\frac{\sin 50°}{\sin 40°}\right) v_1 = \left(\frac{0.766}{0.643}\right) v_1 = 1.19 \, v_1 \quad (3)$$

Substituting for v_2 in Eq. 1 and canceling the masses (as above) gives

$$v_{1_0} = v_1 \cos 50° + (1.19 v_1)(\cos 40°)$$

and

$$v_1 = \frac{v_{1_0}}{\cos 50° + (1.19)(\cos 40°)} = \frac{0.95 \, \text{m/s}}{0.643 + (1.19)(0.766)} = 0.61 \, \text{m/s}$$

Using this value in Eq. 3 gives

$$v_2 = 1.19 v_1 = (1.19)(0.61 \, \text{m/s}) = 0.73 \, \text{m/s}$$

The directions of v_1 and v_2 are given by θ_1 and θ_2, respectively.

The conservation of momentum is one of the most important principles in physics. As mentioned previously, it is used to analyze the collisions of objects ranging from subatomic particles to automobiles in traffic accidents. In many instances, external forces may be acting on the objects, which means that the momentum is not conserved. But, as you will learn in the next section, the conservation of momentum often allows a good approximation *over the short* time of a collision because the internal forces, for which momentum is conserved, are much greater than the external forces. For example, external forces such as gravity and friction also act on colliding objects, but are usually relatively small compared to the internal forces. Therefore, if the objects interact for only a brief time, the effects of the external forces are usually negligible compared to those of the large internal forces.

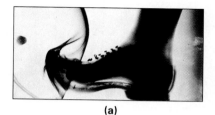

(a)

6.3 ■ Impulse

When two objects—such as a hammer and a nail, a golf club and a golf ball, or even two cars—collide, they exert large forces on one another for a short period of time (see ●Fig. 6.7a). Newton's second law in momentum form is useful for analyzing such situations. Written in this form, the law states that the average force is equal to the time rate of change of momentum, $\overline{\mathbf{F}} = \Delta\mathbf{p}/\Delta t$ (Eq. 6.3). Rewriting the equation to express the change in momentum, we have

$$\overline{\mathbf{F}} \Delta t = \Delta\mathbf{p} = \mathbf{p} - \mathbf{p}_o \qquad (6.5)$$

The term $\overline{\mathbf{F}} \Delta t$ is known as the **impulse** of the force,

$$\boxed{\text{Impulse} = \overline{\mathbf{F}} \Delta t = \Delta\mathbf{p} = m\mathbf{v} - m\mathbf{v}_o} \qquad (6.6)$$

Thus, *the impulse exerted on a body is equal to the change in the body's momentum.* This is referred to as the **impulse-momentum theorem.** Impulse obviously has units of newton-seconds (N-s), which are also units for momentum by Eq. 6.5.

In Chapter 5 you learned that by the work-energy theorem ($W = F \Delta x = \Delta K$), the area under an F versus x curve is equal to the work or the change in kinetic energy. Similarly, the area under an F versus t curve is equal to the impulse or the change in momentum (Fig. 6.7b). An impulse force varies with time and is not a constant force, as Eq. 6.5 might suggest. However, in general, it is convenient to talk about the equivalent *constant* average force $\overline{F}$ acting over a time interval Δt to give the same impulse (same area under the force versus time curve) as shown in Fig. 6.8.

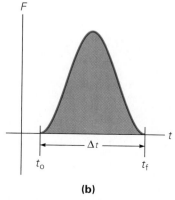

(b)

●**FIGURE 6.7 Collision impulse**
(a) Collision impulse causes the football to be deformed. **(b)** The impulse is the area under the curve of an F versus t graph. Note that the impulse force acting on an object first increases and then decreases.

Compare Fig. 6.7b with Fig. 5.5.

 Figure 6.8

EXAMPLE 6.6 ■ Impulse and Force

A golfer drives a 0.10-kg ball from an elevated tee, giving it an initial horizontal speed of 40 m/s (about 90 mi/h). The club and the ball are in contact for 1.0 ms (millisecond). What is the average force exerted by the club on the ball during this time?

Solution.

Given: $m = 0.10$ kg
$\quad\quad\quad v = 40$ m/s
$\quad\quad\quad v_o = 0$
$\quad\quad\quad \Delta t = 1.0$ ms $= 1.0 \times 10^{-3}$ s

Find: $\overline{F}$ (average force)

Notice that the mass and the initial and final velocities are given, so the change

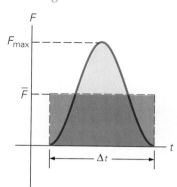

●**FIGURE 6.8 Average impulse force** The area under the average force line ($\overline{F} \Delta t$) is the same as the area under the F versus t curve, which is usually difficult to evaluate.

in momentum can be easily found. Then, the average force may be computed by the impulse-momentum theorem:

$$\bar{F} \, \Delta t = p - p_o = mv - mv_o$$

Thus,

$$\bar{F} = \frac{mv - mv_o}{\Delta t} = \frac{(0.10 \text{ kg})(40 \text{ m/s}) - 0}{1.0 \times 10^{-3} \text{ s}}$$

$$= 4000 \text{ N} \quad (\text{or } 900 \text{ lb})$$

The force is in the direction of the motion and is the *average* force. The instantaneous force is even greater near the midpoint of time interval of the collision (Δt in Fig. 6.8).

Example 6.6 illustrates the large internal forces that colliding objects can exert on one another during short contact times. However, in some instances the impulse may be manipulated to reduce the force. Suppose there is a fixed change in momentum in a given situation. Then, since $\Delta p = \bar{F} \, \Delta t$, if Δt could be made longer, the impulse force $\bar{F}$ would be reduced.

You have probably tried to minimize the impulse force on occasion. For example, when jumping from a height onto a hard surface, you try not to land stiff-legged. The abrupt stop (small Δt) would apply a large impulse force to your leg bones and joints and could do damage. Instead, you bend your knees on landing. Since the impulse is constant in this case ($\bar{F} \, \Delta t = \Delta p = -mv_o$, with the final velocity zero), increasing the time interval Δt makes the impulse force smaller.

Another situation you have probably experienced is discussed in the following example.

EXAMPLE 6.7 ■ Increasing the Contact Time to Reduce the Sting

A boy catches a 0.16-kg baseball coming directly toward him at a speed of 25 m/s in his bare hands with his arms rigidly extended, and emits an audible "ouch!" because the ball stings his hands. He learns quickly that if he moves his hands with the ball as he catches it, the sting is reduced. If the contact time for the collision is increased from 2.5 ms to 7.5 ms in this way, how do the average impulse forces compare? (Assume that the ball has the same initial velocity for each catch.)

Solution.

Given: $m = 0.16$ kg   *Find:* $\bar{F}_1$ and $\bar{F}_2$ (average forces)
$\quad\quad v_o = 25$ m/s
$\quad\quad v = 0$ (when caught)
$\quad\quad t_1 = 2.5$ ms $= 2.5 \times 10^{-3}$ s
$\quad\quad t_2 = 7.5$ ms $= 7.5 \times 10^{-3}$ s

The impulses, $\bar{F}_1 \, \Delta t_1$ and $\bar{F}_2 \, \Delta t_2$, are equal to the same change in momentum for each catch. With $v = 0$, the change in momentum is equal in magnitude to the initial momentum ($\bar{F} \, \Delta t = mv_o$, where the directional minus sign is omitted because only magnitudes are considered).

Thus,

$$\overline{F}_1 \, \Delta t_1 = \overline{F}_2 \, \Delta t_2 = mv_o = (0.16 \text{ kg})(25 \text{ m/s}) = 4.0 \text{ N-s}$$

and the magnitudes of the forces may be computed directly.

$$\overline{F}_1 = \frac{mv_o}{\Delta t_1} = \frac{4.0 \text{ N-s}}{2.5 \times 10^{-3}} = 1.6 \times 10^3 \text{ N}$$

$$\overline{F}_2 = \frac{mv_o}{\Delta t_2} = \frac{4.0 \text{ N-s}}{7.5 \times 10^{-3}} = 5.3 \times 10^2 \text{ N}$$

The force of 1600 N (360 lb) is the average force. The maximum instantaneous force is considerably larger.

Since we are asked only to compare the forces, another approach would be to form a ratio. With

$$\overline{F}_1 \, \Delta t_1 = \overline{F}_2 \, \Delta t_2$$

we have

$$\overline{F}_2 = \left(\frac{\Delta t_1}{\Delta t_2}\right)\overline{F}_1 = \left(\frac{0.25 \text{ s}}{0.75 \text{ s}}\right)\overline{F}_1 = \tfrac{1}{3}\,\overline{F}_1$$

Thus the average force is reduced by two-thirds by increasing the contact time.

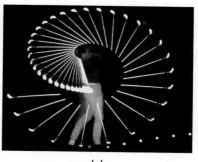

(a)

(b)

FIGURE 6.9 Increasing the contact time (a) A golfer follows through on a drive swing to increase the contact time so that the ball receives greater impulse and momentum. (b) The follow-through on a long putt increases the contact time for greater impulse and momentum, and also improves directional control. Notice that the putter is in contact with the ball for a time equivalent to about 4 flash intervals.

Another example in which the contact time is increased so as to decrease the impulse force is given in the chapter Insight.

In other instances, the *applied* impulse force may be relatively constant, and the contact time (Δt) may be deliberately increased to produce a greater impulse, and thus a greater change in momentum ($\overline{F} \, \Delta t = \Delta p$). This is the principle of "following through" in sports, for example, when hitting a ball with a bat or driving a golf ball. In the latter case (see Fig. 6.9a), assuming that the golfer supplies the same average force with each swing, the longer the contact time, the greater will be the impulse or the momentum the ball receives. That is, with $\overline{F} \, \Delta t = mv$ (since $v_o = 0$), the greater the value of Δt, the greater will be the velocity of the ball. As you learned in Section 3.3, a greater projection velocity gives a projectile a greater range. (What angle of projection should a golfer try to achieve when driving the ball on a level fairway?)

In some instances, a long follow-through that increases the contact time may be used to improve control (Fig. 6.9b).

The word "impulse" implies that the impulse force acts only briefly, and this is true in many instances. However, the definition of impulse places no limit on the time interval over which the force may act. Technically, a comet at its closest approach to the Sun is involved in a collision, because in physics collision forces do not have to be contact forces. Basically, a **collision** is an interaction of objects in which there is an exchange of momentum and energy. As you might expect from the work-energy theorem and the impulse-momentum theorem, momentum and kinetic energy are directly related. Kinetic energy can be expressed in terms of the magnitude of momentum:

Collision—exchange of momentum and energy

$$K = \tfrac{1}{2}mv^2 = \frac{(mv)^2}{2m} = \frac{p^2}{2m} \tag{6.7}$$

Kinetic energy and momentum

Air bags are now installed on most new cars. The bags, along with seatbelts, are safety devices designed to prevent injuries to passengers in the front seat in automobile collisions. (Back seat air bags are also available.)

When a car collides with something immovable such as a bridge abutment, or has a head-on collision with another vehicle, it stops almost instantaneously. If the front-seat passengers have not "buckled up" (and there are no air bags), they keep moving until acted upon by an external force (Newton's first law). For the driver, this force is supplied by the steering wheel and column; for the passenger, by the dashboard and/or windshield.

Even when everyone has buckled up, there can be injuries. Seatbelts absorb energy by stretching, and spread the force over a wide area to reduce the pressure. However, if a car is going fast enough and hits something truly immovable, there may be too much energy for the belts to absorb. This is where the air bag comes in. The bag inflates automatically on hard impact (Fig. 1), cushioning the driver (and front-seat passenger if both sides are equipped with air bags). In terms of impulse, the air bag increases the stopping contact time—the fraction of a second it takes your head to sink into the inflated bag

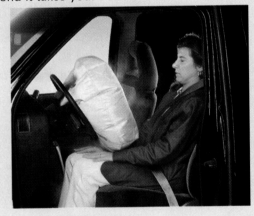

FIGURE 1 Impulse and safety An automobile air bag increases the contact time, thereby decreasing the impulse force that could cause injury.

is many times longer than the instant in which your head is stopped by the dashboard. A longer contact time means a reduced impact force, and thus much less likelihood of an injury. (The total impact force is also spread over a larger area of the body, so the force on any one part of the body is less.)

An interesting point is the inflating mechanism of an air bag. Think of how little time elapses between the front-end impact and the driver hitting the steering column in the collision of a fast-moving automobile. We say such a collision takes place instantaneously, yet during this time the air bag must be inflated! How is this done?

First, the air bag is equipped with sensors that detect the sharp deceleration associated with a head-on collision the instant it begins. If the deceleration exceeds the sensors' threshold settings, an electric current in an igniter in the air bag sets off a chemical explosion that generates gas to inflate the bag. The complete process from sensing to full inflation takes only on the order of 25 *thousandths* of a second (0.025 s).

The sensors' signals go first to a control unit, which determines whether a frontal collision is occurring rather than a system malfunction. (Accidental deployment could be dangerous as well as costly.) Typically, the control unit compares signals from two different sensors for collision verification. The unit is equipped with its own power source, since the car's battery and alternator are usually destroyed in a hard front-end collision. Sensing a collision, the control unit completes the circuit to the air bag igniter. It heats rapidly and initiates a chemical reaction in a sodium azide propellant. Gas (mostly nitrogen) is generated at an explosive rate, which inflates the air bag. The bag itself is made of thin nylon that is covered with cornstarch. The cornstarch acts as a lubricant to help the bag unfold smoothly on inflation.

Air bags offer protection only if the occupants in the front seat are thrown forward, since the bags are designed to deploy only in front-end collisions. They are of little use in side-impact crashes. It is therefore essential that, whether or not the car is equipped with air bags, *all* passengers wear their seat belts at *all* times.

6.4 ■ Elastic and Inelastic Collisions

Taking a closer look at collisions in terms of the conservation of momentum is simpler if an isolated system is considered, such as a system of particles (or balls) involved in head-on collisions. These collisions can also be analyzed in terms of the conservation of energy. On the basis of what happens to the total kinetic energy we can define two types of collisions: elastic and inelastic.

In an **elastic collision**, the total kinetic energy is conserved. That is, the *total* kinetic energy of all the objects of the system after the collision is the same as their *total* kinetic energy before the collision:

Elastic collision—kinetic energy conserved

$$\boxed{\begin{array}{c} \text{total } K \text{ after} \; = \; \text{total } K \text{ before} \\ K_f = K_i \end{array}} \quad \begin{array}{l} \text{condition for an} \\ \text{elastic collision} \end{array} \quad (6.8)$$

During the collision, some of the initial kinetic energy is temporarily converted to potential energy as the objects are deformed. But, after the maximum deformations, the objects elastically spring back to their original shapes, and the system has the same total kinetic energy as it did initially. For example, two steel balls or two billiard balls may have a nearly elastic collision with each ball having the same shape afterwards as before; that is, there is no permanent deformation.

In reality, only atoms and subatomic particles have truly elastic collisions, but some larger hard objects have nearly elastic collisions in which the kinetic energy is approximately conserved.

An elastic collision between two objects can be represented on a graph of force versus displacement as shown in ● Fig. 6.10. (Recall from Chapter 5 that the area under such a curve is equal to the work.) Since the areas bounded by the curves and the x axis are equal in magnitude, they cancel each other so no net work is done, and the change in kinetic energy is zero by the work-energy theorem (Chapter 5.3). As you might guess, objects interact via conservative forces in elastic collisions.

Refer again to Fig. 5.5

In an **inelastic collision**, the total kinetic energy is *not* conserved. For example, one or more of the colliding objects does not spring back to its original shape, so work is done and some kinetic energy is lost.

Inelastic collision—kinetic energy not conserved

$$\boxed{\begin{array}{c} \text{total } K \text{ after} \; < \; \text{total } K \text{ before} \\ K_f < K_i \end{array}} \quad \begin{array}{l} \text{condition for an} \\ \text{inelastic collision} \end{array} \quad (6.9)$$

For example, a hollow aluminum ball that collides with a solid steel ball may be dented, which means work is done in permanently deforming the ball. Inelastic collisions involve nonconservative forces such as friction.

For isolated systems, momentum is conserved in both elastic and inelastic collisions. For an inelastic collision, only an amount of kinetic energy consistent with the conservation of momentum may be lost. It may seem strange that energy can be lost and momentum still be conserved, but this fact provides insight into the difference between scalar and vector quantities.

Energy and Momentum in Inelastic Collisions

To see how momentum can remain constant while the kinetic energy changes (decreases) in inelastic collisions, consider the examples illustrated in ● Fig. 6.11. In the first case, two balls of equal mass approach each other with equal and opposite velocities. Hence, the total momentum before the collision is vectorially zero, but the scalar kinetic energy is *not* zero. After the collision the balls are stuck together and stationary, so the total momentum is unchanged and still zero. The total kinetic energy, however, has decreased—to zero. In this case, the kinetic energy went into the work done in permanently deforming the balls. Some energy may also have gone into doing work against friction (producing heat) or been lost in some other way (for example, in producing sound).

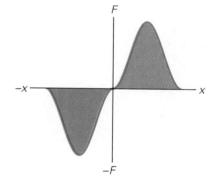

● **FIGURE 6.10 Work in an elastic collision** When two objects collide in an elastic collision, the opposing reaction forces do equal amounts of work. Since the net work is zero, the change in kinetic energy is zero ($W = \Delta K = 0$), and the kinetic energy is conserved.

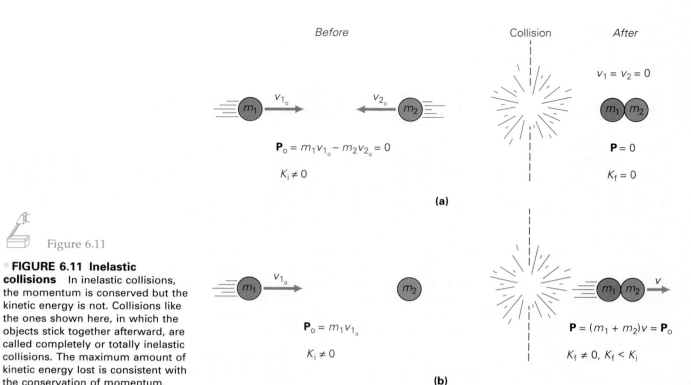

Before Collision After

$v_1 = v_2 = 0$

$\mathbf{P}_o = m_1 v_{1_o} - m_2 v_{2_o} = 0$

$K_i \neq 0$

$\mathbf{P} = 0$

$K_f = 0$

(a)

Figure 6.11

FIGURE 6.11 Inelastic collisions In inelastic collisions, the momentum is conserved but the kinetic energy is not. Collisions like the ones shown here, in which the objects stick together afterward, are called completely or totally inelastic collisions. The maximum amount of kinetic energy lost is consistent with the conservation of momentum.

$\mathbf{P}_o = m_1 v_{1_o}$

$K_i \neq 0$

$\mathbf{P} = (m_1 + m_2)v = \mathbf{P}_o$

$K_f \neq 0, K_f < K_i$

(b)

It should be noted that the balls need not come to rest after collision. In a less inelastic collision, the balls may recoil in opposite directions at reduced but equal speeds. The momentum would be conserved (still equal to zero—why?), but the kinetic energy would again not be conserved.

In Fig. 6.11b, the balls stick together after collision, but are still in motion. Both of these cases are examples of a **completely inelastic collision**, in which the objects stick together and have the same velocity after colliding.

Assume that the balls in Fig. 6.11b have different masses. Since the momentum is conserved even in inelastic collisions,

$$\overset{\text{before}}{m_1 v_o} = \overset{\text{after}}{(m_1 + m_2)v}$$

and

$$v = \left(\frac{m_1}{m_1 + m_2}\right)v_o \tag{6.10}$$

Thus, v is less than v_o, since $m_1/(m_1 + m_2)$ must be less than 1. Now let us consider how much kinetic energy has been lost. Initially, $K_i = \frac{1}{2}m_1 v_o^2$, and finally, or after the collision,

$$K_f = \frac{1}{2}(m_1 + m_2)v^2 = \frac{1}{2}(m_1 + m_2)\left(\frac{m_1 v_o}{m_1 + m_2}\right)^2$$

where Eq. 6.10 was used to substitute for v. Then

$$K_f = \frac{\frac{1}{2}m_1^2 v_o^2}{m_1 + m_2} = \left(\frac{m_1}{m_1 + m_2}\right)\frac{1}{2}m_1 v_o^2 = \left(\frac{m_1}{m_1 + m_2}\right)K_i$$

and

$$\frac{K_f}{K_i} = \frac{m_1}{m_1 + m_2} \qquad (6.11)$$

Eq. 6.11 gives the fractional amount of the initial kinetic energy that the system has after a completely inelastic collision. For example, if the masses of the balls are equal ($m_1 = m_2$), then $m_1/(m_1 + m_2) = 1/2$ and $K_f/K_i = 1/2$ or $K_f = K_i/2$. That is, half of the initial kinetic energy is lost (consistent with the conservation of momentum).

Note that all the kinetic energy cannot be lost in this case no matter what the masses of the balls are. The momentum (velocity) after collision cannot be zero, since it was not zero initially. Thus, the balls must be moving and must have kinetic energy.

EXAMPLE 6.8 ■ Completely Inelastic Collision—Stuck Together

A 1.0-kg ball with a speed of 4.5 m/s strikes a 2.0-kg stationary ball. If the collision is completely inelastic, (a) what are the speeds of the balls after the collision? (b) What percent of the initial kinetic energy do they have after the collision? (c) What is the total momentum after the collision?

Solution. Using the labeling as in the preceding discussion, we have

Given: $m_1 = 1.0$ kg   *Find:* (a) v (speed after collision)

$\qquad m_2 = 2.0$ kg

$\qquad v_o = 4.5$ m/s

(b) $\dfrac{K_f}{K_i}$ ($\times$ 100%)

(c) $\mathbf{P}_f$ (total momentum after collision)

(a) The balls stick together and have the same speed after collision. This is given by Eq. 6.10:

$$v = \left(\frac{m_1}{m_1 + m_2}\right)v_o = \left(\frac{1.0 \text{ kg}}{1.0 \text{ kg} + 2.0 \text{ kg}}\right)(4.5 \text{ m/s})$$

$$= 1.5 \text{ m/s}$$

(b) The fractional part of the initial kinetic energy that the balls have after the completely inelastic collision is given by Eq. 6.11. Notice that this fraction, as given by the masses, is the same as that for the speeds (Eq. 6.10) in this special case. By inspection, we can write

$$\frac{K_f}{K_i} = 1/3 = 0.33 \,(\times\, 100\%) = 33\%$$

(c) The momentum is conserved in all collisions (in the absence of external forces), so the total momentum after collision is the same as before, which is simply the momentum of the incident ball, with a magnitude of

$$P_f = p_{1_o} = m_1 v_o = (1.0 \text{ kg})(4.5 \text{ m/s}) = 4.5 \text{ kg-m/s}$$

and in the same direction.

EXAMPLE 6.9 ■ Collision Impulse

A ball with a large mass M and a ball with a small mass m have a head-on, *elastic* collision. Which ball receives the greater impulse? (a) The large mass ball M, (b) the small mass ball m, (c) the impulses are equal. *Clearly establish the reasoning and physical principle(s) used in determining your answer. That is, why did you select your answer?*

Reasoning and Answer. Impulse is defined as $\overline{F}\Delta t$. In a collision, the contact time is the same for both balls ($\Delta t_1 = \Delta t_2$). The force on one ball at any instant is equal and opposite to the force on the other (Newton's third law), and the average magnitudes of these *internal* forces are equal ($\overline{F}_1 = \overline{F}_2$), so the answer is (c). This means that each ball experiences the same change in momentum. The motion of one ball, however, will be affected more than that of the other. (Why?)

Follow-up Exercise. Suppose that the balls had an *inelastic* collision. Would this make a difference in the impulse? (*Reasoning and answer may be found in the Answers to Exercises section at the back of the book.*)

Energy and Momentum in Elastic Collisions

For a general elastic collision of two objects, the conditions are as follows:

$$\overset{\text{before}}{} \qquad \overset{\text{after}}{}$$

Conservation of **P**: $\quad m_1\mathbf{v}_{1_0} + m_2\mathbf{v}_{2_0} = m_1\mathbf{v}_1 + m_2\mathbf{v}_2 \qquad (6.12a)$

Conservation of K: $\quad \frac{1}{2}m_1v_{1_0}^2 + \frac{1}{2}m_2v_{2_0}^2 = \frac{1}{2}m_1v_1^2 + \frac{1}{2}m_2v_2^2 \qquad (6.12b)$

Knowing the masses of the objects and the initial velocities allows you to find the final velocities (there are two equations and two unknowns).

A common collision situation is that in which one of the objects is initially stationary (● Fig. 6.12). Assume that the motion of both of the balls after the collision is to the right, or in the positive direction, as shown in the figure. If this is not the case, the math will tell you.

The conditions for an elastic collision in this situation are

$$m_1v_{1_0} = m_1v_1 + m_2v_2 \qquad (6.13a)$$

and

$$\frac{1}{2}m_1v_{1_0}^2 = \frac{1}{2}m_1v_1^2 + \frac{1}{2}m_2v_2^2 \qquad (6.13b)$$

Figure 6.12

● **FIGURE 6.12 General elastic collision** For an elastic collision between two bodies, one of which is initially at rest, the velocities after the collision depend on the relative masses of the bodies.

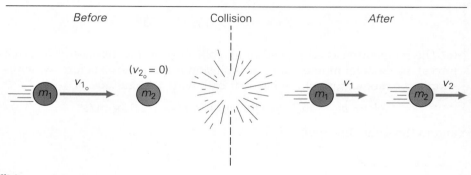

| Before | Collision | After |

v_{1_0} $(v_{2_0} = 0)$ v_1 v_2

Rearranging these equations gives

$$m_1(v_{1_o} - v_1) = m_2 v_2 \tag{1}$$

and

$$m_1(v_{1_o}^2 - v_1^2) = m_2 v_2^2 \tag{2}$$

Then, using the relationship $x^2 - y^2 = (x + y)(x - y)$ and dividing the second equation by the first

$$\frac{m_1(v_{1_o} + v_1)(v_{1_o} - v_1) = m_2 v_2^2}{m_1(v_{1_o} - v_1) = m_2 v_2} \quad \frac{(2)}{(1)} \text{ (divide)}$$

we obtain

$$v_{1_o} + v_1 = v_2 \tag{6.14}$$

Eq. 6.14 can be used to eliminate v_1 or v_2 from either of the preceding two equations. Thus, each of these final velocities may be expressed in terms of the initial velocity (v_{1_o}):

$$v_1 = \left(\frac{m_1 - m_2}{m_1 + m_2}\right) v_{1_o} \qquad \text{(final velocities} \tag{6.15}$$
$$\text{for special case:}$$
$$v_2 = \left(\frac{2m_1}{m_1 + m_2}\right) v_{1_o} \qquad \text{elastic collision with}$$
$$m_2 \text{ initially stationary)} \tag{6.16}$$

Note that the final velocities depend on the masses of the objects. Looking at Eq. 6.15, you can see that if m_1 is greater than m_2, then v_1 is positive, or in the same direction as the initial velocity of the incoming ball, as was assumed in Fig. 6.16. However, if m_2 is greater than m_1, then v_1 is negative, and the incoming ball recoils in the opposite direction after collision. Eq. 6.16 shows that v_2 is always in the same direction as the velocity of the incoming ball.

You can also get some general ideas about what happens after such a collision by considering three possible situations for the relative masses of the objects, illustrated in (Fig. 6.13).

> Consider a large truck moving at 100 km/h (about 60 mph) and striking a stationary ping-pong ball elastically.

Case 1. $m_1 = m_2$ (Fig. 6.13a). From Eqs. 6.15 and 6.16

$$v_1 = 0 \quad \text{and} \quad v_2 = v_{1_o}$$

$m_1 = m_2$ collision

That is, if the masses of the colliding objects are equal, the objects simply exchange momentum and energy. The incoming ball is stopped on collision, and the originally stationary ball moves off with the same velocity that the incoming ball had.

Case 2. $m_1 \gg m_2$ (m_1 very much greater than m_2, Fig. 6.13b). In this case, m_2 can be ignored in the addition and subtraction with m_1 in Eqs. 6.15 and 6.16, and

$$v_1 \cong v_{1_o} \quad \text{and} \quad v_2 \cong 2v_{1_o}$$

$m_1 \gg m_2$ collision

This tells you that if a very massive object collides with a stationary light object, the massive object is slowed down only slightly by the collision, and the light object is knocked away with a velocity almost twice that of the initial velocity of the massive object.

Case 3. $m_1 \ll m_2$ (m_1 very much less than m_2, Fig. 6.13c). Here, m_1 can be ignored in the addition and subtraction with m_2 in Eqs. 6.15 and 6.16, and

$$v_1 \cong -v_{1_o} \quad \text{and} \quad v_2 \cong 0$$

$m_1 \ll m_2$ collision

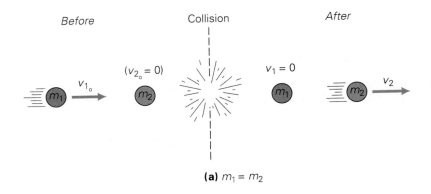

(a) $m_1 = m_2$

Figure 6.13

●**FIGURE 6.13 Special cases of elastic collisions** **(a)** When a moving object collides elastically with a stationary object of equal mass, there is a complete exchange of momentum and energy. **(b)** When a very massive moving object collides elastically with a much less massive stationary object, the very massive object continues to move essentially as before, and the less massive object is given a velocity almost twice the initial velocity of the large mass. **(c)** When a moving object of small mass collides elastically with a very massive stationary object, the incoming object recoils in the opposite direction with approximately the same speed and the very massive object remains essentially stationary.

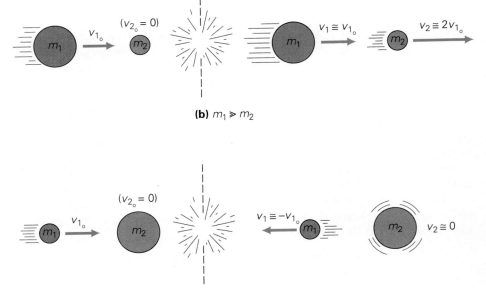

(b) $m_1 \gg m_2$

(c) $m_1 \ll m_2$

(In the second equation, the approximation $m_1/m_2 \cong 0$ is made.) Thus, if a light object collides with a massive stationary one, the massive object remains *almost* stationary, and the light object recoils backward with aproximately the same speed it had before collision. An extreme case of this type seems similar to a particle striking a solid, immovable wall (● see Fig. 6.3). But, for the case in Fig. 6.13c, the massive ball must move a bit after the collision since the momentum is conserved.

EXAMPLE 6.10 ■ Elastic Collision

A 0.30-kg object with a speed of 2.0 m/s in the positive x direction has a head-on elastic collision with a stationary 0.70-kg object located at $x = 0$. What is the distance separating the objects 2.5 s after the collision?

Solution. Using the previous notation, we have

Given: $m_1 = 0.30$ kg $\qquad$ *Find:* $\Delta x = x_2 - x_1$ (separation distance)
$\qquad\quad v_{1_0} = 2.0$ m/s
$\qquad\quad m_2 = 0.70$ kg
$\qquad\quad t = 2.5$ s

From Eqs. 6.15 and 6.16, the velocities after collision are

$$v_1 = \left(\frac{m_1 - m_2}{m_1 + m_2}\right)v_{1_0} = \left(\frac{0.30 \text{ kg} - 0.70 \text{ kg}}{0.30 \text{ kg} + 0.70 \text{ kg}}\right)(2.0 \text{ m/s}) = -0.80 \text{ m/s}$$

$$v_2 = \left(\frac{2m_1}{m_1 + m_2}\right)v_{1_0} = \left[\frac{2(0.30 \text{ kg})}{0.30 \text{ kg} + 0.70 \text{ kg}}\right](2.0 \text{ m/s}) = 1.2 \text{ m/s}$$

Here m_1 is less than m_2, but not *very much* less, as it was in case (c) of Fig. 6.12. The objects are separating after collision and their positions are

$$x_1 = v_1 t = (-0.80 \text{ m/s})(2.5 \text{ s}) = -2.0 \text{ m}$$
$$x_2 = v_2 t = (1.2 \text{ m/s})(2.5 \text{ s}) = 3.0 \text{ m}$$

and

$$\Delta x = x_2 - x_1 = 3.0 \text{ m} - (-2.0 \text{ m}) = 5.0 \text{ m}$$

The objects are 5.0 m apart at that time.

EXAMPLE 6.11 ■ Another Elastic Collision

A 7.5-kg bowling ball with a speed of 6.0 m/s has a head-on collision with a stationary 0.50-kg billiard ball. (a) What are the velocities of the balls after the collision? (b) What is the total momentum after the collision?

Solution. Listing the data as before, we have

Given: $m_1 = 7.5$ kg $\qquad$ *Find:* (a) $\mathbf{v}_1$ and $\mathbf{v}_2$ (velocities after collision)
$\qquad\quad v_{1_0} = 6.0$ m/s $\qquad\qquad\quad$ (b) $\mathbf{P}$ (total momentum after collision)
$\qquad\quad m_2 = 0.50$ kg

(a) The velocities are given directly by Eqs. 6.15 and 6.16, where the direction of the incoming ball is taken as positive.

$$v_1 = \left(\frac{m_1 - m_2}{m_1 + m_2}\right)v_{1_0} = \left(\frac{7.5 \text{ kg} - 0.50 \text{ kg}}{7.5 \text{ kg} + 0.50 \text{ kg}}\right)(6.0 \text{ m/s}) = 5.3 \text{ m/s}$$

$$v_2 = \left(\frac{2m_1}{m_1 + m_2}\right)v_{1_0} = \left[\frac{2(7.5 \text{ kg})}{(7.5 \text{ kg} + 0.50 \text{ kg})}\right](6.0 \text{ m/s}) = 11 \text{ m/s}$$

So both balls move in the same direction. Here m_1 is greater than m_2, but not *very much* greater such that m_2 could be ignored, as in the case (b) of Fig. 6.13.

(b) Again, momentum is conserved in elastic (and inelastic) collisions, so the total momentum afterwards is the same as that before the collision, which is that of the incident ball, with a magnitude of:

$$P_f = P_{1_0} = m_1 v_{1_0} = (7.5 \text{ kg})(6.0 \text{ m/s}) = 45 \text{ kg-m/s}$$

and in the same direction.

6.5 ■ Center of Mass

The conservation of total momentum gives us a method for analyzing a "system of particles." Such a system may be virtually anything—a volume of gas, water in a container, or a baseball. Another important concept allows us to analyze the overall motion of a system of particles. It involves representing the whole system as a single particle. This concept will be introduced here and applied in more detail in the upcoming chapters.

If no net external force acts on a particle, its linear momentum is constant. Similarly, if no net external force acts on a system of particles, the linear momentum of the system is constant. This similarity implies that a system of particles might be represented by an *equivalent* single particle. Moving objects, such as balls, automobiles, and so forth, are essentially systems of particles and can be effectively represented by equivalent single particles when analyzing motion. Such representation is done through the concept of the **center of mass (CM)**.

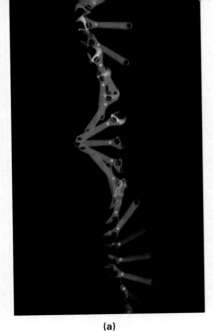

(a)

(b)

● FIGURE 6.14 Center of mass
(a) The center of mass of the wrench moves as though it were a particle. Note the dot on the wrench that marks the center of mass. **(b)** Even after exploding, the center of mass of a fireworks projectile follows a parabolic path.

Consider an artillery shell heading on a true path toward its target that explodes prematurely. Does the center of mass still hit the target? Where is the center of mass of the pieces after they hit the ground?

The center of mass is the point at which all of the mass of an object or system may be considered to be concentrated.

Even if the object is rotating, the center of mass moves as though it were a particle (● Fig. 6.14). The center of mass is sometimes descriptively said to be at the balance point of a solid object. For example, if you balance a meterstick on your finger, the center of mass of the stick is located directly above your finger and all of the mass (or weight) seems to be concentrated there.

Newton's second law applies to a system when the center of mass is used:

$$\mathbf{F} = M\mathbf{A}_{CM} \tag{6.17}$$

where $\mathbf{F}$ is the net external force, M is the total mass of the system or the sum of the masses of the particles of the system ($M = m_1 + m_2 + m_3 + \cdots + m_n$), where the system has n particles), and $\mathbf{A}_{CM}$ is the acceleration of the center of mass. In words, Eq. 6.17 says that the center of mass of a system of particles moves as though all the mass of the system were concentrated there and acted on by the resultant of the external forces.

Also, if the net external force on a system of particles is zero, the total linear momentum of the center of mass is conserved (stays constant) since $\mathbf{F} = M\Delta\mathbf{V}_{CM}/\Delta t$ as for a particle. This means that the center of mass either moves with a constant velocity or remains at rest. Although you may more readily visualize the center of mass of a solid object, the concept of the center of mass applies to any system of particles or objects, even a quantity of gas.

For a system of n particles arranged in one dimension, along the x axis (● Fig. 6.15), the location of the center of mass is given by

$$\mathbf{X}_{CM} = \frac{m_1\mathbf{x}_1 + m_2\mathbf{x}_2 + m_3\mathbf{x}_3 + \cdots + m_n\mathbf{x}_n}{m_1 + m_2 + m_3 + \cdots + m_n} \tag{6.18}$$

That is, X_{CM} is the x coordinate of the center of mass of a system of particles. In shorthand notation (using signs to indicate vector directions),

$$\boxed{X_{CM} = \frac{\Sigma_i m_i x_i}{M}} \tag{6.19}$$

Center of mass coordinates

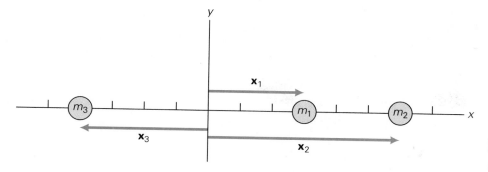

•FIGURE 6.15 System of particles in one dimension Where is the system's center of mass? Example 6.13.

where Σ_i indicates the summation of the products $m_i x_i$ for i particles ($i =$ 1, 2, 3, ..., n). If $\Sigma_i m_i x_i = 0$, then $X_{CM} = 0$, and the center of mass of the one-dimensional system is located at the origin.

Other coordinates of the center of mass for systems of particles are similarly defined. For a two-dimensional distribution of masses, the coordinates of the center of mass are (X_{CM}, Y_{CM}).

EXAMPLE 6.12 ■ Finding the Center of Mass

Three masses, 2.0 kg, 3.0 kg, and 6.0 kg, are located at positions (3.0, 0), (6.0, 0), and (−4.0, 0), respectively, in meters from the origin (●Fig. 6.15). Where is the center of mass of this system?

Solution.

Given: $m_1 = 2.0$ kg   *Find:* X_{CM} (CM coordinate)
$m_2 = 3.0$ kg
$m_3 = 6.0$ kg
$x_1 = 3.0$ m
$x_2 = 6.0$ m
$x_3 = -4.0$ m

Then, simply performing the summation as indicated in Eq. 6.19,

$$X_{CM} = \frac{\Sigma_i m_i x_i}{M}$$

$$= \frac{(2.0 \text{ kg})(3.0 \text{ m}) + (3.0 \text{ kg})(6.0 \text{ m}) + (6.0 \text{ kg})(-4.0 \text{ m})}{2.0 \text{ kg} + 3.0 \text{ kg} + 6.0 \text{ kg}}$$

$$= 0$$

The center of mass is at the origin.

EXAMPLE 6.13 ■ Center of Mass and Frame of Reference

A dumbbell (●Fig. 6.16) has a connecting bar of negligible mass. Find the location of the center of mass (a) if m_1 and m_2 are each 5.0 kg, and (b) if m_1 is 5.0 kg and m_2 is 10.0 kg.

Solution.

Given: (a) $m_1 = m_2 = 5.0$ kg **Find:** (a) (X_{CM}, Y_{CM}) (CM coordinates)
$x_1 = 0.25$ m   (b) (X_{CM}, Y_{CM})
$x_2 = 0.75$ m
$y_1 = y_2 = 0.25$ m
(b) $m_1 = 5.0$ kg
$m_2 = 10.0$ kg

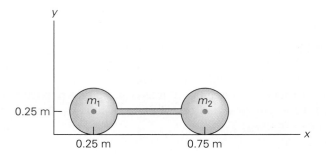

● **FIGURE 6.16 Location of the center of mass** Example 6.14.

Note that each mass is considered to be a particle located at the center of the sphere (its center of mass).

(a) Finding X_{CM} gives

$$X_{CM} = \frac{m_1 x_1 + m_2 x_2}{m_1 + m_2}$$

$$= \frac{(5.0 \text{ kg})(0.25 \text{ m}) + (5.0 \text{ kg})(0.75 \text{ m})}{5.0 \text{ kg} + 5.0 \text{ kg}} = 0.50 \text{ m}$$

Similarly, it is easy to find that $Y_{CM} = 0.25$ m. (You might have seen this right away, since each center of mass is at this height.) The center of mass of the dumbbell is then located at $(X_{CM}, Y_{CM}) = (0.50 \text{ m}, 0.25 \text{ m})$, or midway between the end masses.

(b) With $m_2 = 10.0$ kg,

$$X_{CM} = \frac{m_1 x_1 + m_2 x_2}{m_1 + m_2}$$

$$= \frac{(5.0 \text{ kg})(0.25 \text{ m}) + (10.0 \text{ kg})(0.75 \text{ m})}{5.0 \text{ kg} + 10.0 \text{ kg}} = 0.58 \text{ m}$$

which is $\frac{1}{3}$ of the length of the bar away from m_2. (You might expect the balance point of the dumbbell in this case to be closer to m_2.)

That the location of the center of mass does not depend on the frame of reference can be shown by putting the origin at the point where the 5.0-kg mass touches the x axis. In this case, $x_1 = 0$ and $x_2 = 0.50$ m, and

$$X_{CM} = \frac{(5.0 \text{ kg})(0) + (10.0 \text{ kg})(0.50 \text{ m})}{5.0 \text{ kg} + 10.0 \text{ kg}} = 0.33 \text{ m}$$

The y coordinate of the center of mass is again $Y_{CM} = 0.25$ m, as you can easily prove for yourself.

In Example 6.13, when the value of one of the masses changed, the x coordinate of the center of mass changed. You might have expected the y coordinate to change also. However, the centers of the end masses were still at the same height, and Y_{CM} remained the same. To increase Y_{CM}, one or both of the end masses would have had to rise, which would have required work against gravity and resulted in an increase in potential energy.

As you know, mass and weight are directly related. Closely associated with the center of mass is the **center of gravity (CG)**, the point where all of the weight of an object may be considered to be concentrated in representing the object as a particle. Taking the acceleration due to gravity to be constant, as it is generally done near the Earth's surface, allows Eq. 6.19 to be rewritten as

$$MgX_{CM} = \Sigma_i\, m_i g x_i \qquad (6.20)$$

Then, all of the weight, Mg, is concentrated at X_{CM}, and the center of mass and the center of gravity coincide.* As you may have noticed, the location of the center of gravity was implied in some figures where the vector arrows for weight ($m\mathbf{g}$) were drawn from a point at or near the center of an object.

In some cases, the center of mass or the center of gravity of an object may be located by symmetry. For example, for a spherical object which is homogeneous (the mass is distributed evenly throughout), the center of mass is at the geometrical center (or center of symmetry). In Example 6.14a, where the end masses of the dumbbell were equal, it was probably apparent that the center of mass was midway between them.

The location of the center of mass of an irregularly shaped object is not so evident and is usually difficult to calculate (even with advanced mathematical methods that are beyond the scope of this book). In some instances, the center of mass may be located experimentally. For example, the center of mass of a flat, irregularly shaped object may be determined experimentally by suspending it freely from different points (Fig. 6.17). A moment's thought should convince you that the center of mass (or center of gravity) always lies vertically below the point of suspension. Since the center of mass is defined as the point at which all the mass of a body can be considered to be concentrated, this is analogous to a particle of mass suspended from a string. Suspending the object from two or more points and marking the vertical lines on which the center of mass must lie locates the center of mass as the intersection of the lines.

The center of mass of an object may lie outside the body of the object. Examples are shown in Fig. 6.18. The center of mass of a homogeneous ring is at its center. The mass in any section of the ring is canceled out by the mass in an equivalent section directly across the ring, and by symmetry the center of mass is at the center. For an L-shaped object with equal legs, the center of mass lies on a line that makes a 45° angle with both legs. Its location can easily be determined by suspending the L from a point on one of the legs and noting where a vertical line from that point intersects the diagonal line.

Keep in mind that the location of the center of mass or center of gravity of an object depends on the distribution of mass. Therefore, for a flexible object such as the human body, the position of the center of gravity changes as the object changes configuration (mass distribution). For example, when a person

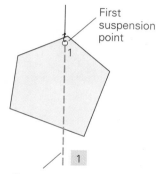

Center of mass lies along this line

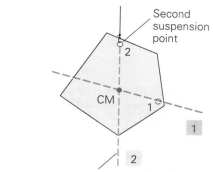

Center of mass also lies along this line

FIGURE 6.17 Location of the center of mass by suspension The center of mass of a flat, irregularly shaped object may be found by suspending the object from two or more points. The CM (and CG) lies on a vertical line under any point of suspension, so the intersection of two such lines marks its location midway through the thickness of the body. The sheet could be balanced horizontally at this point. Why?

*This is a special case to which we will limit our discussion. If the acceleration due to gravity is not constant (Chapter 7), the center of mass and center of gravity in general do not coincide.

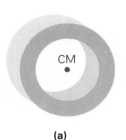

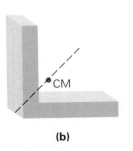

(a)

(b)

● **FIGURE 6.18 The center of mass may be located outside of a body** The center of mass may lie either inside or outside of a body, depending on its mass distribution. **(a)** For a uniform ring, the center of mass is at its center. **(b)** For an L-shaped object, if the mass distribution is uniform and the legs are of equal length, the center of mass lies on the diagonal between the legs.

● **FIGURE 6.19 Center of gravity** By arching her body, this high jumper can get over the bar even though her center of gravity passes beneath it. Work or effort is required to raise the center of gravity and only the jumper's body need clear the bar.

raises both arms overhead, his or her center of gravity is raised several centimeters. For a high jumper going over a bar, the center of gravity lies outside the arched body (● Fig. 6.19). In fact, the center of gravity passes *beneath* the bar. This is done purposefully because work must be done to raise the center of gravity, and only the jumper's body has to clear the bar, not the CG.

6.6 ■ Jet Propulsion and Rockets ◇

The word "jet" is sometimes used to refer to a stream of liquid or gas emitted at a high speed, for example, a jet of water from a fountain or a jet of air from an automobile tire. **Jet propulsion** is the application of such jets to the production of motion. This usually brings to mind jet planes and rockets, but squids and octopuses propel themselves by squirting jets of water. You have probably tried the simple application of blowing up a balloon and releasing it. Lacking any guidance or rigid exhaust system, the balloon zig-zags around, driven by the escaping air. In terms of Newton's third law, the air is forced out by the contraction of the stretched balloon—that is, the balloon exerts a force on the air. Thus there must be an equal and opposite reaction force exerted by the air on the balloon. It is this force that propels the balloon on its erratic path.

Jet propulsion is explained by Newton's third law, and in the absence of external forces, the conservation of momentum also applies. You may understand this better by considering the recoil of a rifle, taking the rifle as an isolated system (● Fig. 6.20). Initially, the total momentum of this system is zero. When

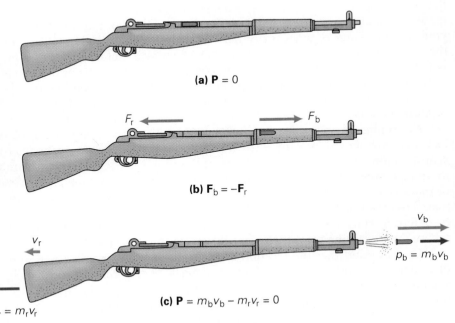

(a) P = 0

(b) F$_b$ = −**F**$_r$

(c) P = $m_b v_b - m_r v_r$ = 0

Figure 6.20

● **FIGURE 6.20 Conservation of momentum** **(a)** Before it is fired, the total momentum of the rifle (as an isolated system) is zero. **(b)** During firing, there are equal and opposite internal forces, and the instantaneous total momentum is zero (neglecting external forces). **(c)** When the bullet leaves the barrel, the total momentum is still zero.

the rifle is fired (by remote control to avoid external forces), the expansion of the gases from the exploding charge accelerates the bullet down the barrel. These gases push backward on the rifle as well, producing a recoil force (the "kick" experienced by the person firing the weapon). Since the initial momentum of the system is zero, and the force of the expanding gas is an internal force, the momenta of the bullet and of the rifle must be equal and opposite at any instant. After the bullet leaves the barrel, there is no propelling force, so the bullet and the rifle move with constant velocities.

Similarly, the thrust for a rocket is created by exhausting the gas from burning fuel out the rear of the rocket. The rocket exerts a force on the exhaust gas, and the gas exerts an equal and opposite force on the rocket. The latter force propels the rocket forward (● Fig. 6.21). If the rocket is at rest when the engines are turned on and there are no external forces (as in space, where friction is zero and gravitational forces are negligible), then the instantaneous momentum of the exhaust gas is equal and opposite to that of the rocket. The numerous exhaust gas molecules have small masses and high velocities, and the rocket has a much larger mass and a smaller velocity. (See Demonstration 4.)

Unlike a rifle firing a single shot, a rocket engine continually loses mass when burning fuel (it is more like a machine gun). Thus, the rocket is a system for which the mass is not constant. As the mass of the rocket decreases, it accelerates more easily. Multistage rockets take advantage of this fact. The hull of a burnt-out stage is jettisoned to give an in-flight reduction in mass (● Fig. 6.22). The payload is actually a very small part of the initial mass of rockets for space flights.

Suppose that the purpose of a space flight is to land a payload on the moon. At some point on the journey, the gravitational attraction of the moon will become greater than that of Earth, and the spacecraft will accelerate toward the moon. A soft landing is desirable, so the spacecraft must be slowed down enough to go into orbit about the moon. This is done by using the rocket engines to apply a **reverse thrust**, or braking thrust. The spacecraft is maneuvered through a 180° angle, or turned around, which is quite easy to do in space. The rocket engines are then fired, expelling the exhaust gas toward the moon and supplying a braking action.

You have experienced a reverse thrust effect if you have flown in a commercial jet. (In this instance, however, the craft is not turned around.) After landing, the jet engines are revved up and a braking action can be felt. Ordinarily, revving up the engines accelerates the plane forward. The reverse thrust is accomplished by placing something in the stream of exhaust gas that deflects the gas forward. The gas experiences an impulse force and a change in momentum in the forward direction (see Fig. 6.3b), and the engine and the aircraft have an equal and opposite momentum change and braking impulse force.

(a)

(b)

● **FIGURE 6.22 Multistage rockets** (a) The space shuttle makes use of a multistage rocket. Both the two booster rockets and the huge external fuel tank are jettisoned in flight. (b) Separation of the first and second stages of a Saturn V rocket after 148 seconds of burn time.

● **FIGURE 6.21 Jet propulsion and mass reduction** A rocket burning fuel is continually losing mass, and so becomes easier to accelerate. The resulting force on (or thrust of) the rocket can be shown to depend on the product of the rate of change of its mass with time and the velocity of the exhaust gases: $(\Delta m/\Delta t)\mathbf{v}_{ex}$. Since the mass is decreasing, $(\Delta m/\Delta t)$ is negative and the thrust is opposite $\mathbf{v}_{ex}$.

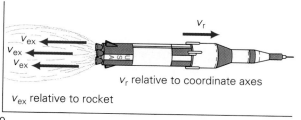

v_r relative to coordinate axes

v_{ex} relative to rocket

0

(a) Ignition and blast off!

(b) Away we go—equal and opposite forces

A demonstration of Newton's third law and conservation of momentum. For a rocket or jet engine, thrust is obtained by burning fuel and exhausting the gas out the rear of the engine. Here a fire extinguisher exhausting CO_2 takes the place of the engine.

Important Concepts

You should be able to define and explain these chapter concepts clearly.

linear momentum	impulse	inelastic collision	center of gravity (CG)
total linear momentum	impulse-momentum	completely inelastic	◇ jet propulsion
conservation of linear	theorem	collision	◇ reverse thrust
momentum	elastic collision	center of mass (CM)	

Important Relationships for Review

These relationships are mathematical statements of the concepts and principles presented in the chapter. You should be able to identify the symbols and to explain the relationships before proceeding to the Exercises. In-text equation reference numbers are given for convenience.

Linear Momentum:

$$\mathbf{p} = m\mathbf{v} \tag{6.1}$$

Total Linear Momentum of a System:

$$\mathbf{P} = \mathbf{p}_1 + \mathbf{p}_2 + \mathbf{p}_3 + \cdots \tag{6.2}$$

Newton's Second Law in Terms of Momentum:

$$\mathbf{F} = \frac{\Delta \mathbf{p}}{\Delta t} \tag{6.3}$$

Impulse-Momentum Theorem:

$$\text{impulse} = \overline{\mathbf{F}} \, \Delta t = \Delta \mathbf{p} = m\mathbf{v} - m\mathbf{v}_0 \tag{6.6}$$

Condition for an Elastic Collision:

$$K_f = K_i \tag{6.8}$$

Condition for an Inelastic Collision:

$$K_f < K_i \tag{6.9}$$

Final Velocities in Head-On Two-Body Collisions:
$(v_{2_0} = 0)$:

$$v_1 = \left(\frac{m_1 - m_2}{m_1 + m_2} \right) v_{1_0} \tag{6.15}$$

$$v_2 = \left(\frac{2m_1}{m_1 + m_2} \right) v_{1_0} \tag{6.16}$$

Coordinate of the Center of Mass:

$$X_{CM} = \frac{\Sigma_i \, m_i x_i}{M} \tag{6.19}$$

Exercises

6.1 ■ Linear Momentum

1 Linear momentum has units of (a) N/m, (b) N-s, (c) kg-m/s, (d) both (b) and (c). (d)

2 Momentum is (a) always conserved, (b) a scalar quantity, (c) a vector quantity that can be resolved into components, (d) unrelated to force. (c)

3 A fan boat of the type used in swampy and marshy areas is shown in ● Fig. 6.23. Explain the principle of its propulsion. see ISM

● **FIGURE 6.23 Fan propulsion** See Exercise 3.

4 ■ Find the magnitude of the momentum of (a) a 0.50-kg ball traveling at 8.0 m/s and (b) a 1500-kg automobile traveling at 100 km/h. (a) 4.0 kg-m/s (b) 4.17 × 10⁴ kg-m/s

● **5** ■ What is the magnitude of the linear momentum of a 7.3-kg bowling ball going down the alley with a speed of 20 m/s? 1.5 × 10² kg-m/s

6 ■ The magnitude of the instantaneous momentum of a runner who is moving at 20.0 km/h is 479 kg-m/s. What is the runner's mass? 86.2 kg

7 ■■ The fastest time for the straight 200-m dash is about 19.8 s. What is the average momentum of a 68-kg runner who finishes the dash in that time? 6.9 × 10² kg-m/s

8 ■ How fast would an automobile having a mass of 1000 kg have to be going to have the same linear momentum as a truck having a mass of 2.00 metric tons and traveling along a straight road with a constant speed of 30.0 km/h? 60.0 km/h = 16.7 m/s

9 ■■ A 0.15-kg baseball traveling with a horizontal speed of 3.40 m/s is hit by a bat and is then moving with a speed of 44.7 m/s in the opposite direction. What is the change in the ball's momentum? 7.2 kg-m/s in direction opposite v_o

● **10** ■■ If two electrons aproach each other with speeds of 3.4 × 10² m/s and 4.5 × 10² m/s, respectively, what is the total momentum of the two-particle system? [*Hint:* Find the mass of an electron in the Table inside the back cover.] 1.0 × 10⁻²⁸ kg-m/s in direction of faster electron

11 ■■ What is the total momentum of a system consisting of two particles, $m_1 = 0.015$ kg and $m_2 = 0.025$ kg, if (a) $v_1 = 8.5$ m/s and $v_2 = 9.0$ m/s in the positive x direction, and (b) $v_1 = -6.4$ m/s and $v_2 = 7.5$ m/s along the y axis? (a) 0.35 kg-m/s x (b) 9.2 × 10⁻² kg-m/s y

12 ■■ A loaded tractor-trailer with a total mass of 5.0 × 10³ kg traveling at 3.0 km/h coasts into a loading dock and collides, coming to a stop in 0.64 s. What is the magnitude of the average force exerted on the truck by the dock? 6.5 × 10³ N

13 ■■ A loaded school bus with a total mass of 1.95 metric tons travels down a straight road with a speed of 60 km/h. The driver sees a speed limit sign reading 45 km/h and slows down to that speed. What is the change in the momentum of the bus? 8.2 × 10³ kg-m/s

14 ■■ Taking the density of air to be 1.29 kg/m³, what is the magnitude of the momentum of a liter of air moving with a wind speed of (a) 30 km/h, and (b) 74 mi/h (the wind speed at which a tropical storm becomes a hurricane). (a) 1.1 × 10⁻² kg-m/s (b) 4.3 × 10⁻² kg-m/s

15 ■■ Two balls of equal mass (0.45 kg) approach the origin along the positive x and y axes at the same speed (3.3 m/s). (a) What is the total momentum of the system? (b) Will the balls necessarily collide at the origin? What is the total momentum of the system after they have both passed through the origin? (a) 2.1 kg-m/s 45° above the +x axis (b) a collision does not have to occur; momentum would be the same

● **16** ■■ A 0.20-kg billiard ball traveling at a speed of 15 m/s strikes the side rail of the table at an angle of 60° (● Fig. 6.24). If the ball rebounds at the same speed and angle, what is the change in its momentum? $\Delta p_x = 0$, $\Delta p_y = -3.0$ kg-m/s

17 ■■■ Suppose that the billiard ball in Fig. 6.24 approaches the rail at a speed of 15 m/s and an angle of 60° as shown, but rebounds at a speed of 10 m/s and an angle of 50°. What is the change in momentum in this case? [*Hint:* Use components.] $\Delta p = 2.8$ kg-m/s x − 1.1 kg-m/s y

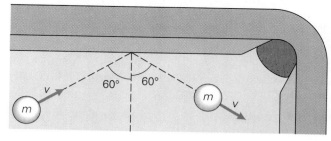

● **FIGURE 6.24 Glancing collision** See Exercises 16 and 17.

18 ■■■ Four particles have masses $m_1 = 10$ g, $m_2 = 15$ g, $m_3 = 13$ g, and $m_4 = 18$ g and velocities $v_1 = (2.0$ m/s)x, $v_2 = (-0.50$ m/s) x, $v_3 = (0.70$ m/s) y, and $v_4 = (-1.9$ m/s) y. What is the total momentum of the four-particle system? p = 1.3 × 10⁻² kg-m/s x −2.5 × 10⁻² kg-m/s y

19 ■■■ At a basketball game, a 120-lb cheerleader is tossed vertically upward with a speed of 2.8 m/s by another cheerleader. (a) What is the cheerleader's change in momentum from the time of release to just before being caught if she is caught at the same height? (b) Would there be any difference if she were caught 0.25 m below the point of release? If so, what is it?   (a) -3.1×10^2 kg-m/s   (b) yes; -3.5×10^2 kg-m/s

6.2 ■ The Conservation of Linear Momentum

20   The momentum of an object is conserved if (a) the force acting on the object is conservative, (b) there is a single, unbalanced internal force acting on the object, (c) the mechanical energy is conserved, (d) none of these.   (d)

● **21**   Internal forces do not affect the conservation of momentum because (a) they cancel each other, (b) their effects are cancelled by external forces, (c) they can never produce a change in velocity, (d) Newton's second law is not applicable to them.   (a)

22   The momenta of the particles of an isolated system can continually change, and the total momentum of the system can remain constant. Explain why this is so.   internal forces do not affect total linear momentum

23   The component of an object's momentum in one direction may be constant, while the component in another direction is not—for example, for a projectile. What does this imply? Is the total momentum conserved?   no, net force is present

24   ■ A 70-kg man and his 40-kg daughter on skates stand together on a frozen lake. If they push apart and the father has a velocity of 0.50 m/s eastward, what is the velocity of the daughter? (Neglect friction.)   0.88 m/s westward

25   ■ An 80-kg hunter jumps to shore from a stationary 35-kg canoe. The hunter has an initial horizontal speed of 0.75 m/s (relative to the shore). At what speed does the canoe initially move away from the shore?   1.7 m/s in opposite direction

● **26**   ■■ A 100-g bullet is fired horizontally into a 14.9-kg block of wood resting on a horizontal surface, and the bullet becomes embedded in the block. If the muzzle velocity of the bullet is 250 m/s, what is the velocity of the block containing the embedded bullet immediately after the impact? (Neglect surface friction and assume an elastic collision.)   1.67 m/s in original direction of bullet

27   ■■ A 70-kg athlete achieves a height of 2.25 m in a standing high jump. Considering the jumper and the Earth as an isolated system, with what speed does the Earth initially move? 7.7×10^{-23} kg-m/s

● **28**   ■■ An isolated 3.0-kg object moves along the y axis between the third and fourth quadrants at a speed of 2.0 m/s. When it reaches the origin, an internal explosion fractures the object into two pieces sending a 1.0-kg fragment in the positive x direction at a speed of 3.7 m/s. What is the velocity of the remainder of the object after the explosion? [*Hint:* Use components.]   3.6 m/s 58° (first quadrant)

29   ■■ A 10-kg cannon ball in flight with a speed of 50 m/s in the $+x$ direction explodes and breaks into three pieces. A 2.5-kg piece goes off at an angle of $+30°$ to the x-axis with a speed of 60 m/s and a 4.5-kg piece goes off at

an angle of $-45°$ to the x axis with a speed of 75 m/s. What is the momentum of the third piece?   p = 1.3×10^2 kg-m/s x + 1.6×10^2 kg-m/s y

30   ■■ A 90-kg astronaut is stranded in space at a point 12 m from his spaceship. In order to get back, he throws a 0.50-kg piece of equipment so that it moves at a speed of 4.0 m/s directly away from the spaceship. With what speed will the astronaut reach the ship? 2.2×10^{-2} m/s

31   ■■ A 65-kg astronaut in an orbiting spacecraft pushes a 2.0-kg instrument package away from herself. If a constant force of 20 N is applied for a time of 0.80 s in doing this, describe mathematically what happens to the astronaut if she is initially floating at rest relative to the spacecraft.   0.25 m/s in direction opposite instrument pushed

32   ■■ A toy gun fired vertically sends a 0.10-kg ball into a 0.050-kg holder sitting on the open end of the barrel. The ball lodges in the holder, and both rise to a height of 0.76 m above the end of the barrel. What was the magnitude of the muzzle velocity of the ball, the velocity just before it hits the holder?   5.9 m/s

☞ **33**   ■■ Identical railroad freight cars hit each other and couple together. In each of the following cases, what are the velocities of the cars immediately after coupling? (a) A moving car approaches a stationary one with a velocity of $+10$ km/h. (b) Two cars approach each other with velocities of $+20$ km/h and -15 km/h, respectively. (c) Two cars travel in the same direction with velocities of $+20$ km/h and $+15$ km/h.   (a) 5.0 km/h in direction of moving car   (b) 2.5 km/h in direction of faster car   (c) 18 km/h in same direction as initial motion

● **34**   ■■ A 1600-kg empty hopper car rolls under a loading bin with a speed of 2.5 m/s, and a 3500-kg load is deposited in the car. What is the magnitude of the velocity of the car immediately after being loaded?   0.78 m/s

☞ **35**   ■■ For a movie scene, an 80-kg stunt man drops from a tree onto a 25-kg sled moving with a velocity of 4.0 m/s toward the shore on a frozen lake. (a) What is the speed of the sled after the stunt man is on board? (b) If the sled hits the bank and stops but the stunt man keeps on going, with what speed does he leave the sled? (Neglect friction.)   (a) 0.95 m/s   (b) 0.95 m/s

36   ■■■ A projectile that is fired from a gun has an initial velocity of 90.0 km/h at an angle of 60.0° to the horizontal. When the projectile is at the top of its trajectory, an internal explosion causes it to separate into two fragments that have equal masses. One of the fragments falls straight downward as though it has been released from rest. How far does the other fragment land from the gun?   82.9 m

37   ■■■ A ballistic pendulum is a device used to measure the velocity of a projectile, for example, the muzzle velocity of a rifle bullet. The projectile is shot horizontally into and becomes embedded in the bob of a pendulum as illustrated in (Fig. 6.25). The pendulum swings upward to some height (h), which is measured. The masses of the block and the bullet are known. Using the laws of momentum and energy, show that the initial velocity of the projectile is given by $v_o = [(m + M)/m]\sqrt{2gh}$. (Neglect rotational considerations.)   see ISM

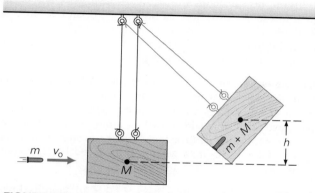

• **FIGURE 6.25 A ballistic pendulum**   See Exercise 37.

6.3 ■ Impulse

6.4 ■ Elastic and Inelastic Collisions

38   Impulse is equal to (a) $F\Delta x$, (b) the change in kinetic energy, (c) the change in momentum, (d) $\Delta p/\Delta t$.   (c)

39   Which of the following is *not* conserved in an inelastic collision: (a) momentum, (b) mass, (c) kinetic energy, (d) total energy?   (c)

• **40**   If an elastic collision mass m_1 strikes a stationary mass m_2, there is a complete transfer of energy if (a) $m_1 = m_2$, (b) $m_1 \gg m_2$, (c) $m_1 \ll m_2$, (d) the masses stick together.   (a)

41   Explain the difference for each of the following pairs of actions in terms of impulse: (a) a golfer's drive and chip shot, (b) a boxer's jab and knock-out punches, (c) a baseball player's bunting action and home-run swing.   see ISM

42   Explain the principle behind (a) using styrofoam as packing to prevent objects from breaking and (b) padding dashboards in cars to prevent injurious bumps.
greater contact time—less force

43   If $K = p^2/2m$, how can kinetic energy be lost in an inelastic collision, yet the total momentum still be conserved? Explain.   momentum is a vector and kinetic energy is a scalar

44   Is it possible for a nonconservative force to act during a collision and still have the total momentum be conserved? Explain.   see ISM

45   A neutron (an electrically neutral subatomic particle) moving at a high speed collides elastically with a stationary (a) carbon atom and (b) uranium atom. Generally describe what happens in each case. [*Hint:* Find the relative masses of these particles in Appendix V.]   uranium would have far less speed due to much larger mass

46   ■ Show that the units for impulse (N-s) are equivalent to those for momentum (kg-m/s).   see ISM

47   ■ When tossed upward and hit horizontally by a batter, a 0.20-kg softball receives an impulse of 10 N-s. With what speed does the ball move away from the bat?   50 m/s

48   ■ An automobile with a linear momentum of 3.2×10^4 kg-m/s is brought to stop in 3.0 s. What is the average braking force? 1.1×10^4 N

49   ■ A pool player imparts with an impulse of 7.0 N-s to a 0.25-kg cue ball with a cue stick. What is the initial speed of the ball?   28 m/s

• **50**   ■■ During a snowball fight, a 0.15-kg snowball traveling at a speed of 8.0 m/s hits a student in the back of the head. (a) What is the impulse? Is this an elastic collision? (b) If the contact time is 0.10 s, what is the average impulse force on the student's head?   (a) -1.2 N-s   (b) -12 N

51   ■■ A basketball with a mass of 0.30 kg is thrown horizontally against a wall with a velocity of 20 m/s. If the ball rebounds with a velocity of 15 m/s, what is the impulse of the collision? (Was the collision elastic?)   11 N-s

52   ■■ For a typical drive, the golf club and the ball are in contact for about 0.50 ms, and the ball leaves the tee with a speed of 70 m/s. What is the average force exerted by the club on the ball? Express your final answer in pounds. (The official weight of a golf ball is 1.620 oz.) 1.4×10^3 lb

53   ■■ A 4.0-kg ball with a velocity of 4.0 m/s has a head-on elastic collision with a stationary 2.0-kg ball. Describe the motions of the balls after the collision. $v_1 = 1.3$ m/s; $v_2 = 5.3$ m/s

• **54**   ■■ A neutron (an electrically neutral subatomic particle) with a mass of 1.67×10^{-27} kg and traveling with a speed of 4.00×10^5 m/s collides elastically head-on with a stationary nucleus with a mass of 6.64×10^{-27} kg. (a) What are the velocities of the neutron and nucleus after collision? (b) What percentage of the neutron's initial kinetic energy is given to the nucleus?   (a) $v_{neutron} = -2.39 \times 10^5$ m/s; $v_{nucleus} = 1.61 \times 10^5$ m/s   (b) 64.4%

55   ■■ A ball with a mass of 100 g is traveling with a velocity of 50 cm/s in the x direction and collides head-on with a 5.0-kg ball that was at rest. Find the velocities of the balls after the collision, assuming that it is elastic. $v_1 = -48$ cm/s; $v_2 = 2.0$ cm/s

56   ■■ For the apparatus in Fig. 6.13, show that one ball swinging in with velocity v_o will not cause two balls to swing out with velocity $v_o/2$.   see ISM

57   ■■ A one-dimensional impulse force acts on a 2.0-kg object as diagrammed in the graph of • Fig. 6.26. Find (a)

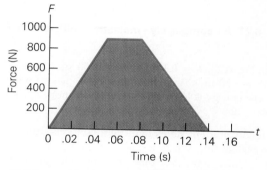

• **FIGURE 6.26 Force versus time graph**   See Exercise 57.

the impulse given to the object, (b) the average force, and (c) the final velocity if the object had an initial velocity of -6.0 m/s.   (a) 77 N-s   (b) 5.5×10^2 N   (c) 33 m/s

58 ■■ A 0.35-kg piece of putty is dropped from a height of 1.5 m above a flat surface. When the putty hits the surface, it comes to rest in 0.10 s. What is the average force exerted on the putty by the surface?   19 N upward

59 ■■ If the billiard ball in Fig. 6.24 is in contact with the rail for 0.01 s, what is the magnitude of the average force exerted on the ball? (See Exercise 16.) 3.0×10^2 N

60 ■■ Suppose that the block and surface in Exercise 26 have a coefficient of kinetic friction of 0.35. After collision, how far will the block with the embedded bullet slide before coming to rest?   0.41 m

61 ■■ Two balls with masses of 2.0 kg and 6.0 kg travel toward each other at speeds of 12 m/s and 4.0 m/s, respectively. If the balls have a head-on, inelastic collision and the 2.0-kg ball recoils with a speed of 8.0 m/s, how much kinetic energy is lost in the collision? $K_{lost} = 1.1 \times 10^2$ J

62 ■■ Two balls with small and large masses (m and M) approach each other as shown in ● Fig. 6.27. If they collide and stick together at the origin, what is the velocity v' of the balls after collision in terms of the other parameters?
$v'_x = mv/(m + M)$; $v'_y = MV/(m + M)$

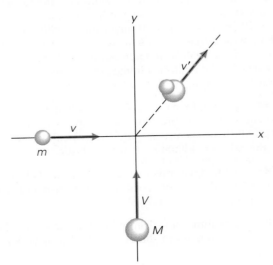

● **FIGURE 6.27 A completely inelastic collision** See Exercise 62.

63 ■■ If in Fig. 6.25, $m = 0.100$ kg, $M = 4.00$ kg, $v_o = 25.0$ m/s, and $V = 3.50$ m/s (immediately after collision) what percent kinetic energy is lost in the collision? (Ignore significant figures.)   19.6%

64 ■■ A fellow student states that the total momentum of a 3-particle system ($m_1 = 0.25$ kg, $m_2 = 0.20$ kg, and $m_3 = 0.33$ kg) is initially zero, and that after an inelastic triple

collision, he calculates the particles have velocities of 4.0 m/s in the $+x$ direction, 6.0 m/s at 120°, and 2.5 m/s at 230°, respectively. Do you agree with his calculations? If not, assuming the first two answers to be correct, what should be the momentum of the third particle? $P \neq P_o$; see ISM

65 ■■ In nuclear reactors, subatomic particles called neutrons are slowed down by allowing them to collide with the atoms of a moderator material, such as carbon. (a) In a head-on, elastic collision with a carbon atom, what percentage energy is lost by a neutron? (b) If the neutron has an initial speed of 1.5×10^7 m/s what will its speed be after collision?   (a) 28%   (b) 1.3×10^7 m/s

● **66** ■■ A freight car with a mass of 2.0×10^4 kg rolls down an inclined track through a vertical distance of 3.3 m. At the bottom of the incline, on a level track, the car collides and couples with an identical freight car that was at rest. What percentage of the initial kinetic energy is lost in the collision?   50%

67 ■■ A gondola car has a mass of 4.5 metric tons when empty. While it is moving at a speed of 2.0 m/s along a level track under a grain elevator, 7.5 metric tons of wheat are loaded into it from directly above. Is this an elastic collision? If not, where did the kinetic energy go?   Since $K_o \neq K$, collision is not elastic; primarily heat

68 ■■ In an elastic, head-on collision with a stationary target particle, a moving particle recoils at $\frac{1}{3}$ of its incident speed. (a) What is the ratio of the particle masses (m_1/m_2)? (b) What is the speed of the target particle after the collision in terms of the initial speed of the incoming particle?
(a) $m_1/m_2 = 1/2$ (b) $v_2 = (2/3)v_o$

69 ■■■ Show that the fraction of kinetic energy lost in the collision in Fig. 6.10b is equal to $m_2/(m_1 + m_2)$.
see ISM

● **70** ■■■ Show that the fraction of kinetic energy lost in a ballistic pendulum collision (as in Fig. 6.25) is equal to $M/(m + M)$.   see ISM

71 ■■■ A 10-g bullet is fired horizontally into and becomes embedded in a suspended block of wood whose mass is 0.890 kg (see Fig. 6.25). (a) What is the magnitude of the velocity of the block with the embedded bullet immediately after the collision in terms of the initial velocity (v_o)? (b) If the block with the embedded bullet swings upward and its center of mass is raised 0.40 m, what was the initial velocity of the bullet? (c) Was the collision elastic? If not, what percentage of the initial kinetic energy was lost?   (a) $v = v_o/90$
(b) 2.5×10^2 m/s   (c) 98.9%

72 ■■■ A moving billiard ball collides with an identical stationary one, and the incoming ball is deflected at an angle of 45° from its original direction. Show that if the collision is elastic, both balls will have the same speed afterward and will move at a right angle (90°) relative to each other.
see ISM

73 ■■■ (a) For an elastic, two-body collision, show that in general $v_2 - v_1 = -(v_{2_o} - v_{1_o})$. That is, the relative speed of recession after the collision is the same as the relative

speed of approach before it. (b) In general, a collision is either completely inelastic, completely elastic, or somewhere in between. The degree of elasticity is sometimes expressed as the *coefficient of restitution* (e), which is defined as the ratio of the relative velocities of recession and approach: $v_2 - v_1 = -e(v_{2_0} - v_{1_0})$. What are the values of e for an elastic collision and a completley inelastic collision?   (a) see ISM
(b) $e_{elastic} = 1.0$; $e_{inelastic} = 0$

74 ■■■ The coefficient of restitution (see Exercise 73) for steel colliding with steel is 0.95. If a steel ball is dropped from a height h_o above a steel plate, to what height will the ball rebound? $h_1 = (0.90)h_o$

6.5 ■ Center of Mass

75 The center of mass of an object (a) always lies at the midpoint of the object, (b) is at the location of the most massive particle in the object, (c) always lies within the object, (d) none of these.   (d)

76 The center of mass and center of gravity of an object coincide if (a) the object is flat, (b) there is a uniform mass distribution, (c) they both lie inside the object, (d) the acceleration due to gravity is constant.   (d)

77 Could the suspension method be used to locate the center of mass of a three-dimensional object? Explain.   yes; hang from three different locations

78 A spacecraft is initially at rest in a space, and then its rocket engines are fired. Describe the motion of the center of mass of the system.   momentum zero, CM does not move

79 ■ The center of mass of a system consisting of two 100-g particles is located at the origin. If one of the particles is at (0.35 m, 0) where is the other?   −0.35 m

80 ■ The centers of a 4.0-kg sphere and a 7.5-kg sphere are separated by a distance of 1.5 m. Where is the center of mass of the system? $x_{CM} = 0.98$ m from less massive sphere

81 ■■ (a) Find the center of mass of the Earth-moon system. [*Hint:* Use data from Appendix I and consider the distance between the two to be measured from their centers.] (b) Where is that center of mass relative to the surface of the Earth?   (a) 4.6 × 10⁶ m from the center of Earth
(b) 1.8 × 10⁶ m below the surface of Earth

82 ■■ Find the center of mass of a system composed of three spherical objects with masses of 3.0 kg, 2.0 kg, and 4.0 kg and centers located at (−6.0 m, 0), (1.0 m, 0), and (3.0 m, 0), respectively.   (−0.44 m, 0)

83 ■■ For the system described in Exercise 82, where would a fourth sphere with a mass of 0.50 kg have to be located for the center of mass of the system to be at the origin?   (8.0 m, 0)

84 ■■ Suppose that the mass of dumbbell bar in Example 6.14 is not negligible, that the bar has a uniform mass of 1.0

kg. How does this affect the location of the center of mass of the dumbbell? (Consider both cases.)   does not change in either case

85 ■■ A uniform, flat piece of metal is shaped like an equilateral triangle with sides that are 30 cm long. What are the coordinates of the center of mass in the xy plane if one apex is at the origin and one side along the y axis?
(8.7 cm, 15 cm)

86 ■■ Three particles, each with a mass of 0.25 kg, are located at (−4.0 m, 0), (2.0 m, 0), and (0, 3.0 m) and are acted on by forces $F_1 = (−3.0 N) y$, $F_2 = (5.0 N) y$, and $F_3 = (4.0 N) x$, respectively. Find the acceleration (magnitude and direction) of the center of mass. [*Hint:* Consider the components of that acceleration.] $a = 5.3$ m/s² x + 2.7 m/s² y

87 ■■■ A system of two masses has a center of mass given by X_{CM_1}. Another system of three masses has a center of mass given by X_{CM_2}. Show that if all five masses are considered to be one system, the center of mass of that combined system is *not* $X_{CM} = X_{CM_1} + X_{CM_2}$.   see ISM

88 ■■■ Two skaters with masses of 90 kg and 60 kg, respectively, stand 8.0 m apart; each holds one end of a piece of rope. (a) If they pull themselves along the rope until they meet, how far does each skater travel? (Neglect friction.) (b) If only the 60-kg skater pulls herself along the rope until she meets her friend on the opposite end (who just holds onto the rope), how far does each skater travel?   see ISM

89 ■■■ A 75-kg man stands in the far end of a 50.0-kg boat 100 m from the shore, as pictured in Fig. 6.28. If he walks to the other end of the 6.00-m boat, how far is the man from the shore? Neglect friction and assume that the center of mass of the boat is at its center. [*Hint:* The answer is *not* 94 m, because the boat moves as the man walks. Why? With no external force, the acceleration of the center of mass of the man-boat system is zero, and thus it is stationary. Calculate the location of the center of mass initially, and then apply to the final configuration.]   97.6 m from the shore

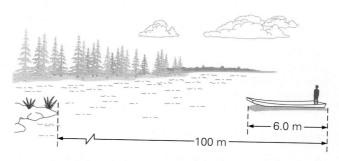

● **FIGURE 6.28 Walking toward shore**   See Exercise 89.

■ **Additional Exercises**

90 Another way of describing an elastic collision between two objects is to say that their relative velocities are the same before and after collision. Apply this statement to the

collision between a tennis racket and ball when the ball is being served (the ball is thrown upward and hit just at its maximum height). Show that the ball leaves the racket with a speed approximately twice that of the initial speed of the racket. see ISM

• **91** A 2.0-kg sphere with a velocity of 6.0 m/s collides head-on and elastically with a stationary 10-kg sphere. Describe the motions of the spheres after the collision.
$v_1 = -4.0$ m/s; $v_2 = 2.0$ m/s

92 Four particles, m_1, m_2, m_3, and m_4, are located at (0, 0), (0, 4.0 m), (4.0 m, 4.0 m), and (4.0 m, 0), respectively. Locate the center of mass of the system (a) if all the masses are equal; (b) if $m_2 = m_4 = 2m_1 = 2m_3$; and (c) if $m_1 = 100$ g, $m_2 = 200$ g, $m_3 = 300$ g, and $m_4 = 400$ g. [*Hint:* Make a sketch of the system.] (a) (2.0 m, 2.0 m) (b) (2.0 m, 2.0 m) (c) (2.8 m, 2.0 m)

93 Two identical billiard balls approach each other at the same speed (2.0 m/s). At what speeds do they rebound after a head-on elastic collision? $v_1 = v_2 = 2.0$ m/s

94 If a 0.25-kg ball is dropped from a height of 50 m, what is the momentum of the ball (a) 1.8 s after being released and (b) just before it hits the ground? (a) −4.5 kg-m/s (b) −7.8 kg-m/s

95 A truck with a mass of 25 metric tons travels at a constant speed of 90 km/h. (a) What is the magnitude of the truck's instantaneous linear momentum? (b) What average force would be required to stop the truck in 4.0 s?
(a) 6.3 × 10⁵ kg-m/s (b) 1.6 × 10⁵ N

96 A 15,000-N automobile travels at a speed of 45 km/h northward along a street, and a 7500-N sports car travels at a speed of 60 km/h eastward along an intersecting street. (a) If neither driver brakes and the cars collide at the intersection and stick together, what will the velocity of the cars be immediately after the collision? (b) How much of the initial kinetic energy will be lost in the collision?
(a) 36 km/h at 56° (b) 50% lost

97 Once in a while, there is a so-called grand alignment of planets; that is, all nine of the (known) planets are located along a straight line running through the Sun as viewed from above. Assuming that there is such an alignment, with all the planets on one side of the Sun, compute the approximate location of the center of mass of the planets' system at that time. (Obtain the necessary data from an introductory astronomy book or other reference.) Where is that center of mass relative to the surface of the Sun? When the planets move out of alignment, what effect does this have on the location of the center of mass? 1.1 × 10¹² m; none

98 A piece of uniform sheet metal measures 25 cm by 25 cm. If a circular piece with a radius of 5.0 cm is cut from the center of the sheet, where is the center of mass of the resulting shape? at center of sheet

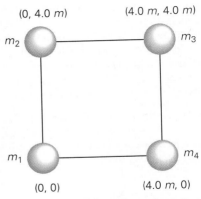

(0, 4.0 m)   (4.0 m, 4.0 m)

m_2 m_3

m_1 m_4

(0, 0)   (4.0 m, 0)

• **FIGURE 6.29 Where's the center of mass?** See Exercise 99.

99 Locate the center of mass of the system shown in Fig. 6.29 (a) if all of the masses are equal; (b) if $m_2 = m_4 = 2m_1 = 2m_3$; (c) if $m_1 = 1.0$ kg, $m_2 = 2.0$ kg, $m_3 = 3.0$ kg, and $m_4 = 4.0$ kg. (Coordinates are in meters.) (a) (2.0 m, 2.0 m) (b) (2.0 m, 2.0 m) (c) (2.8 m, 2.0 m)

100 A 2.5-kg block sliding on a frictionless horizontal surface with a constant velocity of 6.0 m/s approaches a stationary 6.5-kg block. (a) If the blocks have a completely inelastic collision, what is their velocity after it? (b) How much mechanical energy is lost in the completely inelastic collision? (a) 1.7 m/s (b) 32 J

101 A constant force gives a 2.5-kg block that is initially at rest on a horizontal surface a linear speed of 6.0 m/s in 4.5 s. What is the magnitude of the force? 3.3 N

102 An isolated 3.0-kg object initially at rest explodes and splits into three fragments. One fragment has a mass of 0.50 kg and flies off along the negative x axis at a speed of 2.8 m/s, and another has a mass of 1.3 kg and flies off along the negative y axis at a speed of 1.5 m/s. What is the speed and direction of the third fragment? 2.0 m/s 53°

103 Suppose that the 3.0-kg object in Exercise 102 is initially traveling at a speed of 2.5 m/s in the positive x direction. What will the speed and direction of the third fragment be in this case? 7.7 m/s 12°

104 Four coal cars roll from rest down an inclined ramp from a vertical height of 3.50 m onto a straight level track. The first car is inadvertently stopped at the bottom of the ramp. The next car that comes down couples to it, and both cars move off together. Finally, the last two cars come down the ramp and successively catch up to and couple with the already joined cars. Neglecting external forces, what is the final velocity of the four-car system? 6.21 m/s

7

Circular Motion and Gravitation

Circular motion is everywhere, from atoms to galaxies, from flagella to ferris wheels. As shown in the opening time-exposure photograph, the stars appear to follow circular paths. This, of course, is due to the Earth's rotation on its axis. At the same time, the Earth is revolving in a nearly circular orbit about the Sun. In general, we say that an object *rotates when the axis of rotation lies within the body, and that it* revolves *when the axis is outside it. Thus, the Earth rotates on its axis and revolves about the Sun.*

When a body rotates on its axis, all the particles of the body revolve—that is, they move in circular paths about the body's axis of rotation. For example, the particles that make up a car's wheel travel in circles about the wheel's axle and particles in a playing compact disc (CD) rotate about its hub. In fact, as a "particle" on Earth, you are continually in circular motion.

Circular motion is motion in two dimensions, and so can be described by rectangular components as used in Chapter 3. However, it is usually more convenient to describe

207

circular motion in terms of angular quantities that will be introduced in this chapter. This kind of description is particularly useful for rotational motion that will be studied in Chapter 8. Being familiar with the description of circular motion will make the study of rotating rigid bodies much easier.

Gravity plays a large role in determining the motions of the planets, since it supplies the force necessary to maintain their nearly circular orbits. This chapter will consider Newton's law of gravitation, which describes this fundamental force, and will analyze planetary motion in terms of certain basic laws. As you will learn, the orbits of the planets are not perfectly circular, and this fact affects their orbital speeds. Knowledge of circular motion will help you understand the motions of these solar satellites, as well as the motions of Earth satellites, of which there is one natural one (the moon) and many artificial ones.

Chapter 7 includes topics of study in rotational kinematics, Newton's law of gravitation, and Kepler's laws of motion.

7.1 ■ Angular Measure

Motion is described as a time rate of change of position. As you might guess, angular speed and velocity also involve a time rate of change of position, expressed using an angle. Consider a particle traveling in a circular path, as shown in ● Fig. 7.1. At a particular instant, the particle's position (*P*) may be designated by the Cartesian coordinates *x* and *y*. However, the position may also be designated by the polar coordinates *r* and *θ*. The distance *r* extends from the origin, and the angle *θ* is commonly measured counterclockwise from the *x* axis. The transformation equations that relate one set of coordinates to the other are

$$x = r \cos \theta \tag{7.1a}$$

$$y = r \sin \theta \tag{7.1b}$$

Note that *r* is the same for any point on a given circle. As a particle travels in a circle, the value of *r* is constant and only *θ* changes with time. Thus, circular motion can be described using only one polar coordinate (*θ*) that changes with time, instead of two Cartesian coordinates (*x* and *y*).

Analogous to linear distance is **angular distance**, which is given by

$$\Delta\theta = \theta - \theta_0 \tag{7.2}$$

or simply $\Delta\theta = \theta$ when $\theta_0 = 0°$. A unit commonly used to express angular distance (or angles) is the degree (°); there are 360° in one complete circle, or revolution. Each degree is divided into 60 minutes and each minute into 60 seconds (no relationship to time units).

It is important to be able to relate the angular description of circular motion to the orbital or tangential description, that is, to relate the angular distance *θ* to the arc length *s* (● Fig. 7.2a). The arc length is the distance traveled along the circular path, and the angle *θ* is said to be *subtended* by the arc length. A unit that is very convenient for relating the angular distance to the arc length is the **radian (rad)**, which is defined as the angle subtended by an arc length (*s*) that is equal to the radius (*r*). By this definition (see Fig. 7.2b),

$$1 \text{ rad} = 57.3°$$

To get a general relationship between radians and degrees, let's consider the distance around a complete circle (360°). Basically, a radian is the angle

Polar coordinates, *r* and *θ*

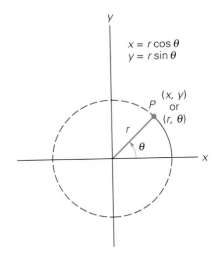

FIGURE 7.1 Polar coordinates
A point, or location, may be described using polar coordinates instead of Cartesian coordinates—that is, as (*r, θ*) instead of (*x, y*). For a circle, *θ* is called the angular distance and *r* is the radius. The two types of coordinates are related by the transformation equations $x = r \cos \theta$ and $y = r \sin \theta$.

subtended when you mark off an arc equal in length to the radius. Since the circumference $c = 2\pi r$, one can do this 2π times along a circle. There are thus 2π radians in a complete circle, or

$$2\pi \text{ rad} = 360°$$

This relationship can be used to obtain convenient conversions of common angles (Table 7.1). Thus, 1 rad = 360°/2π = 57.3°. Notice in the table how the angles in radians are expressed using π explicitly. This is done for convenience.

Similarly, the number of radians subtended by an arbitrary arc length s is equal to the number of radii that will go into s, or the number of radians (θ) = s/r. Thus can write

$$s = r\theta \tag{7.3}$$

which is an important relationship between the arc length s of a circle and its radius r. Notice that since $\theta = s/r$, the angle in radians is the ratio of two lengths. This means that a radian measure is a pure number—that is, it is dimensionless or has no units.

EXAMPLE 7.1 ■ Arc Length

A spectator standing at the center of a circular running track observes a runner start a practice race 256 m due east of her position (Fig. 7.3). The runner runs on the same track to the finish line, which is located due north of the observer's position. (a) What is the distance of the run? (b) What is the distance of one complete lap around the track?

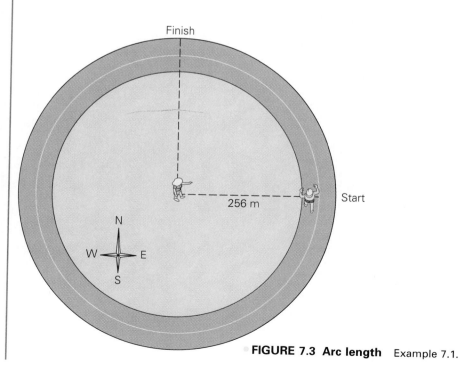

FIGURE 7.3 **Arc length** Example 7.1.

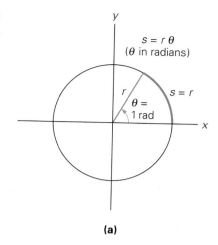

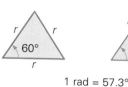

1 rad = 57.3°

(b)

FIGURE 7.2 Radian measure
(a) Angular distance may be measured either in degrees or in radians (rad). An angle θ is subtended by an arc length s. When $s = r$, the angle subtending s is defined to be 1 radian. More generally, $s = r\theta$ with θ in radians.
(b) A radian is equal to 57.3°, and the circular sector marked off by this angle is similar to an equilateral with sides r.

TABLE 7.1 ■ **Equivalent Degree and Radian Measures**

Degrees	Radians
360°	2π
180°	π
90°	$\pi/2$
60°	$\pi/3$
57.3°	1
45°	$\pi/4$
30°	$\pi/6$

Solution. We are given the radius of the track and the angles involved in each case. Listing these and what is to be found, we have

Given: $r = 256$ m
(a) $\theta = 90° = \pi/2$ rad
(b) $\theta = 360° = 2\pi$ rad

Find: (a) s (for 90°)
(b) $s = c$ (circumference)

(a) Using Eq. 7.3 to find the arc length,

$$s = r\theta = (256 \text{ m})\left(\frac{\pi}{2}\right) = 402 \text{ m}$$

Note that rad is omitted, and the equation is dimensionally correct.
(b) Similarly, with $\theta = 2\pi$ rad,

$$s = r\theta = (256 \text{ m})(2\pi) = 1610 \text{ m} = 1.61 \text{ km} \qquad \text{(about a mile)}$$

Warn students not to measure the angular quantities in revolutions or degrees when using the expression $s = r\theta$.

Does it seem odd that the circle is divided into 360 degrees? Ask students if they know why that value was selected.

EXAMPLE 7.2 ■ Angular Distance

A sailor measures the length of a distant tanker as an angular distance of 1° 9′ (1 degree, 9 minutes) with a divided circle, as illustrated in ● Fig. 7.4. He knows from the shipping charts that the tanker is 150 m in length. Approximately how far away is the tanker?

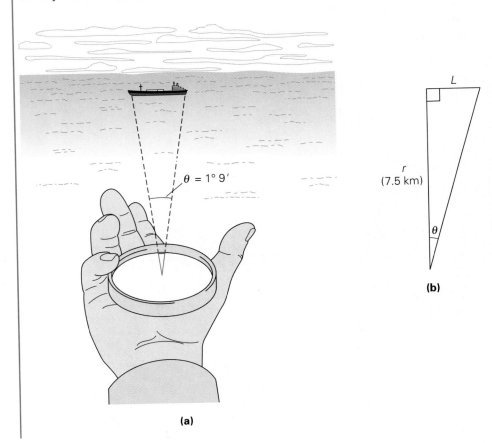

● **FIGURE 7.4 Angular distance** For small angles, the arc length is approximately a straight line, or the chord length. Knowing the length of the tanker, one can find how far away it is by measuring its angular size. Example 7.2.

Solution. To approximate the distance, we take the ship's length to be roughly equal to the arc length subtended by the measured angle. The data is then

> **Given:** $\theta = 1°9'$ **Find:** r (radial distance)
> $s = 150$ m

Eq. 7.3 can be used to find r, but first we convert θ to radians. Changing the minutes to degrees: $9'\left(\dfrac{1°}{60'}\right) = 0.15°$, and $\theta = 1.15°$. Then,

$$\theta = 1.15°\left(\frac{2\pi\,\text{rad}}{360°}\right) = 0.020\,\text{rad}$$

(Note that any of the equivalent relationships in Table 7.1 could be used for the conversion, such as π rad/180° or even 1 rad/57.3°.)
Then,

$$r = \frac{s}{\theta} = \frac{150\,\text{m}}{0.020} = 7500\,\text{m} = 7.5\,\text{km}$$

As pointed out, this is an approximation because s is a *circular* arc length. For small angles, however, the length of the linear chord is very nearly equal to the arc length. To illustrate this, we can consider the known distance to the ship to be part of a right triangle as shown in Fig. 7.4b and find the length of the ship L using trigonometry. We see that $\tan\theta = L/r$ (side opposite over side adjacent), so

$$L = r\tan\theta = (7.5\,\text{km})\tan(0.020\,\text{rad}) = 0.15\,\text{km} = 150\,\text{m}$$

Thus the arc length and chord are equal within the number of significant figures we are working with here.

■ Problem-Solving Hint

In computing trigonometric quantities such as $\tan\theta$ or $\sin\theta$, the angle may be expressed in degrees or radians, that is, $\sin 30° = \sin \pi/6$ rad $= \sin(0.524\,\text{rad}) = 0.500$. When finding trig functions with a calculator, note that there is usually a way to change the angle entry between "deg" and "rad" functions. Calculators commonly are set in the degree function mode, so if you want to find the value of, say, $\sin(1.22\,\text{rad})$, first change to the "rad" mode and enter 1.22, and $\sin(1.22\,\text{rad}) = 0.940$.

7.2 ■ Angular Speed and Velocity

The description of circular motion in angular form is analogous to the description of linear motion. In fact, you'll notice that the equations are almost identical. Different symbols are used to indicate that the quantities have different meanings. The Greek letter omega is used to represent **average angular speed** ($\overline{\omega}$), the angular distance divided by the total time to travel the distance:

$$\overline{\omega} = \frac{\Delta\theta}{\Delta t} \quad \text{(average angular speed)} \quad (7.4)$$

The units of angular speed are rad/s (actually s^{-1} since rad is unitless).

Another common descriptive unit for angular speed is rpm (revolutions per minute); for example, a turntable speed is $33\frac{1}{3}$ rpm. However, this unit is readily converted to rad/s, since 1 revolution $= 2\pi$ rad.

The instantaneous angular speed is given for an extremely small time interval (Δt approaches zero). Also, $\overline{\omega} = \theta/t$ or $\theta = \overline{\omega}t$, where θ_0 and t_0 are taken to be zero. If the angular speed is *constant*, $\overline{\omega} = \omega$, and

$$\boxed{\theta = \omega t} \quad \text{(angular distance, constant angular speed)} \quad (7.5)$$

The **average and instantaneous angular velocities** are analogous to their linear counterparts. Angular velocity is associated with angular displacement. Both are vectors, and so have direction; however, this is specified in a special way. In one-dimensional, or linear, motion, a particle can go only in one direction or the other (+ or −), so the displacement and velocity vectors can have only these two directions. In the angular case, a particle moves one way or the other along its circular path. Thus, the angular displacement and angular velocity vectors of a particle in circular motion can have only two directions, which correspond to going around the circular path with either increasing or decreasing angular distance ($\Delta\theta$). Let's focus on the angular velocity vector, **ω**. (The direction of the angular displacement will be the same. Why?)

The direction of the angular velocity vector is given by a right-hand rule, illustrated in ● Fig. 7.5a. When the fingers of your right hand are curled in the direction of circular motion, your extended thumb points in the direction of **ω**. Note that circular motion can be in only one of two circular *senses*, clockwise or counterclockwise. Plus and minus signs can be used to distinguish circular rotation directions relative to angular velocity vector. It is customary to take a counterclockwise rotation as positive (+) since the measurement of positive angular distance (and displacement) is conventionally done counterclockwise from the positive x axis.

QUESTION: Why not just designate the direction of the angular velocity vector to be either clockwise or counterclockwise?

ANSWER: Clockwise (cw) and counterclockwise (ccw) are really directional senses or indications rather than actual directions. The rotational senses are like right and left. Stand facing another person and ask if some object is on the right or left. (You'll disagree.) Similarly, if you held this book up toward a person facing you and rotated it, would it be rotating cw or ccw?

We can use cw and ccw to indicate rotational "directions" when specified relative to a reference, for example, the positive x axis as in the preceding discussion. Referring to Fig. 7.5, imagine yourself being first on one side of one of the rotating disks and then on the other. Which way is the disk rotating from each vantage point, cw or ccw? Then, apply the right-hand rule on both sides. You should find that the angular velocity vector direction is the same for both locations (because it is referenced to the right hand). Relative to this

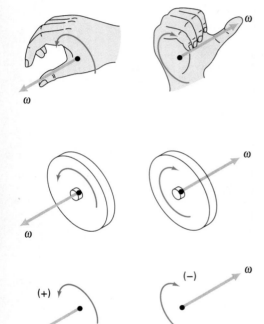

● **FIGURE 7.5 Angular velocity** The direction of the angular velocity vector for an object in rotational motion is given by a right-hand rule: when the fingers of the right hand are curled in the direction of the rotation, the extended thumb points in the direction of the angular velocity vector. Circular senses or directions are commonly indicated by plus and minus signs, as shown in parts (a) and (b), respectively.

vector—for example, looking at the tip—there is no ambiguity in using + and − to indicate rotational senses or directions.

Relationship Between Tangential and Angular Speeds

A particle moving in a circle has an instantaneous velocity tangential to its circular path. For a constant angular velocity and speed, the particle's orbital or **tangential speed** v (the magnitude of the tangential velocity) is also constant (●Fig. 7.6). How the angular and tangential speeds are related is revealed by starting with Eq. 7.3 ($s = r\theta$) and Eq. 7.5 ($\theta = \omega t$),

$$s = r\theta = r(\omega t)$$

The arc length, or distance, is also given by

$$s = vt$$

Combining the equations for s gives

$$\boxed{v = r\omega} \tag{7.6}$$

where ω is in rad/s. Eq. 7.6 holds in general for instantaneous tangential and angular speeds.

Note that all the particles of an object rotating with constant angular velocity have the same angular speed, but the tangential speeds are different at different distances from the axis of rotation (Fig. 7.6 and Demonstration 5).

$v = r\omega$ (ω in rad/s)

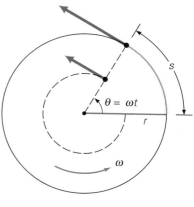

(a)

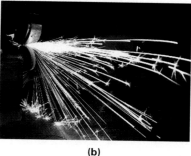

(b)

●**FIGURE 7.6 Tangential and angular speeds** (a) The tangential and angular speeds are related by $v = r\omega$, with ω in rad/s. Note that all of the particles of a rotating object travel in circles. For uniform angular motion, all the particles have the same angular speed (ω), but particles at different distances from the axis of rotation have different tangential speeds. (b) Sparks from a grinding wheel provide a graphic illustration of instantaneous tangential velocity. (Can you explain why the paths curve slightly?)

Figure 7.6

DEMONSTRATION 5 ■ Constant Angular Velocity, but not Tangential Velocity

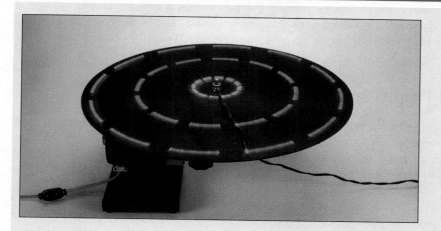

This demonstration of uniform circular motion shows that tangential velocities are different at different radii. Neon bulbs are placed at different radii on a spinning wheel. The bulbs light up every 1/120 second for the same period of time. As seen in the photo, the bulbs at the greater radii have greater tangential speeds since they have longer lighted path lengths for the same time periods.

EXAMPLE 7.3 ■ Different Radii, Different (Tangential) Speeds

On a 45-rpm record, the beginning of the track is 8.0 cm from the center, and the end 5.0 cm from the center. What are (a) the angular speeds and (b) the tangential speeds of a point on the record at these distances when it is spinning at 45 rpm?

Solution

Given: ω = 45 rpm   *Find:* (a) ω_1 and ω_2 (angular speeds)
r_1 = 8.0 cm = 0.080 m   (b) v_1 and v_2 (tangential speeds)
r_2 = 5.0 cm = 0.050 m

(a) It should be apparent that $\omega_1 = \omega_2$; that is, the record as a whole rotates with the same angular speed. So every point on the record has the same angular speed, 45 rpm (rev/min), which can be converted to rad/s:

$$\omega = \left(\frac{45 \text{ rev}}{\text{min}}\right)\left(\frac{2\pi \text{ rad}}{1 \text{ rev}}\right)\left(\frac{1 \text{ min}}{60 \text{ s}}\right) = 4.7 \text{ rad/s}$$

(A convenient conversion factor is 1 rev/min = $\pi/30$ rad/s. Can you see how this is obtained?)

(b) The tangential speeds at different locations on the record are different. All of the particles that make up the record go through one revolution in the same time. Therefore, the farther a point is from the center of the record, the longer its orbital path will be, and its tangential speed must be greater (see Fig. 7.5), as Eq. 7.6 indicates. Thus,

$$v_1 = r_1\omega = (0.080 \text{ m})(4.7 \text{ rad/s}) = 0.38 \text{ m/s}$$

$$v_2 = r_2\omega = (0.050 \text{ m})(4.7 \text{ rad/s}) = 0.24 \text{ m/s}$$

where the unitless rad is included in the angular quantities for clarity.

A point on the outer part of the record has a greater tangential speed than one on the inner part, as you might expect. The outer point must travel faster than the inner point to make one revolution in the same time.

Period and Frequency

Some other quantities commonly used to describe circular motion are period and frequency. The **period** (T) is the time it takes for an object in circular motion to make one complete revolution, or cycle. For example, the period of revolution of the Earth about the Sun is 1 year, and the period of the Earth's axial rotation is 24 hours. The standard unit for the period is the second (s). Descriptively, the period is sometimes given in s/rev or s/cycle.

Closely related to the period is the **frequency** (f), which is the number of revolutions, or cycles, made in a given time, generally a second. For example, if a particle traveling uniformly in a circular orbit makes 5.0 revolutions in 1.0 s, the frequency (of revolution) is f = 5.0 rev/1.0 s = 5.0 rev/s, or 5.0 cycle/s (cps). Revolution and cycle are merely descriptive terms and not part of the frequency unit. The unit of frequency is 1/s, or s^{-1}, which is called the **hertz** (Hz) in the SI.

The hertz (Hz), a unit of frequency, is named for Heinrich Hertz (1857–1894), a German physicist and pioneering investigator of electromagnetic waves, which are also characterized by frequency.

Since the descriptive units for frequency and period are inverses of one another (cycles/s and s/cycle) it follows that the two quantities are related by

$$f = \frac{1}{T}$$ (7.7)

The frequency can also be related to the angular speed. For uniform circular motion, the orbital speed may be written as $v = 2\pi r / T$, which gives the distance traveled in one revolution per the time for one revolution (1 period). Similarly, for the angular case, when a distance of 2π rad is traveled in 1 period,

$$\omega = \frac{2\pi}{T} = 2\pi f$$ (7.8)

EXAMPLE 7.4 ■ Frequency and Period

A phonograph record rotates at a constant speed of $33\frac{1}{3}$ rpm (revolutions per minute). What is the record's (a) frequency and (b) period of rotation?

Solution. The angular speed is not in standard units, and so must be converted. Using the convenient conversion factor from Example 7.3 (1 rev/min = $\pi/30$ rad/s), we can write

Given: $\omega = 33.3$ rpm $\left(\dfrac{\pi/30 \, \text{rad/s}}{\text{rpm}} \right)$ *Find:* (a) f (frequency)
                 $= 3.49$ rad/s   (b) T (period)

(a) Rearranging Eq. 7.8 and solving for f,

$$f = \frac{\omega}{2\pi} = \frac{3.49 \, \text{rad/s}}{2\pi} = 0.555 \, \text{Hz}$$

(The units of 2π are rad/cycle or revolution, so the result is in cycles/s or 1/s, which is hertz.)
(b) Eq. 7.8 could be used to find T, but Eq. 7.7 is a bit simpler,

$$T = \frac{1}{f} = \frac{1}{0.555 \, \text{Hz}} = 1.80 \, \text{s.}$$

Thus, it takes 1.80 s for the record to make one revolution. (Notice that with Hz = 1/s, the equation is dimensionally correct.)

7.3 ■ Uniform Circular Motion and Centripetal Acceleration

A simple, but important, type of circular motion is **uniform circular motion,** which occurs when an object moves at a constant speed in a circular path. An example of this is a car going around a circular track (● Fig. 7.7). The motion of the moon around the Earth and of some electrons around the nucleus of an

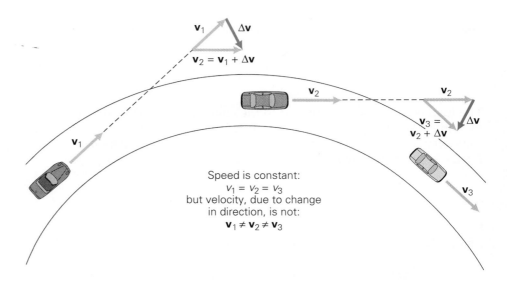

Speed is constant:
$v_1 = v_2 = v_3$
but velocity, due to change
in direction, is not:
$\mathbf{v}_1 \neq \mathbf{v}_2 \neq \mathbf{v}_3$

●FIGURE 7.7 Uniform circular motion The speed of an object in uniform circular motion is constant, but its velocity changes because of changes in the direction of motion. Thus, there is an acceleration.

atom are approximated by uniform circular motion. Such motion is curvilinear, so you know from the discussion in Chapter 3 that there must be an acceleration. But what is its magnitude and direction?

Centripetal Acceleration

Obviously, the acceleration is not in the same direction as the instantaneous velocity (which is tangent to the circular path at any point). If it were, the object would speed up and the motion wouldn't be uniform. Recall that acceleration is

Review the discussion of curvilinear motion in Section 3.1.

the time rate of change of velocity, and velocity has both magnitude and direction. In uniform circular motion, the direction of the velocity is continually changing, which is a clue to the direction of the acceleration. Notice in the vector triangles in Fig. 7.7 that the $\Delta \mathbf{v}$'s generally point inward toward the center of the circular path.

This is shown more explicitly in ●Fig. 7.8. The velocity vectors at the beginning and end of a time interval give the change in velocity, or $\Delta \mathbf{v}$, via vector addition (or subtraction). The vector triangles for several instantaneous velocities are shown in the figure. All of the instantaneous velocity vectors have the same magnitude or length (constant speed), but differ in direction. Note that since $\Delta \mathbf{v}$ is not zero, there must be an acceleration ($\mathbf{a} = \Delta \mathbf{v} / \Delta t$).

Uniform circular motion—constant speed, but not constant velocity

> *The speed is constant but the velocity is changing because of the changing direction.*

As illustrated in the figure, as Δt (or $\Delta \theta$) becomes smaller, $\Delta \mathbf{v}$ points more toward the center of the circular path. As Δt approaches zero, the instantaneous change in the velocity, and therefore the acceleration, is toward the center of the circle. As a result, the acceleration in uniform circular motion is called **centripetal acceleration**, which means center-seeking acceleration (in Latin *centri* means "center" and *petere* means "to move toward").

Centripetal (center-seeking) acceleration

> *Centripetal acceleration is always directed toward the center of the circle.*

Without the centripetal acceleration, the motion would not be in a curved path, but in a straight line. The centripetal acceleration must be directed radially inward, that is, with no component in the direction of the (tangential) velocity, or else the magnitude of that velocity would change (●Fig. 7.9). Note that for

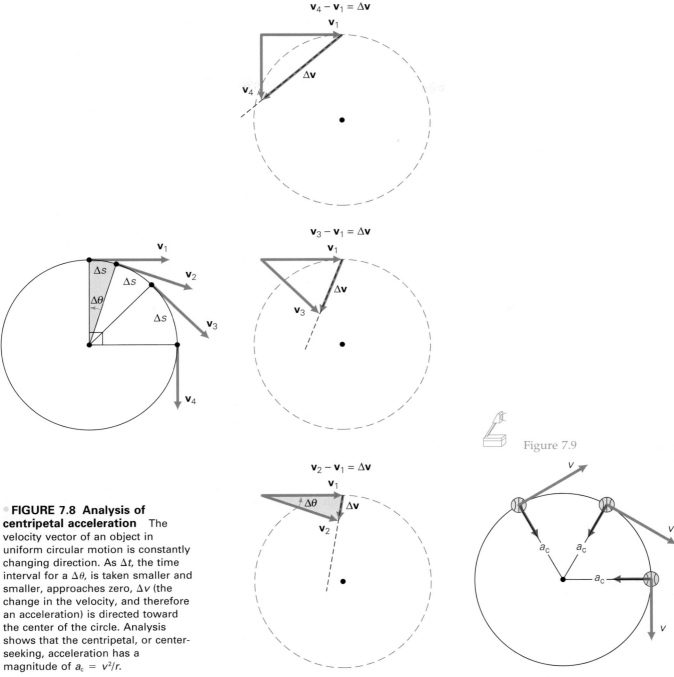

FIGURE 7.8 Analysis of centripetal acceleration The velocity vector of an object in uniform circular motion is constantly changing direction. As Δt, the time interval for a $\Delta\theta$, is taken smaller and smaller, approaches zero, Δv (the change in the velocity, and therefore an acceleration) is directed toward the center of the circle. Analysis shows that the centripetal, or center-seeking, acceleration has a magnitude of $a_c = v^2/r$.

Figure 7.9

FIGURE 7.9 Centripetal acceleration For an object in uniform circular motion, the centripetal acceleration is directed radially inward. There is no acceleration component in the tangential direction, otherwise, the magnitude of the velocity would change.

an object in uniform circular motion, the direction of the centripetal acceleration is continually changing. In terms of x and y components, a_x and a_y are not constant. Can you describe how this differs from projectile motion?

The magnitude of the centripetal acceleration may be deduced from the small shaded triangles in Fig. 7.8. (For very short time intervals, the arc length of Δs is almost a straight line.) These two triangles are similar because each has a pair of equal sides surrounding the same angle $\Delta\theta$. Thus, Δv is to v as Δs is to r, which may be written as

$$\frac{\Delta v}{v} = \frac{\Delta s}{r}$$

For the approximation using short time intervals, $\Delta s \cong v\,\Delta t$, so

and

$$\frac{\Delta v}{v} \cong \frac{v\,\Delta t}{r}$$

$$\frac{\Delta v}{\Delta t} \cong \frac{v^2}{r}$$

Then, as Δt approaches zero, $\Delta v / \Delta t = a_c$, the magnitude of the instantaneous centripetal acceleration, and

Centripetal acceleration depends on tangential speed (v) and radius (r)

$$\boxed{a_c = \frac{v^2}{r}} \qquad \text{(centripetal acceleration)} \qquad (7.9)$$

Video Demonstration 3 relates to this material.

Using Eq. 7.6 ($v = r\omega$), the centripetal acceleration equation may be written in terms of the angular speed:

$$a_c = \frac{v^2}{r} = \frac{(r\omega)^2}{r} = r\omega^2 \qquad (7.10)$$

EXAMPLE 7.5 ■ Centripetal Acceleration

A communications satellite is in a circular orbit about the Earth at an altitude h of 5.0×10^2 km. If the satellite makes one revolution every 90 min, what are its (a) orbital speed and (b) centripetal acceleration?

Solution.

Given: $h = 5.0 \times 10^2$ km   **Find:** (a) v (tangential speed)
$\qquad\quad = 5.0 \times 10^5$ m   (b) $\mathbf{a}_c$ (centripetal acceleration)
$\qquad t = 90$ min $= 5.4 \times 10^3$ s

(a) The radius of the circular orbit is not h, but $R_e + h$, where R_e is the radius of the Earth, 6.4×10^6 m (see Table 6 in Appendix I):

$$r = R_e + h = (6.4 \times 10^6 \text{ m}) + (0.50 \times 10^6 \text{ m}) = 6.9 \times 10^6 \text{ m}$$

The satellite travels the circumference of its circular orbit ($2\pi r$) in the given time, so the tangential speed is

$$v = \frac{2\pi r}{t} = \frac{2\pi(6.9 \times 10^6 \text{ m})}{5.4 \times 10^3 \text{ s}} = 8.0 \times 10^3 \text{ m/s}$$

(b) Then the centripetal acceleration is (Eq. 7.9),

$$a_c = \frac{v^2}{r} = \frac{(8.0 \times 10^3 \text{ m/s})^2}{6.9 \times 10^6 \text{ m}} = 9.3 \text{ m/s}^2$$

directed toward the center of the orbit.

Alternately, we could have computed the angular speed, $\omega = 2\pi/t$ and used Eq. 7.10, $a_c = r\omega^2$.

The concept of centripetal acceleration may be clearer if you look at the displacement components of uniform circular motion for a short time interval, Δt (●Fig. 7.10). The rectangular components in this case are directed tangentially and radially. An object travels a distance $v\Delta t$ tangent to the circle, and at the same time it is displaced a distance $\frac{1}{2}a_c(\Delta t^2)$ radially toward the center of the circle. As Δt approaches zero, a circular path is described. Because of its centripetal acceleration, the moon in a sense is continually falling toward the Earth. Fortunately, it never gets here because it is traveling tangentially at the same time.

Suppose that an automobile is moving into a circular curve. To negotiate the curve, the car must have a centripetal acceleration determined by the equation $a_c = v^2/r$. This acceleration is supplied by the force of friction between the tires and the road. However, this friction has a maximum limiting value. If the speed of the car is high enough, the friction will not be sufficient to supply the necessary centripetal acceleration, and the car will skid outward from the center of the curve. If the car moves onto a wet or icy spot, the friction between the tires and the road may be reduced, allowing the car to skid at an even lower speed. When this occurs, passengers may consider the car to be sliding toward the outside of the curve, but what is really happening is that the car is not being sufficiently accelerated radially toward the center. If the car hit a wet or icy spot and the friction and centripetal acceleration became essentially zero, the car would skid tangentially to the curve in a straight line.

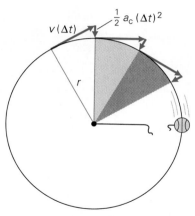

FIGURE 7.10 Components of uniform circular motion In a short time, Δt, an object travels a distance $v\Delta t$ tangentially and a distance $\frac{1}{2}a_c(\Delta t)^2$ radially. Thus, the object is essentially falling toward the center of the circle. As Δt approaches zero, these components describe a circular path. (What happens if the string breaks? See Example 7.6.)

EXAMPLE 7.6 ■ Away We Go!

A ball attached to a string is swung with uniform motion in a horizontal circle above a person's head (●Fig. 7.11a). If the string breaks, which of the trajectories shown in Fig. 7.11b (viewed from above) would the ball follow? *Clearly establish the reasoning and physical principle(s) used in determining your answer before checking it below. That is, **why** did you select your answer?*

Reasoning and Answer. When the string breaks, the centripetal force goes to zero. Newton's first law states that if no force acts on an object in motion, the object will continue to move in a straight line. This rules out (b), (d), and (e). It is evident from Figs. 7.6, 7.8, 7.9, and 7.10 that at any instant (including the instant when the string breaks), the isolated ball has a horizontal, tangential velocity. The downward force of gravity acts on it, but this force affects only its vertical motion, which is not visible in Fig. 7.11b. The ball thus flies off tangentially and is essentially a horizontal projectile (with $v_{x_o} = v$, $v_{y_o} = 0$, and $a_y = g$). Viewed from above, the ball would follow the path labeled (c).

Follow-up Exercise. If you swing a ball in a horizontal circle about your head, can the string be exactly horizontal? (See Fig. 7.11c.) Explain your answer. *Hint:* Analyze the forces acting on the ball. (*Reasoning and answer may be found in the Answers to Exercises section at the back of the book.*)

Centripetal Force

To produce a centripetal acceleration, we must have a **centripetal force**, the magnitude of which in terms of Newton's second law may be expressed as follows:

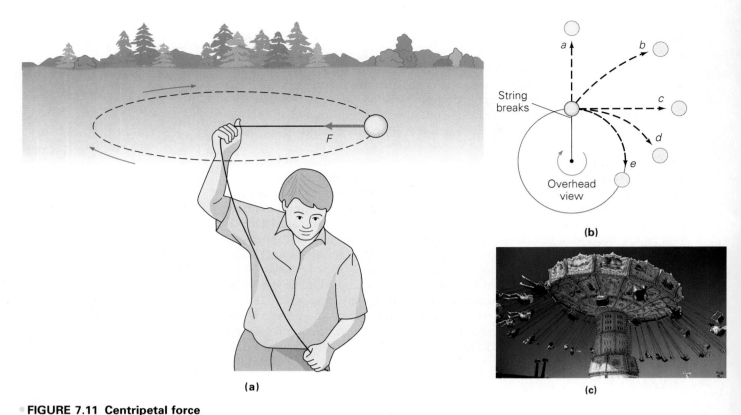

FIGURE 7.11 Centripetal force
(a) A ball is swung in a horizontal circle. (b) If the string breaks, and the centripetal force goes to zero, what happens to the ball? (c) What happens to the passengers on this amusement park ride as the angular velocity increases? See the Follow-up Exercise for Example 7.6.

$$F_c = ma_c = \frac{mv^2}{r} \qquad (7.11)$$

The centripetal force, like the centripetal acceleration, is directed toward the center of the circular path.

Keep in mind that in general a force applied at an angle to the direction of motion of an object produces changes in the magnitude *and* direction of the velocity. However, when a force is continuously applied at an angle of 90° to the direction of motion (as is the case with centripetal force), only the direction of the velocity is changed. Also notice that because it is always perpendicular to the direction of motion, a centripetal force does no work, and so, by the work-energy theorem, does not increase the kinetic energy or speed of the object.

As we have seen in the two preceding Examples, a centripetal force can be provided by gravitational attraction or by a tension force in a string. In other situations, it could be a frictional force, as ● Fig. 7.12 and the following Example show.

Video Demonstration 4 relates to this material.

EXAMPLE 7.7 ■ Frictional Centripetal Force

A car approaches a level, circular curve with a radius of 45.0 m. If the concrete pavement is dry, what is the maximum speed at which the car can negotiate the curve at a constant speed?

Solution. First writing down what is given and what is to be found, we have

> *Given:* $r = 45.0$ m   *Find:* v (maximum speed)

To go around the curve at a particular speed, the car must have a centripetal acceleration and therefore a centripetal force must act on it. As mentioned previously, this force is supplied by static friction between the tires and the road. Recall from Chapter 4 that the maximum frictional force is given by $f_{s_{max}} = \mu_s N$, where N is equal to the weight of the car, mg, on the level road. We may set this equal to the expression for centripetal force ($F_c = mv^2/r$) to find the maximum speed. To find $f_{s_{max}}$ we will need the coefficient of friction between rubber and concrete, and from Table 4.1, $\mu_s = 1.20$. Then, we can write

$$f_{s_{max}} = F_c$$

$$\mu_s N = \mu_s mg = \frac{mv^2}{r}$$

So

$$v = \sqrt{\mu_s rg}$$
$$= \sqrt{(1.20)(45.0 \text{ m})(9.80 \text{ m/s}^2)} = 23.0 \text{ m/s}$$

● **FIGURE 7.12 Frictional centripetal force** The centripetal force necessary for the skaters to round the curve is supplied by frictional force. If there were no friction between the skates and the ice, what would happen?

(about 83 km/h or 52 mi/h).

Note that the speed is independent of the mass (the m's cancel out), so the equation holds for any vehicle. (Is the centripetal force the same for all vehicles?)

The proper safe speed for a highway curve is an important consideration. The coefficient of friction between tires and road may vary, depending on weather, road conditions, the design of the tires, the amount of tread wear, and so on. In designing a curved road, safety may be promoted by banking or inclining the roadway. This reduces the chances of skidding because the normal force exerted on the car by the road then has a component toward the center of the curve that reduces the need for friction. In fact, for a circular curve with a given banking angle and radius, there is one speed for which no friction is required at all (see Exercise 44).

Let's look at one more centripetal force example with two objects in uniform circular motion. It will help you understand the motions of satellites in circular orbits in a later section.

EXAMPLE 7.8 ■ Newton's Second Law and Centripetal Force

Suppose that two masses, $m_1 = 2.5$ kg and $m_2 = 3.5$ kg, are in uniform circular motion on a frictionless surface as illustrated in ● Fig. 7.13, where $r_1 = 1.0$ m and $r_2 = 1.3$ m. The forces acting on the masses are $F = 4.5$ N and $T = 2.9$ N. Find (a) the centripetal accelerations and (b) the magnitudes of the tangential velocities (tangential speeds) of the masses.

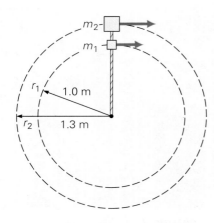

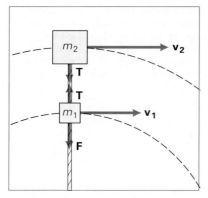

● **FIGURE 7.13 Newton's second law and centripetal force** Example 7.8.

Solution

Given: $r_1 = 1.0$ m and $r_2 = 1.3$ m $\quad$ *Find:* (a) $\mathbf{a}_{c_1}$ and $\mathbf{a}_{c_2}$ (centripetal
$\quad m_1 = 2.5$ kg and $m_2 = 3.5$ kg $\qquad\qquad$ accelerations)
$\quad F = 4.5$ N $\qquad\qquad\qquad\qquad$ (b) v_1 and v_2
$\quad T = 2.9$ N

By isolating m_2 in the figure, you can see that the centripetal force is provided by the tension in the string (**T** is the only force acting on m_2 toward the center of its circular path). Thus,

$$T = m_2 a_{c_2}$$

and

$$a_{c_2} = \frac{T}{m_2} = \frac{2.9 \text{ N}}{3.5 \text{ kg}} = 0.83 \text{ m/s}^2$$

The acceleration is toward the center of the circle. You can find the tangential speed of m_2 from $a_c = v^2/r$:

$$v_2 = \sqrt{a_{c_2} r_2} = \sqrt{(0.83 \text{ m/s}^2)(1.3 \text{ m})} = 1.0 \text{ m/s}$$

The situation is a bit different for m_1. In this case, the centripetal force is *not* simply **F**, because Newton's second law requires that you use the net force. The tension force **T** also acts radially on m_1, and

$$F_{\text{net}} = F - T = m_1 a_{c_1} = \frac{m_1 v_1^2}{r_1}$$

where the radial direction (toward the center of the circular path) is taken to be positive. Then

$$a_{c_1} = \frac{F - T}{m_1} = \frac{4.5 \text{ N} - 2.9 \text{ N}}{2.5 \text{ kg}} = 0.64 \text{ m/s}^2$$

and

$$v_1 = \sqrt{a_{c_1} r_1} = \sqrt{(0.64 \text{ m/s}^2)(1.0 \text{ m})} = 0.80 \text{ m/s}$$

Note that the centripetal accelerations are different; that of m_2 (the outer mass) is greater. This is to be expected since it must travel a greater radial distance to maintain a circular orbit. Also note that v_2 is greater than v_1. This is to be expected as well, since m_2 must travel a longer distance than m_1 in the same time (the circular path of m_2 has a larger radius than that of m_1).

7.4 ■ Angular Acceleration

As you might have guessed, another type of acceleration in angular motion is angular acceleration. This is the time rate of change of angular velocity. In the case of circular motion, if there were an angular acceleration, the motion would not be uniform because the speed would be changing. Analogous to the linear case, the magnitude of the **average angular acceleration** ($\overline{\alpha}$) is

$$\overline{\alpha} = \frac{\Delta \omega}{\Delta t}$$

where the bar over the alpha indicates that it is an average. With $t_o = 0$ and the angular acceleration constant, we have

or

$$\alpha = \frac{\omega - \omega_o}{t}$$

$$\boxed{\omega = \omega_o + \alpha t} \qquad \text{(7.12)}$$

The standard units for angular acceleration are radians per second squared (rad/s^2).

No boldface symbols are used in Eq. 7.12, since in general plus and minus signs will be used to indicate angular directions. As with linear motion, if the angular acceleration increases the angular velocity, both quantities have the same sign, meaning that their vector directions are the same (α is in the same direction as ω as given by the right-hand rule). If the angular acceleration decreases the angular velocity, the two quantities have opposite signs, meaning that their vector directions oppose one another (α is in the direction opposite to ω as given by the right-hand rule, or is an angular deceleration, so to speak).

EXAMPLE 7.9 ■ An Accelerating Record

A $33\frac{1}{3}$-rpm record drops onto a turntable, and accelerates uniformly to its operating speed in 0.42 s. (a) What is the angular acceleration of the record during this time? (b) What is the angular acceleration after this time?

Solution

Given: $\omega_o = 0$
$\qquad \omega = 33.3 \text{ rpm} = 3.49 \text{ rad/s}$
$\qquad t = 0.42 \text{ s}$

Find: (a) α when $t < 0.42$ s
$\qquad$ (b) α when $t > 0.42$ s

(a) Using Eq. 7.12,

$$\alpha = \frac{\omega - \omega_o}{t} = \frac{3.49 \text{ rad/s} - 0}{0.42 \text{ s}} = 8.3 \text{ rad/s}^2$$

in the direction of the angular velocity. [Record turntables rotate clockwise when viewed from above. Can you specify the vector directions of ω and α?]
(b) After the turntable reaches its operating speed of $33\frac{1}{3}$ rpm, the angular speed remains constant at that value, so $\alpha = 0$.

Similar to arc length and angle ($s = r\theta$) and tangential and angular speeds ($v = r\omega$), there is a relationship between the tangential acceleration and the angular acceleration. The **tangential acceleration** is associated with the tangential velocity, and hence continuously changes direction. The magnitudes of the tangential and angular accelerations are related by a factor of r:

$$\boxed{a_t = r\alpha} \qquad \text{(7.13)}$$

The tangential acceleration (a_t) is written with a subscript t to distinguish it from

Students often confuse tangential and centripetal acceleration. It may be helpful to discuss the end of chapter question: Is it possible for a particle in circular motion to have a constant tangential acceleration and a constant centripetal acceleration at the same time?

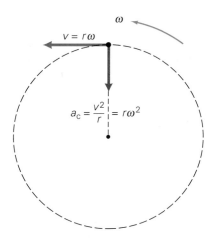

(a) Uniform circular motion
$(a_t = r\alpha = 0)$

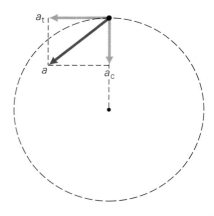

(b) Nonuniform circular motion
$(\mathbf{a} = \mathbf{a_t} + \mathbf{a_c})$

FIGURE 7.14 Acceleration and circular motion (a) In uniform circular motion, there is centripetal acceleration but no angular acceleration ($\alpha = 0$) or tangential acceleration ($a_t = r\alpha = 0$). **(b)** In nonuniform circular motion, there are angular and tangential accelerations, and the total acceleration is the vector sum of the tangential and centripetal components.

the radial, or *centripetal, acceleration* (a_c), which is necessary for circular motion.

For uniform circular motion, there is no angular acceleration ($\alpha = 0$) or tangential acceleration, as can be seen from Eq. 7.13. There is only centripetal acceleration (Fig. 7.14 a).

However, when there is an angular acceleration α (and therefore a tangential acceleration, $a_t = r\alpha$), there is a change in *both* the angular and the tangential velocities ($v = r\omega$). As a result, the centripetal acceleration, $a_c = v^2/r = r\omega^2$, must increase or decrease if the object is to maintain the same circular orbit (if r is to stay the same). When there are both tangential and centripetal accelerations, the total instantaneous acceleration is their vector sum (Fig. 7.14b). The tangential acceleration vector and the centripetal acceleration vector are perpendicular to each other at any instant.

EXAMPLE 7.10 ■ Tangential and Centripetal Accelerations

An object travels with uniform circular motion at an orbital speed of 3.0 m/s and a radius of 1.5 m. The object then experiences an angular acceleration of 0.40 rad/s² for 2.0 s and remains in the same circular orbit. (a) By what factor does the centripetal acceleration increase during the 2.0-s interval? (b) What is the total acceleration at $t = 1.0$ s?

Solution. Listing the data, we have

Given: $v_o = 3.0$ m/s   **Find:** (a) a_c/a_{c_o} (increase in centripetal
$r = 1.5$ m   acceleration)
$\alpha = 0.40$ rad/s²   (b) **a** (total acceleration)
(a) $t = 2.0$ s
(b) $t = 1.0$ s

(a) The magnitude of the initial centripetal acceleration is

$$a_{c_o} = \frac{v_o^2}{r} = \frac{(3.0 \text{ m/s})^2}{1.5 \text{ m}} = 6.0 \text{ m/s}^2$$

and the initial angular speed is

$$\omega_o = \frac{v_o}{r} = \frac{3.0 \text{ m/s}}{1.5 \text{ m}} = 2.0 \text{ rad/s}$$

Thus, by Eq. 7.12, in 2.0 s, the angular speed increases to

$$\omega = \omega_o + \alpha t = 2.0 \text{ rad/s} + (0.40 \text{ rad/s}^2)(2.0 \text{ s}) = 2.8 \text{ rad/s}$$

and the centripetal acceleration increases to

$$a_c = r\omega^2 = (1.5 \text{ m})(2.8 \text{ rad/s}^2)^2 = 12 \text{ m/s}^2$$

The centripetal acceleration has doubled, which means that the centripetal force acting on the object must have doubled; otherwise, the object would spiral outward. [The factor of change could also have been found from the ratio $a_c/a_{c_o} = (v/v_o)^2 = (\omega/\omega_o)^2 = (2.8/2.0)^2 = 2.0$.]
(b) At $t = 1.0$ s, by Eq. 7.12, $\omega = 2.4$ rad/s, so

$$a_c = r\omega^2 = (1.5 \text{ m})(2.4 \text{ rad/s})^2 = 8.6 \text{ m/s}^2$$

The tangential acceleration is

$$a_t = r\alpha = (1.5 \text{ m})(0.40 \text{ rad/s}^2) = 0.60 \text{ m/s}^2$$

Then, at that instant ($t = 1.0$ s), $\mathbf{a} = a_t\mathbf{t} + a_c\mathbf{r}$, where $\mathbf{t}$ and $\mathbf{r}$ are unit vectors directed tangentially and radially inward, respectively. You should be able to find the magnitude of $\mathbf{a}$ and the angle it makes relative to a_t using trigonometry (Fig. 7.14b). However, to find the actual direction of the vector at a given time, you would need to know or be given where the object was at $t_o = 0$.

By now, the direct correspondence between the linear and angular kinematic equations should be apparent to you. The other angular equations can be derived, as was done for the linear ones in Chapter 2. That development will not be shown; the set of angular equations with their linear counterparts for constant accelerations are listed in Table 7.2. A quick review of Chapter 2 (with a change of symbols) will show you how the angular equations are derived.

TABLE 7.2 ■ Equations for Linear and Angular Motion with Constant Acceleration*

Linear	Angular	
$x = \bar{v}t$	$\theta = \bar{\omega}t$	(1)
$\bar{v} = \dfrac{v + v_o}{2}$	$\bar{\omega} = \dfrac{\omega + \omega_o}{2}$	(2)
$v = v_o + at$	$\omega = \omega_o + \alpha t$	(3)
$x = v_o t + \frac{1}{2}at^2$	$\theta = \omega_o t + \frac{1}{2}\alpha t^2$	(4)
$v^2 = v_o^2 + 2ax$	$\omega^2 = \omega_o^2 + 2\alpha\theta$	(5)

*For these equations, $x_o = 0$, $\theta_o = 0$, and $t_o = 0$. The first equation in each column is general, that is, is not limited to situations where the acceleration is constant.

EXAMPLE 7.11 ■ Rotational Kinematics

A child on a merry-go-round selects a horse located 5.5 m from the central axis. When the ride begins, the merry-go-round accelerates at a rate of 0.069 rad/s² for 12 s and then maintains a constant angular speed. (a) How many revolutions does the child make before the constant operating speed is reached? (b) What is the constant angular speed of the ride and the magnitude of the child's tangential velocity? (c) Are the angular speed and the tangential speed the same for all the riders on the merry-go-round?

Solution

Given: $r = 5.5$ m
$\alpha = 0.069$ rad/s²
$t = 12$ s

Find: (a) θ (in revolutions)
(b) ω and v (angular and tangential speeds)
(c) whether ω and v are the same for all values of r

(a) Using Eq. 4 from Table 7.2 gives

$$\theta = \omega_o t + \tfrac{1}{2}\alpha t^2 = 0 + \tfrac{1}{2}(0.069 \text{ rad/s}^2)(12 \text{ s})^2 = 5.0 \text{ rad}$$

Since 2π rad = 1 rev (or 360°),

$$(5.0 \text{ rad})\left(\frac{1 \text{ rev}}{2\pi \text{ rad}}\right) = 0.80 \text{ rev}$$

(b) Using Eq. 7.12 (Eq. 3 in Table 7.2) gives

$$\omega = \omega_o + \alpha t = 0 + (0.069 \text{ rad/s}^2)(12 \text{ s}) = 0.83 \text{ rad/s}$$

Then, by Eq. 7.6,

$$v = r\omega = (5.5 \text{ m})(0.83 \text{ rad/s}) = 4.6 \text{ m/s}$$

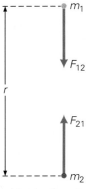

(a) Point masses

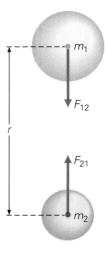

(b) Homogeneous spheres

$$F = \frac{Gm_1 m_2}{r^2}$$

• **FIGURE 7.15 Universal law of gravitation** **(a)** Any two particles or point masses are gravitationally attracted to each other with a force that has a magnitude given by Newton's universal law of gravitation. **(b)** For homogeneous spheres, the masses may be considered to be concentrated at their centers.

Universal gravitational constant— "big G"

(c) The angular speed (ω) is the same for each rider on the merry-go-round. The tangential speed (v) varies, depending on the rider's radial distance from the center of the merry-go-round ($v = r\omega$). All riders at the same distance from the central axis (on a circle with the same radius) have the same tangential speed.

7.5 ■ Newton's Law of Gravitation

Another of Isaac Newton's many accomplishments was the formulation of what is called the **universal law of gravitation.** This law is very powerful and fundamental. Without it, for example, we would not understand the cause of tides or know how to put satellites into particular orbits around the Earth. This law allows us to analyze the motions of planets and comets, stars and galaxies. The word "universal" in the name indicates that we believe it to apply every-where in the universe. (This term highlights the importance of the law, but for brevity it is common to refer simply to Newton's law of gravitation.)

Newton's law of gravitation in mathematical form gives a simple relation-ship for the gravitational interaction between two particles, or point masses, m_1 and m_2, separated by a distance r (• Fig. 7.15). Basically, every particle in the universe has an attractive interaction with every other particle. The forces of mutual interaction are equal and opposite, a force pair as described by Newton's third law (Chapter 4).

The gravitational attraction or force (F) decreases as the distance (r) between two point masses increases; that is, the magnitude of the gravitational force and the distance separating the two particles are related as follows:

$$F \propto \frac{1}{r^2} \tag{7.14}$$

(This type of relationship is called an *inverse square law.*)

Newton's law also correctly postulates that the gravitational force or at-traction of a body depends on its mass—the greater the mass, the greater the attraction. However, since gravity is a mutual interaction between masses, it should be directly proportional to both masses—i.e., to their product ($F \propto m_1m_2$).

Hence, Newton's law of gravitation has the form $F \propto m_1m_2/r^2$. Expressed as an equation, the magnitude of the mutually attractive gravitational forces between two masses is given by

$$F = \frac{Gm_1m_2}{r^2} \tag{7.15}$$

where G is a constant called the **universal gravitational constant:**

$$G = 6.67 \times 10^{-11} \text{ N-m}^2/\text{kg}^2$$

This constant is often referred to as "big G" to distinguish it from "little g," the acceleration due to gravity. Note from Eq. 7.15 that F approaches zero only when r is infinitely large. Thus, the gravitational force has, or acts over, an infinite range.

You might wonder how Newton came to his conclusions about the force

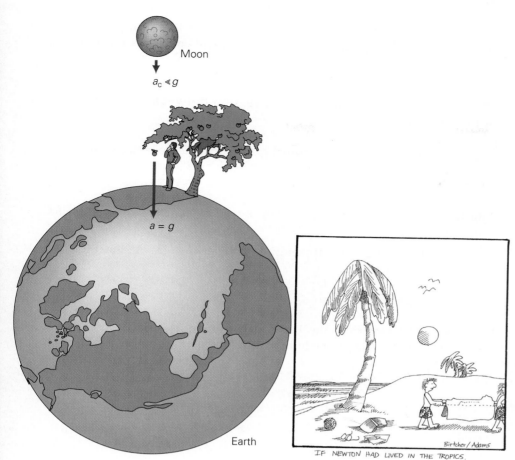

Moon

$a_c \ll g$

$a = g$

Earth

(a)

IF NEWTON HAD LIVED IN THE TROPICS.

Birtcher/Adams

(b)

●**FIGURE 7.16 Gravitational insight?** **(a)** Newton developed his law of gravitation while studying the orbital motion of the moon. According to legend, his thinking was spurred when he observed an apple falling from a tree. He supposedly wondered whether the force causing the apple to accelerate toward the ground could extend to the moon and cause it to fall toward Earth, that is, supply its orbital centripetal acceleration. **(b)** A popular version of the Newton legend—that he was actually hit on the head by a falling apple while sitting under the tree—inspired this cartoon.

of gravity. Legend has it that his insight came after he observed an apple fall from a tree to the ground. Newton had been wondering what supplied the centripetal force to keep the moon in orbit and might have had this thought: "If gravity attracts an apple toward the Earth, perhaps it also attracts the moon, and the moon is falling, or accelerating toward the Earth, under the influence of gravity" (●see Fig. 7.16).

Whether or not the legendary apple did the trick, Newton assumed that the moon and the Earth were attracted to each other and could be treated as point masses, with their total masses concentrated at their centers. The inverse square relationship had been speculated on by some of his contemporaries. Newton's achievement was demonstrating that that relationship could be deduced using one of Johannes Kepler's laws of planetary motion (see Section 7.6).

Newton expressed Eq. 7.14 as a proportion ($F \propto m_1 m_2 / r^2$) because he did not know the value of G. It was not until 1798 (71 years after Newton's death) that the value of the universal gravitational constant was experimentally determined by an English physicist, Henry Cavendish. He used a sensitive balance to measure the gravitational force between separated spherical masses. Knowing F, r, and the m's from measurement, big G can be computed from Eq. 7.15.

As mentioned earlier, Newton considered the nearly spherical Earth and moon to be point masses located at their respective centers. It took him some years, using mathematical methods he developed, to prove that this is the

(a)

(b)

• FIGURE 7.17 Uniform spherical masses (a) Gravity acts between any two (or more) particles. The resultant gravitational force exerted on an object outside a homogeneous sphere by two particles at symmetric locations within the sphere is directed toward the center of the sphere. (b) Because of the sphere's symmetry and uniform mass distribution, the net effect is as though all the mass of the sphere were concentrated as a particle at its center. For this special case, the gravitational center of force and center of mass coincide, but this is not generally true for other objects.

Remind students that r is measured from the Earth's center, not the surface.

case for spherical, *homogeneous* objects.* The general concept is illustrated in • Fig. 7.17.

EXAMPLE 7.12 ■ Gravitational Attraction between the Earth and the Moon

Estimate the magnitude of the mutual gravitational force between the Earth and the moon. (You can assume that the Earth and the moon are homogeneous spheres.)

Solution. No data are given, so they must be available from reference sources.

Given: (from Table 1 in Appendix I)   *Find:* F (gravitational force)
$$M_e = 6.0 \times 10^{24} \text{ kg}$$
$$m_m = 7.4 \times 10^{22} \text{ kg}$$
$$r_{em} = 3.8 \times 10^8 \text{ m}$$

The average distance from the Earth to the moon (r_{em}) is taken to be the distance from the center of one to the center of the other, and using Eq. 7.14,

$$F = \frac{Gm_1m_2}{r^2} = \frac{GM_e m_m}{r_{em}^2}$$

$$= \frac{(6.67 \times 10^{-11} \text{ N-m}^2/\text{kg}^2)(6.0 \times 10^{24} \text{ kg})(7.4 \times 10^{22} \text{ kg})}{(3.8 \times 10^8 \text{ m})^2}$$

$$= 2.1 \times 10^{20} \text{ N}$$

This is the magnitude of the centripetal force that keeps the moon revolving in its orbit around the Earth. It is a very large force, but the moon is a very massive object, with a correspondingly large inertia to overcome.

An interesting relationship that is revealed when Newton's second law and gravitational attraction are used to measure mass is explained in the Insight feature. This feature shows how things are questioned in science.

The acceleration due to gravity can also be investigated using Newton's second law of motion and his law of gravitation. In general, the magnitude of g is found by setting mg, an object's weight, equal to the gravitational attraction on the surface of a spherical mass M with radius r:

$$mg = \frac{GmM}{r^2}$$

Then,

$$g(r) = \frac{GM}{r^2} \tag{7.16}$$

The expression $g(r)$ is read "g as a function of r." It is understood that $g(r)$ is the magnitude of a vector directed toward the center of the spherical mass M.

The above equation can be applied to the moon or any planet, but taking

---

*For a homogeneous sphere, the equivalent point mass is located at the center of mass. However, this is a special case. The center of gravitational force and the center of mass of a configuration of particles or an object do not generally coincide (see Exercise 80).

M_e as the mass of the Earth and R_e as its radius gives g at the Earth's surface:

$$g = \frac{GM_e}{R_e^2}$$

(7.17)

This equation has several interesting implications. First, it reveals that taking g to be constant everywhere on the surface of the Earth involves assuming that the Earth has a homogeneous mass distribution and that the distance from the center of the Earth to any location on its surface is the same. Since these two assumptions are not true, taking g to be a constant is only an approximation, but one that works pretty well for most situations.

Also, you can see why the acceleration due to gravity is the same for all free-falling objects, that is, independent of the mass of the object. This mass doesn't appear in Eq. 7.17, so all objects in free fall accelerate at the same rate.

Finally, if you're observant, you'll notice that Eq. 7.17 can be used to compute the mass of the Earth. All of the other quantities in the equation are measurable and their values are known, so M_e can readily be calculated. This is what Cavendish did after he determined the value of G experimentally. Then he also found the average density of the Earth.

■ **Problem-Solving Hint**

When comparing accelerations due to gravity or gravitational forces, you will often find it convenient to work with ratios. For example, comparing $g(r)$ to g (Eqs. 7.16 and 7.17) for the Earth gives

$$\frac{g(r)}{g} = \frac{GM_e/r^2}{GM_e/R_e^2} = \frac{R_e^2}{r^2} = \left(\frac{R_e}{r}\right)^2 \quad \text{or} \quad \frac{g(r)}{g} = \left(\frac{R_e}{r}\right)^2$$

Note how the constants cancel out. Taking $r = R_e + h$ (Eq. 7.18), you can easily compute g'/g, or the acceleration due to gravity at some altitude above the Earth compared to g on the Earth's surface (9.80 m/s²).

Because R_e is very large compared to readily-attainable altitudes above the Earth's surface, the acceleration due to gravity does not decrease very rapidly as we ascend. At an altitude of 16 km (10 mi), $g'/g = 0.99$, or g' is still 99% of the value of g at the Earth's surface. At an altitude of 160 km (100 mi), g' is 95% of g.

The acceleration due to gravity does vary slightly with altitude. At a distance h above the Earth's surface, the acceleration is given by

$$g' = \frac{GM_e}{(R_e + h)^2} \tag{7.18}$$

The value of g at the Earth's surface is referred to as the standard acceleration and is sometimes used as a unit. For example, when a spacecraft lifts off, astronauts are said to experience several g's. This means that their acceleration is several times the standard acceleration, g. Since $g = F/m$, we can also think of g as the *gravitational force per unit mass*. Thus the unit **g's of force** is often used for the force corresponding to a given acceleration. For example, on the surface of the Earth you experience an *acceleration* of 1 g $= g$, and the corresponding *force* of 1 g is your weight: $F = m(1 \text{ g}) = mg = w$. If you experience a force of 2 g's, the force has a magnitude of twice your weight, and so on. (See the Insight on g's of force.)

INSIGHT ■ g's of Force

To help better understand the use of the gravitational unit g as a unit of force, let's look at some examples. During the takeoff of a jet airliner, you experience an average horizontal force of about 0.20 g. This means that as the plane accelerates down the runway, the seat back exerts a force of about one-fifth of your weight against you. The reaction force (Newton's third law) is the force you exert against the seat, and you experience a feeling of being pushed back into the seat. On takeoff at an angle of 30°, this increases to about 0.70 g with a component of gravity helping push you into the seat.

A jet pilot diving in a circular arc is said to "pull" so many g's at the bottom of the arc. This is the magnitude of the force on the pilot, which supplies the centripetal acceleration necessary for circular motion. At about 10 g's, a pilot experiences a blackout, or loss of conscious-

ness, because this force causes a lack of blood flow to the brain.

Similarly, an astronaut experiences as much as 8 g's of force on blast-off because of the upward acceleration (Fig. 1). This force (experienced as apparent weight) is exerted on the astronaut by the seat. If an astronaut were standing upright, such a force would cause the blood to go to the legs, where the blood vessels would be distended and the capillaries would rupture. Lack of blood circulation to the eyes and brain would cause temporary blindness and loss of consciousness. Consequently, astronauts are in a reclining position for blast-off.

You may have experienced several g's of force on a roller coaster. This occurs when the coaster is climbing after a dip. Usually, the force is limited to less than 3.5 g's. Otherwise, people become frightened.

FIGURE 1 *g*'s of force   Lt Col. John Stapp is shown before and during a high-speed rocket-propelled sled run. He reached Mach 1.7 (1.7 times the speed of sound or about 1200 mi/h) in just seconds and then quickly decelerated to rest.

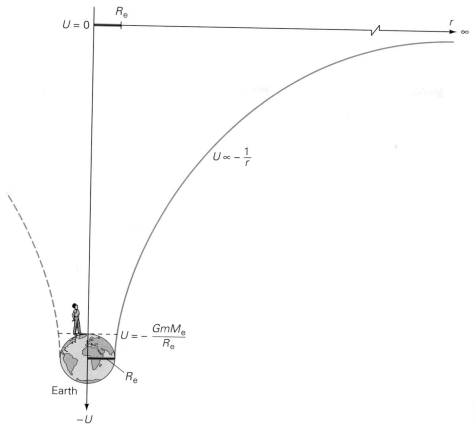

Figure 7.18

● **FIGURE 7.18 Gravitational potential energy well**   On Earth, we are in a negative gravitational potential energy well. As with an actual well or hole in the ground, work must be done against gravity to get higher in the well. The potential energy of an object increases as it moves higher in the well. This means that the value of U becomes less negative. The top of the Earth's gravitational well is at infinity, where the gravitational potential energy is zero.

Another aspect of the change of g with altitude concerns potential energy. In Chapter 5, you learned that $U = mgh$ for an object at a height h above some zero reference point, since g is essentially constant near the Earth's surface. This potential energy is equal to the work done in raising the object a distance h above the Earth's surface in a *uniform* gravitational field. But what if the change in altitude is so large that g cannot be considered constant while work is done in moving an object? In this case, the equation $U = mgh$ doesn't apply. In general, it can be shown (using mathematical methods that are beyond the scope of this book) that the **gravitational potential energy** of two point masses separated by a distance r is given by

Gravitational potential energy

$$U = -\frac{Gm_1m_2}{r} \qquad (7.19)$$

The minus sign in Eq. 7.19 arises from the choice of the zero reference point (the point where $U = 0$), which is $r = \infty$. In terms of the Earth and some altitude h,

$$U = -\frac{Gm_1m_2}{r} = -\frac{GmM_e}{R_e + h} \qquad (7.20)$$

where r is the distance separating the Earth's center and the mass m which is at an altitude h above the Earth's surface. What this means is that on Earth we are in a negative gravitational potential energy well (● Fig. 7.18) that extends

Potential energy wells are discussed in Section 5.4

to infinity because the force of gravity has an infinite range. (Recall that finite potential energy wells arise using the approximation $U = mgh$.)

Thus, when work is done against gravity (an object moves higher in the well) or gravity does work (an object falls lower in the well), there is a *change* in potential energy. As with finite potential energy wells, this change in energy is usually what's important in analyzing situations.

EXAMPLE 7.13 ■ Gravitational Potential Energy

Two 50-kg satellites are in circular orbits about the Earth at altitudes of 1000 km (about 620 mi) and 35,000 km (about 22,000 mi). (The lower one monitors particles about to enter the atmosphere, and the higher one, synchronous with the Earth's rotation, takes weather pictures from its stationary position with respect to the Earth's surface.) How much more work was done on the second (higher) satellite than on the first to place them in these orbits?

Solution. Listing the data so we can better see what's given, we have

Given: $m = 50$ kg *Find:* W (work)
$h_1 = 1000$ km $= 1.0 \times 10^6$ m
$h_2 = 35{,}000$ km $= 35 \times 10^6$ m
$M_e = 6.0 \times 10^{24}$ kg (from table)
$R_e = 6.4 \times 10^6$ m

The extra work associated with the difference in orbital altitudes is simply the difference in their gravitational potential energies at these altitudes. (This is not the difference in the total energy. Why? There is also a difference in kinetic energy.) Air resistance is not a factor because both satellites are above the atmosphere. The path the upper satellite took to reach the higher altitude is not a factor either, since gravity is a conservative force (see Chapter 5). Thus,

$$W = \Delta U = U_2 - U_1 = -\frac{GmM_e}{R_e + h_2} - \left(-\frac{GmM_e}{R_e + h_1}\right)$$

$$= GmM_e\left(\frac{1}{R_e + h_1} - \frac{1}{R_e + h_2}\right)$$

$$= (6.67 \times 10^{-11}\ \text{N-m}^2/\text{kg}^2)(50\ \text{kg})(6.0 \times 10^{24}\ \text{kg})$$

$$\times \left[\frac{1}{(6.4 \times 10^6\ \text{m}) + (1.0 \times 10^6\ \text{m})} - \frac{1}{(6.4 \times 10^6\ \text{m}) + (35 \times 10^6\ \text{m})}\right]$$

$$= 2.2 \times 10^9\ \text{J}$$

This is the amount of work done in raising the satellite higher in the Earth's negative potential energy well. Note that U_2 has a smaller negative value than U_1 has.

Using the gravitational potential energy (Eq. 7.19) gives the equation for the total mechanical energy a different form than it had in Chapter 5. For example, the total mechanical energy of a mass m_1 moving near a stationary mass m_2 is

$$E = K + U = \tfrac{1}{2}m_1v^2 - \frac{Gm_1m_2}{r} \qquad (7.21)$$

Explain the significance of the minus sign in the equation for total mechanical energy.

(Note that the potential energy $U = -Gm_1m_2/r$ is negative in this case, and *not* equal to *mgh*.) This equation and the principle of the conservation of energy can be applied to the Earth moving about the Sun, by neglecting other gravitational forces. The Earth's orbit is not quite circular, but slightly elliptical. At perihelion (the point of the Earth's closest approach to the Sun), the mutual gravitational potential energy is less than it is at aphelion (the point farthest from the Sun).

Therefore, as can be seen from Eq. 7.21 in the form $\tfrac{1}{2}mv^2 = E + Gm_1m_2/r$, where E is constant, the Earth's kinetic energy and orbital speed are greatest at perihelion (the smallest value of r) and least at aphelion (the greatest value of r). Or, in general, the Earth's orbital speed is greater when it is nearer the Sun than when it is farther away.

Mutual gravitational potential energy also applies to a group, or configuration, of more than two masses. That is, there is potential energy due to the masses being in a configuration. This is because work was done in bringing them together. Suppose that there is a single fixed mass m_1, and another mass m_2 is brought close to it, from an infinite distance (where $U = 0$). The work done is equal to the mutual potential energy of the masses, which are now separated by a distance r_{12}, that is, $U_{12} = -Gm_1m_2/r_{12}$.

If a third mass m_3 is brought close to the other two fixed masses, there are then two forces of gravity acting on m_3, so $U_{13} = -Gm_1m_3/r_{13}$ and $U_{23} = -Gm_2m_3/r_{23}$. The total gravitational potential energy of the configuration is therefore

$$U = U_{12} + U_{13} + U_{23}$$

$$= -\frac{Gm_1m_2}{r_{12}} - \frac{Gm_1m_3}{r_{13}} - \frac{Gm_2m_3}{r_{23}} \qquad (7.22)$$

A fourth mass could be brought in, but this development should be sufficient to suggest that the total gravitational potential energy of a configuration of particles is equal to the sum of the individual potential energies for all pairs of particles.

EXAMPLE 7.14 ■ Total Gravitational Potential Energy

Three masses are in a configuration as shown in ● Fig. 7.19. What is their total gravitational potential energy?

Solution. From the figure we have,

Given: $m_1 = 1.0$ kg
$m_2 = 2.0$ kg
$m_3 = 2.0$ kg
$r_{12} = 3.0$ m
$r_{13} = 4.0$ m
$r_{23} = 5.0$ m (3-4-5-triangle)

Find: U (total gravitational potential)

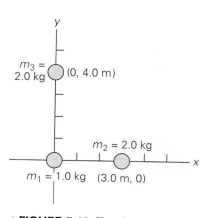

FIGURE 7.19 Total gravitational potential energy Example 7.14.

We can use Eq. 7.22 directly since only three masses are used in this example. (Note that Eq. 7.22 can be extended to four, five, . . . any number of masses.)

$$U = -\frac{Gm_1m_2}{r_{12}} - \frac{Gm_1m_3}{r_{13}} - \frac{Gm_2m_3}{r_{23}}$$

$$= (6.67 \times 10^{-11} \text{ N-m}^2/\text{kg}^2)$$

$$\times \left[-\frac{(1.0 \text{ kg})(2.0 \text{ kg})}{3.0 \text{ m}} - \frac{(1.0 \text{ kg})(2.0 \text{ kg})}{4.0 \text{ m}} - \frac{(2.0 \text{ kg})(2.0 \text{ kg})}{5.0 \text{ m}} \right]$$

$$= -1.3 \times 10^{-10} \text{ J}$$

This is the amount of *negative* work done in bringing the masses together. To separate the masses by infinite distances, an equal amount of positive work (against gravity) would have to be done.

7.6 ■ Kepler's Laws and Earth Satellites

The force of gravity determines the motions of the planets and of Earth satellites and holds the solar system (and galaxy) together. A general description of planetary motion had been set forth shortly before Newton's time by the German astronomer and mathematician Johannes Kepler (1571–1630). Kepler was able to formulate three empirical laws from observational data gathered during a 20-year period by the Danish astronomer Tycho Brahe (1546–1601). Brahe's extensive observations of stars and planets were done without the benefit of a telescope, which had not yet been invented. But the observations Brahe made were more accurate than those of most of his contemporaries because of better instrumentation he had developed. As a result, he is considered to have been one of the greatest practical astronomers.

Kepler went to Prague to assist Brahe, who was the official mathematician at the court of the Holy Roman Emperor. Brahe died the next year and Kepler succeeded him, inheriting his records of the positions of the planets. Analyzing this data, Kepler announced the first two of his three laws in 1609 (the year Galileo built his first telescope). These laws were applied initially only to Mars. Kepler's third law came 10 years later.

Interestingly enough, Kepler's laws of planetary motion, which took him about 15 years to deduce from observed data, can now be derived theoretically with a page or two of calculations. These three laws apply not only to planets, but to any system composed of a body revolving about a larger body, to which the inverse square law of gravitation applies (for example, the moon and solar-bound comets obey these laws).

Kepler's first law (the law of orbits) says that

Kepler's first law—the law of orbits

Planets move in elliptical orbits with the Sun at one of the focal points.

An ellipse, shown in ● Fig. 7.20 (a), has, in general, an oval shape, resembling a flattened circle. In fact, a circle is a special case of an ellipse in which the focal points, or foci (plural of focus), are at the same point (at the center of the circle). Although the orbits of the planets are elliptical, most do not deviate very much from a circle (Mercury and Pluto are notable exceptions; see Table

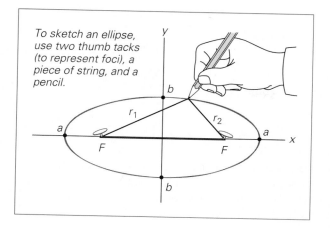

To sketch an ellipse, use two thumb tacks (to represent foci), a piece of string, and a pencil.

Figure 7.20

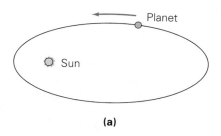

(a)

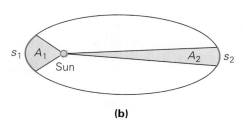

(b)

FIGURE 7.20 Kepler's first and second laws of planetary motion (a) In general, an ellipse has an oval shape. The sum of the distances from the focal points to any point on the ellipse is constant: $r_1 + r_2 = 2a$, where a is the length of the line joining the two points on the ellipse at the greatest distance from its center, called the semimajor axis. (The line joining the two points closest to the center is $2b$, the semiminor axis.) Planets revolve about the Sun in elliptical orbits for which the Sun is at one of the focal points. (b) A line joining the Sun and a planet sweeps out equal areas in equal times. Since $A_1 = A_2$, a planet travels faster along s_1 than along s_2.

6 in Appendix I). For example, the difference between the perihelion and aphelion of the Earth (its closest and farthest distances from the Sun) is about 3 million miles. This may sound like a lot, but it is only a little over 3% of 93 million miles, which is the average distance between the Earth and the Sun.

Kepler's second law (the law of areas) says that

A line from the Sun to a planet sweeps out equal areas in equal lengths of time.

Kepler's second law—the law of areas

This law is illustrated in Fig. 7.20b. Since the time to travel the different orbital distances (s_1 and s_2) is the same, this law tells you that the orbital speed of a planet varies in different parts of its orbit. Because a planet's orbit is elliptical, its orbital speed is greater when it is closer to the Sun than when it is farther away. (This was deduced for the Earth using the conservation of energy in Section 7.5.)

Kepler's third law (the law of periods) says that

The square of the period of a planet is directly proportional to the cube of the average distance of the planet from the Sun; that is, $T^2 \propto r^3$.

Kepler's third law—the law of periods

Kepler's third law is easily derived for the special case of a circular orbit, using Newton's law of gravitation. Since the centripetal force is supplied by the force of gravity, the expressions for these forces can be set equal:

$$\underset{\substack{\text{centripetal} \\ \text{force}}}{\frac{m_p v^2}{r}} = \underset{\substack{\text{gravitational} \\ \text{force}}}{\frac{G m_p M_s}{r^2}}$$

Emphasize this relationship for orbiting bodies. Also show this to be a statement of Newton's second law $F = ma$.

Where m_p and M_s are the masses of the planet and the Sun, respectively, and v is the orbital speed. But $v = 2\pi r/T$ (circumference/period), so

$$\left(\frac{m_p}{r}\right)\left(\frac{2\pi r}{T}\right)^2 = \frac{Gm_pM_s}{r^2}$$

Solving for T^2 gives

$$T^2 = \left(\frac{4\pi^2}{GM_s}\right)r^3$$

The law of periods

or

$$\boxed{T^2 = Kr^3}$$ (7.23)

The constant K is easily evaluated from orbital data (T and r) for the Earth: $K = 2.97 \times 10^{-19}\ \text{s}^2/\text{m}^3$. (As an exercise, you might wish to convert K to the more useful units of y^2/km^3.) Knowing the value of K, notice from Eq. 7.23 that the mass of the Sun can be determined.

Earth Satellites

We are still less than half a century into the space age. Since the 1950s, numerous unmanned satellites have been put into orbit about the Earth, and now astronauts regularly spend days or weeks in orbiting space laboratories.

Putting a spacecraft into orbit about the Earth (or the moon) is an extremely complex task. However, you can get a basic understanding of the problem from fundamental principles. First, suppose that a projectile could be given the initial velocity required to take it just to the top of the Earth's potential energy well. At the exact top of the well, which is an infinite distance away ($r = \infty$), the potential energy is zero. By the conservation of energy and Eq. 7.19,

Newton's law of gravitation suggests that the gravitational force diminishes with distance, but never goes to zero. Nevertheless, we describe the ability to "escape the Earth's gravitational attraction." Be prepared to discuss this in terms of the conservation of energy.

$$\overset{\text{initial}}{K_o + U_o} = \overset{\text{final}}{K + U}$$

$$\tfrac{1}{2}mv_e^2 - \frac{GmM_e}{R_e} = 0 + 0$$

Escape speed—initial speed needed to escape from a planet.

where v_e is the **escape speed**. The final energy is zero since the projectile stops at the top of the well and $U = 0$ there. Solving for v_e gives

$$\tfrac{1}{2}mv_e^2 = \frac{GmM_e}{R_e}$$

and

$$v_e = \sqrt{\frac{2GM_e}{R_e}}$$ (7.24)

Since $g = GM_e/R_e^2$ (Eq. 7.16), it is convenient to write

$$v_e = \sqrt{2gR_e}$$ (7.25)

Although derived for Earth, this equation may be used generally to find the escape speeds for other planets and our moon.

EXAMPLE 7.15 ■ Escape Speed

What is the escape speed for the Earth?

Solution. Eq. 7.25 can be used with known data:

$$v_e = \sqrt{2gR_e} = \sqrt{2(9.80 \text{ m/s}^2)(6.4 \times 10^6 \text{ m})} = 11 \times 10^3 \text{ m/s}$$

$$= 11 \text{ km/s} \qquad (\text{about } 7 \text{ mi/s})$$

Thus, if a projectile or spacecraft could be given an initial speed of 11 km/s (about 40,000 km/h or 25,000 mi/h), it would leave the Earth and not fall back, or return. Note from the equation that the escape speed is independent of the mass of the spacecraft; this is the initial speed required to send *any* object to the top of the Earth's potential energy well. Such an initial speed is not now attainable, but it does provide an upper limit for orbital speeds.

It is clear that a tangential speed smaller than the escape speed is required to orbit a satellite. Consider the centripetal force for a satellite in circular orbit about the Earth. Since the centripetal force on the satellite is supplied by the gravitational attraction between the satellite and the Earth, we may write:

$$F = \frac{mv^2}{r} = \frac{GmM_e}{r^2}$$

Then

$$v = \sqrt{\frac{GM_e}{r}} \tag{7.26}$$

where $r = R_e + h$. For example, suppose that a satellite is in a circular orbit at an altitude of 500 km (about 300 mi); its tangential speed is

$$v = \sqrt{\frac{GM_e}{r}} = \sqrt{\frac{GM_e}{R_e + h}}$$

$$= \sqrt{\frac{(6.67 \times 10^{-11} \text{ N-m}^2/\text{kg}^2)(6.0 \times 10^{24} \text{ kg})}{(6.4 \times 10^6 \text{ m}) + (0.50 \times 10^6 \text{ m})}} = 7.6 \times 10^3 \text{ m/s}$$

$$= 7.6 \text{ km/s} \qquad (\text{about } 4.7 \text{ mi/s})$$

This is about 27,000 km/h or 17,000 mi/h. As can be seen from Eq. 7.26, the orbital speed decreases with altitude.

In practice, a satellite is given a tangential speed by a component of the thrust from a rocket stage (●Fig. 7.21a). The inverse square relationship of Newton's law of gravitation means that the satellite orbits that are possible about a large mass are ellipses, of which a circular orbit is a special case. This is illustrated in Fig. 7.21b for Earth, using the previously calculated values. If a satellite is not given a sufficient tangential speed, it will fall back to Earth (and possibly be burned up while falling through the atmosphere). If the tangential speed reaches the escape speed, the satellite will leave its orbit and go off into space.

Finally, the energy of an orbiting satellite is

$$E = K + U = \tfrac{1}{2}mv^2 - \frac{GmM_e}{r} \tag{7.27}$$

Substituting the expression for v from Eq. 7.26 in the kinetic energy term gives

$$E = \frac{GmM_e}{2r} - \frac{GmM_e}{r}$$

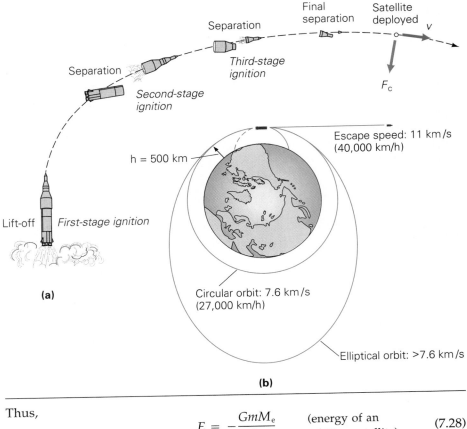

FIGURE 7.21 Satellite orbits
(a) A satellite is put into orbit by giving it a tangential speed sufficient for maintaining an orbit at a particular altitude—the higher the orbit, the smaller the tangential speed. (b) At an altitude of 500 km, a tangential speed of 7.6 km/s is required for a circular orbit. With a greater tangential speed, the satellite would move out of the circular orbit. Since it would not have the escape speed, it would "fall" around the Earth in an elliptical orbit with the Earth at one focal point. A tangential speed less than 7.6 km/s would also give an elliptical path, but a certain minimum speed is needed to keep the satellite from striking the Earth.

Thus,

$$E = -\frac{GmM_e}{2r} \qquad \text{(energy of an orbiting satellite)} \qquad (7.28)$$

Note that the total energy of the satellite is negative. To understand this, recall that on Earth we are in a negative potential energy well (see Fig. 17.18). More work is required to put a satellite into a higher orbit, where it has more energy because it is higher in the well. (The situation is analogous to throwing a ball upward from the bottom of a real well.) The total energy E increases as its *numerical value* becomes smaller—less negative—in going to a higher orbit toward the zero potential at the top of the well. That is, the farther the distance of a satellite from Earth, the greater its total energy.

Also note from the development of Eq. 7.28 that the kinetic energy of an orbiting satellite is equal to the absolute value of its total energy:

$$K = \frac{GmM_e}{2r} = |E| \qquad (7.29)$$

EXAMPLE 7.16 ■ Forward Thrust

The engines of a spacecraft in circular orbit about the Earth are fired briefly so as to give the craft a forward thrust. How will the spacecraft's energy be changed after the firing? (a) K increased and U decreased, (b) K decreased and U increased, (c) both K and U increased, (d) the total energy decreased. *Clearly establish the reasoning and physical principle(s) used in determining your answer before checking it below. That is, **why** did you select your answer?*

Reasoning and Answer: You might think that a forward thrust would cause the spacecraft to speed up and increase its kinetic energy. However, an increase in v would require an increase in the centripetal force. This force is supplied by gravitational attraction and so can increase only if r is reduced (i.e., if the spacecraft moves to a smaller orbit). Applying a forward thrust, however, will send the craft into a larger rather than a smaller orbit, because the forward thrust does positive work on the spacecraft, causing it to rise higher in its potential energy well and increasing its potential energy. With an increase in r, the kinetic energy *decreases* (Eq. 7.29) while the total energy increases by becoming less negative (Eq. 7.28). Hence, the answer is (b).

To help understand why we say that the total energy increases when its value becomes less negative, think of a change in energy from, say, 5.0 J to 10 J. Certainly this would be considered an increase in energy. Similarly, a change from -10 J to -5.0 J would be an increase in energy, even though the *numerical* value has decreased:

$$\Delta U = U - U_\text{o} = -5.0 \text{ J} - (-10 \text{ J}) = +5.0 \text{ J}.$$

(After the brief engine "burn" the spacecraft would go into an elliptical orbit with its lowest altitude equal to the radius of the initial circular orbit. To move a spacecraft from one circular orbit to a higher or lower circular orbit, two thrusts at different times are necessary.)

Follow-up Exercise: How could a spacecraft in orbit about the Earth use its rockets to increase its speed? Explain. (*Reasoning and answer may be found in the Answers to Exercises section at the back of the book.*)

EXAMPLE 7.17 ■ Applying the Brakes to Speed Up

A spacecraft is in a circular orbit about the Earth well outside the atmosphere. Describe what happens to the spacecraft if its retrorockets are fired. (Retrorockets point in the direction of the spacecraft's motion, and so produce a reverse thrust.)

Solution. The retrorockets applying a braking force to the spacecraft, *negative* work is done, and the spacecraft loses some energy (the work-energy theorem). That is, the total energy of the spacecraft,

The work-energy theorem is discussed in Section 5.3

$$E = -\frac{GmM_\text{e}}{2r}$$

is reduced. If E decreases (greater negative value), r must also decrease, and the spacecraft will spiral inward toward the Earth. From Eq. 7.29,

$$K = \tfrac{1}{2}mv^2 = \frac{GmM_\text{e}}{2r} \quad \text{or} \quad v^2 = \frac{GM_\text{e}}{r}$$

So, if r decreases, v increases, or the spacecraft speeds up.

The spacecraft would continue to spiral inward as long as the retrorockets were fired. When it reached the atmosphere, they could be shut off because air resistance would cause the spacecraft to lose further energy and continue its downward spiral.

"Space—the final frontier." Someday instead of brief stays in Earth-orbiting spacecraft, we may have permanent space colonies (see the chapter-closing Insight).

FIGURE 1 Space colony and artificial gravity It has been suggested that a space colony could be housed in a huge, rotating wheel as in this artist's conception. The rotation would supply the "artificial gravity" for the colonists.

Space colonies, long a staple of science fiction stories and films, are now actually being planned. However, humans cannot live for long periods of time in a "zero-gravity" environment (see Exercise 64) without detrimental physical effects. It has been suggested that this problem could be avoided if the colony were housed in a huge rotating wheel, which would supply "artificial gravity" (Fig. 1).

As you know, centripetal force is necessary to keep an object in rotational circular motion. On the rotating Earth, that force is supplied by gravity, and we refer to it as weight. Because of our weight, we exert a force on the ground, and the reaction force (by Newton's third law) on our feet gives us the feeling of "having our feet on solid ground." In a rotating space colony, the situation is somewhat reversed. The rotating colony would supply the centripetal force on the inhabitants, which would be perceived as a weight sensation, or artificial gravity (Fig. 2a). Rotation at the proper speed would produce a simulation of normal gravity ($g = 9.80$ m/s^2) within the colony wheel. Note that in the colonists' world, "down" would be outward, toward the periphery of the space station, and "up" would always be inward, toward the axis of rotation.

The inhabitants would experience normal gravitational effects. For example, in Fig. 2a, suppose that the person simply let go of the ball. Viewed from outside the rotating wheel, you would say the ball has a tangential velocity v as a result of the rotation, and would go off in a straight line (Newton's first law) until it hits the side of the wheel (as indicated by the dashed line in the figure). However, the inhabitant rotates with the wheel, and from that perspective, the ball merely falls to the floor as it would on Earth. Think about it.

Interestingly, the very same principle that could be used to provide artificial gravity for space colonists is the basis of a device that has long been used here on Earth. This is the centrifuge, a machine used to separate particles of different sizes and densities suspended in a liquid (or a gas). For example, cream is separated from milk by centrifuging, and blood components are separated in centrifuges in medical laboratories (Fig. 7. 22).

Blood components will eventually settle toward the bottom of a tube in layers—a process called sedimenta-

■ **FIGURE 7.22 Centrifuge** Centrifuges are used to separate particles of different sizes and densities suspended in liquids. For example, red and white blood cells can be separated from each other as well as the plasma that makes up the liquid portion of the blood.

EXAMPLE 7.18 ■ Acceleration in a Centrifuge

A laboratory centrifuge like that shown in ● Fig. 7.22 operates at a rotational speed of 12,000 rpm. (a) What is the magnitude of the centripetal acceleration on a red blood cell at a radial distance of 8.00 cm from the centrifuge's axis of rotation? (b) How does this acceleration compare to g?

Solution. The data are as follows,

Given: $\omega = 1.20 \times 10^4$ rpm
$\quad\quad = 1.26 \times 10^3$ rad/s
$\quad\quad r = 8.00$ cm $= 0.0800$ m

Find: (a) a_c
(b) How a_c compares to g

(a) The centripetal acceleration is easily found using Eq. 7.10:

$$a_c = r\omega^2 = (0.0800 \text{ m})(1.26 \times 10^3 \text{ rad/s})^2 = 1.27 \times 10^5 \text{ m/s}^2$$

(b) Using the relationship $1\,g = 9.80$ m/s^2 to express a_c in terms of g,

$$a_c = (1.27 \times 10^5 \text{ m/s}^2)\left(\frac{1g}{9.80 \text{ m/s}^2}\right) = 13,000\,g$$

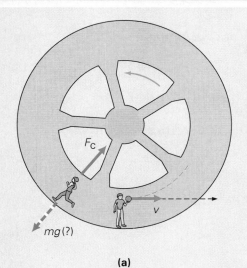

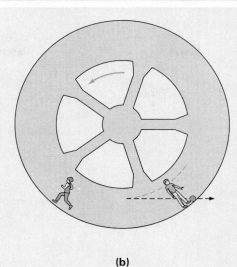

(a)

(b)

FIGURE 2 Rotating space colony (a) In the frame of reference of someone in a rotating space colony, the centripetal force on the person would be perceived as weight sensation, or artificial gravity. Rotation at the proper speed would simulate normal gravity. From the point of view of an outside observer, a dropped ball would follow a tangential straight-line path. (b) A colonist, on the other hand, would observe the ball to fall downward as in a normal gravitational situation.

tion—under the influence of normal gravity alone. The viscous drag of the plasma on the particles is analogous to (but much greater than) the air resistance that determines the terminal velocity of falling objects (Section 4.6). Red blood cells settle in the bottom layer of the plasma because they reach a greater terminal velocity than the white blood cells and platelets, and so get to the bottom sooner. However, gravitational sedimentation is generally very slow.

Since clinicians cannot afford to wait a long time to see the fractional volume of red cells in the blood, centrifugation is used to speed up the sedimentation process. Tubes are spun horizontally. Particles in the ro-

tating, accelerating system are in effect thrust outward, just as the space colonists would be pressed against the "floor" of their rotating space station. The resistance of the fluid medium on the particles supplies the centripetal force that keeps them moving in slowly widening circles as they settle toward the bottom of the tube. The bottom of the tube itself must exert a strong force on the contents as a whole, and must be strong enough so as not to break.

The centrifuge thus supplies an "artificial gravity" many times that of normal gravity. Laboratory centrifuges commonly operate at speed sufficient to produce thousands of g's (see Example 7.18.)

Important Concepts

You should be able to define and explain these chapter concepts clearly.

angular distance	period (T)	g's of force	Kepler's third law
radian (rad)	frequency (f)	gravitational potential	(the law of periods)
average angular speed	hertz (Hz)	energy	escape speed
angular velocity	uniform circular motion	Kepler's first law	tangential acceleration
average	centripetal acceleration	(the law of orbits)	universal law of gravitation
instantaneous	centripetal force	Kepler's second law	universal gravitational
tangential speed	average angular	(the law of areas)	constant, G
	acceleration		

Important Relationships for Review

These relationships are mathematical statements of the concepts and principles presented in the chapter. You should be able to identify the symbols and to explain

the relationships before proceeding to the Exercises. In-text equation reference numbers are given for convenience.

Arc Length (angle in radians):

$$s = r\theta \tag{7.3}$$

Angular Kinematic Equations (see Table 7.2 for linear analogues):

$$\theta = \omega t \tag{7.5}$$

$$\overline{\omega} = \frac{\omega + \omega_\circ}{2} \tag{2, Table 7.2}$$

$$\omega = \omega_\circ + \alpha t \tag{7.12}$$

$$\theta = \omega_\circ t + \tfrac{1}{2}\alpha t^2 \tag{4, Table 7.2}$$

$$\omega^2 = \omega_\circ^2 + 2\alpha\theta \tag{5, Table 7.2}$$

Tangential and Angular Speeds:

$$v = r\omega \tag{7.6}$$

Angular Speed (with uniform circular motion):

$$\omega = \frac{2\pi}{T} = 2\pi f \tag{7.8}$$

Frequency and Period:

$$f = \frac{1}{T} \tag{7.7}$$

Centripetal Acceleration:

$$a_c = \frac{v^2}{r} = r\omega^2 \tag{7.10}$$

Tangential and Angular Accelerations:

$$a_t = r\alpha \tag{7.13}$$

Centripetal Force and Acceleration:

$$F_c = ma_c = \frac{mv^2}{r} \tag{7.11}$$

Newton's Law of Gravitation:

$$F = \frac{Gm_1m_2}{r^2} \tag{7.15}$$

$$G = 6.67 \times 10^{-11} \text{ N-m}^2/\text{kg}^2$$

Acceleration Due to Gravity at an Altitude h:

$$g' = \frac{GM_e}{(R_e + h)^2} \tag{7.18}$$

Gravitational Potential Energy:

$$U = -\frac{Gm_1m_2}{r} \tag{7.19}$$

Kepler's Law of Periods:

$$T^2 = Kr^3 \tag{7.23}$$

$$K = 2.97 \times 10^{-19} \text{ s}^2/\text{m}^3$$

Escape Speed:

$$v_e = \sqrt{\frac{2GM_e}{r_e}} = \sqrt{2gR_e} \tag{7.24}$$

Energy of Orbiting Satellite:

$$E = -\frac{GmM_e}{2r} \tag{7.27}$$

$$K = |E| \tag{7.28}$$

Exercises

7.1 ▪ Angular Measure

1 The radian unit is equivalent to (a) degree/time, (b) length, (c) length/length, (d) length/time. (c)

2 The arc length s of a circle is directly related to the radius r by the subtended angle in (a) degrees, (b) radians, (c) revolutions, (d) rotations. (b)

3 ▪ The Cartesian coordinates of a point on a circle are $(\sqrt{2}$ m, $\sqrt{2}$ m$)$. What are the polar coordinates (r, θ) of this point? (2m, 45°)

4 ▪ The equation for a circle using Cartesian coordinates is $x^2 + y^2 = a^2$, where a is the radius of the circle. What is the equation of a circle using polar coordinates? $r = a$

5 ▪ Convert the following angles from degrees to radians: (a) 7.5°, (b) 270°, (c) 45°, and (d) 540°. (a) 0.13 rad (b) 4.71 rad (c) 0.79 rad (d) 9.42 rad

6 ▪ Convert the following angles from radians to degrees: (a) $\pi/25$ rad, (b) $\pi/10$ rad, (c) 1.25 rad, and (d) 4π rad. (a) 7.2° (b) 18° (c) 72° (d) 720°

7 ▪▪ What is the arc length subtended by an angle of $\pi/6$ rad on a circle with a radius of 8.0 cm? 4.2 cm

●8 ▪▪ A jogger on a circular track that has a radius of 0.25 km jogs a distance of 1.0 km. What angular distance does the jogger cover in (a) radians and (b) degrees? (a) 4.0 rad (b) 229°

9 ▪▪ Assuming that the Earth's orbit around the Sun is circular, what is the approximate orbital distance the Earth travels in 3 months? 2.4×10^8 km

10 ■■ Two race cars start from rest and travel around a circular track whose diameter is 1.0 km. (a) One car develops engine trouble and stops after having traveled 300° around the track. What distance did it travel? (b) The other car makes four complete laps. How far did it travel? (a) 2.6 km
(b) 13 km
11 ■■ At sunset, the Sun has an angular width of about 0.50°. From the time the lower edge of the Sun just touches the horizon, about how long does it take it to disappear?
120 s
● **12** ■■ A garden hose wound on a rotating storage cylinder has an outer layer at a distance of 0.45 m from the center of the cylinder. As this layer of wound hose is pulled out, the cylinder makes three revolutions. What is the length of the unwound hose? 8.5 m

13 ■■ Electrical wire with a diameter of 0.75 cm is wound on a spool with a radius of 30 cm and a length of 24 cm. (a) Through how many radians must the spool be turned to wrap one even layer of wire? (b) What is the length of this wound wire? (a) 2.0×10^2 rad (b) 60 m

14 ■■ The Cartesian coordinates of a point on a circle with its center at the origin are (0.40 m, 0.30 m). What is the arc length measured counterclockwise on the circle from the positive x axis to this point? 0.33 m

7.2 ■ Angular Speed and Velocity

15 The angular velocity vector of an object in uniform circular motion is (a) tangential to the circle, (b) clockwise as viewed from above, (c) perpendicular to the plane of the circle, (d) given by a left-hand rule. (c)

● **16** The frequency unit of hertz is equivalent to (a) that of the period, (b) cycle, (c) radian/s, (d) s^{-1}. (d)

17 A turntable rotates clockwise as viewed from above. What is the direction of the angular velocity vector?
into the turntable
18 When clockwise or counterclockwise is used to describe rotational motion, why is a phrase such as "viewed from above" added? viewed from opposite sides would give different circular senses
19 ■ A race car makes two laps around a circular track in 2.5 min. What is the car's average angular speed?
0.084 rad/s
20 ■ A satellite in a circular orbit has a period of 10 hours. What is its frequency in revolutions per day? 2.4 rev/d

21 ■■ Determine which has the greater angular speed: particle A, which travels 160° in 2.0 s, or particle B, which travels 3π rad in 7.0 s. A is faster

● **22** ■■ A runner running at a constant pace gets halfway around a circular track that has a diameter of 500 m in 2.5 min. What are the runner's (a) angular speed and (b) tangential speed? (a) 2.1×10^{-2} rad/s (b) 5.3 m/s

23 ■■ For the second hand and the hour hand of a (running) clock, find (a) the period, (b) the frequency, (c) the angular speed, and (d) the angular velocity. (a) $T_h = 4.3 \times 10^4$ s; $T_s = 60$ s (b) $f_s = 1.7 \times 10^{-2}$ s^{-1}; $f_h = 2.3 \times 10^{-5}$ s^{-1} (c–d) $\omega_h = 1.4 \times 10^{-4}$ rad/s; into clock face

● **24** ■■ Jupiter, the largest planet in our solar system, has a period of rotation of 10 hours and a period of revolution of 12 years. (a) What are its frequencies of rotation and revolution? (b) If the planet's diameter is 1.38×10^4 km and its mean distance from the Sun is 7.78×10^8 km, what are Jupiter's tangential speed at its equator and its angular speed of revolution? (Assume a circular orbit.) (a) $f_{rot} = 2.8 \times 10^{-5}$ s^{-1}; $f_{rev} = 2.6 \times 10^{-9}$ s^{-1} (b) $v_{rot} = 1.2$ km/s; $\omega_{rev} = 1.7 \times 10^{-8}$ rad/s
25 ■■ In riding a bicycle, it is noted that the 26.0-inch diameter wheel makes 15.0 revolutions in a time of 8.50 s. (a) What is the angular speed of the wheel? (b) What distance does the bicycle travel during this time?
(a) 11.1 rad/s (b) 102 ft
26 ■■ The driver of a motorboat sets the throttle and ties the wheel, making the boat travel at a uniform speed of 12.5 m/s in a circle with a diameter of 115 m. (a) Through what angular distance does the boat move in 4.00 min? (b) What arc distance does it travel in this time? (a) 52.0 rad
(b) 3.0×10^3 m
27 ■■■ What are (a) the frequency of the Earth's rotation and (b) its angular *velocity*? (c) Express the tangential and angular speeds of a person at a latitude of 45° N (or S) as percentages of those of a person at the Equator. (a) 1.16×10^{-5} s^{-1} (b) 7.29×10^{-5} rad/s, north (c) $v_{45°}/v_{eq} = 70.7\%$

7.3 ■ Uniform Circular Motion and Centripetal Acceleration

28 In uniform circular motion, there is a (a) constant velocity, (b) constant angular velocity, (c) zero acceleration, (d) net tangential acceleration. (b)

● **29** If the centipetal force on a particle in uniform circular motion is increased, (a) the motion will remain uniform, (b) the tangential speed will decrease, (c) the radius of the circular path will increase, (d) the tangential speed will increase and/or the radius will decrease. (d)

30 A level curve on an interstate highway is a segment of a large circle and has a speed limit of 65 mi/h. A cloverleaf exit which feeds directly into an interstate highway is also a level, circular segment, but has a speed limit of only 25 mi/h. Why the difference? different radii

31 It is possible for a particle in circular motion to have a constant tangential acceleration and a constant centripetal acceleration at the same time? Explain. only if $a_t = 0$

32 An apparatus as illustrated in Fig. 7.23 is used to dem-

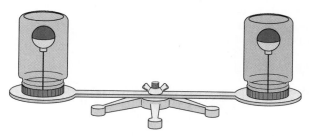

● **FIGURE 7.23 A rotating system (when set into motion)** See Exercise 32.

onstrate forces in a rotating system. The floats are in jars of water. When the arm is rotated, which way will the floats move? (Compare with the accelerometer in Fig. 4.26.) Does it make a difference which way the arm is rotated?
in direction of acceleration, inward

33 When rounding a curve in a fast-moving car, we experience a feeling of being thrown outward (●Fig. 7.24). It is sometimes said that this is because of an outward *centrifugal*

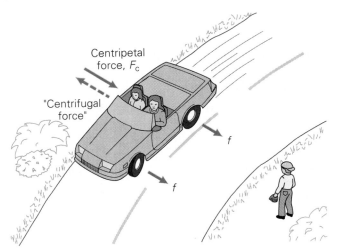

●**FIGURE 7.24 A center-fleeing force?** See Exercise 33.

(center-fleeing) force. However, in terms of Newton's laws, this is called a pseudo or false force because it doesn't really exist. Analyze the situation in the figure and show that this is the case. [*Hint:* Start with Newton's first law.] see ISM

34 ■ A ball is swung at a uniform speed in a horizontal circle with a radius of 1.25 m. If the ball has a centripetal acceleration of 5.35 m/s², what is its orbital speed?
2.59 m/s

35 ■ A race car goes around a level, circular track with a radius of 1.0 km at a speed of 1.2×10^2 km/h. What is the centripetal acceleration of the car? 1.1 m/s²

●**36** ■■ The moon revolves around the Earth in 29.5 days in a nearly circular orbit with a radius of 3.80×10^5 km. Assume that the moon's motion is uniform. With what acceleration is it falling toward the Earth? 2.31×10^{-3} m/s²

37 ■■ (a) Using the data given in Exercise 24, compute the centripetal acceleration for the revolution of the planet Jupiter, assuming a circular orbit. (b) Compare the centripetal acceleration of the moon in its revolution about the Earth to that of Jupiter about the Sun. Comment on any major difference. (a) 2.2×10^{-4} m/s² (b) moon 10 times greater

38 ■■ A car enters a circular curve with a radius of curvature of 0.400 km at a constant speed of 83.0 km/h. If the friction between the road and the car's tires can supply a centripetal acceleration of 1.25 m/s², does the car negotiate the curve smoothly? Justify your answer.
no; 1.32 m/s² > 1.25 m/s²

●**39** ■■ Suppose that you swing an object attached to the end of a string at a constant speed in a horizontal circle with a radius of 1.5 m. It takes 1.6 s for the object to make one revolution. (a) What is the magnitude of the tangential velocity of the object? (b) What centripetal acceleration are you imparting to the object via the string? (Is the string exactly horizontal?) (a) 5.9 m/s (b) 23 m/s²

40 ■■ At a homecoming, one competition is to swing a bucket of water in a vertical circle without spilling any. If the distance from a person's shoulder to the center of mass of the bucket of water is 0.95 m, what is the minimum angular speed required to keep the water from coming out of the bucket at the top of the swing? 3.2 rad/s

41 ■■ The Earth revolves around the Sun in a nearly circular orbit; its average distance from the Sun is about 1.5×10^8 km. Assume that the Earth's motion is uniform. (a) Compute the orbital speed in SI units. (Convert the answer to mi/h to get a more familiar measure of how fast you are traveling through space.) (b) Is the centripetal acceleration required to keep a person at the Equator rotating with the Earth the same as g, the acceleration due to gravity? Explain. (Justify your answer with actual values.)
(a) 3.0×10^4 m/s (b) no; $a_c = 3.4 \times 10^{-2}$ m/s²

●**42** ■■■ A daredevil stunt involves riding a motorcycle around the vertical inside wall of a cylindrical structure (●Fig. 7.25). Assume that the cylinder has a radius of 15 m

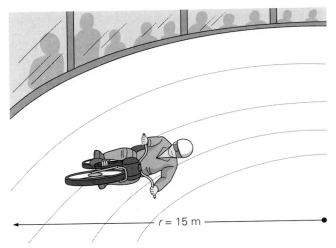

●**FIGURE 7.25 A daredevil stunt** See Exercise 42.

and that the coefficient of static friction between the motorcycle tires and the wall is 1.1. (a) What is the minimum speed that will keep the motorcycle and the rider from sliding down the wall? Express this speed in km/h and mi/h. (b) Would the speed be different for a bigger (heavier) motorcycle? Explain. (a) 12 m/s; 43 km/h; 27 mi/h (b) no

43 ■■■ A pendulum swinging in a circular arc under the influence of gravity, as shown in ●Fig. 7.26, has both centripetal and tangential components of acceleration. (a) If the pendulum bob has a speed of 2.7 m/s when the cord makes an angle of $\theta = 15°$ with the vertical, what are the magnitudes of the components at this time? [*Hint:* Consider the acceleration due to gravity.] (b) When is the centripetal acceleration a maximum? What is the value of the tangential acceleration at that time? (a) $a_t = 2.5$ m/s²; $a_c = 9.7$ m/s² (b) at the lowest point of the swing; $a_t = 0$

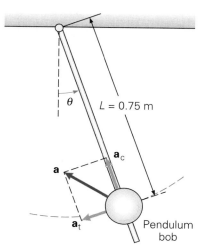

● **FIGURE 7.26 A swinging pendulum** See Exercise 43.

44 ■■■ In the design of a highway, curves are usually banked to improve safety (● Fig. 7.27). The horizontal component of the normal force toward the center of the curve reduces the need for friction to prevent skidding. In fact, for a circular curve banked at a given angle, there is one speed for which no frictional force is required—the centripetal force is supplied completely by the inward component of the normal force. (a) Show that the banking angle θ for which no friction is required for a car to safely negotiate a circular curve is given by tan $\theta = v^2/gr$, where v is the car's speed and r is the radius of curvature. (b) What is the proper banking angle for a speed of 35 km/h on a curve with a radius of 50 m for which friction does not contribute to the centripetal force? (a) see ISM (b) 11°

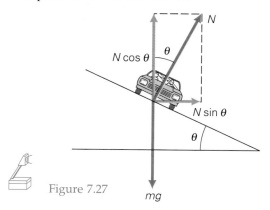

Figure 7.27

● **FIGURE 7.27 Banking safety** See Exercise 44.

7.4 ■ Angular Acceleration

45 The angular acceleration in circular motion (a) is equal in magnitude to the tangential acceleration divided by the radius, (b) increases the angular velocity if in the same direction, (c) has units of s^{-2}, (d) all of the preceding. (d)

46 The angular acceleration vector and the angular velocity vector in circular motion are in opposite directions. This means that (a) the centripetal acceleration will decrease, (b) there is no tangential acceleration, (c) the angular velocity will increase, (d) the vector sum of the accelerations is constant. (a)

● **47** A rotating phonograph turntable is turned off. What happens to the angular velocity and angular acceleration vectors during the time it takes the turntable to stop? see ISM

48 ■■ A phonograph turntable sets for 45 rpm is turned on and reaches its operating speed in 1.25 s. What is the average angular acceleration of the turntable? [*Hint:* Turntables rotate clockwise as viewed from above.] 3.8 rad/s² into turntable

49 ■■ A particle initially at rest accelerates in a circular path at a rate of 2.5 rad/s² for 7.0 s. The circle has a radius of 15 cm. What are (a) the magnitude of the tangential acceleration during this time and (b) the tangential speed at the end of this time? (a) 0.38 m/s² (b) 2.7 m/s

50 ■■ A merry-go-round accelerating uniformly from rest achieves its operating speed of 2.5 rpm in four revolutions. What is the magnitude of its angular acceleration? 1.4×10^{-3} rad/s²

51 ■■ A $33\frac{1}{3}$-rpm record on a turntable uniformly reaches its operating speed in 2.45 s once the record player is turned on. (a) What is the angular distance traveled during this time? (b) What is the corresponding arc length on the circumference of a 12-in. record? (a) 4.26 rad (b) 4.26 ft

● **52** ■■ In a factory, a machine has a large flywheel with a diameter of 1.50 m, which operates with an angular speed of 7.65 rad/s. When the machine is shut down, it takes 24.8 s for the wheel to come to rest. How many revolutions does the wheel go through during this time? 15.1 rev

53 ■■ A Ferris wheel with a diameter of 35.0 m starts from rest and achieves its maximum operational tangential speed of 2.20 m/s in a time of 15.0 s. (a) What is the magnitude of the wheel's angular acceleration? (b) What is the magnitude of the tangential acceleration after reaching the maximum operational speed? (a) 8.40×10^{-3} rad/s² (b) 0

54 ■■ The blades of a fan running at low speed turn at 250 rpm. When the fan is switched to high speed, the rotation rate increases uniformly to 350 rpm in 5.75 s. (a) What is the magnitude of the angular acceleration of the blades? (b) How many revolutions do the blades go through while the fan is accelerating? (a) 1.82 rad/s² (b) 28.8 rev

55 ■■ A 200-kg satellite in a circular orbit about the Earth at an altitude of 500 km makes one revolution in 95 min. (a) What is the magnitude of the centripetal force acting on the satellite? (b) What supplies this force? (a) 1.7×10^3 N (b) gravity

● **56** ■■ When a motorized potter's wheel is turned on, it accelerates uniformly at a rate of 2.6 rad/s² for 5.0 s to achieve its operating speed. What is that speed in rpm? 1.2×10^2 rpm

57 ■■ A phonograph record initially moving at $33\frac{1}{3}$ rpm accelerates uniformly to 45 rpm in 2.5 s. What is the angular distance traveled by the record during this time? 10 rad

58 ■■ A car on a circular track with a radius of 0.30 km accelerates from rest with a constant angular acceleration whose magnitude is 4.5×10^{-3} rad/s². (a) How long does it take the car to make one lap around the track? (b) What is the total acceleration of the car when it has completed half of a lap? (a) 53 s (b) $a = 8.4$ m/s² **r** $+ 1.4$ m/s² **t**

● **59** ■■ A student investigating circular motion places a dime on a $33\frac{1}{3}$-rpm record on a turntable at a distance of 10 cm from the center. Switching on the turntable, the student notes that the dime slides outward when the record is turning at 90% of its operating speed. (a) Why does the dime slide outward? (b) What is the coefficient of friction between the dime and the record? (a) friction not enough to provide the centripetal force (b) 9.8×10^{-2}

60 ■■■ A 60-kg girl prepares to swing out over a pond on a rope, as shown in ● Fig. 7.28. What tension in the rope is required if she is to maintain a circular arc (a) when she starts her swing by stepping off the platform, (b) when she is 5.0 m above the pond surface, and (c) at the lowest point of the swinging motion? [*Hint:* Treat the girl and the rope as a simple pendulum.] (a) 0 (b) 1.2×10^3 N (c) 1.8×10^3 N

61 ■■■ Suppose that the girl swinging on the rope in Fig. 7.28 (see Exercise 60) starts her swing at a lower point and has a tangential speed of 8.5 m/s when the rope is at an angle of 30° to the vertical. (a) What is her tangential acceleration at that time? (b) What is the magnitude of the vector sum of the tangential and centripetal accelerations at that time? (c) From what height above the pond surface did she start her swing? (a) 4.9 m/s² (b) 8.7 m/s² (c) 5.0 m

7.5 ■ Newton's Law of Gravitation

62 Newton's law of gravitation (a) has the same form as the expression for gravitational potential energy, (b) applies only on Earth, (c) is an inverse square law, (d) all of the preceding. (c)

63 The acceleration due to gravity of an object on the Earth's surface (a) is a universal constant like G, (b) does not depend on the Earth's mass, (c) is directly proportional to the Earth's radius, (d) does not depend on the object's mass. (d)

64 Astronauts in a spacecraft orbiting the Earth or out for a "spacewalk" (● Fig. 7.29) are seen to "float" in midair. This is sometimes referred to as *weightlessness* or *zero gravity* (*zero g*). Are these terms correct? Explain why an astronaut appears to float in or near an orbiting spacecraft. see ISM

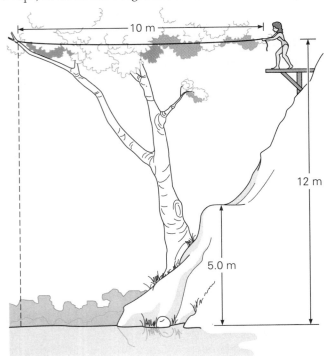

● **FIGURE 7.28 A swinging time** See Exercises 60 and 61.

● **FIGURE 7.29 Out for a walk**
Why does this astronaut seem to "float"? See Exercise 64.

● **65** Take a look at ● Fig. 7.30. If the cup were dropped, describe what would be observed as it fell. see ISM

66 ■ Using Eq. 7.17, compute the value of *g* at the surface of the Earth. see ISM

FIGURE 7.30 **Let it go** See Exercise 65.

67 ■ Compute the mass of the Sun using Eq. 7.23.
2.0×10^{30} kg

68 ■■ (a) Two objects, each with a mass of 1 metric ton, are in contact with each other. How far apart are the centers of gravitational force, if there is a mutual gravitational attraction of 0.00010 N between them? (b) If the objects are homogeneous spheres of the same size, what is their minimum uniform density? [*Hint:* Remember that they are in contact.] (a) 0.82 m (b) 3.4×10^3 kg/m^3

69 ■■ (a) What is the acceleration due to gravity on the top of Mt. Everest? The summit is about 8.8 km above sea level. (Ignore significant figures.) (b) How far above the Earth's surface would a 75-kg person have to go to "reduce" his weight by 10%?
(a) 9.74 m/s^2 (b) 3.4×10^5 m

● **70** ■■ Four identical masses of 2.5 kg each are located at the corners of a square with 1.0-m sides. What is the net force on one of the masses? 8.1×10^{-10} N, $\theta = 45°$

71 ■■ For a spacecraft going directly from the Earth to the moon, beyond what point will lunar gravity begin to dominate; that is, where will the lunar gravitational force be equal to the Earth's gravitational force? Are the astronauts on board truly weightless at this point? Explain. 3.4×10^8 m from Earth; there are other gravitational forces present

72 ■■ Which is greater, the gravitational force exerted on the Earth by the Sun or by the moon? (Compare these attractions by forming a ratio and giving a factor of how many times greater or smaller.) $F_{E\text{-}S} = (1.7 \times 10^2) F_{E\text{-}M}$

73 ■■ Plans are being made for a colony on Mars. Compute the approximate value of the acceleration due to gravity the colonists would experience on the Martian surface. (See Appendix I for data and compute explicitly.) 3.7 m/s^2

74 ■■ Assuming that we could get more earth (dirt) from somewhere, estimate how thick would an added uniform outer layer on the Earth have to be to have $g = 10.0$ m/s^2 exactly? (Take the average density of the Earth to be 5.52 g/cm^3 and $R_e = 6.40 \times 10^3$ km.) 90 km

75 ■■ A plane catapulted from an aircraft carrier goes from rest to a speed of 54 m/s in a distance of 50 m. How many g's of force does the pilot experience because of the catapult action? 3.0 g's

76 Shortly after blast-off, the upward acceleration of a Saturn V rocket is about 80 m/s^2. (a) How many g's of force are experienced by an astronaut on board as a result of this acceleration? (b) What is a 75-kg astronaut's apparent weight because of this acceleration? (a) 9.2 g's (b) 6.8×10^3 N

77 ■■ (a) What is the mutual gravitational potential energy of the configuration shown in Fig. 7.31 if all the masses are 1.0 kg? (b) What is the gravitational force per unit mass at the center of the configuration? (a) -2.5×10^{-10} J (b) 0

78 ■■ Find the mutual gravitational potential energy for the configuration of masses shown in Fig. 6.30. (Use the mass values given in part c of the related Exercise.)
5.29×10^{-10} J

● **79** ■■■ Show that the value of the acceleration due to gravity on the moon's surface is about one-sixth of that on the Earth's surface. [*Hint:* Use a ratio.] see ISM

80 ■■■ For the configuration of 1.0-kg masses in Fig. 7.31, suppose that another 1.0-kg mass (m_4) is placed on the y axis the same distance below the origin of m_2 is above. (a) Find the net gravitational force on m_4. (b) Show that the result from part (a) is *not* the same as the gravitational force on m_4 calculated by considering all of the mass of the initial triangular configuration to be located at its center of mass. [*Hint:* Recall that a homogeneous sphere is a special case; in

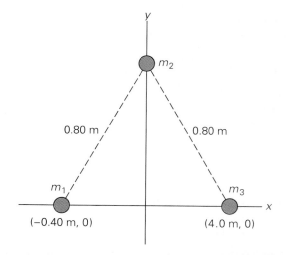

FIGURE 7.31 **Gravitational potential, gravitational force, and center of mass** See Exercises 77 and 80.

general, the center of mass of a body is not the same as the center of gravitational force.] (a) 2.2×10^{-10} N y (b) 2.4×10^{-10} N

7.6 ■ Kepler's Laws and Earth Satellites

81 A new planet is discovered and its period determined. The new planet's distance from the Sun could then be found using Kepler's (a) first law, (b) second law, (c) third law. (c)

82 For a satellite to orbit the Earth, it must (a) have a tangential velocity, (b) be acted on by a centripetal force, (c) have greater gravitational potential energy than on Earth, (d) all of these. (d)

83 How much work does the centripetal force do on a satellite in circular orbit about the Earth in one revolution? none

84 Must the Earth's rotation be taken into account when determining the direction in which a rocket is launched to put a satellite into orbit? Explain. yes; rocket has an initial velocity due to motion of the Earth

85 A shuttlecraft transporting a replacement crew to a space station in circular orbit about a planet is in a circular orbit at a lower altitude than that of the space station. (a) Explain how the shuttlecraft can change its orbit in order to dock with the space station. What effects would this change of orbit have on the shuttlecraft's energy and speed. (b) If the shuttlecraft moves to the altitude of the space station and establishes the same circular orbit, moving behind the space station, will it be able to overtake the space station and dock with it? Explain. (a) fire rockets for forward thrust (b) no; would have same speed

86 ■■ An instrument package is projected vertically upward to collect data at the top of the Earth's atmosphere (at an altitude of about 800 km). (a) What initial speed is required at the Earth's surface for the package to reach this height? (b) What percentage of the escape speed is this? (a) 3.7×10^3 m/s (b) 34%

87 ■■ A 50-kg instrument package is put into a circular orbit about the Earth at an altitude of 1000 km. What is the orbital kinetic energy of the satellite? 1.4×10^9 J

● **88** ■■ Compute the constant K of Kepler's third law for (a) Earth and (b) Venus. (See Appendix I for data.) (a) 3.0×10^{-19} s²/m (b) same

89 ■■ The asteroid belt that lies between Mars and Jupiter may be the debris of a planet that broke apart or did not form. The asteroid belt has a period of 5.0 years. How far from the Sun would this "fifth" planet have been? 4.4×10^{11} m

90 ■■ Two identical satellites are in circular orbits around the Earth at altitudes of 500 km and 850 km. Which satellite has the (a) greater kinetic energy and (b) greater potential energy? (c) How many times greater is each of these values than that for the other satellite? (Take $R_e = 6.40 \times 10^3$ km.) (a) K_{500} (b) U_{850} (c) $K_1 = (1.05) K_2$; $U_2 = (1.05) U_1$

91 ■■ (a) Suppose that the wheel space colony as shown in Fig. 1 of the chapter Insight had a diameter of 0.75 km. With what angular speed would the wheel have to rotate to give normal Earth gravity conditions at the wheel's circumference? (b) Suppose that the diameter were 1.5 km (almost a mile). Would the rotational speed have to be increased or decreased, and by what factor? (a) 0.16 rad/s (b) decreased by a factor of 0.71

92 ■■ An ultracentrifuge operates at a speed of 500,000 rpm. (a) What is the centripetal acceleration on a virus at a radial distance of 4.00 cm from the centrifuge's axis of rotation? (b) How does this acceleration compare with g, the acceleration due to gravity? (a) 1.1×10^8 m/s² (b) 1.1×10^7 g's

93 ■■■ In 1610, Galileo discovered four of the sixteen moons of Jupiter, the largest of which is Ganymede. This Jovian moon revolves around the planet in a nearly circular orbit whose radius is about 1.07×10^6 km in 7.16 days. Using this data, find the mass of Jupiter. 1.9×10^{27} kg

94 ■■■ Syncom satellites are communications satellites whose orbits are synchronized with the Earth's rotation. That is, they remain relatively stationary above some location on the Earth's surface because the period of satellite revolution equals that of Earth's rotation. What is the altitude of the satellites? 3.56×10^7 m or 2.2×10^4 mi

■ Additional Exercises

95 Ocean tides are primarily produced by the gravitational attraction of the moon. Explain how this attraction gives rise to two tidal "bulges" on opposite side of the Earth, resulting in two daily high tides. [*Hint:* Consider the inverse square relation of the distance and the gravitational force acting on the water on opposite sides of the Earth *and* on the Earth.] (You may want to go to the library for a little help on this one.) see ISM

96 ● Fig. 7.32, part of the depiction of a battle that took place nearly 2700 years ago, shows Assyrian warriors armed with slings and stones. Explain the physical principles involved in the use of this weapon. What factors would have limited its effectiveness? see ISM

97 The same side of the moon always faces the Earth, because the period of the moon's revolution around the Earth is the same as that of the moon's rotation. Prove that these periods being equal means that the same hemisphere of the moon is always visible from Earth. see ISM

98 Assuming that the Earth revolves about the Sun in uniform circular motion, find (a) the frequency of revolution in hertz and (b) the angular speed of revolution. (Use data given in Appendix I.) (a) 3.2×10^{-8} Hz (b) 2.0×10^{-7} rad/s

FIGURE 7.32 Sling, sling, sling! See Exercise 96.

99 For a scene in a movie, a stunt driver drives a 1500-kg pickup truck with a length of 4.25 m around a circular curve with a radius of curvature of 0.333 km (Fig. 7.33). The truck is to curve off the road, jump across a gully, and land on the other side 2.96 m below and 10.0 m away. What is the minimum centripetal acceleration the truck must have going around the circular curve so that it will clear the gully and land on the other side? 1.02 m/s²

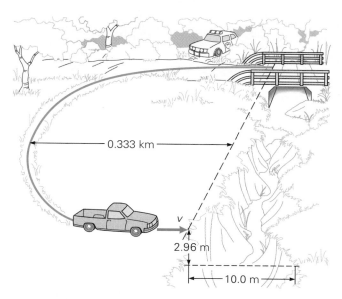

FIGURE 7.33 Over the gully See Exercise 99.

100 What is the escape speed for a satellite in a circular orbit about the Earth at an altitude of 750 km?
1.1 × 10³ m/s
101 Show that the escape speed for Earth is about 41% greater than the orbital speed of a satellite in a circular orbit near the surface of the Earth ($h \cong 0$), that is, that $v_e = 1.41v$.
see ISM
102 An astronaut has a weight of 735 N on Earth. What would be the astronaut's weight in a spacecraft in a circular orbit at an altitude of 450 km? 640 N

103 Compute the centripetal acceleration of a person on the rotating Earth (a) at the Equator, (b) at a latitude of 40°N, and (c) at the North Pole. (a) 3.4 × 10⁻² m/s²
(b) 2.6 × 10⁻² m/s² (c) 0
104 The value of the Kepler constant is $K = 2.97 \times 10^{-19}$ s²/m³, and $T^2 = (2.97 \times 10^{-19} \text{ s}^2/\text{m}^3)r^3$. With nonstandard units, K can be made equal to 1, and the equation may be written $T^2 = r^3$. Find the units of K that make it equal to 1, using Earth data. [*Hint:* The average, or mean, distance of the Earth from the Sun is used as a unit in astronomy and is called an astronomical unit (AU).] 1 y²/AU³

105 Two point masses, $m_1 = 0.50$ kg and $m_2 = 0.75$ kg, are located at the positions (0, 0) and (0.60 m, 0), respectively. (a) What is the gravitational force on a third point mass, $m_3 = 0.25$ kg, located midway between the first two? (b) What is the gravitational force per unit mass midway between m_1

and m_2? (c) At what point on the x axis would the net force on m_3 be zero? Are there any other zero points? see ISM
● **106** An automobile traveling at 60 km/h on a circular track with a diameter of 1.0 km speeds up to 90 km/h in 10 s. What is the magnitude of the average angular acceleration?
1.7 × 10⁻³ rad/s²
107 If an arc length is twice the radius of a circle, what angle is subtended by that arc? 115°

108 Using Kepler's law of periods, show that the centripetal force acting on the moon as it moves in its nearly circular orbit is an inverse square force. [*Hint:* Use the general form of the force and recall that $v = 2\pi r/T$.] see ISM

109 A particle traveling with uniform circular motion at a speed of 0.480 m/s in a path with a radius of 1.75 m experiences a tangential acceleration of 0.766 m/s². If the particle stays on the same circular path, (a) what is its centripetal acceleration and (b) its total acceleration at the end of 3.00 s?
(a) 4.42 m/s² (b) **a** = 4.42 m/s² **r** + 0.766 m/s² **t**
110 An object initially at rest undergoes an angular acceleration of 3.5 rad/s² and travels in a circular orbit that has a radius of 0.18 m. (a) What is the centripetal acceleration of the object when it has been moving for 3.0 s? (b) What is the instantaneous total acceleration at this time? (a) 20 m/s²
(b) **a** = 20 m/s² **r** + 0.63 m/s² **t**
111 A ball on a rope is swung in a horizontal circle 2.0 m above the ground. The ball has a centripetal acceleration of 24 m/s², and the radius of its circular path is 1.5 m. If the ball suddenly breaks free of the rope, where will it hit the level ground? 3.8 m

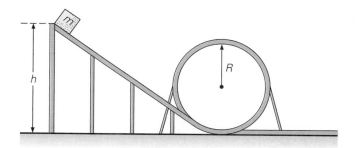

FIGURE 7.34 Loop-the-loop See Exercise 113.

● **112** (a) What is the angular width of a full moon in degrees and radians as viewed from Earth? (b) What is the angular width of a full Earth in degrees and radians as viewed from the moon? [*Hint:* Use data from Appendix I.]
(a) 0.53° = 9.2 × 10⁻³ rad (b) 1.9° = 3.4 × 10⁻² rad
113 A block of mass m slides down an inclined plane into a loop-the-loop of radius R (Fig. 7.34). (a) Neglecting friction, what is the minimum speed the block must have at the highest point of the loop to stay in the loop? [*Hint:* What force must act on the block at the top of the loop to keep it on a circular path?] (b) At what vertical height on the inclined plane (in terms of the radius of the loop) must the block be released if it is to have the required minimum speed at the top of the loop? (a) $v = [Rg]^{1/2}$ (b) $h = (5/2)R$

8

Rotational Motion and Equilibrium

This chapter deals with rotational dynamics. Topics include torques, mechanical equilibrium, stability, center of gravity, rotational work and power, rotational kinetic energy, and angular momentum.

Rotational motion is very important in physics because rotating objects are all around us: wheels on vehicles; gears and pulleys in machinery; carousels, like the one in the photo above; planets in our solar system; and even many bones in the human body. (Can you think of a few bones that rotate in sockets?)

In Chapter 7, the circular motions of particles in rotating bodies were considered, But the description of the rotational motion of a body as a whole must take into account the motions of all its particles. As you may have noticed in previous chapters, the study of physics proceeds in a stepwise manner, going from simple, ideal descriptions to more complex, more realistic ones, which better describe actual physical events. This chapter takes another step into the real world of things, going beyond the circular motions of particles to the rotations of real objects.

In addition, the conditions of equilibrium and stability are considered. In many situations we don't want things to rotate, or perhaps to move at all. (Think of a stepladder—or a bridge.)

Fortunately, the equations describing rotational motion can be written as almost direct analogs of those for translational (linear) motion. In Chapter 7, this similarity was pointed out for the kinematic equations. With the addition of equations describing rotational dynamics, you will be able to analyze the general motions of real objects.

8.1 ■ Rigid Bodies, Translations, and Rotations

It was convenient initially to consider motions of objects or particles with the understanding that the motion of an object may be represented by a particle located at its center of mass. Rotation, or spinning, was not a consideration then because a point mass has no physical dimensions. Rotational motion, on the other hand, is generally applied to solid extended objects or rigid bodies, on which this chapter will focus.

> A *rigid body* is an object or system of particles in which the interparticle distances (distances between particles) are fixed and remain constant.

Definition of a rigid body

A quantity of liquid water is obviously not a rigid body, but the ice that would be formed if the water were frozen would be. The discussion of rigid body rotation is thus conveniently restricted to solids. Actually, the concept of a rigid body is an idealization. In reality, the particles (atoms and molecules) of a solid vibrate constantly. Also, solids can undergo elastic (and inelastic) deformations (Chapter 6). Even so, most solids can be considered to be rigid bodies for purposes of analyzing rotational motion.

A rigid body may be subject to either or both of two different types of motions: translational and rotational. Translational motion is basically the linear motion we studied in previous chapters. If an object has only **translational motion** (● Fig. 8.1a), every particle has the same instantaneous velocity, which means there is no rotation. (Why?)

An object may have only **rotational motion**, about a fixed axis (Fig. 8.1b). In this case, all the particles of an object have the same instantaneous angular velocity and travel in circles about the axis of rotation. Although the axis of rotation of an object is commonly taken through its center of mass, this is not always the case. For example, you might pivot a meterstick through one end and have rotational motion about an axis through that end.

General rigid body motion is a combination of translational and rotational motions. When you throw a ball, the translational motion is described by the motion of its center of mass (as in projectile motion). But the ball may also spin, or rotate, and usually does.

A common example of rigid body motion involving both translation and rotation is rolling, as illustrated in Fig. 8.1c. The combined motion of any point or particle is given by the vector sum of its instantaneous velocity vectors. (Three points or particles are shown in the figure—top, middle, and bottom). At each instant, a rolling object rotates about an **instantaneous axis of rotation** through its point of contact with the surface (for a sphere) or along its line of contact with the surface (for a cylinder). The location of this axis changes with

> Demonstration/Activity: To distinguish between translational and rotational motion, mount an object such as a wheel or a stick so that it can rotate *about an axis*. Then set the support on which it is mounted on a lab cart and translate *it across the room*.

The words "rotation" and "revolution" are commonly used synonymously. In general, this book uses rotation when the axis of rotation goes through the body (the Earth's rotation on its axis) and revolution when the axis is outside the body (the revolution of the Earth about the Sun).

Review the discussion of tangential velocity in Section 7.2

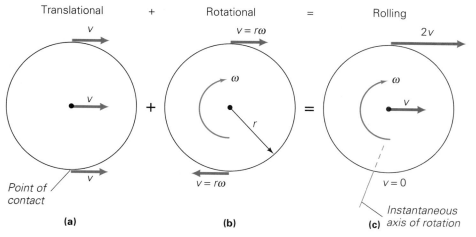

 Figure 8.1

● **FIGURE 8.1 Rolling—a combination of translational and rotational motions** As shown here with vectors, the center of mass of a rolling object translates at a speed *v* while rotating at an angular speed *ω*. The point (or line) of contact is instantaneously at rest and is on a line called the instantaneous axis of rotation. Note that the center of the object moves linearly and remains over the point of contact.

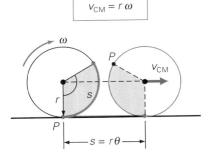

$$v_{CM} = r\omega$$

FIGURE 8.2 Rolling without slipping As an object rolls without slipping, the length of the arc between two points of contact on the circumference is equal to the linear distance traveled. This distance is $s = r\theta$. The speed of the center of mass is $v = r\omega$.

Figure 8.2

time. However, note in Fig. 8.1 that the point or line of contact of the body with the surface is instantaneously at rest (zero velocity), as can be seen from the vector addition of the combined motions at that point.

When an object rolls without slipping, the translational and rotational motions are related simply. For example, when a uniform ball (or cylinder) rolls in a straight line on a flat surface (Fig. 8.2), it turns through an angle θ, and a point on the object which was initially in contact with the surface moves through an arc distance s. From Chapter 7, you know that $s = r\theta$. The center of mass of the ball is directly over the point of contact and moves a linear distance s. Then

$$v = \frac{\Delta s}{\Delta t} = \frac{r\Delta\theta}{\Delta t} = r\omega$$

In terms of the speed of the center of mass and the angular speed, the *condition for rolling without slipping* is

$$v = r\omega \qquad (8.1)$$

The condition is also expressed by $s = r\theta$, where s is the distance an object rolls, or the distance the center of mass moves. Carrying Eq. 8.1 one step further ($\Delta v/\Delta t = r\Delta\omega/\Delta t$) gives an equation for accelerated rolling without slipping:

$$a = \frac{\Delta v}{\Delta t} = \frac{r\Delta\omega}{\Delta t} = r\alpha$$

8.2 ■ Torque, Equilibrium, and Stability

As with translational motion, a force is necessary to produce a change in rotational motion. But rotational motion is not always produced when a force acts on a rigid body. The motion or angular acceleration depends on *where* the force is applied. If a force acts through the axis of rotation, no rotation is produced (Fig. 8.3a). But when the line of action of the force does not go through the axis of rotation, the body rotates (Fig. 8.3b and c). The line of action of a force is simply a line extending through the force vector arrow, that is, the line along which the force acts.

As a practical example, think of applying a force to a heavy glass door that swings in and out. Where you apply the force makes a great difference in how easily the door opens, or rotates on its axis (through the hinges). Have you ever tried to open such a door and inadvertently pushed on the side near the hinges?

Torque

The rate of change of rotation depends not only on the magnitude of the force, but also on the perpendicular distance of its line of action from the axis of rotation $r_{\perp}$ (Fig. 8.3). Note from the figure that $r_{\perp} = r\sin\theta$, *where θ is the angle between a radial vector r and the force F*. This perpendicular distance is called the **moment arm** or **lever arm**.

The factors of force and distance are expressed in terms of their product, which is called **torque**, τ (from the Latin *torquere* meaning "to twist"). The

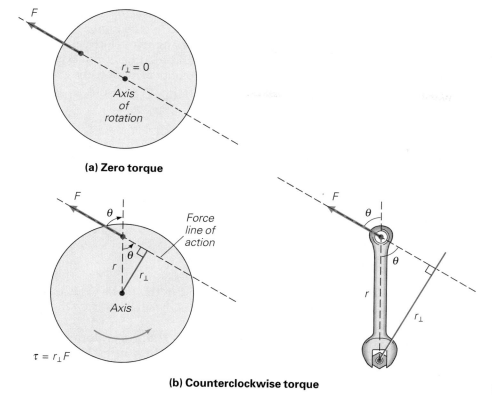

(a) Zero torque

(b) Counterclockwise torque

Figure 8.3

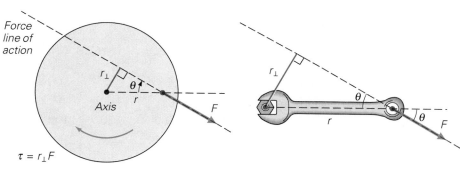

(c) Smaller clockwise torque

magnitude of the torque is

$$\tau = r_\perp F = rF \sin \theta \qquad (8.2)$$

The SI units for torque are meters times newtons (m-N). These are the same as the units for work, $W = Fd$ (N-m, or J). However, we will always write the units for torque as m-N to avoid confusion. Notice from Eq. 8.2 that $\tau = 0$ when $\theta = 0°$, or when the force acts through the axis of rotation. When $\theta = 90°$, the torque is maximum and the force acts perpendicularly to the lever arm, as when you push on a door to open it. Also note that $\mathbf{r}_\perp F = rF_\perp$.

Torque in rotational motion may be thought of as the analog of force in translational motion. An unbalanced or net force changes translational motion, and an unbalanced or net torque changes rotational motion. The product of a

8.2 Torque, Equilibrium, and Stability **253**

The right-hand rule for angular velocity is given in Section 7.2.

force and moment arm, both vectors, torque is also a vector. Its direction is always perpendicular to the plane of the force and momentum arm vectors and is given by a right-hand rule similar to that for angular velocity in Chapter 7. If the fingers of the right hand are curled around the axis of rotation in the direction that the torque would tend to produce rotational motion, the extended thumb points in the direction of the torque.

EXAMPLE 8.1 ■ Human Torque

In our bodies, torques produced by the contraction of our muscles cause some bones to rotate at joints. For example, in lifting something with the forearm, a torque is applied by the biceps muscle on the lower arm (● Fig. 8.4). With the axis of rotation through the elbow joint and the muscle attached 4.0 cm from the joint, what are the magnitudes of the applied torques for the cases in Fig. 8.4 if the muscle exerts a force of 600 N?

Solution. First we list the data given here and in the figure. This example demonstrates an important point. Recall that θ is the angle *between* the radial vector **r** and the force **F**.

Given: $r = 4.0 \text{ cm} = 0.040 \text{ m}$
$\qquad F = 600 \text{ N}$
(a) $\theta = 30° + 90°$
(b) $\theta = 90°$

Find: (a) τ (torque values)
(b) τ

(a) In this case, **r** is directed along the forearm, so the angle between the **r** and **F** vectors is $\theta = 30° + 90°$. Thus, $r \sin \theta = r \sin (30° + 90°) = r \cos 30°$. (See Appendix I for this and other useful trigonometric relationships.) Or, notice from the figure inset that $r_{\perp} = r \cos 30°$.
    Using Eq. 8.2, we have

$$\tau = r_{\perp} F = rF \sin (30° + 90°) = rF \cos 30°$$
$$= (0.040 \text{ m})(600 \text{ N})\cos 30° = 21 \text{ m-N}$$

at that instant.

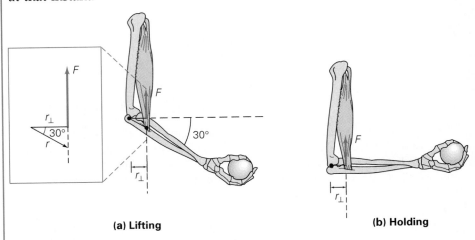

(a) Lifting

(b) Holding

● **FIGURE 8.4 Human torque** Example 8.1.

(b) Here, the lever arm and the line of action for the force are perpendicular ($\theta = 90°$), and $r_\perp = r \sin 90° = r$. Then,

$$\tau = r_\perp F = rF = (0.040 \text{ m})(600 \text{ N}) = 24 \text{ m-N}$$

In (a), since the ball was lifted by a rotation of the forearm, then there must have been a net torque. In (b), the ball is just being held and there is no rotational motion, then the net torque is zero. Where is the other balancing torque? As you might guess, the weights of the ball and the forearm produce torques that tend to cause a rotation in the direction opposite that of the applied torque.

Before considering rotational dynamics with net torques and rotational motions, let's first take a look at the situation where the forces and torques acting on a body are balanced.

Equilibrium

In general, equilibrium means that things are in balance or are stable. This definition applies in the mechanical sense to forces and torques. Unbalanced forces produce translational motion, but *balanced* forces produce the condition we call **translational equilibrium**. Similarly, unbalanced torques produce rotational motion, and *balanced* torques produce **rotational equilibrium**.

According to Newton's first law of motion, when the sum of the forces acting on a body is zero, it remains either at rest (static) or in motion with a constant velocity. In either case, the body is said to be in translational equilibrium. Stated another way, the *condition for translational equilibrium* is that the net force on a body is zero, $\Sigma F_i = 0$. In other words, the vector sum of all forces equals zero: $\Sigma F_i = F_1 + F_2 + F_3 + \cdots = 0$. It should be apparent that this condition is satisfied for the situations illustrated in (Fig. 8.5a and b). Forces with lines of action through the same point are called **concurrent forces**, and when these vectorially add to zero as in (a) and (b), the body is in translational equilibrium.

But what about the situation pictured in Fig. 8.5c? Here $\Sigma F_i = 0$, but the opposing forces will cause the object to rotate, and it will clearly not be in a state of static equilibrium. (Such a pair of equal and opposite forces not having the same line of action is called a *couple*.) Thus, the condition $\Sigma F_i = 0$ is a necessary, but not sufficient, condition for static equilibrium.

Since $\Sigma F_i = 0$ is the condition for translational equilibrium, you might predict (and correctly so) that $\Sigma \tau_i = 0$ is the *condition for rotational equilibrium*. That is, if the sum of the *torques* acting on an object is zero, then the object is rotationally at rest or rotates with a constant angular velocity.

Thus we see that there are two equilibrium conditions. Taken together, they define what is called **mechanical equilibrium**.

A body is said to be in mechanical equilibrium when the conditions for both translational and rotational equilibrium are satisfied:

Conditions for mechanical equilibrium

$$\begin{array}{ll} \Sigma F_i = 0 & \textbf{(for translational equilibrium)} \\ \Sigma \tau_i = 0 & \textbf{(for rotational equilibrium)} \end{array} \qquad (8.3)$$

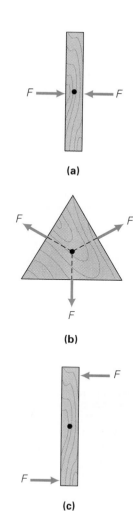

(a)

(b)

(c)

FIGURE 8.5 Equilibrium and forces   Forces with lines of action through the same point are said to be concurrent. The resultants of the concurrent forces acting on the objects in (a) and (b) are zero, and the objects are in equilibrium. In (c), the object is in translational equilibrium, but it will rotate.

Static refers to bodies that are at rest.

Video demonstration 5

A review of free-body diagrams may be necessary.

A rigid body in mechanical equilibrium may be either at rest or moving with a constant linear and/or angular velocity. An example of the latter is an object rolling without slipping on a level surface with the center of mass having a constant velocity. Of course, this is an ideal condition because there is always some friction in reality. Of greater practical interest is **static equilibrium**, the condition that exists when a rigid body is at rest. There are many instances in which we do not want things to move, and this absence of motion can occur only if the equilibrium conditions are satisfied. It is particularly comforting to know, for example, that a bridge you are crossing is in static equilibrium, and not subject to translational or rotational motions.

Let's consider examples of static translational equilibrium and static rotational equilibrium separately, and then one in which both apply.

EXAMPLE 8.2 ■ Translational Static Equilibrium

A picture hangs motionless on a wall as shown in (Fig. 8.6). If it has a mass of 5.0 kg, what are the tension forces in the wires?

Solution. Note that all the forces are concurrent—that is, their lines of action pass through a common point, the nail. Because of this, the condition for rotational equilibrium ($\Sigma\tau_i = 0$) is automatically satisfied—with respect to an axis of rotation along the nail, the moment arms of the forces are zero. Thus, we need consider only translational equilibrium.

You will find it helpful to isolate the forces acting on the picture in a free-body diagram as was done in Chapter 4 for force problems (see Fig. 8.6). In this case, the diagram shows the concurrent forces acting through their common point. Note that we have moved all the force vectors to the common point which is taken as the origin of the coordinate axes. The weight force mg acts downward.

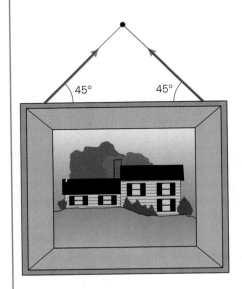

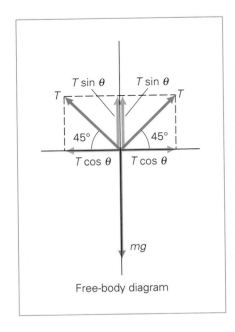

Free-body diagram

● **FIGURE 8.6 Translational static equilibrium** Since the picture hangs motionless on the wall, the sum of the forces acting on it must be zero. The forces are concurrent, with their lines of action passing through a common point at the nail, as shown in the free body diagram. Example 8.2.

With the system in static equilibrium, the net force is zero; that is, $\Sigma \mathbf{F}_i = 0$. Thus, the sums of the rectangular components are also zero: $\Sigma \mathbf{F}_{x_i} = 0$ and $\Sigma \mathbf{F}_{y_i} = 0$. Then

$$\Sigma \mathbf{F}_{x_i}: \qquad T \cos 45° - T \cos 45° = 0$$

This tells you that the x components of the forces are equal and opposite (by symmetry). For the y components, with two upward components of $T \sin 45°$, we have

$$\Sigma \mathbf{F}_{y_i}: \qquad T \sin 45° + T \sin 45° - mg = 0$$

or

$$2T \sin 45° - mg = 0$$

Thus,

$$T = \frac{mg}{2 \sin 45°} = \frac{(5.0\ \text{kg})(9.8\ \text{m/s}^2)}{2(0.707)} = 35\ \text{N}$$

EXAMPLE 8.3 ■ Rotational Static Equilibrium

Three masses are suspended from a meterstick as shown in (Fig. 8.7). How much mass must be suspended on the right side to have the system be in static equilibrium? (Neglect the mass of the meterstick.)

Solution. From the figure, we have

Given: $m_1 = 25$ g   **Find:** m_3 (unknown mass)
$\qquad r_1 = 50$ cm
$\qquad m_2 = 75$ g
$\qquad r_2 = 30$ cm
$\qquad r_3 = 35$ cm

The condition for translational equilibrium ($\Sigma \mathbf{F}_i = 0$) is automatically satisfied by the upward vertical reaction force of the stand on the stick. This force balances the stick's weight at the center of mass; thus, $R = Mg$, or $R - Mg = 0$, where M is the total mass. However, unless the proper mass for m_3 is placed on the right side, the stick will experience a net torque and rotate.

Notice that the masses on the left side produce torques that tend to rotate the stick counterclockwise, and the mass on the right side produces a torque that tends to rotate it clockwise. As pointed out earlier, torque is a vector and directional difference is important. As in the linear case in which + and − were used to express opposite directions (e.g., $+x$ and $-x$), we may designate torques as + and −, depending on the rotational motion they tend to produce. The rotational "directions" are taken as clockwise and counterclockwise around the axis of rotation, and we will use the sign convention shown in Fig. 8.7: A torque that tends to produce counterclockwise rotation will be taken as positive (+), and a torque that tends to produce a clockwise rotation will be taken as a negative (−).

This convention is arbitrary. The key thing to remember is that the magnitude of the sum of the counterclockwise torques must equal that of the clockwise torques to have rotational equilibrium. Thus, we apply the condition for

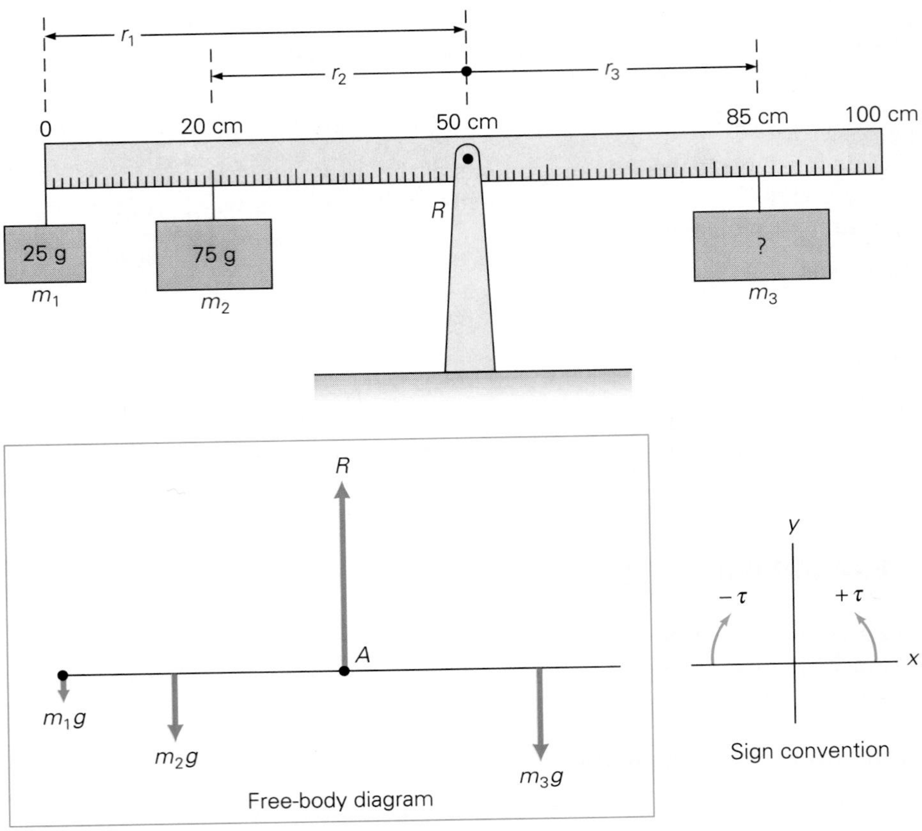

FIGURE 8.7 Rotational static equilibrium For the meterstick to be in rotational equilibrium, the sum of the torques acting about any selected axis must be zero. Example 8.3.

rotational equilibrium by summing the torques about an axis. Let's take this to be through the center of the stick, point A in Fig. 8.7. Then,

$$\Sigma\tau_i: \quad \tau_1 + \tau_2 - \tau_3 = r_1F_1 + r_2F_2 - r_3F_3 \qquad \text{(using our}$$
$$= r_1(m_1g) + r_2(m_2g) - r_3(m_3g) = 0 \quad \text{sign convention)}$$

Noting that the g's cancel out, and solving for m_3, we have

$$m_3 = \frac{m_1r_1 + m_2r_2}{r_3}$$

$$= \frac{(25 \text{ g})(50 \text{ cm}) + (75 \text{ g})(30 \text{ cm})}{35 \text{ cm}} = 100 \text{ g}$$

where it was convenient not to convert to standard units. (The mass of the stick was neglected. If the stick is uniform, however, its mass will not affect the equilibrium as long as the pivot point is at the 50-cm mark. Why?)

It should be noted that the axis of rotation could have been taken through any point along the stick. (Confirm this for yourself by repeating these calculations for another axis—for instance, one passing through m_1.) If a system is in static rotational equilibrium, the condition $\Sigma\tau_i = 0$ holds for any axis of rotation.

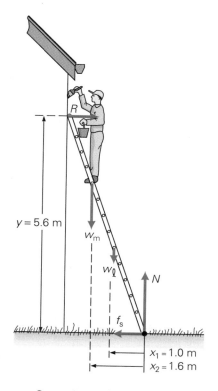

$y = 5.6$ m

w_m

w_ℓ

N

f_s

$x_1 = 1.0$ m

$x_2 = 1.6$ m

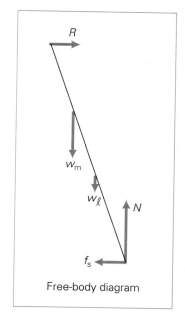

R

w_m

w_ℓ

N

f_s

Free-body diagram

• FIGURE 8.8 Static equilibrium
For the painter's sake, the ladder has to be in static equilibrium; that is, both the sum of the forces and the sum of the torques must be zero. See Example 8.4.

Sometimes the conditions for both translational and rotational equilibrium need to be written explicitly to solve the problem. The next example is one such case.

EXAMPLE 8.4 ■ Static Equilibrium

A ladder with a mass of 15 kg rests against a smooth wall • (Fig. 8.8). A painter, who has a mass of 78 kg, stands on the ladder as shown in the figure. What frictional force must act on the bottom of the ladder to keep it from slipping?

Solution.

    Given: $m_l = 15$ kg   *Find:* f_s (force of static friction)
            $m_m = 78$ kg
              Distances given in figure

Since the wall is smooth, there is negligible friction between it and the ladder, and only the normal reaction force of the wall acts on the ladder at this point.

    In applying the conditions for static equilibrium, you are free to choose any axis of rotation for the rotational condition. (The conditions must hold for any part of a system for static equilibrium; that is, there can't be motion in any part of the system.) Note that choosing an axis at the end of the ladder where it touches the ground eliminates the torques due to f_s and N since the moment arms are zero. Then you can write three equations (using mg for w):

> Discuss the problem of a ladder sliding down because of too little friction. Does the angle at which sliding tends to occur depend on how high the painter is working?

$$\Sigma_i F_{x_i}: \qquad R - f_s = 0$$

$$\Sigma_i F_{y_i}: \qquad N - m_m g - m_l g = 0$$

and

$$\sum_i \tau_i : \quad Ry - (m_l g)x_1 - (m_m g)x_2 = 0$$

The weight of the ladder is considered to be concentrated at its center of gravity. Solving the third equation for R and substituting the given values for the masses and distances gives

$$R = \frac{(m_l g)x_1 + (m_m g)x_2}{y}$$

$$= \frac{(15 \text{ kg})(9.8 \text{ m/s}^2)(1.0 \text{ m}) + (78 \text{ kg})(9.8 \text{ m/s}^2)(1.6 \text{ m})}{5.6 \text{ m}}$$

$$= 2.4 \times 10^2 \text{ N}$$

From the first equation, then,

$$f_s = R = 2.4 \times 10^2 \text{ N}$$

■ Problem-Solving Hint

As may be seen from the preceding examples, a good procedure to follow in working problems involving static equilibrium is as follows:

1. Sketch a space diagram of the problem.
2. Draw a free-body diagram, showing and labeling all external forces, and normally resolving the forces into x and y components.
3. Apply the equilibrium conditions. Sum the forces. ($\Sigma \mathbf{F}_i = 0$, or usually in component form $\Sigma \mathbf{F}_{x_i} = 0$ and $\Sigma \mathbf{F}_{y_i} = 0$.) Sum the torques. ($\Sigma \tau_i = 0$), remembering to select an appropriate axis of rotation to reduce the number of terms as much as possible.
4. Solve for the unknown quantities.

Stability and Center of Gravity

The equilibrium of a particle or a rigid body can also be described as being stable or unstable in a gravitational field. For rigid bodies, these categories of equilibria are conveniently analyzed in terms of the center of gravity. Recall from Chapter 6 that the **center of gravity** is the point at which all the weight of an object may be considered to be concentrated in representing it as a particle. When the acceleration due to gravity is constant, the center of gravity and the center of mass coincide.

If an object is in **stable equilibrium**, any small displacement results in a restoring force or torque, which tends to return the object to its original equilibrium position. As illustrated in • Fig. 8.9a, a ball in a bowl is in stable equilibrium. Analogously, the center of gravity of an extended body in stable equilibrium is essentially in a potential energy bowl. Any slight displacement raises its center of gravity, and a restoring force tends to return it to the position of minimum potential energy. This force is actually a torque that is due to a component of the weight force and that tends to rotate the object about a pivot point back to its original position.

The center of gravity is discussed in Section 6.5

Stable equilibrium

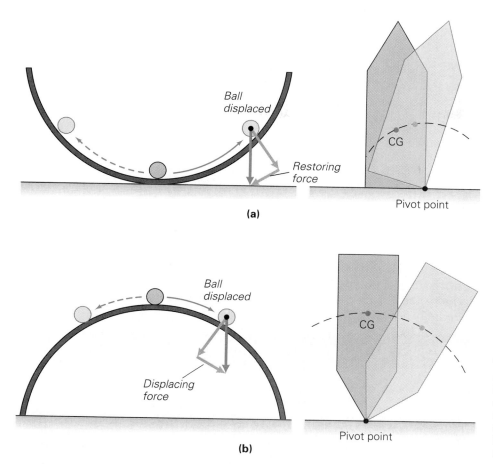

Figure 8.9

FIGURE 8.9 Stable and unstable equilibrium **(a)** When an object is in stable equilibrium, any small displacement from an equilibrium position results in a force or torque that tends to return the object to that position. A ball in a bowl (left) returns to the bottom after being displaced. Analagously, the center of gravity of an extended object (right) can be thought of as being in a potential energy bowl—a small displacement raises the cg, increasing the object's potential energy. **(b)** For an object in unstable equilibrium, any small displacement from its equilibrium position results in a force or torque that tends to take it further away from that position. The ball on top of an overturned bowl (left) is in unstable equilibrium. For an extended object (right), the center of gravity can be thought of as being on an inverted potential energy bowl—a small displacement lowers the cg, decreasing the object's potential energy.

For an object in **unstable equilibrium**, any small displacement from equilibrium results in a force or torque that tends to take the object further away from its equilibrium position. This situation is illustrated in Fig. 8.9b. Note that the center of gravity is at the top of an overturned, or inverted, potential energy bowl, that is, the potential energy is at a maximum in this case.

Small displacements or slight disturbances have profound effects on objects in unstable equilibrium—it doesn't take much to cause such an object to change its position. On the other hand, the angular displacement of an object in stable equilibrium can be quite substantial, and the object will still be restored to its equilibrium position. An object lying on a long side can be rotated through quite a distance and will still fall back to that original position. This is another way of expressing the *condition of stable equilibrium*:

> An object is in stable equilibrium as long as its center of gravity lies above and inside its original base of support.

When this is the case, there will always be a restoring torque (Fig. 8.10a). However, when the center of gravity or center of mass falls outside the base of support, over goes the object—because of a gravitational torque that rotates it away from its equilibrium position (Fig. 8.10b).

Rigid bodies with wide bases and low centers of gravity are therefore more stable and less likely to tip over. This relationship is evident in the design of

Unstable equilibrium

Demonstration/activity: Try to stand a meterstick on end on the desk. If it stands, it is in stable equilibrium. Why is it considered stable when many students will think it is unstable?

Condition for stable equilibrium

Compare the center of gravity for humans. What happens to the center of gravity for women when they are pregnant?

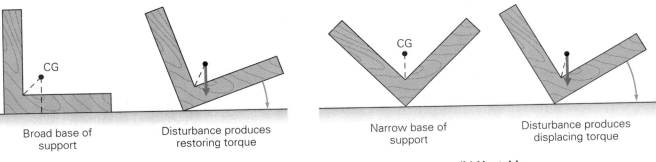

Broad base of support   **Disturbance produces restoring torque**

(a) Stable

Narrow base of support   **Disturbance produces displacing torque**

(b) Unstable

●**FIGURE 8.10 Examples of stable and unstable equilibrium** (a) When the center of gravity is above and inside an object's base of support, the object is in stable equilibrium (there is a restoring torque). (b) When the center of gravity lies outside the base of support, the object is unstable (there is a displacing torque).

high-speed race cars, which have wide wheel bases and centers of gravity close to the ground (●Fig. 8.11). Also, the location of the center of gravity of the human body has an effect on certain physical abilities. For example, women can generally bend over and touch their toes or touch their palms to the floor more easily than can men, who often fall over trying. On the average, men have higher centers of gravity (larger shoulders) than do women (larger pelvises), so it is more likely that a man's center of gravity will be outside his base of support when he bends over. (See Demonstration 6.)

DEMONSTRATION 6 ■ Stability and Center of Gravity

A demonstration to show that women generally have more stability than men—because of mass distribution and location of the center of gravity.

(a) An object is placed just beyond the fingertips of a man kneeling with elbows against his knees. Maintaining his balance, he can topple the object with his nose.

(b) With hands now clasped behind his back, the man's center of gravity shifts beyond his base of support (knees) as he bends over to topple the object, and he topples instead.

(c) A woman can generally do it both ways.

(d) Because her center of gravity is positioned lower in the body, it does not move outside the base of support (past her knees) as she bends with hands clasped behind her back.

(a)   (b)   (c)

FIGURE 8.11 Stable, stabler, stablest (a) The acrobat's base of support is very narrow—the small area of head-to-head contact. As long as his center of gravity remains above this area, he is in equilibrium, but a displacement of only a few inches would probably be enough to topple him. (Why he is in a spread-eagle position will become clearer in the next section.) (b) This balancing rock in Idaho has a slightly larger base of support; still, a displacement of its center of gravity by a few feet would probably send it crashing down. (c) Race cars are very stable because of their wide wheel bases and low centers of gravity.

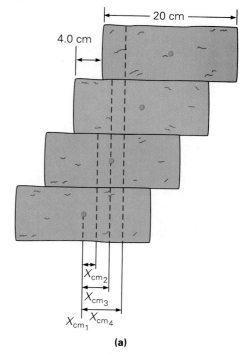

(a)

EXAMPLE 8.5 ■ Stack Them Up!

Uniform, identical bricks 20 cm long are stacked so that 4.0 cm of a brick extends beyond the brick beneath, as shown in ● Fig. 8.12a. How many bricks can be stacked in this way before the stack falls over?

Solution. The stack will fall over when its center of mass is no longer above its base of support, which is the bottom brick. All of the bricks have the same mass, and the center of mass of each is located at its midpoint.

    Taking the origin to be at the center of the bottom brick, the horizontal coordinate of the center of mass (or center of gravity) for the first two bricks in the stack is given by Eq. 6.19, where $m_1 = m_2 = m$ and x_2 is the displacement of the second brick:

$$X_{CM_2} = \frac{mx_1 + mx_2}{m + m}$$

$$= \frac{m(x_1 + x_2)}{2m} = \frac{x_1 + x_2}{2} = \frac{0 + 4.0 \text{ cm}}{2} = 2.0 \text{ cm}$$

Note that the masses of the bricks cancel out (since they are all the same). For three bricks,

$$X_{CM_3} = \frac{m(x_1 + x_2 + x_3)}{3m} = \frac{0 + 4.0 \text{ cm} + 8.0 \text{ cm}}{3} = 4.0 \text{ cm}$$

For four bricks,

$$X_{CM_4} = \frac{m(x_1 + x_2 + x_3 + x_4)}{4m} = \frac{0 + 4.0 \text{ cm} + 8.0 \text{ cm} + 12.0 \text{ cm}}{4} = 6.0 \text{ cm}$$

and so on.

    This series of results shows that the center of mass of the stack moves horizontally 2.0 cm for each brick added to the bottom one. For a stack of six bricks, the center of mass is 10 cm from the origin and directly over the edge of the bottom brick (2.0 cm × 5 *added* bricks = 10 cm, which is $\frac{1}{2}$ of the length

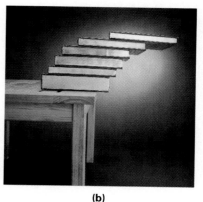

(b)

● **FIGURE 8.12 Stack them up!** (a) How many bricks can be stacked like this before the stack falls over? See Example 8.5. (b) You may want to try this experiment with books.

of the bottom brick), so the stack is in unstable equilibrium. The stack may not topple if the sixth brick is positioned very carefully, but it is doubtful that this could be done in practice. In any case, the seventh brick would definitely cause the stack to fall.

INSIGHT ■ Stability in Action

When riding a bicycle and going around a curve or making a circular turn on a level surface, a rider instinctively leans into the curve. Why is this? One would think that leaning over, rather than remaining upright, is more likely to cause a spill. However, leaning really does increase stability—it's all a matter of torques.

When a bicycle or vehicle goes around a level circular curve, a centripetal force is needed, as we learned in Chapter 7. This is generally supplied by the force of static friction between the tires and the road. As illustrated in Fig. 1a, the reaction force R of the ground on the bicycle is the sum of the required centripetal force F_c to round the curve and the normal force N.

Suppose the rider tried to go around the curve with these forces while remaining upright as shown in the figure. Note that the line of action of R does not go through the system's center of gravity (indicated by a dot). Considering an axis of rotation through this point, there would be a counterclockwise torque that would tend to rotate

the bicycle in such a way that the wheels would slide inward underneath the rider. However, if the rider leans inward at the proper angle (Fig. 1b), the line of action of R and the weight force both go through the center of gravity and there is no rotational instability (as the gentleman on the bicycle well knew).

(Being alert, you might observe there is still a torque on the rider. Indeed, when the rider leans into the curve, the weight force gives rise to a torque about an axis through the point of contact with the ground. This, along with the turning of the handle bars, causes the bicycle to turn. If the bicycle were not moving, there would be a rotation about this axis and the bicycle and rider would fall over.)

The need to lean into a curve is readily apparent in bicycle and motorcycle races on level tracks. Things can be made easier for the riders if tracks or roadways are banked to provide a natural lean (recall Section 7.3 and Exercise 7.44).

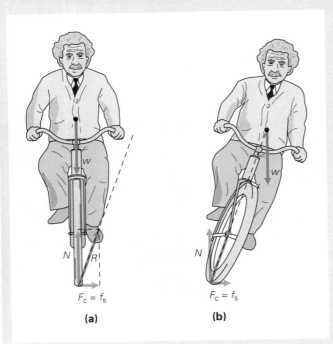

$F_c = f_s$

(a)

$F_c = f_s$

(b)

FIGURE 1 Leaning into a curve When rounding a curve or making a turn, a bicycle rider must lean into the curve. (This rider could have told you why.)

8.3 ■ Rotational Dynamics

Moment of Inertia

Torque is the rotational analog of force in linear motion, and a net torque produces rotational motion. To analyze this relationship, consider a constant force acting on a particle of mass m. The magnitude of the torque on the particle is

$$\tau = r_\perp F = rF_\perp = rma_\perp = mr^2\alpha \quad \textit{(torque on a particle)} \qquad (8.4)$$

Torque on a particle

where $a_\perp = r\alpha$ is the tangential acceleration (a_t). For the rotation of a system of fixed particles (a rigid body) about a fixed axis, this equation can be applied to each particle and the results summed over the entire body to find the total torque. Since, as you recall, all the particles of a rotating body have the same angular acceleration,

$$\begin{aligned}\tau &= \tau_1 + \tau_2 + \tau_3 + \cdots + \tau_n \\ &= m_1 r_1^2\alpha + m_2 r_2^2\alpha + m_3 r_3^2\alpha + \cdots + m_n r_n^2\alpha \\ &= (m_1 r_1^2 + m_2 r_2^2 + m_3 r_3^2 + \cdots + m_n r_n^2)\alpha\end{aligned}$$

or

$$\tau = \left(\sum_{i=1}^{n} m_i r_i^2\right)\alpha \qquad (8.5)$$

But for a rigid body, the masses (m_i's) and the distances from the axis of rotation (r_i's) are constant. Therefore, the quantity in the parentheses is constant and called the **moment of inertia**, I:

$$\boxed{I = \sum m_i r_i^2} \quad \textit{(moment of inertia)} \qquad (8.6)$$

Moment of inertia

The magnitude of the torque is then

$$\boxed{\tau = I\alpha} \quad \textit{(torque on a rigid body)} \qquad (8.7)$$

This is the rotational form of Newton's second law ($\mathbf{F} = m\mathbf{a}$ and $\boldsymbol{\tau} = I\boldsymbol{\alpha}$, in vector form). Keep in mind that *net* forces and torques ($\mathbf{F}_{net}$ and $\boldsymbol{\tau}_{net}$) are necessary to produce motions, although this is not indicated explicitly.

Rotational form of Newton's second law

As you might infer from a comparison of these two forms of Newton's law, the moment of inertia I is a measure of *rotational inertia*, or a body's tendency to resist change in its rotational motion. Although I is said to be constant for a rigid body and is the rotational analog of mass, you must keep in mind that, unlike the mass of a particle, the moment of inertia is referenced to a particular axis and can have different values for different axes. The moment of inertia also depends on the mass distribution *relative* to the axis of rotation, as the following example illustrates.

Demonstration/Activity: Use two ½-inch wooden dowels 36 inches long. Tape 500-gram masses to each end of one dowel and two other 500-gram masses about 12 inches in from the ends of the other dowel. Have a student grab the sticks in the centers, one in each hand, and attempt to twist them back and forth at the same rate.

Can you calculate how much "harder" it is to twist one than the other?

EXAMPLE 8.6 ■ Changes in Rotational Inertia

Find the moment of inertia for each of the simple one-dimensional dumbbell configurations in ● Fig. 8.13. (Consider the mass of the connecting bar to be negligible and ignore significant figures.)

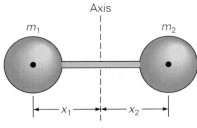

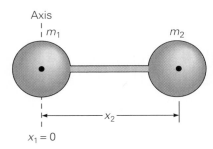

(a) $m_1 = m_2 = 30$ kg
    $x_1 = x_2 = 0.50$ m

(b) $m_1 = 40$ kg, $m_2 = 10$ kg
    $x_1 = x_2 = 0.50$ m

(c) $m_1 = m_2 = 30$ kg
    $x_1 = x_2 = 1.5$ m

Axis

(d) $m_1 = m_2 = 30$ kg
    $x_1 = 0$, $x_2 = 3.0$ m

(e) $m_1 = 40$ kg, $m_2 = 10$ kg
    $x_1 = 0$, $x_2 = 3.0$ m

FIGURE 8.13 Moment of inertia
The moment of inertia depends on the distribution of mass relative to a particular axis of rotation. Example 8.6.

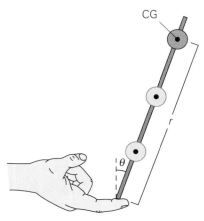

FIGURE 8.14 Greater stability with a higher center of gravity?
Example 8.7.

Solution.

Given: Values of m and r from figure   *Find:* $I = \Sigma m_i r_i^2$

With $I = m_1 r_1^2 + m_2 r_2^2$,

(a) $I = (30 \text{ kg})(0.50 \text{ m})^2 + (30 \text{ kg})(0.50 \text{ m})^2 = 15 \text{ kg-m}^2$

(b) $I = (40 \text{ kg})(0.50 \text{ m})^2 + (10 \text{ kg})(0.50 \text{ m})^2 = 12.5 \text{ kg-m}^2$

(c) $I = (30 \text{ kg})(1.5 \text{ m})^2 + (30 \text{ kg})(1.5 \text{ m})^2 = 135 \text{ kg-m}^2$

(d) $I = (30 \text{ kg})(0 \text{ m})^2 + (30 \text{ kg})(3.0 \text{ m})^2 = 270 \text{ kg-m}^2$

(e) $I = (40 \text{ kg})(0 \text{ m})^2 + (10 \text{ kg})(3.0 \text{ m})^2 = 90 \text{ kg-m}^2$

Example 8.6 clearly shows how the moment of inertia depends on the mass distribution relative to a particular axis of rotation. In general, the moment of inertia is larger the farther the mass is from the axis of rotation. This principle is important in the design of flywheels, which are used in automobiles to keep the engine running smoothly between cylinder firings. The mass of a flywheel is concentrated near the rim, giving a large moment of inertia, which resists changes in motion.

EXAMPLE 8.7 ■ Balancing Act

A rod with a movable ball, like that shown in Fig. 8.14, is more easily balanced if the ball is in a higher position. This is because with the ball in a higher position, (a) the system has a higher center of gravity and more stability, (b) when the center of gravity is off the vertical there is less torque and smaller angular acceleration, (c) the moment of inertia about the axis of rotation is larger, (d) the center of gravity is closer to the axis of rotation. *Clearly establish the reasoning and physical principle(s) used in determining your answer before checking it below. That is, **why** did you select your answer?*

Reasoning and Answer. With the ball in any position and the rod vertical, the system is in unstable equilibrium. We saw in Section 8.2 that rigid bodies with wide bases and *low* centers of gravity are more stable, so answer (a) isn't correct. Any slight movement will cause the rod to rotate about an axis through its point of contact. With the center of gravity (cg) at a higher position and off the vertical, there would be a greater lever arm (and thus a *greater* torque), so (b) too is incorrect. Also with the ball in a higher position, the center of gravity is *farther* from the axis of rotation, which eliminates (d). However, moving the cg farther from the axis of rotation has an interesting consequence—a greater moment of inertia or resistance to change in rotational motion.

With the ball in a higher position, as the rod starts to rotate there is a greater torque, but the increased moment of inertia produces a greater resistance to rotational motion. With the rod moving more slowly, the person has more time to adjust his or her hand and bring the finger and the axis of rotation under the center of gravity. The torque is then zero and the rod is again in equilibrium, albeit unstable. Thus, the answer is (c).

We can understand this result more clearly if we consider the angular

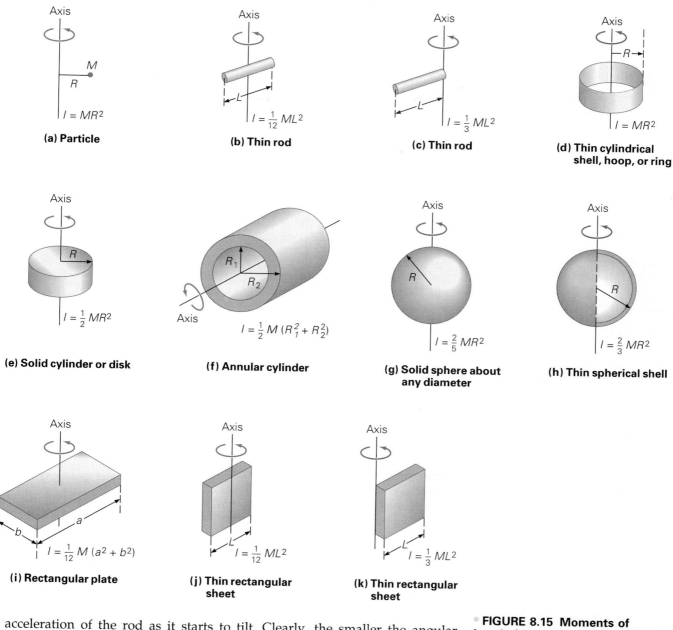

(a) Particle
$$I = MR^2$$

(b) Thin rod
$$I = \frac{1}{12} ML^2$$

(c) Thin rod
$$I = \frac{1}{3} ML^2$$

(d) Thin cylindrical shell, hoop, or ring
$$I = MR^2$$

(e) Solid cylinder or disk
$$I = \frac{1}{2} MR^2$$

(f) Annular cylinder
$$I = \frac{1}{2} M (R_1^2 + R_2^2)$$

(g) Solid sphere about any diameter
$$I = \frac{2}{5} MR^2$$

(h) Thin spherical shell
$$I = \frac{2}{3} MR^2$$

(i) Rectangular plate
$$I = \frac{1}{12} M (a^2 + b^2)$$

(j) Thin rectangular sheet
$$I = \frac{1}{12} ML^2$$

(k) Thin rectangular sheet
$$I = \frac{1}{3} ML^2$$

acceleration of the rod as it starts to tilt. Clearly, the smaller the angular acceleration, the more time is available to adjust one's hand under the rod to balance it. The angular acceleration of the rod is given by Eq. 8.7: $\alpha = \tau/I$. Now let us look at how the torque, τ, and moment of inertia, I, vary with r, the distance of the system's cg from the axis of rotation (through the balancing finger). Neglecting the mass of the rod and considering the cg to be simply that of the balls, $\tau = rF = rmg \sin \theta$ (Eq. 8.2) and $I = mr^2$ (Eq. 8.5). In other words, τ increases as r while I increases as r^2. Combining these relationships, we can see that α is proportional to $1/r$. Thus we have our paradoxical result: the farther the rod's cg from the axis of rotation, the more slowly it tilts and the easier it is to balance.

• **FIGURE 8.15 Moments of inertia for some uniform objects with common shapes**

Figure 8.15

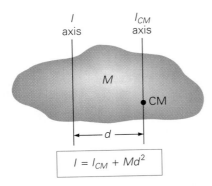

$$I = I_{CM} + Md^2$$

FIGURE 8.16 Parallel axis theorem The moment of inertia about an axis parallel to one through the center of mass of a body is given by $I = I_{CM} + Md^2$, where M is the total mass of the body and d is the distance between the two axes.

Demonstration/Activity: Place a coin on the end of a meterstick and allow the meterstick to undergo an angular acceleration by holding the opposite end stationary and allowing the stick to rotate downward. The coin will accelerate at a rate different from that of the meterstick. Using torques, determine the tangential acceleration of a point on the end of the meterstick.

Follow-up Exercise. When walking on a thin bar or rail, such as a railroad rail, you have probably found that it helps to hold your arms outstretched. Similarly, tightrope walkers often carry long poles. How does this help a performer to maintain his or her balance? (*Reasoning and answer may be found in the Answers to Exercises section at the back of the book.*)

The calculations for the moments of inertia of extended rigid bodies require math that is beyond the scope of this book. The results for some common shapes are given in Fig. 8.15. The rotational axes are generally taken along axes of symmetry, that is, running through the center of mass so as to give a symmetrical mass distribution. An exception is the rod with an axis of rotation through one end (Fig. 8.15c). This axis is parallel to an axis of rotation through the center of mass of the rod (Fig. 8.15b). The moment of inertia about such a parallel axis is given by a useful theorem called the **parallel axis theorem**:

$$I = I_{CM} + Md^2 \qquad (8.8)$$

where I is the moment of inertia about an axis that is parallel to one through the center of mass, I_{CM}, and at a distance d from it (Fig. 8.16), and M is the total mass of the body. For the axis through the end of the rod (Fig. 8.15c), the moment of inertia is obtained by applying the parallel axis theorem to the thin rod in Fig. 8.15b:

$$I = I_{CM} + Md^2 = \tfrac{1}{12}ML^2 + M\left(\frac{L}{2}\right)^2 = \tfrac{1}{12}ML^2 + \tfrac{1}{4}ML^2 = \tfrac{1}{3}ML^2$$

Applications of Rotational Dynamics

The rotational form of Newton's second law allows us to analyze dynamic rotational situations. The following examples illustrate how this is done. In such situations, it is very important to make certain that all the data are properly listed to help with the increasing number of variables.

FIGURE 8.17 Torque in action Example 8.8.

EXAMPLE 8.8 ■ Torque In Action

A student opens a 12-kg door by applying a constant force of 40 N at a perpendicular distance of 0.90 m from the hinges (Fig. 8.17). If the door is 2.0 m in height and 1.0 m wide, what is the magnitude of the angular acceleration of the door? (Assume that the door rotates freely on its hinges.)

Solution. From the problem, we can list the following:

Given: $M = 12$ kg   *Find:* α (magnitude of angular acceleration)
$F = 40$ N
$r_\perp = r = 0.90$ m
$h = 2.0$ m
$w = 1.0$ m

From the data and what is to be found, it should be apparent that we need to apply the rotational form of Newton's second law (Eq. 8.7), that is, $\tau = r_\perp F = I\alpha$. As can be seen, the problem boils down to finding the moment of inertia of the door.

Looking at Fig. 8.15, we see that (k) applies to a door rotating on hinges, and $I = \frac{1}{3}ML^2$, where L is the width of the door. Essentially, the problem is now solved. We must simply do the calculations:

$$\tau = I\alpha$$

or

$$\alpha = \frac{\tau}{I} = \frac{r_\perp F}{\frac{1}{3}ML^2} = \frac{3rF}{Mw^2}$$

$$= \frac{3(0.90\ \text{m})(40\ \text{N})}{(12\ \text{kg})(1.0\ \text{m})^2}$$

$$= 9.0\ \text{rad}/\text{s}^2$$

In problems involving pulleys in Chapter 4, the mass (and inertia) of the pulley was always neglected to simplify things. Now you know how to include this and can treat pulleys more realistically.

EXAMPLE 8.9 ■ Pulley with Inertia

A block of mass m hangs from a string wrapped around a frictionless pulley of mass M and radius R, as shown in Fig. 8.18. If the block descends from rest under the influence of gravity, what is the magnitude of the angular acceleration of the pulley? (Neglect the mass of the string.)

Solution. The pulley is a disk and thus has a moment of inertia of $I = \frac{1}{2}MR^2$ (Fig. 8.15e). It experiences a torque due to the tension force in the string (T). With $\tau = I\alpha$ (considering only the upper dashed box in Fig. 8.18),

$$\tau = r_\perp F = RT = I\alpha = \frac{1}{2}MR^2\alpha$$

so

$$\alpha = \frac{2T}{MR} \tag{1}$$

But T is unknown. Looking at the descending mass (the lower dashed box) and summing the forces gives

$$mg - T = ma$$

The linear acceleration of the block and the angular acceleration of the pulley are related by $a = R\alpha$, so

$$T = mg - mR\alpha \tag{2}$$

Using Eq. 2 to eliminate T from Eq. 1 gives

$$\alpha = \frac{2T}{MR} = \frac{2(mg - mR\alpha)}{MR}$$

and

$$\alpha = \frac{2mg}{(2m + M)R} \tag{3}$$

Note that you could use the same equations to find the linear acceleration of the block, in case you want to know how long it takes to fall a certain distance.

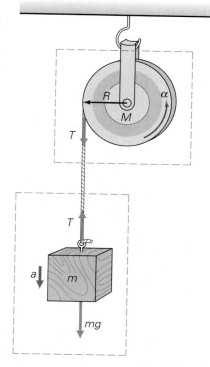

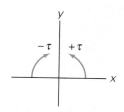

FIGURE 8.18 Pulley with inertia Taking the mass, or rotational inertia, of a pulley into account allows a more realistic description of the motion. Example 8.9.

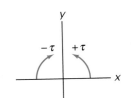

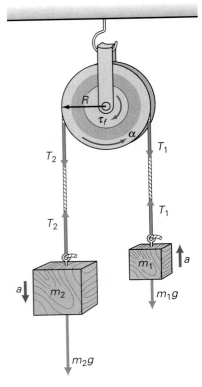

FIGURE 8.19 The Atwood
FIGURE 8.19 The Atwood machine revisited For this pulley, rotational inertia and frictional torque are taken into account. Example 8.10.

Problem-Solving Hint

For problems like those of Examples 8.8 and 8.9, dealing with coupled rotational and translational motions, keep in mind that the magnitude of the accelerations are usually related by $a = r\alpha$, while $v = r\omega$ relates the magnitude of the velocities if the angular velocity is constant. Applying Newton's second law (in rotational or linear form) to different parts of the system gives equations that can be combined using such relationships. Also, for rolling without slipping, $a = r\alpha$ and $v = r\omega$ relate the angular quantities to the linear motion of the center of mass.

Pulleys can be analyzed even more accurately. In Example 8.9, friction was neglected, but the next example includes it. (Neglecting the mass of the string will still give a good approximation because the string is relatively light. Taking the mass of the string into account would give a continuously varying mass, and such a problem is beyond the scope of this study.)

EXAMPLE 8.10 ■ One Up, One Down

For a pulley with two suspended masses, as shown in ● Fig. 8.19, the pulley itself has a mass of 0.20 kg, a radius of 0.15 m, *and* a constant torque of 0.35 m-N due to friction between the rotating pulley and its axle. What is the magnitude of the acceleration of the suspended masses if $m_1 = 0.40$ kg and $m_2 = 0.80$ kg? (Neglect the mass of the string.)

Solution.

Given: $M = 0.20$ kg   *Find:* a (magnitude of acceleration)
         $R = 0.15$ m
         $\tau_f = 0.35$ m-N
         $m_1 = 0.40$ kg
         $m_2 = 0.80$ kg

As before, Newton's second law is applied: the rotational form with the net torque acting on the pulley and the translational form with the net forces acting on the masses. Summing the torques on the pulley using our sign convention gives

$$\Sigma\tau = \tau_2 - \tau_1 - \tau_f = I\alpha$$

or $$RT_2 - RT_1 - \tau_f = (\tfrac{1}{2}MR^2)\left(\frac{a}{R}\right) \quad (1)$$

where the frictional torque opposes the rotation. (Note that the relationship $a = r\alpha$ was used to rewrite the expression for the torque in terms of the linear acceleration a, which is what we want to find.)

Then, summing the forces on the masses gives

$$m_2g - T_2 = m_2a \quad (2)$$

and

$$T_1 - m_1g = m_1a \quad (3)$$

In terms of the physics, the problem is solved. There are three equations (Eqs. 1, 2, and 3) and three unknowns (a, T_1, and T_2). The rest of the problem (computing the value of a) is simple algebra. The result is

$$a = \frac{(m_2 - m_1)gR - \tau_f}{(m_2 + m_1 + M/2)R} = 1.2 \text{ rad/s}^2$$

as you can show.

What would have been the case if m_2 had been on the right side of the pulley? By our torque sign convention, we would have $\Sigma \tau_i = -\tau_2 + \tau_1 + \tau_f = I(-\alpha)$. In this case, τ_2 and α are both in the minus (clockwise) direction, but the result would have been the same with the force summations written as above. (If you did not know the values of the masses or which way the pulley would rotate, you would simply assume a direction. If the result came out with the opposite sign, this would indicate that you assumed the wrong direction.)

Another application of rotational dynamics is the analysis of motion on an inclined plane. Previously, you worked with objects sliding down planes. Now you can let them roll.

EXAMPLE 8.11 ■ Rolling Down

A solid, rigid spherical ball of mass M and radius R is released at the top of a hard-surfaced inclined plane. It rolls without slipping, with only static friction between it and the plane (Fig. 8.20). What is the acceleration of the ball's center of mass?

Solution. Let's analyze this situation closely. Since the ball rolls without slipping, there is no relative motion between it and the plane at the point of contact (through which the instantaneous axis of rotation passes). At this

Figure 8.20

FIGURE 8.20 **Rolling down an inclined plane** For a rigid ball rolling down a hard-surfaced inclined plane, the force of static friction acts on the ball in response to the parallel component of the weight force. Example 8.11.

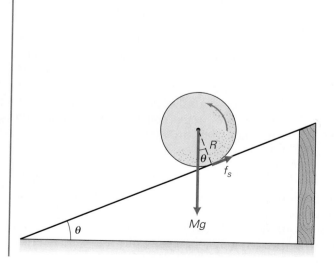

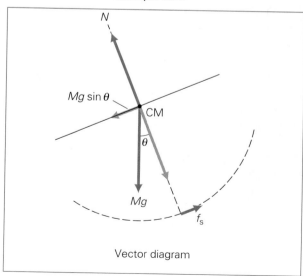

Vector diagram

point the force of static friction (f_s) acts on the ball. Without this force, the ball would slide down the plane. This force is perpendicular to the instantaneous axis of rotation, or directed up the plane.

You can analyze the motion using two different axes of rotation. First, let's consider an axis through the center of mass of the ball (the central dot in the figure) and parallel to the plane so that we can see the effect of the frictional torque. The acceleration of the ball's center of mass is down the plane, and the ball's angular acceleration is about an axis through that center. Since the reaction force (the normal force N) and the weight force act through the axis of rotation (see the vector diagram in the figure), they produce no torques, and the angular acceleration is produced entirely by the frictional torque due to f_s.

Then applying Newton's second law, $\tau = I\alpha$, and using the moment of inertia about an axis through the ball's center of mass as found in Fig. 8.15g ($I = \frac{2}{5}MR^2$), we have

$$\tau = Rf_s = I\alpha = (\tfrac{2}{5}MR^2)\left(\frac{a}{R}\right)$$

where $a = R\alpha$ was used. Then, solving for f_s,

$$f_s = \tfrac{2}{5}Ma$$

Looking at the translational motion of the center of mass and summing the forces acting parallel to the plane (see vector diagram) gives

$$Mg\sin\theta - f_s = Ma$$

or

$$Mg\sin\theta - \tfrac{2}{5}Ma = Ma$$

Thus,

$$a = \tfrac{5}{7}g\sin\theta$$

Here is another situation where the acceleration does not depend on M or R. Therefore, uniform spheres of any size or density will roll down the same plane with the same acceleration. But the acceleration will be *different* for objects with different shapes (for example, cylinders or hoops) because they have different moments of inertia. This reflects the dependence of rotational motion on mass distribution.

Now let's look at the situation from another perspective. Consider the rotational motion about the instantaneous axis of rotation, which runs through the point of contact of the sphere and the plane's surface. This point is instantaneously at rest, as noted earlier. As can be seen from Fig. 8.20, the normal force and the frictional force both act through this point, and so produce no torque. The torque on the ball from this perspective comes from the component of the weight force parallel to the plane, $Mg\sin\theta$ (see the vector diagram). The torque is then

$$\tau = r_\perp F = R(Mg\sin\theta) = I_i\alpha \tag{1}$$

where I_i is the moment of inertia *about the instantaneous axis of rotation*. Note that this is *not* equal to the moment of inertia used above, $I_{CM} = \frac{2}{5}MR^2$, which is the moment of inertia about an axis *through the center of mass* of the ball. You can find the moment of inertia about the instantaneous axis using the parallel axis theorem.

$$I_i = I_{CM} + Md^2 = \tfrac{2}{5}MR^2 + MR^2 = \tfrac{7}{5}MR^2 \qquad (2)$$

Substituting for I_i from Eq. 2 into Eq. 1 gives

$$(R \sin \theta)(Mg) = I_i \alpha = (\tfrac{7}{5}MR^2)\left(\frac{a}{R}\right)$$

and

$$a = \tfrac{5}{7}g \sin \theta$$

This is the same as the acceleration found using the other axis of rotation (as it should be).

8.4 ■ Rotational Work and Kinetic Energy

This section gives the rotational analogs associated with work and kinetic energy for constant torques. Because their development is very similar to that given for their linear counterparts, detailed discussion is not needed.

Rotational Work. For rotational motion, $W = Fs$ for a force acting along an arc length s:

$$W = Fs = F(r\theta) = \tau\theta$$

Thus,

$$\boxed{W = \tau\theta} \qquad (8.9) \quad \text{Rotational work}$$

In this book, both torque (τ) and angular displacement (θ) vectors are almost always along the axis of rotation, so you will not need to be concerned about parallel components as you were for translational work. The torque and angular displacement may be in opposite directions, in which case the applied torque decreases, or slows down, the rotation of the body. Then the rotational work is negative, which is analogous to F and d being in opposite directions for translational motion.

Rotational Power. An expression for power, or the time rate of doing work, is easily obtained from Eq. 8.9:

$$\boxed{P = \frac{W}{t} = \tau\left(\frac{\theta}{t}\right) = \tau\omega} \qquad (8.10) \quad \text{Rotational power}$$

The Work-Energy Theorem and Kinetic Energy

The relationship between rotational work and rotational kinetic energy may be derived as follows, starting with the equation for rotational work:

$$W = \tau\theta = I\alpha\theta$$

For a constant angular acceleration, $\omega^2 = \omega_o^2 + 2\alpha\theta$, and therefore

$$W = I\left(\frac{\omega^2 - \omega_o^2}{2}\right) = \tfrac{1}{2}I\omega^2 - \tfrac{1}{2}I\omega_o^2$$

Thus,

$$W = \tfrac{1}{2}I\omega^2 - \tfrac{1}{2}I\omega_o^2 = K - K_o = \Delta K \tag{8.11}$$

where K is the **rotational kinetic energy**.

Rotational kinetic energy

$$K = \tfrac{1}{2}I\omega^2 \tag{8.12}$$

Thus, the rotational work is equal to the change in the rotational kinetic energy. Consequently, in order to change the rotational energy of an object, a torque must be applied, since Eq. 8.9 shows that rotational work involves a torque.

It is possible to derive the kinetic energy of a rotating rigid body directly. Summing the instantaneous tangential velocities and kinetic energies of the individual particles of the body gives

$$K = \tfrac{1}{2}\Sigma m_i v_i^2 = \tfrac{1}{2}(\Sigma m_i r_i^2)\omega^2 = \tfrac{1}{2}I\omega^2$$

where $v_i = r_i\omega$. Thus, Eq. 8.12 doesn't represent a new form of energy. It is simply another expression for kinetic energy, in a form that is more convenient for rigid body rotation.

A summary of the analogous equations for translational and rotational motion is given in Table 8.1. (The table contains momentum, which is discussed in the next section.)

When an object has both translational and rotational motions, its total kinetic energy may be divided into parts to reflect this. For example, for a cylinder rolling without slipping on a level surface, the motion is purely rotational relative to the instantaneous axis of rotation, which is instantaneously at rest (see Example 8.11). The kinetic energy is

$$K = \tfrac{1}{2}I_i\omega^2$$

where I_i is the moment of inertia about the instantaneous axis. This moment of inertia is given by the parallel axis theorem (Eq. 8.8): $I_i = I_{CM} + MR^2$, where R is the radius of the cylinder. Then

$$K = \tfrac{1}{2}I_i\omega^2 = \tfrac{1}{2}(I_{CM} + MR^2)\omega^2 = \tfrac{1}{2}I_{CM}\omega^2 + \tfrac{1}{2}MR^2\omega^2$$

TABLE 8.1 ■ Translational and Rotational Quantities and Equations

Translational		Rotational	
Force:	F	Torque (moment of force):	$\tau = rF \sin\theta$
Mass (inertia):	m	Moment of inertia:	$I = \Sigma m_i r_i^2$
Newton's second law:	$F = ma$	Newton's second law:	$\tau = I\alpha$
Work:	$W = Fd$	Work:	$W = \tau\theta$
Power:	$P = Fv$	Power:	$P = \tau\omega$
Kinetic energy:	$K = \tfrac{1}{2}mv^2$	Kinetic energy:	$K = \tfrac{1}{2}I\omega^2$
Work-energy theorem:	$W = \tfrac{1}{2}mv^2 - \tfrac{1}{2}mv_o^2 = \Delta K$	Work-energy theorem:	$W = \tfrac{1}{2}I\omega^2 - \tfrac{1}{2}I\omega_o^2 = \Delta K$
Momentum:	$p = mv$	Momentum:	$L = I\omega$

But since there is no slipping, $R\omega = v_{cm}$, and

$$\boxed{\underset{\substack{\text{total} \\ \text{KE}}}{K} = \underset{\substack{\text{rotational} \\ \text{KE}}}{\tfrac{1}{2}I_{CM}\omega^2} + \underset{\substack{\text{translational} \\ \text{KE}}}{\tfrac{1}{2}Mv_{CM}^2}}$$

(8.13)   **Rolling kinetic energy**

Note that although a cylinder was used as an example here, this is a general result and applies to any object that is rolling without slipping. Thus, the total kinetic energy of such an object is the sum of two contributions: the translational kinetic energy of the center of mass and the rotational kinetic energy relative to a horizontal axis through the center of mass.

> *A rolling body has both translational kinetic energy and rotational kinetic energy.*

EXAMPLE 8.12 ■ A Division of Energy

A uniform solid 1.00-kg cylinder rolls without slipping on a flat surface at a speed of 2.40 m/s. (a) What is the total kinetic energy of the cylinder? (b) What percentage of this is rotational kinetic energy?

Solution.

Given: $M = 1.00$ kg   *Find:* (a) K (total kinetic energy)
  $v_{CM} = 2.40$ m/s
  $I_{CM} = \tfrac{1}{2}MR^2$ (from   (b) $\dfrac{K_r}{K}$ ($\times$ 100%) (percentage of
      Fig. 8.15e)   rotational energy)

(a) The cylinder rolls without slipping, so the condition $v_{CM} = R\omega$ applies. Then

$$K = \tfrac{1}{2}I_{CM}\omega^2 + \tfrac{1}{2}Mv_{CM}^2 = \tfrac{1}{2}(\tfrac{1}{2}MR^2)\left(\frac{v_{CM}}{R}\right)^2 + \tfrac{1}{2}Mv_{CM}^2$$

$$= \tfrac{3}{4}Mv_{CM}^2 = \tfrac{3}{4}(1.00 \text{ kg})(2.40 \text{ m/s})^2 = 4.32 \text{ J}$$

(b) The rotational kinetic energy K_r of the cylinder is

$$K_r = \tfrac{1}{2}I_{CM}\omega^2 = \tfrac{1}{2}(\tfrac{1}{2}MR^2)\left(\frac{v_{CM}}{R}\right)^2 = \tfrac{1}{4}Mv_{CM}^2$$

Then

$$\frac{K_r}{K} = \frac{\tfrac{1}{4}Mv_{CM}^2}{\tfrac{3}{4}Mv_{CM}^2} = 0.33 \ (\times \ 100\%) = 33\%$$

Thus, the total kinetic energy of the cylinder is made up of rotational and translational parts, with 33% being rotational.

Don't think that this division of energy is a general result. It is easy to show that the percentage is different for objects with different moments of inertia. For example, you should expect a rolling sphere to have a smaller percentage of rotational kinetic energy than a cylinder has since it has a smaller moment of inertia. Note that the radius of the cylinder was not needed at all, and that neither the mass nor any numerical values were needed in part (b).

Potential energy can be brought into the act by applying the conservation of energy to a cylinder rolling down an inclined plane.

EXAMPLE 8.13 ■ Rolling Down versus Sliding Down

A uniform solid cylinder is released from rest at a height of 0.45 m near the top of an inclined plane (Fig. 8.21). If the cylinder rolls down the plane without slipping and there is no energy loss due to friction, what is the linear speed of the cylinder's center of mass at the bottom of the incline?

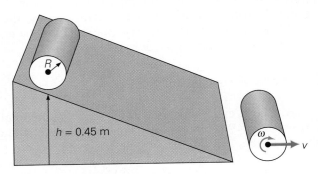

FIGURE 8.21 Rolling motion and energy When an object rolls down an inclined plane, potential energy is converted to translational *and* rotational kinetic energy. This makes the rolling slower than frictionless sliding. Example 8.13.

Solution.

Given: $h = 0.45$ m **Find:** v_{CM} (speed of CM)
$I_{CM} = \frac{1}{2}MR^2$ (from Fig. 8.15e)

Since the total mechanical energy of the cylinder is conserved, you can write

$$E_o = E$$

or

$$\underset{\text{at rest}}{Mgh} = \underset{\text{at bottom of incline}}{\frac{1}{2}I_{CM}\omega^2 + \frac{1}{2}Mv_{CM}^2}$$

Using the condition $v_{CM} = R\omega$ gives

$$Mgh = \frac{1}{2}(\frac{1}{2}MR^2)\left(\frac{v_{CM}}{R}\right)^2 + \frac{1}{2}Mv_{CM}^2$$

which, solved for v_{CM}, is

$$v_{CM} = \sqrt{\tfrac{4}{3}gh} = \sqrt{\tfrac{4}{3}(9.80 \text{ m/s}^2)(0.45 \text{ m})} = 2.4 \text{ m/s}$$

Again, not much data were needed here. If the surface of the inclined plane were frictionless and the cylinder slid down, the speed of the center of mass at the bottom would be $v_{CM} = \sqrt{2gh}$, since $Mgh = \frac{1}{2}Mv_{CM}^2$ in that case. Thus, v_{CM} would be greater if the cylinder slid down without friction. Can you see why?

As Example 8.13 shows, v_{CM} is independent of M and R. The masses and radii cancel out, so all objects of a particular shape (with the same formula for the moment of inertia) roll with the same speed, regardless of size or density.

But the rolling speed does vary with the moment of inertia, which varies with the shape. Therefore, rigid bodies with different shapes roll with different speeds. For example, if you released a cylindrical hoop, a solid cylinder, and a uniform sphere at the same time from the top of an inclined plane, the sphere would win the race to the bottom, followed by the cylinder, with the hoop coming in last—every time!

You can try this experiment using a couple of food cans or other cylindrical containers—one full of some solid material (a rigid body) and one empty with the ends cut out—and a smooth ball of some sort. Remember that the masses and the radii make no difference. You might think that an annular cylinder (a hollow cylinder with inner and outer radii that vary appreciably—Fig. 8.15f) would be a possible front-runner, or front-roller, in such a race, but it wouldn't win. The rolling race down an incline is fixed even when you vary the masses and the radii. (See Exercise 70.)

Another aspect of rolling is discussed in the Insight feature.

INSIGHTS ■ Slide or Roll to a Stop?—Antilock Brakes

In an emergency situation, you may instinctively jam on the brakes of a car, trying to come to a quick stop, that is, to stop in the shortest distance. But with the wheels locked, the car skids, or slides, to a stop, often out of control. In this case, the force of sliding friction is acting on the wheels.

In order to prevent this, you learn to pump the brakes in order to roll rather than slide to a stop, particularly on wet or icy roads. Many newer automobiles have computerized antilock braking systems (ABS) that do this automatically. When the brakes are applied firmly and the car begins to slide, sensors in the wheels note this and a computer takes over control of the braking system. It momentarily releases the brakes and then varies the brake-fluid pressure with a pumping action so that the wheels will continue to roll without slipping.

In the absence of sliding, both rolling friction and static friction act. In many cases, however, the force of rolling friction is small and only static friction need be taken into account, as will be done here.

Does sliding instead of rolling make a big difference in the stopping distance for an automobile? It is easy to calculate the difference by assuming that rolling friction is negligible. Although the external force of static friction does no work to dissipate energy in slowing a car (this is done internally by friction on the brake pads), it does determine whether the wheels roll or slide.

In Example 2.7, a vehicle stopping distance was given by

$$x = \frac{v_0^2}{2a}$$

By Newton's second law,

$$F = f = \mu N = \mu mg = ma$$

and the stopping acceleration is then

$$a = \mu g$$

Thus,

$$x = \frac{v_0^2}{2\mu g} \qquad (1)$$

But, as was noted in Chapter 4, the coefficient of sliding (kinetic) friction is generally less than that of static friction; that is, $\mu_k < \mu_s$. The general difference between rolling and sliding stops can be seen by using the same initial velocity (v_0 for both cases). Then, using Eq. 1 to form a ratio,

$$\frac{x_{roll}}{x_{slide}} = \frac{\mu_k}{\mu_s} \quad \text{or} \quad x_{roll} = \left(\frac{\mu_k}{\mu_s}\right) x_{slide}$$

The value of μ_k for rubber on wet concrete from Table 4.1 is 0.60, and the value of μ_s for these surfaces is 0.80. Using these values for a comparison of the stopping distances gives

$$x_{roll} = \left(\frac{0.60}{0.80}\right) x_{slide} = (0.75)\, x_{slide}$$

The car comes to a rolling stop in 75 percent of the distance required for a sliding stop, for example, 15 m instead of 20 m. Although this may vary for different conditions, it could be an important, perhaps life-saving, difference.

8.5 ■ Angular Momentum

Another important quantity in rotational motion is angular momentum. Recall from Chapter 6 how linear momentum is associated with force. Similarly, angular momentum is associated with torque. As we have seen, torque is the product of a moment arm and a force. In a similar manner, **angular momentum (L)** is the product of a moment arm and linear momentum. For a particle of mass m, the momentum is $p = mv$ and $v = r\omega$. The magnitude of the angular momentum is

Particle angular momentum

$$L = r_\perp p = mr_\perp v = mr_\perp^2 \omega \qquad (8.14)$$

where v is the tangential speed of the particle, $r_\perp$ the moment arm, and ω the angular speed. For circular motion, $r_\perp = r$, since $\mathbf{v}$ is perpendicular to $\mathbf{r}$.

For a system of particles or a rigid body, the total magnitude of the angular momentum is

$$L = (\Sigma m_i r_i^2)\omega = I\omega \qquad (8.15)$$

which, in vector notation, is

Rigid body angular momentum

$$\boxed{\mathbf{L} = I\boldsymbol{\omega}} \qquad (8.16)$$

Thus, $\mathbf{L}$ is in the direction of the angular velocity vector ($\boldsymbol{\omega}$) as given by the right-hand rule.

Review Eq. 6.3, Section 6.1

For linear motion, momentum is related to force by $\mathbf{F} = \Delta \mathbf{p}/\Delta t$. Angular momentum is analogously related to torque, as can be easily shown:

$$\boldsymbol{\tau} = I\boldsymbol{\alpha} = \frac{I\,\Delta\boldsymbol{\omega}}{\Delta t} = \frac{\Delta(I\boldsymbol{\omega})}{\Delta t} = \frac{\Delta \mathbf{L}}{\Delta t}$$

That is,

Torque—the time rate of change of angular momentum

$$\boxed{\boldsymbol{\tau} = \frac{\Delta \mathbf{L}}{\Delta t}} \qquad (8.17)$$

Thus, torque is equal to the time rate of change of angular momentum.

Conservation of Angular Momentum

Eq. 8.17 was derived using $\tau = I\alpha$, which applies to a rigid system of particles or a rigid body having a constant moment of inertia. However, Eq. 8.17 is a general one that also applies to a nonrigid system of particles. In such a system, there may be a change in the mass distribution and a change in the moment of inertia. As a result, there may be an angular acceleration even in the absence of a torque. How can this be?

If the net torque on a system is zero, then, by Eq. 8.17, $\tau = \Delta \mathbf{L}/\Delta t = 0$, and

$$\Delta \mathbf{L} = \mathbf{L} - \mathbf{L}_o = I\boldsymbol{\omega} - I_o\boldsymbol{\omega}_o = 0$$

or

$$\boxed{I\boldsymbol{\omega} = I_o\boldsymbol{\omega}_o} \qquad (8.18)$$

Thus, the condition for the **conservation of angular momentum** is as follows:

> In the absence of an external, unbalanced torque, the total angular momentum of a system is conserved (remains constant).

As with total linear momentum, the internal torques arising from internal forces cancel out.

For a rigid body with a constant moment of inertia (that is, $I = I_o$), the angular speed remains constant ($\omega = \omega_o$) in the absence of a net torque. But it is possible for the moment of inertia to change in some systems, giving rise to a change in the angular velocity, as the following example illustrates.

EXAMPLE 8.14 ■ Conservation of Angular Momentum

A small ball at the end of a string that passes through a tube is swung in a circle, as illustrated in (Fig. 8.22). When the string is pulled downward through the tube, the angular speed of the ball increases. (a) Is this caused by a torque due to the pulling force? (b) If the ball is initially swung in a circle with a radius of 0.30 m at a speed of 2.8 m/s, what will be its tangential speed if the string is pulled down far enough to reduce the radius of the circle to 0.15 m?

Solution.

Given: $r_1 = 0.30$ m   *Find:* (a) Cause of the increase in angular speed
$\quad\quad\quad r_2 = 0.15$ m   (b) v_2 (tangential speed)
$\quad\quad\quad v_1 = 2.8$ m/s

(a) The change in the angular velocity, or the angular acceleration, is not caused by a torque due to the pulling force. The force on the ball as transmitted

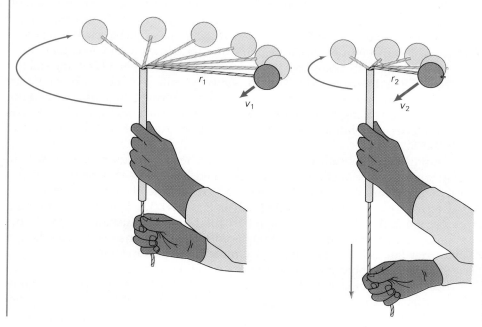

FIGURE 8.22 Conservation of angular momentum When the string is pulled downward through the tube, the revolving ball speeds up. Example 8.14.

(a)

(b)

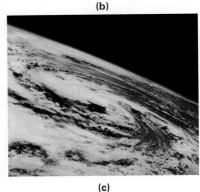

(c)

FIGURE 8.23 Moment of inertia change (a) Spinning slowly with masses in outstretched arms, the man's moment of inertia is relatively large (*r* of masses is large). Note that he is isolated, with no external torques (neglecting friction), so the angular momentum, $L = I\omega$, is conserved. (b) Pulling the arms inward decreases the moment of inertia (*r* of masses decreases). Consequently ω must increase, and he goes into a dizzy spin. (c) The same principle helps to explain the violence of the winds that spiral around the center of a hurricane. As air rushes in toward the center of the storm, where the pressure is low, its rotational velocity must increase for angular momentum to be conserved.

by the string acts through the axis of rotation, and therefore the torque is zero. As the rotating portion of the string is shortened, the moment of inertia of the ball ($I = mr^2$, from Fig. 8.15) decreases. This causes the ball to speed up, since in the absence of an external torque, the angular momentum of the ball is conserved.

(b) Since the angular momentum is conserved,

$$I_o\omega_o = I\omega$$

Then, using $I = mr^2$ and $\omega = v/r$ gives

$$mr_1v_1 = mr_2v_2$$

and

$$v_2 = \left(\frac{r_1}{r_2}\right)v_2 = \left(\frac{0.30 \text{ m}}{0.15 \text{ m}}\right)2.8 \text{ m/s} = 5.6 \text{ m/s}$$

When the radial distance is shortened, the ball speeds up.

You can also look at this situation in terms of work and energy. The pulling force produces no torque, but it does do work, and the kinetic energy of the ball increases.

Example 8.14 should help you understand Kepler's law of areas. When a planet is closer to the Sun in its elliptical orbit, and so has a shorter moment arm, its speed is greater by the conservation of angular momentum. Similarly, when an orbiting satellite changes the altitude of its orbit, it speeds up or slows down in accordance with the same principle.

A popular lecture demonstration of the conservation of angular momentum is shown in ●Fig. 8.23. A person sitting on a stool that rotates holds weights with his arms outstretched and is started slowly rotating. An external torque to start this rotation must be supplied by someone else, because the person on the stool cannot initiate the motion by himself. (Why not?) Once rotating, if the person brings his arms inward, the angular speed increases and he spins much faster. Extending his arms again slows him down. Can you explain this?

Refer to Eq. 8.15. If L is constant, what happens to ω when I is made smaller by reducing r? Ice skaters do dizzy toe spins by pulling in their arms to reduce their moments of inertia. Similarly, a diver spins during a high dive by tucking in, greatly decreasing his or her moment of inertia. The enormous wind speeds of tornadoes and hurricanes represent another example of the same effect (Fig. 8.23c).

Angular momentum **L** is a vector, and when it is conserved or constant, its magnitude *and* direction remain unchanged. Thus, when no external torques act, the direction of **L** is fixed in space. This is the principle of the gyrocompass, as well as of passing a football accurately (●Fig. 8.24). The **L** vector of a spinning gyroscope in the compass is set in a particular direction (usually north). In the absence of external torques, the compass direction remains fixed, even though its carrier (an airplane or ship, for example) changes directions.

A football is normally passed with a spiraling rotation. This spin, or gyroscopic action, stabilizes the ball's motion in the direction of its spin axis. Similarly, rifle bullets are set spinning by the rifling in the barrel for directional stability.

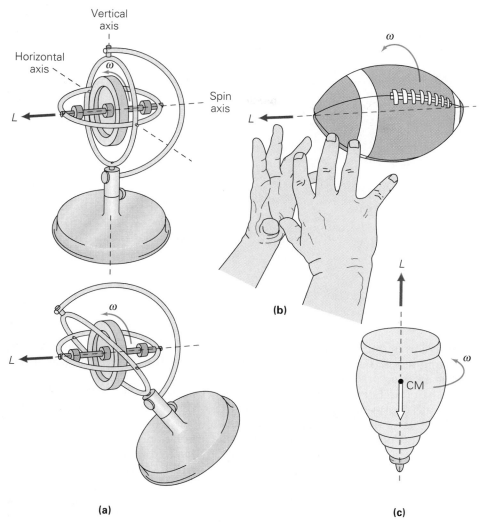

(a)

(b)

(c)

FIGURE 8.24 Constant direction of angular momentum When the angular momentum is conserved, its direction is constant in space. **(a)** This may be demonstrated with a gyroscope, a rotating wheel that is universally mounted on gimbals (rings) so that it is free to turn about any axis. This is the principle of the gyrocompass. **(b)** Gyroscopic action also occurs when a football is passed. **(c)** The axis of rotation of a sleeping top passes through its center of gravity. Thus, there is no torque due to the top's weight.

The motion of a "sleeping" top is also illustrated in Fig. 8.24. A spinning top will stand straight up with its angular momentum vector fixed in space for some time. Note in the figure that the axis of rotation passes through the top's center of gravity, and thus there is no torque due to the top's weight.

However, you know that a sleeping top—or a gyroscope—eventually slows down because of frictional torques. In doing so, you have probably noticed how the spin axis revolves, or *precesses*, about the vertical axis. It revolves tilted over, so to speak (Fig. 8.25). Since the gyroscope precesses, the angular momentum vector **L** is no longer constant in direction, indicating that a torque must be acting to give a change (Δ**L**) with time. As can be seen from the figure, the torque arises from the vertical component of the weight force, since the center of gravity no longer lies directly above the point of support or on the vertical axis of rotation. The instantaneous torque is such that the gyroscope's axis moves or precesses about the vertical axis.

In a similar manner, the Earth's rotational axis precesses. The Earth's rotational spin axis is tilted $23\frac{1}{2}°$ with respect to a line perpendicular to the plane of its revolution about the Sun, and precesses about this line (Fig. 8.25b). The

Demonstration/Activity: Have a student sit on a friction-free platform or stool holding a rotating bicycle wheel with its axis about 45 degrees from the vertical. Then start rotating the student while asking him to explain the tendency of the wheel. If the student allows the rotating wheel to do what it "wants" to do, it will tend to rotate so that its axis is vertical. In what direction will a rotating wheel attached to the earth tend to point?

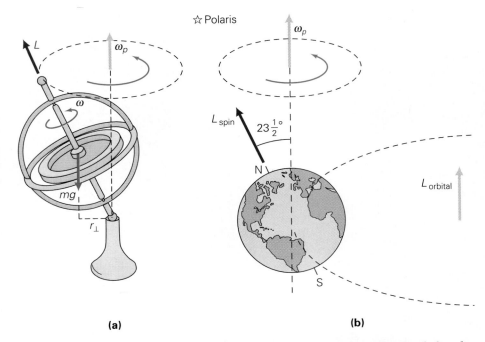

FIGURE 8.25 Precession An external torque causes a change in angular momentum. **(a)** For a spinning gyroscope, this change is directional, and the axis of rotation precesses about a vertical line. (The torque due to the weight force would point out of the page as drawn here, as would **ΔL**.) Note that although there is a torque that would topple a nonspinning gyroscope, a spinning gyroscope doesn't fall. **(b)** Similarly, the Earth's axis precesses because of gravitational torques caused by the Sun and the moon. We don't notice this motion since the period of precession is about 26,000 years.

(a)

(b)

precession is due to slight gravitational torques exerted on the Earth by the Sun and moon. The torques arise because the Earth is not perfectly spherical—it has an equatorial bulge.

The period of the precession of the Earth's axis is about 26,000 years, so it has little day-to-day effect. However, it does have an interesting long-term effect. Polaris will not always be (nor has it always been) the North Star, that is, the star "pointed at" by the Earth's axis of rotation. About 5000 years ago, Alpha Draconis was the North Star, and 5000 years from now, it will be Alpha Cephei, which is at an angular distance of about 68° away from Polaris on the circle described by the precession of the Earth's axis.

There are some other long-term torque effects on the Earth and the moon. Did you know that the Earth is slowing down and hence the days are getting longer? Also, that the moon is receding or getting farther away from the Earth? This is primarily due to ocean tidal friction, which gives rise to a torque. As a result, the Earth's spin angular momentum, and therefore its rate of rotation, is changing. The slowing rate of rotation causes the average day to be longer; this century will be about 25 seconds longer than last. But this is an average rate. At times, the Earth's rotation speeds up for relatively short periods. This is thought to be associated with the rotational inertia of the liquid layer of the Earth's core (see the Insight feature in Chapter 13).

The tidal torque on the Earth results chiefly from the moon's gravitational attraction, which is the main cause of ocean tides. This torque is *internal* to the Earth-moon system, so the total angular momentum of that system is conserved. Since the Earth is losing angular momentum, the moon must be gaining angular momentum to keep the total angular momentum constant. The Earth loses rotational (spin) angular momentum and the moon gains orbital angular momentum. As a result, the moon drifts slightly farther from Earth and its orbital speed decreases. The moon moves away from the Earth at about 1 cm per lunar revolution, or lunar month (calculated from the rate of the slowing down of the Earth's rotation). Thus, the moon moves in a slowly widening spiral.

Important Concepts

You should be able to define and explain these chapter concepts clearly.

rigid body

translational motion

rotational motion

instantaneous axis of rotation

moment (lever) arm

torque

translational equilibrium

concurrent forces

rotational equilibrium

mechanical equilibrium

static equilibrium

center of gravity

equilibrium

   stable

   unstable

moment of inertia

parallel axis theorem

rotational

   work

   power

   kinetic energy

angular momentum

   conservation of

Important Relationships for Review

These relationships are mathematical statements of the concepts and principles presented in the chapter. You should be able to identify the symbols and to explain the relationships before proceeding to the Exercises. In-text equation reference numbers are given for convenience.

Condition for Rolling without Slipping:

$$v = r\omega \tag{8.1}$$
$$\text{(or } s = r\theta \quad \text{or } a = r\alpha)$$

Torque (magnitude):

$$\tau = r_\perp F = rF \sin\theta \tag{8.2}$$

Conditions for Translational and Rotational Mechanical Equilibrium:

$$\Sigma \mathbf{F}_i = 0 \quad \text{and} \quad \Sigma \boldsymbol{\tau}_i = 0 \tag{8.3}$$

Torque on a Particle (magnitude):

$$\tau = mr^2\alpha \tag{8.4}$$

Moment of Inertia:

$$I = \Sigma m_i r_i^2 \tag{8.6}$$

Rotational Form of Newton's Second Law:

$$\boldsymbol{\tau} = I\boldsymbol{\alpha} \tag{8.7}$$

Parallel Axis Theorem:

$$I = I_{CM} + Md^2 \tag{8.8}$$

Rotational Work:

$$W = \tau\theta \tag{8.9}$$

Rotational Power:

$$P = \tau\omega \tag{8.10}$$

Work-Energy Theorem:

$$W = \Delta K = \tfrac{1}{2}I\omega^2 - \tfrac{1}{2}I\omega_0^2 \tag{8.11}$$

Rotational Kinetic Energy:

$$K = \tfrac{1}{2}I\omega^2 \tag{8.12}$$

Kinetic Energy of a Rolling Object:

$$K = \tfrac{1}{2}I_{CM}\omega^2 + \tfrac{1}{2}Mv_{CM}^2 \tag{8.13}$$

Angular Momentum of a Particle in Circular Motion:

$$L = mr^2\boldsymbol{\omega} \tag{8.14}$$

Angular Momentum of a Rigid Body:

$$L = I\omega \tag{8.16}$$

Torque in Terms of Angular Momentum:

$$\boldsymbol{\tau} = \frac{\Delta\mathbf{L}}{\Delta t} \tag{8.17}$$

Conservation of Angular Momentum (with $\tau = 0$):

$$I\boldsymbol{\omega} = I_o\boldsymbol{\omega}_o \tag{8.18}$$

Exercises

8.1 ■ Rigid Bodies, Translations, and Rotations

1 In pure translational motion of a rigid body, (a) all the particles of the body have different angular velocities, (b) a centripetal acceleration is required, (c) the acceleration is always zero, (d) none of the preceding. (d)

2 The condition for rolling without slipping is (a) $a = r\omega^2$, (b) $v = r\omega$, (c) $F = ma$, (d) $a_c = v^2/r$. (b)

3 Suppose someone in your physics class said that it is possible for a rigid body to have translational motion and rotational motion at the same time. Would you agree? Why? yes; see ISM

● **4** For a rolling cylinder, what would happen if v were less than $R\omega$? Is it possible for v to be greater than $R\omega$? Explain. slipping and deformation

5 ■ A circular disk with a radius of 0.15 m rolls without slipping on a level surface with an angular speed of 3.0 rad/s. (a) What is the linear speed of the center of mass of the disk? (b) What is the instantaneous tangential speed of the top of the disk? (a) 0.45 m/s (b) 0.90 m/s

6 ■ Show that $a = r\alpha$ is a condition for rolling without slipping. see ISM

7 ■■ A rope goes over a circular pulley with a radius of 6.0 cm. If the pulley makes four revolutions without the rope slipping, what length of rope passes over the pulley? 1.5 m

● **8** ■■ A cylinder with a radius of 20 cm rolls with an angular speed of 2.4 rad/s. If the center of the cylinder moves 96 cm in 2.0 s, does the cylinder roll without slipping? no

9 ■■ A ball with a radius of 15 cm rolls on a level surface with a translational speed of the center of mass of 0.25 m/s. What is the angular speed about the center of mass if the ball rolls without slipping? 1.7 rad/s

10 ■■ A disk with a radius of 0.20 m rotates through 270° as it travels 0.47 m. Does the disk roll without slipping? no

11 ■■ A friend tells you that he measured the translational speed of the center of mass of a sphere whose diameter is 0.182 m to be 0.465 m/s when the sphere was rolling with an angular speed of 4.98 rad/s. Is his measurement accurate? no

12 ■■■ A cylinder with a diameter of 20 cm rolls on a level surface with an angular speed of 0.50 rad/s. If the cylinder experiences a uniform tangential acceleration of 0.018 m/s² without slipping until its angular speed is 1.25 rad/s, how many rotations does the cylinder go through during the acceleration time? 0.57 rotations

8.2 ■ Torque, Equilibrium, and Stability

● **13** It is possible to have a net torque when (a) all forces act through the axis of rotation, (b) $\Sigma\mathbf{F}_i = 0$, (c) a body is in rotational equilibrium, (d) a body remains in unstable equilibrium. (b)

14 If an object in stable equilibrium is slightly displaced, (a) its potential energy is increased, (b) the center of gravity is above the axis of rotation, (c) no gravitational work is done, (d) unstable equilibrium follows. (a)

15 (a) Why does a lower center of gravity give an object greater stability? (b) Does it make any difference how a tractor-trailer is loaded? [*Hint:* Think about an eighteen-wheeler going around a banked curve.] see ISM

16 Explain the balancing acts in ● Fig. 8.26. Where are the centers of gravity? see ISM

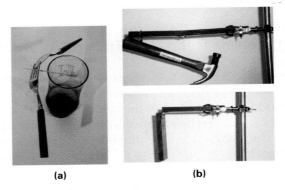

(a)   (b)

● **FIGURE 8.26 Balancing acts** See Exercise 16.

17 (a) How many different positions of stable equilibrium and unstable equilibrium are there for a cube? Consider each surface, edge, and corner to be a different position. (b) A wooden letter L is 3 cm thick. How many different positions of stability and instability does the L have? Consider each surface, edge, and corner to be a different position. (a) 6 stable, 20 unstable (b) see ISM

18 The effects of torque and center of gravity on potential energy for slight displacements of bodies in stable and unstable equilibrium are discussed in the text. For a body in *neutral equilibrium*, a slight displacement produces no restoring or toppling torques, and the potential energy remains constant. Give some examples of bodies in neutral equilibrium. cylinder or sphere on level surface

19 ■ A student opens a glass door in the science building by pushing perpendicularly with a force of 45 N on a metal plate located 90 cm from the hinge side of the door. Another student opens the door by pushing perpendicularly at a distance of 50 cm from the hinge side. What is the magnitude of the force applied by the second student that produces the same torque as the first? 81 N

20 ■ A 35-kg child sits on a uniform seesaw 2.0 m from the pivot point (or fulcrum). How far from the pivot point on the other side will her 30-kg playmate have to sit for the seesaw to be in equilibrium? 2.3 m

21 ■■ The pedals of a bicycle rotate in a circle with a diameter of 40 cm. What is the maximum torque a 55-kg rider can apply by putting all of her weight on a pedal?
1.1 × 10² N-m

● 22 ■■ A uniform meterstick pivoted at its center, as in Example 8.3, has a 100-g mass suspended at the 25.0-cm position. (a) At what position should a 75.0-g mass be suspended to put the system in equilibrium? (b) What mass would have to be suspended at the 90.0-cm position for the system to be in equilibrium? (a) 83.3 cm (b) 62.5 g

23 ■■ Show that the balanced meterstick in Example 8.3 is in static rotational equilibrium about a horizontal axis through the zero end of the stick. see ISM

24 ■■ A force of 36 N is applied to a particle located 0.15 m from its axis of rotation. What is the magnitude of the torque about this axis if the angle between the direction of the applied force and the radius vector is (a) 60°, (b) 90°, and (c) 120°? (a) 4.7 m-N (b) 5.4 m-N (c) 4.7 m-N

25 ■■ If *r* = 0.25 m in Exercise 24, what force would have to be applied to have the same torque in each case? 22 N

● 26 ■■ In Fig. 8.27, what is the force F_m supplied by the deltoid muscle so as to hold up the outstretched arm if the mass of the arm is 3.0 kg? (F_j is the joint force on the bone of the upper arm—the humerus.) 3.0 × 10² N

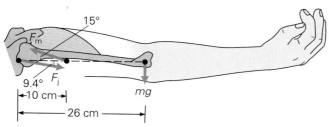

FIGURE 8.27 Arm in static equilibrium See Exercise 26.

27 ■■ A variation of Russell traction (Fig. 8.28) helps support the lower leg when in a cast. Suppose that the patient's

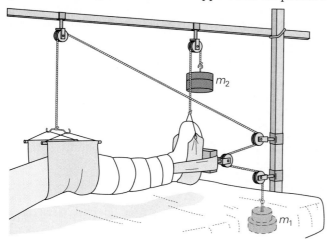

FIGURE 8.28 static traction See Exercise 27.

leg and cast have a combined mass of 15.0 kg and m_1 is 4.50 kg. (a) What is the reaction force of the leg muscles to the traction? (b) What must m_2 be to keep the leg horizontal? (a) 88.2 N (b) 10.5 kg

28 ■■ Telephone and electrical lines are intentionally allowed to sag between poles so that the tension will not be too great when something hits or sits on the line. Suppose that a line were stretched perfectly horizontally between two poles, which are 30 m apart. If a bird with a mass of 0.25 kg perched on the wire midway between the poles and the wire sagged 1.0 cm, what would the tension in the wire be?
1.8 × 10³ N (> 400 lb)

29 ■■ A truck is on a flat 40-m bridge supported by end piers. The truck weighs 2.00 × 10⁵ N, and the bridge weighs 7.50 × 10⁷ N. If the center of mass of the truck is 15.0 m from one end of the bridge, what are the forces exerted on the bridge by the end piers? F_L = 4.26 × 10⁷ N;
F_R = 4.26 × 10⁷ N

30 ■■ A uniform meterstick weighing 5.0 N is pivoted so it can rotate about a horizontal axis through one end. If a 0.15-kg mass is suspended 75 cm from the pivoted end, what force must be applied to the free end of the meterstick to maintain it in a horizontal position? 3.6 N

31 ■■ (a) For a bicycle rider making a circular turn on a level surface as in Fig. 1 of the Insight feature, show that the proper angle of lean is given by $\theta = \tan^{-1} \mu_s$, where μ_s is the coefficient of static friction between the tires and the road. (b) If μ_s = 0.57, what is the angle of lean? (Does the angle increase with an increasing coefficient? Explain.)
(a) tan θ = f_s/N = μ_s (b) 30°, yes

● 32 ■■ A mass is suspended by two cords as shown in Fig. 8.29. What are the tensions in the cords?
T_1 = 50 N; T_2 = 41 N

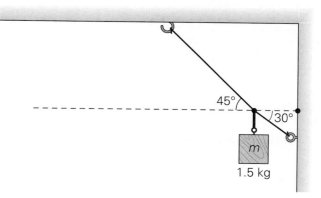

FIGURE 8.29 Strung-out equilibrium See Exercises 32 and 33.

33 ■■ If the cord attached to the vertical wall in Fig. 8.29 were horizontal (instead of at a 30° angle), what would the tensions in the cords be? T_1 = 15 N; T_2 = 21 N

34 ■■ A 10.0-kg solid uniform cube with 0.250-m sides rests on a level surface. What is the minimum amount of work necessary to put the cube in unstable equilibrium?
5.01 J

35 ■■ A uniform L-shaped object has a leg that is 0.50 m long and a shorter leg that is 0.30 m long (both measured along the outside edge). The L has a width of 6.0 cm and a thickness of 2.0 cm. Locate the center of gravity. [*Hint:* Divide the L into two rectangles and locate the center of gravity of each. Express mass in terms of density.] $x_{cm} = 7.9$ cm; $y_{cm} = 16$ cm; $z_{cm} = 1.0$ cm

36 ■■ (a) How many uniform, identical textbooks with a width of 25 cm can be stacked on top of each other on a level surface without the stack falling over if each successive book is displaced 3.0 cm in the width direction relative to the next lower book? (b) If the books are 5.0 cm thick, what will be the height of the center of mass of the stack above the level surface? (a) 9 books (b) 22.5 cm

37 ■■■ The outer wheels of an oversize tractor-trailer rig are 3.66 m apart. When the trailer is loaded, its center of gravity is equidistant from its sides and 3.58 m above the road surface. What banking angle of the road will put the trailer in unstable equilibrium? 27°

● **38** ■■■ A 70-kg painter paints the side of a house while standing on a long board resting on a scaffold, as shown in ● Fig. 8.30. If the board has a mass of 15 kg, how close to the end can the painter stand without tipping the board over? 0.27 m left of support

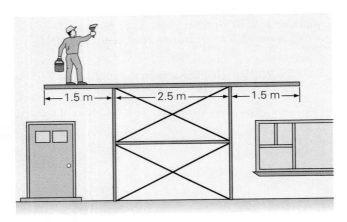

● **FIGURE 8.30 Not too far!** See Exercise 38.

39 ■■■ Suppose that the board in Fig. 8.30 were suspended from vertical ropes attached to each end instead of resting on scaffolding. If the painter stood 1.5 m from one end of the board, what would the tensions in the ropes be? (See Exercise 38 for additional data and ignore significant figures.) $T_1 = 261$ N; $T_2 = 572$ N

40 ■■■ An artist wishes to construct a birds-and-bees mobile, as shown in ● Fig. 8.31. If the bee on the lower left has a mass of 0.10 kg and each vertical support string has a length of 30 cm, what are the masses of the other birds and bees? (Neglect the masses of the bars and strings.) $m_2 = 0.20$ kg, $m_3 = 0.50$ kg, $m_4 = 0.40$ kg

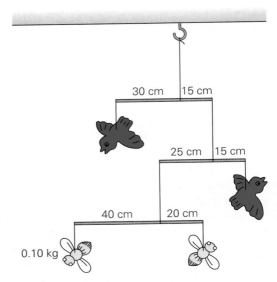

● **FIGURE 8.31 Birds and bees** See Exercise 40.

8.3 ■ Rotational Dynamics

41 The moment of inertia of a rigid body (a) depends on the axis of rotation, (b) cannot be zero, (c) depends on mass distribution, (d) all of the preceding. (d)

42 Which of the following best describes the physical quantity called torque: (a) rotational analog of force, (b) energy due to rotation, (c) rate of change of linear momentum, (d) force that is tangent to a circle? (a)

43 (a) Is the moment of inertia of a rigid body related in any way to the center of mass? Explain. (b) Can a moment of inertia have a negative value? If so, explain what this would mean. (a) yes; (b) no

44 Why does the moment of inertia of a rigid body have different values for different axes of rotation? What does this mean physically? depends on how mass is distributed about an axis

45 The string of a stationary yo-yo is pulled as shown in ● Fig. 8.32. (a) Which way does the yo-yo roll (without slip-

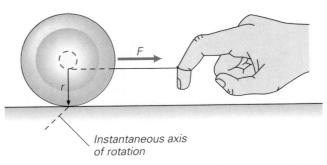

Instantaneous axis of rotation

● **FIGURE 8.32 Pulling the yo-yo's string** See Exercise 45.

ping), and why? [*Hint:* It may not be as you expect. Think about torque and axis of rotation. If you don't have a yo-yo, a spool of thread also works.] (b) Explain what happens if the string is pulled at a greater and greater angle to the horizontal. see ISM

46 (a) Why is it easier to balance a meterstick vertically on your finger than a pencil? The pencil has a lower center of mass. (b) A softball, a volleyball, and a basketball are released at the same time from the top of an inclined plane. Give the results of the race down the plane. (a) meterstick has larger I and smaller α (b) tie

47 The release of vast amounts of carbon dioxide may give rise to an increase in the Earth's average temperature through the so-called greenhouse effect and cause a melting of the polar ice caps. If this occurred and the ocean level rose substantially, what effect would this have on the Earth's rotation and length of the day? lengthen day

48 ■ Show that $r_\perp F = rF_\perp$ for a torque. see ISM

49 ■ A fixed 0.15-kg pulley with a radius of 0.050 m is acted on by a net torque of 6.4 m-N. What is the angular acceleration of the pulley? 3.4×10^4 rad/s²

50 ■■ A light meterstick is loaded with masses of 2.0 kg and 4.0 kg at the 30-cm and 75-cm positions, respectively. (a) What is the moment of inertia about an axis through the 0-cm end of the meterstick? (b) What is the amount of inertia about an axis through the center of mass of the system? (c) Use the parallel axis theorem to find the moment of inertia about the axis through the 0-cm end of the stick, and compare the result with the result of part (a). (a) 2.4 kg-m² (b) 0.27 kg-m² (c) 2.4 kg-m²

51 ■■ Using the parallel axis theorem, show that the moment of inertia for a thin rectangular sheet with an axis of rotation along one side is $I = \frac{1}{3}ML^2$ (see Fig. 8.15k). see ISM

52 ■■ Four 0.25-kg point masses connected by rods of negligible mass form a square rigid body with 0.10-m sides. Find the moment of inertia of the body about each of the following axes: (a) an axis perpendicular to the plane of the square and through its center, (b) an axis perpendicular to the plane and through one of the masses, (c) an axis in the plane and along one side through two of the masses, and (d) an axis in the plane running diagonally through two of the masses. see ISM

53 ■■ If the Earth rotated about a N–S axis tangent to the Earth at the Equator, how much more rotational inertia would it have? 2.5×10^{38} kg-m², or twice as much

54 ■■ A Ferris wheel accelerates from rest to an angular speed of 1.5 rad/s in 12 s. Approximating the Ferris wheel as a circular disk with a radius of 30 m and a mass of 2.0 metric tons, what is the net torque on the wheel? 1.2×10^5 m-N

55 ■■ A 15-kg uniform sphere with a radius of 15 cm rotates about an axis tangent to its surface at a rate of 3.0 rad/s. A constant torque of 10 m-N then increases the

rotational speed to 7.5 rad/s. Through what angle does the sphere rotate while accelerating? 1.1 rad

56 ■■ A disk-shaped 0.25-kg machine part with a radius of 6.0 cm rotates eccentrically about an axis that is normal to its flat surface and located $\frac{2}{3}$ of the distance from the center to the circumference. (a) What torque is required to rotate the part about this off-center axis with an angular acceleration of 2.0 rad/s²? (b) How much greater is this than the torque 1required to rotate the part with the same angular acceleration about a parallel axis through the center? (a) 1.7×10^{-3} m-N (b) 1.3×10^{-3} m-N

57 ■■ For the system shown in ● Fig. 8.33, $m_1 = 8.0$ kg, $m_2 = 3.0$ kg, $\theta = 30°$, and the radius and mass of the pulley are 0.10 m and 0.10 kg, respectively. (a) What is the acceleration of the masses? (Neglect axle friction and the string's mass.) (b) If the pulley has a constant frictional torque of 0.050 m-N when the system is in motion, what is the acceleration? [*Hint:* Isolate the forces. The tensions in the strings are different. Why?] (a) 0.89 m/s² (b) 0.44 m/s²

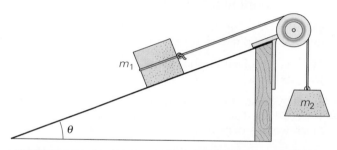

FIGURE 8.33 Inclined plane and pulley See Exercise 57.

58 ■■ What is the magnitude of the tangential force that must be applied to the rim of a 2.0-kg wheel that is disk-shaped and has a radius of 0.75 m in order to give it an angular acceleration of 4.8 rad/s²? (Neglect friction.) 3.6 N

59 ■■ A man starts his lawn mower by applying a constant tangential force of 150 N to the flywheel using a starter rope. The flywheel is a cylindrical ring with a mass of 0.30 kg and a diameter of 18 cm. What is the angular speed of the wheel after it has turned through 1 revolution? (Neglect friction and motor compression.) 3.9×10^2 rad/s

60 ■■■ A uniform hoop rolls without slipping down a plane with an incline of 15° above the horizontal. What is the acceleration of the hoop's center of mass? 1.3 m/s²

61 ■■■ For the ball rolling down the incline in Fig. 8.20, what would be the maximum angle of incline in terms of the coefficient of static friction for it to roll without slipping? $\theta = \tan^{-1}(7\mu/2)$

62 ■■■ A uniform cylinder with a mass of 2.0 kg and a radius of 0.15 m is suspended by two strings wrapped around it (● Fig. 8.34). As the cylinder descends, the strings unwind

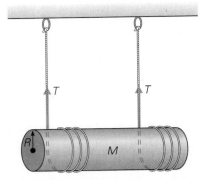

FIGURE 8.34 Unwinding See Exercise 62.

from it. What is the acceleration of the center of mass of the cylinder? (Neglect the mass of the string.) 6.5 m/s²

8.4 ▪ Rotational Work and Kinetic Energy

63 With $W = \tau\theta$, the unit(s) of rotational work is (a) watt, (b) m-N, (c) kg-rad/s², (d) N-rad. (b)

64 The rotational kinetic energy of a body is equal to (a) $\frac{1}{2}I\upsilon^2$, (b) I/τ, (c) $L\alpha$, (d) $L^2/2I$. (d)

65 ▪ A constant retarding torque of 20 m-N stops a rolling wheel whose diameter is 0.80 m in a distance of 15 m. How much work is done by the torque? 7.5 × 10² J

66 ▪ A person opens a door by applying a force of 15 N perpendicular to it at a distance of 0.90 m from the hinges. The door is pushed wide open (to 90°) in 2.0 s. (a) How much work was done on the average? (b) What was the average power delivered? (a) 22 J (b) 11 W

67 ▪▪ Use the conservation of mechanical energy to find the linear speed of the descending mass ($m = 1.0$ kg) of Fig. 8.18 after it has descended a vertical distance of 2.0 m from rest. (For the pulley, $M = 0.30$ kg and $R = 0.15$ m. Neglect friction and mass of the string.) 5.9 m/s

68 ▪▪ A sphere with a radius of 15 cm rolls on a level surface with a constant angular speed of 8.0 rad/s. How high on a 30° inclined plane will the sphere roll before coming to rest? (Neglect frictional losses.) 0.10 m

69 ▪▪ A flywheel with a moment of inertia of 4.50 × 10² kg-m² rotates with a speed of 7500 rpm. How much work is required to bring it to rest? (b) If this is done uniformly in 1.5 min, how much power is expended? (a) 1.39 × 10⁸ J (b) 1.54 × 10⁶ W

70 ▪▪ A cylindrical hoop, a cylinder, and a sphere of equal radius and mass are released at the same time from the top of an inclined plane. Using kinetic energy, show that the sphere always gets to the bottom of the incline first and the hoop last. hoop: $K = mv^2$; cylinder: $K = (3/4)mv^2$; sphere: $K = (7/10)mv^2$

71 ▪▪ For the following objects, which are rolling without slipping, determine the rotational kinetic energy about the center of mass as a percentage of the total instantaneous kinetic energy: (a) solid sphere, (b) a thin spherical shell, (c) a thin cylindrical shell. (a) 29% (b) 40% (c) 50%

72 ▪▪ Which is greater and by what factor: the Earth's kinetic energy due to its rotation or its kinetic energy due to its revolution about the Sun? (Assume a circular orbit.) $K_{rev} = (1.1 \times 10^4) K_{rot}$

73 ▪▪ A 1.5-hp motor drives a machine disk having a mass of 4.5 kg and a diameter of 0.40 m from rest to a speed of 750 rpm. (a) What is the average work done by the motor? (b) How long does it take to reach the operating speed? (a) 2.8 × 10² J (b) 0.25 s

74 ▪▪ A solid 5.0-kg sphere with a diameter of 0.60 m starts from rest at a height of 1.2 m above the base of an inclined plane and rolls down under the influence of gravity. What is the linear speed of the sphere's center of mass just as it leaves the incline and rolls onto a horizontal surface? (Neglect friction.) 4.1 m/s

75 ▪▪ A 0.05-kg phonograph record with a radius of 0.15 m drops onto a turntable and is soon rotating at $33\frac{1}{3}$ rpm. How much work must be supplied to get the record to rotate at this speed, and what supplies it? 3.4 × 10⁻³ J; the work is supplied by the motor

76 ▪▪ A pencil 20 cm long is initially held so that it stands vertically on its sharpened point on a table by a person with a finger on the eraser end. When the finger is removed, the pencil falls over onto the table. Assuming that the point does not slip, at what speed is the eraser end of the pencil moving when it strikes the horizontal surface? 2.4 m/s

77 ▪▪▪ A steel ball rolls down an incline into a loop-the-loop of radius R (Fig. 8.35a). (a) What is the minimum

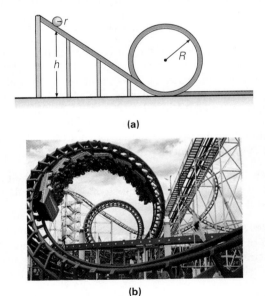

(a)

(b)

FIGURE 8.35 Loop-the-loop and rotational speed See Exercise 77.

speed the ball must have at the top of the loop in order to stay on the track? (b) At what vertical height (*h*) on the incline, in terms of the radius of the loop, must the ball be released in order for it to have the required minimum speed at the top of the loop? (Neglect frictional losses.) (c) Fig. 8.35b shows a loop-the-loop for a roller coaster. What are the sensations of the riders if the roller coaster has the minimum speed or a greater speed at the top of the loop? [*Hint:* In case the speed is below the minimum, seat and shoulder straps hold the riders in.]   (a) $v = (Rg)^{\frac{1}{2}}$ (b) $h = (2.7)R$ (c) weightlessness

8.5 ■ Angular Momentum

78 The units of angular momentum are (a) N-m, (b) kg-m/s², (c) m-N-s, (d) J-m.   (c)

● **79** Angular momentum is conserved when (a) a couple exists, (b) a body is in translational equilibrium, (c) **L** does not change with time, (d) none of the preceding.   (c)

80 A classroom demonstration is illustrated in ● Fig. 8.36. A person on a rotating stool holds a rotating bicycle wheel (by handles attached to the wheel). When the wheel is held horizontally, she rotates one way (clockwise as viewed from above). When the wheel is turned over, she rotates in the opposite direction. Explain why this occurs. [*Hint:* Consider angular momentum vectors.]   see ISM

81 An ice skater goes into a fast spin by tucking her arms in, and a ballerina does a pirouette with her arms over her head. Why are the arms put into these positions in each case?   see ISM

82 Cats always seem to land on their feet when they fall, even if held upside down and dropped. While a cat is falling, there is no external torque and its center of mass falls as a particle. How can cats turn themselves over while falling? see ISM

83 For a non-sleeping top or gyroscope, the precessional motion is around the vertical axis. Explain why this is, and whether the precession is clockwise or counterclockwise (as viewed from above).   see ISM

84 (a) Large helicopters have two overhead rotors that rotate in opposite directions (● Fig. 8.37), and small helicopters have a single overhead rotor and an "anti-torque" tail rotor. Explain the reason(s) for the differences. (b) Unlike helicopters, a single-engine airplane has only one propeller, or set of rotor blades. Discuss the stability of such an aircraft in light of part (a).   see ISM

(a)

(b)

● **FIGURE 8.37 Different rotors**   See Exercise 84.

85 ■ A 0.010-kg ball in an arrangement like that shown in Fig. 8.22 swings initially in a circle having a 15-cm radius at an angular speed of 3π rad/s. If the force on the string is reduced enough that the ball swings in a circle with a radius of 25 cm, what is its angular speed?   3.4 rad/s

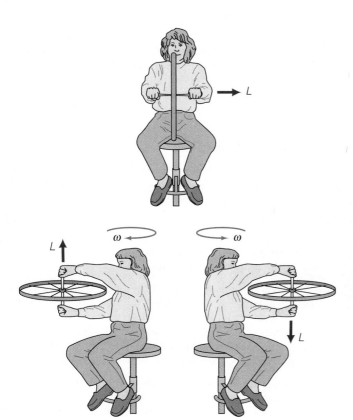

● **FIGURE 8.36 A double rotation**   See Exercise 80.

86 ■ What is the angular momentum of a 2.0-g particle moving in a horizontal circle of 15-cm radius counter-clockwise as viewed from above and with an angular speed of 6π rad/s? (Give magnitude and direction.)
8.5×10^{-4} kg-m/s upward

87 ■■ The Earth's orbit about the Sun is slightly elliptical. The points on the orbit at maximum and minimum distances from the Sun are called the aphelion and the perihelion, respectively. (a) Show that the Earth's orbital speed is greater at the perihelion than at the aphelion. (b) If $r_a = 1.52 \times 10^8$ km and $r_p = 1.47 \times 10^8$ km, how many times greater is the Earth's speed at the perihelion than its speed at the aphelion? (c) What is the difference between r_a and r_p in miles?
(a) $v_p > v_a$ (b) 1.03 (c) 3.1×10^6 mi

● **88** ■■ Compute the ratio of the Earth's orbital angular momentum and its spin angular momentum. Are these momenta in the same direction? $L_{orbital}/L_{spin} = 3.8 \times 10^6$ (not in the same direction)

89 ■■ The period of the moon's rotation is the same as the period of its revolution—$29\frac{1}{2}$ days. What is the angular momentum for each motion? (Because the periods are equal, we see only one side of the moon from Earth.) $L_{rev} = 2.6 \times 10^{34}$ m-N-s; $L_{rot} = 2.2 \times 10^{30}$ m-N-s

90 ■■ A rotating spherical star in part of its life cycle expands to six times its normal volume. Assuming the mass to remain constant, how is the period of rotation affected?
$t_2 = (5.8) t_1$

91 ■■ A person on a rotating stool with arms outstretched as in Fig. 8.23a rotates at a speed of 0.75 rev/s. On bringing in his arms (Fig. 8.23b), the speed increases to 1.25 rev/s. By what factor does the moment of inertia change?
$I = (0.60) I_0$

● **92** ■■ A skater has a moment of inertia of 100 kg-m² when his arms are outstretched and a moment of inertia of 75 kg-m² when his arms are tucked in close to his chest. If he starts to spin at an angular speed of 2.0 rps with his arms outstretched, what will his angular speed be when they are tucked in?
2.7 rps

93 ■■ An ice skater doing a toe spin with outstretched arms has an angular speed of 4.0 rad/s. The skater then tucks in her arms, decreasing her moment of inertia by 7.5%. (a) What is the resulting angular speed? (b) By what factor does the skater's kinetic energy change? (Neglect any frictional effects.) (a) 4.3 rad/s (b) $K = (1.1) K_0$

94 ■■■ In an ice show, two 70-kg skaters, both traveling at 8.0 m/s, approach each other along straight-line parallel paths that are separated by a distance of 4.0 m. When directly opposite each other, the skaters grab the ends of a light rod 4.0 m in length. What is the initial rotational speed of the joined skaters? [*Hint:* Use the conservation of angular momentum.] 4.0 rad/s

95 ■■■ A comet approaches the Sun as illustrated in ● Fig. 8.38 and is deflected by the Sun's gravitational attraction. This may be considered a collision, and b is called the impact parameter. Find the distance of closest approach (d) in terms of the impact parameter and the velocities. Consider the radius of the Sun to be negligible compared to d. (As the

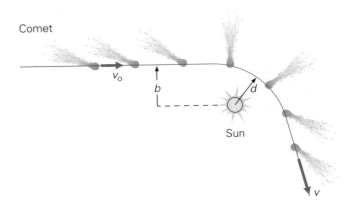

● **FIGURE 8.38 A comet collision**   See Exercise 95.

figure shows, the tail of a comet always "points" away from the Sun.) $d = b (v_0/v)$

96 ■■■ A 0.50-kg kitten stands on the edge of a lazy susan (a kind of turntable) that has a mass of 1.5 kg and a radius of 0.30 m. Assume that the lazy susan has frictionless bearings and is initially at rest. (a) What will happen if the kitten starts walking around the edge of the lazy susan? (b) If the kitten walks at a speed of 0.25 m/s relative to the ground, what will the angular speed of the lazy susan be? (c) When the kitten has walked completely around the edge and is back at its starting point, will that point be above the same point on the ground as it was at the start? If not, where is the kitten relative to the starting ground point? (Speculate on what might happen if everyone on Earth suddenly started to run eastward. What effect might this have on the length of a day?) (a) turntable would turn in the opposite direction (b) 0.55 rad/s (c) 2.2 rad

■ **Additional Exercises**

97 A solid 0.75-kg ball rolls without slipping on a level surface. If the ball rolls in a straight line at a speed of 0.50 m/s, what is its kinetic energy? 0.13 J

● **98** Show that if two rigid bodies with the same axis of rotation are stuck together, the moment of inertia of the combined rigid body about the common axis is equal to the sum of the individual moments, or $I = I_1 + I_2$.
see ISM

99 A 75.0-kg person stands at the outer edge of a round platform whose radius is 8.00 m and whose mass is 450 kg. The platform is rotating at a speed of 0.625 rad/s. Assuming the platform to be a freely rotating disk (no axle friction), what would be its angular speed if the person walked to its center? [*Hint:* See Exercise 98.] 0.833 rad/s

100 A spring is twisted 60° about its linear axis (axis of symmetry) by a torque of 100 m-N. How much torsional, or rotational, energy is stored in the spring? [*Hint:* Let the restoring torque of the spring be $\tau = k\theta$, the rotational analog of $F = kx$ (Hooke's law—see Chapter 5) with the minus sign omitted.]   52 J

101   A thin 1.0-m rod pivoted at one end falls (rotates) frictionlessly from a vertical position, starting from rest. What is the angular speed of the rod when it is horizontal? [*Hint:* Consider the center of mass and use the conservation of mechanical energy.]   5.4 rad/s

102   Prove that a thin hoop or ring starting from rest will roll more slowly down an inclined plane than an annular cylinder. (Remember that masses and radii do not have to be taken into account.)   see ISM

● **103**   A bicycle wheel has a diameter of 0.80 m and a mass of 2.0 kg. (a) If the bicycle travels with a linear speed of 1.5 m/s, what is the angular speed of the wheel? (b) What is the angular momentum of the wheel? (Consider the wheel to be a thin ring.)   (a) 3.8 rad/s   (b) 1.2 m-N-s

104   A uniform sphere and a uniform cylinder with the same mass and radius roll side by side on a level surface without slipping and at the same velocity. If the sphere and the cylinder approach an inclined plane and roll up it without slipping, will they be at the same height on the plane when they come to a stop? If not, what will be the percent height difference?   cylinder goes higher by 7%

105   A piece of machinery with a moment of inertia of 2.6 kg-m² rotates with a constant angular speed of 4.0 rad/s when experiencing a frictional torque of 0.56 m-N. What is the net torque acting on the piece?   net torque is zero

106   The location of a person's center of gravity relative to his or her height can be found using the arrangement shown in ● Fig. 8.39. The scales are initially adjusted to read zero with the board alone. Locate the center of gravity of the person relative to the horizontal dimension. Would you expect the location of the center of gravity in other dimensions to be exactly at the midway points? Explain.   0.87 m from the feet

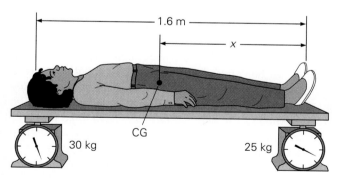

● **FIGURE 8.39 Locating the center of gravity**   See Exercise 106.

107   Using bricks identical to those in Example 8.5, you are asked to create a stack of nine bricks (maximum) without its

toppling, with each added brick displaced an equal distance horizontally in the same direction. (a) What is the maximum displacement that will allow this? (b) What is the height of the center of mass of the stack if the uniform bricks are 8.0 cm thick?   (a) 2.5 cm   (b) 36 cm

108   The *radius of gyration* (k) is defined as the distance from an axis of rotation at which all the mass (M) of an object would have to be concentrated to give the same moment of inertia as would be calculated with the actual mass distribution, that is, $I = Mk^2$. This represents an object as a rotating particle of mass M. For the axes shown in Fig. 8.15, determine the radius of gyration for (a) a uniform sphere, (b) a uniform cylinder, and (c) a particle.   (a) $k = R (2/5)^{1/2}$ (b) $k = R (1/2)^{1/2}$ (c) $k = R$

109   A 0.25-kg pulley has a diameter of 14 cm. Two masses, 0.10 kg and 0.15 kg, are suspended from the ends of a light string passing over the pulley. (a) Neglecting axle friction, what will the acceleration be when the masses are moving under the influence of gravity? (b) What is the magnitude of the combined kinetic frictional torque (due to both friction of the pulley bearings and friction between the string and the pulley) that would keep the masses moving at a constant speed? [*Hint:* Isolate the forces. The tensions in the strings are different. Why?]   (a) 1.3 m/s²   (b) 1.1 × 10⁻² m-N

110   A thin rectangular plate has dimensions of 30 cm by 40 cm and a mass of 0.50 kg. For which axis of rotation does the plate have the greater rotational inertia, an axis tangent to a smaller side at its midpoint or an axis tangent to a larger side at its midpoint? How many times more? (Both axes are perpendicular to the plate surface.)   smaller, 1.6

111   The initial setup for a classroom demonstration is shown in ● Fig. 8.40. A hinged board is propped up with a stick. In a depression on the upper end of the board is a steel ball, and a plastic cup is fixed to the board at a proper place near the hinge. When the stick is removed, the board and the ball fall and the ball lands in the cup. (a) Show that this happens only when the stick is propped at an angle less than or equal to a critical angle of $\theta_o = 35°$. (b) If the hinged board is 1.0 m long and the stick props it up at an angle of 35°, where should the cup be located on the stick?   (a) 35°   (b) 0.18 m from the end

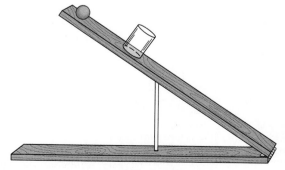

● **FIGURE 8.40 A ball-and-cup trick**   See Exercise 111.

9

Solids and Fluids

Chapter 9 topics include elastic moduli, fluids at rest (pressure, buoyancy, and surface tension), and fluids in motion (Bernoulli's equation, viscosity, and flow rate).

On the basis of general physical distinctions, matter is commonly divided into three phases: *solid, liquid, and gas. A* solid *has a definite shape and volume. A* liquid *has a definite volume, but assumes the shape of its container. A* gas *takes on the shape and volume of its container. Two of these phases can be seen in the photograph above. The liquid magma pouring from this Hawaiian volcano is actually formed by the melting of solid rock. (We will learn about how such phase changes occur in Chapter 11.) You may think that the third phase is also visible, but this is deceptive—the "steam" in the picture is not really a gas but a suspension of tiny liquid droplets, like those that make up a cloud or fog.*

Solids and liquids are sometimes called condensed matter. In this chapter, however, we will utilize a different classification scheme and consider matter in terms of solids and fluids. Liquids and gases are referred to collectively as fluids. A **fluid** is a substance that can flow, so liquids and gases qualify, but solids do not.

A simplistic description of solids is that they are made up of particles called atoms that are held rigidly together by interatomic forces. This concept of an ideal rigid body was used in Chapter 8 to describe rotational motion. Real solid bodies are not absolutely rigid and can be elastically deformed by external forces. Elasticity usually brings to mind a rubber band or spring that will resume its original dimensions even after being greatly deformed. In fact, all materials are elastic to some degree, even very hard steel. But, as you will learn, there's a limit to such deformation—an elastic limit.

Fluids, on the other hand, have little or no elastic response to a force—it merely causes the fluid to flow. Fluids are important in everyday life. You are surrounded by a fluid (air) and, for the most part, are composed of fluids. Most of this chapter will focus on the behavior of fluids.

Because of their fluidity, liquids and gases have many properties in common, and it is convenient to study them together. Of course, liquids and gases have some important differences, a major one being compressibility. Liquids are not very compressible, whereas gases are easily compressed. You should keep this distinction in mind.

9.1 ■ Solids and Elastic Moduli

As stated previously, all solid materials are elastic to some degree. That is, a body that is slightly deformed by an applied force will return to its original dimensions when the force is removed. The deformation may not be noticeable for many materials, but it's there. (See Demonstration 7.)

You may be able to visualize why materials are elastic if you think in terms of the simplistic model of a solid in ● Fig. 9.1. The atoms of the solid substance are imagined to be held together by springs. The elasticity of the springs represents the resilient nature of the interatomic forces. The springs resist permanent deformation, as do the forces between atoms.

The elastic properties of solids are commonly discussed in terms of stress and strain. **Stress** is a measure of the force causing a deformation. **Strain** is a relative measure of the deformation a stress causes. Quantitatively, *stress is the applied force per unit cross-sectional area:*

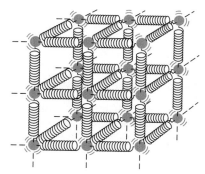

●**FIGURE 9.1 A springy solid**
The elastic nature of interatomic forces is indicated by simplistically representing them as springs, which similarly resist deformation.

$$\text{stress} = \frac{F}{A} \qquad (9.1)$$

DEMONSTRATION 7 ■ Stress, Strain, and Elasticity

(a) A glass bottle with colored water is fitted with a capillary tube in a rubber stopper.

(b) Squeezing the bottle on the flat sides raises the water level in the tube because the bottle's cross section is reduced.

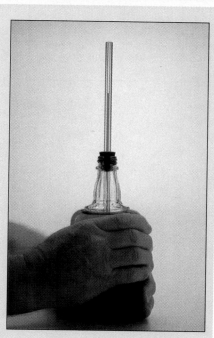

(c) Squeezing the bottle on its narrow sides lowers the water level because the bottle's cross section is increased.

A demonstration to show a material property—the elasticity of glass.

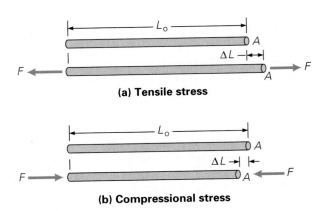

FIGURE 9.2 Tensile and

(a) Tensile stress

(b) Compressional stress

FIGURE 9.2 Tensile and compressional stress and strain Tensile and compressional stresses are due to forces applied normally to the surface area of the ends of bodies, as shown here. **(a)** A tension, or tensile stress, tends to increase the length of an object. **(b)** A compressional stress tends to shorten the length.

Here F is the magnitude of the applied force normal (perpendicular) to the cross-sectional area. The equation shows that the SI units for stress are newtons per square meter (N/m^2).

As illustrated in ● Fig. 9.2, a force applied to the ends of a rod gives rise to either an elongating tension, or **tensile stress**, or a **compressional stress**, depending on the direction of the force. In these cases, the **tensile strain** is the ratio of the change in length to the original length:

$$\text{strain} = \frac{\text{change in length}}{\text{original length}} = \frac{\Delta L}{L_o} \qquad (9.2)$$

where $\Delta L = L - L_o$. Note that strain is a unitless quantity (length/length). It is the *fractional change* in length. For example, if the strain is 0.05, the material has changed in length by 5% of its original length. (See the Insight feature about stresses and strains on some materials on p. 297.)

As might be expected, the strain is proportional to the stress; that is, strain ∝ stress. For relatively small stresses, this is a direct or linear proportion. The constant of proportionality depends on the nature of the material and is called the **elastic modulus**. Thus,

$$\text{stress} = \text{elastic modulus} \times \text{strain}$$

Elastic modulus   or

$$\text{elastic modulus} = \frac{\text{stress}}{\text{strain}} \qquad (9.3)$$

That is, the elastic modulus is the stress divided by the strain, or the ratio of stress to strain.

Three general types of elastic moduli (plural of modulus) are associated with stresses that produce changes in length, shape, or volume. These are called Young's modulus, shear modulus, and bulk modulus, respectively.

Change in Length: Young's Modulus

● Fig. 9.3 is a graph of the tensile stress versus the strain for a typical metal rod. The curve is a straight line up to a point called the *proportional limit*. Beyond this point, the strain begins to increase more rapidly to another critical point called the **elastic limit**. If the tension is removed at this point, the material

Demonstration/activity: A student's weight of approximately 150 lb can be supported by an aluminum drink can. Have a student stand on a can. Describe the mechanical properties present. Lightly tap the can with a stick to cause it to collapse. Discuss the mechanical properties this demonstrates.

Use the micrometer to measure the thickness of the can and calculate the stress applied by the 150 lb student.

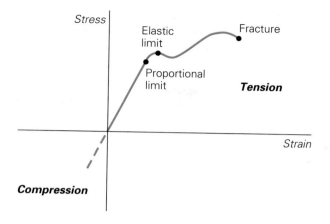

FIGURE 9.3 **Stress versus strain** A plot of stress versus strain for a typical metal rod is a straight line up to the proportional limit. Then elastic deformation continues until the elastic limit is reached. Beyond that, the rod will be permanently deformed and will eventually fracture or break.

will return to its original length. If the tension is applied beyond the elastic limit and then removed, the material will recover somewhat but will retain some permanent deformation.

The straight-line part of the graph shows a direct proportionality between stress and strain. This relationship was first formalized by Robert Hooke in 1678 and is now known as *Hooke's law*. (It is the same general relationship as that given for a spring in Chapter 5.) The elastic modulus for a tension or a compression is called **Young's modulus** (Y):

$$\frac{F}{A} = Y\left(\frac{\Delta L}{L_o}\right)$$

or

$$Y = \frac{F/A}{\Delta L/L_o} \qquad (9.4)$$

The units for Young's modulus are the same as for stress (N/m^2). Some typical values of Young's modulus are given in Table 9.1.

The elastic limit is an important point for construction and/or engineering applications.

Compare Eq. 5.3

Thomas Young (1773–1829) was a British physicist who investigated the mechanical properties of materials and optical phenomena.

Young's modulus

Demonstration/activity: Suspend a spring from a support and apply weights to demonstrate Hooke's law.

TABLE 9.1 ■ Elastic Moduli for Various Materials (in N/m²)

Substance	Young's Modulus (Y)	Shear Modulus (S)	Bulk Modulus (B)
Solids			
Aluminum	7.0×10^{10}	2.5×10^{10}	7.0×10^{10}
Brass	9.0×10^{10}	3.5×10^{10}	7.5×10^{10}
Copper	11×10^{10}	3.8×10^{10}	12×10^{10}
Glass	5.7×10^{10}	2.4×10^{10}	4.0×10^{10}
Iron	15×10^{10}	6.0×10^{10}	12×10^{10}
Steel	20×10^{10}	8.2×10^{10}	15×10^{10}
Liquids			
Alcohol, ethyl			1.0×10^{9}
Glycerin			4.5×10^{9}
Mercury			26×10^{9}
Water			2.2×10^{9}

EXAMPLE 9.1 ■ Tensile Stress and Young's Modulus

What mass would have to be suspended from a steel wire with a diameter of 0.20 cm to increase its length by 0.10%?

Solution. Listing the data, along with Young's modulus for steel from Table 9.1, we have

Given: $d = 0.20$ cm, so $r = 0.10$ cm   *Find:* m (mass)
$\qquad\qquad\qquad\qquad = 1.0 \times 10^{-3}$ m
$\qquad$ strain $= \Delta L/L_o = 0.10\%$
$\qquad\qquad\qquad\qquad = 0.0010 = 1.0 \times 10^{-3}$
$\qquad Y_{steel} = 20 \times 10^{10}$ N/m^2
$\qquad\qquad$ (from Table 9.1)

The force required to produce the elongation is, by Eq. 9.4,

$$F = Y\left(\frac{\Delta L}{L_o}\right)A = Y\left(\frac{\Delta L}{L_o}\right)\pi r^2$$

$$= (20 \times 10^{10}\text{ N/m}^2)(1.0 \times 10^{-3}\text{ m})\pi(1.0 \times 10^{-3}\text{ m})^2 = 6.3 \times 10^2\text{ N}$$

This force is produced by the weight of the suspended mass, or $F = mg$. Thus,

$$m = \frac{F}{g} = \frac{6.3 \times 10^2\text{ N}}{9.8\text{ m/s}^2} = 64\text{ kg}$$

Change in Shape: Shear Modulus

Another way an elastic body can be deformed is by a **shear stress**. In this case, the deformation is due to an applied force that is tangential to the surface area (●Fig. 9.4). A change in shape results without a change in volume. The **shear strain** is given by x/h, where x is the relative displacement of the faces and h is the distance between them.

The shear strain is sometimes defined in terms of the **shear angle** (ϕ). As can be seen from the figure, $\tan \phi = x/h$. But the shear angle is usually quite small, so a good approximation is $\tan \phi \cong \phi \cong x/h$, where the angle ϕ is in radians. (For $\phi = 10°$, there is only 1.0% difference between ϕ and $\tan \phi$.) The **shear modulus** (sometimes called the modulus of rigidity) is then

$$S = \frac{F/A}{x/h} \cong \frac{F/A}{\phi} \qquad (9.5)$$

As you may note in Table 9.1, the shear modulus is generally less than Young's modulus; S is approximately $Y/3$ for many materials, which indicates that there is a greater response to a shear stress than to a tensile stress.

A shear stress may also be of the torsional type, resulting from the twisting action of a torque. For example, a torsional shear stress may shear off the head of a bolt that is being tightened.

Note in Table 9.1 that liquids do not have shear moduli (or Young's moduli). A shear stress cannot be effectively applied to a liquid or a gas. It is said that *fluids cannot support a shear*. Why?

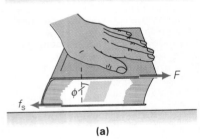

(a)

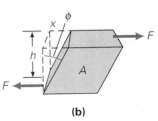

(b)

●**FIGURE 9.4 Shear stress and strain** A shear stress is produced when a force is applied tangentially to a surface area. The strain is measured in terms of the relative displacement of the object's faces or the shear angle ϕ.

The breaking of wooden boards, concrete blocks, or similar materials with a bare hand or foot is an impressive demonstration often done by karate experts (Fig. 1). The physics of this feat can be analyzed in terms of properties of the materials involved. In delivering the blow, the expert imparts a large impulse force (Chapter 6) to the top board or block, which bends under the pressure. (Note that the objects are struck midway between the end supports.) At the same time, the bones of the hand are being compressed as well. Fortunately, human bone can withstand more compressive force than can wood, tile, or concrete, which is why the expert's bones aren't damaged. (The ultimate compressive strength of bone is at least four times greater than that of concrete.*)

When the board or block is hit, the upper surface is compressed and the lower surface is elongated, or subjected to a tension force. Wood, tile, and concrete are weaker under tension than under compression. (The ultimate tensile strength of concrete is only about a twentieth of its ultimate compressive strength.) Thus, the board or block begins to crack at the bottom surface first. The crack propagates from the underside *toward* the hand, widens, and becomes a complete break. Thus the hand never actually "cuts through" the board or block.

The amount of force required to break a board or block in this way depends on several factors. For a wooden board, these are the type of wood, the width and thickness, and the distance between the end supports. Also, the edge of the hand must strike the board parallel to the grain of the wood. Similar considerations apply for a tile or a concrete block. Because tile and concrete are more rigid, more force is generally required to break these substances.

Some karate experts are able to break through a stack

FIGURE 1 Feat of strength or knowledge of materials? Breaking wooden boards, concrete blocks, or roofing tiles (as shown here) with a karate blow depends on both the physical strength of the expert and the strength of the material. Wood, concrete, and tile have different maximum tensile and compressional stresses, but the maximum compressional strength of bone is greater than any of these. (This photo, from a high-speed photography sequence, was made towards the end of the blow, when 3 of the 4 tiles were broken.)

of boards or tiles and more than one concrete block. The force required does not increase by a factor equal to the number of objects. High-speed photography shows that the hand makes contact with only one or two boards or tiles at the top of the stack. Then, as each object breaks, it collides with and breaks through the one below it.

Even though the properties of the materials seem to guarantee the success of this demonstration, you should not attempt the feat unless you are an expert and know what you're doing. The board or block might win.

*Ultimate strength is the maximum strength a material can stand before it breaks or fractures.

Change in Volume: Bulk Modulus

Suppose that a force directed inward acts over the entire surface of a body (Fig. 9.5). Such a **volume stress** is often applied by pressure transmitted by a fluid (Section 9.2). An elastic material will be compressed by a volume stress, that is, will show a change in volume but not in general shape. The (change in) pressure is equal to the volume stress, or $\Delta p = F/A$. The **volume strain** is the ratio of the volume change (ΔV) to the original volume (V_o). The **bulk modulus** (B) is then

$$B = \frac{F/A}{-\Delta V/V_o} = -\frac{\Delta p}{\Delta V/V_o}$$

(9.6)   Bulk modulus

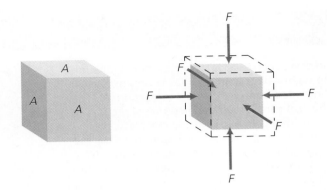

The minus sign is introduced to make B a positive quantity, since $\Delta V = V - V_\circ$ is negative for an increase in external pressure.

Bulk moduli are listed for solids *and* liquids in Table 9.1. Gases have bulk moduli too, since they can be compressed. For a gas, it is common to talk about the reciprocal of the bulk modulus, which is called the **compressibility** (k):

Compressibility

$$k = \frac{1}{B} \qquad (9.7)$$

Solids and liquids are relatively incompressible and have small values of compressibility. Gases, on the other hand, are easily compressed and have large compressibilities, which vary with pressure and temperature.

EXAMPLE 9.2 ■ Volume Stress and Bulk Modulus

How much pressure must be exerted on a liter of water to compress it by 0.10%?

Solution.
Listing the data, we have

Given: $-\Delta V/V_\circ = 0.0010$ *Find:* $\Delta p = F/A$
$V_\circ = 1.0 \text{ L} = 1000 \text{ cm}^3$
$B_{\text{H}_2\text{O}} = 2.2 \times 10^9 \text{ N/m}^2$
(from Table 9.1)

Note that $-\Delta V/V_\circ$ is the *fractional* change in the volume, and for a compression of 0.10%, this is 0.0010 (with no units since it expresses a ratio of volumes). With $V_\circ = 1000 \text{ cm}^3$, the volume reduction is

$$-\Delta V = (0.0010) \, V_\circ = (0.0010)(1000 \text{ cm}^3) = 1.0 \text{ cm}^3$$

However, the change in volume is not needed. The fractional change may be used directly in Eq. 9.6 to find the pressure (or change or increase in pressure):

$$\Delta p = B\left(\frac{-\Delta V}{V_\circ}\right) = (2.2 \times 10^9 \text{ N/m}^2)(0.0010) = 2.2 \times 10^6 \text{ N/m}^2$$

(This is about 22 times normal atmospheric pressure.)

9.2 ■ Fluids: Pressure and Pascal's Principle

A force can be applied to a solid at a point of contact, but this won't work with a fluid, as we saw in the preceding discussion on bulk moduli. With fluids, a force must be applied over an area. Such an application of force is expressed in terms of **pressure, or the force per unit area:**

$$p = \frac{F}{A}$$

(9.8)

Pressure—force per unit area

The force in this equation is understood to be acting normally (perpendicularly) to the surface area. F may be the perpendicular component of a force that acts at an angle to the surface (● Fig. 9.6). Pressure is a scalar quantity (with magnitude only), but the force producing it or the force produced within a fluid by a pressure does have direction and so is a vector.

Pressure has SI units of newtons per square meter (N/m^2). This combined unit is given the special name **pascal (Pa)** in honor of the French scientist and philosopher Blaise Pascal (1623–1662):

$$1 \text{ Pa} \equiv 1 \text{ N/m}^2$$

(One of Pascal's major contributions has to do with fluids and will be discussed shortly.) In the British system, a common unit of pressure is lb/in² (pounds per square inch, or psi). Other units, some of which will be introduced later, are used in special applications.

Pressure and Depth

If you have done any diving, you know that pressure increases with depth, because you have felt increased pressure on your eardrums. An opposite effect is commonly felt when flying in a plane or riding in a car going up a mountain. With increasing altitude, your ears may "pop" because of reduced air pressure.

How the pressure in a fluid varies with depth can be demonstrated by considering a container of liquid at rest, like the one in ● Fig. 9.7. With the liquid at rest (in static equilibrium) all points at the same depth must experience the same pressure; otherwise, there would be a pressure difference and horizontal movement of the liquid. For the rectangular column shown in the figure, the force on the surface at the bottom of the container is equal to the weight of the liquid making up the column:

$$F = mg = \rho V g = \rho g A h$$

where ρ (the Greek letter rho) is the density of the liquid ($\rho = m/V$). Since the liquid is assumed incompressible, ρ is constant. The volume V of the liquid is equal to $V = Ah$, and with $p = F/A$, the pressure at a depth h due to the weight of the column is

$$p = \rho g h$$

(9.9)

The pressure is the same everywhere on a horizontal plane at a depth h. Note that the total cross-sectional area of the bottom of the container can be taken as the base of a circular column with the same result.

Why do:
—ladies' high-heel shoes do more damage to floors than flat shoes?
—wheel tractors pack the dirt more than track tractors?
—sharp thumbtacks push in more easily than dull thumbtacks?

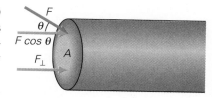

$$p = \frac{F_\perp}{A} = \frac{F \cos \theta}{A}$$

● **FIGURE 9.6 Pressure** Pressure is usually written $p = F/A$, where it is understood that F is the force or component of force normal to the surface. The normal or perpendicular component of force would be $p = (F \cos \theta)/A$.

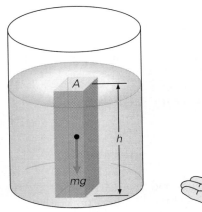

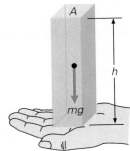

• FIGURE 9.7 Pressure and depth
The pressure at a depth h in a liquid
is due to the weight of the liquid
above: $p = \rho gh$, where ρ is the
density of the liquid (assumed to be
constant). This is shown here for an
arbitrary column of liquid.

The derivation of Eq. 9.9 did not take into account pressure being applied
to the open surface of the liquid. This adds to the pressure at a depth h to give
a *total* pressure of

Pressure-depth equation

$$\boxed{p = p_{\text{o}} + \rho gh}$$ (9.10)

pressure-depth equation

where p_{o} is the pressure applied to the liquid surface (that is, at $h = 0$).

> *Demonstration/activity:*
> *Demonstrate the effect of*
> *atmospheric pressure using*
> *Magdeburg Spheres. Recall that*
> *the area involved in this*
> *demonstration is the cross section*
> *of the sphere, not the surface*
> *area. Why?*

For an open container, p_{o} is atmospheric pressure, or the weight (force) per
area due to the gases in the atmosphere above the liquid's surface. The average
atmospheric pressure at sea level is sometimes used as a unit, called an
atmosphere (atm):

$$1 \text{ atm} \equiv 101.325 \text{ kPa} = 1.01325 \times 10^5 \text{ N/m}^2 = 14.7 \text{ lb/in}^2$$

How atmospheric pressure is measured will be described shortly.

EXAMPLE 9.3 ■ Pressure and Force

> *Video demonstration 7 relates to*
> *this material.*

(a) What is the total pressure on the back of a scuba diver in a lake at a depth
of 8.00 m? (b) What is the force on the diver's back due to the water alone,
taking the surface of the back to be a rectangle 60.0 cm by 50.0 cm?

Solution.

Given: $h = 8.00$ m

         $A = 0.600 \text{ m} \times 0.500 \text{ m}$

           $= 0.300 \text{ m}^2$

        $\rho_{\text{H2O}} = 1.00 \times 10^3 \text{ kg/m}^3$

           (known from original
definition of the kilogram
or from Table 9.2 in the
next section)

        $p_{\text{a}} = 1.01 \times 10^5 \text{ N/m}^2$

Find: (a) p (total pressure)

       (b) F (force due to water)

> *Adding atmospheric pressure is*
> *confusing to some students. A*
> *tire pressure of 35 psi as*
> *measured with a standard gauge*
> *does not include the atmospheric*
> *pressure.*

(a) The total pressure is the sum of the pressure due to the water and atmospheric pressure (p_{a}), by Eq. 9.10 is:

$$p = p_a + \rho g h$$
$$= (1.01 \times 10^5 \, \text{N/m}^2) + (1.00 \times 10^3 \, \text{kg/m}^3)(9.80 \, \text{m/s}^2)(8.00 \, \text{m})$$
$$= (1.01 \times 10^5 \, \text{N/m}^2) + (0.784 \times 10^5 \, \text{N/m}^2) = 1.79 \times 10^5 \, \text{N/m}^2 \ (\text{or Pa})$$

(b) The pressure p_w due to the water alone is the $\rho g h$ portion of the preceding equation, so $p_w = 0.784 \times 10^5 \, \text{N/m}^2$. Then, $p_w = F/A$, and

$$F = p_w A = (0.784 \times 10^5 \, \text{N/m}^2)(0.300 \, \text{m}^2)$$
$$= 2.35 \times 10^4 \, \text{N} \quad (\text{or } 5.29 \times 10^3 \, \text{lb—about 2.6 tons!})$$

You might think this answer is wrong—how could the diver support such a force? Look at the force on the diver's back from just atmospheric pressure:

$$F_a = p_a A = (1.01 \times 10^5 \, \text{N/m}^2)(0.300 \, \text{m}^2)$$
$$= 3.03 \times 10^4 \, \text{N} \quad (\text{or } 6.82 \times 10^3 \, \text{lb—about 3.4 tons!})$$

This is roughly the force on your back right now. Our bodies don't collapse under atmospheric pressure because cells are filled with fluids that react with an equal outward pressure (equal and opposite forces). As with forces, it is a pressure *difference* that gives rise to dynamic effects.

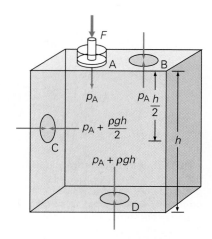

FIGURE 9.8 Pascal's principle
The pressure applied at A is fully transmitted to all parts of the fluid and to the walls of the container. There is also pressure due to the weight of the fluid above at different depths.

Pascal's Principle

When the pressure (for example, air pressure) is increased on the open surface of an incompressible liquid at rest, the pressure at any point in the liquid or on the boundary surfaces increases by the same amount. The effect is the same if pressure is applied by means of a piston to any surface of an enclosed fluid (Fig. 9.8). The transmission of pressure in fluids was studied by Blaise Pascal (after whom the SI pressure unit is named), and the observed effect is called **Pascal's principle**:

> Pressure applied to an enclosed fluid is transmitted undiminished to every point in the fluid and to the walls of the container.

Pascal's principle

For an incompressible liquid, the pressure change is transmitted instantaneously. For a gas, the pressure change is transmitted throughout the fluid, and after equilibrium has been reestablished (following changes in volume and/or temperature), Pascal's principle is valid.

Common practical applications of Pascal's principle include the hydraulic braking systems used on automobiles. A relatively small force on the brake pedal transmits a large force to the wheel brake cylinder. Also, hydraulic lifts and jacks are used to raise automobiles and other heavy objects (see Fig. 9.9). The input pressure p_i, supplied by compressed air for a garage lift, gives an input force F_i on a small piston area A_i. The full magnitude of the pressure is transmitted to the output piston, which has an area A_o. Since $p_i = p_o$,

$$\frac{F_i}{A_i} = \frac{F_o}{A_o}$$

and

$$F_o = \left(\frac{A_o}{A_i}\right) F_i \quad \text{(force multiplication)} \qquad (9.11)$$

Cars with power brakes have an intermediate system powered with a difference in air pressure to amplify the force exerted by the driver. Where does that difference in air pressure come from?

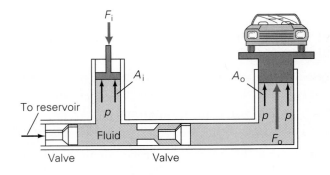

$$F_o = \left(\frac{A_o}{A_i}\right)F_i$$

Figure 9.9

With A_o larger than A_i, then F_o will be larger than F_i. The input force is greatly multiplied.

EXAMPLE 9.4 ■ The Hydraulic Lift

A garage lift has input and lift pistons with diameters of 10 cm and 30 cm, respectively. The lift is used to hold up a car with a weight of 1.4×10^4 N. (a) What is the force on the input piston? (b) What pressure is applied to the input piston?

Solution.

Given: $d_i = 10$ cm   *Find:* (a) F_i (input force)
          $d_o = 30$ cm   (b) p_i (input pressure)
          $F_o = 1.4 \times 10^4$ N

(a) Rearranging Eq. 9.11 and using $A = \pi r^2 = \pi d^2/4$ for the circular piston ($r = d/2$) gives

$$F_i = \left(\frac{A_i}{A_o}\right)F_o = \left(\frac{\pi d_i^2/4}{\pi d_o^2/4}\right)F_o = \left(\frac{d_i}{d_o}\right)^2 F_o$$

$$F_i = \left(\frac{10\text{ cm}}{30\text{ cm}}\right)^2 F_o = \frac{F_o}{9} = \frac{1.4 \times 10^4\text{ N}}{9} = 1.6 \times 10^3\text{ N}$$

The input force is $\frac{1}{9}$ of the output force, or there was a force multiplication of 9 ($F_o = 9F_i$).
(b) Then

$$p_i = \frac{F_i}{A_i} = \frac{F_i}{\pi r_i^2} = \frac{1.6 \times 10^3\text{ N}}{\pi(0.050\text{ m})^2}$$

$$= 2.0 \times 10^5\text{ N/m}^2 \quad (= 200\text{ kPa})$$

This pressure is about 30 lb/in², a common pressure used in automobile tires, and about twice atmospheric pressure (about 100 kPa or 15 lb/in²).

As the example shows, we can relate the force directly to the diameters of the pistons, $F_o = (d_o/d_i)^2 F_i$. By making $d_i \ll d_o$, we can get huge factors of force multiplication, as is typical for hydraulic presses and jacks. (Inversely, we can get a force reduction by making $d_i > d_o$.)

However, don't think that you are getting something for nothing with large force multiplications. Energy is still a factor, and this can never be multiplied by a machine. (Why not?) Looking at the work involved, and assuming the work output is equal to the work input, $W_o = W_i$ (an ideal condition—why?), we have from Eq. 5.1

$$F_o x_o = F_i x_i \quad \text{or} \quad F_o = \left(\frac{x_i}{x_o}\right) F_i$$

where x_o and x_i are the output and input distances moved by the respective pistons. Thus, the output force can be much greater than the input force only if the input distance is much greater than the output distance. For example, if $F_o = 10 F_i$, then $x_i = 10 x_o$, and the input piston must travel ten times the distance of the output piston. We say that force is multiplied at the expense of distance.

Pressure Measurement

Pressure can be measured with a variety of mechanical devices that are often spring-loaded. Another type of instrument uses a liquid, usually mercury, and is called a manometer. An **open-tube manometer** is illustrated in ● Fig. 9.10a. One end of the U-shaped tube is open to the atmosphere, and the other is connected to the container of gas whose pressure is to be measured. The liquid in the U-tube acts as a reservoir through which pressure is transmitted according to Pascal's principle.

The pressure of the gas (p) is balanced by the weight of the column of liquid (of height h measured from the lower surface level) and the atmospheric

Figure 9.10

● **FIGURE 9.10 Pressure measurement** **(a)** For an open-tube manometer, the pressure of the gas in the container is balanced by the pressure due to the liquid column and atmospheric pressure acting on the open surface of the liquid. The absolute pressure of the gas equals the sum of the atmospheric pressure and $\rho g h$, the gauge pressure. **(b)** A tire gauge measures gauge pressure, that is, the difference between the pressure in the tire and atmospheric pressure: $p_{gauge} = p - p_a$. Thus, if a tire gauge reads 200 kPa (30 lb/in²), the actual pressure within the tire is 1 atm higher, or 300 kPa. **(c)** A closed-tube manometer is a closed system. It measures absolute pressure since here atmospheric pressure is not a consideration. **(d)** A barometer is a closed-tube manometer that is exposed to the atmosphere and thus reads atmospheric pressure.

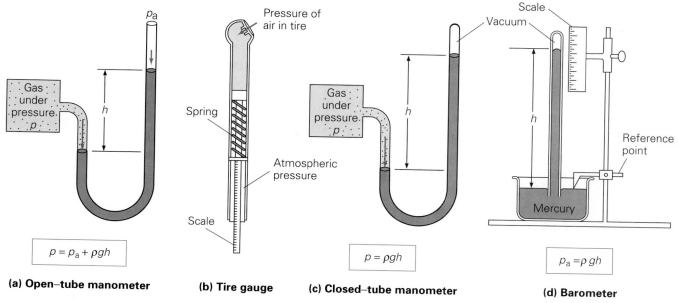

(a) Open–tube manometer $\quad$ **(b) Tire gauge** $\quad$ **(c) Closed–tube manometer** $\quad$ **(d) Barometer**

(a) $p = p_a + \rho g h$

(c) $p = \rho g h$

(d) $p_a = \rho g h$

pressure (p_a) on the open liquid surface:

$$p = p_a + \rho g h \qquad (9.12)$$

The pressure p is called the **absolute pressure**, and $p - p_a = \rho g h$ is called the **gauge pressure**. The gauge pressure is the pressure that registers on many common pressure-measuring devices such as a tire pressure gauge. This instrument basically compares the pressure inside a tire with the pressure outside and records the pressure difference: $p_{gauge} = p - p_a$ (Fig. 9.10b).

For the manometer illustrated in Fig. 9.10a, the absolute pressure of the gas is greater than the atmospheric pressure. Thus, the gauge pressure and the height (h) are positive ($p - p_a = \rho g h$). But if the absolute pressure of the gas in the container were equal to the atmospheric pressure, the gauge pressure would be zero. In this case, the column height would be zero ($h = 0$), or the liquid levels in the tubes would be at the same height. If the absolute pressure of the gas were less than atmospheric pressure (a partial vacuum), the gauge pressure and the column height would be negative.

If the open end of the U-shaped tube is evacuated and sealed, the device is a **closed-tube manometer** (Fig. 9.10c). Atmospheric pressure then has no effect on the liquid in the closed system, so the gauge pressure registered by this manometer is equal to the absolute pressure: $p_{gauge} = \rho g h = p - p_a = p - 0 = p$. A practical application of a closed-tube manometer is the **barometer** (Fig. 9.10d), used to measure atmospheric pressure. Essentially, the container of gas whose pressure the barometer measures is the atmosphere.

The mercury barometer was invented by Evangelista Torricelli (1608–1647), who was Galileo's successor as professor of mathematics at the university in Florence. A tube filled with mercury is inverted into a reservoir. Some mercury runs out, but as much as is supported by the air pressure on the surface of the reservoir pool remains in the tube. (You may be wondering about the space in the tube above the liquid column. Since the tube was initially full, there is no air above the liquid surface. Some of the mercury will evaporate and the mercury vapor will exert some pressure on the surface, but this is taken to be negligibly small.) The atmospheric pressure is then equal to the pressure due to the weight of the column of mercury, or

$$\boxed{p_a = \rho g h} \qquad (9.13)$$

A **standard atmosphere** is defined as the pressure supporting a column of mercury exactly 76 cm in height at sea level and 0°C.

Demonstration/activity:
Demonstrate the heights of two liquids in an open-tube manometer and of a liquid in a closed-tube manometer. Also display a mercury barometer.

Barometric pressure

EXAMPLE 9.5 ■ Standard Atmospheric Pressure

If a standard atmosphere supports a column height of exactly 76 cm of mercury (chemical symbol Hg), what is the standard atmospheric pressure in pascals? (The density of mercury is 13.5951×10^3 kg/m³ at 0°C, and $g = 9.80665$ m/s².)

Solution.

Given: $h = 76$ cm $= 0.76$ m (exact)   *Find:* p_a (atmospheric pressure)
$\rho_{Hg} = 13.5951 \times 10^3$ kg/m³
$g = 9.80665$ m/s²

Using Eq. 9.13

$$p_a = \rho g h = (13.5951 \times 10^3 \text{ kg/m}^3)(9.80665 \text{ m/s}^2)(0.760000 \text{ m})$$
$$= 101{,}325 \text{ N/m}^2 = 1.01325 \times 10^5 \text{ Pa} \quad \text{(or } 101.325 \text{ kPa)}$$

Changes in atmospheric pressure are observed as changes in the height of the mercury column. Atmospheric pressure is commonly reported in terms of the height of the barometer column, and weather forecasters say that the barometer is rising or falling. As noted previously, 1 standard atmosphere is a unit for air pressure expressed in height units as

1 atm = 76 cm Hg = 760 mm Hg = 29.92 in. Hg (about 30 in. Hg)

In honor of Torricelli, a pressure supporting 1 mm of mercury is given the name **torr**:

1 mm Hg ≡ 1 torr   and   1 atm = 760 torr

Since mercury is very toxic, it is sealed inside a barometer. A safer and less expensive device widely used to measure atmospheric pressure is the aneroid (without fluid) barometer. In an aneroid barometer, a sensitive metal diaphragm on an evacuated container (something like a drum head) responds to pressure changes, which are indicated on a dial. This is the kind of barometer you frequently find in homes in decorative wall mountings.

Since air is compressible, the density and pressure of the atmosphere increase with depth as measured from the top of the atmosphere. Atmospheric pressure is greatest at the Earth's surface and decreases with altitude. The rate of decrease is fairly uniform near the Earth. The pressure is approximately halved for each 5 km increase in altitude up to about 20 km. Above this, the atmospheric pressure decreases rapidly to about 6 Pa at an altitude of 70 km. The result is that half of the entire atmosphere is below 11 km (about 7 mi), and 99% is below 30 km (about 19 mi). The air is quite thin in the upper part of the atmosphere, which extends for several hundred kilometers.

We live at the bottom of the atmosphere but don't notice its pressure very much in our ordinary daily activities. Remember that our bodies are composed largely of fluids, which exert a matching outward pressure. Indeed, the external pressure of the atmosphere is so important to our normal functioning that we take it with us wherever we can. The pressurized suits worn by astronauts in space or on the moon are needed, not only to supply oxygen, but also to provide an external pressure similar to that on the Earth's surface (Fig. 9.11).

9.3 ■ Buoyancy and Archimedes' Principle

When an object is placed in a fluid, it will either sink or float. This is most commonly observed with liquids, for example, objects floating or sinking in water. But the same effect occurs in gases. A falling object is sinking in the atmosphere, and other bodies float (Fig. 9.12).

Things float because they are buoyant, or are buoyed up. For example, if you immerse a cork in water and release it, the cork will be buoyed up to the surface and float there. From your knowledge of forces, you know that such

 FIGURE 9.11 Dressed for the occasion Astronauts need pressurized suits not only to supply oxygen but also to protect them from the vacuum of space, where there is no air pressure to oppose our body's internal pressure.

Another unit sometimes used in weather reports is the millibar (mb). By definition, 1 atm = 1.01325×10^5 N/m² = 1.01325 bar = 1,013.25 mb. Normal atmospheric pressures are around 1000 mb.

 FIGURE 9.12 Fluid buoyancy The air is a fluid in which objects such as these hot-air balloons float. The hot air inside the balloons is lighter or less dense than the surrounding cooler air. A balloon is buoyed upward by a force opposite that of the balloon's weight.

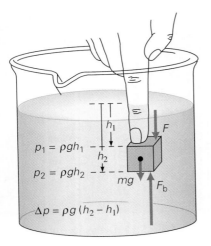

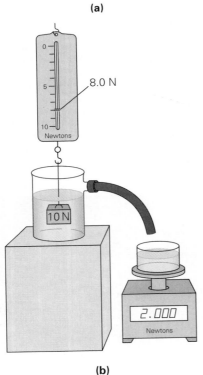

8.0 N

10 N

2.000

Newtons

(b)

FIGURE 9.13 Buoyancy and Archimedes' principle (a) A buoyant force arises from the pressure difference between different depths. The pressure on the bottom of the submerged block is greater than that on the top, so there is a (buoyant) force directed upward. (b) Archimedes' principle: The buoyant force on the object is equal to the weight of the volume of fluid it displaces. (The scale is set to read zero with an empty container.)

Figure 9.13

action requires an upward net force on the object. That is, there must be an upward force acting on the object that is greater than its downward weight force. The forces are equal when the object floats or stops sinking and is stationary. The upward force resulting from an object being wholly or partially immersed in a fluid is called the **buoyant force**.

How the buoyant force arises can be seen by considering a buoyant object being held under the surface of a fluid, as shown in ● Fig. 9.13a. The pressures on the upper and lower surfaces of the block are $p_1 = \rho_f g h_1$, and $p_2 = \rho_f g h_2$, where ρ_f is the density of the fluid. Thus, there is a pressure difference, $\Delta p = p_2 - p_1 = \rho_f g(h_2 - h_1)$, between the top and bottom of the block, which gives rise to an upward force (the buoyant force), F_b. This force is balanced by the applied force and the weight of the block.

It is not difficult to derive an expression for the magnitude of the buoyant force. We know that pressure is force per unit area. Thus, if the top and bottom areas of the block are A, the magnitude of the net buoyant force in terms of the pressure difference is

$$F_b = p_2 A - p_1 A = (\Delta p)A = \rho_f g(h_2 - h_1)A.$$

Since $(h_2 - h_1)A$ is the volume of the block, and hence the volume of fluid displaced by the block, V_f, we can write this expression as

$$F_b = \rho_f g V_f$$

But $\rho_f V_f$ is simply the mass of the fluid displaced by the block, m_f. Thus we can write the expression for the buoyant force $F_b = m_f g$, or the magnitude of the buoyant force is equal to the weight of the water displaced by the block (Fig. 9.13b). This general result is known as **Archimedes' principle**:

A body immersed wholly or partially in a fluid is buoyed up by a force equal in magnitude to the weight of the volume of fluid it displaces.

$$F_b = m_f g = \rho_f g V_f \qquad (9.14)$$

Archimedes (287–212 B.C.) was given the task of determining whether a crown made for a certain king was pure gold or contained some other, cheaper metal. Legend has it that the solution to the problem came to him when he was bathing, perhaps from seeing the water level rise when he got into the tub and experiencing the buoyant force on his limbs. In any case, it is said that he was so excited that he ran through the streets of the city shouting "Eureka!" (Greek for "I have found it"). Although Archimedes' solution to the problem involved density and volume (see Exercise 56), it presumably got him thinking about buoyancy.

EXAMPLE 9.6 ■ Lighter Than Air

What is the buoyant force on a helium balloon with a radius of 30 cm in air if $\rho_{air} = 1.3 \text{ kg/m}^3$?

Solution.

Given: $r = 30 \text{ cm} = 0.30 \text{ m}$ *Find:* F_b (buoyant force)
$\rho_{air} = 1.3 \text{ kg/m}^3$

The volume of the balloon is

$$V = \tfrac{4}{3}\pi r^3 = \tfrac{4}{3}\pi (0.30 \text{ m})^3 = 0.11 \text{ m}^3$$

Then, by Eq. 9.14, the weight of the air displaced by the balloon's volume, or the magnitude of the upward buoyant force, is

$$F_b = m_{air}g = (\rho_{air}V)g = (1.3 \text{ kg/m}^3)(0.11 \text{ m}^3)(9.8 \text{ m/s}^2) = 1.4 \text{ N}$$

Note that the buoyant force depends on the *density of the fluid and the volume of the body*. Shape makes no difference.

EXAMPLE 9.7 ■ Weight and Buoyant Force

An overflow container filled with water, like that shown in Fig. 9.13b, sits on a scale, which reads 40 N. (The water level is just below the exit tube in the side of the container.) A 5.0-N object that is less dense than water (e.g., a block of wood), is placed in the container. The water it displaces runs out the exit tube into another container that is not on the scale. Will the scale reading then be (a) exactly 45 N, (b) between 40 N and 45 N, (c) exactly 40 N, (d) less than 40 N? *Clearly establish the reasoning and physical principle(s) used in determining your answer before checking it below. That is, **why** did you select your answer?*

Reasoning and Answer. By Archimedes' principle, the block is buoyed upward with a force equal in magnitude to the weight of the water displaced. Since the block floats, the upward buoyant force must balance the weight of the block, and has a magnitude of 5.0 N. Thus a volume of water weighing 5.0 N is displaced from the container as a 5.0-N weight is added to the container. The scale still reads 40 N, so the answer is (c).

Note that the upward buoyant force and the block's weight force act *on the block*. The reaction force (pressure) of the block *on the water* is transmitted to the bottom of the container (Pascal's principle) and is registered on the scale.

Follow-up Exercise. Would the scale reading still be 40 N if the object had a density greater than that of water? (*Reasoning and answer may be found in the Answers to Exercises section at the back of the book.*)

We commonly say that helium and hot-air balloons float because they are lighter than air. However, technically, they are *less dense than air*. An object's density will tell you whether it will sink or float in a fluid, as long as you also know the density of the fluid. Consider an object totally immersed in a fluid. The weight of the object is

$$w_o = m_o g = \rho_o V_o g$$

The weight of the volume of fluid the object displaces, or the magnitude of the buoyant force, is

$$F_b = w_f = m_f g = \rho_f V_f g$$

But if the object is completely submerged, $V_o = V_f$, and dividing the second equation above by the first gives

$$\frac{F_b}{w_o} = \frac{\rho_f}{\rho_o} \qquad or \qquad F_b = \left(\frac{\rho_f}{\rho_o}\right)w_o \qquad (9.15)$$

Thus, if ρ_o is less than ρ_f, then F_b will be greater than w_o, and the object will be buoyed to the surface and float. If ρ_o is greater than ρ_f, then F_b will be less than w_o, and the object will sink. Also, if ρ_o equals ρ_f, then F_b will be equal to w_o, and the object will remain in equilibrium at any submerged depth (as long as the density of the fluid is constant).

These three conditions expressed in words are as follows:

An object will float in a fluid if the density of the object is less than the density of the fluid.

An object will sink in a fluid if the density of the object is greater than the density of the fluid.

An object will be in equilibrium at any submerged depth in a fluid if the densities of the object and the fluid are equal.

(See Demonstration 8.)

The densities of some solids and fluids are given in Table 9.2. A quick look will tell you if an object will float in a fluid, regardless of the shape or volume of the object. The conditions stated above also apply to a fluid in a fluid, provided the two are immiscible (do not mix).

In general, the densities of objects or fluids will be assumed to be uniform

Float or sink?

DEMONSTRATION 8 ■ Buoyancy and Density

(a) Unopened cans of Coke are dropped into a container of water.

(b) The can of Classic Coke sinks and the can of Diet Coke floats.

A demonstration of buoyancy that shows that the overall density of a can of Diet Coke is less than that of water, while the density of a can of Classic Coke is greater.

Consider the following questions: Does one can have a greater volume of metal? higher gas pressure inside? more fluid volume? Do calories make a difference? You can investigate the possibilities yourself to determine the reason(s) for the different densities.

TABLE 9.2 ■ Densities of Some Common Substances (in kg/m³)

Substance	Density (ρ)	Substance	Density (ρ)	Substance	Density (ρ)
Solids		*Liquids*		*Gases**	
Aluminum	2.7×10^3	Alcohol, ethyl	0.79×10^3	Air	1.29
Brass	8.7×10^3	Alcohol, methyl	0.82×10^3	Helium	0.18
Copper	8.9×10^3	Blood, whole	1.05×10^3	Oxygen	1.43
Glass	2.6×10^3	Blood plasma	1.03×10^3	Water vapor (100°C)	0.63
Gold	19.3×10^3	Gasoline	0.68×10^3		
Ice	0.92×10^3	Kerosene	0.82×10^3		
Iron	7.9×10^3	Mercury	13.6×10^3		
Lead	11.4×10^3	Sea water (4°C)	1.03×10^3		
Silver	10.5×10^3	Water, fresh (4°C)	1.00×10^3		
Steel	7.8×10^3				
Wood, oak	0.81×10^3				

*At 0°C and 1 atm unless otherwise specified.

and constant in this book. Of course, the density of the atmosphere varies with altitude but is relatively constant near the surface of the Earth. In some instances, the overall density of an object may be purposefully varied. For example, a submarine submerges by flooding its tanks with sea water (called taking on ballast), which increases its overall density. When the sub is to surface, the water is pumped out of the tanks, and the density of the sub becomes less than that of the surrounding sea water.

EXAMPLE 9.8 ■ Float or Sink?

A cube of material 10 cm on a side has a mass of 700 g. (a) Will the cube float in water? (b) If so, how much of its volume will be submerged?

Solution.

Given: $m = 700$ g
$L = 10$ cm
$\rho_w = 1.00$ g/cm³ (density of water from Table 9.2)

Find: (a) Whether the cube will float in water
(b) The percentage of the volume submerged

It is sometimes convenient to work in cgs units in comparing quantities, particularly when working with ratios. For densities in g/cm³, drop the "$\times 10^3$" from the values given in Table 9.2 for solids and liquids and add "$\times 10^{-3}$" for gases.

(a) The density of the cube material is

$$\rho_c = \frac{m}{V_c} = \frac{m}{L^3} = \frac{700 \text{ g}}{(10 \text{ cm})^3} = 0.70 \text{ g/cm}^3$$

Since ρ_c is less than ρ_w, the cube will float.

(b) The weight of the cube is $w_c = \rho_c g V_c$. When the cube is floating (in equilibrium), its weight is balanced by the buoyant force. That is, $F_b = \rho_w g V_w$, where V_w is the volume of water the submerged part of the cube displaces. Equating the expressions for weight and buoyant force gives

> Demonstration/activity: Place ice in a beaker of water filled to the rim. Ask students to predict what will occur when the ice melts. Start this demo at the beginning of a class period.

FIGURE 9.14 The tip of the iceberg The vast majority of an iceberg's bulk is underneath the water. This large "berg" was photographed off the Antarctic Peninsula.

$$\rho_w g V_w = \rho_c g V_c$$

or

$$\frac{V_w}{V_c} = \frac{\rho_c}{\rho_w} = \frac{0.70 \text{ g/cm}^3}{1.00 \text{ g/cm}^3} = 0.70$$

Thus, $V_w = (0.70)V_c$, and 70% of the cube is submerged.

Similarly, most of an iceberg floating in the ocean is submerged (Fig. 9.14). All that is visible is the proverbial tip of the iceberg. Now you can show what percentage this is. Icebergs are frozen fresh water, and finding the densities of ice (ρ_i) and sea water (ρ_{sw}) in Table 9.2, we have

$$\frac{V_{sw}}{V_i} = \frac{\rho_i}{\rho_{sw}} = \frac{0.92 \text{ g/cm}^3}{1.03 \text{ g/cm}^3} = 0.89$$

Thus, 89% of an iceberg is submerged and only 11% is seen above the surface.

Specific gravity

You may have heard of a quantity called specific gravity, which is related to density. It is commonly used for liquids but also applies to solids. Basically, it is a comparison of the weight of a volume of a substance with the weight of an equal volume of water. The **specific gravity** of a substance is equal to the ratio of the density of the substance (ρ_s) to the density of water (ρ_w, at 4°C):

$$\text{sp. gr.} = \frac{\rho_s}{\rho_w}$$

Because it is a ratio of densities, specific gravity has no units. Since $\rho_w = 1.00$ g/cm^3,

$$\text{sp. gr.} = \frac{\rho_s}{1.00} = \rho_s$$

That is, the specific gravity of a substance is equal to the numerical value of its density in cgs units. For example, if a liquid has a density of 1.5 g/cm^3, its specific gravity is 1.5, which tells you that it is 1.5 times denser than water.

Measuring the specific gravity of a liquid involves an application of Archimedes' principle and a device called a **hydrometer**. A common type of hydrometer consists of a sealed glass tube with a weighted bulb (to make the tube float upright) and an enclosed calibrated scale in the upper stem (Fig. 9.15). The hydrometer is calibrated so that its scale reads 1.000 at the surface level when it is floating in water. When the hydrometer is placed in another liquid with a different density, it floats higher or lower, depending on the liquid's density (why?), and the specific gravity of the liquid is read from the scale.

Hydrometers have many applications. They are used medically to measure the specific gravities of body fluids and commercially to measure the specific gravities of acids, alcohols, gasoline, and various solutions. A common use of a hydrometer is as a battery-acid tester. The liquid in a lead storage battery, such as those used in automobiles, is a solution of sulfuric acid and water. When the battery is charging, the concentration of sulfuric acid, which is denser than water, increases. When the battery is discharging, water is formed and sulfuric acid is depleted. Checking the specific gravity of the liquid in the battery is thus an easy way to determine how charged the battery is.

Since sealed, maintenance-free batteries have become more common recently, perhaps a more frequent application of the hydrometer is for determining the concentration of antifreeze in an automobile's cooling system.

FIGURE 9.15 A hydrometer The specific gravity of a liquid in which the hydrometer is immersed is read from the calibrated scale on the upper stem. This hydrometer is constructed for measuring specific gravities greater than that of water, or 1.00000.

9.4 ■ Surface Tension and Capillary Action

A fluid will not support a shear stress, yet you have probably seen water striders walking on the surface of a pond (● Fig. 9.16). Also, a razor blade or a needle placed carefully on the surface of water will float, even though steel is denser than water (about eight times denser according to Table 9.2). These things are possible because of an interesting property of liquids. A free liquid surface acts like a thin membrane that can be placed under a slight tension.

Surface Tension

The molecules of a liquid exert small attractive forces on each other. Even though molecules are electrically neutral overall, there is often some slight asymmetry of charge that gives rise to attractive forces between molecules (called van der Waals forces). Within a liquid, where any molecule is completely surrounded by other molecules, the net force is zero (● Fig. 9.17a). However, for molecules at the surface of the liquid, there is no attractive force acting from above the surface. (The effect of air molecules is small and considered negligible.) As a result, the molecules of the surface layer experience net forces due to the attraction of neighboring molecules just below the surface. This inward pull on the surface molecules causes the surface of the liquid to contract and to resist being stretched or broken, a property called **surface tension**.

If a sewing needle is carefully placed on the surface of a bowl of water, the surface acts like an elastic membrane under tension. There is a slight depression in the surface, and molecular forces along the depression are at an angle to the surface (Fig. 9.17b). The vertical components of these forces balance the weight (*mg*) of the needle and it "floats" on the surface. Similarly surface tension supports the weight of a water strider (Fig. 9.17c).

The net effect of surface tension is to make the surface area of a liquid as small as possible. That is, a given volume of liquid tends to assume the shape

FIGURE 9.16 Surface tension Insects like this water strider can walk on water because of surface tension, much as you might walk on a large trampoline. Note the depressions in the surface of the liquid where the legs touch it.

Surface tension

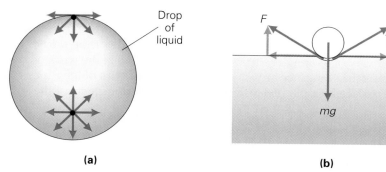

(a)　　　　(b)　　　　(c)

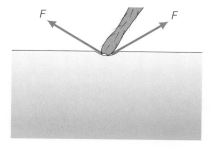

FIGURE 9.17 Surface tension **(a)** The net force on a molecule in the interior of a liquid is zero, since it is surrounded by other molecules. However, a molecule at the surface experiences a nonzero net force due to the attractive forces of the neighboring molecules just below the surface. **(b)** To form a surface depression, work must be done, since more interior molecules must be brought to the surface to increase the area. As a result, the surface area acts like a stretched elastic membrane, and the weight force of an object such as a needle is supported by the upward components of the surface tension. **(c)** Insect legs make a similar depression, and the resulting upward force components allow the insect to walk on water. **(d)** Because of surface tension, water droplets tend to assume the shape that minimizes their surface area—that of a sphere.

(d)

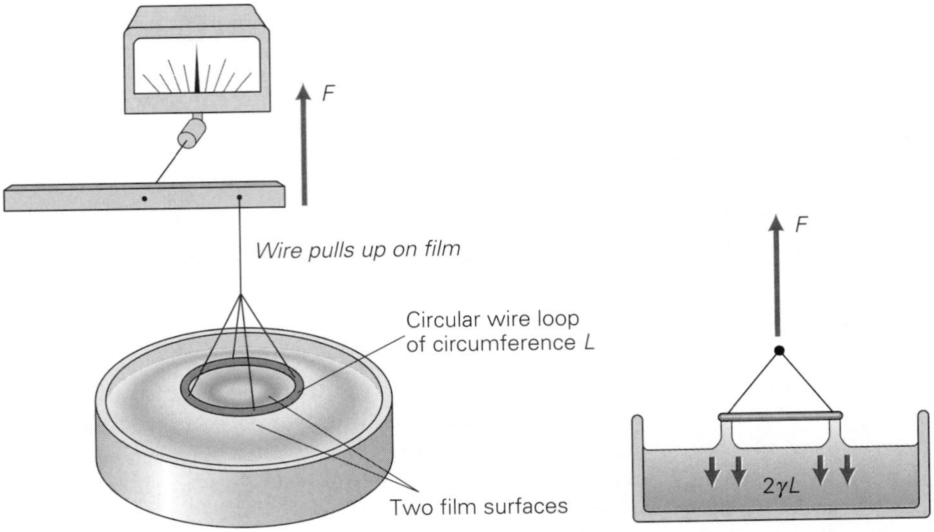

FIGURE 9.18 **Measuring surface tension** An experimental method of measuring the force required to just overcome surface tension. It is equal to $2\gamma L$, since there are two film surfaces.

Measuring surface tension

that has the least surface area. As a result, drops of water and soap bubbles have spherical shapes because a sphere has the smallest surface area for a given volume (Fig. 9.17d). In forming a drop or bubble, surface tension pulls the molecules together to minimize the surface area.

Quantitatively, the surface tension (γ) in a liquid film is defined as the force per unit length acting along a line (for example, on a length of wire) when stretching the surface:

$$\gamma = \frac{F}{L} \quad \text{(surface tension)} \quad (9.16)$$

The SI units for surface tension are newtons per meter (N/m), as can be seen from the equation. Some surface tensions for liquids are given in Table 9.3. As might be expected, surface tension is highly dependent on temperature.

An apparatus used to measure surface tension is shown in Fig. 9.18. Basically, the device measures the force required to overcome the surface tension. For a circular wire loop, L is the length of the circumference, and $\gamma = F/2L$, since there are two film surfaces (one on each side of the wire).

Another way of looking at surface tension is in terms of the work or energy needed to stretch the surface area. If a straight piece of wire of length L is used to stretch a surface by a parallel distance Δx, the work done against the surface tension is

$$\Delta W = F\Delta x = \gamma L \Delta x = \gamma \Delta A$$

since $F = \gamma L$ and $\Delta A = L\Delta x$ (the change in surface area). Thus,

$$\gamma = \frac{\Delta W}{\Delta A}$$

The surface tension, or force per unit length, is equivalently the work per unit change in the surface area, with units of J/m^2. (Is J/m^2 equivalent to N/m?)

Adhesion, Cohesion, and Capillary Action

Note the relatively small surface tension given in Table 9.3 for soapy water. Soaps and detergents have the effect of lowering the surface tension of water. Such substances are called *surfactants*. The relatively high surface tension of

TABLE 9.3 ■ **Surface Tensions for Some Liquids (in N/m)**

Liquid		Surface tension (γ)
Alcohol, ethyl	(20°C)	0.022
Blood, whole	(37°C)	0.058
Blood plasma	(37°C)	0.072
Mercury	(20°C)	0.45
Soapy water	(20°C)	0.025
Water	(0°C)	0.076
Water	(20°C)	0.073
Water	(100°C)	0.059

plain water tends to prevent it from getting into small places, such as between the fibers of clothing. (You can also see from the table why warm water is generally used for cleaning.)

Soaps and detergents also act as *wetting agents*. Whether or not a liquid "wets" or adheres to a surface depends on the relative strengths of the adhesive and cohesive forces between molecules. **Adhesive forces** (or adhesion) are attractive forces between unlike molecules. **Cohesive forces** (or cohesion) are attractive forces between like molecules. Cohesive forces hold a substance together, and adhesive forces hold different substances together. (Adhesives such as glue hold things together.)

If the adhesive forces between the molecules of the liquid and those of the surface are greater than the cohesive forces among the molecules of the liquid, the liquid wets the surface. On the other hand, if the cohesive forces are greater than the adhesive forces, the liquid does not wet the surface. Water beading up on a recently waxed car is a good example of the latter situation (see Fig. 9.17d). Water does not adhere well to waxes or oils on a surface. (This phenomenon, as well as surface tension, helps water striders to walk on water. The insects' feet are covered with a waxlike substance that prevents wetting.)

Although adhesive and cohesive forces are difficult to analyze, a relative measure of their effects is the **contact angle (ϕ)**. This is the angle between the surface and a line drawn tangent to the liquid (●Fig. 9.19). Note from the figure that ϕ is less than 90° if the liquid wets the surface and greater than 90° if it does not. Water on clean glass has a contact angle of approximately 0° (it spreads out in a thin layer), and water on paraffin has a contact angle of 107° (Table 9.4). A little detergent on the paraffin will cause the water to spread out, or wet the surface more, and the contact angle will decrease. The cleansing action of soaps and detergents is due in large part to their enhancing of the capability of water to wet dirt particles so they can be washed away.

In a container, the free liquid surface curves upward (is concave) if the liquid wets the container wall and curves downward (is convex) if it does not (●Fig. 9.20). The curved shape of the liquid surface is referred to as the *meniscus* (from a Greek word meaning "crescent moon"). If a tube with a small diameter is positioned vertically with one end submerged in a liquid that wets the walls of the tube, the liquid in the tube will rise some distance above the surface of the surrounding liquid. This is called **capillary action** (or capillarity) and is a consequence of both surface tension and adhesion. Essentially, adhesion draws the water up the sides of the tube, and cohesion (surface tension) then pulls the column upward.

As might be expected, the height to which a liquid rises in a capillary tube depends on the diameter. (Capillary comes from a Latin word meaning "hairlike." Your smallest blood vessels are called capillaries and are so narrow that blood cells must pass through them in single file.) At equilibrium, the upward component of the surface tension force and the downward weight force of the liquid column must be equal in magnitude. The surface tension force is

$$F = \gamma L = \gamma(2\pi r)$$

where $L = 2\pi r$ since the liquid is in contact with the tube at all points around its circumference. The vertical component of this force has a magnitude of

$$F \cos \phi = \gamma(2\pi r)(\cos \phi)$$

The weight of the liquid column is given by

$$w = mg = \rho Vg = \rho(\pi r^2 h)g$$

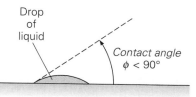

(a) Wetting condition

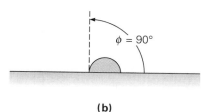

(b)

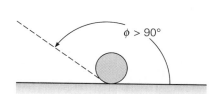

(c) Nonwetting condition

●**FIGURE 9.19 Contact angle** A relative measure of adhesive and cohesive forces is given by the contact angle. **(a)** An angle ϕ less than 90° means that the liquid wets the surface, indicating that the adhesive forces are greater than the cohesive forces. **(b)** When ϕ is equal to 90°, the drop forms a hemisphere. **(c)** An angle ϕ greater than 90° means that the liquid does not wet the surface, indicating that the cohesive forces are greater than the adhesive forces.

TABLE 9.4 ■ Contact Angles for Some Liquids on Solids

Liquid-solid	Contact angle (ϕ) (approximate)
Alcohol-glass	0°
Kerosene-glass	26°
Mercury-glass	140°
Water-glass	0°
Water-silver	90°
Water-paraffin	107°

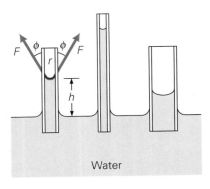

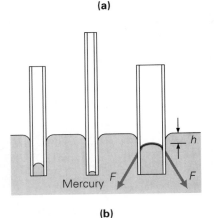

Water

(a)

Mercury

(b)

● **FIGURE 9.20 Capillary action**
(a) Liquids rise in small (capillary) tubes because of adhesion (wetting) and surface tension. The upward force (balanced by the weight of the liquid column) and the contact angle are shown in one tube. **(b)** If a liquid does not wet a capillary tube, there is a column depression.

Demonstration/activity: Show the rise of water in capillary tubes. Add food coloring to the water for better visibility.

Demonstration/activity: Demonstrate how capillary action changes by using different brands of paper towels.

where mass in terms of density is $m = \rho V$ and the volume of the cylinder of liquid is $V = \pi r^2 h$. (Atmospheric pressure is not a consideration because it is the same on both surfaces.) Equating these force magnitudes ($F \cos \phi = w$) and solving for h gives

$$h = \frac{2\gamma \cos \phi}{\rho g r} \qquad (9.17)$$

A similar analysis for capillary depression, which occurs when the liquid does not wet the tube surface, gives the same equation. In this case, ϕ is greater than 90°, and h is negative.

EXAMPLE 9.9 ■ Soap Makes a Difference

(a) How high will plain water at 20°C rise in a 1.0-mm-diameter glass capillary tube? (b) Will soapy water rise more or less in the same capillary tube?

Solution.

Given: $d = 1.0$ mm $= 1.0 \times 10^{-3}$ m,   *Find:* (a) h_w (water height)
so $r = 5.0 \times 10^{-4}$ m   (b) h_{sw} (soapy water
$\rho_w = \rho_{sw} = 1.0 \times 10^3$ kg/m³   height compared
(from Table 9.2)   to h_w)
$\gamma_w = 0.073$ N/m
(from Table 9.3)
$\gamma_{sw} = 0.025$ N/m
(from Table 9.3)

(a) From Table 9.4, the contact angle for water and glass is $\phi = 0°$. With $\cos \phi = \cos 0° = 1$, Eq. 9.17 becomes $h = 2\gamma/\rho g r$. For water,

$$h_w = \frac{2\gamma_w}{\rho_w g r}$$

$$= \frac{2(0.073 \text{ N/m})}{(1.0 \times 10^3 \text{ kg/m}^3)(9.8 \text{ m/s}^2)(5.0 \times 10^{-4} \text{ m})}$$

$$= 0.030 \text{ m} \quad \text{(or 3.0 cm)}$$

(b) The height of the soapy water could be calculated using the same equation, but using a proportion saves some time.

$$h_{sw} = \left(\frac{\gamma_{sw}}{\gamma_w}\right) h_w = \left(\frac{0.025 \text{ N/m}}{0.073 \text{ N/m}}\right) h_w = (0.34) h_w$$

Hence, the rise of soapy water in the capillary is less than that of pure water—only about one-third as great, or $h_{sw} \approx 0.010$ m (or 1.0 cm).

Capillary action is important in liquid transport. Common examples are the absorption of water by paper towels, the wicking of oil in oil lamps, the distribution of water and nutrients in plants, and the holding of water in the soil. If it were not for the latter, the ground would be pretty dry down to the water table. Less desirable results of capillary action are the wetting of concrete blocks in building walls and the rise of water in cracks and joints.

9.5 ■ Fluid Dynamics and Bernoulli's Equation

In general, fluid motion is difficult to analyze. For example, think of trying to describe the motion of a particle (a molecule) of water in a rushing stream. The overall motion of the stream may be apparent, but a mathematical description of the motion of some particle of it may be virtually impossible because of eddy currents (small whirlpool motions), the gushing of water over rocks, frictional drag on the stream bottom, and so on. A basic description of fluid flow is conveniently obtained by ignoring such complications and considering an ideal fluid. Actual fluid flow can then be approximated in reference to this more simple theoretical model.

In this simple approach to fluid dynamics, it is customary to consider four characteristics of ideal fluids: steady, irrotational, nonviscous, and incompressible flow.

> Steady flow means that all the particles of the fluid have the same velocity as they pass a given point.

Steady flow might be called smooth or regular flow. The path of steady flow can be depicted in the form of **streamlines** (●Fig. 9.21).

Every particle that passes a particular point moves along a streamline. That is, every particle moves along the same path (streamline) as particles that passed by earlier. Streamlines never cross. If they did, a particle would have alternate paths and abrupt changes in its velocity, in which case the flow would not be steady. Steady flow requires low velocities. For example, it is approximated by the flow relative to a canoe that is gliding slowly through still water. When the flow velocity is high, eddies tend to appear, especially near boundaries, and the flow becomes turbulent. (The conditions for turbulent flow are discussed more fully at the end of the next Section.)

Streamlines also indicate the relative magnitude of the fluid velocity. The velocity is greater where the streamlines are closer together. Notice this effect in Fig. 9.23a. The reason for this will be explained shortly.

> Irrotational flow means that a fluid element (a small volume of the fluid) has no net angular velocity, which eliminates the possibility of whirlpools and eddy currents (the flow is nonturbulent).

Consider the small paddle wheel in Fig. 9.21a. With a zero net torque, it does not rotate. Thus, the flow is irrotational (Fig. 9.21b).

> Nonviscous flow means that viscosity is neglected. Viscosity refers to a fluid's internal friction, or resistance to flow.

A truly nonviscous fluid would flow freely with no energy lost within it. Also, there would be no frictional drag between the fluid and the walls containing it. In reality, when a liquid flows through a pipe, a velocity is less near the walls because of frictional drag and is greater toward the center of the pipe. (Viscosity is discussed in more detail in Section 9.6.)

Incompressible flow means that the fluid's density is constant. Liquids

Ideal fluid flow

Streamlines

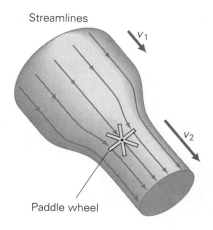

(a)

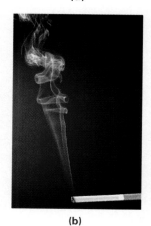

(b)

● **FIGURE 9.21 Streamline flow** **(a)** Streamlines never cross and are closer together in regions of greater fluid velocity. The stationary paddle wheel indicates that the flow is irrotational, or without whirlpools and eddy currents. **(b)** The smoke begins to rise in nearly streamline flow, but quickly becomes rotational and turbulent.

can usually be considered incompressible. Gases, on the other hand, are quite compressible. Sometimes, however, gases approximate incompressible flow, for example, in air flowing relative to the wings of an airplane traveling at low speeds.

Theoretical or ideal fluid flow is not characteristic of most real situations. But the analysis of ideal flow provides results that approximate, or generally describe, a variety of applications. This analysis is not done in terms of Newton's laws but in terms of two basic principles: the conservation of mass and the conservation of energy.

Equation of Continuity

If there are no losses of fluid within a uniform tube, the mass of fluid flowing into the tube in a given time must be equal to the mass flowing out of the tube in the same time (by the conservation of mass). For example, in ● Fig. 9.22, the mass (Δm_1) entering the tube during a short time (Δt) is

$$\Delta m_1 = \rho_1 \, \Delta V_1 = \rho_1 (A_1 v_1 \Delta t)$$

where A_1 is the cross-sectional area of the tube at the entrance, and in a time Δt, a fluid particle moves a distance equal to $v_1 \, \Delta t$. Similarly, the mass leaving the tube in the same time interval is

$$\Delta m_2 = \rho_2 \, \Delta V_2 = \rho_2 (A_2 v_2 \Delta t)$$

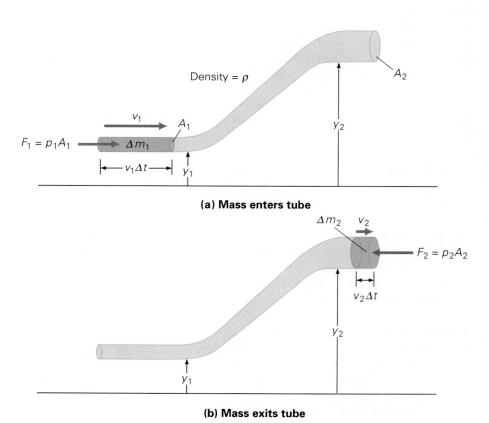

●FIGURE 9.22 Flow continuity
Ideal fluid flow can be described in terms of the conservation of mass by the equation of continuity. See text for description.

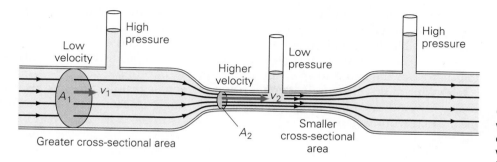

FIGURE 9.23 Flow rate and work-energy By the flow rate equation, Av = constant, and by the work-energy theorem, $W = \Delta K$. See text for description.

Since the mass is conserved, $\Delta m_1 = \Delta m_2$, and

$$\rho_1 A_1 v_1 = \rho_2 A_2 v_2$$

or
$$\rho A v = \text{constant} \qquad (9.18) \qquad \text{Equation of continuity}$$

This general result is called the **equation of continuity**.
 For an incompressible fluid, the density ρ is constant, so

$$\boxed{\begin{array}{l} A_1 v_1 = A_2 v_2 \\[4pt] Av = \text{constant} \end{array}} \quad \text{(for an incompressible fluid)} \qquad (9.19) \qquad \text{Flow rate equation}$$

or

This is sometimes called the **flow rate equation**, since Av has the standard units of cubic meters per second (m³/s, or volume/time). (In the British system, gal/min. is often used.) Note that the flow rate equation shows that the fluid velocity is greater where the cross-sectional area of the tube is smaller. That is,

$$v_2 = \left(\frac{A_1}{A_2}\right) v_1$$

and v_2 is greater than v_1 if A_2 is less than A_1 (see ● Fig. 9.23). (The streamlines will be closer together in a smaller tube.) This effect is evident in the common experience that the velocity of water is greater from a hose fitted with a nozzle than from the same hose without a nozzle.
 The flow rate equation can be applied to the flow of blood in your body. Blood flows from the heart into the aorta. It then makes a circuit through the circulatory system, passing through arteries, arterioles (small arteries), capillaries, and back to the heart through veins. The velocity is lowest in the capillaries—a contradiction? No. The *total* area of the capillaries is much greater than that of the arteries or veins, so the flow rate equation holds.

Bernoulli's Equation

The conservation of energy or the general work-energy theorem leads to another relationship that has great generality for fluid flow. This relationship was first derived in 1738 by the Swiss mathematician Daniel Bernoulli (1700–1782) and is named for him.
 The derivation of Bernoulli's equation is somewhat involved, so let's look at a qualitative "derivation" for better understanding. Consider the streamline liquid flow illustrated in Fig. 9.23. A constriction in the pipe forces the stream-

lines closer together, which means that the velocity of the liquid is greater. This can be understood in several ways. First, since liquids are practically incompressible, if a certain mass of liquid goes into one end of the tube in a given time, an equal mass must come out the other end. The only way for this to occur is for the liquid to speed up in the smaller part of the tube.

Looking at this effect from another perspective, note that the pressure also changes, as indicated by the standpipe pressure gauges. A change in pressure means that work was done. The pressure decreases in the constricted part of the tube, so work was done on the liquid (in forcing it into the smaller tube). Then, by the work-energy theorem ($W = \Delta K$), the kinetic energy (and thus the speed) of the liquid increases.

In applying the work-energy theorem and the conservation of energy to fluid flow, there are three aspects to be considered: kinetic energy, potential energy, and the work associated with pressure differences. Since we are dealing with fluid flow, these terms are expressed as energy (or work) per unit volume. To do this, the kinetic and potential energies are written in terms of density, $\rho = m/V$. Thus, $\frac{1}{2}mv^2/V$(energy/volume) $= \frac{1}{2}\rho v^2$ and mgh/V (energy/volume) $= \rho gh$. Work is given by $W = Fx$, and in terms of pressure, $p = F/A$, we have $W = pAx$. In a pipe of cross-sectional area A and length x, the product Ax is the volume of a quantity of fluid. So, $W = pV$, and $p = W/V$ (work/volume).

Compare the derivation of Eq. 5.9 in Section 5.5

Point out the similarities between Bernoulli's equation and the law of conservation of energy.

The work-energy theorem is then applied to ideal fluid flow, similar to the application in Chapter 5 that resulted in an expression for the conservation of mechanical energy. In the fluid case, we have $W = \Delta K + \Delta U$. In the general case, some work is assumed to go into the gravitational potential energy of the fluid (as in Fig. 9.22). Thus ΔU has two components, a pressure-related term and a gravity-related term. With the work expressed in terms of pressure, the result is expressed by what is known as **Bernoulli's equation**:

Bernoulli's equation for ideal fluid flow

or

$$p_1 + \tfrac{1}{2}\rho v_1^2 + \rho g y_1 = p_2 + \tfrac{1}{2}\rho v_2^2 + \rho g y_2$$

$$p + \tfrac{1}{2}\rho v^2 + \rho g y = \text{constant}$$

(9.20)

Bernoulli effects

Bernoulli's equation can be applied to many situations (see Demonstration 9). If there is horizontal flow ($y_1 = y_2$), then $p + \frac{1}{2}\rho v^2 = $ constant, which indicates that the pressure decreases if the fluid velocity increases (and vice versa). Chimneys and smokestacks are tall in order to take advantage of the more consistent and higher wind speeds at greater heights. The faster the wind blows over the top of a chimney, the lower the pressure, and the greater the pressure difference between the bottom and top of the chimney. Thus, the chimney draws better. Bernoulli's equation and the continuity equation ($Av = $ constant) also tell you that if the cross-sectional area of a pipe is reduced, so that the velocity of the fluid passing through it is increased, the pressure is reduced.

Suppose a fluid is at rest ($v_2 = v_1 = 0$). Bernoulli's equation becomes

$$p_2 - p_1 = \rho g(y_1 - y_2)$$

This should look familiar; it is the pressure-depth relationship ($p = \rho gh$) derived earlier.

(a) The demonstrator prepares to place the ball in an inverted funnel and to blow downward through the funnel.

(b) With air blowing through, the ball remains "trapped" in the funnel.

(c) When the demonstrator is out of breath, the ball falls.

How does blowing downward through a funnel keep a Ping Pong ball from falling? In this demonstration, seeing is believing and Bernoulli.

The air stream above and around the ball (along the wall of the funnel) is moving faster than the air below the ball, and the faster air moves, the less pressure it exerts. Thus the greater pressure of the slower air below the ball holds it up. Without an air stream, the pressures are equalized (both atmospheric), and the ball falls.

EXAMPLE 9.10 ■ Flow Rate from a Tank

Water escapes through a small hole in a tank as depicted in ● Fig. 9.24. What is the flow rate of the water from the tank?

Solution.

By Bernoulli's equation,

$$p_1 + \tfrac{1}{2}\rho v_1^2 + \rho g y_1 = p_2 + \tfrac{1}{2}\rho v_2^2 + \rho g y_2$$

where $y_2 - y_1$ is the height of the liquid surface above the hole. The atmospheric pressures acting on the open surface and at the hole, p_1 and p_2, respectively, are essentially equal and cancel from the equation, as does the density, giving

$$v_1^2 - v_2^2 = 2g(y_2 - y_1)$$

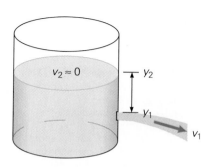

● **FIGURE 9.24 Fluid flow from a tank** The flow rate is given by Bernoulli's equation. Example 9.10.

The equation of continuity says that $A_1v_1 = A_2v_2$, where A_2 is the cross-sectional area of the tank and A_1 is that of the hole. Since A_2 is much greater than A_1, then v_1 is much greater than v_2. So a good approximation is

$$v_1^2 = 2g(y_2 - y_1) \qquad \text{or} \qquad v_1 = \sqrt{2g(y_2 - y_1)}$$

The flow rate (volume/time) is then

$$\text{flow rate} = A_1v_1 = A_1\sqrt{2g(y_2 - y_1)}$$

Given the area of the hole and the height of the liquid above it, you can find the instantaneous speed of the water coming from the hole and the instantaneous flow rate.

Another example of the Bernoulli effect (as it is sometimes called) is given in the Insight feature.

INSIGHT ■ Throwing a Curve

In baseball, a pitcher can cause a baseball to curve as it moves toward the catcher by giving it an appropriate spin. Why a curve ball curves can be understood in terms of Bernoulli's equation and the viscous properties of air.

If air were an ideal fluid, the spin of a pitched ball would have no effect in changing the direction of its motion. (See Fig. 1a; the streamlines are those that would be seen by the ball or an observer moving with the ball.) However, since air is viscous, friction between it and the ball causes a thin layer of air to be dragged around by the spinning ball. The speed of the ball relative to the air is thus greater on one side than on the other because the velocities are in the same direction on one side and in opposite directions on the other (Fig. 1b). By Bernoulli's equation, the low-velocity side (v') has a greater pressure than the high-velocity side (v''). Thus, the ball experiences a net force

toward the low-pressure side and is deflected (curves). (The Bernoulli effect is important at relatively low velocities. However, baseballs are usually pitched at speeds high enough for another factor, known as the Magnus effect, to predominate. This effect, which involves the turbulent wake created by the ball, alters the ball's flight path in much the same way as the Bernoulli effect that we have described.)

A Bernoulli effect takes place in your body. Looking at the blood vessels again, we find that, because of frictional drag, the blood velocity is smaller near the vessel walls than in the center. (This phenomenon is described more fully in Section 9.6.) Hence, the pressure is higher at the walls than near the center of a vein or artery. As a result, the blood cells experience a "Bernoulli force" and are concentrated toward the center of the vessel.

FIGURE 1 Curve ball **(a)** If air were an ideal fluid (without viscosity), or if no spin were applied, air flow would be the same on both sides of the ball. **(b)** Because air has viscosity, some air is dragged around the ball, so the speed of the ball with respect to the air is greater on one side. Pressure on that side is lower, resulting in a net force in that direction (see text).

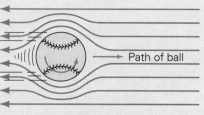

(a) If air were an ideal fluid, air flow would be the same on both sides.

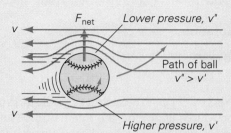

(b) Since air has viscosity, air flow is faster on one side of the ball, creating a pressure difference.

9.6 ■ Viscosity, Poiseuille's Law, and Reynolds Number ◇

All real fluids have an internal resistance to flow, which is described as **viscosity**. Viscosity may be considered to be friction between the molecules of a fluid. In liquids, it is caused by short-range cohesive forces, and in gases, by collisions between molecules (see the discussion of air resistance in Chapter 4). The viscous drag for both liquids and gases depends on velocity and may be directly proportional to it in some cases. However, the relationship varies depending on conditions; for example, the drag is approximately proportional to v^2 or v^3 for turbulent flow.

Viscosity—resistance to flow

Air resistance is discussed in Section 4.6

Internal friction causes fluid layers to move relative to each other in response to a shear stress (Fig. 9.25a). This layered motion, called *laminar flow*, is characteristic of steady flow for viscous liquids at low velocities. At greater velocities, the flow becomes rotational, or turbulent. As Fig. 9.25a indicates, the magnitude of the shear stress can be used as a measure of viscosity. Viscosity is characterized by a **coefficient of viscosity**, η (the Greek letter eta):

Many of the newer home-built airplanes, including the Voyager that flew non-stop around the world without refueling, have what is called a laminar flow airfoil. For these airfoils, the air flow from front to back remains laminar for a greater distance, hence the skin friction is reduced.

$$\eta = \left(\frac{F}{A}\right)\left(\frac{h}{v}\right) = \frac{Fh}{Av} \tag{9.21}$$

where F/A is the shear stress needed to maintain laminar flow between two parallel planes separated by a distance h at a relative velocity of magnitude v. Actually, η (commonly referred to as simply the viscosity) is the ratio of the shear stress to the rate of change of the shear strain: $(\Delta x/h)\Delta t = v/h$. Viscosities of some fluids are given in Table 9.5.

The SI units for viscosity are pascal-seconds:

$$(N/m^2)(m/[m/s]) = (N/m^2)(s) = Pa\text{-}s$$

This combined unit is called the **poiseuille (Pl)**, in honor of the French scientist Jean Poiseuille (1799–1869), who studied the flow of liquids, particularly blood. The cgs unit for viscosity is the **poise (P)**. A smaller multiple, the centipoise (cP), is widely used because of its convenient size.

As might be expected, viscosity, and thus fluid flow, vary with temperature, which is evident from the old saying "slow as molasses in January." A familiar application is the viscosity grade given to motor oil used in automobiles. In winter, a low-viscosity, or relatively thin, oil should be used (such as SAE grade 10W or 20W) because it will flow more readily, particularly when the

 FIGURE 9.25 Laminar flow
(a) A shear stress causes fluid layers to move over each other in laminar flow. The shear force and the flow rate depend on the viscosity of the fluid. **(b)** For laminar flow through a pipe, the fluid velocity is less nearer the walls of the pipe because of frictional drag between the walls and the fluid.

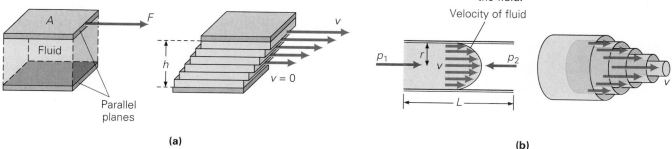

(a)   (b)

TABLE 9.5 ■ Viscosities of Various Fluids*

Fluid	Viscosity (η) Pl	cP
Liquids		
Alcohol, ethyl	1.2×10^{-3}	1.2
Blood, whole (37°C)	1.7×10^{-3}	1.7
Blood plasma (37°C)	2.5×10^{-3}	2.5
Glycerin	1.5	1.5×10^{3}
Mercury	1.55×10^{-3}	1.55
Oil, light machine	0.11	1.1×10^{3}
Water	1.00×10^{-3}	1.00
Gases		
Air	1.9×10^{-5}	1.9×10^{-2}
Oxygen	2.2×10^{-5}	2.2×10^{-2}

*At 20°C unless otherwise indicated. 1 poiseuille (Pl) $= 10^{3}$ centipoise (cP).

SAE stands for Society of Automotive Engineers, an organization that designates the grades of motor oils based on flow rate or viscosity.

engine is cold at start-up. In summer, a higher-viscosity, or thicker, oil is used (SAE 30, 40, or even 50).

Of course, seasonal changes in the grade of motor oil are not necessary if you use the multigrade year-round oils. These contain additives called viscosity improvers, which are polymers whose molecules are long, coiled chains. A temperature increase causes the molecules to uncoil and intertwine with each other. Thus, the normal decrease in viscosity is counteracted. The action is reversed on cooling, and the oil maintains a relatively constant viscosity over a temperature range. Such motor oils are graded as, for example, SAE 10W-40 (or 10W-40 for short).

Poiseuille's Law

When a fluid flows through a pipe, there is frictional drag between the liquid and the walls, and the fluid velocity is greater toward the center of the pipe (Fig. 9.25b). The average *flow rate Q* is given by

$$Q = A\bar{v} = \Delta V / \Delta t \tag{9.22}$$

where ΔV is the volume passing a given point in the time Δt. The flow rate depends on properties of the fluid and dimensions of the pipe, as well as on the pressure difference between the ends of the pipe. Jean Poiseuille studied flow in pipes and tubes, assuming constant viscosity and steady or laminar flow. He derived a relationship for the flow rate, which is known as **Poiseuille's law:**

Poiseuille's law—viscous flow

$$Q = \frac{\pi r^{4} \Delta p}{8 \eta L} \tag{9.23}$$

where r is the radius of the pipe and L is its length.

As should be expected, the flow rate (m³/s) is inversely proportional to the viscosity and the length of the pipe. Also as expected, the flow rate is directly proportional to the pressure difference. Somewhat surprisingly, however, the flow rate is proportional to r^4, which makes it highly dependent on the radius of the tube, more so than might have been thought.

EXAMPLE 9.11 ■ Fourth-Power Dependence

The radius of a length of pipe carrying a liquid is decreased by 5.0% because of deposits on the inner surface. By how much would the pressure difference between the ends of the constricted pipe have to be increased to maintain a constant flow rate?

Solution.

> *Given:* $r_2 = 95\% \times r_1$, or $\qquad$ *Find:* Δp (pressure difference)
> $\qquad r_2 = (0.95)r_1$
> $\qquad Q_2 = Q_1$

With $Q_2 = Q_1$, by Eq. 9.23

$$\frac{Q_2}{Q_1} = \frac{r_2^4 \Delta p_2}{r_1^4 \Delta p_1}$$

and

$$\Delta p_2 = \left(\frac{r_1}{r_2}\right)^4 \Delta p_1 = \left(\frac{1}{0.95}\right)^4 \Delta p_1 = (1.23)\,\Delta p_1$$

The pressure difference must be increased by 23% to maintain the same flow rate.

Thus, the blood pressure has to increase to maintain the normal flow of blood in constricted arteries. As pointed out in the Insight feature, this makes the heart work harder. A similar application of Poiseuille's law concerns administering gases in respiratory therapy for conditions that reduce air flow to the lungs because the obstructions in the air passages.

INSIGHT ■ Blood Pressure

Basically, a pump is a machine that transfers mechanical energy to a fluid, causing it to flow. There are a wide variety of pumps, but one of interest to everyone is the human heart. The heart is a muscular pump that drives blood through the body's circulatory network of arteries, capillaries, and veins.

At the start of each pumping cycle, the heart's interior chambers enlarge and fill with freshly oxygenated blood arriving from the lungs (see Fig. 1). On contraction, the blood is forced out through the aorta into the arterial network. Smaller and smaller arteries branch off from the

main ones, until the very small capillaries are reached. There food and oxygen being carried by the blood are exchanged with surrounding tissues, and wastes are picked up. The blood then flows into the veins to complete its circuit back to the heart.

The walls of the arteries have considerable elasticity and expand and contract with each pumping cycle. When the heart contracts, the blood pressure in the arteries increases. When the heart relaxes, the blood pressure decreases. The maximum pressure is called the systolic pressure, and the minimum pressure is called the dia-

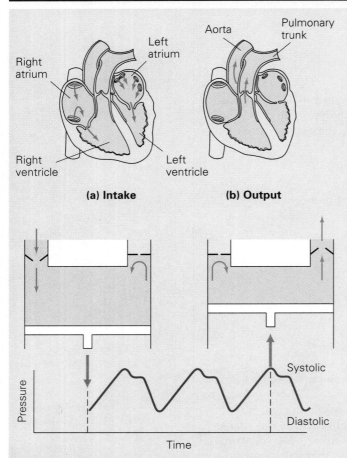

FIGURE 1 The heart as a pump The human heart is analogous to a mechanical force pump. Its pumping action, consisting of (a) intake and (b) output, gives rise to variations in blood pressure.

stolic pressure (after the two parts of the pumping cycle, the systole and diastole).

Taking a person's blood pressure involves measuring the pressure of the blood on the arterial walls. This is done with a sphygmomanometer (the Greek word *sphygmo* means "pulse"). An inflatable cuff is used to shut off the blood flow temporarily. The cuff pressure is slowly released, and the artery is monitored with a stethoscope (Fig. 2). When the cuff pressure is slightly below the maximum blood pressure, blood will begin to flow through the artery with each beat of the heart. This produces a distinct beating sound that can be heard with the stethoscope. As soon as the sound is heard, the pressure is

noted, normally about 120 mm Hg. This is the systolic pressure. As more air is released from the cuff, the blood flows more steadily, until at some pressure the distinct beating sound is no longer heard. The pressure at this point, normally about 80 mm Hg, corresponds to the diastolic pressure. Blood pressure is commonly reported by giving the systolic and diastolic pressures separated by a slash, for example, 120/80 (which is read as "120 over 80"). Normal blood pressure ranges between 100 and 140 for the systolic pressure and between 70 and 90 for the diastolic pressure.

High blood pressure is a common health problem. The elastic walls of the arteries expand under the hydraulic force of the blood pumped from the heart. Their elasticity may diminish with age, however. Fatty deposits (of cholesterols) can narrow and roughen the arterial passageways, impeding the blood flow and giving rise to a form of arteriosclerosis, or hardening of the arteries. Because of these defects, the driving pressure must increase to maintain a normal blood flow. The heart must work harder, which places a greater demand on its muscles. A relatively slight decrease in the effective cross-sectional area of a blood vessel has a rather large effect on the flow rate. (See the discussion of Poiseuille's law.)

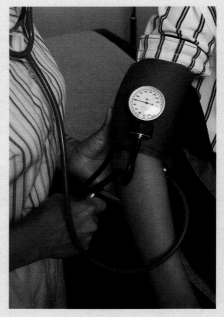

FIGURE 2 Measuring blood pressure The pressure is indicated on the gauge in mm Hg or torr.

Reynolds Number

When the flow rate of a fluid exceeds a certain velocity, the flow ceases to be laminar and becomes turbulent. Poiseuille's law then no longer applies. Analyzing turbulent flow is a difficult task, but a value that tells when the onset of turbulence will occur has been determined experimentally. This is expressed in terms of a dimensionless number called the **Reynolds number** (R_n):

$$R_n = \frac{\rho \bar{v} d}{\eta}$$

(9.24)   Reynolds number—turbulent flow

where ρ is the density of the fluid, $\bar{v}$ the average flow speed, d the diameter of a cylindrical pipe or tube, and η the viscosity.

In a pipe with smooth walls, the flow is laminar if R_n is below 2000. Turbulence generally sets in when R_n is about 2000 or more ($R_n \geq 2000$). It is possible to have laminar flow if R_n is above 2000, but that flow will be very unstable. Any slight disturbance will cause it to become turbulent. Interestingly, flow rate is then often greater than if Poiseuille's law applied. Thus engineers deliberately design pipelines for turbulent flow, and much of the flow in our circulatory system is turbulent.

Important Concepts

You should be able to define and explain these chapter concepts clearly.

fluid	volume strain	torr	viscous flow
stress	bulk modulus	buoyant force	incompressible flow
strain	compressibility	Archimedes' principle	equation of continuity
tensile stress	pressure	specific gravity	flow rate equation
tensile strain	pascal (Pa)	hydrometer	Bernoulli's equation
elastic moduli	atmosphere (atm)	surface tension	◇ viscosity
elastic limit	Pascal's principle	adhesive forces	◇ coefficient of viscosity
Young's modulus	open-tube manometer	cohesive forces	◇ poiseuille (Pl)
shear stress	absolute pressure	contact angle	◇ poise (P)
shear strain	gauge pressure	capillary action	◇ Poiseuille's law
shear angle	closed-tube manometer	steady flow	◇ Reynolds number
shear modulus	barometer	streamlines	
volume stress	standard atmosphere	irrotational flow	

Important Relationships for Review

These relationships are mathematical statements of the concepts and principles presented in the chapter. You should be able to identify the symbols and to explain the relationships before proceeding to the Exercises. In-text equation reference numbers are given for convenience.

Stress:

$$\text{stress} = \frac{F}{A}$$

(9.1)

Strain (tensile):

$$\text{strain} = \frac{\Delta L}{L_0} = \frac{L - L_0}{L_0}$$

(9.2)

Young's Modulus:

$$Y = \frac{F/A}{\Delta L/L_o}$$ (9.4)

Shear Modulus:

$$S = \frac{F/A}{x/h} \simeq \frac{F/A}{\phi}$$ (9.5)

Bulk Modulus:

$$B = \frac{F/A}{-\Delta V/V_o} = -\frac{\Delta p}{\Delta V/V_o}$$ (9.6)

Compressibility:

$$k = \frac{1}{B}$$ (9.7)

Pressure:

$$p = \frac{F}{A}$$ (9.8)

Pressure-Depth Equation:

$$p = p_o + \rho g h$$ (9.10)

Barometric (Atmospheric) Pressure:

$$p_a = \rho g h$$ (9.13)

Archimedes' Principle:

$$F_b = m_f g = \rho_f g V_f$$ (9.14)

Surface Tension:

$$\gamma = \frac{F}{L}$$ (9.16)

Capillary Height:

$$h = \frac{2\gamma \cos \phi}{\rho g r}$$ (9.17)

Equation of Continuity:

$$\rho_1 A_1 v_1 = \rho_2 A_2 v_2 \quad \text{or} \quad \rho A v = \text{constant}$$ (9.18)

Flow Rate Equation (for an incompressible fluid):

$$A_1 v_1 = A_2 v_2 \quad \text{or} \quad A v = \text{constant}$$ (9.19)

Bernoulli's Equation:

$$p_1 + \tfrac{1}{2}\rho v_1^2 + \rho g y_1 = p_2 + \tfrac{1}{2}\rho v_2^2 + \rho g y_2$$
$$\text{or} \quad p + \tfrac{1}{2}\rho v^2 + \rho g h = \text{constant}$$ (9.20)

◇ *Coefficient of Viscosity:*

$$\eta = \frac{Fh}{Av}$$ (9.21)

◇ *Flow Rate:*

$$Q = A\bar{v} = \frac{\Delta V}{\Delta t}$$ (9.22)

◇ *Poiseuille's Law:*

$$Q = \frac{\pi r^4 \Delta p}{8\eta L}$$ (9.23)

◇ *Reynolds Number:*

$$R_n = \frac{\rho \bar{v} d}{\eta}$$ (9.24)

Exercises

9.1 ▪ Solids and Elastic Moduli
(Ignore significant figures for very small changes.)

1 The deformation of an elastic body is described by (a) a modulus, (b) work, (c) stress, (d) strain.   (d)

●**2** Shear moduli exist for (a) solids, (b) liquids, (c) gases, (d) all of these.   (a)

3 One material has a greater Young's modulus than another. What does this tell you?   less strain for a given stress

4 Why are scissors sometimes called shears? Is this a descriptive name?   see ISM

5 Is it possible for the bulk modulus to be negative? Explain, and give an example if possible.   no; ΔV is $(-)$ for a decrease in volume

6 Write the general form of Hooke's law and find the units of the "spring constant" for elastic deformation. N/m^2

7 ▪ A force of 800 N is applied at an angle of 30° to the cross-sectional area at the end of a bar. That surface is 4.0

cm on a side. What are (a) the compressional stress and (b) the shear stress on the bar? (a) 2.5×10^5 N/m² (b) 4.3×10^5 N/m²

8 ■ A 100-kg mass is suspended using a cable with a diameter of 2.0 cm. What is the stress in the cable? 3.2×10^6 N/m²

9 ■ A carpenter applies a tangential force of 125 N to the upper surface of a block of wood. If the dimensions of the surface are 20 cm by 30 cm, what is the shear stress on the block? 2.1×10^3 N/m²

10 ■■ A metal wire 1.0 mm in diameter and 2.0 m long hangs vertically with a 6.0-kg mass suspended from it. If the wire stretches 1.4 mm under the tension, what is the value of Young's modulus for the metal? 1.1×10^{11} N/m²

11 ■■ A cable initially 130.00 cm long and 2.00 mm in diameter is stretched to a length of 130.26 cm by a force of 600 N. What is Young's modulus for the material of the cable? 9.5×10^{10} N/m²

12 ■■ A rectangular steel column (20.0 cm × 15.0 cm) supports a load of 12.0 metric tons. If the column was 4.00 m in length before being stressed, what is its loaded length? 3.99992 m

13 ■■ A bimetallic rod as illustrated in ● Fig. 9.26 is composed of brass and copper. If the rod is subjected to a compressive force of 5.00×10^4 N, which way will the rod bow or bend? (Justify your answer mathematically.) bends toward copper, see ISM

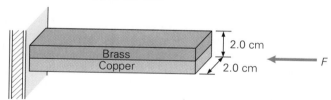

●FIGURE 9.26 Bimetallic rod and mechanical stress See Exercise 13.

14 ■■ Two metal posts of aluminum and copper of the same size are subjected to equal shear stresses. Which post will show the larger deformation and by what factor? $\phi_{Al} = (1.5)\,\phi_{Cu}$

15 ■■ A rectangular block of Jell-O with length, width, and height of 10 cm, 8.0 cm, and 4.0 cm, respectively, is subjected to a shear force of 0.40 N on its upper surface. If the top surface is displaced 0.30 mm relative to the bottom surface, what is the shear modulus of the gelatin? 6.7×10^3 N/m²

16 ■■ A shear force of 500 N is applied to one face of a cube of aluminum measuring 10 cm on a side. What is the relative displacement of the opposite face? 2.0×10^{-7} m

17 ■■ Two metal plates are held together by two rivets with diameters of 0.40 cm. If the maximum shear stress a single rivet can withstand is 5.0×10^8 N/m², how much force must be applied parallel to the plates to shear off both rivets? 1.3×10^4 N

18 ■■ What pressure difference is required to compress a volume of water by 0.10%? 2.2×10^6 N/m²

19 ■■ (a) Which of the liquids in Table 9.1 has the greatest compressibility? (b) For equal volumes of ethyl alcohol and water, which would require more pressure to be compressed by 0.10%, and how many times more? (a) ethyl alcohol has the smallest β (b) $\rho_W/\rho_{Al} = 2.2$

●20 ■■ A brass cube 6.0 cm on a side is placed in a pressure chamber and subjected to a pressure of 1.2×10^7 N/m² on all its surfaces. What is the volume of the cube under this pressure? 2.1996×10^{-4} m³

21 ■■ A cube of aluminum 10 cm on a side receives equal pressure on all its faces. What would be the magnitude of the force required to compress the cube's volume by 0.01%? 7.0×10^4 N

22 ■■■ A 45-kg traffic light is suspended with two steel cables of equal length and radii of 0.50 cm. If the cables make an angle of 15° with the horizontal, what is the fractional increase in their length due to the weight of the light? 5.4×10^{-5}

23 ■■■ A length of steel wire with a diameter of 0.10 cm is lengthened by 0.40% by a tension force. If a piece of copper wire with the same initial length is lengthened by the same percentage by a tension force of the same magnitude, what is the diameter of the copper wire? 0.13 cm

9.2 ■ Fluids: Pressure and Pascal's Principle

24 The pressure at a depth h in a liquid column is independent of (a) the liquid's density, (b) the acceleration due to gravity, (c) the height of the column, (d) the shape of the liquid's container. (d)

25 Pressure applied to an enclosed fluid is transmitted (a) only in liquids, (b) in all directions, (c) only to the bottom of the container, (d) only to the sides of the container. (b)

●26 Two dams form artificial lakes of equal depth. However, one lake backs up 5 km behind the dam and the other backs up 10 km. What effect does the difference in length have on the pressures on the dams? none—same depth, same pressure

27 The water in each of the containers in ● Fig. 9.27 exerts

●FIGURE 9.27 Same pressure and same force? See Exercise 27.

the same pressure *and* the same total force on the bottom of the container (equal base areas). But it is obvious that the weight of the water in each container is different. (a) What is the explanation of this so-called hydrostatic paradox? (b) Suppose that water could be added to the containers until the depth in each was doubled. How would this affect the weight of the contained water and the pressure on the bottom of each container?   see ISM

28   A water dispenser for pets has an inverted plastic bottle, as shown in ● Fig. 9.28. (The water is dyed blue for contrast.) When a certain amount of water is drunk from the bowl, more water flows automatically from the bottle into the bowl. The bowl never overflows. Explain the operation of the dispenser. Does the height of the water in the bottle depend on the surface area of the water in the bowl?   see ISM

● **FIGURE 9.28 Pet barometer**   See Exercise 28.

29   (a) Liquid storage cans, such as gasoline cans, generally have capped vent holes. What is the purpose of the vent, and what happens if you forget to remove the cap before pouring the liquid? (b) Explain how a medicine dropper works. (c) Explain how we breathe, i.e., inhalation and exhalation.   see ISM

30   ■ A 100-mL graduated cylinder is filled halfway with ethyl alcohol, and then the other half is filled with gasoline. What is the total mass of the combined liquids?   74 g

31   ■ A 60-kg woman balances herself on the heel of one of her high-heeled shoes. If the heel is a square with sides of 1.5 cm, what is the pressure exerted on the floor? (Express your answer in lb/in.² and atm.) 3.8×10^2 lb/in² or 26 atm

● **32**   ■ Which has the greater volume and how many times greater, 10 kg of aluminum or 2.4 kg of lead? $V_{Al} = (18) V_{Pb}$

33   ■■ Show that a water barometer would be impractical.
10 m high

34   ■■ A pure gold nugget with a volume of 15 cm³ is placed on a pan balance. What would be the volume of the brass weights needed to balance the nugget?   33 cm³

35   ■■ What is the pressure exerted on a 15° ski slope by a 90-kg skier going down it? (Assume that the area of contact of the skis with the snow is 0.40 m².) 2.1×10^3 Pa

● **36**   ■■ A scuba diver dives to a depth of 12 m in a lake. If the circular glass plate on the diver's face mask has a diameter of 18 cm, what is the force on it? 3.0×10^3 N

37   ■■ A water tank sits on a hill above a town. If the water level in the tank is 20 m above the town's water plant, what is the static water pressure at the plant if the water is delivered through an 0.80-m diameter pipe? 2.0×10^5 Pa

38   ■■ The gauge pressure in both tires of a bicycle is 400 kPa. If the bicycle and the child riding it have a combined mass of 30 kg, what is the area of contact of each tire with the ground? 3.8×10^{-4} m²

39   ■■ In a sample of seawater taken from an oil spill region, it is found that an oil layer 4.0 cm thick floats on 55 cm of water. If the density of the oil is 0.75×10^3 kg/m³, what is the absolute pressure on the bottom of the container?
1.07×10^5 Pa

40   ■■ The pressure exerted by a person's lungs can be measured by having the person blow as hard as possible into one side of a manometer. If a person blowing into one side of an open-tube manometer produces a 70-cm difference in the heights of the columns of water in the manometer arms, what is the lung pressure? 1.07×10^5 Pa

41   ■■ In 1960, the U.S. Navy's bathyscaphe *Trieste* descended to a depth of 10,912 m (about 35,000 ft) in the Mariana Trench in the Pacific Ocean. (a) What was the pressure at that depth? (Assume that sea water is incompressible.) (b) What was the force on a circular observation window with a diameter of 15 cm?   (a) 1.1×10^8 Pa   (b) 2.0×10^6 N

● **42**   ■■ Water is poured into one arm of an open U-tube containing mercury. What column height of water will cause the mercury level in the other arm to rise 5.0 cm?   68 cm

43   ■■ Oil is poured into one side of an open tube manometer containing mercury. What is the density of the oil if a 7.0-cm column of mercury supports a 40-cm column of oil?   2.4 g/cm³

44   ■■ The output piston of a hydraulic garage lift has a cross-sectional area of 0.20 m². (a) How much pressure on the input piston is required to support a car with a mass of 1.4 metric tons? (b) What force is applied to the input piston if it has a diameter of 5.0 cm?   (a) 7.0×10^4 Pa   (b) 140 N

45   ■■■ A syringe with a plunger diameter of 2.0 cm is attached to a hypodermic needle with a diameter of 1.5 mm. What minimum force must be applied to the plunger to inject a fluid into a vein, where the blood pressure is 12 torr?
3.7×10^{-2} N

46 ■■■ An open-tube mercury manometer has a height difference between the columns of −10 cm. What are (a) the gauge pressure and (b) the absolute pressure?
(a) −1.3 × 10⁴ Pa   (b) 8.8 × 10⁴ Pa

47 ■■■ A hydraulic balance used to detect small mass changes is shown in ● Fig. 9.29. If a mass of 0.25 g is placed on the balance platform, by how much will the height of the water in the smaller cylinder have changed when the balance comes to equilibrium?   1.6 cm

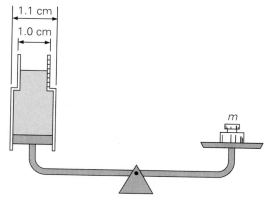

● **FIGURE 9.29 A hydraulic balance**   See Exercise 47.

9.3 ■ Buoyancy and Archimedes' Principle

48 If a barometer were constructed using a liquid less dense than mercury, (a) a tube of larger diameter would be needed (b) the liquid column would be unchanged for a given pressure, (c) a longer tube may be necessary (d) none of these.   (c)

49 If an object displaces an amount of liquid of greater weight than its own, the object will (a) float, (b) sink, (c) remain in equilibrium at any submerged position.   (a)

50 (a) What is the criterion for constructing a life jacket that will keep a person afloat? (b) Why is it so easy to float in the Great Salt Lake in Utah?   see ISM

● **51** An ice cube floats in a glass of water. As the ice melts, how does the level of the water in the glass change? Would it make any difference if the ice cube were hollow? Explain.   see ISM

52 A person sitting in a boat in the middle of a swimming pool notes the water level on the side of the pool. He then drops several concrete blocks over the side of the boat into the water. Does the water level of the pool change? If so, how?   water level falls

53 ■■ A cube 8.5 cm on a side has a mass of 0.65 kg. Will the cube float in water?   no

54 ■■ A helium-filled balloon supports a load of 1 metric ton. (a) If the load includes the balloon's mass, what is its volume? (b) If the balloon were spherical, what would be its radius?   (a) 7.8 × 10² m³   (b) 5.7 m

55 ■■ A submarine has a mass of 10,000 metric tons. How much water must be displaced for the sub to be in equilibrium just below the ocean surface?   10,000 metric tons

● **56** ■■ Suppose that Archimedes found that the king's crown had a mass of 0.750 kg and a volume of 3.980 × 10⁻⁵ m³. (a) How did he determine the volume? (b) Was the crown pure gold?   (a) water displacement   (b) no

57 ■■ An 0.80-kg crown is submerged in water and its apparent weight is measured to be 7.30 N. Is the crown pure gold?   no

58 ■■ A crane lifts a rectangular iron bar whose dimensions are 0.25 m by 0.20 m by 10 m from the bottom of a lake. What is the minimum upward force the crane must supply when the bar is (a) in the water and (b) out of the water?
(a) 3.4 × 10⁴ N   (b) 3.9 × 10⁴ N

59 ■■ A rectangular boat as illustrated in ● Fig. 9.30 is overloaded so that the water level is just 1.0 cm below the top of the side of the boat. What is the combined mass of the people and the boat?   2.6 × 10³ kg

● **FIGURE 9.30 An overloaded boat**   See Exercise 59.

60 ■■ Ice will float in water, but sink in ethyl alcohol. In what proportion should these liquids be mixed so that an ice cube will be able to float at any depth?
1 part water to 0.62 parts alcohol, or 38% alcohol

61 ■■ An irregularly shaped piece of metal has a mass of 90 g in air. It is suspended from a scale and the scale reads 75 g when the piece is submerged in water. What are the volume and density of the piece of metal?   1.5 × 10⁻⁵ m³,
6.0 × 10³ kg/m³

● **62** ■■ A flat-bottomed rectangular boat has a length of 4.0 m and a width of 1.5 m. If the load is 2000 kg (including the mass of the boat), how much of the boat will be submerged when it floats in a lake?   0.33 m

63 ■■ A block of iron quickly sinks in water, but ships constructed of iron float. A solid cube of iron 1.0 m on a side is made into sheets. To make an open cube that will not sink from these sheets, what should the minimum length of the sides be?   2.0 m

64 ■■ A girl floats in a lake with 90% of her body beneath the water. What are (a) her mass density and (b) her weight density?   (a) 9.0 × 10² kg/m³   (b) 8.8 × 10³ N/m³

65 ■■■ A bar of 24-karat (pure) gold is supposed to be completely solid. To check this, a scientist takes the bar into the lab, and using a scale finds that it has a mass of 1.00 kg in air and a measured mass of 0.93 when completely immersed in water. Is the bar solid?   no

9.4 ■ Surface Tension and Capillary Action

66 Surface tension (a) depends on cohesion, (b) is a factor in capillary action, (c) causes liquids to form spherical drops, (d) all of these.   (d)

●**67** Capillary action depends on (a) surface tension only, (b) cohesive forces only, (c) adhesive forces only, (d) all of the preceding.   (d)

68 Can you suggest (a) why the sand is wet well above the water line on a beach (assume there are no waves), and (b) why wet sand is firmer and easier to walk on than dry sand?   see ISM

69 The speed of blood flow is greater in arteries than in capillaries. However, the flow rate equation (Av = a constant) seems to predict that the speed should be greater in the smaller capillaries. Can you resolve this apparent inconsistency?   see ISM

70 ■ A vertical force of 0.014 N is required to lift a wire ring with a radius of 1.5 cm away from the surface of a liquid. What is the surface tension of the liquid? (Neglect the diameter of the wire.) 7.4×10^{-2} N/m

71 ■■ If a circular wire loop with a diameter of 3.0 cm is used to measure the surface tension of (a) soapy water and (b) mercury at room temperature (20°C), what would be the vertical force required to lift the ring from the liquid in each case?   (a) 4.7×10^{-3} N   (b) 8.4×10^{-2} N

●**72** ■■ To what height will water rise in a glass capillary tube with a diameter of 2.0 mm at room temperature?   0.015 m

73 How high will water rise in a 0.75-mm-diameter capillary tube made of silver? $\phi = 90°$, $h = 0$

74 ■■ In order to have water at 20°C rise a distance of 5.0 cm in a glass capillary tube, what would the diameter of the tube have to be? 6.0×10^{-4} m

75 ■■ (a) Would water or blood rise higher in a glass capillary tube having a diameter of 0.80 mm? (b) How many times greater would the height difference be? (Assume a contact angle of 0° for each and water at 20°C.)   (a) water (b) $h_w = (1.3) h_b$

76 Given capillary tubes of the same diameter, will water or alcohol rise higher in the tubes, and how many times higher? $h_{water} = (3.3) h_{alcohol}$

77 ■■■ A glass capillary tube with a diameter of 1.0 mm is placed vertically in a dish of mercury. How far is the mercury in the tube depressed below the surface level in the dish? $h = -0.010$ cm

78 ■■■ Water rises in plants through tiny capillaries (called xylem), which have diameters that range from about 0.01 mm to 0.30 mm. What is the maximum height to which water can rise in a xylem system under capillary action? (Obviously, this does not account for moisture getting to the tops of tall trees. Atmospheric pressure will support a column of water only about 10 m in height. It is believed that water is raised to great heights in trees by a negative pressure created by evaporation. Water evaporates from leaves, and the cohesive forces between water molecules pull up more molecules from below.)   3.0 m

79 ■■■ A measurement instrument like that shown in Fig. 9.18 is used to apply a vertical force to a 4.0-cm-diameter wire loop to determine the surface tension of water. If the same force were used to determine the surface tension of mercury, what would be the diameter of the loop?   0.65 cm

9.5 ■ Fluid Dynamics and Bernoulli's Equation

80 If the velocity at some point changes with time, the fluid flow is *not* (a) steady, (b) irrotational, (c) incompressible, (d) nonviscous.   (a)

81 According to Bernoulli's equation, if the pressure on a liquid is increased, (a) the flow velocity necessarily increases, (b) the height of a liquid always increases, (c) both the flow velocity and the height of the liquid may increase, (d) none of these.   (c)

●**82** When a car is overtaken and passed by a fast-moving semitrailer truck, the driver of the car notices a force that tends to push the car toward the truck. What causes this? (This also occurs with ships, which is why they do not pass each other too closely.)   see ISM

83 (a) The shape of an airplane wing causes the air to move faster over its upper surface than across its lower surface. Explain how this supplies lift to the plane, in terms of Bernoulli's equation. (b) What supplies lift for a helicopter?   see ISM

84 (a) Why must an airplane reach a certain take-off speed before leaving the runway? (b) Why do large commercial planes take off with their flaps down? (A flap is a movable curved section on the trailing edge of the wing that may be extended to change the curvature of the wing.)   see ISM

85 If you hold a narrow strip of paper in front of your mouth and blow over the top surface, the strip will rise. (Try it.) Explain why.   see ISM

86 Two Bernoulli effects are shown in ● Fig. 9.31. One photo shows a faucet aspirator commonly used to provide suction in chemistry labs, and the other shows a plastic egg trapped by a Bernoulli effect. Explain these effects.   see ISM

FIGURE 9.31 Bernoulli effects See Example 86.

87 ■ Show that the pressure-depth equation can be derived from Bernoulli's equation. $\Delta p = \rho g (y_2 - y_1)$, with $v = 0$

88 ■ What is the flow rate of water that moves at an average speed of 2.8 m/s through a pipe with a diameter of 15 cm? 5.0×10^{-2} m³/s

89 ■■ Fluid A flows three times as fast as fluid B through the same horizontal pipe. Which has the greater density, and how many times greater is it? $\rho_b = 3\,\rho_a$

90 ■■ Water flows from a 2.0-cm-diameter pipe at a speed of 0.35 m/s. How long will it take to fill a 10-L container? 9.2×10^3 s

91 ■■ The speed of blood in a main artery, which has a diameter of 1.0 cm, is 4.5 cm/s. (a) What is the flow rate in the artery? (b) If the capillary system has a total cross-sectional area of 2500 cm², the average speed of blood through the capillaries is what percentage of that through the main artery? (c) Why is there a need for a low blood speed through the capillaries? (a) 3.5 cm³/s (b) 0.031%

● **92** ■■ A room measures 3.0 m by 4.5 m by 6.0 m. If the heating and air conditioning ducts to and from the room have a diameter of 30 cm, what is the average flow rate of air necessary for all the air in the room to be exchanged every 10 min? (Assume that the density of the air is constant.) 0.14 m³/s

93 ■■ In an industrial cooling process, water is circulated through a system. If water is pumped from the first floor through a 6.0-cm diameter pipe with a speed of 0.45 m/s under a pressure of 400 torr, what will be the pressure on the next floor 4.0 m above in a pipe with diameter of 2.0 cm? 5.7×10^3 Pa

94 ■■ The water level in a 15-cm-diameter cylinder like that in Fig. 9.32 is maintained at a constant height of 0.45 m. If the diameter of the spout pipe is 0.50 cm, how high is the vertical stream of water? (Assume the water to be an ideal fluid.) 0.46 m

● **FIGURE 9.32 How high a fountain?** See Exercise 94.

95 ■■■ Water flows through a horizontal 7.0-cm-diameter pipe under a pressure of 6.0 Pa with a flow rate of 25 L/min. At one point, calcium deposits reduce the cross-sectional area of the pipe to 30 cm². What is the pressure at this point? (Consider the water to be an ideal fluid.) 2.2 Pa

96 ■■■ A Venturi meter can be used to measure the flow speed of a liquid. A simple one is shown in ● Fig. 9.33. Show that the flow speed for an ideal fluid is given by

see ISM

$$v_1 = \sqrt{\frac{2gh}{(A_1^2/A_2^2) - 1}}$$

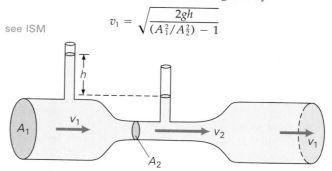

● **FIGURE 9.33 A flow speed meter** See Exercise 96.

9.6 ■ Viscosity, Poiseuille's Law, and Reynolds Number ◇

97 For a given flow rate of a viscous liquid, the required pressure difference varies as (a) r^2, (b) $1/r^2$, (c) r^4, (d) $1/r^4$. (c)

98 The Reynolds number (a) is part of Poiseuille's law, (b) has units of poise, (c) gives an indication of turbulent flow, (d) is usually negative. (c)

99 ■ Show that the units for viscosity are N-s/m² or Pa-s. see ISM

100 ■ Show that the Reynolds number is a dimensionless quantity. see ISM

101 ■■ A horizontal pipe with an inside diameter of 3.0 cm and a length of 6.0 m carries water, which has a flow rate of 40 L/min. What is the required pressure difference between the ends of the pipe? 2.0×10^2 Pa

102 ■■ A hospital patient receives a 500-cc blood transfusion through a needle with a length of 5.0 cm and an inner diameter of 1.0 mm. If the blood bag is suspended 0.85 m above the needle, how long does the transfusion take? (Neglect the viscosity of the blood flowing in the plastic tube between the bag and the needle.) 2.0×10^2 s

103 ■■ What is the maximum flow rate of water for laminar flow in a 5.0-cm-diameter pipe? 7.9×10^{-5} m³/s

104 Blood goes through a vessel with a diameter of 1.5 cm. What would be the minimum average flow speed if the flow were turbulent?   0.22 m/s

105 ■■■ (a) Show that the average flow rate for turbulent flow is given by

$$Q = \frac{\pi d R_n \eta}{4\rho}$$

where d is the diameter of the pipe. (b) What is the flow rate of water through a 3.0-cm-diameter pipe?   (a) see ISM
(b) 4.7×10^{-5} m³/s

■ **Additional Exercises**

106 A steel machine part has a volume of 0.36 m³. If it is suspended from a scale and immersed in gasoline, what will the scale read? 2.5×10^4 N

107 A 75-kg athlete does a single handstand. If the area of the hand in contact with the floor is 125 cm², what pressure is exerted on the floor? 5.9×10^4 Pa

108 A container 100 cm deep is filled with water. If a small hole is punched in its side 25 cm from the top, with what initial velocity will the water flow from the hole?
2.2 m/s

109 Ancient stonemasons sometimes split huge blocks of rock by inserting wooden pegs into holes drilled in the rock and then pouring water on the pegs. Can you explain the physics that underlies this technique? [*Hint*: Think about sponges and paper towels.]   see ISM

110 The spout heights for the container shown in ● Fig. 9.34 are 10 cm, 20 cm, and 30 cm. The water level is maintained at a height of 35 cm. (a) What is the speed of the water coming from each hole? (b) Which stream of water has the greatest range relative to the base of the container? Justify your answer. (You may be surprised.)
(a) $v_1 = 0.99$ m/s; $v_2 = 1.7$ m/s; $v_3 = 2.2$ m/s
(b) $x_1 = 0.31$ m; $x_2 = 0.34$ m; $x_3 = 0.25$ m
111 A hypodermic needle with an inner diameter of 0.10 cm and a length of 3.0 cm is attached to a syringe whose plunger has a cross-sectional area of 5.0 cm². The syringe is filled with saline (salt) solution. If a force of 100 N is applied to the plunger, what is the flow rate from the needle when the solution is squirted into the air? (Assume that the solution has the same density and viscosity as water.)
8.2×10^{-5} m³/s

● **FIGURE 9.34 Streams as projectiles**   See Exercise 111.

112 What is the force exerted on a person's body by the atmosphere if the surface area is 1.30 m²? (Express your answer in both newtons and pounds.)
1.32×10^5 N or 2.97×10^4 lb
113 A copper wire 100.00 cm long is stretched to a length of 100.02 cm when it supports a certain load. If an aluminum wire of the same diameter is used to support the same load, what should its initial length be if its stretched length is to be 100.02 cm also?   99.97 cm

114 What is the fractional decrease in pressure when a barometer is raised 30 m to the top of a tall building? (Assume that the density of air remains constant.)   0.38%

115 An oceangoing barge is 50.0 m long and 20.0 m wide and has a mass of 145 metric tons. Will the barge clear a reef 1.50 m below the surface of the water?   yes

116 If water rises to a height of 4.5 cm in a glass capillary tube, what is the diameter of the tube? 6.6×10^{-4} m

117 If the atmosphere had a constant density equal to its density at the Earth's surface, how high would it extend?
8.0×10^3 m
118 A vertical steel beam with a rectangular cross-sectional area of 24 cm² is used to support a sagging floor in a building. If the beam supports a load of 10,000 N, by what percentage is the beam compressed? 2.1×10^{-3}%

119 Show that specific gravity is the ratio of densities, given that by definition it is the ratio of the weight of a given volume of a substance and the weight of an equal volume of water.   see ISM

120 The flow rate of water through a garden hose is 66 cm³/s, and the hose and nozzle have cross-sectional areas of 6.0 cm² and 1.0 cm², respectively. (a) If the nozzle is held 10 cm above the spigot, what are the flow speeds through the spigot and the nozzle? (b) What is the pressure difference between these points? (Consider the water to be an ideal fluid.)   (a) $v_s = 11$ cm/s; $v_n = 66$ cm/s (b) 1.2×10^3 Pa

10 Temperature

10.1	**Temperature and Heat**
10.2	**The Celsius and Fahrenheit Temperature Scales**
10.3	**Gas Laws and Absolute Temperature**
10.4	**Thermal Expansion**
10.5	**The Kinetic Theory of Gases**

INSIGHT

■ **Physiological Diffusion**

Temperature and heat are frequent subjects of conversation, but if you had to state what the words really mean, you might find yourself at a loss. Certainly temperature and heat are related. You know from experience about how hot or cold something will be if you know its temperature. Thus we use thermometers of all sorts to record temperatures: not only within our homes and outside them (as with the window thermometer in the photo above), but inside car engines, home freezers, ovens—even the human body. You know, too, that a temperature change generally results from the application or removal of heat, and that temperature is therefore some sort of indication, or measure, of heat. But what is heat?

An early theory of heat considered it to be a fluidlike substance called caloric (from the latin word calor, *meaning "heat") that could be made to flow in and out of a body. Even though this theory has been abandoned, we still speak of heat as flowing from one object to another. Heat is now characterized as energy, and temperature and thermal properties are explained by considering the atomic and molecular behavior of substances. This and the next two chapters examine the nature of temperature and heat in terms of microscopic (molecular) theory and macroscopic observations.*

Chapter 10 includes the temperature scale, the gas laws, thermal expansion, and the kinetic theory of gases.

10.1 ■ Temperature and Heat

A good way to begin studying thermal physics is with definitions of temperature and heat. **Temperature** is a relative measure, or indication, of hotness or coldness. A hot stove is said to have a high temperature, and an ice cube to

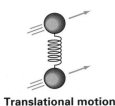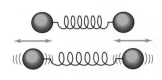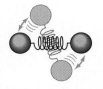

FIGURE 10.1 Molecular motions
(a) A molecule may move as a whole in translational motion and/or have linear and rotational vibrations. **(b)** Temperature is associated with random translational motion; the internal energy of a system is the total energy.

Translational motion   **Linear vibration**   **Rotational vibration**
(a)

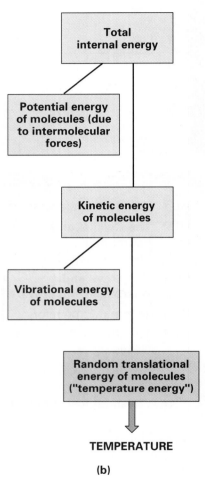

TEMPERATURE

(b)

Average random kinetic energy is sometimes referred to as thermal energy, but that term is used to mean different things. When this energy is associated with random translational motion, it is the "temperature energy." An object moving as a whole, such as a thrown ball, is in translational motion, but this is ordered, not random, motion.

have a low temperature. An object that has a higher temperature than another object does is said to be hotter. Note that hot and cold are relative terms, like tall and short.

We can perceive temperature by touch. However, this temperature sense is somewhat unreliable, and its range is too limited to be useful for scientific purposes. Methods and scales have been devised to measure temperature and temperature differences quantitatively. Temperature measurements tell *how much* hotter or colder one thing is than another (as length measurements tell how much taller or shorter something is than something else). Temperature is measured with thermometers, devices that will be discussed in the next section.

Heat is related to temperature and describes the process of energy transfer from one object to another. **Heat** describes *energy that is transferred from one object to another because of a temperature difference.* Thus, heat is energy in motion, so to speak. Once transferred, the energy becomes part of the total energy of the molecules of the object or system, its **internal energy**.

Kinetic theory (to be covered in Section 10.5) shows that the temperature of a substance is a measure of the average random translational kinetic energy of its molecules (Fig. 10.1a). Besides this "temperature" energy, molecules may also have kinetic energy due to linear and rotational vibrations, as well as potential energy due to the attractive forces between molecules. These energies do not contribute directly to temperature, but are part of the internal energy, which is the sum total of all such energies (Fig. 10.1b).

However, a higher temperature does not necessarily mean that one system has a greater internal energy than another. For example, in a classroom on a cold day, the air temperature is relatively high compared to that of the outdoor air. But all that cold air outside the classroom has far more internal energy as a whole than does the warm air inside, simply because there is so much *more* of it. If this were not the case, heat pumps would not be practical (see Chapter 12). The internal energy of a system also depends on its mass, or the number of molecules in the system.

When heat is transferred between two objects, whether or not they are touching, they are said to be in **thermal contact.** Heat will be transferred between them as long as there is a temperature difference. When there is no longer a net heat transfer between objects in thermal contact, they are at the same temperature and are said to be in **thermal equilibrium.**

10.2 ▪ The Celsius and Fahrenheit Temperature Scales

A measure of temperature is obtained using a **thermometer**, a device constructed to make evident some property of a substance that changes with

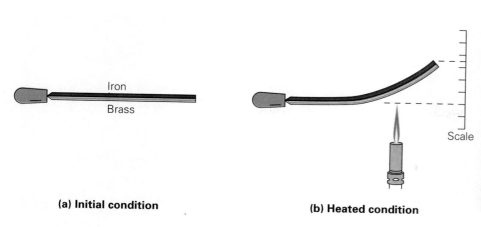

(a) Initial condition   (b) Heated condition

temperature. Fortunately, many physical properties of materials change suffi-ciently with temperature to be used as the bases for thermometers. By far the most obvious and commonly used property is **thermal expansion**, a change in the dimensions or volume of a substance that occurs when the temperature changes. Galileo is said to have constructed one of the first thermometers, a "thermoscope" that used the expansion of air in a bulb to measure temperature.

Almost all substances expand with increasing temperature, but to different extents. Most substances also contract with decreasing temperature. (Thermal expansion is used to mean both expansion and contraction; contraction is considered to be a negative expansion.) Because some metals expand more than others, a bimetallic strip (strips of two different metals bonded together) can be used to measure temperature changes. As heat is applied, the composite strip will bend away from the side made of the metal that expands more (Fig. 10.2). Helical coils formed from such strips are used in dial thermometers and in common household thermostats (Fig. 10.3).

A common thermometer is the liquid-in-glass type, which is based on the thermal expansion of a liquid. A liquid in a glass bulb expands into a glass stem, rising in a capillary bore. Mercury and alcohol (usually colored with a red dye to make it more visible) are the liquids used in most liquid-in-glass thermometers. These substances were chosen because of their relatively large thermal expansion and because they remain liquids over normal temperature ranges.

Thermometers are calibrated so that a numerical value may be assigned to a given temperature. For the definition of any standard scale or unit, two fixed reference points are needed. Since all substances change dimensions with temperature, an absolute reference for expansion is unavailable. However, the necessary fixed points may be correlated to physical phenomena which always occur at the same temperatures.

The ice point and the steam point of water are two convenient fixed points. Also commonly known as the freezing and boiling points, these two points are the temperatures at which pure water freezes and boils under a pressure of 1 atm (standard pressure).

The two familiar temperature scales are the **Fahrenheit temperature scale** and the **Celsius temperature scale**, named for their originators.* As shown in

*Daniel Gabriel Fahrenheit (1686–1736), a German scientist and instrument maker; and Anders Celsius (1701–1744), a Swedish astronomer.

FIGURE 10.2 Thermal expansion (a) A bimetallic strip is made of two strips of metal bonded together. (b) When such a strip is heated, it bends because of unequal expansions of the two metals. Here brass expands more than iron and the deflection is toward the iron. The deflection of the end of a strip could be used to measure temperature.

FIGURE 10.3 Bimetallic coil Helical bimetallic coils are used in dial thermometers.

Other temperature-dependent properties on which thermometers are based include the change in electrical resistance (these devices are called resistance thermometers), the variation of electric voltage at the junction of two wires made of different metals (thermocouples), and the wavelength of emitted light or electromagnetic waves (pyrometers).

The Celsius scale is based on the boiling and freezing points of water at standard pressure.

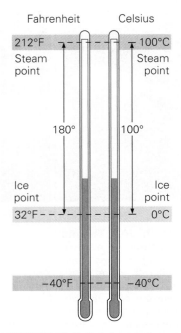

Fahrenheit   Celsius

212°F ⌐ ⌐ ⌐ ⌐ ⌐ 100°C
Steam   Steam
point   point

180°   100°

Ice   Ice
point   point
32°F ⌐ ⌐ ⌐ ⌐ ⌐ 0°C

−40°F ⌐ ⌐ ⌐ ⌐ −40°C

FIGURE 10.4 Celsius and Fahrenheit temperature scales
Between the ice and steam fixed points, there are 100 degrees on the Celsius scale and 180 degrees on the Fahrenheit scale. Thus, a Celsius degree is 1.8 times larger than a Fahrenheit degree.

Figure 10.4

> *Construct this graph and show that the equation of the line is the conversion formula from Celsius to Fahrenheit.*

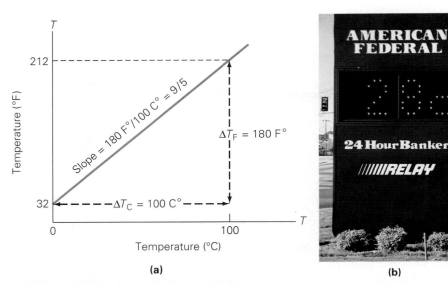

(a)   (b)

FIGURE 10.5 Fahrenheit versus Celsius **(a)** A plot of Fahrenheit temperature versus Celsius temperature gives a straight line of the general form $y = mx + b$, where $T_F = \frac{9}{5} T_C + 32$. **(b)** The temperature is given here in degrees Celsius. What is the equivalent temperature in degrees Fahrenheit?

    Figure 10.5

● Fig. 10.4, the ice and steam points have values of 32°F and 212°F, respectively, on the Fahrenheit scale and 0°C and 100°C on the Celsius scale. On the Fahrenheit scale, there are 180 equal intervals, or degrees, (F°), between the two reference points, and on the Celsius scale, there are 100 (C°)*. Therefore, a Celsius degree is larger than a Fahrenheit degree; since $180/100 = 9/5 = 1.8$, it is almost twice as large.

A relationship for converting between the two scales may be obtained from a graph of Fahrenheit temperature (T_F) versus Celsius temperature (T_C), like the one in ● Fig. 10.5. The equation of the straight line (in general slope-intercept form, $y = mx + b$) is

$$T_F = \left(\frac{180}{100}\right) T_C + 32$$

and
$$\boxed{\begin{array}{l} T_F = \frac{9}{5} T_C + 32 \\ T_F = 1.8\, T_C + 32 \end{array}}$$
(Celsius to Fahrenheit conversion)   (10.1)

where $\frac{9}{5}$ is the slope of the line and 32 is the intercept on the vertical axis. Thus, to change from a Celsius temperature (T_C) to its equivalent Fahrenheit temperature (T_F), you simply multiply the Celsius reading by $\frac{9}{5}$ and add 32.

The equation may be solved for T_C to convert from Fahrenheit to Celsius:

$$\boxed{T_C = \frac{5}{9}(T_F - 32)}$$
(Fahrenheit to Celsius conversion)   (10.2)

*For distinction, a particular temperature, e.g., $T = 20°C$ is written with °C (20 degress Celsius), whereas a temperature *interval*, e.g., $\Delta T = 80°C-60°C = 20\ C°$, is written with C° (20 Celsius degrees).

EXAMPLE 10.1 ■ Changing Temperatures

What are (a) 68°F on the Celsius scale, (b) normal body temperature, 98.6°F on the Celsius scale, and (c) −40°C on the Fahrenheit scale?

Solution. We wish to make the following conversions:

Given: (a) $T_F = 68°F$ *Find:* (a) T_C
         (b) $T_F = 98.6°F$ (b) T_C
         (c) $T_C = −40°F$ (c) T_F

(a) Eq. 10.2 is for changing Fahrenheit readings to Celsius:

$$T_C = \tfrac{5}{9}(T_F − 32) = \tfrac{5}{9}(68 − 32) = \tfrac{5}{9}(36) = 20°C$$

This temperature is commonly taken to be room temperature, and is a good one to remember.

(b) Again, Eq. 10.2 can be used:

$$T_C = \tfrac{5}{9}(T_F − 32) = \tfrac{5}{9}(98.6 − 32) = \tfrac{5}{9}(66.6) = 37.0°C$$

On the Celsius scale, normal body temperature has a whole-number value. Keep in mind that a Celsius degree is 1.8 times (almost twice) as large as a Fahrenheit degree, so a temperature elevation of several degrees on the Celsius scale makes a big difference. For example, a temperature of 39.5°C represents an elevation of 2.5 C° over normal body temperature. This is an elevation of $2.5 \times 1.8 = 4.5°$ on the Fahrenheit scale, or a temperature of $98.6 + 4.5 = 103.1°F$.

(c) Finally, we go the other way (Celsius to Fahrenheit) using Eq. 10.1:

$$T_F = \tfrac{9}{5}T_C + 32 = \tfrac{9}{5}(−40) + 32 = −72 + 32 = −40°F$$

Hence, − 40°C = − 40°F, and the temperature scales have the same numerical value at − 40°.

> *Point out that a change of 1 degree Celsius is not equivalent to a change of 1 Fahrenheit degree. Is a body temperature of 40°C serious?*

■ Problem-Solving Hint

Because of the similarity between Eqs. 10.1 and 10.2, it is easy to miswrite them. Since they are equivalent, you need to know only one of the equations, say Celsius to Fahrenheit (Eq. 10.1, $T_F = \tfrac{9}{5}T_C + 32$). Solving this for T_C algebraically gives Eq. 10.2. A good way to make sure that you have written the conversion equation correctly is to test it with a known temperature, such as the boiling point of water. For example, $T_C = 100°$, and

$$T_F = \tfrac{9}{5}T_C + 32 = \tfrac{9}{5}(100) + 32 = 212°F$$

and we know the equation is correct.

Liquid-in-glass thermometers are adequate for many temperature measurements, but problems arise when very accurate determinations are needed. A material may not expand uniformly over a wide temperature range. When calibrated to the ice and steam points, an alcohol thermometer and a mercury thermometer have the same readings at these points. But because of different expansion properties, the thermometers will not have exactly the same reading at an intermediate temperature, such as room temperature. For very sensitive temperature measurements and to define intermediate temperatures precisely, some other type of thermometer must be used. One such thermometer is discussed in the next section.

10.3 ■ Gas Laws and Absolute Temperature

Demonstration/activity: Place marshmallows and/or a partially filled balloon in a vacuum chamber and pump out the air.

Ask students whether they can explain why their ears "pop" when driving in the mountains or going up and down in an airplane.

Different liquid-in-glass thermometers show slightly different readings for temperatures other than the fixed points because of differing expansion properties. A thermometer that uses a gas, however, gives the same readings regardless of which gas is used. Experiments show that all gases at very low densities exhibit the same expansion behavior.

The variables that describe the behavior of a given quantity (mass) of gas are pressure, volume, and temperature (p, V, and T). When temperature is held constant, the pressure and volume of a quantity of gas are related as follows:

$$pV = \text{constant} \quad \text{or} \quad p_1V_1 = p_2V_2$$
$$\text{(at constant temperature)}$$

(10.3)

That is, the product of the pressure and volume is a constant. This relationship is known as **Boyle's law**, after Robert Boyle (1627–1691), the English chemist who discovered it. (See Demonstration 10.)

DEMONSTRATION 10 ■ Boyle's Shaving Cream

(a) A swirl of shaving cream is placed in a transparent vacuum chamber.

(b) A vacuum pump begins to evacuate the chamber.

(c) As the pressure is reduced, the swirl of cream grows in volume from the expansion of the air bubbles trapped in the cream.

(d) In a dramatic volume reduction, the shaving cream cannot stand up to the sudden increase in pressure when the chamber is vented to the atmosphere.

A demonstration that shows the inverse relationship of pressure (p) and volume (V) of Boyle's law, $p \propto 1/V$. Note: A similar effect occurs for a Boyle's marshmallow.

When the pressure is held constant, the volume of a quantity of gas is related to the *absolute* temperature (to be defined shortly):

$$\frac{V}{T} = \text{constant} \quad \text{or} \quad \frac{V_1}{T_1} = \frac{V_2}{T_2} \qquad (10.4)$$

Note in discussing the gas laws that temperatures must be in kelvins and not in Celsius degrees.

That is, the ratio of the volume and the temperature is a constant. This relationship is known as **Charles' law**, after the French scientist Jacques Charles (1747–1823), who made early hot-air balloon flights and was therefore quite interested in the relationship between the volume and temperature of a gas. A popular demonstration of Charles' law is shown in ● Fig. 10.6.

Low density gases obey these laws, which may be combined into a single relationship. Notice that since $pV = \text{constant}$ and $V/T = \text{constant}$ for a given quantity of gas, then pV/T must equal a constant also ($pV/T = \text{constant}$). This is known as the **ideal gas law**:

Students should know how to write both expressions for the ideal gas law. They may have trouble understanding how the $p_1V_1/T_1 = p_2V_2/T_2$ form is obtained. Explain that since $pV/T = \text{constant}$, then pV/T is the same at any time: $p_1V_1/T_1 = p_2V_2/T_2 = p_3V_3/T_3$, etc., for t_1, t_2, t_3, etc.

$$pV = NkT \quad \text{or} \quad \frac{pV}{T} = Nk \qquad (10.5)$$

Alternatively, we can write this as

$$p_1V_1/T_1 = p_2V_2/T_2$$

The constant is written as Nk because N, the number of molecules in the quantity of gas, reflects the constant mass of the gas. The constant k is called Boltzmann's constant: $k = 1.38 \times 10^{-23}$ J/K. The ideal gas law (sometimes called the perfect gas law) applies to all gases at low densities and describes fairly accurately the behavior of gases at normal densities.

The important point for the purpose of this section is that the pressure and volume are directly proportional to temperature: $pV \propto T$. This relationship allows a gas to be used to measure temperature in a *constant-volume gas thermometer*. Holding the volume of a gas constant, which can be done easily in a rigid container (● Fig. 10.7), gives $p \propto T$. Thus, with a constant-volume gas

(a) (b)

● **FIGURE 10.6 Charles' law in action** Demonstrations of the relationship between the volume and temperature of a quantity of gas. A weighted-down balloon, initially at room temperature, is placed in a beaker of water. **(a)** When ice is placed in the beaker and the temperature lowered, the balloon's volume is reduced. **(b)** When the water is heated and the temperature is raised, the balloon's volume increases.

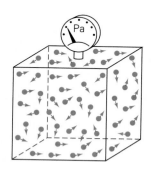

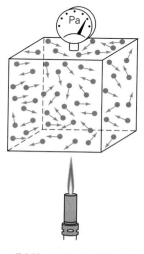

(a) Initial temperature   **(b) Heated condition**

● **FIGURE 10.7 Constant-volume gas thermometer** Such a thermometer indicates temperature as a function of pressure, since for a low-density gas, $p \propto T$. **(a)** At some intitial temperature, the pressure reading has a certain value. **(b)** When the gas thermometer is heated, the pressure (and temperature) reading is higher.

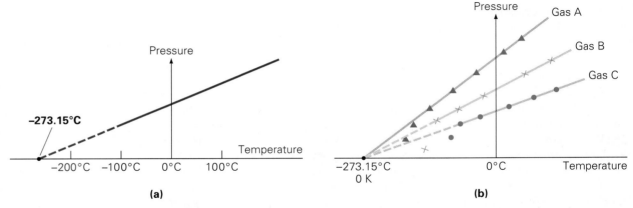

(a)

(b)

● **FIGURE 10.8 Pressure versus temperature** **(a)** A low-density gas kept at a constant volume gives a straight line on a *p* versus *T* graph: $p = (Nk/V)T$. When the line is extended to the zero pressure value, a temperature of $-273.15°C$ is obtained, which is taken to be absolute zero. **(b)** Extrapolation of lines for all low-density gases indicate the same absolute zero temperature. The actual behavior of gases deviate from the straight-line relationship at low temperatures because they start to liquefy.

A demonstration device consisting of a pressure gauge attached to a spherical bulb can be used to measure points on this line. The bulb is immersed in baths of different temperatures and the resulting pressure is read from the gauge. Easily obtained temperatures are:
    boiling water
    ice water
    freon (if still available)
    liquid nitrogen

Caution: always use Kelvin (absolute) temperatures with the ideal gas law.

There is also an absolute temperature scale whose basic unit is the same size as the Fahrenheit degree. This is called the Rankine scale, after William Rankine (1820–1872), a Scottish engineer, and $T_R = T_F + 459.67$, where T_R is in degrees Rankine (°R).

thermometer, temperature is read in terms of pressure. A plot of pressure versus temperature gives a straight line in this case, as shown in ● Fig. 10.8(a).

As can be seen in Fig. 10.8(b), measurements of real gases (plotted data points) deviate from the values predicted by the ideal gas law at low temperatures. This is because the gases liquefy at low temperatures. However, the relationship is linear over a large temperature range, and it looks as though the pressure might reach zero with decreasing temperature if the gas continued to be a gas (ideal or perfect).

The absolute minimum temperature for an ideal gas may therefore be inferred by extrapolating, or extending the straight line to the axis, as in Fig. 10.8. This temperature is found to be $-273.15°C$ and is designated as **absolute zero**. Absolute zero is believed to be the lower limit of temperature, but it has never been attained. In fact, there is a physical law that says it never can be (Chapter 12).

There is no known upper limit to temperature. The temperatures at the center of stars are estimated to be greater than 100 million degrees.

Absolute zero is the foundation of the **Kelvin temperature scale**, named after the British scientist Lord Kelvin.* On this scale, $-273.15°C$ is taken as the zero point, that is, as 0 K (● Fig. 10.9). The size of the unit for Kelvin temperature is the same as the Celsius degree, so temperatures on these scales are related by

$$\boxed{T_K = T_C + 273.15} \qquad (10.6)$$

where T_K is the temperature in **kelvins** (*not* degrees Kelvin). The kelvin is abbreviated as K (not °K). In many instances, a rounded value may be used in Eq. 10.1, that is, $T_K = T_C + 273$.

The absolute Kelvin scale is the official SI temperature scale; however, the Celsius scale is usually used for everyday temperature readings. The absolute temperature in kelvins is used primarily in scientific applications. **Caution:** Keep in mind that Kelvin temperatures must be used with the ideal gas law. It is a common mistake to use Celsius or Fahrenheit temperatures in that equation. Also note that there can be no negative temperatures on the Kelvin scale if absolute zero is the lowest possible temperature.

*Lord Kelvin, born William Thomson (1824–1907), developed devices to improve telegraphy and the compass and was involved in the laying of the first transatlantic cable. When he became a lord, it is said that he considered choosing Lord Cable or Lord Compass as his title, but decided on Lord Kelvin after a river that runs near the University of Glasgow in Scotland where he was a professor of physics for 50 years.

EXAMPLE 10.2 ■ Absolute Temperature

What is absolute zero on the Fahrenheit scale?

Solution.

Given: $T_K = 0$ K   *Find:* T_F

Temperatures on the Kelvin scale are related directly to Celsius temperatures by $T_K = T_C + 273.15$, so first you convert 0 K to a Celsius value:

$$T_C = T_K - 273.15 = 0 - 273.15 = -273.15°C$$

Then converting to Fahrenheit (Eq. 10.1) gives

$$T_F = \tfrac{9}{5}T_C + 32 = \tfrac{9}{5}(-273.15) + 32 = -459.67°F$$

Thus, absolute zero is about $-460°F$.

Initially, gas thermometers were calibrated using the ice and steam points. The Kelvin scale uses absolute zero and a fixed point adopted in 1954 by the International Committee on Weights and Measures. The second fixed point is the **triple point of water**, which represents a unique set of conditions where water coexists simultaneously in equilibrium as a solid, and a gas. The conditions for the triple point are a pressure of 4.58 mm Hg, or about 610 Pa (760 mm Hg is normal atmospheric pressure), and a temperature taken to be 273.16 K. The kelvin is then defined as 1/273.16 of the temperature at the triple point of water.

Now let's use the ideal gas law, which requires absolute temperatures.

EXAMPLE 10.3 ■ Absolute Temperature and the Ideal Gas Law

A quantity of low-density gas in a rigid container is initially at room temperature (20°C) and a particular pressure (p_1). If the gas is heated to a temperature of 60°C, by what factor does the pressure change?

Solution.

Given: $T_1 = 20°C$ *Find:* p_2/p_1 (pressure ratio or factor)
       $T_2 = 60°C$

Since we want the factor of pressure change, we write p_2/p_1 as a ratio. For example, if $p_2/p_1 = 2$, then $p_2 = 2p_1$, or the pressure would change (increase) by a factor of 2. The ratio should also indicate that we will make use of the ideal gas law in ratio form. The gas law requires *absolute* temperatures, so we first change the Celsius temperatures to kelvins:

$$T_1 = 20°C + 273 = 293 \text{ K}$$
$$T_2 = 60° + 273 = 333 \text{ K}$$

where a rounded value of 273 was used in Eq. 10.6 for convenience. Then using the ideal gas law (Eq. 10.5) in the form $p_2V_2/T_2 = p_1V_1/T_1$, and

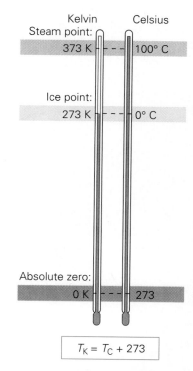

FIGURE 10.9 The absolute Kelvin temperature scale The lowest temperature on the Kelvin scale (corresponding to $-273.15°C$) is absolute zero. A unit interval on the Kelvin scale, called a kelvin and abbreviated K, is equivalent to a temperature change of a Celsius degree; thus $T_K = T_C + 273.15$. (The constant is usually rounded to 273 for convenience.)

since $V_1 = V_2$,

$$p_2 = \left(\frac{T_2}{T_1}\right)p_1 = \left(\frac{333 \text{ K}}{293 \text{ K}}\right)p_1 = (1.14)p_1$$

Thus, the pressure increases by a factor of 1.14, or p_2 is 1.14 times p_1. (What would the factor be if the Celsius temperatures were *incorrectly* used? It would be much larger, as you can readily show.)

Another Form of the Ideal Gas Law

The ideal gas law is sometimes written in another form:

$$pV = nRT \tag{10.7}$$

the constants nR rather than Nk are used for convenience. Here, n is the number of moles (mol) of the gas, which reflects the mass (and defined below), and R is called the **universal gas constant**:

$$R = 8.31 \text{ J/mol-K}$$

or

$$R = 0.0821 \text{ L-atm/mol-K}$$

The latter form is convenient because in working with gases, volumes are often measured in liters (L) and pressure in atmospheres (atm).

You may have learned in a chemistry course that a mole of a substance contains **Avogadro's number** (N_A) of molecules,

$$N_A = 6.02 \times 10^{23} \text{ molecules/mole}$$

Thus, the n and N in the two forms of the ideal gas law are related by $N = nN_A$. Also, a mole of any gas occupies 22.4 L at STP (standard temperature and pressure, 0°C and 1 atm).

Eq. 10.7 is a very practical form of the gas law, because we generally work with measured quantities or moles (n) of gases, rather than the number of molecules (N). To use Eq. 10.7, we need to know the number of moles of a gas. One **mole** of any substance is simply its formula mass expressed in grams. The formula mass is determined from the chemical formula and the atomic masses of the atoms (these are listed in Appendix IV and on the periodic table on p. 838). For example, water, H_2O, with two hydrogen atoms and one oxygen atom, has a formula mass of $2.0 + 16.0 = 18.0$; thus one mole of water has a mass of 18.0 grams. As steam, 18.0 g, or 1.0 mol, of water at STP would occupy 22.4 L.

EXAMPLE 10.4 ■ Applying the Ideal Gas Law

What volume in liters does one mole of helium occupy at one atmosphere of pressure and a temperature of (a) 0°C and (b) 20°C?

Solution.

> **Given:** $n = 1.00$ mol   **Find:** (a) V_1 (volume)
> $p = 1.00$ atm   (b) V_2
>   (a) $T_1 = 0°C$
>   (b) $T_2 = 20°C$
> $R = 0.0821$ L-atm/mol-K (known)

Note that we use the liter-atm form of the universal gas constant. (Why?) Eq. 10.7 can be used to find the volumes, but first we must change the temperatures to kelvins. *Probably the most common error in working with the gas laws is forgetting to use absolute temperatures.*

$$\text{(a) } T_1 = T_C + 273 = 0°C + 273 = 273 \text{ K}$$
$$\text{(b) } T_2 = T_C + 273 = 20°C + 273 = 293 \text{ K}$$

(a) Then, rearranging Eq. 10.7, we have

$$V_1 = \frac{nRT_1}{p} = \frac{(1.00 \text{ mol})(0.0821 \text{ L-atm/mol-K})(273 \text{ K})}{1 \text{ atm}} = 22.4 \text{ L}$$

You may have noticed that the gas was at STP and knew the answer already, since 22.4 L is the volume that one liter of *any* gas occupies at STP. This example shows it. The gas in the example was helium; what would be the difference in the calculation for one mole of, say, Ar, or O_2, or N_2?

(b) Computing the volume at T_2,

$$V_2 = \frac{nRT_2}{p} = \frac{(1.00 \text{ mol})(0.0821 \text{ L-atm/mol-K})(293 \text{ K})}{1 \text{ atm}} = 24.1 \text{ L}$$

and the volume increases as one might expect. Essentially, the gas is heated to 20°C at constant pressure.

10.4 ■ Thermal Expansion

Changes in the dimensions and volumes of materials are common thermal effects. As you learned earlier, thermal expansion provides a means of temperature measurement. The thermal expansion of gases is generally described by the ideal gas law, and is very obvious. Less dramatic, but by no means less important, is the thermal expansion of solids and liquids.

Thermal expansion results from a change in the average distance separating the atoms of a substance. The atoms are held together by bonding forces, which can be simplistically represented as springs in a simple model of a solid. The atoms vibrate back and forth, and with increased temperature (more internal energy) become increasingly active and vibrate over greater distances. The solid as a whole expands as a result.

Solids are discussed in Section 9.1

The change in one dimension of a solid (length, width, or thickness) is called linear expansion. For small temperature changes, linear expansion is approximately proportional to ΔT, or $T - T_o$ (● Fig. 10.10a). The fractional change in length is $(L - L_o)/L_o$, or $\Delta L/L_o$, where L_o is the original length (at

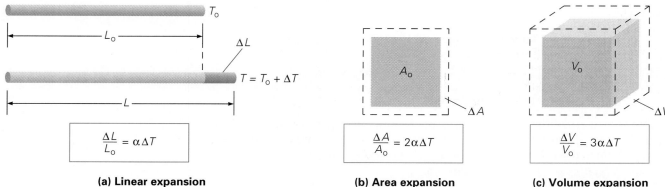

$$\frac{\Delta L}{L_o} = \alpha \Delta T$$

(a) Linear expansion

$$\frac{\Delta A}{A_o} = 2\alpha \Delta T$$

(b) Area expansion

$$\frac{\Delta V}{V_o} = 3\alpha \Delta T$$

(c) Volume expansion

●**FIGURE 10.10 Thermal expansion** (a) Linear expansion is proportional to the temperature change; that is, the change in length ΔL is proportional to ΔT, and $\Delta L = \alpha L_o \Delta T$, where α is the thermal coefficient of linear expansion. (b) For isotropic expansion, the thermal coefficient of area expansion is approximately 2α. (c) The thermal coefficient of volume expansion for solids is about 3α.

Figure 10.10

Demonstration/activity: Use the standard ring and ball apparatus to demonstrate linear expansion.

the initial temperature). This is related to the change in temperature by

$$\frac{\Delta L}{L_o} = \alpha \Delta T \qquad \text{or} \qquad \Delta L = \alpha L_o \Delta T \tag{10.8}$$

where α is the **thermal coefficient of linear expansion**. Note that the unit of α is inverse temperature: $1/\text{C}°$, or $\text{C}^{°-1}$. Values of α for some materials are given in Table 10.1.

A solid may have different coefficients of linear expansion for different directions, but for simplicity this book will assume that the same coefficient applies to all directions (that solids show *isotopic* expansion). Also, the coefficient of expansion may vary slightly for different temperature ranges. Since this variation is negligible for most common applications, α will be considered to be constant and independent of temperature.

Eq. 10.8 may be rewritten to give the final length (L) after a temperature change:

$$\Delta L = \alpha L_o \Delta T$$
$$L - L_o = \alpha L_o \Delta T$$
$$L = L_o + \alpha L_o \Delta T$$

or

$$L = L_o(1 + \alpha \Delta T) \tag{10.9}$$

TABLE 10.1 ■ Values of Thermal Expansion Coefficients (in $\text{C}^{°-1}$) for Some Materials at 20°C

Material	Coefficient of linear expansion (α)	Material	Coefficient of volume expansion (β)
Aluminum	24×10^{-6}	Alcohol, ethyl	1.1×10^{-4}
Brass	19×10^{-6}	Gasoline	9.5×10^{-4}
Brick or concrete	12×10^{-6}	Glycerin	4.9×10^{-4}
Copper	17×10^{-6}	Mercury	1.8×10^{-4}
Glass, window	9.0×10^{-6}	Water	2.1×10^{-4}
Glass, Pyrex	3.3×10^{-6}		
Ice	52×10^{-6}	Air (and most other gases) at 1 atm	3.5×10^{-3}
Iron and steel	12×10^{-6}		

Since area (A) is length squared (L^2),

$$A = L^2 = L_0^2(1 + \alpha\Delta T)^2 = A_0(1 + 2\alpha\Delta T + \alpha^2\Delta T^2)$$

Since the values of α for solids are much less than 1 ($\sim 10^{-6}$, as shown in Table 10.1), the second-order term (containing α^2) may be dropped with negligible error. As a first-order approximation, then,

$$A = A_0(1 + 2\alpha\Delta T) \qquad \text{or} \qquad \frac{\Delta A}{A_0} = 2\alpha\Delta T \qquad\qquad (10.10)$$

Thus, the **thermal coefficient of area expansion** (Fig. 10.10b) is twice as large as the coefficient of linear expansion (that is, is equal to 2α).

Similarly, a first-order expression for volume expansion is

$$V = V_0(1 + 3\alpha\Delta T) \qquad \text{or} \qquad \frac{\Delta V}{V_0} = 3\alpha\Delta T \qquad\qquad (10.11)$$

The **thermal coefficient of volume expansion** (Fig. 10.10c) is equal to 3α.

The following example illustrates that the equations for thermal expansions are approximations. Even though an equation is a description of a physical relationship, you should always keep in mind that it may be only an approximation of physical reality and/or may only apply in certain situations. Because calculations of thermal expansion involve such small numbers, you may assume that all quantities are exact, that is, that they have any number of significant figures you wish.

Demonstration/activity: Fill a flask with water to the brim. Heat it. The water level will rise and water will spill over as the temperature increases.

EXAMPLE 10.5 ■ A Paradox or an Approximation?

An aluminum rod has a length of 1.0 m at 30°C. The temperature is decreased to 10°C and then raised back to 30°C again. Is the length of the rod still 1.0 m?

Solution.

Given: $L_0 = 1.0$ m $= 100$ cm $\qquad$ *Find:* Final length of the rod
$T_0 = 30°C$
$T = 10°C$
$\Delta T_1 = T - T_0 = 10°C - 30°C = -20$ C°
$\Delta T_2 = T_0 - T = 30°C - 10°C = 20$ C°
$\alpha = 24 \times 10^{-6}$ C°$^{-1}$ (from Table 10.1)

Intuitively, you know the rod should have the same length after its temperature is lowered and then raised by the same number of degrees. But what result does Eq. 10.9 give? With the first temperature change, the rod will contract to a length L_1:

$$L_1 = L_0(1 + \alpha\Delta T_1) = (100 \text{ cm})[1 + (24 \times 10^{-6} \text{ C}°^{-1})(-20 \text{ C}°)]$$

$$= (100 \text{ cm})(1 - 0.00048) = 99.952 \text{ cm}$$

The minus from the negative temperature difference indicates a thermal contraction. Using the smaller units of centimeters better indicates the small effect. Any units of length can be used in the equation.

With the temperature rise, the rod with an initial length of L_i will expand to a length L_2:

$$L_2 = L_1(1 + \alpha\Delta T_2) = (99.952 \text{ cm})[1 + (24 \times 10^{-6}\, \text{C}^{\circ-1})(20\, \text{C}^\circ)]$$
$$= (99.952 \text{ cm})(1 + 0.00048) = 99.999 \text{ cm}$$

Thus, the result obtained by applying the equation for linear expansion to this situation does *not* reflect reality—the rod would not be shorter when it came back to its starting temperature. If this were true, putting the rod through numerous cycles would make it shorter and shorter, and it would eventually disappear!

The problem arises because Eq. 10.9 is an approximation, and as a result is not symmetric with respect to the temperature change ΔT. This may be easily seen using Eq. 10.8. For a cooling, $\Delta L = L_o\, \alpha\, \Delta T$, whereas for heating, $\Delta L' = L_1\, \alpha\, \Delta T$. Thus in cooling, ΔL is proportional to L_o, which is greater than L_1 to which the heating increase $\Delta L'$ is proportional. In other words, there is a change in basis or reference. To help understand this, consider a monetary cycle analogy. Suppose you invested some money, and in one day your investment increased by 100 percent. Then, the next day you lost 100 percent of your investment. The percentages of increase and decrease were equal, but would you have the same amount of money as you started with—or be broke? Think about it.

The thermal expansion of materials is an inportant consideration in construction. Seams are put in concrete highways and sidewalks to allow room for expansion and prevent cracking. Expansion gaps in large bridge structures and between railroad rails are necessary to prevent damage (●Fig. 10.11). The thermal expansion of steel beams and girders can produce tremendous pressures, as the following example shows.

● **FIGURE 10.11 Expansion gaps** Expansion gaps can be found built into bridge roadways. They are designed to prevent contact stresses produced by thermal expansion.

EXAMPLE 10.6 ■ Thermal Expansion and Stress

A steel I-beam is 5.0 m long at a temperature of 20°C (68°F). On a hot day, the temperature rises to 40°C (104°F). (a) What is the change in the beam's length due to thermal expansion? (b) Suppose that the ends of the beam are initially in contact with rigid vertical supports. How much force will the expanded beam exert on the supports if it has a cross-sectional area of 60 cm²?

Solution.

Given: $L_o = 5.0$ m
$T_o = 20°C$
$T = 40°C$
$\alpha = 12 \times 10^{-6}\, \text{C}^{\circ-1}$
(from Table 10.1)

Find: (a) ΔL (change in length)
(b) F (force)

$$A = 60 \text{ cm}^2\left(\frac{1 \text{ m}}{100 \text{ cm}}\right)^2 = 6.0 \times 10^{-3}\, \text{m}^2$$

(a) Using Eq. 10.8 to find the change in length with $\Delta T = T - T_o = 40°C - 20°C = 20\ C°$,

$$\Delta L = L_o \alpha \Delta T = (5.0\ \text{m})(12 \times 10^{-6}\ C°^{-1})(20\ C°)$$
$$= 1.2 \times 10^{-3}\ \text{m} = 1.2\ \text{mm}$$

This may not seem like much of an expansion, but it can give rise to a great deal of force if the beam is constrained and kept from expanding.

(b) The force exerted by the beam as it tries to expand by a length ΔL is the same as the force that would be required to compress the beam by that length. Using the Young's modulus form of Hooke's law (Chapter 9) with $Y = 20 \times 10^{10}\ \text{N/m}^2$ (Table 9.1), the stress on the beam is

$$\frac{F}{A} = \frac{Y \Delta L}{L_o} = \frac{(20 \times 10^{10}\ \text{N/m}^2)(1.2 \times 10^{-3}\ \text{m})}{5.0\ \text{m}}$$

and the force is
$$= 4.8 \times 10^7\ \text{N/m}^2$$

$$F = (4.8 \times 10^7\ \text{N/m}^2)A = (4.8 \times 10^7\ \text{N/m}^2)(6.0 \times 10^{-3}\ \text{m}^2)$$

$$= 2.9 \times 10^5\ \text{N} \qquad (\text{about } 65{,}000\ \text{lb, or } 32\tfrac{1}{2}\ \text{tons!})$$

EXAMPLE 10.7 ■ A Larger or Smaller Hole?

A circular piece is cut from a flat metal sheet (●Fig. 10.12a). If the sheet is then placed in an oven and heated, the size of the hole will (a) become larger, (b) become smaller, (c) be unchanged. *Clearly establish the reasoning and physical principle(s) used in determining your answer. That is,* **why** *did you select your answer?*

Reasoning and Answer. It is a common misconception to think that the area of the hole will shrink because of expansion of the metal around it. Think of the piece of metal removed from the hole, rather than of the hole itself. This piece would expand with increasing temperature. The metal in the heated sheet reacts as if the piece removed were still part of it. (Think of putting the piece of metal back into the hole and heating, as in Fig. 10.12b. Or consider drawing a circle on an uncut metal sheet and heating it.) So, the answer is (a).

Follow-up Exercise. A spherical cavity is cut inside a metal block. If the block is placed in a freezer, how would the cavity be affected? (*Reasoning and answer may be found in the Answers to Exercises section at the back of the book.*)

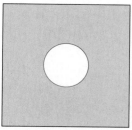

(a) Metal plate with hole

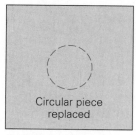

Circular piece replaced

(b) Metal plate without hole

●**FIGURE 10.12 A larger or smaller hole?** Example 10.7.

EXAMPLE 10.8 ■ The Size of a Hole

A circular piece with a diameter of 8.00 cm is cut from an aluminum sheet at room temperature (Fig. 10.12a). If the sheet is then placed in an oven and heated to 150°C, what will the area of the hole be?

> Watch the reading on a mercury thermometer when a hot flame is passed near the bulb. The reading will first drop slightly while the glass bulb heats up, then the mercury will start expanding as expected.

10.4 Thermal Expansion   **347**

Solution.

Given: $d_o = 8.00$ cm $\qquad\qquad$ **Find:** A (area)
$T_o = 20°C$
$T = 150°C$
$\alpha = 24 \times 10^{-6}\, C^{°-1}$
(from Table 10.1)

We may use Eq. 10.10 directly, with

$$\Delta T = 150°C - 20°C = 130\ C°$$

and

$$A_o = \frac{\pi d_o^2}{4} = \frac{\pi(8.00\ \text{cm})^2}{4} = 50.3\ \text{cm}^2$$

The area of the hole (or the metal piece) after heating will therefore be

$$A = A_o(1 + 2\alpha\Delta T) = (50.3\ \text{cm}^2)[1 + 2(24 \times 10^{-6}\, C^{°-1})(130\ C°)]$$
$$= (50.3\ \text{cm}^2)[1 + 0.00624] = 50.6\ \text{cm}^2$$

Liquids, like solids and gases, normally expand with increasing temperature. Because fluids (liquids and gases) have no definite shape, only volume expansion can be analyzed. The expression is

$$\boxed{\frac{\Delta V}{V_o} = \beta\Delta T} \qquad \text{(fluid volume expansion)} \qquad (10.12)$$

where β is the coefficient of volume expansion for fluids. Note in Table 10.1 that the values for β for fluids are typically larger than the values of 3α for solids.

Water exhibits an anomalous volume expansion near its freezing point. The volume of a given amount of water decreases as it is cooled from room temperature, until its temperature reaches 4°C (●Fig. 10.13). Below 4°C, the

Figure 10.13

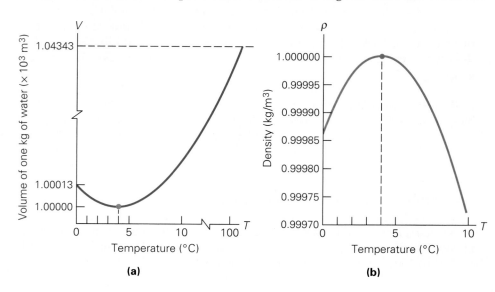

(a)

(b)

●**FIGURE 10.13 Thermal expansion of water** Water exhibits nonlinear expansion behavior near its freezing point. (a) Above 4°C (actually 3.98°C), water expands with increasing temperature, but from 4°C down to 0°C, it expands with decreasing temperature. (b) As a result, water has its maximum density near 4°C.

volume increases, and therefore the density decreases. This means that water has a maximum density ($\rho = m/V$) at 4°C (actually 3.98°C).

When water freezes, the molecules form a hexagonal (six-sided) lattice pattern. (This is why snowflakes have hexagonal shapes.) It is the open structure of this lattice that gives water its almost unique property of expanding on freezing and being less dense as a solid than as a liquid. (This is why ice floats in water and frozen water pipes burst.) The variation of the density of water over the temperature range of 4°C to 0°C indicates that the open lattice structure is beginning to form at about 4°C, rather than arising instantaneously at the freezing point.

This property has an important environmental effect: Bodies of water, such as lakes and ponds, freeze at the top first. As a lake cools toward 4°C, water near the surface loses energy to the atmosphere, becomes denser, and sinks. The warmer, less dense water near the bottom rises. However, once the colder water on top reaches temperatures below 4°C, it becomes less dense and remains at the surface, where it freezes. If water did not have this property, lakes and ponds would freeze from the bottom up, which would destroy much of their animal and plant life (and would make ice skating a lot less popular). There would also be no oceanic ice caps at the polar regions. Rather, there would be a thick layer of ice at the bottom of the ocean, covered by a layer of water.

10.5 ■ The Kinetic Theory of Gases

The notion that matter consists of tiny particles dates back to the Greek philosopher Democritus, who lived around 450 B.C. For centuries, this concept of the nature of matter was purely speculative. There was no experimental evidence that matter was made up of particles.

In the early nineteenth century, the English chemist John Dalton developed explanations of several chemical phenomena using atomic theory. Not all of the explanations were satisfactory because Dalton considered atoms to be indivisible particles, and mistakenly thought gas molecules were atoms. (Molecules often divide in chemical reactions.) Although there were some errors in his approach, Dalton's work helped give atomic theory a firm foundation. Today, atoms are no longer considered to be indivisible; many subatomic particles have been observed (Chapter 29). However, in common reactions and interactions, atoms maintain their identity as separate particles.

More direct evidence for the atomic theory was accidentally discovered in 1827 by Robert Brown, a Scottish botanist. Brown observed under a microscope that pollen grains suspended in water move about in a jerky fashion. This so-called Brownian motion is readily explained by atomic theory. The tiny pollen grains are being pushed here and there because of collisions with randomly moving water molecules. Oddly enough, this simple explanation was not suggested until 1905—by Albert Einstein.

If atoms and molecules are viewed as colliding particles, the laws of mechanics can be applied to a quantity (system) of gas in describing its characteristics in terms of molecular motion, pressure (force/area), energy, and so on. Because of the large number of particles involved, statistical analysis is necessary.

Figure 10.14

Elastic collisions are discussed in Section 6.4

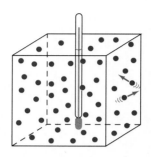

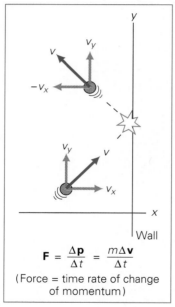

$$\mathbf{F} = \frac{\Delta \mathbf{p}}{\Delta t} = \frac{m\Delta \mathbf{v}}{\Delta t}$$

(Force = time rate of change of momentum)

FIGURE 10.14 Kinetic theory of gases The pressure a gas exerts on the walls of a container is due to the force resulting from the change in momentum of the gas molecules that collide with the wall. The force exerted by an individual molecule is equal to the time rate of change of momentum: $\mathbf{F} = \Delta \mathbf{p}/\Delta t = m\Delta \mathbf{v}/\Delta t$, where $\mathbf{p} = m\mathbf{v}$. The sum of the instantaneous normal components of the collision forces gives rise to the average pressure on the wall.

One of the major accomplishments of theoretical physics was to derive the ideal gas law from mechanical principles by interpreting the temperature in terms of the kinetic energy of the gas molecules. Theoretically, the molecules of an ideal gas are viewed as essentially point masses in random motion with relatively large distances separating them. As point masses, the molecules have no internal structure, so vibrational and rotational motions are not a consideration.

The molecules make perfectly elastic collisions with each other and with the walls of the container. The forces between molecules are considered to act only over a short range, so the molecules interact with each other only during collisions. Since the molecules of an ideal gas interact only during collisions, such a gas would remain a gas at low temperatures and would not liquefy as real gases do. This gives the theoretical ideal gas its idealness. Colliding molecules lose and gain kinetic energy, with corresponding changes in their potential energy. This potential energy is ignored in summing the total energy of the system because molecules spend a negligible fraction of the time in collisions.

From Newton's laws of motion, the force on the walls of the container can be calculated from the change in momentum of the gas molecules when they collide with the walls (Fig. 10.14). If this force is expressed in terms of pressure (force/area), the following equation is obtained (see Appendix II for derivation):

$$pV = \tfrac{1}{3}Nm\bar{v}^2 \qquad (10.13)$$

where V is the volume of the container or gas, N the number of gas molecules in the closed container and m the mass of a gas molecule. In this expression $\bar{v}$ is the average speed, but a special kind of average. It is obtained by averaging the squares of the speeds and then taking the square root of the average—that is, $\sqrt{\bar{v^2}} = \bar{v}$. As a result, $\bar{v}$ is called the root-mean-square (rms) speed.

Solving Eq. 10.5 for pV and equating that expression with Eq. 10.13 shows how the temperature came to be interpreted as a measure of the kinetic energy:

$$pV = NkT = \tfrac{1}{3}Nm\bar{v}^2$$
$$\text{or} \qquad \tfrac{1}{2}m\bar{v}^2 = \tfrac{3}{2}kT \qquad (10.14)$$

Thus, the temperature of a gas (and that of the walls of the container or a thermometer bulb in thermal equilibrium with the gas) is directly proportional to its average random kinetic energy (per molecule): $K = \tfrac{1}{2}m\bar{v}^2 = \tfrac{3}{2}kT$. (Don't forget that T is the absolute temperature in kelvins.)

EXAMPLE 10.9 ■ Molecular Speed

What is the average (rms) speed of a helium atom (He) in a helium balloon at room temperature? (Take the mass of the helium atom to be 6.65×10^{-27} kg.)

Solution.

Given: $m = 6.65 \times 10^{-27}$ kg
$\qquad T = 20°C$ (room temperature)
$\qquad k = 1.38 \times 10^{-23}$ J/K (known)

Find: $\bar{v}$ (rms speed)

It should be evident that we will use Eq. 10.13, so we list k in the given quantities.

As we already know, we must change the Celsius temperature to kelvins, but note that k has units of J/K, which is a good reminder. So,

$$T_K = T_C + 273 = 20°C + 273 = 293 \text{ K}$$

where the 273.15 in Eq. 10.6 was rounded to 273 for convenience.

Rearranging Eq. 10.14, we have

$$\bar{v} = \sqrt{\frac{3kT}{m}} = \left[\frac{3(1.38 \times 10^{-23} \text{ J/K})(293 \text{ K})}{6.65 \times 10^{-27} \text{ kg}} \right]^{\frac{1}{2}}$$

$$= 1.35 \times 10^3 \text{ m/s} \quad (= 1.35 \text{ km/s})$$

This is over 3000 mi/h—pretty fast!

Follow-up Exercise. An oxygen molecule (O_2) has about 16 times the mass of a helium atom. How many times faster, on the (rms) average, does an atmospheric oxygen molecule move under the same conditions? (*Reasoning and answer may be found in the Answers to Exercises section at the back of the book.*)

The average kinetic energy of gas molecules is dependent only upon the temperature. Compare the rms speeds of molecules of different masses at the same temperature. Also compare the temperatures of molecules having different masses but the same speed.

There are a couple of interesting points concerning Eq. 10.13. First, it predicts that at absolute zero ($T = 0$ K) all molecular motion of a gas would cease. According to classical theory, this would correspond to absolute zero energy. However, modern theory says that there would still be some zero-point motion, and a corresponding minimum zero-point energy.

Also, since the "particles" in an ideal gas interact only through collisions as explained previously, the total average kinetic energy is the total internal energy of the gas. That is, its internal energy is all "temperature" energy as discussed in Section 10.1.

Thus, the internal energy of an ideal gas is directly proportional to its absolute temperature. This means that if the absolute temperature of a gas is doubled (by heat transfer), for example, from 200 K to 400 K, then its internal energy is also doubled. This does not apply to the Celsius and Fahrenheit temperatures, since their zero points are not referenced to zero energy. If a gas is at a temperature of 0° C (or 0°F) and its internal energy is doubled, the Celsius (or Fahrenheit) temperature is not doubled—doubling zero gives zero, which is not a new temperature.

Diffusion

We depend on our sense of smell to detect odors, such as the smell of smoke from something burning. That you can smell something from a distance implies that molecules get from one place to another in the air, from the source to your nose. This process of random molecular mixing in which particular molecules move from a region where they are present in higher concentration to one where they are in lower concentration is called **diffusion**. Diffusion also occurs readily in liquids; think about what happens to a drop of ink in a glass of water (Fig. 10.15). It even occurs to some degree in solids.

FIGURE 10.15 Diffusion
Diffusion occurs in liquids. Random molecular motion would eventually distribute the dye throughout the water. Here there is some distribution due to mixing and the ink colors the water after a few minutes. The distribution would take some time with diffusion only.

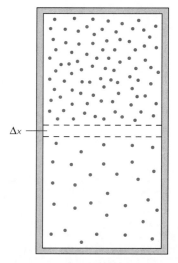

Region of high
concentration, C_2

Δx

Region of low
concentration, C_1

FIGURE 10.16 Gaseous diffusion With regions of different concentrations, there is a net movement from the region of higher concentration C_2 to the region of lower concentration C_1. The rate of diffusion is proportional to $(C_2 - C_1)/\Delta x$. Once the concentrations are equal ($C_2 = C_1$), the distributions are uniform and no further net movement occurs.

The random molecular motion that produces gaseous diffusion can be understood in terms of kinetic theory. Consider the hypothetical situation of a container of gas in which most of the gas molecules are concentrated on one side (Fig. 10.16). Although the motions of the molecules are entirely random, there will be a net movement of molecules across the central section (of length Δx) of the container, from the region of higher concentration to the region of lower concentration. Molecules from both regions enter that central section, but more enter on the average from the region of higher concentration because there are more molecules there. Since only one gas is present, this is called self-diffusion.

In general, the diffusion rate of gases and liquids is found to be proportional to $D(C_2 - C_1)/\Delta x$, where D is a diffusion coefficient and depends on the substance that is diffusing, and $(C_2 - C_1)/\Delta x$ is called the concentration gradient [concentration, e.g., kg/m^3 per length]. Note that C_1 and C_2 are the concentrations (kg/m^3) in adjoining regions, and when $C_2 = C_1$, the concentrations are equal, or the mixture is uniform, and the diffusion ceases.

Gases can also diffuse through porous materials or permeable membranes. Energetic molecules enter the material through the pores (openings), and, colliding with the pore walls, they slowly meander through the material. Such gaseous diffusion can be used to physically separate different gases from a mixture.

The kinetic theory of gases says that the average kinetic energy (per molecule) is proportional to the absolute temperature of a gas, $\frac{1}{2}m\bar{v}^2 = \frac{3}{2}kT$. So, on the average, the molecules of different gases (having different masses) move at different speeds at a given temperature. For example, at a particular temperature, molecules of oxygen (O_2) move faster on the average than the more massive molecules of carbon dioxide (CO_2). Because of this difference in molecular speed, oxygen can diffuse through a barrier faster than carbon dioxide can. Suppose that a mixture of equal volumes of oxygen and carbon dioxide is contained on one side of a porous barrier, as illustrated in Fig. 10.17. After a while, some O_2 molecules and CO_2 molecules will have diffused through the barrier. Of course, the diffused gases would still be mixed. After another diffusion stage, however, the oxygen concentration would become even greater. Almost pure oxygen can be obtained by repeating the process many times.

Separation by gaseous diffusion is a key process in obtaining enriched uranium, which was used in the atomic bomb and is currently used in nuclear reactors that generate electricity (Chapter 29).

The rate of diffusion for a particular gas depends on its rms speed. This is the case for diffusion through a porous barrier, or the diffusion of one gas through another. Even though the gas molecules have large average speeds

FIGURE 10.17 Separation by gaseous diffusion The molecules of both gases diffuse through the porous barrier. But, since oxygen molecules have the greater average speed, more of them pass through. Thus, there is a greater concentration of oxygen molecules on the other side of the barrier with time.

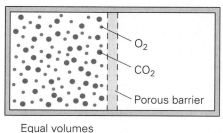
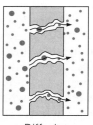

Equal volumes
of O_2 and CO_2

Diffusion
through barrier

O_2

CO_2

Porous barrier

(Example 10.9), the molecules (of, say, the air in a room) do not fly from one wall to another. There are frequent collisions, and as a result the molecules "drift" rather slowly. For example, suppose someone opened a bottle of ammonia on the other side of a closed room. It would take some time for the ammonia to diffuse across the room for you to smell. (Much of the movement that people commonly attribute to diffusion is actually due to air currents.)

The rate of diffusion depends on the average speed of the molecules ($R \propto \bar{v}$), which in turn depends, as we have seen, both on the temperature and on the mass of the gas' molecules. Let's form a ratio and investigate:

$$\frac{R_1}{R_2} = \frac{\bar{v}_1}{\bar{v}_2}$$

where R_1 and R_2 are the diffusion rates (mass/time) of two gases, and the $\bar{v}$'s are the average rms speeds of the molecules. Then, by Eq. 10.14 for constant temperature (and pressure),

$$\frac{R_1}{R_2} = \frac{\sqrt{\frac{3kT}{m_1}}}{\sqrt{\frac{3kT}{m_2}}} = \frac{\sqrt{\frac{1}{m_1}}}{\sqrt{\frac{1}{m_2}}} = \sqrt{\frac{m_2}{m_1}} \qquad (10.15)$$

Thus, *the rate of diffusion of gas molecules is inversely proportional to the square root of the molecular mass.* This result is known as **Graham's law** and was experimentally determined by Thomas Graham (1805–1869), a Scottish chemist.

EXAMPLE 10.10 ■ Diffusion and Graham's Law

Which gas will diffuse faster, nitrogen (N_2) or oxygen (O_2), and how many times faster? (The molecular masses of N_2 and O_2 are about 4.65×10^{-26} kg and 5.31×10^{-26} kg, respectively.)

Solution.

Given: $m_N = 4.65 \times 10^{-26}$ kg   *Find:* Which gas diffuses faster, and
      $m_O = 5.31 \times 10^{-26}$ kg   how many times faster

By Graham's law, the diffusion rate is inversely proportional to the molecular mass, and since nitrogen has the smaller molecular mass, it will have the greater diffusion rate and diffuse faster.

The ratio of the diffusion rates is given by Eq. 10.14, and

$$\frac{R_N}{R_O} = \sqrt{\frac{m_O}{m_N}} = \sqrt{\frac{5.31 \times 10^{-26} \text{ kg}}{4.65 \times 10^{-26} \text{ kg}}} = \sqrt{1.14}$$

$$= 1.07$$

Thus nitrogen diffuses about 7 percent faster than oxygen: $R_N = (1.07)R_O$.

Follow-up Exercise. At what temperature would oxygen diffuse as fast as nitrogen does at room temperature? (*Reasoning and answer may be found in the Answers to Exercises section at the back of the book.*)

Diffusion plays a central role in many life processes. Let's take a look at a couple of important examples. Consider a cell membrane in the lung. Such a membrane is permeable to a number of substances, any of which will diffuse through the membrane from a region where its concentration is high to a region where its concentration is low. Most importantly, the lung membrane is permeable to oxygen (O_2), and the transfer of O_2 across the membrane occurs because of a concentration gradient.

The blood carried to the lungs is low in O_2, having given up the oxygen to tissues for metabolism during its circulation through the body. The air in the lungs, on the other hand, is high in O_2 because there is a continual exchange of fresh air in the breathing process. As a result of this concentration difference, or gradient, O_2 diffuses from the lung volume into the blood as it flows through the lung tissue, and the blood leaving the lungs is high in O_2.

Exchanges between the blood and the tissues occur across capillary walls, and diffusion again is a major factor. The chemical composition of arterial blood is regulated to maintain the proper concentrations of particular solutes (substances dissolved in the blood solution), so that diffusion takes place in the appropriate directions across capillary walls. For example, as cells take up O_2 and glucose (blood sugar), the blood continuously brings in fresh supplies of the substances to maintain the concentration gradient needed for diffusion to the cells. The

continual production of carbon dioxide (CO_2) and metabolic wastes in the cells produces concentration gradients in the opposite direction for these substances. They therefore diffuse out of the cells into the blood, to be carried away from the tissue by the circulatory system.

During periods of work and physical exercise, there is an increase in cellular activity. More O_2 is used up and more CO_2 is produced, thereby increasing the concentration gradients and the diffusion rates. With an increased demand for O_2, how do the lungs respond to provide this to the blood? As you might expect, the rate of diffusion depends on the amount of surface area of the lung membrane and its thickness. Deeper breathing during exercise causes the aveoli (small air sacs in the lungs) to increase in volume. Such stretching increases the aveolar surface area and decreases the thickness of the membrane wall.

Also, the heart works harder during exercise, and the blood pressure is raised. The increased pressure forces open capillaries that are normally closed during rest or normal activity. As a result, total exchange area between the blood and cells is increased. Each of these changes help expedite the exchange of gases during exercise.

Now you can understand why people with emphysema (a disease involving the breakdown of aveoli walls) or pneumonia (characterized by fluid accumulations within or around the lungs) have difficulty providing enough oxygen to their tissues.

Fluid diffusion is very important to plants and animals. In plant photosynthesis, carbon dioxide from the air diffuses into leaves, and oxygen and water (vapor) diffuse out. The diffusion of liquid water across a permeable membrane because of a concentration gradient is called **osmosis** and is vital in living cells. Osmotic diffusion is also important to kidney functioning: tubules in the kidneys concentrate waste matter and toxins from the blood in much the same way oxygen is removed from mixtures. (See the Insight feature for some other vital examples of diffusion.)

Important Concepts

You should be able to define and explain these chapter concepts clearly.

temperature	Boyle's law	mole
heat	Charles' law	thermal coefficient of linear expansion
internal energy	ideal (perfect) gas law	thermal coefficient of area expansion
thermal contact	absolute zero	thermal coefficient of volume expansion
thermal equilibrium	Kelvin temperature scale	kinetic theory of gases
thermometer	kelvins	diffusion
thermal expansion	triple point of water	Graham's law
Fahrenheit temperature scale	universal gas constant	osmosis
Celsius temperature scale	Avogadro's number	

Important Relationships for Review

These relationships are mathematical statements of the concepts and principles presented in the chapter. You should be able to identify the symbols and to explain the relationships before proceeding to the Exercises. In-text equation reference numbers are given for convenience.

Celsius-Fahrenheit Conversion:

$$T_F = \tfrac{9}{5}T_C + 32 \tag{10.1}$$

$$T_C = \tfrac{5}{9}(T_F - 32) \tag{10.2}$$

Boyle's Law:

$$pV = \text{constant} \quad \text{or} \quad p_1V_1 = p_2V_2 \tag{10.3}$$

Charles' Law:

$$\frac{V}{T} = \text{constant} \quad \text{or} \quad \frac{V_1}{T_1} = \frac{V_2}{T_2} \tag{10.4}$$

Ideal (or perfect) Gas Law:

$$pV = NkT \quad \text{or} \quad \frac{p_1V_1}{T_1} = \frac{p_2V_2}{T_2} \tag{10.5}$$

or $\quad pV = nRT \quad$ (always absolute temperature)

$$k = 1.38 \times 10^{-23}\,\text{J/K}$$
$$R = 0.0821\,\text{L-atm/mol-K}$$
$$= 8.31\,\text{J/mol-K}$$

Kelvin-Celsius Conversion:

$$T_K = T_C + 273.15 \tag{10.6}$$

or $\quad T_K = T_C + 273 \quad$ (general calculations)

Thermal Expansion of Solids:

$$\text{linear:}\quad \frac{\Delta L}{L_o} = \alpha\Delta T \quad \text{or} \quad L = L_o(1 + \alpha\Delta T) \tag{10.8}$$

$$\text{area:}\quad \frac{\Delta A}{A_o} = 2\alpha\Delta T \quad \text{or} \quad A = A_o(1 + 2\alpha\Delta T) \tag{10.10}$$

$$\text{volume:}\quad \frac{\Delta V}{V_o} = 3\alpha\Delta T \quad \text{or} \quad V = V_o(1 + 3\alpha\Delta T) \tag{10.11}$$

Volume: Thermal Expansion of Fluids:

$$\frac{\Delta V}{V_o} = \beta\Delta T \tag{10.12}$$

Kinetic Theory of Gases:

$$pV = \tfrac{1}{3}Nm\bar{v}^2 \tag{10.13}$$

$$\tfrac{1}{2}m\bar{v}^2 = \tfrac{3}{2}kT \tag{10.14}$$

Diffusion Rate Ratio: (Graham's Law):

$$\frac{R_1}{R_2} = \sqrt{\frac{m_2}{m_1}} \tag{10.15}$$

Exercises

10.2 ■ The Celsius and Fahrenheit Temperature Scales

1 Which of the following is the hottest temperature: (a) 10°C, (b) 10°F, (c) 10 K, (d) −10 K? (a)

2 Which of the following temperature scales has the smallest unit interval: (a) Celsius, (b) Fahrenheit, (c) Kelvin? (b)

3 Heat always flows spontaneously from a body at a higher temperature to one at a lower temperature that is in thermal contact with it. Does it always flow from one with more internal energy to one with less internal energy? Explain. not necessarily; internal energy depends on mass in addition to temperature

4 Suppose you didn't know that −40°C = −40°F (see Example 10.1) and you were asked if the Celsius and Fahrenheit temperatures were ever equal. How could you find out algebraically? $T_F = T_C$; $(9/5)T_C + 32 = (5/9)(T_F - 32)$; solve

5 ■ (a) If the temperature drops by 10°C, what is the corresponding temperature change on the Fahrenheit scale? (b) If the temperature rises by 10°F, what is the corresponding change on the Celsius scale? (a) 18 F° (b) 5.6 C°

● **6** ■ Convert the following to Celsius readings: (a) 0°F, (b) 1500°F, (c) −20°F, (d) −40°F. (a) −18°C (b) 816°C (c) −29°C (d) −40°C

7 ■ Convert the following to Fahrenheit readings: (a) 32°C, (b) 125°C, (c) −15°C, (d) −273.16°C. (a) 90°F (b) 257°F (c) 5°F (d) −460°F

8 ■ To conserve energy, thermostats in an office building are set at 78°F in the summer and 65°F in the winter. What would the settings be if the thermostats had a Celsius scale? 26°C and 18°C

9 ■ The highest and lowest recorded air temperatures in the United States are 134°F (Death Valley, California, 1913) and −80°F (Prospect Creek, Alaska, 1971), respectively. What are these temperatures on the Celsius scale? 57°C and −62°C

10 ■ When the humidity is high we feel hotter, and this subjective response is sometimes quantified by referring to the *apparent* temperature. For example, when the actual temperature is 90°F and the relative humidity is 70%, the apparent temperature we feel is 106°F, or a difference of 16 degrees. What is the degree difference on the Celsius scale?   8.9 C°

● **11** ■ A person running a fever has a body temperature of 38.8°C. What is the person's temperature on the Fahrenheit scale?   102°F

12 ■■ In the troposphere (the lowest part of the atmosphere), the temperature decreases rather uniformly with altitude at a lapse rate of about 6.5°C/km. What are the temperatures (a) near the top of the troposphere (which has an average thickness of 11 km) and (b) outside a commercial aircraft flying at a cruising altitude of 34,000 ft? (Assume that the ground temperature is normal room temperature.)
(a) −51.5°C   (b) −47.6°C

10.3 ■ Gas Laws and Absolute Temperature

13 The temperature used in the ideal gas law is (a) Fahrenheit, (b) Celsius, (c) Kelvin, (d) it doesn't matter.   (c)

14 When the temperature of a quantity of gas is increased, (a) the pressure must increase, (b) the volume must increase, (c) both the pressure and volume must increase, (d) none of the preceding.   (d)

● **15** The temperature of a quantity of gas is increased. (a) How is the density affected if the pressure is held constant? (b) How is the density affected if the volume is held constant?   (a) decreases   (b) constant

16 A type of constant-volume gas thermometer is shown in ● Fig. 10.18. Describe how it operates.   see ISM

17 Describe how a constant-pressure gas thermometer might be constructed.   see ISM

18 In terms of the ideal gas law, what would a temperature of absolute zero imply? A negative absolute temperature?   (a) zero volume;   (b) negative volume

19 ■ The pressure on a low-density gas in a cylinder is kept constant as its temperature is increased from 10°C to 40°C. (a) Does the piston in the cylinder advance or recede? (b) What is the fractional change in the volume of the gas?   (a) advances   (b) 0.11

● **20** ■ Convert the following temperatures to absolute temperatures in kelvins: (a) 0°C, (b) 100°C, (c) 20°C, and (d) −35°C.   (a) 273 K   (b) 373 K   (c) 293 K   (d) 238 K

21 ■ Convert the following temperatures to degrees Celsius: (a) 0 K, (b) 250 K, (c) 88 K, and (d) 273.16 K.
(a) −273°C   (b) −23°C   (c) −185°C   (d) 0°C

22 ■■ The surface temperature of the Sun is about 6000 K. (a) What is this temperature on the Fahrenheit and Celsius scales? (b) The surface temperature is sometimes reported

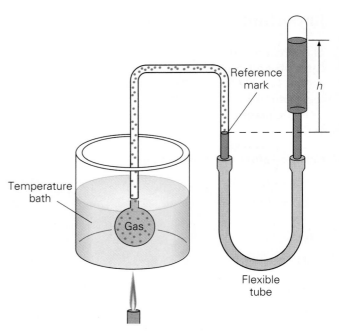

● **FIGURE 10.18 A type of constant-volume gas thermometer**   See Exercise 16.

to be 6000°C. Assuming that 6000 K is correct, what is the percentage error of this Celsius value?   (a) 10,340°F; 5727°C
(b) 4.6%

23 ■■ The core of the Sun is estimated to have a temperature of about 15 million kelvins. What is this temperature on the Fahrenheit and Celsius scales? 27×10^6 °F, 1.5×10^7 °C

● **24** ■■ How many moles are there in (a) 40 g of water, (b) 245 g of H_2SO_4 (sulfuric acid), (c) 138 g of NO_2 (nitrogen dioxide), (d) 56 L of SO_2 (sulfur dioxide) at STP?
(a) 2.2 moles   (b) 2.5 moles   (c) 3.0 moles   (d) 2.5 moles

25 ■■ How many molecules are there in each of the quantities in the preceding problem?   see ISM

26 ■■ A constant-volume gas thermometer has a pressure of 1000 Pa at 15°C. If the pressure increases to 2000 Pa, what is the new Celsius temperature?   303°C

27 ■■ A constant-pressure gas thermometer is initially at a temperature of 25°C. If the volume of gas in the thermometer increases by 5.0%, what is the final Celsius temperature?   40°C

28 ■■ On a warm day (92°F), the air in a balloon occupies a volume of 0.20 m³ and has a pressure of 20 lb/in². If the balloon is placed in a refrigerator and cooled to 32°F, the pressure decreases to 14.7 lb/in². What is the volume of the balloon? (Assume that the air behaves as an ideal gas.)
0.24 m³

29 ■■ The temperature of a quantity of ideal gas is doubled and its volume is decreased by one-half. How is the pressure affected? $p_2 = 4p_1$

30 ■■ If 2.4 m³ of a gas initially at STP is compressed to 1.6 m³ and its temperature raised 30°C, what is the final pressure?   1.7 atm

31 ■■ A quantity of gas in a 5.0 L container has a pressure of 2.5 atm at room temperature. How many molecules are in the container? 3.1×10^{23} molecules

32 ■■ An ideal gas occupies a container with a volume of 0.75 L at STP (standard temperature and pressure, 0°C and 1 atm). Find (a) the number of moles and (b) the number of molecules of the gas. (c) If the gas is carbon monoxide (CO), what is its mass?   (a) 3.3×10^{-2} moles
(b) 2.0×10^{22} molecules   (c) 0.92 g

33 ■■ A quantity of an ideal gas at 10°C occupies 4.0 L and has a pressure of 150 kPa. (a) What will the volume be if the temperature is kept constant and the pressure is decreased to 120 kPa? (b) What will the pressure be if the temperature is kept constant and the volume is compressed to 2.5 L? (c) What will the Celsius temperature be at a pressure of 120 kPa and 2.5 L?   (a) 5.0 L   (b) 2.4×10^{2} kPa   (c) -132°C

34 ■■■ The geothermal gradient, the rate at which temperature increases with depth in the Earth's crust, is determined in deep mines and wells to be about 1 F°/150 ft. Assuming that this rate is uniform to any depth, what is the absolute temperature at the center of the Earth? Is this a reasonable assumption? (See Exercise 70.)
7.8×10^{4} K, hotter than the Sun's surface

35 ■■■ Is there a temperature that has the same numerical value on the Kelvin and Fahrenheit scales? Justify your answer.   574.61 K, see ISM

10.4 ■ Thermal Expansion

36 The units of the thermal coefficient of linear expansion are (a) m/C°, (b) m²/C°, (c) m-C°, (d) C°⁻¹.   (d)

37 The thermal coefficient of volume expansion for a liquid is (a) α, (b) 2α, (c) 3α, (d) α^3.   (c)

38 Consider a cube sitting on a bimetallic strip at room temperature, as depicted in ● Fig. 10.19. What will happen

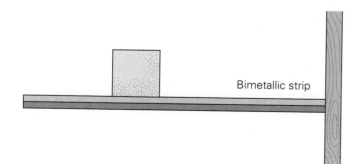

Bimetallic strip

● **FIGURE 10.19 Which way will the cube go?**   See Exercise 38.

● **FIGURE 10.20**   Ball-and-ring expansion See Exercise 39.

if the cube is ice and (a) the upper strip aluminum and the lower strip brass, or (b) the upper strip iron and the lower strip copper? (c) If the two strips are brass and copper, which should be on top to keep a hot metal cube from falling off?   (a) upward   (b) downward   (c) Cu

39 A demonstration of thermal expansion is shown in ● Fig. 10.20. Initially, the ball goes through the ring. (a) When the ball is heated, it does not go through the ring. (b) If both the ball and the ring are heated, the ball again goes through the ring. Explain what is being demonstrated.   see ISM

40 A solid metal disk rotates freely, so the conservation of angular momentum applies (Chapter 8). If the disk is heated while it is rotating, will there be any effect on the rate of rotation (the angular speed)?   see ISM

41 ■ A copper wire has a length of 0.500 m at 20°C. If the temperature is increased to 100°C, what is the change in the wire's length? 6.8×10^{-4} m

42 ■ A rectangular steel plate whose area is 0.060 m² is cooled from 350°C to room temperature. By what percentage does the area decrease?   0.79%

43 ■ What temperature change would cause a 0.10% increase in the volume of a quantity of water that was initially at room temperature?   4.8 C°

● **44** ■■ A piece of copper tubing used in plumbing has a length of 60 cm and an inner diameter of 1.50 cm at room temperature. When hot water at 85°C flows through the tube, what are (a) its new length and (b) the change in its cross-sectional area? Does the latter affect the flow rate?
(a) 60.07 cm   (b) 3.91×10^{-3} cm²; yes

45 ■■ Steel rails 7.50 m long are laid end to end on a cold day when the temperature is 0°C. The engineer on the project knows that the highest recorded summer temperature for that region is 40°C. If the engineer adds 20% to this value as a safety factor, what minimum expansion space should be left between the rails to avoid contact stress?
4.3×10^{-3} m

46 ■■ Show that a general expression for thermal stress is given by $F/A = Y\alpha\Delta T$, where Y is Young's modulus and α is the coefficient of linear expansion.   see ISM

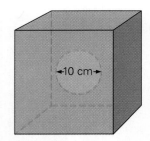

● FIGURE 10.21 A hole in a block   See Exercise 48.

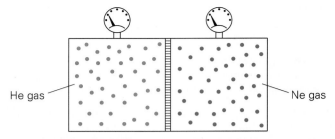

● FIGURE 10.22 What happens as time passes?   See Exercise 56.

47 ■■ Show that the density of a substance varies with temperature as $\rho = \rho_o(1 - 3\alpha\Delta T)$ or $\rho = \rho_o(1 - \beta\Delta T)$. see ISM

48 ■■ A copper block has an internal spherical cavity with a diameter of 10 cm (● Fig. 10.21). The block is placed in an oven and heated from room temperature to 500 K. (a) Does the cavity get larger or smaller? (b) What is the change in the cavity's volume?   (a) larger   (b) 1.3 × 10⁻⁵ m³

49 ■■ A brass rod has a circular cross section with a radius of 0.500 cm. It fits into a circular hole in a copper sheet with a clearance of 0.010 mm completely around it when both it and sheet are at 20°C. (a) At what temperature will the clearance be zero? (b) Would such a tight fit be possible if the sheet were brass and the rod were copper?
(a) 1.02 × 10³ °C   (b) no

● **50** ■■ Mercury has a density of 13.59 g/cm³ at room temperature. What is its density at 100°C?   13.4 g/cm³

51 ■■ Using the perfect gas law, express the coefficient of volume expansion in terms of temperature and pressure. see ISM

52 ■■ One morning when the temperature is 10°C, an employee at a rent-a-car company fills the 25-gal gas tank of an automobile to the top and then parks the car a short distance away. That afternoon, when the temperature is 20°C, gasoline drips from the tank onto the pavement. How much gas will be lost? (Neglect the expansion of the tank.)   0.24 gal

53 ■■■ A Pyrex beaker that has a capacity of 1000 cm³ at 20°C contains 990 cm³ of mercury at that temperature. Is there some temperature at which the mercury will completely fill the beaker? Justify your answer. (Assume no mass loss by vaporization.)   76°C

10.5 ■ The Kinetic Theory of Gases

54 In the kinetic theory of gases, the particles (a) experience gravitational attractions, (b) interact only through collisions, (c) move at very slow speeds, (d) have negligible masses.   (b)

● **55** If the temperature of a quantity of ideal gas is raised from 20°C to 40°C, its internal energy is (a) doubled, (b) quadrupled, (c) unchanged, (d) none of the preceding.   (d)

56 Equal volumes of helium gas (He) and neon (Ne) gas at the same temperature (and pressure) are on opposite sides

of a porous membrane, as shown in ● Fig. 10.22. Describe what happens after a period of time, and why.   see ISM

57 An aqueous saline (salt) solution with a concentration of 0.85% is said to be isotonic. There is no osmosis through the cell membranes of red blood cells placed in an isotonic solution. Explain what would happen and why if red blood cells were placed in hypertonic and hypotonic solutions. [*Hint:* Think in terms of *water* concentrations.]   see ISM

58 ■■ What is the average kinetic energy per molecule of a low-density gas at (a) 20°C and (b) 100°C?   (a) 6.1 × 10⁻²¹ J   (b) 7.7 × 10⁻²¹ J

59 ■■ If the temperature of an ideal gas is raised from 25°C to 100°C, what will the percentage change in the average (rms) velocity of the gas molecules be?   12% greater

60 ■■ (a) What is the average kinetic energy of molecules in a volume of gas at a temperature of 27°C? (b) What is the average (rms) speed of the molecules if the gas is helium? (A helium molecule is a single atom, which has a mass of 6.65 × 10⁻²⁷ kg.)   (a) 6.2 × 10⁻²¹ J   (b) 1.4 × 10³ m/s

61 ■■ What is the average speed of the molecules in low-density oxygen gas at room temperature? (The mass of an oxygen molecule, 0_2, is 5.31 × 10⁻²⁶ kg.)   4.8 × 10² m/s

● **62** ■■ Heat is added to a quantity of an ideal gas that is initially at 25°C until its internal energy is doubled. What is the final Celsius temperature of the gas?   323°C

63 ■■ The temperature of an ideal gas is doubled from 20°C to 40°C. What is the percentage increase in the internal energy of the gas?   7% increase

64 ■■ At a given temperature, which would be greater, the rms speed of oxygen (O_2) or of ozone (O_3), and how many times greater? $v_{oxygen} = (1.2) v_{ozone}$

65 ■■ Which has the faster diffusion rate, oxygen (O_2) or ozone (O_3), and how many times faster? $R_{oxygen} = (1.2) R_{ozone}$

66 ■■ The rms speed of one gas is 2.5 times that of another. How do their diffusion rates vary?   2.5:1

67 ■■■ A quantity of an ideal gas has a temperature of 0°C. An equal quantity of another ideal gas is twice as hot. What is its temperature?   273°C

■ Additional Exercises

● 68 Compute the final length of the rod in Example 10.5 with the thermal cycle reversed, that is, heating first and then cooling. 0.99999 cm

69 Use algebraic formulas (not numbers) to demonstrate the discrepancy shown in Example 10.5. *see ISM*

70 Assuming that the interior temperature of the Sun decreases uniformly from the central core to the surface, what is the Kelvin temperature gradient (K/km) if the core temperature is 15 million kelvins? (Use data given inside back cover.) 21 K/km

71 Which is the lower temperature, $-45°C$ or $-45°F$, and how much lower? $-45°C$ by 4 F°

72 A mercury thermometer has a uniform capillary bore whose cross-sectional area is 0.012 mm². The volume of the mercury in the thermometer bulb at 10°C is 0.130 cm³. If the temperature is increased to 50°C, how much will the height of the mercury column in the capillary bore change? (Neglect the expansion of the mercury in the bore and of the glass of the thermometer.) 7.8 cm

73 What temperature increase would produce a stress of 8.0×10^7 N/m² on a rigidly held steel beam? 33°C

74 A solid aluminum sphere has a diameter of 8.00 cm at room temperature. If the sphere is heated to 360°C, what is the change in its diameter? 0.06 cm

75 The largest temperature drop recorded in the United States is 100F° in 1 day (from 44°F to $-56°F$, in Browning, Montana, 1916). The largest rise is 49F° in 2 min (from $-4°F$ to 45°F, in Spearfish, South Dakota, 1943). (a) What are the corresponding temperatures and temperature changes on the Celsius scale? (b) What are the rates of both temperature changes on both scales? *see ISM*

76 Concrete highway slabs are poured in lengths of 10.0 m. How wide should the expansion gaps between the slabs be to ensure that there will be no contact stress over a temperature range of $-25°C$ to 45°C? 8.4×10^{-3} m

77 Suppose that a copper cube with sides 5.0 cm long undergoes a 200°C temperature change. What is the cube's new volume? 126 cm³

78 Equal quantities of an ideal gas are at temperatures of 10°C and 10°F. (a) If the temperatures of the gases are raised to 60°C and 60°F, respectively, which one has a greater increase in internal energy? (b) How much greater is that increase than the other one? (a) the one initially at 10°C (b) 7% greater

79 A new metal alloy is made into a 50-cm rod to determine its coefficient of linear expansion. After being heated from room temperature to 250°C, the rod is found to have a length of 50.044 cm. Is this a potentially valuable alloy? Justify your answer. $\alpha = 3.8 \times 10^{-6}$ C° $^{-1}$; yes, low coefficient of linear expansion

80 A square aluminum sheet 25 cm on a side is heated from room temperature to 300°C. What is the change in the length of a side? 0.17 cm

81 The highest and lowest recorded air temperatures in the world are 58°C (Libya, 1922) and $-89°C$ (Antarctica, 1983). What are these temperatures on the Fahrenheit scale? 136°F; $-128°F$

82 Show that the coefficient of volume expansion for solids is approximately equal to 3α. *see ISM*

83 A mercury thermometer has a bulb volume of 0.200 cm³ and a capillary bore diameter of 0.65 mm. How far up the bore will the column of mercury move if the overall temperature of the thermometer is increased by 30 C°? 3.1 cm

84 Steel train rails are 12.0 m long when the track is laid, on a day when the temperature is 15°C. To prevent contact stress enough space must be left between the ends of adjacent rails so they will not touch up to a temperature of 45°C. What is the width of the required gap between the rails? 4.3×10^{-3} m

85 An automobile tire is inflated to 200 kPa when the air temperature is at the freezing point. It later warms up to 12.0°C and the air pressure in the tire is found to be 204 kPa. Does the volume of the tire change? If so, by what percentage? (Assume that the air is an ideal gas and that atmospheric pressure is constant.) increases by 2.4%

86 Derive a single formula that may be used to directly convert Kelvin temperatures to Fahrenheit (and vice versa). $K = (5/9) T_F + 255$

87 A constant-volume gas thermometer has a pressure of 2.0×10^5 Pa at 10°C. (a) What is the pressure for a temperature of $-10°C$? (b) What is the Celsius temperature for a pressure of 1 atm? (a) 1.9×10^5 Pa (b) $-130°C$

88 A quantity of an ideal gas initially at atmospheric pressure is maintained at a constant temperature while it is compressed to half of its volume. What is the final pressure of the gas? 2.026×10^5 Pa

11

Heat

INSIGHTS

- ■ **Phase Changes and Ice Skating**

- ■ **The Greenhouse Effect and Global Warming**

- ■ **Heat Transfer by Radiation: The Microwave Oven**

Topics of study in Chapter 11 include specific heat, latent heat, and heat transfer.

Think for a moment how crucial heat is in our existence. Our bodies must precisely balance heat loss and heat gain to stay within the narrow temperature range necessary for life. In cold weather, we insulate our bodies to prevent the loss of vital heat by putting on layers of clothing. Without them, we might quickly freeze to death. The same strategy can be used to keep life-threatening extremes of heat out, as demonstrated by the insulating suit for fire-fighters in the photo above. (Fortunately, we don't often have to resort to such extreme measures.)

On a larger scale, the average temperature of the Earth, so critical to our environment and the survival of the organisms that inhabit it, is maintained through a similar balance. We are warmed by energy that comes to us across a 93-million mile void. Each day, vast quantities of heat reach our planet's surface and atmosphere from the Sun, only to be radiated away into the cold of space.

These balances are delicate, and any disturbance can have serious consequences. In an individual, sickness can disrupt the balance, producing a chill or fever. Similarly, we worry that pollution of the atmosphere by "greenhouse" gases, the product of our industrial society, could give our whole planet a "fever" that could affect all of us.

Thus, heat, heat transfer, and phase changes not only affect the quality of our lives but are some of the most important topics in the study of physics. An understanding of them will allow you to explain many everyday things, as well as providing a basis for understanding the conversion of thermal energy into useful mechanical work.

11.1 ▪ Units of Heat

Like work, heat involves a transfer of energy. **Heat** is energy that is transferred from one object or system to another because of a temperature difference. It is commonly said that heat is a form of energy, but this is not true in the strictest sense. References to "heat energy" in this book refer to the addition or removal of internal energy of a body or system.

Even though heat is energy in transit, it is measurable as energy losses or gains and is described by standard energy units like any other quantity of energy. Recall that the SI standard unit of energy is the joule (J), or newton-meter (N-m). Thus, it is correct to say, for example, that 20 J of heat energy is transferred from one body to another. However, there are other commonly used units of heat. A chief one is the **calorie (cal)** [● Fig. 11.1a]:

> One calorie is defined as the amount of heat needed to raise the temperature of 1 g of water 1 C° (from 14.5°C to 15.5°C).

Because the calorie is a small unit, its larger multiple, the **kilocalorie (kcal)**, is often used (1 kcal = 1000 cal). *One kilocalorie is the amount of heat needed to raise the temperature of 1 kg of water by 1 C°* (*from 14.5°C to 15.5°C*) [Fig. 11.1b].

A familiar use of the kilocalorie is for specifying the energy values of foods. However, in this context the word is usually shortened to Calorie (Cal). That is, people on diets really count kilocalories; for example, a piece of cake may contain 400 Cal, or 400,000 cal. A capital C is used to distinguish the larger kilogram-Calorie, or kilocalorie, from the smaller gram-calorie. They are sometimes referred to as "big Calorie" and "little calorie." (In some countries, the joule is used for food values—see ● Fig. 11.2.)

A unit of heat that is commonly used in industry is the **British thermal unit (Btu)**. *One Btu is the amount of heat needed to raise the temperature of 1 lb of water by 1 F°* (*from 63°F to 64°F*) [Fig. 11.1c]. A Btu is more than 250 times larger than a calorie, but only about a fourth of a kilocalorie (1 Btu = 252 cal = 0.252 kcal). If you buy an air conditioner or an electric heater, you will find it rated in Btu's; for example, window air conditioners range from 5000 to 25,000 Btu's. This number is actually Btu's *per hour* and specifies how much heat the unit will transfer (or deliver in the case of a heater) during that time.

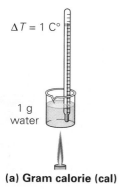

$\Delta T = 1$ C°

1 g water

(a) Gram calorie (cal)

$\Delta T = 1$ C°

1 kg water

(b) Kilocalorie (kcal) or Calorie (Cal)

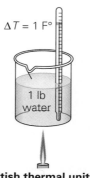

$\Delta T = 1$ F°

1 lb water

(c) British thermal unit (Btu)

●**FIGURE 11.1 Units of heat** **(a)** A calorie raises the temperature of 1 g of water by 1C°. **(b)** A kilocalorie raises the temperature of 1 kg of water by 1 C°. **(c)** A Btu raises the temperature of 1 lb of water by 1 F°. (Drawings not to scale.)

●**FIGURE 11.2 It's a joule** In Australia, diet drinks are labeled as being "low joule." In Germany, the labeling is a bit more specific. The label reads: "Less than 4 kilojoules (1 kcal) in 0.3 Liter." How does this compare to our diet drinks?

Benjamin Thompson (1753–1814) was born in New England, but went to England at the time of the American Revolution. He later worked in Bavaria, where he was made a count. He took the name Count Rumford after his birthplace, now known as Concord, New Hampshire.

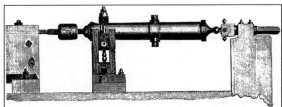

FIGURE 11.3 Benjamin Thompson, Count Rumford (1753–1814) Engravings of the statesman and physicist Count Rumford and a cannon borer. Rumford's experience with the boring of cannon barrels led to his experiments on the nature of heat.

The Mechanical Equivalent of Heat

The current idea that heat is a transfer of energy is the result of work by many scientists on the relationship of heat and energy. Some early observations were made by Benjamin Thompson, Count Rumford while he was supervising the boring of cannon barrels in Germany (see Fig. 11.3). Rumford noticed that water put into the bore of the cannon to prevent overheating during drilling boiled away and had to be frequently replenished. He did several experiments, including ones in which he tried to detect "caloric fluid" by changes in the weights of heated substances. He eventually concluded that mechanical work was responsible for the heating of the water.

This conclusion was later proven quantitatively by James Joule, the English scientist after whom the SI unit for work and energy is named. Using the apparatus shown in Fig. 11.4, Joule demonstrated that when a given amount of mechanical work was done, the water was heated, as indicated by an increase in temperature. He found that for every 4.19 J of work done the temperature of the water rose 1 C° per gram, or that 4.19 J was equivalent to 1 cal:

$$1 \text{ cal} = 4.19 \text{ J} \quad \text{(actually 4.186 J)}$$

or

$$1 \text{ kcal} = 4.19 \times 10^3 \text{ J}$$

This relationship is called the **mechanical equivalent of heat** and provides a conversion factor between calories and joules.

FIGURE 11.4 Joule's apparatus for measuring the mechanical equivalent of heat As the weights descend, the paddle wheels churn the water and mechanical energy or work is converted into heat energy. It was found that for every 4.19 J of work done, the temperature of the water rose 1 C° per gram, or that 4.19 J was equivalent to 1 calorie.

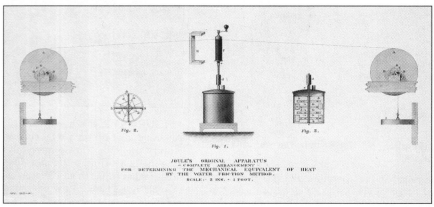

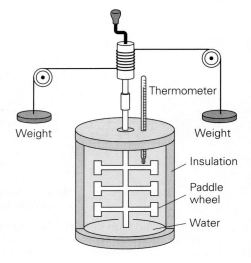

Thermometer

Weight

Weight

Insulation

Paddle wheel

Water

EXAMPLE 11.1 ■ Mechanical Equivalent of Heat

How much mechanical work in joules would have to be done in Joule's apparatus to raise the temperature of a liter of water from room temperature to 25°C?

Solution. Listing the given data and what we want to find, we have

Given: $T_i = 20°C$ Find: W (work in joules)
$\quad\quad\quad T_f = 25°C$
$\quad\quad\quad V = 1.0\ L$

where T_i and T_f are the initial and final temperatures, respectively.

From the definition of the kilocalorie (or calorie) we know how much heat (Q) it takes to raise the temperature of a certain amount of water one Celsius degree. Here, the amount or mass of water is given in terms of its volume. However, we know from Chapter 1 that 1.0 L of water has a mass of 1.0 kg.

One kcal raises the temperature of one kg of water one Celsius degree. So, to raise the temperature of one kilogram of water by five Celsius degrees ($\Delta T = T_f - T_i = 25°C - 20°C = 5\ C°$) would require 5.0 kcal, assuming no heat loss. Then,

$$Q = \underset{\substack{\text{heat}\\\text{energy}}}{5.0\ \text{kcal}} \underset{\substack{\text{mechanical}\\\text{equivalent}\\\text{of heat}}}{\left(\frac{4.19 \times 10^3\ J}{1\ \text{kcal}}\right)} = \underset{\substack{\text{mechanical}\\\text{energy}}}{2.1 \times 10^4\ J}$$

To get an idea of the mechanical equivalent of 5.0 kcal of heat, this would be the energy used in lifting a mass of 10 kg (22 lb) a height of 21 m, or about the work done in carrying a 10 kg object to the top of a 5-story building.

Heat of Combustion

Fossil fuels are burned to obtain energy for heating homes and for a variety of industrial applications. In a similar sense, our bodies use foods as fuels. The intrinsic energy values of foods and fuels are expressed in terms of the **heat of combustion** (H), which is the heat produced per unit mass of substance when it is burned in oxygen. That is,

$$H = \frac{Q}{m} \quad \text{or} \quad Q = mH \tag{11.1}$$

where Q is the amount of heat released and m is the mass. The common units for heat of combustion are cal/g, kcal/kg, and Btu/lb. (See Table 11.1 for some typical values of the heat of combustion.)

For example, the heat of combustion of soft (bituminous) coal is approximately 7500 kcal/kg. This means that when 1 kg of coal burns completely (undergoes complete combustion), 7500 kcal of heat energy on the average are liberated. Compared to fuel oil, which has a heat of combustion of 10,300

TABLE 11.1 ■ Heats of Combustion for Some Substances

Substance	J/kg	kcal/kg
Fuels		
Alcohol	2.7×10^7	6,400
Coal, hard	3.4×10^7	8,000
Coal, soft	3.2×10^7	7,500
Coke	2.5×10^7	6,000
Gasoline	4.8×10^7	11,400
Natural gas*	4.4×10^7	(10,500 kcal/m³)
Oil, diesel	4.4×10^7	10,500
Oil, fuel	4.3×10^7	10,300
Wood (average)	2.1×10^7	5,000
Foods		
Bread, white	0.84×10^7	2,000
Butter	3.4×10^7	8,000
Eggs, boiled	0.67×10^7	1,600
Eggs, scrambled	0.88×10^7	2,100
Ice cream	0.88×10^7	2,100
Meat, lean	0.50×10^7	1,200
Potatoes, boiled	0.38×10^7	970
Sugar, white	1.7×10^7	4,000

*At STP (standard temperature and pressure, 0°C and 1 atm)

kcal/kg, coal has less intrinsic heat value per kilogram. The heat of combustion of ice cream is 2100 kcal/kg. In a food chart, this would be listed in terms of the "calories" (really Calories) in an average-size serving, for example, 480 Cal per 8-oz serving.

Heats of combustion are ordinarily measured using a device called a bomb calorimeter. The bomb is a heavy steel cylinder that is leakproof. A known mass of a substance is placed in a cup inside the bomb in an atmosphere of pure oxygen. Electric current is used to ignite the sample, and combustion takes place in the form of an explosion. The heat of combustion is transmitted to a quantity of water surrounding the bomb. Knowing the temperature rise and the mass of the substance and calorimeter cup, the amount of heat released can be computed (see Section 11.2).

11.2 ■ Specific Heat

Review Section 10.1 and Fig. 10.1.

Recall from Chapter 10 that when heat is added to a substance, the energy may go to increase the random molecular motion, which results in a temperature change, and to increase the potential energy associated with the molecular bonds. Different substances have different molecular configurations and bonding. Thus, if equal amounts of heat are added to equal masses of different substances, the resulting temperature changes will not generally be the same. For example, suppose that you had 1 kg of water and 1 kg of aluminum and added 1 kcal of heat to each. From the definition of the kilocalorie, you know that the temperature of the water would rise 1 C°. You would find that the temperature of the aluminum would increase by 4.5 C°. The reason for this

Students should be able to state the specific heat for water from the definition of a calorie or a kilocalorie.

difference is that more of the added energy goes into the nonkinetic part of the internal energy of the water than of the aluminum.

The amount of heat (Q) required to change the temperature of a substance is proportional to the mass (m) of the substance and to the change in the temperature (ΔT). That is, $Q \propto m\Delta T$, or, in equation form,

$$\boxed{Q = mc\Delta T}$$

(11.2)

Here $\Delta T = T_f - T_i$ is the temperature change, or the difference between the initial temperature (T_i) and the final (T_f) temperature, and c is called the specific heat capacity. The constant c is commonly referred to as simply the **specific heat**. It is characteristic of, or *specific* for, a given substance and gives an indication of its internal molecular configuration and bonding. The specific heats for some common substances are given in Table 11.2.

Writing Eq. 11.2 as $c = Q/m\Delta T$ shows us that the units of specific heat are J/kg-K or kcal/kg-C° (or cal/g-C° in cgs units). The standard SI unit for specific heat is the one with the temperature in kelvins. However, the use of Celsius temperature in this unit (giving J/kg-C°) is commonly accepted because the size of a Celsius degree is the same as the size of a kelvin. Note that *the specific heat is the amount of energy required to raise the temperature of 1 kg of a substance by 1 C° (or 1 g of a substance by 1 C°)*. The specific heat depends somewhat on temperature (and pressure), but you can consider this effect to be negligible.

The greater the specific heat of a substance, the more energy must be transferred to it to change the temperature of a given mass. That is, a substance with a greater specific heat has a greater heat capacity, or accepts or yields more heat for a given temperature change (and mass).

Water has relatively large specific heat of 1.00 kcal/kg-C°. The value is exactly 1.00 because the definition of the kilocalorie states that 1 kcal raises the temperature of 1 kg of water by 1 C°.

You have been the victim of the large specific heat of water if you have ever burned your mouth on a baked potato or the cheese of a pizza. These foods have a high water content and a large heat capacity, so they don't cool off as quickly as some other foods do.

Specific heat

Heat equal masses of lead, steel, and aluminum to the same temperature in a hot water bath. Then place them on a block of paraffin. What does this explain about the specific heats of the elements?

Water, because of its high specific heat, is used to store heat energy in solar homes. It is also used as a coolant in cars because it can store, and then transfer, a great deal of heat energy away from the engine—and is inexpensive.

TABLE 11.2 ■ Specific Heats of Various Substances (20°C and 1 atm)

	Specific heat (c)	
Substance	*J/kg-C°*	*kcal/kg-C° (or cal/g-C°)*
Air (50°C)	1050	0.25
Alcohol, ethyl	2430	0.58
Aluminum	920	0.22
Copper	390	0.093
Glass	840	0.20
Ice (-5°C)	2100	0.50
Iron or steel	460	0.11
Lead	130	0.031
Mercury	140	0.033
Soil (average)	1050	0.25
Steam (110°C)	2010	0.48
Water (15°C)	4190	1.00
Wood (average)	1680	0.40

EXAMPLE 11.2 ■ Specific Heat

How much heat is required to raise the temperature of 0.20 kg of water from 15°C to 45°C?

Solution

Given: $m = 0.20$ kg   *Find:* Q (heat)
$\Delta T = T_f - T_i = 45°C - 15°C = 30\ C°$
$c = 4190\ \text{J/kg-C}°$
(from Table 11.2)

Eq. 11.2 can be used directly with the quantities given:

$$Q = mc\Delta T = (0.20\ \text{kg})(4190\ \text{J/kg-C}°)(30\ C°) = 2.5 \times 10^4\ \text{J}$$

$+Q$ = heat added
$-Q$ = heat removed

Note that when there is a temperature increase, ΔT and Q are positive. This corresponds to energy being *added to* a system. Conversely, ΔT and Q are negative when energy is *removed from* a system.

EXAMPLE 11.3 ■ Finding Temperature

A half-liter of water at 30°C is cooled, with the removal of 15 kcal of heat. What is the final temperature of the water?

Solution.

Given: $m = 0.50$ kg   *Find:* T_f (final temperature)
(since 1 L of water has a mass of 1 kg)
$T_i = 30°C$
$Q = -15$ kcal
(negative because energy is removed)
$c = 1.00\ \text{kcal/kg-C}°$
(from Table 11.2)

Writing out the ΔT term of Eq. 11.2 gives

$$Q = mc\Delta T = mc(T_f - T_i)$$

Solving for T_f gives

$$T_f = \frac{Q}{mc} + T_i = \frac{-15\ \text{kcal}}{(0.50\ \text{kg})(1.00\ \text{kcal/kg-C}°)} + 30°C$$

$$= 0°C$$

The (liquid) water is at its freezing point; the removal of more heat would cause the water to freeze. However, Eq. 11.2 *does not apply* when the temperature change (ΔT) is an interval that includes a change of phase, as you will learn in the next section.

EXAMPLE 11.4 ■ Greater Capacity

Equal masses of aluminum (Al) and copper (Cu) are at the same temperature. Which will require the greater heat to raise its temperature by a given amount, and how many times greater is this than the heat that would have to be added to the other metal?

Solution.

$$\text{Given: } \begin{aligned} m_{Al} &= m_{Cu} \\ \Delta T_{Al} &= \Delta T_{Cu} \\ c_{Al} &= 0.22 \text{ kcal/kg-°C} \\ c_{Cu} &= 0.093 \text{ kcal/kg-°C} \\ &\text{(from Table 11.2)} \end{aligned}$$

Find: The greater value of Q and how many times greater it is

The first part is easy. Since aluminum has a larger specific heat, it has a larger heat capacity, and more heat is required to raise its temperature by a given amount.

The question "how many times greater" generally implies a ratio. That is, if we find $x = Q_{Al}/Q_{Cu}$, we can write $Q_{Al} = xQ_{Cu}$, or Q_{Al} is x times greater than Q_{Cu}. To find the ratio we use Eq. 11.2 for each metal and divide one by the other.

$$\frac{Q_{Al}}{Q_{Cu}} = \left(\frac{m_{Al}}{m_{Cu}}\right)\left(\frac{c_{Al}}{c_{Cu}}\right)\left(\frac{\Delta T_{Al}}{\Delta T_{Cu}}\right) = \frac{c_{Al}}{c_{Cu}}$$

where common quantities have been associated so it can be easily seen that the equal masses and temperature changes cancel out.

Then,

$$Q_{Al} = \left(\frac{c_{Al}}{c_{Cu}}\right)Q_{Cu} = \left(\frac{0.22 \text{ kcal/kg-C°}}{0.093 \text{ kcal/kg-C°}}\right)Q_{Cu}$$

$$Q_{Al} = (2.4)Q_{Cu}$$

That is, 2.4 times more heat is required for the aluminum than for the copper.

■ Problem-Solving Hint

When the quantities given in a problem are equal, like the masses and temperature intervals in Example 11.4, it is a good indication that a ratio can be used to cancel out the equal quantities. Also keep in mind that phrases like "how many times more" imply a factor derived from a ratio.

Calorimetry

The specific heat of a substance is determined by measuring the quantities in Eq. 11.2 ($Q = mc\Delta T$) other than c. A simple laboratory apparatus for measuring specific heats is shown in ● Fig. 11.5. A substance of known mass and temperature is put into a quantity of water in a calorimeter. The water is at a different temperature, usually a lower one. The calorimeter is an insulated container that allows little (ideally, no) heat loss. The principle of the conservation of energy is then applied to determine c. This procedure is sometimes called the *method of mixtures*.

● **FIGURE 11.5 Calorimetry apparatus** The calorimeter cup (center, with black insulating ring) goes into the larger container. The cover with the thermometer and stirrer is seen at the right. Metal shot or pieces of metal are heated in the small cup with handle that is inserted into the hole at the top of the steam generator on the tripod.

EXAMPLE 11.5 ■ Calorimetry

Students in a physics lab are to determine the specific heat of copper experimentally. They heat 0.150 kg of copper shot to 100°C and then carefully pour the hot shot into a calorimeter cup (Fig. 11.5) containing 0.200 kg of water at 20°C. The final temperature of the mixture in the cup is measured to be 25°C. If the aluminum cup has a mass of 0.037 kg, what is the specific heat of copper? (Assume that there is no heat loss.)

Solution. In calorimetry problems, it is important to identify and label all of the quantities so as to keep them straight. We will use subscripts m, w, and c to refer to the metal, water, and calorimeter cup, respectively, and the subscripts h, i, and f to refer to the temperatures of the hot metal shot, the water (and cup) initially at room temperature, and the final temperature of the system, respectively. With this notation, we have,

$$Given: \quad m_m = 0.150 \text{ kg} \qquad\qquad Find: \quad c_m \text{ (specific heat)}$$
$$m_w = 0.200 \text{ kg}$$
$$c_w = 1.00 \text{ kcal/kg-C°}$$
$$\text{(from Table 12.2)}$$
$$m_c = 0.037 \text{ kg}$$
$$c_c = 0.22 \text{ kcal/kg-C°}$$
$$\text{(from Table 12.2)}$$
$$T_h = 100°C, T_f = 25°C,$$
$$\text{and } T_i = 20°C$$

Then, assuming no heat is lost from the system, its energy is conserved and the heat lost by the metal $(-Q_m)$ must equal the heat gained by the water and cup (Q_{w+c}):

$$\text{heat lost (by metal)} = \text{heat gained (by water and cup)}$$
$$-Q_m = Q_{w+c}$$

Substituting for these heats from Eq. 11.2 $(Q = mc\Delta T)$ gives

$$m_m c_m (T_h - T_f) = m_w c_w (T_f - T_i) + m_c c_c (T_f - T_i)$$

where the metal initially at T_h cools to T_f and $-\Delta T = (T_h - T_f)$; the water and the cup at T_i are heated to T_f and $\Delta T = (T_f - T_i)$.
Solving for c_m gives

$$c_m = \frac{(m_w c_w + m_c c_c)(T_f - T_i)}{m_m (T_h - T_f)}$$

$$= \frac{[(0.200 \text{ kg})(1.00 \text{ kcal/kg-C°}) + (0.037 \text{ kg})(0.22 \text{ kcal/kg-C°})] \times (25°C - 20°C)}{0.150 \text{ kg}(100°C - 25°C)}$$

$$= 0.093 \text{ kcal/kg-C°}$$

The experimentally determined value would actually be slightly less than this calculated value, because some heat is lost in transferring the shot to the cup and from the calorimeter while the mixture comes to thermal equilibrium.

11.3 ■ Phase Changes and Latent Heat

Matter normally exists in three *phases*: solid, liquid, and gas (see ●Fig. 11.6). The phase that a substance is in depends on its internal energy (as indicated by its temperature) and the pressure on it. You probably think more readily of adding or removing heat to change the phase of a substance because most of your experience with phase changes has been at normal atmospheric pressure, which is relatively constant.

In the **solid phase**, molecules are held together by attractive forces, or bonds. (Simplistically, these bonds can be represented as springs, as was done in Section 9.1.) Adding heat causes increased motion about the molecular equilibrium positions. If enough heat is added to provide sufficient energy to break the intermolecular bonds, the solid undergoes a phase change and becomes a liquid. The temperature at which this occurs is called the **melting point**. The temperature at which a liquid becomes a solid is called the **freezing point**. In general, these temperatures are the same, but they can differ slightly.

Ice, table salt, and most metals are crystalline solids. That is, they have orderly molecular or atomic arrangements. Other substances, however, such as glass, are noncrystalline, or amorphous. Instead of melting at a particular temperature, these substances melt over a temperture range. This discussion will be concerned primarily with substances that have definite melting points.

In the **liquid phase**, molecules of a substance are relatively free to move, and a liquid assumes the shape of its container. In certain liquids, there may be some ordered structure, giving rise to so-called liquid crystals, such as are used in LCD's (liquid crystal displays) of calculators and clocks (Chapter 23). Adding heat increases the motion of the molecules of a liquid, and when they have enough energy to become separated by large distances (compared to their diameters), the liquid changes to the **gaseous phase**, or **vapor phase**. (The distinction between a gas and a vapor will be made shortly.) This change may occur slowly by the process of evaporation or rapidly at a particular temperature called the **boiling point**. The temperature at which a gas condenses and becomes a liquid is the **condensation point**.

Some solids, such as dry ice (solid carbon dioxide), moth balls, and certain air fresheners, change directly from the solid to the gaseous phase. This is called **sublimation**. Like the rate of evaporation, the rate of sublimation increases with temperature. A phase change from a gas to a solid is called *deposition*. Frost, for example, is solidified water vapor deposited directly on grass, car windows, and other objects. Frost is not frozen dew, as is sometimes mistakenly assumed.

Latent Heat

In general, when heat energy is transferred to a substance, its temperature increases. However, when added (or removed) heat causes only a phase change, the temperature of the substance does *not* change. For example, if heat is added to a quantity of ice ($c = 0.50$ kcal/kg-C°) at $-10°C$, the temperature of the ice increases until it reaches its melting point of $0°C$. At this point, the addition of more heat does not increase the temperature but causes the ice to melt, or change phase. (Of course, the heat must be added slowly so that the ice and melted water remain in thermal equilibrium.) Once the ice is melted, adding more heat will cause the temperature of the water to rise. A similar situation

Solid, liquid, and gas are sometimes referred to as states of matter, but the state of a system has a different meaning in physics, as you will learn in Chapter 12.

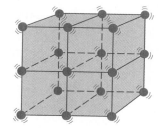

(a) Solid

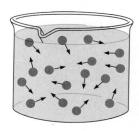

(b) Liquid

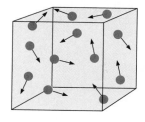

(c) Gas

●**FIGURE 11.6 Three phases of matter** (a) The molecules of a solid are held together by bonds; consequently, a solid has definite shape and volume. (b) The molecules of a liquid are relatively free to move, so a liquid has a definite volume but assumes the shape of its container. (c) The molecules of a gas are separated by relatively large distances; thus, a gas has no definite shape or volume.

TABLE 11.3 ■ Temperatures of Phase Changes and Latent Heats for Various Substances (1 atm)

Substance	Melting point	L_f		Boiling point	L_v	
		J/kg	kcal/kg		J/kg	kcal/kg
Alcohol, ethyl	−114°C	1.0×10^5	25	78°C	8.5×10^5	204
Gold	1063°C	0.645×10^5	15.4	2660°C	15.8×10^5	377
Helium*	—	—		−269°C	0.21×10^5	5
Lead	328°C	0.25×10^5	5.9	1744°C	8.67×10^5	207
Mercury	−39°C	0.12×10^5	2.8	357°C	2.7×10^5	65
Nitrogen	−210°C	0.26×10^5	6.1	−196°C	2.0×10^5	48
Oxygen	−219°C	0.14×10^5	3.3	−183°C	2.1×10^5	51
Tungsten	−3410°C	1.8×10^5	44	5900°C	48.2×10^5	1150
Water	0°C	3.3×10^5	80	100°C	22.6×10^5	540

*Not a solid at 1 atm pressure; melting point −272°C at 26 atm.

occurs during the liquid-gas phase change at the boiling point. Adding more heat to boiling water only causes more vaporization, not a temperature increase.

From the earlier description of the molecular nature of the different phases of matter, you can see that during a phase change the heat energy goes into the work of breaking bonds and separating molecules, rather than into increasing the temperature. The heat involved in a phase change is called the **latent heat** (L), and

Latent heat and phase changes

$$Q = mL \qquad (11.3)$$

where m is the mass of the substance. As you can see from this equation, the latent heat has units of joules per kilogram (J/kg) in the SI, or kcal/kg in other common units. The latent heat for a solid-liquid phase change is called the **latent heat of fusion** (L_f), and that for a liquid-gas phase change is called the **latent heat of vaporization** (L_v). These are often referred to as simply the heat of fusion and the heat of vaporization. The latent heats of some substances, along with their freezing and boiling points, are given in Table 11.3. (The latent heat for the less common solid-gas phase change is called the latent heat of sublimation and symbolized by L_s.)

It is helpful to focus on the fusion and vaporization of water. A plot of temperature versus heat energy for a quantity of water is shown in ● Fig. 11.7.

Figure 11.7

● **FIGURE 11.7 Temperature versus heat for water** As heat is added to the various phases of water, the temperature increases. During a phase change, however, the heat energy does the work of separating the molecules and the temperature remains constant. Note the different slopes of the phase lines, which indicate different values of specific heat.

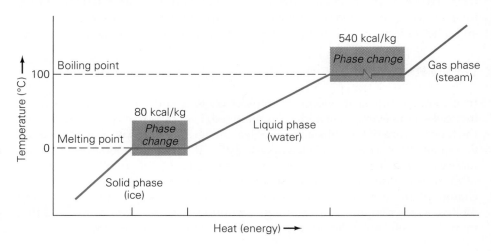

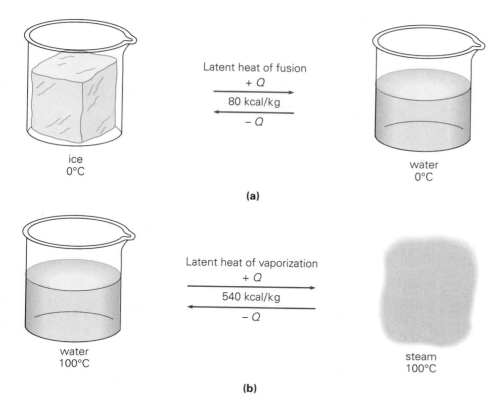

ice
0°C

Latent heat of fusion
+ Q
80 kcal/kg
– Q

water
0°C

(a)

water
100°C

Latent heat of vaporization
+ Q
540 kcal/kg
– Q

steam
100°C

(b)

FIGURE 11.8 Phase changes and latent heats (a) At 0°C, 80 kcal must be added to 1 kg of ice or removed from 1 kg of water to change its phase. (b) At 100°C, 540 kcal must be added to 1 kg of water or removed from 1 kg of steam to change its phase.

Note that when heat is added (or removed) at the temperature of a phase change, 0°C or 100°C, the temperature remains constant. Once the phase change is complete, adding more heat causes the temperature to increase. Note in Fig. 11.7 that the slopes of the phase lines are not all the same, which indicates that the specific heats of the various phases are not all the same. (Why?)

For water, the latent heats of fusion and vaporization are

$$L_f = 3.3 \times 10^5 \text{ J/kg} \quad \text{(or 80 kcal/kg or 80 cal/g)}$$
$$L_v = 22.6 \times 10^5 \text{ J/kg} \quad \text{(or 540 kcal/kg or 540 cal/g)}$$

latent heats for water

That is, 3.3×10^5 J (or 80 kcal) of heat are needed to melt 1 kg of ice at 0°C, and 22.6×10^5 J (or 540 kcal) of heat are needed to convert 1 kg of water to steam at 100°C (● Fig. 11.8). Note that the latent heat of vaporization is almost 7 times the latent heat of fusion. This indicates that more energy is needed to separate the molecules in going from water to steam than to break up the lattice structure in going from ice to water.

The word "latent" means "hidden," and its use in this context may be understood by considering a situation involving human skin. Since 540 kcal of heat energy are required to convert 1 kg of water into steam, the conservation of energy tells you that when 1 kg of steam condenses into water, 540 kcal of energy must be given up. As a result, burns from steam are usually more serious than those from boiling water. The condensing of the steam on the skin provides an additional 540 kcal/kg of heat that was seemingly hidden.

Emphasize that there is no change in temperature in the equation Q = mL. In other words, while ice is melting, it does not change its temperature. The melt water immediately after melting is the same temperature as the ice immediately prior to melting.

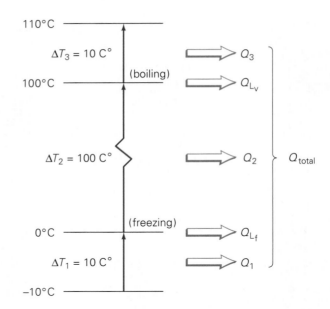

FIGURE 11.9 Steps in changing phases In going from ice at $-10°C$ to steam at $110°C$, five heat-adding steps are involved. See Example 11.6.

Demonstration/activity: Perform a standard calorimetry experiment using water and ice. Use a small enough piece of ice so it all melts. Crushed ice works better because it melts more quickly. Keep the ice as dry as you can prior to putting it into the water.

When solving phase-change problems, students often forget to take into account the changes in specific heat. For example, if steam condenses and there is a change in the temperature of water, they will sometimes still use the specific heat of steam— not the specific heat of water.

EXAMPLE 11.6 ■ Latent Heat

Heat is added to 0.50 kg of ice at $-10°C$. How many kilocalories are required to change the ice to steam at $110°C$? (Ignore significant figures.)

Solution.

Given: $m = 0.50$ kg
$T_i = -10°C$, $T_f = 110°C$
$L_f = 80$ kcal/kg,
$L_v = 540$ kcal/kg (from Table 11.3)
(specific heats from Table 11.2)

Find: Q (heat in kcal)

The temperature intervals and the heats required in the process are shown in ● Fig. 11.9. The Q's for the various steps are

$$Q_1 = mc_i\Delta T_1 = (0.50 \text{ kg})(0.50 \text{ kcal/kg-C°})(10 \text{ C°}) = 2.5 \text{ kcal}$$
$$Q_{L_f} = mL_f = (0.50 \text{ kg})(80 \text{ kcal/kg}) = 40 \text{ kcal}$$
$$Q_2 = mc_w\Delta T_2 = (0.50 \text{ kg})(1.0 \text{ kcal/kg-C°})(100 \text{ C°}) = 50 \text{ kcal}$$
$$Q_{L_v} = mL_v = (0.50 \text{ kg})(540 \text{ kcal/kg}) = 270 \text{ kcal}$$
$$Q_3 = mc_s\Delta T_3 = (0.50 \text{ kg})(0.48 \text{ kcal/kg-C°})(10 \text{ C°}) = 2.4 \text{ kcal}$$

Then

$$Q_{total} = \Sigma_i Q_i = 2.5 + 40 + 50 + 270 + 2.4 = 364.9 \text{ kcal}$$

■ Problem-Solving Hint

Note that you must compute the latent heat at each phase change. It is a common error to use the specific heat equation with a temperature interval that includes a phase change.

EXAMPLE 11.7 ■ Thermal Equilibrium

A 0.30-kg piece of ice at 0°C is placed in a liter of water at room temperature (20°C) in an insulated container. Assuming that no heat is lost to the container, what is the final temperature of the water?

Solution.

Given: $m_i = 0.30$ kg   **Find:** T_f (final temperature)
$\quad\quad\quad T_i = 0$°C
$\quad\quad\quad V_w = 1.0$ L, so $m_w = 1.0$ kg
$\quad\quad\quad T_w = 20$°C

The subscripts i and w refer to ice and water, respectively. You know that 1.0 L of water has a mass (m_w) of 1.0 kg and that the water supplies the heat to melt the ice. If all of the ice melts, you could then view the system as being two masses of water at different temperatures, which come to equilibrium at some intermediate temperature. Since there is no heat loss, it is tempting to write an equation equating the amount of heat lost by the water to the amounts of heat used to melt the ice and to warm up the ice water.

But does all the ice melt? To melt 0.30 kg of ice requires

$$Q_i = m_i L_f = (0.30 \text{ kg})(80 \text{ kcal/kg}) = 24 \text{ kcal}$$

Then, looking at the water, you must ask how much heat it can supply to the ice. The maximum amount would be that given up in lowering the temperature of the water to 0°C, which would be a temperature decrease of $\Delta T = -20$C°. This maximum amount of heat would be

$$Q_w = m_w c_w \Delta T = (1.0 \text{ kg})(1.0 \text{ kcal/kg-C°})(-20\text{C°}) = -20 \text{ kcal}$$

Thus, all of the ice does not melt since the water does not have enough energy. The final temperature of the water is therefore 0°C.

The heat given up by the water (20 kcal) will melt a mass of ice equal to

$$m_i = \frac{Q_i}{L_f} = \frac{20 \text{ kcal}}{80 \text{ kcal/kg}} = 0.25 \text{ kg}$$

Thus, the final mixture would be 0.30 kg -0.25 kg $= 0.05$ kg of ice in thermal equilibrium with 1.25 kg (or 1.25 L) of water at $T_f = 0$°C.

Information about phase changes is represented on graphs called **phase diagrams**. An example is the p-T (pressure-temperature) diagram for water shown in ● Fig. 11.10. The curves are formed of the points (p, T), or the pressure-temperature combinations at which different phases are in equilibrium. For example, the point at 1 atm and 100°C corresponds to the normal boiling point, at which liquid water and steam are in equilibrium.

The triple point is the point at which all three phases coexist. This is the unique point used as a reference for the Kelvin scale (Chapter 10). The three curves branching out from this point separate the phase regions. Note that if you started heating a quantity of ice that was below 0°C at 1 atm, the plotted state of the system would first cross the fusion curve (along the horizontal dashed line extending from 1.0 atm in the graph) into the liquid phase region

It is interesting to note from the phase diagram of water (Figure 11.10) that the fusion curve slopes upward to the left and the vaporization curve slopes upward to the right. These slopes indicate that the freezing point of water decreases slightly with increasing pressure and the boiling point increases with increasing pressure. The latter fact is what makes possible the operation of the pressure cooker, which cooks foods faster because boiling temperatures greater than 100°C can be obtained at pressures above atmospheric pressure.

Such a slope of the fusion curve is characteristic of only a very few substances that, like water, expand on freezing. It was once thought that the lowering of the freezing point of water by pressure provided the mechanism that made ice skating possible. Supposedly, the pressure of the narrow skate blade on the ice would lower the melting point below the ambient temperature. The skater would thus glide on a thin film of water, which quickly refroze when the pressure was removed. How-

ever, from the phase diagram for water, it is found that a pressure of about 140 atm is needed to lower the freezing point from 0°C to −1°C. Only a very heavy person could produce a pressure on this order, and the outdoor ice skating temperature is usually well below −1°C.

Frictional heating between the skate blade and the ice does contribute to melting. For high-friction surfaces, such as skis on snow, this is the main mechanism. However, for ice skating there is another factor that contributes to the low coefficient of friction of ice. This is called *surface melting*, proposed by the English scientist Michael Faraday (1791–1867; see Chapter 19). He suggested that a thin layer of liquid normally exists on the surface of a solid even at temperatures well below the solid's melting point. Modern techniques have shown this to be the case for most solids. For ice, the thickness of the water film is about 40 nm near 0°C and about 0.50 nm near −35°C.

Explain how a change in pressure can affect the phases of matter, for example in a pressure cooker. For a thought question, ask students what should happen to the boiling and freezing points of water at high altitudes? Do eggs really have to be boiled longer in Denver than in Detroit?

As a general rule, the increase of pressure favors the liquid phase of matter.

and, with continued heating, would cross the vaporization curve into the vapor phase region at 100°C.

"Vapor" is another term commonly used for "gas." Water vapor is water in the gas phase. (Vapor is also used in a nontechnical sense to mean visible droplets of water, such as condensed steam or clouds.) The distinction between vapor and gas is often made relative to the critical point at the end of the vaporization curve. At temperatures less than the critical temperature (374°C for water), a gas will change to a liquid if sufficient pressure is applied. If a

Figure 11.10

● **FIGURE 11.10 Phase diagram for water** The curves branch out from the triple point, where water exists in all three phases. On the transition curves, it exists in two phases.

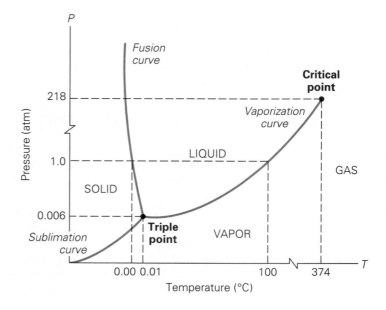

gas is above its critical temperature, no amount of pressure will cause it to become a liquid. It becomes denser and denser with increasing pressure, but never quite becomes a liquid. A substance that is gaseous and has a temperature above its critical temperature is called a gas, and a substance that is gaseous but with a temperature below its critical temperature is known as a vapor.

A phase change consideration is given in the Insight feature.

11.4 ▪ Heat Transfer

Since heat is defined as energy in transit, how the transfer takes place is important. Heat moves from place to place (from a higher-temperature region to a lower-temperature region) by three mechanisms: conduction, convection, and radiation.

Conduction

You can keep a pot of coffee hot on a stove because heat is conducted through the coffee pot from the hot burner. The process of **conduction** is visualized as resulting from molecular interactions. Molecules in one part of a body at a higher temperature vibrate faster. They collide with and transfer some of their energy to less energetic molecules located toward the cooler part of the body. In this way, energy is conductively transferred from a higher-temperature region to a lower-temperature region.

Solids can be divided into two general categories: metals and nonmetals. Metals are good conductors of heat, or **thermal conductors**. Modern theory views metals as having a large number of electrons that are free to move around (not permanently bound to a particular molecule or atom). These free electrons are believed to be primarily responsible for the heat conduction of metals. Nonmetals, such as wood or cloth, have relatively few free electrons and are poor heat conductors. A poor heat conductor is called a **thermal insulator**.

In general, the ability of a substance to conduct heat depends on its phase. Gases are poor thermal conductors because their molecules are relatively far apart, and collisions are therefore infrequent. Liquids are better thermal conductors than gases are because their molecules are closer together and can interact more readily.

Heat conduction may be described quantitatively as the time rate of heat flow ($\Delta Q/\Delta t$) in a material for a given temperature difference (ΔT), as illustrated in ●Fig. 11.11. It has been established through experiments that the rate of heat flow through a substance depends on the temperature difference between its boundaries. Heat conduction also depends on the size and shape of an object. Thus, the analysis of heat flow is generally done using a uniform slab of the material.

A moment's thought should convince you that the heat flow through a slab of material is directly proportional to its surface area (A) and inversely proportional to its thickness (d). That is,

$$\frac{\Delta Q}{\Delta t} \propto \frac{A\Delta T}{d}$$

The term $\Delta T/d$ is called *thermal gradient* (the change in temperature per unit of length). Using a constant of proportionality allows the relation to be written

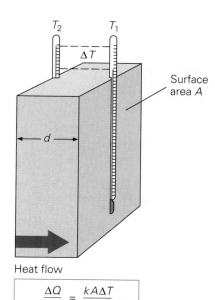

$$\frac{\Delta Q}{\Delta t} = \frac{kA\Delta T}{d}$$

●**FIGURE 11.11 Thermal conduction** Heat conduction is characterized by the time rate of flow of heat ($\Delta Q/\Delta t$) in a material with a temperature difference ΔT. For a slab of material, $\Delta Q/\Delta t$ is directly proportional to the cross-sectional area (A) and the thermal conductivity of the material (k), and is inversely proportional to the slab thickness (d).

as an equation:

$$\frac{\Delta Q}{\Delta t} = \frac{kA\Delta T}{d} \qquad (11.4)$$

The constant k is called the **thermal conductivity** and characterizes the heat-conducting ability of a material. The greater the value of k for a material, the more heat it will conduct.

The units of k are J/m-s-C° (or kcal/m-s-C°). The thermal conductivities of various substances are listed in Table 11.4. These values actually vary slightly over different temperature ranges, but can be considered to be constant over normal temperature ranges and differences. Compare the relatively large thermal conductivities of the good thermal conductors, the metals, with the relatively small thermal conductivities of some good thermal insulators, such as Styrofoam and wood. Plastic foams are good insulators mainly because they contain pockets of air. Recall that gases are poor conductors, and note the low thermal conductivity of air in the table.

> Stress that heat spontaneously flows from a high temperature region to a lower temperature region.

EXAMPLE 11.8 ■ Thermal Conductivity and Insulation

A room with a pine ceiling that measures 3.0 m × 5.0 m × 2.0 cm thick has a 6.0 cm layer of glass wool insulation above it (Fig. 11.12a). On a cold day, the temperature inside the room is 20°C and the temperature in the attic is 8.0°C. Assuming that the temperatures remain constant, how much energy is saved in one hour by having installed the layer of insulation?

Solution. Computing some of the quantities and making conversions as we list the data, we have

Given: $A = 3.0\,\text{m} \times 5.0\,\text{m} = 15\,\text{m}^2$
$d_1 = 6.0\,\text{cm} = 0.060\,\text{m}$
$d_2 = 2.0\,\text{cm} = 0.020\,\text{m}$
$\Delta T = T_2 - T_1 = 20°\text{C} - 8.0°\text{C} = 12\,\text{C}°$
$\Delta t = 1.0\,\text{h} = 3.6 \times 10^3\,\text{s}$
$\left. \begin{array}{l} k_1 = 0.042\,\text{J/m-s-C}° \\ k_2 = 0.12\,\text{J/m-s-C}° \end{array} \right\}$ (from Table 11.4)

Find: Energy saved in 1.0 h

(In working such problems, it is important to label all the data correctly.)

Then, to find the energy saved in one hour, we need to compute how much heat is conducted in this time with and without the layer of insulation. First, let's consider how much heat would be conducted in one hour through the wooden ceiling with no insulation present.

Eq. 11.4 gives the rate of heat flow ($\Delta Q/\Delta t$). Since we know Δt, we can rearrange the equation to find ΔQ_c (heat conducted through the wooden ceiling).

$$\Delta Q_c = \left(\frac{k_2 A \Delta T}{d_2}\right)\Delta t$$

$$= \left[\frac{(0.12\,\text{J/m-s-C}°)(15\,\text{m}^2)(12\,\text{C}°)}{0.020\,\text{m}}\right](3.6 \times 10^3\,\text{s})$$

$$= 3.9 \times 10^6\,\text{J}$$

TABLE 11.4 ■ Thermal Conductivities of Some Substances

Substance	Thermal conductivity (k)	
	J/m-s-C°	kcal/m-s-C°
Metals		
Aluminum	240	5.7×10^{-2}
Copper	390	9.4×10^{-2}
Iron and steel	46	1.1×10^{-2}
Silver	420	10×10^{-2}
Liquids		
Transformer oil	0.18	4.2×10^{-5}
Water	0.57	14×10^{-5}
Gases		
Air	0.024	0.57×10^{-5}
Hydrogen	0.17	4.0×10^{-5}
Oxygen	0.024	0.58×10^{-5}
Other materials		
Brick	0.71	17×10^{-5}
Concrete	1.3	31×10^{-5}
Cotton	0.075	1.8×10^{-5}
Fiberboard	0.059	1.4×10^{-5}
Floor tile	0.67	16×10^{-5}
Glass (typical)	0.84	20×10^{-5}
Glass wool	0.042	1.0×10^{-5}
Ice	2.2	53×10^{-5}
Styrofoam	0.042	1.0×10^{-5}
Wood		
Wood, oak	0.15	3.5×10^{-5}
Wood, pine	0.12	2.8×10^{-5}
Vacuum	0	0

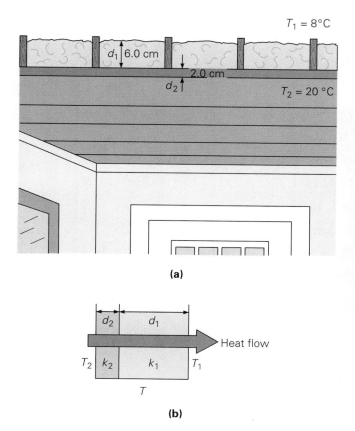

$T_1 = 8°C$

d_1 6.0 cm

2.0 cm

d_2

$T_2 = 20 °C$

(a)

d_2 d_1

Heat flow

T_2 k_2 k_1 T_1

T

(b)

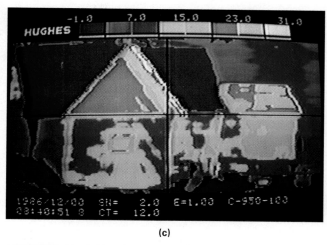

(c)

FIGURE 11.12 Insulation and thermal conductivity **(a, b)** Attics are insulated to prevent the loss of heat through conduction. See Example 11.8. **(c)** This thermogram of a house allows us to visualize heat loss. (Blue represents the coolest areas, white, pink, and red the hottest.) What conclusions can you draw from it? (Compare Fig. 11.16.)

Now we need to find the heat conducted through the ceiling *and* the insulation layer together. Let T be the temperature at the interface of the materials, and T_2 and T_1 refer to the warmer and cooler temperatures, respectively (see Fig. 11.12b). Then,

$$\frac{\Delta Q_1}{\Delta t} = \frac{k_1 A(T - T_1)}{d_1} \quad \text{and} \quad \frac{\Delta Q_2}{\Delta t} = \frac{k_2 A(T_2 - T)}{d_2}$$

The problem is that we don't know T. However, when the conduction is steady, the flow rates are the same for both, that is, $\Delta Q_1 / \Delta t = \Delta Q_2 / \Delta t$. This relationship enables us to eliminate T. We equate the two preceding equations, finding an expression for T, and then substitute this expression into either equation, which will give us a general expression for $\Delta Q / \Delta t$ for the combined layers.

$$\frac{\Delta Q_1}{\Delta t} = \frac{\Delta Q_2}{\Delta t}$$

or

$$\frac{k_1 A(T - T_1)}{d_1} = \frac{k_2 A(T_2 - T)}{d_2}$$

The A's cancel, and solving for T gives

$$T = \frac{k_1 d_2 T_1 + k_2 d_1 T_2}{k_1 d_2 + k_2 d_1} \tag{11.5*}$$

Substituting into either of the flow rate equations, and rearranging, we get

$$\frac{\Delta Q}{\Delta t} = \frac{A(T_2 - T_1)}{(d_1/k_1) + (d_2/k_2)}$$

$$= \frac{(15 \text{ m}^2)(12 \text{ C}°)}{(0.060 \text{ m})/(0.042 \text{ J/m-s-C}°) + (0.020 \text{ m})/(0.12 \text{ J/m-s-C}°)}$$

$$= 1.1 \times 10^2 \text{ J/s}$$

and in $1.0 \text{ h} = 3.6 \times 10^3 \text{ s}$,

$$\Delta Q = (1.1 \times 10^2 \text{ J/s}) \, \Delta t = (1.1 \times 10^2 \text{ J/s})(3.6 \times 10^3 \text{ s})$$

$$= 4.0 \times 10^5 \text{ J}$$

Thus, the heat loss is decreased by

$$\Delta Q_c - \Delta Q = 39 \times 10^5 \text{ J} - 4.0 \times 10^5 \text{ J} = 35 \times 10^5 \text{ J}$$

Ideally, this represents a savings of

$$\frac{35 \times 10^5 \text{ J}}{39 \times 10^5 \text{ J}} (\times 100\%) = 90\%$$

*This equation may be extended to any number of layers or slabs of materials by adding additional d/k terms to the denominator: $\Delta Q/\Delta t = A(T_2 - T_1)/\Sigma(d_i/k_i)$.

Convection

In general, compared to solids, liquids and gases are not good thermal conductors. The mobility of molecules in fluids permits heat transfer by another process—**convection**. Heat transfer by convection involves mass transfer. For example, when cold water is run over a hot object, the object transfers heat to the water by conduction, and the water carries the heat away with it by convection.

Natural convection cycles occur in liquids and gases. Such cycles are important in atmospheric processes, as illustrated in ● Fig. 11.13. During the day, the ground heats up more quickly than do large bodies of water, as anyone who has been to the beach knows. This is because the water has a high specific heat, as well as mixing currents that disperse the absorbed heat throughout the great volume of water. The air in contact with the warm ground is heated by conduction. That air expands, becoming less dense than in surrounding cooler air. As a result, the warm air rises (air currents) and other air moves

> Convection is the only form of heat transfer that involves mass transfer.

● **FIGURE 11.13 Convection cycles** During the day, natural convection cycles give rise to sea breezes near large bodies of water. At night, the pattern of circulation is reversed and land breezes blow.

horizontally (winds) to fill the space—creating a sea breeze near a large body of water. Cooler air descends, and a thermal convection cycle is set up, which transfers heat away from the Earth. At night, the ground loses its heat more quickly, and the water surface is warmer than the land. As a result, the cycle is reversed.

You can see convection currents in the air above a hot road surface in the summer and in transparent liquids, such as heated water in a glass container. This is because regions of different temperatures have different densities, which cause a bending, or refraction, of light (see Chapter 21).

Convection can also be forced, which means that the medium of heat transfer is moved mechanically. Common examples of forced convection systems are forced-air heating systems in homes (Fig. 11.14), the human circulatory system, and the cooling system of an automobile engine. The human body does not use all of the energy obtained from food; a great deal is lost. [Keep in mind that there's usually a temperature difference between your body and the surroundings.] So that body temperature will stay normal, the internally generated heat energy is transferred close to the surface by blood circulation. From the skin, it is conducted to the air or lost by radiation (the other heat-transfer mechanism, to be discussed shortly).

Water or some other coolant is circulated (pumped) through most automobile cooling systems (some engines are air-cooled). The fluid medium carries heat to the radiator (a heat exchanger), where forced air flow produced by the fan carries it away. The radiator of an automobile is actually misnamed—most of the heat is transferred from it by convection rather than radiation.

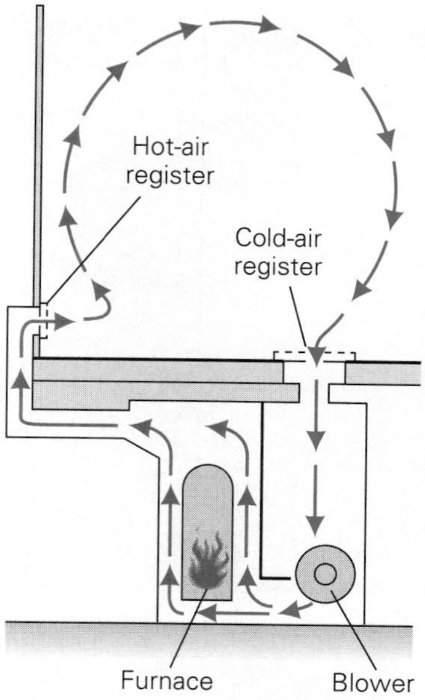

FIGURE 11.14 Forced convection Houses are commonly heated by forced convection. In older homes, natural convection was (and still is) used. Registers or gratings in the floors or walls allow heated air to enter and cooler air to return to the heat source.

EXAMPLE 11.9 ■ More Insulation?

Polymer foam insulation is sometimes blown into the space between the inner and outer walls of a house. Since air is a good thermal insulator, why is the foam insulation needed? (a) To prevent cold air from occupying the space, (b) to prevent loss of heat by conduction, (c) to prevent loss of heat by convection, (d) for fireproofing. *Clearly establish the reasoning and physical principle(s) used in determining your answer. That is, **why** did you select your answer?*

Reasoning and Answer. Polymer foams are porous and contain a lot of air. Also, they will generally burn, so (a) and (d) aren't good answers. Air is a poor thermal conductor, even poorer than plastic foam (Table 11.4). However, as a gas, it is subject to convection within the wall space. In the winter, the air near the warm inner wall is heated and rises, thus setting up a convection cycle in the space and transferring heat to the cold outer wall. In the summer, with air conditioning, the heat-loss cycle is reversed. Hence, the answer is (c).

Follow-up Exercise. Thermal underwear and thermal blankets are loosely knitted with lots of small holes. Wouldn't it be better if the material were denser? (*Reasoning and answer may be found in the Answers to Exercises section at the back of the book.*)

(a)

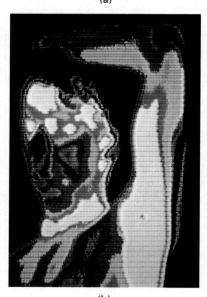

(b)

● **FIGURE 11.16 Thermograms**
(a) A thermogram of a saw cutting through a board. The saw blade is heated by friction. The warmest colors, yellow and red, appear in the area of the blade that has just passed through the board. The handle of the saw where a gloved hand grips it also (white). (b) A thermogram of a man holding his arm above his head. The skin temperature varies about 1C° for each color, from white (hottest) to black (coldest). Note the high temperature of the armpit, due to the proximity of blood vessels to the skin, and the cooler nose and hair.

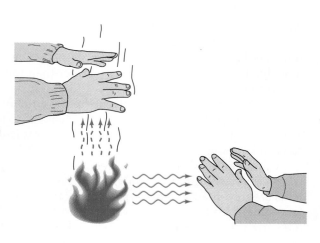

● **FIGURE 11.15 Convection and radiation heating** The hands of the person at left are warmed by the convection of rising hot air (and some radiation). The hands of the person at right are warmed by radiation.

Radiation

Conduction and convection require some material as a transport medium. The third mechanism for heat transfer needs no medium and is called **radiation**, which refers to energy transfer by electromagnetic waves (Chapter 19). This is how heat is transferred to the Earth from the Sun through empty space. Visible light and other forms of electromagnetic radiation are commonly referred to as radiant energy.

You have experienced heat transfer by radiation if you've ever stood near an open fire (● Fig. 11.15). You can feel the heat on your exposed hands and face. This heat transfer is not due to convection or conduction, since heated air rises and air is a poor conductor. Visible radiation is emitted from the burning material, but most of the heating effect comes from the invisible **infrared radiation** emitted by the glowing embers or coals. You feel this radiation because it is absorbed by water molecules in your skin. (Body tissue is about 85% water.) The water molecule has an internal vibration whose frequency coincides with that of infrared radiation, which is therefore absorbed readily. [This is called *resonance absorption*. The electromagnetic wave drives the molecular vibration, and energy is transferred to the molecule, somewhat like pushing a swing; see Chapter 13.]

Infrared radiation is sometimes referred to as "heat rays." You may have noticed the red infrared lamps used to keep food warm in cafeterias. Heat transfer by infrared radiation is also important in maintaining our planet's warmth by a mechanism known as the *greenhouse effect*. This important environmental topic, along with another common example of heating by radiation absorption, is discussed in the Insight feature.

Although infrared radiation is invisible to the human eye, it can be detected by other means. For example, you can buy special infrared film for some cameras. A picture taken with this film will be an image consisting of contrasting light and dark areas corresponding to regions of higher and lower temperatures. Special instruments that apply such thermography are used in industry and medicine (● Fig. 11.16). The frequency of the infrared radiation is proportional to the temperature of its source. This fact is the basis of infrared thermometers, which can measure temperature remotely.

The rate at which an object radiates energy has been found to be proportional to the fourth power of the absolute temperature (T^4). This is expressed in an equation known as **Stefan's law**:

$$P = \sigma A e T^4 \qquad (11.6)$$

where P is the power radiated in watts (W) or joules/s (J/s). [The calorie is not generally used to measure radiation. Radiated power can be written $\Delta Q/\Delta t$ to indicate a rate of heat loss if desired.] The symbol σ (the Greek letter sigma) is called the *Stefan-Boltzmann constant*: $\sigma = 5.67 \times 10^{-8}$ W/m²-K⁴. The radiated power is also proportional to the surface area (A) of the object. The **emissivity** (e) is a number between 0 and 1 that is characteristic of the material (e is unitless). Dark surfaces have emissivities close to 1, and shiny surfaces have emissivities close to 0. The emissivity of human skin is about 0.70.

Dark surfaces are not only better emitters of radiation, they are also good absorbers. In general, *a good emitter is also a good absorber.* An ideal, or perfect, absorber (and emitter) is referred to as a **black body** ($e = 1.0$). Shiny surfaces are poor absorbers, since most of the incident radiation is reflected. This fact may be demonstrated easily as shown in ● Fig. 11.17. (You should see why it is better to wear light-colored clothes in the summer and dark-colored clothes in the winter.)

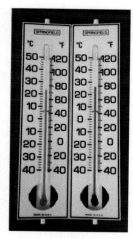

● **FIGURE 11.17 Good absorber**
Black objects are good absorbers. The bulb of the thermometer on the right has been painted black. Note the difference in the temperatures.

INSIGHTS ■ I. The Greenhouse Effect and Global Warming

Recently, rising concern about global warming caused by the so-called greenhouse effect has made headlines all over the world. Normally, the greenhouse effect helps regulate the Earth's long-term average temperature, which is fairly constant. A portion of the solar radiation we receive reaches the Earth's surface and warms it. The Earth in turn reradiates energy in the form of infrared radiation. The balance between absorption and radiation is a major factor in stabilizing the Earth's temperature.

It is this balance that is affected by the concentration

of "greenhouse gases"—primarily water vapor and carbon dioxide (CO_2)—in the atmosphere.

As the infrared radiation passes through the atmosphere, some of it is absorbed by the greenhouse gases. These gases are selective absorbers; that is, they absorb radiation at certain wavelengths but not at others (Fig. 1). If radiant infrared energy is absorbed, the atmosphere warms, warming the Earth. This rise in surface temperature shifts the wavelength of the reemitted radiation so that it no longer corresponds to the absorption wave-

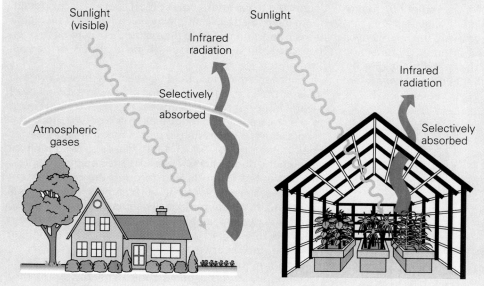

Figure 1

FIGURE 1 The greenhouse effect The "greenhouse" gases of the atmosphere, particularly water vapor and carbon dioxide, are selective absorbers with absorption properties similar to those of glass used in greenhouses. Visible light is transmitted and heats the surface, while much of the infrared radiation that is reemitted is absorbed.

length of the gases. The radiation therefore passes through the atmosphere into space, so the Earth loses energy and cools. Cooling shifts the radiation back to a wavelength that is absorbed, and the process starts again. Hence, the selective absorption of the greenhouse gases provides a thermostatic action that helps regulate the Earth's temperature.

You may be wondering why this phenomenon is called the greenhouse effect. The reason is that the atmosphere functions like the glass in a greenhouse. That is, the absorption and transmission properties of glass are similar to those of the atmospheric greenhouse gases—in general, visible radiation is transmitted, but infrared radiation is selectively absorbed (Fig. 1). We have all observed the warming effect of sunlight passing through glass, for example, in a closed car on a sunny, cold day. Similarly, a greenhouse heats up by absorbing sunlight and trapping the reradiated infrared radiation. Thus it is quite warm inside on a sunny day, even in winter. (Of course, the glass enclosure also keeps warm air from escaping upward, as it normally would. In practice, this elimination of heat loss by convection is the chief factor

in maintaining an elevated temperature. The temperature of a greenhouse in the summer is controlled by painting the glass panels white so that some of the sunlight is reflected, and opening panels to let some hot air escape.)

The problem is that on Earth, human activities may accentuate the greenhouse warming. With all the combustion of fuels for heating and industrial processes, vast amounts of CO_2 and other greenhouse gases are vented into the atmosphere. It is feared that the result of this trend will be global warming: an increase in the Earth's average temperature that could dramatically affect the environment. For example, the climate in many parts of the globe might be altered, with effects that are very difficult to predict. Changes in average temperatures and rainfall patterns could affect the great crop-growing regions of the world, disrupting agricultural production and reducing the world's food supplies. It has also been suggested that a general rise in temperature might cause partial melting of the polar ice caps. Sea levels would rise, flooding low-lying regions and endangering coastal ports and population centers.

II. Heat Transfer by Radiation: The Microwave Oven

The microwave oven has quickly become a common kitchen appliance. It is both time-saving and energy-saving, since the oven doesn't have to be warmed up as does a conventional oven. The principle of operation of the microwave oven is heat transfer by radiation.

Microwaves are a form of electromagnetic radiation; they have a frequency range just below that of infrared radiation (Chapter 21). Like infrared radiation, microwaves are absorbed chiefly by water molecules (in a molecular resonance). In a microwave oven, the microwaves are generated electronically and distributed by reflection from a metal stirrer or fan and the metal walls (Fig. 2). Because the walls reflect the radiant energy, they do not get hot.

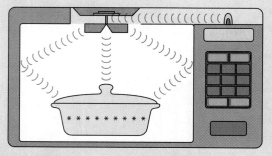

FIGURE 2 Heat transfer by radiation In a microwave oven, microwaves (a type of electromagnetic radiation) are absorbed by molecules, chiefly water molecules, raising their temperature.

Microwaves pass through plastic wrap, glass, or dishes made of other "microwave-safe" materials and are absorbed by water molecules in the food, which causes it to be heated. (Metal utensils or objects can become electrically charged, causing sparking and possibly damage.) The microwaves do not penetrate the food completely but are absorbed near the surface. Heat is then conducted to the interior of the food, just as it is in conventional oven heating. This is why it is advisable to let large items or portions sit for a time after the microwave oven has shut off, so that they will be warmed or cooked throughout.

Since microwaves could be absorbed by water molecules in the skin, causing burns, microwave ovens have several important safety features. The door is tight-fitting so that microwaves cannot leak out. You will note that the glass in the door is fitted with a metal shield that has small holes through which food can be viewed without opening the door. Microwaves are reflected by this shield and prevented from coming through the glass (essentially, the waves are larger than the holes). Also, there is a mechanism that automatically shuts off the oven when the door is opened, so a person cannot get into the oven while it is running. In fact, the oven cannot be turned on when the door is open.

You may be hearing more about microwaves soon. Developmental work is being done on microwave clothes dryers, in which clothes are dried by microwaves instead of hot air.

When an object is in thermal equilibrium with its surroundings, its temperature is constant. Thus, it must be emitting and absorbing radiation at the same rate. If the temperatures of the object and its surroundings are different, there must be a net flow of radiant energy. If an object is at a temperature T and its surroundings are at a temperature T_s, the net rate of energy loss or gain power is given by

Demonstration/activity: Wrap one of two identical cans with masking tape and leave the surface of the other can shiny. Fill them both with boiling water at the start of a lecture period and measure their temperatures about every five minutes. The can wrapped with tape will cool faster. Explain why.

$$P_{net} = \sigma Ae(T_s^4 - T^4) \qquad (11.7)$$

Note that if T_s is less than T, then P (or $\Delta Q/\Delta t$) will be negative, indicating a net energy loss. *Keep in mind that the temperatures used in calculating radiated power are the absolute temperatures in kelvins.*

It is sometimes convenient to consider the **intensity** (I) of radiation. This is simply the power per area ($I = P/A$) or the energy per area per time ($I = E/At$). In terms of Eq. 11.7,

$$I = \frac{P_{net}}{A} = \sigma e(T_s^4 - T^4) \qquad (11.8)$$

EXAMPLE 11.10 ■ Radiant Heat Transfer

Suppose that your skin has an emissivity of 0.70 and its exposed area is 0.27 m². (a) How much net energy will be radiated per second from this area if the air temperature is 20°C? (Assume that your body temperature is normal. Recall from Chapter 10 that this is 37°C.) (b) What is the intensity of this radiation?

Solution.

Given: $T_s = 20°C + 273 = 293$ K   *Find:* (a) P_{net} (net power)
$\quad\quad\quad T = 37°C + 273 = 310$ K $\quad\quad\quad$ (b) I (intensity)
$\quad\quad\quad e = 0.70$
$\quad\quad\quad A = 0.27$ m²
$\quad\quad\quad \sigma = 5.67 \times 10^{-8}$ W/m²-K⁴ (known)

(a) Using Eq. 11.7,

$P_{net} = \sigma Ae(T_s^4 - T^4)$
$\quad\quad = (5.67 \times 10^{-8}$ W/m²-K⁴$)(0.27$ m²$)(0.70)[(293$ K$)^4 - (310$ K$)^4]$
$\quad\quad = -20$ W   (or -20 J/s)

Since $P_{net} = \Delta Q/\Delta t$, 20 J of energy is radiated or *lost* (as indicated by the minus sign) each second.
(b) Since $I = P/A$,

$$I = \frac{P_{net}}{A} = \frac{-20 \text{ W}}{0.27 \text{ m}^2} = -74 \text{ W/m}^2$$

■ Problem-Solving Hint

Note that in part (a) of Example 11.10 the fourth powers of the temperatures were found first and then their difference. It is *not* correct to find the temperature difference and then raise it to the fourth power: $T_s^4 - T^4 \neq (T_s - T)^4$.

(a)

(b)

FIGURE 11.18 Vapor pressure and boiling (a) In a closed container, vapor fills the space above a liquid until equilibrium is reached—that is, the same number of molecules are reentering the liquid as are leaving it. The space is then said to be saturated, and the pressure of the vapor is termed the saturated vapor pressure. (b) A pan of boiling water. When the saturated vapor pressure inside the bubbles formed near the heated surface equals or exceeds the external pressure, the bubbles increase in size and rise to the surface. With continued heating, the water will boil vigorously.

11.5 ■ Evaporation and Relative Humidity ◇

The evaporation of water from an open container becomes evident only after a relatively long period of time. This phenomenon can be explained in terms of kinetic theory (Section 10.5). The molecules in a liquid are in motion, at different speeds. A faster-moving molecule that is near the surface may momentarily leave the liquid. If its velocity is not too large, it will return to the liquid because of the attractive forces exerted by the other molecules. Occasionally, however, a molecule has a large enough velocity that it leaves the liquid entirely and becomes part of the air. The higher the temperature of the liquid, the more likely this is to occur. (Why?)

The escaping molecules take their energy with them. Since those molecules with greater than average energy are the ones most likely to escape, the energy and temperature of the remaining liquid will be reduced. Thus, *evaporation is a cooling process*. You have probably noticed this when drying off after a bath or shower. About 600 kcal are needed to evaporate 1 kg of water from the skin. This energy requirement can be estimated by considering a different process. The amount of heat per kilogram (Q/m) needed to raise the temperature of water from 35°C (approximately skin temperature) to 100°C ($\Delta T = 100°C - 35°C = 65\ C°$) is $c\Delta T$ (since $Q = mc\Delta T$). Adding the latent heat of vaporization, which is also the heat per unit mass ($Q = mL_v$ and $L_v = Q/m$), gives

$$\frac{Q}{m} = c\Delta T + L_v = (1\ \text{kcal/kg-C°})(65\ C°) + 540\ \text{kcal/kg} = 605\ \text{kcal/kg}$$

Of course, evaporation doesn't involve heating to the boiling point, but this estimate gives a limiting approximation.

Although evaporation is a relatively slow process, it is often important in preventing our bodies from overheating. Usually, radiation and conduction are sufficient to maintain a rate of heat loss to the air that keeps us comfortable, given the temperature difference between our bodies and the surroundings. However, when the air gets really hot and the temperature difference narrows (or disappears), we start to perspire. The evaporation of perspiration helps to cool our bodies. On a hot summer day, a person may stand in front of a fan and remark how cool the blowing air feels. But the fan is merely blowing hot air from one place to another. The air feels cool because its flow promotes evaporation, which removes heat energy. Of course, evaporation depends on the humidity (the amount of moisture already in the air). We feel less comfortable on hot humid days because evaporation is reduced.

Air normally contains water vapor (water in the gaseous phase), mainly from evaporation. Consider an evacuated container partially filled with water or some other liquid (● Fig. 11.18a). Energetic molecules escape into the space above the liquid. In bouncing around, some strike the liquid surface and again become part of the liquid phase. Eventually, equilibrium will be reached, when the average number of molecules in the space is constant, with the same number of molecules entering the liquid as leaving it. The region above the liquid is then said to be saturated, and the pressure of the vapor is called the **saturated vapor pressure** (also known as equilibrium vapor pressure). As you might expect, the saturated vapor pressure of any liquid depends on temperature. The higher the temperature, the more molecules have enough kinetic energy to escape into the vapor phase and the greater the pressure.

The same general description applies for a liquid evaporating into air. Evaporation takes place until an equilibrium is reached between escaping and returning molecules. The effect of the air molecules is that collisions with them may lengthen the time to reach equilibrium. The "container" is not closed in this case, so the air does not generally become saturated. If it does hold all of the water it can at *a given temperature*, it usually gives up the moisture as rain or some other form of precipitation.

Heating a liquid speeds up the evaporation process. As mentioned above, the saturated vapor pressure increases with temperature. When the saturated vapor pressure equals the external pressure (generally atmospheric pressure), the liquid boils. Tiny bubbles form in the heated region as the boiling point is approached. When the saturated vapor pressure inside these bubbles equals or exceeds the external pressure, they increase in size and rise to the surface, giving evidence that boiling has begun. With continued heating, the liquid will boil vigorously (Fig. 11.18b).

As you can see from Fig. 11.10 (in Section 11.3), the boiling point of water decreases with decreasing pressure. In fact, a container of water in a vacuum chamber will boil at room temperature. The cooling effect of the boiling (the removal of latent heat) will eventually cause the remaining water to freeze if the vacuum is maintained. At high altitudes, where there is lower atmospheric pressure, the boiling point of water is lowered. For example, at Pike's Peak, Colorado, at an elevation of about 4300 m, the atmospheric pressure is about 600 torr, and water boils at about 94°C rather than at 100°C. The lower temperature lengthens the cooking time of food. A pressure cooker may be used to reduce the cooking time (at Pike's Peak or at sea level)—by increasing the pressure, a pressure cooker raises the boiling point.

Another example of vapor pressure is provided by the operation of spray cans (Fig. 11.19). We use spray cans a great deal—for hair sprays, deodorants, paints, etc. Let's look at what causes the spray. In addition to the active ingredients, a spray can contains a liquid propellant. A major requirement for propellant is that it have an equilibrium or saturated vapor pressure greater than atmospheric pressure at room temperature. As a result, the space above the can's contents is at high pressure from the propellant vapor.

When the nozzle button of the can is depressed, the propellant pressure forces the liquid up through a tube and out the nozzle. When the nozzle button is released, its opening is resealed (usually by spring action). The propellant vapor then builds up again to its equilibrium pressure, and the can is ready for another spray.

Of course, there are other propellant requirements. It must be inert, or not react chemically with the can's contents. Also, it must be environmentally safe. At one time chlorofluorocarbon compounds (CFCs) are used as propellants, but these are being discontinued because of the effect of CFCs on the Earth's atmospheric ozone layer (discussed in the Insight in Chapter 19).

Relative Humidity

The vapor pressure of water in the air is commonly expressed in terms of **relative humidity**. In a mixture of gases such as air, the pressure of the individual gases (called the partial pressures) make up the total pressure. The relative humidity is defined as the ratio of the partial pressure of the water vapor to the saturated water vapor pressure at a given temperature. Relative humidity

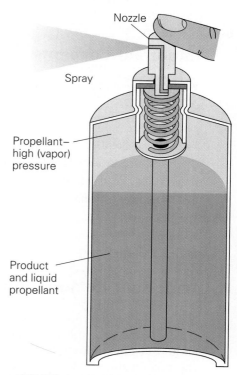

● **FIGURE 11.19 Spray cans and saturated vapor pressure** The propellant in a spray can must have an equilibrium or saturated vapor pressure greater than atmospheric pressure at room temperature. (See text for description of operation.)

is usually expressed as a percentage. Thus, a relative humidity of 60% means that a volume of air is 60% full of water, so to speak, at a particular temperature.

At the dew point, the relative humidity = 100%

If the air temperature falls, the saturated vapor pressure, or maximum moisture capacity, decreases, and the relative humidity increases. When the air is completely saturated, the relative humidity is 100%. This occurs at a temperature called the **dew point**. If the temperature drops further, the air is said to be supersaturated. When air is at the dew point or below, condensation occurs, and the result is precipitation such as dew, rain, or snow. Since particles, such as dust or ice crystals, are needed for the formation of raindrops, air may be supersaturated and no precipitation will occur. The principle involved in rain-making is that the clouds are seeded with crystals of silver iodide, which have a structure similar to that of ice, or with dry ice pellets that sublime and cause the freezing of ice crystals. Ideally, the moisture in supersaturated air in the clouds will condense on these particles and produce rain.

Demonstration/activity: Take a fan to class and let the students explain its purpose. Have students dip their hands in water and then hold them in the air flow from the fan.

The optimum range of relative humidity for human comfort and health is 40–50%. High humidity makes us feel hot and uncomfortable because perspiration does not evaporate, so cooling is lessened. Low humidity can cause dry skin, eyes, and nasal membranes. Many people use dehumidifiers in the summer to remove moisture from the air and humidifiers in the winter.

Important Concepts

You should be able to define and explain these chapter concepts clearly.

heat	freezing point	phase diagrams	emissivity
calorie (cal)	liquid phase	conduction	black body
kilocalorie (kcal)	gaseous (vapor) phase	thermal conductors	intensity
British thermal unit (Btu)	boiling point	thermal insulators	saturated vapor pressure
mechanical equivalent of heat	condensation point	thermal conductivity	relative humidity
heat of combustion	sublimation	convection	dew point
specific heat	latent heat	radiation	
solid phase	latent heat of fusion	infrared radiation	
melting point	latent heat of vaporization	Stefan's law	

Important Relationships for Review

These relationships are mathematical statements of the concepts and principles presented in the chapter. You should be able to identify the symbols and to explain the relationships before proceeding to the Exercises. In-text equation reference numbers are given for convenience.

Heat of Combustion:

$$Q = mH \tag{11.1}$$

Mechanical Equivalent of Heat:

1 kcal = 4.19 × 10³ J
1 cal = 4.19 J

Specific Heat (capacity):

$$Q = mc\Delta T \tag{11.2}$$

Latent Heat:

$$Q = mL \tag{11.3}$$

(for water)

$L_f = 3.3 \times 10^5 \text{ J/kg}$ (or 80 kcal/kg)

$L_v = 22.6 \times 10^5 \text{ J/kg}$ (or 540 kcal/kg)

Thermal Conduction:

$$\frac{\Delta Q}{\Delta t} = \frac{kA\Delta T}{d} \tag{11.4}$$

Stefan's Law:

$$P = \sigma A e T^4 \tag{11.6}$$

Radiant Power Loss or Gain:

$$P_{net} = \sigma A e (T_s^4 - T^4) \tag{11.7}$$

Intensity:

$$I = \frac{P_{net}}{A} = \sigma e(T_s^4 - T^4) \tag{11.8}$$

Exercises*

11.1 ■ Units of Heat

1 Which of the following is the largest unit of heat energy: (a) calorie, (b) kilocalorie, (c) Btu, (d) joule?   (b)

2 The SI unit of heat energy is the (a) calorie, (b) kilocalorie, (c) Btu, (d) joule.   (d)

3 What is the source of the heat of combustion?   see ISM

•**4** Gasohol is gasoline that contains about 10% ethyl alcohol by volume. Does gasoline or gasohol have a greater intrinsic heat value?   see ISM

5 ■ An apple is listed on a food chart as having 95 Cal. What is its energy value in joules? 4.0×10^5 J

6 ■ A person goes on an 1800-Cal diet to lose weight. What is the equivalent daily allowance in joules? 7.5×10^6 J

7 ■ A window air conditioner has a rating of 15,000 Btu (per hour). What would this rating be in watts? 4.4×10^3 W

8 ■■ Suppose that an 80-kg person wants to "work off" a 600-Cal hot fudge sundae (including nuts, whipped cream, and a cherry) by climbing up a rope. How high would the person have to climb to do an equivalent amount of work? 3.2×10^3 m

9 ■■ Suppose food energy values were listed in Btu. What would be the listed values of (a) a less-than-one-Calorie soft drink and (b) a 210-Calorie candy bar?   (a) < 4.0 Btu (b) 8.3×10^2 Btu

10 ■■ What is the mechanical equivalent of heat in Btu? 9.48×10^{-4} Btu/J

11 ■■ On the average, how much wood would have to be burned to give the amount of heat that is obtained from the burning of 100 m³ of natural gas?   210 kg

•**12** ■■ How much energy is released by the complete combustion of 1 gal of gasoline? [*Hint:* See Table 9.2.] 1.2×10^8 J

13 ■■■ Gasohol is about 90% gasoline and 10% ethyl alcohol by volume. How does the energy content of a gallon of gasohol compare to that of a gallon of gasoline? That is, what is the energy difference? [*Hint:* See Table 11.1.]   4% difference

14 ■■ A person has a meal of 150 g of lean meat, 125 g of boiled potatoes with a 10-g pat of butter, and two 25-g slices of plain white bread. What was the person's caloric intake? 4.7×10^2 Cal

11.2 ■ Specific Heat

15 The amount of heat necessary to change the temperature of 1 kg of a substance by one degree Celsius is the (a) specific heat, (b) latent heat, (c) heat of combustion, (d) mechanical equivalent of heat.   (a)

•**16** Twelve kilocalories would raise the temperature of three liters of water (a) 2 C°, (b) 3 C°, (c) 4 C°, (d) 5 C°.   (c)

17 When you swim in the ocean or a lake at night, the water may feel pleasantly warm even when the air is quite cool. Why?   water has a high specific heat and does not cool as quickly

18 Is it possible to have a negative specific heat? Explain.   no

19 ■ When 50 cal of heat is removed from 10 g of a substance, its temperature is observed to decrease from 40°C to 15°C. What is the specific heat of the substance?   0.20 cal/g-C°

20 ■ If 175 J of energy is added to 10 g of water at 25°C, what is the final temperature of the water?   29°C

21 ■■ A 100-g piece of aluminum at 90°C is immersed in 100 mL of water at 20°C. Assuming no heat is lost to the surroundings or container, what is the temperature of the metal and water when they reach thermal equilibrium?   32.6°C

•**22** ■■ A student mixes 1 L of water at 40°C with 1 L of ethyl alcohol at 20°C. Assuming that there is no heat loss to the container or the surroundings, what is the final temperature of the mixture? [*Hint:* See Table 11.1.]   34°C

23 ■■ A 0.25-kg coffee cup at room temperature is filled with 250 cc of boiling coffee. The cup and the coffee come to thermal equilibrium at 80°C. If no heat is lost, what is the specific heat of the cup? [*Hint:* Consider the coffee to be essentially boiling water.]   0.33 cal/g-C°

•**24** ■■ A gallon of ice tea at 25°C is placed in a refrigerator to cool. How much energy must be removed from the tea to cool it to 10°C? (Take the specific heat of the sweetened tea to be 1.05 kcal/kg-C°.)   60 kcal

25 ■■ A block of aluminum 10 cm on a side is cooled from 100°C to room temperature. If the energy removed from the aluminum block were added to a copper block of similar dimensions at room temperature, what would the final temperature of the copper block be?   77°C

*Neglect heat losses in the exercises unless instructed otherwise.

26 ■■ A camper heats 20 L of water to boiling to take a bath. How much water from a 15°C stream must he add to have the bath water at 45°C? (Neglect any losses.) 37 L

27 ■■ Equal amounts of heat are added to different quantities of copper and lead. The temperature of the copper increases by 10 C°, and the temperature of the lead by 5 C°. Which piece of metal has the greater mass and how much greater? $m_{Pb} = 6\ m_{Cu}$

● **28** ■■ In a calorimetry experiment, 0.35-kg of aluminum shot at 100°C is carefully poured into 50 mL of water that has been chilled to 10°C. Neglecting any losses to the container (or otherwise), what is the final equilibrium temperature of the mixture? 65°C

29 ■■ In a calorimetry experiment, 0.50 kg of a metal at 100°C is added to 0.50 L of water at room temperature in an aluminum calorimeter cup. The cup has a mass of 250 g. If the final temperature of the mixture is 25°C, what is the specific heat of the metal? What would be the effect if some water splashed out of the cup when the metal was added? 0.074 kcal/kg-C°, final temperature higher

30 ■■ An electric immersion heater has a power rating of 1200 W. If the heater is placed in 0.75 of water at room temperature, how long will it take to bring the water to a boil? (Assume that there is no heat loss.)
2.1×10^2 s = 35 min

31 ■■ At what average rate would heat have to be removed from 1.5 L of (a) water and (b) mercury to reduce the liquid's temperature from room temperature to its freezing point in 3.0 min? (a) 7.2×10^2 W (b) 9.4×10^2 W

● **32** ■■■ A 0.030-kg lead bullet hits a steel plate and melts and splatters on impact. (This action has been photographed.) Assuming that the bullet receives 80% of the collision energy, at what minimum speed must it be traveling to melt on impact? (Assume that the bullet and the steel plate are initially at room temperature.) 4.0×10^2 m/s

33 ■■■ An aluminum rod has a diameter of 4.0 cm and a length of 15 cm. The rod, initially at room temperature, absorbs 24 kcal of heat. What is the change in the rod's length? 7.6×10^{-2} cm

11.3 ■ Phase Changes and Latent Heat

34 The units of latent heat are (a) $1/C°$, (b) kcal/kg-C°, (c) kcal/C°, (d) kcal/kg. (d)

35 Latent heat is (a) part of the specific heat, (b) the same as the mechanical equivalent of heat, (c) involved in a phase change, (d) the same as the heat of combustion. (c)

36 Why do different substances have different freezing and boiling points? Would you expect the latent heats to be different for different substances? Explain. see ISM

● **37** Does it make any sense to talk about the value of the specific heat for a phase change? Explain.
no, no change in temperature

38 Ice cubes left in a refrigerator freezing compartment for a long time become smaller. Explain why. [*Hint:* newer refrigerators have circulating fans.] see ISM

39 Automobile cooling systems operate under pressure. What is the purpose of this? What would happen if you removed the radiator cap shortly after turning off a hot engine? (*Don't do this*—it is quite dangerous!) see ISM

40 ■ How much heat is required to boil away 1.0 L of water that is at 100°C? 5.4×10^2 kcal

41 ■ How much more heat is needed to convert 1.0 kg of ice at 0°C to steam at 100°C than to raise the temperature of 1.0 kg of water from 0°C to 100°C? 6.2×10^2 kcal greater

● **42** ■■ How much heat must be removed from 2.0 L of water at room temperature to form ice whose temperature is −10°C? 2.1×10^2 kcal

43 ■■ How much heat must be added to 0.75 kg of lead at room temperature to cause it to melt? 12 kcal

44 ■■ (a) A 0.25-kg piece of ice at −20°C is converted to steam at 115°C. How much heat in kcal must be supplied to do this? (b) To convert the steam back to ice at −5°C, how much heat would have to be removed? (a) 184 kcal
(b) −182 kcal

45 ■■ Heat is added to 40 mL of water and to 40 mL of ethyl alcohol, both initially at room temperature, at a rate of 0.27 kcal/s. How long will it take for each liquid to change completely to a gas? water, 92 s; alcohol, 28 s

● **46** ■■ A 0.20-kg piece of ice at −10°C is placed in 0.10 L of water at 50°C. How much liquid is there when the system reaches thermal equilibrium? 0.15 L

47 ■■ Steam at 100°C is added to 250 cm³ of water at room temperature in a calorimeter cup. How much steam will have been added when the water in the cup is at 60°C? (Ignore the effect of the cup.) 1.7×10^{-2} kg

48 ■■ Ice is added to 0.75 L of tea at room temperature to make iced tea. If enough ice is added to bring ice and liquid to thermal equilibrium, how much liquid is in the pitcher when this occurs? 0.94 kg

49 A mercury diffusion vacuum pump contains 0.015 kg of mercury vapor at a temperature of 630 K. Suppose that you wanted to condense the vapor. How much heat would have to be removed? 0.98 kcal

● **50** ■■ An ice cube tray holds one-half liter of water at room temperature. If it is placed in a refrigerator freezer compartment at −5°C, how much heat has been removed from the water when it is ice at the freezer temperature?
51 kcal

51 ■■ If the ice cube tray in the preceding exercise is made of aluminum and has a mass of 0.250 kg, what would be the total heat removed from the tray of water in making ice cubes? 52.4 kcal

52 ■■ Suppose 2.5 cm of rain is received over a region that has equivalent rectangular dimensions of 2.0 km × 3.0 km. How much energy would be released when water vapor condensed to form this amount of rain? Is it a large amount of energy? 8.1×10^{10} kcal

53 ■■■ In an experiment, a 150-g piece of ceramic superconductor material at room temperature is placed in liquid nitrogen (N_2) to cool. Assuming that this is done in a perfectly insulated flask, how many liters of liquid nitrogen will be boiled away in doing this? (Take the specific heat of the ceramic material to be the same as that of glass, and the density of liquid nitrogen to be 0.80×10^3 kg/m³.) 0.17 L

11.4 ■ Heat Transfer

54 Solid thermal conductors commonly contain (a) free electrons, (b) air cavities, (c) metallic additives, (d) black bodies. (a)

55 The warming of the atmosphere involves (a) conduction, (b) convection, (c) radiation, (d) all of these. (d)

56 Underground water pipes sometimes freeze during extended cold spells. Why don't they freeze in a day or two? the ground insulates the water pipes, reducing the heat flow

● **57** Why is the warning shown on the highway road sign in ● Fig. 11.20 so often necessary? see ISM

● **FIGURE 11.20 A cold warning** See Exercise 57.

58 What is the purpose of the fins on the radiator shown in ● Fig. 11.21? to increase the surface area for better conduction

● **FIGURE 11.21 Radiator with fins** See Example 58.

59 Newton's law of cooling (Sir Isaac was a busy man) states that, in general, the heat loss from an object is proportional to the temperature difference between it and the surroundings: $\Delta Q / \Delta t = K \Delta T$, where K is a constant that includes losses by conduction, convection, and radiation. Apply this law to the following question: Would a cup of hot coffee stay hotter longer if you put cream in it right away or waited until you were ready to drink it, at a later time? see ISM

60 A Thermos bottle, shown in ● Fig. 11.22, keeps cold beverages cold and hot ones hot. It consists of a double-walled, partially evacuated container with silvered walls (mirrored interior). The bottle is constructed to counteract all three mechanisms of heat transfer. Explain how. see ISM

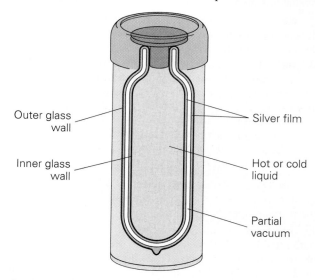

Outer glass wall

Inner glass wall

Silver film

Hot or cold liquid

Partial vacuum

● **FIGURE 11.22 Thermal insulation** The Thermos bottle is an application involving all three methods of heat transfer. See Exercise 60.

61 ■ How many times faster would heat be conducted from your bare feet by a tile floor than by an oak floor? (Assume that both floors have the same temperature and thickness.) $t_o = (4.5) \, t_t$

62 ■ If an object has an emissivity of 0.65 and an area of 0.20 m², how much energy does it radiate at room temperature? 54 W

63 ■■ Of the metals listed in Table 11.4, for which two do the thermal conductivities form the greatest ratio? What does this tell you? $k_{silver}/k_{iron} = 9.1$

64 ■■ A copper teakettle with a circular bottom that is 30 cm in diameter and has a uniform thickness of 2.5 mm sits on a burner whose temperature is 180°C. (a) If the teakettle is full of boiling water, what is the rate of heat conduction through its bottom? (b) Assuming that the heat from the burner is the only heat input, how much water is boiled away in 5.0 min? Is your answer reasonable? If not, explain why. (a) 2.1×10^2 kcal/s (b) 1.2×10^2 kg

65 ■■ An aluminum bar and a copper bar of identical cross-sectional area have the same temperature difference between their ends and conduct heat at the same rate. Which bar is longer and how much longer? $d_{Cu} = (1.6)\, d_{Al}$

● **66** ■■ The thermal insulation used in building is commonly rated in terms of its R-value, defined as L/k, where L is the thickness of the insulation in inches and k is the thermal conductivity. For example, 3.0 in. of foam plastic would have an R-value of 3.0 in./0.30 = 10, where, in British units, k = 0.30 Btu-in./ft²-h-°F. This value is expressed as R-10. (a) What does the R-value tell you, that is, how does thermal insulation vary with R-value? (b) What thicknesses of (1) fiberboard and (2) brick would give an R-value of R-10? (a) greater the R-value, the greater the insulation value. (b) (1) 4.2 in. (2) 51 in.

67 ■■ Pine wood 14 in. thick has an R-value of 19. What thickness of (a) glass wool and (b) foam plastic would have the same R-value? (See Exercise 66.) (a) 5.1 in. (b) 5.7 in.

68 ■■ A large picture window measures 2.0 m by 3.0 m. At what rate will heat be conducted through the window when the room temperature is 20°C and the outside temperature is 0°C (a) if the window has a single pane of glass 4.0 mm thick and (b) if the window has a double pane of glass (a thermopane), each pane 2.0 mm thick with an intervening air space of 1.0 mm? Do thermopanes save energy? (Assume that there is a constant temperature difference.)
(a) 2.5×10^4 J/s (b) 2.6×10^3 J/s

69 ■■ The emissivity of an object is 0.70. How many times greater would the radiation be from a similar black body at the same temperature? 1.4

● **70** ■■ A house wall is composed of a solid concrete block with outside brick veneer and faced on the inside with fiberboard as illustrated in Fig. 11.23. If the outside temperature on a cold day is −5.0 C and it is room temperature on the inside, how much energy in joules is conducted through a wall with dimensions of 3.5 m × 5.0 m in one hour?
3.3×10^3 J

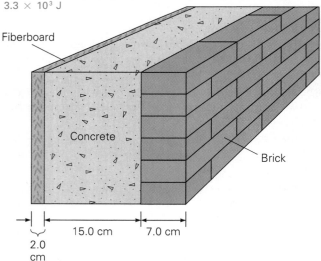

FIGURE 11.23 Thermal conductivity and heat loss See Exercise 70.

71 ■■ Suppose you wished to cut the heat loss through the wall in the preceding problem by 50% by installing insulation. What thickness of Styrofoam would have to be placed between the fiberboard and concrete block to do this?
2.3 cm

72 ■■ If the temperature of an object at room temperature is doubled to 40°C, how are its (a) emissivity and (b) radiation rate affected? (c) What are the effects if the object's temperature (room temperature) in kelvins is doubled? (a) same (b) (1.3) P_o (c) (16) P_o

73 ■■ A metal sphere with a diameter of 15 cm has an emissivity of 0.58. If it is heated to a temperature of 300°C when the temperature of the surroundings is 20°C, approximately how much radiant energy does the sphere lose in 1.0 min? (Why is this value approximate without further heating?) 1.4×10^4 J

● **74** ■■ The average intensity of solar radiation received at the top of the atmosphere is called the solar constant and is about 0.33 kcal/m²-s. How much radiant energy is emitted by the Sun each second? [Hint: The Earth intersects a cross-sectional area of the radiation at the Earth's average distance from the Sun. Think of the Sun's emitted energy as being spread evenly over a sphere with a radius equal to the distance of the Earth from the Sun.] 9.2×10^{22} kcal/s

75 ■■ Solar heating takes advantage of solar collectors such as the type shown in Fig. 11.24. About 50% of the solar

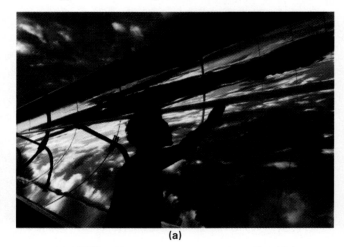

(a)

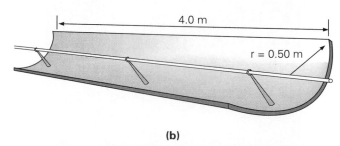

(b)

FIGURE 11.24 Solar collector and solar heating See Exercise 75.

radiation received at the top of the atmosphere reaches the Earth. (The other part is reflected, scattered, absorbed, etc.) Assuming this is the case, how much heat energy would be collected, on the average, by the cylindrical collector shown in the figure during 8.0 h of collection? 2.0×10^4 kcal

76 ■■ A steel cylinder with a radius of 5.0 cm and a length of 4.0 cm is placed in end-to-end thermal contact with a copper cylinder of the same dimensions. If the free ends of the two cylinders are maintained at constant temperatures of 95°C (steel) and 15°C (copper), how much heat will flow through the cylinders in 10 min? 4.0×10^5 J

77 ■■ For the metal cylinders in Exercise 76, what is the temperature at the interface of the cylinders?   23°C

11.5 ■ Evaporation and Relative Humidity ◇

78 The cooling mechanism of our bodies involves (a) latent heat of fusion, (b) heat of combustion, (c) latent heat of vaporization, (d) the mechanical equivalent of heat.   (c)

79 The relative humidity (a) increases with increasing temperature, (b) is the same as the saturated vapor pressure, (c) is 100% at the dew point temperature, (d) none of the preceding.   (c)

80 Spray cans carry warnings such as "Store in a cool area away from heat or open flame." Why is this?
heat causes pressure to increase; the can could explode.

● **81** Fogs are basically low-lying clouds. Why do fogs sometimes form in valleys overnight? What do people mean when they say that the Sun burns off the fog?   see ISM

82 ■■ How much heat would be removed from a body by the evaporation of 10 cm³ of perspiration (water)? (Assume that $Q/m = 600$ kcal/kg.) 2.5×10^4 J

83 ■■ The body generates heat at a rate of about 60 kcal/h. (a) How much water would have to evaporate each hour to remove all of this heat? (b) Assuming that 20% of this heat goes into body functions and 10% is lost by other heat transfer mechanisms, how much water evaporation would be required in one day?   (a) 100 g   (b) 1.7 kg

● **84** ■■ How many milliliters of perspiration (water) would have to be evaporated to remove one Calorie of heat from one's body?   1.7 mL

■ Additional Exercises

85 If 50 g of ice at 0°C is added to 300 cm³ of water at 25°C in a 100-g aluminum calorimeter cup, what is the final temperature of the water?   11°C

86 A lamp filament radiates 100 W of power when the temperature of the surroundings is 20°C and 99.5 W of power when the temperature of the surroundings is 30°C. If the temperature of the filament is the same in each case, what is that temperature on the Celsius scale?   411° C

87 What is the Calorie content of a slice of bread (15 g) spread with 4.0 g of butter?   62 Cal

88 A student doing an experiment pours 150 g of heated copper shot into a 375-g aluminum calorimeter cup containing 200 mL of water at 25°C. The mixture (and the cup) comes to thermal equilibrium at 28°C. What was the initial temperature of the shot?   89°C

89 What is the Celsius temperature of a black body that radiates with an intensity of 462 W/m²?   27°C

90 Determine the ratio of the intensities of radiation from the exposed skin of a person when the air temperature is 55°F and 95°F? $I_2/I_2 = 11$

91 A Styrofoam ice chest is filled with a block of ice at 0°C. The dimensions of the chest are 0.75 m by 0.40 m by 0.40 m, with walls 2.0 cm thick. (a) How much heat is conducted through the walls (including top and bottom) in 30 min if the outside temperature is 25°C? (Assume that the temperature difference remains constant.) (b) How much ice will melt in this time? (Assume that ice and water are in thermal equilibrium.) (c) Will water overflow from the chest?
(a) 34 kcal   (b) 0.43 kg   (c) no, since ice is less dense than water

92 After a barrel jump, a 65-kg ice skater traveling at 25 km/h glides to a stop. If 40% of the frictional heat generated by the skate blades goes into melting ice at 0°C, how much ice is momentarily melted? Where does the other 60% of the energy go?   1.7 g; energy heats skates and some is lost to the environment

● **93** A waterfall is 75 m high. If all of the gravitational potential energy of the water were converted into heat energy, by how much would the temperature of the water increase in going from the top to the bottom of the falls? [*Hint:* Consider a gram or kilogram of water going over the falls.]   0.18°C

94 The label on a container of ice cream says that there are 270 Cal per serving and 4 servings per container. How much ice cream does the container hold?   0.514 kg

95 A 5.0-g pellet of aluminum at 30°C gains 88 cal of heat. What is its final temperature?   110°C

96 Initially at room temperature, 0.50 kg of aluminum and 0.50 kg of iron are heated to 100°C. Which metal gains more heat and how much more?   Al, 4.4 kcal

97 A 0.025-kg lead bullet traveling with a speed of 250 m/s hits a 4.0-kg wooden block, penetrates, and stops inside the block. Assuming no heat loss, what is the temperature increase of the bullet and the block when they come to thermal equilibrium? (Assume that both had the same initial temperature.)   0.12°C

● **98** How much heat is required to convert 0.75 kg of ice at −10°C to water at 20°C?   79 kcal

99 How many joules of energy are required to boil away 1.0 L of liquid nitrogen? (Take the density of liquid nitrogen to be 0.80×10^3 kg/m³.)   38 kcal or 1.6×10^5 J

12

Thermodynamics

Topics for study in Chapter 12 include the laws of thermodynamics, entropy, heat engines and heat pumps.

*As the word implies, **thermodynamics** deals with the transfer or the actions (dynamics) of heat (the Greek word for "heat" is* therme*). The development of thermodynamics started about 200 years ago and grew out of efforts to develop heat engines. The steam engine was one of the first of such devices, which convert heat energy to mechanical work. Steam engines, in factories and in locomotives (such as the one shown above, still in service in Alaska), powered the industrial revolution that changed the world. Although this course of study is primarily concerned with heat and work, thermodynamics is a broad and comprehensive science that includes a great deal more than heat engine theory.*

Classical thermodynamics was not based on hypotheses about the structure of matter, but on experimental observations. However, it is possible to gain deeper insight into the principles of thermodynamics by applying modern molecular and kinetic theory and statistical mechanics, which deals with large numbers of particles. In this chapter, you will learn about the two general laws on which thermodynamics is based, as well as the concept of entropy.

12.1 ■ Thermodynamic Systems, States, and Processes

Thermodynamics is a field that makes use of many special terms, or everyday terms with special meanings, and you will find it useful to become familiar with some of them at the onset. As you know, a **system** (thermodynamic or other) is a definite quantity of matter enclosed by boundaries, either real or imaginary. A system may be open or closed. An **open system** is one that mass can be transferred into or out of. A **closed system** is one for which there can be no transfer of mass across its boundaries, or a system with constant mass.

An important thermodynamic occurrence is the interchange of energy between a system and its surroundings through a transfer of heat and/or the

Thermodynamic systems

performance of mechanical work. For example, an expanding system does work on its surroundings; if the air in a balloon is heated, the balloon expands and work is done against the surrounding atmospheric pressure (and in stretching the balloon). But if the transfer of heat is excluded, the system is said to be a **thermally isolated system** (no heat flows into or out of the system). Work may be done by or on a thermally isolated system. For example, a thermally isolated balloon can be compressed by an external force or pressure (work is done on the system). If there is no interaction at all between a system and its surroundings, the system is said to be a **completely isolated system** (or simply an isolated system).

When heat does enter or leave a system, it is absorbed from or given up to the surroundings or to theoretical heat reservoirs, which have constant temperatures. A **heat reservoir** is a system with an unlimited heat capacity. This implies that any quantity of heat may be withdrawn from or added to the reservoir without appreciably changing its temperature. By analogy, this is like taking a cup of water from or adding a cup of water to the ocean—you don't change its level appreciably.

State of a System

Just as there are kinematic equations to describe aspects of the motion of an object, there are **equations of state** to describe the conditions of thermodynamic systems. Such an equation expresses a mathematical relationship of the thermodynamic variables of a system. The ideal or perfect gas law, $pV = NkT$, is a simple equation of state. This expression establishes a relationship involving the pressure (p), volume (V), absolute temperature (T), and mass (N, the number of molecules) of a gas, which are state variables. Different states have different sets of values for these variables. A set of these variables that satisfies the perfect gas law specifies a state of a system of ideal gas completely. Of course, the system must be in thermal equilibrium and have a uniform temperature.

The state of a given mass of gas in a closed system can be specified by the variables p, V, and T. It is convenient to plot the states using the thermodynamic coordinates (p, V, T), much as we plot graphs with Cartesian coordinates (x, y, z). The three-dimensional pVT plot for carbon dioxide is shown in Fig. 12.1. Each point (p, V, T) on the surface represents an equilibrium state.

Note that the two-dimensional projections in the figure are phase diagrams, like those introduced in Chapter 11. The fusion curve between the solid and liquid phase regions on the p-T diagram for carbon dioxide slopes slightly upward and to the right. (Recall that the slope of the fusion curve for water was upward and to the left.) As you can see from the pVT surface, when a quantity of carbon dioxide goes from liquid to solid, its volume is reduced; that is, it contracts on freezing (unlike water, which expands on freezing). The pVT plot of a substance is very helpful for understanding and predicting its behavior when the state variables change.

Review Fig. 11.10.

Processes

A **process** is a change in the state, or the thermodynamic coordinates, of a system. That is, when a system undergoes a process, the set of coordinates on the pVT plot describing it changes. Processes are said to be either reversible or irreversible.

Note that a change of state (of a system) does not necessarily mean a change of phase, such as from liquid to solid. To avoid confusion, refer to solid, liquid, and gas as phases rather than states of matter.

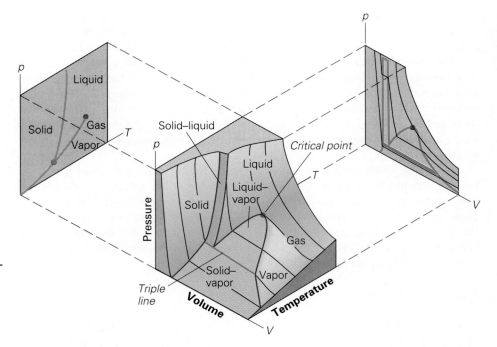

Figure 12.1

● **FIGURE 12.1 The *pVT* surface for carbon dioxide** The thermodynamic states are points (*p, V, T*) on a three-dimensional *pVT* surface. The two-dimensional projections on the *pT* and *pV* planes illustrate the behavior of the substance at constant volume and constant temperature, respectively.

Suppose that a system of gas in equilibrium is allowed to expand quickly. The state of the system will change rapidly and unpredictably, but will eventually return to equilibrium with another set of thermodynamic coordinates, in another state. On a graph, such as a *p-V* diagram, one could show the initial equilibrium point, or state, and the final state, but not what happened in between. Since the intermediate states changed so fast, there would be no data describing them. This is called an **irreversible process**, that is, one for which the intermediate states are unknown (the path is not known). "Irreversible" does not mean that the system can't be taken back to the initial state; it only means the process path can't be retraced, because it isn't known.

If, however, the gas expands very, very slowly, passing from one known equilibrium state to the neighboring one and eventually arriving at the final state, then the process path between the initial and final states would be known. This is called a **reversible process**, that is, one whose path is known. In fact, a perfectly reversible process cannot be achieved. All real thermodynamic processes are irreversible to some degree because they follow complicated paths with many intermediate states. However, the concept of an ideal reversible process is useful.

These thermodynamic terms and ideas are applied in the laws of thermodynamics.

12.2 ■ The First Law of Thermodynamics

The conservation of energy is considered to be valid for any system, and the **first law of thermodynamics** is simply a statement of the conservation of energy for thermodynamic systems. Heat, internal energy, and work are the quantities involved in a thermodynamic system. Suppose that some heat (Q) is added to a system. Where does it go? Into increasing the system's internal energy (ΔU) is certainly a possibility. Another possibility is that it could result in work (W) being done by the system. For example, when a heated gas expands, it does work on its surroundings (think of expanding air in a balloon).

Thus, it is possible for the added heat to go into internal energy or work, or both. Writing this as an equation gives a mathematical expression for the first law:

The ideal gas law is very important to an understanding of the first law of thermodynamics.

$$\boxed{Q = \Delta U + W} \qquad (12.1)$$

First law of thermodynamics—conservation of energy

For this equation, a positive value of heat ($+Q$) means that heat is *added to* the system and a positive value of work ($+W$) means that work is *done by* the system. Negative quantities mean that heat *is removed* ($-Q$) from the system and work is *done on* ($-W$) the system.

The first law can be applied to several processes for a closed system of an ideal gas in which one of the thermodynamic variables is kept constant. These processes have names beginning with iso- (from the Greek *isos*, meaning "equal").

Isobaric Process. A constant-pressure process is called an **isobaric process**. One for an ideal gas is illustrated in ● Fig. 12.2. On a p-V diagram, the path of an isobaric process is along a horizontal line called an **isobar**. When heat is added to the gas in the cylinder, the ratio V/T must remain constant ($pV = NkT$, so $V/T = Nk/p = $ a constant when p is constant). The heated gas expands; there is an increase in its volume. The temperature must also increase, which means the internal energy of the gas increases. (Recall from Chapter 10 that kinetic theory says that the internal energy of an ideal gas is directly proportional to its absolute temperature.)

Demonstration/activity: Fill a plastic garbage bag partially full of air and close the end. Then heat the bag over an electric heater to observe the change in volume with little or no change in pressure. (Don't melt the plastic.)

Review Sections 10.1 and 10.5.

Work is done by the gas as it expands, in moving the piston, and

$$W = F\Delta x$$

Explain what happens in the diagram if the pressure is not constant.

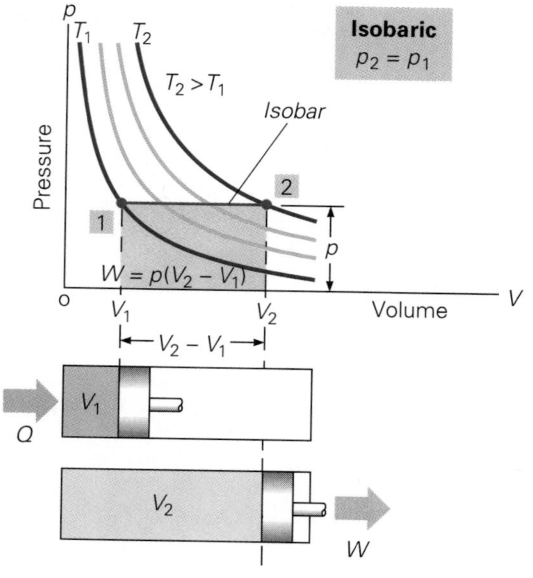

Figure 12.2

●**FIGURE 12.2 Isobaric (constant pressure) process** The heat added to the gas goes into increasing the internal energy and doing work (the expanding gas moves the piston): $Q = \Delta U + W$. The work is equal to the area under the process path (from state 1 to state 2 here) on the p-V diagram.

In terms of pressure ($p = F/A$), the force may be written $F = pA$, where A is the area of the piston. Then,

$$W = pA\Delta x$$

But $A\,\Delta x$ is simply the change in the volume of the gas, $A\Delta x = \Delta V = V_2 - V_1$. Thus,

$$\boxed{W = p\Delta V = p(V_2 - V_1)} \quad \text{(for an isobaric process)} \quad (12.2)$$

In Fig. 12.2, you can see that $p\Delta V$ is the area under the isobar on the p-V diagram. For a nonisobaric process (one in which the pressure changes), the work is also equal to the area under the line showing the process path. Thus, the work depends on the process path as well as on the initial and final states (there will be different areas under different paths).

On the other hand, since the internal energy of a quantity of an ideal gas depends only on its (absolute) temperature, a change in this internal energy is *independent* of the process path and depends only on the initial and final states, or the temperatures of these states ($\Delta U = U_2 - U_1 \propto T_2 - T_1$).

Since V_2 is greater than V_1 for an expanding gas, work is done by the system ($+W$). In terms of the first law, then,

$$Q = \Delta U + W = \Delta U + p\Delta V \quad (12.3)$$

That is, the heat added to the system goes into both increasing the internal energy ($+\Delta U$, since $\Delta U = Q - W$, which is greater than zero) and into work done by the system. If the process were reversed and the gas were being compressed by an external force doing work *on* the system, all the quantities would be negative. Heat would go out of the system ($-Q$), and the internal energy, or the temperature, of the gas would decrease ($-\Delta U$).

Isometric Process. An **isometric process** (short for isovolumetric process) is a constant-volume process, sometimes called an isochoric process. As illustrated in ● Fig. 12.3, the process path on a p-V diagram is along a vertical line, commonly called an **isomet**. No work is done ($W = p\,\Delta V = 0$, since $\Delta V = 0$), so all the added heat goes into increasing the internal energy, and therefore

Figure 12.3

●**FIGURE 12.3 Isometric (constant volume) process** All of the heat added to the gas goes into increasing the internal energy when the volume is held constant: $Q = \Delta U$. This causes an increase in temperature.

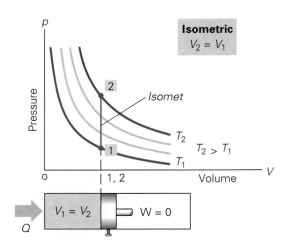

the temperature, of the gas. By the first law,

$$Q = \Delta U + W = \Delta U + 0$$

and

$$Q = \Delta U \quad \text{(for an isometric process)}$$

Isothermal Process. An **isothermal process** is a constant-temperature process (Fig. 12.4). In this case, the process path is along an **isotherm**, or a line of constant temperature. Since $p = (NkT)/V = \text{(constant)}/V$ for an isothermal process, an isotherm is a hyperbola on a p-V diagram (the general form of the equation for a hyperbola is $y = a/x$).

In going from state 1 to state 2 in Fig. 12.4, heat is added to the system, and both the pressure and volume change in order to keep the temperature constant (pressure decreases and volume increases). The work done by the expanding system ($+W$) is again equal to the area under the process path.

For an isothermal process, the internal energy of the ideal gas remains constant ($\Delta U = 0$), since the temperature is constant. By the first law,

$$Q = \Delta U + W = 0 + W$$

and

$$Q = W \quad \text{(for an isothermal process)}$$

Thus, for an ideal gas an isothermal process is one in which heat energy is converted to mechanical work (or vice versa for the reverse path).

Adiabatic Process. There's one more type of process in which a thermodynamic condition remains constant (Fig. 12.5). An **adiabatic process** is one in which no heat is transferred into or out of the system; that is, $Q = 0$ (the Greek

> *pV is a constant for an isothermal process. Discuss the problems one might have in attempting to demonstrate this process in class.*

Adiabatic process: no heat transfer

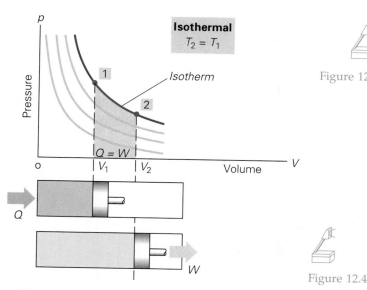

FIGURE 12.4 Isothermal (constant temperature) process All of the heat added to the gas goes into doing work (the expanding gas moves the piston): $Q = W$. The work is equal to the area under the process path on the p-V diagram.

Figure 12.5

Figure 12.4

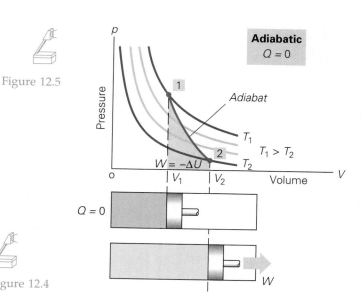

FIGURE 12.5 Adiabatic (no heat transfer) process In an adiabatic process, no heat is added or removed from the system: $Q = 0$. Work is done at the expense of the internal energy: $W = -\Delta U$. The pressure, volume, and temperature all change in the process.

word *adiabatos* means "impassable.") The condition $Q = 0$ is satisfied for a thermally isolated system. Of course, this is an ideal or theoretical situation, since there is always some heat transfer in actual system processes. Under actual conditions, we can only approximate adiabatic processes. For example, processes that are nearly adiabatic can take place in systems that are not thermally isolated if they occur rapidly enough that there isn't time for much energy to be transferred into or out of the system.

As the system follows the process path, a curve called an **adiabat**, all three thermodynamic coordinates change. For example, suppose a quantity of ideal gas were compressed in a thermally isolated piston-cylinder. If the piston were suddenly released, the gas would expand (i.e., there would be changes in p and V). Work would be done at the expense of the internal energy of the gas, so its temperature would change. By the first law,

$$Q = 0 = \Delta U + W$$

and

$$W = -\Delta U \quad \text{(for an adiabatic expansion)}$$

Thus, in an adiabatic expansion, work (the area under the process path) is done by the system with a corresponding decrease in its internal energy. The decrease in internal energy is evidenced by a decrease in temperature in going from state 1 to state 2.

EXAMPLE 12.1 ■ Joule's Free Expansion Experiment

In 1843, Joule performed a simple experiment that produced some important information about the internal energy of gases. His experimental apparatus is shown in ●Fig. 12.6. Two containers connected by a valve are submerged in a water bath. Container A is filled with a gas at high pressure, and container B is empty (evacuated). If the valve is opened quickly, the gas in A undergoes *free expansion* into the evacuated container.

With free expansion into a vacuum or void, no work is done by the gas on the surroundings, so $W = 0$. During the process, the gas remaining in A does do work on the gas that has already flowed into B. This is work done on one part of the system by another, however, not by the system as a whole on the surroundings.

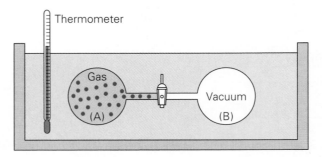

●**FIGURE 12.6 Joule's free expansion experiment**   Joule used this type of apparatus to show that the internal energy of a gas is not an independent function of either pressure or volume, but of temperature. See text for description.

The temperature of the water bath is the same after the gas has come to equilibrium as it was before the valve was opened. Thus, there is no heat transfer between the system and its surroundings (the water bath), so $Q = 0$. By the first law,

$$Q = \Delta U + W$$

or

$$\Delta U = Q - W$$

Since $Q = W = 0$,

$$\Delta U = 0$$

The internal energy of the gas remains constant, which indicates that this energy is not an independent function of either the pressure or volume, since both these quantities changed during the process. The fact that the temperature of the gas is the only other variable that, like the internal energy, remained unchanged suggests that these quantities are proportional to each other.

EXAMPLE 12.2 ■ Isobaric Expansion

A quantity of an ideal gas has a volume of 22.4 L at STP (standard temperature and pressure). While absorbing 315 cal of heat from the surroundings, the gas expands isobarically to 32.4 L. (a) What is the change in the internal energy of the gas? (b) What is the equilibrium temperature of the gas after the expansion?

Solution. Listing the data for inspection, along with appropriate conversions, we have

Given: $p_1 = p_2 = 1$ atm
　　　$= 1.01 \times 10^5 \, N/m^2$ (Pa)
　　$V_1 = 22.4 \, L = 22.4 \times 10^{-3} \, m^3$
　　$V_2 = 32.4 \, L = 32.4 \times 10^{-3} \, m^3$
　　$T_1 = 0°C = 273$ K
　　$Q = 315 \, cal = 1.32 \times 10^3 \, J$

Find: (a) ΔU (change in internal energy)
　　(b) T_2 (temperature after expansion)

(a) As you might expect, the change in internal energy is given by the first law. $\Delta U = Q - W$. We know Q, but we must first compute W, the work done by the expanding gas, to find ΔU. Using Eq. 12.2,

$$W = p\Delta V = p(V_2 - V_1)$$
$$= (1.01 \times 10^5 \, N/m^2)[(32.4 \times 10^{-3} \, m^3) - (22.4 \times 10^{-3} \, m^3)]$$
$$= 1.01 \times 10^3 \, J$$

Then, by the first law,

$$\Delta U = Q - W = (1.32 \times 10^3 \, J) - (1.01 \times 10^3 \, J) = 310 \, J$$

(b) We know the pressure, temperature, and volume for the initial state, as well as the pressure and volume for the final state. We can therefore use the ideal gas law (Eq. 10.5) to find the final temperature, T_2.

Since $p_1V_1/T_1 = p_2V_2/T_2$ and $p_1 = p_2$,

$$T_2 = \left(\frac{V_2}{V_1}\right)T_1 = \left(\frac{32.4 \times 10^{-3}\,\text{m}^3}{22.4 \times 10^{-3}\,\text{m}^3}\right)(273\,\text{K}) = 395\,\text{K}$$

or

$$T_2 = 395\,\text{K} - 273 = 122°\text{C}$$

12.3 ■ The Second Law of Thermodynamics and Entropy

Suppose that a piece of hot metal is placed in an insulated container of cool water. Heat will be transferred from the metal to the water, and the two will come to thermal equilibrium at some intermediate temperature. For a thermally isolated system, the total energy remains constant, in accordance with the first law of thermodynamics. Could there be a reverse process in which heat would be transferred from the cooler water to the hot metal? Everyone realizes that this would not happen naturally. But if it did, the total energy of the system would remain constant, and the energy transfer would not violate the first law.

Clearly, there must be another principle, not expressed in the first law of thermodynamics, that specifies the *direction* in which a process can take place. This principle is embodied in the **second law of thermodynamics**, which says that certain processes do not take place, or have never been observed to take place, even though they are consistent with the first law.

There are many equivalent statements of the second law, which are worded differently according to their application. One of these, which is applicable to the above situation, is

Heat will not flow spontaneously from a colder body to a warmer body.

Another common statement of the second law is this:

Heat energy cannot be transformed completely into mechanical work (or vice versa).

This statement applies to heat engines, which are discussed in more detail in the next section.

In general, the second law applies to all forms of energy. It is considered to be true because no one has ever found an exception to it (regardless of the form of its statement). If it were not valid, a perpetual motion machine could be built. Such a machine could transform heat from a reservoir completely into work and motion (mechanical energy), with no energy loss. The mechanical energy could then be transformed back into heat and be used to reheat the reservoir (again with no loss). Since the processes could be repeated indefinitely, the machine would run perpetually. All of the energy is accounted for, so this situation does not violate the first law. However, it is obvious that real machines are always less than 100% efficient, that the work output is always less than the energy input. Another statement of the second law is therefore:

It is impossible to construct an operational perpetual motion machine.

Many attempts have been made to construct perpetual motion machines, but there have been no known successes.

It would be convenient to have some way of expressing the direction of a process in terms of the thermodynamic properties of a system. One such property is temperature. In analyzing a conductive heat transfer process, you need to know the temperatures of the system and its surroundings. Knowing the temperature difference allows you to state the direction in which the heat transfer will spontaneously take place, into or out of the system (indicated by $+Q$ or $-Q$, respectively).

A more general property that indicates the direction of a process was first described by Rudolf Clausius (1822–1888), a German physicist. This property is called **entropy**, a name coined by Clausius. Physically, entropy is a particularly rich, multifaceted concept. Its interpretations are diverse and intriguing. One might draw an analogy to the fable of the Blind Men and the Elephant, in which three blind men describe what each thinks an elephant to be after touching different parts. Thus, a discussion of entropy is likely to resemble a debate:

"Entropy is a measure of a system's ability to do useful work. As a system loses the ability to do work, its entropy increases."

"No, I say that entropy is what determines the direction of time. It's 'time's arrow' that points out the forward flow of events, thereby distinguishing the past from the future."

"You're both wrong. Entropy is really a measure of disorder. A system naturally moves toward a state of greater disorder or disarray. The more order, the less entropy you have."

As you see, entropy has several physical interpretations, and these are discussed more fully in the Insight feature. Here, however, we will be concerned with the mathematical definition of entropy in terms of thermodynamic properties.

Entropy is related to temperature and heat. The change in entropy (ΔS) when an amount of heat (Q) is added to (or removed from) a system by a reversible process at a constant temperature is given by

$$\Delta S = \frac{Q}{T} \quad \text{(change in entropy)} \quad (12.4)$$

where T is the Kelvin temperature. [Note what would happen if you used the Celsius temperature in this equation for a phase change from ice to water at $0°C$.] Thus, the units for entropy are joules per kelvin (J/K).

If the temperature changes during the process, the change in entropy can be calculated using advanced mathematics. This discussion will be limited to isothermal processes or ones with relatively small temperature changes. For the latter, reasonable approximations of entropy changes may be obtained using average temperatures.

Perpetual motion machines are distinguished as being of the first kind or the second kind. A perpetual motion machine of the first kind would violate the first law of thermodynamics, the law of the conservation of energy. That is, energy would be created, or the machine would have an efficiency *greater* than 100%. A perpetual motion machine of the second kind would not create energy but would violate the second law, which specifies the direction of spontaneous heat flow. In this case, 100% efficiency would be enough.

Entropy can be explained as a measure of disorder (see the Insight feature).

Students sometimes leave the temperature in Celsius. Warn students not to make this mistake.

In developing the concept of entropy, Clausius realized that when heat flows from one body to another, there is more involved than energy transfer. Energy could be transferred from either body to the other without violating the first law of thermodynamics, but heat will flow spontaneously only from a hotter body to a colder one. The temperature difference between the bodies that is necessary for heat flow corresponds to different "levels" of energy. When the bodies reach thermal equilibrium their temperatures (and thus their energy levels) are the same, and the heat flow ceases.

Even though there is a transfer of heat, the total energy of such an isolated two-body system is the same before and after the bodies come to equilibrium. But something is lost or reduced through the heat transfer— *the ability to do work.* When a temperature difference exists, there can be useful work output, as in a heat engine (Section 12.4). With no temperature difference, there is no thermal ability to do work. We can make an analogy here with different levels of gravitational potential energy represented by different heights. As the water in a stream flows down a mountainside, its descent can be made to do useful work (turning a mill wheel, for example). Once the water reaches sea level, however, it will fall no further and so cannot be made to do more work.

This loss of the ability to do useful work is one aspect of entropy. *As a system loses its ability to do useful work, its entropy increases.*

A closely related aspect of entropy is that it is a mea-sure of disorder: *The more disordered the system, the greater is its entropy.* This is consistent with the previous discussion. Useful work can generally be extracted from highly ordered systems. As the degree of order decreases, so does the ability of the system to do work. For example, consider Joule's free expansion experiment (Example 12.1 and Fig. 12.6). This system is more ordered when all of the gas is in one container. When the expansion occurs, the molecules are distributed randomly between two chambers, so the disorder (and thus the entropy) is increased. If you put a paddle wheel between the chambers, you could force the gas to perform work as it expanded into the second chamber. Once the gas has filled both chambers, however, and the system is in a state of maximum disorder, no more work can be extracted from it in this manner.

According to the second law of thermodynamics, the entropy of the universe increases in every natural process. This means that systems naturally move toward states of greater disorder or disarray. You know from experience that highly ordered systems—the neatly arranged items on the shelves of a stockroom, or the neatly piled papers on your desk—do not generally stay that way for long. As time passes, they tend to become disordered. (Eventually, they may become totally random, so that you can't predict where *anything* will be found.) Similarly, when ice cubes melt in a tray, the orderly crystalline structure of the ice is replaced by the more disorderly arrangement of molecules in the liquid.

EXAMPLE 12.3 ■ Change in Entropy for an Isothermal Process

What is the change in entropy when 0.25 kg of ethyl alcohol ($L_v = 1.0 \times 10^5$ J/kg) vaporizes at its boiling point of 78°C?

Solution. From the problem, we have

Given: $m = 0.25$ kg *Find:* ΔS (change in entropy)
$T = 78°C + 273 = 351$ K
$L_v = 1.0 \times 10^5$ J/kg

Notice that the temperature was converted directly to kelvins. Then, to compute the entropy change, we need to know the amount of heat involved in the process. This is a phase change and we are given the latent heat, so

$$Q = mL_v = (0.25 \text{ kg})(1.0 \times 10^5 \text{ J/kg}) = 2.5 \times 10^4 \text{ J}$$

Then

$$\Delta S = \frac{Q}{T} = \frac{2.5 \times 10^4 \text{ J}}{351 \text{ K}} = 71 \text{ J/K}$$

Note that Q is positive because heat is added to the system. The change in entropy, then, is also positive, and the entropy of the alcohol increases.

Students should be reminded that Q can be either positive or negative, therefore entropy may be either positive or negative.

It is important to realize that in applying the second law, one must take a very broad view. Within a given system, order can be created, restored, or even increased through the expenditure of energy. For example, you could work all afternoon tidying up the shelves of the stockroom or organizing your papers. And your freezer, using electrical energy, could perform the work necessary to refreeze the tray of ice cubes. As we shall see in a later Insight, however, such work merely purchases local order at the expense of greater disorder elsewhere. The entropy of the universe *as a whole* always increases.

Another interesting and important point about the second law is that it is based on probability. Consider again Joule's expansion experiment. You know that the expansion will not spontaneously reverse, with all the gas returning to the original container, thereby increasing the order. Statistically, the probability of this occurring is very small. However, statistics does not say that it is impossible, only that the likelihood is virtually zero. The validity of the second law is based not on the impossibility of certain processes (although they tend to be viewed as such), but on the fact that they have never been observed.

Like the simple process of gas expansion, more complex processes also have a natural direction specified by the second law. Suppose that you filmed a match being lit, a car rolling down a hill, or cream being poured into a cup of coffee. If the developed film were run backward, you would see the match "unburn," the car roll uphill, and the cream separate itself out from the coffee—processes that you would immediately recognize as unnatural. Such impossible processes would involve the creation of a more highly ordered state from a less highly ordered one. They would thus correspond to negative changes in the entropy of the universe.

The increase in entropy provides direction not only for processes but also for time. This is necessarily the case because all our measurements of time involve some physical process, whether it be the fall of sand in an hour-glass, the swing of a pendulum, or the vibration of a quartz crystal. Thus, entropy is sometimes referred to as *time's arrow*. It points out the forward direction of the flow of events. Basically, entropy distinguishes the past from the future. Overall there will be more entropy in the future, and there was less entropy in the past.

As time goes on and natural processes occur, the total entropy of the universe increases, and differences in energy levels or temperatures are reduced. For example, if a system is at a high temperature, heat will naturally be transferred to its surroundings until thermal equilibrium is reached. In a universe of ongoing natural processes, there is a continual net entropy increase as heat is continually being transferred from systems at higher temperatures to those at lower temperatures. Over a long period of time, this will lead to the so-called *heat death of the universe:* the condition when the entropy of the universe has reached a maximum, and everything is at the same temperature. Not a very appealing future—but fortunately not too imminent either.

EXAMPLE 12.4 ■ An Increase and Decrease in Entropy

A piece of metal at 24°C is placed in 1.00 L of water at 18°C. The thermally isolated system comes to equilibrium at a temperature of 20°C. Find the approximate change in the entropy of the system.

Solution. Using subscripts m and w for metal and water, and i and f for initial and final, respectively, we have

Given: $T_{m_i} = 24°C$ *Find:* ΔS (change in entropy
$T_{w_i} = 18°C$ of the system)
$m_w = 1.00$ kg
$T_f = 20°C$
$c_w = 4190$ J/kg-C°
(from Table 11.2)

As in the preceding example, we need to find the amount of heat (Q) to solve for ΔS. However, in this system there are two Q's—Q_w, the heat gained by the water and Q_m, the heat lost by the metal.

With $\Delta T = T_f - T_{w_i} = 20°C - 18°C = 2\,C°$, the heat gained by the water is

$$Q_w = m_w c_w \Delta T = (1.00 \text{ kg})(4190 \text{ J/kg})(2 \text{ C°}) = 8.38 \times 10^3 \text{ J}$$

403

This is also the amount of heat *lost* by the metal, since by the conservation of energy, heat gained must equal heat lost. Therefore, $Q_m = -8.38 \times 10^3$ J, where the minus sign indicates a loss of heat.

In this case, we have small temperature changes, so we can approximate the entropy changes using average temperatures:

$$\overline{T}_w = \frac{T_f + T_{wi}}{2} = \frac{20°C + 18°C}{2} = 19°C = 292 \text{ K}$$

$$\overline{T}_m = \frac{T_{mi} + T_f}{2} = \frac{24°C + 20°C}{2} = 22°C = 295 \text{ K}$$

We can then use these average temperatures to compute the entropy changes for both the water and the metal.

$$\Delta S_w = \frac{Q_w}{\overline{T}_w} = \frac{8.38 \times 10^3 \text{ J}}{292 \text{ K}} = 28.7 \text{ J/K}$$

$$\Delta S_m = \frac{Q_m}{\overline{T}_m} = \frac{-8.38 \times 10^3 \text{ J}}{295 \text{ K}} = -28.4 \text{ J/K}$$

The change in the entropy of the whole system is the sum of these individual changes:

$$\Delta S = \Delta S_w + \Delta S_m = 28.7 \text{ J/K} - 28.4 \text{ J/K} = +0.3 \text{ J/K}$$

The entropy of the metal decreased because heat flowed out of it. The entropy of the water increased by a greater amount, so the total entropy change of the system was positive.

Note that the total entropy change for the system in Example 12.4 is positive. In general, the direction of any process is toward an increase in entropy. That is, *the entropy of an isolated system never decreases.*

Another way to state this observation about entropy is to say that *the entropy of an isolated system increases for every natural process* ($\Delta S > 0$). The water and metal in coming to an intermediate temperature in Example 12.4 are undergoing a natural process, so-called because this is what is always observed to occur.

The entropy of a closed system increases for every natural process.

If the metal in coming to thermal equilibrium with the water somehow spontaneously became hotter while the water became cooler, this would certainly be an *unnatural* process ($\Delta S < 0$). Similarly, water at room temperature in an isolated ice cube tray will not naturally turn into ice.

If a system is not isolated, there may be a decrease in the entropy *of the system*. For example, if an ice cube tray filled with water is put into a freezer compartment, the water will freeze, with a decrease in entropy. But there will be a larger increase in entropy somewhere else in the environment or the universe. Work has to be done or energy expended to bring about the transfer of heat involved in the change from water to ice. Thus, a general statement of the second law of thermodynamics in terms of entropy is:

The total entropy of the universe increases in every natural process.

Energy can neither be created nor destroyed; entropy can be created but not destroyed.

The entropy of a system is a function of its state. Each state of a system has a particular value of entropy, and a change in entropy depends only on the initial and final states for a process, or $\Delta S = S_f - S_i$. (This is analogous to the change in internal energy for an ideal gas, $\Delta U = U_f - U_i$.) An entropy

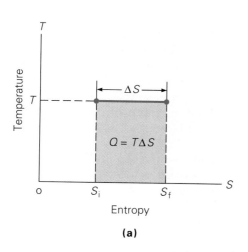

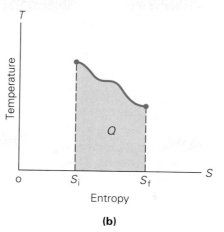

(a)

(b)

● FIGURE 12.7 *T-S* (temperature-entropy) diagrams When the temperature (T) is plotted versus entropy (S), the area under the process path is equal to the heat (Q) transferred in the process. This is similar to work being equal to the area under a process path on a p-V diagram. **(a)** This *T-S* diagram is for an isothermal process. Note that the area under the path is that of a rectangle. **(b)** This *T-S* diagram is for a general process.

change can be represented on a *T-S* diagram, as shown in ● Fig. 12.7. Just as the area under a curve on a *p-V* diagram for a gas is equal to the work ($W = p\Delta V$), the area under a *T-S* curve is equal to the change in the heat energy ($Q = T\Delta S$), which is easily seen for the isothermal process in the figure. Like work for a process with variable pressure, heat transferred in a process with variable temperature is also equal to the area under the *T-S* curve (which generally requires advanced mathematics to calculate).

There is a theoretical process for which the entropy is constant. Recall for an adiabatic process $Q = 0$. Since in this case $\Delta S = Q/T = 0$, there is no change in entropy. Hence, an adiabatic process is a constant-entropy process, or an **isentropic process**. With the inclusion of an ideal adiabatic process, we may say that the entropy can only increase or remain constant ($\Delta S \geq 0$). How would an isentropic process appear on a *T-S* diagram?

EXAMPLE 12.5 ■ Entropy Changes

Which of the following involves a *decrease* in entropy: (a) a hot fudge sundae is left uneaten too long, so that the ice cream melts and the fudge solidifies, (b) a green plant combines water and carbon dioxide molecules in photosynthesis to make a larger, more complex molecule of sugar, (c) you drop your term paper on the way up the library stairs and find the pages are no longer in sequence after picking them up, (d) all the perfume in an opened bottle evaporates and fills the room with scent, (e) a shortage of library personnel makes it increasingly difficult to find the books you want in their proper locations on the shelves? *Clearly establish the reasoning and physical principle(s) used in determining your answer before checking it below. That is,* **why** *did you select your answer?*

Reasoning and Answer. It is not difficult to see that cases (c) and (e) both involve a decrease in order, and thus an increase in entropy. The same is true, in a slightly less obvious way, of case (d). When the perfume molecules are no longer confined within the restricted space of the bottle but instead are distributed at random throughout a larger volume of air in the room, there is a loss of order—the positions of individual perfume molecules cannot be specified or predicted as accurately. Similarly in case (a), the system (fudge plus ice cream) is less highly ordered when the speeds of all the molecules

FIGURE 12.8 Work from heat At this geothermal energy facility in California, natural heat is exploited to do useful work (generating electricity) without the need to "burn" chemical or nuclear fuel. Water heated by thermal processes beneath the Earth's surface serves as the hot reservoir. What is the cold reservoir?

Heat engine: heat → work

are randomized than when they are divided into two distinct populations, one hot and one cold. (You may also recall that we showed in another way in Example 12.4 how the flow of heat from a warmer body to a colder one entails an increase in entropy.) Hence the answer is (b). In this instance a more complex, highly ordered structure was created from simpler components—a process that will not occur naturally without an input of energy (see the Insight feature on Life, Order, and the Second Law).

Follow-up Exercise. You look through a microscope and find that your pet amoeba has died. Does this represent an increase or decrease in the total entropy of the amoeba? Of the universe? (*Reasoning and answer may be found in the Answers to Exercise section at the back of the book.*)

12.4 ■ Heat Engines and Heat Pumps

A **heat engine** is any device that converts heat energy to work. Since the second law says that perpetual motion machines are impossible, some of the heat supplied to a heat engine will necessarily be lost. In studying thermodynamics, there is no need to be concerned with the usual mechanical components of an engine, such as pistons, cylinders, and gears. For theoretical purposes, a heat engine is simply a device that takes heat from a high-temperature source (a hot reservoir), converts some of it to useful work, and transfers the rest to its surroundings (a cold, or low-temperature, reservoir). For example, most of the turbines that generate the electricity we use (Chapter 19) are heat engines, using heat from various sources: the burning of chemical fuels (oil, gas, or coal); nuclear reactions (Chapter 29); or the heat already present beneath the Earth's surface (●Figure 12.8). A generalized heat engine is usually represented as shown in ●Fig. 12.9a.

Simply adding heat to a gas in a piston-cylinder arrangement will produce work in a single process. But since a continual output is usually wanted, practical heat engines usually operate in a cycle, or a series of processes, which

Figure 12.9

FIGURE 12.9 Heat engine
(a) In this diagram of energy flow for a generalized cyclic heat engine, note that the width of the arrow representing Q_{hot} is equal to the combined widths of the arrows representing W and Q_{cold}, reflecting the conservation of work-energy.
(b) A cyclic process consisting of two isobars and two isomets is shown here. The work output is the area of the rectangle formed by the process paths.

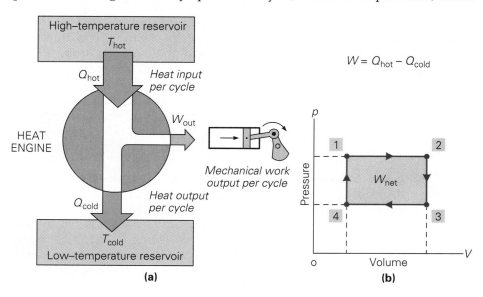

The second law of thermodynamics is one of the foundations of physics and operates universally—the total entropy of the universe increases in every natural process. As we saw in the previous Insight, this means that systems naturally move toward states of greater disorder or disarray. Certainly life is a natural process. Yet clearly, there is an increase in order during the development of the human embryo into a mature adult. Moreover, in virtually all life processes larger, more complex molecules and cellular structures are continuously being synthesized from simpler components. Does this mean that living things are exempt from the second law?

To answer this question, we must realize that cells are not closed systems. They are involved in continual exchanges of matter and energy with their surroundings. Living cells are constantly occupied in the exchange and conversion of energy from one form to another, which is governed by the first law of thermodynamics. For example, cells convert chemical potential energy into kinetic energy (movement) and electrical energy (the basis of nerve impulses).

Many cells, such as those of plants, can also convert sunlight (radiant energy) into chemical energy through photosynthesis (Fig. 1). In all these processes, biological systems maintain or even increase their organization by causing a larger decrease in the order of their surroundings. The second law is not violated by a *local* decrease in entropy, as long as the *total* entropy of the universe increases in the process.

As an example of how cells can create order while still remaining within the confines of the second law, consider how a cell uses chemical raw materials such as the sugar glucose ($C_6H_{12}O_6$). Cells use carbon atoms from glucose (and other organic compounds) to build up more complex molecules—a decrease in entropy. But they get the energy for these reactions by "burning" a great many glucose molecules as well—oxidizing them to produce many smaller, simpler molecules of water (H_2O) and carbon dioxide (CO_2). This process involves an increase in entropy. For one thing, the product molecules, being less highly structured than glucose, have greater entropy. Moreover, heat is released by the oxidation process, raising the entropy of the cell's surroundings. Thus a cell "pays" for its increase in order by increasing the disorder of the rest of the universe.

But, you may ask, where did the glucose molecule, with its complex structure and conveniently available chemical potential energy, come from in the first place? As we mentioned above, photosynthetic cells use the energy of sunlight to make glucose and related compounds. Thus all life on Earth is dependent on the Sun's energy—the plants that use it directly, the animals that eat the plants, the animals that eat those animals, and so on down the line.

We can look at the flow of energy through the living world in a more abstract and general way. As we will see in the Chapter 19 Insight on the greenhouse effect, life on Earth is dependent on receiving radiant energy in the form of sunlight and reradiating infrared energy back into space. In effect, the Sun is a high-temperature source and space is a low-temperature sink. The flow of energy from one to another allows living things to decrease entropy locally, creating this island of order that we call the biosphere. Thus biological organisms are not exempt from the second law—they merely take advantage of a loophole in it.

(a)

(b)

FIGURE 1 Solar collectors (a) Human beings have developed devices to capture some of the energy of sunlight, although the technology is not yet in widespread use. (b) Green plants capture trillions of joules of solar energy every day, and have been doing so for billions of years. The chemical energy that they store in the form of sugars and other complex molecules is used by nearly all organisms on Earth (animals as well as plants) to maintain and increase the highly organized state that we call life.

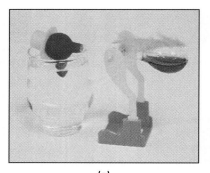

(a)

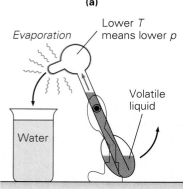

Evaporation

Lower T means lower p

Volatile liquid

Water

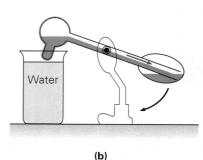

Water

(b)

FIGURE 12.10 A heat engine? See Example 12.6.

brings the engine or system back to its original condition. Such cyclic heat engines include steam engines and internal combustion engines, such as automobile engines. An idealized, rectangular thermodynamic cycle is shown in Fig. 12.9b. It consists of two isobars and two isomets. When these processes occur in the sequence indicated, the system goes through a cycle (1-2-3-4-1), returning to its original condition. Recall that for an isobaric process, the work done is equal to the area under the isobar, $W = p\Delta V$, on a p-V diagram (see Fig. 12.2). Here, the work in the 1-2 process is positive and the work in the 3-4 process is negative ($-\Delta V$, a compression). Hence, the net work is the area of the rectangle formed by the isobars and isomets.

EXAMPLE 12.6 ■ The Drinking Bird—A Heat Engine

An unusual example of a cyclic heat engine is a novelty item called a drinking bird, shown in ● Fig. 12.10. The glass bird contains a volatile liquid (usually ether). When the absorbent material on the head and beak is wet, evaporation occurs. Heat is removed from the head, which lowers its temperature and internal vapor pressure. With a temperature difference, there is a pressure difference between the head and body, and the liquid rises to the head. This raises the bird's center of gravity above the pivot point, and a torque causes a forward rotation. The bird then rewets its beak in the glass of water.

The liquid from the body runs into the head, warming it. With the tube not completely filled with liquid and partially open, the pressures in the head and body equalize so that there is no longer a pressure difference. The liquid then drains back into the body. With the lowering of its center of gravity below the pivot point, the bird swings back to the vertical position, and the cycle begins again. The net effect is that heat is transferred from a high-temperature reservoir (the body) to a low-temperature reservoir (the head), with work being done (evidenced by the motion). Of course, gravity and the atmosphere also play roles in the operation of this specialized heat engine.

The most common cyclic heat engine is the internal combustion engine used in automobiles and for a variety of other applications. Two different cycles of operation are in general use: a two-stroke cycle and a four-stroke cycle. (The number of strokes tells how many up and down motions a piston makes in a cycle; for example, a two-stroke cycle has one upstroke and one downstroke per cycle.) Engines with a two-stroke cycle are used in motorcycles, outboard motors, chain saws, and some lawn mowers.

Most automobile engines have a four-stroke cycle. The steps in this cycle are shown in ● Fig. 12.11, along with a p-V diagram of the theoretical thermodynamic processes that the cycle approximates. The theoretical cycle is called the **Otto cycle**, for the German engineer Nikolas Otto (1832–1891), who built one of the first successful gasoline engines.

During the intake stroke (1–2), an isobaric expansion, the air-fuel mixture is admitted through the open valve. This mixture is adiabatically compressed on the compression stroke (2–3). On ignition, there is a quick isometric heating (3–4), which is followed by an adiabatic expansion during the power stroke (4–5). Next is an isometric cooling of the system when the piston is at its lowest position (5–2). The final, exhaust stroke is along the isobaric leg of the Otto cycle (2–1).

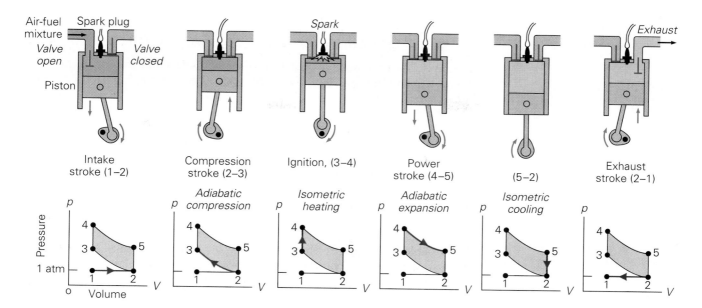

FIGURE 12.11 The four-stroke cycle for a heat engine The process steps for the four-stroke Otto cycle. The piston moves up and down two times each cycle, making four strokes each cycle.

Automobile engines typically have four, six, or eight cylinders operating in alternating cycles, so there is a continual work output. There are four strokes (two up and two down) as the cylinder goes through a cycle, which corresponds to two revolutions of the crankshaft. Since there is one power stroke per cycle, a four-cylinder engine is timed to have a cylinder fire every $\frac{2}{4}$, or $\frac{1}{2}$, revolution of the crankshaft; a six-cylinder engine has a cylinder firing every $\frac{2}{6}$, or $\frac{1}{3}$, revolution; and so on.

A two-stroke engine does not have an isobaric leg on its theoretical cycle. The air-fuel mixture is added to the cylinder at the same time as the previously combusted gas is expelled, during the isometric cooling process (5–2). This results in some loss of fuel (with the exhaust gas), which is the major disadvantage of the two-stroke engine. The advantage is that there is one power stroke for every crankshaft revolution for a two-stroke engine, compared to only one power stroke for two crankshaft revolutions for a four-stroke engine.

The diesel engine may have a two-stroke or four-stroke cycle. The latter cycle is similar to the Otto cycle, but there are other differences between diesel and gasoline engines. One is the type of fuel—the diesel engine runs on oil, not gasoline. Another is that the diesel engine lacks an ignition system or sparkplugs. During the compression stroke, oil is injected into the cylinder. Combustion occurs spontaneously when the temperature of the air-fuel mixture gets high enough (because it is being compressed).

The higher compression and pressure present in a diesel engine give it a higher efficiency than a gasoline engine, about 40% as opposed to 30%. Thermal efficiency is used to rate heat engines. The **thermal efficiency** (ε_{th}) is

$$\varepsilon_{th} = \frac{\text{work out}}{\text{heat in}} = \frac{W}{Q_{in}} \qquad (12.5) \qquad \text{Thermal efficiency}$$

The diesel engine was developed by Rudolf Diesel (1858–1913), a German engineer.

The efficiency tells you what you get out for what you put in, so to speak. For one cycle of a cyclic heat engine, the work output per cycle is

$W = Q_{hot} - Q_{cold}$ (see Fig. 12.9a), or, by the first law,

$$Q = \Delta U + W \quad \text{and} \quad Q_{hot} - Q_{cold} = 0 + W$$

where ΔU is zero, since the system returns to its original state in completing a cycle. Thus, with $Q_{in} = Q_{hot}$, the thermal efficiency (per cycle) of a cyclic heat engine is

$$\varepsilon_{th} = \frac{W}{Q_{in}} = \frac{Q_{hot} - Q_{cold}}{Q_{hot}} = 1 - \frac{Q_{cold}}{Q_{hot}} \ (\times \ 100\%) \tag{12.6}$$

In general, an engine is expected to have the same efficiency each cycle.

Like mechanical efficiency, thermal efficiency is a fraction and is commonly expressed as a percentage. From Eq. 12.6, you can see that a heat engine could have 100% efficiency if Q_{cold} were zero. This would mean that no heat energy would be lost and all the heat input would be converted to useful work, which is impossible, according to the second law of thermodynamics. In 1851, Lord Kelvin stated this observation, another form of the second law:

No heat engine operating in a cycle can convert its heat input completely to work.

The second law applied to heat engines

EXAMPLE 12.7 ■ Thermal Efficiency

With each cycle a heat engine removes 1500 J of heat energy from a high-temperature reservoir and exhausts 1000 J to a low-temperature reservoir. (a) What is the engine's thermal efficiency? (b) What is its work output?

Solution.

Given: $Q_{hot} = 1500 \text{ J}$ *Find:* (a) ε_{th} (thermal efficiency)
$Q_{cold} = 1000 \text{ J}$ (b) W (work output)

(a) We may use Eq. 12.6 directly, and

$$\varepsilon_{th} = 1 - \frac{Q_{cold}}{Q_{hot}}$$
$$= 1 - \frac{1000 \text{ J}}{1500 \text{ J}} = 0.33 \ (\times \ 100\%) = 33\%$$

(b) The work is the difference between the heat input and output:

$$W = Q_{hot} - Q_{cold} = 1500 \text{ J} - 1000 \text{ J} = 500 \text{ J}$$

There are 500 J of work output each cycle.

Heat Pumps

As used here, the term "heat pump" is general and does not refer specifically to the devices that are commonly used for heating and cooling of buildings. These will be discussed shortly.

The function performed by a heat pump is basically the reverse of that of a heat engine. That is, a **heat pump** is a device that transfers heat energy from a low-temperature reservoir to a high-temperature reservoir (Fig. 12.12). To do this, there must be work input, since the second law says that heat will not spontaneously flow from a cold body to a hot body. An example of a heat pump is an air conditioner.

Another familiar example of a heat pump is a refrigerator. With work input (from electrical energy), heat is transferred from inside the refrigerator (low-temperature reservoir) to the surroundings (high-temperature reservoir). Have you ever wondered how a refrigerator (or an air conditioner, another heat pump) operates? The knowledge you have gained about thermodynamic processes will allow you to understand the basic operation (see Fig. 12.13).

The refrigerant, or heat-transferring medium, is a substance with a relatively low boiling point. Ammonia (boiling point $-33.3°C$ at 1 atm), sulfur dioxide ($-10.1°C$), and Freon ($-29.8°C$) can be used as refrigerants. Freon is

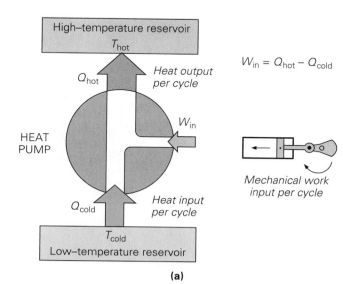

$$W_{in} = Q_{hot} - Q_{cold}$$

Mechanical work
input per cycle

Figure 12.12a

FIGURE 12.12 Heat pump (a) An energy flow diagram for a generalized cyclic heat pump. Note that the width of the arrow representing Q_{hot} is equal to the combined widths of the arrows representing W and Q_{cold}, reflecting the conservation of work-energy. (b) An air conditioner is an example of heat pump. With work input, it transfers heat (Q_{cold}) from a low-temperature reservoir (inside the house) to a high-temperature reservoir (the outside).

(a)

(b)

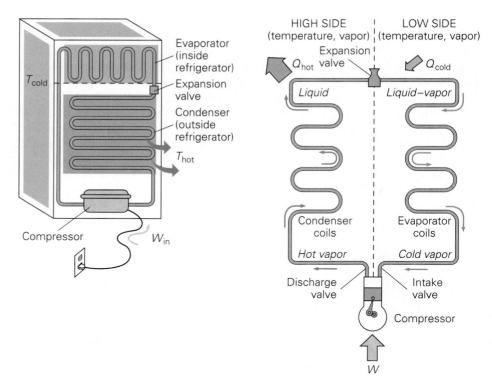

FIGURE 12.13 Refrigerator operation Heat (Q_{cold}) is carried away from the interior by the refrigerant as latent heat. This heat energy and that of the work input (W) are discharged from the condenser to the surroundings (Q_{hot}). See text for detailed description.

Freon, a chlorofluorocarbon (CFC), is involved in the depletion of the atmospheric ozone (O_3) layer that absorbs and protects us from harmful ultraviolet rays from the sun. Substitutes are now being developed and used as refrigerants (see the Insight in Chapter 19).

Ask students to see if they can find where the condenser coils are located on their refrigerator or on their car air conditioner.

used in most domestic refrigerators. The other compounds were once used, and their smells were quite evident if the system happened to leak.

It is convenient to think of the system diagrammed in Fig. 12.13 as having high and low (temperature and pressure) sides. On the low side, heat is transferred to the evaporator coils from inside the refrigerator. The heat causes the refrigerant to boil and is carried away as latent heat of vaporization. The gaseous vapor is drawn into the compressor chamber on the downstroke of the piston and is compressed on the upstroke (work input by compressor motor). The compression increases the temperature of the gas, which is discharged from the compressor as a superheated vapor (on the high side). This vapor condenses in the cooler condenser unit, where circulating air (or water in larger units) carries away the latent heat of condensation and the heat of compression. The condensed liquid collects in a receiver.

An expansion valve maintains a balance between the high and low sides, thereby controlling the rate at which the refrigerant passes back to the low side. On being admitted to the low side, the liquid immediately boils and vaporizes because of the low pressure. This is a cooling process because the heat of vaporization is supplied by the internal energy of the liquid. The cooled refrigerant is drawn into the evaporator, and the cycle begins again.

The cooling efficiency of a refrigerator or heat pump depends on the heat output extracted from the low temperature reservoir, Q_{out}, and the work input, W_{in}. This is expressed by what is called the **coefficient of performance (cop):**

$$\text{cop} = \frac{Q_{out}}{W_{in}} \tag{12.7}$$

Since a practical heat pump operates in a cycle to provide a continuous removal of heat, $\Delta U = 0$ for the cycle, as in the case of a heat engine. Then, by the conservation of energy (or first law),

$$W_{in} = Q_{hot} - Q_{cold}$$

The *cop per cycle* is then

$$\text{cop} = \frac{Q_{out}}{W_{in}} = \frac{Q_{cold}}{Q_{hot} - Q_{cold}} \qquad (12.8)$$

For normal refrigerator operation, Q_{hot} is between 1 and 2 times greater than Q_{cold}. As a result, the cop is greater than 1 (or 100%). The cop's of actual refrigerator cycles range from 1.5 to 4.0, depending on the operating conditions.

Commercial heat pumps are commonly used to cool homes and offices in the summer and to heat them in the winter. In the summer, the interior of the building is the low-temperature reservoir. The heat pump cools the air inside by transferring heat to the outside (the high-temperature reservoir), thus operating somewhat like a refrigerator. In the winter, these conditions are reversed—the interior is the high-temperature reservoir. Operating in reverse, the heat pump removes heat from the air outside and transfers it into the building. This may seem hard to believe since it is cold outside. However, keep in mind that air has internal energy regardless of its temperature.

If the outside temperature is very low, a heat pump may not be efficient enough to heat a building adequately. Then it must be supplemented by some conventional heating system, such as an electrical one. Some heat pumps use water from underground reservoirs, wells or in buried loops of pipe as a heat source and sink. These are more efficient than the ones that use the outside air because water has a larger specific heat than air, and the average temperature difference between the water and the inside air is smaller.

12.5 ▪ The Carnot Cycle and Ideal Heat Engines

Lord Kelvin's statement of the second law of thermodynamics says that any *cyclic* heat engine, regardless of its design, will always lose some heat energy. But how much heat must be lost? In other words, what is the maximum possible efficiency of a heat engine? In designing heat engines, engineers strive to make them as efficient as possible, but there must be some theoretical limit, and, according to the second law, it must be less than 100%.

Sadi Carnot (1796–1832), a French engineer, solved this problem. The first thing he considered was the thermodynamic cycle an ideal heat engine would use, that is, the most efficient cycle. A heat engine absorbs heat from a *constant* high-temperature reservoir and exhausts it to a *constant* low-temperature reservoir. These are ideally reversible isothermal processes and may be represented as two isotherms on a *p-V* diagram. But what are the processes that complete the cycle and make it the most efficient cycle? Carnot showed that these are reversible adiabatic processes, called adiabats when represented on a graph (Fig. 12.14a).

Thus, the ideal **Carnot cycle** consists of two isotherms and two adiabats and is conveniently represented on a *T-S* diagram, where it forms a rectangle (Fig. 12.14b). The area under the upper isotherm (1–2) is the heat added to the system from the high-temperature reservoir: $Q_{hot} = T_{hot}\Delta S$. Similarly, the area

Figure 12.14

FIGURE 12.14 The Carnot cycle (a) The Carnot cycle consists of two isotherms and two adiabats. Heat is absorbed during the isothermal expansion and exhausted during the isothermal compression. (b) On a *T-S* diagram, the Carnot cycle forms a rectangle, the area of which is equal to *Q*.

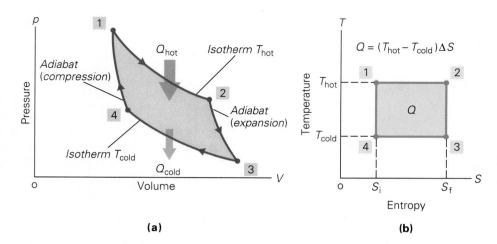

(a)   **(b)**

under the lower isotherm (3–4) is the heat exhausted: $Q_{cold} = T_{cold}\Delta S$. The difference between these is the work output, which is equal to the area of the rectangle enclosed by the processes (the shaded area on the diagram):

$$W = Q_{hot} - Q_{cold} = (T_{hot} - T_{cold})\Delta S$$

Since ΔS is the same for the areas under both isotherms, these expressions can be used to relate the temperatures and heats. That is, since

$$Q_{hot} = T_{hot}\Delta S \quad \text{and} \quad Q_{cold} = T_{cold}\Delta S$$

then

$$\frac{Q_{hot}}{T_{hot}} = \frac{Q_{cold}}{T_{cold}} \quad \text{or} \quad \frac{Q_{hot}}{Q_{cold}} = \frac{T_{hot}}{T_{cold}} \tag{12.9}$$

This equation can be used to express the efficiency of an ideal heat engine in terms of temperature. From Eq. 13.6, this ideal **Carnot efficiency** (ε_C) is

$$\varepsilon_C = 1 - \frac{Q_{cold}}{Q_{hot}} = 1 - \frac{T_{cold}}{T_{hot}}$$

or

$$\boxed{\varepsilon_C = 1 - \frac{T_{cold}}{T_{hot}} = \frac{T_{hot} - T_{cold}}{T_{hot}}} \tag{12.10}$$

where the fractional efficiency is often expressed as a percentage. Note that T_{cold} and T_{hot} are Kelvin temperatures.

The Carnot efficiency expresses the theoretical upper limit on thermodynamic efficiency for a cyclic heat engine, which can never be achieved. It corresponds to the theoretical mechanical advantage for a simple machine. A true Carnot engine cannot be built because the necessary reversible processes can only be approximated. Although reversible adiabatic processes can be approached, reversible isothermal processes are virtually impossible to approximate during the heat transfer processes in a real engine.

However, the Carnot efficiency does show that the greater the difference in the temperatures of the heat reservoirs, the greater the efficiency. For example, if T_{hot} is two times T_{cold}, or $T_{cold}/T_{hot} = \frac{1}{2} = 0.50$, the Carnot efficiency will be

$$\varepsilon_C = 1 - \frac{T_{cold}}{T_{hot}} = 1 - 0.50\,(\times 100\%) = 50\%$$

Students often leave the temperature in Celsius. Warn them that temperatures must be in kelvins.

The Carnot efficiency, which can never be attained, gives an ideal upper limit.

But if T_{hot} is four times T_{cold}, or $T_{cold}/T_{hot} = \frac{1}{4} = 0.25$, then

$$\varepsilon_C = 1 - \frac{T_{cold}}{T_{hot}} = 1 - 0.25 \, (\times \, 100\%) = 75\%$$

EXAMPLE 12.8 ■ Carnot Efficiency

An engineer is designing a cyclic heat engine to operate between temperatures of 150°C and 27°C. What is the maximum theoretical efficiency that can be achieved?

Solution.

Given: $T_{hot} = 150°C = 423$ K   *Find:* ε_C (Carnot efficiency)
     $T_{cold} = 27°C = 300$ K

Using Eq. 12.10

$$\varepsilon_C = 1 - \frac{T_{cold}}{T_{hot}} = 1 - \frac{300 \text{ K}}{423 \text{ K}} = 0.291 \, (\times \, 100\%) = 29.1\%$$

The Third Law of Thermodynamics

Another interesting conclusion can be drawn from the expression for the Carnot efficiency (Eq. 12.10). To have ε_C equal to 100%, T_{cold} would have to be (absolute) zero. Since a heat engine with 100% efficiency is impossible by the second law, the **third law of thermodynamics** is

It is impossible to reach a temperature of absolute zero.

Absolute zero has never been observed experimentally. If it were, this would in effect violate the second law. In cryogenic (low-temperature) experiments, scientists have come close to absolute zero—to about 0.000001 (one-millionth) of a kelvin—but have never reached it. Near absolute zero, reducing the temperature by an order of magnitude becomes more difficult at each step.

Important Concepts

You should be able to define and explain these chapter concepts clearly.

thermodynamics	process	isothermal process	Otto cycle
system	irreversible process	isotherm	thermal efficiency
open system	reversible process	adiabatic process	heat pump
closed system	first law of	adiabat	coefficient of performance
thermally isolated system	thermodynamics	second law of	(cop)
completely isolated	isobaric process	thermodynamics	Carnot cycle
system	isobar	entropy	Carnot (ideal) efficiency
heat reservoir	isometric process	isentropic process	third law of thermodynamics
equations of state	isomet	heat engine	

Important Relationships for Review

These relationships are mathematical statements of the concepts and principles presented in the chapter. You should be able to identify the symbols and to explain the relationships before proceeding to the Exercises. In-text equation reference numbers are given for convenience.

First Law of Thermodynamics:

$$Q = \Delta U + W$$

$+Q$, heat *added* to system
$-Q$, heat *removed* from system $\qquad$ (12.1)
$+W$, work done *by* system
$-W$, work done *on* system

Work Done by an Expanding Gas (Constant Pressure):

$$W = p\Delta V = p(V_2 - V_1) \qquad (12.2)$$

Change in Entropy:

$$\Delta S = \frac{Q}{T} \qquad (12.4)$$

Thermal Efficiency of a Heat Engine:

$$\varepsilon_{\text{th}} = \frac{W}{Q_{\text{in}}} = \frac{Q_{\text{hot}} - Q_{\text{cold}}}{Q_{\text{hot}}} = 1 - \frac{Q_{\text{cold}}}{Q_{\text{hot}}} \qquad (12.6)$$

Coefficient of Performance:

$$\text{cop} = \frac{Q_{\text{out}}}{W_{\text{in}}} = \frac{Q_{\text{cold}}}{Q_{\text{hot}} - Q_{\text{cold}}} \qquad (12.8)$$

Carnot Efficiency of an Ideal Heat Engine:

$$\varepsilon_{\text{C}} = \frac{T_{\text{hot}} - T_{\text{cold}}}{T_{\text{hot}}} = 1 - \frac{T_{\text{cold}}}{T_{\text{hot}}} \qquad (12.10)$$

Exercises

12.1 ■ Thermodynamic Systems, States, and Processes

1 A change in thermodynamic state (a) occurs only when all the variables of the system change, (b) occurs only for an open system, (c) requires a heat reservoir, (d) cannot take place in a completely isolated system in equilibrium. (d)

● **2** Only initial and final states are known for irreversible processes on (a) p-V diagrams, (b) p-T diagrams, (c) V-T diagrams, (d) all of these. (d)

3 The pVT diagram for water is shown in ● Fig. 12.15 (compare this with Fig. 11.10). (a) Explain the processes on the liquid-vapor and solid-vapor surfaces. (b) Why is there a triple-point line? see ISM

4 Explain how you can show from Fig. 12.15 that a quantity of water has a greater volume when it is frozen than when it is melted. see ISM

5 ■■ On a p-T diagram, sketch the general paths for the following reversible processes for an ideal gas: (a) isothermal, (b) isobaric, (c) isometric, and (d) adiabatic. see ISM

6 ■■ On a V-T diagram, sketch the general paths for the following reversible processes for an ideal gas: (a) isothermal, (b) isobaric, (c) isometric, and (d) adiabatic. see ISM

7 ■■ On a p-V diagram, sketch a general cyclic process that consists of (a) an isothermal expansion, (b) an isobaric compression, and (c) an isometric process, in that order. see ISM

8 ■■ Sketch the projection of the pVT surface for water in Fig. 12.15 on the pT plane and compare it with that for CO_2 in Fig. 12.1. What do the pT diagrams tell you about the freezing points of the substances as a function of temperature? see ISM

12.2 ■ The First Law of Thermodynamics

9 According to the first law of thermodynamics, if heat is added to a system, then (a) the internal energy of the system must change, (b) work must be done on the system, (c) the internal energy of the system changes and/or work is done by the system, (d) none of the preceding. (c)

● **10** If a system of ideal gas has heat removed by an isothermal process, (a) work is done on the system, (b) the internal

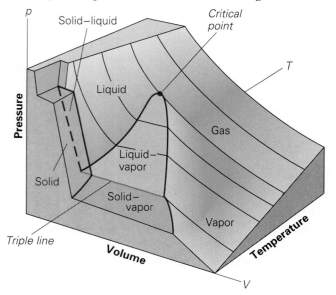

● **FIGURE 12.15 The pVT surface for water** See Exercises 3, 4, and 8.

energy decreases, (c) the effect is the same as for an isometric process, (d) the process is also adiabatic. (b)

11 (a) Why does the body of a hand pump become hot when a tire is pumped up? (Neglect friction. The plunger is usually lubricated to prevent air leakage.) (b) Why does the valve of a tire become cold when air is released from the tire? see ISM

12 Apply the first law to the stretching of a rubber band to determine the effect on its internal energy. (*Hint:* Stretch a wide rubber band and immediately hold it to your forehead; note its temperature.) see ISM

13 ■ An ideal gas in a cylinder expands adiabatically while doing 2.5×10^3 J of work in moving the piston. What is the change in the internal energy of the gas? Does the temperature of the gas increase or decrease?
-2.5×10^3 J; temperature decreases

14 ■ A quantity of ideal gas goes through a cyclic process and does 200 J of work. (a) Is heat added or removed from the system, and how much heat is involved? (b) Does the temperature of the gas increase or decrease? (a) 200 J
(b) $\Delta U = 0$ for a complete cycle, ΔT is 0

15 ■■ A quantity of ideal gas is taken through processes that are depicted as straight-line paths on a p-V diagram: from (1 atm, 1 m³) to (4 atm, 1 m³), then to (4 atm, 3 m³), and back to (1 atm, 1 m³). What is the net work done?
3.0×10^5 J

● **16** ■■ A system of gas at low density has an initial pressure of 1.28×10^4 Pa and occupies a volume of 0.25 m³. The slow addition of 200 cal of heat to the system causes it to expand isobarically to a volume of 0.30 m³. (a) How much work is done by the system in the process? (b) Did the internal energy of the system change? If so, by how much?
(a) 640 J   (b) 198 J

17 ■■ A rigid container contains 1 mole of nitrogen gas that slowly receives 2.0 kcal of heat. What is the change in the internal energy of the gas? 8.4×10^3 J

18 An olympic weight lifter lifts 150 kg a vertical distance of 2.0 m. In doing so, his internal energy decreases by 5.0×10^3 J. Considering the weight lifter to be a thermodynamic system, how much heat flows and in what direction? -2.1×10^3 J out

19 One mole of ideal gas at STP expands isobarically to 2.5 times its original volume. (a) Is work done by or on the system? How much work? (b) Does heat flow into or out of the system? (a) 1.13×10^3 J   (b) heat in

20 ■■ During an isobaric process at 1 atm of pressure, a quantity of gas absorbs 125 cal of heat while expanding its volume from 1.0 L to 1.2 L. What is the change in the internal energy of the system? 5.0×10^2 J

21 ■■ A gram of water (1.00 cm³) at 100°C is converted to 1671 cm³ of steam at atmospheric pressure. What is the change in the internal energy of the system? 2.09×10^3 J

22 ■■ A quantity of gas undergoes the reversible changes

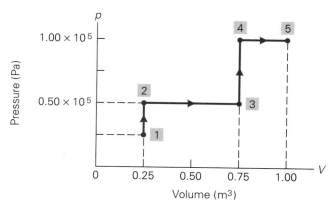

● **FIGURE 12.16 A p-V diagram and work** See Exercises 22 and 23.

illustrated in the p-V diagram in ● Fig. 12.16. How much work is done in each process? (1–2) 0; (2–3) 2.5×10^4 J; (3–4) 0; (4–5) 2.5×10^4 J

23 ■■ Suppose that after the final process shown in Fig. 12.16 (see Exercise 22) the pressure of the gas is decreased isometrically from 1.0×10^5 Pa to 0.70×10^5 Pa, and then the gas is compressed isobarically from 1.0 m³ to 0.80 m³. What is the total work done in all of these processes?
3.6×10^4 J

● **24** ■■ A quantity of gas is compressed as shown on the p-V diagram in ● Fig. 12.17. How much work is done on the system? -1.75×10^5 J

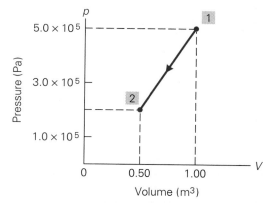

● **FIGURE 12.17 A variable p-V process and work** See Exercises 24 and 25.

25 ■■■ Suppose that after the process shown in Fig. 12.17 (see Exercise 24) the gas undergoes an isobaric expansion to 1.0 m³ and then the pressure is increased isometrically to 5.0×10^5 Pa. (a) How much work is done in each of these processes? (b) What is the total work done in taking the gas from its initial state [(1) in the figure] to its final state? (c) Is heat added or removed from the system? How much heat?
(a) 1.0×10^5 J; 0   (b) -7.5×10^4 J   (c) -7.5×10^4 J

26 ■■■ One mole of ideal gas is taken through the cyclic

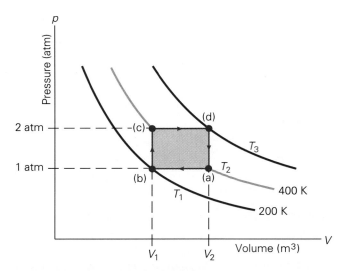

● FIGURE 12.18 A cyclic process See Exercise 26.

process shown in ● Fig. 12.18. (a) Compute the work involved (*W*) for each of the four processes. (b) Find ΔU, *W*, and *Q* for the complete cyclic process. (c) What is T_3? (a) ab: $W =$ -1.66×10^3 J; bc: $W = 0$; cd: $W = 3.31 \times 10^3$ J; da: $W = 0$ (b) $\Delta U = 0$; $W = Q = 1.66 \times 10^5$ J   (c) 799 K

12.3 ■ The Second Law of Thermodynamics and Entropy

27 The second law of thermodynamics (a) precludes perpetual motion machines, (b) applies only when the first law is satisfied, (c) gives an equation of state for a system, (d) does not apply to a closed system. (a)

28 Overall in a natural process, the change in the entropy of a system is (a) positive, (b) negative, (c) zero, (d) the same as the change in the internal energy. (a)

29 In Example 12.4, what would be the implication if the total change in entropy had been negative? energy created

● **30** When a quantity of hot water is mixed with a quantity of cold water, the combined system comes to thermal equilibrium at some intermediate temperature. How does the entropy of the system change? positive

31 ■ What change in entropy is associated with the reversible phase change of 1.0 kg of ice to water at 0°C? 1.2×10^3 J/K
32 ■ What is the change in entropy when 0.35 kg of steam condenses to water at 100°C? -2.1×10^3 J/K

33 ■ Which process has the greater change in entropy— 0.75 kg of water changing to ice at 0°C or 0.25 kg of water changing to steam at 100°C? Comment on the entropy changes in terms of order and disorder.
$S_i = -9.2 \times 10^2$ J/K; $S_s = 1.5 \times 10^3$ J/K; ice has more order
34 ■■ A heat reservoir supplies heat to melt 0.75 kg of ice

at 0°C. What is the change in entropy of the ice once it has all melted? 9.2×10^2 J/K

35 ■■ A quantity of an ideal gas initially at STP undergoes a reversible isothermal expansion and does 3.0×10^3 J of work on its surroundings in the process. What is the change in the entropy of the gas? 11 J/K

● **36** ■■ An ice tray is filled with 250 mL of water at 10°C from a cold-water tap and placed in a freezer, where the temperature is -6.0°C. Estimate the change in entropy that will have occurred when the water freezes and the ice comes to thermal equilibrium with its surroundings. -4.6×10^2 J/K
37 ■■ Suppose that two automobiles with equal masses of 1.5×10^3 kg are both traveling at 60 km/h when they have a head-on collision. Estimate the change in entropy for this process. 1.5×10^3 J/K

38 ■■ A certain system undergoes an isothermal process at 27°C, and there is an entropy change of 35 J/K. If 2.0×10^3 J of work is done on the system in the process, what is the change in the system's internal energy? 8.5×10^3 J
39 ■■ What is the change in entropy when 0.50 kg of mercury vapor ($L_v = 2.7 \times 10^5$ J/kg) condenses to a liquid at its boiling point of 357°C? -2.2×10^2 J/K

40 ■■ Two heat reservoirs at temperatures of 200°C and 60°C are brought into thermal contact and 1500 J of heat spontaneously flows from one to the other. What is the change in the entropy of the universe? What would a negative entropy change imply? 1.1 J/K

41 ■■ One mole of ideal gas goes through an isothermal compression at room temperature. If 5.0×10^3 J of work is done in compressing the gas, what is the change in entropy of the system? -17 J/K

● **42** ■■ (a) How much heat is transferred in the processes shown on the *T-S* diagram in ● Fig. 12.19? (b) What type of process takes place as the system goes from state b to state c? (a) 2.73×10^4 J/K (b) isentropic since $W = 0$

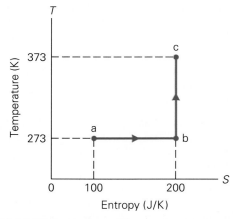

● **FIGURE 12.19 Entropy and heat** See Exercises 42 and 43.

43 ■■ Suppose that the system described by the *T-S* diagram in Fig. 12.19 is returned to its original state (state a) by a reversible process that is depicted by a straight line from c to a. What is the change in entropy for the total cycle? How much heat is transferred in the cyclic process? zero, -5.0×10^3 J

44 ■■■ A 50-g ice cube at 0°C is placed in 500 mL of water at room temperature. Estimate the change in entropy when all the ice has melted (a) for the ice, (b) for the water, and (c) for the universe. (a) 68 J/K (b) −65 J/K (c) 3.0 J/K

45 ■■■ A half-liter of ethyl alcohol at 18.0°C is mixed with a half-liter of water at 25.0°C. What is the change in entropy once the mixture has reached thermal equilibrium? (Consider the system to be thermally isolated.) 0.2 J/K

12.4 ■ Heat Engines and Heat Pumps

46 For a cyclic heat engine, (a) $\varepsilon_{th} > 1$, (b) $Q_{hot} = W$, (c) $\Delta U = W$, (d) $Q_{hot} > Q_{cold}$. (d)

47 A heat pump (a) is rated by thermal efficiency, (b) requires work input, (c) is not consistent with the second law, (d) violates both the first and second laws. (b)

48 Is any external work done on the drinking bird in Example 12.6? How would the bird be represented in a generalized heat engine diagram? external work is done by gravity

49 Lord Kelvin's statement of the second law as applied to heat engines refers to their operating *in a cycle*. Why is this phrase included? see ISM

● **50** In atmospheric convection cycles, colder air from a higher altitude is transferred to a lower, warmer level. Does this violate the second law? Explain. no

51 Is leaving a refrigerator door open a practical way to air condition a room? Explain. no

52 ■ A heat engine with an efficiency of 40% does 800 J of work each cycle. How much heat is rejected to the surroundings (the low-temperature reservoir)? 1.2×10^3 J

53 ■ In each cycle, a heat engine absorbs 150 kcal of heat from a high-temperature reservoir and exhausts 100 kcal. (a) What is the engine's thermal efficiency? (b) What is its work output? (a) 33% (b) 2.1×10^5 J or 50 kcal

● **54** ■ A heat engine has a thermal efficiency of 30%. It absorbs 750 J from a high-temperature reservoir each cycle. (a) What is the work output of the engine? (b) How much of the heat input per cycle is not converted to work? (a) 2.25×10^2 J (b) 70% or 5.25×10^2 J

55 ■■ A steam engine does 8.5×10^3 J of useful work each cycle but loses 5.5×10^2 J to friction and exhausts 2.0 kcal of heat energy. What is the engine's efficiency? 49%

56 ■■ A theoretical cyclic heat engine absorbs 1800 kcal of heat from a high-temperature reservoir and exhausts 600 kcal to a low-temperature reservoir. When run in reverse

as a heat pump with the same reservoirs, 5870 J of work must be input to deliver 1800 kcal of heat energy to the high-temperature reservoir. (a) What is the engine's thermal efficiency? (b) Is the heat engine reversible in the sense that the work input and output are the same when the engine is run between the same reservoirs? (a) 67% (b) not the same

57 ■■ A gasoline engine consumes 10 L of fuel per hour. (a) What is the energy input during a period of 2.0 h? (b) If the engine delivers 25 kW of power during this time, what is its efficiency? [*Hint:* Recall the heat of combustion from Chapter 11.] (a) 6.6×10^8 J (b) 27%

● **58** ■■ A refrigerator takes 250 kcal from a low-temperature reservoir and exhausts 400 kcal to a high-temperature reservoir each cycle. (a) What is its coefficient of performance? (b) What is the work input each cycle? (a) 1.67 (b) 6.29×10^5 J

59 ■■ A refrigerator with a cop of 2.2 removes 100 kcal of heat from its storage area each cycle. (a) How much heat is exhausted each cycle? (b) What is the total work input in joules for 10 cycles? (a) 145 kal (b) 1.89×10^6 J

60 ■■ An air conditioner has a cop of 1.75. What is the wattage input required for the unit to remove 2.50×10^5 kcal of heat in 20 minutes? 500 kW

61 ■■■ A heat engine has a thermal efficiency of 28%. If its heat input each cycle is supplied by the condensation of 7.5 kg of steam at 100° C, (a) what is the work output per cycle? (b) How much heat is rejected to the surroundings each cycle? (a) 4.7×10^6 J (b) 1.3×10^7 J

62 ■■■ A coal-fired power plant produces 900 MW of electricity and operates at an overall thermal efficiency of 35%. (a) What is the thermal input to the plant? (b) What is the rate of heat discharge from the plant? (c) Why is the water heated by the discharged heat cooled in a cooling tower before being discharged into a nearby river? (a) 2.6×10^3 MW (b) 1.7×10^9 W (c) to minimize thermal pollution

63 ■■■ Show that the change in entropy for a cycle of a heat engine is

see ISM
$$\Delta S = \frac{Q_{cold}}{T_{cold}} - \frac{Q_{hot}}{T_{hot}}$$

12.5 ■ The Carnot Cycle and Ideal Heat Engines

64 The Carnot cycle consists of (a) two isobars and two isotherms, (b) two isomets and two adiabats, (c) two adiabats and two isotherms, (d) four arbitrary processes which return the system to its initial state. (c)

65 The Carnot efficiency of a heat engine (a) may be greater than one, (b) would be 100% if the low temperature reservoir were at absolute zero, (c) decreases when the difference in the reservoir temperatures increases, (d) none of the preceding. (b)

66 Automobile engines can either be air-cooled or water-cooled. Which type of engine would you expect to be more efficient and why?   efficiency depends on ΔT; water can have higher temperature differences

67 There is also a Carnot coefficient of performance (cop_C) for an ideal or Carnot heat pump:

$$\text{cop}_C = \frac{T_{\text{cold}}}{T_{\text{hot}} - T_{\text{cold}}}$$

What does this tell you about the operating parameters of a heat pump?   when the outside temperature is low, the heat pump is less efficient

68 ■ A heat engine operates between a reservoir at 27°C and one at 250°C. What is the maximum theoretical efficiency of the engine? Is this ever achieved?   43%

69 ■ An ideal heat engine takes in heat from a high-temperature reservoir at 175°C and exhausts heat to a low-temperature reservoir at 0°C. What is its Carnot efficiency?   39%

70 ■■ The exhaust temperature of a Carnot engine is 95°C, and the engine has a heat output of 100 kcal per cycle at this temperature. If the engine has an efficiency of 30%, what is its heat input? 6.0×10^5 J

71 ■■ An ideal heat engine with a Carnot efficiency of 35% takes in heat from a high-temperature reservoir at 147°C. What is the temperature of the low-temperature reservoir?   273 K

● **72** ■■ An ideal heat engine takes 6.5 kcal of heat from a high-temperature reservoir at 320°C and exhausts some of it to a low-temperature reservoir at 120°C. How much work is done by the engine? 9.2×10^3 J

73 ■■ A Carnot engine with an efficiency of 40% operates with a low-temperature reservoir at 50°C and exhausts 1200 J of heat each cycle. What is the heat input and the temperature of the high-temperature reservoir? 2.00×10^3 J; 265°C

74 ■■ An ideal heat engine takes in heat from a reservoir at 327°C and has an efficiency of 30%. If the exhaust temperature did not vary and the efficiency increased to 40%, what would be the increase in the temperature of the hot reservoir?   100C° change in temperature

75 ■■ An inventor claims to have developed a heat engine that, on each cycle, takes in 120 kcal of heat from a high-temperature reservoir at 400°C and exhausts 48 kcal to the surroundings at 125°C. Would you invest your money in the production of this engine? Explain.   no; not possible

● **76** ■■ Which has the greatest theoretical efficiency—a heat engine operating between reservoirs at 300°C and 100°C or one operating between reservoirs at 300 K and 100 K? What are the efficiencies in each case? $\varepsilon_1 = 35\%$; $\varepsilon_2 = 67\%$

77 ■■ An engineer wants to run a heat engine with an efficiency of 40% between a high-temperature reservoir at 300°C and a low-temperature reservoir. Below what temperature must the low-temperature reservoir be for practical operation of the engine?   71°C

78 ■■ A Carnot engine operates between temperature reservoirs at 250°C and 80°C. During each cycle, it absorbs 3.0 $\times 10^4$ J of heat from the high-temperature reservoir. (a) What is the power output of this ideal engine? (b) How much heat does it reject to the low-temperature reservoir each cycle?   (a) 1.0×10^4 J   (b) 2.0×10^4 J

79 ■■ Assume a heat engine has a thermal efficiency of 100%. Prove that this engine would also have a Carnot efficiency of 100%.   see ISM

80 ■■ The working substance of a cyclic heat engine is 0.75 kg of an ideal gas. The cycle consists of two isobaric processes and two isometric processes as shown in ● Fig. 12.20. What would be the efficiency of a Carnot engine operating with the same high-temperature and low-temperature reservoirs?   53%

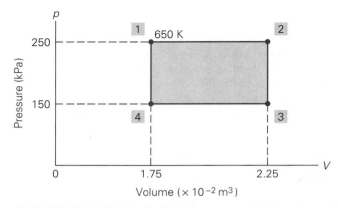

● **FIGURE 12.20 Thermal efficiency**   See Exercise 80.

81 ■■ In each cycle, a Carnot heat engine takes 774 J of heat from a high-temperature reservoir and discharges 258 J to a low-temperature reservoir. (a) What is the Carnot efficiency of the engine? (b) How many times greater is the temperature of the high-temperature reservoir than that of the low-temperature reservoir?   (a) 67%   (b) $T_h = 3T_c$

82 ■■ A Carnot engine operating between reservoirs at 27°C and 227°C does 1500 J of work each cycle. (a) What is the efficiency of the engine? (b) What is the change in entropy each cycle? [*Hint:* See Exercise 63.]   (a) 40%   (b) cycle comes back to original entropy value; $\Delta S = 0$

83 ■■ A heat engine has an efficiency that is half that of a Carnot engine that operates between temperatures of 100°C and 375°C. If the real engine absorbs heat at a rate of 50 kW per cycle, at what rate is heat exhausted each cycle?   40 kW/cycle

84 ■■■ It has been proposed that the temperature difference in the ocean could be utilized to run a heat engine to generate electricity. In tropical regions, the water temperature is about 25°C at the surface and about 5°C at a low depth. (a) What would be the maximum theoretical efficiency of such an engine be? (b) Do you think a heat engine with such a relatively low efficiency would be practical? Explain. (a) 6.7% (b) probably not, due to the poor return; fossil fuels are much better

85 ■■■ An ideal heat pump is equivalent to a Carnot engine running in reverse. Show that the work required for an ideal heat pump is given by

see ISM
$$W = Q_{cold}\left[\left(\frac{T_{hot}}{T_{cold}}\right) - 1\right]$$

■ Additional Examples

86 Show that the coefficient of performance of a Carnot engine is given by

see ISM
$$cop_C = \frac{T_{cold}}{T_{hot} - T_{cold}}$$

87 A thermally isolated quantity of gas has an initial volume of 10 L and is compressed isobarically at a pressure of 300 kPa. If 6.0×10^2 J of work is done on the system, what is the final volume of the gas? 8.0 L

88 Use the general definition of efficiency, $\varepsilon = W/Q_{in}$, to derive the Carnot efficiency. [*Hint:* Consider the cycle in terms of entropy, and see Fig. 12.14b.] see ISM

89 A gas is enclosed in a piston-cylinder arrangement with a diameter of 24.0 cm. Heat is slowly added to the system while the pressure is maintained at 1.00 atm. During the process, the piston moves 6.00 cm. (a) What type of process is this? (b) If the quantity of heat transferred to the system during the expansion is 0.10 kcal, what is the change in the internal energy of the gas? (a) isobaric (b) 145 J

90 In a piston-cylinder arrangement 2.0 L of water at room temperature slowly receive 3.5 kcal of heat with the piston locked in place. (a) What is the change in the internal energy of the water? (b) What is the final temperature of the water? (Neglect pressure effects.) (a) 1.5×10^4 J (b) 22°C

91 The energy or useful work lost as a result of a change in the entropy of the universe is given by $W = T_{cold}\Delta S_u$, where T_{cold} is the temperature of the cold reservoir for the process. (a) Show that for heat (Q) transferred from a high-temperature reservoir to a low-temperature reservoir $W = Q[1 - (T_{cold}/T_{hot})]$. (b) What is the quantity within the brackets? see ISM

92 For the two-step reversible process illustrated in ● Fig. 12.21, the same amount of heat is added to and removed from the system. Prove that the total entropy (of the universe) increases in the process. see ISM

93 In a thermodynamic process, 2.50×10^3 J of heat is transferred to a system, and 6.00×10^2 J of work is done on

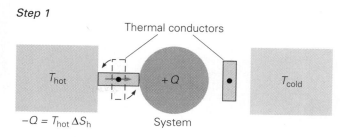

Step 1

Thermal conductors

T_{hot} $+Q$ T_{cold}

$-Q = T_{hot}\Delta S_h$ System

Step 2

T_{hot} $-Q$ T_{cold}

System $+Q = T_{cold}\Delta S_c$

● **FIGURE 12.21 Does the entropy increase?** See Exercise 92.

the system. What is the change in the internal energy of the system? 3.1×10^3 J

94 The maximum temperature for the superheated steam used in a turbine for electrical generation is about 540°C because of material limitations. (a) If the steam condenser operated at room temperature, what would the ideal efficiency be? (b) The actual efficiency is about 35–40%. What does this tell you? (a) 64% (b) a lot more energy is lost than ideally predicted

95 It can be shown that the work done in an isothermal process with an ideal gas is

$$W = nRT\left[\ln\left(\frac{V_2}{V_1}\right)\right]$$

If 2.0 moles of an ideal gas expand isothermally from 3.8×10^{-2} m^3 to 7.6×10^{-2} m^3 at room temperature, (a) how much heat is added to the system? (b) What is the change in the internal energy of the gas? (c) What is the change in the entropy of the system? (a) 3.4×10^3 J (b) $\Delta U = 0$ since the process is isothermal (c) 12 J/K

96 The change in entropy for a liquid that undergoes a temperature change can be shown to be

$$\Delta S = mc\left[\ln\left(\frac{T_f}{T_i}\right)\right]$$

where m is the mass, c the specific heat, and T_f and T_i the final and initial temperatures. If 1.0 kg of water at 0°C and 1.0 kg of water at 100°C are mixed together in a thermally isolated container, will there be a net change in entropy? Justify your answer. 102 J/K

13

Vibrations and Waves

INSIGHTS

- ■ **Earthquakes and Seismology**
- ■ **Unwanted Resonance**

A vibration or oscillation involves back-and-forth motion. There are many familiar examples. You may have felt a power jig saw or an electric toothbrush vibrate in your hand. A swinging pendulum oscillates back and forth, as does the dial on a bathroom scale as it comes to equilibrium to give a weight reading. Another example is a small ball in a bowl with a rounded bottom. If the ball is displaced from its equilibrium position, it will roll back and forth, or oscillate, and finally come to rest at the equilibrium position, where its potential energy is lowest.

Recall from Chapter 8 that the ball must be oscillating about a point of stable equilibrium, for which there is a restoring force or torque. This is true in general for particles that undergo vibrating or oscillating motions. However, in a material medium, the oscillation of one particle affects that of its neighbors. A simplistic model of a solid pictures the intermolecular (bonding) forces as springs (Fig. 9.1). The molecules joined by these elastic forces thus influence each other. For example, suppose that a molecule is disturbed. Then there is a restoring force that tends to return it to its original position, and it begins to oscillate. In so doing, it affects the adjacent particle, which is also set into oscillation. That is, something happens at point A at time t_1, and this causes a similar happening at point B at t_2, and so on. This is referred to as propagation. *(By analogy, think of how a rumor is propagated.)*

But what is propagated by the molecules in a material? A moment's thought should tell you it is energy. Suppose that energy is added to a material mechanically, such as by a blow or (in the case of a gas) by compression. This sets the molecules vibrating, and energy is propagated through the medium by molecular interactions. Such propagation of energy resulting from a disturbance is called a wave. *A single disturbance, such as you produce when you give the end of a stretched rope a quick shake, gives rise to what is referred to as a* wave pulse. *A continuous, repetitive disturbance gives rise to a continuous propagation of energy, or wave motion.*

Topics studied in Chapter 13 include simple harmonic motion, equations of motion, wave motion and phenomena, and standing waves.

In this chapter, you will learn how to describe vibrations and wave motions. This will be very important in the study of future topics, for wave phenomena are everywhere. The ripples on the surface of a pond or tank (like those shown in the opening photograph) or the waves of the ocean are only the most familiar examples. Indeed, without waves, we would know very little about our world, for sound (Chapter 14) is a type of wave, and light (Chapter 19) is a type of wave. In fact, all electromagnetic radiations are waves—radio waves, microwaves, X-rays, and so on. Also, in Chapter 27, you'll learn how moving particles have wave-like properties. But first, we need to look at the basic descriptions of waves.

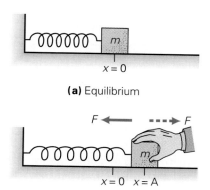

(a) Equilibrium

(b) $t = 0$

13.1 ■ Simple Harmonic Motion

The motion of an oscillating particle depends on the restoring force producing it. It is convenient to begin to study such motion by considering the simplest type of force, which is one that is directly proportional to the displacement. One such force is the spring force, described by **Hooke's law**,

$$F = -kx \qquad (13.1)$$

where k is the spring constant. Recall from Chapter 5 that the minus sign indicates that the force is always in the opposite direction to the displacement; that is, it always tends to restore the spring to its equilibrium position.

Suppose that a mass on a horizontal frictionless surface is connected to a spring as shown in ● Fig. 13.1. When the mass is displaced to one side of its equilibrium position, it will move back and forth, will vibrate or oscillate. Here, we consider an oscillation or a vibration to be **periodic motion**, that is, a motion that repeats itself again and again along the same path. For linear oscillations, like those of a mass attached to a spring, the path may be back and forth or up and down. For the angular oscillation of a pendulum, the path is back and forth along a circular arc.

Motion under the influence of the type of force described by Hooke's law is called **simple harmonic motion (SHM)**, because the force is the simplest force and because the motion can be described by harmonic functions (sines and cosines), as you will see later in the chapter. Descriptions of simple harmonic motion refer to the directional distance of the object from its equilibrium position as its **displacement**. Note in the figure that the displacement can be either positive or negative ($+x$ or $-x$), which indicates direction. The maximum displacements are $+A$ and $-A$ [(b) and (d) in Fig. 13.1]. The magnitude of the maximum displacement, or the maximum distance of a mass or particle from its equilibrium position is called the **amplitude** (A). It is a scalar quantity that expresses the distance for both extreme displacements.

Besides the amplitude, other important quantities for describing an oscillation are its period and frequency. The **period** (T) is the time required for one complete cycle of motion. A cycle is a complete round trip, or motion through a complete oscillation. For example, if an object starts at $x = A$ (Fig. 13.1b), when it returns to position A (as in Fig. 13.1f), it will have completed one cycle in a time of one period. If an object were initially at $x = 0$ when disturbed, then its second return to this point would mark a cycle. In either case, the object would travel a distance of $4A$ in a cycle. Do you agree?

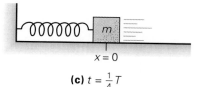

(c) $t = \frac{1}{4}T$

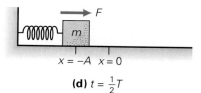

(d) $t = \frac{1}{2}T$

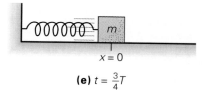

(e) $t = \frac{3}{4}T$

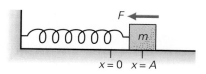

(f) $t = T$

● **FIGURE 13.1 Simple harmonic motion (SHM)** When displaced from its equilibrium position $x = 0$ and released (**a**, **b**), the mass on a spring undergoes SHM (assuming no frictional losses). The time it takes to complete one cycle is the period of oscillation (T). At $t = T/4$ (**c**), the mass is back at its equilibrium position; and at $t = T/2$ (**d**), it is at $x = -A$. During the next half cycle, the motion is to the right; and at $t = T$, the mass is back at its initial ($t = 0$) starting position.

Figure 13.1

The **frequency** (f) is the number of cycles per second. The frequency and the period are related by

$$f = \frac{1}{T}$$ (frequency and period)   (13.2)

The inverse relationship is reflected in the units: The period is the number of seconds per cycle, and the frequency is the number of cycles per second. For example, when $T = \frac{1}{2}$ s/cycle, $f = 2$ cycles/s.

hertz (Hz)—the unit of frequency

The standard unit for frequency is the **hertz (Hz)**, which is one cycle per second (cps).* From Eq. 13.2, frequency has the unit $1/s$, or s^{-1}, since the period is a measure of time. Although cycle is not really a unit, you might find it convenient at times to express frequency in cycles/s to help with dimensional analysis. This is similar to the way the radian (rad) is used in the description of circular motion.

See Sections 7.1 and 7.2

Energy and Speed in SHM

Compare Fig. 5.5, Section 5.2

Recall from Chapter 5 that the potential energy stored in a stretched or compressed spring is given by

$$U = \frac{1}{2}kx^2$$ (13.3)

This discussion will be limited to *light* springs, the mass of which can be considered negligible.

This is equal to the work done on the spring, which is equal to the area under the curve on a graph of force versus displacement, such as the one in ● Fig. 13.2. A mass m oscillating on a spring has kinetic energy. Thus, the kinetic and potential energies together give the total mechanical energy of the system:

$$E = K + U = \frac{1}{2}mv^2 + \frac{1}{2}kx^2$$ (13.4)

When the mass is at one of its maximum displacements, $+A$ or $-A$, it is instantaneously at rest ($v = 0$). Thus, all the energy is potential energy at this time; that is,

$$E = \frac{1}{2}m(0)^2 + \frac{1}{2}kA^2 = \frac{1}{2}kA^2$$

or

$$E = \frac{1}{2}kA^2$$ (total energy in SHM of a spring)   (13.5)

Demonstration/activity: Demonstrate simple harmonic motion with a mass hung on a spring or a light spring connected to an air track glider. Motion on a horizontal surface such as the air track is better at this time to avoid discussing gravitational potential energy.

Discuss points where the potential energy is a maximum and points where the kinetic energy is a maximum. Discuss the location where the potential and kinetic energies are equal by considering the shaded area on the force-displacement graph.

where A is the magnitude of the maximum displacement or the amplitude. Neglecting any losses, the total energy is conserved. This being the case, we can say in general that

The total energy of an object in simple harmonic motion is proportional to the square of the amplitude.

Equation 13.5 allows the velocity to be expressed as a function of position:

$$E = K + U \qquad \text{or} \qquad \frac{1}{2}kA^2 = \frac{1}{2}mv^2 + \frac{1}{2}kx^2$$

Then

$$v^2 = \frac{k}{m}(A^2 - x^2)$$

*The unit is named for Heinrich Hertz (1857–1894), a German physicist and early investigator of electromagnetic waves.

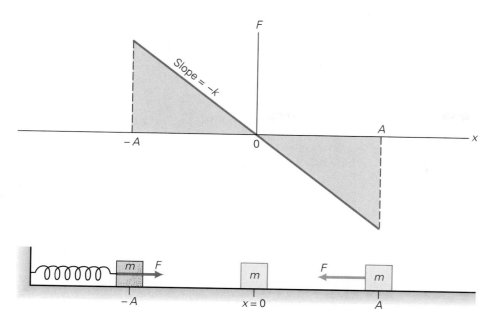

FIGURE 13.2 A plot of $F = -kx$ and work The work done by a force in stretching a spring is equal to the area under the curve on a graph of force (F) versus displacement (x). Here the spring force is plotted ($F = -kx$), and the work is negative. Note that at the amplitude positions, $W = U = \frac{1}{2}kA^2$.

and

$$v = \pm\sqrt{\frac{k}{m}(A^2 - x^2)}$$

(13.6)   **Velocity of a particle in SHM**

where the $\pm$ indicates direction. Note that at $x = \pm A$, the velocity is zero and the mass is instantaneously at rest at its maximum displacement positions.

Also note that when the oscillating mass passes through the origin, or equilibrium position ($x = 0$), the potential energy is zero. At that instant, all the energy is kinetic and the mass is traveling at its maximum speed, v_{max}. (See Fig. 13.1c and e.) The energy expression for this case is

$$E = \tfrac{1}{2}kA^2 = \tfrac{1}{2}mv_{max}^2$$

and

$$v_{max} = \sqrt{\frac{k}{m}}(A)$$

(13.7)

EXAMPLE 13.1 ■ Motion—Simple and Harmonic

A block with a mass of 0.25 kg sitting on a frictionless surface is connected to a light spring that has a spring constant of 180 N/m (see Fig. 13.1). If the block is displaced 15 cm from its equilibrium position and released, what are (a) the total energy of the system, (b) the maximum speed of the block, and (c) the speed of the block when it is 10 cm from its equilibrium position?

Solution. First we sort out and list the given data, as usual, along with determining what it is we are to find.

Given: $m = 0.25$ kg
$k = 180$ N/m
$A = 15$ cm $= 0.15$ m
$x = 10$ cm $= 0.10$ m

Find: (a) E (total energy)
(b) v_{max} (maximum speed)
(c) v (speed)

(a) The total energy is given directly by Eq. 13.5:

$$E = \tfrac{1}{2}kA^2 = \tfrac{1}{2}(180 \text{ N/m})(0.15 \text{ m})^2 = 2.0 \text{ J}$$

(b) The block has its maximum speed when it passes through the equilibrium position ($x = 0$), where all of its energy is kinetic energy. From Eq. 13.7,

$$v_{max} = \sqrt{\frac{k}{m}}\,(A) = \sqrt{\frac{180 \text{ N/m}}{0.25 \text{ kg}}}\,(0.15 \text{ m}) = 4.0 \text{ m/s}$$

(c) The instantaneous speed of the block at a distance of 10 cm from the equilibrium position is given by Eq. 13.6 without directional signs:

$$v = \sqrt{\frac{k}{m}}\,(A^2 - x^2) = \sqrt{\frac{180 \text{ N/m}}{0.25 \text{ kg}}\,[(0.15 \text{ m})^2 - (0.10 \text{ m})^2]}$$

$$= \sqrt{9.0 \text{ m}^2/\text{s}^2}$$

$$= 3.0 \text{ m/s}$$

EXAMPLE 13.2 ■ Determination of the Spring Constant

When a 0.50-kg mass is suspended from a spring, the spring stretches 10 cm (● Fig. 13.3). The mass is then pulled down another 5.0 cm and released. (a) What is the total energy of the oscillating system? (Neglect gravitational potential energy.) (b) What is the highest position of the oscillating mass?

Given: $m = 0.50$ kg
$y_o = 10$ cm $= 0.10$ m
$-A = 5.0$ cm $= 0.050$ m

Find: (a) E (total energy)
(b) A (amplitude)

Since this oscillation is vertical, the displacement is designated by y, the coordinate commonly used for this direction.

(a) We are asked to find the total energy. Eq. 13.5 ($E = \tfrac{1}{2}kA^2$) gives the energy in terms of the spring constant and amplitude. We know the latter, and we can find the former by considering the spring at its equilibrium position. (Suspending a mass on a spring is a common method of determining k.) When the suspended mass and the stretched spring are in equilibrium (Fig. 13.3a), the weight force of the mass and the spring force are equal and opposite. That is,

$$F = w$$

or

$$ky_o = mg$$

Thus,

$$k = \frac{mg}{y_o} = \frac{(0.50 \text{ kg})(9.8 \text{ m/s}^2)}{0.10 \text{ m}} = 49 \text{ N/m}$$

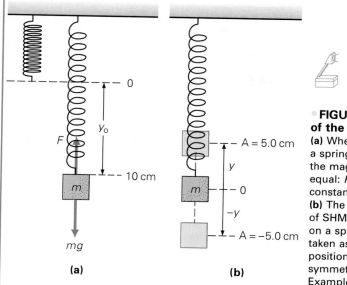

Figure 13.3

FIGURE 13.3 Determination of the spring constant (a) When a mass suspended on a spring is in equilibrium, the magnitudes of the forces are equal: $F = ky_0 = mg$. The spring constant k can then be computed. (b) The zero reference point of SHM of a mass suspended on a spring is conveniently taken as the equilibrium position, as the motion is symmetric about this point. Example 13.2

Then, the magnitude of the initial downward displacement from the equilibrium position (5.0 cm) is the amplitude of the oscillation, and

$$E = \tfrac{1}{2}kA^2 = \tfrac{1}{2}(49 \text{ N/m})(0.050 \text{ m})^2 = 0.061 \text{ J}$$

(b) Once set into motion, the mass oscillates up and down through the equilibrium position. Since the motion is symmetric about this point, it is designated as the zero reference point of the oscillation (Fig. 13.3b). The amplitude of the oscillation is 5.0 cm, so the highest position of the mass will be 5.0 cm above the equilibrium position.

13.2 ■ Equations of Motion

We refer to the **equation of motion** for an object or particle as that equation which gives its position as a function of time. For example, the equation of motion with a constant linear acceleration is $x = v_0 t + \tfrac{1}{2}at^2$, where v_0 is the initial velocity (Chapter 2). However, the acceleration is not constant for simple harmonic motion, so the kinematic equations of Chapter 2 do not apply to this case.

The equation of motion for an object in simple harmonic motion can be derived using a relationship between simple harmonic and uniform circular motions. SHM can be simulated by a component of uniform circular motion, as illustrated in ● Fig. 13.4. Note that as the illuminated object moves in uniform circular motion (with constant angular speed ω) in a horizontal plane, its shadow moves back and forth horizontally, following the same path as the mass on the spring, which is in simple harmonic motion. Since the shadow and the mass have the same position at any time, it follows that the equation of *horizontal* motion for the object in circular motion is the same as the equation of motion for horizontally oscillating mass on the spring.

Figure 13.4

FIGURE 13.4 Reference circle for horizontal motion (a) The shadow of the object in uniform circular motion has the same horizontal motion as the mass on the spring in simple harmonic motion. (b) The motion can thus be described by $x = A \cos \theta = A \cos \omega t$.

Eq. 7.5, Section 7.2

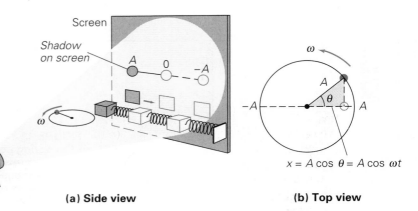

(a) Side view   (b) Top view

$x = A \cos \theta = A \cos \omega t$

From the reference circle in Fig. 13.4b the x coordinate (position) of the object is given by

$$x = A \cos \theta$$

But the object moves with a constant angular velocity with a magnitude ω, and in terms of angular distance θ we have $\theta = \omega t$, so

$$x = A \cos \omega t \qquad (13.8)$$

The angular speed (in rad/s) is sometimes called the angular frequency, since $\omega = 2\pi f$, where f is the frequency of revolution or rotation. (Note in the figure that this is the same as the frequency of oscillation of the mass on the spring.) Thus,

$$x = A \cos 2\pi f t \qquad (13.9)$$

Also, $f = 1/T$, so

$$x = A \cos \frac{2\pi t}{T} \qquad (13.10)$$

Different forms of an equation of motion

Equations 2.4 and 2.7–2.10, Section 2.4

Equations 13.8 through 13.10 are three equivalent forms of the equation of motion for a mass in simple harmonic motion. Any one of them may be used for convenience, depending on the known parameters. (Recall how the kinematic equations in Chapter 2 were expressed in different or combined forms for convenience.) For example, suppose you are given the time t in terms of the period T—say $t_0 = 0$, $t_1 = T/2$, and $t_3 = T$—and asked to find the position of an object in SHM at these times.

In this case, it is convenient to use Eq. 13.10, and

$$t_0 = 0 \qquad x_0 = A \cos 2\pi(0)/T = A \cos 0 = A$$

$$t_1 = \frac{T}{2} \qquad x_1 = A \cos 2\pi(T/2)/T = A \cos \pi = -A$$

$$t_2 = T \qquad x_2 = A \cos 2\pi T/T = A \cos 2\pi = A$$

Hence, the results tell us that the object was initially at $x = A$. One-half period later it was at $x = -A$ or the opposite extreme of its oscillation; and at a time of one period (T), it was back where it started, which is to be expected since the motion is periodic.

The period of a mass oscillating on a spring can be expressed in terms of the system's parameters m and k. These are general parameters—that is, we can experimentally select different masses (m) and springs with different constants (k). Our algebraic equation of motion describes the motion of any combination once you know or select the numerical values. It can be shown using a reference circle as in Fig. 13.4b that the period for a mass oscillating in SHM on a spring is given by

$$T = 2\pi \sqrt{\frac{m}{k}} \quad \text{(period of mass oscillating on a spring)} \quad (13.11)$$

Thus, the greater the mass, the greater the period; and the greater the spring constant (the stiffer the spring), the smaller the period.

Since $f = 1/T$,

$$f = \frac{1}{2\pi} \sqrt{\frac{k}{m}} \quad \text{(frequency of mass oscillating on a spring)} \quad (13.12)$$

Thus, the greater the spring constant (the stiffer the spring), the faster or more frequently the spring vibrates, as you might expect.

Also, note that since $\omega = 2\pi f$, we may write

$$\omega = \sqrt{\frac{k}{m}} \quad (13.13)$$

As another example, a simple pendulum (with no energy losses) undergoes simple harmonic motion for small angles of oscillation. The equation for the period of a pendulum in simple harmonic motion is similar in form to that for a mass oscillating on a spring. It can be shown that the period of a simple pendulum oscillating through a small angle ($\theta \leq 10°$) is given by

$$T = 2\pi \sqrt{\frac{L}{g}} \quad \text{(period of a pendulum)} \quad (13.14)$$

where L is the length of the pendulum and g is the acceleration due to gravity. Note that the pendulum period is independent of the mass of the bob.

EXAMPLE 13.3 ■ Applying the Equation of Motion

A mass on a spring oscillates horizontally on a frictionless surface with an amplitude of 15 cm, a frequency of 0.20 Hz, and an equation of motion as given in Eq. 13.8. (a) What is the displacement of the mass at $t = 3.1$ s? (b) How many oscillations does it make during this time?

Solution.

Given: $A = 15$ cm $= 0.15$ m
$\quad\quad\quad f = 0.20$ Hz
$\quad\quad\quad x = A \cos \omega t$ (Eq. 13.8)
$\quad\quad\quad t = 3.15$

Find: (a) x (displacement)
$\quad\quad\quad$ (b) n (number of oscillations)

Demonstration/activity: The deflection of most materials is approximately linearly proportional to the force for small deflections. Clamp one end of a meterstick to the desk and put weights of 20, 40 etc. grams on the end to observe this linear deflection. Measure its "spring constant." Then tape a larger weight of about 500 grams to the end of the stick and start it oscillating up and down. Measure its period and compare it with the calculated period. Close?

Demonstration/activity: Set up several pendulums and show how the period is independent of the mass and amplitude for small amplitudes.

(a) First, since we are given the frequency f, it is convenient to use the equation of motion in the form $x = A \cos 2\pi ft$ (Eq. 13.9), where $\omega = 2\pi f$. At $t = 0$, we have $x = A$, so initially the mass is released from its positive amplitude distance. Then, at $t = 3.1$ s,

$$x = A \cos 2\pi ft$$
$$= (0.15 \text{ m}) \cos [2\pi(0.20 \text{ s}^{-1})(3.1 \text{ s})] = (0.15 \text{ m}) \cos (3.9) = -0.11 \text{ m}$$

and the mass is 0.11 m on the other side of its equilibrium position.
(b) The number of oscillations (cycles) is equal to the product of the frequency (cycle/s) and the elapsed time (s), which are both given:

$$n = ft = (0.20 \text{ cycle/s})(3.1 \text{ s}) = 0.62 \text{ cycle}$$

Thus, at $t = 3.1$ s, the mass has traveled to its maximum negative displacement position (in 0.50 cycle) and is back at $x = -0.11$ m approaching its equilibrium position ($x = 0$).

■ Problem-Solving Hint

Note that in the part (a) calculation of Example 13.3, where we have $\cos (3.9)$, the angle is in radians, *not* degrees. Don't forget to set your calculator to radians (instead of degrees) when finding the value of a trigonometric function in equations for simple harmonic or circular motion.

EXAMPLE 13.4 ■ Period and Length of a Pendulum

Thomas Jefferson once proposed that a standard reference for time be a simple pendulum with a period of one second. What would be the length of such a pendulum?

Solution.
   Given: $T = 1.00$ s   *Find:* L (pendulum length)

The length of the simple pendulum can be found directly from Eq. 13.14. Squaring both sides of the equation and solving for L gives

$$L = \frac{T^2 g}{4\pi^2} = \frac{(1.00 \text{ s})^2(9.80 \text{ m/s}^2)}{4\pi^2} = 0.248 \text{ m}$$

This standard was not adopted. (Can you think of any reasons?)

A vertical reference circle can also be used to describe simple harmonic motion, as illustrated in ● Fig. 13.5 for a suspended mass oscillating on a spring. In this case, the vertical displacement is given by $y = A \sin \theta$. A development similar to that for the horizontal case leads to this equation of motion:

$$y = A \sin \omega t$$

Since $y = 0$ at $t = 0$, the mass initially started its upward motion from its equilibrium point ($y = 0$).

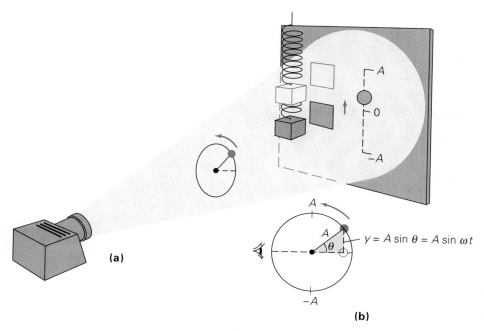

Figure 13.5

FIGURE 13.5 Reference circle for vertical motion **(a)** The shadow of the object in uniform circular motion has the same vertical motion as the mass oscillating on the spring in simple harmonic motion. **(b)** The motion can thus be described by $y = A \sin \theta = A \sin \omega t$.

However, if the mass starts initially from its maximum positive displacement position, as shown in ● Fig. 13.6, the resulting curve is a cosine, and

$$y = A \cos \omega t$$

Note that in this case the mass is released from $y = +A$ at $t = 0$.

Thus, the equation of motion for an oscillating mass may be either a sine or a cosine function. Both of these functions are referred to as being *sinusoidal*. That is, simple harmonic motion is described by a sinusoidal function of time.

Initial Conditions and Phase

You may be wondering how to decide whether to use a sine or cosine function to describe a particular case of simple harmonic motion. In general, the form of the function is determined by the initial displacement of the mass, or the initial condition of the system. This initial condition is the value of the displacement at $t = 0$ and tells how the system is initially set into motion.

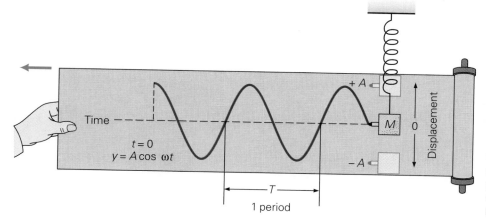

Figure 13.6

FIGURE 13.6 Sinusoidal equation of motion As time passes, the oscillating mass traces out a sinusoidal curve on the moving paper. In this case, $y = A \cos \omega t$.

If an object in simple harmonic motion has an initial displacement of $y = 0$ at $t = 0$, then the equation of motion is given by $y = A \sin \omega t$. That is, a sine curve satisfies this initial condition since $y = A \sin \omega t = A \sin \omega(0) = 0$. Note that $y = A \cos \omega t$ does *not* satisfy the initial condition: Since $\cos 0 = 1$, the initial displacement at $t = 0$ would be $y = A$ rather than $y = 0$. A situation for which a cosine function is appropriate in the equation of motion is shown in Fig. 13.6. Note that $y = A$ at $t = 0$, so $y = A \cos \omega t$ satisfies the initial condition.

For the general case we may write

General equation for SHM

$$y = A \sin(\omega t + \delta)$$

(13.15)

where $(\omega t + \delta)$ is the phase angle and the δ is the **phase constant**. The phase constant essentially matches the appropriate sinusoidal funtion to the motion. The phase constant conditions for four special cases are shown in Fig. 13.7.

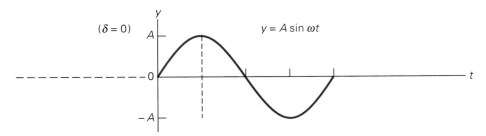

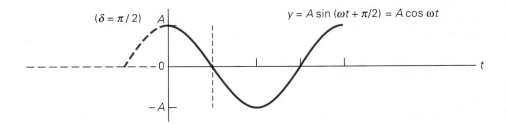

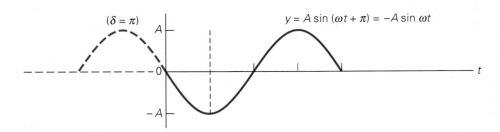

Figure 13.7

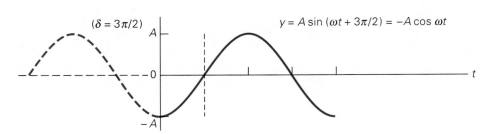

FIGURE 13.7 Phase differences With the general equation $y = A \sin (\omega t + \delta)$, motions are described by either sine and cosine terms for the phase constants shown. The initial displacement determines δ. Each curve is 90° (or $\pi/2$ rad) out of phase with the preceding one. Note that this is equivalent to shifting the wave a quarter of a cycle.

For $\delta = 0$, the equation of motion is $y = A \sin \omega t$, and the object is set into upward motion at its equilibrium position. For $\delta = 90°$ (or $\pi/2$ rad), the equation of motion is $y = A \cos \omega t$, and a mass oscillating up and down on a spring would be initially released at the $+A$ position. You probably recognize that the curve for the $\delta = 90°$ is a cosine. This can be shown algebraically using the trigonometric double angle formula: $\sin (a + b) = \sin a \cos b + \cos a \sin b$. Then, with $\delta = 90°$,

$$y = A \sin(\omega t + 90°) = A[\sin \omega t \cos 90° + \cos \omega t \sin 90°] = A \cos \omega t$$

since $\cos 90° = 0$ and $\sin 90° = 1$.

Notice that the sine and cosine curves have the *same shape*. In fact, they are actually the same curve shifted in phase. The curves for $\delta = 0$ and $\delta = 90°$ (or $\pi/2$ rad) are said to be 90° out of phase, or shifted by a quarter cycle, with respect to one another. (Notice in Fig. 13.7 that the curve for $\delta = 90°$ has essentially been "pulled over" or shifted to the left by a quarter cycle (90°) from the curve for $\delta = 0$.) The cases for $\delta = 180°$ (or π rad) and $\delta = 270°$ (or $3\pi/2$ rad) are each shifted an additional 90° out of phase.

Two masses oscillating in phase (having the same δ) will oscillate together. Two that oscillate completely out of phase (180° difference in δ) will always be going in opposite directions or be at opposite maximum displacements.

For each of the four special cases in Fig. 13.7, the equation of motion is either a sine or cosine curve. If the initial displacement is not zero or $\pm A$, then δ is not a multiple of 90°, and things get a bit more complicated. Eq. 13.15 still applies, but the double-angle formula given above gives an equation of motion with both sine and cosine terms.

Keep in mind that the phase angle $(\omega t + \delta)$ is usually expressed in radians. For example, if $y = \sin [0.50t + \pi]$ then at $t = 20$ s, we have $y = \sin [(0.50)(20) + \pi] = \sin 13$, or the sine of 13 rad. If you are using a calculator for such computations, again make sure it is in the "rad" mode and not the "deg" mode.

See Demonstration 11 for a pictorial view of a sinusoidal wave and phase.

Velocity and Acceleration in SHM

Expressions for the velocity and acceleration of an object in SHM can be easily derived using energy and force considerations. We have already derived an expression (Eq. 13.6) for the velocity of a mass oscillating on a spring in terms of the displacement. We can now express the velocity as a function of time. For the case of vertical simple harmonic motion, with the displacement $y = A \sin \omega t$, we can rewrite Eq. 13.6 as

$$v = \sqrt{\frac{k}{m}(A^2 - y^2)} = \sqrt{\frac{k}{m}(A^2 - A^2 \sin^2 \omega t)} = A\sqrt{\frac{k}{m}}\sqrt{1 - \sin^2 \omega t}$$

Since $\omega = \sqrt{k/m}$ (Eq. 13.13) and $\cos \theta = \sqrt{1 - \sin^2\theta}$ (see Appendix I), we can write this expressions as

$$v = A\omega \cos \omega t \qquad (13.16)$$

Max velocity $v = \pm \omega A$.

where the directional signs are given by the cosine function. Using Newton's second law to find the acceleration with the spring force $F = -ky$, we have

$$a = \frac{F}{m} = \frac{-ky}{m} = \frac{-k}{m} A \sin \omega t$$

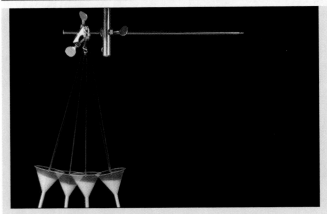

(a) A salt-filled funnel oscillates from a bi-filar suspension.

(b) The salt falls on a black-painted poster board that will be pulled in a direction perpendicular to the plane of the funnel's oscillation.

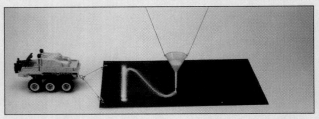

(c) Away we go.

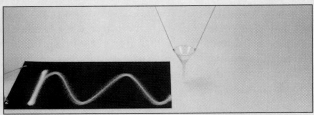

(d) The salt trail traces out a plot of displacement versus time, or $y = A \sin(\omega t + \delta)$. Note that in this case the phase constant is about $\delta = 90°$, and $y = A \cos \omega t$. (Why?)

A demonstration to show that SHM can be represented by a sinusoidal wave function. A "graph" of the wave function is generated with an analogue of a strip chart recorder.

and since $\omega = \sqrt{k/m}$,

$$a = -\omega^2 A \sin \omega t = -\omega^2 y \qquad (13.17)$$

Max acceleration $a = \pm \, \omega^2 A$.

Note that the functions for the velocity and acceleration are out of phase with that for the displacement (see Fig. 13.7). Since the velocity is 90° out of phase with the displacement, the velocity is greatest when the oscillating mass is at (passing through) its equilibrium position. The acceleration is 180° out of phase with the displacement (as indicated by the minus sign on the right-hand side of Eq. 13.17). Therefore, the acceleration is a maximum when the displacement is a maximum, or when the mass is at an amplitude position. At any position except the equilibrium position, the directional sign of the acceleration is the opposite of that of the displacement, as it should be for an acceleration resulting from a restoring force. At the equilibriuim position, the displacement and acceleration are both zero. (Can you see why?)

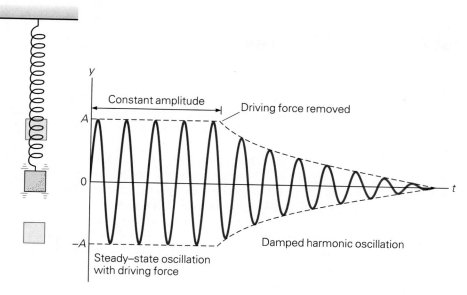

• FIGURE 13.8 Damped harmonic motion When a driving force adds energy to a system equal to its losses, the oscillation is steady with a constant amplitude. When the driving force is removed, the oscillations decay (that is, are damped), and the amplitude decreases exponentially with time.

Damped Harmonic Motion

Simple harmonic motion with a constant amplitude implies that there are no losses of energy, but in practical applications there are always some frictional losses. Therefore, to maintain a constant-amplitude motion, energy must be added to the system by some driving force (• Fig. 13.8). Without a driving force, the amplitude and the energy of an oscillator decrease with time, giving rise to **damped harmonic motion**. If the frictional force is proportional to the velocity of the oscillator, the motion changes over time as illustrated in Fig. 13.8. The time required for the oscillations to cease, or be damped out, depends on the magnitude of the damping force.

In many applications involving continuous periodic motion, damping is unwanted and necessitates energy input. However, in some instances, damping is desirable. For example, the dial in a spring-operated bathroom scale oscillates briefly before stopping at the weight. If not properly damped, these oscillations would continue for some time and you would have a wait before you could read your weight. Damping is also required for shock absorbers on automobiles and needle indicators on instruments measuring electrical quantities.

13.3 ■ Wave Motion

The world is full of waves of various types—some examples are water waves, shock waves, sound waves, waves generated by earthquakes, and light waves. Any type of wave results from a disturbance. In this chapter we will be concerned with mechanical waves, or those that are propagated in some medium. (Light waves, which do not require a propagating medium, will be considered in more detail in later chapters.)

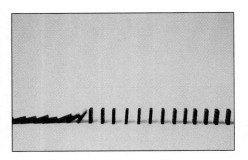

FIGURE 13.9 Energy transfer The propagation of a disturbance or a transfer of energy is seen in a row of falling dominos.

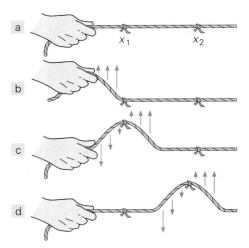

FIGURE 13.10 Wave pulse The hand disturbs the stretched rope in a quick up and down motion, and a wave pulse propagates along the rope. Note that the rope "particles" move up and down as the pulse passes.

When a medium is disturbed, energy is imparted to it. This energy propagates, or spreads, by means of interactions between the particles of the medium, as described at the beginning of the chapter. An analogy for this process is shown in Fig. 13.9, where the "particles" are dominoes. As each domino falls, it topples the one next to it. Thus energy is transferred from domino to domino, and the disturbance propagates through the medium.

In this case, there is no restoring force between the dominoes, so they do not oscillate as do particles in a continuous material medium. Therefore the disturbance moves in space, but it does not repeat itself in time at any one location.

Similarly, if the end of a stretched rope is given a quick shake, the disturbance transfers energy from the hand to the rope, as illustrated in Fig. 13.10. The forces acting between the rope "particles" cause them to move in response to the motion of the hand, and a **wave pulse** travels down the rope. Each "particle" goes up and then back down as the pulse passes by. This motion and that of the wave pulse propagation can be observed by tying pieces of ribbon onto the rope (at x_1 and x_2 in Fig. 13.10). As the disturbance passes point x_1, the ribbon rises and falls, as do the rope "particles." Later, the same occurs for the ribbon at x_2, which indicates that the disturbance energy is propagating or traveling along the rope.

In a continuous material medium, particles interact with their neighbors, and restoring forces cause them to oscillate when they are disturbed. Thus any disturbance not only propagates through space but may be repeated over and over in time at each place along the way. Such a regular, rhythmic disturbance in both time and space is called a **wave**, and the transfer of energy is said to take place by means of **wave motion**.

A continuous wave motion, or **periodic wave**, requires a constant disturbance from an oscillating source (Fig. 13.11). In this case, the particles move up and down continuously. If the driving source is such that a constant amplitude is maintained (and the restoring force has the form of Hooke's law), the particle motion can be described as simple harmonic motion.

Such wave motion will have sinusoidal forms (sine or cosine) in both time and space. Being sinusoidal in space means that if you took a photograph of the wave at any instant (freezing it in time), you would see a sinusoidal waveform (such as one of the curves in Fig. 13.11). However, if you looked at a single point in space as a wave passed by, you would see a particle of the medium oscillating up and down sinusoidally with time, like the mass on a spring discussed in Section 13.2. (For example, imagine looking through a thin slit at the moving paper in Fig. 13.6. The wave trace would be seen rising and falling like a particle.)

FIGURE 13.11 Periodic wave A continuous disturbance can set up a sinusoidal wave in stretched rope or string, and the wave travels down the rope with a wave speed *v*. Note that the rope "particles" oscillate vertically in simple harmonic motion. The distance between two adjacent points in phase, (e.g., at two crests) on the waveform is the wavelength of the wave.

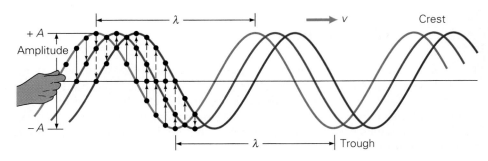

Wave Characteristics

Specific characteristics of sinusoidal waves are used to describe them. As for a particle in simple harmonic motion, the *amplitude* (*A*) of a wave is the magnitude of the maximum displacement or the maximum distance from the particle equilibrium position (Fig. 13.11). This corresponds to the height of a crest or the depth of a trough. As in SHM, the total energy transported by a wave is proportional to the square of its amplitude ($E \propto A^2$).

Wave amplitude

The distance between two successive crests (or troughs) is called the **wavelength** (λ). Actually, it is the distance between any two successive particles that are in phase (at identical points on the wave form). The crest and trough positions are usually used for convenience. Note that a wavelength corresponds spatially to one cycle.

The *frequency* (*f*) of a wave is the number of cycles per second, that is, the number of complete waveforms, or wavelengths, that pass by a given point during a second. Keep in mind that the wave is moving. The period $T = 1/f$ is the time for one complete waveform (a wavelength) to pass by a given point.

Since a wave moves, it has a **wave speed** (or velocity if direction is specified). You should be able to convince yourself that the wave (or a particular point, such as a crest) travels a distance of one wavelength λ in a time of one period T. Then, since $v = d/t$,

$$v = \frac{\lambda}{T} = \lambda f$$

(13.18)   **Wave speed**

Note that the dimensions are correct (length/time). In general, the wave speed depends on the nature of the medium.

EXAMPLE 13.5 ■ Wave Speed

A person on a pier observes incoming waves that have a sinusoidal form with a distance of 1.6 m between the crests. If a wave laps against the pier every 4.0 s, what are (a) the frequency and (b) the speed of the waves?

Solution. The distance between crests is the wavelength, so we have

Given: $\lambda = 1.6$ m   *Find:* (a) f (frequency)
$\quad\quad\quad T = 4.0$ s $\quad\quad\quad$ (b) v (wave speed)

(a) The lapping indicates the arrival of a wave crest, so 4.0 s is the period of the wave (time it takes to travel one wavelength, or the crest-to-crest distance). Then

$$f = \frac{1}{T} = \frac{1}{4.0 \text{ s}} = 0.25 \text{ s}^{-1} = 0.25 \text{ Hz}$$

(b) The frequency or the period can be used in Eq. 13.18 to find the wave speed:

$$v = \lambda f = (1.6 \text{ m})(0.25 \text{ s}^{-1}) = 0.40 \text{ m/s}$$

or

$$v = \frac{\lambda}{T} = \frac{1.6 \text{ m}}{4.0 \text{ s}} = 0.40 \text{ m/s}$$

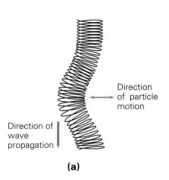
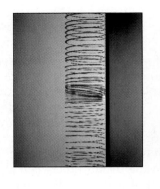
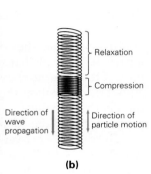

Relaxation

Compression

Direction of particle motion

Direction of wave propagation

Direction of wave propagation

Direction of particle motion

(a)　　　　　　　　　　　　　　　**(b)**

• FIGURE 13.12 Transverse and longitudinal waves (a) For a transverse wave, the particle motion is perpendicular to the direction of the wave velocity, as shown here in a spring for a wave moving toward the bottom of the page. (b) For a longitudinal wave, the particle motion is parallel to (or *along*) the direction of the wave velocity. Here a wave pulse moves toward the bottom of the page.

Demonstration/activity: Use a slinky to demonstrate both types of waves.

• FIGURE 13.13 Water waves Water waves are a combination of longitudinal and transverse motions. (a) At the surface, the water particles move in circles, but their motions become more longitudinal with depth. (b) When a wave approaches shore, the lower particles are forced into steeper paths until finally the wave breaks or falls over to form a surf.

Types of Waves

Waves may be divided into two types based on the direction of the particles' oscillations relative to the wave velocity. A **transverse wave** is one for which the particle motion is perpendicular to the direction of the wave velocity. The wave produced in a stretched string (Fig. 13.11) is an example of a transverse wave, as is the wave shown in • Fig. 13.12a. A transverse wave is sometimes called a shear wave because the disturbance supplies a force that tends to shear the medium. Shear waves can propagate only in solids, since a liquid or a gas cannot support a shear. That is, a liquid or a gas does not have sufficient restoring forces between its particles to propagate a transverse wave.

In a **longitudinal wave**, the particle oscillation is parallel to the direction of the wave velocity. A longitudinal wave may be produced in a stretched spring by moving the coils back and forth along the spring axis (Fig. 13.12b). Alternating pulses of compressions and relaxations move along the spring. A longitudinal wave is sometimes called a *compressional* wave.

Sound waves are another example of longitudinal waves. A periodic disturbance produces compressions in the air. The intervening relaxations are called rarefactions because the density of the air in these regions is reduced, or rarefied. Sound waves will be discussed in detail in Chapter 14.

Longitudinal waves can propagate in solids, liquids, and gases. All phases of matter can be compressed to some extent. The propagations of transverse and longitudinal waves in different media give information about the Earth's interior structure, as discussed in the Insight feature.

The sinusoidal profile of water waves might make you think that these are transverse waves. Actually, they reflect a combination of longitudinal and transverse motions (• Fig. 13.13). The particle motion may be nearly circular at the surface but becomes more elliptical with depth, eventually becoming longitudinal. A hundred meters or so below the surface of a large body of

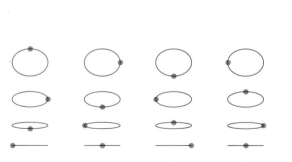

(a)

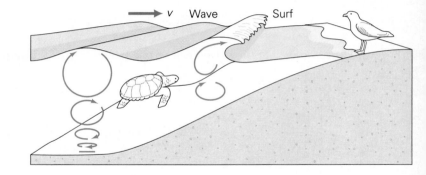

v　Wave　　Surf

(b)

The Earth's interior structure is still something of a mystery. The deepest mine shafts and drillings extend only a few kilometers into the Earth. Using waves to probe the Earth's structure is one way to investigate it further. Appropriately, waves generated by earthquakes have proved to be especially useful for this purpose. Seismology is the study of these so-called seismic waves.

Earthquakes are caused by the sudden release of built-up stress along cracks and faults, such as the famous San Andreas Fault in California. The geological theory of plate tectonics views the outer layer of the Earth as being a series of rigid plates, or huge slabs of rock, that are in very slow motion relative to one another. Stresses are continually being built up, particularly along boundaries between plates.

The energy from a stress-relieving disturbance propagates outward (seismic) waves. These are of two general types: surface waves and body waves. The surface waves, which move along the Earth's surface, account for most earthquake damage. Body waves, as the name implies, travel through the Earth. There are both longitudinal and transverse vibrations. The compressional (longitudinal) waves are called P waves, and shear (transverse) waves are called S waves (see Fig. 1). The P and S stand for primary and secondary and indicate the waves' relative speeds (actually, their arrival times at monitoring stations). Primary waves travel through materials faster than do secondary waves and are detected first. An earthquake's rating on the Richter scale is related to the amplitude or energy of the seismic waves.

Seismic stations around the world monitor these

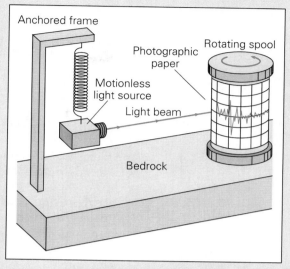

FIGURE 2 A simple seismograph The amplitude of the ground vibrations is proportional to the energy of the waves, and the amplitude of these waves is recorded on a seismograph. (See text for description.)

waves with sensitive detecting instruments called seismographs (Fig. 2). With this data, the paths of the waves through the Earth can be mapped, giving knowledge of the interior structure. The Earth's interior seems to be divided into three general regions: the crust, the mantle, and the core, which has a solid inner part and a liquid outer part.* The locations of the boundaries of these regions are determined by the refraction, or bending, of the waves (see Section 13.4).

Longitudinal waves can travel through solids or liquids, but transverse waves can travel only through solids. When an earthquake occurs at a particular location, P waves are detected on the other side of the Earth and S waves are not (see Fig. 1). The absence of S waves in a shadow zone leads to the conclusion that the Earth must have a region near its center that is in the liquid phase. This region is a highly viscous metallic liquid, but definitely a liquid since it does not support a shear (transverse waves are not propagated). When the transmitted P waves enter and leave the liquid region, they are refracted (bent). This gives rise to a P wave shadow zone, which indicates that only the outer part of the core is liquid.

As you will learn in Chapter 18, the combination of a liquid outer core and rotation may be responsible for the Earth's magnetic field.

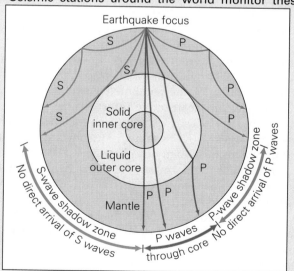

FIGURE 1 Earth waves Earthquakes produce waves that travel through the Earth. Because transverse S waves are not detected on the opposite side of the Earth, scientists believe that part of the Earth's interior is a viscous liquid.

*The crust is about 24–30 km (15-20 mi) thick; the mantle is 2900 km (1800 mi) thick; and the core has a radius of 3450 km (2150 mi). The solid inner core has a radius of about 1200 km (750 mi).

water, the wave disturbances have little effect. For example, a submarine at these depths is undisturbed by large waves on the ocean's surface. As a wave approaches shallower water near shore, the water particles have difficulty completing their elliptical paths. Finally, when the water becomes too shallow, the particles can no longer move through the bottom parts of their paths and the wave breaks. Its crest falls forward to form surf.

13.4 ▪ Wave Phenomena

Among the phenomena common to all waves are interference, superposition, reflection, refraction, and diffraction.

Interference and Superposition

Strange as it may seem, when two or more waves meet, or pass through the same region of a medium, they pass through each other and proceed without being altered. While they are in the same region, the waves are said to be interfering.

What happens during interference, or what does the combined waveform look like? The relatively simple answer to this is given by the **principle of superposition**:

> At any time, the combined waveform of two or more interfering waves is given by the sum of the displacements of the individual waves at each point in the medium.

This principle is illustrated in ● Fig. 13.14. The displacement of the combined waveform at any point is given by $y = y_1 + y_2$, where y_1 and y_2 are the displacements of the individual pulses at that point (directions are indicated by plus and minus signs). Interference, then, is the physical addition of waves.

In Fig. 13.14, the vertical displacements of the two pulses are in the same direction, and the amplitude of the combined waveform is greater than that of either pulse. This is called **constructive interference**. On the other hand, if two pulses tend to cancel each other when they overlap (one pulse has a negative displacment), the amplitude of the combined waveform is smaller than that of either pulse. This is called **destructive interference**.

Special cases of constructive and destructive interference are shown in ● Fig. 13.15. These occur when waves with the same frequency and amplitude meet. When these interfering waves are exactly in phase (crest coincides with crest), the amplitude of the combined waveform is twice that of either individual wave. This is sometimes referred to as **total constructive interference**. When these interfering waves are completely out of phase (180° difference, or crest coinciding with trough), the waveforms disappear; that is, the amplitude of the combined wave is zero. This is called **total destructive interference**.

The word "destructive" unfortunately tends to imply that the energy as well as the form of the waves is destroyed. This is not the case. At the point of total destructive interference, the wave energy is stored in the medium as potential energy. Also, keep in mind that the total destructive interference of waves is an instantaneous condition. As time progresses, the waveforms reappear.

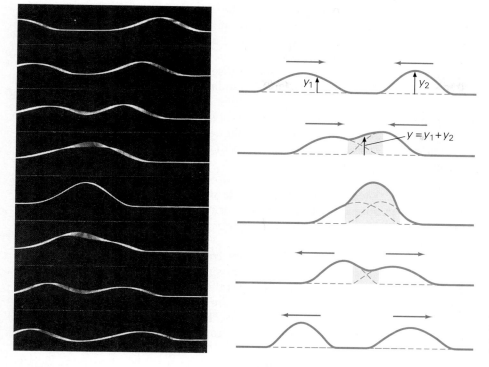

FIGURE 13.14 Principle of superposition When two waves meet, they interfere. The displacement at any point on the combination wave is equal to the sum of the displacements on the individual waves, $y = y_1 + y_2$.

Reflection, Refraction, and Diffraction

Besides meeting other waves, waves can (and do) meet objects or a boundary with another medium. In such cases, several things may occur. One of these is reflection. **Reflection** occurs when a wave strikes an object or comes to a boundary of another medium and is at least partly diverted backward. An echo is an example of reflection (sound waves), and mirrors reflect light waves.

Two cases of reflection are illustrated in ●Fig. 13.16. If the end of the string is fixed, the reflected pulse is inverted, or undergoes a 180° phase shift (Fig. 13.16a). This is because the pulse causes the string to exert an upward force

Have a student hold one end of the slinky fixed to show how the wave pulse reflects from a fixed end. Then tie a 3-foot string to the end and have a student hold the end of the string to demonstrate reflection from a loose end. This difference in reflection will come up again in optics.

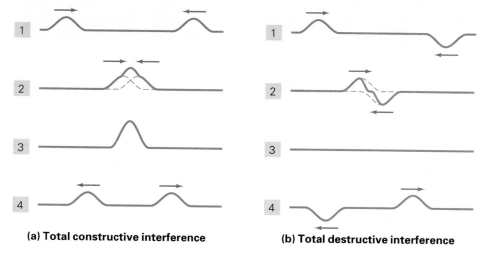

(a) Total constructive interference

(b) Total destructive interference

FIGURE 13.15 Interference (a) When two waves of the same frequency and amplitude meet and are in phase, they interfere constructively (shown here for wave pulses). When the waves are exactly superimposed (3), total constructive interference occurs. **(b)** When the interfering waves are completely out of phase (180°) and are exactly superimposed (3), total destructive interference occurs.

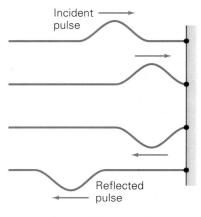

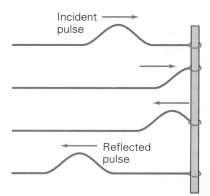

(a) 180° Phase shift

(b) No phase shift

FIGURE 13.16 Reflection
(a) When a wave (pulse) in a string is reflected from a fixed boundary, the reflected wave is inverted, or undergoes a 180° phase shift. **(b)** If the string is free to move at the boundary, there is no phase shift of the reflected wave.

on the wall, and the wall exerts an equal and opposite downward force on the string (by Newton's third law). Thus, the reflected pulse is downward, or inverted. If the end of the string is free to move, then the reflected pulse is not inverted (zero or no phase shift). This is illustrated in Fig. 13.16b, where the string is attached to a light ring that can move freely on a smooth pole. The ring is accelerated upward by the incoming pulse and is then brought downward.

In general, when a wave strikes a boundary, some of its energy is reflected and some is transmitted or absorbed. When a wave crosses a boundary into another medium, its velocity changes because the new material has different characteristics. Entering the medium obliquely (at an angle), the transmitted wave moves in a direction different from that of the incident wave. This phenomenon is called **refraction**.

Diffraction refers to the bending of waves around an edge of an object. For example, if you stand along an outside wall of a building near the corner, you can hear people talking around the corner. Assuming there are no reflections or air motion (wind), this would not be possible if the sound waves traveled in a straight line (Fig. 13.17a).

In general, the effects of diffraction are greater when the size of the diffracting object or opening is about the same or smaller than the wavelength of the waves. For example, water waves in a pond or lake pass by blades of grass or reeds with little noticeable effect because the widths of these objects are much smaller than the wavelength of the waves. The dependence of diffraction on wavelength and size of the object is illustrated in Fig. 13.17b using a ripple tank.

Reflection, refraction, and diffraction will be considered in more details for light waves in Chapter 21.

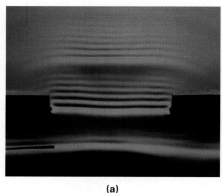

(a)

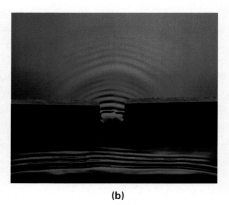

(b)

FIGURE 13.17 Diffraction   Diffraction effects are greater when the opening (or object) is about the same size as or smaller than the wavelength of the waves, as shown here for water waves. **(a)** With an opening much larger than the wavelength of the plane water waves, diffraction is noticeable only near the edges. **(b)** With an opening about the same size as the wavelength of the waves, diffraction produces circular waves.

13.5 ■ Standing Waves and Resonance

If you shake one end of a stretched rope, waves travel down it to the fixed end and are reflected back. The waves going down and back interfere. In most cases, the combined waveforms will have a changing, jumbled appearance. But if the rope is shaken at just the right frequency, a waveform appears to stand in place along the rope. Appropriately, this phenomenon is called a **standing wave** (see ● Fig. 13.18a and Demonstration 12, p. 445).

Some points on the rope are stationary and are called **nodes**. By the principle of superposition, the interfering waves must cancel each other completely at these points; that is, a crest exactly coincides with a trough, and the rope does not undergo a displacement. At all other points, the rope oscillates back and forth with the same frequency. The points of maximum amplitude, where constructive interference is greatest, are called **antinodes**. As you can see in Fig. 13.18b, adjacent antinodes are separated by half of a wavelength ($\lambda/2$), or one loop; adjacent nodes are also separated by half of a wavelength. The wavelength is that of the interfering waves that produce the standing wave.

Standing waves can be generated in a rope by more than one driving frequency; the higher the frequency, the greater the number of oscillating half-wavelength loops in the rope. The frequencies at which large-amplitude standing waves are produced are called **natural frequencies**, or **resonant frequencies**. The standing wave patterns are called normal, or resonant, modes of vibration. In general, all things have one or more natural frequencies, which depend on such factors as mass, elasticity or restoring force, and geometry (boundary conditions). The natural frequencies of a system are sometimes called its characteristic frequencies.

A stretched string or rope can be analyzed to determine its natural frequencies (● Fig. 13.19). The boundary conditions are that the ends are fixed; thus, there must be a node at each end. The number of closed segments or loops of a standing wave that will fit between the nodes at the ends (along the length of the string) is equal to an integral number of half-wavelengths. Note that $L = \lambda_1/2$, $L = 2\lambda_2/2$, $L = 3\lambda_3/2$, $L = 4\lambda_4/2$, and so on. In general,

$$L = n\frac{\lambda_n}{2} \quad \text{or} \quad \lambda_n = \frac{2L}{n} \quad \text{(for } n = 1, 2, 3, \ldots)$$

The natural frequencies of oscillation, where v is the wave speed, are

$$f_n = \frac{v}{\lambda_n} = \frac{nv}{2L} \quad \text{(for } n = 1, 2, 3, \ldots) \qquad \text{(natural frequencies for a stretched string)} \qquad (13.19)$$

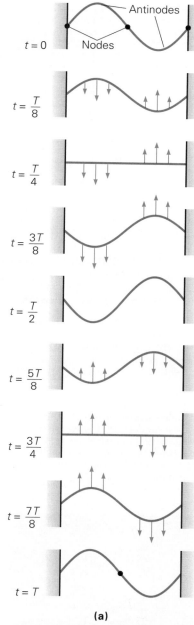

●FIGURE 13.18 Standing waves Standing waves are formed by interfering waves traveling in opposite directions. **(a)** Conditions of destructive and constructive interference recur as each wave travels a distance of $\lambda/4$ in a time of $t = T/4$. The motions of the rope particles are indicated by the arrows. **(b)** This gives rise to standing waves with stationary nodes and maximum amplitude antinodes.

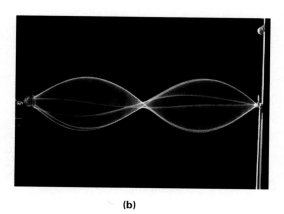

(b)

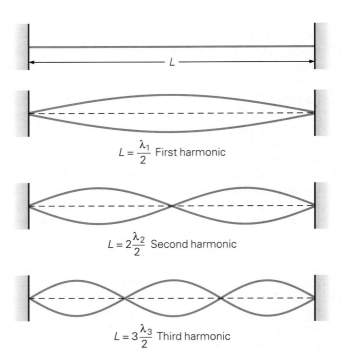

$L = \dfrac{\lambda_1}{2}$ First harmonic

$L = 2\dfrac{\lambda_2}{2}$ Second harmonic

$L = 3\dfrac{\lambda_3}{2}$ Third harmonic

Demonstration/activity: A 1/4-inch rubber rope several feet long works well for demonstrating standing waves. Attach a variable-speed electric motor to one end.

•**FIGURE 13.19 Resonant frequencies**   A stretched string can have standing waves only at certain frequencies. These correspond to the numbers of half-wavelength loops that will fit along the length of string between the nodes at the fixed ends.

Harmonics are whole-number multiples of the fundamental frequency.

A musical note or tone is referenced to the fundamental vibrational frequency. In musical terms, the first overtone is the second harmonic, the second overtone is the third harmonic, and so on.

•**FIGURE 13.20 Fundamental frequencies**   Performers on string instruments such as the violin or guitar use their fingers to stop or fret the strings. By pressing a string against the fingerboard the player alters the amount of its length that is free to vibrate. This changes the resonant frequency of the string and the pitch of the tone it produces.

The lowest natural frequency ($f_1 = v/2L$) is called the **fundamental frequency**. All of the other natural frequencies are integral multiples of the fundamental frequency: $f_n = nf_1$ ($n = 1, 2, 3, \ldots$). The set of frequencies $f_1, f_2 = 2f_1, f_3 = 3f_1$, and so on, is called a **harmonic series**: f_1 is the *first harmonic*, f_2 the *second harmonic*, and so on.

Strings fixed at each end are found in musical instruments such as violins and guitars. When such a string is excited, the resulting vibration will include several harmonics in addition to the fundamental frequency. The number of harmonics depends on how the string is excited, that is, plucked or bowed. In any case, it is the combination of harmonic frequencies that gives a particular instrument its characteristic sound quality. (More on this in Chapter 14.) As Eq. 13.19 shows, the fundamental frequency of a stretched string, as well as the other harmonics, depends on the length of the string. Think of how different notes are obtained on a particular string on a violin or a guitar (•Fig. 13.20).

Natural frequencies also depend on other parameters, such as mass and force, which affect the wave speed. For a stretched string, the wave speed (v) can be shown to be

$$v = \sqrt{\dfrac{F}{\mu}} \qquad (13.20)$$

where F is the tension force in the string and μ is the linear mass density (mass per unit length, $\mu = m/L$). Thus Eq. 13.19 can be written

$$f_n = \dfrac{nv}{2L} = \dfrac{n}{2L}\sqrt{\dfrac{F}{\mu}} \qquad \text{(for } n = 1, 2, 3, \ldots) \qquad (13.21)$$

Note that the greater the linear mass density of a string, the lower its natural frequencies are. As you may know, the low-note strings on a violin or guitar are thicker, or more massive, than the high-note strings.

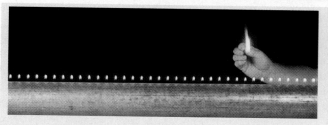

(a) Regularly spaced holes in a piece of downspout allow a line of uniformly sized flames when (natural) gas is supplied to the pipe. One end of the pipe is closed by a metal plate and the other end is fitted with a rubber diaphragm and loudspeaker.
(b) When an audio oscillator driving the speaker is tuned to give standing waves in the pipe, the presence

of pressure maxima (antinodes) and minima (nodes) are indicated by higher and lower flames, respectively. [See *The Physics Teacher*, 17, 307 (1979).]
Note: Some fast, contemporary music with good bass instead of an oscillator output produces a rapidly varying, dancing flame pattern.

A demonstration of a different type of standing wave form, where nodes and antinodes correspond to pressure minima and maxima, respectively, in the gas column.

EXAMPLE 13.6 ■ Fundamental Frequency

A particular string in a piano has an effective length of 1.15 m and a mass of 20.0 g and is under a tension of 6.30×10^3 N. (a) What is the fundamental frequency of the tone produced when the string is struck? (b) What are the frequencies of the next two harmonics?

Solution

Given: $L = 1.15$ m
$\quad\quad m = 20.0$ g $= 0.0200$ kg
$\quad\quad F = 6.30 \times 10^3$ N

Find: (a) f_1 (fundamental frequency)
$\quad\quad$ (b) f_2 and f_3 (frequencies of next two harmonics)

(a) The linear mass density of the string is

$$\mu = \frac{m}{L} = \frac{0.0200 \text{ kg}}{1.15 \text{ m}} = 0.0174 \text{ kg/m}$$

Then, using Eq. 13.21,

$$f_1 = \frac{1}{2L}\sqrt{\frac{F}{\mu}} = \frac{1}{2(1.15 \text{ m})}\sqrt{\frac{6.3 \times 10^3 \text{ N}}{0.0174 \text{ kg/m}}} = 262 \text{ Hz}$$

This is approximately the frequency of middle C (C_4) on a piano (261.6 Hz).
(b) Since $f_2 = 2f_1$ and $f_3 = 3f_1$,

$$f_2 = 2f_1 = 2(262 \text{ Hz}) = 524 \text{ Hz}$$

and

$$f_3 = 3f_1 = 3(262 \text{ Hz}) = 786 \text{ Hz}$$

The second harmonic corresponds approximately to C_5 on a piano, since the frequency doubles with each octave (or every eighth white key).

Demonstration/activity: Stretch a 1/4-inch rubber rope the length of the lecture room. Shake a pulse in the rope and measure the time it takes to make 3 or 4 round trips. Calculate its speed. Then measure the mass per unit length of the rope and the tension, and calculate the speed from expression given in the text. Do they agree?

EXAMPLE 13.7 ■ Changing the Pitch

You wish to raise the pitch of a violin or guitar string from the A note below middle C (220 Hz) to the A note above middle C (440 Hz). Would you (a) loosen the string so that its tension is halved, (b) tighten the string so as to double the tension, (c) use another string of the same material with half the diameter at the same tension, (d) use another string of the same material with twice the diameter at the same tension? *Clearly establish the reasoning and physical principle(s) used in determining your answer. That is, **why** did you select your answer?*

Reasoning and Answer. The fundamental frequency or pitch of a stretched string is given by Eq. 13.21:

$$f = \frac{1}{2L} \sqrt{\frac{F}{\mu}} \qquad (n = 1)$$

Since the frequency is proportional to the *square root* of the tension force F, loosening the string or decreasing F would not increase the frequency. Nor would doubling the tension double the frequency ($\sqrt{2F} \neq 2\sqrt{F}$). Thus, neither (a) nor (b) is the correct answer.

The question then is, how does the frequency vary with the diameter of the string? Recall that μ is the mass per unit length ($\mu = m/L$), and the mass can be written in terms of the volume density (ρ) and the geometry of the string, $m = \rho V$, where V is the volume of the string. The volume can be written in terms of its circular cross-sectional area (assumed constant) and the length of the string:

$$V = AL = \frac{\pi d^2 L}{4}$$

where d is the diameter of the string ($A = \pi r^2 = \pi d^2/4$).

Writing the frequency in terms of these parameters, we have

$$f = \frac{1}{2L} \sqrt{\frac{F}{\mu}} = \frac{1}{2L} \sqrt{\frac{FL}{m}} = \frac{1}{2L} \sqrt{\frac{4F}{\rho \pi d^2}}$$

So, with constant F and L (the active length of the string between the bridge and neck of a stringed instrument is constant), the frequency is inversely proportional to the diameter of the string,

$$f \propto \frac{1}{d}$$

Hence, using a string with half the diameter ($d_2 = d_1/2$) would double the frequency and the answer is (c).

Forming a ratio shows this explicitly,

$$\frac{f_2}{f_1} = \frac{d_1}{d_2} \quad \text{or} \quad f_2 = \left(\frac{d_1}{d_2}\right) f_1$$

For $f_2 = 2f_1$ (that is, 440 Hz = 2 × 220 Hz), we must have $d_1/d_2 = 2$ or $d_2 = d_1/2$.

Follow-up Exercise. The pitch of a violin string is A below middle C (220 Hz). Could you tune this string to middle C (264 Hz)? If so, how? (*Reasoning and answer may be found in the Answers to Exercises section at the back of the book.*)

When an oscillating system is driven at one of its natural, or resonant, frequencies, maximum energy transfer to the system occurs. The system is physically suited to a natural frequency, or wants to vibrate at it, so to speak. The condition of vibrating at a natural frequency is referred to as **resonance**.

A common example of a system in mechanical resonance is someone being pushed on a swing. Basically, a swing is a simple pendulum and has only one resonant frequency for a given length $[f = (1/2\pi)\sqrt{g/L}]$. If you push the swing with this frequency, the amplitude and energy increase. (What happens if you push at a slightly different frequency? The energy transfer is no longer a maximum.)

Unlike a simple pendulum, a stretched string has many natural frequencies. A standing wave will be set up in such a string by a disturbance at any frequency. However, if the frequency of the driving force is not equal to one of the natural frequencies, the amplitude of the standing wave will be relatively small. When the frequency of the driving force is at one of the natural frequencies, more energy is transferred to the string, and the amplitude of the antinodes is relatively large.

Mechanical resonance is not the only type. When you tune a radio, you are changing the resonance frequency of an electrical circuit (Chapter 20) so that it will be driven by, or pick up, the frequency signal of the station you want. A classic example of undesirable mechanical resonance is described in the Insight feature.

INSIGHT ■ Unwanted Resonance

FIGURE 1 Galloping Gertie The collapse of the Tacoma Narrows Bridge on November 7, 1940.

When a large number of soldiers march over a small bridge, they are often ordered to break step, because it is possible that the marching frequency will correspond to one of the natural frequencies of the bridge and set it into resonant vibration. One famous incidence of bridge vibration wasn't due to marching soldiers, however, but to the driving force of the wind.

On the morning of November 7, 1940, winds gusting to 40–45 mi/h started the main span of the Tacoma Narrows Bridge (in Washington state) vibrating. The bridge, 2800 ft (855 m) long and 39 ft (12 m) wide, had been opened to traffic only 4 months earlier.

During the first month of use, small transverse modes of vibration had been observed. But on November 7, special wind effects drove the bridge in near resonance, and the main span vibrated with a frequency of 36 vib/min and an amplitude of 1.5 ft.* At 10 a.m., the main span began to vibrate in a torsional (twisting) mode in two segments with a frequency of 14 vib/min. The wind continued to drive the bridge in resonance and the vibrational amplitude increased. Shortly after 11 a.m., the main span collapsed (Fig. 1).

"Galloping Gertie" (the nickname given to the bridge) was rebuilt using the same tower foundations. However, the new design made the structure stiffer to increase its resonant frequency so that high winds could not produce unwanted resonance.

* It is doubtful that the gusting of the wind set the bridge into vibration. The wind velocity was moderately steady and gust fluctuations are normally random. One explanation for the driving source of the oscillations involves the formation of vortices as the wind blew past the bridge. Vortices are like the eddies that form in the water at the end of the oars when a boat is rowed. The wind blowing over and under the bridge formed vortices that rotated in opposite directions. The formation and "shedding" of the vortices (like eddies coming off oars) would impart energy to the bridge, and if the frequency of this action were near a natural frequency, a standing wave would be set up.

Important Concepts

You should be able to define and explain these chapter concepts clearly.

Hooke's law	wave pulse	total destructive interference
periodic motion	wave motion	reflection
simple harmonic motion (SHM)	wave	refraction
displacement	periodic wave	diffraction
amplitude	wavelength (λ)	standing wave
period	wave speed	node
frequency	transverse wave	antinode
hertz (Hz)	longitudinal wave	natural (resonant) frequencies
equation of motion	principle of superposition	fundamental frequency
initial conditions	constructive interference	harmonic series
phase constant	destructive interference	resonance
damped harmonic motion	total constructive interference	

Important Relationships for Review

These relationships are mathematical statements of the concepts and principles presented in the chapter. You should be able to identify the symbols and to explain the relationships before proceeding to the Exercises. In-text equation reference numbers are given for convenience.

Hooke's Law:

$$F = -kx \tag{13.1}$$

Frequency and Period for SHM:

$$f = \frac{1}{T} \tag{13.3}$$

Total Energy of a Spring and Mass in SHM:

$$E = \tfrac{1}{2}kA^2 = \tfrac{1}{2}mv^2 + \tfrac{1}{2}kx^2 \tag{13.4–5}$$

Velocity of Oscillating Mass on a Spring:

$$v = \pm\sqrt{\frac{k}{m}(A^2 - x^2)} \tag{13.6}$$

Maximum Speed of Oscillating Mass on a Spring:

$$v_{\text{max}} = \sqrt{\frac{k}{m}}\,(A) \tag{13.7}$$

Period of Mass Oscillating on a Spring:

$$T = 2\pi\sqrt{\frac{m}{k}} \tag{13.11}$$

Frequency of a Mass Oscillating on a Spring:

$$f = \frac{1}{2\pi}\sqrt{\frac{k}{m}} \tag{13.12}$$

Angular Frequency:

$$\omega = 2\pi f = \sqrt{\frac{k}{m}} \tag{13.13}$$

Period of a Simple Pendulum (small-angle approximation):

$$T = 2\pi\sqrt{\frac{L}{g}} \tag{13.14}$$

Displacement of a Mass in SHM:

$$y = A\sin(\omega t + \delta) \tag{13.15}$$

Velocity of a Mass in SHM ($\delta = 0$):

$$v = A\omega\cos\omega t \tag{13.16}$$

Acceleration of a Mass in SHM ($\delta = 0$):

$$a = -A\omega^2\sin\omega t = -\omega^2 y \tag{13.17}$$

Wave Speed:

$$v = \frac{\lambda}{T} = \lambda f \tag{13.18}$$

Natural Frequencies in a Stretched String:

$$f_n = \frac{nv}{2L} = \frac{n}{2L}\sqrt{\frac{F}{\mu}} \quad (\text{for } n = 1,2,3, \ldots) \tag{13.19–20}$$

Exercises

13.1 ■ Simple Harmonic Motion

1 A particle in SHM (a) has variable amplitudes, (b) has a restoring force in the form of Hooke's law, (c) has a frequency directly proportional to its period, (d) may be represented graphically by $y = ax + b$.   (b)

●2 The total energy of a particle in SHM is directly proportional to (a) f, (b) T, (c) A^2, (d) none of these.   (c)

3 If the amplitude of a mass in SHM is doubled, how are (a) the energy and (b) the maximum speed affected?
(a) four times greater   (b) twice as large

4 How is the speed of a mass in SHM affected the further it gets from its equilibrium position? Explain.   decreases

5 ■ A mass oscillating on a spring completes a cycle every 0.050 s. What is the frequency of the oscillation?   20 Hz

6 ■ A particle in simple harmonic motion has a frequency of 20 Hz. What is the period of oscillation? 5.0×10^{-2} s

7 ■ The frequency of a simple harmonic oscillation is doubled from 0.25 Hz to 0.50 Hz. What is the change in the period of oscillation?   2.0 s

●8 ■ Show that the total energy of a system in simple harmonic motion is given by $\frac{1}{2}m\omega^2 A^2$.   see ISM

9 ■ Show that for a pendulum to oscillate with the same frequency as a mass on a spring, the pendulum's length must be $L = mg/k$.   see ISM

10 ■■ What is the spring constant of a spring that stretches 4.0 cm when a 0.25-kg mass is suspended from it?   61 N/m

11 ■■ Atoms in a solid are in continual vibration due to thermal energy. At room temperature, the amplitude of the atomic vibrations is about 10^{-9} cm, and the frequency of oscillation is about 10^{12} Hz. (a) What is the approximate period of oscillation of an atom? (b) What is the magnitude of its velocity?   (a) 10^{-12} s;   (b) 63 m/s

12 ■■ A 0.350-kg mass resting on a horizontal frictionless surface is attached to a spring with a spring constant of 150 N/m. If the mass is pulled 0.100 m from its equilibrium position and released, what is the force on the mass and its acceleration at (a) $t = 0$, (b) $x = 0.050$ m, and (c) $x = 0$?
(a) $F = 15.0$ N; $a = 42.9$ m/s^2 (b) $F = 7.50$ N; $a = 21.4$ m/s^2
(c) $F = 0$; $a = 0$

13 ■■ A 0.20-kg mass is oscillating on a spring that has a spring constant of 40 N/m. The instantaneous speed of the mass is 0.95 m/s as it passes through its equilibriuim position. What is the total energy of the system? 9.0×10^{-2} J

●14 ■■ A 0.25-kg object suspended on a light spring is released from a position 15 cm above the equilibrium position. The spring has a spring constant of 75 N/m. (a) What is the total energy of the system? (Neglect gravitational potential energy.) (b) Does this energy depend on the mass of the object? Explain.   (a) 0.84   (b) no

15 ■■ What is the speed of the object in Exercise 14 when it is (a) 5.0 cm above its equilibrium position and (b) 5.0 cm below its equilibrium position? (c) What is the object's maximum speed, and where does this occur?   (a) 2.4 m/s
(b) 2.4 m/s   (c) 2.6 m/s, equilibrium position

16 ■■ A mass of 0.25 kg oscillates on a light spring with a spring constant of 15 N/m. If the mass was initially displaced 6.0 cm from its equilibrium position, what is (a) its total energy, and (b) the difference between its maximum speed and its speed at $x = 2.0$ cm?   (a) 2.7×10^{-2} J
(b) $\Delta v = 0.02$ m/s

17 ■■ For the waves in ●Figs. 13.21 and 13.22, which has more energy and how many times more?   figure 13.22 by a factor of 4

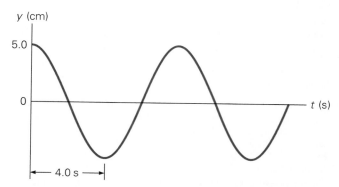

FIGURE 13.21 Wave energy and equation of motion
See Exercises 17 and 32.

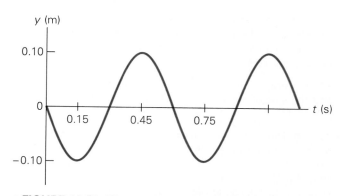

FIGURE 13.22 Wave energy and equation of motion
See Exercises 17 and 39.

18 ■■■ Two objects oscillate on springs in simple harmonic motion. One has twice the mass of the other and oscillates with half the amplitude. The more massive object's spring has a spring constant that is $\frac{2}{3}$ of that of the other spring. How do the total energies of the two systems compare?
$E_2/E_1 = 1/6$

19 ■■■ A 75-kg circus performer jumps from a height of 5.0 m onto a trampoline and stretches it a distance of 0.30 m. Assume that the trampoline obeys Hooke's law. (a) How far will it stretch if the performer jumps from a height

of 8.0 m? (b) How far will it be stretched when the performer stands still on it while taking a bow?   (a) 0.38 m
(b) 8.4 × 10⁻³ m

13.2 ■ Equations of Motion

20 An equation of motion for a particle in SHM (a) is always a cosine function, (b) reflects damping action, (c) is independent of the initial conditions, (d) may be represented by a sine and/or cosine function.   (d)

● **21** The angular frequency (ω) of an object oscillating in SHM on a light spring is directly proportional to (a) the object's mass, (b) the period of oscillation, (c) the phase constant, (d) the square root of the spring constant.   (d)

22 The apparatus in Fig. 13.6 demonstrates that the motion of a mass on a spring can be described by a sinusoidal function of time. How could this be demonstrated for a pendulum?   trace out path on scrolling vertical paper

23 Could simple harmonic motion be described using a tangent function? Explain.   no; tangent goes to ∞

24 ■ A mass of 0.30 kg oscillates in simple harmonic motion on a spring with a spring constant of 1.5×10^2 N/m. What are (a) the period and (b) the frequency of the oscillation?
(a) 0.28 s   (b) 3.6 Hz

25 ■ Write the general equation of motion for a mass that is on a horizontal frictionless surface and connected to a spring at equilibrium (a) if the mass is initially given a quick push away from the spring and (b) if the mass is pulled away from the spring and released.   (a) $x = A \sin \omega t$
(b) $x = \pm A \cos \omega t$

26 ■ Make sketches showing two masses oscillating on springs in equilibrium (a) in phase, (b) 90° out of phase, (c) 180° out of phase, (d) 270° out of phase, and (e) 30° out of phase.   see ISM

27 ■ Compute the percentage between the angle θ in rad and the sine of θ for (a) $\theta = 10°$, (b) $\theta = 20°$, and (c) $\theta = 25°$.   (a) 0.50%   (b) 2%   (c) 3%

● **28** ■■ The equation of motion for a 10-g mass in simple harmonic motion is given by $y = 7.0 \cos 10t$, where y is in centimeters. What are (a) the period and (b) the phase constant for this motion? (c) What is the total energy of the system?   (a) 0.63 s   (b) 90°   (c) 2.4×10^{-3} J

29 ■■ The equation of motion of a particle in simple harmonic motion is given by $y = 10 \sin 0.50t$ (where y is in centimeters). What are the particle's (a) displacement, (b) velocity, and (c) acceleration at $t = 1.0$ s?   (a) 4.8 cm
(b) 4.4 cm/s   (c) -1.2 cm/s²

30 ■■ A 0.50-kg mass oscillates in simple harmonic motion on a spring with a spring constant of 2.4×10^3 N/m. The spring was initially compressed 8.0 cm. (a) Write the equation of motion. (b) Determine the phase constant. (c) What is the displacement of the mass at $t = 0.25$ s?
(a) $y = (-8.0$ cm$) \cos [(69$ rad/s$) t]$ (b) 90°   (c) -0.23 cm
31 ■■ The simple harmonic motion of a 0.20-kg mass on

a spring is described by $y = 20 \sin 2\pi t$ (where y is in centimeters). At $t = \frac{1}{8}$ s, what are the particle's (a) displacement and (b) velocity? (c) What is the total energy of the system?   (a) 14 cm   (b) 89 cm/s   (c) 0.17 J

● **32** ■■ The motion of a particle is described by the curve in Fig. 13.21. Write the equation of motion in three equivalent forms (see Eqs. 13.8–13.10). $y = (5.0$ cm$) \cos \pi t/4$,
all the same
33 ■■ If the phase constant for the motion of the particle in Exercise 33 is $\delta = 90°$, what are the magnitudes (in general terms), the times, and the locations of (a) the maximum velocity and (b) the maximum acceleration?   see ISM

34 ■■ A 0.20-kg mass is suspended on a spring, which stretches a distance of 5.0 cm. The mass is then pulled down an additional distance of 10 cm and released. What are (a) the period of oscillation, (b) the equation of motion of the mass, (c) the total energy of the system, and (d) the displacement of the mass at $t = T/6$ s?   (a) 0.45 s
(b) $x = (-10$ cm$) \cos (14 t)$ (c) 0.20 J   (d) -5.0 cm
35 ■■ The equation of motion for a particle in simple harmonic motion is $x = 15 \sin 2\pi t$ (where x is in centimeters). At $t = 2.5$ s, what are the particle's (a) displacement, (b) velocity, and (c) acceleration?   (a) $x = 0$ (b) $v = 94$ cm/s
(c) $a = 0$
36 ■■ The motion of an object is described by $x = A \cos \omega t$. What are the general equations for the object's velocity and acceleration? $v = -A\omega \sin (\omega t)$;
$a = -A\omega^2 \cos (\omega t)$
37 ■■ Two equal masses m_1 and m_2 oscillate on light springs, the second with a spring constant twice that of the first. Which mass will have the greater period and how many times greater? $T_1 = (1.4) T_2$

38 ■■.A mass of 0.075 kg oscillates on a light spring, and another 0.075-kg mass oscillates as the bob of a simple pendulum in SHM. If the pendulum length is 0.30 m, what would be the spring constant of the spring if the periods of the oscillations are equal?   2.5 N/m

39 ■■ The motion of a 0.35-kg mass oscillating on a light spring is described by the curve in Fig. 13.22. (a) Write the equation for its displacement as a function of time. (b) What is the spring constant of the spring?
(a) $y = -(0.10$m$) \sin (10\pi/3)t$ (b) $k = 38$ N/m
40 ■■■ A grandfather clock has a pendulum that is 75.00 cm long. It is accidentally broken, and when repaired the length is shorter by 2.0 mm. (a) Will the repaired clock gain or lose time? (b) By how much will the repaired clock differ from the correct time (taken to be the time determined by the original pendulum in 24 h)? (c) If the pendulum were metal, would temperature make a difference in the timekeeping of the clock? Explain.   (a) gain time   (b) 1.8 min   (c) yes, because of linear expansion

13.3 ■ Wave Motion

41 Wave motion in a material medium involves (a) the propagation of a disturbance, (b) interparticle interactions, (c) the transfer of energy, (d) all of the preceding.   (d)

42 The wave speed in a medium (a) is inversely proportional to the period of particle oscillation, (b) has the same value as the angular frequency, (c) is directly proportional to the frequency of a transverse wave but not a longitudinal wave, (d) is directly proportional to the frequency of a compressional wave but not a shear wave. (a)

●43 What type(s) of wave(s) will propagate in (a) solids, (b) fluids, and (c) stretched strings? (a) transverse and longitudinal (b) longitudinal (c) transverse

44 ■ A longitudinal sound wave has a speed of 340 m/s in air. If this produces a tone having a frequency of 6.0 × 10³ Hz, what is the wavelength? 5.7 × 10⁻² m

45 ■ A transverse wave has a wavelength of 1.5 m and a frequency of 30 Hz. What is the wave speed? 45 m/s

●46 ■■ Light waves travel in a vacuum at a speed of 300,000 km/s. The frequency of visible light is about 10¹⁴ Hz. What is the approximate wavelength of visible light? 3.0 × 10⁻⁶ m

47 ■■ Red light has a frequency of 3.8 × 10¹⁴ Hz. What is the wavelength of this light in a vacuum? 7.9 × 10⁻⁷ m

48 ■■ A sonar generator on a submarine produces periodic ultrasonic waves at a frequency of 2.50 MHz. The wavelength of the waves in sea water is 4.80 × 10⁻⁴ m. When the generator is directed downward, an echo reflected from the ocean floor is received 16.7 s later. How deep is the ocean at that point? (Assume wave + length is constant at all depths.) 1.00 × 10⁴ m

49 ■■ In watching a transverse wave go by, a person notes that 12 crests go by in a time of 3.0 s. If the distance between two successive crests is measured to be 0.80 m, what is the speed of the wave? 3.2 m/s

50 ■■ A wave traveling in the $+x$ direction is shown in Fig. 13.23a. The particle displacement in the medium through which the wave travels is shown in Fig. 13.23b. (a) What is the amplitude of the traveling wave? (b) What is the wave speed? (a) 0.15 m (b) 0.15 m/s

51 ■■ A wave with a frequency of 60 Hz has a velocity of 12 m/s in a particular medium. (a) What is the wavelength? (b) If the wave is transmitted into another medium, in which it is propagated at a speed of 20 m/s, by how much will the wavelength change? (The frequency remains the same.) (a) 0.20 m (b) 13 cm

●52 ■■ The speed of longitudinal waves traveling in a long solid rod is given by $v = \sqrt{Y/\rho}$, where Y is Young's modulus and ρ is the density of the solid. If a disturbance has a frequency of 40 Hz, what is the wavelength of the waves it produces in (a) an aluminum rod and (b) a copper rod? [*Hint:* See Tables 9.1 and 9.2.] (a) 1.3 × 10² m (b) 88 m

53 ■■ As noted in Problem 52, the speed of longitudinal waves in a solid rod is given by $v = \sqrt{Y/\rho}$. Someone strikes a steel train rail with a hammer at a frequency of 0.50 Hz, and someone else puts his or her ear to the rail 1.0 km away. (a) How long after the first strike does the observer hear

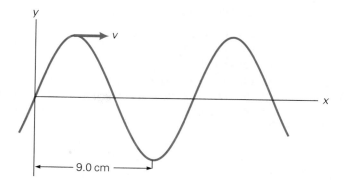

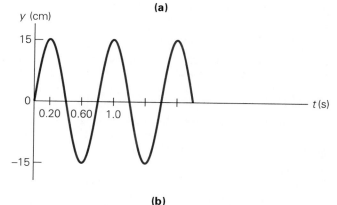

(b)

●FIGURE 13.23 How high and how fast? See Exercise 50.

the sound? (b) What is the time interval between successive sound pulses heard by the observer? [*Hint:* See Tables 9.1 and 9.2] (a) 0.20 s (b) 2.0 s

54 ■■■ The amplitude of a damped harmonic oscillation is given as a function of time by $A = A_0 e^{-t/\tau}$, where e is the base of the natural logarithm, A_0 is the maximum amplitude at $t = 0$, and τ is a constant that depends on the damping forces. (a) How long does it take for the amplitude to be halved if $\tau = 0.12$ s? (b) How long does it take for the amplitude to go to zero? (a) 8.3 × 10⁻³ s (b) theoretically it will never reach zero

55 ■■■ The general form of the equation of motion for a traveling wave, expressing both spatial and temporal dependence, is

$$y = A \sin(kx - \omega t)$$

where k is the *wave number* (*not* the spring constant) and is equal to $2\pi/\lambda$. (a) Show mathematically or graphically that this equation represents a wave traveling in the positive x direction. (b) Show that

$$y = A \sin(kx + \omega t)$$

represents a wave traveling in the negative x direction. see ISM

13.4 ■ Wave Phenomena

56 When waves meet each other and interfere, the resultant waveform is determined by (a) reflection, (b) refraction, (c) diffraction, (d) superposition. (d)

57 Refraction (a) involves constructive interference, (b) refers to a "bending" or change in direction at media interfaces, (c) is synonomous with diffraction, (d) occurs only for media or mechanical waves. (b)

58 What is destroyed when destructive interference occurs? only waveform; energy is not destroyed

13.5 ■ Standing Waves and Resonance

59 For a standing wave, a node has a displacement of (a) 0, (b) $+A$, (c) $-A$, (d) either (b) or (c). (a)

● **60** When a stretching string or cord is oscillated at the third harmonic of its natural frequencies, then the standing wave in the string will exhibit (a) 3 wavelengths, (b) 1/3 wavelength, (c) 3/2 wavelengths, (d) 2 wavelengths. (c)

61 A child's swing (a pendulum) has only one natural frequency f_o. Yet, it can be driven or pushed smoothly at frequencies of $f_o/2, f_o/3, 2f_o, 3f_o$. How is this possible? see ISM

62 Give an example or two of physical objects that can vibrate in resonance with $L = \lambda/4$ or $3\lambda/4$. [*Hint:* Think of some linear, metal objects.] see ISM

63 ■ The fundamental frequency of a stretched string is 220 Hz. What is the frequency of (a) the third harmonic and (b) the fourth harmonic? (a) 660 Hz (b) 880 Hz

64 ■ A standing wave is formed in a stretched string that is 4.0 m long. What is the wavelength of (a) the first harmonic and (b) the third harmonic? (a) 8.0 m (b) 2.7 m

65 ■ A father pushes a 25 kg child on a swing. If the length of the swing is 4.0 m, how often does the father have to push to drive the system in resonance? 4.0 s

● **66** ■■ A piece of rubber tubing with a linear mass density of 0.125 kg/m is stretched by a force of 9.0 N. (a) What will be the wave speed in the tubing? (b) If the stretched tubing has a length of 10 m, what are its natural frequencies?
(a) 8.5 m/s (b) $f_n = (0.43)n$ Hz; $n = 1, 2, 3, \ldots$

67 ■■ Find the first four harmonics for a string that is 1.5 m long, has a linear mass density of 2.3×10^{-3} kg/m, and is under a tension of 30 N. 38 Hz; 76 Hz; 114 Hz; 152 Hz

68 ■■ Will a standing wave be formed in a 4.0-m length of stretched string that transmits waves with a speed of 12 m/s if it is driven at a frequency of (a) 15 Hz or (b) 20 Hz? (a) yes (b) no

69 A standing wave has nodes at $x = 0$ cm, $x = 6$ cm, $x = 12$ cm, and $x = 18$ cm. (a) What is the wavelength of the waves that are interfering to produce this standing wave? (b) At what positions are the antinodes? (a) 12 cm
(b) 3 cm, 9 cm, 15 cm

70 ■■ Two stretched strings A and B have the same tension and linear mass density. Are any of the first six harmonics of the strings equal if the string lengths are (a) 1.0 m and 3.0 m or (b) 1.5 m and 2.0 m respectively? (a) the third harmonic of 3.0 m = first harmonic of 1.0 m (b) the third harmonic of 2.0 m = fourth harmonic of 1.5 m

71 ■■ A stretched string with a length of 2.0 m is driven by a variable frequency osciallator. It is observed that a standing wave with six closed loop segments occurs when the string is driven at 720 Hz. What frequency will produce a standing wave with four loop segments? 480 Hz

● **72** ■■ A "whip" CB antenna on a car is 3.0 m long. When the car is moving along a highway, a standing wave is observed in the antenna that has 1.5 loop segments in it. What is the wavelength of the fundamental frequency of the antenna? 12 m

73 ■■■ A thin, flexible metal rod is 1.0 m long. It is clamped at one end to a table, and the other end can vibrate freely. What are the natural frequencies of the rod if the wave speed in the material is 3.5×10^3 m/s?
$(2n + 1) (8.8 \times 10^2$ Hz$), n = 0, 1, 2, \ldots$

74 ■■■ In a common laboratory experiment on standing waves, the waves are produced in a stretched string by an electrical vibrator that oscillates at 60 Hz (● Fig. 13.24). The string runs over a pulley, and a hanger is suspended from the end. The tension in the string is varied by adding weights to the hanger. If the active length of string (the part that vibrates) is 1.5 m and this length of string has a mass of 0.1 g, what weights must be suspended to produce the first four harmonics in that length? 2.2 kg; 0.55 kg; 0.24 kg; 0.14 kg

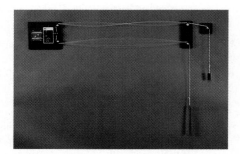

● **FIGURE 13.24 Standing waves in strings** Twin vibrating strings with standing waves. This demonstration model allows variations in string tension and length (linear density) of the string. Also, the vibration frequency may be adjusted. See Exercise 74 for a given set of conditions.

■ Additional Exercises

75 The forces acting on a simple pendulum are shown in ● Fig. 13.25. (a) Show that for the small-angle approximation $(\sin \theta \cong \theta)$ the force producing the motion has the same form as Hooke's law. (b) Show by analogy with a mass on a spring that the period of a simple pendulum is given by $T = 2\pi\sqrt{L/g}$. see ISM

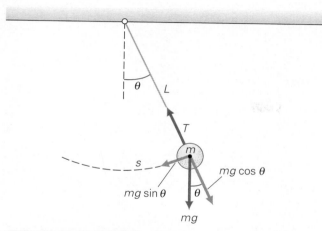

FIGURE 13.25 SHM of a pendulum See Exercise 75.

76 A 0.10-kg mass suspended on a spring is pulled to 8.0 cm below its equilibrium position and released. When the mass passes through the equilibrium position, it has a speed of 0.40 m/s. What is the speed of the mass when it is 3.0 cm from the equilibrium position? 0.37 m/s

77 A pendulum makes 6.0 complete cycles in a time of 10 s. What are the frequency and period of the pendulum's oscillation? 1.7 s; 0.60 Hz

78 The range of sound audible to the human ear has frequencies from about 20 Hz to 20 kHz. The speed of sound in air is 345 m/s. What are the limits of this audible range in wavelengths? 1.7 cm to 17 m

79 A steel piano wire is 60 cm long and has a mass of 3.0 g. If the tension in the wire is 550 N, what are the fundamental frequency and wavelength? 2.8×10^2 Hz; 1.2 m

80 A 0.15-kg mass oscillates on a spring that has a spring constant of 500 N/m. (a) What is the energy of the system if the mass oscillates with an amplitude of 0.10 m? (b) What is the energy of the system if the 0.15-kg mass is replaced with a 0.30-kg mass that oscillates with the same amplitude? (a) 2.5 J (b) 2.5 J

81 A stretched string is observed to have four equal loops in a standing wave when driven at a frequency of 420 Hz. What driving frequency will set up a wave with two equal segments? 210 Hz

82 On a violin, a correctly tuned A string has frequency of 440 Hz. If an A string is found to produce a sharp note of 450 Hz when under a tension of 500 N, to what should the tension be adjusted to produce the correct frequency? 478 N

83 An object in simple harmonic motion is described by $y = 0.20 \sin 1.8\pi t$ (where y is in centimeters). What is the speed of the object at $t = 10$ s? 1.1 cm/s

84 Radio broadcasting frequencies (of radio waves) have magnitudes in the kHz and MHz ranges. What are the corresponding wavelength and period ranges for radio waves? (The speed of radio waves is about 10^8 m/s.) 10^5 m to 10^2 m; 10^{-6} s to 10^{-3} s

85 A spring-loaded toy gun with a spring constant of 30 N/m is cocked through a distance of 5.0 cm. With what speed is a 0.065-kg ball propelled by the spring? (Neglect friction.) 1.1 m/s

86 For an object in simple harmonic motion with a frequency of 3.0 Hz and an amplitude of 7.5 cm, what are the magnitudes of (a) the maximum displacement, (b) the maximum velocity, and (c) the maximum acceleration? (d) Where is the object when each of these conditions occurs? (a) 7.5 cm (b) 1.4 m/s (c) -27 m/s^2 (d) ±7.5 cm

87 A mass resting on a horizontal frictionless surface is connected to a fixed spring. The mass is displaced 16 cm from its equilibrium position and released. At $t = 0.50$ s, the mass is 8.0 cm from its equilibrium position (and has not passed through it yet). What is the period of oscillation of the mass? 3.0 s

88 The speed of longitudinal waves in liquids is given by $v = \sqrt{B/\rho}$, where B is the bulk modulus and ρ is the density of the liquid. (a) What is the speed of longitudinal waves in water? (b) What is the speed of transverse waves in water? [*Hint:* See Tables 9.1 and 9.2.] (a) 1.5×10^3 m/s (b) 0, since fluids cannot support a shear

89 Assume that P and S waves from an earthquake with a focus near the Earth's surface travel through the Earth at average speeds of 8.0 km/s and 6.0 km/s, respectively. Assume that there is no deflection or refraction of the waves. (a) How long is the delay between the arrivals of successive waves at a seismic monitoring station located 90° latitude from the epicenter of the quake? (b) Do the waves cross the boundary of the mantle? (c) How long does it take for the waves to arrive at a monitoring station on the opposite side of the Earth? (a) 4×10^2 s (b) 1.9×10^3 km (c) 1.6×10^3 s; shear waves not transmitted through outer liquid core

90 Since the energy of a wave is proportional to the square of its amplitude ($E \propto A^2$), two waves of equal amplitude A would have a total energy $E = E_1 + E_2 = A^2 + A^2 = 2A^2$. Yet, when the waves interfere and overlap each other exactly, by the principle of superposition, $y = y_1 + y_2 = A + A = 2A$, and $E \propto (2A)^2 = 4A^2$. Is this correct? Can you explain your answer? [*Hint:* Think of what happens to the kinetic energy and the potential energy when two wave pulses are superimposed.] see ISM

14

Sound

Chapter 14 discusses details of wave motion, sound intensity, and sound characteristics and acoustics.

Demonstration/activity: Show how sound is produced by striking the desk, activating a tuning fork, blowing across a bottle, whistling, talking, swinging a flexible corrugated pipe around in circles, using an oscillator and small speaker, etc.

Sound waves provide us with our major form of communication (speech) and a favorite source of enjoyment (music)—as this painting by Caravaggio (c. 1600) so eloquently suggests. But sound waves can also constitute a highly irritating distraction (noise). Sound waves become speech, music, or noise only when our ears perceive them as disturbances (usually in the air). Physically, sound waves are longitudinal waves that are propagated in solids, liquids, and gases. Without a medium, there can be no sound. That is, in a vacuum, there would be no sound.

This distinction between the sensory and physical meanings of sound gives you a way to answer this old philosophical question: If a tree falls in the forest and there is no one to hear it, will there be a sound? The answers are no in terms of sensory hearing and yes in terms of physical waves. That is, the answer depends on how sound is defined. The definition of sound covers three aspects: its source, the medium of its propagation (in the form of longitudinal sound waves), and its detector, which may be human ears.

Since sound waves are all around us most of the time, we are exposed to many interesting and important sound phenomena. A few of these, which often have distinct effects on the sounds we perceive, are covered in this chapter.

14.1 ■ Sound Waves

For there to be sound waves, there must be a disturbance or vibrations in some medium. This disturbance may be the clapping of hands, or the skidding of tires as a car comes to a sudden stop. Underwater you can hear the click of rocks against one another. If you put your ear to a thin wall, you can hear

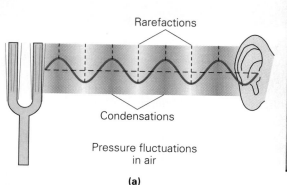

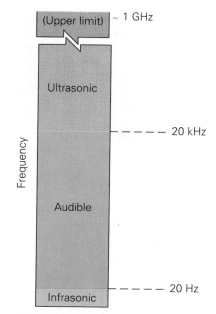

FIGURE 14.1 Vibrations makes waves (a) A vibrating tuning fork disturbs the air, producing alternating high-pressure regions (condensations) and low-pressure regions (rarefactions), which form sound waves. (b) After being picked up by a microphone, the pressure variations are converted to electrical signals. When these are displayed on an oscilloscope, the sinusoidal waveform is evident.

Rarefactions

Condensations

Pressure fluctuations in air

(a)

(b)

sounds from the other side through it. **Sound waves** are longitudinal waves that are propagated in solids, liquids, and gases (Section 13.3).

The characteristics of sound waves can be visualized by considering those produced by a tuning fork (see Fig. 14.1). A tuning fork is essentially a metal bar bent into a U shape. The prongs, or tines, vibrate when struck. The fork vibrates at its fundamental frequency (with an antinode at the end of each tine), so a single tone is heard. The vibrations disturb the air, producing alternating higher-density (compressed) regions called *condensations* and lower-density regions called *rarefactions*. As the fork vibrates, these disturbances propagate outward, and a series of them can be described by a sinusoidal wave.

As the disturbances traveling through the air reach the ear, the eardrum (a thin membrane) is set into vibration by the pressure variations. On the other side of the eardrum, tiny bone structures (the hammer, anvil, and stirrup) carry the vibrations to the inner ear, where they are picked up by the auditory nerve (see the Insight feature).

Characteristics of the human ear limit the perception of sound. Only sound waves with frequencies between about 20 Hz and 20 kHz (kilohertz) initiate nerve impulses that are interpreted by the brain as sound. This frequency range is called the **audible region** of the **sound frequency spectrum** (Fig. 14.2). Frequencies lower than 20 Hz are in the **infrasonic region**. The longitudinal waves generated by earthquakes have infrasonic frequencies. Above 20 kHz is the **ultrasonic region**. Ultrasonic waves can be generated by high-frequency vibrations in crystals.

Ultrasonic waves, or ultrasound, cannot be detected by humans but can be by other animals. The audible region for dogs extends beyond that of humans, so ultrasonic whistles can be used to call dogs without disturbing people. A more recent application of this idea is a cone-shaped whistle that is attached to a car or truck to generate ultrasonic waves when the vehicle travels above a certain speed. The intention is to scare deer, which can apparently hear these sounds, away from the road so that they won't run in front of a car or truck. Some success has been reported: however, the overall effectiveness of the whistles has been seriously questioned.

There are many other practical applications of ultrasound. Since ultrasound can travel for kilometers in water, it is used in sonar. Sonar is the ultrasound counterpart of radar, which uses radio waves for ranging and detection. Sound pulses generated by the sonar apparatus are reflected by underwater objects, and the resulting echoes are picked up by a detector. The time required for a sound pulse to make the round trip, together with the speed of sound in water,

Show how sound can be analyzed with an oscilloscope by using a microphone. The same small speaker used to produce sound can also be used as a transducer to display the sound on the oscilloscope. See who can whistle a note clearly enough to produce a sine wave on the scope.

(Upper limit)   ~ 1 GHz

Ultrasonic

—— —— 20 kHz

Frequency

Audible

—— —— 20 Hz

Infrasonic

FIGURE 14.2 Sound frequency spectrum The audible region of sound for humans lies between about 20 Hz and 20 kHz. Below this is the infrasonic region, and above it is the ultrasonic region.

Sound is one of our most important means of interpersonal communication, and our most important sound source is the human voice. Let's take a look at the anatomy and physics of the voice and of the ear, which serves as our receiver for sound.

The waveforms of voice sounds are quite specific for individuals and provide a "voice print" (see Fig. 14.14, Section 14.5) that can be used for identification, just as fingerprints are. The validity of voice prints for legal identification, however, is still highly controversial.

The Human Voice

The energy for sounds associated with the human voice originates in the muscle action of the diaphragm, which forces air up from the lungs. To produce variations in sound, this steady stream of air must be periodically disturbed or "modulated." The fundamental modulating organ is the larynx (the "voice box"), across which are stretched membrane-like bands called the vocal cords. The opening and closing of the vocal cords modulates the air stream to produce sounds.

The effect of the vocal cords is similar to that of the reed or reeds of a woodwind instrument such as a clarinet or oboe. In these instruments the induced vibrations of the reeds convert a steady air flow into a periodic one. For the voice, the fundamental frequency of the modulated sound waves is determined by the tension of the vocal cords, which can be controlled voluntarily. The extent of this control is a primary factor in determining the basic frequency range of the speaking and singing voice.

The sound waves from the larynx are further modulated in the numerous resonance cavities in the throat, mouth, and nose, where standing waves are set up. Some of the vocal cavities can be altered by means of controllable structures, such as the tongue and lips, so as to produce a wide variety of sounds. (Sound out the vowel letters *a*, *e*, *i*, *o*, and *u*, and notice the positions of the tongue and lips.)

Hearing

The anatomy of the human ear is illustrated in Fig. 1. Sound enters the outer ear and travels through the ear (or auditory) canal to the *eardrum* (the tympanum), which separates the outer ear from the middle ear. The eardrum is a membrane that vibrates in response to the impinging sound waves. The vibrations are transmitted through the middle ear, which contains an intricate set of connected bones, commonly called the *hammer* (malleus), *anvil* (incus), and *stirrup* (stapes), because of their resemblance to these objects.

The bones of the middle ear form a delicate set of levers with a force multiplication factor (mechanical advantage) of about 2. However, the amplification of the pressure in a wave is much greater because the area of the eardrum is about 20 times that of the *oval window*, the membrane-covered opening to the inner ear. (The stirrup is in contact with the membrane covering.) Thus, pressure is amplified by a factor of about 40.

The inner ear includes the semicircular canals, which are important in controlling balance, and the *cochlea*. It is in the cochlea that sound waves are translated into nerve impulses and that pitch or frequency discrimination is made. The cochlea consists of a series of liquid-filled tubes, coiled into a spiral shape that resembles a snail shell. Coiled up between two of the tubes is a structure called the *basilar membrane* that is supported and stiff-

gives the distance of the reflecting object. Sonar can be used not only for ranging but also to create images of the seafloor and objects resting on it, such as shipwrecks (Fig. 14.3a, page 458).

Interestingly, sonar appeared in the animal kingdom long before it was developed by human engineers. Bats use a kind of natural sonar to navigate in and out of their caves and to locate and catch flying insects on their nocturnal hunting flights. The bats emit pulses of ultrasound and track their prey by means of the reflected echoes (Fig. 14.3b). Certain species of cave-dwelling birds have evolved the same capability.

In medicine, ultrasound is used to examine internal tissues and organs that are nearly invisible to X-rays. Perhaps the best known medical application of ultrasound is its use to view a fetus without exposing it to the dangerous effects of X-rays (Fig. 14.3c). Ultrasonic generators (transducers) made of quartz

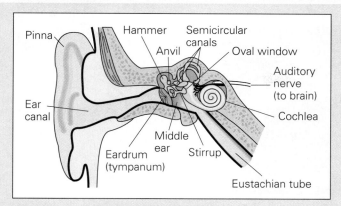

FIGURE 1 Anatomy of the human ear The ear converts pressure variations in the air into electrical nerve impulses that are interpreted by the brain as sound. (See text for details.)

ened by reed-like *basilar fibers*. On the surface of the basilar membrane rest thousands of special receptor cells called *hair cells*.

The response of the basilar membrane to sounds of different frequency depends largely on resonance. Different sound frequencies transmitted by vibrations of the oval window to the fluid in the cochlea cause different sections of the basilar membrane to vibrate in resonance. As in the resonance of a violin string (Section 13.5), the resonant frequencies depend on mass and tension. The relevant mass in this case is partly that of the fluid in the ducts, which vibrates back and forth, and partly that of the membrane itself, which varies along its length (it is narrower near the windows and broader near the tip). The tension is developed mainly by the bending of the basilar fibers, which vary in length and rigidity in such a

way that the membrane is most taut near the windows and loosest near the tip. As we would expect, the broad, loose tip of the membrane vibrates most strongly in response to sounds of low frequency, while higher-frequency tones produce the most vibration in the narrower, tauter portion of the membrane nearest the windows. Thus frequency information is translated into positional information (different regions of the basilar membrane).

The hair cells that rest on the basilar membrane are specialized nerve receptors. When a region of the basilar membrane is set into vibration by sound of a particular frequency, hair cells in that region are stimulated and nerve impulses are sent to the brain, where they are interpreted as sound. The brain translates this positional information (impulses from particular fibers originating in specific regions of the basilar membrane) back into frequency information (the subjective physiological sensation of pitch).

The sensation of loudness is determined by the amplitude of vibration of the basilar membrane. The greater the amplitude, the more strongly the hair cells are stimulated and the greater the number of impulses they send to the brain in a given period of time. The brain translates this information into degrees of loudness.

Incidentally, the middle ear is connected to the throat by the Eustachian tube, the end of which is normally closed. It opens during swallowing and yawning to permit air to enter and leave, so that internal and external pressures are equalized. You have probably experienced a "stopping up" of the ears with a sudden change in atmospheric pressure (for example, during rapid ascents or descents in elevators or airplanes). Swallowing opens the Eustachian tubes and relieves the excess pressure difference on the middle ear.

crystals produce high-frequency waves that are used to scan a designated region of the body from several angles. Reflections from the scanned areas are monitored, and a computer constructs an image from the reflected signals. Images are recorded several times each second. The series of images provides a "moving picture" of an internal structure, such as the heart of a fetus. A still shot, or echogram, is shown in the figure.

In industrial and home applications, ultrasonic baths are used to clean metal machine parts and jewelry. The high-frequency (short-wavelength) ultrasound vibrations loosen particles in otherwise inaccessible places.

Ultrasonic frequencies extend into the megahertz (MHz) range, but the sound frequency spectrum does not continue indefinitely. There is an upper limit of about 10^9 Hz, or 1 GHz (gigahertz), which is determined by the upper limit of the elasticity of materials.

Let students "see" ultrasonic waves. Blow a dog whistle in front of a microphone connected to an oscilloscope to demonstrate the presence of waves and the lack of detection by the human ear.

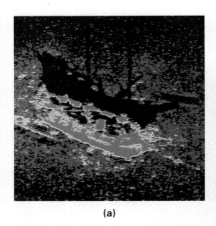

(a)

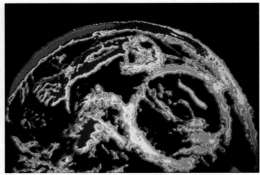

(c)

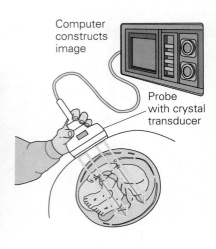

Computer constructs image

Probe with crystal transducer

(b)

FIGURE 14.3 Some uses of ultrasound (a) This computer-enhanced false-color image, constructed from sonar signals, shows the wreck of the American sailing ship Hamilton. The schooner, sunk in Lake Ontario during the War of 1812, lies under 300 feet of water. (b) Bats hunt flying insects with the aid of their own natural sonar systems. They emit pulses of ultrasonic waves, which lie within their audible region, and use the echoes reflected from their prey to guide their attack. (Notice the size of the animal's ears.) (c) Ultrasound generated by crystal transducers is transmitted through tissue and is reflected from internal structures. The reflected waves are detected by the transducers and the signals are used to construct an image, or echogram, as shown here for a well-developed fetus.

14.2 ■ The Speed of Sound

In general, the speed at which a disturbance moves through a medium depends on the elasticity of the medium and its density. For example, as you learned in Chapter 13, the wave speed in a stretched string is given by $v = \sqrt{F/\mu}$, where F is the tension in the string and μ is the linear mass density.

Similar expressions describe wave speeds in solids and liquids. Here, the elasticity is expressed in terms of moduli (Chapter 9). In general, the speeds of sound in solids and liquids are given by $v = \sqrt{Y/\rho}$ and $v = \sqrt{B/\rho}$, respectively, where Y is Young's modulus, B is the bulk modulus, and ρ is the density.)

Solids are generally more elastic than liquids, which in turn are more elastic than gases. In a highly elastic material, the restoring forces between the atoms or molecules cause a disturbance to propagate faster. Thus, the speed of sound is generally greater in solids than in liquids, and greater in liquids than in gases (Table 14.1). Notice from the table that, in general, sound travels about 3–4 times faster in solids than in liquids and about 10–15 times faster in solids than in gases (such as air).

Although not expressed explicitly in the preceding equations, the speed of sound also depends on the temperature of the medium. We will consider a common example of this aspect, the speed of sound in air.

In air, the speed of sound is 331 m/s (about 740 mi/h) at a temperature of 0°C. As the temperature increases, so do the speeds of the gas molecules. As a result, the molecules collide with each other more frequently, and a disturbance is transmitted more quickly. Thus the speed of sound in air increases with increasing temperature. For *normal environmental temperatures*, the speed of sound in air increases by about 0.6 m/s for each degree Celsius above 0°C. Thus, a good approximation of the speed of sound in air for a particular (environmental) temperature is given by

$$v = (331 + 0.6\, T_C)\ \text{m/s} \qquad (14.1)$$

where T_C is the air temperature in degrees Celsius.* (Although not written explicitly, the units associated with the factor 0.6 are m/s -C°.)

Let's take a comparative look at the speeds of sound in different media.

EXAMPLE 14.1 ■ Speeds of Sound in Different Media

Find the general values of the speeds of sound in (a) a solid copper rod, (b) liquid water, and (c) air at room temperature (20°C).

Solution. To find the general speeds of sound in copper and water, we will need the moduli and densities. These are given in Tables 9.1 and 9.2.

$Given:$ $Y_{Cu} = 11 \times 10^{10}$ N/m² $Find:$ (a) v_{Cu} (speed in copper)
$B_{H_2O} = 2.2 \times 10^9$ N/m²   (b) v_{H_2O} (speed in water)
$\rho_{Cu} = 8.9 \times 10^3$ kg/m³   (c) v_{air} (speed in air)
$\rho_{H_2O} = 1.0 \times 10^3$ kg/m³

$T_C = 20°C$ (for air)

(a) To find the general value for the speed of sound in a copper rod, we use the expression $v = \sqrt{Y/\rho}$, and

$$v_{Cu} = \sqrt{Y/\rho} = \sqrt{(11 \times 10^{10} \text{ N/m}^2)/(8.9 \times 10^3 \text{ kg/m}^3)}$$
$$= 3.5 \times 10^3 \text{ m/s}$$

(b) For water, $v = \sqrt{B/\rho}$, and

$$v_{H_2O} = \sqrt{B/\rho} = \sqrt{(2.2 \times 10^9 \text{ N/m}^2)/(1.0 \times 10^3 \text{ kg/m}^3)}$$
$$= 1.5 \times 10^3 \text{ m/s}$$

(c) For air at 20°C, by Eq. 14.1, we have

$$v_{air} = 331 + 0.6 \, T_C \text{ m/s} = 331 + 0.6 \,(20) = 343 \text{ m/s}$$

A generally useful figure for the speed of sound in air is $\frac{1}{3}$ km/s (or $\frac{1}{5}$ mi/s). Using this figure, you can, for example, estimate how far away lightning has struck by counting the number of seconds between the time the flash is observed and the time the associated thunder is heard. Because of the very fast speed of light, the lightning flash is seen almost instantaneously. The sound waves of the thunder travel relatively slowly, at about $\frac{1}{3}$ km/s. For example, if the interval between the two is measured to be 6 s (often by counting "one thousand one, one thousand two, . . ."), the lightning stroke was approximately 2 km away (that is, $\frac{1}{3}$ km/s × 6 s).

You may also have noticed the delay of sound relative to light at a baseball game. If you're sitting in the outfield stands, you see the batter hit the ball before you hear the crack of the bat.

---

*A better approximation for these and higher temperatures is given by the expression

$$v = 331 \sqrt{1 + \frac{T_C}{273}} \text{ m/s}$$

See v for air at 100°C in Table 14.1, which is outside the normal environmental temperature range.

TABLE 14.1 ■ The Speed of Sound in Various Media (typical values)

Medium	Speed (m/s)
Solids	
Aluminum	5100
Copper	3500
Iron	4500
Glass	5200
Polystyrene	1850
Liquids	
Alcohol, ethyl	1125
Mercury	1400
Water	1500
Gases	
Air (0°C)	331
Air (100°C)	387
Helium (0°C)	965
Hydrogen (0°C)	1284
Oxygen (0°C)	316

Demonstration/activity: Show that air is necessary for sound to be produced and transmitted by placing a sound source in a bell jar and removing the air.

Demonstration/activity: Measure the speed of sound in aluminum, steel, brass, etc. Hold a lab equipment/support rod lightly at its center and strike the end with a small hammer to produce a longitudinal standing wave with a node in the center. Many students can quickly match the resulting frequency with a variable oscillator and speaker. The speed is then the product of the frequency and twice the length of the rod. Don't confuse the longitudinal wave with the lower-frequency transverse wave that is also produced.

EXAMPLE 14.2 ■ Safe or Out?

On a cool October afternoon (air temperature = 15°C), from your seat in the center-field stands 113 m from first base, you witness the play that will decide the 2001 World Series. You see the runner's foot touch the bag; half a second later, straining your ears, you hear the faint thud of the ball in the first baseman's glove. The umpire signals safe; half the fans boo loudly. As a student of physics, you make the call—did the ump blow it?

Solution. Listing the data, we have

> *Given:* $d = 113$ m   *Find:* t (sound travel time)
> $T = 15°C$

The visual observation takes place almost instantaneously, but the sound generated by the ball striking the glove travels more slowly than light and takes more time to reach you. With a temperature of 15°C, the speed of sound is (Eq. 14.1):

$$v = (331 + 0.6\, T_C) \text{ m/s} = 331 + 0.6\,(15°C) = 340 \text{ m/s}$$

For a constant speed, the general distance-time relation is $d = vt$, so the time for the sound to travel the distance to your seat is

$$t = \frac{d}{v} = \frac{113 \text{ m}}{340 \text{ m/s}} = 0.332 \text{ s}$$

or just under 1/3 of a second. Thus the runner actually reached the base about 1/2 s − 1/3 s = 1/6 s before the ball arrived. The ump was right!

To show why it is justified to say that the visual observation takes place almost instantaneously, let's see how long it takes for light to travel from first base to your seat. For practical purposes, the speed of light in air is the same as that in a vacuum, $c = 3.00 \times 10^8$ m/s, which is on the order of 10^6 (a million) times greater than the speed of sound. Then, the time for light to travel 113 m is

$$t = \frac{d}{c} = \frac{113 \text{ m}}{3.00 \times 10^8 \text{ m/s}} = 3.77 \times 10^{-7} \text{ s}$$

or 0.377 μs—not a very long time.

14.3 ■ Sound Intensity

Wave motion involves the propagation of energy. The rate of the energy transfer is expressed in terms of **intensity**, which is the energy transported per unit time across a unit area. Since energy/time is power, intensity is power/area:

$$\text{intensity} = \frac{\text{energy/time}}{\text{area}} = \frac{\text{power}}{\text{area}}$$

The standard units for intensity (power/area) are watts per meter squared (W/m²).

In the definition of sound intensity we use the component of force perpen-

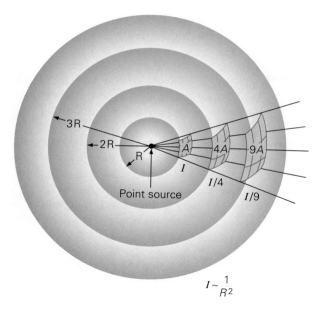

$$I \sim \frac{1}{R^2}$$

● **FIGURE 14.4 Intensity of a point source** The energy emitted from a point source spreads out equally in all directions. Since intensity is power/area, $I = P/A = P/4\pi R^2$, where the area is that of a spherical surface. The intensity is then related to the distance from the source by $1/R^2$.

dicular to the surface area in the direction of wave propagation (just as we used the component of force perpendicular to an area in the definition of pressure in Chapter 9).

For example, consider a point source that sends out spherical sound waves, as shown in ● Fig. 14.4. If there are no losses, the sound intensity at a distance R from the source is

$$I = \frac{P}{A} = \frac{P}{4\pi R^2} \qquad (14.2)$$

where P is the power of the source and $4\pi R^2$ is the area of a sphere with a radius R, through which the sound energy passes perpendicularly.

The intensity for a point source is therefore *inversely proportional to the square of the distance from the source* (an inverse square relationship). Two intensities at different distances from a source of constant power may be compared using a ratio:

$$\frac{I_2}{I_1} = \frac{P/4\pi R_2^2}{P/4\pi R_1^2} = \frac{R_1^2}{R_2^2}$$

or

$$\frac{I_2}{I_1} = \left(\frac{R_1}{R_2}\right)^2 \qquad (14.3)$$

Suppose that the distance from a point source is doubled; that is, $R_2 = 2R_1$, or $R_1/R_2 = \frac{1}{2}$. Then

$$\frac{I_2}{I_1} = \left(\frac{R_1}{R_2}\right)^2 = \left(\frac{1}{2}\right)^2 = \frac{1}{4}$$

and

$$I_2 = \frac{I_1}{4}$$

Since the intensity decreases by a factor of $1/R^2$, doubling the distance decreases the intensity to a quarter of its original value.

A good way to understand this inverse-square relationship intuitively is to look at the geometry of the situation. As you can see in Fig. 14.4, the greater the distance from the source, the larger the area over which a given amount of sound energy is spread, and thus the lower its intensity. (As an analogy, imagine having to paint two walls of different areas. If you had the same amount of paint to use on each, you'd obviously have to spread it more thinly over the larger wall.) Since this area increases as the square of the radius r, the intensity decreases accordingly—that is, as $1/r^2$.

Sound intensity is perceived by the ear as loudness. On the average, the human ear can detect sound waves (at 1 kHz) with an intensity as low as 10^{-12} W/m^2. This intensity is referred to as the *threshold of hearing*. Thus, for us to hear a sound, it must not only have a frequency in the audible range, but also be of sufficient intensity. As the intensity is increased, the perceived sound becomes louder. At an intensity of 1.0 W/m^2, the sound is uncomfortably loud and may be painful to the ear. Thus, this intensity is called the *threshold of pain*.

Note that the thresholds of pain and hearing differ by a factor of 10^{12}:

Threshold of pain: $I = 1.0$ W/m^2

Threshold of hearing: $I = 10^{-12}$ W/m^2

$$\frac{I_p}{I_h} = \frac{1.0\ \text{W/m}^2}{10^{-12}\ \text{W/m}^2} = 10^{12}$$

That is, the intensity at the threshold of pain is a *trillion* times greater than that at the threshold of hearing. Within this enormous range, the perceived loudness is not directly proportional to the intensity. That is, if the intensity is doubled, the perceived loudness does not double. A doubling of perceived loudness corresponds approximately to an increase in intensity by a factor of 10. For example, a sound with an intensity of 10^{-5} W/m^2 would be perceived to be twice as loud as one with an intensity of 10^{-6} W/m^2 (the smaller the negative exponent, the larger the number).

It is convenient to compress the large range of sound intensities by using a logarithmic scale (base 10) to express intensity levels. The intensity level of a sound must be referenced to a standard intensity, which is taken to be that of the threshold of hearing, $I_o = 10^{-12}$ W/m^2. Then, for any intensity I, the intensity level is the log of the ratio of I to I_o: $\log I/I_o$. For example, if a sound has an intensity of $I = 10^{-6}$ W/m^2,

$$\log \frac{I}{I_o} = \log \frac{10^{-6}\ \text{W/m}^2}{10^{-12}\ \text{W/m}^2} = \log 10^6 = 6$$

(Recall that $\log_{10} 10^x = x$.) Thus, a sound with an intensity of 10^{-6} W/m^2 has an intensity level of 6 bel (B) on this scale. In this way, the intensity range from 10^{-12} W/m^2 to 1.0 W/m^2 is compressed into a scale of intensity levels running from 0 B to 12 B.

The bel was named in honor of Alexander Graham Bell, the inventor of the telephone.

A finer intensity scale is obtained by using a smaller unit, the **decibel (dB)**, which is a tenth of a bel. The 0–12 B range corresponds to 0–120 dB. In this case, the equation for the relative **sound intensity level**, or **decibel level (β)**, is

Sound level intensity in decibels

$$\beta = 10 \log \frac{I}{I_o} \qquad \text{where } I_o = 10^{-12}\ \text{W/m}^2 \qquad (14.4)$$

The decibel intensity scale and familiar sounds at some intensity levels are shown in ● Fig. 14.5.

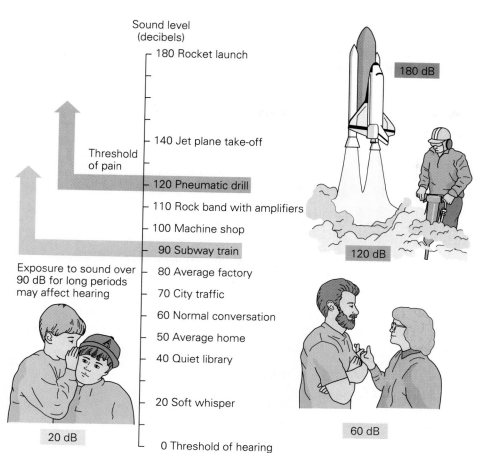

Sound level
(decibels)

— 180 Rocket launch

— 140 Jet plane take-off

Threshold
of pain

— 120 Pneumatic drill

— 110 Rock band with amplifiers

— 100 Machine shop

— 90 Subway train

Exposure to sound over
90 dB for long periods
may affect hearing

— 80 Average factory

— 70 City traffic

— 60 Normal conversation

— 50 Average home

— 40 Quiet library

— 20 Soft whisper

— 0 Threshold of hearing

20 dB

180 dB

120 dB

60 dB

Figure 14.5

FIGURE 14.5 Sound intensity levels and the decibel scale The intensity levels of some common sounds on the decibel (dB) scale.

EXAMPLE 14.3 ■ Sound Intensity Levels

What are the intensity levels for sounds with intensities of (a) 10^{-12} W/m² and (b) 5.0×10^{-6} W/m²?

Solution.

Given: (a) $I = 10^{-12}$ W/m²   *Find:* (a) β (sound level intensity)
(b) $I = 5.0 \times 10^{-6}$ W/m²   (b) β

Using Eq. 14.4 we have,

(a)
$$\beta = 10 \log \frac{I}{I_o} = 10 \log \frac{10^{-12} \text{ W/m}^2}{10^{-12} \text{ W/m}^2}$$

$$= 10 \log 1 = 0 \text{ dB}$$

The intensity is the same as that at the threshold of hearing. (Recall that $\log 1 = 0$, since $1 = 10^0$ and $\log 10^0 = 0$.)

(b)
$$\beta = 10 \log \frac{I}{I_o} = 10 \log \left(\frac{5.0 \times 10^{-6} \text{ W/m}^2}{10^{-12} \text{ W/m}^2} \right)$$

$$= 10 \log (5.0 \times 10^6) = 10(\log 5.0 + \log 10^6)$$

$$= 10(0.70 + 6.0) = 67 \text{ dB}$$

Note that 5.0×10^{-6} W/m² is half way between 10^{-6} and 10^{-5} W/m² (or 60 and 70 dB). However, this does not correspond to 65 dB, since the dB scale is logarithmic, not linear.

EXAMPLE 14.4 ■ **Intensity Level Factors**

(a) What is the difference in the intensity level if the intensity of a sound is doubled? (b) By what factors does the intensity increase for intensity level *differences* of 10 dB and 20 dB?

Solution. Note in (b) that these values are differences, $\Delta\beta = \beta_2 - \beta_1$, not intensity levels. Then,

Given: (a) $I_2 = 2I_1$ *Find:* (a) $\Delta\beta$ (intensity level difference)
          (b) $\Delta\beta = 10$ dB   (b) I_2/I_1 (factors of increase)
          $\Delta\beta = 20$ dB

(a) The intensity is doubled, so $I_2/I_1 = 2$. This ratio can be used directly to find the intensity level *difference* because of the properties of logarithms (see Exercise 45):

$$\Delta\beta = 10 \log \frac{I_2}{I_1} = 10 \log 2 = 3 \text{ dB}$$

Thus, doubling the intensity increases the intensity level by 3 dB (for example, an increase from 60 dB to 63 dB).
(b) For a 10-dB difference,

$$\Delta\beta = 10 \text{ dB} = 10 \log \frac{I_2}{I_1}, \quad \text{and} \quad \log \frac{I_2}{I_1} = 1.0$$

Since $\log 10^1 = 1$, $\dfrac{I_2}{I_1} = 10^1$ and $I_2 = 10 I_1$

Similarly, for a 20-dB difference,

$$\Delta\beta = 20 \text{ dB} = 10 \log \frac{I_2}{I_1} \quad \text{and} \quad \log \frac{I_2}{I_1} = 2.0$$

Since $\log 10^2 = 2$, $\dfrac{I_2}{I_1} = 10^2$ and $I_2 = 100 I_1$

Thus, an intensity level difference of 10 dB corresponds to increasing (or decreasing) the intensity by a factor of 10. An intensity level difference of 20 dB corresponds to increasing (or decreasing) the intensity by a factor of 100. You should be able to guess the factor that corresponds to an intensity level difference of 30 dB. In general, the factor of the intensity change is $10^{\Delta B}$, where ΔB is the level difference in bels. Since 30 dB = 3 B and $10^3 = 1000$, the intensity changes by a factor of 1000 for an intensity level difference of 30 dB.

Sound intensities can have detrimental effects on hearing, and because of this the government has set occupational noise-exposure limits.

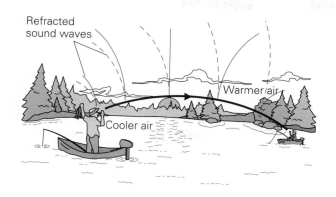

FIGURE 14.6 Sound refraction
Sound is refracted (bent) more in the overlying warmer air. This increases the intensity of the sound at a distance where it otherwise might not be heard.

14.4 ■ Sound Phenomena

There are many interesting sound phenomena. Several you may have experienced are discussed in this section.

Reflection, Refraction, and Diffraction

A sound wave can be reflected. An echo is a familiar example of sound *reflection*.

Sound *refraction* is less common, but you may have experienced this effect on a calm summer evening. Then it is sometimes possible to hear distant voices or other sounds that ordinarily would not be audible. This effect is due to the refraction, or bending, of the sound waves as they pass into a region where the air density is different. The effect is similar to what would happen if the sound passed into another medium.

The required conditions are a layer of cooler air near the ground with a layer of warmer air above it. These conditions occur frequently over bodies of water after sunset because of cooling (●Fig. 14.6). The sound waves spread out and upward from their source and would ordinarily be too faint to be heard by someone at a distance. But, as the waves pass into the overlying layer of warm air, they travel faster and are bent, or refracted, toward the distant person. This increases the intensity of the sound received by that person. If the intensity is above the threshold of hearing, the distant sound can be heard.

Sound is also *diffracted* or bent around corners, as you know from experience. Reflection, refraction, and diffraction of light are important phenomena that will be considered in more detail in Chapters 21 and 23.

Interference

Like waves of any kind, sound waves *interfere* when they meet. Suppose that two loudspeakers separated by some distance emit sound waves in phase at the same frequency (see ●Fig. 14.7a). Consider the speakers to be point sources—the waves spread out spherically and interfere. The lines from a particular speaker represent wave crests (or condensations), and the troughs (or rarefactions) lie in the intervening white areas.

In particular regions of space, there will be constructive or destructive

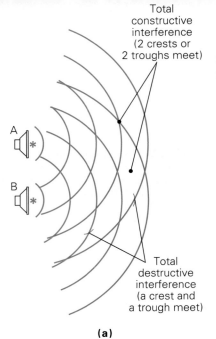

Total constructive interference (2 crests or 2 troughs meet)

A

B

Total destructive interference (a crest and a trough meet)

(a)

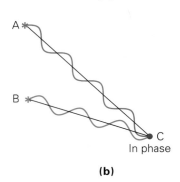

A

B

C

In phase

(b)

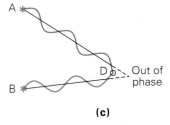

A

D Out of phase

B

(c)

• **FIGURE 14.7 Interference (a)** Sound waves from two point sources spread out and interfere. At points where the waves arrive in phase (zero phase difference), such as point C in **(b)**, constructive interference occurs. At points where the waves arrive completely out of phase (phase difference of 180°), such as point D in **(c)**, destructive interference occurs. The phase difference at a particular point depends on the path lengths the waves travel to reach that point.

Figure 14.7

interferences. For example, if the waves meet in a region where they are exactly in phase (i.e., where two crests or two troughs coincide), there will be total **constructive interference**. This situation is illustrated in Fig. 14.7b. Notice that the waves have the same motion at point C. Conversely, if the waves meet so that the crest of one coincides with the trough of the other, the two waves will cancel each other out. The result will be total **destructive interference**.

It is convenient to describe the path lengths traveled by the waves in terms of wavelength (λ) to determine whether they arrive in phase. In such analysis we work with points of zero displacement rather than crests or troughs for convenience. Consider the waves arriving at point C in Fig. 14.7b. The path lengths in this case are AC = 4λ and BC = 3λ. The **phase difference** ($\Delta\theta$) is related to the **path difference** (PD) by the simple relationship:

$$\Delta\theta = \frac{2\pi}{\lambda}\,(PD) \qquad \text{(phase difference and path difference)} \qquad (14.5)$$

Since there are 2π rad per wavelength ($2\pi/\lambda$), multiplying that ratio by the path difference in *wavelength units* gives the phase difference.

For the example illustrated in Fig. 14.7b, we have

$$\Delta\theta = \frac{2\pi}{\lambda}\,(AC - BC) = \frac{2\pi}{\lambda}\,(4\lambda - 3\lambda) = 2\pi \text{ rad}$$

A $\Delta\theta = 2\pi$ rad means that the waves are shifted by one wavelength. This is the same as $\Delta\theta = 0°$, so the waves are in phase. Thus, the waves interfere constructively in the point C region, increasing the intensity, or loudness, of the sound detected there.

From Eq. 14.5, it should be clear that the sound waves are in phase at any point where the path difference is zero or an integral multiple of the wavelength. That is,

$$PD = n\lambda \qquad \text{(for } n = 0,1,2,3,\ldots)$$

condition for constructive interference

$$(14.6)$$

A similar analysis for the situation in Fig. 14.7c, where $AD = 2\frac{3}{4}\lambda$ and $BD = 2\frac{1}{4}\lambda$, gives

$$\Delta\theta = \frac{2\pi}{\lambda}(2\tfrac{3}{4}\lambda - 2\tfrac{1}{4}\lambda) = \pi \text{ rad}$$

or $\Delta\theta = 180°$. At point D, the waves are completely out of phase, and destructive interference occurs in this region.

Sound waves will be out of phase at any point where the path difference is an odd number of half-wavelengths ($\lambda/2$), or

$$PD = m\left(\frac{\lambda}{2}\right) \qquad \text{(for } m = 1,3,5,\ldots)$$

condition for destructive interference

$$(14.7)$$

At these points, a softer or less intense sound will be heard or detected. If the amplitudes of the waves are equal, the destructive interference is total, and no sound is heard!

EXAMPLE 14.5 ■ Interference and Path Difference

At an open-air concert on a hot day (air temperature of 25°C), a person sits at a location that is 7.0 m and 9.1 m, respectively, from speakers at each side of the stage. A musician, warming up, plays a single 494-Hz tone. What does the spectator hear? (Consider the speakers to be point sources.)

Solution. Since the sound waves interfere, what the person hears depends on the phase difference at that location. Hence, we have

Given: $d_1 = 7.0$ m and $d_2 = 9.1$ m   *Find:* $\Delta\theta$ (phase difference)
$f = 494$ Hz
$T = 25°C$

The phase difference between the waves arriving at the spectator's location is found by expressing the path lengths in wavelengths. To do this, we first need to know the wavelength. Given the frequency, this can be found from the relationship $\lambda = v/f$ provided we know the value of the speed of sound v at the given temperature. We calculate v by means of Eq. 14.1:

$$v = 331 + 0.6T_C = 331 + 0.6(25) = 346 \text{ m/s}$$

The wavelength of the sound waves is then

$$\lambda = \frac{v}{f} = \frac{346 \text{ m/s}}{494 \text{ Hz}} = 0.700 \text{ m}$$

Thus, the distances in terms of wavelength are

$$d_1 = (7.0 \text{ m})\left(\frac{\lambda}{0.700 \text{ m}}\right) = 10\lambda$$

and

$$d_2 = (9.1 \text{ m})\left(\frac{\lambda}{0.700 \text{ m}}\right) = 13\lambda$$

The path difference is

$$PD = d_2 - d_1 = 13\lambda - 10\lambda = 3\lambda$$

This is an integral number of wavelengths ($n = 3$), so constructive interference occurs, and a loud sound is heard by the spectator.

If the path difference had been 2.5λ, for example, $2.5\lambda = 5(\lambda/2)$, which is a condition for destructive interference ($m = 5$), no sound would be heard if the waves from the speakers had equal amplitudes. Of course, during a concert the sound would not be single-frequency tone, but would have a variety of frequencies, wavelengths, and amplitudes. Spectators at certain locations might not hear certain parts of the audio spectrum, but probably wouldn't notice.

> Constructive interference will occur when the path difference is an integral multiple of the wavelength. Destructive interference will occur when the path difference is an odd number of half-wavelengths.

Another interesting interference effect occurs when two tones of nearly the same frequency ($f_1 \cong f_2$) are sounded simultaneously. The ear senses pulsations in loudness known as **beats**. (The human ear can detect as many as 7 beats per second.)

For example, suppose that two sinusoidal waves with the same amplitude have similar frequencies. As these waves interfere, the total displacement at some location is (by the principle of superposition)

$$y = y_1 + y_2 = A \sin 2\pi f_1 t + A \sin 2\pi f_2 t$$

The trigonometric identity

$$\sin a + \sin b = 2 \cos\left(\frac{a-b}{2}\right) \sin\left(\frac{a+b}{2}\right)$$

allows the equation for y to be rewritten as

$$y = \left[2A \cos 2\pi t\left(\frac{f_1 - f_2}{2}\right)\right]\left[\sin 2\pi t\left(\frac{f_1 + f_2}{2}\right)\right] \quad (14.8)$$

● Fig. 14.8 represents the resulting sound wave. The red curve is a sinusoidal wave whose frequency is the average of the two frequencies producing it: $(f_1 + f_2)/2$. This wave is described by the sine term in Eq. 14.8. The amplitude of the wave varies sinusoidally, as shown by the black curve (known as an *envelope*). This variation in amplitude, expressed by the bracketed cosine term in Eq. 14.8, has a frequency of $(f_1 - f_2)/2$. The maximum amplitude is $2A$ (at those points where maxima of the two original tones interfere constructively).

What does this mean in terms of perception? A listener will hear a sound of frequency $(f_1 + f_2)/2$ that pulsates with a frequency of $(f_1 - f_2)/2$. You can see from Fig. 14.8 that a full cycle of the envelope representing this pulsation comprises two maximum amplitudes of the heard tone (red curve). This means that two maximum amplitudes of $2A$ are heard each cycle. Thus, the **beat frequency** (f_b) that is perceived is twice the envelope (cosine) frequency, or

$$\boxed{f_b = |f_1 - f_2|} \quad (14.9)$$

(The absolute value is taken because there cannot be a negative frequency.)

Beats may be produced when tuning forks of nearly the same frequency are vibrating at the same time. For example, using forks with frequencies of 516 Hz and 513 Hz, the beat frequency is $f_b = 516$ Hz $- 513$ Hz $= 3$ Hz, and 3 pulsations, or beats, are heard each second. Musicians tune two stringed instruments to the same note by adjusting the strings until the beats disappear ($f_1 = f_2$).

Figure 14.8

● **FIGURE 14.8 Beats** Two traveling waves of equal amplitude and slightly different frequencies interfere and give rise to pulsating tones called beats. The beat frequency is given by $f_b = |f_1 - f_2|$.

The Doppler Effect

If you stand along a highway and a car or truck approaches you with the horn blowing, the pitch (the perceived frequency) of the sound is higher as the

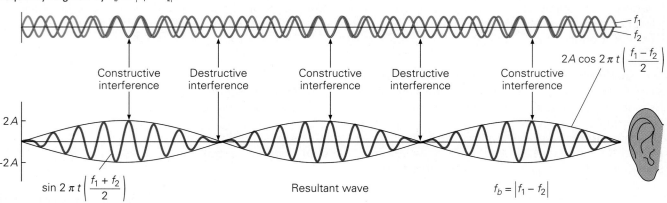

Constructive interference	Destructive interference	Constructive interference	Destructive interference	Constructive interference

$2A$
$-2A$

$\sin 2\pi t\left(\frac{f_1 + f_2}{2}\right)$

Resultant wave

$f_b = |f_1 - f_2|$

$2A \cos 2\pi t\left(\frac{f_1 - f_2}{2}\right)$

f_1
f_2

vehicle approaches and lower as it recedes. You can also hear variations in the frequency of the motor noise when watching a racing car going around a track. A variation in the perceived sound frequency due to the motion of the sound source is an example of the **Doppler effect**.

As Fig. 14.9 shows, the sound waves emitted by a moving source tend to bunch up in front of the source and spread out in back. The Doppler shift in frequency can be found by assuming that the air is at rest in a reference frame such as that depicted in Fig. 14.10. The speed of sound in air is v and the speed of the moving source is v_s. The frequency of the sound produced by the source is f_s. In one period, $T = 1/f_s$, a wave crest moves a distance $d = vT = \lambda$. (The sound wave would travel this distance in still air in any case, whether or not the source were moving.) But, in one period, the source travels a distance $d_s = v_sT$ before emitting another wave crest. The distance between the successive wave crests is thus shortened to a wavelength λ':

$$\lambda' = d - d_s = vT - v_sT$$
$$= (v - v_s)T = \frac{v - v_s}{f_s}$$

The Austrian physicist Christian Doppler (1803–1853) first described what we now call the Doppler effect.

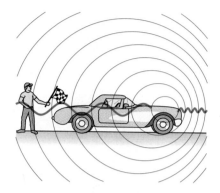

The frequency heard by the observer (f_o) is related to the shortened wavelength by $f_o = v/\lambda'$, and substituting v/λ' gives

$$f_o = \frac{v}{\lambda'} = \left(\frac{v}{v - v_s}\right)f_s$$

or

$$f_o = \left(\frac{1}{1 - v_s/v}\right)f_s \qquad (14.10)$$

source moving toward a stationary observer

● **FIGURE 14.9 The Doppler effect** The sound waves bunch up in front of a moving source, giving a higher frequency. They trail out behind the source, giving a lower frequency. [Drawing not to scale. Why?]

Since $(1 - v_s/v)$ is less than 1, f_o is greater than f_s in this situation. For example, suppose that the speed of the source is a tenth of the speed of sound: $v_s = v/10$, or $v_s/v = \frac{1}{10}$. Then, by Eq. 14.10, $f_o = \frac{10}{9}f_s$.

Similarly, when the source is moving away from the observer ($\lambda' = d + d_s$), the observed frequency is given by

$$f_o = \left(\frac{v}{v + v_s}\right)f_s = \left(\frac{1}{1 + v_s/v}\right)f_s \qquad (14.11)$$

source moving away from a stationary observer

Demonstration/activity: Use an oscillator and a small speaker connected to a cord. Swing the speaker in a circular path around your head parallel to the floor to demonstrate the Doppler effect. Use small wires taped to the cord to transmit the oscillator signal to the speaker. Try swinging the speaker as a simple pendulum.

Here f_o is less than f_s. (Why?)

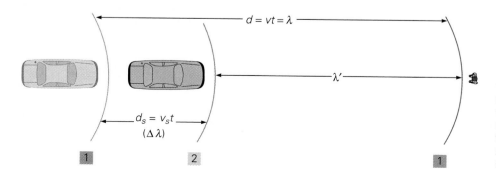

● **FIGURE 14.10 The Doppler effect and wavelength** Sound from a moving car's horn travels a distance d in a time t. During this time, the car (the source) travels a distance d_s, thereby shortening the observed wavelength of the sound.

Combining Eqs. 14.10 and 14.11 yields a general equation for the observed frequency with a moving source and a stationary observer.

$$f_o = \left(\frac{v}{v \mp v_s}\right) f_s = \left(\frac{1}{1 \mp v_s/v}\right) f_s \qquad (14.12)$$

$$\begin{cases} - \text{ for source moving toward stationary observer} \\ + \text{ for source moving away from stationary observer} \end{cases}$$

As you might expect, the Doppler effect also occurs with a moving observer and a stationary source. This situation is a bit different. As the observer moves toward the source, the distance between successive wave crests is the normal wavelength (or $\lambda = v/f_s$), but the measured wave speed is different. Relative to the approaching observer, the sound from the stationary source has a wave speed of $v' = v + v_o$, where v_o is the speed of the observer and v is the speed of sound in still air. (The observer moving toward the source is moving in the opposite direction to the propagating waves and thus meets more wave crests in a given time.) The observed frequency (f_o) is then (with $\lambda = v/f_s$):

$$f_o = \frac{v'}{\lambda} = \left(\frac{v + v_o}{v}\right) f_s$$

or

$$f_o = \left(1 + \frac{v_o}{v}\right) f_s \qquad (14.13)$$

observer moving toward a stationary source

Similarly, for an observer moving away from a stationary source, the perceived wave speed is $v' = v - v_o$, and

$$f_o = \frac{v'}{\lambda} = \left(\frac{v - v_o}{v}\right) f_s$$

or

$$f_o = \left(1 - \frac{v_o}{v}\right) f_s \qquad (14.14)$$

observer moving away from a stationary source

Eqs. 14.13 and 14.14 can be combined into another general equation for a moving observer and a stationary source.

$$f_o = \left(\frac{v \pm v_o}{v}\right) f_s = \left(1 \pm \frac{v_o}{v}\right) f_s \qquad (14.15)$$

$$\begin{cases} + \text{ for observer moving toward stationary source} \\ - \text{ for observer moving away from stationary source} \end{cases}$$

There are also the cases when both the source and the observer are moving, either toward one another or away from one another. However, these will not be considered mathematically.

EXAMPLE 14.6 ■ The Doppler Effect

As a truck traveling at 96 km/h approaches and passes a person standing along the highway, the driver sounds the horn. If the horn has a frequency of 400 Hz, what are the frequencies of the sound waves heard by the person (a) as the truck approaches and (b) after it has passed? (Assume that the speed of sound is 346 m/s.)

Solution.

Given: $v_s = 96$ km/h $= 27$ m/s   *Find:* (a) f_o (observed frequency)
        $f_s = 400$ Hz   (b) f_o
        $v = 346$ m/s

(a) Using Eq. 14.12 with a minus sign (source approaching stationary observer),

$$f_o = \left(\frac{v}{v - v_s}\right)f_s = \left(\frac{343 \text{ m/s}}{343 \text{ m/s} - 27 \text{ m/s}}\right)(400 \text{ Hz}) = 434 \text{ Hz}$$

(b) A plus sign is used when the source is moving away:

$$f_o = \left(\frac{v}{v + v_s}\right)f_s = \left(\frac{343 \text{ m/s}}{343 \text{ m/s} + 27 \text{ m/s}}\right)(400 \text{ Hz}) = 371 \text{ Hz}$$

EXAMPLE 14.7 ■ Moving Source and Moving Observer

Suppose a sound source and an observer were moving away in opposite directions, each at one half the speed of sound in air. Then, the observer would (a) receive sound with a frequency greater than the source frequency, (b) receive sound with a frequency less than the source frequency, (c) receive sound with the same frequency as the source frequency, (d) receive no sound from the source. *Clearly establish the reasoning and physical principle(s) used in determining your answer before checking it below. That is, why did you select your answer?*

Reasoning and Answer. As we know, when a source moves away from a stationary observer, the observed frequency is lower (Eq. 14.12). Similarly, when an observer moves away from a stationary source, the observed frequency is also lower (Eq. 14.14). With both source and observer moving away from each other in opposite directions, the combined effect would make the observed frequency even less, so the answer is not (a) or (c).

Looking at (b) and (d), it would appear that (b) is the correct answer, but we must logically eliminate (d) for completeness. It should be remembered that the speed of sound relative to the air is constant. Therefore, (d) would be correct *only if the observer is moving faster than the speed of sound* relative to the air. Since the observer is moving at only one half the speed of sound, (b) is the correct answer. Think about it this way. The sound from the source is moving at the speed of sound through the air toward the observer regardless of how fast the source is moving. The observer is moving at only one half the speed of sound through the air, so the sound from the source can easily reach the observer.

Follow-up Exercise. What would be the situation if the source and the observer were both traveling in the same direction with the same subsonic speed? (Subsonic, as opposed to supersonic, refers to a speed that is less than the speed of sound in air.) (*Reasoning and answer are given in the Answers to Exercises section at the back of the book.*)

■ Problem-Solving Hint

You may find it difficult to remember whether a plus or minus sign is used in the general equations for the Doppler effect. Let your experience help you. For the commonly experienced case of being a stationary observer, the sound frequency increases when the source approaches, so the denominator in Eq. 14.12 must be smaller than the numerator. For this case, you use the minus sign. When the source is receding, the frequency is lower. The denominator in Eq. 14.12 must then be larger than the numerator, and you use the plus sign for this case. Similar thinking will help you choose a plus or minus sign for the numerator in Eq. 14.15.

The Doppler effect also occurs for light waves, although the formulas describing it are different from those given above. It has been found that the wavelengths of light from distant galaxies have increased (frequencies decreased) when the light reaches Earth (shifted toward the red end of the visible spectrum). This indicates that the galaxies are moving away from the Earth. This so-called Doppler red shift is evidence for the theory of an expanding universe. The Doppler shift of light from stars in our galaxy, the Milky Way, indicates that the galaxy is rotating.

You have been subjected to a practical application of the Doppler effect if you have ever been caught speeding in your car by police radar, which uses reflected radio waves. (Radar stands for *ra*dio *d*etecting *a*nd *r*anging and is similar to underwater sonar, which uses ultrasound.) If the radio waves are reflected from a parked car, the reflected waves return to the source with the same frequency. But for a car that is moving toward a patrol car, the reflected waves have a higher frequency, or are Doppler-shifted. Actually, there is a double Doppler shift. The moving car acts like a moving observer in receiving the wave (first Doppler shift), and in reflecting it, the car acts like a moving source emitting a wave (second Doppler shift). The magnitudes of the shifts depend on the speed of the car. A computer quickly calculates this speed and displays it for the police officer.

Sonic Booms

Consider a jet plane that can travel at supersonic speeds. As the speed of a moving source of sound approaches the speed of sound, the waves ahead of the source come close together (• Fig. 14.11). When a plane is traveling at the speed of sound, the waves can't outrun it, and they pile up in front. At supersonic speeds, the waves overlap. This overlapping of a large number of waves produces many points of constructive interference, forming a large pressure ridge, or shock wave. This is sometimes called a bow wave because it is analogous to the wave produced by the bow of a boat moving through water at a speed greater than the speed of the water waves. (Fig. 14.11c).

From aircraft traveling at supersonic speed, the shock wave trails out to the sides and downward. When this pressure ridge passes over an observer on the ground, the large concentration of energy produces what is known as a **sonic boom**. There is really a double boom because shock waves are formed at both ends of the aircraft. Under certain conditions, the shock waves can break windows and cause other damage. (Sonic booms are no longer heard as frequently as in the past. Pilots are now instructed to fly supersonically only at high altitudes and away from populated areas.)

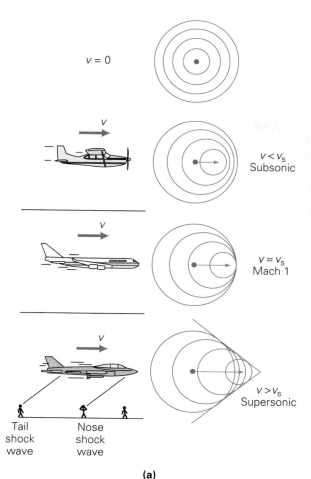

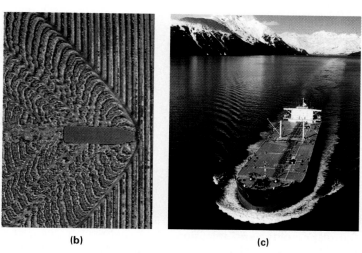

(b)   (c)

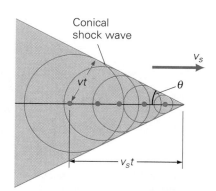

• FIGURE 14.11 **Bow waves and sonic booms** (a) When an aircraft exceeds the speed of sound in air, the sound waves form a pressure ridge, or shock wave. As the trailing shock wave passes over the ground, observers hear a sonic boom—actually two booms, because shock waves are formed at the front and tail of the plane. (b) A bullet traveling at a speed of 500 m/s. Note the shock waves produced (and the turbulence behind the bullet). The image was made using interferometry with polarized light and a pulsed laser, with an exposure time of 20 ns. (c) A shock wave in water. A bow wave like the one shown here is produced whenever a boat travels faster than the speed of the water waves that it creates.

On a smaller scale, you have probably heard a "mini" sonic boom. The "crack" of a whip is actually a sonic boom created by the transonic speed of the whip's tip.

A common misconception is that a sonic boom occurs only when the plane breaks the sound barrier. As an aircraft approaches the speed of sound, the pressure ridge in front of it is essentially a barrier that must be overcome with extra power. However, once supersonic speed is reached, this barrier is no longer there, and the trailing shock wave produces booms along its ground path.

Ideally, the sound waves produced by a supersonic aircraft form a cone-shaped shock wave (• Fig. 14.12). The waves travel outward with a speed v, and the speed of the plane (the source) is v_s. Note from the figure that the angle between a line tangent to the spherical waves and the line along which the plane is moving is given by

$$\sin \theta = \frac{vt}{v_s t} = \frac{v}{v_s} \qquad (14.16)$$

The inverse ratio is called the **Mach number** (M):

$$\boxed{M = \frac{1}{\sin \theta} = \frac{v_s}{v}} \qquad (14.17)$$

• FIGURE 14.12 **Shock wave cone and Mach number** When the speed of the source (v_s) is greater than the speed of sound in air (v), the interfering spherical sound waves form a V-shaped pressure ridge (side view), or a conical shock wave. The angle θ is given by $\sin \theta = v/v_s$, and the inverse ratio v_s/v is called the Mach number.

Mach number, not surprisingly, was named for a physicist, Ernst Mach (1838–1916), also an Austrian.

If v equals v_s, the plane is flying at the speed of sound, and the Mach number is 1 (that is, $v_s/v = 1$). Therefore, a Mach number less than 1 indicates a subsonic speed, and a Mach number greater than 1 indicates a supersonic speed. In the latter case, the Mach number tells the speed of the aircraft in terms of a multiple of the speed of sound.

14.5 ■ Sound Characteristics

Perceived sounds are described by terms whose meanings are similar to those used to describe the physical properties of sound waves. Physically, a wave is generally characterized by intensity, frequency, and waveform (harmonics). The corresponding terms used to describe the sensations of the ear are loudness, pitch, and quality (or timbre). These general correlations are shown in Table 14.2. However, the correspondence is not perfect. The physical properties are objective and can be measured directly. The sensory effects are subjective and vary from person to person. (As an analogy, think of temperature as measured by a thermometer and by the sense of touch.)

Sound intensity and its measurement on the decibel scale were covered in Section 14.3. **Loudness** is related to intensity, but the human ear responds differently to sounds of different frequencies. For example, two tones with different frequencies but the *same* intensities (in W/m²) may be judged by the ear to have different loudnesses.

Frequency and **pitch** are often used synonymously, but again there is an objective-subjective difference. If the same low-frequency tone is sounded at two intensity levels, most people will say that the more intense sound has a lower pitch, or perceived frequency.

The curves in the graph of intensity level versus frequency shown in ● Fig. 14.13 are called equal loudness contours. They join points representing intensity-frequency combinations that the average person judges to be equally loud. The top curve shows that the decibel level of the threshold of pain does not vary much from 120 dB, regardless of the frequency of the sound. However, the threshold of hearing (0 dB at 1000 Hz) varies widely with frequency, as indicated by the lowest curve in the graph. A tone with a frequency of 100 Hz

TABLE 14.2 ■ General Correlation between Perception and Physical Characteristics of Sound

Sensory Effect	Physical Wave Property
Loudness	Intensity
Pitch	Frequency
Quality (timbre)	Waveform (harmonics)

Distinguish between loudness and intensity, and between frequency and pitch.

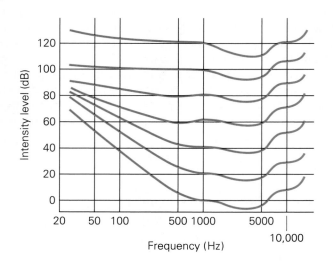

●**FIGURE 14.13 Equal loudness contours** The curves indicate tones that are judged to be equally loud, though they have different frequencies and intensity levels. For example, on the next-to-lowest contour, a 1000-Hz tone at 20 dB sounds as loud as a 100-Hz tone at 50 dB. Note that the frequency scale is logarithmic to compress the large frequency range.

must have an intensity of almost 40 dB to be heard. The dips (or minima) in the curves indicate that the human ear is most sensitive to sounds whose frequencies are between 1000 Hz and 5000 Hz. Note that a tone with a frequency around 3300 Hz can be heard at intensity levels *below* 0 dB.

The **quality** of a tone is the characteristic that enables it to be distinguished from another of basically the same intensity and frequency. Tone quality depends on the waveform, specifically, on the number of harmonics (overtones) present (● Fig. 14.14). The tone of a voice depends in large part on the vocal resonance cavities. One person can sing a tone with the same basic frequency and intensity as another, but different combinations of overtones give the two voices different qualities.

Sound quality is highly subjective. Some sounds are pleasing to some people and cultures, but not to others. Certain music may seem discordant to some people, but others find it quite pleasing.

The notes of a musical scale correspond to certain frequencies; for example, middle C (C_4) has a frequency of 262 Hz. When a note is played on an instrument, its assigned frequency is that of the first harmonic, the fundamental frequency. This frequency is dominant over the accompanying overtones that determine the sound quality of the instrument. Recall from Chapter 13 that the overtones produced depend on how an instrument is played in some cases. Whether a violin string is plucked or bowed can be discerned from the quality of identical notes.

Musical instruments provide good examples of standing waves and boundary conditions. On some stringed instruments, different notes are produced by varying the lengths of the strings using finger pressure (● Fig. 14.15). As you learned in Chapter 13, the natural frequencies of a stretched string (fixed at each end, as is the case for the strings on an instrument) are $f_n = nv/2L$, where $v = \sqrt{F/\mu}$. Initially adjusting the tension in a string tunes it to a particular (fundamental) frequency. Then the effective length of the string is varied by finger pressure.

Standing waves are also set up in wind instruments. For example, consider a pipe organ with fixed pipe lengths, which may be open or closed (● Fig. 14.16). An open pipe is open at both ends, and a closed pipe is closed at one end and open at the other (the antinode end). Analysis similar to that done in Chapter 13 for a stretched string with the proper boundary conditions shows that the natural frequencies for the pipes (where v is the speed of sound in air) are

$$f_n = \frac{v}{\lambda_n} = \frac{nv}{2L} \quad \text{for } n = 1, 2, 3, \ldots$$
natural frequencies for an open pipe
(14.18)

and

$$f_m = \frac{v}{\lambda_m} = \frac{mv}{4L} \quad \text{for } m = 1, 3, 5, \ldots$$
natural frequencies for a closed pipe
(14.19)

Note that the natural frequencies depend on the length of a pipe. This is an important consideration in a pipe organ (Fig. 14.16c), particularly in selecting

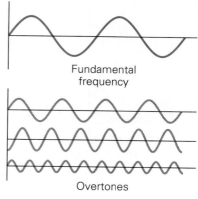

Fundamental frequency

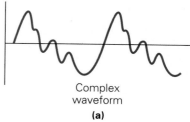

Overtones

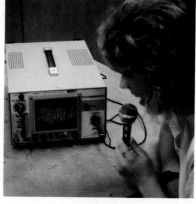

Complex waveform

(a)

(b)

● **FIGURE 14.14 Waveform and quality** **(a)** The superposition of sounds of different frequencies and amplitudes gives a complex waveform. The overtones determine the quality of the sound. **(b)** A voice print. The waveform of a student's voice tone is electronically displayed on an oscilloscope.

Video demonstration 8 relates to this material.

Different notes are
produced on stringed instruments
such as guitars and violins by
placing a finger on a string to
change its effective, or vibrating
length. The painting is "Angelo
Musicante" by Rosso Fiorentino
(1494–1540).

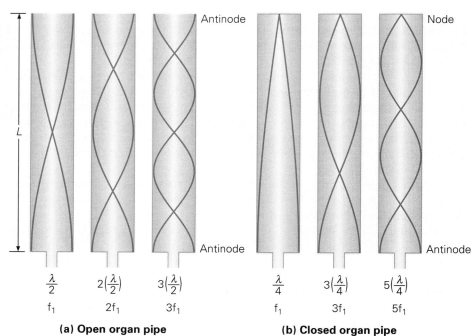

Antinode

Node

Antinode

Antinode

$\dfrac{\lambda}{2}$ $2\left(\dfrac{\lambda}{2}\right)$ $3\left(\dfrac{\lambda}{2}\right)$

f_1 $2f_1$ $3f_1$

(a) Open organ pipe

$\dfrac{\lambda}{4}$ $3\left(\dfrac{\lambda}{4}\right)$ $5\left(\dfrac{\lambda}{4}\right)$

f_1 $3f_1$ $5f_1$

(b) Closed organ pipe

Figure 14.16a,b

●FIGURE 14.17 Wind
instruments   The fife, a close
relative of the flute and piccolo, is
essentially an open tube. The
effective length of the tube is varied
by opening and closing holes with
the fingers. This changes the length
of the vibrating column of air, and
thus the pitch. This painting by
Edouard Manet (1832–1883) is "Le
Fifre" (The Piper).

(c)

●FIGURE 14.16 Organ pipes   Longitudinal standing waves are formed in vibrating air
columns in pipes (illustrated here with sinusoidal curves). **(a)** An open pipe has antinodes
at both ends. **(b)** A closed pipe has a closed (node) end and an open (antinode) end. **(c)** A
modern pipe organ. The pipes can be open or closed.

the dominant or fundamental frequency. (Pipe diameter is also a factor, but is
not considered in this simple analysis.)

The same physical principles apply to wind and brass instruments. In all
of these, the human breath is used to create standing waves in an open tube.
Most such instruments allow the player to vary the effective length of the tube,
and thus the pitch produced—either by opening and closing holes in the tube,
as in woodwinds (●Fig. 14.17), or with the help of slides or valves that vary
the actual length of tubing in which the air can resonate, as in most brasses.

Important Concepts

You should be able to define and explain these chapter concepts clearly.

sound waves	intensity	path difference	sonic boom
sound frequency spectrum	decibel (dB)	destructive interference	Mach number
audible region	sound intensity level	beats	loudness
infrasonic region	constructive interference	beat frequency	pitch
ultrasonic region	phase difference	Doppler effect	quality

Important Relationships for Review

These relationships are mathematical statements of the concepts and principles presented in the chapter. You should be able to identify the symbols and to explain the relationships before proceeding to the Exercises. In-text equation reference numbers are given for convenience.

Speed of Sound (in m/s):

$$v = (331 + 0.6\, T_C) \text{ m/s} \tag{14.1}$$

Intensity of a Point Source:

$$I = \frac{P}{4\pi R^2} \quad \text{and} \quad \frac{I_2}{I_1} = \left(\frac{R_1}{R_2}\right)^2 \tag{14.2–3}$$

Intensity Level (in dB):

$$\beta = 10 \log \frac{I}{I_o} \quad \text{where } I_o = 10^{-12} \text{ W/m}^2 \tag{14.4}$$

Phase Difference (where PD is path difference):

$$\Delta\theta = \frac{2\pi}{\lambda}(\text{PD}) \tag{14.5}$$

Condition for Constructive Interference:

$$\text{PD} = n\lambda \quad \text{where } n = 0, 1, 2, 3, \ldots \tag{14.6}$$

Condition for Destructive Interference:

$$\text{PD} = m\left(\frac{\lambda}{2}\right) \quad \text{where } m = 1, 3, 5, \ldots \tag{14.7}$$

Beat Frequency:

$$f_b = |f_1 - f_2| \tag{14.9}$$

Doppler Effect:

$$f_o = \left(\frac{v}{v \mp v_s}\right)f_s = \left(\frac{1}{1 \mp v_s/v}\right)f_s \tag{14.12}$$

$$\begin{cases} - \text{ for source moving toward stationary observer} \\ + \text{ for source moving away from stationary observer} \end{cases}$$

$$f_o = \left(\frac{v \pm v_o}{v}\right)f_s = \left(1 \pm \frac{v_o}{v}\right)f_s \tag{14.15}$$

$$\begin{cases} + \text{ for observer moving toward stationary source} \\ - \text{ for observer moving away from stationary source} \end{cases}$$

Mach Number:

$$M = \frac{1}{\sin\theta} = \frac{v_s}{v} \tag{14.17}$$

Natural Frequencies of Open Organ Pipe:

$$f_n = \frac{nv}{2L} \quad \text{for } n = 1, 2, 3, \ldots \tag{14.18}$$

Natural Frequencies of Closed Organ Pipe:

$$f_m = \frac{mv}{4L} \quad \text{for } m = 1, 3, 5, \ldots \tag{14.19}$$

Exercises

14.1 ■ Sound Waves

14.2 ■ The Speed of Sound

1 A sound wave is a (a) longitudinal wave, (b) transverse wave, (c) compressional wave, (d) both (a) and (c). (d)

2 A sound wave with a frequency of 25 kHz is in what region of the sound spectrum: (a) audible, (b) infrasonic, (c) ultrasonic, (d) subsonic? (c)

3 The speed of sound is generally greatest in (a) solids, (b) liquids, (c) gases, (d) vacuum. (a)

4 The speed of sound in air (a) is about 1/3 km/s, (b) is about 1/5 mi/s, (c) depends on temperature, (d) all of these. (d)

● **5** Suggest a possible explanation of why some flying insects produce buzzing sounds and some do not. sounds not all in the audible range

6 Is the speed of sound in a medium the same for all frequencies? Explain. see ISM

7 In Table 14.1, the speed of sound in polystyrene (a polymer) is noticeably smaller than it is in other solids, and the speed of sound of hydrogen is noticeably larger than it is in other gases. Can you explain why?   depends on density

8 In a popular lecture demonstration, the instructor breathes helium gas. This gives the instructor's voice a high-pitched sound, like that of Donald Duck. Why? [*Hint:* The resonant wavelengths in the vocal cavities remain the same.]   see ISM

9 ■ What is the speed of sound in air at (a) 10°C and (b) 25°C?   (a) 337 m/s   (b) 346 m/s

10 ■ The thunder from a lightning flash is heard by an observer 6.0 s after she sees the flash. What is the approximate distance to the lightning strike in (a) kilometers and (b) miles?   (a) 2.0 km   (b) 1.3 mi

11 ■ What is the air temperature if the speed of sound is 0.340 km/s?   15°C

12 ■■ Particles that are about 2.9×10^{-2} cm in diameter are to be scrubbed loose in an aqueous ultrasonic cleaning bath. Above what frequency should the bath be operated to produce wavelengths of this size and smaller? 5.1×10^6 Hz

13 ■■ Calculate the difference in the speed of sound in mercury and water. 1.3×10^2 m/s

14 ■■ Brass is an alloy of copper and zinc. Does the addition of zinc to copper cause an increase or decrease in the speed of sound in brass rods as compared to copper rods? If so, by what factor? $v_{Br} = (0.91) v_{Cu}$

15 ■■ How long does it take sound to travel 3.5 km in air if the temperature is 30°C?   10 s

16 ■■ A tuning fork vibrates at a frequency of 512 Hz. What is the wavelength of the sound coming from the fork when the air temperature is (a) 0°C and (b) 20°C?   (a) 0.646 m   (b) 0.670 m

17 ■■ The speed of sound in steel is about 4500 m/s. A steel rail is struck with a hammer, and there is an observer 0.30 km away with one ear to the rail. (a) How much time will elapse from the time the sound is heard through the rail to the time it is heard through the air? Assume that the air temperature is 20°C and there is no wind blowing. (b) How much time would elapse if the wind were blowing toward the observer at 36 km/h from where the rail was struck?   (a) 0.81 s   (b) 0.78 s

● **18** ■■ At a baseball game on a cool day (air temperature of 16°C), a fan hears the crack of the bat 0.40 s after observing a batter hit a ball. How far is the fan from home plate? 1.7×10^2 m

19 ■■ One hunter sees another who is 1.5 km away fire his rifle (sees smoke come from the barrel). If the air temperature is 5°C, how long will it be until the first hunter hears the report of the shot?   4.5 s

20 ■■ A hiker gives a shout and hears the echo reflected from a rock wall 5.0 s later. If the air temperature is 12°C, how far is the hiker from the wall? 8.5×10^2 m

21 ■■ The driver of a truck parked on a mountain road gives the air horn a quick blast toward a rock cliff 0.760 km away. If he hears the echo 4.50 s later, what is the approximate air temperature at this mountain elevation?   12°C

● **22** ■■ The value of the speed of sound in air at 100°C is listed in Table 14.1 as 387 m/s. Is this the value given by Eq. 14.1? If not, explain.   no; see ISM

23 ■■■ The speed of sound in air for normal *environmental* temperatures to a good approximation is given by $v = 331 + 0.6T_C$ m/s (Eq. 14.1). Yet, a better approximation is given by $v = (331) \sqrt{1 + (T_C/273)}$ m/s. Show that the environmental approximation is justified. [*Hint:* Consider a series expansion, $(1 \pm x)^n = 1 \pm nx + \dfrac{n(n-1)}{2}x^2 \pm \ldots\ldots$ $x^2 \ll 1$.]   see ISM

24 ■■■ (Here's an old one.) A person drops a stone into a deep well and hears the splash from its hitting the water 3.16 s later. How deep is the well? (Assume the air temperature in the well to be 10°C.)   45 m

25 ■■■ Sound propagating through air at 20°C passes through a vertical cold front into air that is 4.0°C. If the sound has a frequency of 2400 Hz, by what percentage does its wavelength change as it crosses the boundary?   2.8%

14.3 ■ Sound Intensity

26 Intensity is (a) energy/time-area, (b) energy/time, (c) power/area, (d) both (a) and (c).   (d)

27 The decibel scale is referenced to a standard intensity of (a) 1.0 W/m², (b) 10^{-12} W/m², (c) normal conversation, (d) the threshold of pain.   (b)

28 The Richter scale used to measure the intensity levels of earthquakes is a logarithmic scale, as is the decibel scale. Why are such scales used?   to compress a large range into a smaller numerical scale

● **29** Can there be negative decibel levels, such as −10 dB? If so, what would these mean?   yes, an intensity below the threshold intensity

30 ■ By how many times must the distance from a point source be increased to *reduce* the sound intensity by $\frac{1}{3}$? $R_2 = (1.7) R_1$

31 ■ Find the intensity levels in decibels for sounds with intensities of (a) 10^{-2} W/m², (b) 10^{-4} W/m², and (c) 10^{-13} W/m².   (a) 100 dB   (b) 80 dB   (c) −10 dB

32 ■ What is the intensity of a sound that has an intensity level of (a) 50 dB and (b) 100 dB?   (a) 10^{-7} W/m²   (b) 10^{-2} W/m²

33 ■■ An observer 9.0 m from a stationary point source moves 5.0 m closer to the source. (a) By what factor does the intensity of the sound change? (b) How much farther away from the source would the observer then have to move to reduce the intensity by $\frac{1}{2}$?   (a) 5.1   (b) 5.7 m

34 ■■ If the intensity of one sound is 10^{-4} W/m² and the intensity of another is 10^{-2} W/m², what is the difference in their intensity levels?   20 dB

35 ■■ A tape player has a signal-to-noise ratio of 53 dB. How many times larger is the intensity of the signal than that of the background noise? 2.0×10^5 times greater

36 ■■ A person standing 4.0 m from a wall shouts so that the sound strikes the wall with an intensity of 2.5×10^{-5} W/m². Assuming that the wall absorbs 10% of the incident energy and reflects the rest, what is the sound intensity level just after the sound is reflected?　74 dB

37 ■■ The sound intensity levels for a machine shop and a quiet office are 90 dB and 40 dB, respectively. (a) How many times greater is the intensity of the sound in the machine shop than that in the office? (b) What is each intensity?
(a) $I_s/I_o = 10^5$　(b) 10^{-3} W/m², 10^{-8} W/m²

● **38** ■■ Two identical balloons burst simultaneously, producing a sound with an intensity level of 100 dB. What would the intensity level have been if only one balloon had burst?　97 dB

39 ■■ At a rock concert, the sound intensity level for a person in a front-row seat is 110 dB for a single band. If all the bands scheduled to play produce sound of that same intensity, how many of them would have to play simultaneously for the sound level to be at or above the threshold of pain?　10 bands

40 ■■ At a 4th of July celebration, a firecracker explodes as illustrated in ● Fig. 14.18. Considering the firecracker to be a point source, what are the intensities heard by observers at B, C, and D, relative to that for the observer at A? $I_A = (1.8) I_B$; $I_A = (4.0) I_C$; $I_A = (5.8) I_D$

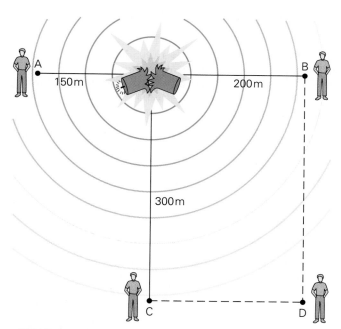

● **FIGURE 14.18 A big bang**　See Exercise 40.

41 ■■ An orchestra plays a movement pianissimo (very softly) at an average intensity of 7.5×10^{-6} W/m² and another movement is played fortissimo (very loudly) at 2.5×10^{-4} W/m². What is the difference in the average sound levels for the movements?　15 dB

42 ■■ A person has lost 30 dB of the normal range of hear-ing response. How many times must the intensity of a sound with an intensity level of 60 dB be amplified by a hearing aid to allow the person to hear it at a normal intensity?　10^3

43 ■■ An engine-powered lawnmower is rated at 95 dB. (a) What is the sound intensity for this particular mower? (b) How many times more intense is the sound of this mower than that of an electric-powered mower rated at 83 dB?
(a) 3.2×10^{-3} W/m²　(b) 16

● **44** ■■ Show that $10 \log I/I_o$ is $\frac{1}{10}$ of a bel.　see ISM

45 ■■ Show that the difference in the intensity levels for intensities I_2 and I_1 is given by $10 \log I_2/I_1$.　see ISM

46 ■■ What would be the intensity level of a 33-dB sound after being amplified a million times?　93 dB

47 ■■■ A 1000-Hz tone issuing from a loudspeaker has an intensity level of 100 dB at a distance of 2.5 m. If the speaker is assumed to be a point source, how far from the speaker will the sound have intensity levels of (a) 60 dB and (b) just barely enough to be heard?　(a) 2.5×10^2 m　(b) 2.5×10^5 m

48 ■■■ A flying bee produces a buzzing sound that is just barely audible to a person 3.0 m away. How many bees would have to be buzzing at that distance to produce a sound with an intensity level of 40 dB?　10^4 bees

14.4 ■ Sound Phenomena

49 The interference of waves from two point sources de-pends primarily on (a) reflection, (b) refraction, (c) diffrac-tion, (d) path difference.　(d)

50 The Doppler effect is used in (a) radar, (b) sonar, (c) determining galactic red shifts, (d) all of these.　(d)

51 Do interference beats have anything to do with the beat of music? Explain.　no (see ISM)

52 Is there a Doppler effect if a sound source and an ob-server are moving (a) with the same velocity or (b) at right angles? (c) What would be the effect if a moving source accelerated toward a stationary observer?　(a) no Doppler effect—same velocity　(b) no Doppler effect—perpendicular　(c) higher frequency

53 How fast would a "jet fish" have to swim to create an aquatic sonic boom?　faster than the speed of sound in water

● **54** ■■ Two sound waves with the same wavelength, 1.5 m, arrive at a point after having traveled (a) 13.50 m and 16.50 m and (b) 9.00 m and 11.25 m. What type of interference occurs in each case?　(a) constructive　(b) destructive

55 ■■ Two adjacent point sources, source A and source B, are in front of an observer and emit identical 600-Hz tones. How far directly behind source B would source A have to be moved for the observer to hear no sound? (Assume that the air temperature is 20°C.)　0.286 m

56 ■■ A violinist and a pianist simultaneously sound notes with frequencies of 352 Hz and 355 Hz, respectively. What beat frequency will be heard by the musicians?　3 Hz

57 ■■ A violinist tuning an instrument to a piano note of 256 Hz detects three beats per second. What are the possible frequencies of the violin tone?   259 Hz and 253 Hz

● **58** ■■ While standing near a railroad crossing, a person hears a train horn. The frequency emitted by the horn is 400 Hz. If the train is traveling at 90 km/h and the air temperature is 25°C, what is the frequency heard by the bystander (a) when the train is approaching and (b) when it has passed by?   (a) 429 Hz   (b) 371 Hz

59 ■■ How fast must a sound source be moving toward you to make the observed frequency 5.0% greater than the true frequency? (Assume that the speed of sound is 340 m/s.)
16 m/s

60 ■■ What is the frequency heard by a person driving 50 km/h toward a blowing factory whistle (f = 800 Hz) if the air temperature is 0°C?   834 Hz

61 ■■ A bystander hears a siren vary in frequency from 476 Hz to 404 Hz as a fire truck approaches, passes by, and moves away on a straight street. What is the speed of the truck? (Take the speed of sound in air to be 343 m/s.)
28 m/s

● **62** ■■ What is the half-angle of the shock wave of a jet aircraft just as it breaks the sound barrier?   90°

63 ■■ The supersonic transport (SST) *Concorde* on transAtlantic flights flies at a speed of Mach 1.5. What is the half-angle of the conical shock wave formed by the *Concorde* at this speed?   42°

64 ■■ The half-angle of the conical shock wave formed by a supersonic jet is 30°. What are (a) the Mach number of the aircraft and (b) the actual speed of the aircraft if the air temperature is 0°C?   (a) 2   (b) 663 m/s

65 ■■■ Two point-source loudspeakers are a certain distance apart and a person stands 12.00 m in front of one of them on a line perpendicular to the base line of the speakers. If the speakers emit identical 1000-Hz tones, what is their minimum nonzero separation so the observer hears no sound? (Let the speed of sound be exactly 340 m/s.)
2.027 m

66 ■■■ A stationary submerged submarine tracks an approaching submarine with sonar. A short pulse of ultrasound with a frequency of 400 kHz is sent toward the sub and returns 10 s later with an observed frequency of 412 kHz. (a) What is the speed of the approaching sub? Take the speed of sound in seawater to be 1200 m/s. [*Hint:* The ultrasound received by the moving sub is Doppler-shifted, and the moving sub acts as a moving source of sound with this shifted frequency. That is, the stationary sub receives a reflected pulse that is doubly Doppler-shifted.] (b) What is the approximate range of the approaching sub at the time of reflection?
(a) 17.7 m/s   (b) 6.0 km

67 ■■■ Show that the general equation for the Doppler effect for a moving source and a moving observer is given by

$$f_\text{o} = f_\text{s} \left(\frac{v \pm v_\text{o}}{v \mp v_\text{s}} \right)$$

using the sign convention of Eqs. 14.12 and 14.15.   see ISM

68 ■■■ A fire engine travels at a speed of 90 km/h with its siren emitting sound at a frequency of 500 Hz. What is the frequency heard by a passenger in a car traveling at 65 km/h in the opposite direction to the fire engine, (a) approaching it and (b) moving away from it? [*Hint:* See Exercise 67 and take the speed of sound to be 354 m/s.]
(a) 567 Hz   (b) 442 Hz

14.5 ■ Sound Characteristics

69 The loudness of a sound as perceived by the ear depends on (a) intensity, (b) overtones, (c) frequency, (d) both (a) and (c).   (d)

● **70** The quality of sound depends on its (a) waveform, (b) frequency, (c) speed, (d) intensity.   (a)

71 (a) After a snowfall, why does it seem particularly quiet? (b) Why do sounds in empty rooms sound hollow? (c) Why do people's voices sound fuller or richer when they sing in the shower?   see ISM

72 ■■ The first three natural frequencies of an organ pipe are 136 Hz, 408 Hz, and 680 Hz. (a) Is the pipe an open or closed one? (b) Taking the speed of sound in air to be 340 m/s, find the length of the pipe.   (a) closed pipe   (b) 0.63 m

73 It is possible for an open and a closed organ pipe of the same length to produce notes of the same frequency? Justify your answer.   not possible, see ISM

74 ■■ A closed organ pipe has a fundamental frequency of 528 Hz (a C note) at room temperature. What is the fundamental frequency of the pipe when the temperature is 0°C?   509 Hz

75 ■■ An open organ pipe has a length of 0.75 m. What would be the length of a closed organ pipe whose third harmonic (m = 3) is the same as the fundamental frequency of the open pipe?   1.1 m

76 ■■ Derive Eqs. 14.18 and 14.19.   see ISM

● **77** ■■ A closed organ pipe has a length of 0.80 m. At room temperature, what are the frequencies of (a) the second harmonic and (b) the third harmonic?   (a) f_2 does not exist—only odd harmonics   (b) 322 Hz

78 ■■■ A resonance tube is an apparatus used to determine the speed of sound in air (see ● Fig. 14.19). The length of the empty part of the tube (L) is adjusted by raising and lowering the water level (by moving the reservoir can). If, when a vibrating tuning fork is held over the tube, the length of the empty part is such that there is a node at the water boundary and an antinode at the top of the tube, a loud resonance sound is heard. The wavelength can be determined by measuring the length of the empty tube. When a fork with a frequency of 512 Hz is used, the first resonance is heard when the water drops 16.5 cm below the top of the tube. (a) What is the speed of sound? (b) What is the air temperature? (c) If the overall tube length is 1.00 m, how many different resonances could be produced?   (a) 338 m/s   (b) 12°C   (c) three

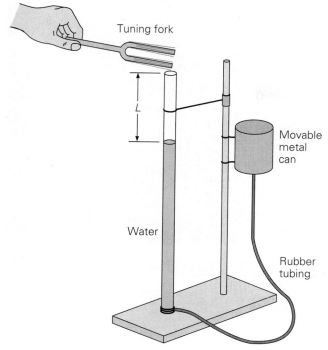

FIGURE 14.19 A resonance tube apparatus See Exercises 78–80.

Tuning fork

L

Movable metal can

Water

Rubber tubing

79 ■■■ In a lab experiment using a resonance tube apparatus (see Exercise 78), the second resonance is heard when the water level is 36 cm below the top of the tube. If the air is at room temperature, what is the frequency of the tuning fork? 716 Hz

80 ■■■ A tuning fork with a frequency of 440 Hz is held above a resonance tube partially filled with water. Assuming that the speed of sound in air is 342 m/s, for what heights of the air column will resonances occur? 0.190 m; 0.582 m; 0.970 m

81 ■■■ A closed organ pipe is filled with helium. The pipe has a fundamental frequency of 660 Hz in air at 0°C. What is the fundamental frequency with the helium?
1.92 × 10³ Hz

■ Additional Exercises

82 What is the beat frequency of two tones that have equal amplitudes and frequencies of 256 Hz and 260 Hz? How could the beat frequency be made zero? 4.0 Hz; frequencies should be equal

83 A jet flies at a speed of Mach 2.0. What is the half-angle of the conical shock wave formed by the aircraft? 30°

84 Two identical strings on different cellos are tuned to the 440-Hz A note. The peg holding one of the strings slips, so its tension is decreased by 1.5%. What is the beat frequency heard when the strings are then played together? 3 Hz

85 The takeoff and landing noise levels for some common commercial jet aircraft are given in Table 14.3. What are the lowest and highest intensities for (a) takeoff and (b) landing? (a) 3.72 × 10⁻⁴ W/m²; 1.00 × 10⁻¹ W/m²
(b) 9.55 × 10⁻³ W/m²; 6.03 × 10⁻² W/m²

TABLE 14.3 ■ Takeoff and Landing Noise Levels for Some Common Commercial Jet Aircraft*

Aircraft	Takeoff noise (dB)	Landing noise (dB)
737	85.7–97.7	99.8–105.3
747	89.5–110.0	103.8–107.8
DC-10	98.4–103.0	103.8–106.6
L-1011	95.9–99.3	101.4–102.8

*Noise level readings taken from 650 ft. Range depends on the aircraft model and the type of engine used.

86 Two sources of 440-Hz tones are located 6.97 m and 8.90 m from an observation point. If the air temperature that day is 15°C, how do the waves interfere at the point? destructively

87 An open organ pipe has a fundamental frequency of 440 Hz at room temperature. (a) What is the length of the pipe? (b) If the air temperature rose to 25°C, what would the change in the fundamental frequency be? (a) 0.39 m
(b) 4 Hz

88 If a person standing 30 m from a 550-Hz point source walks 5.0 m toward the source, by what factor does the sound intensity change? $I_2 = (1.4) I_1$

89 What is the decrease in intensity if the distance from a point source is tripled? $I_2 = (1/9) I_1$

90 Two people speak at the same time with intensity levels of 60 dB and 65 dB. What is the intensity level of the combined sounds heard by another person? 66.2 dB

91 The intensity level of a sound is 90 dB at a certain distance from its source. How much energy falls on a 1.5-m² area in 5.0 s? 7.5 × 10⁻³ J

92 An engineer who uses units of feet and degrees Fahrenheit wants an equation for the speed of sound in air. Rewrite Eq. 14.1 to meet this need. $v = (1075.1 + 0.33 T_F)$ ft/s

93 The note A (440 Hz) is sounded on a stringed musical instrument. The air temperature is 20°C. Approximately how many vibrations does the string make before the sound reaches a person 30 m away? 38 vibrations

94 A stereo speaker is rated at 60 W of output at 1000 Hz. At the lower limit of the audible sound range, the output decreases by 1.5 dB. What is the power output of the speaker at the lower frequency? 43 W

95 A factory whistle is heard by one worker at an intensity level of 60 dB and by another at 80 dB. How many times farther away from the whistle is the first worker?
$R_{60} = 10 (R_{80})$

96 A person fires a rifle at a target. The bullet has a muzzle velocity of 460 m/s, and the person hears the bullet strike the target 2.00 s after firing it. The air temperature is 72°F. What is the distance to the target? 394 m

97 If the room temperature of the air increases by 10 C°, what will be the percentage change in the speed of sound? 1.7%

15

Electric Charge, Force, and Energy

Topics for study in Chapter 15 include electric charge and charging, electric force and electric field, electric energy and electric potential, and capacitance and dielectrics.

Electricity seems to have a flair for the dramatic. To most of us, the word conjures up striking images: a blaze of floodlights, crackling live wires, the skyline of a great city at night, a flash of lightning—or a demonstration such as the one pictured above. There is often a hint of anxiety in our associations, for we all know that electricity can be dangerous.

We also know, however, that electricity can be tamed, even domesticated. In the home or the office, we have come to take it almost for granted. Indeed, the extent to which we depend on it becomes evident only when the power goes off unexpectedly, giving us a dramatic reminder of the role that it plays in our daily lives. Yet less than a century ago there were no power lines crossing the land, no electric lights or appliances—none of the seemingly endless electrical applications surrounding us today.

The word "electricity" is derived from elecktron, the Greek word for amber. The early Greeks were familiar with the attraction a piece of amber rubbed with cloth or fur has for certain other materials. For centuries, such electrical properties were not understood. Only during the last 400 years have theories of electricity been developed.

Less than 200 years ago, the discovery was made that electricity and magnetism are in fact related phenomena. Today, the electromagnetic force is recognized as being one of the four fundamental forces (along with gravity and the strong and weak nuclear forces, the latter of which is believed to be related to the electromagnetic force. See Chapter 29.) It is customary in introductory physics courses to consider the electrical part of the electromagnetic force before the magnetic part and only then to combine electricity and magnetism into electromagnetism. This will be the approach followed in this book.

15.1 ▪ Electric Charge

What is electricity? Perhaps the broadest answer is that electricity is a collective term describing phenomena associated with the interaction (or force) between *electric charges.*

Like mass, **electric charge** is a fundamental property of matter. Electric charge is associated with atomic particles, the electron and the proton. The simplistic solar system model of the atom likens its structure to planets orbiting the Sun (Fig. 15.1). The electrons are viewed as orbiting a nucleus, a core containing protons and another type of electrically neutral particle. The centripetal force that keeps the planets in their orbits is supplied by gravity; the force that keeps the electrons orbiting the nucleus is supplied by electrical attraction. However, there is an important difference between the gravitational and electrical forces.

Mass particles are apparently of one type and give rise only to attractive gravitational forces. Electric charges, on the other hand, are of two types, distinguished as being positive (+) or negative (−). A positive charge is associated with the proton and a negative charge with the electron. Different combinations of the two kinds of charges produce *both* attractive and repulsive forces.

The directions of the electrical forces when charges interact with one another are given by the **law of charges:**

Like charges repel, and unlike charges attract.

That is, two negatively charged particles or two positively charged particles experience mutually repulsive forces, whereas particles with opposite charges are mutually attracted (Fig. 15.2).

The charge of an electron and the charge of a proton are equal in magnitude, though opposite in sign. The charge on the electron (e) is taken as the fundamental unit of charge, since it is the smallest charge that has been observed in nature.* An electric charge q is a charge with an integral multiple of fundamental

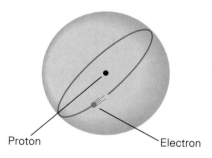

Proton ———— Electron

(a) Hydrogen atom

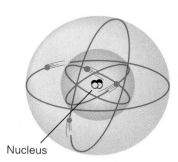

Nucleus ————

(b) Beryllium atom

 FIGURE 15.1 Simplistic model for atoms The so-called solar system model of **(a)** a hydrogen atom and **(b)** a beryllium atom views the electrons as orbiting the positively charged nucleus, analogous to the planets orbiting the Sun. (The electronic structure of atoms is actually much more complicated than this.)

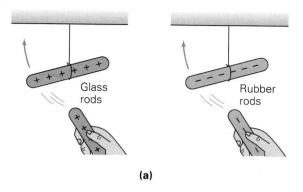

Glass rods

Rubber rods

(a)

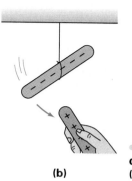

(b)

 FIGURE 15.2 The law of charges **(a)** Like charges repel. **(b)** Unlike charges attract.

*According to a recent theory, protons may be made up of particles called quarks that carry charges of $\frac{1}{3}$ and $\frac{2}{3}$ of the electronic charge. There is experimental evidence for the existence of quarks within the nucleus, but not outside the nucleus in common electrical interactions. See Chapter 29.

TABLE 15.1 ■ Particles and Electric Charge

Particle	Electric Charge	Mass
Electron	-1.6×10^{-19} C	$m_e = 9.11 \times 10^{-31}$ kg
Proton	$+1.6 \times 10^{-19}$ C	$m_p = 1.672 \times 10^{-27}$ kg
Neutron*	0	$m_n = 1.674 \times 10^{-27}$ kg

*The neutron is an electrically neutral particle found in the nucleus of an atom.

charges (either positive or negative). That is,

$$q = ne \qquad (15.1)$$

where n is an integer. It is sometimes said that charge is quantized, which means that it occurs only in integral multiples of the fundamental electronic charge. Mass, on the other hand, is not quantized.

The standard unit of charge is the **coulomb (C)**, named for the French physicist Charles A. de Coulomb (1736–1806), who discovered a relationship between electrical force and charge (to be considered later in this chapter). The charges of the electron and the proton in terms of the coulomb are given in Table 15.1.

There are some other terms frequently used in discussing electrical properties. Saying that an object has a **net charge** means that it has an excess of either positive or negative charges. As you will learn in the next section, excess charges are obtained by a transfer of electrons. For example, if an object has a net charge of $+1.6 \times 10^{-19}$ C, it could be a charged atom, or ion, that has had one electron removed. This means that it has one proton whose charge is not offset by that of an electron, so it is not electrically neutral.

EXAMPLE 15.1 ■ Quantized Charge

An object has a net charge of -1.0 C. How many excess electrons does this represent?

Solution.

  Given: $q = -1.0$ C   *Find:* n (number of excess electrons)

The net charge is made up of an integral number of electronic charges. Using Eq. 15.1,

$$n = \frac{q}{e} = \frac{-1.0\ \text{C}}{-1.6 \times 10^{-19}\ \text{C/electron}} = 6.3 \times 10^{18}\ \text{electrons}.$$

In dealing with any electrical phenomena, an important principle is the **conservation of charge**.

The net charge of an isolated system remains constant.

The net charge may be other than zero, but it remains constant. Suppose, for example, that a system consists initially of two electrically neutral objects, and electrons are transferred from one to the other. The object with added electrons will then have a net negative charge and the object with fewer electrons will have a net positive charge of equal magnitude. However, the net charge of the *system* is still zero. If we assume the universe as a whole to be electrically neutral, the conservation of charge requires that the summation of the electric charge on all positively and negatively charged particles add up to zero. In any case, the conservation of charge requires that the net charge of the universe be constant, but no one knows for sure what the net charge is.

This principle does not mean that charged particles cannot be created or destroyed. Physicists have done this in this century, as you will learn in later chapters on modern physics. However, by the conservation of charge, charged particles are created or destroyed only in pairs with equal and opposite charges. Like the other conservation laws of physics, the concept of the conservation of charge is called a law because no violation of it has ever been observed.

15.2 ■ Electrostatic Charging

That there are two types of electrical charges and attractive and repulsive forces can be demonstrated easily. Before learning how this is done, you need to be able to distinguish between electrical conductors and insulators. What distinquishes these broad groups of substances is their ability to conduct, or transmit, electrical charge. Some materials, particularly metals, are good **conductors** of electrical charge. Others, such as glass, rubber, and most plastics, are **insulators**, or poor electrical conductors. A comparison of the relative magnitudes of the conductivities of some materials is given in ● Fig. 15.3.

Distinguish between conductors, insulators, and semiconductors.

A general picture is that in conductors, valence electrons of atoms—the ones in the outermost orbits—are relatively free; that is, they are not permanently bound to a particular atom. (This electron mobility also plays a major role in thermal conduction.) In insulators, on the other hand, the valence electrons are more tightly bound. The conduction of electrical charge can create an electrical current, which will be considered in more detail in Chapter 16.

As Fig. 15.3 shows, there is an intermediate class of materials called **semiconductors**. Their ability to conduct charge is much less than that of metals, though much greater than that of insulators. The conductivity of a semiconductor can be adjusted by adding certain types of atomic impurities in varying concentrations. Semiconductors form the basis of transistors and solid-state circuits, which have almost completely replaced vacuum tubes in electronic applications.

The electroscope is a simple device that can be used to demonstrate the characteristics of electrical charge (● Fig. 15.4). When charged objects are brought close to the insulated metal bulb, electrons in the bulb are either attracted or repelled, according to the law of charges. For example, if a negatively charged rod is brought near the bulb, electrons are repelled, and the bulb is left with a positive charge. The electrons are conducted to the metal foil leaves, which diverge because of repulsive electrical forces (Fig. 15.4b). Similarly, if a positively charged rod is brought near the bulb, electrons are attracted to the bulb, leaving the leaves positively charged, and they diverge. (Why?)

Notice that the net charge of the electroscope remains zero in these in-

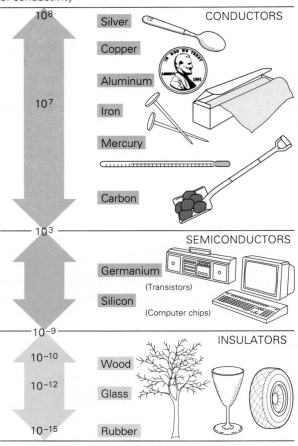

Relative magnitude   Material
of conductivity

10^8 — Silver — CONDUCTORS

10^7 — Copper

Aluminum

Iron

Mercury

Carbon

10^3 — SEMICONDUCTORS

Germanium
(Transistors)

Silicon
(Computer chips)

10^{-9} — INSULATORS

10^{-10} — Wood

10^{-12} — Glass

10^{-15} — Rubber

•**FIGURE 15.3 Conductors, semiconductors, and insulators** A comparison of the relative magnitudes of the electrical conductivities of various materials. (Not to scale.)

Demonstration/activity: Use a Van de Graaff generator to produce large amounts of electric potential. Demonstrate attraction and repulsion using such things as aluminum pie pans, ping pong balls suspended on strings, etc.

Show how to bend a stream of water using a charged rubber rod. This effect depends on the polarization of the water molecule.

stances—only the *distribution* of charge is altered. However, it is possible to give an electroscope (and other objects) a nonzero net charge by electrostatic charging.

In general, **electrostatic charging** is a process by which an insulator or an isolated conductor receives a net charge. In one such charging process, when certain insulator materials are rubbed with cloth or fur, they become electrically charged. For example, if a hard rubber rod is rubbed with fur, the rod will acquire a net negative charge; and rubbing a glass rod with silk will give the rod a net positive charge. This is called **charging by friction**, although the transfer of charge is due to the contact and nature of the materials and is not merely rubbed off by friction.

You have almost certainly experienced a result of electrostatic charging when, after walking across a carpet on a dry day, you get "zapped" by a spark when you reach for a metal object, such as a doorknob. This happens because you have become electrostatically charged. That is, through charging by friction you have picked up a net charge. The charge produces an electric force great enough to *ionize* (free electrons from) the air molecules when your hand comes close to the metal knob. The resulting flow of charges between hand and metal gives rise to a spark discharge. This doesn't occur on humid days. With adequate

Bulb

(a) Neutral electroscope has charges evenly distributed; leaves are close together.

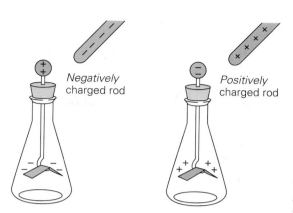

Negatively charged rod

Positively charged rod

(b) Electrostatic forces cause leaves to diverge.

● **FIGURE 15.4 The electroscope** An electroscope can be used to determine whether an object is electrically charged. When a charged object is brought near the bulb, the leaves diverge.

humidity, a thin film of moisture on objects prevents the build-up of charge by conducting it away. A great deal of money is spent to prevent electrostatic charging and the resulting "static cling" (● Fig. 15.5).

As with heat, when a charge moves in a conductor, it is common to talk about a flow. Like heat flow, the flow of electricity was once thought to be due to some type of fluid transfer. Ben Franklin proposed a single-fluid theory of electricity. He assumed that all bodies had some normal amount of "electrical fluid." When some of this was transferred, for example, by rubbing two bodies together, one body would then have an excess of it and the other a deficiency. Franklin indicated these conditions by plus and minus signs, respectively, which is the origin of the sign convention for charge.

Bringing a charged rod close to an electroscope bulb will reveal that the rod is charged, but won't tell you *how* the rod is charged (positively or negatively). This distinction can be made, however, if the electroscope is first given a known type of charge. For example, electrons can be transferred from one object to another if the two objects touch, as illustrated in ● Fig. 15.6a for an electroscope bulb and negatively charged rod. This is called **charging by contact**. If a negatively charged rod is brought close to the now negatively charged

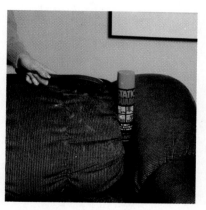

● **FIGURE 15.5 Static cling** Commercial products are available that control static cling by eliminating static electricity. How do you think this is done?

(a) Neutral electroscope is touched with negatively charged rod; charges transferred to bulb.

(b) Electroscope has net negative charge.

(c) Positively charged rod attracts electrons; leaves collapse.

● **FIGURE 15.6 Charging by contact** Charge is transferred to the electroscope when the charged rod touches the bulb. Then, when an oppositely charged rod is brought near the bulb, the leaves collapse or come closer together.

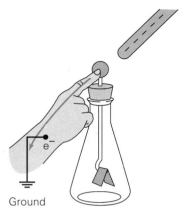

Ground

(a) Electrons are transferred to ground.

(b) Electroscope is left positively charged.

● **FIGURE 15.7 Charging by induction** **(a)** Touching the bulb provides a path to ground for charge transfer. **(b)** When the finger is removed, the electroscope has a net charge.

Electrical ground refers to earth (hence "ground") or some other object that can receive or supply electrons without significantly changing its own electrical condition.

Video demonstration 9 relates to this material.

● **FIGURE 15.8 Polarization** **(a)** Some molecules are polar in nature—that is, they have regions of positive and negative charge. But even some molecules that are not normally polar can be polarized by the presence of a nearby charged object. The electric field induces a separation of charge, making the molecules into induced molecular dipoles. **(b)** When the balloons are charged by friction and placed in contact with the wall, an opposite charge is induced on the wall's surface, to which the balloons then stick by the force of electrostatic attraction.

bulb, the leaves will diverge further (Fig. 15.6b). An oppositely (positively) charged rod will cause the leaves to collapse, or come closer together (Fig. 15.6c).

Since it is electrons that are transferred, you might wonder how an electroscope can be positively charged. This can be done by **charging by induction** (● Fig. 15.7). Touching the bulb with a finger grounds the electroscope, that is, provides a path by which electrons can escape from the bulb. Then when a negatively charged rod is brought close to the bulb, it repels electrons from the bulb. Removing the finger leaves the electroscope with a net positive charge.

Charging by induction does not have to involve a removal of charge from an object. Charge can be moved *within* an object to give different regions of charge. In this case, induction brings about **polarization**, or separation of charge (● Fig. 15.8). Now you can understand why a balloon will stick to the wall or ceiling after being rubbed on someone's hair or sweater. The balloon is charged by friction, and the charged balloon induces an opposite charge on the surface, creating an attractive electrical force. Polarization of molecules is an important consideration in the storing of electrical energy.

Electrostatic charging can be annoying (even dangerous), but it can also be beneficial in a variety of practical applications. Clothes and papers often stick together because of static cling, and an electrostatic spark discharge can start a fire or cause an explosion in the presence of a flammable gas. On the

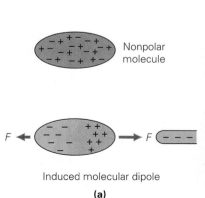

Nonpolar molecule

$F \leftarrow$ $\rightarrow F$

Induced molecular dipole

(a)

(b)

Xerography, coined from the Greek words *xeros* (meaning "dry") and *graphein* (meaning "to write"), refers to a dry process by which almost any printed material can be copied. This process makes use of a photoconductor, which is a light-sensitive semiconductor, such as selenium. When kept in darkness, a photoconductor is a good insulator that can be electrostatically charged. However, when light strikes the material, it becomes conductive and the electrical charge can be removed.

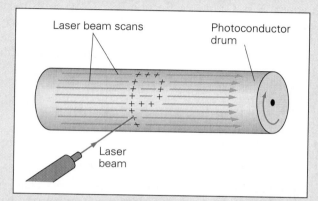

FIGURE 1 Laser printer A computer-controlled laser scans the charged photoconductor drum, causing the charge to bleed off where the beam strikes. When the laser beam is turned off, charged regions remain, which can be reproduced as in the xerography process.

In transfer xerography, a photoconductor-coated plate, drum, or belt is electrostatically charged, and then receives a projected image of the page to be copied. The illuminated portions of the photoconductor coating become conducting and are discharged, but the areas corresponding to the dark print remain charged. Essentially, this creates an electric charge copy of the original sheet. Then the photoconductor copy comes into contact with a negatively charged powder called toner, or dry ink. The toner is attracted to and adheres on the charged regions. Paper is placed over the inked photoconductor and given a positive charge. The toner is then attracted to the paper and heating causes it to be permanently fused to the paper. All of this takes place very quickly—out comes your copy!

The laser printer used with computers is basically a xerographic machine. In this case, there is no "original copy" per se; the information to be printed is stored in the computer. In the printer a laser (Chapter 26) scans back and forth across a rotating, charged drum (Fig. 1). The laser beam passes through a device called a modulator, which allows the beam to reach the drum or blocks it, according to the signals received from the computer. Wherever the fine light beam strikes the drum, that point is discharged, while unilluminated areas, corresponding to the letters, remain charged. In this way a charged image of what is to be printed is produced. The rest of the printing process then takes place as described previously.

other hand, the air we breathe is cleaner because of electrostatic precipitators used in smokestacks. In these devices, electrical discharges cause the particles that are a byproduct of fuel combustion to be charged; then this particulate matter is removed from the flue gases through the use of electrical force. On a smaller scale, electostatic air cleaners are available for the home. Another, almost indispensable, application is the electrostatic copier (see the Insight feature).

15.3 ■ Electric Force

The relative directions of the electric forces on mutually interacting changes are given by the law of charges. However, what about the *magnitude* or strength of the electric force? This was investigated by Charles de Coulomb (after whom the unit of electric charge is named), using a delicate balance to measure the force. He found that the magnitude of the electric force between two charges (q_1 and q_2) depended on the product of the charges and varied inversely as the square of the distance between them; that is $F \propto q_1 q_2 / r^2$. (Note that this is

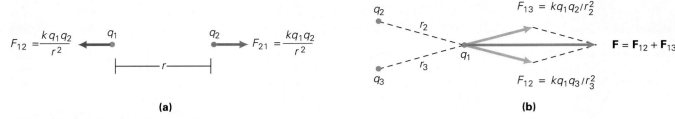

(a)

(b)

• **FIGURE 15.9 Coulomb's law** **(a)** The mutual electrostatic forces on two point charges are equal and opposite. **(b)** For a configuration of more than two charges, the force on a particular charge is the vector sum of the forces on it due to all the other charges.

Coulomb's law: the force between two charges

another inverse-square relationship. You can see the similarity between this expression and the one for the force of gravity, $F \propto m_1 m_2 / r^2$.)

Like Cavendish's measurements for the determination of the universal gravitational constant G, where $F = Gm_1 m_2 / r^2$ (Section 7.4), Coulomb's measurements provided a constant of proportionality so that the electric force may be represented in equation form. Thus, the magnitude of the electric force between two electrical charges is described by an equation called **Coulomb's law**:

$$F = \frac{kq_1 q_2}{r^2}$$ (15.2)

where r is the distance between the charges (● Fig. 15.9a) and k is a constant:

$$k = 9.0 \times 10^9 \text{ N-m}^2/\text{C}^2$$

By Newton's third law, there are equal and opposite forces on the charges (Fig 15.9a). These forces are sometimes referred to as *electrostatic* forces, emphasizing the fact that Coulomb's law applies to fixed or static charges.

In some instances, we are concerned with the force on a particular charge in a configuration of two or more charges. The net electric force on a particular charge is simply the vector sum of the forces on it due to all the other charges (Fig. 15.9b).

Students may be rusty on operations with vectors. Electric forces are non-contact forces and may be more difficult for some students to comprehend. Problems at the end of the chapter allow students to practice finding resultant forces.

EXAMPLE 15.2 ■ Coulomb's Law

(a) Two point charges of $-1.0\ \mu C$ and $2.0\ \mu C$ are separated by a distance of 0.30 m, as illustrated in ● Fig. 15.10a. [1 μC (1 microcoulomb) $= 10^{-6}$ C.] What is the electrostatic force on each particle? (b) A configuration of three charges is shown in Fig. 15.10b. What is the electrostatic force on q_3?

Demonstration/activity: Paint two ping-pong balls with metallic paint and suspend them from 50 to 100 cm of silk thread. Charge them with an electrostatic generator and observe them to stand apart due to the repulsive force. Measure their weight and the angle of separation, and use Coulomb's law to calculate the size of charge on the balls. (This activity requires dry weather.)

Solution. Listing the data and converting microcoulombs directly to coulombs, we have

Given: (a) $q_1 = -1.0\ \mu C \left(\dfrac{10^{-6}\ C}{1 \mu C} \right) = -1.0 \times 10^{-6}\ C$ *Find:* (a) F_{12} and F_{21}

(b) F_{net} on q_3

$q_2 = 2.0\ \mu C \left(\dfrac{10^{-6}\ C}{1 \mu C} \right) = 2.0 \times 10^{-6}\ C$

$r = 0.30\ \text{m}$

$k = 9.0 \times 10^9\ \text{N-m}^2/\text{C}^2$ (known)

(b) Data given in Figure 15.10b.

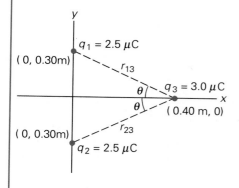

(a)

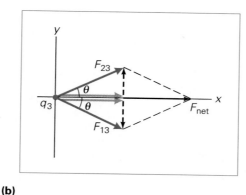

(b)

(Note that values in μC have been changed to C. Don't forget this.)

(a) Eq. 15.2 gives the magnitude of the force acting on each particle.

$$F_{12} = F_{21} = F = \frac{kq_1q_2}{r^2} = \frac{(9.0 \times 10^9 \text{ N-m}^2/\text{C}^2)(1.0 \times 10^{-6} \text{ C})(2.0 \times 10^{-6} \text{ C})}{(0.30 \text{ m})^2}$$

$$= 0.20 \text{ N}$$

Since the charges are unlike, the force is attractive. By Newton's third law, F_{21} is equal to F_{12} but in the opposite direction (see the Problem-Solving Hint on p. 492).

(b) Here, we must vectorially add the forces F_{13} and F_{23}. Since all the charges are positive, the forces are repulsive as shown in the vector diagram in Fig. 15.10b. Since $q_1 = q_2$ and the charges are equidistant from q_3, then F_{13} and F_{23} are of equal magnitude.

Note from the figure that $r_{13} = r_{23} = 0.50$ m (why?). With data from the figure we again use Equation 15.2.

$$F_{13} = F_{23} = \frac{kq_2q_3}{r_{23}^2}$$

$$= \frac{(9.0 \times 10^9 \text{ N-m}^2/\text{C}^2)(2.5 \times 10^{-6} \text{ C})(3.0 \times 10^{-6} \text{ C})}{(0.50 \text{ m})^2}$$

$$= 0.27 \text{ N}$$

Taking into account the directions of F_{13} and F_{23}, we see by symmetry that the y components of the vectors cancel. Thus F_{net} is along the x axis:

$$F_{net} = F_{13} \cos \theta + F_{23} \cos \theta = 2F_{23} \cos \theta$$

since $F_{13} = F_{23}$. The angle θ may be determined from the distance triangles;

that is,

$$\tan \theta = \frac{y}{x} = \frac{0.30 \text{ m}}{0.40 \text{ m}} = 0.75$$

and
$$\theta = \tan^{-1}(0.75) = 37°$$

So
$$F_{\text{net}} = 2F_{23} \cos \theta = 2(0.27 \text{ N}) \cos 37° = 0.43 \text{ N}$$

in the $+x$ direction.

■ Problem-Solving Hint

The signs of the charges may be used explicitly in Eq. 15.2; if they are, a positive value for F indicates a repulsive force and a negative value an attractive force. However, it is usually less confusing to calculate the magnitude of the force without using the charge signs (as was done in Example 15.2) and apply the law of charges to determine whether the force is attractive or repulsive.

The forces in Example 15.2 may not seem very large, but keep in mind that the forces are generally on *very* small particles. Also, the inverse-square relationship makes a difference, as the following example shows.

EXAMPLE 15.3 ■ Repulsive Force in the Nucleus

What is the magnitude of the repulsive electrostatic force between two protons in a nucleus? The distance between nuclear protons is about 10^{-13} cm.

Solution.

Given: $r \cong 10^{-13} \text{ cm} = 10^{-15} \text{ m}$ *Find:* F (magnitude of force)

$q_1 = q_2 = p = 1.6 \times 10^{-19} \text{ C}$

$k = 9.0 \times 10^9 \text{ N-m}^2/\text{C}^2$

Using Coulomb's law (Eq. 15.2), we have

$$F = \frac{kq_1q_2}{r^2} \cong \frac{(9.0 \times 10^9 \text{ N-m}^2/\text{C}^2)(1.6 \times 10^{-19} \text{ C})^2}{(10^{-15} \text{ m})^2}$$

$$\cong 230 \text{ N}$$

This is more than 50 lb of force. Large atoms contain many protons in their nuclei, so the repulsive force on any of these protons would be even larger, which would tend to cause the nucleus to fly apart. Since this doesn't generally happen, there must be a stronger attractive force holding the nucleus together (the nuclear force, which will be discussed in Chapter 28).

As pointed out previously, there is a striking similarity between the expressions for the electrical and gravitational forces. However, there is a huge difference in the relative strengths of these forces, as shown in Example 15.4.

EXAMPLE 15.4 ■ Electrical Force versus Gravitational Force

How do the strengths, or magnitudes, of the electrical and gravitational forces between a proton and an electron compare? Express your answer in terms of a ratio so that the difference can be represented as a factor.

Solution. The charges and masses of the particles are known (Table 15.1), and we also know the values of the constants in the equations for the electrical and gravitational forces. Thus, we have

Given: $q_1 = e = -1.6 \times 10^{-19}$ C   *Find:* $\dfrac{F_e}{F_g}$ (ratio of forces)

$q_2 = p = 1.6 \times 10^{-19}$ C
$m_e = 9.11 \times 10^{-31}$ kg
$m_p = 1.67 \times 10^{-27}$ kg
$k = 9.0 \times 10^9$ N-m^2/C^2
$G = 6.67 \times 10^{-11}$ N-m^2/kg^2

We don't have to worry about the separation distance, since this is the same for each equation and will cancel.

The expressions for the forces are

$$F_e = \frac{kq_1q_2}{r^2} \quad \text{and} \quad F_g = \frac{Gm_1m_2}{r^2}$$

Then, forming a ratio for comparison (and to eliminate r) gives

$$\frac{F_e}{F_g} = \frac{kq_1q_2}{Gm_1m_2}$$

$$= \frac{(9.0 \times 10^9 \text{ N-m}^2/\text{C}^2)(1.6 \times 10^{-19} \text{ C})^2}{(6.67 \times 10^{-11} \text{ N-m}^2/\text{kg}^2)(9.11 \times 10^{-31} \text{ kg})(1.67 \times 10^{-27} \text{ kg})}$$

$$= 0.23 \times 10^{40}$$

or $F_e = (2.3 \times 10^{39})F_g$

The magnitude of the electrostatic force between a proton and an electron is more than 10^{39} times greater than the gravitational force. For this reason, the gravitational force between charged particles is usually neglected. (10^{39} is 1000 trillion trillion trillion. By comparison, our national debt is less than 10 trillion dollars.)

> Electric forces are different from gravitational forces in several ways. One way is that electrical forces may be either attractive or repulsive whereas gravitational forces are only attractive forces. Compare the two expressions and note the magnitude of the forces.
>
> Discussions such as "What would the world be like if both attractive and repulsive gravitational forces did exist?" may be helpful in understanding electric forces.

15.4 ■ Electric Field

The electrical force, like the gravitational force, acts over a distance; hence it is termed an action-at-a-distance force. In fact, we say that the range of the electrical force is infinite: $F_e \propto 1/r^2$ approaches 0 as r approaches ∞. Thus, a particular arrangement, or configuration, of charges will have an effect on an additional charge placed anywhere nearby (or indeed anywhere in space).

The idea of a force acting across space was a difficult one for early investiga-

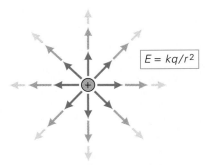

$E = kq/r^2$

(a) Electric field vectors

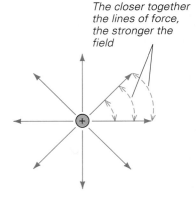

The closer together the lines of force, the stronger the field

(b) Electric field lines (lines of force)

FIGURE 15.11 Electric field
(a) Electric field vectors are determined using a positive test charge. Note that the magnitude of the field (lengths of vectors) becomes smaller as the distance from the charge increases (an inverse square relationship). **(b)** The vectors are connected to give electric field lines, or lines of force.

494

tors, and the concept of a force field, or simply a field, was introduced. Conceptually, an *electric field* extends outward from every electrical charge, permeating all space. This conception allows us to see another charge as interacting with the electrical field rather than with the originating charge itself.

The electric field tells you what force a charge experiences at a particular position in space. The strength of the electric field is expressed as force per unit charge (F/q_o). We map an electric field by placing a *test charge* at various locations and measuring the force per unit charge at points throughout space. With these values, the electric field can be plotted. Knowing the *effect* of a charge or charge configuration throughout space, you can then ignore the charge(s) and talk solely in terms of the electric field.

In investigating the electric field around a charge or configuration of charges, it is important for the test charge to be small enough that the force it exerts on the field-producing charge or charges is negligible. Since the direction of the force on the charge depends on whether it is positive or negative, this too must be specified in a field representation. By convention, a *positive* test charge (q_o) is used for plotting the field. The **electric field** ($\mathbf{E} = \mathbf{F}/q_o$) is a vector field, and a vector at any point in it has the direction of the force that would be experienced by a positive charge placed there. The magnitude of the electric field (E), or the force per unit charge, at a distance of r meters from a charge of q coulombs is given by

$$E = \frac{F}{q_o} = \frac{kq_o q}{q_o r^2} = \frac{kq}{r^2}$$

That is,

$$\boxed{E = \frac{kq}{r^2}} \quad \text{(electric field, charge } q) \qquad (15.3)$$

The direction of the field is given by the law of charges, with q_o taken as positive. The units for the electric field strength can be seen to be newtons per coulomb (N/C).

The test charge basically specifies or determines the direction of the electric field at a point. Note in the equation that the q_o's cancel. Thus, the magnitude of the electric field is independent of the magnitude of the test charge. A *unit* test charge could be used ($q_o = 1.0$ C) or any other size positive charge, as long as it didn't alter the field-producing configuration as mentioned above.

Some electric field vectors in the vicinity of a positive charge are illustrated in ● Fig. 15.11. Note that the vectors point away from the positive charge, which is the direction of the force a positive test charge experiences, and that the magnitude of the vectors decreases with distance from the charge (an inverse-square relationship). Connecting the vectors allows us to represent the electric field graphically by *lines of force*. Notice that the electric field is stronger nearer a charge, where the lines of force are closer together. In general, for a single charge or a configuration of charges, the closer together the lines of force, the stronger the electric field. For a configuration of charges, the total electric field at any point is the vector sum of the electric fields due to the individual charges.

EXAMPLE 15.5 ■ Electric Field

What is the electric field at the origin for the three-charge configuration shown in ● Fig. 15.12?

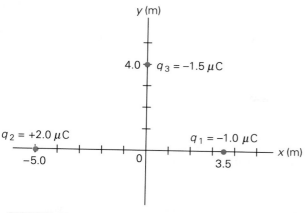

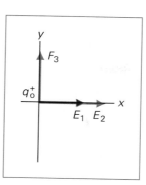

FIGURE 15.12 Finding the electric field Example 15.5

Solution. From the figure, we have

Given: $q_1 = -1.0\ \mu C = -1.0 \times 10^{-6}\ C$
$q_2 = 2.0\ \mu C = 2.0 \times 10^{-6}\ C$
$q_3 = 1.5\ \mu C = 1.5 \times 10^{-6}\ C$
$r_1 = 3.5$ m
$r_2 = 5.0$ m
$r_3 = 4.0$ m

Find: $\mathbf{E}_t$ (total electric field at origin)

By the law of charges, a *positive* test charge q_o at the origin would experience an electric field for each charge in the directions indicated in the figure inset. Then, using Eq. 15.3 to find the magnitude of each of the electric field vectors,

$$E_1 = \frac{kq_1}{r_1^2} = \frac{(9.0 \times 10^9\ \text{N-m}^2/\text{C}^2)(1.0 \times 10^{-6}\ \text{C})}{(3.5\ \text{m})^2}$$

$$= 7.3 \times 10^2\ \text{N/C}$$

$$E_2 = \frac{kq_2}{r_2^2} = \frac{(9.0 \times 10^9\ \text{N-m}^2/\text{C}^2)(2.0 \times 10^{-6}\ \text{C})}{(5.0\ \text{m})^2}$$

$$= 7.2 \times 10^2\ \text{N/C}$$

$$E_3 = \frac{kq_3}{r_3^2} = \frac{(9.0 \times 10^9\ \text{N-m}^2/\text{C}^2)(1.5 \times 10^{-6}\ \text{C})}{(4.0\ \text{m})^2}$$

$$= 8.4 \times 10^2\ \text{N/C}$$

The magnitudes of the x and y components of the total field are then

$$E_x = E_1 + E_2 = 7.3 \times 10^2\ \text{N/C} + 7.2 \times 10^2\ \text{N/C}$$
$$= 14.5 \times 10^2\ \text{N/C}$$

$$E_y = E_3 = 8.4 \times 10^2\ \text{N/C}$$

So, is component form

$$\mathbf{E}_t = E_x\ \mathbf{x} + E_y\ \mathbf{y} = (14.5 \times 10^2\ \text{N/C})\ \mathbf{x} + (8.4 \times 10^2\ \text{N/C})\ \mathbf{y}$$

You should be able to show that in magnitude-angle form this is

$$E_t = 1.7 \times 10^3\ \text{N/C at } \theta = 30° \text{ relative to the } x \text{ axis}$$
$$\text{(1st quadrant)}$$

The concept of an electric field is difficult for most beginning students. An electric field exists in a region in which a stationary charge experiences a force. The intensity of that field is a measure of the ratio of the force to the charge. The direction of that field is in the direction of the force on a positive charge.

Demonstration/activity: Bring a charged pith ball near a Van de Graaff generator to illustrate electric fields. The direction will be obvious. Using a charged ping-pong ball as described previously, one can actually get an approximate value of the electric field. Students are surprised at the large number, indicating a field strength in thousands of newtons per coulomb.

Electric fields are vector quantities. When showing fields from two charges, draw two vectors and show how their vector sum is in the direction of the field.

495

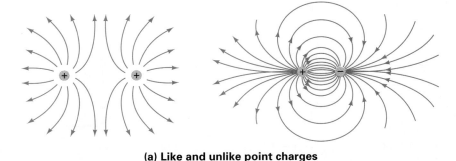

(a) Like and unlike point charges

 Figure 15.13

FIGURE 15.13 Electric fields
Electric fields for various charge
configurations: **(a)** Like and unlike
point charges. **(b)** Charged metal
sphere. The electric field outside the
sphere is as though all the charge on
the sphere were concentrated at its
center. The electric field inside the
sphere is zero. **(c)** Metal parallel
plates. The field is relatively uniform
between the plates. **(d)** An irregularly
shaped charged conductor. Charge
tends to accumulate at sharp points,
or locations of greatest curvature,
producing large electric fields.

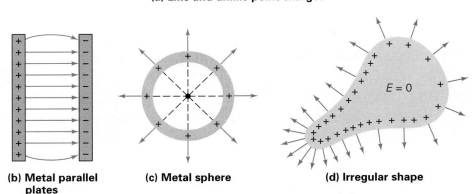

(b) Metal parallel plates　　**(c) Metal sphere**　　**(d) Irregular shape**

The electric fields for some charge configurations and charged conductors
are depicted in ● Fig. 15.13. Note that the electric field lines begin on positive
charges and end on negative charges (or at infinity where there is no nearby
negative charge). Also note that the lines do not cross. Figure 15.13a illustrates
the electric fields associated with pairs of like and unlike charges. Notice that
the fields lines are directed away from a positive charge and toward a negative
charge. (Why?) As a result, the field is diminished in the region between two
like charges and concentrated in the region between two unlike charges. The
electric field between two oppositely charged metal parallel plates (Fig. 15.13b)
is uniform, except for the fringe fields near the ends of the plates.

The electric fields associated with charged conductors (that are isolated or
insulated) have several interesting properties. Suppose that a metal sphere is
given a charge $+Q$, as shown in Fig. 15.13c. Because of the mutually repelling
forces between them, the like charges will be distributed evenly over the sphere
and will eventually come to rest. At this point, the conductor is in electrostatic
equilibrium. Because of symmetry, the electric field outside a charged sphere
is as though all the excess charge on the sphere were concentrated at its center;
that is, $E = kQ/r^2$, where r is greater than the radius of the sphere. Inside the
actual charged sphere, the electric field is zero.

Notice the analogy to the gravitational case, in which all the mass of a
uniform sphere can be thought of as concentrated at its center (Section 7.5).
Like an electric field, a gravitational field can be mapped. The field strength
is the force per unit mass, F/m_o, which is equal to $F/m_o = Gm/r^2 = g(r)$, or
the acceleration due to gravity.

As you might expect, charges on a conductor try to get as far away from each other as possible. Thus,

> Any excess charge on an isolated conductor resides entirely on the surface of the conductor.

Also, it is easy to show that

> The electric field is zero everywhere inside a charged conductor.

If this were not the case, free charges inside the conductor would experience a force, and the conductor wouldn't be in electrostatic equilibrium.

Another property of the electric field of a charged conductor is

> The electric field at the outer surface of a charged conductor is perpendicular to the surface.

Again, if this were not the case, there would be a component of the force tangential to the surface of the conductor, which would cause charges to move and would contradict the condition of electrostatic equilibrium.

EXAMPLE 15.6 ■ An Ice Pail Experiment

A positively charged rod is held inside an isolated metal container which has uncharged electroscopes conductively attached to its inside and outside surfaces (Fig. 15.14a). With the charged rod held inside the container, (a) neither electroscope would show a deflection, (b) only the outside-connected electroscope would be deflected, (c) only the inside-connected electroscope would be deflected, (d) both electroscopes would be deflected. *Clearly establish the reasoning and physical principle(s) used in determining your answer. That is, why did you select your answer?*

Reasoning and Answer. The positively charged rod would attract negative charges, causing the inside of the metal container to become negatively charged. The outside would thus acquire a positive charge. Hence, both electroscopes would be charged (though with opposite signs) and would show deflections, so the answer is (d). A similar experiment was performed by the English scientist Michael Faraday using ice pails, and this arrangement is commonly called Faraday's ice pail experiment.

Follow-up Exercise. Suppose that the positively charged rod touched the metal container as illustrated in Fig. 15.14b. What would be the effect on the electroscopes? (*Reasoning and answer may be found in the Answers to Exercises section at the back of the book.*)

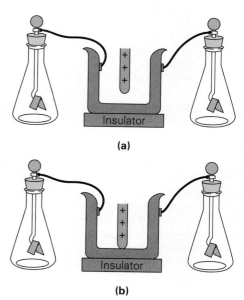

(a)

(b)

FIGURE 15.14 An ice pail experiment Example 15.6

The preceding general characteristics apply to all charged conductors, not just the spherical one used for illustration. For irregularly shaped, or asymmetric, charged conductors, however, the electric field is not symmetric. As Fig. 15.13d shows, the field lines are not distributed evenly around an irregular shape. There is an accumulation of charge in certain places.

Charge tends to accumulate at sharp points, or locations of greatest curvature, on asymmetric charged conductors.

What do you think would happen if all the people on a raft in the middle of a lake tried to get as far from each other as possible? If the raft were irregularly shaped (not round), who would fall off first? In what direction relative to the edge of the raft would that person fall?

To help understand why, consider the forces acting *between* charges on the surface of the conductor. Where the surface is slightly curved, these forces will be directed nearly parallel to the surface (they would be parallel for a completely flat surface). At a sharp end, the intercharge forces will be directed more nearly perpendicular to the surface, and there will be less tendency for the charges to move or be forced apart. Thus, a concentration of charge results. (Draw some surfaces and force vectors to prove this to yourself.)

If there is a large concentration of charge on a charged conductor with a sharp point, the molecules of air near the point may be ionized and a spark discharge may occur. More charge can be placed on a gently curved conductor, such as a sphere, before a spark discharge will occur. The concentration of charge at the sharp point of a conductor is one reason for the effectiveness of lightning rods. (See the Insight feature on lightning.)

INSIGHT ■ Lightning and Lightning Rods

We are all familiar with the violent release of electrical energy in the form of lightning. Although it is a common occurence, we still have a lot to learn about the formation of lightning. We do know that during the development of a cumulonimbus or storm cloud, a separation of charge occurs. The cloud acquires regions of different charge, with the bottom of the cloud generally negatively charged. As a result, an opposite charge is induced on the surface of the Earth (Fig. 1a). Eventually, lightning may equalize these potential differences by ionizing the air, allowing a flow of charge between cloud and ground. However, air is a good insulator and the electrical field must be quite strong for this to occur.

How the separation of charge takes place in a cloud is not fully understood, but it must be associated somehow with the rapid vertical movement of air and moisture within storm clouds. Water is a polar molecule—it has regions of charge, and under certain circumstances water molecules can break apart to produce positively charged and negatively charged ions. It is thought that some ionization may occur as a result of frictional forces between water droplets. However, a more plausible theory describes the separation as taking place during ice pellet formation. It has been shown experimentally that as water droplets freeze, positively charged ions are concentrated in the colder, outer regions of the droplets, whereas negatively charged ions are concentrated in the warmer, interior regions. Thus, a freezing droplet has a positively charged outer ice shell and a negatively charged liquid interior.

As the interior of a droplet begins to freeze, it expands and shatters the outer shell. This gives rise to positively charged ice fragments, which are carried upward by the internal cloud turbulence. This occurs on a large scale, and the remaining, relatively heavy droplets with their negative charge eventually settle to the base of the cloud.

Most lightning occurs entirely within a cloud (intracloud discharges) where it cannot be seen directly. However, familiar visible discharges do take place between clouds (cloud-to-cloud discharges) and between a cloud and the Earth (cloud-to-ground discharges). Pictures of cloud-to-ground discharges taken with special high-speed cameras reveal a nearly invisible downward ionization path. This occurs in a series of jumps or steps and so is called a *stepped leader*. As the leader nears the ground, positively charged ions in the form of a *streamer* rise from trees, tall buildings, or the ground to meet it.

When a streamer and leader make contact, the electrons along the leader channel begin to flow downward. The initial flow is near the ground, and as it continues, electrons positioned successively higher begin to migrate downward. Hence, the path of electron flow is continuously extended upward in what is called a *return stroke*. The surge of charge flow in the return stroke causes the conductive path to be illuminated, producing the bright flash seen by the eye and recorded in time-exposure photographs of lightning (Fig. 1b). Most lightning flashes have a duration of less than 0.50 s. Usually after the initial discharge, ionization again takes place along the original channel and another return stroke occurs. Most lightning events have 3 or 4 return strokes.

Ben Franklin is often said to have been the first to demonstrate the electrical nature of lightning. In 1750 he suggested an experiment using a metal rod on a tall building. However, a Frenchman named d'Alibard set up the experiment and drew sparks from a rod during a

15.5 ■ Electrical Energy and Electric Potential

As with any force, work and energy are associated with an electrical force. When two or more charges are brought closer together or moved further apart, work is done and energy is expended or stored. Recall how intermolecular (electrical) forces in matter can be seen as analogous to springs. Bringing charges closer together or separating them is like compressing or extending an electrical "spring," so to speak.

The expressions for the electrical force and the gravitational force are mathematically identical, and so are those for the potential energies, except for the use of charge instead of mass and the sign associated with the charge. The **electrostatic potential energy** for two charges separated by a distance r is mutual and is given by

$$U = \frac{kq_1q_2}{r} \tag{15.4}$$

Electric potential and electric potential energy are both scalar quantities. Students often get these equations mixed up. Students also sometimes try to make potential and energy vectors.

Potential energy is associated with position.

thunderstorm. Franklin later performed a similar experiment with a kite he flew during a thunderstorm. He drew sparks to his knuckles from a key hanging at the end of a conductive kite cord. Ben was extremely lucky that he wasn't electrocuted. *Under no circumstances should you try to duplicate this experiment.* On the average, lightning kills 200 people a year in the United States and injures another 550.

A practical outcome of Franklin's work with lightning was the lightning rod, which was described in *Poor Richard's Almanac* in 1753. It consists simply of a pointed metal rod connected by a wire to a metal rod that has

been driven into the Earth, or grounded (Fig. 1c). Franklin wrote that the rod "either prevents the stroke from the cloud or, if the stroke is made, conducts the stroke to Earth with safety of the building."

The latter is the general principle of operation of a lightning rod. The idea is for the elevated rod to intercept the downward ionized stepped leader from a cloud, and to harmlessly discharge it to the ground before it reaches the structure or makes contact with an upward streamer. This prevents the formation of the damaging electrical surge associated with the return stroke.

(a)

(b)

(c)

FIGURE 1 Lightning and lightning rods (a) Cloud polarization induces a charge on the Earth's surface. (b) When the field becomes large enough, an electrical discharge results, which we call lightning. See text for details. (c) A lightning rod or arrestor provides a path to ground so as to prevent damage.

Review the discussion of gravitational potential energy in Section 7.5

The electrical potential energy can be either positive or negative, since the force between two charges can be either attractive or repulsive, depending on the signs of the charges. For *unlike* charges, the electrical force is attractive. In that case, by direct analogy with an attractive gravitational force, the electrostatic potential energy is taken to be negative. For *like* charges, the electrical force is repulsive, and the potential energy is positive. The signs of the charges can be used to keep things straight mathematically, as shown in the following example. Remember that energy is a scalar quantity, and in the case of a static configuration of charges, the *total* potential energy is the algebriac sum of the mutual potential energies of all pairs of charges. For example, for a configuration of three charges, the total potential energy (U_t) is

$$U_t = U_{12} + U_{23} + U_{13} \qquad (15.5)$$

This notation simply means to sum all of the different charge pairs algebraically (using the appropriate charge signs).

In general, using summation notation

$$\boxed{U_t = \Sigma_{ij} U_{ij}} \quad (i \neq j) \qquad (15.6)$$

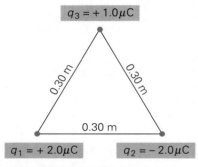

$q_3 = +1.0 \mu C$

0.30 m   0.30 m

0.30 m

$q_1 = +2.0 \mu C$ $q_2 = -2.0 \mu C$

FIGURE 15.15 Electrostatic potential energy A charge configuration has electric potential energy due to the work done in bringing the charges together. This energy can be easily computed. Example 15.7.

EXAMPLE 15.7 ■ Electrostatic Potential Energy

What is the total electrostatic potential energy of the charge configuration shown in ● Fig. 15.15?

Solution. From the figure, we have

Given: $q_1 = +2.0 \ \mu C$ *Find:* U_t (total potential energy)
$\quad\quad = +2.0 \times 10^{-6} C$
$\quad q_2 = -2.0 \ \mu C = -2.0 \times 10^{-6} C$
$\quad q_3 = +1.0 \ \mu C = +1.0 \times 10^{-6} C$
$\quad r = 0.30$ m (same for all in this case)
$\quad k = 9.0 \times 10^9 \ \text{N-m}^2/\text{C}^2$

The total potential energy is the algebraic sum of the mutual potential energies of all pairs of charges. Here, with three charges, Eq. 15.5 applies directly.

$$U_t = U_{12} + U_{23} + U_{13} = k\left(\frac{q_1 q_2}{r_{12}} + \frac{q_2 q_3}{r_{23}} + \frac{q_1 q_3}{r_{13}}\right)$$

$$= (9.0 \times 10^9 \ \text{N-m}^2/\text{C}^2)\left[\frac{(+2.0 \times 10^{-6} \ C)(-2.0 \times 10^{-6} \ C)}{0.30 \ \text{m}}\right.$$
$$\frac{(-2.0 \times 10^{-6} \ C)(+1.0 \times 10^{-6} \ C)}{0.30 \ \text{m}}$$
$$\left.\frac{(+2.0 \times 10^{-6} \ C)(+1.0 \times 10^{-6} \ C)}{0.30 \ \text{m}}\right]$$

$$= -0.12 \ \text{J}$$

This configuration of charge has potential energy because work had to be done in bringing the charges together.

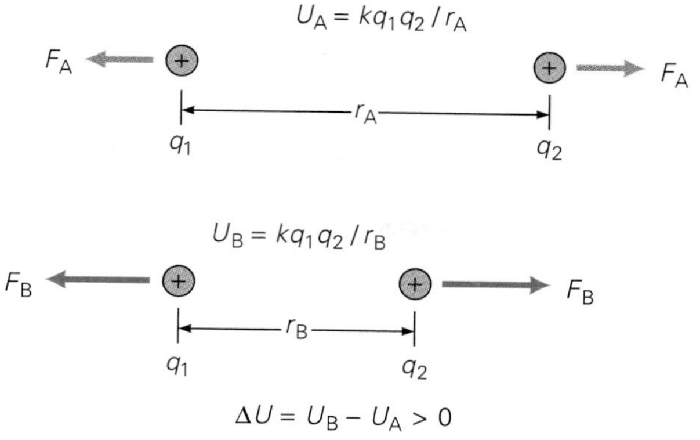

$$U_A = kq_1q_2 / r_A$$

$$U_B = kq_1q_2 / r_B$$

$$\Delta U = U_B - U_A > 0$$

FIGURE 15.16 Change in potential energy If two charges are brought together (or moved apart), a change in potential energy occurs as a result of work being done. In the case illustrated here, there is an increase in potential energy because work is done against the mutually repulsive forces between the charges.

When charges are moved, it is important to consider the *change* in potential energy. For example, consider two positive charges separated by a distance r_A (see ● Fig. 15.16). Then mutual potential energy is

$$U_A = \frac{k(+q_1)(+q_2)}{r_A} = \frac{+kq_1q_2}{r_A}$$

If the charges are brought closer together (by work that is done by an external force—compressing the electrical spring, so to speak) so their separation is r_B (where $r_B < r_A$), then the mutual potential energy is

$$U_B = \frac{+kq_1q_2}{r_B}$$

The change in potential energy is

$$\Delta U = U_B - U_A = \frac{kq_1q_2}{r_B} - \frac{kq_1q_2}{r_A}$$

Since r_B is less than $r_A (r_B < r_A)$, ΔU is greater than zero, or the change is positive, which indicates an increase in potential energy. As you can easily show, if two charges are moved farther apart, the change in their mutual potential is negative, indicating a decrease in the potential energy of the system.

Suppose that after the two charges are brought closer together, they are released. They will move apart because of the mutually repulsive force and there will be a decrease (a negative change) in their mutual potential energy. When the distance separating them is again r_A, the change in their potential energy is manifested as the kinetic energy of the particles. By conservation of energy, for the system in this case,

$$\underset{\text{initial energy}}{U_B} = \underset{\text{final energy}}{K_A + U_A}$$

$$\frac{kq_1q_2}{r_B} = \tfrac{1}{2}m_1v_1^2 + \frac{kq_1q_2}{r_A}$$

Electric Potential

The energy per unit charge is more commonly useful for electrical applications than the mutual potential energy. This quantity is determined by using a positive test charge, as for mapping an electric field. For such a test charge (q_o) located at a distance r from a charge q, the mutual electrostatic potential energy is

$$U = \frac{kq_\circ q}{r}$$

The **electric potential** or **voltage** (V) is the energy per unit charge, or

$$V = \frac{U}{q_\circ} = \frac{kq}{r} \qquad 15.7$$

> Compare the definitions of electric field and electric force with those of electric potential and electric potential energy.

Equivalently,

$$V = \frac{W}{q_\circ} \qquad 15.8$$

where W is the work done in bringing the test charge in from infinity, which is the zero reference point for the electrostatic potential energy and the electrical potential. As you can see, the units for electrical potential are joules per coulomb (J/C). This combination of units has been given the special name **volt** (**V**). That is, $1V \equiv 1 \, J/C$. The electrical potential is commonly referred to as *voltage*. Keep in mind that V is not potential energy but electric *potential* (energy/charge). The potential energy for a charge with potential V is $U = qV$.

The volt unit was named for Alessandro Volta (1745–1827), an Italian scientist who developed one of the first practical batteries.

What is usually of practical interest is the voltage *difference* between two points. This difference is equal to the negative of the work that must be done to move a test charge between the points ($-W_{AB}$)

$$\Delta V = V_B - V_A = -\frac{W_{AB}}{q_\circ} \qquad (15.9)$$

Recall that the change in potential energy is equal to the negative of the work done by a conservative force. For example, the work done in lifting an object against the force of gravity is negative, with a corresponding increase in potential energy. The electrical force is conservative, and $\Delta V = \Delta U / q_\circ = -W/q_\circ$. The work in Eq. 15.9 may be positive, negative, or zero. For these three possibilities, the potential at point B will be lower than, higher than, or the same as the potential at A, respectively. Later, when you study electrical circuit applications, you will learn that the potential of the Earth, or ground, is taken as a reference and assigned a value of zero since only voltage changes or differences are of interest. (This is similar to taking $h = 0$ for the reference position of the gravitational potential.)

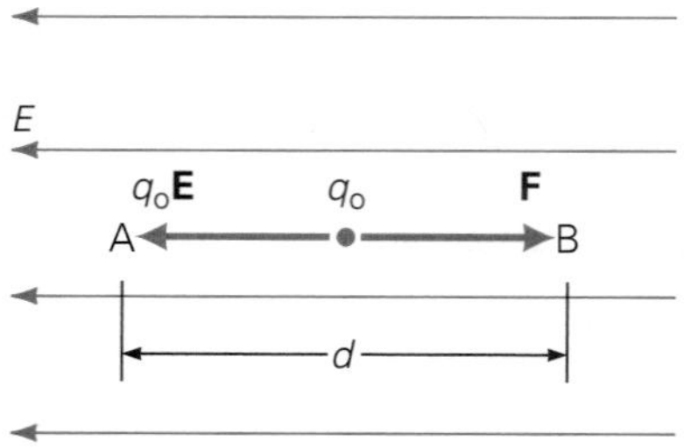

The Relationship of Electric Potential to a Uniform Electric Field

● **FIGURE 15.17 Electric field and electric potential** An external force (**F**) does work against the electric field, or electric force ($q_\circ \mathbf{E}$), in moving the charge a distance d. The change in potential is $+ \Delta V = Ed$. Note that $q_\circ$ is a positive charge, since it experiences an electrical force in the direction of the field. The change in the potential is positive as the charge is moved to a higher electric potential. If the charge were negative, there would be a $-\Delta V$. Why?

Although electronic potential can be defined for any electric field, we will consider their relationship only for a special case. The relationship between the electric potential and a *uniform* electric field can be explored by moving a positive test charge, as illustrated in ● Fig. 15.17. This is done by an external force **F**, which acts against an opposing electrical force $q_\circ \mathbf{E}$. The work done in moving the charge through a distance d (from A to B) is

$$W_{AB} = Fd = -q_\circ Ed$$

By Eq. 15.9, the potential difference is

$$\Delta V = V_B - V_A = \frac{-W_{AB}}{q_o} = Ed$$

The potential at B is higher than that at A because work was done against the field. This is analogous to lifting an object to a greater height in the Earth's gravitational field, and is valid only for the special case of a uniform electric field.

The potential difference, or voltage, ΔV, is commonly written simply as V (not to be confused with the electric potential *at* a point). Recall that this was also done in earlier chapters for length and time intervals Δx and Δt for convenience. From the equation $V = Ed$, you can see that units for an electric field are volts per meter (V/m). Since E also has units of newtons per coulomb (N/C), as can be seen from $E = F/q_o$, then N/C must be equivalent to V/m.

When the displacement of the test charge is not parallel to the field, the potential difference is found using the *component* of the electric field parallel to the displacement. For a uniform electric field, this is expressed as

$$\boxed{V = -Ed \cos \theta} \qquad (15.10)$$

where θ is the angle between the electric field and the displacement. In Fig. 15.17, **E** and **d** are in opposite directions ($\theta = 180°$), so

$$V = -Ed \cos 180° = -Ed(-1) = Ed$$

(If the displacement were from B to A, can you show that $V = -Ed$? Also, what does this mean with regard to energy?)

Keep in mind that there are two types of electrical charge ($+$ and $-$), and this fact affects the change of potential. A positive charge will gain potential energy (and electric potential) when moved in the direction opposite that of the field. If a positive charge were released from rest, it would move in the direction of the field, gaining kinetic energy (and losing an equivalent amount of potential energy)—down an electrical "hill," so to speak. A moment's thought should tell you that a negative charge would gain potential energy if it were moved in the direction of the field. (Why?) If released from rest, which way would a negative charge move?

As mentioned above, the potential difference, rather than the electric field, is usually specified for practical applications. For example, chemical energy produces a potential difference between the terminals of a battery. A D-cell flashlight battery has a terminal voltage of 1.5 V, that is, a potential difference of 1.5 V between its terminals.

You should also keep in mind that since the electrical force is conservative and that the work done in moving a charged particle between two points (and the potential difference between those points) is independent of path. That is, the potential difference depends only on the end points, not on how the charge is moved between them. Some points in space may have the same (equal) potentials. Lines connecting such points are called **equipotentials**. Some equipotentials for a point charge are shown in ● Fig. 15.18. The lines are at right angles to the field lines. Thus, no work is done in moving a charge along an equipotential since the displacement is at an angle of 90° to the electric field ($W = -Ed \cos \theta$ and $\cos 90° = 0$).

The concept of electric potential provides a convenient unit of energy that

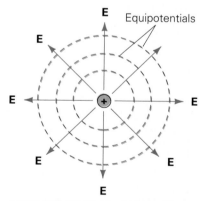

● **FIGURE 15.18 Equipotentials** Lines connecting positions at which there are equal electric potentials are called equipotentials. These are at right angles to the field lines, and no work is done in moving a charge along an equipotential.

A method of generating large electrostatic potentials was developed by the American physicist Robert Van de Graaff around 1930. A diagram illustrating the operation of a simple Van de Graaff generator is given in Fig. 1a. Such generators are commonly used for classroom demonstrations and can develop potential differences of more than 50,000 V.

A continuous, motor-driven, insulating (rubber) belt runs vertically around two pulleys. When the generator is turned on, frictional contact with the lower pulley transfers electrons from the belt to the pulley. The positively charged belt moves upward to the upper pulley, where electrons flow from the pulley to the belt. With continuous running, charges build up on both pulleys.

After a short time, the built-up charge is sufficient to ionize the air in the vicinity of the metal electrodes, or combs (C_1 and C_2 in the figure). When this occurs at the upper electrode (C_2), electrons are drawn from the metal sphere and transferred to the belt, which takes them to the lower pulley. When the charge on the lower pulley is sufficient to cause an ionizing discharge, the excess electrons are transferred to the lower electrode (C_1), which is grounded. The net effect is that electrons are transferred from the metal sphere to a ground, leaving the sphere positively charged. Since the electric field is virtually zero inside the charged sphere (cf. Fig. 15.13b),

electrons are continuously removed from it and a build-up of positive charge results, giving a high electric potential. (Another method of charging the belt is through electric arcing, or corona discharge, from a high-voltage source.)

An electric potential between 50,000 V and 100,000 V is great enough to cause a so-called corona discharge through the ionization of air molecules around the sphere, and small electric arcs can be seen (and heard) between the sphere and a nearby object (Fig. 1b). In a demonstration of this, an instructor may arc the discharge to his or her knuckles. Another common demonstration is to have an insulated person touch the generator. The person becomes negatively charged, and repulsive forces cause his or her hair to stand on end, as shown in the chapter opening photograph. In effect, the person becomes an electroscope, with hair serving as the leaves.

A voltage of 50,000–100,000 V may sound impressive, but there is no shock hazard for a healthy person because of the small amount of charge involved. The corona discharge can be suppressed and higher voltages achieved by surrounding the sphere with a gas that has a greater ionization potential than air. Because of this possibility of producing large voltages, Van de Graaff generators are used to accelerate charged particles in particle accelerators (Chapter 29).

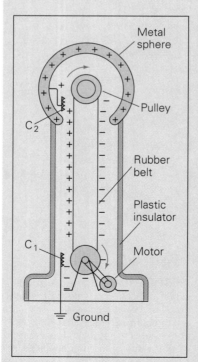

(a)

(b)

FIGURE 1 The Van de Graaff generator (a) A charged, continuously moving belt transports charge from the metal sphere, giving rise to large potential differences between the sphere and ground. (See text for description.) (b) A small, classroom model generator. Such machines are capable of voltages of 50–100 kV and can produce corona discharges of 5–10 cm in length. The discharges to the grounded bulb that the person is holding are difficult to photograph in a lighted room, though they can be seen easily in the dark. (c) The discharges from these large demonstration Van de Graaff generators in the Boston Science Museum are a good deal more conspicuous. They reach the ground through the wire cage in which the operator sits.

(c)

is used in calculating the kinetic energy of charged particles that have been accelerated through potential differences. An **electron volt (eV)** is the energy acquired by an electron in moving through a potential difference of 1 V, and

$$\Delta U = e\Delta V = (1.6 \times 10^{-19}\,C)(1.0\,V)$$
$$= 1.6 \times 10^{-19}\,J$$

That is,

$$1\,eV = 1.6 \times 10^{-19}\,J$$

The energy of any charged particle accelerated through any potential difference can be expressed in electron volts. For example, if an electron is accelerated through a potential difference of 1000 V, its kinetic energy (K) is

$$K = eV = (1\,e)(1000\,V) = 1000\,eV = 1\,keV$$

This is a kiloelectron volt. Although an electron volt is defined for an electron, any integral number of units of electronic charge can be considered.

Similarly, if a particle with a $+2$ charge were accelerated through this voltage, it would have $K = (2e)(1000\,V) = 2000\,eV = 2\,keV$. In this case, a charge of $+2$ is the equivalent of two units of electronic charge ($2e$). Note how easy it is to compute the kinetic energy of an accelerated charged particle in eV.

Larger units for energy are sometimes needed, and mega-electron volts (MeV) and giga-electron volts (GeV) are used (1 MeV $= 10^6$ eV, and 1 GeV $= 10^9$ eV). At one time, a billion electron volts was referred to as BeV, but this was abandoned because confusion arose. In some countries, such as Great Britain and Germany, a billion means 10^{12} (which Americans call a trillion).

15.6 ■ Capacitance and Dielectrics

The electric field that exists between charged conductors provides a means for storing electrical energy. Consider the parallel metal plates shown in ● Fig. 15.19a. Such an arrangement of conductors is called a **capacitor**. Work must be done to put charge on the plates. Suppose that one electron was moved to a plate. Putting the next electron on would be more difficult since it would be repelled by the original charge. The transfer of more charge requires more and more work as the charge on the plate accumulates. (This is analogous to pump-

> Capacitors store energy in an electric field.

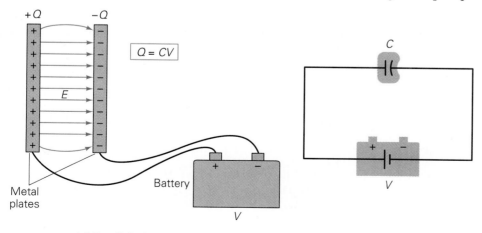

(a) **Parallel-plate capacitor**   (b) **Circuit diagram**

● **FIGURE 15.19 Capacitor and circuit diagram** **(a)** Two parallel metal plates are charge by a battery to a charge Q, and $Q = CV$, where C is the capacitance. Work is done in charge the capacitor, and energy is stored in the electric field. **(b)** This circuit diagram shows the symbols used for a battery and a capacitor. The longer line of the battery symbol is taken as the positive terminal and the short line as the negative terminal. The symbol for a capacitor is similar, but with a curved line for the negative terminal. (At one time, it was common to represent a capacitor with two equal parallel lines, but the curved-line symbol is preferred to avoid confusion with the symbol for the battery.)

Discuss charging parallel plates by theoretically carrying charge from one plate to the other. The first load of charge went easily. Each successive load of charge requires more and more work to carry it from the first plate to the second plate. Explain why.

ing up a tire with a hand pump: the more air there is in the tire, the harder the pumping gets.)

The work done in charging parallel plates may be very quickly done by a battery. The battery charges the plates up to its terminal voltage, and a relatively uniform electric field builds up between the plates. Then the capacitor is disconnected from the battery and becomes a vehicle of stored electrical energy (stored in the electric field), which can be used to do work.

The potential difference across the plates of a capacitor is proportional to the charge Q:

$$Q \propto V$$

In equation form, the relation is

The net charges on the plates are $+Q$ and $-Q$, but it is customary to refer in general to the charge Q on a capacitor.

$$\boxed{Q = CV} \qquad (15.11)$$

The constant of proportionality C is called the **capacitance** and expresses the charge per voltage, $C = Q/V$. In this sense, capacitance is a measure of the charge capacity (Q) of the parallel plates (at a given potential difference V).

The units of capacitance are coulombs per volt (C/V), and this combined unit is called a **farad (F)**. The farad is a rather large unit, so the microfarad (μF) is frequently used (1 μF $= 10^{-6}$ F).

The farad was named for the English scientist Michael Faraday (1791–1867), an early investigator of electrical phenomena who first introduced the concept of the electric field.

The capacitance depends on the geometry of the conductors. For parallel plates of equal surface areas (A), which are large compared to the distance (d) between them, the capacitance is given by

$$\boxed{C = \frac{\varepsilon_0 A}{d}} \qquad \text{(for a parallel-plate capacitor)} \qquad (15.12)$$

Here ε_0 is a constant called the *permittivity of free space* (vacuum), which has a value of 8.85×10^{-12} C^2/N-m^2. ε_0 is a fundamental constant and is contained in the constant in Coulomb's law: $k = 1/4\pi\varepsilon_0$.

EXAMPLE 15.8 ■ How Large Is a Farad?

What would be the area of the plates of a 1.0-F parallel-plate capacitor with a plate separation of 1.0 mm?

Solution.

Given: $C = 1.0$ F   *Find:* A (area of plates)
$d = 1.0$ mm $= 10^{-3}$ m
$\varepsilon_0 = 8.85 \times 10^{-12}$ C^2/N-m^2

Solving Eq. 15.12 for the area (A) gives

$$A = \frac{Cd}{\varepsilon_0} = \frac{(1.0 \text{ F})(10^{-3} \text{ m})}{8.85 \times 10^{-12} \text{ C}^2/\text{N-m}^2}$$

$$= 1.1 \times 10^8 \text{ m}^2$$

This is about 100 km^2, or a square 10 km on a side (about 36 mi^2, or a square 6 mi on a side). A 1.0-F capacitor is therefore a very large capacitor! Most capacitors in common use are in the microfarad or picofarad range (1 pF $= 10^{-12}$ F).

The expression for the energy stored in a charged capacitor may be obtained through graphical analysis. A plot of voltage versus charge for charging a capacitor gives a straight line with a slope of $1/C$ (see Fig. 15.20). The graph represents the charging of an initially uncharged capacitor ($V_0 = 0$) to some final voltage (V). The total work done is equivalent to transferring the charge through the average voltage $\overline{V}$, and

$$\overline{V} = \frac{V_{final} + V_{initial}}{2} = \frac{V + 0}{2} = \frac{V}{2}$$

(Compare this to the analysis of the work done in compressing a spring, Section 5.2.) Thus, the energy (or work) is expressed as

$$U = W = Q\overline{V} = \tfrac{1}{2}QV$$

(Recall that voltage is energy, or work/charge.) Since $Q = CV$, this equation may be written in several forms:

$$U = \tfrac{1}{2}QV = \frac{Q^2}{2C} = \tfrac{1}{2}CV^2 \qquad (15.13)$$

The form $U = \tfrac{1}{2}CV^2$ is usually the most practical, since the capacitance and the applied voltage are often known.

Dielectrics

In most capacitors, a sheet of insulating material, such as paper or plastic, is placed between the plates. This insulating material is called a **dielectric** and serves several purposes. It keeps the plates from coming into contact, which would short the capacitor. It also allows flexible plates of aluminum foil to be rolled into a cylinder, giving the capacitor a practical size. Finally, it increases the energy storage capacity of the capacitor. This capability varies with materials and is characterized by the **dielectric constant** (K). Values of the dielectric constant for some materials are given in Table 15.2.

How a dielectric affects the electrical properties of a capacitor is illustrated in Fig. 15.21. The capacitor is charged, disconnected from the battery, and then a dielectric is inserted. In the capacitor, work is done on molecular dipoles as they experience electrical forces that tend to align them with the field. (The molecular polarization of charge may be permanent or temporarily induced

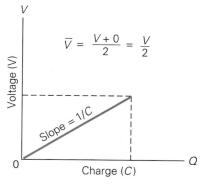

FIGURE 15.20 Voltage versus charge A plot of voltage versus charge for a charging capacitor is a straight line with a slope of $1/C$ (since $Q = CV$). The average voltage is $\overline{V} = V/2$, and the total work done is equivalent to transferring the charge through $\overline{V}$. Thus, $U = W = \tfrac{1}{2}QV$, which is the area under the curve.

Figure 15.21

FIGURE 15.21 The effects of a dielectric (a) A charged capacitor and a dielectric with randomly oriented permanent molecular dipoles. (b) If the dielectric is put into the capacitor, the dipoles are oriented (or partially oriented) with the field, giving rise to an opposing electric field. The dipole field effectively cancels some of the field due to the plate charges. (c) The net effects are decreases in the electric field and voltage by a factor of $1/K$, where K is the dielectric constant of the material.

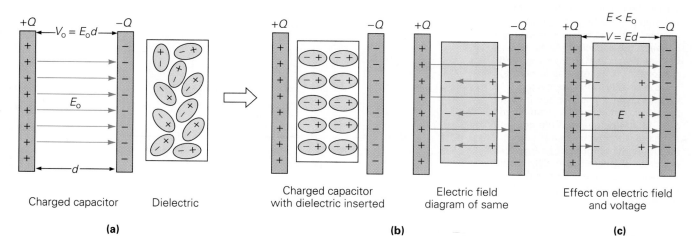

Charged capacitor · Dielectric

(a)

Charged capacitor with dielectric inserted · Electric field diagram of same

(b)

Effect on electric field and voltage

(c)

by the electric field.) Note in Fig. 15.21c how this effectively cancels some of the field lines between the plates. The field between the plates and the voltage across the plates are reduced by a factor of K (●Fig. 15.22a):

$$V = \frac{V_o}{K} \tag{15.14}$$

Here V_o is a given voltage on the plates without the dielectric, or the voltage on the plates before the dielectric was added. The charge Q is unaffected, so the capacitance is

$$C = \frac{Q}{V} = \frac{Q}{V_o/K} = KC_o \qquad \text{or} \qquad \boxed{C = KC_o} \tag{15.15}$$

If, on the other hand, the dielectric is inserted with the battery still connected, the original voltage is maintained and the battery must supply more charge. The total charge on the plates is then $Q = KQ_o$, and again, $C = KC_o$, as can be easily shown. Thus, a dielectric *increases* the capacitance by a factor of K. (Note that this equation shows that the dielectric constant is unitless.) For a

Figure 15.22

●**FIGURE 15.22 Dielectric and capacitance** **(a)** A parallel-plate capacitor in air is charged by a battery to a charge Q and a voltage V_o (left). If the battery is disconnected and the potential across the capacitor measured by a voltmeter, a reading of V_o is obtained (center). But if a dielectric is inserted between the capacitor plates, the voltage drops to $V = V_o/K$ (right). **(b)** A capacitor is charged as in part (a), but the battery is left connected. When a dielectric is inserted into the capacitor, the voltage is maintained at V_o by the battery, but the amount of charge that the capacitor accepts (and thus the amount of energy that it stores) is greater: $Q = KQ_o$. Both of these observations reflect the increase in capacitance produced by the dielectric.

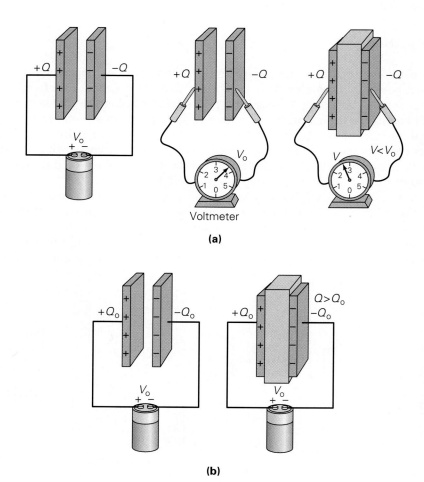

Voltmeter

(a)

(b)

parallel-plate capacitor with a dielectric,

$$C = KC_o = \frac{K\varepsilon_o A}{d} \qquad (15.16)$$

capacitance of a parallel-plate capacitor with a dielectric

This is sometimes written as $C = \varepsilon A/d$, where $\varepsilon = K\varepsilon_o$ is the permittivity of the dielectric material.

Since a dielectric increases the capacitance, a capacitor with a dielectric can store more charge (Fig. 15.22b), and hence energy, than one of the same dimensions without a dielectric (assuming both are charged to the *same* voltage V). This can be shown as follows:

$$U = \tfrac{1}{2}CV^2 = \tfrac{1}{2}KC_oV^2 = KU_o \qquad (15.17)$$

The energy is increased by a factor of K for a given voltage.

An example of the use of a capacitor for electrical energy storage is shown in ● Fig. 15.23.

EXAMPLE 15.9 ■ Dielectric Effects

A parallel-plate capacitor having a plate area of 0.70 m² and a plate separation of 1.0 mm is connected to a source with a voltage of 50 V. Find the capacitance, the charge on the plates, and the energy of the capacitor (a) when there is air between the plates and (b) when the capacitor contains a material with a dielectric constant of 2.5.

Solution.

Given: $A = 0.70$ m²   *Find:* for (a) and (b)
    $d = 1.0$ mm $= 10^{-3}$ m C (capacitance)
    $V = 50$ V Q (charge)
    $K = 2.5$ U (energy)

(a) For air, K is essentially 1.0, and

$$C_o = \frac{\varepsilon_o A}{d} = \frac{(8.85 \times 10^{-12}\,\text{C}^2/\text{N-m}^2)(0.70\,\text{m}^2)}{10^{-3}\,\text{m}}$$

$$= 6.2 \times 10^{-9}\,\text{F} = 0.0062\;\mu\text{F}\;(= 6.2\,\text{nF})$$

Then

$$Q = C_oV = (6.2 \times 10^{-9}\,\text{F})(50\,\text{V}) = 3.1 \times 10^{-7}\,\text{C} = 0.31\;\mu\text{C}$$

and

$$U = \tfrac{1}{2}C_oV^2 = \tfrac{1}{2}(6.2 \times 10^{-9}\,\text{F})(50\,\text{V})^2 = 7.8 \times 10^{-6}\,\text{J}$$

(b) With a dielectric constant of 2.5,

$$C = KC_o = (2.5)(0.0062\;\mu\text{F}) = 0.016\;\mu\text{F}$$

$$Q = KQ_o = (2.5)(0.31\;\mu\text{C}) = 0.78\;\mu\text{C}$$

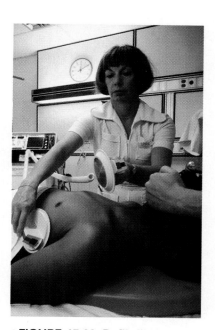

● **FIGURE 15.23 Defibrillator**
A burst of electric current from a defibrillator can often restore normal heartbeat in persons who have suffered cardiac arrest. Capacitors are used to store the electrical energy that the device depends on.

and

$$U = KU_0 = (2.5)(7.8 \times 10^{-6} \text{ J}) = 2.0 \times 10^{-5} \text{ J}$$

Note that the energy is increased from 7.8 μJ to 20 μJ.

Series and Parallel Capacitor Circuits

For so-called sandwiched dielectric parallel-plate capacitors, there is no head or tail distinction between the leads. Some types of capacitors do have particular positive and negative sides, and then the distinction must be made (+ to − in series).

Capacitors can be connected in two basic ways: in series or in parallel. In series (Fig. 15.24a), the capacitors are connected head to tail, so to speak. When they are connected in parallel, all the leads on one side of the capacitors have a common connection (all the tails hooked together, and all the heads hooked together).

When capacitors are in series, the charge is the same on all the plates: $Q = Q_1 = Q_2 = Q_3$. The sum of the individual voltage drops across all the capacitors is the voltage of the source:

$$V = V_1 + V_2 + V_3$$

Since $V = Q/C$,

$$\frac{Q}{C_s} = \frac{Q}{C_1} + \frac{Q}{C_2} + \frac{Q}{C_3}$$

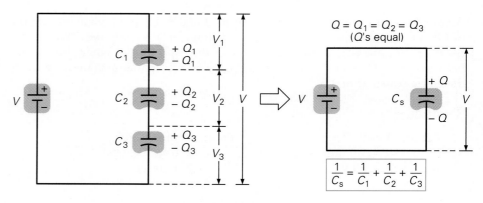

(a) Capacitors in series

Figure 15.24

FIGURE 15.24 Capacitors in series and parallel **(a)** Capacitors connected in series all have the same charge, and the sum of the voltage drops is equal to the voltage of the battery. The total capacitance is equivalent to C_s. **(b)** When capacitors are connected in parallel, the voltage drops across the capacitors are the same, and the total charge is equal to the sum of the charges on the individual capacitors. The total capacitance is equivalent to C_p.

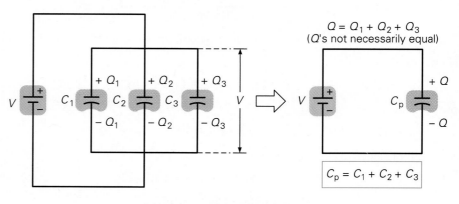

(b) Capacitors in parallel

All the Q's have the same value and cancel out. Thus, for three capacitors in series,

$$\frac{1}{C_s} = \frac{1}{C_1} + \frac{1}{C_2} + \frac{1}{C_3} \qquad (15.18)$$

The value of C_s is the equivalent series capacitance. That is, the three capacitors in series could be replaced with one capacitor whose capacitance is C_s. It is interesting to note that the value of C_s is always smaller than the smallest capacitance in the combination.

The general form of Eq. 15.18 applies to any number of capacitors in series.

$$\boxed{\frac{1}{C_s} = \frac{1}{C_1} + \frac{1}{C_2} + \frac{1}{C_3} + \cdots} \qquad (15.19)$$

equivalent capacitance for capacitors in series

With a parallel hook-up (Fig. 15.24b), the voltages across the capacitors are the same: $V = V_1 = V_2 = V_3$. The total charge is equal to the sum of the charges on the individual capacitors.

$$Q = Q_1 + Q_2 + Q_3$$

Since $Q = CV$,

$$C_p V = C_1 V + C_2 V + C_3 V$$

All the voltages are the same, so for the three capacitors in parallel,

$$C_p = C_1 + C_2 + C_3 \qquad (15.20)$$

In this case, the equivalent capacitance C_p is simply the sum of the individual capacitances.

The general form of Eq. 15.20 applies to any number of capacitors in parallel.

$$\boxed{C_p = C_1 + C_2 + C_3 + \cdots} \qquad (15.21)$$

equivalent capacitance for capacitors in parallel

EXAMPLE 15.10 ■ Series-Parallel Combination of Capacitors

Three capacitors are connected in a circuit as shown in Fig. 15.25a. What is the voltage across each capacitor?

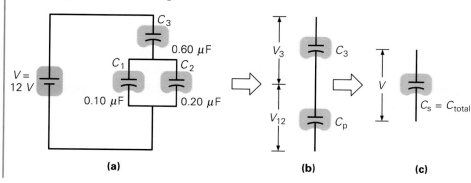

(a)　　　　(b)　　　　(c)

FIGURE 15.25 Circuit reduction By combining capacitances, the capacitor combination may be reduced to a single equivalent capacitance. Example 15.10.

Solution.

Given: Values of capacitance and voltage from figure

Find: V_1, V_2, and V_3 (voltages across capacitors)

The voltage across each capacitor could be found from $V = Q/C$ if the charge on each capacitor were known. The total charge transferred from the battery in charging the capacitors is found by reducing the series-parallel combination to a single equivalent capacitance. Starting with the parallel combination,

$$C_p = C_1 + C_2 = 0.10 \ \mu\text{F} + 0.20 \ \mu\text{F} = 0.30 \ \mu\text{F}$$

and the circuit is reduced as shown in Fig. 15.25b. For the series combination of C_p and C_3,

$$\frac{1}{C_s} = \frac{1}{C_3} + \frac{1}{C_p} = \frac{1}{0.60 \ \mu\text{F}} + \frac{1}{0.30 \ \mu\text{F}} = \frac{1}{0.20 \ \mu\text{F}}$$

and

$$C_s = 0.20 \ \mu\text{F}$$

(Don't forget to take the inverse of $1/C_s$.) Note that C_s is less than either of the capacitances. Using the total equivalent capacitance,

$$Q = C_s V = (0.20 \times 10^{-6} \ \text{F})(12 \ \text{V}) = 2.4 \times 10^{-6} \ \text{C} = 2.4 \ \mu\text{C}$$

This is the charge on C_3, since it is in series with the battery, and

$$V_3 = \frac{Q}{C_3} = \frac{2.4 \ \mu\text{F}}{0.60 \ \mu\text{F}} = 4.0 \ \text{V}$$

The voltage drops across the capacitors equal the voltage rise of the battery: $V = V_{12} + V_3$ [see Fig. 15.25]. Thus,

$$V_{12} = V - V_3 = 12 \ \text{V} - 4.0 \ \text{V} = 8.0 \ \text{V}$$

There is 8.0 V across both C_1 and C_2. Could you find the charges on these capacitors?

Important Concepts

You should be able to define and explain these chapter concepts clearly.

electric charge	semiconductors	electric field	electron volt (eV)
law of charges	electrostatic charging	electrostatic (electric) potential energy	capacitor
coulomb (C)	charging by friction		capacitance
net charge	charging by contact	electric potential (voltage, V)	farad (F)
conservation of charge	charging by induction		dielectric
conductors	polarization	volt (V)	dielectric constant (K)
insulators	Coulomb's law	equipotentials	

Important Relationships for Review

These relationships are mathematical statements of the concepts and principles presented in the chapter. You should be able to identify the symbols and to explain the relationships before proceeding to the Exercises. In-text equation reference numbers are given for convenience.

Quantization of Electric Charge:

(where $e = \pm 1.6 \times 10^{-19}$ C):

$$q = ne \tag{15.1}$$

Coulomb's Law (where $k = 9.0 \times 10^9$ N-m²/C²):

$$F = \frac{kq_1 q_2}{r^2} \quad \text{(magnitude)} \tag{15.2}$$

Electric Field (due to a charge q):

$$E = \frac{F}{q_o} = \frac{kq}{r^2} \quad \text{(magnitude)} \tag{15.3}$$

Electrostatic Potential Energy:

$$U = \frac{kq_1 q_2}{r} \quad \text{(two charges)} \tag{15.4}$$

$$U_t = \sum_{\substack{i,j \\ i \neq j}} U_{ij} \quad \text{(multiple charges)} \tag{15.6}$$

Electric Potential (due to a charge q):

$$V = \frac{U}{q_o} = \frac{kq}{r} = \frac{W}{q_o} \tag{15.7–8}$$

Potential (voltage) Difference:

$$\Delta V = V_B - V_A = \frac{-W_{AB}}{q_o} \tag{15.9}$$

Voltage in a Uniform Electric Field:

$$V = -Ed \cos \theta \tag{15.10}$$

Capacitance:

$$Q = CV \tag{15.11}$$

Capacitance of a Parallel-Plate Capacitor:

(where $\varepsilon_o = 8.85 \times 10^{-12}$ C²/N-m²):

$$C = \frac{\varepsilon_o A}{d} \tag{15.12}$$

Energy in a Charged Capacitor:

$$U = \tfrac{1}{2}QV = \frac{Q^2}{2C} = \tfrac{1}{2}CV^2 \tag{15.13}$$

Dielectric Effects:

$$V = \frac{V_o}{K} \tag{15.14}$$

$$C = KC_o \tag{15.16}$$

$$U = KU_o \tag{15.17}$$

Equivalent Capacitance for Capacitors in Series:

$$\frac{1}{C_s} = \frac{1}{C_1} + \frac{1}{C_2} + \frac{1}{C_3} + \cdots \tag{15.19}$$

Equivalent Capacitance for Capacitors in Parallel:

$$C_p = C_1 + C_2 + C_3 + \cdots \tag{15.21}$$

Exercises

15.1 ▪ Electric Charge

•**1** A combination of two electrons and one proton would have a net charge of (a) $+1$, (b) -1, (c) 1.6×10^{-19} C, (d) -1.6×10^{-19} C. (d)

2 The directions of the interacting electric forces on two charges is given by (a) the conservation of charge, (b) the law of charges, (c) the magnitude of the charges, (d) none of these. (b)

3 (a) How do we know that there are two types of electric charge? (b) What would be the effect of designating the charge on the electron as positive and the charge on the proton as negative? see ISM

4 An electrically neutral object can be given a net charge by several means. Does this violate the conservation of charge? Explain. no (see ISM)

5 ▪ What net electric charge would 10,000 protons have? 1.6×10^{-15} C

•**6** ▪▪ A glass rod rubbed with silk acquires a charge of $+8.0 \times 10^{-10}$ C. (a) What is the charge on the silk? (b) How many electrons have been transferred to the silk?
(a) -8.0×10^{-10} C   (b) 5.0×10^9 electrons

7 ▪▪ A rubber rod rubbed with fur acquires a charge of -2.4×10^{-9} C. (a) What is the charge on the fur? (b) How much mass is transferred to the rod? (a) $+2.4 \times 10^{-9}$ C
(b) 1.4×10^{-20} kg

8 ▪▪ In walking across a carpet, a person acquires a net negative charge of -50 μC. How many excess electrons does the person have? 3.1×10^{14} electrons

15.2 ▪ Electrostatic Charging

9 When a negatively charged rod is brought near the bulb of a negatively charged electroscope, the leaves (a) diverge farther, (b) collapse, (c) remain unchanged. (a)

10 Electrostatic charging may be done by (a) friction, (b) contact, (c) induction, (d) all of these. (d)

•**11** How could an electroscope be negatively charged by induction? How could you prove it was negatively charged? see ISM

12 Two metal spheres mounted on insulated supports are in contact. How could both spheres be electrically charged without touching the metal? Would the spheres be charged positively or negatively?   charge by induction; spheres would have opposite charges

15.3 ▪ Electric Force

13 The magnitude of the electric force between two point charges is given by (a) the law of charges, (b) the conservation of charge, (c) Coulomb's law, (d) both (a) and (b).   (c)

14 Compared to that of the electric force, the strength of the gravitational force between two electrons is (a) about the same, (b) somewhat larger, (c) very much larger, (d) very much smaller.   (d)

● **15** An electron is a certain distance from a proton. How would the electric force be affected if the electron were moved (a) half that distance toward the proton, and then (b) three times that distance away from the proton?
(a) 4 times larger   (b) 1/9 as large

16 We often experience the effect of the gravitational force, for example, in picking up a heavy object, but we don't normally experience the electrical force. Explain why. see ISM

17 ▪ An electron and a proton are separated by 10 cm. (a) What is the magnitude of the force on the electron? (b) What is the net force on the system?   (a) 2.3×10^{-26} N   (b) zero

18 ▪ Two charges originally separated by 30 cm are moved farther apart, until the force between them has decreased by a factor of 10. How far apart are the charges then?   95 cm

19 ▪▪ Two charges are attracted by a force of 25 N when separated by 20 cm. What is the force between the charges when the distance between them is 50 cm?   4.0 N

● **20** ▪▪ Two charges of -1.0 C are placed at opposite ends of a meterstick. Where could (a) a free electron and (b) a free proton be placed on the meterstick and be in electrostatic equilibrium?   (a) 50 cm   (b) 50 cm

21 ▪▪ Charges of -1.0 C and $+1.0$ C are placed at opposite ends of a meterstick. Where could (a) a free electron and (b) a free proton be placed and be in electrostatic equilibrium?   (a) no place   (b) no place

22 ▪▪ Two charges, q_1, and q_2, are located at the origin and at (0.50 m, 0). Where on the x axis must a third charge, q_3, of arbitrary sign be placed to be in electrostatic equilibrium if (a) q_1 and q_2 are like charges of equal magnitude, (b) q_1 and q_2 are unlike charges of equal magnitude, and (c) $q_1 = +3.0 \mu C$ and $q_2 = -7.0 \mu C$?   (a) $x = 0.25$ m
(b) no location   (c) $x = -0.94$ m for either $\pm q_3$

23 ▪▪ What are the magnitude and the direction of a vertically oriented electric field that would just support the weight of an electron? 5.6×10^{-11} N/C downward

● **24** ▪▪ The electron and proton in a hydrogen atom are separated on the average by a distance of 5.3×10^{-11} m (● Fig. 15.26). Assuming the orbit of the electron to be circular, (a) what is the electric force on the electron? (b) What is the electron's orbital speed?   (a) 8.2×10^{-8} N   (b) 2.2×10^6 m/s

25 ▪▪ Compute the magnitude of the gravitational force

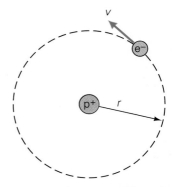

● **FIGURE 15.26 Hydrogen atom**   See Exercises 24, 25, and 65.

between the electron and proton in the hydrogen atom (Fig. 15.26), and compare it to the electric force between them.
$F_e = (2.3 \times 10^{39}) F_g$

26 ▪▪ Three charges are located at the corners of an equilateral triangle, as depicted in ● Fig. 15.27. What are the magnitude and the direction of the force on q_1? $F_x = 3.6$ N in the $+x$ direction

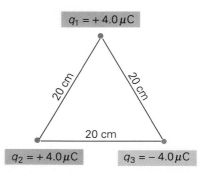

● **FIGURE 15.27 A charge triangle**   See Exercises 26, 42, 43, 57, and 60.

27 ▪▪ Four charges are located at the corners of a square as illustrated in ● Fig. 15.28. What are the magnitude and the direction of the force on (a) q_2 and (b) q_4?   (a) 96 N, 39° below the $-x$ axis; (b) 61 N, 84° above $-x$ axis

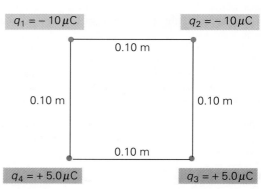

● **FIGURE 15.28 A charge rectangle**   See Exercises 27, 44, 47, 58, 66, 67, and 71.

28 ■■■ Two 0.10-g pith balls are suspended from the same point by threads 30 cm long. (Pith is a light insulating material once used to make helmets worn in tropical climates.) When the balls are given equal charges, they come to rest 18 cm apart. What is the magnitude of the charge on each ball? (Neglect the mass of the thread.) 3.2×10^{-8} C

15.4 ■ Electric Field

29 The electric field of a negative charge (a) varies as $1/r$, (b) points toward the charge, (c) has a finite range, (d) is the same as that of a positive charge.   (b)

● **30** The units of electric field are (a) V/m, (b) N/C, (c) both (a) and (b), (d) neither (a) or (b).   (d)

31 Electric field diagrams usually indicate only the direction of the field. How can the relative magnitudes of the field in different regions be determined from a diagram?
by the relative density of the field lines

32 Why do electric field lines never cross?   see ISM

33 A positive charge is inside an isolated metal sphere as shown in ●Fig. 15.29. Describe the situation in terms of the electric field and the charge on the sphere. How would the situation change if the charge were negative?   see ISM

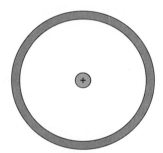

Metal conductor

●**FIGURE 15.29 A point charge in a metal sphere** See Exercise 33.

34 Explain how a region may be shielded from electrical influences (electric fields). [*Hint:* Think of a metal enclosure.] (Do you think gravitational shielding is possible? Explain.)   by enclosing the region in a metal container

35 Sketch the electric field between a positively charged flat plate and a negatively charged pointed object near the plate and pointing toward it.   see ISM

36 ■ An isolated electron experiences an electrical force of 3.2×10^{-4} N. What is the magnitude of the electric field at the electron's location? 2.0×10^{15} N/C

37 ■ What is the electric field at a point 0.25 cm away from a charge of 4.0 μC? 5.8×10^9 N/C

● **38** ■■ Two charges, -4.0 μC and -5.0 μC, are separated by a distance of 20 cm. What is the electric field halfway between the charges?
9.0×10^5 N/C toward the -5.0 μC charge

39 ■■ Two charges, -3.0 μC and -4.0 μC, are located at $(-0.50$ m 0$)$ and $(0.75$ m, 0$)$, respectively. (a) Where on the x axis is the electric field zero? (b) Is there a position (or

positions) where the electric field has only a y component? Explain.   (a) (0.080 m, 0)   (b) any point above and below
x axis where $E_x = 0$

40 Three charges, $+ 2.5$ μC, $- 4.8$ μC, and -6.3 μC, are located at $(-0.20$ m, 0.15 m$)$, $(0.50$ m, -0.35 m$)$, and $(-0.42$ m, -0.32 m$)$ respectively. What is the electric field at the origin?   E = 2.3×10^5 N/C x $-$ 4.1×10^5 N/C y

41 ■■ Two charges of $+ 4.0$ μC and $+9.0$ μC are 30 cm apart. Where on the line joining the charges is the electric field zero?   12 cm from the 4.0 μC charge (between the charges)

42 ■■ What is the electric field at the center of the triangle in Fig. 15.27? 5.4×10^6 N/C toward $-$ 4.0 μC charge

43 ■■ Compute the electric field at a point midway between charges q_1 and q_2 in Fig. 15.27 1.2×10^6 N/C toward the -4.0 μC charge

● **44** ■■ What is the electric field at the center of the square in Fig. 15.28? 3.8×10^7 N/C in the $+y$ direction

45 ■■ A particle with a mass of 1.0×10^{-5} kg and a charge of $+2.0$ μC is released in a uniform electric field of 12 N/C. How far does it travel in 0.50 s?   0.30 m

46 ■■ A thin, hollow spherical conductor with a radius of 0.20 m has a uniform charge density of -10 C/m^2 on its surface. What is the electric field (a) inside the sphere and (b) at the surface of the sphere?   (a) 0   (b) 1.1×10^{12} N/C toward the center

47 ■■■ Compute the electric field at a point 4.0 cm from q_2 along a line running toward q_3 in Fig. 15.28.
E = $(-4.2 \times 10^6$ N/C) x $+ (7.3 \times 10^7$ N/C) y

15.5 ■ Electrical Energy and Electric Potential

48 The electrostatic potential energy of two point charges (a) is inversely proportional to their separation distance, (b) is a vector quantity, (c) is always positive, (d) has units of N/C.   (a)

● **49** Voltage is (a) the same as potential energy, (b) equivalent to the electric field, (c) the energy per unit charge, (d) none of these.   (c)

50 What is the difference (a) between electrostatic potential energy and electric potential, and (b) between electric potential and voltage?   (a) see ISM   (b) no difference

51 When is a charge said to be at a higher electric potential relative to some other position in an electric field?   see ISM

52 Explain why an electron volt is a unit of energy. Which is larger, a GeV or a MeV?   see ISM

53 When a person touches the metal sphere of a charged Van de Graaff generator, his or her hair may stand on end, as shown in the chapter introductory photo. Explain why. Why must the person be insulated?   see ISM

● **54** ■■ A charge of -4.0 μC is initially 40 cm from a fixed charge of -6.0 μC and is then moved to a position 90 cm from the fixed charge. (a) What is the change in the mutual potential energy of the charges? (b) Does this change depend on the path through which the one charge is moved?
(a) -0.30 J   (b) no

55 ■■ How much work is done in moving an electron along two legs of an equilateral triangle with sides 0.25 m long in an electric field of 15 V/m parallel to the other side of the triangle? (Take the electron as being initially at one end of the side of the triangle parallel to the electric field.)
6.0×10^{-19} J against the field

56 ■■ (a) What is the electrical potential energy of an electron located 15 cm from a charge of $+6.0$ μC? (b) How much work is required to move the electron to an infinite distance from the charge? (a) -5.8×10^{-14} J (b) 5.8×10^{-14} J

57 ■■ Compute the energy necessary to bring together the charges in the configuration shown in Fig. 15.27. -0.72 J

58 ■■ Compute the energy necessary to bring together the charges in the configuration shown in Fig. 15.28. -4.1 J

59 ■■ What is the electric potential at the center of the triangle in Fig. 15.27? 3.1×10^5 V

60 ■■ Compute the electric potential at a point midway between q_2 and q_3 in Fig. 15.27. 2.1×10^5 V

61 ■■ What speed will an electron reach if it is accelerated from rest through a potential difference of 100 V?
5.9×10^6 m/s

62 ■■ Two electrons that are initially at rest 10 cm apart are released. What will be their speed when they are 20 cm apart?   36 m/s

63 ■■ A proton moving directly toward another fixed proton has a speed of 0.50 m/s when the two are 1.0 m apart. How close to the stationary proton will the moving proton be when it stops and reverses course?   0.52 m

64 ■■ The potential difference involved in a typical lightning discharge may be up to 100 million volts. What would be the kinetic energy of an electron after moving through this potential difference in (a) eV and (b) joules? (Assume no collisions.) (a) 1.00×10^8 eV (b) 1.6×10^{-11} J

65 ■■ In the simplified model of the hydrogen atom (Fig. 15.26), (a) what is the electric potential due to the proton at the electron's distance? (b) What is the electric potential energy of the electron? (c) Is this its total energy? Explain.
(a) 27 V (b) -4.3×10^{-18} J (c) no, also has kinetic energy

66 ■■ What is the electric potential at the center of the square in Fig. 15.28? -1.3×10^6 V

67 ■■ Compute the electric potential at a point midway between q_1 and q_4 in Fig. 15.28. -1.3×10^6 V

68 ■■ A proton is moved 15 cm on a path parallel to the field lines of a uniform electric field of 2.0×10^5 V/m. (a) What is the change in the proton's potential? Consider both cases, of moving the proton with and against the field. (b) What is the change in energy in electron volts? (c) How much work would be done if the proton were moved perpendicularly to the electric field? (a) against: 3.0×10^4 V; with: -3.0×10^4 V (b) $\pm 3.0 \times 10^4$ eV (c) zero

69 ■■ Two charged parallel plates separated by a distance of 5.0 cm have an electric field of 2.0×10^3 V/m between them (Fig. 15.30). (a) What is the potential difference between the plates? (b) What would be the change in the poten-

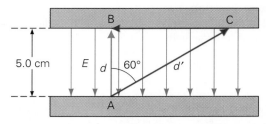

● **FIGURE 15.30 Potential and energy**   See Exercise 69.

tial energy of a charge of 1.0 μC if it were moved from A to B? From A to C to B? (a) 1.0×10^2 V (b) 1.0×10^{-4} J (both situations)

70 ■■ In a television picture tube, electrons are accelerated from rest through a potential difference of 10 kV in an electron gun. (a) What is the "muzzle velocity" of the electrons emerging from the gun? (b) If the gun is directed at a screen 35 cm away, how long does it take the electrons to reach the screen? (a) 5.9×10^7 m/s (b) 5.9×10^{-9} s

71 ■■■ Compute the electric potential at a point 4.0 cm from q_4 along the line joining it to q_1 in Fig. 15.28.
-7.4×10^5 V

15.6 ■ Capacitance and Dielectrics

72 Capacitance has units of (a) farads, (b) joules, (c) coulomb/volt, (d) both (a) and (c). (d)

73 Putting a dielectric in a charged, parallel-plate capacitor (a) decreases the capacitance, (b) decreases the voltage, (c) increases the charge, (d) causes a discharge because the dielectric is a conductor. (b)

74 The dielectric constant of a conductor is said to be infinite. Explain why. see ISM

● **75** Explain why a stream of water is deflected, or bent, when an electrically charged object is brought close to it. (Try this yourself with a plastic pen, hard rubber comb, or balloon.) see ISM

76 As can be seen from the formulas given in this chapter, the capacitance and the energy-storing capability of a parallel-plate capacitor can be increased by reducing the plate separation distance. This fact seems to imply that any amount of capacitance or energy storage could be had. Why is this not the case? see ISM

77 ■ How much charge flows through a 12-V battery when a 1.0-μF capacitor is connected across its terminals?
1.2×10^{-5} C

78 ■ What is the capacitance of a capacitor that acquires a charge of 0.40 μC when connected to a 250-V source?
1.6×10^{-9} F

79 ■ A capacitor has a capacitance of 50 pF, which increases to 175 pF with a dielectric material between its plates. What is the dielectric constant of the material? $K = 3.5$

● **80** ■■ A 12-V battery is connected to a parallel-plate air capacitor with plate areas of 0.20 m² and a plate separation

of 5.0 mm. (a) What is the resulting charge on the capacitor? (b) How much energy is stored in the capacitor?
(a) 4.2×10^{-9} C   (b) 2.5×10^{-8} J

81 ■■ If the capacitor in Exercise 80 is discharged and immersed in silicone oil ($K = 2.6$) while still connected to the battery, what will be the charge on it and the amount of stored energy? $q = 1.1 \times 10^{-8}$ C; $U = 6.5 \times 10^{-8}$ J

82 ■■ How much more energy is stored at a given voltage in a capacitor with a paper dielectric ($K = 3.5$) than in a capacitor with the same dimensions containing a polyethylene dielectric ($K = 2.3$)? $U_{pap} = (1.5)\ U_{poly}$

83 ■■ A parallel-plate capacitor has a capacitance of 1.5 μF with air between the plates. The capacitor is connected to a 12-V battery and charged. When a dielectric is placed between the plates, a potential difference of 5.0 V is measured across the plates. What is the dielectric constant of the material?   2.4

84 ■■ A parallel-plate capacitor has rectangular plates with dimensions of 6.0 cm × 8.0 cm. If the plates are separated by a sheet of Teflon ($K = 2.1$) 1.5 mm thick, how much energy is stored in the capacitor when it is connected to a 12-V battery? 4.3×10^{-9} J

● **85** ■■ What is the equivalent capacitance of two capacitors of 0.60 μF and 0.80 μF when they are connected (a) in series and (b) in parallel?   (a) 0.34 μF   (b) 1.40 μF

86 ■■ When a series combination of two uncharged capacitors is connected to a 12-V battery, 173 μJ of energy is drawn from the battery. If one of the capacitors has a capacitance of 4.0 μC, what is the capacitance of the other?   6.0 μF

87 ■■ For the arrangement of three capacitors in Fig. 15.22, with $C_2 = 0.20$ μF and $C_3 = 0.30$ μF, what value of C_1 will give a total equivalent capacitance of 0.17 μF?   0.19 μF

88 ■■ Three capacitors of 0.25 μF each are connected in parallel to a 12-V battery. (a) What is the charge on each capacitor? (b) How much charge is drawn from the battery?   (a) 3.0 μC   (b) 9.0 μC

● **89** ■■ Two capacitors, C_1 and C_2, are connected in parallel, and this combination is connected in series with another capacitor, C_3. If $C_1 = C_3 = 1.0$ μF and the total equivalent capacitance is $\frac{3}{4}$ μF, what is the capacitance of C_2?   2.0 μF

90 ■■ What is the charge on each of the capacitors in the circuit in   Fig. 15.31?   0.10 μF : 0.60 μC; 0.20 μF : 1.2 μC; 0.30 μF : 1.8 μC

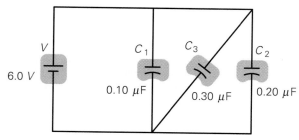

FIGURE 15.31 A capacitor triad   See Exercise 90.

91 ■■ What are the maximum and minimum equivalent capacitances that can be obtained by combinations of three capacitors of 1.5 μF, 2.0 μF, and 3.0 μF?   6.5 μF; 0.67 μF

92 ■■■■ Four capacitors are connected in a circuit as illustrated in   Fig. 15.32. Find (a) the charge on and (b) the voltage difference across each of the capacitors.
$C_1 : Q = 2.4$ μC; $V = 6.0$ V $C_2 : Q = 2.4$ μC; $V = 6.0$ V
$C_3 : Q = 1.2$ μC; $V = 6.0$ V $C_4 : Q = 3.6$ μC; $V = 6.0$ V

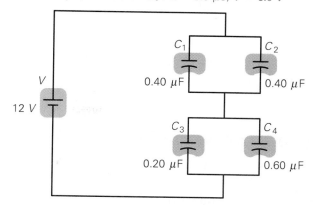

FIGURE 15.32 Double parallel in series   See Exercise 92.

■ **Additional Exercises**

93 An electron is placed in a constant electric field of 2.5×10^3 N/C directed along the $+y$ axis. (a) What is the force on the electron? (b) If the electron is released, how much kinetic energy does it have after moving 10 cm?
(a) $4.0 - 10^{-16}$ N in the $-y$ direction   (b) 4.0×10^{-17} J

94 Three capacitors with values $C_1 = 0.15$ μF, $C_2 = 0.25$ μF, and $C_3 = 0.30$ μF are connected in a circuit as shown in Fig. 15.25. (a) What is the equivalent capacitance of the arrangement? (b) If the circuit is connected to a 12-V battery, how much charge will be drawn from the battery? (c) What is the voltage difference across each capacitor?
(a) 0.17 μF   (b) 2.0 μC   (c) $V_1 = V_2 = 5.3$ V, $V_3 = 6.7$ V

● **95** How much work is required to bring two charges of $+2.5$ μC and $+4.5$ μC that are separated by an infinite distance to a distance of 0.50 m?   0.20 J

96 How many electrons are required to make up a net charge of -0.10 μC? 6.3×10^{11} electrons

97 Two charges, -3.0 μC and -5.0 μC, are 0.40 m apart. (a) Where should a charge of -1.0 μC be placed to put it in electrostatic equilibrium? (b) Where should a charge of $+1.0$ μC be placed to be in electrostatic equilibrium with the two original charges?   (a) 0.17 m   (b) 0.17 m

98 Two horizontal, conductive parallel plates are separated by a distance of 1.5 cm. What voltage across the plates would be required to suspend (a) an electron or (b) a proton in midair between them? (c) What would be the signs of the charges on the plates in each case?   (a) 8.4×10^{-13} V
(b) 1.5×10^{-9} V   (c) in (a), top plate is $+$; in (b), top plate is $-$
99 Show that N/C is equivalent to V/m.   see ISM

16 Electric Current and Resistance

INSIGHTS

- ■ **Superconductivity**
- ■ **Joule Heat and Battery Testing**
- ■ **The Cost of Portable Electricity**

Chapter 16 topics include batteries and dc current, drift velocity, Ohm's law, and electrical power.

Most of the practical applications of electricity involve electric current. When you turn on the headlights of your car, a current of electrons flows through the circuit of which the filaments of the headlights are a part. Because of the resistance of the filaments, electrical energy is converted into light, or radiant energy. In homes and offices, electrons flow in the wiring of electrical applicances when they are plugged in and turned on. In this case, the electrons surge back and forth 60 times a second, in an alternating current.

For sustained currents, voltage sources are required. These are devices such as batteries and generators, which produce electrical energy. That is, potential differences are produced by the conversion of other forms of energy into electrical energy. This chapter considers direct current, which is produced by batteries. The source of alternating current will be discussed in Chapter 20. However, the general relationships for voltage, current, and resistance developed here apply to both types of current and are useful in analyzing a variety of everyday practical applications.

16.1 ■ Batteries and Direct Current

After studying electric force and energy in Chapter 15, you can probably guess what is required to produce an **electric current**, or a flow of charge. A flow of fluid (mass flow) may occur when there is a gravitational potential energy difference. Spontaneous heat flow occurs only when there is a temperature difference. Similarly, a flow of electric charge is dependent on a *difference* in electric potential, or voltage.

Discuss the law of conservation of energy.

In solid conductors, particularly metals, outer atomic electrons of atoms are relatively free to move. The positively charged protons of the nuclei are fixed in the lattice structure of the material. (In liquid conductors, both positive and negative ions can move.) Energy is required to do the work of moving electric charge. Electrical energy is produced, or generated, through the conversion of other forms of energy into electrical energy, giving rise to a potential, or voltage, difference.

A **battery** is a device that converts chemical energy into electrical energy. The Italian scientist Allesandro Volta (1745–1827) is credited with constructing one of the first practical batteries. Basically, a battery consists of two electrodes in an electrolyte, which is a solution that conducts electricity (●Fig. 16.1). With the appropriate electrodes and electrolyte (Volta used zinc and copper electrodes in dilute sulfuric acid), a potential difference develops across the electrodes as a result of chemical action. One electrode becomes negatively charged and this is the **cathode**, or negative (−) terminal of the battery. The other electrode develps a deficiency of electrons, becoming positively charged; this is the **anode**, or positive (+) terminal of the battery.

Because of the positive and negative charges on the battery terminals, a potential difference exists between them. The maximum potential difference across the terminals of a battery is called the **electromotive force (emf)**. This is a somewhat misleading name because the electromotive force is not a force but a voltage difference, and, as such, is measured in volts. The emf ($\mathcal{E}$) is determined when the battery is not connected to an external circuit (●Fig. 16.2a). When a battery is connected to a circuit and charge flows, the voltage across the battery is slightly less than the emf because of internal resistance. The operating voltage (V) of a battery is called its **terminal voltage** (Fig. 16.2b). In this chapter, we will be chiefly concerned with terminal voltage.

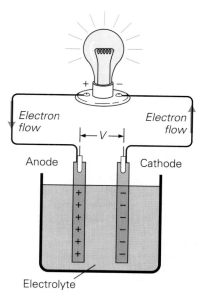

●**FIGURE 16.1 The chemical battery or cell** Chemical processes involving the electrolyte and two unlike metal electrodes cause one electrode to become positively charged (the anode) and the other to become negatively charged (the cathode). The electric potential, or voltage, developed across the electrodes can cause a current, or a flow of charge in the circuit.

Figure 16.1

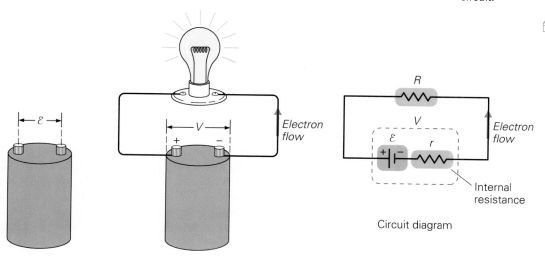

(a) Electromotive force (emf)

(b) Terminal voltage

●**FIGURE 16.2 Electromotive force (emf) and terminal voltage** **(a)** The emf of a battery is the maximum potential difference across its terminals. This occurs when the battery is not connected to an external circuit. **(b)** Because of internal resistance (r), the terminal voltage when the battery is in operation is less than the emf. As will be learned, $V = \mathcal{E} - I(r + R)$, where I is the current and R the resistance of the circuit element.

Figure 16.2

Figure 16.3

●**FIGURE 16.3 Batteries in series and in parallel** **(a)** When batteries are connected in series, their voltages add and the voltage across the resistance R is the sum of the voltages. **(b)** When batteries of the same voltage are connected in parallel, the voltage across the resistance is the same, as if only a single battery were present, and each battery supplies a fraction of the total current.

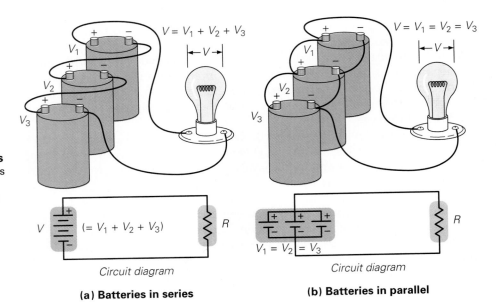

$V = V_1 + V_2 + V_3$

$V = V_1 = V_2 = V_3$

$V \quad (= V_1 + V_2 + V_3) \qquad R$

$V_1 = V_2 = V_3 \qquad R$

Circuit diagram

Circuit diagram

(a) Batteries in series

(b) Batteries in parallel

Direct current—flow of charge directed one way only.

The internal resistance r of the battery in Fig. 16.2b is represented explicitly in the circuit diagram. This resistance or opposition to charge flow results from the resistance of the materials from which the battery is made. Internal resistances are typically small, and the terminal voltage of a battery is usually only slightly less than its emf. However, when a large current is supplied to a circuit from a battery, the terminal voltage V may drop appreciably below the maximum emf value. As you will learn in our discussion of electricity, the emf for the circuit in Fig. 16.2b is given by $\mathcal{E} = V + I(r + R)$ or $V = \mathcal{E} - I(r + R)$, where I is the electric current. Note that when $I = 0$, we have $V = \mathcal{E}$.

So long as the internal chemical action maintains a potential difference across its terminals, current is supplied to a circuit as in Fig. 16.2b. The battery is said to deliver current to a circuit. Alternatively, we say that the circuit (or its components) draws current from the battery. Since electrons can flow only in one direction in such a circuit, from the negative ($-$) terminal (cathode) to the positive ($+$) terminal (anode), this is called **direct current (dc)**. Notice the circuit diagram in the figure. The symbol for a battery is two parallel lines, the longer of which represents the positive ($+$) terminal and the shorter the negative ($-$) terminal. A circuit element, such as the lamp, may be generally represented by the symbol —⋀⋀—, which stands for a resistance (R). (Electrical resistance—resistance to charge flow—will be considered in detail in later sections of this chapter. Here, we merely introduce the circuit symbol.)

There are a wide variety of batteries now in use. One of the most common batteries is the 12-V automobile battery. It actually consists of six 2-V cells connected in series. That is, the positive terminal of each cell is connected to the negative terminal of the next cell (shown for three cells in ● Fig. 16.3a). When batteries or cells are connected in this fashion, their voltages add.*

---

*Chemical energy is converted to electrical energy in a chemical *cell*. The term *battery* generally refers to a "battery" of cells.

If cells are connected in parallel, all their positive terminals have a common connection, as do their negative terminals (Fig. 16.3b). When identical batteries are connected in this way, the potential difference is the same across all of them, and each one supplies a fraction of the current to the circuit (for three batteries with equal voltages, each one supplies $\frac{1}{3}$ of the current). If different voltage sources are connected in parallel, the situation is more complex.

Other sources of voltage, such as generators and photocells, will be considered in later chapters.

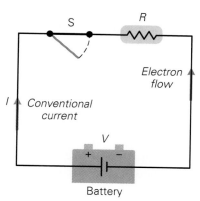

16.2 ■ Current and Drift Velocity

A battery or some other voltage source connected to a continuous conducting path forms a complete circuit. In ordinary applications, *a sustained electric current requires a voltage source and a complete circuit.* As shown in Fig. 16.4, a circuit may have a switch (symbolized by S), which is used to open or close the circuit. When the switch is open, the current or circuit component is turned off; when the switch is closed, there is current in the circuit.

Since it is electrons that move in the wires of an external circuit, the net flow of charge is in the direction in which an electron experiences a force—away from the negative terminal of the battery and toward the positive terminal (recall the law of charges). However, historically, circuit analysis has generally been done in terms of the **conventional current**, which is in the direction in which positive charges would flow, or the direction opposite to the electron flow (see Fig. 16.4).

Conventional current—in direction opposite that of electron flow

Early investigators thought that electrical phenomena were the result of positive and negative fluids. (Recall from Chapter 12 that heat was also considered to be a fluid.) Benjamin Franklin advanced a single-fluid theory of electricity in which he postulated the existence of a special fluid in all matter. Objects could contain a certain quantity of this fluid without showing any electrical properties, but did show such properties when they contained either a surplus or a deficit of it. An object with an excess of the fluid was considered to be positively excited, and one with a deficit to be negatively excited. The designation of conventional (positive) current flow is apparently a consequence of Franklin's single-fluid theory.

An electric current is then a flow of charge. Quantitatively, the *electric current* (I) in a wire is defined as the net amount of charge (q) passing through a cross-sectional area of the wire per unit time. (Fig. 16.5). For a constant rate of flow, the current is given by

$$I = \frac{q}{t} \qquad (16.1)$$

As you can see, the units of current are coulombs per second (C/s). This combination is called an **ampere (A)** in honor of the French physicist André Ampère (1775–1836), an early investigator of electrical and magnetic phenomena. A value such as 10 A is usually read as "ten amps." Small currents are expressed in milliamperes (mA, 10^{-3} A) or microamperes (μA, or 10^{-6} A), which are shortened to milliamps and microamps.

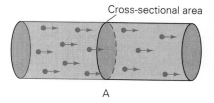

FIGURE 16.5 Electric current Electric current (I) in a wire is defined as the net amount of charge (q) that passes through a cross-sectional area of the wire per time, that is, $I = q/t$. I has the units of amperes (A), or "amps" for short.

EXAMPLE 16.1 ■ Current and Charge

A steady current of 0.50 A flows in a circuit for 2.0 min. How much charge passes through a cross-sectional area of one of the connecting wires during this time? How many electrons does this represent?

Solution.

Given: $I = 0.50$ A   *Find:* q (amount of charge)
        $t = 2.0$ min $= 120$ s

By Eq. 16.1, $I = q/t$, so the charge is given by $q = It$.

$$q = It = (0.50 \text{ A})(120 \text{ s}) = 60 \text{ C}$$

The amount of charge q is made up of n electrons or electronic charges, that is, $q = ne$ (Eq. 15.1). Solving for n, we have

$$n = \frac{q}{e} = \frac{60 \text{ C}}{1.6 \times 10^{-19} \text{ C}} = 3.8 \times 10^{20} \text{ electrons}$$

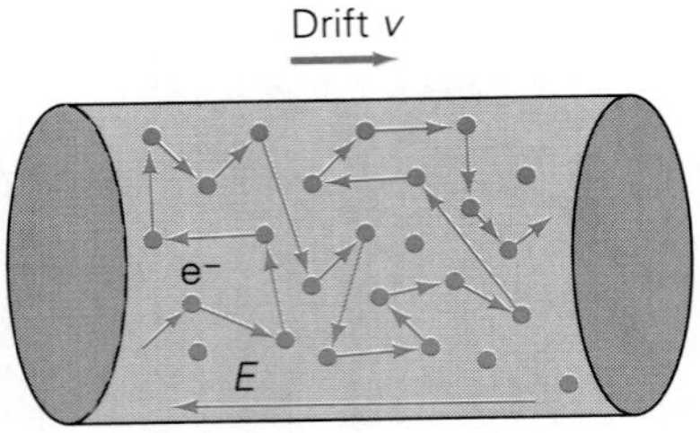

Drift v

● **FIGURE 16.6 Drift velocity**
Because of collisions with the atoms of the conductor, electron motion is random. However, when the conductor is connected in a circuit, there is a relatively small net motion in the direction opposite the electric field (or toward the positive terminal of the battery or other voltage source). This net motion is characterized by a drift velocity.

Demonstration/activity: Connect a DC lab power supply to a small, flashlight-type bulb. Turn the bulb on at the beginning of a lecture period. Then stop at some time during the period (depending on wire length) and point out that the electrons that left the power supply at the beginning of the period have not yet reached the bulb.

Electric fields are introduced in Section 15.4

Although charge flow is frequently mentioned, you should be aware that electric charge does not flow in a conductor exactly like mass flow (such as water through a pipe). In the absence of a potential difference, the free electrons in a metal wire move about randomly at very high speeds. As a result, there is no net flow of charge one way or the other in the wire. (On the average, the same amounts of charge pass through a cross-sectional area in opposite directions in a given time.) When a potential difference is applied across the ends of the wire (for example, by a battery), this creates an electrical bias in one direction and thus a net flow of charge in that direction. This does *not* mean electrons are moving directly from one end of the wire to the other. They still move about in all directions because they collide with the atoms of the conductor (● Fig. 16.6). But their motions are less random, and more of them move toward the positive terminal of the battery than away from it.

The net electron flow is characterized by an average velocity called the **drift velocity**, which is much smaller than the random velocities of the electrons themselves. The magnitude of the drift velocity is on the order of 0.10 cm/s. Motion at this speed is relatively slow. A quick calculation using this drift velocity would show that on the average it would take an electron about 17 minutes to travel 1 m along a wire. Yet a lamp comes on almost instantaneously when you turn the switch (close the circuit), and the electronic signals carrying telephone conversations travel almost instantaneously over miles of wire.

Evidently, something must be moving faster than the net electron motion. This is the electric field. When a potential difference is applied, the associated electric field that drives the electrons in their net motion travels along the conductor at a speed close to the speed of light (on the order of 10^8 m/s). The electric field therefore influences the motion of electrons throughout the conductor almost instantaneously. Thus there is current everywhere in the circuit almost simultaneously. You don't have to wait for electrons to "get there" from somewhere else.

16.3 ■ Ohm's Law and Resistance

Given that an applied voltage causes an electrical current in a conductor, how much charge flows for a given voltage? As you might expect, the current is directly proportional to the voltage:

$$I \propto V$$

That is, the greater the voltage, the greater the current. This is analogous to more water flowing through a pipe when there is a greater gravitational potential or larger pressure difference.

However, another factor besides voltage affects current flow. Just as internal friction (viscosity) affects fluid flow, the internal resistance of materials affects the flow of electric charge. The current is inversely proportional to the **resistance** (R). For many materials, the relationship between current, voltage, and resistance is given by

$$I = \frac{V}{R} \quad \text{or} \quad \boxed{V = IR} \tag{16.2}$$

This equation is called **Ohm's law**, after Georg Ohm (1789–1851), a German physicist who investigated the relationship between current and voltage. The units for resistance can be seen to be volts per ampere (V / A). This combined unit is called an **ohm (Ω)**. A plot of voltage versus current for an ohmic conductor (one for which Ohm's law holds) gives a straight line with a slope equal to R (●Fig. 16.7).

Although commonly referred to as Ohm's law, Eq. 16.2 is not a fundamental law in the sense that Newton's law of gravitation or the first and second laws of thermodynamics are. Eq. 16.2 essentially defines ohmic resistance as the constant of proportionality between voltage and current. The voltage-current relationship expressed by the equation applies to a wide range of materials, particularly metals. However, keep in mind that the relationship is not linear (that is, Ohm's law does not apply) for other materials, such as semiconductors.

EXAMPLE 16.2 ■ Electrical Resistance

When a 12-V battery is connected across an unknown resistor, there is a current of 8.0 mA in the circuit. What is the value of the resistor?

Solution.

Given: $V = 12\,\text{V}$ *Find:* R (resistance)
$I = 8.0\,\text{mA} = 8.0 \times 10^{-3}\,\text{A}$

Resistors are used in circuits to limit current and are manufactured in a wide variety of resistance values (see ●Fig. 16.8). In this case, using Ohm's law,

$$R = \frac{V}{I} = \frac{12\,\text{V}}{8.0 \times 10^{-3}\,\text{A}} = 1.5 \times 10^3\,\Omega$$

That is, the resistor is a 1.5-kΩ (kilohm) resistor. Resistors in the kΩ and MΩ (megohm) ranges are quite common.

> Sketch a graph showing V versus I for a common conductor and a diode. Point out the property of the graph that indicates the diode is not ohmic.

Ohm's law—voltage, resistance and current.

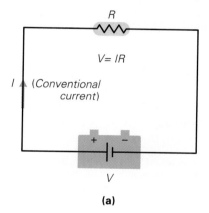

(a)

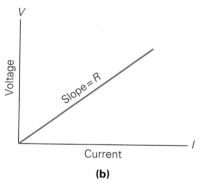

(b)

●**FIGURE 16.7 Ohm's law**
(a) Resistive materials that show a voltage-current relationship of $V = IR$ are said to obey Ohm's law, or to be ohmic. **(b)** A plot of voltage versus current gives a straight line with a slope equal to R.

FIGURE 16.8 Resistors (and capacitors) Various types of resistors of different values on a circuit board. The resistance (in ohms) of a particular resistor is indicated by a 3- or 4-band color code. The capacitors (Chapter 15) have their value and voltage rating indicated by a 5-band color code.

Resistivity—a material property

You should be careful not to confuse resistivity with mass density, which has the same symbol (rho).

On the atomic level, resistance arises from collisions of electrons with the atoms or ions making up a material. Thus, resistance is a material property; that is, it depends on the type of material. The resistance of a particular conductor of uniform cross section, such as a length of wire, depends on several factors (Fig. 16.9). These factors directly affecting resistance are

1. Type of material
2. Length
3. Cross-sectional area
4. Temperature

As you might expect, the resistance of a conductor is directly proportional to its length (L) and inversely proportional to its cross-sectional area (A).

$$R \propto \frac{L}{A}$$

For example, a uniform metal wire 4 m long offers twice as much resistance as a similar wire 2 m long. Also, the larger the cross-sectional area of a conductor, the more current (less resistance) there is in it. These geometrical conditions are analogous to those for liquid flow in a pipe. The longer the pipe, the more resistance (drag) there is; but the larger the diameter (or cross-sectional area) of the pipe, the greater the amount of liquid it can carry.

The resistance of a material also depends on instrinsic properties, which are characterized by the **resistivity** (ρ). This acts as a constant of proportionality for the relationship of resistance to length and area.

$$R = \frac{\rho L}{A} \quad \text{or} \quad \rho = \frac{R A}{L} \tag{16.3}$$

You can see that ρ has units of ohm-meters (Ω-m).

The resistivities of some common conductors (and insulators) are given in Table 16.1. The values given are for a particular temperature (20°C, or room

TABLE 16.1 ■ Resistivities (at 20°C) and Temperature Coefficients of Resistivity for Various Materials*

	ρ (Ω-m)	α (C$^{\circ-1}$)		ρ (Ω-m)	α (C$^{\circ-1}$)
Conductors			*Semiconductors*		
Aluminum	2.82×10^{-8}	4.29×10^{-3}	Carbon	3.6×10^{-5}	-5.0×10^{-4}
Copper	1.70×10^{-8}	6.80×10^{-3}	Germanium	4.6×10^{-1}	-5.0×10^{-2}
Iron	10×10^{-8}	6.51×10^{-3}	Silicon	2.5×10^{2}	-7.0×10^{-2}
Mercury	98.4×10^{-8}	0.89×10^{-3}	*Insulators*		
			Glass	10^{12}	
Nichrome alloy of Nickel and Chromium	100×10^{-8}	0.40×10^{-3}	Rubber	10^{15}	
			Wood	10^{10}	
Nickel	7.8×10^{-8}	6.0×10^{-3}			
Platinum	10×10^{-8}	3.93×10^{-3}			
Silver	1.59×10^{-8}	6.1×10^{-3}			
Tungsten	5.6×10^{-8}	4.5×10^{-3}			

*Values for semiconductors are general values, and resistivities for insulators are typical orders of magnitude.

temperature), because resistivity is somewhat temperature-dependent. *For most metallic conductors, the resistivity increases as the temperature rises.* Higher temperatures cause increased atomic or molecular vibrations in a conductor. Thus, more collisions with electrons take place, which hinder the net flow of charge.

A quantity that is sometimes mentioned is the electrical conductivity of a material. As you might guess, the **conductivity** (σ) and the resistivity (ρ) are inversely proportional.

$$\sigma = \frac{1}{\rho}$$

(16.4) Conductivity—the reciprocal of resistivity

The unit of conductivity is $(\Omega\text{-m})^{-1}$.

EXAMPLE 16.3 ■ Resistance of a Wire

A total length of 1.5 m of insulated copper wire is used to connect the components of a circuit. If the wire has a diameter of 2.3 mm (equivalent to American Wire Gauge, or AWG, No. 12), what is its resistance at room temperature?

Solution.

Given: $L = 1.5$ m   Find: R (resistance)
$d = 2.3$ mm $= 2.3 \times 10^{-3}$ m
$\rho_{copper} = 1.70 \times 10^{-8}$ Ω-m
    (at room temperature, Table 16.1)

The cross-sectional area of the wire is

$$A = \pi r^2 = \frac{\pi d^2}{4} = \frac{\pi(2.3 \times 10^{-3} \text{ m})^2}{4} = 4.2 \times 10^{-6} \text{ m}^2$$

Then, using Eq. 16.3,

$$R = \frac{\rho L}{A} = \frac{(1.70 \times 10^{-8} \text{ }\Omega\text{-m})(1.5 \text{ m})}{4.2 \times 10^{-6} \text{ m}^2}$$
$$= 0.0061 \text{ } \Omega$$

This result demonstrates why the resistances of the connecting wires in an electrical circuit are usually considered to be negligible. The resistances of the circuit components are generally much larger than this.

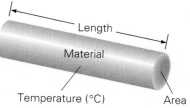

● **FIGURE 16.9 Resistance factors**   Factors directly affecting the electrical resistance of uniform conductor are the type of material, the length (L), the cross-sectional area (A), and the temperature.

The temperature dependence of resistivity is nearly linear if the temperature change is not too great. An expression similar to the one for thermal linear expansion (Chapter 10) may be written for this relationship. That is, the resistivity (ρ) at a temperature T after a temperature change $\Delta T = T - T_o$ is given by

Temperature dependence of resistivity and resistance

$$\rho = \rho_o(1 + \alpha\Delta T)$$

(16.5) Compare Eq. 10.9, Section 10.4

Here α is a constant (over a small temperature range) called the **temperature coefficient of resistivity**, and ρ_o is a reference resistivity at T_o (usually 20°C or 0°C). Eq. 16.5 may also be written like this:

$$\Delta\rho = \rho_o\alpha\Delta T$$

(16.6)

Over a large temperature range, high-order terms may be needed to accurately describe the resistivity, for example, $\rho = \rho_o[1 + \alpha\Delta T + \beta(\Delta T)^2 + \gamma(\Delta T)^3]$. The coefficients β and γ are very small, but the higher-order terms become significant for large ΔT's.

Demonstration/activity: Use a
variable laboratory power supply
with demonstration meters to plot
I vs V for a typical 110-V light
bulb. Observe that the curve is
not linear as the bulb heats up,
showing how the resistance
increases with temperature. If
possible, do the same with an old
carbon-filament bulb and observe
how its resistance decreases with
temperature.

Video demonstration 11 relates to
this material.

where $\Delta\rho = \rho - \rho_0$ is the change in resistivity for a given change in temperature (ΔT). Since the ratio $\Delta\rho/\rho_0$ is dimensionless, α must have the unit of $C^{\circ-1}$ (or $1/C^\circ$). The temperature coefficients of resistivity for some materials are listed in Table 16.1. This discussion will consider these coefficients to be constant over the temperature ranges used in the examples and problems.

Resistance is directly proportional to the resistivity, and Eqs. 16.3, 16.5, and 16.6 can be used to derive an expression for the resistance of a conductor of uniform cross section, where R_0 is the resistance of the conductor at the reference temperature.

$$R = R_0(1 + \alpha\Delta T) \quad \text{or} \quad \Delta R = R_0\alpha\Delta T \tag{16.7}$$

The variation of resistance with temperature provides a means of measuring temperature in the form of an electrical resistance thermometer.

EXAMPLE 16.4 ■ Variation in Resistance with Temperature

What is the variation (as a percentage) in the resistance of platinum wire over the range of 0°C to 100°C? (Assume that α is constant over this temperature range.)

Solution.

Given: $T_0 = 0°C$
$T = 100°C$
$\alpha = 3.93 \times 10^3\, C^{\circ-1}$
(from Table 16.1)

Find: $\Delta R/R_0$ (variation in resistance as a percentage)

The ratio $\Delta R/R_0$ is the fractional change in the initial resistance R_0 (at 0°C). Using Eq. 16.7 with the values given for α, T, and T_0,

$$\frac{\Delta R}{R_0} = \alpha(T - T_0) = (3.93 \times 10^{-3}\, C^{\circ-1})(100°C - 0°C)$$

$$= 0.393\,(\times\,100\%) = 39.3\%$$

EXAMPLE 16.5 ■ Electrical Resistance Thermometer

The resistance of a coil of platinum wire is measured as 250 mΩ at room temperature (20°C). When the coil is placed in a hot oven, its resistance is measured as 496 mΩ. What is the termperature of the oven?

Solution.

Given: $R = 496$ mΩ
$R_0 = 250$ mΩ
$T_0 = 20°C$
$\alpha = 3.93 \times 10^{-3}\, C^{\circ-1}$
(from Table 16.1)

Find: T (temperature)

First, find the change in temperature ($\Delta T = T - T_0$) from the change in resistance using Eq. 16.7:

$$\Delta T = \frac{R - R_o}{R_o \alpha}$$

$$= \frac{496 \text{ m}\Omega - 250 \text{ m}\Omega}{(250 \text{ m}\Omega)(3.93 \times 10^{-3} \text{C}^{\circ -1})} = 250 \text{ C}^\circ$$

Then

$$T = \Delta T + T_o = 250 \text{C}^\circ + 20^\circ \text{C} = 270^\circ \text{C}$$

Note that it was dimensionally correct and convenient to work directly with the quantities in milliohms rather than converting them to ohms.

Carbon and other semiconductors have negative temperature coefficients of resistivity (see Table 16.1). This means that the resistance of a semiconductor decreases with increasing temperature or increases with decreasing temperature. Most materials have positive temperature coefficients of resistivity, and their resistances decrease with decreasing temperature. You might be wondering how far electrical resistance can be reduced by lowering the temperature. In certain cases, the resistance can be reduced to zero! This condition of superconductivity is discussed in the Insight feature.

INSIGHT ■ Superconductivity

The electrical resistance of metals and alloys generally decreases with decreasing temperature. At relatively low temperatures, some materials exhibit what is called **superconductivity**; that is, the electrical resistance vanishes, or goes to zero.

Superconductivity was discovered in 1911 by Heike Kamerlingh Onnes, a Dutch physicist. The phenomenon was first observed in solid mercury at a temperature of about 4 K (−269°C). The mercury was cooled to this temperature using liquid helium. The boiling point of helium (the temperature at which it condenses to a liquid) is about −267°C at 1 atm. Lead also exhibits superconductivity when cooled to such a temperature.

An electrical current established in a superconducting loop should persist indefinitely with no resistive losses. Currents introduced in superconducting loops have been observed to remain constant for several years. In 1957, a theory was presented by the American physicists John Bardeen, Leon Cooper, and Robert Schrieffer in an attempt to explain various aspects of metallic superconductors. Known as the BCS theory for obvious reason, it presents a quantum-mechanical model in which electrons are viewed as waves traveling through a material. According to the theory, lattice vibrations and imperfections, which scatter electrons and give rise to resistance and energy loss in metals under normal conditions, have no effect on the electrons in superconductors. In the absence of this electron scattering, the resistance is zero,

and a current can persist as long as the metal is in a superconducting state.

Keeping a superconductor at the low temperatures of liquid helium is somewhat difficult and costly. Thus, there has been an ongoing search for materials that become superconducting at higher temperatures. Other superconducting metals and alloys were found for which the critical temperature was about 18 K (−255°C). In 1973, a material with a critical temperature of 23 K (−250°C) was discovered.

In 1986, there was a major breakthrough—a new class of superconductors was discovered. These were the so-called ceramic alloys of rare earth elements, such as lanthanum and yttrium. These superconductors were prepared by grinding the mixture of metallic elements together and heating it to a high temperature to produce a ceramic material. For example, one such mixture consisted of barium, yttrium, and copper oxide. The critical temperature for these ceramic mixtures was about 57 K (−216°C).

In 1987, a ceramic material with a critical temperature of 98 K (−175°C) was discovered. This was a major advance because it meant that superconductivity could be obtained using liquid nitrogen, which has a boiling point of 77 K (−196°C). (See Fig. 1.) Liquid nitrogen is relatively plentiful (nitrogen is the chief constituent of air) and inexpensive—it costs cents per liter compared to dollars per liter for liquid helium.

FIGURE 1 Magnetic levitation A magnetic cube levitates above a piece of "high-temperature" superconducting material that is cooled with liquid nitrogen. When the material becomes cold enough to superconduct, it acquires certain magnetic properties that cause a magnet to be repelled and levitate.

There have been more recent reports of higher critical temperatures, and even suggestions that certain copper oxide compounds may lose their resistance to electrical current at room temperature and above. As scientists study the new superconductors, better understanding will be gained, and no doubt the critical temperature will rise. One problem is the difficulty of confirming a superconducting state in small samples or regions of samples. Another is that the BCS theory does not appear to apply to the new high-temperature superconductors. The mechanism for their superconductivity is not well understood.

There are many possible applications for superconductors. One is superconducting magnets. The strength of an electromagnet depends on the magnitude of the current in the windings (Chapter 18). If there were no resistance, there would be greater current and no losses. Used in motors or engines, superconducting electromagnets would provide more power. (Superconducting magnets cooled by liquid helium have been used in ships' engines for some time.) Such magnets could be used to levitate and propel trains and electric cars. Another application of superconductors might be underground transmission cables with no resistive losses. However, among the many technological problems still to overcome is how to form wires out of ceramic materials (which are generally brittle). The application most likely to be realized soon will be in making faster computer circuits.

Imagine what it would mean if someone discovered a material that was superconducting at refrigerator or room temperature. The absence of electrical resistance opens many possibilities. You're likely to hear more about superconductor applications in the near future.

16.4 ■ Electric Power

With a sustained current in a circuit, the electrons are given energy by the voltage source, such as a battery. As these charge carriers pass through a circuit component, they collide with the atoms of the material (experience resistance) and lose energy because of these collisions (Chapter 6). The energy transfer results in a temperature increase of the component, and electrical energy is transformed into thermal energy. Overall, all of the energy given to the charge carriers by the battery must be lost in the circuit. That is, a charge carrier transversing the circuit must have the same (zero) electric potential (or potential energy) when it returns to the negative terminal of the battery as it had when starting. Otherwise, a charge carrier could gain energy with each round trip, which would violate the law of the conservation of energy.

Energy changes in collisions are discussed in Section 6.4

The energy gained by a charge q from a voltage source that has an electric potential V is qV (this is the work done by the voltage source on the charge, $W = qV$). Over a period of time, the rate at which energy is obtained or expended as thermal energy in a circuit is given in terms of **electric power** (P):

$$P = \frac{W}{t} = \frac{qV}{t}$$

Since $I = q/t$ by Eq. 16.1,

$$\boxed{P = IV} \qquad (16.8)$$

As you learned in Chapter 5, the SI unit of power is the watt (W).

Using Ohm's law ($V = IR$), power can be expressed in two other convenient forms:

$$P = IV = \left(\frac{V}{R}\right) V = \frac{V^2}{R}$$

and

$$P = IV = I(IR) = I^2 R$$

That is,

$$\boxed{P = \frac{V^2}{R} = I^2 R} \qquad (16.9)$$

The thermal energy expended in a current-carrying conductor is sometimes referred to as **joule heat**, or **I squared R (I^2R) loss.** In many instance, joule heat loss is undesirable, for example, in electrical transmission lines. However, in other instances, the conversion of electrical energy to thermal energy is of practical use. These applications include the heating elements (burners) of electric stoves, hair dryers, and toasters. A novel application of joule heat in battery testers is discussed in the Insight feature.

Electric light bulbs are rated directly in watts (power), for example, 100-W or 60-W (●Fig. 16.10a). In such bulbs, electrical energy excites the atoms of the filament, and the excited atoms then radiate away the extra energy in the form of light (Chapter 26). The greater the wattage of a bulb, the greater the energy consumption per unit time (joules per second). Incandescent lamps are relatively inefficient light sources. Only about 5% of the electrical energy is converted to visible light (most of the radiant energy produced is invisible infrared radiation; see Section 19.4).

Electrical appliances are tagged or stamped with their power ratings. Either the voltage and power requirements or the voltage and current requirements are given (Fig. 16.10b). In either case, the current, power, and effective resistance may be found using Eqs. 16.8 and 16.9. The power requirements of some household appliances are given in Table 16.2. You should keep in mind that

(a)

(b)

●**FIGURE 16.10 Power ratings**
(a) Light bulbs are rated directly in watts. This 60-W bulb uses 60 J of energy each second. **(b)** Appliance ratings list either voltage and power or voltage and current. In either case, the current, power, and effective resistance may be found. Here, one appliance is rated at 120 V and 88 W and the other at 120 V and 1.50 A.

TABLE 16.2 ■ Typical Power and Current Requirements for Household Appliances (120 V)

Appliance	Power	Current	Appliance	Power	Current
Air conditioner, room	1500 W	12.5 A	Radio-cassette player	14 W	0.12 A
Air conditioning, central	5000 W	41.7 A	Refrigerator, regular	400 W	3.3 A
Blender	800 W	6.7 A	Refrigerator, frost-free	500 W	4.2 A
Coffee maker	1625 W	13.5 A	Stove, top burners	6000 W	50.0 A
Dishwasher	1200 W	10.0 A	Stove, oven	4500 W	37.5 A
Electric blanket	180 W	1.5 A	Television, black-and-white	50 W	0.42 A
Hair dryer	1200 W	10.0 A	Television, color	100 W	0.83 A
Heater, portable	1500 W	12.5 A	Toaster	950 W	7.9 A
Microwave oven	625 W	5.2 A	Water heater	4500 W	37.5 A

You may have noticed and used the individual battery testers on battery packages (Fig. 1). A battery is placed between the terminals of the tester and a rising yellow stripe indicates the condition of the battery. Joule heat is one of the ingredients of the operation.

The tester consists of a polyester strip with a printed circuit on one side and a liquid crystal on the other. (Many organic substances exhibit an "intermediate" phase of matter reminiscent of both a solid and a liquid, hence the name liquid crystal. These substances have unique optical properties that are affected by heat and voltage. See LCDs—liquid crystal displays—in Chapter 23.)

The printed circuit is a combination of silver and graphite that appears as a dark film. It is designed so as to produce graduated resistances. For example, for the AA battery tester, the circuit film is tapered, being narrow at the bottom and wide at the top. The resistance per unit length ($R/L = \rho/A$) varies accordingly, with the greatest resistance at the bottom.

You test a battery by inserting it into the back of the package so as to make contact with the ends of the tapered resistor. The printed circuit strip then carries a current I, the magnitude of which depends upon the condition of the battery, and joule heat raises the temperature of the resistor. Since $P = I^2R$, the heating for a given current is greatest at the bottom, where the resistance is highest, and least at the top, where the resistance is lowest (Fig. 2).

The heating causes a reaction in the liquid crystal, which results in the thermometer effect on the display side of the tester. The black liquid crystal becomes transparent at a critical temperature. The "color change" occurs when the underlying yellow strip becomes visible through the transparent overcoat. The scale rates the battery as "good" if it supplies enough current to heat the complete liquid crystal strip above the critical temperature, and indicates "replace" if the critical heating occurs only at the bottom of the strip.

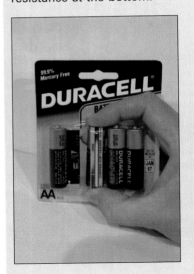

FIGURE 1 Battery tester The test battery is inserted into the back of the package and contact is made by pressing with the fingers. The rising yellow stripe indicates the condition of the battery—the longer the stripe, the better the battery.

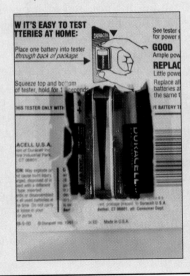

FIGURE 2 Resistance and joule heat As seen in the back of the test package, a printed circuit provides a calibrated resistance. In this AA battery tester, the circuit film is tapered, being narrow at the bottom and wide at the top. It thus has the greatest resistance at the bottom.

lamps and appliances generally operate on alternating current, which will be discussed in Chapters 19 and 20. However, the current-voltage equations for alternating current have the same form as those for direct current.

Using different light bulbs, measure the current for different wattages. Explain why high-current devices such as stoves and dryers require a 240-V source.

EXAMPLE 16.6 ■ Power and Current

A 60-W light bulb operates on 120-V household voltage. (Household voltage varies between 110 V and 120 V and is commonly given as 110 V, 115 V, or 120 V.) (a) How much current does the light bulb draw? (b) What is the resistance of the bulb?

Solution.

Given: $P = 60$ W
$\qquad\quad V = 120$ V

Find: (a) I (current)
$\qquad\quad$ (b) R (resistance)

(a) The current may be found using $P = IV$:

$$I = \frac{P}{V} = \frac{60 \text{ W}}{120 \text{ V}} = 0.50 \text{ A}$$

(b) Using Ohm's law ($V = IR$), the resistance is

$$R = \frac{V}{I} = \frac{120 \text{ V}}{0.50 \text{ A}} = 240 \; \Omega$$

(R could also be found using the relationship $P = V^2/R$.)

EXAMPLE 16.7 ■ Joule Heat

An electric bathroom heater is constructed using a coil of wire. For efficient operation, the heater coil should have relatively (a) low resistance, (b) high resistance, or (c) the resistance doesn't matter. *Clearly establish the reasoning and physical principle(s) used in determining your answer. That is,* **why** *did you select your answer?*

Reasoning and Answer. This problem is simplified by choosing the most convenient form of Eq. 16.9. If we use the form $P = I^2R$, we might be fooled into thinking that *increasing* the resistance would increase the power consumption, and thus the joule heating. However, note that power also increases with the square of the current—and current, being inversely proportional to resistance, increases with *lower* resistance.

We might conclude (rightly) that because of the I^2 dependence, this second effect will outweigh the first. However, a simpler approach is to use the form of Eq. 16.9 that eliminates the current (which is a variable here) and gives the power in terms of the resistance and the voltage (which we can assume to be household voltage and therefore constant): $P = V^2/R$. From this we can see clearly that, for a circuit with constant voltage, power is inversely proportional to resistance. Thus, to promote joule heating, the heating element should have a relatively low resistance, and the answer is (a).

Follow-up Exercise. Two heater coils are made out of the same wire but one is 95 percent as long as the other. How do their power outputs compare when connected to the same voltage source? (*Reasoning and answer may be found in the Answers to Exercises section at the back of the book.*)

EXAMPLE 16.8 ■ More Joule Heat

A radiant heater is rated at 1500 W for 120 V. The uniform wire filament breaks near one end, and the owner repairs it by unlooping and reconnecting it. The filament then has a length 10% shorter than its original length. What effect will this have on the heater's power output?

Solution.

> *Given:* $P_o = 1500$ W   *Find:* P (new power output)
> $V = 120$ V
> $R = (0.90) R_o$

Since 10% of the filament's length is gone, it will have 90% of its original resistance (R_o) after being reconnected or $R = (0.90) R_o$. (Why?) This means that the current will increase. With the same applied voltage before and after, $(V = V_o)$,

$$IR = I_o R_o$$

and

$$I = \left(\frac{R_o}{R}\right) I_o = \left(\frac{1}{0.90}\right) I_o = (1.11) I_o$$

Under the initial conditions, the current was

$$I_o = \frac{P_o}{V} = \frac{1500 \text{ W}}{120 \text{ V}} = 12.5 \text{ A}$$

Then

$$I = (1.11) I_o = (1.11)(12.5 \text{ A}) = 13.9 \text{ A}$$

The power output could be computed directly from $P = IV$ or by computing the resistance and then using Eq. 16.9. By the latter method, using Ohm's law,

$$R_o = \frac{V}{I_o} = \frac{120 \text{ V}}{12.5 \text{ A}} = 9.60 \text{ } \Omega$$

and

$$R = (0.90) R_o = (0.90)(9.60 \text{ } \Omega) = 8.60 \text{ } \Omega$$

For the repaired heater, then,

$$P = I^2 R = (13.9 \text{ A})^2 (8.60 \text{ } \Omega) = 1670 \text{ W}$$

The power output of the heater has been increased.

You should not attempt such a repair job. With reduced resistance and increased current, the filament could overheat, melt, and possibly start a fire.

People often complain about their electric bills. But, what do we buy from the electric or power company? As you may know, we buy electricity in units of kilowatt-hours (kWh). Let's look at our defining equation of work and power to see what this is a unit of. Recall that power is the time rate of doing work, $P = W/t$ or $W = Pt$, so work has the units of watt-second (power × time). Converting this to a larger derived unit of kilowatt-hour, we see that the kWh is a unit of work (or energy). Thus, we pay the "power" company for electrical energy which we use to do work. Let's now take a look at the cost of electricity.

EXAMPLE 16.9 ■ Electrical Cost

If a frost-free refrigerator runs 15% of the time, how much does it cost to operate (a) per day and (b) per month if the power company charges 11¢ per kilowatt-hour?

Solution.

(a) Since the refrigerator operates 15% of the time, in one day it runs

$$t = (0.15)(24 \text{ h}) = 3.6 \text{ h}$$

Taking the power requirement of the refrigerator to be 500 W (see Table 16.2), the electrical work done, or the energy expended, per day is, since $P = W/t$,

$$W = Pt = (500 \text{ W})(3.6 \text{ h}) = 1800 \text{ Wh} = 1.8 \text{ kWh}$$

Then

$$(1.8 \text{ kWh})(\$0.11/\text{kWh}) = \$0.20$$

It costs 20¢ a day to operate the refrigerator.
(b) For a 30-day month,

$$(\$0.20/\text{day})(30 \text{ day/month}) = \$6.00/\text{month}$$

The cost of electricity varies around the country. On the average, it ranges from about 8¢ to 18¢ per kilowatt-hour. Do you know the price of electricity in your locality? Check an electric bill to find out. The Insight feature on electrical costs gives a comparison you might find interesting.

● **FIGURE 16.11 All lit up** A satellite image of the Americas at night. Can you identify the major population centers in the United States and elsewhere? The red spots across part of South America indicate large-scale burning of vegetation. The small yellow spot in Central America shows burning gas flares at oil production sites. At the top right edge you can just glimpse a few of the white city lights of Europe. The image was recorded by a visible/infrared system.

Electrical Efficiency

In our electrical society, use of and demand for electricity are continually increasing. About 25% of the electricity generated in the United States goes into lighting (● Fig. 16.11). This is roughly equivalent to the output of 100 electrical generating (power) plants. Refrigerators consume about 7% of the electricity produced in the United States (the output of about 25 power plants).

This huge consumption of electricity has prompted the federal government to set minimum efficiency limits for refrigerators, freezers, air conditioners, water heaters, dishwashers, heat pumps, and so forth (● Fig. 16.12). It is estimated that these new standards will save over 30 terawatt-hours (10^9 kilowatt-hours) of electricity by 1995.

Also, more efficient fluorescent lighting is being developed. The most efficient fluorescent lamp now in use consumes about 25–30% less energy than the average fluorescent lamp and roughly 75% less energy than incandescent lamps with an equivalent illumination output. Researchers are using new techniques in an effort to develop fluorescent lamps with increased efficiencies.

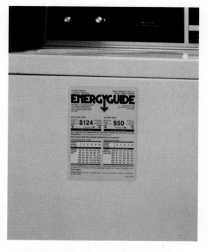

● **FIGURE 16.12 Energy guide** Consumers are made aware of appliance efficiences in terms of the average yearly cost of operation. The yearly cost is given for different kilowatt-hour (kWh) rates, which vary around the country.

As mentioned in Example 16.9, the cost of household electricity ranges from about 8¢ to 18¢ per kilowatt-hour, depending on where one lives. The example also shows that electrical energy prices are still a bargain, considering the work that electricity does and the convenience it provides.

As noted earlier, batteries are widely and increasingly used to supply electrical energy. They power toys, radios, TV remote controls, and a variety of other portable devices. Have you ever thought about the cost of this "portable" electricity, as compared to that of household electricity? Let's take a look and make a general cost comparison. To do this, we need some data on batteries.

For one thing, we need to know the battery cost. This varies a great deal with location and quantity purchased. Focusing on just the AA-cell and the D-cell, a brief survey in the author's area showed average costs to be about those listed in Table 1.

TABLE 1 ■ Battery Data

	Cost	Rating	Voltage
AA-cell	$0.95	2450 mA-h = 2.45 A-h	1.5 V
D-cell	$1.15	14,250 mA-h = 14.24 A-h	1.5 V

Then, we need to know how much "electricity" we get out of a battery. Batteries are rated in ampere (amp)-hours (A-h). This allows you to figure the lifetime of a battery for a given current output, for example, that needed to operate a flashlight: battery life (h) = amp-h rating/current output. Thus, the greater the current out-

put, the shorter the battery life, as one would expect. The amp-hour ratings are not usually listed on batteries or their packages, but they are available from manufacturers, and some typical values are given in Table 1. (The ratings were furnished in units of mA-h.)

Now, notice that amp-hour is current × time (It), and recall that $q = It$. Hence, the battery rating is a measure of the charge supplied over the chemical lifetime of a battery. Converting the ratings to amp-s, which is equivalent to coulombs (C), we have

AA-cell: $q = It = 2.45$ A-h $(3.60 \times 10^3 \text{ s/h}) = 8.82 \times 10^3$ C
D-cell: $q = It = 14.25$ A-h $(3.60 \times 10^3 \text{ s/h}) = 5.13 \times 10^4$ C

The energy supplied by a battery is given by $W = qV$, so the energy outputs over the lifetimes of the batteries are

AA-cell: $qV = (8.82 \times 10^3 \text{ C})(1.5 \text{ V})$
$$= 1.3 \times 10^4 \text{ J} \left(\frac{1 \text{ kWh}}{3.6 \times 10^6 \text{ J}} \right) = 0.0036 \text{ kWh}$$

D-cell: $qV = (5.13 \times 10^4 \text{ C})(1.5 \text{ V})$
$$= 7.7 \times 10^4 \text{ J} \left(\frac{1 \text{ kWh}}{3.6 \times 10^6 \text{ J}} \right) = 0.021 \text{ kWh}$$

where the conversions to kWh are made for comparison purposes. The cost of portable (battery) electricity is then about

AA-cell: price/kWh = $0.95/0.0036 kWh = $260/kWh
D-cell: price/kWh = $1.15/0.021 kWh = $55/kWh

Compared to $0.08–$0.18/kWh for household electrical energy costs, we pay a great deal for battery convenience.

Important Concepts

You should be able to define and explain these chapter concepts clearly.

electric current	terminal voltage	resistance	temperature coefficient of resistivity
battery	direct current (dc)	Ohm's law	superconductivity
cathode	conventional current	ohm (Ω)	electric power
anode	ampere (A)	resistivity	joule heat (I^2R loss)
electromotive force (emf)	drift velocity	conductivity	

Important Relationships for Review

These relationships are mathematical statements of the concepts and principles presented in the chapter. You should be able to identify the symbols and to explain the relationships before proceeding to the Exercises. In-text equation reference numbers are given for convenience.

Current:

$$I = \frac{q}{t} \tag{16.1}$$

Ohm's law:

$$V = IR \tag{16.2}$$

Resistivity:

$$\rho = \frac{RA}{L} \tag{16.3}$$

Conductivity:

$$\sigma = \frac{1}{\rho} \tag{16.4}$$

Temperature-dependence of Resistivity:

$$\rho = \rho_o(1 + \alpha \Delta T)] \quad \text{or} \quad \Delta\rho = \rho_o \alpha \Delta T \tag{16.5}$$

Temperature-dependence of Resistance: (for a conductor with a uniform cross section):

$$R = R_o(1 + \alpha \Delta T) \quad \text{or} \quad \Delta R = R_o \alpha \Delta T \tag{16.6}$$

Electrical Power:

$$P = IV \quad \text{and} \quad P = \frac{V^2}{R} = I^2 R \tag{16.8–9}$$

Exercises

16.1 ■ Batteries and Direct Current

16.2 ■ Current and Drift Velocity

1 When three 6-V batteries are connected in parallel, the output voltage of the combination is (a) 6 V, (b) 2 V, (c) 18 V, (d) none of these. (a)

2 The unit of current is (a) C, (b) C/s, (c) A, (d) both (b) and (c). (d)

3 Is the electromotive force really a force? Explain. no (see ISM)

● **4** A battery has a small internal resistance r (● Fig. 16.13). Explain why the emf of a battery is the terminal voltage for an open circuit condition. see ISM

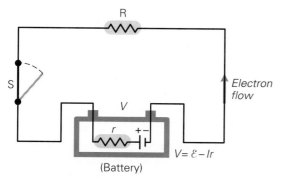

● **FIGURE 16.13 Emf and terminal voltage** See Exercise 4.

5 ■ (a) Two 1.5-V dry cells are connected in series. What is the total voltage of the combination? (b) What would be the total voltage if the cells were connected in parallel? (a) 3.0 V; (b) 1.5 V

6 ■ What is the voltage across six 1.5-V batteries when they are connected (a) in series and (b) in parallel? (a) 9.0 V; (b) 1.5 V

7 ■ A net charge of 15 C passes through a cross-sectional area of a wire in 1.0 min. What is the current carried by the wire? 0.25 A

8 ■■ Two 6.0-V batteries and one 12-V battery are connected in series. (a) What is the voltage across the whole arrangement? (b) What arrangement of these three batteries would give a total voltage of 12 V? (a) 24 V; (b) 6.0 V in series, combined in parallel with 12 V

9 ■■ Given three batteries with voltages of 1.5 V, 6.0 V, and 12 V, how many different voltages could be obtained using one or more of the batteries and what are these voltages? see ISM

● **10** ■■ How long would it take for a net charge of 1.8 C to pass through a cross-sectional area of wire so as to produce a steady current of 3.0 mA? 6.0×10^2 s

11 ■■ A hand calculator draws 0.15 mA of current from a 3.0-V nicad battery. In 20 minutes of operation, (a) how much net charge flows in the circuit and (b) how much energy is delivered to the circuit? (a) 0.18 C; (b) 0.54 J

12 ■■ A large motor draws 30 A of current when starting up. If the start-up time is 0.75 s, what is the net charge passing through a cross-sectional area of the circuit in this time? 23 C

13 ■■ A net charge of 20 C passes through a wire with a cross-sectional area of 0.30 cm^2 in 1.25 min; a net charge of 30 C passes through another wire with a cross-sectional area of 0.450 cm^2 in 1.52 min. Which wire carries more current and how much more? 2nd wire by 0.060 A

14 ■■ A battery is rated at 60 A-h at 3.0 A. Thus the life of the battery is 20 h (60 A-h/3.0 A) when fully charged. How much charge will the battery deliver in this time?
2.2 × 10⁵ C

15 ■■ What is the net number of electrons that move past a point in a wire carrying 500 mA of current in 4.0 min?
7.5 × 10²⁰ electrons

● **16** ■■■ ● Fig. 16.14 shows charge carriers having a charge *q* and moving with a speed v_d (drift speed) in a conductor of cross-sectional area *A*. Let *n* be the number of free charge carriers per unit volume. (a) Show that the total charge (ΔQ) free to move in the volume element shown is given by $\Delta Q = (nAx)q$. (b) Show that the current in the conductor is given by $I = nqv_d A$.   see ISM

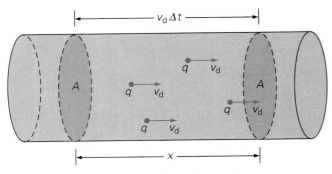

● **FIGURE 16.14 Total charge and current**   See Exercise 16.

17 ■■■ A copper wire with a cross-sectional area of 13.3 mm² (AWG No. 6) carries a current of 1.2 A. If there are 8.5 × 10²² free electrons per cubic centimeter, what is the drift velocity of the electrons? [*Hint:* See Exercise 16.]
6.6 × 10⁻⁶ m/s

16.3 ■ Ohm's Law and Resistance*

18 According to Ohm's law, current and resistance (a) vary with temperature, (b) are directly proportional, (c) are independent of voltage, (d) none of these.   (d)

19 Electrical conductivity (a) is the reciprocal of the resistivity, (b) has units of Ω-m, (c) is equal to 1/*R*, (d) does not vary with temperature.   (a)

● **20** If the voltage (*V*) were plotted versus current (*I*) for two conductors with different resistances on the same graph, what would be the general results and why?
slopes give resistances—Ohm's law

21 (a) If a connecting wire in a circuit is replaced with one of the same material that is twice as long and has twice the cross-sectional area, how will the current in the wire be affected? Assume that the potential difference remains constant. (b) How will the current be affected if the new wire has the same length but half the diameter of the old one?
(a) same;   (b) one quarter the current

22 What does a negative temperature coefficient of resisti-

*Assume that the temperature coefficients of resistivity given in Table 16.1 apply over large temperature ranges.

vity for a semiconductor imply about the conducting electrons in such a material?   see ISM

23 ■ What voltage must a battery have to produce 0.50 A of current through a 2.0-Ω resistor?   1.0 V

24 ■ How much current is drawn from a 12-V battery when a 150-Ω resistor is connected across its terminals?
8.0 × 10⁻² A

25 ■ A circuit component with a resistance of 14 Ω draws 1.5 A of current when connected to a dc power supply. What is the voltage of the power supply?   21 V

26 ■ A copper wire is 1.2 m long and has a diameter of 0.10 cm. What is the resistance of the wire?   2.6 × 10⁻² Ω

● **27** ■■ A 100-Ω resistor is placed in a circuit with two 9.0-V batteries connected in series. How much current flows through the resistor?   0.18 A

28 ■■ An automobile starter is connected to a 12-V battery. If the starter has an effective resistance of 2.5 Ω, what is the net rate at which electrons move in the circuit?
3.0 × 10¹⁹ electrons

29 ■■ A certain material is formed into a long rod with a square cross section 0.50 cm on a side. When a voltage of 100 V is applied across a length of the rod 2.0 m long, 5.0 A of current flows. What are (a) the resistivity and (b) the conductivity of the material? Is it a good conductor?
(a) 2.5 × 10⁻⁴ Ω-m   (b) 4.0 × 10³ Ω-m

● **30** ■■ Two copper wires have equal cross-sectional areas and lengths of 2.0 m and 0.50 m. What is the ratio of the resistance of the shorter to that of the longer wire?
R_s/R_l = 0.25

31 ■■ Two copper wires have equal lengths, but the diameter of one is twice that of the other. What is the ratio of the resistance of the thicker to that of the thinner wire?
R_{thick}/R_{thin} = 1/4

32 ■■ The wire in a heating element of an electric stove burner has an effective length of 0.75 m and a cross-sectional area of 2.0 × 10⁻⁶ m². If the wire is made of iron, what is its resistance when operating at a temperature of 380°C?
0.13 Ω

33 ■■ What is the percentage variation of the resistivity of copper over the range from room temperature to 100°C?   54%

● **34** ■■ A copper wire has a resistance of 25 mΩ at 20°C. When the wire is carrying a current, joule heat causes its temperature to increase 27 C°. What is the change in the wire's resistance?   4.6 mΩ

35 ■■ What is the fractional change in the resistance of an aluminum wire 1.0 m long over a temperature range from the freezing point to the boiling point of water?
$\Delta R/R_o$ = 0.43

36 ■■ When a resistor is connected to a 12-V source, it draws 185 mA of current. When the same resistor is connected to a 90-V source, it draws 1.38 A of current. Is the resistor ohmic? Justify your answer mathematically.
65 Ω; ohmic

37 ■■ What length of wire is required to make a 20-Ω resistor by winding AWG (American Wire Gauge) No. 10

nichrome wire in a coil? (AWG No. 10 wire has a diameter of 2.588 mm.)   105 m

38 ■■ A 1.0-m length of copper wire with a diameter of 2.0 mm is stretched out; its length increases by 25% while its cross-sectional area decreases but remains uniform. Is the resistance of the wire the same before and after? Justify your answer mathematically. $R_2/R_1 = 1.6$; not the same

39 ■ An iron conductor with a resistance of 1.75 Ω at 20°C is connected to a 12-V source. If the conductor is placed in an oven and heated to 300°C, what is the change in the current flowing through it?   4.4 A less

40 ■■ A particular application requires a 20-m length of aluminum wire to have a resistance of 1.0 mΩ. What diameter should the wire have? 2.7×10^{-2} m

41 ■■ A 10-m length of 1.0-mm-diameter nichrome wire is wound in a coil and connected to a 1.5-V flashlight battery. How much current will initially flow in the coil?   0.12 A

42 ■■ A carbon resistor has a resistance of 150 Ω at room temperature. In a circuit with a voltage of 12 V across it, its operational temperature is 90°C. What is the difference in current through the resistor at 90°C and at room temperature when the circuit is initially closed?   3.0 mA

●**43** ■■ A rod of silicon is used in a circuit that carries a current of 0.50 A with a potential difference of 6.0 V at 20°C. If the temperature of the rod is increased to 25°C, how much current flows through it?   0.77 A

44 ■■■ Pieces of carbon and copper have uniform cross sections and the same resistance at room temperature. (a) If the temperature of each piece is increased by 10C°, which will have the greater resistance? (b) What is the ratio of the resistances?   (a) copper   (b) $R_{Cu}/R_C = 1.1$

45 ■■■ An electrical resistance thermometer made of platinum has a resistance of 5.0 Ω at 20°C and is in a circuit with a 1.5-V battery. What is the change in the current in the circuit when the thermometer is heated to 2020°C? (Assume α constant over temperature range.)   0.27 A decrease

46 ■■■ Pieces of aluminum and copper wire are identical in length and diameter. At some temperature, one of the wires will have the same resistance as the other has at room temperature. What is the temperature? (Is there more than one temperature?)   copper 117°C or aluminum − 72.8°C

16.4 ■ Electric Power

47 Electric power has units of (a) A²-Ω, (b) J/s, (c) V²/Ω, (d) all of these.   (d)

●**48** If the voltage of a circuit with a constant resistance is doubled, the power expended in the circuit (a) increases by a factor of 2, (b) increases by a factor of 4, (c) decreases by one-half, (d) none of these.   (b)

49 ■ An applicance is rated for 2.5 A at 120 V. How much electric power does it require? 3.0×10^2 W

50 ■ How much power will be expended in a 10-kΩ resistor when it is connected across a 120-V voltage source?   1.44 W

51 ■■ A hair dryer is rated for 1200 W at 120 V. (a) How much current does the dryer use? (b) What is its resistance?   (a) 10 A   (b) 12 Ω

52 ■■ Show that V^2/R has the same units as power.   see ISM

53 ■■ An electric heater is designed to produce 50 kW of heat when connected to a 240-V source. What must the resistance of the heater be?   1.2 Ω

54 ■■ An electric toy rated at 2.5 W is operated by four 1.5-V batteries connected in series. (a) How much current does the toy draw? (b) What is its effective resistance?   (a) 0.42 A;   (b) 14 Ω

55 ■■ A welding machine draws 18 A of current at 240 V. (a) How much energy does the machine use per second? (b) What is its effective resistance?   (a) 4.3×10^3 J/s;   (b) 13 Ω

56 ■■ An electric water heater automatically operates for 2.0 h each day. (a) If the cost of electricity is 12¢/kWh, what is the cost of operating the heater during a 30-day month? (b) What is the effective resistance of a typical water heater? [*Hint:* See Table 16.2.]   (a) $32,40;   (b) 3.2 Ω

57 ■■ What resistor should be used to generate 10 kJ of heat per minute when connected to a 120-V source?   85 Ω

●**58** ■■ A 1500-W hair dryer is used for 10 min every day. If the cost of electricity is 12¢/kWh, what is the cost of using the dryer for a month (30 days)? $0.90

59 ■■ A 240-V air conditioner unit draws 15 A of current. If it operates for a period of 20 min, (a) how much energy in kilowatt hours does it use? (b) If the cost of electricity is 12¢/kWh, what is the cost of operating the unit for this time?   (a) 1.2 kWh   (b) $0.14

60 ■■ Two resistors of 100 Ω and 25 kΩ are rated for maximum wattages of 1.5 W and 0.25 W, respectively. What is the maximum voltage that can be safely applied to each resistor?   100 Ω: 12 V; 25 Ω: 79 V

61 ■■ A wire 5.0 m long and 3.0 mm in diameter has a resistance of 100 Ω. A potential difference of 15 V is applied across the wire. Find (a) the current in the wire, (b) the resistivity of its material, and (c) the rate at which heat is being produced in it.   (a) 0.15 A   (b) 1.4×10^{-4} Ω-m   (c) 2.3 W

62 ■■ A coil of tungsten wire initially dissipates 500 W of power when connected to a voltage source. In a short time, the temperature of the coil increases by 150 C° because of joule heat. What is the corresponding change in the power?   − 185 W

63 ■■ A toaster oven is rated for 1600 W at 120 V. When the filament ribbon of the oven burns out, the owner makes a repair that reduces the length of the filament by 10%. How does this affect the power output of the oven, and what is the new value? 1.8×10^3 W

64 ■■ A 0.50-kΩ resistor is in a circuit with three 12-V batteries. What is the joule heat loss of the resistor (a) if the batteries are connected in series and (b) if the batteries are connected in parallel? (a) 2.6 W (b) 0.29 W

● **65** ■■ A 150-W light bulb operates on 120-V household voltage. What is the resistance of the bulb? 96 Ω

66 ■■ A 5500-W water heater operates on 240 V. (a) Should the heater circuit have a 20-A or a 30-A circuit breaker? (A circuit breaker is a safety device that opens the circuit at its rated current.) (b) Assuming that there is no energy loss, how long will the heater take to heat the water in a 55-gal tank from room temperature to 80°C? (a) 30 A breaker (b) 9.5 $\times$ 10^3 s = 2.6 h

67 ■■ By what factor is the power expended in a 150-Ω resistor increased when the voltage across it is increased from 15 V to 45 V? $P_2 = 9 P_1$

68 ■■■ What is the efficiency of a 350-W motor that draws 4.0 A from a 120-V line? 73%

● **69** ■■■ An immersion heater with a resistance of 9.0 Ω operates on 120 V. The heater is placed in a cup of water (0.25 L) at room temperature (20°C) to heat it up to make coffee. Assuming that no heat loss occurs, how long will the heater have to operate to bring the water to a boil? 53 s

70 ■■■ How long will the heater in Exercise 69 have to operate to boil all the water away? 406 s

71 ■■■ A 500-W heating element made of nichrome wire whose diameter is 4.0 mm operates on 120 V. (a) What is the length of the wire? (b) What power would be dissipated if the voltage were reduced to 105 V? (a) 3.6 $\times$ 10^2 m (b) 3.8 $\times$ 10^2 W

72 ■■■ Find the total charged on a monthly (30-day) electric bill for the following appliance usage if the utility rate is 9.5¢/kWh: central air conditioning runs 30% of the time; a blender is used 0.50 h/month; a dishwasher is used 8.0 h/month; a microwave oven is used 15 min/day; a frostfree refrigerator runs 15% of the time; a stove (range and oven) is used a total of 10 h/month; and a color television is operated 120 h/month. (Use the information given in Table 16.2) $120.24

■ Additional Exercises

73 A resistance thermometer made of platinum shows a 25% change in resistance over a certain temperature range. What is the change in temperature for this range? 64°C

74 An electric heater that operates on 120 V is rated at 1500 W. (a) How much current does the heater draw? (b) What is the resistance of its heating element? (a) 12.6 A; (b) 9.6 Ω

75 If the price of electricity is 8¢/kWh, how much would it cost to operate ten 75-W light bulbs for 6 hours? $0.36

76 A 20-Ω, 5-W resistor is used in a circuit. (a) What is the maximum voltage that should be applied across the resistor? (b) Could the resistor safely carry a current of 0.60 A? (a) 10 V; (b) no

77 An aluminum wire has a mass of 10 g and a length of 50 cm. Find the current in the wire when it is connected to a 10-V source. 5.3 $\times$ 10^3 A

78 The tungsten filament of an incandescent lamp has a resistance of 200 Ω at room temperature. What will the filament's resistance be at an operating temperature of 1600°C? 1.6 $\times$ 10^3 Ω

79 Find the resistance of a piece of copper wire that is 75 cm long and has a diameter of 2.0 mm. 4.1 $\times$ 10^{-3} Ω

80 An electric iron with a 14-Ω heating element operates on 120 V. How much heat does the iron produce in 30 min? 1.9 $\times$ 10^6 J

81 When a wire 1.0 mm in diameter and 2.0 m in length is connected to a 3.00-V source, a current of 11.8 A flows. (a) What is the resistivity of the wire? (b) What material is the wire made of? (a) 1.0 $\times$ 10^{-7} Ω-m; (b) iron or platinum

82 A 100-Ω resistor is rated at $^1/_4$ W. (a) What is the maximum current the resistor can carry? (b) What is the maximum operating voltage that should be applied across the resistor? (a) 5.0 $\times$ 10^{-2} A; (b) 5.0 V

83 A wire carries a current of 75 mA. How many electrons have to pass through a cross-sectional area of the wire in 2.0 s to produce this current? 9.4 $\times$ 10^{17} electrons

84 When a 1.0-m length of wire is connected to a 6.0-V source, 0.25 A of current flows in it. Calculate the current in a 2.0-m length of the same type of wire connected to the same voltage source. 0.13 A

85 What is the resistance of a 5.0-km length of aluminum cable with an effective diameter of 1.0 cm? 1.8 Ω

86 A current of 0.10 A flows in a resistor in a circuit with a 40-V voltage source. (a) What is the resistance of the resistor? (b) How much power is dissipated by the resistor? (c) How much energy is dissipated in 2.0 min? (a) 4.0 $\times$ 10^2 Ω; (b) 4.0 W; (c) 4.8 $\times$ 10^2 J

87 Find the ratio of the resistance of a copper wire to the resistance of an aluminum wire (a) if the wires have the same length and diameter and (b) if the copper wire is twice as long as the aluminum wire and has a diameter that is half that of the aluminum wire. (a) R_c/R_{al} = 0.61 (b) R_c/R_{al} = 4.9

88 A portable radio draws 150 mA of current from a battery. If the radio is played for a half-hour, what is the net number of electrons that move through it? 1.7 $\times$ 10^{21} electrons

89 A copper bus bar used to conduct large amounts of current at a power station is 1.5 m long and 8.0 cm by 4.0 cm in cross section. What voltage difference across the ends of the bar will cause a current of 3000 A to flow through it? 2.4 $\times$ 10^{-2} V

Basic Electric Circuits

Electric circuits are of many kinds and can be designed for many puposes. Some are very complex, with numerous components. Others, like the one shown above, are extremely simple. This circuit incorporates a saline (salt water) solution and electrodes of pencil "lead"—actually a form of carbon called graphite. Since the bulb is lit, the saline and graphite must be good conductors of electricity. The setup in the photograph is designed as a demonstration, but circuits of this type do have practical uses in the laboratory and in industry: for example, synthesizing or purifying chemical substances and electroplating metals (silver plate).

Armed with the principles learned in Chapters 15 and 16, you are now ready to analyze some relatively simple circuits. This will give you an appreciation of how electricity actually works.

Circuit analysis most often deals with voltage, current, and power requirements. For example, a circuit component with a given resistance may be designed to operate at a particular voltage. The voltage source used then depends on specific current and power requirements. Also, how two or more components are connected in a circuit affects the voltage across those components and/or current through them. A circuit may be analyzed theoretically before actually being hooked up. The analysis might show that the circuit will not function properly as designed or that there is a safety problem (such as overheating due to joule heat).

Circuit diagrams are used to visualize circuits in order to understand their functioning. A few of these diagrams were included in figures in Chapter 16. In actuality, the wires of a circuit are not placed in the rectangular pattern of a circuit diagram. The rectangular form is simply a convention that provides a neater presentation and easier visualization of the circuit arrangement.

Let's begin our analysis of circuits by looking at arrangements of resistive elements.

Topics in Chapter 17 include series, parallel, and series-parallel combinations of resistors, multiloop circuits, RC circuits, ammeter and voltmeter circuits, and a discussion of household currents and electrial safety.

Figure 17.1

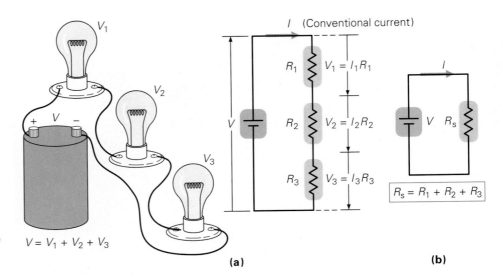

$V = V_1 + V_2 + V_3$

(a)

$$R_s = R_1 + R_2 + R_3$$

(b)

● **FIGURE 17.1 Resistors in series** **(a)** When resistors are connected in series, the current through each of them is the same. Also, the sum of the voltage drops across the resistors is equal to the voltage (rise) of the batttery. **(b)** The equivalent resistance R_s of the resistors in series.

17.1 ■ Resistances in Series, Parallel, and Series-Parallel Combinations

The resistance symbols in a circuit diagram can represent any resistive element—a commercial resistor, a light bulb, an appliance, and so on. This discussion will consider all of these to be ohmic and to have a constant operating resistance, unless otherwise stated. Also, the resistance of the connecting wires in a circuit will be considered to be negligible.

In analyzing a circuit, you need to keep in mind that the sum of the voltages around a circuit loop is zero. For example, for the circuit in ●Fig. 17.1a the sum of the voltage drops (decreases) across the circuit components, or resistances (V_i), equals the voltage "rise" of the battery, V (the voltage across the circuit elements):

or
$$V = \Sigma_i\, V_i = \Sigma_i\, IR_i$$
$$V - \Sigma_i\, IR_i = 0 \tag{17.1}$$

where the i summation is over the individual resistances. (The subscript is often omitted as being understood.) The sum of the voltages around the circuit loop is zero. This is essentially a statement of the conservation of energy. We saw in Chapter 16 that for the charge carriers traversing the circuit, the energy given to them by a voltage source must be lost in the circuit. That is, they must have the same (zero) electric potential on returning to the negative terminal as they had when starting. Otherwise, the charge carriers would gain energy with each round trip of the circuit, which would violate the law of the conservation of energy.

The resistances or resistors in Fig. 17.1 are in series, or connected end-to-end, so to speak. *When resistors are connected in series, the current is the same through all of them*, as required by the conservation of charge. As an analogy, think of an incompressible fluid flowing through a series of pipe sections. The amount of fluid leaving a section must be equal to the amount going into it by the conservation of mass. Thus, the volume of fluid flow must be the same

Recall from Section 16.1 that the operating voltage (V) of a battery is related to its emf ($\mathscr{E}$) by $V = \mathscr{E} - Ir$, where r is the internal resistance of the battery. Note that V varies with the amount of current being drawn from the battery and that $V = \mathscr{E}$ for an open circuit ($I = 0$). In general, this book uses the operating voltages of batteries in circuits.

through all the pipe sections. Otherwise, there would be build-up of mass somewhere along the way. Similarly, the current, or charge flow, is the same through each resistor connected in series. Otherwise, there would be a build-up of charge somewhere along the way.

The total voltage drop around a circuit is equal to the sum of the individual voltage drops across the resistors. With a current I, Eq. 17.1 can be written explicitly (see Fig. 17.1):

$$V = V_1 + V_2 + V_3$$
$$= IR_1 + IR_2 + IR_3$$
$$= I(R_1 + R_2 + R_3)$$

Thus, since $V = IR$, three resistors in series have an equivalent resistance R_s:

$$R_s = R_1 + R_2 + R_3$$

That is, the equivalent resistance of resistors in series is simply the sum of the individual resistances. The three resistors of Fig. 17.1 could be replaced with a single resistor with a resistance value of R_s without having any effect on the voltage source. For example, if each of the resistors in Fig. 17.1 had a value of 10 Ω, then R_s would be 30 Ω.

This result may be extended to any number of resistors in series:

$$\boxed{R_s = R_1 + R_2 + R_3 + \cdots} \qquad (17.2)$$

equivalent resistance for resistors in series

QUESTION: Suppose that one of the light bulbs in the circuit in Fig. 17.1a blew out. What would happen?

ANSWER: All of the bulbs would go out because the circuit would no longer be complete. If the filament of one of the bulbs is broken or blows out, the circuit is open and no current can flow.

Another basic way of connecting resistors in a circuit is in parallel (●Fig. 17.2a). In this case, all the resistors have a common connection (one side of all

Demonstration/activity: Construct circuits using large demonstration ammeter(s) and voltmeter(s) so students can see the current and potential readings. Elements such as 10-V Christmas tree bulbs or flashlight bulbs work well because of the visual effect. Keep in mind the fact that the resistance changes with temperature.

Build a series circuit and show how the total resistance is the sum of the individual resistances.

$$V = V_1 = V_2 = V_3$$

$$I = I_1 + I_2 + I_3$$

$$V_1 = V_2 = V_3$$

(a)

$$\frac{1}{R_p} = \frac{1}{R_1} + \frac{1}{R_2} + \frac{1}{R_3}$$

(b)

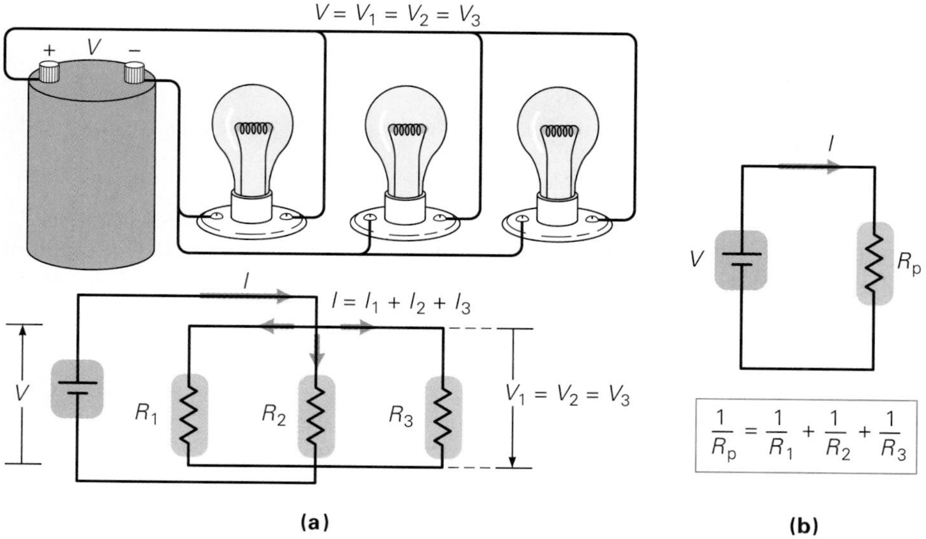

Figure 17.2

●**FIGURE 17.2 Resistors in parallel** **(a)** When resistors are connected in parallel, the voltage drop across each of the resistors is the same. The current from the battery divides proportionally among the resistors. **(b)** The equivalent resistance R_p of the resistors in parallel.

of them connected together). *When resistors are connected in parallel, the voltage drop across each resistor is the same and equal to the voltage of the battery.* However, the current from the battery divides among the different paths, as shown in Fig. 17.2a,

$$I = I_1 + I_2 + I_3 \tag{17.3}$$

If the individual resistances are equal, the current divides equally. But if the resistances are not equal, the current divides among the resistors proportionately; that is, the greatest current flows through the path of least resistance. (It may be helpful to think of how a liquid flowing in a pipe would divide at a junction into pipes with different cross-sectional areas.)

Since the voltage drop (V) across each resistor is the same, by Ohm's law,

$$I = I_1 + I_2 + I_3$$

$$= \frac{V}{R_1} + \frac{V}{R_2} + \frac{V}{R_3}$$

$$= V\left(\frac{1}{R_1} + \frac{1}{R_2} + \frac{1}{R_3}\right) = \frac{V}{R_p}$$

Demonstration/activity: Build a parallel circuit and show how the addition of each successive element in parallel causes the current in the power supply to increase. Note that the resistance of elements in parallel is less than the resistance of the individual elements.

Discuss how the power changes when resistors are placed in series and in parallel.

Thus, the equivalent resistance R_p of three resistors in parallel is given (in reciprocal form) by

$$\frac{1}{R_p} = \frac{1}{R_1} + \frac{1}{R_2} + \frac{1}{R_3}$$

That is, the reciprocal of the equivalent resistance is equal to the sum of the reciprocals of individual resistors connected in parallel. The three resistors could be replaced with a single resistor with a resistance value of R_p (see Fig. 17.2b).

This result may also be generalized to include any number of resistors in parallel.

$$\boxed{\frac{1}{R_p} = \frac{1}{R_1} + \frac{1}{R_2} + \frac{1}{R_3} + \cdots} \tag{17.4}$$

equivalent resistance for resistors in parallel

■ Problem-Solving Hint

Eq. 17.4 gives $1/R_p$. Don't forget to take the reciprocal of this value to get R_p.

If there are only two resistors in parallel, the formula for the equivalent resistance may be written in a nonreciprocal form, which is sometimes more convenient.

Encourage students to do the necessary algebra, and show the best way of using the calculator for these types of problems. Some students will use the ⅟ₓ function on their calculators and then round to one significant figure prior to adding the next term.

$$\frac{1}{R_p} = \frac{1}{R_1} + \frac{1}{R_2} = \frac{R_1 + R_2}{R_1 R_2}$$

or

$$\boxed{R_p = \frac{R_1 R_2}{R_1 + R_2}} \tag{17.5}$$

equivalent resistance for two resistors in parallel

An interesting fact to note is that the equivalent resistance of resistors in parallel is always less than the smallest individual resistance. You can easily demonstrate this for two resistors in parallel by putting some values in Eq. 17.5. More generally, you can see that

$$\frac{1}{R_p} = \Sigma_i \frac{1}{R_i}$$

where $\Sigma_i 1/R_i$ is the sum of the reciprocals of the individual resistances. This sum is greater than the reciprocal of any individual resistance, and

$$\frac{1}{R_p} = \Sigma_i \frac{1}{R_i} > \frac{1}{R_i}$$

Therefore R_p must be less than any R_i, where R_i may be the smallest individual resistance.

Thus, series connections provide a way to increase resistance, and parallel connections a way to reduce resistance.

EXAMPLE 17.1 ■ Resistors in Series and in Parallel

What is the equivalent resistance of three resistors (1.0 Ω, 2.0 Ω, and 3.0 Ω) when they are connected (a) in series (Fig. 17.1a), and (b) in parallel (Fig. 17.2a)? (c) What current will be delivered from a 12-V battery for each of these arrangements?

Solution. Listing the data, we have

> *Given:* $R_1 = 1.0 \, \Omega$ *Find:* (a) R_s (series resistance)
> $R_2 = 2.0 \, \Omega$ (b) R_p (parallel resistance)
> $R_3 = 3.0 \, \Omega$ (c) I (current for each case)
> $V = 12 \, V$

(a) The equivalent series resistance is simply the sum of the resistors (Eq. 17.2).

$$R_s = R_1 + R_2 + R_3 = 1.0 \, \Omega + 2.0 \, \Omega + 3.0 \, \Omega = 6.0 \, \Omega$$

(b) For resistors in parallel, Eq. 17.4 applies.

$$\frac{1}{R_p} = \frac{1}{R_1} + \frac{1}{R_2} + \frac{1}{R_3} = \frac{1}{1.0 \, \Omega} + \frac{1}{2.0 \, \Omega} + \frac{1}{3.0 \, \Omega}$$

With a common denominator,

$$\frac{1}{R_p} = \frac{6.0}{6.0 \, \Omega} + \frac{3.0}{6.0 \, \Omega} + \frac{2.0}{6.0 \, \Omega} = \frac{11}{6.0 \, \Omega}$$

and

$$R_p = \frac{6.0 \, \Omega}{11} = 0.55 \, \Omega$$

Note that R_p is found by inverting the value of $1/R_p$ and it is less than the smallest resistance. Another way to find R_p would be to apply Eq. 17.5 twice. That is, you could combine two of the resistances to get R_{p_1} and then combine that value with the third resistance to get R_p. This avoids working with a reciprocal.

(c) Using Ohm's law with the equivalent resistance for the series arrangement gives the current.

$$I = \frac{V}{R_s} = \frac{12 \text{ V}}{6.0 \ \Omega} = 2.0 \text{ A}$$

Note that the voltage drop across each resistor is

$$V_1 = IR_1 = (2.0 \text{ A})(1.0 \ \Omega) = 2.0 \text{ V}$$
$$V_2 = IR_2 = (2.0 \text{ A})(2.0 \ \Omega) = 4.0 \text{ V}$$
$$V_3 = IR_3 = (2.0 \text{ A})(3.0 \ \Omega) = 6.0 \text{ V}$$

and the sum of the voltage drops equals the battery voltage.

Similarly, for the parallel arrangement, the current is

$$I = \frac{V}{R_p} = \frac{12 \text{ V}}{0.55 \ \Omega} = 22 \text{ A}$$

Note that the current for the parallel combination is large relative to that for the series combination. The current through each of the resistors in parallel is

$$I_1 = \frac{V}{R_1} = \frac{12 \text{ V}}{1.0 \ \Omega} = 12 \text{ A}$$

$$I_2 = \frac{V}{R_2} = \frac{12 \text{ V}}{2.0 \ \Omega} = 6.0 \text{ A}$$

$$I_3 = \frac{V}{R_3} = \frac{12 \text{ V}}{3.0 \ \Omega} = 4.0 \text{ A}$$

The current divides proportionally according to the individual resistances.

The power dissipated by each resistor ($P = IV = I^2R$) could easily be found for either arrangement.

As a wiring application, consider strings of Christmas tree lights. In old strings of lights with large bulbs, when one bulb blew out, all the others on that string also went out, leaving you to hunt for the faulty bulb—and you had a real problem if more than one bulb blew out at the same time. With newer strings having smaller bulbs, one or more bulbs may burn out, but the others remain lit. Are the bulbs now wired in parallel? No, this would make the total equivalent resistance very small, giving a large current in the circuit, which is not desirable. Also, more costly wire would be required.

Instead, an insulated jumper or shunt is wired in parallel with each bulb filament (●Fig. 17.3). When a bulb is in operation, the shunt does not conduct because it is insulated from the filament wires. When the filament breaks and the bulb "burns out," there is momentarily no current in the string. The voltage difference across the open circuit at the broken filament will then be the full 120-V household voltage—the high side of the string at 120 V and the low side at 0 V. This causes a sparking that burns off the shunt insulation material. Making contact with the filament wires, the shunt completes the circuit and the rest of the lights in the string continue to glow. (The shunt, a wire with very little resistance compared to a bulb filament, is indicated by the small symbol in Fig. 17.3.)

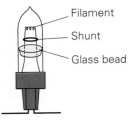

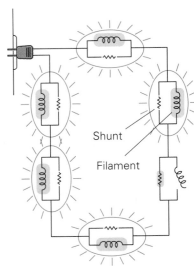

●**FIGURE 17.3 Shunt-wired Christmas tree lights** A shunt or jumper in parallel with the bulb filament reestablishes a complete circuit when one of the bulb filaments burns out. (See text for description.) Without the shunt, if one bulb burned out, all of the bulbs would go out.

EXAMPLE 17.2 ■ Oh, Tannenbaum!

In a string of Christmas tree lights composed of bulbs with jumper shunts, if the filament of one bulb burns out will the other bulbs (a) glow a little more brightly, (b) glow a little more dimly, or (c) be unaffected? *Clearly establish the reasoning and physical principle(s) used in determining your answer. That is, **why** did you select your answer?*

Reasoning and Answer. If one bulb filament burns out and its shunt completes the circuit, there will be less total resistance in the circuit because the shunt resistance is much less than the filament resistance. Note that the filaments of the good bulbs and the shunt of the burned out bulb are in series, so the resistances add.

With less total resistance, there will be more current in the circuit ($I = V/R$) and the bulbs will all glow a little brighter, so the answer is (a). For example, suppose a string of lights had 18 bulbs. Then the voltage drop across each would be $V = V_t/18 = 120\ \text{V}/18 = 6.7\ \text{V}$. (The voltage drops across the bulbs must add up to the input voltage, $V_t = 120\ \text{V}$.) If one bulb is burned out, the voltage across each lighted bulb would be $V = V_t/17 = 120\ \text{V}/17 = 7.1\ \text{V}$. Hence, the current through each bulb increases.

Follow-up Exercises. (a) Suppose that two or more bulbs in the same string burned out. What would be the effect and possible safety ramifications? (b) Suppose that the shunt resistors in Christmas tree bulbs were wired in parallel with the filaments so as always to be part of the circuit—that is, to conduct current at *all* times rather than just when the filament burned out. If one bulb in a string of such lights burned out, would the result be different from that described above? Explain. *(Reasoning and answer may be found in the Answers to Exercises section at the back of the book.)*

Series-Parallel Combinations

Resistors may be connected in a circuit in a variety of series-parallel combinations. As shown in ● Fig. 17.4, circuits with only one voltage source can usually be reduced or collapsed, theoretically into a single loop containing the voltage source and one equivalent resistance, by applying the equations given above.

A general procedure for analyzing circuits with different series-parallel combinations of resistors is to find the voltage drops across and the currents through the various resistors as follows:

> Students must master series and parallel circuits before beginning series-parallel combinations.

1. Starting with the resistor combination farthest from the voltage source, find the equivalent series and parallel resistances.

2. Reduce the circuit until there is a single loop with one total equivalent resistance.

3. Find the current delivered to the reduced circuit using Ohm's law.

4. Expand the circuit by reversing the reduction steps, and use the current for the reduced circuit to find the currents and voltage drops for the resistors in each step.

The following example illustrates this procedure. Analyze the various steps carefully in terms of the series = parallel relationships you have learned.

EXAMPLE 17.3 ■ **Series-Parallel Combination of Resistors**

(a) What is the voltage drop across and the current through each of the resistors R_1 through R_5 in Fig. 17.4a? (b) How much power is dissipated in R_4?

Solution.

Given: Values in figure (any number of significant figures may be assumed) *Find:* V's, I's, and P_4

(a) The parallel combination at the right side of the circuit diagram (farthest from the battery) is first reduced to the equivalent resistance R_{P_1} (see Fig. 17.4b).

$$R_{P_1} = \frac{R_3 R_4}{R_3 + R_4} = \frac{(6.0\ \Omega)(2.0\ \Omega)}{6.0\ \Omega + 2.0\ \Omega} = 1.5\ \Omega$$

This leaves a series combination along that side, which is reduced to R_{s_1} (Fig. 18.4c).

$$R_{s_1} = R_{P_1} + R_5 = 1.5\ \Omega + 2.5\ \Omega = 4.0\ \Omega$$

Then, R_2 and R_{s_1} are in parallel and can be reduced to R_{P_2} (Fig. 17.4d):

$$R_{P_2} = \frac{R_2 R_{s_1}}{R_2 + R_{s_1}} = \frac{(4.0\ \Omega)(4.0\ \Omega)}{4.0\ \Omega + 4.0\ \Omega} = 2.0\ \Omega$$

This leaves two resistances in series, which can be combined to give the *total* equivalent resistance (R_t) of the circuit (Fig. 17.4e),

$$R_t = R_1 + R_{P_2} = 6.0\ \Omega + 2.0\ \Omega = 8.0\ \Omega$$

Then, by Ohm's law, the battery delivers to the reduced circuit a current of

$$I = \frac{V}{R_t} = \frac{24\ V}{8.0\ \Omega} = 3.0\ A$$

This is the current through R_1 and R_{P_2} since they are in series (Fig. 17.4d). Their voltage drops are

$$V_1 = IR_1 = (3.0\ A)(6.0\ \Omega) = 18\ V$$

and

$$V_{P_2} = IR_{P_2} = (3.0\ A)(2.0\ \Omega) = 6.0\ V$$

Since R_{P_2} is made up of R_2 and R_{s_1} (Fig. 17.4c), there is a 6.0-V drop across each of these resistors. With $R_2 = R_{s_1}$, the current divides equally:

$$I_2 = 1.5\ A \qquad \text{and} \qquad I_{s_1} = 1.5\ A$$

Then I_{s_1} is the current through R_{P_1} and R_5 in series (Fig. 17.4b). Their voltage drops are therefore

---

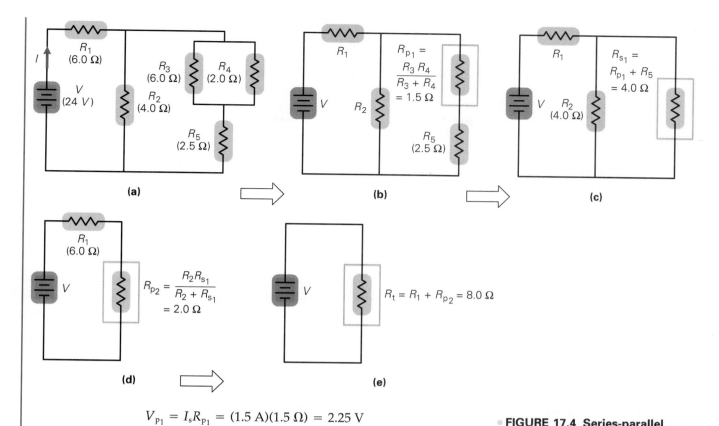

(a)

(b)

(c)

(d)

(e)

$$V_{P_1} = I_s R_{P_1} = (1.5\ \text{A})(1.5\ \Omega) = 2.25\ \text{V}$$
$$V_5 = I_s R_5 = (1.5\ \text{A})(2.5\ \Omega) = 3.75\ \text{V}$$

Note that these add up to 6.0 V.

Finally, the voltage drop across R_3 and R_4 is the same as V_{P_1}.

$$V_3 = V_4 = 2.25\ \text{V}$$

The current of 1.5 A (I_{s_1}) divides at the R_3-R_4 junction, and

$$I_3 = \frac{V_3}{R_3} = \frac{2.25\ \text{V}}{6.0\ \Omega} = 0.375\ \text{A}$$

$$I_4 = \frac{V_4}{R_4} = \frac{2.25\ \text{V}}{2.0\ \Omega} = 1.125\ \text{A}$$

Note that $I_3 + I_4 = I_{s_1}$.

To see how the current divides proportionally between R_3 and R_4, consider these equations:

$$V_3 = V_4 \quad \text{and} \quad I_3 R_3 = I_4 R_4$$

or

$$I_3 = \left(\frac{R_4}{R_3}\right)I_4 = \left(\frac{2.0\ \Omega}{6.0\ \Omega}\right)I_4 = \frac{I_4}{3}$$

That is, I_3 is $\frac{1}{3}$ of I_4. That means that when the current I_{s_1} divides at the junction, $\frac{1}{4}$ of it goes through R_3 and $\frac{3}{4}$ goes through R_4, according to their relative resistances. For R_3, the relative resistance expressed as a fraction of the total resistance is $R_3/(R_3 + R_4) = \frac{6}{8} = \frac{3}{4}$, and for R_4 it is $R_4/(R_3 + R_4) = \frac{2}{8} = \frac{1}{4}$.
(b) The power expended in R_4 is

$$P_4 = I_4 V_4 = (1.125\ \text{A})(2.25\ \text{V}) = 2.53\ \text{W}$$

● **FIGURE 17.4 Series-parallel combinations and circuit reduction** Reducing series combinations and parallel combinations to equivalent resistances reduces the circuit with one voltage source to a single loop with a single equivalent resistance. Example 17.3.

Figure 17.4

Circuit analysis like that done in Example 17.3 may seem involved, but it is simply repeated applications of equations for equivalents of series and parallel combinations and Ohm's law, both of which are quite simple.

17.2 ■ Multiloop Circuits and Kirchhoff's Rules

Try not to give the impression that Kirchhoff's rules apply only to more complex circuits. Show how they also apply to simple series and/or parallel circuits with only one voltage source.

Figure 17.5

Kirchhoff's rules were developed by the German physicist Gustav Kirchhoff (1824–1887).

●**FIGURE 17.5 Multiloop circuit** In general, a circuit that contains voltage sources in more than one loop may not be able to be further reduced by series and parallel reductions. However, some reductions within each loop may be possible, such as from **(a)** to **(b)** here. At a circuit junction, where three or more wires come together, the current divides or currents come together, as at junctions A and B in part **(b)**, respectively. The path between two junctions is called a branch. There are three branches in diagram **(b)**, that is, three different paths between junctions A and B.

Simple series-parallel circuits with a single voltage source that can be reduced to a single loop can be analyzed using Ohm's law, as you learned in the preceding section. However, in general, circuits may contain several loops, with each one having voltage sources and/or resistances (collectively referred to as circuit elements or components). A simple multiloop circuit is shown in ● Fig. 17.5a. Some combinations of resistors may be replaced by equivalent resistances, but this circuit can be reduced only so far by using the procedures illustrated in Section 17.1

A **junction** is a point in a circuit at which three or more connecting wires are joined together, for example, at point A in Fig. 17.5b. The current either divides or comes together at a junction. A path connecting two junctions is called a **branch** and may contain one or more elements.

A general method for analyzing multiloop circuits is by applying Kirchhoff's rules. These rules embody the conservation of charge and the conservation of energy. (Although they were not stated specifically, Kirchhoff's rules were applied to the simple circuits analyzed in Section 17.1.)

Kirchhoff's first rule, or **junction theorem,** is that the algebraic sum of the currents at any junction is zero.

$$\boxed{\sum_i I_i = 0} \quad \text{(sum of currents at junction)} \quad (17.6)$$

This simply means that the sum of the currents going into a junction (taken as positive) is equal to the sum of the currents leaving the junction (taken as negative), or that charge is conserved. For the junction at A in Fig. 17.4b, the algebraic sum of the currents is $I_1 - I_2 - I_3 = 0$, or

$$I_1 = I_2 + I_3$$
$$\text{current in} = \text{current out}$$

(This rule was applied in analyzing parallel resistances in Section 17.1).

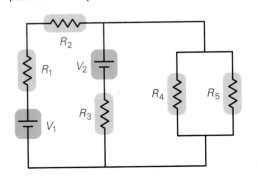

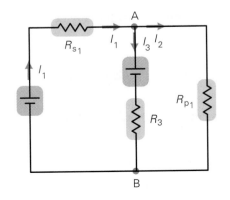

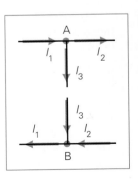

(a)

(b)

Of course, you cannot tell whether a particular current is directed into or out of a junction simply by looking at a multiloop circuit diagram. You have to *assume* current directions at a particular junction. If these assumptions are wrong, you will find out later from the mathematical results. A negative current will indicate the wrong choice of direction. (You will see this in Example 17.4.)

Kirchhoff's second rule, or **loop theorem**, is that the algebraic sum of the voltage differences across all of the elements of any closed loop is zero.

$$\Sigma_i V_i = 0 \quad \text{(sum of voltages around loop)} \quad (17.7)$$

This means that the sum of the voltage rises equals the sum of the voltage drops across the resistances around a closed loop, which must be true if energy is conserved. That is, Σ Voltage rises $= \Sigma$ IR drops. (This rule was used in analyzing series resistances in Section 17.1.)

Since traversing a circuit loop in either a clockwise or a counterclockwise sense will give rise to either a voltage rise or a voltage drop across each circuit element, respectively, it is important to establish a sign convention for voltage changes. This book uses the one illustrated in ● Fig. 17.6. The voltage change across a battery is taken to be positive (a voltage rise) if the loop is traversed toward the positive terminal (in the direction of conventional current flow from the battery), as shown in Fig. 17.6a. The voltage change across a battery is taken to be negative if the loop is traversed in the opposite direction, that is, toward the negative terminal. (Note that the assigned branch currents have nothing to do with determining the sign of the voltage change across a battery. The sign of the change depends only on the direction in which the loop is traversed.)

The voltage change across a resistor is taken to be negative (a voltage drop) if the loop is traversed in the direction of the assigned current in that branch (Fig. 17.6b) and positive if the loop is traversed in the opposite direction.

This sign convention allows you to go around a loop either clockwise or counterclockwise. Either way, Eq. 17.7 remains mathematically the same.

The general steps in applying Kirchoff's rules are as follows:

1. Assign a current and current direction for each branch in the circuit. This is done most conveniently at junctions.

2. Indicate the loops and the arbitrarily chosen directions in which they are to be traversed (see ● Fig. 17.7). Every branch *must* be in at least one loop.

3. Apply Kirchhoff's first rule and write the equations for the currents, one for each junction that gives a different equation. (In general, this gives a set of equations that includes all branch currents.)

4. Traverse the number of loops necessary to include all branches. In traversing a loop, apply Kirchhoff's second rule and write the equations using the adopted sign convention.

Steps 3 and 4 give a set of N equations with N unknowns (the currents), which may be solved for the unknowns. If more loops are traversed than necessary, you will have redundant equations. Only the number of loops that includes all the branches is needed.

The junction theorem is generally accepted by most students. However, the concept that the current flowing into a light bulb equals the current flowing out of the bulb is more difficult for some students. Using two identical ammeters, one on each side of a circuit element, helps to illustrate this important point.

It is extremely important for students to get the correct signs in their expressions when applying Kirchhoff's rules.

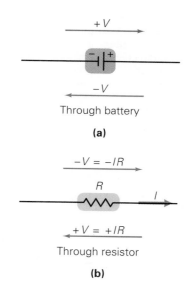

FIGURE 17.6 Sign convention for Kirchhoff's rules **(a)** When Kirchhoff's rules are applied in going around a circuit loop, the voltage change is taken to be positive (+) if a loop is traversed going toward the positive terminal of the battery and negative (−) if the loop is traversed going toward the negative terminal. **(b)** The voltage change across a resistance is taken to be negative (−) if the resistance is traversed in the direction of the assigned branch current and positive (+) if the resistance is traversed in the direction opposite that of the assigned branch current.

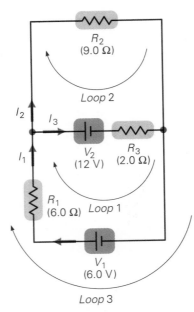

Figure 17.7

FIGURE 17.7 Application of Kirchhoff's rules A current and current direction are assigned for each branch in the circuit (most conveniently done at junctions), and then loops are traversed to write equations. Every branch must be in at least one loop; sign conventions must be used. Example 17.4.

This procedure may seem complicated, but it's really straightforward, as the following example shows.

EXAMPLE 17.4 ■ Kirchhoff's Rules

For the circuit diagrammed in Fig. 17.7, find the branch currents.

Solution. The branch currents and their directions as well as the loops have been arbitrarily assigned in the diagram. Note that there is a current in every branch and that every branch is in at least one loop (some branches are in more than one loop but that is acceptable).

Applying Kirchhoff's first rule at the upper junction gives

$$I_1 = I_2 + I_3 \tag{1}$$

For the lower junction, you could write $I_2 + I_3 = I_1$ (currents in = currents out), but this is the same equation. (If there were another branch, it would have a different current equation for its junctions.)

Going around loop 1 as indicated in the figure and applying Kirchhoff's second rule with the sign convention gives

$$V_1 - I_1 R_1 - V_2 - I_3 R_3 = 0$$

Then, putting in the numerical values from the figures gives

$$6 - I_1(6) - 12 - I_3(2) = 0$$

and

$$6I_1 + 2I_3 = -6 \quad \text{or} \quad 3I_1 + I_3 = -3 \tag{2}$$

(Units are omitted for convenience, and any number of significant figures is assumed.)

Similarly, for loop 2,

$$V_2 - I_2 R_2 + I_3 R_3 = 0$$

and

$$12 - I_2(9) + I_3(2) = 0$$

Thus,

$$9I_2 - 2I_3 = 12 \tag{3}$$

Equations 1, 2, and 3 form a set of three equations with three unknowns. Physically, the problem is solved. The rest is math. You can solve for the I's in several ways. For example, first substitute from Eq. 1 into Eq. 2 to eliminate I_1.

$$3(I_2 + I_3) + I_3 = -3$$

This simplifies to

$$3I_2 + 4I_3 = -3 \tag{4}$$

Then, substituting from Eq. 4 into Eq. 3 eliminates I_2.

$$9(-1 - \tfrac{4}{3}I_3) - 2I_3 = 12$$

That is,

$$-14 I_3 = 21 \quad \text{or} \quad I_3 = -1.5 \text{ A}$$

The minus sign on the results tells you that the wrong direction was assumed for I_3.

Putting the value of I_3 into Eq. 4 gives I_2.

$$3I_2 + 4(-1.5) = -3$$
$$I_2 = 1.0 \text{ A}$$

Then, by Eq. 1,

$$I_1 = I_2 + I_3 = 1.0 \text{ A} - 1.5 \text{ A} = -0.5 \text{ A}$$

The minus sign here indicates that I_1 was also assigned the wrong direction. So, at the upper junction in Fig. 17.7, I_3 really goes into the junction and I_1 and I_2 go out. (You might have suspected this to be the case because of the larger 12-V battery in the I_3 branch.)

Note that loop 3 was not used in this analysis. The equation for this loop would be redundant, giving four equations and three unknowns. However, loop 3 could have been used with either loop 1 or loop 2 in solving the problem. (Why?)

This application of Kirchhoff's rules is called the *branch current method*. A similar loop current method, which some consider to be simpler mathematically, is described in Exercise 48.

17.3 ■ RC Circuits

The previous sections dealt with circuits having constant currents. In some dc circuits, the current may vary with time. This is the case with **RC circuits**, which have a resistor (R) (or other resistive component) and a capacitor (C) in series (● Fig. 17.8). You may be quick to notice that even when the switch is closed, the circuit is still open, or incomplete, because of the plate separation of the capacitor. This is true, but charge does flow for a short time when the switch is closed while the capacitor is charging.

The maximum amount of charge built up (Q_o) on the capacitor depends on the capacitance (C) and the voltage of the battery (V_o). Recall from Section 15.6 that this charge is $Q_o = CV_o$. Immediately after the switch is closed, at $t = 0$, there is an initial current in the circuit, which by Ohm's law is $I_o = V_o/R$ through the resistor. Also recall that as charge accumulates on the capacitor's plates, it takes more work to add charge because of the repulsion between like charges. Eventually, the capacitor is charged to the maximum, and the current diminishes to zero.

The resistance helps determine how fast the capacitor is charged, since the larger its value, the greater the resistance to charge flow. The voltage across the capacitor is found to vary exponentially with time according to this equation.

$$V = V_o(1 - e^{-t/RC}) \tag{17.8}$$

charging voltage for a capacitor in an RC circuit

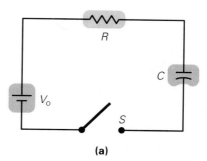

(a)

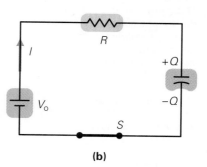

(b)

● **FIGURE 17.8 A series RC circuit** (a) Even though the circuit is open because of the space between the plates of the capacitor, **(b)** there is a current in the circuit when the switch is closed until the capacitor is charged to its maximum value. The rate of charging (and discharging) depends on the product of the values of the resistance and capacitance, which is called the time constant (τ) of the circuit: $\tau = RC$.

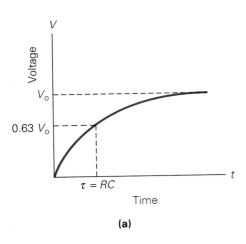

(a)

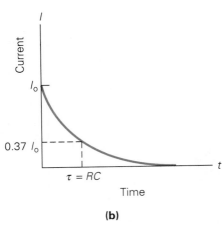

(b)

FIGURE 17.9 Capacitor charging **(a)** In a series RC circuit, the voltage across the capacitor increases exponentially with time, reaching 63% of its maximum voltage (V_o) in one time constant, or $t = \tau = RC$. **(b)** The current in the circuit is initially a maximum ($I_o = V_o/R$) and decays exponentially with time, falling to 37% of its initial value in one time constant.

where $e = 2.718$ is the base of the natural logarithm. A graph of V versus t and a graph of I versus t are presented in Fig. 17.9. The current varies with time according to this equation.

$$I = I_o e^{-t/RC} \qquad (17.9)$$

The current decreases exponentially with time. Note that the current is greatest initially. Why?

According to Eq. 17.8, it would theoretically take an infinite amount of time for the capacitor in an RC circuit to become fully charged (to reach the battery voltage, V_o). However, in practice, such a capacitor charges and discharges in relatively short times. It is customary to use a special value to express the charging and discharging rates. This **time constant** (τ) for an RC circuit is

$$\tau = RC \qquad (17.10)$$

At a time equal to the time constant, $t = \tau = RC$, the voltage across the charging capacitor is

$$V = V_o(1 - e^{-t/\tau}) = V_o(1 - e^{-t/\tau}) = V_o(1 - e^{-1})$$

That is,

$$V = 0.63 V_o$$

After one time constant, the voltage across the capacitor is at 63% of its maximum (or the capacitor is 63% charged, since $Q = CV$). Note that at that time the current has decayed to 37% of its initial maximum value (I_o).

At the end of two time constants, $t = 2\tau = 2RC$, the capacitor is charged to more than 86% of its maximum value, and so on. Practically, the capacitor is considered to be fully charged after only a few time constants.

When a fully charged capacitor is discharged through a resistance, the voltage across the capacitor decays exponentially with time (as does the current).

$$V = V_o e^{-t/RC} \qquad (17.11)$$

discharging voltage for a capacitor in an RC circuit

The voltage is said to "decay" exponentially. In one time constant, the voltage across the capacitor falls to 37% of its original value (see Fig. 17.10).

EXAMPLE 17.5 ■ Time Constant

The capacitance and resistance in the RC circuit in Fig. 17.8 are 6.0 μF and 0.25 MΩ, respectively, and the battery is a 12-V one. (a) What is the voltage across the capacitor at one time constant after the switch is closed if it was initially uncharged? (b) What is the voltage across the capacitor and its charge at $t = 5.0$ s?

Solution.

Given: $C = 6.0\ \mu\text{F} = 6.0 \times 10^{-6}\ \text{F}$
$R = 0.25\ \text{M}\Omega = 2.5 \times 10^5\ \Omega$
$V = 12\ \text{V}$
(a) $t = \tau$
(b) $t = 5.0\ \text{s}$

Find: (a) V (voltage at τ)
(b) V and Q
(voltage and
charge)

(a) In one time constant, the capacitor is charged to 63% of its maximum charge, and

$$V = 0.63\ V_o = (0.63)(12\ \text{V}) = 7.6\ \text{V}$$

(b) When a specific time is given, you also need to know the time constant for the circuit, which in this case is

$$\tau = RC = (2.5 \times 10^5\ \Omega)(6.0 \times 10^{-6}\ \text{F}) = 1.5\ \text{s}$$

Then, for $t = 5.0\ \text{s}$, using Eq. 17.8,

$$V = V_o(1 - e^{-t/\tau}) = (12\ \text{V})\,(1 - e^{-5.0\,\text{s}/1.5\,\text{s}})$$
$$= (12\ \text{V})(1 - e^{-3.3}) = (12\ \text{V})(1 - 0.037)$$
$$= (12\ \text{V})(0.963) = 11.6\ \text{V}$$

Note that after $t = 5.0 = 3.3\tau$, the voltage across the capacitor is more than 96% of its maximum value.

Expressing τ as RC and multiplying both sides of Eq. 17.8 by C gives

$$CV = CV_o(1 - e^{-t/RC})$$

Since $Q = CV$,

$$Q = Q_o(1 - e^{-t/RC})$$

Thus, the charge varies with time just as the voltage does.
The maximum charge (Q_o) for the capacitor is

$$Q_o = CV_o = (6.0 \times 10^{-6}\ \text{F})(12\ \text{V}) = 7.2 \times 10^{-5}\ \text{C}$$

At $t = 5.0\ \text{s} = 3.3\tau$, the charge on the capacitor is 96.3% of the maximum charge (the same percentage as for the voltage):

$$Q_o = (0.963)Q_o = (0.963)(7.2 \times 10^{-5}\ \text{C}) = 6.9 \times 10^{-5}\ \text{C}$$

An application of an RC circuit is diagrammed in ● Fig. 17.11a. This is called a blinker circuit (or, more impressively, a neon-tube relaxation oscillator). The resistor and capacitor are in series, and a small neon tube is connected across (in parallel with) the capacitor. (These neon tubes are about the size of miniature Christmas tree lights.)

When the circuit is closed, the voltage across the capacitor (and the neon tube) rises from 0 to V_b, which is the breakdown voltage of the neon gas in the tube (about 80 V). At that voltage, the gas is ionized and begins to conduct electricity, and the tube lights up. When the tube is in a conducting state, the capacitor discharges through it, and the voltage falls rapidly (Figure 17.11b). When the voltage drops to V_m, called the maintaining voltage, the tube discharge can no longer be sustained, and the tube stops conducting. The capacitor then begins charging again, the voltage rises from V_m to V_b, and the cycle repeats. The continual repetition of this cycle causes the tube to blink on and off. The period of the oscillator, or the time between flashes, depends on the RC time constant.

Other applications of RC circuits are discussed in the Insight feature.

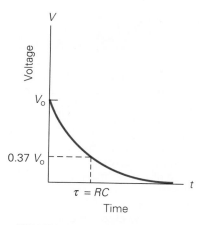

● **FIGURE 17.10 Capacitor discharging** When a charged capacitor is discharged through a resistance, the voltage across the capacitor (and the current in the circuit) decays exponentially with time, falling to 37% of its initial value in one time constant, $t = \tau = RC$.

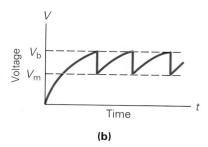

(a)

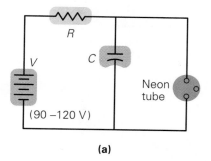

(b)

● **FIGURE 17.11 Neon-tube relaxation oscillator circuit**
(a) When a neon tube is connected across the capacitor in a series RC circuit having the proper voltage source, the voltage across the tube (and capacitor) will relax, or oscillate, with time. As result, the tube periodically flashes or blinks. (b) A graph of voltage versus time shows the oscillating effect.

In order to install shelving or cabinets, or even to hang a picture securely, it is usually necessary to locate the studs in the walls of your house. Studs are the vertical wooden elements to which plasterboard or paneling is nailed, and they can be very elusive. The traditional way of finding them, by tapping on the wall, generally leads to more frustration than success. Commercially available stud finders employ pivoted magnets, which move when they pass over a nail head in the stud behind the wall, but these too are far from infallible.

A new electronic device called Studsensor™, much easier to use, relies on capacitive sensing. The capacitor is in a printed circuit and forms part of a simple RC charging network. However, for this application the metallic plates of the capacitor are arranged end to end rather than facing one another as in a parallel-plate capacitor (Fig.1). In this configuration the fringe electric field extends a considerable distance beyond the plane of the plates, and any nearby material acts as a dielectric in this field.

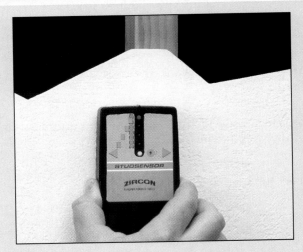

FIGURE 2 Studsensor™ In this cut-away view of the Studsensor in use, the lower green light indicates that the unit is calibrated. Vertical red lights signal the approach to a stud and the top red light indicates that the sensor is over a stud. Some models have both light and sound indicators.

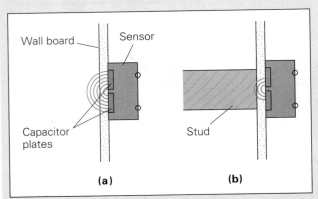

FIGURE 1 RC sensing (a) The fringe electric field of the capacitor plates permeates the wall board, which acts as a dielectric. (b) When over a stud, the capacitance changes. The sensor detects this change as a change in an RC circuit.

The unit is initially calibrated by placing it against the wall and balancing the charging time of the RC circuit against a fixed reference. When the sensor containing the capacitor plates is moved over a stud, the character of the dielectric changes, and the capacitance of the plates increases. The greater the capacitance, the longer the charging time of the circuit relative to the calibration standard. The changes in charging time are sensed electronically and conveyed to the user by lighted indicators (Fig. 2).

You may wonder what would happen if you were unknowingly holding the unit over a stud during the initial calibration. Newer models of the Studsensor™ signal this by being able to detect decreases in the charging time, as well as increases as in normal stud sensing.

17.4 ■ Ammeters and Voltmeters

As the names imply, an **ammeter** measures current (amps) and a **voltmeter** measures voltage (volts). A basic component of both of these meters is a **galvanometer** (● Fig. 17.12). The galvanometer operates on magnetic principles that will be covered in Chapter 19. In this chapter, it will simply be considered to be a circuit element having an internal resistance (r).

The deflection of the needle on the galvanometer's dial is directly proportional to the current through its wire coil. The resistance of the coil (r) is relatively small. In effect, a galvanometer measures current, but because of the

(a)

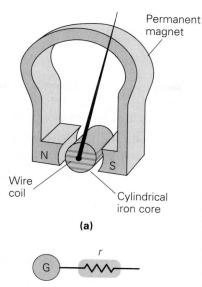

Permanent magnet

Wire coil

Cylindrical iron core

(a)

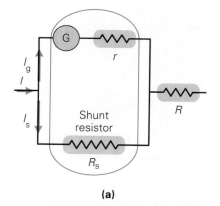

r

G ——/\/\/\——

(b)

• **FIGURE 17.12 The galvanometer** (a) A galvanometer is a current-sensitive device whose needle deflection is proportional to the current through its coil. (b) The circuit symbol is a circle containing a G. The internal resistance (r) of the meter is indicated explicitly.

small coil resistance, only currents in the microamp range can be measured without burning out the coil. An ammeter that can be used to measure larger currents has a small shunt resistor in parallel with a galvanometer (• Fig. 17.13). This provides an alternate path by which part of a large current (I) can bypass the galvanometer ($I = I_g + I_s$, at the left junction in the figure).

The voltages across the galvanometer and the shunt resistor in parallel are equal,

$$V_g = V_s$$

By Ohm's law, then,

$$I_g r = I_s R_s$$

Using $I = I_g + I_s$,

$$I_g r = (I - I_g)R_s$$

and

$$\boxed{I_g = \frac{IR_s}{r + R_s}}$$ (17.12)

galvanometer current for dc ammeter.

This equation allows you to select the proper shunt resistance for a given current range and galvanometer. The following example illustrates how to do this.

EXAMPLE 17.6 ■ A dc Ammeter

A galvanometer that can safely carry a maximum coil current of 200 μA (called the full-scale sensitivity) has a coil resistance of 50 Ω. It is to be used in an ammeter designed to read currents up to 3.0 A (at full scale). What is the required shunt resistance? (Ignore significant figures.)

G ——/\/\/\——
r

I_g

I

I_s

Shunt resistor

/\/\/\——
R_s

——/\/\/\——
R

(a)

———→——(A)——/\/\/\——
R

(b)

• **FIGURE 17.13 dc ammeter** (a) A galvanometer in parallel with a shunt resistor (R_s) is an ammeter capable of measuring various current ranges, depending on the value of R_s. (b) The circuit symbol for an ammeter is a circle with an A inside it.

Solution.

Given: $I_g = 200\ \mu A = 2.0 \times 10^{-4}\ A$ *Find:* R_s (shunt resistance)
$r = 50\ \Omega$
$I_{max} = 3.0\ A$

A full-scale reading of 3.0 A means that when a current of 3.0 A enters the ammeter, I_g should be 200 μA. Solving Eq. 17.12 for the shunt resistance and plugging in the values given (assumed to be exact) gives

$$R_s = \frac{I_g r}{I_{max} - I_g}$$

$$= \frac{(2.0 \times 10^{-4}\ A)(50\ \Omega)}{3.000\ A - 0.0002\ A} \approx 0.0033\ \Omega$$

Note the small size of the shunt resistance as compared to r (50 Ω). This allows most of the current (2.9998 A at full scale) to pass through the shunt resistor branch. This resistor is made of a material that does not burn out as readily as the thin wire of the galvanometer coil. The ammeter will read currents linearly up to 3.0 A. For example, if a current of 1.5 A flowed into the ammeter, there would be a current of 100 μA in the coil of the galvanometer, which would give a half-scale reading.

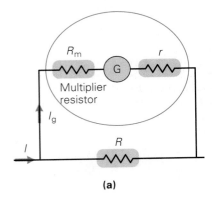

(a)

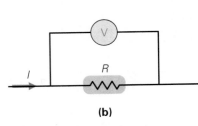

(b)

◦ FIGURE 17.14 dc voltmeter
(a) A galvanometer in series with a multiplier resistor (R_m) is a voltmeter capable of measuring various voltage ranges, depending on the value of R_m. **(b)** The circuit symbol for a voltmeter is a circle with a V inside it.

QUESTION: When an ammeter is used to measure the current through a circuit element, it is connected in series so that the current through the element flows through the ammeter. Since the ammeter has resistance, wouldn't the instrument itself affect the current?

ANSWER: Yes, but the resistance of the ammeter is the equivalent parallel resistance for r and R_s, which is very small. In fact, it may usually be considered negligible compared to the circuit or component resistance. Recall that the equivalent resistance for a parallel combination is less than any of the individual resistances. (What would it be for the resistances in Example 17.6?)

A voltmeter that is capable of reading voltages higher than the microvolt range is constructed by connecting a large multiplier resistor in series with a galvanometer (◦ Fig. 17.14). Because the voltmeter has a large internal resistance due to the multiplier resistor, it draws little current from the main circuit. When connected across a circuit element, the voltmeter (or galvanometer branch) experiences a voltage drop of $V = V_g + V_m$, and most of this drop is across the multiplier resistor rather than the galvanometer coil. By Ohm's law,

$$V = V_g + V_m$$
$$= I_g r + I_g R_m = I_g(r + R_m)$$

and

$$\boxed{I_g = \frac{V}{r + R_m}} \qquad (17.13)$$

galvanometer current for dc voltmeter

This voltage is also the potential difference across the circuit element having a resistance R because of the parallel connection (see Fig. 17.14). If the galvanometer scale is calibrated in volts (instead of amps), the voltage drop across that circuit element can be read from the meter.

EXAMPLE 17.7 ■ A dc Voltmeter

Suppose that the galvanometer described in Example 17.6 is to be used instead in a voltmeter with a full-scale reading of 3.0 V. What is the required value of the multiplier resistor? (Ignore significant figures.)

Solution.

Given: $I_g = 2.0 \times 10^{-4}\,A$
    (from Example 17.6)
$r = 50\,\Omega$ (from Example 17.6)
$V_{max} = 3.0$ V

Find: R_m (multiplier resistance)

Eq. 17.13 can be used to find R_m.

$$R_m = \frac{V_{max} - I_g r}{I_g}$$

$$= \frac{3.0\,V - (2.0 \times 10^{-4}\,A)(50\,\Omega)}{2.0 \times 10^{-4}\,A}$$

$$= 1.495 \times 10^4\,\Omega = 14.95\,k\Omega$$

For versatility, ammeters and voltmeters may be multiranged. This is accomplished by having several shunt or multiplier resistors (●Fig. 17.15).

Also, they may be combined into *multimeters*. Electronic digital multimeters are now common (Fig. 17.15c). In place of mechanical galvanometers, these use electronic circuits which analyze the voltage drops across internal resistors.

●**FIGURE 17.15 Multirange meters** (a) An ammeter or (b) a voltmeter can be used to measure several ranges of current and voltage by switching among different shunt and multiplier resistors, respectively. (Instead of a switch, there may be an exterior terminal for each range.) (c) Both functions may be combined into a single multimeter. Multimeters are available with electronic digital readouts.

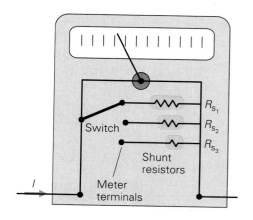

(a) Multirange ammeter

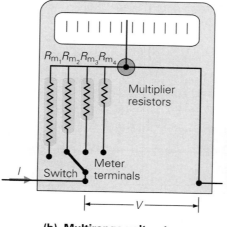

(b) Multirange voltmeter

(c)

17.5 ■ Household Circuits and Electrical Safety

Although household circuits generally use alternating current and that has not yet been discussed, they include practical applications of some of the principles already studied.

For example, would you expect the elements in a household circuit (lamps, appliances, and so on) to be connected in series or in parallel? From the discussion of Christmas tree lights in Section 17.1, it should be apparent that household elements must be connected in parallel. For example, when the bulb in a lamp blows out, other elements in the circuit continue to work. This would not be the case for a series circuit. Moreover, household appliances and lamps are generally rated for 120 V. If these elements were connected in series, the voltage drops across them would add up to a *total* of 120 V. None of the circuit elements would have adequate voltage, and the more that were connected in series, the less voltage each would have.

Power is supplied to a house by a three-wire system (● Fig. 17.16). There is a potential difference of 240 V between the two hot, or high-potential, wires, and each of these has a 120-V potential difference with the ground. The third wire is grounded at the point where the wires enter the house, usually by a metal rod driven into the ground. This wire has a zero potential and is the ground, or neutral, wire.

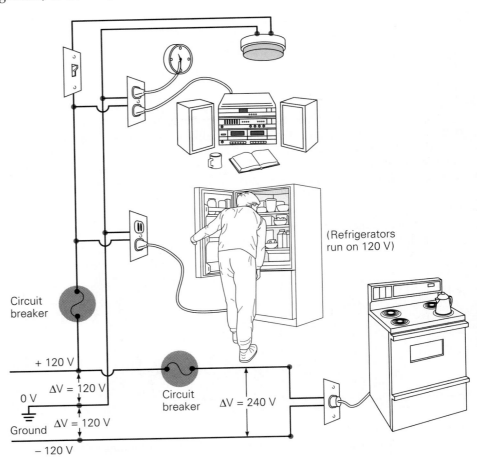

(Refrigerators run on 120 V)

Circuit breaker

Circuit breaker

+ 120 V

$\Delta V = 120\ V$

0 V

Ground

$\Delta V = 120\ V$

$\Delta V = 240\ V$

− 120 V

● **FIGURE 17.16 Household wiring** A 120-V circuit is obtained by connecting between the + 120-V (or − 120-V) line and the ground line. A potential difference of 240 V for large appliances such as electric stoves, central air conditioners, and hot water heaters is obtained by connecting between the + 120-V line and the − 120-V line.

The potential difference of 120 V needed for most household appliances is obtained by connecting them between the grounded wire and either of the high-potential wires: $\Delta V = 120 \text{ V} - 0 \text{ V} = 120 \text{ V}$ or $\Delta V = 0 \text{ V} - (-120 \text{ V}) = 120 \text{ V}$. (See Fig. 17.16.) Even though the grounded wire has a zero potential, it is a *current-carrying* wire because it is part of a circuit. Large appliances such as central air conditioners, ovens, and hot water heaters need 240 V for operation, and this is obtained by connecting them between the two hot wires: $\Delta V = 120 \text{ V} - (-120 \text{ V}) = 240 \text{ V}$.

As you learned in Chapter 16, the amount of current an appliance draws (uses) may be given on a rating tag but can also be determined from the power rating (using $P = IV$). For example, a stereo rated at 180 W at 120 V would draw 1.5 A ($I = P/V$). There are limitations to the number of elements that can be put in a circuit and the total amount of current drawn. In particular, the joule heat (or $I^2 R$ loss) must be considered. Generally, the more elements (resistances) connected in parallel, the smaller the equivalent resistance. In any case, the equivalent resistance is smaller than the resistance of the smallest element. As we saw in Example 16.7 in the previous chapter, for a constant voltage the joule heat produced is inversely proportional to the resistance of the circuit: $P = V^2/R$. Thus by adding too many current-carrying elements it is possible to overload a household circuit so that it draws too much current and produces too much heat. The joule heat in the circuit wires could melt the insulation and start a fire.

Overloading is prevented by limiting the current in a circuit by means of two types of devices: fuses and circuit breakers. **Fuses** are common in older homes. An Edison-base fuse has threads like those on the base of a light bulb (see ● Fig. 17.17a). Inside the fuse is a metal strip that melts because of joule heat when the current is larger than the rated value (which might be 15 A). The melting of the strip opens the circuit, and electrically everything comes to a halt.

One problem with Edison-base fuses is that they are interchangeable. For example, a 30-A fuse can be put in a circuit that is rated for 15 A. This would allow a current flow twice as great as that for which the circuit was designed. Such a problem is avoided with another type of fuse, called a Type-S fuse (Fig. 17.17b). A nonremovable adapter is installed in the socket for a fuse; the adapter is specific for a particular rated value (the fuses and adapters for different ratings have different threads). That is, a 30-A Type-S fuse will not screw into a 15-A Type-S adapter, for example, so the wrong fuse can't be used.

Circuit breakers are now used exclusively in wiring new homes. One type of circuit breaker, shown in ● Fig. 17.18, uses a bimetallic strip (see Chapter 10). As the current through the strip increases, it becomes warmer and bends. At the rated current value, the strip will be bent sufficiently to cause the circuit to open mechanically. The strip quickly cools, so the breaker may be reset. However, when a fuse blows or a circuit breaker trips, this is an indication that the circuit is drawing or attempting to draw too much current. *You should always investigate to find and correct the problem before replacing the fuse or resetting the circuit breaker.*

Switches, fuses, and circuit breakers are placed in the hot (high-potential) side of the line. They would work in the grounded side, but even when the switch was open, the fuse blown, or the breaker tripped, the circuit and any elements in it would still be connected to a high potential, which could be dangerous if a person made electrical contact. And even with fuses or circuit

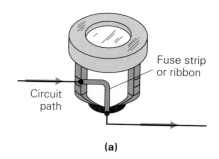

(a)

(b)

(c)

● **FIGURE 17.17 Fuses** (a) A fuse contains a metallic strip, or ribbon, that melts when the current exceeds a rated value. This opens the circuit and prevents overheating. (b) Edison-base fuses (left) have threads similar to those on light bulbs; fuses with different ampere ratings can be interchanged. Type-S fuses (right) have different threads for different ratings that fit specific adapters in the fuse sockets and so cannot be interchanged. (c) A type of small fuse commonly found in electrical equipment. Note that the fuse on the right is "blown."

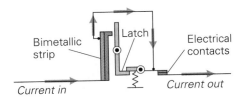

Bimetallic strip   Latch   Electrical contacts

Current in   Current out

Thermal trip   Circuit broken

After current overload

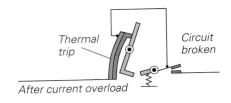

(a)

• **FIGURE 17.18 Circuit breaker** (a) A diagram of a thermal trip element. With increased current and joule heating, the element bends until it opens the circuit at some preset current value. Magnetic trip elements are also used. **(b)** A bank of 20-A household circuit breakers with one switch open. **(c)** These heavy-duty circuit breakers at an electrical substation must be capable of cutting off current almost instantaneously in power lines operating at very high voltages.

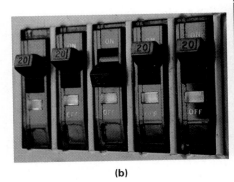

(b)

(c)

breakers in the hot side of the line, there is still a possibility of getting an electrical shock from a defective appliance that has a metal casing, such as a hand drill. A wire may come loose inside and make contact with the casing, which would then be hot, or at a high potential. As illustrated in ● Fig. 17.19, a person could provide a path to the ground and become part of the circuit, thus getting a shock.

To prevent this from happening, a third dedicated grounding wire is added to the circuit that grounds the metal casing (● Fig. 17.20). Note that this wire is not normally a current-carrying wire. If a hot wire comes into contact with the casing, the circuit is completed to this grounded wire, and the fuse is blown or the circuit breaker tripped.

On three-prong **grounded plugs**, the large round prong connects with the dedicated grounding wire. Adapters can be used between a three-prong plug and a two-prong socket. Such an adapter has a grounding lug or grounding wire (● Fig 17.21a). This should be fastened to the receptacle box by the plate-fastening screw or some other means. The receptacle box is grounded by means of the grounding wire. If the adapter lug or wire is not connected, the system is left unprotected, which defeats the purpose of the dedicated grounding safety feature.

Three-prong plugs—dedicated grounding wire

• **FIGURE 17.19 Electrical safety** (a) Switches and fuses or circuit breakers should always be wired in the hot side of the line, *not* in the grounded side as shown here. If these elements are wired in the grounded side, the line is at a high potential even when the fuse is blown or a switch is open. **(b)** Even if the fuse or circuit breaker is in the hot side of the line, a potentially dangerous situation exists. If an internal wire comes into contact with the metal casing of an appliance or power tool, a person touching the casing, which is at high voltage, can get a shock.

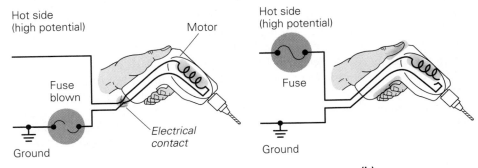

Hot side (high potential)   Motor

Fuse blown

Ground

Electrical contact

(a)

Hot side (high potential)

Fuse

Ground

(b)

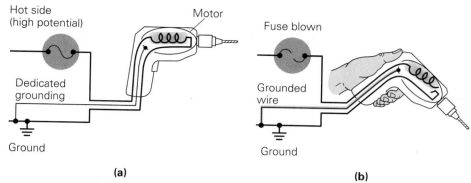

Hot side (high potential)

Motor

Dedicated grounding

Ground

(a)

Fuse blown

Grounded wire

Ground

(b)

● **FIGURE 17.20 Dedicated grounding** **(a)** For electrical safety, a third wire is connected from an appliance or power tool to ground. This dedicated grounding wire normally carries no current (as opposed to the grounded wire of the circuit). **(b)** If a loose internal wire comes into contact with the grounded casing, the shorting to ground through the dedicated grounding wire blows the protective fuse or circuit breaker, and the appliance is not at a high potential as in Fig. 17.19b.

You have probably noticed that there is another type of plug, a two-prong plug that only fits in the socket one way because one prong is larger than the other and one of the slits of the receptacle is also larger (Fig. 17.21b). This is called a polarized plug. *Polarizing* in the electrical sense is a method of identifying the hot and grounded sides of the line (or the voltages) so that particular connections may be made.

Polarized plugs and sockets are an older safety feature. Wall receptacles are wired so that the small slit connects to the hot side and the large slit connects to the neutral, or ground, side. Having the hot side identified in this way makes two safeguards possible. First, the manufacturer of an electrical appliance can design it so that the switch is always in the hot side of the line. Thus, all of the wiring of the appliance beyond the switch is safely neutral when the switch is open and the appliance off. Moreover, the casing of an appliance is connected by the manufacturer to the ground side by means of a polarized plug. Should a hot wire inside the appliance come loose and make contact with the metal casing, the effect would be similar to that with a dedicated grounding system. The hot side of the line would be shorted to the ground, which would blow a fuse or trip a circuit breaker.

However, the three-wire dedicated grounding system is preferred over polarized plugs and sockets. There could be a wiring error in an appliance or receptacle, which might cause a problem even with a polarized plug. (What would the problem be?) Another type of electrical safety device, the ground fault interrupter, is discussed in Chapter 19.

(a)

(b)

● **FIGURE 17.21 Plugging into ground** **(a)** A three-prong plug and adapter for a two-prong socket. Note the grounding lug on the adapter. This should be connected to the plate-fastening screw on the grounded receptacle box—otherwise, the safety feature is lost. **(b)** A polarized plug on an adapter. The differently-sized prongs permit prewired identification of the high and ground sides of the line.

INSIGHT ■ Electricity and Personal Safety

Safety precautions are necessary to prevent injuries when people work with and use electricity. Electricity conductors (such as wires) are coated with insulating materials so that they can be handled safely. However, when a person comes into contact with a charged conductor, a potential difference may exist across part of the body. A bird can sit on a high-voltage line without any problem because the bird and the line are at the same potential,

and there is no potential difference or circuit path. But if a person carrying an aluminum (conducting) ladder touches it to an electrical line, a potential difference exists from the line to the ground, and the ladder and the person are part of the circuit.

The extent of personal injury in such a case depends on the amount of electric current that flows through the body and on the circuit path. The current is given by

$$I = \frac{V}{R_{body}}$$

where R_{body} is the resistance of the body. Thus, for a given voltage, the current depends on the body resistance.

Body resistance varies. If the skin is dry, the resistance may be 0.50 MΩ (0.50×10^6 Ω) or more. For a potential difference of 120 V, there would be a current of

$$I = \frac{V}{R_{body}} = \frac{120 \text{ V}}{0.50 \times 10^6 \text{ Ω}} = 0.24 \times 10^{-3} \text{ A} = 0.24 \text{ mA}$$

This current is almost too weak to be felt.

Suppose, however, that the skin is wet with perspiration. Then R_{body} is about 5.0 kΩ (5.0×10^3 Ω). In this case, the current is

$$I = \frac{V}{R_{body}} = \frac{120 \text{ V}}{5.0 \times 10^3 \text{ Ω}} = 24 \times 10^{-3} \text{ A} = 24 \text{ mA}$$

which could be very dangerous.

A basic precaution to take is to avoid coming into contact with an electrical conductor that might cause a potential difference to exist across the body or part of it.

The effect of such contact depends on the path of the current. If that path is from the little finger to the thumb on one hand, probably only a burn would result from a large current. However, if the path is from hand to hand through the chest, the effect can be much worse. Some of the possible effects of this circuit path are given in Table 1.

Injury results because the current interferes with muscle functions and/or causes burns. Muscle functions are regulated by electrical impulses through the nerves, and these can be influenced by currents from external voltages. Muscle reaction and pain can occur from a current of a few milliamps. At about 10 mA, muscle paralysis can prevent a person from releasing the conductor. At about 20 mA, contraction of the chest muscles occurs, which may cause breathing to be impaired or to stop. Death can occur in a few minutes. At 100 mA, there are rapid uncoordinated movements of the heart muscles (called ventricular fibrillation), which prevent the proper pumping action.

Working safely with electricity requires a knowledge of fundamental electrical principles *and* common sense. Electricity must be treated with respect.

TABLE 1 ■ **Effects of Electric Current on the Human Body***

Current (approximate)	Effect
1.0 mA (0.001 A)	Mild shock or heating
10 mA (0.01 A)	Paralysis of motor muscles
20 mA (0.02 A)	Paralysis of chest muscles, causing respiratory arrest; fatal in a few minutes
100 mA (0.1 A)	Ventricular fibrillation, preventing coordination of the heart's beating; fatal in a few seconds
1000 mA (1 A)	Serious burns; fatal almost instantly

*The effect on the human body of a given amount of current depends on a variety of conditions. This table gives only general and relative descriptions.

Important Concepts

You should be able to define and explain these chapter concepts clearly.

resistances in series	branch	RC circuit	fuse
resistances in parallel	Kirchhoff's first rule	time constant	circuit breaker
resistances in series-parallel combinations	(junction theorem)	ammeter (dc)	grounded plug
junction	Kirchhoff's second rule (loop theorem)	voltmeter (dc)	polarized plug
		galvanometer	

Important Relationships for Review

These relationships are mathematical statements of the concepts and principles presented in the chapter. You should be able to identify the symbols and to explain the relationships before proceeding to the Exercises. In-text equation reference numbers are given for convenience.

Equivalent Series Resistance:

$$R_s = R_1 + R_2 + R_3 + \ldots \qquad (17.2)$$

Equivalent Parallel Resistance:

$$\frac{1}{R_p} = \frac{1}{R_1} + \frac{1}{R_2} + \frac{1}{R_3} + \ldots \qquad (17.4)$$

Equivalent Parallel Resistance For Two Resistors:

$$R_p = \frac{R_1 R_2}{R_1 + R_2} \qquad (17.5)$$

Kirchhoff's Rules:

(1) $\sum_i I_i = 0$ (junction theorem) $\qquad$ (17.6)

(2) $\sum_i V_i = 0$ (loop theorem) $\qquad$ (17.7)

Time Constant:

$$\tau = RC \qquad (17.10)$$

Charging Voltage for an RC Circuit:

$$V = V_o(1 - e^{-t/RC}) = V_o(1 - e^{-t/\tau}) \qquad (17.8)$$

Discharging Voltage for an RC Circuit:

$$V = V_o e^{-t/RC} = V_o e^{-t/\tau} \qquad (17.11)$$

Current (Charging and Discharging) in an RC Circuit:

$$I = I_o e^{-t/RC} = I_o e^{-t/\tau} \qquad (17.9)$$

Ammeter Current:

$$I_g = \frac{IR_s}{r + R_s} \qquad (17.12)$$

Voltmeter Current:

$$I_g = \frac{V}{r + R_m} \qquad (17.13)$$

Exercises

17.1 ■ Resistances in Series, Parallel, and Series-Parallel Combinations

1 The same amount of current goes through each resistor in a circuit when they are connected in (a) series, (b) parallel, (c) series-parallel, (d) parallel-series. (a)

2 The potential difference across each resistor in a circuit is the same when they are connected in (a) series, (b) parallel, (c) series-parallel, (d) parallel-series. (b)

3 Are the voltage drops across resistors in series generally the same? If not, could they be so in certain cases?
no; only if all resistors are equal

● **4** Are the currents in resistors in parallel generally the same? If not, could they be so in certain cases?
no; only if all resistors are equal

5 ■ Three resistors that have values of 20 Ω, 30 Ω, and 40 Ω are to be connected. (a) What is the maximum equivalent resistance? (b) What is the minimum equivalent resistance? (a) 90 Ω (b) 9.2 Ω

6 ■ A combination of two 10-Ω resistors in parallel is connected in series to a 40-Ω resistor. What is the total equivalent resistance? 45 Ω

7 ■■ Three 4.0-Ω resistors can be connected in different ways to produce different equivalent resistances. Draw a circuit diagram showing each way, and find the total resistance in each case. parallel: 1.3 Ω; series: 12 Ω; series-parallel: 2.7 Ω; parallel-series: 6.0 Ω

8 ■■ Three resistors with values of 5.0 Ω, 10 Ω, and 15 Ω are connected in series in a circuit with a 9.0-V battery. (a) What is the total equivalent resistance? (b) What is the current

through each resistor? (c) What is the rate at which energy is dissipated in the 15-Ω resistor? (a) 30 Ω (b) 0.30 A
(c) 1.4 W

9 ■■ Find the equivalent resistances for all possible combinations of two or more of the three resistors in Exercise 8.
series: 15 Ω, 20 Ω, 25 Ω, 30 Ω; parallel: 3.3 Ω, 3.8 Ω, 6.0 Ω, 2.7 Ω; series-parallel: 11 Ω, 13.8 Ω, 18.3 Ω, 4.2 Ω, 6.7 Ω, 7.5 Ω

10 ■■ How many different integral (whole-number) values of resistance can be obtained by using one or more of three resistors with values R, 2R, and 3R? 6

● **11** ■■ A 4.0-Ω resistor and a 6.0-Ω resistor are connected in series. A third resistor is connected in parallel with the 6.0-Ω resistor. This gives a total equivalent resistance of 7.0 Ω for the whole arrangement. What is the value of the third resistor? 6.0 Ω

12 ■■ A length of wire with a resistance of $18\mu\Omega$ is cut into three equal segments. The segments are then wrapped together to form a conductor a third as long as the original wire. What is the resistance of the shortened conductor?
2.0 μΩ

13 ■■ You are given four resistors with values of 2.5 Ω, 3.5 Ω, 4.5 Ω, and 5.5 Ω. Is it possible to connect all of the resistors in a combination so as to produce an effective total resistance of 5.0 Ω? If so, describe the arrangement. no

● **14** ■■ Three resistors with values of 2.0 Ω, 4.0 Ω, and 6.0 Ω are connected in series and put in a circuit with a 12-V battery. (a) How much current is delivered to the circuit by the battery? (b) What is the current through each resistor? (c) How much power is dissipated by each resistor? (d) How does this compare with the power dissipated by the total equivalent resistance? (a) 1.0 A (b) 1.0 A (c) $P_{2\Omega} = 2.0$ W;
$P_{4\Omega} = 4.0$ W; $P_{6\Omega} = 6.0$ W (d) $P_{sum} = P_{total} = 12$ W

15 ■■ Suppose that the resistors in Exercise 14 are connected in parallel. (a) How much current is drawn from the battery? (b) How much current flows through each resistor? (c) How much power is dissipated by each resistor? (d) How does this compare with the power dissipated by the total equivalent resistance? (a) 11 A (b) 2.0 Ω: 6.0 A; 4.0 Ω: 3.0 A; 6.0 Ω: 2.0 A (c) 2.0 Ω: 72 W; 4.0 Ω: 36 W; 6.0 Ω: 24 W (d) $P_{sum} = P_{total} = 132$ W

16 ■■ Two 8.0-Ω resistors are connected in parallel, and two 4.0-Ω resistors are connected in parallel. These combinations are then connected in series and put in a circuit with a 12-V battery. What are the current through and the voltage across each resistor? *I* (for all) = 1.0 A $V_{8\Omega} = 8.0$ V; $V_{4.0\Omega} = 4.0$ V

17 ■■ What is the equivalent resistance for the resistors in ● Fig. 17.22? 0.80 Ω

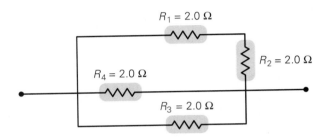

FIGURE 17.22 Series-parallel combination See Exercises 17 and 28.

18 ■■ What is the equivalent resistance between points A and B in ● Fig. 17.23? 2.7 Ω

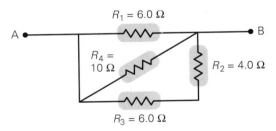

FIGURE 17.23 Series-parallel combination See Exercises 18 and 30.

19 ■■ Find the equivalent resistance for the arrangement of resistors shown in ● Fig. 17.24. 6.9 Ω

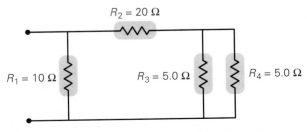

FIGURE 17.24 Series-parallel combination See Exercise 19.

20 ■■ Two 60-W and one 100-W light bulbs are connected across (in parallel with) a 120-V source. (a) What is the current through each bulb? (b) How much power is dissipated by each bulb? (a) 60 W: 0.50 A; 100 W: 0.83 A (b) 60 W: 60 W; 100 W: 100 W

21 ■■ How many 60-W light bulbs can be connected in parallel with a 120-V source without blowing a 15-A fuse in the circuit? (The fuse is in series with the parallel bulb arrangement. Why?) 30 bulbs

22 ■■ Use three 50-Ω resistors and a 120-V source in a circuit in any arrangement. (a) What is the maximum power for the circuit? (b) What is the minimum power? (a) 862 W (b) 96 Ω

23 ■■ Find the current and the voltage drop for the 2.0-Ω resistor in ● Fig. 17.25. 2.6 V; 1.25 A

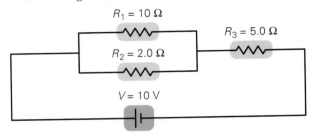

FIGURE 17.25 Current and voltage drop for a resistor See Exercise 23.

24 ■■ A three-way light bulb can produce 50 W, 100 W, or 150 W of power at 120 V. (a) Find the current for each power setting. (b) Find the resistance for each power setting. (a) 50 W: 0.42 A; 100 W: 0.83 A; 150 W: 1.25 A (b) 50 W: 288 Ω; 100 W: 144 Ω; 150 W: 96 Ω

25 ■■ If a combination of three 30-Ω resistors dissipates 3.2 W of energy when connected to a 12-V battery, how are the resistors connected in the circuit? two in parallel and in series with other resistor

26 ■■ For the circuit shown in ● Fig. 17.26, find (a) the current through each resistor, (b) the voltage across each resistor, and (c) the total power dissipated. (a) $I_1 = 1.0$ A; $I_2 = I_3 = 0.50$ A (b) $V_1 = 20$ V; $V_2 = V_3 = 10$ V (c) 30 W

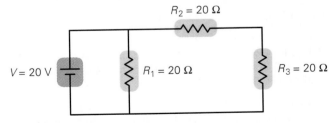

FIGURE 17.26 Circuit reduction See Exercises 26 and 43.

27 ■■ A 120-V circuit has a circuit breaker rated to trip at 20 A. How many 30-Ω resistors could be connected in parallel in the circuit before the circuit breaker opens? 5

28 ■■ Suppose that the resistor arrangement in Fig. 17.22 is connected to a 12-V battery. (a) What will the current through each resistor be? (b) What will the voltage drop across each resistor be? (c) What will the total power dissipated be? (a) $I_1 = I_2 = 3.0$ A; $I_3 = 6.0$ A; $I_4 = 6.0$ A (b) $V_1 = V_2 = 6.0$ V; $V_3 = V_4 = 12$V (c) 1.8×10^{-2} W

● **29** ■■ Three 100-Ω immersible resistors and a 120-V source are used to heat 0.50 kg of water initially at 10°C as rapidly as possible. (Assume that there is no heat loss.) (a) How much time is required to heat the water to 100°C? (b) How much additional time is required to boil away all of the water? (a) 4.3×10^2 s = 7.2 min (b) 2.5×10^3 s = 42 min

30 ■■ The terminals of a 6.0-V battery are connected to points A and B in Fig. 17.23. (a) How much current flows through each resistor? (b) How much power is dissipated by each resistor? (c) Compare the sum of the individual power dissipations to the power dissipation of the equivalent resistance for the circuit. (a) $I_1 = 1.0$ A; $I_2 = 0.60$ A; $I_3 = 0.60$ A; $I_4 = 0.60$ A (b) $P_1 = 6.0$ W; $P_2 = 1.4$ W; $P_3 = 2.2$ W; $P_4 = 3.6$ W (c) $P = 13$ W

31 Light bulbs with the wattage ratings shown in ● Fig. 17.27 are connected in a circuit. (a) What current does the voltage source deliver to the circuit? (b) Find the power dissipated by each bulb. (Assume normal operating resistances of the bulbs to be constant at any voltage.) see ISM

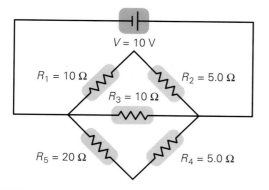

● **FIGURE 17.27 Watt's up?** See Exercise 31.

32 For the circuit in ● Fig. 17.28, find (a) the current through each resistor and (b) the voltage drop across each resistor. (a) $I_1 = 0.67$ A; $I_2 = 0.67$ A; $I_3 = 1.0$ A; $I_4 = 0.40$ A; $I_5 = 0.40$ A (b) $V_1 = 6.7$ V; $V_2 = 3.3$ V; $V_3 = 10$ V; $V_4 = 2.0$ V; $V_5 = 8.0$ V

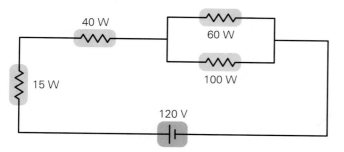

● **FIGURE 17.28 Resistors and currents** See Exercise 32.

33 ■■■ How many different values of resistance can be obtained using four 10-Ω resistors by connecting two or more of them in series, parallel, and series-parallel arrangements? 12

● **34** ■■■ What is the total power dissipated in the circuit shown in ● Fig. 17.29? 35 W

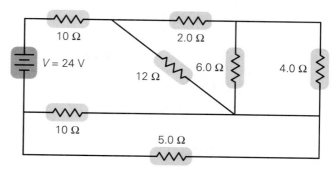

● **FIGURE 17.29 Power dissipation** See Exercise 34.

35 ■■■ What is the equivalent resistance of the arrangement in ● Fig. 17.30? 8.1 Ω

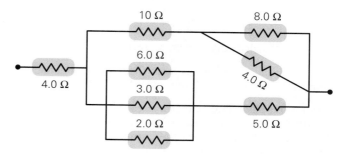

● **FIGURE 17.30 Equivalent resistance replacement** See Exercise 35.

36 ■■■ Find the current through each of the resistors in the circuit in ● Fig. 17.31? see ISM

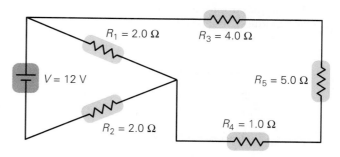

● **FIGURE 17.31 How much current?** See Exercise 36.

37 ■■■ The circuit shown in ● Fig. 17.32 is called a Wheatstone bridge after Sir Charles Wheatstone (1802–1875) and is used to make accurate measurements of resistance without the errors of ammeter-voltmeter measurements (see Exercises 72 and 73). The resistances R_1, R_2, and R_s are all known, and R_x is the unknown resistance to be measured; R_s is variable and is adjusted until the bridge circuit is balanced, when the galvanometer (G) shows a zero reading (no current through the galvanometer branch). Show that when the bridge is balanced R_x is given by

$$R_x = \left(\frac{R_2}{R_1}\right)R_s$$

see ISM

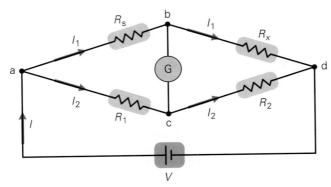

● FIGURE 17.32 Wheatstone bridge   See Exercise 37.

17.2 ■ Multiloop Circuits and Kirchhoff's Rules

38 A multiloop circuit has more than one (a) junction, (b) branch, (c) current, (d) all of these.   (d)

39 By our sign convention, if a resistance is traversed in the direction opposite to that of the conventional current, (a) the current is negative, (b) the current is positive, (c) the potential difference is negative, (d) the potential difference is positive.   (d)

● **40** Traverse loop 3 in Fig. 17.7, and show that the equation for this loop is not needed or could be used with the equation for one of the other loops.   see ISM

41 ■ Find the current in each resistor in ● Fig. 17.33 using Kirchhoff's rules. $I_1 = I_2 = 0.33$ A

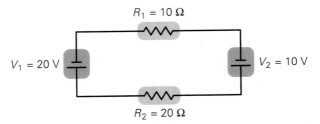

● FIGURE 17.33 Single-loop circuit   See Exercise 41.

42 ■ For the circuit in Fig. 17.7, reverse the directions of the interior loops and show that equations equivalent to those in Example are obtained.   see ISM

43 ■ Apply Kirchhoff's rules to the circuit in Fig. 17.26 to find the current through each resistor.
$I_1 = 1.0$ A; $I_2 = I_3 = 0.50$ A

44 ■■ Apply Kirchhoff's rules to the circuit in ● Fig. 17.34, and find (a) the current in each resistor and (b) the rate at which energy is being dissipated in the 8-Ω resistor.
(a) $I_1 = -0.75$ A; $I_2 = -0.50$ A; $I_3 = I_4 = I_5 = 1.25$ A   (b) 13 W

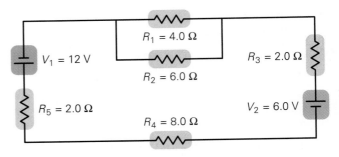

● FIGURE 17.34 A loop in a loop   See Exercise 44.

45 ■■ Find the current through each resistor in the circuit in ● Fig. 17.35. $I_1 = 3.75$ A; $I_2 = -1.25$ A; $I_3 = 5.00$ A

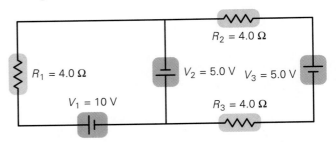

● FIGURE 17.35 Double-loop circuit   See Exercise 45.

● **46** ■■ Find the currents in the circuit branches in ● Fig. 17.36. $I_1 = 6.24$ A; $I_2 = 3.05$ A; $I_3 = 3.19$ A

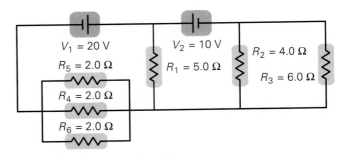

● FIGURE 17.36 How many loops?   See Exercise 46.

47 ■■■ For the multiloop circuit shown in ● Fig. 17.37, what is the current through each branch? $I_1 = 0.664$ A; $I_2 = 0.786$ A; $I_3 = 1.450$ A; $I_4 = 0.770$ A; $I_5 = 0.016$ A

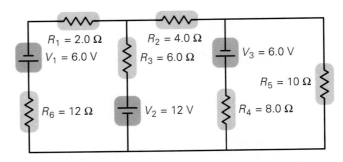

● FIGURE 17.37 Triple-loop circuit   See Exercise 47.

48 ■■■ A procedure called the loop current method is considered by some to be mathematically simpler than Kirchhoff's branch current method. In the loop current method, a particular current is assigned to each loop (in ● Fig. 17.38, I_A and I_B). Kirchhoff's second rule is then used to write an equation for each loop, incorporating the sign convention. The equations are solved for the loop currents, and then Kirchhoff's first rule is applied to each junction in the circuit to find the branch currents by comparison with the loop currents. Show that this loop current method gives the same currents for the circuit in Fig. 17.38 as the branch current method. (Remember that all the loop currents in a branch must be taken into account when applying the loop and junction theorems.)   see ISM

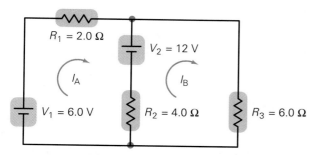

● **FIGURE 17.38 The loop-current method**   See Exercise 48.

17.3 ■ RC Circuits

49 In charging a capacitor in series with a resistance, in one time constant (a) the capacitor is 63% charged, (b) the voltage across the capacitor is 37% of its maximum value, (c) the current in the circuit decreases to 37% of its maximum value, (d) both (a) and (c).   (d)

● **50** When charged with a battery and then discharged, an RC circuit has its maximum current (a) at the beginning of the charging process, (b) near the end of the charging process, (c) at the end of the discharging process, (d) after one time constant.   (a)

51 Why doesn't the voltage across the neon tube in an oscillator circuit fall below the maintaining voltage and go to zero?   when tube ceases to conduct, voltage rises again

52 ■ Show that the unit of the time constant is the second.   see ISM

53 ■ A capacitor in a single-loop RC circuit is charged to 63% of its final voltage in 1.5 s. Find the time constant for the circuit.   1.5 s

54 ■ For an RC circuit with a 2.5 MΩ resistance, what capacitance will give a time constant of 1.0 s?   0.40 µF

55 ■■ An RC circuit has a time constant of 2.25 s. The capacitor is in the microfarad range and has the same numeri-

cal prefix as the resistor. What are the values of the capacitor and the resistor?   1.5 µF, 1.5 MΩ

56 ■■ How many time constants will it take for an initially charged capacitor to be discharged to half of its initial voltage?   0.693 s

57 ■■ A 2.0 µF capacitor in series with a resistor is connected to a 12-V source and charged. (a) What resistance is necessary to cause the capacitor to have only 37% of its initial charge 1.5 s after starting to discharge? (b) What is the voltage across the capacitor at $t = 3\tau$ when it is charging?   (a) 7.5×10^5 Ω   (b)11.4 V

● **58** ■■ An RC circuit with $C = 40$ µF and $R = 6.0$ Ω has a 12-V source. With the capacitor initialy uncharged, an open switch in the circuit is closed. (a) What is the potential difference across the 6-Ω resistor immediately afterward? (b) What is the potential difference across the capacitor at that time? (c) What is the current in the resistor at that time?   (a) 12 V   (b) 0   (c) 2.0 A

59 ■■ (a) In the circuit in Exercise 58 after the switch has been closed for $t = 4\tau$, what is the charge on the capacitor? (b) After a long time, what are the voltages across the capacitor and the resistor?   (a) 4.7×10^{-4} C   (b) $V_C = 12$ V; $V_R = 0$

60 ■■ An RC circuit with a resistance of 5.0 MΩ and a capacitance of 0.40 µF is connected to a 12-V source. If the capacitor is initially uncharged, what is the voltage change across it between $t = 2\tau$ and $t = 4\tau$?   1.4 V

61 ■■ A 3.0-MΩ resistor is connected in series with a 0.28-µF capacitor. If this arrangement is hooked across four 1.5-V batteries connected in series, (a) what is the initial voltage across the capacitor and the initial current in the circuit? (b) How much charge is on the capacitor after 4.0 s?   (a) 0 V, 2.0 µA   (b) 1.7×10^{-6} C

62 ■■■ The time between flashes of the neon tube, or the period of the oscillator circuit, in Fig. 17.11 is the time it takes for the voltage to rise from V_m to V_b. Show that the period of such an oscillator is given by

$$T = t_b - t_m = RC \ln\left(\frac{V_o - V_m}{V_o - V_b}\right)$$

where V_o is the maximum voltage, or the voltage of the battery in the circuit.   see ISM

17.4 ■ Ammeters and Voltmeters

63 An ammeter has a (a) large shunt resistance, (b) large multiplier resistance, (c) small shunt resistance, (d) small multiplier resistance.   (c)

64 To measure the voltage across a circuit element, a voltmeter is connected (a) in series with the element, (b) in parallel with the element, (c) between the high potential side of the element and ground, (d) none of these.   (b)

65 What would happen if (a) an ammeter were connected in parallel in a circuit, and (b) a voltmeter were connected in series?   see ISM

● 66 ■■ A galvanometer with a full-scale sensitivity of 2000 μA has a coil resistance of 100 Ω. If it is to be used in an ammeter with a full-scale reading of 3.0 A, what is the necessary shunt resistance? 67 mΩ

67 ■■ If the galvanometer in Exercise 66 were to be used in a voltmeter with a full-scale reading of 1.5 V, what would the required value of the multiplier resistor be?
6.5 × 10² Ω

68 ■■ A galvanometer with a full-scale sensitivity of 600 μA and a coil resistance of 50 Ω is to be used in an ammeter designed to read 5.0 A at full scale. What is the required value of the shunt resistor? 6.0 mΩ

69 ■■ A galvanometer has a coil resistance of 20 Ω. A current of 200 μA deflects the needle through 10 divisions full-scale. What is needed to convert the galvanometer to a full-scale 10-V voltmeter? 49.98 kΩ

● 70 ■■ An ammeter has a resistance of 1.0 mΩ. Find the current in the ammeter when it is properly connected to a 10-Ω resistor and a 6.0-V source. 0.59994 A

71 ■■ A voltmeter has a resistance of 20 kΩ. What is the current through the meter when it is connected across a 10-Ω resistor that is hooked to a 6.0-V source? 0.30 mA

72 ■■■ An ammeter and a voltmeter can be used to measure the value of a resistor in a circuit. Suppose that the ammeter is connected in series with the resistor, and the voltmeter is placed across the resistor *only*. Show that the value of the resistance when measured like this is given by

$$R = \frac{V}{I - (V/R_v)}$$

where I is the current measured by the ammeter, V is the voltage measured by the voltmeter, and R_v is the resistance of the voltmeter. [*Hint:* Draw a circuit diagram to help analyze the problem.] see ISM

73 ■■■ An ammeter and a voltmeter are used to measure the value of a resistor in a circuit. The ammeter is connected in series with the resistor, and the voltmeter is placed across *both* the ammeter and the resistor. Show that the value of the resistance when measured like this is given by

$$R = (V/I) - R_a$$

where V is the voltage measured by the voltmeter, I is the current measured by the ammeter, and R_a is the resistance of the ammeter, [*Hint:* Draw a circuit diagram to help analyze the problem.] see ISM

17.5 ■ Household Circuits and Electrical Safety

74 The so-called neutral or ground wire in household wiring (a) is a current-carrying wire, (b) is at a voltage difference of 240 V from a "hot" wire, (c) carries no current, (d) is not part of house circuits. (a)

75 A dedicated grounding wire (a) is the basis for the polarized plug, (b) is necessary for a circuit breaker, (c) normally carries no current, (d) none of these. (c)

76 What is wrong with the circuit in ● Fig. 17.39 in terms of electrical safety, and why? see ISM

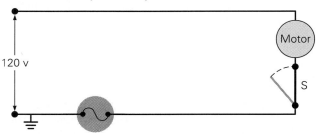

● **FIGURE 17.39 A safety problem?** See Exercise 76.

77 Bodily injury depends on the magnitude of the current and its path, yet you commonly see signs that warn "Danger: High Voltage" (● Fig. 17.40). Shouldn't this refer to high current? Explain. high voltage could produce harmful current

● **FIGURE 17.40 Danger—High Voltage** Shouldn't it be high current? See Exercise 77.

■ **Additional Exercises**

78 Five 30-Ω resistors are connected in parallel. What is the total equivalent resistance? 6.0 Ω

79 Two 1.0-kΩ resistors are connected in parallel. That arrangement is connected in series to another 1.0-kΩ resistor in a circuit containing a battery. If each resistor is rated at 0.25 W, find the maximum voltage of the battery that can be safely used in the circuit. 23.7 V

80 Find the current through each resistor in the circuit in ● Fig. 17.41. I_1 = 2.6 A; I_2 = 1.7 A; I_3 = 0.85 A

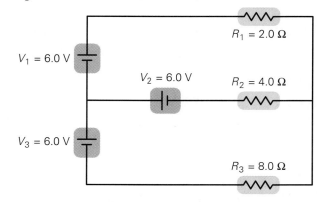

● **FIGURE 17.41 Kirchhoff's rules** See Exercise 80.

81 Resistors of 20 Ω and 15 Ω are connected in parallel in a circuit with a 9.0-V battery. How much energy is dissipated each second by the resistors?   9.5 J/s

82 Four resistors are connected to a 90-V source as shown in ● Fig. 17.42. (a) Which resistor dissipates the most power and how much? (b) What is the total power dissipated in the circuit?   (a) 8.1 × 10² Ω   (b) 2.2 × 10³ Ω

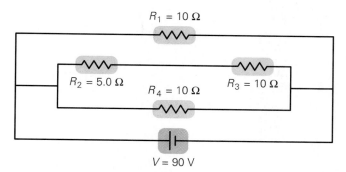

FIGURE 17.42 How much power is dissipated? See Exercise 82.

83 Four resistors are connected in a circuit with a 110-V source, as shown in ● Fig. 17.43. (a) What is the current through each resistor? (b) How much power is dissipated by each resistor?   (a) I_1 = 1.0 A, I_2 = I_4 = 0.40 A, I_3 = 0.20 A   (b) P_1 = 100 W, P_2 = P_4 = 4.0 W, P_3 = 2.0 W

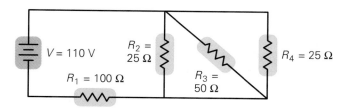

FIGURE 17.43 Joule heat losses See Exercise 83.

84 An RC circuit has a time constant of 0.75 s. If the capacitor has a capacitance of 0.25 μF, what is the value of the resistance?   3.0 MΩ

85 Two 75-W light bulbs are connected to a 110-V power source. What is the current in each bulb and the total power dissipated if the bulbs are connected (a) in series and (b) in parallel?   (a) I = 0.34 A for each; P = 36 W
(b) I = 0.69 A for each; P = 150 W

86 If a 120-V source were connected across the leads (open ends) of the circuit shown in ● Fig. 17.44, how much current would be drawn from it?   11.8 A

87 A battery has three cells, each with an internal resistance of 0.02 Ω and an emf of 1.5 V. The battery is connected in parallel with a 10-Ω resistor. (a) Determine the voltage across the resistor. (b) How much current passes through each cell? (The cells in a battery are connected in series.)   (a) 4.47 V
(b) 0.447 A

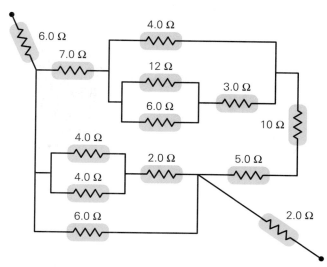

FIGURE 17.44 Connect it to a voltage source See Exercise 86.

88 A galvanometer with an internal resistance of 50 Ω and a full-scale sensitivity of 200 μA is used to construct a multi-voltmeter with the circuit diagram as shown in ● Fig. 17.45. What are the values of the multiplier resistors that will allow the full-scale readings indicated for the various connections? (Ignore significant figures.)   10 V: 4.995 × 10⁴ Ω;
50 V: 24.995 × 10⁴ Ω; 250 V: 24.995 × 10⁴ Ω

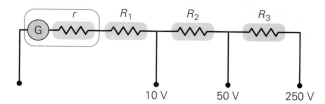

FIGURE 17.45 A multivoltmeter See Exercise 88.

89 A galvanometer with an internal resistance of 100 Ω and a full-scale sensitivity of 100 μA is used to construct a multiammeter with a circuit diagram as shown in ● Fig. 17.46. What are the values of the shunt resistors that will allow the full-scale readings indicated for the various connections? (Ignore significant figures.)   0.30 A: 0.03334 Ω;
3.0 A: 0.003334 Ω; 30 A: 0.0003334 Ω

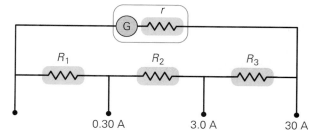

FIGURE 17.46 A multiammeter See Exercise 89.

Magnetism

Children, and many adults, are fascinated by the behavior of magnets. Experiencing the attraction and repulsion between magnets or between a magnet and a piece of metal gives a hands-on demonstration of a magnetic force. Magnets are readily available today and are produced commercially, but they were once quite scarce and existed only as natural magnets (magnetized stones found in nature). It was known as early as about 600 B.C. that a certain type of rock, called lodestone, could attract pieces of iron as well as other pieces of the same kind of rock. Today, we know that lodestone is a type of iron ore called magnetite (Fe_3O_4, iron oxide).

The early history of human use of magnetism is largely undocumented. It is generally believed that magnetic rocks were first found in a region called Magnesia (in what is now Turkey), from which the name "magnet" is derived. For centuries, the attractive properties of magnets were attributed to supernatural forces. Early Greek philosophers believed that a magnet had a soul that caused it to attract a piece of iron. We now associate magnetism with electricity (electromagnetism) as both turn out to be aspects of a single fundamental force or interaction (the electromagnetic force). We have put electromagnetism to use in motors, generators, radios, telephones, and so many other familiar applications that ours might be called an electromagnetic society.

*And there may be many more to come. The development of high-temperature supercon-
ducting magnets (Chapter 16) would open the way for practical exploitation of many
devices that are now found only in the laboratory or as experimental prototypes—for
example, high-speed trains that are levitated and propelled by magnetic fields.*

*Although electricity and magnetism are manifestations of the same fundamental
force, it is instructive to consider them individually and then put them together, so to
speak. This and the next chapter will investigate magnetism and its intimate relationship
to electricity.*

18.1 ■ Magnets and Magnetic Poles

One of the first things anyone notices in examining a common bar magnet is
that it has two poles, or "centers" of force, one at or near each end (● Fig. 18.1).
Rather than being distinguished as positive and negative, these poles are called
north (N) and south (S). This terminology comes from the early use of the
magnetic compass. The north pole of a compass magnet is the *north-seeking*
end. With two bar magnets, we notice a pattern of attraction and a repulsion
between their various ends: Each end is attracted to one end of a second magnet
and repelled by the other.

The attraction and repulsion between poles of magnets is similar to the
behavior of like and unlike electric charges. That is, analogous to the law of
charges is the **law of poles.**

Like magnetic poles repel, and unlike magnetic poles attract.

That is, north and south poles (N-S) attract each other, and north and north
poles (N-N) or south and south (S-S) poles repel each other (● Fig. 18.2).

Magnetic poles always occur in pairs, as a so-called magnetic dipole. You
might think you could break a bar magnet in half and get two single isolated
poles. However, you would find that the pieces are two shorter magnets, both
with north and south poles. Although an isolated, single magnetic pole, a
magnetic monopole, has been postulated to exist, it has yet to be verified
experimentally. Always having two poles may make you wonder about the
nature of magnetism. As will be discussed shortly, magnetism is due to electric
charges in motion, which occurs with electric currents, orbiting atomic electrons,
and so on. This understanding of the source of magnetism and some knowledge
of magnetic materials (to be covered in a later section) will allow you to see
why pieces of a magnet are themselves dipole magnets.

From your knowledge of electrostatics, you may be tempted to think of a
pole as a magnetic "charge" and to wonder if the magnetic force can be ex-

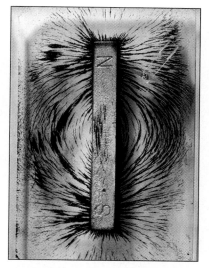

●**FIGURE 18.1 Bar magnet** The
iron filings indicate the poles or
centers of force of a common bar
magnet. We designate these poles
as north (N) and south (S).

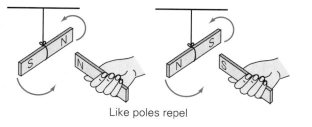

Like poles repel

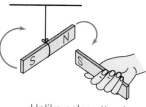

Unlike poles attract

●**FIGURE 18.2 The law of
poles** Like poles (N-N or S-S)
repel, and unlike poles (N-S) attract.

• FIGURE 18.3 Magnetic fields Magnetic field lines may be outlined using iron filings. The filings become tiny magnets and line up with the field. The closer together the field lines, the stronger the magnetic field. **(a)** Iron filing pattern for the magnetic field of a single bar magnet. **(b)** Iron filing pattern for the magnetic field between unlike poles. **(c)** Iron filing for the magnetic field between like poles. Notice how the field lines run between the poles in (b) and diverge in (c).

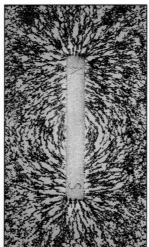

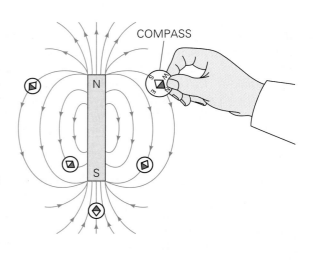

(a)

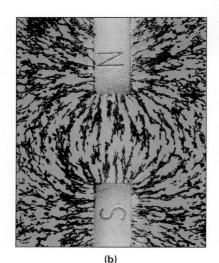

(b)

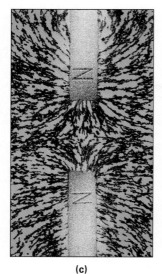

(c)

pressed in a form similar to Coulomb's law for electric charges. In fact, such a law was developed by Coulomb using the magnetic pole strengths in place of the electric charges. However, this law is now used rarely because it doesn't match present understanding of magnetism and because it is more convenient to work with magnetic fields.

In Section 15.4, you learned how convenient it is to describe the interaction of electrically charged objects in terms of electric fields. There is an electric field surrounding any electric charge, and this is represented by electric field lines, or lines of force ($\mathbf{E} = \mathbf{F}/q$). Similarly, it is convenient to describe magnetic interactions in terms of magnetic fields. Hence, we consider the magnetic field that surrounds any magnet. Like an electric field, a **magnetic field** is a vector quantity and is represented by the symbol **B**. The pattern of the magnetic field lines surrounding a magnet can be visually demonstrated by sprinkling iron filings over a magnet covered with a piece of paper or glass (• Fig. 18.3). As you will learn, the iron filings become induced magnets and line up with the field.

To describe any vector field, you must specify both the magnitude, or strength, and direction of the field at various points. The direction of a magnetic field, or B field, can be defined using another north magnetic pole.

> The direction of a B field at any location is in the direction that the north pole of a compass would point with the compass at that location.

This definition provides another method of mapping a B field by placing a small compass near a magnet. The torque on the compass needle (a small magnet) due to the magnetic force causes the needle to line up with the field. If the compass is moved in the direction in which the needle points, the path of the needle traces out a field line, as illustrated in Fig. 18.3a.

Thus, *the direction of the magnetic field at any point is in the direction of the force on a magnetic north pole.* The field, then, by the law of poles, is away from the north pole of a magnet and toward the south pole. As with an electric field, the closer together the field lines, the stronger the field. Note how this is indicated by the concentration of iron filings in the pole regions in Fig. 18.3.

Since a magnetic north pole can be used to map out a magnetic field, you might think that B would be the magnetic force per unit pole (as E is the electric force per unit charge). The magnetic field was once thought of in this manner. However, the magnetic field is now defined in terms of the force on a moving electric charge, as will be discussed in the next section.

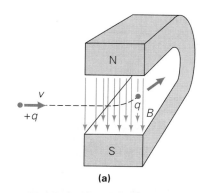

(a)

18.2 ▪ Electromagnetism and the Source of Magnetic Fields

An example of an *electomagnetic* interaction occurs when a particle with a positive charge (q) moving with a constant velocity enters a region with a uniform magnetic field in a path such that the velocity and the magnetic field are at right angles (● Fig. 18.4a). When the charge enters the field, it is deflected into a curved path.

From the study of dynamics, you know that this deflection could be caused by a force perpendicular to the particle's velocity. But what gives rise to this force? No electric field is present (other than that of the charge itself). And the force of gravity is very weak (because the particle is so small), but it would deflect the particle downward rather than sideways. Evidently, the force is due to the interaction of the moving charge and the magnetic field. *A charged particle moving in a magnetic field may experience a force.*

Varying the magnitude of the charge, its velocity, and the magnetic field shows that the magnitude of the deflecting force is directly proportional to each of these quantities. That is,

$$F \propto qvB$$

In equation form,

$$F = qvB \qquad (18.1)$$

(The constant of proportionality has a value of 1 by the selection of the magnetic field unit.) This gives an expression for the strength (magnitude) of the magnetic field in terms of familiar quantities.

$$B = \frac{F}{qv} \qquad (18.2)$$

That is, B is the magnetic force per moving charge. From Eq. 18.2, you can see that a magnetic field has SI units of N/C-(m/s) or N/A-m where A=C/s. This combination of units is given the name **tesla (T)**. The magnetic field is sometimes given in webers per square meter (Wb/m²), and 1 T = 1 Wb/m².

If the direction of the charged particle's velocity is not perpendicular to the magnetic field, the magnitude of the magnetic force on the particle is not given by Eq. 18.1. It is known that the magnitude of this force depends on the angle (θ) between the velocity and the field vectors—more specifically, on the sine of that angle (sin θ). That is, the force is zero when **v** and **B** are parallel and a maximum when those two vectors are perpendicular. In general,

$$\boxed{F = qvB \sin \theta} \qquad (18.3)$$

(b)

● **FIGURE 18.4 Force on a moving charged particle** (a) When a charged particle enters a magnetic field, it experiences a force which is evident because the charge is deflected from its original path. (b) The electron beam in a cathode ray tube (made visible by fluorescent paper) is normally horizontal between the end electrodes, but is deflected here because of the magnet.

Another unit for the magnetic field is the gauss (G); 1 T = 1 Wb/m² = 10⁴ G. Like the tesla, the weber and the gauss are named after early investigators of magnetic phenomena.

Point out which quantities in this equation are vectors and which are scalars.

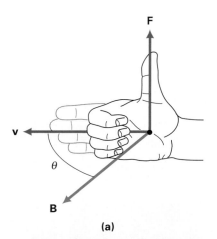

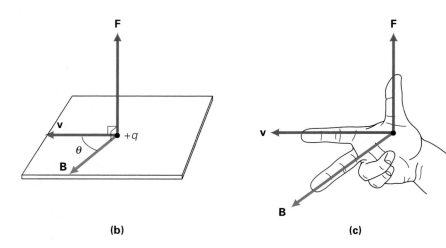

(a)   (b)   (c)

●**FIGURE 18.5 Right-hand rule
for magnetic force (a)** When the
fingers of the right hand are pointed
in the direction of **v** and then turned
or curled toward the direction of **B**,
the extended thumb points in the
direction of the force **F** on a *positive*
charge. **(b)** The force is always
perpendicular to the plane of **B** and
v, or always perpendicular to the
direction of the particle's motion. **(c)**
Another method. When the extended
forefinger of the right hand points in
the direction of **v** and the middle
finger points in the direction of **B**,
the extended right thumb points in
the direction of **F**.

Figure 18.5

There is also a right-hand "palm"
method for determining the
direction of the magnetic force on
a moving positive charge. If the
extended thumb of the right hand
is in the direction of v and the
extended fingers in the direction of
B, the force F is directed out of the
palm or in the direction the palm
would push. Take your choice of
methods. Common to all is that F is
perpendicular to the plane of v and B.

*Compare electrical forces with
magnetic forces. Some students
start thinking of poles as positive
and negative. Try to avoid this.*

When **v** and **B** are perpendicular ($\theta = 90°$), the equation reduces to Eq. 18.1, $F = qvB \sin 90° = qvB$. (maximum force). However, when the **v** and **B** vectors are parallel ($\theta = 0°$ or $180°$), the force on the moving charge is zero, $F = qvB \sin 0° = 0$. You should be able to prove for yourself that in Eq. 18.3, $B \sin \theta$ is the magnitude of the component of **B** perpendicular to the velocity.

The direction of the magnetic force on a positively charged particle can also be determined by considering the velocity and field vectors. This method is stated in the form of a right-hand rule (● Fig. 18.5a).

When the fingers of the right hand are pointed in the direction of v and then turned or curled toward the vector B, the extended thumb points in the direction of F for a *positive* charge. (For a negative charge, the force is in the opposite direction.)

You might imagine the fingers of the right hand to be physically turning or pushing the **v** vector into **B** so that they are both aligned in the same direction. The magnetic force is always *perpendicular to the plane* of the vectors **v** and **B** (Fig. 18.5b). If the moving charge is negative, the right-hand thumb points in the direction opposite to that of the force. (You can simply assume that the negative charge is positive, apply the right-hand rule, and then reverse the direction.)

Some people prefer a right-hand "three finger" rule as illustrated in Fig. 18.5c: With the forefinger of the right hand pointing in the direction of **v** and the middle finger flexed in the direction of **B**, the extended right thumb points in the direction of **F**. Use whichever method you find easier.

EXAMPLE 18.1 ■ The Right-Hand Rule

A beam of protons travels horizontally northward. In order to have the beam deflected east of north by a uniform magnetic field which it enters, the field would have to be directed in which direction? (a) vertically downward, (b) west, (c) vertically upward, (d) south. *Clearly establish the reasoning and physical principle(s) used in determining your answer. That is, why did you select your answer?*

Reasoning and Answer. Since the force on a moving charged particle is perpendicular to the plane of **v** and **B**, the magnetic field could not be in the horizontal plane, which eliminates (b) and (d). The field must be either up or down. Using the right-hand rule for a positively charged particle to see if **B** is downward (answer a), you should find that the initial force would be to the west in this case. Hence, the answer is (c) for an easterly deflection. Prove this to yourself using the right-hand rule.

Follow-up Exercise. In what direction would a beam of *electrons* have to travel to experience the same initial force from the same magnetic field as in Example 18.1? (*Reasoning and answer may be found in the Answers to Exercises section at the back of the book.*)

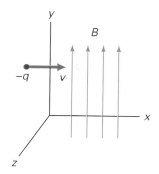

(a) Side view

EXAMPLE 18.2 ■ Force on a Moving Charge

A negatively charged particle with a charge of -5.0×10^{-4} C moves at a speed of 1.0×10^2 m/s in the $+x$ direction toward a uniform magnetic field of 2.0 T in the $+y$ direction (Fig. 18.6a). (a) What is the force on the particle when it enters the magnetic field? (b) Describe the path of the particle while it is in the field.

Solution.

Given: $q = -5.0 \times 10^{-4}$ C $\qquad$ *Find:* (a) F (force)
$\qquad v = 1.0 \times 10^2$ m/s $\qquad\qquad$ (b) Path of the particle in the field
$\qquad B = 2.0$ T

(a) Eq. 18.3 can be used to find the magnitude of the force.

$$F = qvB \sin \theta = (5.0 \times 10^{-4} \text{ C})(1.0 \times 10^2 \text{ m/s})(2.0 \text{ T}) = 0.10 \text{ N}$$

since $\sin \theta = \sin 90° = 1$.

The direction of the force just as the particle enters the magnetic field is in the $-z$ direction (into the page) in Fig. 18.6a. By the right-hand rule, the force on a positive charge would be in the $+z$ direction (out of the page). But since the charge is negative, the force is in the opposite direction.
(b) Since the magnetic force is always perpendicular to the velocity of the particle, it is a centripetal force that causes the particle to move in a circular path (Fig. 18.6b). The radius of the path (r) may be found using Newton's second law:

$$F = ma_c$$

This is, in this case,

$$qvB = \frac{mv^2}{r}$$

and

$$r = \frac{mv}{qB}$$

If the mass of the particle were known, the radius of its circular path could be computed easily.

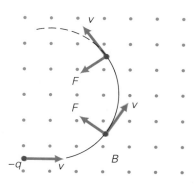

(b) Top view

 FIGURE 18.6 Path of a charged particle in a magnetic field (a) A charged particle entering a uniform magnetic field will be deflected, here into the page by the right-hand rule. Note that the charge is negative. (b) In the field, the force is always perpendicular to the particle's motion and it moves in a circular path. Example 18.2.

 Figure 18.6

A review of circular motion may be necessary.

Demonstration/activity: Use an old oscilloscope or television to demonstrate how a beam of electrons will be deflected by a magnetic field. DO NOT use a good color TV. Serious problems could result (see p. 589).

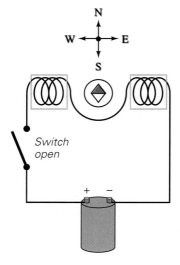

(a) No current

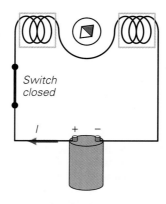

(b) Current

FIGURE 18.7 Electric current and magnetic field (a) With no current in the wire, the compass needle points north. (b) With a current in the wire, the compass needle is deflected, indicating the presence of a magnetic field other than that of the Earth.

Note that since there is no tangential acceleration, only the direction of the particle's motion changes—its speed, and thus its kinetic energy, remains constant. Looked at from another point of view, the fact that the force on the particle is always perpendicular to the direction of motion means that no work is done by the force. Thus the particle's kinetic energy doesn't change.

Electric Currents and Magnetic Fields

Electric and magnetic phenomena are closely related. In fact, magnetic fields are produced by electric currents. This fact was discovered by the Danish physicist Hans Christian Oersted in 1820. He noted that an electric current could produce a deflection of a compass needle. This can be demonstrated using an arrangement like that in ● Fig. 18.7. When the circuit is open and there is no current in it, the compass needle points in the northerly direction due to the Earth's magnetic field (to be discussed in Section 18.6). However, when the switch is closed and there is current in the circuit, the compass needle is deflected, indicating that another magnetic field (other than that of the Earth) is affecting the needle. When the switch is opened, the compass needle goes back to pointing north.

Developing expressions for the magnitude of the magnetic field around a current-carrying wire requires mathematics beyond the scope of this book. This section will simply present the resulting formulas for the magnitude of the B field for several arrangements that are used in many practical applications.

Magnetic Field Around a Long, Straight, Current-carrying Wire. At a perpendicular distance d from a long, straight wire carrying a current I (● Fig. 18.8), the magnitude of **B** is

$$B = \frac{\mu_o I}{2\pi d} \qquad (18.4)$$

where

$$\mu_o = 4\pi \times 10^{-7} \text{ T-m/A} \quad \text{(or Wb/A-m)}$$

is a constant called the *permeability of free space*. The field lines are closed circles around the wire.

Note in Fig. 18.8 that the direction of **B** is given by a right-hand rule.

If a current-carrying wire is grasped with the right hand with the extended thumb pointing in the direction of the conventional current, the curled fingers indicate the circular sense of the magnetic field. (The field vector is tangent to a circular field line at any point on the circle.)

Magnetic Field at the Center of Circular Current-carrying Wire Loop. At the *center* of a circular loop of wire of radius r carrying a current I (● Fig. 18.9), the magnitude of **B** is

$$B = \frac{\mu_o I}{2r} \qquad (18.5)$$

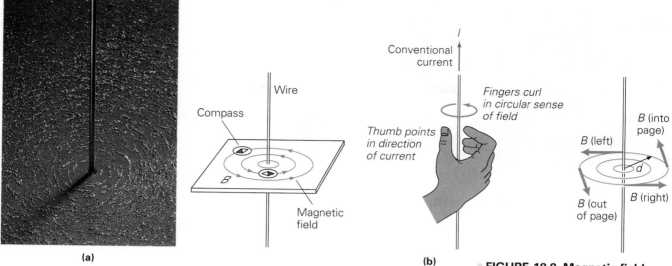

(a)

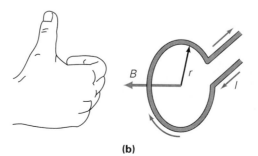

(b)

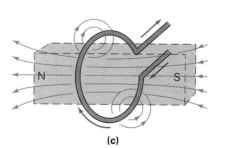

(c)

This equation is valid only at the center of the loop. The direction of **B** is given by the right-hand rule, and is perpendicular to the plane of the loop at its center (Fig. 18.9b). Notice that the total field of the loop (Fig. 18.9c) is similar to that of a bar magnet.

Magnetic Field in a Current-carrying Solenoid. A solenoid is constructed by winding a long wire in a tight coil, or helix (many circular loops, as shown in (•Fig. 18.10). If the radius of the loops is small compared to the length (L) of the coil, the magnitude of **B** *along a longitudinal axis* through the center of a solenoid of N turns (loops) carrying a current I is

$$B = \frac{\mu_o NI}{L} \qquad (18.6)$$

The direction of **B** is given by the right-hand rule as applied to one of the loops of the coil.

The expression $n = N/L$, or the number of turns per meter, is called the *linear turn density*. Using it allows Eq. 18.6 to be written in simpler form.

$$B = \mu_o nI \qquad (18.7)$$

•FIGURE 18.8 Magnetic field around a straight current-carrying wire **(a)** The field lines form concentric circles around the wire, as revealed by the iron filing pattern. **(b)** The circular sense of the field lines is given by a right-hand rule, and the field vector is tangent to the circular field line at any point.

Figure 18.8

•FIGURE 18.9 Magnetic field at the center of a current-carrying loop **(a)** Iron filing pattern for a current-carrying loop. The magnetic field is perpendicular to the plane of the loop. **(b)** The direction of the field is given by the right-hand rule. With the thumb of the right hand in the direction of the conventional current, the curled fingers indicate the direction of **B**. **(c)** The overall magnetic field of a current-carrying circular loop is similar to that of a bar magnet.

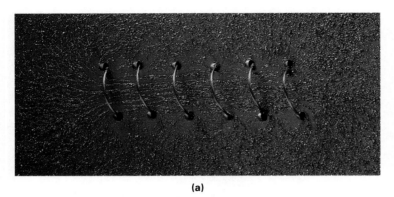

(a)

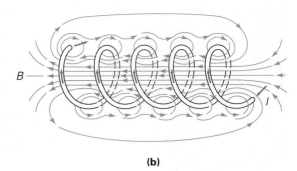

(b)

• **FIGURE 18.10 Magnetic field of a solenoid** **(a)** The magnetic field of a current-carrying solenoid is fairly uniform near the central axis of the solenoid, as seen in this iron filing pattern. **(b)** The direction of the field may be determined by applying the right-hand rule to one of the loops.

The longer the solenoid, the more uniform the magnetic field across the cross-sectional area within its coil. An ideal, or infinitely long, solenoid would have a uniform internal field. Of course, this ideal condition can only be approached in reality, but a long solenoid does produce a relatively uniform magnetic field. Note how the pattern of the field lines for a loop (Fig. 18.9) and especially that for a solenoid (18.10) are similar to that for a bar magnet.

• **FIGURE 18.11 Magnetic field** Finding the magnetic field produced by a straight current-carrying wire. Example 18.3.

EXAMPLE 18.3 ■ Magnetic Field Vector

A long, straight, horizontal wire carries a conventional current of 50 A. Considered as a conventional current, it flows in a west-to-east direction (● Fig. 18.11). What is the magnitude and direction of the magnetic field 1.0 m directly below the wire?

Solution

Given: $I = 50$ A   *Find:* **B** (magnitude and direction)
$d = 1.0$ m

We use Eq. 18.4 to find the magnitude of the field at the point 1.0 m directly below the line:

$$B = \frac{\mu_o I}{2\pi d} = \frac{(4\pi \times 10^{-7} \text{ T-m/A})(50 \text{ A})}{2\pi(1.0 \text{ m})} = 1.0 \times 10^{-5} \text{ T}$$

By the right-hand rule, the vector points northward. (Note that in Fig. 18.11 north is into the paper and south is outward.)

■ Problem-Solving Hint

In Example 18.3, with $\mu_o = 4\pi \times 10^{-7}$ T-m/A, the 4π conveniently canceled. This is not the case in the following example. Note that π may be left in symbol form unless a numerical answer is needed.

Magnetohydrodynamics refers to the use of magnetic fields (*magneto*) in a liquid (*hydro*) propulsion system. Such systems are now being developed for use in seagoing ships and for other applications. For ships, the idea is to use MHD thrusters instead of the conventional propeller drive for propulsion. Arrays of thrusters would be used on each side of a ship like jet engines on an aircraft, or perhaps around the hull of a submarine.

The principle of an MHD thruster is illustrated in Fig. 1. The thruster uses the simple $F = qvB$ relationship in producing a jet of seawater. A powerful magnet sur-rounds a thruster tube through which seawater flows. The magnet is fashioned so as to produce a magnetic field perpendicular to the tube's length (see Fig. 1). Inside the tube, electrodes produce a current across the tube, composed of ions from the dissolved salts in seawater (Video Demonstration 11). Hence, we have charged particles moving in a magnetic field, which you know gives rise to a force on the particles. By the right-hand rule, the force is along the length of the tube, so seawater is ejected from the thruster. By Newton's third law, this jet action would cause a boat to move in the opposite direction (again analogous to a jet aircraft).

The intense magnetic fields needed for an MHD system are produced by superconducting magnets. Recall from Chapter 16 that certain materials become superconducting, or lose all electrical resistance, at some low critical temperature. With no resistance, large currents are possible without joule heat (I^2R losses), and large magnetic field can be produced ($B \propto I$).

MHD propulsion systems are currently in the development stage, and at present can only propel small boats very slowly. It is predicted, however, that some day MHD will be able to propel ships and submarines at speeds in excess of 160 km/h (100 mi/h). The speed of propeller-driven vessels is limited by what is known as cavitation. At higher propeller speeds, areas of low pressure form in front of the blades, cuasing the water to vaporize and form bubbles or cavities. This reduces the propeller's efficiency, and at very high speeds can cause damage. Because MHD systems have no propellers, they are not subject to this limitation. For the same reason they are relatively quiet—there is no propeller noise or noise resulting from cavitation. Another advantage of MHD is that, because the MHD thrusters have no mechanical moving parts, they should require less maintenance than conventional propulsion systems.

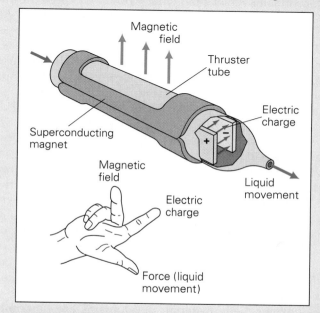

FIGURE 1 MHD thruster The thruster uses an $F = qvB$ interaction to produce a jet of sea water. See text for description.

EXAMPLE 18.4 ■ Magnetic Field in a Solenoid

A solenoid 0.30 m long with 10^3 turns per meter carries a current of 5.0 A. What is the magnitude of the magnetic field through the center of the solenoid?

Solution.

> ***Given:*** $I = 50$ A   ***Find:*** B (magnitude of magnetic field)
> $n = 10^3$ turns/m

Here we are given the turn density, so using Eq. 18.7,

$$B = \mu_o nI = (4\pi \times 10^{-7} \text{ T-m/A})(10^3 \text{ m}^{-1})(5.0 \text{ A}) = 2\pi \times 10^{-3} \text{ T}$$

Notice that we can increase the magnetic field in a solenoid by increasing the turn density and/or the current.

18.3 ■ Magnetic Materials

You may wonder why some materials are magnetic or easily magnetized, and others are not. Knowing that a current produces a magnetic field and comparing the magnetic fields of a bar magnet and a current-carrying loop (see Figs. 18.1 and 18.9), you might be tempted to think that the magnetic field of the bar magnet is in some way due to internal currents.

Recall that, according to the simplistic solar system model of the atom, the electrons orbit the nucleus. They are thus charges in motion, and each orbiting electron is in effect a current loop that produces a magnetic field. This model pictures a material as having many internal atomic magnets. However, detailed analysis of atomic structure shows that the magnetic field produced by the orbiting atomic electrons is much smaller than what would be expected from this simplistic model. Also, the atoms are sometimes randomly arranged. As a result, the net magnetic effect due to orbiting electrons for most materials is either zero or very small.

What then is the source of the magnetism of magnetic materials? Modern theory says that magnetism is an atomic effect due to electron spin. The word "spin" is used because of an early idea that an additional contribution to the atomic magnetic field might come from an electron spinning on its axis. The classical picture likens a spinning electron to the Earth rotating on its axis. However, this is *not* the case. Electron spin is strictly a quantum mechanical effect with no direct classical analog.

In atoms with two or more electrons, the electrons usually pair up, in pairs whose spins are opposite. The magnetic fields then cancel each other, and the material is not magnetic. Aluminum is an example. However, in certain strongly magnetic materials, electron spins do not pair, or cancel, completely. These are called **ferromagnetic materials**. In such materials there is a strong coupling or interaction between neighboring atoms, forming large groups of atoms, or **magnetic domains**, in which many of the electron spins are aligned. There aren't very many ferromagnetic materials. The most common are iron, nickel, and cobalt (gadolinium and certain alloys are also ferromagnetic).

In an unmagnetized ferromagnetic material, the domains are randomly oriented and there is no net magnetic effect (● Fig. 18.12a). But, when a ferromagnetic material is placed in an external magnetic field, two things may happen:

1. Domain boundaries change, and the domains with magnetic orientations parallel to the external field grow at the expense of the other domains.

2. The magnetic orientation of some domains may change slightly so as to be more aligned with the field (Fig. 18.12b).

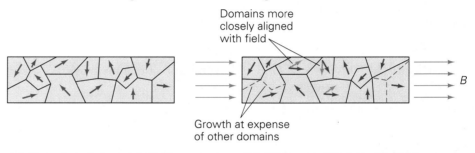

Domains more closely aligned with field

B

Growth at expense of other domains

(a) **No external magnetic field**   (b) **With external magnetic field**

● **FIGURE 18.12 Magnetic domains** **(a)** With no external magnetic field, the magnetic domains of a ferromagnetic material are randomly oriented, and the material is unmagnetized. **(b)** In an external magnetic field, domains with orientations parallel to the field may grow at the expense of other domains, and the orientations of some domains may become more aligned with the field. As a result, the material is magnetized, or exhibits magnetic properties.

Demonstration/activity: Put some iron filings in a Petri dish and place on a magnetic stirrer plate (as used in chemistry labs). Adjust the stirrer frequency and note movements of the filings (called Ampere's ants).

You now can understand why an unmagnetized piece of iron is attracted to a magnet and why iron filings line up with a magnetic field. Essentially, the pieces of iron become induced magnets.

Ferromagnetic materials are used to make electromagnets, usually by wrapping a wire around an iron core (•Fig. 18.13). Turning the current on and off, the magnet (or magnetic field) may be turned on and off at will. The iron used in an electromagnet is called soft iron. It is treated so that when the external field is removed, the magnetic domains quickly become unaligned and the iron is unmagnetized. ("Soft" does not refer to the metal's mechanical hardness but to its magnetic properties.)

When an electromagnet is on (Fig. 18.13a), the iron core becomes magnetized and adds to the field of the solenoid. The total field is expressed as

$$B = \mu n I$$

Notice that this equation is similar to that for the magnetic field of a solenoid (Eq. 18.7), but contains μ instead of μ_o. Here, μ expresses the **permeability** of the core material. The role permeability plays in magnetism is similar to that of permittivity ε in electricity: $\varepsilon = K\varepsilon_o$, where K is the dielectric constant of a material. For magnetic materials, we have

$$\mu = K_m\mu_o \qquad (18.8)$$

K_m, called the *relative permeability*, is the magnetic analog of the dielectric constant.

Both K and K_m are equal to one in a vacuum. Magnetic permeability is a useful material property. A core of a ferromagnetic material with a large permeability in an electromagnet can enhance its field thousands of times (compared to an air core).

Note that the strength of the field of an electromagnet depends on the current ($B = \mu n I$). Large currents produce large fields, but this field generation is accompanied by much greater joule heating (or $I^2 R$ losses). Large electromagnets need water-cooling coils to remove this heat. The problem may be alleviated by using a superconductor, which has zero resistance. However, superconducting magnets now in use also require tremendous cooling just to keep the conductors in the superconducting state. Perhaps someday we will have electromagnets made with new high-temperature superconductors, but currently these are ceramic materials which are not easily made into wires that can be wound in an electromagnet.

Iron that retains some magnetism after being in an external magnetic field is called "hard" iron and is used to make so-called permanent magnets. You may have noticed that a paper clip or a screwdriver blade becomes slightly magnetized after being near a magnet. Permanent magnets are produced by heating pieces of some ferromagnetic material in an oven and then cooling them in a strong magnetic field to get the maximum effect.

A *permanent* magnet is not really permanant, however—its magnetism can be lost. Hitting such a magnet with a hard object or dropping it on the floor may cause it to lose its domain alignment. Also, heating can cause a loss of magnetism because the resulting increase in random motions of atoms tends to unalign the domains. Above a certain critical temperature, called the **Curie temperature** (or Curie point), domain coupling disappears and a ferromagnetic material loses its ferromagnetism. The Curie temperature for iron is 770°C.

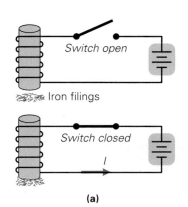

(a)

(b)

•**FIGURE 18.13 Electromagnet** (a) With no current in the circuit, there is no magnetism. However, with a current in the coil, there is a magnetic field and the iron core becomes magnetized, attracting other ferromagnetic materials. (b) An electromagnet in action picking up scrap metal.

See the Insight on superconductivity in Section 16.3.

Pierre Curie (1859–1906), a French physicist, discovered this effect and was the husband of Marie (Madame) Curie.

18.4 ■ Magnetic Forces on Current-Carrying Wires

An electric charge moving in a magnetic field experiences a force unless it is moving along or parallel to the field lines. Thus, a current-carrying wire in a magnetic field should experience such a force because a current is made up of moving charges. That is, the magnetic forces on the moving charges should give a resultant force on the wire conducting them, and this is the case.

As you know, a conventional current can be thought of as positive charges moving in a straight wire, as depicted in ● Fig. 18.14. In a time t, a charge (q) would move on the average through a length given by $\ell = vt$, where v is the drift velocity and is perpendicular to the B field. Since all the moving charges ($\Sigma_i\, q_i$) in this length of wire will experience a force due to the magnetic field, the magnitude of the total force on the length of wire ℓ is

$$F = \left(\Sigma_i\, q_i\right)vB = \left(\Sigma_i\, q_i\right)\left(\frac{\ell}{t}\right)B = \left(\frac{\Sigma_i\, q_i}{t}\right)\ell B$$

But $\Sigma_i\, q_i/t$ is the current (I) since it represents the charge passing through a cross-sectional area of the wire per unit time. Therefore,

$$F_{\text{max}} = I\ell B \tag{18.9}$$

Eq. 18.9 gives the maximum force on the wire (F_{max}) because the direction of the charges' motion (the current) is at an angle of 90° to the magnetic field. If the wire is not perpendicular to the field, but at some angle θ, then the force on the wire will be less. Similar to the force on a moving charge, for the general case, the force on a length of current-carrying wire is given by:

$$\boxed{F = I\ell B \sin \theta} \tag{18.10}$$

Note that if the wire is parallel to the field, the force on it is zero (as is the force on a charge moving parallel to a field).

The direction of the magnetic force on a current-carrying wire is also given by a right-hand rule. When the vector ℓ is turned by the fingers of the right hand toward **B**, the extended thumb points in the direction of **F**. The vector ℓ for the length of wire is in the direction of the *conventional current*. The right-hand rule is often stated in terms of this direction.

●**FIGURE 18.14 Force on a wire segment** The diagram helps visualize and determine the force on a length of current-carrying wire in a magnetic field. Charge carriers are considered to be positive for convenience. See text for description.

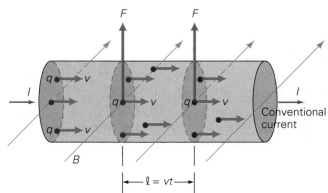

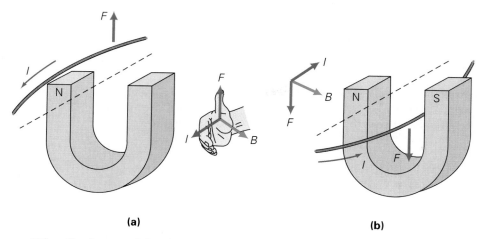

(a)

(b)

Figure 18.15

FIGURE 18.15 Force on a current-carrying wire The direction of the force is given by pointing the fingers of the right hand in the direction of the conventional curent *I* and then turning or curling them toward **B**. The extended thumb points in the direction of **F**. The force is upward for the case shown in (a) and downward for the case shown in (b).

When the fingers of the right hand are pointed in the direction of the conventional current *I* and then turned or curled toward the vector B, the extended thumb points in the direction of the magnetic force on the wire.

Two examples of magnetic forces on current-carrying wires are given in ● Fig. 18.15. Apply the right-hand rule in each case to see how the direction of the force is found.

EXAMPLE 18.5 ■ **Current, Magnetic Field, and Magnetic Force, and Definition of the Ampere**

Two long parallel wires are separated by a distance *d* and have conventional currents I_1 and I_2 going in the same direction, as illustrated in ● Fig. 18.16. There is a force on each wire because of the magnetic field produced by the current in the other wire. By a right-hand rule, the magnetic field due to I_1 at the second wire is into the page, and its magnitude is (Eq. 18.4)

$$B_1 = \frac{\mu_o I_1}{2\pi d}$$

The direction of B_1 is perpendicular to the wire carrying I_2, so from Eq. 18.9 the magnitude of the *force per unit length* (F_2/ℓ) on the wire is

$$\frac{F_2}{\ell} = I_2 B_1 = \frac{\mu_o I_1 I_2}{2\pi d}$$

By a right-hand rule, with the fingers in the direction of I_2 and turning them into $\mathbf{B}_1$, we see that the magnetic force on the wire carrying I_2 is *toward* the other wire.

The magnetic field due to I_2 also exerts a force per unit length on the wire carrying I_1, which by a similar analysis is given by

$$\frac{F_1}{\ell} = \frac{\mu_o I_1 I_2}{2\pi d}$$

Again, by a right-hand rule, this force is *toward* the other wire (and the forces are the equal and opposite forces of Newton's third law). Thus, two parallel wires carrying current in the same direction are attracted toward one another.

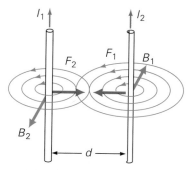

FIGURE 18.16 Mutual interaction For two parallel current-carrying wires, the magnetic field due to each current interacts with the other current, and there is a force on each wire. If the currents are in the same direction, the forces are attractive. See text for description and how this effect is used to define the ampere unit of current.

The magnetic force between parallel wires in an arrangement like that analyzed in Example 18.5 is used to define the ampere. According to the National Institute of Standards and Technology (formerly the National Bureau of Standards),

The ampere is defined as that current which, if maintained in each of two long parallel wires separated by a distance of 1 m in free space, would produce a force between the wires (due to their magnetic fields) of 2×10^{-7} N for each meter of wire.

Force on a Current-Carrying Loop

Another important application of magnetic force on a current is that on a current-carrying loop, for example, the rectangular loop in ● Fig. 18.17a. (The wires connecting the loop to a voltage source are not shown.) Suppose that the loop is free to rotate about an axis passing through opposite sides, as shown in the figure. There are no net forces or torques on the pivot sides of the loop (the sides through which the axis of rotation passes), because the forces on these sides are equal and opposite in the plane of the loop or zero when the sides are parallel to the field. (Why?) However, the equal and opposite forces on the other two sides of the loop (the sides parallel to the axis of rotation) produce torques that rotate the loop.

The magnitude of the magnetic force on each of the nonpivoted sides, whose length is L, if given by

$$F = ILB$$

Torque is discussed in Section 8.2.

since L and B are always at right angles to each other. Recall that the torque produced by a force is given by $\tau = r_\perp F$, where $r_\perp$ is the perpendicular lever arm from the axis of rotation to the line of the force. From Fig. 18.17b, we see that $r_\perp = \frac{1}{2}w \sin \theta$, where w is the width or length of the side of the loop and θ is the angle between the normal to the plane of the loop and the direction of the magnetic field. The total torque on the loop from both forces is

$$\tau = r_\perp F + r_\perp F = (\tfrac{1}{2}w \sin \theta)F + (\tfrac{1}{2}w \sin \theta)F = wF \sin \theta$$
$$= w(ILB) \sin \theta$$

Then, since wL is the area (A) of the loop, we may write for the magnitude of the torque on a pivoted, current-carrying loop,

$$\boxed{\tau = IAB \sin \theta} \tag{18.11}$$

Even though derived for a rectangular loop, Eq. 18.11 is valid for a flat loop of any shape.

The quantity IA is often referred to as the **magnetic moment**, represented by the symbol m (that is, $m = IA$). Thus we can rewrite Eq. 18.11 as

$$\tau = mB \sin \theta \tag{18.12}$$

The magnetic moment is a vector perpendicular to the plane of the loop, and the torque tends to line up this vector with the magnetic field.

A loop in a magnetic field will rotate, or experience a torque, until $\sin \theta = 0$ (when the forces producing the torque are parallel to the plane of

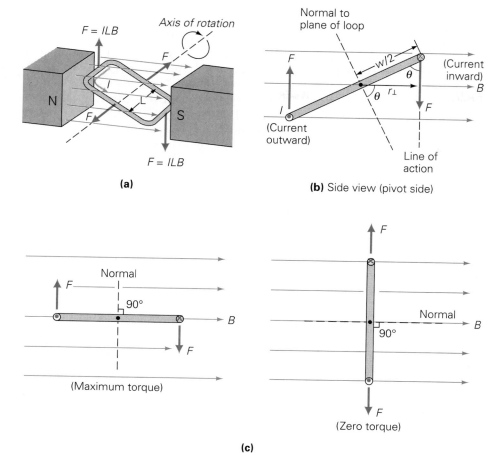

(a)

(b) Side view (pivot side)

(c)

Figure 18.17

FIGURE 18.17 Force and torque on a current-carrying pivoted loop **(a)** A current-carrying rectangular loop oriented in a magnetic field as shown here experiences a force on each of its sides. The forces on the sides parallel to the axis of rotation produce a torque that causes the loop to rotate. **(b)** A side view shows the geometry for determining the torque. See text for description. **(c)** Conditions of maximum and zero torque.

the loop, Fig. 18.17c). This occurs when the plane of the loop is perpendicular to the field. The loop then tends to stop rotating because the net torque becomes zero and the loop stops accelerating.

Current-carrying loops, even ones with such limited motion, have important applications. Since the current is directly related to the force, such a loop provides a means to measure current. Also, there are ways to keep a loop rotating, which allows continuous conversion of electrical energy to mechanical energy and is the operating principle of motors. These and other applications of electromagnetism are considered in the following section.

18.5 ■ Applications of Electromagnetism

With the principles you have learned so far about electromagnetic interactions, you can understand the operation of several widely used scientific instruments.

The Galvanometer

In the discussion of ammeters and voltmeters in Chapter 17, the galvanometer was considered to be simply a circuit element that measured small currents. See Section 17.4.

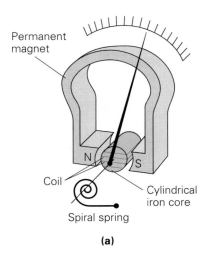

Permanent magnet

Coil

Cylindrical iron core

Spiral spring

(a)

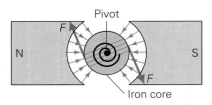

Pivot

N

F

F

S

Iron core

(b)

●**FIGURE 18.18 The galvanometer** (a) The deflection of the needle is proportional to the current in the coil. A galvanometer can therefore be used to detect and measure currents. (b) A magnet with curved pole faces is used so that the field lines are always perpendicular to the core surface and the torque does not vary with angle.

Now you are ready to see how it does this. As ● Fig. 18.18a shows, a galvanometer consists of a coil of wire loops on an iron core that is pivoted between the pole faces of a permanent magnet. When a current flows in the coil, the coil experiences a torque. A small spring supplies a counter (restoring) torque, and in equilibrium, a pointer indicates a deflection ϕ that is proportional to the current in the coil.

As you learned in Section 18.4, the torque on a current-carrying loop in a magnetic field is given by $\tau = IAB \sin \theta$. The dependence of the torque on the angle θ between the field lines and a line normal to the plane of the loop would cause a problem in measuring current with a galvanometer coil. As the coil rotates from its position of maximum torque ($\theta = 90°$, or the B field is perpendicular to the nonpivoted sides of the coil), the torque would become less, and the pointer deflection ϕ would not be proportional to the current alone. This problem is avoided by making the pole faces curved and wrapping the coil on a cylindrical iron core. The core tends to concentrate the field lines so that **B** is always perpendicular to the nonpivoted side of the coil (Fig. 18.18b).

The magnetic force is then always perpendicular to the plane of the coil, and the torque does not vary with the angle θ. Thus, the total torque on the coil is

$$\tau = NIAB \qquad (18.13)$$

where N is the number of loops in the coil.

The restoring torque (τ_s) of the spring in a galvanometer is given by a rotational form of Hooke's law ($F = kx$ for linear restoring force).

$$\tau_s = k\phi \qquad (18.14)$$

Here ϕ is the deflection angle as measured by the pointer. With a current in the coil, the coil and the pointer will rotate until the torque on the coil is balanced by that on the spring. In equilibrium, the magnitudes of the torques are equal, and by combining Eqs. 18.13 and 18.14 we obtain

$$\phi = \frac{NIAB}{k} \qquad (18.15)$$

Thus, the pointer deflection (ϕ) is directly proportional to the current (I).

A properly calibrated galvanometer can be used to measure small currents (in the microamp range). With an appropriate shunt resistor, a galvanometer forms the basis of an ammeter for measuring larger currents. Also, with an appropriate series multiplier resistor, the galvanometer forms a voltmeter.

The dc Motor

In general, *a motor is a device that converts electrical energy into mechanical energy.* An example of such a conversion occurs with a galvanometer: A current causes the coil to rotate mechanically. However, a galvanometer is not considered to be a motor. A practical motor must have continuous rotation for continuous energy output.

A pivoted, current-carrying loop in a magnetic field will rotate freely, but for only a half-cycle, or through a maximum angle of 180°. Recall that $\tau = IAB \sin \theta$, and when the magnetic field is perpendicular to the plane of the loop ($\sin \theta = 0$), the torque is zero and the loop is in equilibrium.

To provide for continuous rotation, the current is reversed every half turn so that the torque forces are reversed. This is done by means of a split-ring commutator, an arrangement of two metal half-rings insulated from each other (●Fig. 18.19a). The ends of the wire that forms the loop are fixed to the half-rings, and the current is supplied to the loop through the commutator by means of contact brushes. Then, with one half-ring electrically positive (+) and the other negative (−), the loop and ring rotate. When they have gone through half a rotation, the half-rings come in contact with the opposite brushes. This reverses their polarity, and the current flows in the loop in the opposite direction. This changes the directions of the torque forces (Fig. 18.19b). The torque is zero at the equilibrium position, but the loop is in unstable equilibrium. The loop has enough inertial motion to move through this point, and it rotates through another half-cycle. The process repeats in continuous operation. Of course, a real motor has many more loops or windings on its rotating shaft or *armature*.

The Cathode Ray Tube (CRT)

The **cathode ray tube (CRT)** is a vacuum tube that is used in an oscilloscope (●Fig. 18.20a). Electrons are "boiled off" a hot filament in an electron gun and accelerated through a potential difference. Because they are emitted from the negative terminal, or cathode, the electrons are called cathode rays. (The posi-

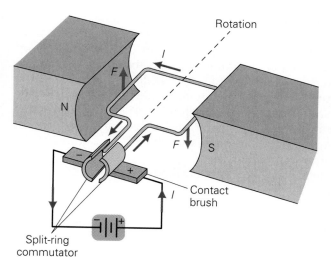

(a)

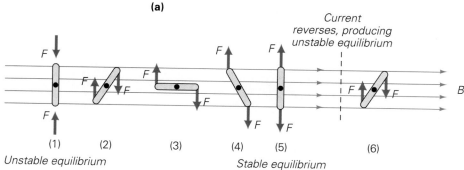

(b) End view of loop showing loop rotation sequence

Figure 18.19

●**FIGURE 18.19 dc motor** (a) A split-ring commutator reverses the polarity and current each half-cycle, so the loop rotates continuously. (b) An end view shows the forces on the loop and its orientation during a half-cycle.

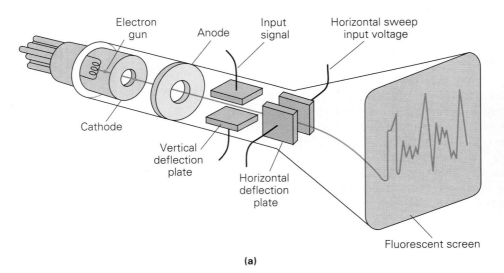
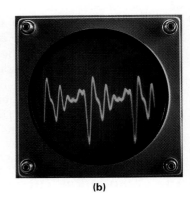

(a)

(b)

tive terminal toward which the electrons are accelerated is the anode.) In 1897, the English physicist J. J. Thomson used a type of cathode ray tube to demonstrate the existence of the electron. Previously, it had not been known that the rays emitted from the hot filament of such a tube were electrons.

The electrons pass through a small hole in the anode, forming a beam. The beam passes through two sets of parallel plates, one for vertical deflections and the other for horizontal deflections. Input voltages (which usually vary with time) are placed across the plates, and the deflected beam makes patterns on a screen. (The fluorescent material coating the inside of the screen glows when struck by the electrons.) In normal operation, the horizontal input voltage from an internal voltage source sweeps the dot illuminated by the beam across the screen periodically at a constant rate. The vertical deflection is proportional to an external voltage input. Thus, the display on a CRT is a graph of the external voltage versus time (Fig. 18.20b).

The picture tube of a television set is also a cathode ray tube (Fig. 18.21).

FIGURE 18.21 Television tube (a) A television picture tube is a CRT. (b) The beam scans every other line on the screen in one downward scan that takes $\frac{1}{60}$s and then scans the lines in between in a second scan of $\frac{1}{60}$s. This yields a complete picture of 525 lines in $\frac{1}{30}$s.

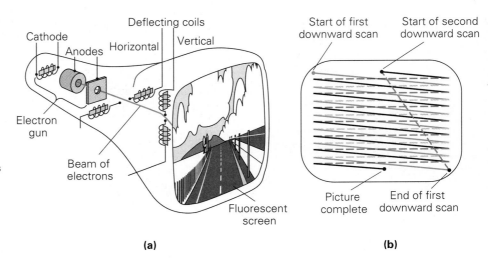

(a)

(b)

In this case, magnetic coils are usually used instead of plates to deflect the electron beam. The beam scans the fluorescent screen in a 525-line pattern in a fraction of a second. For a black-and-white TV, the signals transmitted from a video camera reproduce an image on the screen as a mosaic of light and dark dots, depending on whether or not the intermittent beam is on or off at a particular instant.

Producing the images of color TV is somewhat more involved. A common color picture tube has three beams, one for each of the primary colors (Chapter 24). Phosphor dots on the screen are arranged in groups of three (triads) with one dot for each of the primary colors. The excitation of the appropriate dots and the resulting combination of colors produce a color picture. There are also single-beam color picture tubes in which color dots are excited sequentially in rapid succession to produce the image on the screen.

The fact that a picture tube is a CRT can be demonstrated by bringing a magnet near a black-and-white TV screen. The magnetic field will deflect the electrons and distort the picture. *Do not try this on a color TV screen!* Internal metallic parts may become magnetized, causing the picture to be permanently distorted.

The Mass Spectrometer

Have you ever thought about how the mass of an atom or molecule is measured? Electric and magnetic fields provide a way in one form of a **mass spectrometer** (often called a mass spec for short). Actually, the masses of ions are measured since electric and magnetic fields have motional effects only on charged particles. (Recall that an ion is an atom or molecule with a net electric charge.)

Ions with a known charge $(+q)$ are produced by heating a substance. A beam of ions introduced into the mass spec has a wide distribution of speeds. Ions with a particular velocity are selected by setting the values of the electric and magnetic fields between the plates of a *velocity selector* (● Fig. 18.22). For a positively charged ion, the electric field produces a downward force (magnitude $F_e = qE$), and the magnetic field produces an upward force (magnitude $F_m = qvB_1$). [In the figure, a cross ($\times$) means into the page, and a dot ($\cdot$) means out of the page; you can visualize these symbols as though observing the feathered end and tip of an arrow.] If the beam is not deflected, the resultant

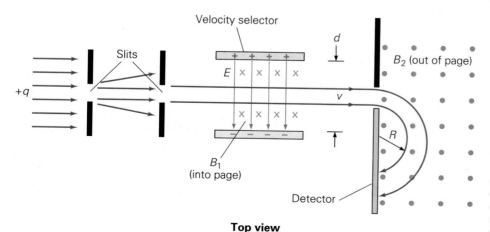

FIGURE 18.22 Principles of the mass spectrometer Ions pass through the velocity selector; those with a particular velocity enter a magnetic field (B_2). The particles are deflected, with the radius of the circular path depending on the mass of the particle. Paths of two different radii indicate that the beam contains particles of two different masses.

Top view

force must be zero, so

$$qE = qvB_1$$

Compare situations where $v > (E/B)$ and $v < (E/B)$

or

$$v = \frac{E}{B_1}$$

If E is written as V/d, since the voltage and the plate separation distance are the controllable or measured quantities,

$$v = \frac{V}{B_1 d} \qquad (18.16)$$

Thus, a beam with a known speed can be selected by varying the values of V, B, and d.

The beam then passes through a slit into another magnetic field ($\mathbf{B_2}$), which is perpendicular to the direction of the beam. The force due to this magnetic field (magnitude $F = qvB_2$) is always perpendicular to the velocity of the ions, which are deflected in a circular arc. The magnetic force supplies the centripetal force for this motion, and

.A thought experiment to help with the concept of a moving charge in a magnetic field: What would the world be like if the Earth's gravitational field acted on mass in the same way that a magnetic field acts on charge?

$$\frac{mv^2}{R} = qvB_2$$

Using the expression for the selected ion velocity (Eq. 18.16),

$$m = \left(\frac{qdB_1B_2}{V}\right)R \qquad (18.17)$$

The greater the mass of an ion, the greater the radius of the circular path. Two circular paths of different radii are shown in Fig. 18.22, which indicates that the beam contains ions of two different masses. If the radius of curvature R for an ion beam is measured (e.g., by recording the position of the beam with a detector), the mass of the ion can be calculated using the other known quantities in Eq. 18.17.

Another type of mass spectrometer is shown in ● Fig. 18.23. It is known as a "time-of-flight" mass spectrometer. As illustrated in the drawing, a gas enters an evacuated tube where an electron beam produces positive ions. The ions are accelerated down the tube by a potential difference V, acquiring a kinetic energy of $qV = \frac{1}{2}mv^2$. The time of flight of the ions down the tube is measured, and, knowing the length of the tube, their speed can be found. The mass of a particular ion can then be determined from the previous equation.

● **FIGURE 18.23 Mass spectrometer** (a) Diagram of a "time-of-flight" mass spectrometer tube. In the evacuated tube, the time for ions to reach the detector is measured. Heavier ions take longer to travel down the tube. (b) Computer display of a mass spectrometer. The molecule being analyzed is myoglobin, a protein that stores oxygen in muscle tissue. Each of the peaks on the display represents the mass of a particular ionized fragment of the molecule. Such mass spectra can help to determine the composition and structure of large molecules. The mass spectrometer can also be used to identify tiny amounts of a particular molecule in a complex mixture.

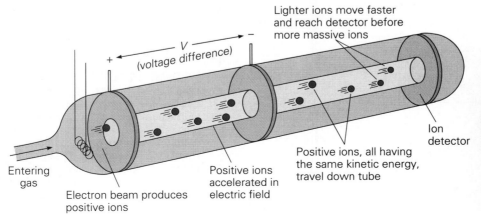

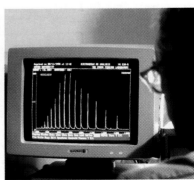

(a)

(b)

Notice that there are different ions represented in the figure. Most naturally occuring elements consist of different species of atoms, called isotopes. Isotopes of an element have the same number of protons in their nuclei but different numbers of neutrons (see Chapter 28). As a result, isotopes of the same element have different masses. As ions, all of the isotopes in a gas receive the same energy (qV) when accelerated down the tube. However, the more massive ions travel more slowly and arrive at the detector later than less massive ions. If the detector is connected to a device that measures the number of ions of a given mass that reach the detector in a given time, the relative abundance of each isotope of an element can be determined.

Mass spectrometers have many functions in modern laboratories. For example, they are used to help establish the age of ancient rocks and more recent human artifacts by measuring the relative abundance of different isotopes that they contain. (This method, called radioactive dating, is discussed in Section 28.3.) Among their other roles are tracking short-lived intermediates in studies of the biochemistry of living organisms; helping to determine the structure of large organic molecules; analyzing the composition of complex mixtures, such as the output of a petroleum refinery; and detecting tiny amounts of impurities in metal alloys and semiconductors.

The Electronic Balance

Traditional laboratory balances measure mass by balancing the weight force of an unknown mass against that of a known mass. The newer, digital electronic balances (Fig. 18.24a) work on a different principle. There is still a suspended beam with a pan on one end that holds the object to be weighed (or massed), but there is no known mass. The balancing downward force is supplied by a current-carrying coil of wire in the field of a permanent magnet (Fig. 18.24b). The coil moves up and down in the cylindrical gap of the magnet, and the downward force is proportional to the current in the coil. The mass of the object in the pan is determined from the coil current that produces a force just sufficient to balance the beam.

The current required to produce balance is controlled automatically. This is done by means of light photosensing and a feedback loop. When the beam

FIGURE 18.24 Electronic balance (a) A digital electronic balance. (b) Diagram of the principle of an electronic balance. The balance force is supplied by electromagnetism. See text for description.

(a)

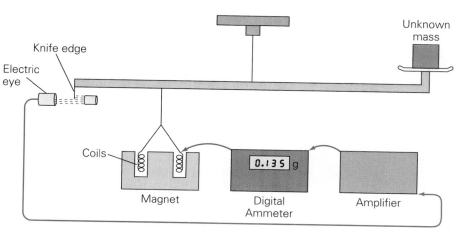

Electric eye

Knife edge

Coils

Magnet

Digital Ammeter

0.135 g

Amplifier

Unknown mass

(b)

is horizontal or balanced, a knife-edge obstruction cuts off part of the light from a source that falls on a photosensitive "electric eye" (see Chapter 26). The resistance of the electric eye is a function of the amount of light falling on it. This resistance controls the current that an amplifier sends through the coil. For example, if the beam tilts so that the knife-edge raises and more light strikes the eye, the current in the coil is increased to counterbalance the tilting. In this manner, the beam is electronically maintained in nearly horizontal equilibrium. The current that keeps the beam in the horizontal position is read out on a digital ammeter that is calibrated in grams or milligrams instead of amperes.

18.6 ■ The Earth's Magnetic Field ◇

The magnetic field of the Earth has been used for centuries. In ancient times, navigators used lodestones or magnetized needles to show them where north was. An early study of magnetism was done by the English scientist Sir William Gilbert around 1600. In investigating the magnetic field of a specially cut spherical lodestone that simulated the Earth, he came to believe that the magnetic field was associated with the entire Earth, or that the Earth as a whole acted as a magnet. Gilbert thought that the field might be produced by a large body of permanently magnetized material within the Earth.

In fact, the Earth's magnetic field has approximately the configuration that would be produced by a large interior bar magnet, as ● Fig. 18.25 shows. The magnitude of the horizontal component of the Earth's magnetic field at the magnetic equator is about 10^{-5} T, and the vertical component at the geomagnetic poles is about 10^{-4} T. It has been calculated that for a ferromagnetic material of maximum magnetization to produce this field, it would have to occupy about 0.01% of the Earth's volume.

The idea of a ferromagnet of this size within the Earth may not seem unreasonable at first, but this cannot be a true model. The interior temperature of the Earth is well above the Curie temperatures of iron and nickel, the ferromagnetic materials believed to be the most abundant in the Earth's interior. For iron, this temperature is 770°C, which is attained at a depth of only 100 km. The temperature of the Earth is higher at greater depths. This means that such a permanent internal magnet is impossible.

The fact that an electric current produces a magnetic field has led scientists to speculate that the Earth's magnetic field is associated with motions in the liquid outer core. (Recall that the magnetic field of a loop of wire or a solenoid is similar to that of a bar magnet—compare Figs. 18.9 and 18.10 to Fig. 18.1.) These motions may be associated in some way with the Earth's rotation. Jupiter and Saturn have magnetic fields much larger than that of Earth. These planets are largely gaseous and rotate rapidly, with periods of 10–11 hours. Mercury and Venus, on the other hand, have very weak magnetic fields. These planets are more like Earth in density and rotate relatively slowly, with periods of 58 and 243 days, respectively.

Several theoretical models have been proposed to explain the Earth's magnetic field. For example, it has been suggested that it arises from currents associated with thermal convection cycles in the liquid outer core caused by heat from the inner core. But the details of the mechanism are not clear.

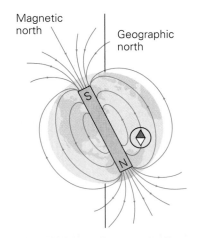

Magnetic north

Geographic north

●FIGURE 18.25 Geomagnetic field The Earth's magnetic field is similar to that of a bar magnet. However, a permanent magnet could not exist within the Earth because of the high temperatures there. The field is believed to be associated with motions in the liquid outer core.

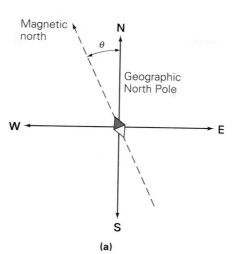

FIGURE 18.26 Magnetic declination
(a) The angular variation between magnetic north and "true" or geographic north is called the magnetic declination. **(b)** The magnetic declination varies with location. The map shows isogonic (same magnetic declination) lines for the conterminous United States. For locations on the 0° line, magnetic north is in the same direction as true (geographic) north. On either side of this line, a compass has an easterly or westerly variation. For example, on a 20°W line, a compass has a westerly declination of 20° (magnetic north is 20° west of true north).

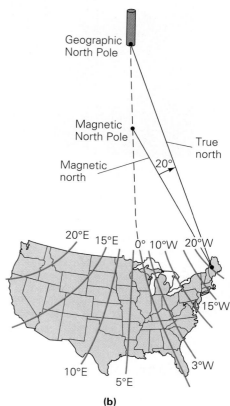

(b)

The Earth's magnetic poles and geographic poles do not coincide. Hence, a compass indicates the direction of magnetic north, not "true" or geographic north. The angular variation in the directions is called the *magnetic declination* (Fig. 18.26). As shown on the map in the figure, the magnetic declination varies for different locations. Knowing the variations in the magnetic declination is particularly important in navigation, as you can imagine.

Charged particles from the Sun and cosmic rays entering the Earth's magnetic field give rise to other phenomena. A charged particle, entering a uniform magnetic field not perpendicular to the field, spirals in a helix (Fig. 18.27a). This is because there is a component of the particle's velocity parallel to the field, and therefore a force perpendicular to the field (by a right-hand rule: **v** turned into **B**). The motions of charged particles in a non-uniform field are

> Demonstration/activity: Turn on a CRT such as a computer display and mark the location of something on the screen with a piece of tape. Then turn the CRT upside down and notice how the printing on the screen has moved to one side. Explain. Do all CRT's behave this way?

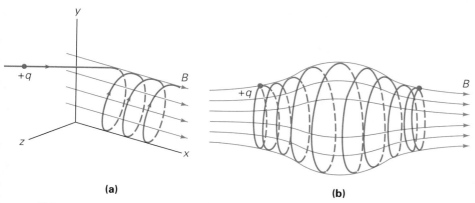

(a) **(b)**

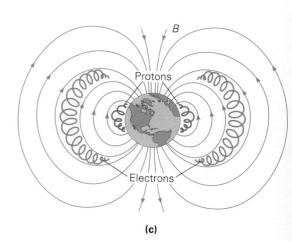

(c)

FIGURE 18.27 Magnetic confinement **(a)** A charged particle entering a uniform magnetic field at an angle moves in a spiraling path. **(b)** In a nonuniform, bulging magnetic field, particles spiral back and forth as though confined in a magnetic bottle. **(c)** Particles are trapped in the Earth's magnetic field, and the regions where they are concentrated are called Van Allen belts.

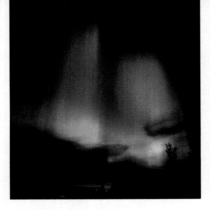

FIGURE 18.28 Aurora borealis—the northern lights This spectacular aurora display is believed to be caused by energetic solar particles trapped in the Earth's magnetic field. The particles excite or ionize air molecules, and on de-excitation (or recombination), light is emitted. Aurorae are most commonly observed in the polar regions, where the solar particles are concentrated by the Earth's magnetic field.

quite complex. However, for a bulging field such as that depicted in Fig. 18.27b, analysis shows that the particles spiral back and forth as though in a magnetic container, a so-called bottle. This is an important consideration in confining a plasma (a gas of charged particles) in fusion research (Chapter 29).

An analogous phenomenon occurs in the Earth's magnetic field, giving rise to regions where there are concentrations of charged particles. Two large donut-shaped regions at altitudes of several thousand kilometers are called the Van Allen belts (Fig. 18.27c). It is in the lower Van Allen belt that light emissions called aurorae occur—the aurora borealis, or northern lights, in the Northern Hemisphere and the aurora australis, or southern lights, in the Southern Hemisphere. These eerie, flickering lights are most commonly observed in the Earth's polar regions, but have been seen at all latitudes (● Fig. 18.28).

It is believed that an aurora is created when charged solar particles are trapped in the Earth's magnetic field. Maximum aurora activity is noted to occur after a solar disturbance, such as a solar flare. Solar flares are violent magnetic storms on the Sun that spew out enormous quantities of charged particles and radiation. Trapped in the Earth's magnetic field, the charged particles are guided toward the polar regions. They excite or ionize oxygen and nitrogen molecules in the atmosphere. When the excited molecules return to their normal state and ions regain their normal number of electrons, light is emitted—the glow of the aurora.

Important Concepts

You should be able to define and explain these chapter concepts clearly.

law of poles	tesla (T)	magnetic domains	Curie temperature	cathode ray tube (CRT)
magnetic field	ferromagnetic materials	permeability	magnetic moment	mass spectrometer

Important Relationships for Review

These relationships are mathematical statements of the concepts and principles presented in the chapter. You should be able to identify the symbols and to explain the relationships before proceeding to the Exercises. In-text equation reference numbers are given for convenience.

Magnitude of the Magnetic Field on a Moving Charge (q):

$$F = qvB \sin \theta \qquad (18.3)$$

Directional right-hand rule: When the fingers of the right hand are pointed in the direction of **v** and then turned or curled toward the vector **B**, the extended thumb points in the direction of the force **F** on a *positive* charge. (**F** is in the opposite direction for a negative charge.)

Magnitude of the Magnetic Field Around a Long, Straight Current-Carrying Wire:

$$B = \frac{\mu_o I}{2\pi d} \qquad (18.4)$$

Magnitude of the Magnetic Field at the Center of Circular Loop of Current-Carrying Wire:

$$B = \frac{\mu_o I}{2r} \qquad (18.5)$$

Magnitude of the Magnetic Field in a Solenoid (along the axis):

$$B = \frac{\mu_o NI}{L} \qquad (18.6)$$

or

$$B = \mu_o nI \quad \text{(where } n = N/L) \qquad (18.7)$$

Directional right-hand rule: When a current-carrying wire is grasped with the right hand with the the extended thumb pointing in the direction of the conventional current, the curled fingers indicate the circular sense of the magnetic field. (The field vector is tangent to the circular field line at any point on the circle.)

Permeability of Free Space:

$$\mu_o = 4\pi \times 10^{-7} \text{ T-m/A}$$

Magnetic Permeability:

$$\mu = K_m \mu_o \qquad (18.8)$$

Magnitude of Force on a Straight Current-Carrying Wire:

$$F = I\ell B \sin\theta \qquad (18.10)$$

Directional right-hand rule: When the fingers of the right hand are pointed in the direction of the conventional current I and then turned or curled toward the **B** vector, the extended thumb points in the direction of the force on the wire.

Magnitude of Torque on a Current-Carrying Loop:

$$\tau = IAB \sin\theta \qquad (18.11)$$

Magnitude of Restoring Spring Torque in a Galvanometer:

$$\tau_s = k\phi \qquad (18.14)$$

Pointer Deflection in a Galvanometer:

$$\phi = \frac{NIAB}{k} \qquad (18.15)$$

Velocity Selector (mass spectrometer):

$$v = \frac{E}{B_1} = \frac{V}{B_1 d} \qquad (18.16)$$

Particle Mass (mass spectrometer):

$$m = \left(\frac{qdB_1B_2}{V}\right)R \qquad (18.17)$$

Exercises

18.1 ▪ Magnets and Magnetic Poles

1 When two bar magnets are near each other, how many separate magnetic force interactions are there? (a) 1, (b) 2, (c) 4, (d) the number depends on the orientation of the magnets. (c)

2 If a compass is placed in a horizontal magnetic field, it will point (a) perpendicular to the field lines, (b) with the south pole in the opposite direction of the field, (c) at a nonzero angle $\theta < 90°$ to the field lines, (d) none of these. (b)

3 Given two identical iron bars, one of which is a permanent magnet and the other unmagnetized, how could you tell the difference using only the two bars? see ISM

18.2 ▪ Electromagnetism and the Source of Magnetic Fields

4 A magnetic field (a) has units of N/(A-m), (b) is effectively the force per moving charge, (c) always produces a force on a moving charge, (d) both (a) and (b). (d)

5 A proton moving along the x axis enters a uniform magnetic field in the y direction as it passes the origin. As it travels in the field, the proton (a) will experience a force in the $-y$ direction, (b) will experience a force in the $-z$ direction, (c) will move in a circular path in the x-z plane, (d) none of these. (a)

● **6** If a negatively charged particle were moving downward along the right edge of this page, which way should a magnetic field be oriented so that the particle would initially be deflected to the left? into the page

7 A magnetic field can be used to determine the sign of charge carriers. Consider a wide conducting strip in a magnetic field oriented as shown in ● Fig. 18.29. The charge carriers are deflected by the magnetic force and accumulate on one side of the strip, giving rise to the measureable voltage across it. (This is known as the Hall effect.) If the sign of the charge carriers is unknown (they are either positive charges moving as indicated by the arrows in the figure, or negative charges moving in the opposite direction), how does the measured voltage allow this to be determined? see ISM

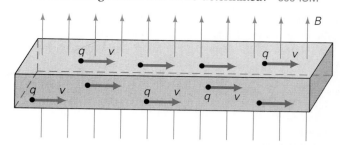

● **FIGURE 18.29 The Hall effect** See Exercise 7.

8 ▪ A positive charge of 0.25 C moves horizontally at a speed of 2.0×10^2 m/s and enters a magnetic field of 0.40 T directed vertically downward. What is the magnitude of the force on the charge? 20 N

9 ▪ A long, straight wire carries a current of 2.5 A. Find the magnitude of the magnetic field 75 cm from the wire. 6.7×10^{-7} T

10 ▪ A loop of wire with a diameter of 40 cm carries a current of 0.50 A in a clockwise sense, as viewed from above. Find the magnetic field at the center of the loop. 1.6×10^{-6} T, downward

11 ■■ An electron traveling at a uniform speed of 100 m/s in the $+x$ direction enters a uniform magnetic field and experiences a maximum force of 5.0×10^{-15} N in the $+y$ direction. What is the magnitude and direction of the magnetic field? 3.1×10^2 T in $+z$ direction

12 ■■ An electron travels at a speed of 2.0×10^4 m/s through a uniform magnetic field whose magnitude is 900 T. What is the magnitude of the force on the electron if its velocity vector and the magnetic field vector (a) are perpendicular, (b) make an angle of $45°$ or (c) are parallel?
(a) 2.9×10^{-12} N   (b) 2.0×10^{-12} N   (c) 0

13 ■■ A proton has a velocity of magnitude v near the equator. In which direction will the proton experience a force due to the Earth's magnetic field if its velocity is directed (a) due south, (b) northwest, (c) upward?
(a) 0   (b) down   (c) west

● **14** ■■ A horizontal proton beam in a particle accelerator is accelerated to a constant speed of 3.0×10^5 m/s. The beam enters a uniform magnetic field of 0.50 T oriented at an upward angle of $37°$ relative to the direction of the beam. (a) What is the initial acceleration experienced by a proton in the beam? (b) If the beam were made up of electrons, what would be the difference in the forces on the particles as the beams enter the magnetic field?   (a) 8.6×10^{12} m/s² (b) same magnitude but in opposite direction

15 ■■ A beam of protons is accelerated to a speed of 5.0×10^6 m/s in a particle accelerator and emerges horizontally from the accelerator into a uniform magnetic field. What B field will keep the beam moving exactly horizontally?
2.0×10^{-14} T, left, looking in direction of beam

16 ■■ The magnetic field at the center of a 250-turn coil of radius 15 cm is 6.0 mT. Find the current in the coil.   5.7 A

17 ■■ What current in a long, straight wire will produce a magnetic field with a magnitude of 5.0μT at a perpendicular distance of 20 cm from the wire?   5.0 A

● **18** ■■ Three particles enter a uniform magnetic field as shown in Fig. 18.30a. Particles 1 and 3 have equal speeds and charges of the same magnitude. What can you say about (a) the charges of the particles and (b) their masses?
see ISM

19 ■■ It is desired to deflect a positively charged particle in an "S" path as shown in Fig. 18.30b using only magnetic fields. (a) Explain how this could be done. (b) How does the magnitude of the emerging velocity v compare to that of the initial velocity v_0?   see ISM

20 ■■ A solenoid 50.0 cm long with 2000 turns/m has a central magnetic field of 0.75 mT. Find the current in the coil.   0.30 A

● **21** ■■ Two long, parallel wires separated by 50 cm carry currents of 8.0 A each in a horizontal direction. Find the magnetic field midway between the wires if the curents are (a) in the same direction and (b) in opposite directions.
(a) 0   (b) 1.28×10^{-5} T

22 ■■ A long, straight wire has a resistance of 2.0 Ω. What potential difference between its ends will produce a magnetic field of 10^{-4} T at a distance of 4.0 cm from the wire?   40 V

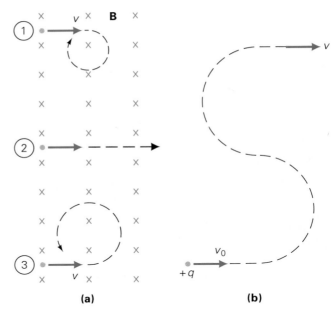

(a)   (b)

● **FIGURE 18.30 Charges in motion**   See Exercises 18 and 19.

23 ■■ Two long, parallel wires separated by 0.20 m carry equal currents of 1.5 A in the same direction. Find the magnetic fields 0.15 m away from each wire on the side opposite the other wire on a line running perpendicularly through both wires. 2.9×10^{-6} T

24 ■■ Two long, parallel wires carry currents of 8.0 A and 2.0 A (● Fig. 18.31). (a) What is the magnitude of the magnetic field at the point midway between the wires? (b) Where on a line perpendicular to and joining the wires is the magnetic field zero?
(a) 2.0×10^{-5} T   (b) 9.6×10^{-2} m from wire 1

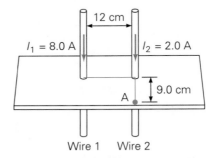

● **FIGURE 18.31 Parallel current-carrying wires**   See Exercises 24, 26, 27, 64, and 65.

25 A proton is accelerated from rest to a velocity of 4.0×10^6 m/s in a horizontal direction. It then enters a vertical magnetic field of 3.0 T. (a) Calculate the radius of the circular path of the particle. (b) Find the electric field that would cause the proton to move in a straight line.
(a) 1.4×10^{-2} m   (b) 1.2×10^7 N/C

● **26** ■■ Find the magnetic field (magnitude and direction) at point A in Fig. 18.31, which is located 9.0 cm away from wire 2 on a line perpendicular to the line joining the wires.
1.4×10^{-5} T at $39°$

27 ■■ Suppose that the current in wire 1 in Fig. 18.31 were in the opposite direction. (a) What would be the magnetic field at a point midway between the wires? (b) Where on a line perpendicular to and joining the wires would the magnetic field be zero?   (a) 3.3×10^{-5} T   (b) no place

28 ■■ A circular loop of wire with a diameter of 12 cm is horizontal and carries a current of 0.60 A in a clockwise sense, as viewed from above. What is the magnetic field (magnitude and direction) at the center of the loop?
6.3×10^{-6} T, down

29 ■■ How much current must flow in a circular loop with a radius 10 cm to produce a magnetic field at its center that is approximately the same magnitude as the horizontal component of the Earth's magnetic field at the Equator?
1.6 A

30 ■■ A wire with four circular loops of radius 5.0 cm carries a current of 2.0 A in a clockwise sense, as viewed along a line perpendicular to the circles formed by the loops. What is the magnetic field at the center of the loops?
1.0×10^{-4} T, away from the observer

31 ■■ A circular loop of wire with a radius of 5.0 cm carries a current of 1.0 A. Another circular loop of wire is concentric with the first (has a common center with it) and carries a current of 2.0 A. The magnetic field at the center of the loops is measured to be zero. What is the radius of the second loop?   10 cm

32 ■■ A solenoid 10 cm long is wound with 3000 turns of wire. How much current must flow through the windings to produce a magnetic field of 2.5×10^{-3} T at the solenoid's central axis? 6.6×10^{-2} A

33 ■■ An electron has a kinetic energy of 20 keV. It enters a uniform magnetic field of 35 T perpendicular to its line of motion. What is the radius of the resulting circular path?
1.4×10^{-5} m

34 ■■ A proton travels in a circular path with a radius of 60 mm in a uniform magnetic field of 12 μT. What is the kinetic energy of the proton? 4.0×10^{-24} J

35 ■■■ Two long, perpendicular wires carry currents of 15A, as illustrated in ● Fig. 18.32. What is the magnetic field at the midpoint of the line joining the wires? 4.2×10^{-5} T, 45° (from page)

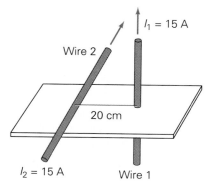

● **FIGURE 18.32 Perpendicular current-carrying wires**   See Exercise 35.

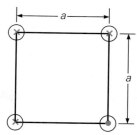

● **FIGURE 18.33 Current-carrying wires in a square array**   See Exercise 36.

36 ■■■ Four wires running through the corners of a square with sides of length a as shown in ● Fig. 18.33 carry equal currents I. Calculate the magnetic field at the center of the square in terms of these parameters. $B = (1.4)\mu_0 I/\pi a$, at 45° (3rd quadrant)

37 ■■■ A solenoid is wound with 200 turns/cm. An outer layer of insulated wire with 180 turns/cm is wound over the first layer of wire. In operation, the inner coil carries a current of 5.0 A, and the outer coil carries a current of 7.5 A in the direction opposite to that of the current in the inner coil. What is the magnitude of the magnetic field at the central axis of this doubly wound solenoid? 4.4×10^{-2} T

38 ■■■ A fast-moving electron moves in a horizontal plane at a right angle to a uniform vertical magnetic field. (a) What is the frequency of the electron's revolution? (This frequency is called the cyclotron frequency.) (b) Show that the time required for the electron (or any charged particle) to make one complete revolution is independent of its speed and radius. (c) Compute the path radius and frequency for an electron if $v = 10^5$ m/s and $B = 10^{-4}$ T.   (a) $f = qB/2\pi m$ (b) $t = 2\pi m/(qB)$ (c) 5.7×10^{-3} m; $f = 2.8 \times 10^6$ Hz

18.3 ■ Magnetic Materials

39 The main source of magnetism in magnetic materials is from (a) electron orbits, (b) electron spin, (c) magnetic poles, (d) nuclear properties.   (b)

40 When a ferromagnetic material is placed in an external magnetic field, (a) the domain orientation may change slightly, (b) the domain boundaries may change, (c) new domains are created, (d) both (a) and (b).   (d)

41 Can a magnetic monopole be obtained by continually breaking a bar magnet in two? Explain.   see ISM

● **42** ■■ The magnitude of the magnetic moment (m) of a current loop is the product of the current (I) and the cross-sectional area of the loop (A): $m = IA$. Show that for an atomic electron in a circular orbit of radius r at an orbital speed v the magnetic moment is $m = evr/2$.   see ISM

43 ■■ What is the magnetic moment of the electron in a hydrogen atom, which travels around the nuclear proton at a radius of 0.053 nm? [Hint: Find the electron's speed by considering the centripetal force and see Exercise 42.]
9.3×10^{-24} A-m^2

44 ■■ What is the magnitude of the magnetic moment of a rectangular loop of wire that measures 10 cm by 15 cm and carries a current of 2.0 A? 3.0×10^{-3} A-m²

45 ■■ What is the magnetic field at the center of the circular orbit of the electron in a hydrogen atom (the orbital radius is 0.053 nm)? [*Hint:* Find the electron's period by considering the centripetal force.]   13 T

46 ■■ A solenoid with 90 turns/cm has an iron core with a relative permeability of 2000. The solenoid carries a current of 1.6 A. (a) What is the magnetic field at the central axis of the solenoid? (b) How much greater is the magnetic field with the iron core than it would be without it?   (a) 36 T (b) $B = (2.0 \times 10^3)B_o$

18.4 ■ Magnetic Forces on Current-Carrying Wires

47 A long straight, horizontal wire has a conventional current directed toward the north. In what direction would the needle point if a compass were placed above the wire: (a) east, (b) west, (c) south, (d) upward?   (a)

48 Hold up your left hand, palm to the right, the index and middle fingers forming a V. If the middle finger pointed in the direction of a magnetic field and the index finger in the direction of a conventional current in a straight wire, in what direction would the force on the wire be: (a) downward, (b) toward the back of the hand, (c) out of the palm, (d) none of these?   (b)

49 Electric fields have equipotentials (Section 15.5), or paths along which no work is done in moving an electric charge from one point to a different point. Are there analogous paths for magnetic fields? Explain.   see ISM

50 Two straight wires are positioned at right angles to each other as in Fig. 18.32. If one wire carries a conventional current toward the north and the other a current toward the east, what is the direction of the force on each wire?   see ISM

51 ■ A straight, horizontal segment of wire is 1.0 m long and carries a current of 5.0 A in the $+x$ direction in a magnetic field of 0.60 T that is directed vertically upward. What is the force on the wire (magnitude and direction)?   3.0 N, $-y$ direction (taking $+z$ as upward)

52 ■ A 2.0-m length of straight wire carries a current of 20 A in a uniform magnetic field of 50 mT whose field lines make an angle of 37° with the direction of the conventional current. Find the force on the wire.   1.2 N perpendicular to the plane of B and I

53 ■ Use a right-hand rule to find the direction of the conventional current in a wire in a magnetic field that results in the force on the wire shown for each case in Fig. 18.34. (a) to the right   (b) toward top of page   (c) into the page (d) to the left   (e) into or out of the page

54 ■■ A straight wire 50 cm long conducts a current of 4.0 A directed vertically upward. If the wire experiences a force of 2.0×10^{-2} N in the eastward direction due to a magnetic field at right angles to its length, what is the magnitude and direction of the magnetic field? 1.0×10^{-2} T north to south

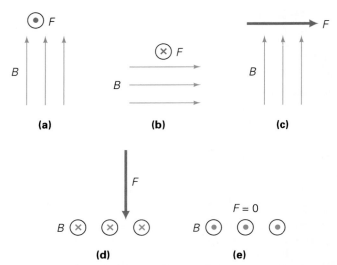

FIGURE 18.34 The right-hand rule   See Exercise 53.

55 ■■ A horizontal magnetic field of 10^{-2} T is at an angle of 30° to the direction of the current in a straight, horizontal wire 75 cm long. If the wire carries a current of 15 A, what is the magnitude of the force on the wire? 5.6×10^{-2} N

56 ■■ A wire carries a current of 10 A in the $+x$ direction in a uniform magnetic field of 40 T. Find the magnitude of the force per unit length and direction of the force on the wire if the magnetic field is directed in (a) the $+x$ direction, (b) the $+y$ direction, (c) the $+z$ direction, (d) the $-y$ direction, and (e) the $-z$ direction.   (a) zero; 4.0×10^2 N/m (b) $+z$ (c) $-y$ (d) $-z$ (e) $+y$

57 ■■ A straight wire 25 cm long is oriented vertically in a uniform horizontal magnetic field of 3.0×10^2 T in the $-x$ direction. What current flowing in what direction in the wire will cause it to experience a force of 5.0 N in the $+y$ direction? 6.7×10^{-2} A; $-z$

58 ■■ A wire carries a current of 10 A in the $+x$ direction. Find the force per unit length on the wire if the magnetic field has components $B_x = 0.020$ T and $B_y = 0.040$ T.   0.40 N/m; $+z$

59 ■■ A set of jumper cables used to start one car from another's battery is connected to the terminals of the cars' batteries. If 15 A of current flows in the cables when the one car is started and the cables are parallel and 15 cm apart, what is the force per unit length on the cables? 3.0×10^{-4} N/m repulsive

● **60** ■■ Find the force per unit length on each of two long, straight, parallel wires that are 24 cm apart when one carries a current of 2.0 A and the other a current of 4.0 A in the same direction. 6.7×10^{-6} N/m attractive

61 ■■ Two long, straight, parallel wires 10 cm apart carry equal currents of 3.0 A in opposite directions. What is the force per unit length on the wires? 1.8×10^{-5} N; repulsive

62 ■■ Two parallel wires are 4.0 m long and 18 cm apart and carry equal currents of 2.5 A in opposite directions. (a) Find the force on each wire. (b) Find the force on each wire if their currents are in the same direction.   (a) 2.8×10^{-5} N; repulsive   (b) 2.8×10^{-5} N; attractive

63 ■■ What is the force per unit length on wire 1 in Fig. 18.31? 2.7 × 10⁻⁵ N/m toward the other wire

64 ■■ What is the force per unit length on wire 2 in Fig. 18.31? 2.7 × 10⁻⁵ N/m toward the other wire

65 ■■ A wire is bent as shown in • Fig. 18.35 and placed in a magnetic field with a magnitude of 10 T in the indicated direction. Find the force on each segment of the wire if $x = 50$ cm and the wire carries a current of 5.0 A. short: 25 N; long: 75 N

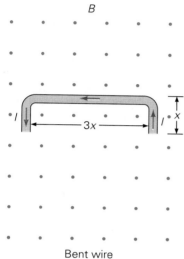

B

3x

Bent wire

• **FIGURE 18.35 Current-carrying wire in a magnetic field** See Exercise 65.

66 ■■ A nearly horizontal dc power line carries a current of 500 A directly eastward. If the only component of the Earth's magnetic field at the location is 5.0×10^{-5} T due north, what is the magnitude and direction of the magnetic force on a 75-m section of the line between two utility poles? 1.9 N upward

• **67** ■■ A rectangular loop of wire whose dimensions are 20 cm by 30 cm carries a current of 1.5 A. (a) What is the magnitude of the magnetic moment of the loop? (b) How should the loop be pivoted and oriented in a uniform magnetic field to obtain the maximum torque on it?
(a) 9.0×10^{-2} A-m² (b) with plane of coil parallel to **B**

68 ■■■ A 50.0-cm length of wire carries a current of 3.0 A in the $+x$ direction in a uniform magnetic field with components of $B_x = B_y = B_z = 100$ mT. Find the force on the wire.
$F = -0.23$ N **y** + 2.3 N **z**

69 ■■■ A rectangular wire loop of a cross-sectional area of 0.20 m² carries a current of 0.75 A. If the loop is free to rotate about an axis perpendicular to a uniform magnetic field of 3.0 T and the plane of the loop is initially at an angle of 30° to the direction of the magnetic field, what is the magnitude of the torque on the loop? 0.39 m-N

70 ■■■ A loop of wire with a magnetic moment of 1.6 A-m² is in a uniform magnetic field of 100 T. Find the magnitude of the torque on the loop if the magnetic field is (a) perpendicular to the plane of the loop and (b) at an angle of 30° to the plane of the loop. (a) 0 (b) 1.4×10^2 m-N

18.5 ■ Applications of Electromagnetism

71 For continuous operation, a dc motor must have (a) a split-ring commutator, (b) brushes, (c) current input, (d) all of these. (d)

72 A mass spectrometer (a) can be used to determine the masses of atoms and molecules, (b) requires charged particles, (c) can be used to determine relative abundances of isotopes, (d) all of these. (d)

73 Explain the operation of the door bell and door chimes illustrated in • Fig. 18.36. see ISM

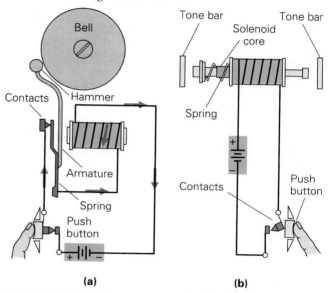

(a)   **(b)**

• **FIGURE 18.36 Electromagnetic applications** (a) A doorbell and **(b)** door chimes both have electromagnets. See Exercise 73.

• **74** ■■ An electron is accelerated from rest through a potential difference of 250 V in a CRT. The electron then passes through a horizontal electric field of 0.10 V/m, moving a horizontal distance of 12 cm. Find the electron's deflection from its original straight-line path. 1.5×10^{-6} m

75 ■■ An ionized deuteron (a particle with a +1 charge) passes through a velocity selector whose magnetic and electric fields have magnitudes of 40 mT and 4.0 kV/m, respectively. Find the speed of the ion. 1.0×10^5 m/s

76 ■■ A 3.5-keV proton passes in an undeflected straight line through a region of crossed electric and magnetic fields. If the magnetic field has a magnitude of 200 mT, what is the magnitude of the electric field? 1.6×10^5 V/m

77 ■■ In a velocity selector, a uniform magnetic field of 1.5 T from a large magnet and two parallel plates with a separation distance of 1.5 cm produce a perpendicular electric field. What voltage should be applied across the plates so that (a) a singly charged ion traveling at a speed of 8.0×10^4 m/s will pass through or (b) a doubly charged ion traveling at the same speed will pass through?
(a) 1.8×10^3 V (b) same speed, independent of the charge

78 ■■ A charged particle travels undeflected through crossed electric and magnetic fields whose magnitudes are

6000 N/C and 300 mT, respectively. Find the speed of the particle if it is (a) a proton or (b) an alpha particle. (An alpha particle is a helium nucleus, a positive ion with a double positive charge.) (a) 2.0×10^4 m/s (b) same v, independent of the charge

79 ■■ An alpha particle (see Exercise 78) is accelerated in the $+x$ direction through a potential of 1000 V. The particle then moves in an undeflected path between two oppositely charged parallel plates in a uniform magnetic field of 50 mT in the $+y$ direction. (a) If the plates are horizontal in the xy plane, what is the magnitude of the electric field between them? (b) If the distance between the plates is 10 cm, what is the potential difference between them? (c) Which plate (top or bottom) is positively charged? Take the mass of the alpha particle to be 6.64×10^{-27} kg. (a) 1.6×10^4 V/m (b) 1.6×10^3 V (c) top is negative

● **80** ■■ In a mass spectrometer, a singly charged ion having a particular velocity is selected using a magnetic field of 0.10 T perpendicular to an electric field of 1.0×10^3 V/m. The same magnetic field is used to deflect the ion, which moves in a circular path with a radius of 1.2 cm. What is the mass of the ion? 1.9×10^{-26} kg

81 ■■ In a mass spectrometer, a doubly charged ion having a particular velocity is selected using a magnetic field of 100 mT perpendicular to an electric field of 1.0 kV/m. The same magnetic field is used to deflect the ion in a circular path with a radius of 15 mm. Find (a) the mass of the ion and (b) the kinetic energy of the ion. (c) Does the kinetic energy of the ion increase in the circular path? Explain. (a) 4.8×10^{-26} kg (b) 2.4×10^{-18} J (c) no, work equals zero

82 ■■■ In a "time-of-flight" mass spectrometer, a beam of protons and a beam of alpha particles (see Exercises 79 and 80) are accelerated through a potential difference of 500 kV. If the length of the flight tube is 1.25 m, what is the time difference for the different particles to arrive at the detector? $\Delta t = 5 \times 10^{-8}$ s

18.6 ■ The Earth's Magnetic Field

83 The Earth's magnetic field (a) has poles that coincide with the geographic poles, (b) is produced by internal ferromagnetic material, (c) reverses polarity every few hundred years, (d) none of these. (d)

84 Aurorae occur (a) only in the Northern Hemisphere, (b) in the lower Van Allen belt, (c) because of pole reversals, (d) predominantly when there are no solar disturbances. (b)

■ Additional Exercises

85 A proton is accelerated through a potential difference of 3.0-keV. It then enters a region between two parallel plates that are separated by 10 cm and have a potential difference of 250 V. Find the magnitude of the magnetic field needed to allow the proton to pass between the plates undeflected. 3.3×10^{-3} T

86 How many turns should a solenoid 30 cm long have to produce a magnetic field of 2.5 mT at its axis when operating with a current of 5.0 A? 1.2×10^2

87 A solenoid 10 cm long has 3000 turns of wire and carries a current of 5.0 A. A 2000-turn coil of wire of the same length

as the solenoid surrounds the solenoid and is concentric with it (shares a common center). If the outer coil carries a current of 10 A in the same direction as that of the current in the solenoid, find the magnetic field at the center of both coils. 0.44 T

88 A horizontal beam of electrons travels at a speed of 10^3 m/s along a north-to-south line in a discharge tube. What is the force on the electrons due to the vertical component of the Earth's magnetic field if it has magnitude of 5.0×10^{-5} T at that location? 8.0×10^{-21} N, east

89 A proton with a speed of 3.5×10^6 m/s enters a uniform magnetic field of 0.80 T in such a way that the proton follows a circular path with a radius of 4.6 cm. What is the kinetic energy of the proton in circular motion? 1.0×10^{-14} J

90 A current of 10 A maintained through a square loop creates a torque of 0.15 m-N about an axis through the loop's center and parallel to one of the sides in a magnetic field of 500 mT that is at an angle of 50° relative to the plane of the loop. Find the length of each side of the square. 0.22 m

91 A proton is accelerated from rest through a potential difference of 1.0 kV. It enters a uniform magnetic field of 4.5 mT that is perpendicular to the direction of its motion. (a) Find the radius of the circular path of the proton. (b) Calculate the period of revolution of the proton. (a) 1.0 m (b) 1.5×10^{-5} s

92 A beam of protons traveling at a speed of 2.0×10^2 m/s passes through the space between two horizontal parallel plates, where a constant electric field and a constant magnetic field are at right angles to one another. If the electric field has a magnitude of 100 V/m and is directed from the bottom to the top plate, what must the magnitude and direction of the magnetic field be to allow the beam to pass undeflected? [*Hint:* Make a sketch of the situation.] 5.0×10^{-1} T; if **v** north and **E** upward, **B** east

93 A horizontal power line carries a current of 2.5 A from west to east. What is the magnetic field 8.0 m below the line? 6.3×10^{-8} T north

● **94** How fast should an electron travel at right angles to a magnetic field of 10^{-4} T so that the magnetic force just balances the gravitational force on it? 5.6×10^{-7} m/s

95 A proton moves at a speed of 1.0×10^4 m/s perpendicularly to a uniform magnetic field of 80 mT. Find the radius of the resulting circular path. 1.3×10^{-3} m

96 A particle with a charge of 4.0×10^{-8} C moves at a speed of 3.0×10^2 m/s through a magnetic field in the direction at which the magnetic force on the particle is maximum. If the force on the particle is 1.8×10^{-6} N, what is the magnitude of the magnetic field? 0.15 T

97 A current-carrying loop is in the form of a rectangle with dimensions of 20 cm by 30 cm. The loop carries a current of 10 A and is in a uniform magnetic field of 50 mT directed parallel to the direction of the current along the 30-cm sides. Find the torque on the loop. 3.0×10^{-2} m-N

98 How much current is flowing in a straight wire 4.0 m long if the force on it is 0.040 N when it is placed in a perpendicular magnetic field of 0.50 T? 2.0×10^{-2} A

19

Electromagnetic Induction

As we saw in Chapter 18, an electric current produces a magnetic field. But the mutual relationship of electricity and magnetism does not stop there. In this chapter we shall see that under certain conditions a magnetic field can be used to produce an electric current.

How is this done? Chapter 18 considered only constant magnetic fields. No current is induced in a loop of wire that is stationary in a constant magnetic field. However, if the magnetic field changes with time, or if the wire loop moves across or is rotated in the field, a current is induced in the wire.

The uses of this interrelationship of electricity and magnetism are legion. One example is of fundamental importance to our whole civilization: the generation of the electric power that we use in so many ways every day. Regardless of the ultimate source of the energy—the burning of oil, coal, or gas, a nuclear reactor, or falling water (as in the turbines shown above)—the conversion to electricity is accomplished by means of magnetic fields.

Another example that will be more familiar to most people is in the playing of a cassette tape. The music you hear was encoded as tiny variations in a magnetic field. These variations produce electrical impulses, which are amplified and drive the speakers. The speakers, in turn, utilize electromagnetic interactions to translate the electrical impulses back into audible sound.

On a larger scale is the generation of electricity. Stop and think how dependent we are on electricity in our everyday lives. Have you ever wondered how this electricity is produced? Electromagnetic phenomena form the basis for the production and transmission of the electricity used in our homes and in industry. This chapter examines the underlying electromagnetic principles.

Topics for study in Chapter 19 include Faraday's Law and Lenz's law, generation and back emf, transformers and power transmission, and electromagnetic waves.

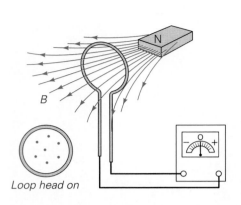

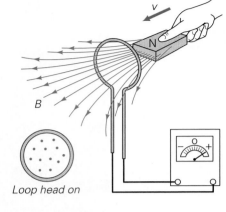

(a) No motion between magnet and loop   **(b)** Magnet is moved toward loop

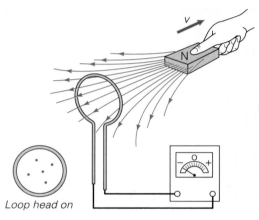

(c) Magnet is moved away from loop

FIGURE 19.1 Electromagnetic induction **(a)** When there is no relative motion between the magnet and the wire loop, the number of field lines through the loop is constant, and the galvanometer needle shows no deflection. **(b)** Moving the magnet toward the loop increases the number of field lines passing through the loop, and an induced current is detected. **(c)** Moving the magnet away from the loop decreases the number of field lines passing through the loop. The induced current is in the opposite direction, as indicated by the needle deflection. **(d)** In an actual experiment of this type, a coil with a number of loops, or solenoid, is used to increase the effect.

19.1 ■ Induced Emf's: Faraday's Law and Lenz's Law

A current can be induced in a conductor in several ways. For example, if a magnet is moved toward a loop of wire, as shown in ● Fig. 19.1b, the deflection of the galvanometer needle indicates that there is a current in the loop. If the magnet is moved away from the loop, as shown in Fig. 19.1c, the galvanometer needle is deflected in the opposite direction, which indicates a reversal of the current's direction. Deflections of the galvanometer needle due to induced currents also occur if the loop is moved toward and away from a stationary magnet. The effect, then, depends on the *relative* motion of the loop and the magnet. Also, the extent of the needle's deflection, or the magnitude of the induced current, depends on the speed of the motion. However, if a loop is moved parallel to a uniform magnetic field (● Fig. 19.2), there is no induced current.

Another way to induce a current in a stationary wire loop is to vary the current in another loop close to it. When the switch in the battery-powered

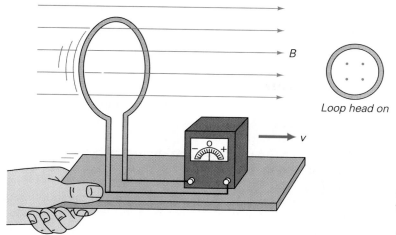

Loop head on

● FIGURE 19.2 Relative motion and no induction
When a loop is moved parallel to a uniform magnetic field, there is no change in the number of field lines passing through the loop and no induced current.

circuit in Fig. 19.3a is closed, the current in its loop goes from zero to a constant value in a short time. During this time, the magnetic field caused by this current and passing through both loops builds up, and the galvanometer needle deflects. When the current in the loop in the powered circuit is at its maximum (constant) value, the resulting magnetic field is also constant, and the galvanometer reads zero. Similarly, when the switch is opened (● Fig. 19.3b), the current and the field quickly decrease to zero, and the galvanometer deflects in the opposite direction. In both cases, the deflection and induced current occur only when the current and the magnetic field are *changing*.

The current induced in a loop is said to be set up by an induced electromotive force (emf) due to **electromagnetic induction**. Recall that an emf represents energy capable of driving charges around a circuit. For example, a battery is a chemical source of emf. In the case of a moving magnet and a stationary loop (Fig. 19.1), mechanical energy is converted to electrical energy. For the two stationary loops (Fig. 19.3), electrical energy is transferred by electromagnetic induction. This effect of a changing current in one circuit inducing an emf in another circuit is called *mutual induction*.

Experiments on electromagnetic induction were done independently around 1830 by Michael Faraday in England and Joseph Henry in the United States. Faraday realized that the important factor in electromagnetic induction was the time rate of change of the magnetic field through the loop. That is,

an induced emf is produced in a loop by a changing magnetic field, or more specifically, by a change in the number of field lines passing through the loop.

(Note that this is the case for the situations in Figs. 19.1 and 19.3, and think about the situation in Fig. 19.2.)

Since the induced emf or current in a loop depends on the change in the number of field lines through it, the ability to quantify the number of field lines through the loop at any time would be useful. Consider a loop of wire in a uniform magnetic field **B** as illustrated in ● Fig. 19.4. The number of field lines through the loop depends on its orientation relative to the B field. To describe this we use a vector **A** normal to the plane of the loop. This is the area vector and its magnitude is equal to the area of the loop. The orientation of the loop may then be described by the angle θ, which is the angle between **B** and **A**.

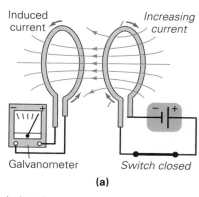

Galvanometer   Switch closed
(a)

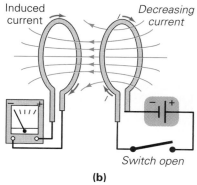

Switch open
(b)

● FIGURE 19.3 Mutual induction (a) When the switch is closed in the right-loop circuit, the current buildup produces a changing magnetic field that passes through the other loop, inducing a current in it. (b) Similarly, when the switch is opened, the magnetic field collapses and there is a decrease in the number of field lines through the second loop. The induced current in this loop is then in the opposite direction.

In general, a relative measure of the number of field lines passing through a particular area (the area within a loop, in our case) is given by the **magnetic flux** (Φ), which is defined as

Magnetic flux—a relative measure of field lines

$$\Phi = BA \cos \theta \qquad\qquad (19.1)$$

The orientation of the loop to the magnetic field affects the number of field lines passing through it, and this is taken into account by including the cosine term in Eq. 19.1. If **B** and **A** are parallel ($\theta = 0$), then the magnetic flux is a maximum ($\Phi_{max} = BA \cos 0° = BA$), and the maximum number of field lines pass through the loop (Fig. 19.4a). If **B** and **A** are perpendicular, there are no field lines through the loop, and the flux is zero ($\Phi = BA \cos 90° = 0$, Fig. 19.4c). In between (Fig. 19.4b), $A \cos \theta$ gives the effective area of the loop open to field lines, so in general, the flux is $\Phi = BA \cos \theta$. Another way of looking at this is that $B \cos \theta$ is the perpendicular component of the field through the loop and $\Phi = (B \cos \theta)A = BA \cos \theta$.

A unit for magentic flux is the weber (Wb). Recall from Section 18.2 that B has the units Wb/m^2, so $\Phi = BA = (Wb/m^2)(m^2) = Wb$, which is a T-m^2 in the SI.

From his experiments, Faraday concluded that the emf induced in a loop depends on the time rate of change of the number of field lines through the loop, or the time rate of change of the magnetic flux.

Faraday's law—a changing flux produces an emf

$$\mathcal{E} = -N \frac{\Delta\Phi}{\Delta t} \qquad\qquad (19.2)$$

That is, $\Delta\Phi$ is the change in flux through N loops of wire in a time Δt. Eq. 19.2 is known as **Faraday's law of induction**. Note that $\mathcal{E}$ is an *average* value over the time interval Δt.

The minus sign is included in Eq. 19.2 to give an indication of the polarity of the induced emf, which is found by considering the induced current and its effect, according to **Lenz's law**:

The law was first stated by Heinrich Lenz (1804–1865), a Russian physicist.

An induced emf gives rise to a current whose magnetic field opposes the change in flux that produced it.

What this means is that the magnetic field due to the induced current is in a

Figure 19.4

FIGURE 19.4 Magnetic flux
Magnetic flux (Φ) is a relative measure of the number of field lines passing through an area. **(a)** When the plane of a rotating loop is perpendicular to the field and $\theta = 0°$, then $\Phi = BA$. **(b)** As the loop is rotated, less area is open to the field lines and the flux decreass. Here $\Phi = BA \cos \theta$. **(c)** When $\theta = 90°$, then $\Phi = 0$.

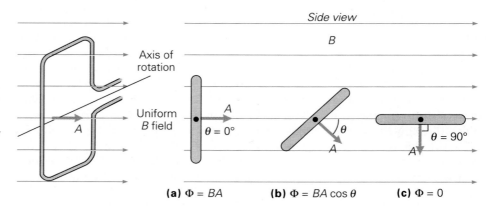

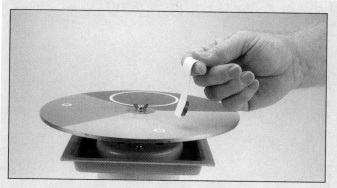

(a) A strong, lightweight magnet is taped to a strip of Manila folder and held above an aluminum disk on a turntable.

(b) When the disk rotates, the magnet levitates above it.

A demonstration of magnetic levitation—not with a superconductor, but with electromagnetic induction and induced currents.

When the disk rotates, it experiences a time-varying magnetic field because of its motion relative to the magnet, and induced currents (called eddy currents) are set up in the disk. By Lenz's law, these currents give rise to a repulsive force. As a result, when the disk is rotating, the magnet levitates above it.

The levitation height depends on the tangential speed of the disk. This can be shown by moving the magnet along a radial axis. Levitation because of air currents is ruled out by using a similar strip of paper without a magnet. (The paper strip is necessary to prevent horizontal movement of the magnet.)

direction that tends to keep the flux through the loop from changing. For example, if the flux increases, the magnetic field due to the induced current will be in the direction opposite to the original field. This tends to cancel the increase in the flux, or oppose the change, since magnetic fields add vectorially.

The current direction is given by a right-hand rule: with the fingers in the direction of the induced field, the extended thumb points in the direction of the conventional current. This is the right-hand rule used to find the direction of a magnetic field produced by a current (Chapter 18), applied in reverse, so to speak. Try applying Lenz's law to the loops in Fig. 19.3. (Also see Demonstration 13.)

Lenz's law incorporates the conservation of energy. Suppose that the magnetic field due to an induced current complemented the original field; that is, added to it, or increased the flux. This would give a greater induced emf and current, which would give a greater magnetic field, which would give a greater induced current, and so on. This would be a something-for-nothing situation that would contradict the law of the conservation of energy.

For the case of inducing an emf in a loop by moving a magnet (Fig. 19.1), recall that a current-carrying loop has a magnetic field similar to that of a bar magnet. The induced current sets up a magnetic field whose effect is such that the loop acts like a bar magnet with a polarity that opposes the motion of the

> *The direction of the induced current is such that energy is conserved. Note the negative sign in Lenz's law.*

real bar magnet. Thus, there is opposition to the motion; work must be done to move the magnet.

Substituting the expression for the magnetic flux (Φ) given by Eq. 19.1 into Eq. 19.2 shows that a change in flux can result from changes other than a change in magnetic field. In general,

$$\mathcal{E} = -\frac{N\Delta\Phi}{\Delta t} = -\frac{N\Delta(BA\cos\theta)}{\Delta t}$$

Hence we see that there are three quantities that can change with time to produce an induced emf: (1) the strength of the magnetic field B, (2) the area of the loop A, and (3) the angle θ. Expanding the equation so as to see these terms explicitly, we have

$$\mathcal{E} = -N\left[\left(\frac{\Delta B}{\Delta t}\right)(A\cos\theta) + B\left(\frac{\Delta A}{\Delta t}\right)(\cos\theta) + BA\left(\frac{\Delta(\cos\theta)}{\Delta t}\right)\right] \qquad (19.3)$$

(1)   (2)   (3)

(1) The first term in the brackets represents the flux change due to a *time-varying magnetic field*, with the area and orientation ($\cos\theta$) constant. A time-varying magnetic field is easily obtained using a time-varying current.

(2) The second term represents a flux change due to a *time-varying loop area*, with a constant magnetic field and a constant orientation. This could occur if a circular loop were being stretched or flattened in a plane, or if it had an adjustable circumference (imagine an adjustable loop around a balloon that was being blown up). Another example will be given shortly.

(3) The third term represents a change in flux resulting from a *change in orientation of the loop with time*, with a constant magnetic field and a constant loop area. This occurs when a loop is rotated in a uniform magnetic field. The change in the number of field lines through a loop in this case is evident in the sequential views of a rotating loop in Fig. 19.4. Another way of looking at this is that there is a change in the *effective* area of the loop, or a change in the area that is open to the field lines.

The induced emf in a rotating loop will be considered in more detail in Section 19.2. The emf's resulting from the first two terms in Eq. 19.3 are analyzed in the following two examples. (Also, see the Insight feature for some applications of electromagnetic induction.)

EXAMPLE 19.1 ■ An Emf Due to a Time-Varying Magnetic Field

The magnitude of a magnetic field passing perpendicularly through a 20-loop wire coil with a radius of 6.0 cm goes from 5.3 T to zero in 0.010 s. (a) What emf is induced in the coil? (b) What is the direction and magnitude of the induced current if the resistance of the coil is 60 Ω?

You have probably used applications of electromagnetic induction many times without knowing it. One of these is magnetic tape—either audio or video. In an audio tape recorder, a plastic tape coated with a film of iron oxide or chromium oxide runs past a recording head, which consists of a coil wound around an iron core with a gap (Fig. 1). Current in the coil produces a magnetic field in the gap which magnetizes the tape's film. The strength and direction of the magnetization in the film are determined by the gap field, which is determined by the current pulses that are generated by sound from a microphone.

To reproduce the sound, the tape is run by the same head or one similar to it. The changing flux due to the magnetization of the film on the tape induces an emf in the coil, matching the original voltage or current fluctuations. These are amplified and converted to sound in a speaker (which also commonly uses electromagnetic induction; see Exercise 4 and Fig. 19.23). A tape may be erased by passing it over the gap in a head that has a high-frequency alternating current (ac) input. The resulting magnetic field diminishes to zero as the tape moves away from the gap, leaving its film in a demagnetized condition.

Electromagnetic induction is also used in an electrical safety device called a ground fault circuit interrupter (GFCI). A GFCI may be plugged into a wall outlet or installed as part of a home circuit. It senses any problem in an electrical appliance plugged into the outlet or circuit and de-energizes the appliance, thus protecting individuals from electrical shock or injury.

The principle of the GFCI is illustrated in Fig. 2. It consists of a sensing coil and a circuit breaker. The coil is wrapped on an iron ring, through which pass the wires carrying current to and from the protected outlet or circuit. This arrangement is sometimes called a *differential transformer* because it can sense a difference in the currents carried by the two wires.

The opposite currents in the two wires produce oppo-

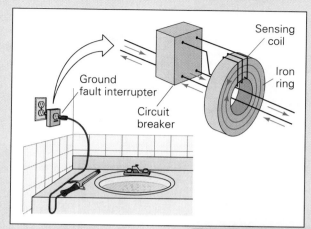

FIGURE 2 Ground fault circuit interrupter (GFCI) A safety device that quickly detects currents to ground and opens the circuit.

site magnetic fields, which are concentrated in the iron ring. With alternating current, the directions of the fields are constantly changing. However, since the fields are always equal and opposite, the total field is normally zero. Hence, the net flux through the coil is zero, and no emf is induced in the sensing coil.

However, suppose that a wire breaks inside an appliance and touches its metal casing (Section 17.5), or a plugged-in curling iron falls into a sink full of water. If someone then touches the appliance or puts a hand in the water, some of the current will pass through the person's body to ground (a fault to ground). The currents in the wires passing through the ring are then not equal, giving rise to a nonzero magnetic flux. A changing flux causes an induced emf in the sensing coil and the resulting current trips the circuit breaker, opening the circuit.

This all happens in about 30 ms (30/1000 of a second) in response to an induced current of 4 to 6 mA. Notice that this is much less than the current that would normally trip a protective circuit breaker, which is usually set at 20 or 30 amps. The GFCI does not protect someone from receiving a shock, but it does limit the time the hazard exists and the potential for serious injury.

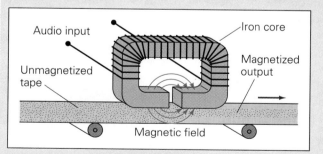

FIGURE 1 Tape recorder head Current pulses in the coil on the iron core produce magnetic fields that magnetize the metallic coating on the tape as it passes by.

Solution. Listing the data carefully, we have

Given: $\theta = 0°$ **Find:** (a) $\mathscr{E}$ (induced emf)
$N = 20$ (b) induced current
$r = 6.0 \text{ cm} = 0.060 \text{ m}$
$\Delta B = B_2 - B_1$
$\quad = 0 \text{ T} - 5.3 \text{ T}$
$\quad = -5.3 \text{ T (or Wb/m}^2)$
$\Delta t = 0.010 \text{ s}$
$R = 60 \ \Omega$

(a) Only the first term in Eq. 19.3 contributes, since A and θ are constant (for example, $\Delta A/\Delta t = 0$), and the induced emf is

$$\mathscr{E} = -N\frac{\Delta \Phi}{\Delta t} = -N\left(\frac{\Delta B}{\Delta t}\right)(A \cos \theta) = -N\left(\frac{\Delta B}{\Delta t}\right)(\pi r^2)(\cos 0°)$$

$$= -(20)\left(\frac{-5.3 \text{ T}}{0.010 \text{ s}}\right)\pi(0.060 \text{ m})^2 = 120 \text{ V}$$

Note that this is an average value.

(b) The magnetic field is decreasing. Thus, by Lenz's law, the induced current is in the direction that sets up a magnetic field in the same direction as that field so as to add to it and oppose the decrease. The magnitude of the current is given by Ohm's law ($\mathscr{E} = IR$), and

$$I = \frac{\mathscr{E}}{R} = \frac{120 \text{ V}}{60 \ \Omega} = 2.0 \text{ A}$$

Again, this is an average value. (Why?)

EXAMPLE 19.2 ■ An Emf Due to a Time-Varying Area

A metal rod is pulled along a metal frame at a constant velocity, as illustrated in ●Fig. 19.5. A uniform magnetic field is normal to the plane of the frame. What is the potential difference, or emf, across the rod?

Solution. In this situation, the magnetic field is constant and $\theta = 0°$, but the flux through the loop formed by the frame and the rod increases because

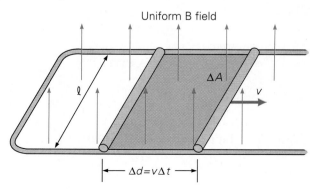

Uniform B field

●**FIGURE 19.5 Motional emf**   As the metal rod moves on the metal frame, the area of the rectangular loop varies with time. A current is induced in the loop as a result of the changing flux. Example 19.2.

its area increases. In a time Δt, the rod moves a distance $\Delta d = v\Delta t$. If ℓ is the length of the rod, the change in the area of the loop is

$$\Delta A = \ell\,\Delta d = \ell v\,\Delta t$$

and

$$\frac{\Delta A}{\Delta t} = \ell v$$

Thus, the emf that develops across the rod is (since $\cos 0° = 1$)

$$\mathscr{E} = -\frac{\Delta\Phi}{\Delta t} = -B\left(\frac{\Delta A}{\Delta t}\right) = -B\ell v$$

This is called *motional emf*. It would set up an induced current in the loop that would produce a magnetic field to oppose the change in flux. Since the moving bar causes an increase in the area of the loop and thus an increase in the number of field lines through the loop, the induced field would be downward in the figure and the current in the bar would be directed toward the near side of the frame.

A potential difference would similarly develop across a bar moving on an insulated frame, but a sustained current would not flow in that case. Similarly, a very small emf develops across the wings of an airplane in flight as they "cut through" the Earth's magnetic field lines.

19.2 ■ Generators and Back Emf

As was pointed out in the preceding section, one method to induce an emf or current in a loop is through a change in the loop's orientation, or a change in its effective area. Note again in Fig. 19.4 that the number of field lines through the loop changes as it *rotates*; the effective area of the loop (the area perpendicular to the field) is $A\cos\theta$. This way of producing a flux change is the principle of operation of a simple electrical generator.

A **generator** is a device that converts mechanical energy into electrical energy. Basically, the function of a generator is the reverse of that of a motor. In some instances, a generator may actually be run "backward" as a motor, and vice versa; however, this is not usually the case.

Motors are discussed in Section 18.5.

A battery supplies direct current (dc). That is, the polarity of the voltage does not change. Generators can produce either direct current or **alternating current (ac)**, for which the polarity of the voltage and the direction of the current periodically change. The electricity used in homes and industry is primarily ac.

Although redundant, the expression ac current is often used, as is ac voltage (usually abbreviated VAC, for volts-ac).

An **ac generator** is sometimes called an *alternator*. The basic elements of a simple ac generator are shown in Fig. 19.6. A wire loop is mechanically rotated in a magnetic field by some external means. The rotation of the loop (called an armature) causes the magnetic flux through it to change, and a current is induced in its wire. The ends of the wire loop are connected to an external circuit by means of slip rings and brushes. (In practice, generators have many loops, or windings, on their armatures.)

Figure 19.6

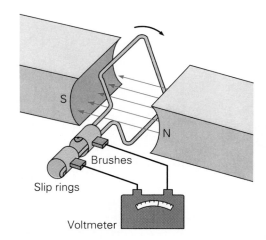

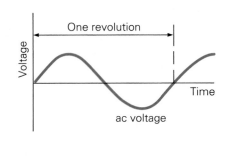

When the loop is rotated with a constant angular speed (ω), the angle (θ) between the magnetic field vector and the area vector of the loop (which is perpendicular to the plane of the loop) changes with time: $\theta = \omega t$. As a result, the cross-sectional area of the loop open to the magnetic field lines changes with time, and from Eq. 19.1 the flux at any time is

$$\Phi = BA\cos\theta = BA\cos\omega t$$

By Faraday's law, the induced emf is then

$$\mathscr{E} = -N\frac{\Delta\Phi}{\Delta t} = -NBA\left(\frac{\Delta(\cos\omega t)}{\Delta t}\right)$$

It can be shown by mathematical methods beyond the scope of this book that as Δt approaches zero,

$$\frac{\Delta(\cos\omega t)}{\Delta t} = -\omega\sin\omega t$$

Thus, the instantaneous value of the emf is

$$\mathscr{E} = NBA\omega\sin\omega t$$

where $NBA\omega$ is the maximum value of the emf, which occurs when $\sin\omega t = \pm 1$. If we designate $NBA\omega$ as $\mathscr{E}_\mathrm{o}$, we can write

Sinusoidal ac emf

$$\boxed{\mathscr{E} = \mathscr{E}_\mathrm{o}\sin\omega t} \tag{19.4}$$

Since the value of the sine function varies between $+1$ and -1, the sign, or polarity, of the emf changes with time (●Fig. 19.7). Note from the figure that the emf has its maximum value when $\theta = 90°$ or $\theta = 270°$, that is, just when the area of the rotating loop becomes closed to field lines. This is because the *change* in flux is greatest at these points.

The direction of the current also periodically changes, and that is why it is called *alternating* current. Since $\omega = 2\pi f$, Eq. 19.4 can be written as

With alternating current, the direction of the current periodically changes. Knowing the drift velocity of the electrons and the rate at which they change direction, can you estimate how far they move along the wire in an ac circuit?

$$\boxed{\mathscr{E} = \mathscr{E}_\mathrm{o}\sin 2\pi ft} \tag{19.5}$$

where f is the rotational frequency of the generator's armature. The ac frequency in the United States is 60 Hz (or 60 cycles/s).

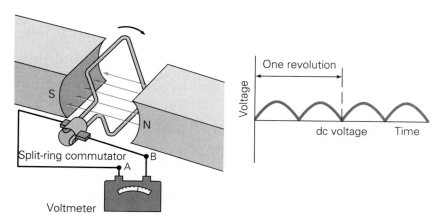

End view of loop
(sequential series of loop rotation)

$+\mathscr{E}_0$

$0°$ $90°$ $180°$ $270°$ $360°$

$-\mathscr{E}_0$

B

FIGURE 19.7 $\mathscr{E} = \mathscr{E}_0 \sin \omega t$ A graph of the sinusoidal generator output, together with an end view of the corresponding orientations of the loop during a cycle.

Keep in mind that Eqs. 19.4 and 19.5 give the instantaneous value of the emf and that $\mathscr{E}$ varies between $+\mathscr{E}_0$ and $-\mathscr{E}_0$ over one period. You will learn how to determine more practical time-averaged values for ac voltage and current in Chapter 20.

EXAMPLE 19.3 ■ Generated Emf

A 60-cycle ac generator produces a maximum emf of 120 V. What is the value of the emf at $\frac{1}{120}$ s after it has a value of zero?

Solution.

    Given: $\mathscr{E}_0 = 120$ V   **Find:** $\mathscr{E}$ (emf)
          $f = 60$ Hz
          $t = \frac{1}{120}$ s

The emf at a given time is given by Eq. 19.5.

$$\mathscr{E} = \mathscr{E}_0 \sin 2\pi f t = (120 \text{ V})[\sin 2\pi (60 \text{ Hz})(\tfrac{1}{120} \text{ s})] = 0$$

Thus, the instantaneous value of the emf at $\frac{1}{120}$ s after it is zero is again zero. Note that the period of rotation of the generator's armature is $T = 1/f = \frac{1}{60}$ s. The emf is zero at $t = 0$, and $\frac{1}{120}$ s is one half-cycle later.

Electromagnetic induction can also be used to generate direct current. A simple **dc generator** is illustrated in ● Fig. 19.8. Here a split-ring commutator

Figure 19.8

FIGURE 19.8 A simple dc generator The polarity of the induced emf reverses every half-cycle, but the split-ring commutator maintains the same polarity on terminals A and B. The output is a fluctuating dc (constant-polarity) voltage. (Compare this dc generator with the dc motor in Fig. 18.19.)

is used to maintain the same polarity. One of the brushes is kept positively charged and the other negatively charged by an exchange of the ends of the armature just as the polarity changes. Note how this occurs from the figure. The output is a pulsating current, but because the polarity does not change, the current is in one direction, or direct. In practice, an armature has many windings, and when the pulses from a large number of loops are superimposed, the resultant direct current is steady and almost free of fluctuations. However, since alternating current is easily rectified, or converted to direct current, direct current is not usually generated in this manner.

A generator that uses a permanent magnet is referred to as a magneto. Magnetos are common in single-cylinder engines, such as those used to power small lawn mowers. (In the gasoline combustion engine of automobiles, the ignition voltage is usually supplied by an induction coil rather than a magneto.) The voltage output of a lawn mower's magneto is used to fire the spark plug that ignites the fuel mixture. In this application, permanent magnets mounted on a flywheel rotate about stationary armature coils. The moving and stationary parts of a generator (or a motor) are called the rotor and stator, respectively. The circuit is opened mechanically or by a solid state device so that a built-up field collapses, producing a large induced voltage. About 15 kV is supplied to the spark plug at the proper time in the engine's cycle.

In most large-scale ac generators (power plants), the armature is stationary and magnets are revolved about it. The revolving magnetic field produces a time-varying flux through the coils of the armature and thus an ac output. The mechanical energy for the generator is supplied by a turbine (Fig. 19.9). Turbines are powered by steam generated from the heat of combustion of fossil fuels, by nuclear reactions, or by water power (hydroelectricity).

● **FIGURE 19.9 Electrical generation** Hydroelectric turbines generate electric power at Hoover dam on the Colorado river.

Back Emf

Motors also generate emf's. Like a generator, a motor has a rotating armature in a magnetic field. The induced emf in this case is called a **back emf** $\mathscr{E}_b$ (or sometimes a counter emf), because its polarity is opposite to that of the line voltage and tends to reduce the current in the armature coils. (What would happen if this were not the case?)

For a motor with a coil or internal resistance R, the current it draws when in operation is given by

$$I = \frac{V - \mathscr{E}_b}{R} \tag{19.6}$$

or $$\mathscr{E}_b = V - IR$$

where V is the line voltage.

The back emf of a motor depends on the rotational speed of the armature and builds up from zero to some maximum value when the armature goes from rest to its normal operating speed. On startup, the back emf is zero, so the starting current is a maximum (Eq. 19.6). Ordinarily, a motor turns something; that is, it has a mechanical load. Without a load, the armature speed will increase until the back emf has built up to a point where it almost equals the line voltage, with just enough current in the coils to overcome friction and joule heat loss. Under normal load conditions, the back emf will be less than the line voltage. The larger the load, the slower the motor will rotate and the smaller the back emf will be. If a motor is overloaded and turns very slowly, the back emf may be reduced so much that the current becomes large enough to burn out the coils. Thus, the back emf is involved in the regulation of a motor's operation.

A back emf in a dc motor circuit can be represented in a circuit diagram as a battery with polarity opposite that of the driving voltage (●Fig. 19.10).

Demonstration/activity: Connect a small 1.5-Vdc motor with flywheel in series with a small flashlight bulb. (Experiment to get the right size motor and bulb.) Immediately after connecting the battery, the bulb is bright, but as the motor speeds up, the bulb gets dimmer. Then grab the flywheel to slow the motor. What happens to the light?

EXAMPLE 19.4 ■ Back Emf

A dc motor with a resistance in its windings of 8.0 Ω operates on 120 V. With a normal load, there is a back emf of 100 V when the motor reaches full speed (see Fig. 19.10). What are (a) the starting current drawn by the motor and (b) the armature current at operating speed under a normal load?

Solution.

Given: $R = 8.0 \ \Omega$ Find: (a) I_s (starting current)
$V = 120$ V   (b) I_a (operational current)
$\mathscr{E}_b = 100$ V

(a) When the motor starts and its armature just begins to turn, the back emf is essentially zero. From Eq. 19.6, the current in the windings at that time is given by

$$I_s = \frac{V}{R} = \frac{120 \text{ V}}{8.0 \ \Omega} = 15 \text{ A}$$

(b) At operating speed, the back emf opposes the driving voltage. The effective voltage is the difference in these two voltages (Eq. 19.6). Thus,

$$I_a = \frac{V - \mathscr{E}_b}{R} = \frac{120 \text{ V} - 100 \text{ V}}{8.0 \ \Omega} = 2.5 \text{ A}$$

Note that, with no back emf, the starting current of the motor is relatively large. (When a big motor, such as that of a central air conditioning unit for a whole building, starts up, you may notice the lights momentarily dim because of the large starting current that it draws.) Here again, there is danger of coil burnout at low operating speeds. In some instances, resistors are temporarily connected in series with a motor's coil to protect the windings from burning out because of large starting currents.

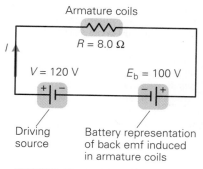

Armature coils

$R = 8.0 \ \Omega$

$V = 120$ V $E_b = 100$ V

Driving source   Battery representation of back emf induced in armature coils

●**FIGURE 19.10 Back emf** The back emf in the armature of a dc motor may be effectively represented as a battery with polarity opposite that of the driving source. See Example 19.4.

Since motors and generators are opposites, so to speak, and a back emf develops in a motor, you may be wondering whether a back force develops in a generator. The answer is yes. When an operating generator is not connected to an external circuit, no current flows and there is no force on the coils of the armature due to the magnetic field. However, when the generator delivers power to a circuit and current flows in the coils (current-carrying wires in a magnetic field), there is a force that produces a **counter torque**, which opposes the rotation of the armature. As more current is drawn, the counter torque becomes greater, and a greater driving force is needed to turn the armature. Therefore, the higher the current output of a generator, the greater will be the energy expended (fuel consumed) in overcoming the counter torque.

Demonstration/activity: Light a bulb using a hand-cranked generator. Show how the generator is easier to turn when the bulb is not connected, and is harder to turn when the bulb is connected. Is there an emf across the ends of the wire when the crank is turned while the bulb is disconnected?

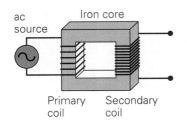

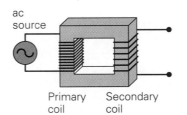

(a) Step-up transformer: high-voltage (low current) output

(b) Step-down transformer: low-voltage (high current) output

(c) Transformer circuit symbol

(d)

● **FIGURE 19.11 Transformers**
(a) A step-up transformer has more turns in the secondary coil than in the primary. **(b)** A step-down transformer has more turns in the primary coil than in the secondary. **(c)** The circuit symbol for a transformer reflects its actual structure to some extent. **(d)** Workers inspect large transformers used in electrical power transmission systems (to be discussed shortly).

For transformers, it is customary to work with voltage rather than emf.

19.3 ▪ Transformers and Power Transmission

Electricity is transmitted by power lines over long distances. Obviously, it is desirable to minimize the I^2R losses through these transmission lines. The resistance of a line is fixed, so reducing the I^2R losses means reducing the current. However, the power output of a generator is determined by its current and voltage outputs ($P = IV$), and for a fixed voltage (such as 120 V), a reduction in current would mean a reduced power output. It might appear that there is no way to reduce the current while maintaining the power supplied. Fortunately, however, electromagnetic induction can be applied to reduce power transmission losses by stepping up the voltage and reducing the current. This is done with a device called a transformer.

As ● Fig. 19.11 shows, a simple **transformer** consists of two coils of insulated wire wound on the same (closed) iron core. When ac voltage is applied to the input coil, or **primary coil**, the alternating current gives rise to an alternating magnetic flux that is concentrated in the iron core. The changing flux passes through the output coil, or **secondary coil**, inducing an alternating voltage and current in it.

The induced secondary voltage differs from the primary voltage depending on the ratio of the numbers of turns in the two coils. By Faraday's law, the secondary voltage is given by

$$V_s = -N_s \frac{\Delta \Phi}{\Delta t}$$

where N_s is the number of turns in the secondary coil. The changing flux in the primary coil produces a back emf equal to

$$V_p = -N_p \frac{\Delta \Phi}{\Delta t}$$

where N_p is the number of turns in the primary coil. If the resistance of the primary coil is neglected, this is equal to the external voltage applied to it. Then, forming a ratio gives

$$\frac{V_s}{V_p} = \frac{-N_s \Delta \Phi / \Delta t}{-N_p \Delta \Phi / \Delta t}$$

or

$$\frac{V_s}{V_p} = \frac{N_s}{N_p} \qquad (19.7)$$

Here it is assumed that the core concentrates the field so the flux through each coil is the same.

If the transformer is assumed to be 100% efficient (no energy losses), the power input is equal to the power output, and since $P = IV$,

$$I_p V_p = I_s V_s \qquad (19.8)$$

Then, using Eq. 19.6, the *transformer currents and voltages are related to the turn ratio* by the relationship

$$\boxed{\frac{I_p}{I_s} = \frac{V_s}{V_p} = \frac{N_s}{N_p}} \qquad (19.9)$$

With these equations, it is easy to see how a transformer affects the voltage and current. In terms of the output,

$$V_s = \left(\frac{N_s}{N_p}\right) V_p \quad \text{and} \quad I_s = \left(\frac{N_p}{N_s}\right) I_p \qquad (19.10)$$

That is, if the secondary coil has a greater number of windings than the primary does ($N_s > N_p$ or $N_s/N_p > 1$) as in Fig. 19.11a, the voltage is stepped up ($V_s > V_p$). However, less current flows in the secondary than in the primary ($N_p/N_s < 1$ and $I_s < I_p$). This type of arrangement is called a **step-up transformer**. For example, if the primary coil of a transformer has 50 turns and the secondary has 100 turns, $N_s/N_p = 2$ and $N_p/N_s = \frac{1}{2}$. Thus, a 220-V input at 10 A will be stepped up to a 440-V output at 5.0 A.

The opposite situation, in which the secondary coil has fewer turns than the primary, characterizes a **step-down transformer** (Fig. 19.11b). In this case, the voltage is stepped down, or reduced, and the current is increased. A step-up transformer may be used as a step-down transformer by simply reversing the output and input connections. Also, it should be apparent that transformers operate only on ac (not on dc).

Although some energy is always lost, this is a good approximation, since a well-designed transformer may have an efficiency of more than 95%. The sources of energy losses will be discussed shortly.

Some students tend to associate high voltage with large amounts of energy, and they may even view transformers as devices that give more energy out than what is put in. Always refer to voltage as a measure of energy per charge.

Step-up transformer: $N_s > N_p$ or $\frac{N_s}{N_p} > 1$ (more turns on secondary than on primary)

Step-down transformer: $N_p > N_s$ or $\frac{N_s}{N_p} < 1$ (more turns on primary than on secondary)

EXAMPLE 19.5 ■ Step-up and Step-down

A transformer has 50 turns on its primary and 150 turns on its secondary. (a) If the primary is connected to a 120-V source, what is the voltage output of the secondary? (b) If the transformer is operated in reverse, and the 120-V input is applied to the coil with 150 turns, what would be the voltage output?

Solution. Here the number of primary and secondary turns are given in part (a). In part (b), when operated in reverse, the secondary becomes the primary and vice versa, so we have

Given: (a) $N_p = 50$ Find: (a) V_s (secondary
        $N_s = 150$ voltage outputs)
        $V_p = 120$ V   (b) V_s
    (b) $N_p = 150$
        $N_s = 50$
        $V_p = 120$ V

(a) The secondary voltage is given by Eq. 19.10. Note that this is a step-up transformer with $N_s > N_p$ and a turn ratio of $N_s/N_p = 150/50 = 3$. Then,

$$V_s = \left(\frac{N_s}{N_p}\right)V_p = (3)(120 \text{ V}) = 360 \text{ V}$$

Looking at the secondary current with $N_p/N_s = 50/150 = 1/3$,

$$I_s = \left(\frac{N_p}{N_s}\right)I_p = \left(\frac{1}{3}\right)I_p$$

Thus we see that the voltage is stepped up or increased by a factor of 3, and the current is stepped down and is 1/3 as large.

(b) In this case, the turn ratio is $N_s/N_p = 50/150 = 1/3$, and we have a step-down transformer with $N_p > N_s$. Then,

$$V_s = \left(\frac{N_s}{N_p}\right)V_p = \left(\frac{1}{3}\right)(120 \text{ V}) = 40 \text{ V}$$

Again looking at the secondary current with $N_p/N_s = 150/50 = 3$,

$$I_s = \left(\frac{N_p}{N_s}\right)I_p = (3)I_p$$

As you might expect, in this case, the voltage is stepped down and is 1/3 as large, while the current is stepped up by a factor of 3. (Since the currents were not given, the actual numerical values cannot be found.)

The preceding equations apply to ideal transformers, but actual transformers do have some energy losses. First, there is always some flux leakage; that is, not all of the flux passes through the secondary coil. In some transformers, one of the insulated coils is wound over (on top of) the other rather than having the two on separate "legs" of a core. This helps avoid flux leakage and reduces the size.

Second, when ac current flows in the primary coil, the changing magnetic flux through the loops gives rise to an induced emf in that coil. This is called *self-induction*. By Lenz's law, the self-induced emf will oppose the change in current and will thus limit the current (similar to a back emf in a motor). Self-induction may be thought of as a kind of electromagnetic inertia—like the inertia of material bodies, it opposes change.

Some energy is also lost due to the resistances of the coil wires (I^2R losses), but this is generally small.

A fourth cause of energy loss is **eddy currents** in the transformer core. A highly permeable material is used for the core to increase the density of the magnetic flux, but such materials are usually good conductors. The changing magnetic flux sets up swirling movements of charge, or eddy currents, in the core material, and these dissipate energy. (See Demonstration 13 earlier in the chapter for an eddy current effect.)

To reduce this effect, transformer cores are made of thin sheets of material (usually iron) laminated with an insulating glue between them. Because of the insulating layers between the sheets, the eddy currents are broken up, or confined to the sheets, which greatly reduces the loss due to them. Well-designed transformers generally have less than 5% internal energy loss.

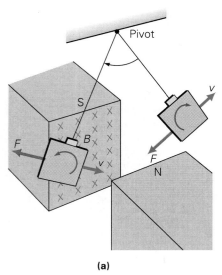

(a)

(b)

FIGURE 19.12 Eddy currents
(a) Eddy currents are induced in a nonmagnetic conductive plate moving in a magnetic field. The induced currents oppose the change in flux, and there is a retarding force that opposes the motion. Note that the currents reverse as the plate swings through the field. (b) If that half of the pendulum plate with the holes swings between the magnet poles, the motion is retarded. However, if the half with the slits is used as shown here, the plate swings relatively freely.

An effect of eddy currents may be demonstrated by swinging a plate made of a nonmagnetic metal, such as aluminum, through a magnetic field, as illustrated in ● Fig. 19.12a. Eddy currents are set up in the plate as a result of its motion in the field and the changing magnetic flux. By Lenz's law, the eddy currents set up opposing fluxes, or effectively opposite magnetic poles on the plate. This gives rise to a repulsive force that retards the swinging motion and brings the plate quickly to rest.

The breaking up of the eddy currents may be demonstrated using a plate with holes and slits (Fig. 19.12b). When the end of the plate with the holes swings between the magnet's poles, the motion is retarded and damped, since eddy currents can have large paths around the holes. However, when the end with the slits swings between the pole faces, the plate swings relatively freely.

The damping effect of eddy currents is applied in the braking systems of rapid-transit cars that travel on rails. When an electromagnet on the car is turned on, it applies a magnetic field to a metal wheel or the rail. The repulsive force due to the induced eddy currents acts as a braking force. As the car slows, the eddy currents decrease, allowing a smooth braking action.

Power Transmission

For power transmission over long distances, transformers provide a means to increase (step up) the voltage and reduce the current of a generator's output in order to cut down the resistive I^2R losses. A schematic diagram of an ac power distribution system is shown in ● Fig. 19.13. The voltage output of the generator is stepped up and transmitted over long distances to an area substation near the consumers. There the voltage is stepped down. There are further voltage step-downs at distributing substations and utility poles before 120–240 V are supplied to homes and businesses.

The following example illustrates the benefits of being able to step up the voltage (and step down the current) for electrical power transmission.

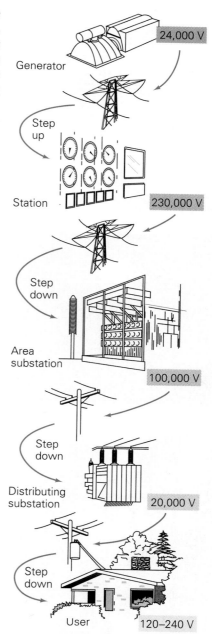

● **FIGURE 19.13 Power transmission system** A diagram of a typical electrical power distribution system.

EXAMPLE 19.6 ■ Power Transmission

A generator produces 10 A at 440 V. The voltage is stepped up to 4400 V (by an ideal transformer) for transmission over 40 km of power line, which has a resistance of 0.50 Ω/km. [A power transmission line has two wires (for a complete circuit); however, for simplicity, you can take the length given to be the total wire length rather than doubling it.] (a) What percentage of the original energy would have been lost in transmission if the voltage had not been stepped up? (b) What percentage of the original energy is lost even with the voltage stepped up?

Given: $V_p = 440$ V

$I_p = 10$ A

$V_s = 4400$ V

$R_o/L = 0.50$ Ω/km

$L = 40$ km

Find: (a) Percentage energy loss without voltage step-up

(b) Percentage energy loss with voltage step-up

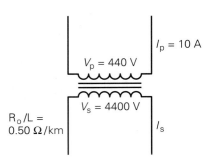

(a) The power output of the generator is

$$P = I_p V_p = (10 \text{ A})(440 \text{ V}) = 4400 \text{ W}$$

The resistance of 40 km of power line is

$$R = \left(\frac{R_o}{L}\right)L = \left(\frac{0.50 \ \Omega}{\text{km}}\right)(40 \text{ km}) = 20 \ \Omega$$

The energy loss (per unit time) in transmitting a current of 10 A is

$$P_{\text{loss}} = I_p^2 R = (10 \text{ A})^2(20 \ \Omega) = 2000 \text{ W}$$

Thus,

$$\% \text{ loss} = \frac{P_{\text{loss}}}{P} (\times 100\%) = \frac{2000 \text{ W}}{4400 \text{ W}} (\times 100\%) = 45\%$$

This is the percentage of energy loss because power is energy per time.
(b) When the voltage is stepped up to 4400 V, the transmitted current is

$$I_s = \left(\frac{V_p}{V_s}\right)I_p = \left(\frac{440 \text{ V}}{4400 \text{ V}}\right)(10 \text{ A}) = 1.0 \text{ A}$$

(Note that the voltage was stepped up by a factor of 10 and the current is stepped down by the same factor.) The power loss in this case is

$$P_{\text{loss}} = I_s^2 R = (1.0 \text{ A})^2(20 \ \Omega) = 20 \text{ W}$$

and

$$\% \text{ loss} = \frac{P_{\text{loss}}}{P} (\times 100\%) = \frac{20 \text{ W}}{4400 \text{ W}} (\times 100\%) = 0.45\%$$

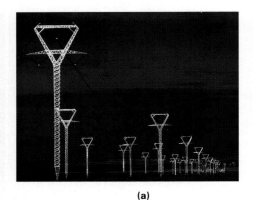

(a)

(b)

● **FIGURE 19.14 Long lines**
(a) Long-distance power lines carry electric power at high voltage across the plains of Wyoming. The use of high voltages in transmission reduces power losses. (b) Workers maintaining insulators at a power substation. Large insulators are needed to reduce leakage losses from high-voltage lines.

Example 19.6 shows the advantage of high-voltage, or high-tension, transmission lines (● Fig. 19.14a). However, there is a practical limit to the degree of voltage step-up. At very high voltages, the molecules in the air surrounding a power line may be ionized, forming a conducting path to nearby trees, buildings, or the ground. This is referred to as a leakage loss. Crews from electric companies are continually clearing foliage from near power lines, and long insulators are used to hold the high-voltage wires away from the metal towers (Fig. 19.14b). Leakage losses are generally greater during wet weather because moist air is more easily ionized. Under certain conditions, you may see the arcing or corona discharge from a high-voltage line and/or hear the accompanying crackling noise.

19.4 ■ Electromagnetic Waves

Electromagnetic waves (or **radiation**) as a means of heat transfer were considered in Section 11.4. Now you are ready to understand more fully the production and characteristics of electromagnetic radiation. As the name implies, these waves have both electric and magnetic properties, which may be described by quantities you have studied.

James Clerk Maxwell (1831–1879), a brilliant Scottish scientist, showed that four fundamental relationships could completely describe all observed electromagnetic phenomena. Maxwell also used this set of equations to predict the existence of waves of an electromagnetic nature. Because of his contributions, the set of equations is known as **Maxwell's equations**, although they were for the most part developed individually by other scientists (for example, Faraday discovered the law of induction).

Maxwell's equations

Essentially, Maxwell's equations combine the electric force and the magnetic force into a single electromagnetic force; the separate forces or fields are shown to be symmetrically related. This symmetry is evident in the equations as presented in their advanced mathematical form. For the purposes of this course of study, a qualitative description is sufficient.

> A time-varying magnetic field produces a time-varying electric field.
> A time-varying electric field produces a time-varying magnetic field.

The first statement reflects the fact that a changing magnetic flux gives rise to an induced voltage in a wire, or to *an electric field in space*. The second statement implies that a changing electric field gives rise to a changing magnetic field. This symmetry is important in the analysis of electromagnetic waves.

A changing B produces a changing E.

A changing E produces a changing B.

Basically, electromagnetic waves are produced by accelerating electric charges, such as an electron oscillating in simple harmonic motion. This could be one of the many electrons in a radio transmitter, which oscillates with frequencies of about 10^6 Hz. As such an electron moves, it accelerates and radiates an electromagnetic wave (•Fig. 19.15). The continually alternating motions of many such charges due to the alternating current in the transmitter rod produce time-varying electric and magnetic fields in the immediate vicinity of the rod. The electric field (in the plane of the paper) continually changes direction, as does the magnetic field (into and out of the paper by a right-hand rule).

Figure 19.15

Both the electric and the magnetic fields store energy and propagate outward with the speed of light (c in a vacuum, 3.00×10^8 m/s). Maxwell's equations show that, except in the immediate vicinity of the source, the electro-

•**FIGURE 19.15 Source of electromagnetic waves** Electromagnetic waves are produced by accelerating electric charges. The wave propagates outward with the electric field vectors (**E**) and magnetic field vectors (**B**) in phase, perpendicular to each other and to the direction of propagation.

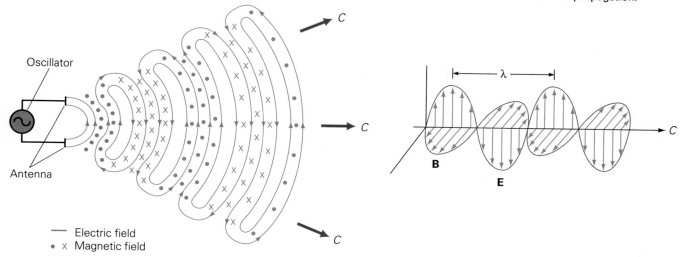

Oscillator

Antenna

— Electric field
• × Magnetic field

magnetic waves at a fixed instant of time are of the form shown in Fig. 19.15. The electric vector of the E field (**E**) is perpendicular to the magnetic vector of the B field (**B**), and each varies sinusoidally with time. Both **E** and **B** are in phase and perpendicular to the direction of propagation. Thus, electromagnetic waves are transverse waves.

EXAMPLE 19.7 ■ The Speed of Light (and other EM Waves)

A *Viking* lander on the planet Mars sends back radio signals to Earth. How much longer does it take for a signal to reach us when Mars is farthest from Earth than when it is closest? The average distances of Mars and Earth from the Sun are 142 million miles and 93 million miles, respectively. Assume circular orbits with these average distances as radii.)

Solution. Listing the data, we have

Given: $d_M = 142 \times 10^6$ mi   *Find:* Δt (time difference)
(average distance of Mars from Sun)
$d_E = 93 \times 10^6$ mi
(average distance of Earth from Sun)

This is a simple time-distance calculation. To find the difference in the travel times of the radio signals, we first find the difference in the distances. The planets are farthest apart when they are on opposite sides of the Sun. At this time, they are separated by a distance $(d_M + d_E)$. The planets are closest when they are aligned on the same side of the Sun; their separation distance is then $(d_M - d_E)$. (Draw yourself a diagram to help visualize these relationships.)

The difference in the separation distances is then

$$\Delta d = (d_M + d_E) - (d_M - d_E) = 2d_E$$

$$= 2(93 \times 10^6 \text{ mi}) \left(\frac{1.61 \times 10^3 \text{ m}}{\text{mi}} \right) = 3.0 \times 10^{11} \text{ m}$$

with a conversion to metric units. Notice that the difference is twice the Earth's distance from the Sun (or the diameter of a circular orbit), so the distance of Mars from the Sun was not needed. This illustrates the advantage of doing operations algebraically before putting in numerical values.

Radio waves are electromagnetic waves and travel at "the speed of light" c. Since $\Delta d = c \, \Delta t$,

$$\Delta t = \frac{\Delta d}{c} = \frac{3.0 \times 10^{11} \text{ m}}{3.0 \times 10^8 \text{ m/s}} = 1.0 \times 10^3 \text{ s} \quad \text{(or 16.7 min)}$$

Not too long to travel 2(93 million mi) = 186 million miles. (Note that this is twice the time it takes light to travel from the Sun to Earth, or twice the time it would take us to learn that the Sun had burned out or exploded.)

Radiation Pressure

An electromagnetic wave carries energy. Consequently, it can do work and can exert a force on a material it strikes. Consider light striking an electron at

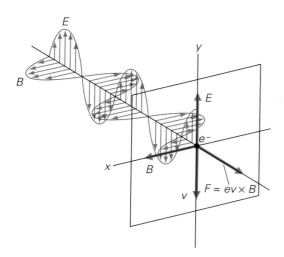

FIGURE 19.16 Radiation pressure The electric field of an electromagnetic wave that strikes a surface acts on an electron, giving it a velocity (**v**). The magnetic field produces a force on the moving charge in the direction of propagation of the incident light. (Check this using the appropriate right-hand rule.)

the surface of a material (Fig. 19.16). The electric field of the electromagnetic wave does work on the electron; assume that this gives the electron a velocity (**v**) as shown in the figure. Recall from Chapter 18 that a moving charged particle in a magnetic field experiences a force. As a result, there is a magnetic force on the electron due to the magnetic field component of the light wave. By the right-hand rule for the magnetic force on a moving charged particle, the force on the electron is in the direction shown in Fig. 19.16. That is, the electromagnetic wave produces a force on the electron in the direction in which the wave is propagating and therefore exerts a force on the material to which the electron is bound.

The radiation force per area is called the **radiation pressure** on an absorbing material. If the electromagnetic wave is re-emitted, the radiation pressure is doubled because of the recoil momentum and force. Radiation pressure is negligible for most common situations but is of importance in some atmospheric and astronomical phenomena. For example, radiation pressure plays a key role in the formation of the tail of a comet, which is made up of fine dust particles. The tail always points away from the Sun.

> Even the light from your desk lamp exerts a very, very small force on your desk.

Types of Electromagnetic Waves

Electromagnetic waves are classified by ranges of frequencies or wavelengths in a spectrum. These are inversely related, $\lambda = c/f$; the greater the frequency, the shorter the wave length and vice versa. The electromagnetic spectrum is continuous, so the limits of the various ranges are approximate. Table 19.1 lists these frequency and wavelength ranges for the general types of electromagnetic waves. (See also Fig. 19.17.)

Power Waves. Electromagnetic waves of 60-cycle frequency result from currents moving back and forth (alternating) in electrical circuits. As Table 19.1 indicates, these power waves have a wavelength of 5.0×10^6 m, or 5000 km (more than 3000 mi). Waves of this low frequency are of little practical use. They may occasionally produce a so-called 60-cycle hum on your stereo. More seriously, there are suspicions about possible health effects of these waves. Some research tends to suggest that very low-frequency fields may have potentially

TABLE 19.1 ■ Classification of Electromagnetic Waves

Type of Wave	Approximate Frequency Range (Hz)	Approximate Wavelength Range (m)	Source
Power waves	60	5×10^6	Electric currents
Radio waves			Electric circuits
AM	$0.53–1.7 \times 10^6$	570–186	
FM	$88–108 \times 10^6$	3.4–2.8	
TV	$54–890 \times 10^6$	5.6–0.34	
Microwaves	$10^9–10^{11}$	$10^{-1}–10^{-3}$	Special vacuum tubes
Infrared radiation	$10^{11}–10^{14}$	$10^{-3}–10^{-7}$	Warm and hot bodies
Visible light	$4.0–7.0 \times 10^{14}$	10^{-7}	Sun and lamps
Ultraviolet radiation	$10^{14}–10^{17}$	$10^{-7}–10^{-10}$	Very hot bodies and special lamps
X-rays	$10^{17}–10^{19}$	$10^{-10}–10^{-12}$	Electron collisions
Gamma rays	above 10^{19}	below 10^{-12}	Nuclear reactions and processes in particle accelerators

Figure 19.17

● FIGURE 19.17 The electromagnetic spectrum
The spectrum of frequencies or wavelengths is divided into regions, or ranges. Note that the visible region is a very small part of the total electromagnetic spectrum. The wavelengths are given in nanometers: 1 nm = 10^{-9} m. (Top illustrative wavelengths not to scale.)

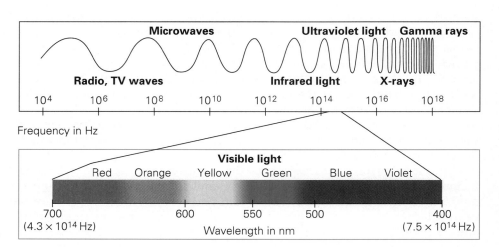

harmful biological effects on cells and tissues. People living near power lines are continually exposed to power frequency radiation, so this is an important concern. At the present time, this is a very controversial issue and much more work needs to be done before a firm conclusion can be reached.

Radio and TV Waves. Radio and TV waves are generally in the frequency range from 200 kHz to about 1000 MHz. The AM (amplitude modulated) band runs from 530 to 1710 kHz (1.71 MHz). Higher frequencies, up to 54 MHz, are used for "short wave" bands. TV bands range from 54 MHz to 806 MHz. The FM (frequency modulated) radio band runs from 88 to 108 MHz, which lies in a gap between channels 6 and 7 of the TV band. Cellular phones use radio waves to transmit voice communication in the ultra-high frequency (UHF) band, with frequencies similar to those of radio waves used for television channels 13 and higher.

Early global communications used the "short wave" bands, as do amateur (ham) radio operators today. But how are the normally straight-line radio waves transmitted around the curvature of the Earth? This is accomplished by reflection from ionic layers in the upper atmosphere. Energetic particles from the Sun ionize gas molecules, giving rise to several ion layers. Certain of these layers reflect radio waves below a certain frequency. By "bouncing" radio waves off these layers, we can send radio transmissions beyond the horizon, to any region of the Earth.

Such reflection of radio waves requires uniformity in the density of the ionic layers. Should a solar disturbance produce a shower of energetic particles that upsets the uniformity, a communications "blackout" would occur. To avoid such disruptions, global communications have relied largely on transoceanic cables. Now we also have communications satellites, which can provide line-of-sight transmission to any point on the globe (●Fig. 19.18).

Microwaves. Microwaves, with frequencies in gigahertz (GHz) range, are produced by special vacuum tubes (called klystrons and megetrons). Micro-waves are used in communications and radar applications. In addition to its many roles in navigation and guidance, radar provides the basis for the speed guns used to time fast balls, tennis serves, and motorists (●Fig. 19.19). When radar waves are reflected from a moving object, their wavelength is shifted by the Doppler effect (Section 14.4). The amount of the shift indicates the velocity of the object toward or away from the observer. Another very common use of microwaves today is in microwave ovens (Section 11.4).

Infrared Radiation. The infrared region of the electromagnetic spectrum lies adjacent to the low-frequency or long-wavelength end of the visible spectrum. The frequency at which a warm body emits radiation depends on its temperature. Actually, such a body emits electromagnetic waves of many different frequencies, but the frequency of the maximum intensity characterizes the radiation. A body at about room temperature emits radiation in the far infrared region (meaning farthest from the visible region).

Recall from Chapter 11 that infrared radiation is sometimes referred to as heat rays. This is because this radiation is readily absorbed by some materials containing water molecules, producing an increase in the material's temperature because of increased molecular motion. Infrared lamps are used in therapeutic applications and to keep food warm in cafeterias. Infrared radiation is also associated with maintaining the Earth's warmth or average temperature through the greenhouse effect.

Visible Light. The visible region occupies a very small portion of the total electromagnetic spectrum. It runs from about 4.3×10^{14} Hz to about 7.5×10^{14} Hz, or a wavelength range of $700 - 400$ nm, respectively (Fig. 19.17). Only the radiation in this region can activate the receptors in our eyes. Visible light emitted or reflected from the objects around us provides us with much information about our world. Visible light will be considered in more detail in Chapters 21–24.

It is interesting to note that not all animals are sensitive to the same range of wavelengths. For example, snakes can detect infrared radiation, and the visible range of many insects extends well into the ultraviolet (see below).

Ultraviolet Radiation. Beyond the violet end of the visible region lies the ultraviolet frequency range. Ultraviolet (or uv) radiation is produced by special lamps and very hot bodies. The Sun emits large amounts of ultraviolet radiation, but fortunately most of it received by the Earth is absorbed in the ozone (O_3)

●**FIGURE 19.18 Global village** Radio and TV transmissions from around the world, relayed by orbiting satellites, can be picked up by back-yard and rooftop antennas.

See the Insight in Section 11.4.

●**FIGURE 19.19 Slow down** Radar speed guns based on the Doppler effect employ radiation in the microwave region of the spectrum.

layer in the atmosphere at an altitude of about 40–50 km. Because the ozone layer plays a protective role, there is concern about its depletion by chlorofluorocarbon gases (such as Freon, used in refrigerators) that drift upward and react with the ozone. (See the Insight feature on the ozone layer.)

The small amount of ultraviolet radiation that reaches the Earth's surface from the Sun can cause human skin to burn or tan. Pigmentation in the skin acts as a protective mechanism against the penetration of ultraviolet radiation. The degree of penetration depends on the amount of a pigment called melanin in the skin and on the thickness of the skin's layers. Persons (except albinos) have varying amounts of melanin in their skin. Exposure to sunlight induces the production of more melanin, causing the skin to tan. Overexposure, especially initially, may cause the skin to burn and turn red (sunburn). Healing thickens the outer skin layer, which then offers greater protection. Creams and lotions containing so-called sunscreens are also available to keep skin from sunburning. The molecules of chemicals in these preparations absorb some of the ultraviolet radiation before it reaches the skin. Doctors now urge people—especially those with fair skins, which contain little protective melanin—to use effective sunscreens or avoid bright sunlight altogether, because uv exposure greatly increases the risk of skin cancer, as mentioned in the Insight feature.

Exposure to sunlight is necessary for the natural production of vitamin D from compounds in the skin. This vitamin is essential for strong bones and teeth. In northern latitudes, where people's exposure to sunlight is relatively seasonal, diets need to be supplemented with synthetic vitamin D. You may have noticed that a vitamin D supplement is commonly provided in milk.

Most ultraviolet radiation is absorbed by certain molecules in ordinary glass. Therefore, you cannot get a tan or a sunburn through glass windows. Sunglasses are now labeled to indicate the uv protection standards they meet in shielding the eyes from this potentially harmful radiation. Welders wear special glass goggles or face masks to protect their eyes from the large amounts of ultraviolet radiation produced by the arcs of welding torches. Similarly, it is important to shield the eyes when using a sun lamp. The ultraviolet component of sunlight reflected from snow-covered surfaces can produce snowblindness in unprotected eyes.

"Black lights" used at discos emit radiation in the violet and near ultraviolet. The ultraviolet radiation causes fluorescent paints and dyes on signs, posters, and performers' clothes to glow with brilliant colors. *Fluorescence* is the process whereby a substance absorbs ultraviolet radiation and emits visible radiation. In fluorescent lamps, an electrically excited mercury vapor in the lamp emits ultraviolet radiation. The white material coating the inside of the tube absorbs the ultraviolet radiation and emits visible light. You may have noticed in the grocery store that many products, such as laundry detergents, are packaged in brightly colored boxes. In some instances, the ink is fluorescent, and appears brighter under the store's fluorescent lamps to get your attention. (A small amount of ultraviolet radiation is emitted from the lamps.)

X-rays. Beyond the ultraviolet region of the electromagnetic spectrum is the important X-ray region. We are all familiar with X-rays, primarily through medical applications. X-rays were discovered accidentally in 1895 by the German physicist Wilhelm Roentgen (1845–1923), when he noted the glow of a piece of fluorescent paper caused by some mysterious (unknown and therefore referred to as X) radiation coming from a cathode ray tube.

The basic elements of an X-ray tube are shown in Fig. 19.20. A potential difference of several thousand volts is applied across the electrodes in a sealed,

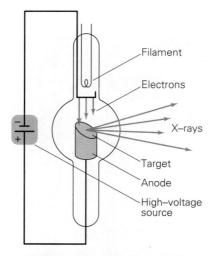

●**FIGURE 19.20 The X-ray tube**
Electrons accelerated through a large potential difference strike a target electrode and interact with the atomic electrons of its material, causing the incoming electrons to be slowed down. Energy is emitted in the form of X-rays.

We hear a lot about the ozone layer these days. What is it, and why should it concern us? Let's take a look.

At an altitude of about 30 km ($\approx$20 mi) in the atmosphere, there exists a concentration of ozone (O_3)—the "ozone layer." Here oxygen molecules (O_2) undergo dissociation by solar ultraviolet radiation ($O_2 + uv \rightarrow O + O$). The resulting atomic oxygen (O) combines with an oxygen molecule to form ozone ($O + O_2 \rightarrow O_3$). However, there is not a continual buildup of ozone. Ozone is very unstable in the presence of sunlight, and when it absorbs ultraviolet radiation, it is again dissociated into atomic and molecular oxygen. Moreover, an oxygen atom may react with an ozone molecule to form two ordinary oxygen molecules ($O + O_3 \rightarrow O_2 + O_2$), thus destroying the ozone. All these processes go on simultaneously in the ozone layer so that a natural balance between ozone production and ozone destruction is maintained.

The ozone layer is vitally important to the Earth's living creatures—ourselves included—because it absorbs most of the uv radiation in sunlight. It thus shields us from these energetic rays, which are potentially harmful to biological organisms. Not only can uv produce severe skin burns and eye damage, it is also a carcinogen, or cancer-causing agent. For example, uv exposure is thought to be responsible for most cases of melanoma, an extremely dangerous form of cancer that originates in the pigment-producing cells of the skin.

This protective blanket of ozone is now threatened by the products of human activity. The cause is the release into the atmosphere of a class of chemicals called chlorofluorocarbons (CFCs for short). CFCs are the most widely used refrigerants (a common commercial example is Freon). We depend on refrigeration to cool our homes, cars, and businesses, and use numerous refrigerators to keep food and beverages cool. But as gases, these CFCs often escape from refrigerator units. CFCs are also used as plastic foam blowing agents and in industrial solvents. They were once used as propellants in spray cans as well, but this use was discontinued in the 1970s after the effects of CFCs on the ozone layers were predicted.

The release of gaseous CFCs into the atmosphere is now a cause for worldwide concern. In the lower atmosphere, there is no known mechanism for their destruction, and the CFCs rise slowly through the atmosphere. In 20 to 30 years, they reach the altitude of the ozone layer. Here, the CFC molecules are broken apart by uv radiation and release reactive chlorine (Cl) atoms. These atoms in turn react with and destroy ozone molecules in a repeating cycle:

$$CFCs \xrightarrow{(uv)} Cl$$
$$Cl + O_3 \longrightarrow ClO + O_2$$
$$ClO + O \longrightarrow Cl + O_2$$

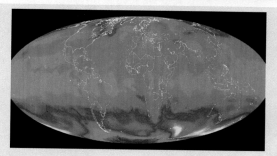

FIGURE 1 Ozone hole The depletion of ozone over the South Pole from the data taken with the Total Ozone Mapping Spectrometer (TOMS) on a NIMBUS-7 satellite. Purple shades indicate the regions of lowest ozone concentrations.

Notice that a chlorine atom is again available for reaction at the end of the sequence. It is estimated that these atoms may remain in the atmosphere for a year or more. During this time, a single Cl atom may destroy as many as 100,000 ozone molecules.

Attention was drawn to the fate of the ozone layer in the late 1980s when an ozone "hole" was discovered over Antarctica (Fig. 1). This prompted companies to seek environmentally safe substitutes for CFCs. Even so, there will be tradeoffs. Currently, the substitutes are much less efficient as refrigerants, and it is estimated that their universal adoption would cause a 3% increase in electrical use. This would both add to consumer cost and increase the amount of carbon dioxide (CO_2) released into the atmosphere, since most electricity is generated by burning fossil fuels. More CO_2 could in turn contribute to the greenhouse effect and global warming (Chapter 11).

It is estimated that for every 1% loss of the ozone layer, an additional 2% of the Sun's uv radiation will reach the Earth's surface. The direct effects of increased uv exposure on human communities are likely to include a higher incidence of melanoma. (In fact, a rise in the number of cases of this deadly cancer has already been observed in Australia and other southern regions where the erosion of the ozone layer has been most marked.) Eye damage, such as cataracts, is also likely to increase. On a global scale, stronger uv radiation could kill many of the oceanic microorganisms that play a key role in the food chain. And an increase in the amount of uv radiation reaching the Earth's surface could contribute to global surface warming, with its many poorly-understood and potentially dangerous consequences (discussed in the Insight feature in Chapter 11).

The problem is real enough to have alarmed political leaders as well as scientists. In 1989, some 24 nations signed the *Montreal Protocol on Substances that Deplete the Ozone Layer*; and in 1990 more than 90 nations agreed to phase out CFCs by the year 2000.

FIGURE 19.21 X-ray CT scan In an ordinary X-ray image, the entire thickness of the body is projected onto the film. Internal structures often overlap, making details hard to distinguish. In computerized tomography (from the Greek words *tomo*, "slice," and *graph*, "picture"), X-ray beams scan across a slice of the body. The transmitted radiation is recorded by a series of detectors and processed by a computer. Using information from multiple slices, the computer can construct a three-dimensional image. Any single slice can also be displayed for study, as on the monitor in the photo. CT scans typically provide doctors with much more information than can be obtained from a conventional X-ray.

evacuated tube. Electrons emitted from the negative electrode are accelerated toward the positive anode, which is called the target. When the electrons strike the target, they interact with the atomic electrons of its material, and the electrical repulsion abruptly decelerates them. This results in a loss of energy in the form of high-frequency X-rays. Similar processes take place in color television picture tubes, which use electron beams and high potential differences. Proper shielding is necessary to protect viewers from the resulting X-rays.

As you will learn in Chapter 26, the energy of electromagnetic radiation depends on its frequency. High-frequency X-rays have very high energies and can cause cancer, skin burns, and other harmful effects. However, at low intensities, X-rays can be used with relative safety to view the internal structure of the human body and other opaque objects.* X-rays can pass through materials that are opaque to other types of radiation. The denser the material, the greater its absorption of X-rays and the less intense the transmitted radiation will be. For example, as X-rays pass through the human body, many more of them are absorbed by bone than by tissue. If the transmitted radiation is directed onto a photographic plate or film, the exposed areas show variations in intensity that form a picture of internal structures.

The combination of the computer with modern X-ray machines permits the formation of three-dimensional images by means of a technique called *computerized tomography,* or CT (● Fig. 19.21).

Gamma Rays. The electromagnetic waves of the upper frequency range of the known electromagnetic spectrum are called gamma rays (γ-rays). This high-frequency radiation is produced in nuclear reactions and in particle accelerators. Gamma rays will be discussed in Chapter 28.

---

*Most health scientists believe that there is no safe "threshold" level for X-rays or other energetic radiation—that is, no level of exposure that is completely risk-free—and that some of the dangerous effects are cumulative over a lifetime. People should therefore avoid unnecessary medical X-rays or any other unwarranted exposure to "hard" radiation (Chapter 28).

Important Concepts

You should be able to define and/or explain these chapter concepts clearly.

electromagnetic induction	back emf	eddy currents	microwaves
magnetic flux	counter torque	electromagnetic waves (radiation)	infrared radiation
Faraday's law of induction	transformer	Maxwell's equations	visible light
Lenz's law	primary coil	radiation pressure	ultraviolet radiation
alternating current	secondary coil	power waves	X-rays
ac generator	step-up transformer	radio and TV waves	gamma rays
dc generator	step-down transformer		

Important Relationships for Review

These relationships are mathematical statements of the concepts and principles presented in the chapter. You should be able to identify the symbols and to explain the relationships before proceeding to the Exercises. In-text equation reference numbers are given for convenience.

Magnetic Flux:

$$\Phi = BA \cos \theta \qquad (19.1)$$

Faraday's Law of Induction:

$$\mathscr{E} = -N \frac{\Delta \Phi}{\Delta t} \qquad (19.2)$$

$$= -N \left[\left(\frac{\Delta B}{\Delta t} \right)(A \cos \theta) + B \left(\frac{\Delta A}{\Delta t} \right)(\cos \theta) \qquad (19.3) \right.$$

$$\left. + BA \left(\frac{\Delta(\cos \theta)}{\Delta t} \right) \right]$$

Generator emf (where $\mathscr{E}_o = NBA\omega$):

$$\mathscr{E} = \mathscr{E}_o \sin \omega t = \mathscr{E}_o \sin 2\pi f t \qquad (19.4\text{–}5)$$

Back emf:

$$\mathscr{E}_b = V - IR \qquad (19.6)$$

Currents, voltage, and turn ratio for a transformer:

$$\frac{I_p}{I_s} = \frac{V_s}{V_p} = \frac{N_s}{N_p} \qquad (19.9)$$

Exercises

19.1 ■ Induced Emf's: Faraday's Law and Lenz's Law

1 A unit of magnetic flux is the (a) Wb, (b) T-m², (c) T-m/A, (d) both (a) and (b). (d)

2 A time rate of change of magnetic flux may be accomplished by varying one or more of how many parameters? (a) 2, (b) 3, (c) 5, (d) any number. (b)

3 A bar magnet is dropped through a coil of wire as shown in ● Fig. 19.22. (a) Describe what is observed on the galvanometer by sketching a graph of $\mathscr{E}$ versus t. (b) Does the magnet fall freely? Explain. (a) see ISM (b) no, Lenz's law

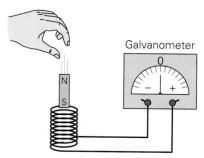

● **FIGURE 19.22 A time-varying magnetic field** Which way will the galvanometer's needle deflect? See Exercise 3.

4 A telephone has both a speaker-transmitter and a receiver (● Fig. 19.23). The transmitter has a diaphragm coupled to a carbon chamber (called the button), which contains loosely packed granules of carbon. As the diaphragm vibrates because of incident sound waves, the pressure on the granules varies, causing them to be more or less closely packed. As a result, the resistance of the button changes. The receiver converts electrical impulses to sound. Applying principles of electricity and magnetism you have learned, explain the basic operation of the telephone. see ISM

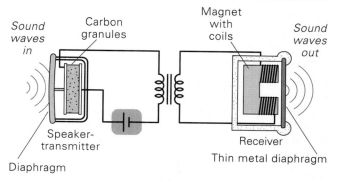

● **FIGURE 19.23 Telephone operation** See Exercise 4.

5 ■ A circular loop of wire that encloses an area of 0.015 m² is in a uniform magnetic field of 0.25 T. What is the flux through the loop if its plane is (a) parallel to the field, (b) at an angle of 37° to the field, and (c) perpendicular to the field? (a) 0 (b) 2.3×10^{-3} T-m² (c) 3.8×10^{-3} T-m²

6 ■ A conductive loop enclosing an area of 2.0×10^{-3} m² is perpendicular to a uniform magnetic field of 3.0 T. If the field goes to zero in 0.0045 s, what is the average emf induced in the loop? 1.3 V

7 ■■ A circular loop with a radius of 10 cm is positioned at various locations in a uniform 0.60-T magnetic field. Find the magnetic flux if the normal to the plane of the loop is (a) perpendicular to the magnetic field, (b) parallel to the magnetic field, and (c) at an angle of 40° to the magnetic field. (a) 0 (b) 1.9×10^{-2} T-m² (c) 1.4×10^{-2} T-m²

● **8** ■■ A loop in the form of an equilateral triangle 40.0 cm on a side is oriented perpendicularly to a uniform magnetic field of 550 mT. What is the flux through the loop? 3.8×10^{-2} T-m²

9 ■■ A square coil of wire with 10 turns is in a magnetic field of 0.25 T. The flux through the loop is 0.50 T-m². Find the effective area of the loop if the field (a) is perpendicular to the plane of the loop and (b) makes an angle of 60° with the plane of the loop.   (a) 0.20 m² (b) 0.17 m²

10 ▪▪ What is the magnetic flux through an ideal solenoid whose windings have a radius of 3.0 cm if the turn density is 250 turns/m and a current of 1.5 A flows through the wire?   1.3 × 10⁻⁶ T-m²

11 ▪▪ A magnetic field perpendicular to the plane of a wire loop with an area of 0.40 m² decreases by 0.10 T in 10⁻³ s. What is the average value of the emf induced in the loop?
40 V

12 ▪▪ A square loop of wire with 50-cm sides experiences a uniform, perpendicular magnetic field of 500 mT. If the field goes to zero in 0.010 s, what is the average emf induced in the loop?   13 V

13 ▪▪ The magnetic flux through a 30-turn coil of wire is reduced from 20 Wb to 5.0 Wb in 0.10 s. Find the average induced current in the coil, which has resistance of 2.5 Ω.
1.8 × 10³ A

14 ▪▪ If the magnetic flux through a loop of wire increases by 30 T-m² and an average emf of 20 V is induced in the wire, over what period of time was the flux increased?
1.5 s

●**15** ▪▪ During a time of 0.20 s, a coil of wire with 50 loops has an average induced emf of +9.0 V due to a changing magnetic field perpendicular to the plane of the coil. If the radius of the coil is 10 cm and the initial value of the magnetic field is 1.5 T, what is the final value of the field?   0.35 T

16 ▪▪ A single strand of wire of adjustable length is wound around the circumference of a round balloon that has a diameter of 20 cm. A uniform magnetic field with a magnitude of 0.15 T is perpendicular to the plane of the loop. If the balloon is blown up so its diameter and the diameter of the wire loop increase to 40 cm in 0.040 s, what is the average value of the emf induced in the loop?   0.35 V

17 The magnetic field perpendicular to the plane of a wire loop with an area of 0.10 m² changes with time as shown in ● Fig. 19.24. What is the average emf induced in the loop for each time interval on the graph (e.g., from 0 to 2.0 ms)?
(a) −50 V   (b) 13 V   (c) 0   (d) 12 V

●**18** ▪▪ A metal rod is pulled at a uniform velocity of magnitude v perpendicularly to a uniform magnetic field of magnitude B. If the rod has a length L, what emf is induced across it? $\varepsilon = BLv$

19 ▪▪ A metal rod 20 cm long moves in a straight line at a speed of 4.0 m/s with its length perpendicular to a uniform magnetic field of 1.2 T. Find the potential difference between the ends of the rod.   −0.96 V

20 ▪▪ Suppose that the metal rod in Fig. 19.5 is 20 cm long and is moving at a speed of 10 m/s in a magnetic field of 1.5 T, but that the metal frame is covered with an insulating material. What would happen? (Give a quantitative answer.)
$V = -3.0$ V across the bar, $I = 0$

21 ▪▪ The flux through a fixed loop of wire changes uniformly from +40 Wb to −20 Wb in 1.5 ms. (a) What is the significance of the negative flux? (b) What is the average induced emf in the loop?   (a) change in direction
(b) 4.0 × 10⁴ V

●**22** ▪▪ A rectangular conductive loop measuring 20 cm by 30 cm has a resistance of 0.10 Ω. At what rate must a magnetic field perpendicular to the plane of the loop change with time to cause a current of 3.0 A to flow in the loop?   −5.0 T/s

23 ▪▪ A coil of wire with 10 turns and a cross-sectional area of 0.055 m² is placed in magnetic field of 1.8 T and oriented so the area is perpendicular to the field. The coil is then flipped in 0.25 s and ends up with the area parallel to the field. What is the average emf induced in the coil?
−4.0 V

24 ▪▪ A uniform magnetic field with a value of 1.5 T passes through a 0.40 m × 0.60 m rectangular loop at an angle of 60° relative to a normal to the plane of the loop. (a) If the field decreases to zero in 0.50 s, what is the average emf developed in the loop? (b) If the field remains constant at its initial value, at what rate would the loop have to shrink in size to generate the same average emf?
(a) 0.36 V   (b) 0.48 m²/s

25 ▪▪▪ A uniform magnetic field of 0.50 T permeates a double incline block as shown in ● Fig. 19.25. Determine the magnetic flux through *each* surface of the block.   see ISM

26 ▪▪▪ A length of 20-gauge copper wire is formed into a circular loop with a radius of 20 cm. (20-gauge wire has a

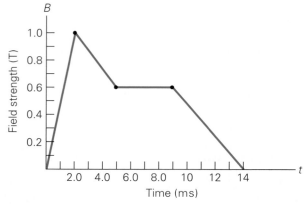

●**FIGURE 19.24 Magnetic field versus time**   See Exercise 17.

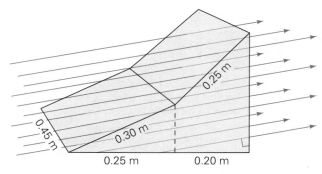

●**FIGURE 19.25 Magnetic flux**   See Exercise 25. (Drawing not to scale.)

diameter of 0.8118 mm). A magnetic field perpendicular to the plane of the loop increases from zero to 7.0 mT in 0.25 s. Find the average electrical energy dissipated in the process. 8.0×10^{-5} J

27 ■■■ A metal airplane with a wing span of 30 m flies horizontally at a constant speed of 320 km/h in a region where the vertical component of the Earth's magnetic field is 5.0×10^{-5} T. What is the induced motional emf across its wing tips? 0.13 V

19.2 ■ Generators and Back Emf

28 A split-ring commutator is used in a(n) (a) ac generator, (b) dc generator, (c) alternator, (d) both (a) and (b). (b)

29 The back emf of a motor depends on (a) the input voltage, (b) the input current, (c) the armature's rotational speed, (d) none of these. (c)

30 What is the orientation of the armature loop in a simple ac generator when the value of the emf is a maximum, and why? see ISM

31 A student has a bright idea for a generator: For the arrangement shown in ● Fig. 19.26, the magnet is pulled down and released. With a highly elastic spring, the inventor thinks there should be a relatively continuous electrical output. What is wrong with this idea? see ISM

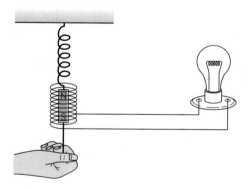

● **FIGURE 19.26 Inventive genius?** See Exercise 31.

32 ■ An ac generator is rated at 60 Hz. At what intervals of the period is the emf (a) a maximum, (b) zero, and (c) a minimum? (a) 1/240 s (b) 1/120 s (c) 1/80 s

33 ■ (a) What is the maximum value of the emf output from a simple ac generator having a single loop with an area of 90 cm² that rotates with a frequency of 60 Hz in a uniform magnetic field of 10^{-2} T? (b) What would the maximum value be if a coil of 15 such loops were used? (a) 3.4×10^{-2} V (b) 0.51 V

34 ■■ A simple ac generator consists of a coil having 10 turns of wire, with each loop having an area of 50 cm². The coil rotates in a uniform magnetic field of 350 mT with a frequency of 60 Hz. (a) Write an equation showing how the emf of the generator varies as a function of time. (b) Compute the maximum emf. (a) $\varepsilon = \varepsilon_0 \sin 120 \pi t$ (b) 6.6 V

35 ■■ A 60-Hz sinusoidal ac voltage has a maximum value of 120 V. What is the value of the voltage at $\frac{1}{180}$ s after it has a value of zero? (Is there more than one such value?) ±104 V

36 ■■ A simple ac generator with a maximum emf value of at least 20 V is to be constructed, using loops of wire with a radius of 0.15 m, a rotational frequency of 60 Hz, and a magnetic field of 10^{-2} T. How many loops of wire will be needed? 75 loops

● **37** ■■ A simple ac generator has a rotational frequency of 60 Hz. What percentage of the maximum emf is the instantaneous emf (a) at 0.50 s after starting up and (b) at $\frac{1}{360}$ s after the emf passes through zero in changing to a negative polarity? (a) 0 (b) 87%

38 ■■ The armature of a simple ac generator has 20 circular loops of wire with a radius of 10 cm. It is rotated with a frequency of 60 Hz in a uniform magnetic field of 800 mT. What is the maximum value of the emf induced in the loops, and how often is this value attained? 1.9×10^2 V every 1/120 s

39 ■■ The armature of an ac generator has 100 turns, which are rectangular loops measuring 8.0 cm by 12 cm. The generator has a sinusoidal voltage output with an amplitude of 24 V. If the magnetic field of the generator is 250 mT, with what frequency does the armature turn? 16 Hz

40 ■■ Which has the greater voltage output—a generator with a loop area of 100 cm² rotating in a magnetic field of 20 mT at 60 Hz or a generator with a loop area of 75 cm² rotating in a magnetic field of 200 mT at 120 Hz? Justify your answer mathematically. the second loop, see ISM

41 ■■ A motor has a resistance of 3.0 Ω and draws a current of 2.5 A when operating at its normal speed on a 115-V line. What is the back emf of the motor? 108 V

● **42** ■■ The starter motor in an automobile has a resistance of 0.40 Ω in its armature windings. The motor operates on 12 V and has a back emf of 10 V when running at normal operating speed. How much current does the motor draw (a) when running at its operating speed and (b) when initially starting up? (a) 5.0 A (b) 30 A

43 ■■ A 240-V dc motor with an armature whose resistance is 1.5 Ω draws a current of 16 A when running at its operating speed. (a) What is the back emf of the motor when it is operating normally? (b) What is the starting current? (Assume that there is no additional resistance.) (c) What series resistance would be required to limit the starting current to 25 A? (a) 216 V (b) 160 A (c) 8.1 Ω

19.3 ■ Transformers and Power Transmission

44 For a step-up transformer, which of the following is *not* correct? (a) $N_p > N_s$, (b) $N_s > N_p$, (c) energy is lost through eddy currents. (d)

45 In order to reduce resistance losses, electric power is transmitted (a) at low voltage, (b) at high current, (c) using only step-down transformers, (d) none of these. (d)

46 The voltage for ignition by a spark plug in an automobile is supplied by what is referred to as the coil, which is actually a pair of induction coils (● Fig. 19.27). Explain how the voltage of the car's battery (12 V) is raised to as high as 25 kV by this device. Explain the function of the distributor. see ISM

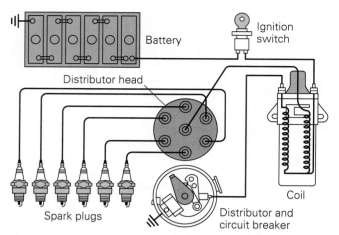

● **FIGURE 19.27 Auto ignition with coil** See Exercise 46.

● **47** ■■ A transformer has 800 turns on its primary coil and 600 turns on its secondary. (a) What type of transformer is this? (b) If the input to the primary coil is 4.0 A at 120 V, what is the output of the secondary?
(a) step down (b) 90 V, 5.3 A

48 ■■ The number of turns on the secondary coil of an ideal transformer is 450 and on the primary it is 75. (a) What type of transformer is this? (b) What is the ratio of the current in the primary coil to the current in the secondary? (c) What is the ratio of the voltage in the primary coil to the voltage in the secondary? (a) step up (b) $V_p/V_s = \frac{1}{6}$

49 ■■ An ideal transfomer steps up 8.0 V to 2000 V, and the 4000-turn secondary coil carries 2.0 A. (a) Find the number of turns on the primary coil. (b) Find the current in the primary. (a) 16 (b) 5.0×10^2 A

50 ■■ The primary coil of an ideal transformer has 720 turns, and the secondary coil has 180 turns. If 15 A flows in the primary with a voltage of 120 V, what are (a) the voltage and (b) the current output of the secondary? (a) 30 V
(b) 60 A

51 ■■ A transformer changes a 120-V input to a 6000-V output. Find the ratio of the number of turns on the primary coil to the number of turns on the secondary coil.
$N_p/N_s = 2.0 \times 10^{-2}$

● **52** ■■ The primary coil of an ideal transformer is connected to a 120-V source and draws 10 A. The secondary coil has 800 turns and a current of 4.0 A. (a) What is the voltage on the secondary? (b) How many turns are on the primary?
(a) 3.0×10^2 V (b) 3.2×10^2 turns

53 ■■ An ideal transformer has 840 turns on its primary coil and 120 turns on its secondary. If the primary draws 1.5

at 120 V, what are (a) the current and (b) the voltage output of the secondary? (a) 11 A (b) 17 V

54 ■■ An electric arc-welding machine requires 200 A of current. The transformer of the machine has 1200 turns on the primary coil, which draws 2.5 A at 240 V. (a) How many turns are on the secondary coil? (b) What is the voltage output of the secondary? (a) 15 turns (b) 3.0 V

55 ■■ A 1.5 kV power line runs to a residential area, where a transformer with 1000 turns on its primary coil steps down the voltage to 240 V. How many turns are on the secondary coil of this transformer? 1.6×10^2 turns

56 ■■ A circuit component operates on 20 V at 0.50 A. A transformer with 300 turns on its primary coil is used to convert 120-V household electricity to the proper voltage. (a) How many turns must the secondary coil have? (b) How much current flows in the primary? (a) 50 turns
(b) 8.3×10^{-2} A

57 ■■ A transformer in a door chime steps down the voltage from 120 V to 10 V and supplies a current of 0.50 A to the chime mechanism. (a) What is the turn ratio of the transformer? (b) What is the current input to the transformer? $N_s/N_p = 1/12$ (b) 4.2×10^{-2} A

● **58** ■■ An ac generator supplies 20 A to 440 V to a 10,000-V power line. If the step-up transformer has 132 turns on its primary coil, how many turns are on the secondary?
3.0×10^3 turns

59 ■■ The electrical energy described in Exercise 58 is transmitted over an 80.0-km line ($R_oL = 0.80$ Ω/km). How many kilowatt-hours are saved in 5.00 h by stepping up the voltage? 128 kWh

60 ■■■ At an area substation, the power-line voltage is stepped down from 100,000 V to 20,000 V. If 1.0 MW of power is delivered to the 20,000-V circuit, what are the primary and secondary currents in the transformer?
10 A, 50 A

61 ■■■ A voltage of 230,000 V in a transmission line is reduced to 100,000 V at an area substation, to 7200 V at a distributing substation, and to 240 V at a utility pole outside a house. (a) What turn ratio is required for each reduction? (b) By what factor is the transmission line current stepped up in each voltage step-down and overall?
(a) $N_s/N_p = \frac{1}{2.3}, \frac{1}{13.9}, \frac{1}{30}$ (b) 2.3, 13.9, 30

62 ■■■ An electric-generating plant produces 50 A at 20 kV. The electricity is transmitted 25 km over transmission lines whose resistance is 1.2 Ω/km. (a) What would the power loss be if the energy is transmitted at 20 kV? (b) To what value should the output voltage of the generator be stepped up to decrease the energy loss by a factor of 15?
(a) 75 kW (b) 77 kW

63 ■■■ Electrical power is transmitted through a 175-km power line with a resistance of 1.2 Ω/km. The generator output is 50 A at its operating voltage of 440 V. This voltage has a single step up for transmission. (a) How much power is lost as joule heat? (b) What is the necessary turn ratio of a transformer at the power delivery point that provides an output voltage of 220 V? (Neglect the voltage drop of the line.)   (a) 53 W   (b) $N_p/N_s = 200$

19.4 ■ Electromagnetic Waves

64 Relative to the red end of the visible spectrum, the blue or violet end has (a) lower frequencies, (b) longer wavelengths, (c) shorter wavelengths, (d) both (a) and (b).   (d)

● **65** Which of the following radiations has the highest frequency: (a) ultraviolet, (b) infrared, (c) X-rays, (d) microwaves?   (c)

66 On a cloudy summer day, you may work outside or go swimming most of the day and feel relatively cool. Yet, that evening, you may find that you have a bad sunburn. Why is this? Explain in terms of infrared and ultraviolet radiations.   see ISM

67 ■ Find the frequencies of electromagnetic waves with wavelengths of (a) 1.0 m, (b) 25 m, (c) 150 m.
(a) 3.0×10^8 Hz   (b) 1.2×10^7 Hz   (c) 2.0×10^6 Hz

68 ■ The frequency ranges of the radio AM and FM bands and the TV band are given in the text. What are the corresponding wavelength ranges for these bands?
AM: 1.8×10^2 m–5.7×10^2 m; FM: 2.8 m–3.4 m;
TV: 3.4 cm–5.6 m

69 ■■ A radar operator notes that a time of 0.24 ms elapses between the sending and return of a radar pulse. How far away was the object that reflected the pulse? 3.6×10^4 m

● **70** ■■ How long does it take for a laser beam to travel from Earth to a reflector on the moon and back? (Take the distance of the moon to be 240,000 mi. This experiment was done to determine the accurate distance of the moon from the Earth.)   2.6 s

71 ■■ Some electromagnetic waves have the following wavelengths: (a) 800 nm and (b) 300 nm. To which spectral regions do these waves belong?   (a) 800 nm > 700 nm;
infrared   (b) 300 nm < 400 nm; ultraviolet

72 ■■ Orange light has a wavelength of 600 nm and green light a wavelength of 510 nm. What is the frequency difference of these colors? $\Delta f = 8.8 \times 10^{13}$ Hz

73 ■■■ A radio antenna made of wire is called a quarter wavelength antenna because of its length. If you were going to make such antennae for the AM and FM radio bands by using the mid frequency of each band, what lengths of wire would you use?   AM: 74 m; FM: 0.77 m

■ Additional Exercises

74 A 120-VAC motor has a resistance of 3.50 Ω. When operating at normal speed, the motor develops a back emf of 115 V. What are (a) the start-up current of the motor, and (b) the current when it is operating at normal speed?
(a) 34.3 A   (b) 1.43 A

75 An electric door bell operates on 9.0 V. If this voltage is obtained from standard 120-V household voltage, which of its windings has more turns and how many times more than the other?   primary, 13

76 A circular wire loop is in a uniform magnetic field of 1.5×10^{-2} T. The flux through the loop is 1.2×10^{-2} T-m². What is the radius of the loop if the normal to the plane of the loop makes an angle of 45° with the field?   0.60 m

77 The efficiency of a transformer is defined as the ratio of the power output to the power input: efficiency = I_sV_s/I_pV_p. Show that in terms of the ratios of currents and voltages given in Eq. 19.9, an efficiency of 100% is obtained. What does this imply?   100%; ideal conditions

78 A pivoted coil of wire is rotated 50 times per second in a uniform magnetic field. How often during each second does the induced emf in the coil have a value of zero?
once every 0.01 s

79 A 120-V dc motor draws a current of 6.0 and has a back emf of 96 V at its operating speed. (a) What starting current is required by the motor? (Assume that there is no additional resistance.) (b) What series resistance would be required to limit the starting current to 15 A?   (a) 30 A   (b) 4 Ω

80 A circular wire loop with a diameter of 28 cm carries a current of 0.50 A. Assume that the magnitude of the magnetic field at the center of the loop is uniform across the whole area. What is the flux through the loop? 1.4×10^{-7} T-m²

81 An ideal transformer has 120 turns on its primary coil and 840 turns on its secondary. If 4.0 A flows in the primary with a voltage of 120 V, what are (a) the voltage and (b) the current output of the secondary?   (a) 840 V   (b) 0.57 A

82 A loop of wire in the form of a square 1.0 m on a side is perpendicular to a uniform magnetic field of 0.75 T. What is the flux through the loop?   0.75 T-m²

83 An ideal transformer has 80 turns on its primary coil and 360 turns on its secondary. If 3.0 A at 12 V is applied to the primary coil, what are (a) the current and (b) the voltage output of the secondary?   (a) 0.67 A   (b) 54 V

84 A square conductive loop 50 cm on a side has a resistance of 200 mΩ. Find the rate at which a magnetic field perpendicular to the plane of the loop must change with time to cause an average of 5.0 J per unit time to be expended in the loop.   4.0 T/s

85 A coil with 150 concentric loops of wire with a radius of 12 cm is used as the armature of a generator with a magnetic field of 0.80 T. With what frequency should the armature be rotated to make the polarity change every 0.010 s?   50 Hz

86 The transformer on a utility pole steps down the voltage from 20,000 V to 220 V for home usage. If a household circuit uses 6.6 kW of power, what are the primary and secondary currents in the transformer? $I_p = 0.33$ A; $I_s = 30$ A

20

AC Circuits

Because alternating voltage and current are used in power transmission systems and in homes and industry, ac circuits are very common. The lights on the television control panel shown above, for example, utilize ac. So do the monitors in the background, the cameras that capture the pictures, and the lights that illuminate the scenes being photographed. You might suppose that ac circuits should be as simple and straight forward as dc circuits. However, the alternating current and the associated time-varying magnetic fields in such circuits add new dimensions to ac circuit analysis.

In common dc circuits, we are concerned only with ohmic resistances, and Ohm's law (V = IR) and expressions for power (P = IV = I²R) applied. There is ohmic resistance in ac circuits, too, but there are other factors that affect the flow of charge. Recall that in a dc circuit, once a capacitor has been fully charged it offers infinite resistance (since in effect it opens the circuit). However, in an ac circuit, this is not the case. Alternating voltage continually charges and discharges a capacitor. As a result, there is current in the circuit, but there is also opposition to the current, and this is expressed in a special way, as will be discussed. Coils of wire also oppose an ac current through induction. An example, the back emf of a motor, was described in Chapter 19.

In this chapter, we will look at some of the basic principles of ac circuits. Forms of Ohm's law and expressions for power will be developed for ac circuits. They will resemble their dc counterparts, but there will be some important differences. In general, we will need to use special average values of the current and voltage. (Keep in mind that with normal 60-Hz voltage, the polarity reverses 120 times per second.)

> *Topics of study in Chapter 20 include resistance in ac circuits, capacitance and inductive reactance, impedance, and circuit resonance.*

Finally, in certain ac circuits there is an electrical analog of mechanical resonance—the condition for maximum energy transfer. How does this occur? We will see that it has to do with the resonance frequency of the circuit, which is determined by the oppositions to ac current mentioned previously.

20.1 ■ Resistance in an AC Circuit

In general, an ac circuit is the one with an ac voltage source and one or more other circuit elements. The circuit diagram for an ac circuit with a single resistive element is shown in ● Fig. 20.1. If the source output is assumed to be sinusoidal, as is the case for a simple generator (Section 19.2), the voltage across the resistor varies with time according to this equation:

$$V = V_o \sin \omega t = V_o \sin 2\pi ft \tag{20.1}$$

where ω is the angular frequency ($\omega = 2\pi f$). The voltage oscillates between maximum instantaneous values of $+V_o$ and $-V_o$, where V_o is the **peak voltage**.

The current through the resistor also oscillates—its direction changes. From Ohm's law, the current can be expressed as a function of time.

$$I = \frac{V}{R} = \frac{V_o}{R} \sin 2\pi ft$$

That is

$$I = I_o \sin 2\pi ft \tag{20.2}$$

where the amplitude $I_o = V_o/R$ is the **peak current**.

Both current and voltage for an ac circuit are plotted versus time in the graph in ● Fig. 20.2. Note that they are in step, or *in phase*, with each other, reaching zero values and maxima at the same times. The current oscillates back and forth and has corresponding positive and negative values during each cycle. Thus, the summation of all the instantaneous current values (positive and negative) over one complete cycle is zero, and the *average current is zero*. Mathematically, this reflects the fact that the time average value of $\sin 2\pi ft$ over one or more *complete* (360°) cycles is zero: in general, $\langle \sin \theta \rangle = \langle \sin 2\pi ft \rangle = 0$. (We use angular brackets instead of overbars here to denote a time average value.) Similarly, $\langle \cos \theta \rangle = 0$.

The fact that the *average* current is zero, however, does not mean that there is no joule heating (I^2R losses). The dissipation of electrical energy depends on the collisions of electrons with the atoms of the material in which there is a current. These collisions occur when the electrons of an alternating current are going in either direction.

The instantaneous power is obtained using the instantaneous current (Eq. 20.2):

$$P = I^2R = I_o^2 R \sin^2 2\pi ft \tag{20.3}$$

Even though the current changes sign each half-cycle, the square of the current, I^2, is always positive. Thus the average value of I^2R is *not* zero, even though the average value of I is zero. The average, or mean, value of I^2 is

$$\bar{I^2} = \langle I_o^2 \sin^2 2\pi ft \rangle = I_o^2 \langle \sin^2 2\pi ft \rangle$$

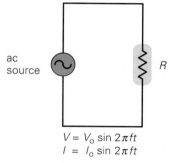

$$V = V_o \sin 2\pi ft$$
$$I = I_o \sin 2\pi ft$$

● **FIGURE 20.1 A purely resistive circuit** The ac source delivers a sinusoidal voltage to the circuit, and the voltage across and current through the resistor are sinusoidal.

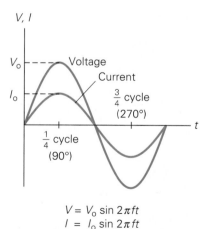

$$V = V_o \sin 2\pi ft$$
$$I = I_o \sin 2\pi ft$$

● **FIGURE 20.2 Voltage and current in phase** In a purely resistive ac circuit, the voltage and current are in step, or in phase, with zero values and maxima occurring at the same respective times.

Using the trigonometric identity $\sin^2\theta = \frac{1}{2}(1 - \cos 2\theta)$, we can write $\langle\sin^2\theta\rangle = \frac{1}{2}(1 - \langle\cos 2\theta\rangle)$, and since $\langle\cos 2\theta\rangle = 0$ (just as $\langle\cos\theta\rangle = 0$), we have $\langle\sin^2\theta\rangle = \frac{1}{2}$. Thus,

$$\bar{I}^2 = I_o^2 \langle\sin^2 2\pi ft\rangle = \frac{1}{2}I_o^2 \qquad (20.4)$$

The average power is therefore

$$\bar{P} = \bar{I}^2 R = \frac{1}{2}I_o^2 R \qquad (20.5)$$

It is customary to write the ac power in the same form as the dc power ($P = I^2 R$). To do this, a special value of current is used,

$$\boxed{I_{rms} = \sqrt{\bar{I}^2} = \sqrt{\frac{1}{2}I_o^2} = \frac{I_o}{\sqrt{2}} = 0.707\,I_o} \qquad (20.6)$$

where $1/\sqrt{2} = 0.707\,I_{rms}$ is called the **rms current**, or **effective current** (rms stands for root-mean-square, in this case indicating the root of the mean value of the square of the current). Then with $I_{rms}^2 = (I_o/\sqrt{2})^2 = \frac{1}{2}I_o^2$, the effective power is

$$\boxed{\bar{P} = \frac{1}{2}I_o^2 R = I_{rms}^2 R} \qquad (20.7)$$

By a similar development, the **rms voltage**, or **effective voltage**, is

$$\boxed{V_{rms} = \frac{V_o}{\sqrt{2}} = 0.707\,V_o} \qquad (20.8)$$

Thus, for power dissipation through a resistance R, an alternating current and voltage with rms values equal to dc current and voltage values will have the same effect, or produce the same $I^2 R$ loss. For alternating current, Ohm's law is written as

$$\boxed{V_{rms} = I_{rms}R} \qquad (20.9)$$

It is customary to measure and specify rms values for ac quantities. For example, the household line voltage of 120 V is an rms value with a peak voltage of

$$V_o = \sqrt{2}\,V_{rms} = \frac{120\text{ V}}{0.707} = 170\text{ V}$$

The various mean and peak values of ac current and voltage are shown in the graphs in ● Fig. 20.3. Another voltage designation sometimes used is the peak-to-peak value:

$$V_{p-p} = 2V_o.$$

Figure 20.3

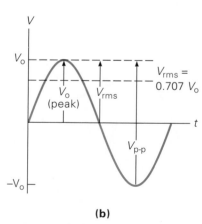

(a)

(b)

● **FIGURE 20.3 Root-mean-square (rms) current and voltage** The rms values of the current **(a)** and the voltage **(b)** are 0.707, or $1/\sqrt{2}$, times the peak or maximum values. The voltage is sometimes described by the peak-to-peak (V_{p-p}) value.

EXAMPLE 20.1 ■ Rms Values

A lamp with a 100-W bulb is plugged into a 120-V outlet. (a) What are the rms and peak currents through the lamp? (b) What is the resistance of the bulb? (Neglect the resistance of the lamp's wiring.)

Solution.

$$\text{Given: } \overline{P} = 100 \text{ W} \qquad \textit{Find: } \text{(a) } I_{rms} \text{ and } I_o \text{ (rms and peak currents)}$$
$$V_{rms} = 120 \text{ V} \qquad \qquad \text{(b) } R \text{ (resistance)}$$

(a) The rms current is

$$I_{rms} = \frac{\overline{P}}{V_{rms}} = \frac{100 \text{ W}}{120 \text{ V}} = 0.833 \text{ A}$$

and the peak current is

$$I_o = \sqrt{2} I_{rms} = \frac{0.833 \text{ A}}{0.707} = 1.18 \text{ A}$$

(b) The resistance of the bulb is

$$R = \frac{V_{rms}}{I_{rms}} = \frac{120 \text{ V}}{0.833 \text{ A}} = 144 \text{ }\Omega$$

Note that the resistance could also be obtained from the peak values (V_o for 120 V was computed above):

$$R = \frac{V_o}{I_o} = \frac{170 \text{ V}}{1.18 \text{ A}} = 144 \text{ }\Omega$$

EXAMPLE 20.2 ■ Big Volts

In European countries, the normal line voltage is 240 V. If you were traveling there and plugged in the hair dryer you use at home, you might expect the dryer to (a) not operate at all, (b) operate normally, (c) operate only for a short period. *Clearly establish the reasoning and physical principle(s) used in determining your answer. That is, why did you select your answer?*

Reasoning and Answer. Our small appliances are designed to operate on 120 V. With double the voltage, an appliance would operate, but not as normally designed. With increased voltage, there would be increased current (Ohm's law). The heating element of the hair dryer would get very hot and would probably burn out, so the answer is (c). Assuming the same element resistance R, with double the voltage, there would be twice the current and four times the power output (recall that $P = V^2/R$). Also, the increased current might cause the windings on the blower motor to burn out (or, depending on the type of motor, the blower motor might run faster than normal).

Fortunately, you would not make the mistake of plugging our 120-V appliances into a 240-V line, because the plugs and sockets are different (as are our 240-V plug and sockets). There are several different types of plugs used in various countries. When on foreign travel with normal appliances, one should carry a converter/adapter kit (●Fig. 20.4). It contains a selection of different plugs into which a voltage converter fits. The converter is a solid-state device that converts 240 V to 120 V. (Some travel appliances are designed for dual voltage and will operate on either 120 V or 240 V.)

● **FIGURE 20.4 Converter and adapters** In foreign countries which have 240-V line voltages, a conversion to 120 V is needed to properly operate our normal appliances. Note the different types of plugs for different countries. The small plugs go into the foreign sockets and the converter prongs then fit into the back of a socket. Our standard two-prong plug fits into the converter which has a 120-V output.

The formulas derived for dc circuits can be used for ac circuits if the rms, or effective values, of the quantities are inserted. *For convenience and simplicity, the rms subscripts will be omitted from now on—but keep in mind that rms values are given for ac quantities unless otherwise specified.*

Both alternating current and direct current are measured in amperes. But how is the ampere physically defined for alternating current? It cannot be derived from the mutual attraction of two parallel wires carrying ac current, as the dc ampere is defined (Section 18.4). An ac current changes direction with the source frequency (for example, 60 Hz), and the attractive force would average to zero. Thus, the ac ampere must be defined in terms of some property that is independent of the direction of the current. Joule heating is such a property, and there is *1 ampere of alternating current in a circuit if the current produces the same average heating effect as 1 ampere of dc current would under the same conditions.* That is, the ampere is defined as an *effective* unit of ac current.

> Review joule heating. If the two bulbs in the previous demonstration shone with the same brightness, they were most likely putting out the same amount of heat.

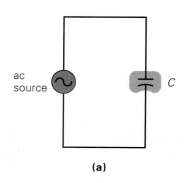

(a)

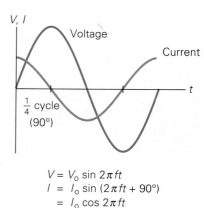

$V = V_0 \sin 2\pi ft$
$I = I_0 \sin (2\pi ft + 90°)$
$\quad = I_0 \cos 2\pi ft$

(b)

● FIGURE 20.5 A purely capacitive circuit In a circuit with only capacitance, the current leads the voltage by 90° or $\frac{1}{4}$ cycle. (Notice that a current zero is 90° or a quarter cycle ahead of a voltage zero.)

20.2 ■ Capacitive Reactance

As you know, when a capacitor is connected to a voltage source in a dc circuit, current will flow for the short time required to charge the capacitor. As charge accumulates on the capacitor's plates, the voltage across them increases, opposing the current. That is, a capacitor in a dc circuit will limit or oppose the current as it charges. When the capacitor is fully charged, the current in the circuit goes to zero.

When a capacitor is in a circuit with an ac source, as shown in ● Fig. 20.5a, it limits, or regulates, the current, but does not completely prevent the flow of charge. The capacitor is alternately charged and discharged as the current and voltage reverse each half-cycle.

Plots of ac current and voltage versus time for a circuit with a capacitor are shown in Fig. 20.5 b. Let's look at the changing conditions of the capacitor. We will choose as our starting point a time (t_o) when the voltage is at its maximum. This means that the capacitor is fully charged ($Q_o = CV_o$). Since the capacitor plates cannot accommodate any additional charge, there is no current in the circuit.

1. As the voltage starts to decrease, the capacitor begins to discharge, giving rise to a current in the circuit. The current reaches its maximum value as the voltage falls to zero and the capacitor plates are completely discharged.*

*Note that although the current is depicted as negative in steps 1 and 2 of Fig. 20.5, this has no special significance. The positive and negative dimensions on the graph merely indicate the *direction* of the current and the *polarity* of the voltage. We could as easily have begun our analysis at step 3, with the voltage negative and the current positive.

2. The voltage then reverses polarity and starts to increase. The capacitor becomes charged again, this time with the opposite polarity. With the plates initially uncharged, there is no opposition to the current (other than the resistance of the connecting wires), so the current is at a maximum. However, the charge accumulating on the capacitor plates opposes the current, which therefore decreases, reaching zero when the voltage is again at a maximum and the capacitor fully charged.

During the next half cycle (steps 3 and 4) the same process is repeated, but with reversed polarities.

3. As the voltage again starts to decline, the capacitor loses its accumulated charge once more. The current is now in the opposite direction, attaining its maximum value as the voltage falls to zero.

4. The voltage rises, the capacitor becomes charged, and the current falls. At the end of this time the circuit has completed a full cycle—the voltage and current are exactly the same as they were at time t_o.

Note that the current and voltage are not in step, or in phase, in this situation. The current reaches a maximum a quarter-cycle ahead of the voltage. The relationship between the current and the voltage is commonly stated in this way:

In a purely capacitive ac circuit, the *current* leads the *voltage* by 90° or $\frac{1}{4}$ cycle.

As in a dc circuit, there is opposition to the charging process in an ac circuit but it is not totally limiting (as in a dc open-circuit condition). The impeding effect of a capacitor on the current in an ac circuit is expressed in terms of a quantity called the **capacitive reactance** (X_c):

$$X_c = \frac{1}{2\pi f C} = \frac{1}{\omega C} \qquad (20.10)$$

Capacitive reactance—opposition to current

where $\omega = 2\pi f$. Here C is the capacitance (in farads), and f is the frequency (in Hz). Like resistance, reactance is measured in ohms (Ω).

Use unit analysis to demonstrate that capacitive reactance is measured in ohms.

As you can see, the reactance is inversely proportional to the capacitance (C) and the voltage frequency (f). Recall that capacitance is the charge per voltage ($C = Q/V$). For a particular voltage, therefore, the greater the capacitance, the more charge the capacitor can accommodate. Putting more charge on the plates requires a larger flow of charge, that is, a greater current, which means the opposition to the current, or the reactance, must be smaller.

Also, the greater the frequency of the ac driving source, the shorter the charging time and the less the charge accumulation on the plates, and therefore the less the opposition to the current. Thus, the capacitive reactance is inversely proportional to the frequency. (Note that if $f = 0$, as is the case for dc, the capacitive reactance is infinite and there is no current flow, which corresponds to an open circuit.)

The capacitive reactance is related to the voltage across the capacitor and the current in the circuit by an equation that has the same form as Ohm's law.

$$\boxed{V = IX_c}$$
(20.11)

(Remember that V and I are rms values.)

EXAMPLE 20.3 ■ Capacitive Reactance

A 15-μF capacitor is connected to a 120-V, 60-Hz source. What are (a) the capacitive reactance and (b) the current in the circuit?

Solution.

Given: $C = 15\ \mu F = 15 \times 10^{-6}\ F$ *Find:* (a) X_c (capacitive reactance)
$V = 120\ V$ (b) I (current)
$f = 60\ Hz$

(a) The capacitive reactance is given by Eq. 20.10,

$$X_c = \frac{1}{2\pi f C} = \frac{1}{2\pi(60\ Hz)(15 \times 10^{-6}\ F)} = 177\ \Omega$$

(b) Then, using Eq. 20.11, the current is

$$I = \frac{V}{X_c} = \frac{120\ V}{177\ \Omega} = 0.678\ A$$

20.3 ■ Inductive Reactance

Another important phenomenon in ac circuits is inductance. A coil of wire (an inductor) in a circuit with a time-varying current has a reverse voltage, or back emf, set up in it as a result of a changing flux. This is an example of self-induction. The induced emf opposes the change in flux (Lenz's law) and thus opposes the current flow in the circuit.

The self-induced emf is given by Faraday's law (Eq. 19.2): $\mathscr{E} = -N\Delta\Phi/\Delta t$. But since the geometry of the coil is fixed, the time rate of change of the flux, or $\Delta\Phi/\Delta t$, is proportional to the time rate of change of the self-induced current in the coil, or $\Delta I/\Delta t$, that is,

$$\mathscr{E} \propto -\frac{\Delta I}{\Delta t}$$

Writing this relationship in equation form with a constant of proportionality, the emf is

$$\mathscr{E} = -L\left(\frac{\Delta I}{\Delta t}\right)$$
(20.12)

Where L is the (self) **inductance** of the coil. As you can see from the equation, inductance has units of volt-second per ampere (V-s/A). This combination is called a **henry** (H). The smaller unit of millihenry (mH) is commonly used (1 mH $= 10^{-3}$ H).

Self-induction is introduced in Section 19.3

Lenz's law is given in Section 19.1

The henry was named for the American physicist Joseph Henry (1797–1879) in honor of his work on electromagnetic induction. Henry discovered this phenomenon independently at about the same time as Michael Faraday did, but Faraday published his findings first.

The impeding effect of an inductor on the current in an ac circuit depends on the inductance, and this is expressed in terms of **inductive reactance** (X_L),

$$X_L = 2\pi f L = \omega L$$ (20.13)

Inductive reactance—opposition to current

where $\omega = 2\pi f$. Here f is the frequency of the driving source and L is the inductance. Like capacitive reactance, inductive reactance is measured in ohms (Ω).

Use unit analysis to demonstrate that inductive reactance is measured in ohms.

Notice that the inductive reactance is directly proportional to the inductance (L) and the frequency (f) of the voltage source. The inductance of a coil is a constant that depends on the number of turns, the coil's diameter, its length, and the material of the core (if any). The greater the inductance, the greater the opposition to current. Also, the higher the frequency of the ac driving source, the greater the inductance opposition to current because of faster time variations.

In terms of X_L, the equation for the voltage across an inductor has an Ohm's law form.

$$V = IX_L$$ (20.14)

The circuit symbol for an inductor and the graphs of the voltage across the inductor and the current in the circuit are shown in ● Fig. 20.6. For this kind of circuit, when the voltage is a maximum, the current is zero; and when the voltage goes to zero, the current is a maximum. The current lags a quarter-cycle behind the voltage, a relationship that is commonly expressed in the following way:

In a purely inductive ac circuit, the *voltage* leads the *current* by 90° or $\frac{1}{4}$ cycle.

The phase relationships of current and voltage for purely inductive and purely capacitive circuits are opposite. A phrase that may help you remember the relationship is

ELI the ICE man

With E representing voltage, ELI indicates that with an inductance (L) the voltage leads the current (I). Similarly, ICE tells you that with a capacitance (C) the current leads the voltage.

EXAMPLE 20.4 ■ Inductive Reactance

A 125-mH inductor is connected to a 120-V, 60-Hz source. What are (a) the inductive reactance and (b) the current in the circuit?

Solution.

Given: $L = 125\,\text{mH} = 125 \times 10^{-3}\,\text{H}$ *Find:* (a) X_L (inductive reactance)
    $V = 120\,\text{V}$ (b) I (current)
    $f = 60\,\text{Hz}$

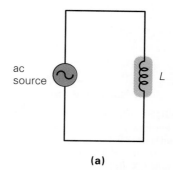

(a)

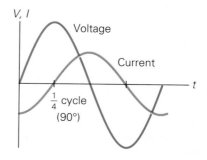

$V = V_0 \sin 2\pi ft$
$I = I_0 \sin(2\pi ft - 90°)$
$\ = I_0 \cos 2\pi ft$

(b)

● **FIGURE 20.6 A purely inductive circuit** In a circuit with only inductance, the voltage leads the current by 90° or $\frac{1}{4}$ cycle. (Notice that a voltage zero is 90° or a quarter cycle ahead of a current zero.)

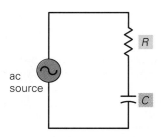

(a) Circuit diagram

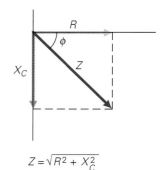

$$Z = \sqrt{R^2 + X_C^2}$$

$$\tan \phi = \frac{X_C}{R}$$

(b) Phase diagram

● **FIGURE 20.7 A series RC circuit**
In a series RC circuit **(a)**, the impedance Z is the phasor sum of the resistance R and the capacitive reactance X_c **(b)**.

Impedance—effective opposition to current

(a) The inductive reactance is given by Eq. 20.13,

$$X_L = 2\pi f L = 2\pi (60 \text{ Hz})(125 \times 10^{-3} \text{ H}) = 47.1 \ \Omega$$

(b) Then, using Eq. 20.14 gives

$$I = \frac{V}{X_L} = \frac{120 \text{ V}}{47.1 \ \Omega} = 2.55 \text{ A}$$

20.4 ■ Impedance: RLC Circuits

The previous sections considered purely capacitive and inductive circuits, in which only *nonresistive* opposition to current flow was found. However, it is impossible to have purely reactive circuits, since there is always some resistance, at minimum that from the connecting wires or the wire in an inductor. Therefore, real ac circuits usually have appreciable resistance, and the current is impeded by both resistances and reactances (capacitive and/or inductive). Analysis of some simple circuits will illustrate these effects.

Series RC Circuit

Suppose that an ac circuit consists of a voltage source and a resistance and a capacitive element connected in series, as illustrated in ● Fig. 20.7a. The phase difference between the current and the voltage is different for each of these circuit elements. As a result, a special graphical method is used to find the effective opposition to the current flow in the circuit. This is conveniently found by means of a **phase diagram**, such as the one shown in Fig. 20.7b.

In a phase diagram, the resistance and reactance of the circuit are given vectorlike properties and their magnitudes are represented as arrows called **phasors**. On a set of x-y coordinate axes, the resistance is plotted on the positive x axis, since the voltage-current phase difference for a resistor is zero ($\phi = 0$). The capacitive reactance is plotted along the negative y axis, to reflect a phase difference of $-90°$. (A negative phase angle implies that the voltage lags behind the current, as is the case for a capacitor.)

The phasor sum is the **impedance (Z)**, or the effective opposition to the current flow. Phasors are added in the same way vectors are, so for the series RC circuit,

$$Z = \sqrt{R^2 + X_c^2} \tag{20.15}$$

The unit for the impedance is the ohm.

The Ohm's law relationship for the RC circuit in terms of impedance is

$$V = IZ \tag{20.16}$$

EXAMPLE 20.5 ■ Capacitive Impedance

A series RC circuit has a 100-Ω resistor and a 15-μF capacitor. What is the current in the circuit when driven by a 120-V, 60-Hz source?

Solution.

    Given: $R = 100\ \Omega$ *Find:* I (current)
          $C = 15\ \mu\text{F} = 15 \times 10^{-6}\ \text{F}$
          $V = 120\ \text{V}$
          $f = 60\ \text{Hz}$

First, the impedance is computed. From Example 20.3, we know the reactance for the capacitor is $X_c = 177\ \Omega$. Then using this directly in Eq. 20.15,

$$Z = \sqrt{R^2 + X_c^2} = \sqrt{(100\ \Omega)^2 + (177\ \Omega)^2} = 203\ \Omega$$

Since $V = IZ$, the current in the circuit is

$$I = \frac{V}{Z} = \frac{120\ \text{V}}{203\ \Omega} = 0.591\ \text{A}$$

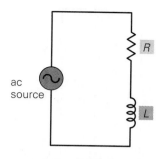

(a) Circuit diagram

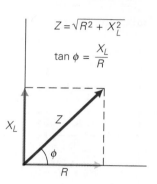

(b) Phase diagram

FIGURE 20.8 A series RL circuit
In a series RL circuit (a), the impedance Z is the phasor sum of the resistance R and the inductive reactance X_L (b).

Series RL Circuit

The analysis of a series RL circuit (Fig. 20.8) is similar to that of a series RC circuit. However, the inductive reactance is plotted along the positive y axis in the phase diagram to reflect a phase difference of $+90°$. (In this case, a positive phase angle implies that the voltage leads the current, as is the case for an inductor.)

    The impedance can be seen to be

$$Z = \sqrt{R^2 + X_L^2} \qquad (20.17)$$

as for an RC circuit, $V = IZ$.

Series RLC Circuit

An ac circuit may contain all three circuit elements—resistor, inductor, and capacitor—as in Fig. 20.9. Again, phasor addition is used to determine the impedance by a component method. Summing the vertical components (the inductive and capacitive reactances) gives the total reactance ($X_L - X_C$, since X_C is defined as a negative phasor). The impedance is then the phasor sum of the resistance and the total reactance. From the phasor diagram, you can see that the impedance is

$$Z = \sqrt{R^2 + (X_L - X_C)^2} = \sqrt{R^2 + \left(2\pi f L - \frac{1}{2\pi f C}\right)^2} \qquad (20.18)$$

Impedance for series RLC circuit

Again,

$$V = IZ \qquad (20.19)$$

    In this case, the **phase angle** (ϕ) between the voltage from the source and the current is given by

$$\tan \phi = \frac{X_L - X_C}{R} \qquad (20.20)$$

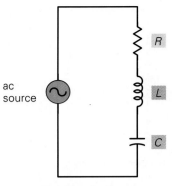

(a) Circuit diagram

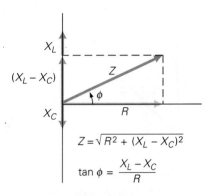

$$Z = \sqrt{R^2 + (X_L - X_C)^2}$$

$$\tan \phi = \frac{X_L - X_C}{R}$$

(b) Phase diagram

FIGURE 20.9 A series RLC circuit In a series RLC circuit **(a)**, the impedance Z is the phasor sum of the resistance R and the total reactance $(X_L - X_C)$, as shown in **(b)**.

Figure 20.9

TABLE 20.1 ■ Impedances and Phase Angles

Circuit element(s)		Impedance, Z (in Ω)	Phase angle, φ
R	—/\/\—	R	$0°$
C	—|(—	X_C	$-90°$
L	—ellell—	X_L	$+90°$
RC	—/\/\—|(—	$\sqrt{R^2 + X_C^2}$	$-\phi$
RL	—/\/\—ellell—	$\sqrt{R^2 + X_L^2}$	$+\phi$
RLC	—/\/\—ellell|(—	$\sqrt{R^2 + (X_L - X_C)^2}$	$+\phi$, if $X_L > X_C$ $-\phi$, if $X_C > X_L$

Note that if X_L is greater than X_C (as is true in Fig. 20.9), the phase angle is positive $(+\phi)$, and the circuit is said to be an **inductive circuit**. If X_C is greater than X_L, the phase angle is negative $(-\phi)$, and the circuit is said to be a **capacitive circuit**.

A summary of impedances and phase angles for the three circuit elements and combinations considered is given in Table 20.1.

EXAMPLE 20.6 ■ Impedance in an RLC Circuit

A series RLC circuit has a resistance of 25 Ω, a capacitance of 50 μF, and an inductance of 0.30 H. If the circuit is driven by a 120-V, 60-Hz source, what are (a) the impedance of the circuit, (b) the current in the circuit, and (c) the phase angle between the current and the voltage supplied?

Solution.

Given: $R = 25\ \Omega$
$C = 50\ \mu\text{F} = 5.0 \times 10^{-5}\ \text{F}$
$L = 0.30\ \text{H}$
$V = 120\ \text{V}$
$f = 60\ \text{Hz}$

Find: (a) Z (impedance)
(b) I (current)
(c) ϕ (phase angle)

(a) To find the impedance (Eq. 20.18), we first compute the reactances.

$$X_C = \frac{1}{2\pi fC} = \frac{1}{2\pi(60\ \text{Hz})(5.0 \times 10^{-5}\ \text{F})} = 53\ \Omega$$

and

$$X_L = 2\pi fL = 2\pi(60\ \text{Hz})(0.30\ \text{H}) = 113\ \Omega$$

Then,

$$Z = \sqrt{R^2 + (X_L - X_C)^2}$$
$$= \sqrt{(25\ \Omega)^2 + (113\ \Omega - 53\ \Omega)^2} = 65\ \Omega$$

(b) Since $V = IZ$, we have

$$I = \frac{V}{Z} = \frac{120\ \text{V}}{65\ \Omega} = 1.8\ \text{A}$$

(c) Solving $\tan\phi = (X_L - X_C)/R$ for the phase angle gives

$$\phi = \tan^{-1}\left(\frac{X_L - X_C}{R}\right) = \tan^{-1}\left(\frac{113\ \Omega - 53\ \Omega}{25\ \Omega}\right) = 67°$$

You should now appreciate the importance of phasor diagrams in finding the impedance and the current in ac circuits. However, you may be wondering what the importance of or the use of the phase angle ϕ might be. This is answered by looking at the power loss for our RLC circuit, which illustrates another important benefit from phasor diagrams.

Power Factor

The power, or energy dissipated per unit time, is an interesting feature of an RLC circuit. *There are no power losses associated with pure capacitances and inductances in an ac circuit.* Capacitors and inductors simply store energy and then give it back. For example, during a half cycle, a capacitor in an ac circuit is charged, and energy is stored in the electric field between the plates. During the next half cycle, the capacitor discharges and returns the charge to the voltage source. Similarly, in an inductor, the energy is stored in the magnetic field associated with the induced current, or back emf, which builds up and reverses each cycle.*

Thus, *the only element that dissipates energy in an RLC series circuit is the resistive element.* As you learned earlier, the average power dissipated by a resistance R is $P = I^2R$ (where I is the rms value). The power can also be expressed in terms of the current and voltage, but the voltage in this case *must be that across the resistive element* (V_R), since this is the only dissipative element. That is, the power dissipated in a series RLC circuit is

$$P = IV_R$$

The voltage across the resistive element is conveniently found from a voltage triangle that corresponds to the phasor triangle (•Fig. 20.10). The voltages across the components of an RLC circuit are given by $V = IZ$, $V_R = IR$, $V_L = IX_L$, and $V_C = IX_C$. Combining the last two voltages, we may write $(V_L - V_C) = I(X_L - X_C)$. Hence, if we multiply each of the phasor triangle legs in Fig. 20.10a by the current I, we obtain an equivalent voltage triangle as in Fig. 20.10b. As you can see from this triangle, the voltage across the resistive element (V_R) depends on the phase angle.

$$V_R = V\cos\phi \qquad (20.21)$$

The term $\cos\phi$ is called the **power factor**, and from Fig. 20.10,

$$\boxed{\cos\phi = \frac{R}{Z}} \qquad (20.22)$$

*In Chapter 15, you saw that the energy stored in a capacitor was given by $U = \frac{1}{2}CV^2$. Similarly, it can be shown that the energy stored in an inductor is given by $U = \frac{1}{2}LI^2$, where L is the inductance and I is the current in the inductor.

In an LC circuit the energy oscillates between the capacitor and the inductor. The case is similar to that of a mass on a spring, where the energy oscillates between the spring and the mass.

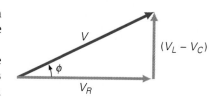

(a) Phasor triangle

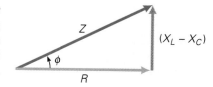

(b) Equivalent voltage triangle

•**FIGURE 20.10 Phasor and voltage triangles** Since the voltages across the components of a series RLC circuit are given by $V = IZ$, $V_R = IR$, and $(V_L - V_C) = I(X_L - X_C)$, and the current is the same through all three, the phasor triangle **(a)** may be converted to a voltage triangle **(b)**. Note that $V_R = V\cos\phi$.

The power in terms of the power factor is

$$P = IV \cos \phi \qquad (20.23)$$

With power dissipated only in the resistance ($P = I^2R$) we can use Eq. 20.22 to express the average power in another form:

$$P = I^2Z \cos \phi \qquad (20.24)$$

Note that $\cos \phi$ varies from 1 to 0. When $\phi = 0$ and $\cos \phi = 1$, the circuit is said to be completely resistive. That is, there is maximum power dissipation (as though the circuit contained only a resistor). The power factor decreases as the phase angle increases [either $+ \phi$ or $- \phi$, since $\cos (- \phi) = \cos \phi$], and the circuit becomes more inductive or capacitive. At $\phi = \pm 90°$, the circuit is completely inductive or capacitive. For these cases, it is as though the circuit contained only an inductor or a capacitor, respectively, and no power is dissipated ($P = IV \cos 90° = 0$).

EXAMPLE 20.7 ■ **Power Factor**

How much power is dissipated in the circuit described in Example 20.6?

Solution. In Example 20.6, the circuit was found to have an impedance of $Z = 65 \, \Omega$, and its resistance is $R = 25 \, \Omega$. The power factor of the circuit is therefore

$$\cos \phi = \frac{R}{Z} = \frac{25 \, \Omega}{65 \, \Omega} = 0.38$$

Using other data from Example 20.6 gives

$$P = IV \cos \phi = (1.8 \text{ A})(120 \text{ V})(0.38) = 82 \text{ W}$$

Note that this is quite different from the power that would be dissipated in a circuit without a capacitor and inductor—that is, with only a resistive element—or when the capacitive and inductive reactances cancel each other out and $\cos \phi = R/R = 1$.

20.5 ■ Circuit Resonance

When the power factor of an RLC circuit is equal to 1 ($\cos \phi = 1$), there is maximum power dissipation. For a particular voltage, the current in the circuit is then a maximum, since the impedance is a minimum. This occurs when the inductive and capacitive reactances effectively cancel each other.

The inductive and capacitive reactances are both frequency-dependent, so the impedance also depends on the frequency. From the expression for the impedance,

$$Z = \sqrt{R^2 + (X_L - X_C)^2}$$

the impedance is a minimum when

$$X_L = X_C$$

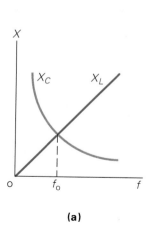

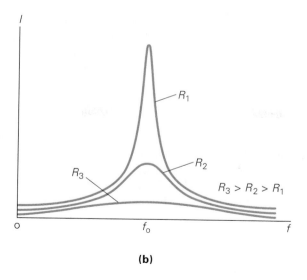

(a)

(b)

FIGURE 20.11 **Resonance frequency** (a) At the resonance frequency (f_o), the reactances are equal ($X_L = X_c$). On a graph of X versus f, this is the frequency at which the X_c and X_L curves intersect. (b) On a graph of I versus f, the current is a maximum at f_o. The curve becomes sharper and narrower as the resistance decreases.

or

$$2\pi f_o L = \frac{1}{2\pi f_o C}$$

Thus,

$$f_o = \frac{1}{2\pi\sqrt{LC}}$$

(20.25)   Resonance frequency

Eq. 20.25 gives the frequency for the condition of minimum impedance, which is called the **resonance frequency**. A graph of reactance versus frequency is shown in ●Fig. 20.11a. The X_c and X_L curves intersect at f_o where their values are equal ($X_L = X_c$).

Thus, when a series RLC circuit is driven in resonance (at its resonance frequency), the impedance is a minimum, and the power transfer from the source to the circuit is a maximum. Since $X_L = X_c$, then $Z = R$, and the impedance is completely resistive. In terms of the power factor, $\cos\theta = R/Z = R/R = 1$; thus the power factor is a maximum and there is maximum power dissipation in the circuit.

A graph of current versus frequency is shown in Fig. 20.11b for several different resistances in an RLC circuit with fixed capacitance and inductance. Notice that the currents have maximum values at f_o. Also, notice how the curve becomes sharper and narrower as the resistance decreases.

Resonant circuits have a wide variety of applications, for example, in the tuning mechanism of a radio. Each radio station has an assigned broadcast frequency at which radio waves are transmitted. The oscillating electric and magnetic fields of the radio waves set electrons in the antenna into regular back-and-forth motion—that is, they produce an alternating current. (The process is just the reverse of the production of electromagnetic waves by oscillating electrons in a transmitter.) Signals broadcast by many different stations generally reach a radio, but the receiver circuit selectively picks up only the one with a frequency at or near its resonant frequency. Most radios allow you to alter this resonant frequency to "tune in" different stations. Variable air capacitors in the tuning circuit have long been used for this purpose (●Fig. 20.12). More compact variable capacitors in some smaller radios employ a polymer dielectric

●FIGURE 20.12 **Variable air capacitor** Rotating the movable plates between the fixed plates changes the overlap area and thus the capacitance. Such capacitors were common in tuning circuits in older radios.

See Section 19.4

between thin plates. The polymer sheets help maintain the plate separation and increase the capacitance, allowing makers to use plates of smaller area. (Recall that $C = \varepsilon_o A/d$.) In other radios, variable capacitors are replaced by solid-state devices.

EXAMPLE 20.8 ■ Electrical Resonance

A series RLC circuit has a 50-Ω resistor, a 0.0060-μF (or 6.0 pF) capacitor, and a 28-mH inductor. The circuit is connected to a wide-range, adjustable-frequency source with an output of 25 V. (a) What is the resonance frequency of the circuit? (b) How much current flows in the circuit when it is in resonance? (c) What is the voltage across each of the circuit elements for this condition?

Solution.

Given: $R = 50\ \Omega$ *Find:* (a) f_o (resonance frequency)
$C = 0.0060\ \mu\text{F} = 6.0 \times 10^{-9}\text{F}$ (b) I (current)
$L = 28\ \text{mH} = 28 \times 10^{-3}\ \text{H}$ (c) V_R, V_C, V_L (voltages)
$V = 25\ \text{V}$

(a) By Eq. 20.25, we have for the resonance frequency,

$$f_o = \frac{1}{2\pi\sqrt{LC}} = \frac{1}{2\pi\sqrt{(28 \times 10^{-3}\ \text{H})(6.0 \times 10^{-9}\ \text{F})}}$$

$$= 12.3 \times 10^3\ \text{Hz} = 12.3\ \text{kHz}$$

(b) At the resonance frequency, $X_L = X_C$, and $Z = R = 50\ \Omega$. The current in the circuit is then

$$I = \frac{V}{Z} = \frac{V}{R} = \frac{25\ \text{V}}{50\ \Omega} = 0.50\ \text{A}$$

(c) The reactances at f_o are

$$X_L = 2\pi f_o L = 2\pi(12.3 \times 10^3\ \text{Hz})(28 \times 10^{-3}\ \text{H}) = 2200\ \Omega$$

and $X_C = X_L = 2200\ \Omega$ at f_o. Then, using Ohm's law of relationships for voltage and current,

$$V_R = IR = (0.50\ \text{A})(50\ \Omega) = 25\ \text{V}$$
$$V_L = IX_L = (0.50\ \text{A})(2200\ \Omega) = 1100\ \text{V}$$
$$V_C = IX_C = (0.50\ \text{A})(2200\ \Omega) = 1100\ \text{V}$$

Note that the voltages across the capacitor and the inductor are significantly higher than the source voltage. Can you explain where the additional voltage comes from?

Important Concepts

You should be able to define and/or explain these chapter concepts clearly.

peak voltage	capacitive reactance (X_C)	phasors	capacitive circuit
peak current	inductance henry (H)	impedance (Z)	power factor
rms (or effective) current	inductive reactance (X_L)	phase angle (ϕ)	resonance frequency
rms (or effective) voltage	phase diagram	inductive circuit	

Important Formulas

These relationships are mathematical statements of the concepts and principles presented in the chapter. You should be able to identify the symbols and to explain the relationships before proceeding to the Exercises. In-text equation reference numbers are given for convenience.

Instantaneous Voltage in an ac Circuit:

$$V = V_o \sin 2\pi ft \qquad (20.1)$$

Instantaneous Current in an ac Circuit:
(where $I_o = V_o/R$):

$$I = I_o \sin 2\pi ft \qquad (20.2)$$

Effective Power in an ac Circuit:

$$\bar{P} = I_{rms}^2 R \qquad (20.7)$$

Ohm's Law for ac:

$$V_{rms} = I_{rms} R \qquad (20.9)$$

Rms (or Effective) Current:

$$I_{rms} = \frac{I_o}{\sqrt{2}} = 0.707\, I_o \qquad (20.6)$$

Rms (or Effective) Voltage:

$$V_{rms} = \frac{V_o}{\sqrt{2}} = 0.707\, V_o \qquad (20.8)$$

Capacitive Reactance and Ohm's Law:

$$X_C = \frac{1}{2\pi fC} = \frac{1}{\omega C} \quad \text{and} \quad V = IX_C \qquad (20.10\text{--}11)$$

Inductive Reactance and Ohm's Law:

$$X_L = 2\pi fL = \omega L \quad \text{and} \quad V = IX_L \qquad (20.13\text{--}14)$$

Impedance for an RLC Circuit:

$$Z = \sqrt{R^2 + (X_L - X_C)^2} \qquad (20.18)$$

$$= \sqrt{R^2 + \left(2\pi fL - \frac{1}{2\pi fC}\right)^2}$$

Phase Angle:

$$\tan \phi = \frac{X_L - X_C}{R} \qquad (20.20)$$

Ohm's Law for RC, RL, and RLC Circuit:

$$V = IZ \qquad (20.19)$$

Power Factor (RLC circuit):

$$\cos \phi = \frac{R}{Z} \qquad (20.22)$$

Power in Terms of Power Factor:

$$P = IV \cos \phi \qquad (20.23)$$

$$\text{or} \quad P = I^2 Z \cos \phi \qquad (20.24)$$

Resonance Frequency:

$$f_o = \frac{1}{2\pi\sqrt{LC}} \qquad (20.25)$$

Exercises

20.1 ■ Resistance in an AC Circuit

1 Which of the following voltages has the greatest magnitude for a sinusoidal ac voltage: (a) V_o, (b) V_{rms}, (c) V_{p-p}, (d) V? (c)

● **2** The ac ampere is defined in terms of (a) Ohm's law, (b) the force between parallel, current-carrying wires, (c) joule heating, (d) the frequency of the ac current. (c)

3 For the circuit in Fig. 20.1, why is the magnitude of V_o greater than that of I_o? Could it be the other way, with the magnitude of I_o greater than that of V_o? Explain. see ISM

4 If the average current in an ac circuit is zero, why isn't the average power zero? see ISM

5 ■ What are the peak voltages of a 120-VAC line and a 240-VAC line? 170 V; 340 V

6 ■ An ac circuit has an rms current of 1.5 A. What is the peak current? 2.1 A

7 ■ The maximum potential difference across a resistor in an ac circuit is 156 V. Find the effective voltage. 110 V

8 ■ Based on the definition of the ac ampere, how much

ac current must flow through a 6.0-Ω resistor to produce 15 J/s of joule heat?   1.6 A

9 ■■ An ac circuit with a resistance of 5.0 Ω has an effective current of 0.75 A. (a) Find the rms voltage and peak voltage. (b) Find the average power dissipated by the resistance.
(a) V_{rms} = 3.8 V; V_o = 5.3 V   (b) 2.8 W

● **10** ■■ A hair dryer rated at 1600 W is plugged into a 120-V outlet. (a) Find the effective current. (b) Find the peak current. (c) Find the resistance of the dryer.   (a) 13 A (b) 18 A   (c) 9.0 Ω

11 ■■ The voltage across a resistor varies according to this equation:
$$V = (120 \text{ V})(\sin 120t)$$
What are (a) the effective voltage across the resistor and (b) the frequency with which the voltage changes polarity?
(a) 85 V   (b) 19 Hz

12 ■■ An ac voltage is applied to a 25-Ω resistor, and it dissipates 500 W of power. Find (a) the effective and peak currents and (b) the effective and peak voltages.
(a) 4.5 A, 6.4 A   (b) 112 V, 158 V

13 ■■ The current through a 60-Ω resistor is given by this equation:
$$I = 2.0 \sin 380t$$
where I is in amps and t is in seconds. (a) What is the frequency of the current? (b) What is the effective current? (c) How much power is dissipated by the resistor? (d) Write an equation expressing the voltage as a function of time.
(a) 60 Hz   (b) 1.4 A   (c) 1.2×10^2 W   (d) $V = 120 \sin (380t)$ V

14 ■■ An ac voltage has an amplitude of 85 V and a frequency of 60 Hz. (a) If $V = 0$ at $t = 0$, what is the instantaneous voltage at $t = 2.0$ s? (b) What is the effective voltage?   (a) 0   (b) 60 V

15 ■■ An ac voltage has an rms value of 120 V. The voltage goes from zero to its maximum value in 1.5 ms. Write an expression for the voltage as a function of time.
$V = 170 \sin (334 \pi t)$ V

● **16** ■■ A circuit has a current (in amps) given by $I = 8.0 \sin 4\pi t$ with an applied voltage (in volts) of $V = 60 \sin 40t$. What is the average power delivered to the circuit? 2.4×10^2 W

17 ■■ The output of an ac generator has a negative peak voltage at $t = 0.075$ s. What is the operational frequency of the generator?   10 Hz

18 ■■ Find the effective and peak currents through a 40-W, 120-V light bulb.   0.33 A; 0.47 A

19 ■■ A 50-kW heater is connected to a 240-V ac source. Find (a) the peak current and (b) the peak voltage.
(a) 3.0×10^2 A   (b) 3.4×10^2 V

20 ■■ The current and voltage outputs of an ac generator have peak values of 1.5 A and 12 V. What is the effective power output of the generator?   9.0 W

21 ■■■ A coil of 200 loops of wire with a radius of 10 cm is used as the armature of a simple generator with a magnetic field of 0.80 T. If the coil has a resistance of 2.0 Ω, how fast must it be rotated to have an effective induced current of 2.7 A?   1.5 rad/s

22 ■■■ Rotating coils are often used to measure magnetic fields. A coil of 40 loops with a radius of 7.5 cm rotates at 60 Hz about an axis along a diameter perpendicular to a uniform magnetic field. An *effective* voltage of 10 V, as read from an ac voltmeter, is induced in the coil. What is the magnitude of the magnetic field? 5.3×10^{-2} T

20.2 ■ Capacitive Reactance

20.3 ■ Inductive Reactance

23 In a purely capacitive ac circuit, (a) the current and voltage are in phase, (b) the current leads the voltage, (c) the voltage lags the current.   (d)

24 The unit of capacitive reactance is (a) F, (b) $(\text{F-s})^{-1}$, (c) Ω, (d) Ω-m.   (c)

● **25** In a purely inductive ac circuit, (a) the current leads the voltage, (b) the current and voltage are in phase, (c) the inductance is zero, (d) none of these.   (d)

26 The unit of inductance is (a) H/s, (b) H, (c) Ω, (d) Ω-s.   (b)

27 In an ac circuit, what can oppose current other than ohmic resistance and why?   see ISM

28 Analyze the graph in Fig. 20.6b and explain why the current is zero when the voltage is a maximum, and vice versa. [*Hint:* $\Delta I \Delta t$ is the slope of the curve for the current.]   see ISM

29 ■ What capacitance is needed in a 60-Hz ac circuit to have a reactance of 45 Ω? 5.9×10^{-8} F

30 ■ Find the frequency at which a 25-μF capacitor will have a reactance of 25 Ω. 2.5×10^2 Hz

31 ■■ A 4.0-μF capacitor is connected across a 60-Hz voltage source and a current of 2.0 mA is measured on an ammeter. What is the capacitive reactance of the circuit?
6.6×10^2 Ω

● **32** ■■ A 50-mH inductor is connected in a circuit with a 120-V, 60-Hz source. (a) What is the inductive reactance of the circuit? (b) How much current is there in the circuit? (c) What is the phase angle between the current and the supplied voltage? (Assume negligible resistance.)   (a) 19 Ω (b) 6.3 A   (c) 90°

33 ■■ How much current is delivered to a circuit with a 50-μF capacitor if it is connected to an ac generator with a 90-V, 60-Hz output?   1.7 A

34 ■■ A variable capacitor in a circuit with a 120-V, 60-Hz source is initially at a value of 0.25 μF and is then increased to 0.40 μF. What is the percentage change in the current in the circuit?   60%

35 ■■ An inductor has a reactance of 90 Ω in a 60-Hz ac circuit. What is the inductance of the inductor?   0.24 H

36 ■■ Find the frequency at which a 450-mH inductor will have a reactance of 200 Ω.   71 Hz

37 ■■ With a 150-mH inductor in a circuit with a 60-Hz voltage source, a current of 1.6 A is measured on an ammeter. (a) What is the voltage of the source? (b) What is the phase angle between the current and that voltage?   (a) 91 V (b) 90°

38 ■■ Using the fact that the self-induced emf in a coil is given by $\mathscr{E} = -L\Delta I/\Delta t$, show that $L = N \Delta\Phi/\Delta I$, where Φ is the flux through the coil and N is the number of turns.   see ISM

● **39** ■■ What inductance will have the same reactance in a 120-V, 60-Hz circuit as a 10-μF capacitor does?   0.70 H

40 ■■ A circuit with a single inductor is connected to a 120-V, 60-Hz source. What is the value of the inductance if there is a current of 10 mA in the circuit?   32 H

20.4 ■ Circuit Impedance: RLC Circuits

20.5 ■ Circuit Resonance

41 The impedance of an RLC circuit depends on (a) frequency, (b) inductance, (c) capacitance, (d) all of these.   (d)

● **42** If the capacitance is increased in a series RLC circuit, (a) the capacitive reactance increases, (b) the impedance increases, (c) the current increases, (d) the power factor increases.   (c)

43 Which of the following has a resonance frequency: (a) RC circuit, (b) RL circuit, (c) RLC circuit, (d) all of these?   (c)

44 When a series RLC circuit is driven at its resonance frequency, (a) energy is dissipated by only the resistive element, (b) the power factor is one, (c) there is maximum energy transfer to the circuit, (d) all of these.   (d)

45 If the resistance in an RLC circuit were zero and the circuit were driven with a variable frequency voltage source, what would a graph of current versus frequency look like? What would this mean physically?   see ISM

● **46** ■■ A series RC circuit has a resistance of 200 Ω and a capacitance of 25 μF and is driven by a 120-V, 60-Hz source. (a) Find the capacitive reactance of the circuit. (b) How much current is drawn from the source?   (a) 1.1 × 10² Ω (b) 0.53 A

47 ■■ To reduce the impedance in the RC circuit in Exercise 46 by 2.0% while maintaining the 120-V source voltage, what would the required source frequency be?   68 Hz

48 ■■ In a series RC circuit with a 100-Ω resistor, what capacitance will give a current of 0.50 A if the circuit is driven by a 120-V, 60-Hz source?   1.2 × 10⁻⁵ F

49 ■■ What is the phase angle between the current and the source voltage in Exercise 48?   −65°

50 ■■ A coil in a 60-Hz circuit has a resistance of 100 Ω and an inductance of 0.45 H. Calculate (a) the coil's reactance and (b) the circuit's impedance.   (a) 1.7 × 10² Ω (b) 2.0 × 10² Ω

51 ■■ A series RLC circuit has a resistance of 25 Ω, an inductance of 0.30 H, and a capacitance of 8.0 μF. At what frequency should the circuit be driven to have the maximum power transferred from the driving source?   1.0 × 10² Hz

52 ■■ In a series RLC circuit, for a particular driving frequency, $R = X_c = X_L = 40$ Ω. If the driving frequency is doubled, what will the impedance of the circuit be?   72 Ω

53 ■■ A resistor, an inductor, and a capacitor have values of 500 Ω, 500 mH, and 3.5 μF, respectively, and are connected in series to a 240-V, 60-Hz power supply. Is the circuit driven in resonance? If not, what values of resistance and inductance would be required for this to occur?   no, 2.0 H for any resistor

54 ■■ How much power is dissipated in the circuit described in Exercise 53?   51 W

55 Referring to Fig. 20.13, find the currents supplied by the ac source for all possible connections.   ab: 1.3 A; ac: 1.2 A; bc: 4.0 A; cd: 1.8 A; bd: 1.7 A; ad: 2.9 A

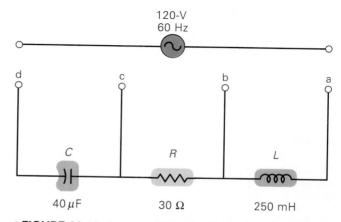

120-V
60 Hz

d   c   b   a

C   R   L

40 μF   30 Ω   250 mH

● **FIGURE 20.13 A series RLC circuit**   See Exercise 55.

● **56** ■■ A tuning circuit in a radio receiver has a fixed 2.5-mH inductance and a variable capacitor. If the circuit is tuned to a radio station broadcasting at 980 kHz on the AM dial, what is the capacitance of the capacitor?   1.1 × 10⁻¹¹ F

57 What would be the range of the variable capacitor in Exercise 56 for tuning over the complete AM band? [*Hint:* See Section 19.5.]   3.6 × 10⁻¹¹ F to 3.4 × 10⁻¹² F

58 ■■ A coil with a resistance of 30 Ω and an inductance of 0.15 H is connected to a 120-V, 60-Hz source. (a) How much current is in the circuit? (b) What is the phase angle between the current and the source voltage?   (a) 1.9 A (b) 62°

59 ■■ A series RLC circuit has a resistance of 25.0 Ω, an inductance of 0.450 H, and a capacitance of 5.00 μF. (a) What is the impedance in a circuit with a driving frequency of 60 Hz? (b) Is the circuit driven in resonance? If not, find the resonance frequency of the circuit.   (a) 4.4 × 10² Ω (b) 1.1 × 10² Hz

60 ■■ A circuit connected to a 220-V, 60-Hz power supply has the following components connected in series: a 10-Ω resistor, a coil with an inductive reactance of 120 Ω, and a capacitor with a reactance of 120 Ω. Compute the potential difference across (a) the resistor, (b) the inductor, and (c) the capacitor. (a) 220 V (b) 2.64 $\times$ 10^3 V (c) 2.64 $\times$ 10^3 V

61 ■■ A series RLC circuit has a 25-Ω resistor, a 0.80-μF capacitor, and a 250-mH inductor. The circuit is connected to a variable-frequency source with a constant output of 12 V. If the supplied frequency is set at the circuit's resonance frequency, what is the voltage across each of the circuit elements? V_R = 12 V V_L = 2.7 $\times$ 10^2 V V_C = 2.7 $\times$ 10^2 V

● **62** ■■ An RL circuit operates on 120 V at 60 Hz and contains a 60.0-Ω resistor. If the wattage rating of the resistor is 10.0 W, what inductance should be used in the circuit so power is expended in the resistor at that rated limit? 764 mH

63 ■■ A series RLC circuit with a resistance of 400 Ω has capacitive and inductive reactances of 300 Ω and 500 Ω, respectively. (a) What is the power factor of the circuit? (b) If the circuit operates at 60 Hz, what additional capacitance should be connected with the original capacitance to give a power factor of one, and how should the capacitances be connected? (a) 0.895 (b) 13 μF

64 ■■ A machine operates on 240 V at 60 Hz. In operation, it uses 550 W of power, and the power factor is 0.75. Find the effective current in the circuit containing the machine. 3.1 A

65 ■■■ The current in a series RLC circuit is 4.0 A when it is connected to a 120-V, 60-Hz source. If the current leads the voltage by 37°, (a) what is the average power dissipated in the circuit? (b) What is the resistance of the circuit? (c) What is the value of $X_L - X_C$? (a) 3.8 $\times$ 10^2 W (b) 24 Ω (c) − 18 Ω

66 ■■■ A series RLC circuit has components with these values: $R = 50\ \Omega$, $L = 0.15$ H, and $C = 20\ \mu$F. The circuit is driven by a 120-V, 60-Hz source. What is the power loss of the circuit as a percentage of its power loss when in resonance? 57%

67 ■■■ If the values of all three circuit components in Exercise 66 were halved, what would be the change in the power loss? − 91 W

■ Additional Exercises

68 A 60-W light bulb operates on 120-V household voltage. What are (a) the peak voltage across the bulb, and (b) the peak current in the circuit? (a) 1.7 $\times$ 10^2 V (b) 0.71 A

69 The rms current in an ac circuit is 0.25 A. What is the maximum instantaneous current? 0.35 A

70 A 120-V, 60-Hz source is connected across a 0.40-H inductor. (a) Find the current through the inductor. (b) Find the phase angle between the current and the supplied voltage. (a) 0.80 A (b) 90°

71 A radio receiver circuit containing an inductor of 150 mH is tuned to an FM station at 98.9 MHz by adjusting a variable capacitor. What is the capacitance of the capacitor when the circuit is tuned to the station? 1.7 $\times$ 10^{-17} F

72 A series RL circuit has a resistance of 50.0 Ω and an inductance of 250 mH. The circuit is driven by a 120-V, 60-Hz source. (a) What is the inductive reactance of the circuit? (b) How much power is dissipated by the circuit? (a) 94 Ω (b) 64 W

73 Prove that $\langle \sin^2 \omega t \rangle = \frac{1}{2}$. see ISM

● **74** A circuit connected to a 110-V, 60-Hz source contains a single coil with an inductance of 100 mH and a resistance of 50 Ω. Find (a) the reactance of the coil, (b) the impedance of the circuit, (c) the current in the circuit, and (d) the power dissipated by the coil. (a) 38 Ω (b) 63 Ω (c) 1.9 A (d) zero

75 Calculate the phase angle between the current and the supplied voltage in Exercise 74. 37°

76 A 1.0-μF capacitor is connected to a 120-V, 60-Hz source. (a) What is the capacitive reactance of the circuit? (b) How much current is there in the circuit? (c) What is the phase angle between the current and the supplied voltage? (a) 2.7 $\times$ 10^3 Ω (b) 4.4 $\times$ 10^{-2} A (c) −90°

77 The circuit in Fig. 20.14a is called a low-pass filter because a large current is delivered to the load (R_L) only by a low-frequency source. The circuit in Fig. 20.14b, on the other hand, is called a high-pass filter because a large current is delivered to the load only by a high-frequency source. Show mathematically why the circuits have these characteristics. see ISM

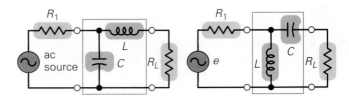

(a) Low–pass filter   **(b) High–pass filter**

● **FIGURE 20.14 Low-pass and high-pass filters** See Exercise 77.

78 In a series RLC circuit with an inductance of 750 mH, what values of resistance and capacitance would give the circuit a resonance frequency of 60 Hz? 9.4 $\times$ 10^{-6} F; any resistor

21

Geometrical Optics: Reflection and Refraction of Light

Optics is the study of light and vision. Human vision, of course, requires light, specifically what is called visible light. The broad definition of light is any radiation in or near the visible region of the electromagnetic spectrum. This definition includes infrared and ultraviolet radiations. Similar optical properties are shared by all of these electromagnetic waves.

In this chapter we will investigate basic optical phenomena, such as reflection and refraction, which we experience every day. The principles that govern reflection explain the behavior of mirrors, while those that govern refraction explain the properties of lenses, such as those found in eyeglasses. With the aid of these and other optical principles, we can understand why rainbows are seen, the cause of the "wet spot" mirage seen on roads in the summer, why a glass prism spreads out light into a spectrum of colors—and why the underwater swimmer in the photo above seems to have a phantom double. We will also explore the fascinating field of fiber optics.

You may have noted that the title of the chapter is geometrical optics. What this means is that a simple geometrical approach, using straight lines and angles, can be used to investigate many aspects of reflection and refraction. For these purposes we do not need to be concerned with the physical nature of electromagnetic waves. The principles of geometrical optics will be introduced here and applied in greater detail in the study of mirrors and lenses in Chapter 22.

Topics of study in this chapter include wave fronts and rays, reflection, refraction, internal reflection and fiber optics, and dispersion.

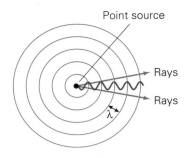

Point source

Rays

Rays

λ

(a) Circular wave fronts

Rays

λ

(b) Plane wave fronts

●**FIGURE 21.1 Wave fronts and rays** A wave front is defined by adjacent points on a wave that are in phase, such as those along wave crests or troughs. **(a)** Near a point source, the wave fronts are circular (two-dimensional) or spherical (three-dimensional). **(b)** Far from a point source, the wave fronts are approximately linear or planar. A line perpendicular to a wave front in the direction of the wave's propagation is called a ray.

Figure 21.1

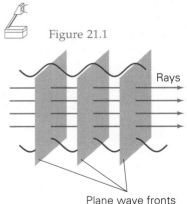

Rays

Plane wave fronts

●**FIGURE 21.2 Light rays** A plane wave front travels in a straight line in the direction of its rays. A beam of light may be represented by a group of parallel rays (or a single ray).

> *Distinguish between the angle of incidence and the angle between the incident ray and the surface.*

21.1 ■ Wave Fronts and Rays

Waves, electromagnetic or other, are conveniently described using wave fronts. A **wave front** is the line or surface defined by adjacent portions of a wave that are in phase. For example, if an arc is drawn along one of the crests of a circular water wave moving out from a point source, all the particles on the line will be in phase (●Fig. 21.1a). A line along a wave trough would work equally well (and would also form a circle). For a three-dimensional spherical wave, such as sound or light emitted from a point source, the wave front is a spherical surface rather than a circle. At a distance far from the source, a short segment of a circular or spherical wave front may be approximated as a **plane wave front** (Fig. 21.1b). A plane wave front may also be produced directly by a linear, elongated source such as a long bulb filament. In a uniform medium, wave fronts propagate outward from the source at a particular wave speed. For light in a vacuum, this speed is $c = 3.00 \times 10^8$ m/s.

The geometrical description of a wave using wave fronts tends to neglect the sinusoidal nature of the wave. This simplification is carried a step further by the concept of a ray. As illustrated in Fig. 21.1, a line drawn perpendicularly to a series of wave fronts and pointing in the direction of propagation is called a **ray**. Note that a ray is in the direction of the energy flow of a wave. A plane wave front is assumed to travel in a straight line in a medium in the direction of its rays (●Fig. 21.2). A beam of light may be represented by a group of parallel rays or simply as a single ray. The representation of light as rays is adequate and convenient for explaining some optical phenomena.

The use of the geometrical representations of wave fronts and rays to explain phenomena such as the reflection and refraction of light is called **geometrical optics**. However, certain other phenomena, such as the interference of light, cannot be treated in this manner and must be explained in terms of actual wave characteristics. These phenomena will be considered in Chapter 23.

21.2 ■ Reflection

The reflection of light is an optical phenomenon of enormous importance—if light were not reflected to our eyes by objects around us, we wouldn't see them at all. **Reflection** involves the absorption and re-emission of light by means of complex electronic vibrations in the atoms of the reflecting medium. However, the phenomenon is easily described using rays.

A light ray incident on a surface is described by an **angle of incidence** (θ_i). This is measured relative to a line perpendicular, or normal, to the reflecting surface, a line commonly referred to as a normal (see ●Fig. 21.3). Similarly, the reflected ray is described by an **angle of reflection** (θ_r). The relationship between these angles is given by the **law of reflection**.

The angle of incidence is equal to the angle of reflection:

$$\theta_i = \theta_r \tag{21.1}$$

Another condition for reflection is that the incident ray, the reflected ray, and the normal all lie in the same plane, which is sometimes called the plane of incidence.

When the reflecting surface is smooth, the reflected rays of a beam of light are parallel (Fig. 21.4). This is called **regular**, or **specular reflection**. Reflection from a flat mirror is regular. If the reflecting surface is rough, on the other hand, the light rays are reflected in nonparallel directions because of the irregular nature of the surface. This is termed **irregular**, or **diffuse reflection**. The reflection of light from this page is an example of diffuse reflection.

Note in Fig. 21.4 that the law of reflection applies to both regular and diffuse reflection. However, the type of reflection involved determines whether we see images from a reflecting surface (Fig. 21.5). In regular reflection, the reflected, parallel rays produce an image. Diffuse reflection does not produce an image because the light is reflected in various directions.

Experience with friction and direct investigations show that all surfaces are rough on a microscopic scale. What then determines whether reflection is specular or diffuse? In general, if the dimensions of the surface irregularities are greater than the wavelength of the light, the reflection is diffuse. Therefore,

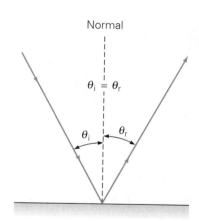

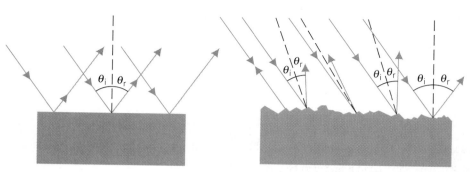

(a) Specular, or regular reflection   (b) Diffuse, or irregular reflection

FIGURE 21.4 Regular (specular) reflection and irregular (diffuse) reflection   **(a)** When a light beam is reflected from a smooth surface, and the reflected rays are parallel, the reflection is said to be regular or specular. **(b)** Reflected rays from a relatively rough surface, such as this page, are not parallel, and the reflection is said to be irregular or diffuse. (Note that the law of reflection applies to each individual ray.)

to make a good mirror, glass (with a metal coating) or metal must be polished until the surface irregularities are about the same size as the wavelength of light. Recall that the wavelength of visible light is on the order of 10^{-5} cm.

It is because of diffuse reflection that we are able to see illuminated objects, for example, the moon. If the moon's spherical surface were smooth, only the reflected sunlight from a small region would come to an observer on Earth,

(a)   (b)

FIGURE 21.5 Reflections
(a) Specular reflection gives a mirror image in Lake Powell in Utah. **(b)** When water is disturbed so that its surface is no longer perfectly smooth, it produces only diffuse reflection, with no clear images.

and only a small illuminated area would be seen. Also, you can see the beam of light from a flashlight or spotlight because of diffuse reflection from dust and particles in the air.

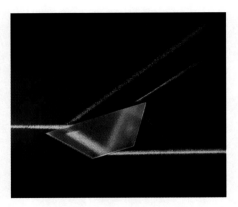

● **FIGURE 21.6 Reflection and refraction** A beam of light is incident on a prism from the left. Part of the beam is reflected and part is refracted. The refracted beam is partially reflected and partially refracted at the bottom glass-air surface. (Can you see what happens to the reflected portion?)

Figure 21.7

● **FIGURE 21.7 Huygens' principle** (a) Each point on a wave front is considered to be the source of a wavelet. The line or surface tangent to all these wavelets defines a new position of the wave front. When a wave enters an optically denser medium, its speed is less, and the wave fronts (or rays) are bent, or refracted. (b) This diagram shows the geometry for the derivation of Snell's law. See text for description.

21.3 ■ Refraction

Refraction refers to the change in direction of a wave at a boundary where it passes from one medium into another. In general, when a wave is incident on a boundary between media, some of its energy is reflected and some is transmitted. For example, when light traveling in air is incident on a transparent material such as glass (● Fig. 21.6), it is partially reflected and partially transmitted. But the direction in which the transmitted light is propagated is different from the direction of the incident light, and the light is said to have been refracted, or bent.

This phenomenon can be analyzed conveniently using a geometrical method developed by the Dutch physicist Christian Huygens (1629–1696). **Huygens' principle** for wave propagation is

Every point on an advancing wave front can be considered to be a source of secondary waves, or wavelets, and the line or surface tangent to all these wavelets defines a new position of the wave front.

Huygens' principle is applied to incident and transmitted wave fronts at a media boundary as shown in ● Fig. 21.7a. The wave speeds are different in the two media. In this case, $v_1 > v_2$. (The speed of light varies in different media, and in general, is less in denser media. Intuitively, you might expect the passage of light to take longer through a medium with more atoms per volume. And, in fact, the speed of light in water is about 75% of that in air or a vacuum.)

The distances the wave fronts travel in a time t are $v_1 t$ in medium 1 and $v_2 t$ in medium 2 (Fig. 21.7b). As a result of the smaller wave speed in the second medium, the direction of the transmitted wave front is different from that of the incident wave front. The particles of the second medium are driven or set in motion by the incident wave disturbance, and thus the waves' frequency is the same in both media. However, the wavelengths are different because of the different wave speeds ($\lambda f = v$).

The change in the direction of wave propagation is described using the **angle of refraction**. In Fig. 21.7b, the angle of incidence is θ_1, and the angle of

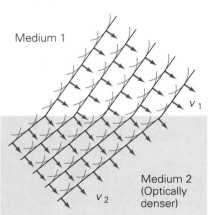

Medium 1

v_1

Medium 2 (Optically denser)

v_2

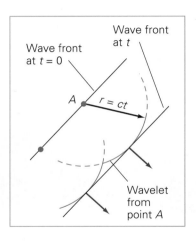

Wave front at $t = 0$

Wave front at t

A $r = ct$

Wavelet from point A

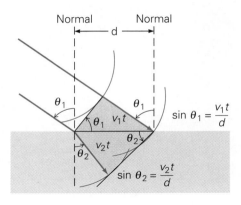

Normal   Normal

d

θ_1 θ_1

θ_1 $v_1 t$

$v_2 t$ θ_2

θ_2

$\sin \theta_1 = \dfrac{v_1 t}{d}$

$\sin \theta_2 = \dfrac{v_2 t}{d}$

refraction is θ_2. From the geometry of two parallel rays, where d is the distance between the normals to the boundary at the points where the rays are incident,

$$\sin \theta_1 = \frac{v_1 t}{d} \qquad \text{and} \qquad \sin \theta_2 = \frac{v_2 t}{d}$$

Combining these two equations in ratio form gives

$$\frac{\sin \theta_1}{\sin \theta_2} = \frac{v_1}{v_2} \qquad\qquad (21.2)$$

If the angle of incidence is zero, what is the angle of refraction?

Snell's law

This expression is known as **Snell's law**.

Thus, light is refracted when passing from one medium into another because the speed of light is different in the two media (see the Insight below). The speed of light is a maximum in a vacuum, and it is convenient to compare the speed of light in other media to this constant value (c). This is done by defining a ratio called the **index of refraction** (n).

This law is named for the Dutch physicist Willebord Snell (1591–1626), who discovered it.

$$n = \frac{c}{v} \frac{\text{(speed of light in a vacuum)}}{\text{(speed of light in a medium)}} \qquad\qquad (21.3)$$

Index of refraction; a ratio of speeds

INSIGHT ■ Understanding Refraction: A Marching Analogy

The refraction of light is not as easy to understand or visualize as reflection. We say that light is bent because waves have different speeds in different media. Intuitively, we might expect light to slow down when it enters a denser medium. The transmission of light through a transparent medium involves complex atomic absorption and emission processes, but it makes sense to suppose that these processes would take longer in a denser medium. This is indeed the case. The speed of light in water is 75% of that in air or vacuum, and in glass it is about 67% or less (depending on the type of glass). What is difficult to visualize is why light is bent or changes direction because of a change in speed.

To give some insight into this phenomenon, let's consider an analogy of a band marching across a field (Fig. 1). Part of the field is wet and muddy, and the marching column enters this region obliquely (at an angle of incidence). As the marchers in a row enter the wet, slippery region, they keep marching with the same cadence (frequency). However, slipping in the mud, the stride (wavelength) of the marchers is shorter, so they are slowed down.

The band members in the other part of the same row are still on dry ground and continue on with their original stride. The effect of the change in speed is a change in direction when the band enters the second medium. (A similar change in direction produced by changes in marching speeds is seen when a marching column turns a corner. The marchers nearest the corner deliberately shorten their stride and slow down, allowing those farther from the corner to swing around and complete their wider turn.) We might think of the marching rows as wavefronts. As in refraction, the frequency (cadence) remains the same, but the wavelength, speed, and direction change on entering another medium.

FIGURE 1 Marching analogy for refraction On obliquely entering a muddy field, the direction of a marching row is changed, analogous to the refraction of a wave front.

TABLE 21.1 ■ Indices of Refraction (at λ = 590 nm)*

Substance	n
Air	1.00029
Water	1.33
Ethyl alcohol	1.36
Fused quartz	1.46
Glycerine	1.47
Polystyrene	1.49
Oil (typical value)	1.50
Glass (depending on type)†	1.45–1.70
crown	1.52
flint	1.66
Zircon	1.92
Diamond	2.42

*A nanometer (nm) is 10^{-9} m.

†Crown glass is a soda-lime silicate glass and flint glass is a lead-alkali silicate glass. Flint glass is more dispersive than crown glass (see Section 21.5).

> When light is refracted, its speed and wavelength are changed and its frequency remains unchanged.

> Demonstration/activity: Use the following methods of demonstrating refraction:
> a) shine a laser into an aquarium which contains water and a small amount of powdered milk.
> b) shine a laser through a piece of smoked glass.
> c) put a ruler into a glass container of water and observe the bending of the image.

As a ratio of speeds, the index of refraction is a unitless quantity. The indices of refraction of several substances are given in Table 21.1. Note that these values are for a specific wavelength of light. This is specified because n is slightly different for different wavelengths ($n = c/v = c/\lambda_m f$, where λ_m is the wavelength of light in a particular material). The values of n given in Table 21.1 will be used in examples and problems in this chapter for all wavelengths of light in the visible region, unless noted otherwise.

Remember that the frequency of light does not change when it enters another medium, but the wavelength of light in a material differs from the wavelength of that light in vacuum, as can be easily shown:

$$n = \frac{c}{v} = \frac{\lambda f}{\lambda_m f}$$

or

$$n = \frac{\lambda}{\lambda_m} \qquad (21.4)$$

The wavelength of light in the medium is then $\lambda_m = \lambda/n$. Note that n is always greater than 1 because the speed of light in a vacuum is greater than the speed of light in any material ($c > v$). Therefore, $\lambda > \lambda_m$.

EXAMPLE 21.1 ■ The Speed of Light in Water

What is the speed of light in water?

Solution. Obviously, there are some known quantities.

Given: $n = 1.33$ (from Table 21.1)   *Find:* v (speed of light in H_2O)
$c = 3.0 \times 10^8$ m/s (known)

Since $n = c/v$,

$$v = \frac{c}{n} = \frac{3.0 \times 10^8 \text{ m/s}}{1.33} = 2.3 \times 10^8 \text{ m/s}$$

Note that $1/n = v/c = 1/1.33 = 0.75$, and v is 75% of the speed of light in a vacuum.

Hence we see that the index of refraction n is a measure of the speed of light in a transparent material, or technically, a measure of the *optical density* of a material. For example, the speed of light in water is less than that in air, and water is said to be optically denser than air. (Optical density does not correlate directly with mass density. In some instances, a material with a greater optical density than another will have a lower mass density.) So, the greater the index of refraction of a material, the greater its optical density and the smaller the speed of light in the material.

For practical purposes, the index of refraction is measured in air rather than in vacuum, since the speed of light in air is very close to c, and

$$n_{air} \cong \frac{c}{c} = 1$$

(From Table 21.1, $n_{air} = 1.00029$.)

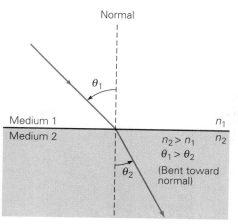

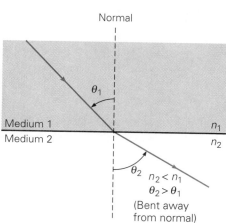

Medium 1 n_1
Medium 2 $n_2 > n_1$
$\theta_1 > \theta_2$
(Bent toward normal)

Medium 1 n_1
Medium 2 n_2
$n_2 < n_1$
$\theta_2 > \theta_1$
(Bent away from normal)

● FIGURE 21.8 **Index of refraction and ray deviation**
(a) When the second medium is more optically dense than the first ($n_2 > n_1$), the refracted ray is bent toward the normal, as in the case of light entering water from air. (b) When the second medium is less optically dense than the first ($n_2 < n_1$), the refracted ray is bent away from the normal. [Note that this is the case if the ray in part (a) is traced in reverse, going from medium 2 to medium 1.]

Figure 21.8

The index of refraction of a material may be determined experimentally using Snell's law.

$$\frac{\sin \theta_1}{\sin \theta_2} = \frac{v_1}{v_2} = \frac{c/n_1}{c/n_2} = \frac{n_2}{n_1}$$

or

$$n_1 \sin \theta_1 = n_2 \sin \theta_2 \qquad (21.5)$$

where n_1 and n_2 are the indices of refraction for the first and second media, respectively.

Eq. 21.5 is a very practical form of Snell's law. If the first medium is air, $n_1 \cong 1$, and $n_2 = \sin \theta_1/\sin \theta_2$. Thus, only the angles of incidence and refraction need to be measured to determine the index of refraction of a material experimentally. If the index of refraction of a material is known, that value can be used in the practical form of Snell's law to find the angle of refraction for a given angle of incidence. In general, the following relationships can be easily deduced from Eq. 21.5:

- If the second medium is more optically dense than the first medium ($n_2 > n_1$), the refracted ray is bent toward the normal ($\theta_2 < \theta_1$), as illustrated in ● Fig. 21.8a.

- If the second medium is less optically dense than the first medium ($n_2 < n_1$), the refracted ray is bent away from the normal ($\theta_2 > \theta_1$), as illustrated in Fig. 21.8b.

Demonstration/activity: Fill a beaker with a solution that has the same index of refraction as glass. (Try microscope immersion oil.) Then place another smaller beaker in the solution and illuminate it with a light. If the solution has the same index of refraction as the glass, the glass beaker should be invisible.

Why is a perfectly clear piece of glass "visible"?

EXAMPLE 21.2 ■ Snell's Law

A beam of light traveling in air strikes a glass plate at an angle of incidence of 45° (● Fig. 21.9). The glass has an index of refraction of 1.5. (a) What is the angle of refraction for the light transmitted into the glass? (b) If the glass plate is 2.0 cm thick, how far is the beam displaced laterally (sideways) from the normal in passing through the glass? (c) Prove that the emergent beam is parallel to the incident beam, that is, that $\theta_4 = \theta_1$.

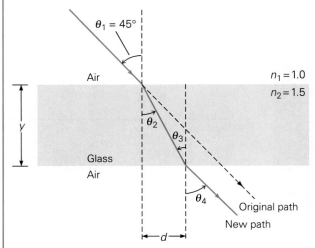

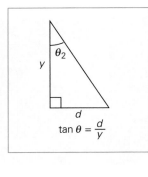

$\tan \theta = \dfrac{d}{y}$

• FIGURE 21.9 Double refraction The refracted ray is displaced laterally (sideways) a distance d, and the emergent ray is parallel to the original ray. Example 21.2.

Solution. Listing the data, we have

> **Given:** $\theta_1 = 45°$ **Find:** (a) θ_2 (angle of refraction)
> $n_1 = 1.0$ (b) d (lateral displacement)
> $n_2 = 1.5$ (c) Show that $\theta_4 = \theta_1$
> $y = 2.0$ cm

(a) Using the practical form of Snell's law, $n_1 \sin \theta_1 = n_2 \sin \theta_2$, with $n_1 = 1.0$ gives

$$\sin \theta_2 = \frac{\sin \theta_1}{n_2} = \frac{\sin 45°}{1.5} = \frac{0.707}{1.5} = 0.47$$

Thus,

$$\theta_2 = \sin^{-1}(0.47) = 28°$$

Note that the beam is bent toward the normal.

(b) You can see from the inset in Fig. 21.9 that the lateral displacement (d) is related to θ_2 by $\tan \theta_2 = d/y$, and

$$d = y \tan \theta_2 = (2.0 \text{ cm})(\tan 28°) = (2.0 \text{ cm})(0.53) = 1.1 \text{ cm}$$

(c) If $\theta_1 = \theta_4$, then the emergent ray is parallel to the incident ray. Applying Snell's law to the beam at both surfaces gives

$$n_1 \sin \theta_1 = n_2 \sin \theta_2$$

and

$$n_2 \sin \theta_3 = n_1 \sin \theta_4$$

From the figure, we see that $\theta_2 = \theta_3$. Therefore,

$$n_1 \sin \theta_1 = n_1 \sin \theta_4$$

or

$$\theta_1 = \theta_4$$

Thus, the emergent beam is parallel to the incident beam but displaced laterally or sideways a distance d. (Can you find the perpendicular distance between the original and emergent ray paths?)

EXAMPLE 21.3 ■ Refraction and Wavelength

A beam of monochromatic (single frequency) light is directed at the side of a fish tank, so that the light passes from air to glass and from glass to water. The wavelength of the light in the water (a) is the same as that in air, (b) is independent of the wavelength in the glass, (c) does not change at the glass-water interface, (d) is shorter than that in the glass. *Clearly establish the reasoning and physical principle(s) used in determining your answer. That is, **why** did you select your answer?*

Reasoning and Answer. First, one needs to have an idea of the relative magnitudes of the optical densities of the materials or their indices of refraction. As can be seen from Table 21.1, $n_{glass} > n_{water} > n_{air}$.

Then, looking at the possible answers, we can eliminate (a) since the wavelength of light in a material medium is less than that in vacuum or air, as shown previously ($\lambda_m = \lambda/n$). Similarly, since the indices of refraction of glass and water are different, the wavelength of the light changes at this interface, so (c) is not the answer. Noting that $n_{glass} > n_{water}$, the wavelength is shorter in the more optically dense medium (glass), so (d) is eliminated.

This leaves only (b), which is the correct answer. Multiple choice questions may sometimes be answered correctly by such a process of elimination, but it is important to understand why a particular answer is correct—in this case, why the wavelength of light in water is independent of that in glass.

Let's look at the interface of two general material media with indices n_2 and n_1. Forming a ratio,

$$\frac{n_2}{n_1} = \frac{c/v_2}{c/v_1} = \frac{v_1}{v_2} = \frac{\lambda_1 f}{\lambda_2 f} = \frac{\lambda_{m_1}}{\lambda_{m_2}}$$

or

$$\lambda_2 = \left(\frac{n_1}{n_2}\right)\lambda_1$$

Then, applying the wavelength-index conditions at the interfaces,

(air–glass) $\qquad \lambda_{glass} = \dfrac{\lambda_{air}}{n_{glass}}$

(glass–water) $\qquad \lambda_{water} = \left(\dfrac{n_{glass}}{n_{water}}\right)\lambda_{glass}$

and combining

$$\lambda_{water} = \left(\frac{n_{glass}}{n_{water}}\right)\frac{\lambda_{air}}{n_{glass}} = \frac{\lambda_{air}}{n_{water}}$$

We can see that the wavelength of the light in water is independent of that in glass, since the latter does not appear in the final equation. This is to be expected, since the speed of light in a transparent medium, and thus the wavelength for a constant frequency, is a property of the material itself.

Follow-up Exercise. A light source with a particular frequency in air is submersed in water in a fish tank. The beam travels in the water, through double plate glass at the side of the tank (each glass plate having a different n), and into air. In general, what are (a) the frequency and (b) the wavelength of the light in air? (*Reasoning and answers may be found in the Answers to Exercises section at the back of the book.*)

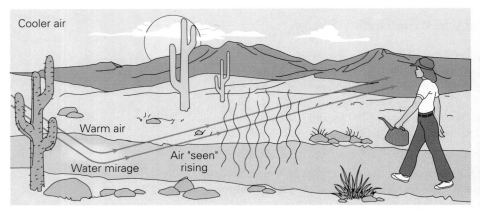

(b)

● **FIGURE 21.10 Refraction in action** (a) The refraction of light from the sky by warmer air near the road gives rise to a "wet spot" mirage. Refraction also accounts for being able to "see" hot air rising from the road. **(b)** Wet spot mirage on road due to refracted skylight. Notice the partial image of the truck on the road in the foreground. This too is a result of refraction, or light being bent.

A couple of other common examples of refraction are shown in ● Fig. 21.10. The refraction of light from the sky by warm air near the surface of a road produces the "wet spot" mirage commonly seen on a road on a hot day. We also "see" hot air rising from a hot road surface. Of course, you cannot really see the air, but gases have a property that makes it possible for you to sense it visually. The index of refraction of a gas is proportional to its density, which is in turn inversely proportional to its temperature. Thus, you can perceive regions of air having different densities (temperatures) because the refraction of light as it passes through the rising air produces distortion of the image, which seems to shimmer. (Ordinarily, the air near a road's surface has a uniform density, and you do not see these effects.)

Refraction of light in air gives rise to other effects. At night, starlight travels to your eyes through air of varying densities, causing the light to be bent. This refraction and the motion of the air are responsible for stars' twinkling effect. Atmospheric refraction also accounts for the fact that the Sun can be seen for a short time before it rises above or after it sinks below the horizon. (See Exercise 19 at the end of the chapter.)

You may have experienced an effect of the refractive bending of light while trying to reach for something that is underwater, for example, a fish or a coin (● Fig. 21.11). You are used to light traveling in straight lines from objects to your eyes, but the light reaching our eyes from the submerged object is bent

● **FIGURE 21.11 Refraction and depth perception** An object immersed in water appears closer to the surface than it actually is because of refraction. See Example 21.4.

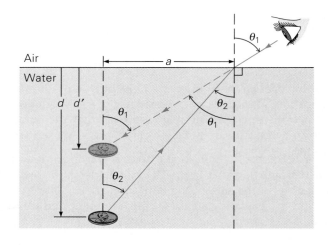

at the water surface. (Note in the figure that the ray is bent away from the normal.) The object appears to be closer to the surface than it actually is, and therefore you tend to miss the object when reaching for it.

EXAMPLE 21.4 ■ Depth Perception

How is the apparent depth (d') of the coin in Fig. 21.11 related to the actual depth d and the angles of incidence and refraction?

Solution. From the figure, we see that the distance a is common to both d and d', and using trigonometry we may write

$$\tan \theta_1 = \frac{a}{d'} \quad \text{and} \quad \tan \theta_2 = \frac{a}{d}$$

Combining these equations to form a ratio,

$$\frac{d'}{d} = \frac{\tan \theta_2}{\tan \theta_1} \quad \text{or} \quad d' = \left(\frac{\tan \theta_2}{\tan \theta_1}\right)d$$

This expression can be simplified if we consider refraction for only small angles. Recall that for small angles ($\theta < 15°$), $\cos \theta \cong 1$ (actually $\cos 15° = 0.966$). Thus

$$\tan \theta = \frac{\sin \theta}{\cos \theta} \cong \sin \theta$$

The previous expression can therefore be approximated by

$$\frac{d'}{d} = \frac{\tan \theta_2}{\tan \theta_1} \cong \frac{\sin \theta_2}{\sin \theta_1}$$

But, by Snell's law, $\sin \theta_2 / \sin \theta_1 = 1/n$, and

$$d' \cong \frac{d}{n}$$

With $n = 1.33$ for water (Table 21.1),

$$d' \cong \frac{d}{1.33} = (0.75)d$$

and the apparent depth of the coin would be $\frac{3}{4}$ of its actual depth.

Figure 21.12a

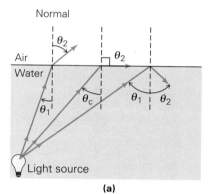

(a)

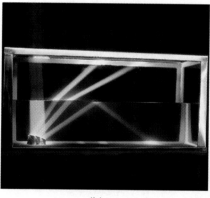

(b)

● **FIGURE 21.12 Internal reflection** (a) When light enters a less optically dense medium, it is refracted away from the normal. At a critical angle (θ_c), the light is refracted along the interface (common boundary) of the media. At an angle greater than the critical angle ($\theta_1 > \theta_c$), there is total internal reflection. (b) Internal reflections in binocular prisms make the instrument short and compact, rather than long like a telescope.

21.4 ■ Total Internal Reflection and Fiber Optics

An interesting phenomenon occurs when light traveling in one medium is incident on the boundary with another medium that is less optically dense, for example, when light goes *from* water *to* air. As you know, in such a case a transmitted beam will be bent away from the normal, and Snell's law tells you that the greater the angle of incidence, the greater the angle of refraction will

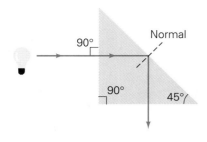

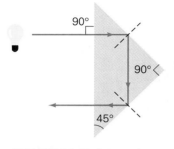

FIGURE 21.13 Internal reflection in a prism Because the critical angle of glass is less than 45°, prisms with 45° and 90° angles can be used to reflect light through 90° and 180°.

Demonstration/activity: Punch a smooth hole in an empty plastic soda bottle and fill it with water and a pinch of dry milk. Then shine a laser through the bottle into the stream of water as it flows out of the bottle. The light beam will bend with the stream of water.

FIGURE 21.14 Panoramic view A "fish-eye" lens provides an extremely wide-angle (but distorted) view. Shown here are replicas of Columbus' Nina, Pinta, and Santa Maria sailing into Miami.

be. That is, the more the angle of incidence increases, the farther the refracted ray diverges from the normal.

However, there is a limit. For a certain angle of incidence called the **critical angle** (θ_c), the angle of refraction is 90°, and the refracted ray is directed along the boundary. But what happens if the angle of incidence is even larger? If the angle of incidence is greater than the critical angle ($\theta_1 > \theta_c$), the light isn't transmitted but is internally reflected (Fig. 21.12). This condition is called **total internal reflection** and the reflection process can be 100% efficient (Fig. 21.13).

The critical angle for total internal reflection may be obtained using Snell's law. If $\theta_1 = \theta_c$ in the optically denser medium, $\theta_2 = 90°$, and

$$\frac{\sin \theta_1}{\sin \theta_2} = \frac{\sin \theta_c}{\sin 90°} = \frac{n_2}{n_1}$$

Since $\sin 90° = 1$,

$$\sin \theta_c = \frac{n_2}{n_1} \quad \text{(where } n_1 > n_2\text{)} \tag{21.6}$$

If the second medium is air, $n_2 \cong 1$, and the critical angle for total internal reflection at the boundary from a medium into air is given by

$$\sin \theta_c = \frac{1}{n} \quad \begin{array}{l}\text{(where } n \text{ is the index of}\\ \text{refraction of the medium)}\end{array} \tag{21.7}$$

Critical angle for total internal reflection at a medium–air boundary

EXAMPLE 21.5 ■ Wide-Angle View

(a) What is the critical angle for light traveling in water and incident on a water-air boundary? (b) If a diver submerged in a pool looked up at the water surface at angles of $\theta < \theta_c$ and $\theta > \theta_c$, what would she see in each case? (Neglect any thermal or motional effects.)

Solution.

Given: $n = 1.33$ (from Table 21.1)   *Find:* (a) θ_c (critical angle)
                                                        (b) Effects of changing θ

(a) The critical angle can be found directly from Eq. 21.7.

$$\theta_c = \sin^{-1}\left(\frac{1}{n}\right) = \sin^{-1}\left(\frac{1}{1.33}\right) = 48.8°$$

(b) For viewing angles of $\theta < \theta_c$, the diver would have a conical view of things above the surface (see Fig. 21.12 and mentally trace the rays in reverse for light coming from all angles in three dimensions). That is, light coming from the above-water 180° panorama could be viewed in a cone with a half-angle of 48.8°. As a result, objects would appear distorted. Such panoramic views are seen in photographs made with "fish-eye" lenses (Fig. 21.14). For a viewing angle greater than θ_c, the diver would see the reflection of something on the bottom of the pool.

Internal reflections enhance the brilliance of cut diamonds. The critical angle for a diamond–air surface is

$$\theta_c = \sin^{-1}\left(\frac{1}{n}\right) = \sin^{-1}\left(\frac{1}{2.42}\right) = 24.4°$$

A so-called brilliant-cut diamond has many facets, or faces (58 in all—33 on the upper face and 25 on the lower), and light entering the main and upper facets (or crown) above the critical angle is internally reflected in the diamond. It then emerges from the upper facets, giving rise to the diamond's brilliance (• Fig. 21.15).

Fiber Optics

When a fountain is illuminated from below, the light is transmitted along the curved streams of water. This phenomenon was first demonstrated in 1870 by the British scientist John Tyndall (1820–1893), who showed how light was "conducted" along the curved path of a stream of water flowing from a hole in the side of a container. As you may have guessed, this phenomenon is observed because the light is internally reflected along the stream of water.

Internal reflection forms the basis of **fiber optics**, a relatively new and fascinating field centered on the use of transparent fibers to transmit light. Multiple internal reflections make it possible to "pipe" light along a transparent rod, even if the rod is curved (• Fig. 21.16). Note from the figure that the smaller the diameter of the light pipe, the greater the number of internal reflections. In a small fiber, there can be as many as several hundred internal reflections per centimeter.

Internal reflection is an exceptionally efficient process. Optical fibers can be used to transmit light over very long distances with losses of only about 25% per kilometer. These losses are primarily due to fiber impurities which scatter the light. Transparent materials have different degrees of transmission. Fibers are made out of certain plastics and special glass for maximum transmission, and the greatest efficiency is achieved with infrared radiation, for which there is less scattering.

The comparatively greater efficiency of multiple internal reflections over multiple external reflections can be illustrated by considering a good reflecting surface, such as a plane mirror, which at best has a reflectivity of about 95%. Suppose that a number of plane mirrors were placed in two parallel rows, and properly spaced so that light incident on the first mirror at one end would be reflected back and forth down the rows (• Fig. 21.17).

After each reflection, the beam intensity would be 95% of the incident beam from the preceding reflection. The intensity I of the reflected beam after n reflections is given by

$$I = (0.95)^n I_o$$

where I_o is the initial light intensity on the first mirror. Then, after 100 reflections,

$$I = (0.95)^{100} I_o = 0.006\, I_o$$

Thus, with multiple reflections on plane mirrors there is only 0.6% transmission. Compare this to about 75% transmission with optical fibers over a kilometer in length even after thousands and thousands of reflections.

• **FIGURE 21.15 Diamond brilliance** Internal reflection gives rise to a diamond's brilliance. The brilliant-cut diamond has 58 facets or faces. Light entering the upper facets is internally reflected and re-emerges through these facets, giving the diamond a sparkling brilliance.

> Shine laser light through a piece of fiber-optic cable to show internal reflection.

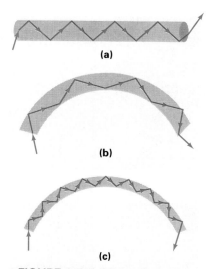

(a)

(b)

(c)

• **FIGURE 21.16 Light pipes** (a) When light is incident on the end of a cylindrical form of transparent material so that the internal angle of incidence is greater than the critical angle of the material, the light is reflected down the length of the light pipe. (b) Light is also transmitted along curved light pipes by internal reflection. (c) As the diameter of the rod or fiber becomes smaller, the number of reflections per unit length becomes greater.

FIGURE 21.17 Multiple reflections Multiple reflections from nearly parallel mirrors.

FIGURE 21.18 Fiber optic bundle Hundreds or even thousands of extremely thin fibers (a) are grouped together to make a bundle (b).

(a)

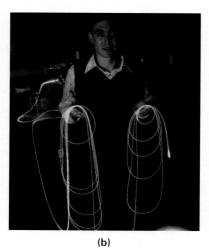

(b)

Fibers whose diameter is about 10 μm (10^{-5} m) are grouped together in flexible bundles that are 4–10 mm in diameter and up to several meters in length, depending on the application (Fig. 21.18). A fiber bundle with a cross-sectional area of 1 cm^2 can contain as many as 50,000 individual fibers. To prevent light from being transmitted between fibers in contact with each other, they are coated with a film.

Fiber optics can be used for purely decorative purposes, such as lamps, but a much more important application involves the piping of light to and images from inaccessible or hard-to-reach places. To do this, the ends of a fiber bundle are cut and polished to form a flexible **fiberscope**. A beam of light can be transmitted along the bundle to illuminate an area, even if the bundle is bent and twisted (Fig. 21.19). Equally important, an image or picture may be tranmitted back by a fiberscope. Light travels through one set of fibers to illuminate an object and, after reflection, travels back in another set. The image has light and dark regions since each fiber has a circular cross section and transmits a dot. The overall image thus has a mosaic pattern, like a newspaper picture. The smaller the elements in the mosaic, the finer the detail. Thus, a fiberscope has a very large number of extremely fine fibers. A transmitted image can be magnified by a lens for viewing.

There are a number of interesting applications of fiber optics. For example, you're probably aware that many telephone transmissions are accomplished by means of fiber optics. Light signals, rather than electrical signals, are transmitted through optical telephone lines. Optical fibers have lower energy losses than current-carrying wires, particularly at higher frequencies. Also, these fibers are lighter than metal wires, have greater flexibility, and are not affected by electromagnetic disturbances (electric and magnetic fields) since they are made of materials that are electrical insulators.

Fiber optics has been widely applied in medicine. For example, endoscopes, or instruments used to view internal portions of the human body, previously consisted of lens systems in long narrow tubes. Some contained a dozen or more lenses and gave relatively poor images. Also, since the lenses had to be aligned in certain ways, the tubes had to have rigid sections, which limited the endoscope's maneuverability. Such an endoscope could be inserted down the throat into the stomach to observe the stomach lining. However, there would be blind spots due to the curvature of the stomach and the inflexibility of the instrument.

The flexibility of fiber bundles has eliminated this problem. Lenses are used at the end of the fiber bundles to focus the light, and a prism is used to change the direction for its return. The incident light is usually transmitted by

FIGURE 21.19 Fibroscopy A technician uses a fiberscope to inspect the interior of a guitar. The label on the inside (image displayed on screen) reveals that it was made by Stradivarius in 1711.

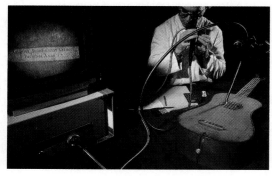

an outer layer of fiber bundles, and the image is returned through a central core of fibers. Mechanical linkages allow maneuverability. The end of a fiber endoscope may be equipped with devices to obtain specimens of the viewed tissues for biopsy (diagnostic examination), or even to perform surgical procedures (Fig. 21.20). For example, you may have heard of arthroscopic surgery being performed on the knees of injured athletes. The arthroscope that is now routinely used for inspecting and repairing damaged joints is simply a fiber endoscope fitted with appropriate surgical implements.

A fiber-optic cardioscope used for direct observation of heart valves typically has a fiber bundle about 4 mm in diameter and 30 cm long. Such a cardioscope passes easily to the heart through the jugular vein, which is about 15 mm in diameter. To displace the blood and provide a clear field of view for observations and photographing, a transparent balloon at the tip of the cardioscope is inflated with saline (salt water) solution.

Another application of fiber optics is the coding and decoding of information. To make a "coded" image of a classified picture, for example, the component fibers of a bundle are deliberately misaligned and randomly interwoven. As a result, a transmitted image is jumbled and unrecognizable unless it is viewed through an identically interwoven bundle that "decodes" it.

FIGURE 21.20 Endoscopy
Surgeons use fiber-optic endoscope to perform internal surgery.

21.5 ■ Dispersion

Light of a single frequency is called monochromatic light (from the Greek, *mono* meaning "one" and *chroma* meaning "color"). Visible light that contains all the component frequencies, or colors, is termed white light. Sunlight is white light. When a beam of white light passes through a glass prism as shown in Fig. 21.21, it is spread out, or dispersed, into a spectrum of colors. This phenomenon led Newton to believe that sunlight is a mixture of component colors. When the beam enters the prism, the component colors, corresponding to different wavelengths, must be refracted at slightly different angles so that they spread out into a spectrum.

The formation of a spectrum indicates that the index of refraction of glass must be slightly different for different wavelengths, and this is true of many transparent media. The explanation has to do with the speed of light. In a vacuum, the speed of light c is the same for all wavelengths, but in a dispersive medium the speed of light is slightly different for different wavelengths. Since the index of refraction n of a medium is a function of the velocity of light in

You can remember the sequence of the colors of the visible spectrum (from the long-wavelength end to the short-wavelength end) using the name ROY G. BIV, which is an acronym for Red, Orange, Yellow, Green, Blue, Indigo, and Violet.

Which color of light is dispersed the most?

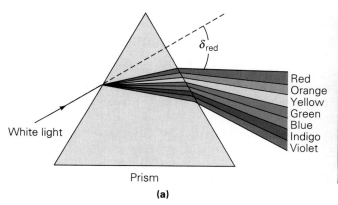

White light

δ_{red}

Red
Orange
Yellow
Green
Blue
Indigo
Violet

Prism

(a)

(b)

FIGURE 21.21 Dispersion
(a) White light is dispersed into a spectrum of colors by a glass prism. **(b)** In a dispersive medium, the index of refraction differs slightly for different wavelengths. Red light, with the longest wavelength, has the smallest index of refraction and is bent least. The angle between the incident beam and a refracted ray is called the angle of deviation (δ) for that ray.

FIGURE 1 Double rainbow Notice that the colors of the rainbows are reversed. In the lower, primary rainbow the colors run vertically from blue to red, whereas in the upper, secondary rainbow they run from red to blue.

"My heart leaps up when I behold a rainbow in the sky."

We have all been fascinated by the beautiful array of colors of a rainbow (Fig. 1). With the optical principles learned in this chapter, we are now in a position to understand the formation of this spectacular display.

A rainbow is produced by refraction, dispersion, and internal reflection of light within water droplets. When millions of water droplets remain suspended in the air after a rainstorm, a multicolored arc is seen, whose colors run from violet along the lower part up the spectrum (in order of wavelength) to red along the upper. Below the arc, the light from the droplets combines to form a region of brightness. Occasionally, more than one rainbow is seen; the main, or primary, rainbow is sometimes accompanied by a fainter and higher secondary rainbow.

The light that forms the primary rainbow is reflected once inside each water droplet (Fig. 2a). Being also refracted and dispersed, the light is spread out into a spectrum of colors. However, because of the conditions for refraction and internal reflection in water, the angles of deviation (between incoming and outgoing rays) for violet to red light lie within a narrow range of 40°–42° (Fig. 2b). This means that you can see a rainbow only when the Sun is behind you, so the dispersed light is reflected to you through these angles.

The less frequently seen secondary rainbow is caused by a double internal reflection (Fig. 2c). This results in an arc whose vertical color sequence is the inverse of the primary rainbow's. The angles of deviation in this case lie between 50.5° for red light and 54° for violet light.

We generally see only rainbow arcs, because the formation by water droplets is cut off at the ground. If you were on a cliff, you might possibly see a complete circular rainbow (Fig. 2c). Also, the higher the Sun is in the sky, the less of a rainbow you will be able to see from the ground. In fact, you won't see a primary rainbow if the Sun's altitude is greater than 42°. (Altitude in this case is the angle above the horizon.) The primary rainbow can still be seen from a height, however. As an observer's elevation increases, more of the arc becomes visible. It is common to see a completely circular rainbow from an airplane. You may also have seen a circular rainbow in the spray from a garden hose.

QUESTION: Can there be a third-order, or tertiary, rainbow?

ANSWER: Yes, a third-order rainbow resulting from three internal reflections in a water droplet is possible.

In a book entitled *Opticks*, Isaac Newton wrote, "The Light which passes through a drop of Rain after two

that medium ($n = c/v = c/\lambda f$), the index of refraction will then be different for different wavelengths. It follows from Snell's law that light of different wavelengths will be refracted at different angles.

We can summarize by saying that in a transparent material with different indices of refraction for different wavelengths of light, there is a separation of the wavelengths, and the material is said to exhibit **dispersion**. Dispersion varies with different media and may be neglected in many instances. Also, since the differences in the indices of refraction for different wavelengths are small, a representative value at some specified wavelength can be used for general purposes. (See Table 21.1.)

A good example of a dispersion material is diamond, which is about five times more dispersive than glass. In addition to the brilliance resulting from internal reflections off many facets, a cut diamond shows a play of colors, or "fire", resulting from the dispersion of the internally reflected light.

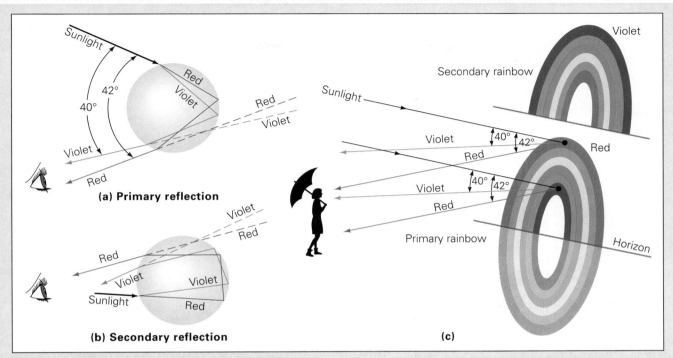

● FIGURE 2 The rainbow Rainbows are due to the refraction, dispersion, and internal reflection of sunlight. **(a)** A single internal reflection gives rise to the primary rainbow. **(b)** A double internal reflection produces a secondary rainbow. **(c)** In the primary rainbow, an observer sees red light at the top of the bow—that is, from the higher droplets—because red light is deviated most. (The other color components from these droplets pass above the observer's eyes.) Similarly, violet or blue is seen from the lower droplets.

Refractions, and three or more Reflexions, is scarcely strong enough to cause a sensible rainbow." (By "sensible," he meant intense enough to be seen.) However, his friend Edmund Halley (for whom the famous comet was named) showed that the reason the tertiary rainbow is not seen is not because the reflected light lacks intensity, but rather because the tertiary arc is formed in the general direction of the Sun and cannot be seen against the brightness of the sky.

Another dramatic example of dispersion is the production of a rainbow, as discussed in the Insight feature.

Important Concepts

You should be able to define and/or explain these chapter concepts clearly.

wave front	angle of reflection	Huygens' principle	total internal reflection
plane wave front	law of reflection	angle of refraction	fiber optics
ray	regular (specular) reflection	Snell's law	fiberscope
geometrical optics	irregular (diffuse) reflection	index of refraction	dispersion
angle of incidence	refraction	critical angle	

Important Relationships for Review

These relationships are mathematical statements of the concepts and principles presented in the chapter. You should be able to identify the symbols and to explain the relationships before proceeding to the Exercises. In-text equation reference numbers are given for convenience.

Law of Reflection:

$$\theta_i = \theta_r \tag{21.1}$$

Snell's Law:

$$\frac{\sin \theta_1}{\sin \theta_2} = \frac{v_1}{v_2} \quad \text{or} \quad n_1 \sin \theta_1 = n_2 \sin \theta_2 \tag{21.2}$$

Index of Refraction:

$$n = \frac{c}{v} = \frac{\lambda}{\lambda_m} \tag{21.3-4}$$

Critical Angle for Boundary Between Two Materials:
(where $n_1 > n_2$):

$$\sin \theta_c = \frac{n_2}{n_1} \tag{21.6}$$

Critical Angle for Material-Air Boundary:
(where n is the index of refraction of the material):

$$\sin \theta_c = \frac{1}{n} \tag{21.7}$$

Exercises

21.1 ■ Wave Fronts and Rays

1 A wave front is (a) always circular, (b) parallel to a ray, (c) described by a surface of equal phase (d) none of these. (c)

2 A ray (a) is perpendicular to the direction of energy flow, (b) is always parallel to other rays, (c) is perpendicular to a series of wave fronts, (d) illustrates the wave nature of light. (d)

21.2 ■ Reflection

3 For regular reflection, (a) the angle of incidence equals the angle of reflection, (b) the rays of a reflected beam are not parallel, (c) the incident ray, the reflected ray, and the normal all lie in the same plane, (d) both (a) and (c). (d)

● **4** For diffuse reflection, (a) the angle of incidence equals the angle of reflection, (b) the rays of a reflected beam are not parallel, (c) the incident ray, the reflected ray, and the local normal all line in the same plane, (d) all of these. (d)

5 Explain why the sunbeams are visible in ● Fig. 21.22.
diffuse reflections by dust particles

6 ■ The angle of incidence of a light ray on a mirrored surface is 38°. What is the angle between the incident and reflected rays? 76°

7 ■■ Light strikes a surface at an angle of 40° relative to the surface. What is the angle of reflection? 50°

8 ■■ A beam of light is incident on a plane mirror at an angle of 36° to the normal. What is the angle between the reflected rays and the surface of the mirror? 54°

9 ■■ Two people stand in a dark room, 3.0 m from a large plane mirror and 4.0 m apart. At what angle of incidence

● **FIGURE 21.22 Sunbeams** See Exercise 5.

should one of them shine a flashlight on the mirror so the reflected beam directly strikes the other person? 34°

10 ■■ Two upright plane mirrors touch along one edge, where their planes make an angle of 70°. If a beam of light is directed onto one of the mirrors at an angle of incidence of 40° and is reflected onto the other mirror, what will be the angle of reflection of the beam from the second mirror? 30°, see ISM

11 ■■ Two plane mirrors are placed 60 cm apart with their mirrored surfaces parallel and facing each other. If the mirrors are 25 cm wide, at what angle should a beam of light be incident at one end of one mirror so that it just strikes the far end of the other? 23°

● **12** ■■ Two plane mirrors M_1 and M_2 are placed together as illustrated in ● Fig. 21.23. (a) If the angle α between the mirrors is 70° and the angle of incidence θ_{i_1} of a light ray incident on M_1 is 35°, what is the angle of reflection θ_{r_2} from M_2? (b) If $\alpha = 115°$ and $\theta_{i_1} = 60°$, what is θ_{r_2}? (a) 35° (b) 55°

13 ■■ For the plane mirrors in Fig. 21.23, what angles α and θ_{i_1} would allow a ray to be reflected back in the same direction from which it came (parallel to the incident ray)? 90°

14 ■■■ Show that for two mirrors placed together at an angle α as in Fig. 21.23, a light ray successively reflected from both mirrors is deflected through an angle of 2α irrespective of the incident angle, where 2α is the obtuse angle between the incident and reflected rays. see ISM

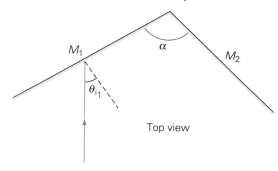

Top view

● **FIGURE 21.23 Plane mirrors together** See Exercises 12, 13, and 14.

21.3 ■ Refraction
21.4 ■ Total Internal Reflection and Fiber Optics

15 Light refracted at the boundary of two media (a) is bent toward the normal when entering the less optically dense medium, (b) is bent away from the normal when entering the more optically dense medium, (c) has the same frequency in both media, (d) always experiences a decrease in speed. (c)

● **16** The index of refraction (a) is always greater than or equal to one, (b) is inversely proportional to the speed of light in a medium, (c) is inversely proportional to the wavelength of light in the medium, (d) all of these. (d)

17 The critical angle for total internal reflection at a material–air boundary (a) is independent of the wavelength of the light in the medium, (b) is smaller for a material with a greater index of refraction, (c) may be greater than 90°, (d) none of these. (b)

18 Total internal reflection will occur if (a) $\theta_1 < \theta_c$, (b) $n_1 = n_2$, (c) $n_2 > n_1$, (d) none of these. (d)

● **FIGURE 21.24 Refraction effects** See Exercise 19.

19 Two refraction phenomena are shown in ● Fig. 21.24. (a) Explain why the pencil appears almost severed. Illustrate with a ray diagram. (b) Explain why the setting Sun appears flattened. see ISM

20 Explain the phenomenon illustrated in ● Fig. 21.25. The pictures are taken with a camera on a tripod at a fixed angle. There is a penny in the container, but only its tip is seen initially. However, when water is added, more of the coin is seen. Why? see ISM

● **FIGURE 21.25 You don't see it, then you do** See Exercise 20.

● **21** ■ The speed of light in a material is determined to be 2.15×10^8 m/s. What is the index of refraction of the material? 1.40

22 ■ Find the speed of light in diamond. 1.24×10^8 m/s

23 ■ Light strikes water perpendicular to the surface. What is the angle of refraction? 0°; not refracted

24 ■■ A beam of light enters water at an angle of 55° with a normal to the water surface. Find the angle of refraction. 38°

25 ■■ A beam of light in air is incident on the surface of a slab of fused quartz. Part of the beam is transmitted into the quartz at an angle of 30° with a normal to the surface, and part is reflected. What is the angle of reflection? 47°

26 ■■ Light passes from air into water. If the angle of refraction is 15°, what is the angle of incidence? 20°

27 ■■ A beam of light is incident on a flat piece of polystyrene at an angle of 55° to a surface normal. What angle does the refracted ray make with the plane of the surface? 57°

● 28 A beam of light traveling in air is incident on a transparent plastic material at an angle of 45° with a normal to the material's surface. The angle of refraction is measured to be 30°. What is the index of refraction of the material? 1.4

29 ■■ A beam of light traveling in water strikes a glass surface at an angle of 43° with a normal to the surface. If the angle of refraction in the glass is 35°, what is the index of refraction of the glass? 1.6

30 ■■ Is the speed of light greater in diamond or zircon? Express the difference as a percentage. 26% greater

31 ■■ Monochromatic blue light having a frequency of 6.5×10^{14} Hz enters a piece of crown glass. What are the frequency and wavelength of the light in the glass?
$f = 6.5 \times 10^{14}$ Hz; $\lambda = 3.0 \times 10^{-7}$ m

32 ■■ Light passes through a piece of flint glass in 20 ps (picoseconds). Find the thickness of the glass. 3.6×10^{-5} m

33 ■■ A He-Ne (helium-neon) laser beam ($\lambda = 632.8$ nm in air) is directed through ethyl alcohol. What are the wavelength and frequency of the light in the alcohol?
465.3 nm, 4.74×10^{14} Hz

● 34 ■■ Light passes from material A, which has an index of refraction of $\frac{4}{3}$, into material B, which has an index of refraction of $\frac{5}{4}$. Find the ratio of the speed of light in material B to the speed of light in material A. 16/15

35 ■■ If the angle of incidence in Exercise 34 is 30°, what is the angle of refraction? 32°

36 ■■ A layer of water 30.0 mm thick floats on a layer of another liquid, which is 24.0 mm thick. The other liquid has an index of refraction of 1.50. How far below the water surface will the bottom of the container appear to be for a small angle of incidence? 38.6 mm

37 ■■ A person in a boat shines a flashlight into the water of a clear lake at an angle of 12° and notes a large fish at an apparent depth of 4.5 m. (a) What are the true x and y coordinates of the fish (with the origin at the point where the light enters the water)? (b) What is the angle of refraction? (a) (0.96 m, −6.0 m) (b) 9.0°

38 ■■ A fish tank is made of glass with an index of refraction of 1.50. A person shines a light beam on the glass at an incident angle of 40° so as to see a fish inside. Is the fish illuminated? Justify your answer. yes, see ISM

39 ■■ A beam of light passes through a 45°-45°-90° prism as shown in ● Fig. 21.26. (a) What can be said about the index

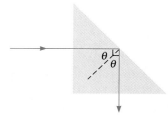

● FIGURE 21.26 Internal reflection in a prism See Exercise 39.

of refraction of the prism for this to occur? (b) What if the prism were under water? (a) 1.41 (b) 1.88

40 ■■ A light ray in air is incident on a glass plate 10.0 cm thick at an angle of 40°. The glass has an index of refraction of 1.65. The emerging ray on the other side of the plate is parallel to the incident ray but laterally displaced. How far is the emerging ray displaced relative to the normal? What is the perpendicular distance between the original direction of the ray and the direction of the emerging ray? 4.2 cm

41 ■■ A 45°–90°–45° prism is made of a material with an index of refraction of 1.85. Can the prism be used to deflect a beam of light by 90° (a) in air or (b) in water? (a) yes; (b) no (see ISM)

● 42 A person lying at poolside looks over the edge of the pool and sees a bottle cap on the bottom directly below, where the pool depth is 3.5 m. How far below the water surface does the bottle cap appear to be to the person? 2.6 m

43 ■■ What percentage of the actual depth is the apparent depth of an object submerged in water? (Assume that the observer is looking almost straight downward.) 75%

44 ■■ How far will a beam of light travel in flint glass in the time it takes light to travel 25 cm in air? 15 cm

45 ■■ The critical angle for a certain type of glass is determined to be 39° (at which no light is seen to emerge from a sample). What is the index of refraction of the glass? 1.6

● 46 ■■ To a submerged diver looking upward through the water, the altitude of the Sun (the angle between Sun and horizon) appears to be 50°. What is the Sun's actual altitude? 31°

47 ■■ At what angle must a diver submerged in a lake look toward the surface to see the setting Sun? >48.8°

48 ■■■ A coin lies on the bottom of a pool under 1.5 m of water and 0.90 m from the side wall (● Fig. 21.27). If a light beam is incident on the water surface at the wall, at what angle (θ) relative to the wall must the beam be directed so it will illuminate the coin? (See Fig. 21.25.) 43°

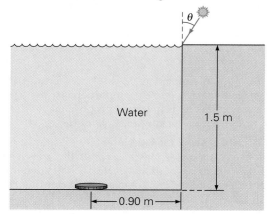

● FIGURE 21.27 Find the coin See Exercise 48. (Drawing not to scale.)

49 ■■■ A crown-glass plate 2.5 cm thick is placed over a newspaper. How far beneath the top surface of the plate

would the print appear to be to a person looking almost vertically downward through the plate?   1.6 cm

50 ■■■ Light travels from a material whose index of refraction is $n_1 = 2.0$ into another material whose index of refraction is $n_2 = \frac{5}{3}$. The opposite surface of the second material is exposed to the air. (a) Find the critical angle at which light will be reflected at the interface of the materials. (b) Find the critical angle at which the light will passs through the interface and then be reflected at the opposite surface of the second material.   (a) 56°   (b) 30°

21.5 ■ Dispersion

● **51** Dispersion occurs for (a) monochromatic light, (b) polychromatic light, (c) both (a) and (b).   (b)

52 A transparent material (a) shows dispersion for $\theta_1 = 90°$, (b) has different n's for different λ's, (c) changes the frequency of a particular light wave, (d) none of these.   (b)

53 A prism disperses white light into a spectrum. Can a second prism be used to recombine the spectral components? Explain.   no

● **54** ■■ The index of refraction of crown glass is 1.515 for red light and 1.523 for blue light. Find the angle separating rays of the two colors in a piece of crown glass if their angle of incidence is 37°.   0.13°

55 ■■ White light passes through a prism made of crown glass and strikes an air interface at an angle of 41.15°. Using the indices of refraction given in Exercise 54, describe what happens.   see ISM

56 ■■ (a) If glass is dispersive, why don't we see a spectrum of colors when sunlight passes through a window pane? (b) Does dispersion occur for polychromatic light incident on a dispersive medium at an angle of 0°? Explain. (Are the speeds of the different colors of light the same in the medium?)
(a) angle is approximately zero   (b) no; (no)
57 ■■ A beam of light with red and blue components having wavelengths of 670 nm and 425 nm, respectively, strikes on a slab of fused quartz at an incident angle of 30°. It is observed that the different components are separated by an angle of 0.00131 rad on refraction. If the index of refraction for the red light is 1.4925, what is the index of refraction for the blue light?   1.4980

58 ■■ Fused quartz is a dispersive medium with an index of refraction of 1.470 for blue light (400 nm in air) and 1.445 for red light (680 nm in air). A beam of light composed of these two colors is incident on a plate of fused quartz at an angle of 45°. (a) What is the angle of separation between the two components of the beam in the quartz? (b) What is the ratio of the wavelengths of the two components in the quartz?
(a) 0.0096 rad   (b) 1.73
59 ■■■ The glass of a prism has an index of refraction of $n_1 = 1.554$ for light with a wavelength of $\lambda_1 = 440$ nm and an index of refraction of $n_2 = 1.538$ for light with a wavelength of $\lambda_2 = 650$ nm. If a beam consisting of light of these two wavelengths is incident on one of the prism's surfaces at an angle of 70° with a normal to the surface, what is the angular separation of the resulting beams in the glass?   0.45°

■ **Additional Exercises**

60 When looking into an opaque, empty container that is 15 cm deep at a viewing angle of 50° relative to the vertical side of the container, an observer sees nothing on the bottom. When the container is filled with water, the observer fully sees the coin on the bottom of and just beyond the side of the container (from the same viewing angle). See Fig. 21.24(b). How far is the coin from the side of the container?   11 cm

61 A light beam is incident on a mirror at an angle θ_i. The mirror is rotated through an angle ϕ about an axis in the plane of the mirror and perpendicular to the plane of incidence. The incident angle is then $\theta_i' = \theta_i - \phi$ (see ● Fig. 21.28). Show that the reflection ray rotates through an angle $\alpha = 2\phi$. (This is the principle of an optical lever, which is used to measure small angles of rotation.)   see ISM

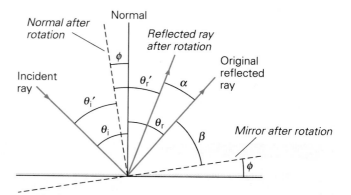

● **FIGURE 21.28 Rotation of reflected ray**   See Exercise 61.

62 Prisms are sometimes used as optical components in place of mirrors. Find the minimum index of refraction for glass that is to be used in a 45°–90°–45° prism that is meant to change the direction of light by 100° on reflection. (Assume that the incident and exit rays are normal to the prism faces.)   1.6

63 Yellow-green light with a wavelength of 550 nm is incident on the surface of a flat piece of crown glass at an angle of 40°. (a) What is the angle of refraction of the light? (b) What is the speed of the light in the glass? (c) What is the wavelength of the light in the glass?   (a) 25°
(b) 1.97 × 10⁸ m/s   (c) 362 nm
64 Light passes from medium A into medium B at an angle of incidence of 30°. The index of refraction of A is 1.5 times that of B. (a) What is the angle of refraction? (b) What is the ratio of the speed of light in B to the speed of light in A? (c) What is the ratio of the frequency of the light in B to the frequency of light in A? (d) At what angle of incidence would the light be internally reflected?   (a) 49°   (b) 1.5
(c) 1.0   (d) 42°
65 A submerged diver shines a light toward the surface of the water at angles of incidence of 40° and 50°. Will a person on the shore see a beam of light emerging from the surface in either case? Justify your answer mathematically.   seen for 40° but not for 50°

22

Mirrors and Lenses

Think of what life would be like if there were no mirrors in bathrooms or on cars, and no one could get glasses. Imagine living in a world without optical images of any kind—no photographs, no movies, no TV. Think about how little we'd know about the universe if there were no telescopes; how little we'd know about biology and medicine if there were no microscopes with which to see cells and bacteria. We sometimes forget how dependent we are on mirrors and lenses.

The first mirror was probably the reflecting surface of a pool of water. Later people discovered that polished metals and glass have reflective properties. People also noticed that when they looked at things through pieces of glass, the objects looked different, depending on the shape of the glass. In some cases, the objects appeared to be enlarged (magnified) or inverted, as in the photo above. Eventually, lenses were constructed to improve people's vision by purposefully shaping pieces of glass.

The optical properties of mirrors and lenses are based on the principles of the reflection and refraction of light. In this chapter, you will learn about these general properties by applying geometrical optics. Our emphasis will be on the formation of images by mirrors and lenses, and on the properties of those images. The practical uses of mirrors and lenses in optical instruments such as microscopes and telescopes will be discussed in Chapter 24.

Topics of study in this chapter include plane and spherical mirrors, lenses, aberrations, and the lens maker's equation.

22.1 ■ Plane Mirrors

Mirrors are smooth reflecting surfaces, usually made of polished metal or glass that has been coated with some metallic substance. As you know, even an uncoated piece of glass such as a window pane can act as a mirror (see Demonstration 14, p. 676). However, when one side of a piece of glass is coated with a compound of tin, mercury, or silver, its reflectivity is increased, and light is not transmitted through the coating. A mirror may be front-coated or back-coated, depending on the application.

When you look directly into a mirror, you see the reflected images of yourself and objects around you. Usually, these images appear to be behind the mirror (on the other side of the surface). The geometry of a mirror's surface affects the size, orientation, and type of image.

A mirror with a flat surface is called a **plane mirror**. How images are formed by a plane mirror is illustrated by the ray diagram in ● Fig. 22.1. As we know, an image appears to be behind or "inside" the mirror. This is because when the mirror reflects a ray from the object to the eye (Fig. 22.1a), it appears to us to originate from behind the mirror. Reflected rays from the top and bottom of an object are shown in Fig. 22.1b. In actuality, light rays coming from all points on the side of the object toward the mirror are reflected, and a complete image is observed.

The image formed in this way *appears* to be behind the mirror. Such an image is called a **virtual image**. Light rays appear to emanate from virtual images, but do not actually do so. However, spherical mirrors (discussed in the next section) can produce images from which light actually emanates. This type of image is called a **real image**. An example of a real image is one that you see on a movie screen.

Notice in Fig. 22.1b the positions or distances of the object and image from the mirror. Quite logically, the distance of an object from a mirror is called the *object distance* (d_o), and the distance its image appears to be behind the mirror is called the *image distance* (d_i). By geometry of similar triangles, it can be shown that $d_o = d_i$, which means that *the image formed by a plane mirror appears to be at a distance behind the mirror that is equal to the distance of the object in front of the mirror.*

We are interested in various characteristics of images. One of these is the size of an image compared to that of its object. This is expressed in terms of the **lateral magnification** (*M*),* which is defined as a ratio:

$$M = \frac{\text{image height}}{\text{object height}} = \frac{h_i}{h_o} \qquad (22.1)$$

Referring to ● Fig. 22.2, you should be able to show by similar triangles that $h_i = h_o$, so $M = 1$ for a plane mirror and there is no magnification. That is, you and your image in a plane mirror are the same size. Note that *M* is a magnification *factor*, since $h_i = Mh_o$.

*Lateral, or height distance, magnification (*M*) is distinguished from angular, or angular distance, magnification (*m*) in Chapter 24. In this chapter, we will refer to *M* simply as the magnification, or magnification factor, for convenience.

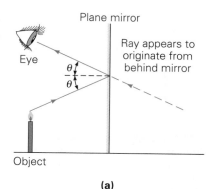

Plane mirror

Ray appears to originate from behind mirror

Eye

θ θ

Object

(a)

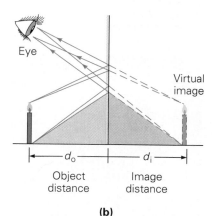

Eye

Virtual image

d_o d_i

Object distance

Image distance

(b)

● **FIGURE 22.1 Image formed by a plane mirror** (a) A ray from a point on the object is reflected in the mirror according to the law of reflection. (b) Rays from various points on the object produce an extended image. Because the two shaded triangles are similar, the image distance d_i (the distance of the image from the mirror) is equal to the object distance d_o. That is, the image is the same distance behind the mirror as the object is in front of the mirror. The rays appear to converge at the image position. In this case, the image is said to be virtual.

It may be necessary to review properties of similar triangles.

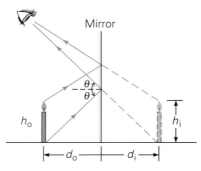

FIGURE 22.2 Magnification
The lateral or height magnification is given by $M = h_i/h_o$. Writing this relationship as $h_i = Mh_o$, the M is then a magnification factor. For a plane mirror $M = 1$, which means that $h_i = h_o$ and the image is the same height as the object.

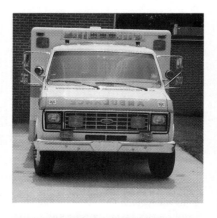

FIGURE 22.4 Practical right–left reversal The letters of the word AMBULANCE are printed backwards and are reversed in sequence so that they appear in the proper orientation and order when seen in a rear-view mirror.

It is common in ray diagrams to represent the object as an arrow and the image as a dashed arrow, as shown in Fig. 22.2. This allows us to address another image characteristic, its orientation—that is, whether it is upright or inverted with respect to the object. For a plane mirror, the image is upright (or erect). This means that the image is oriented in the same vertical direction as the object, which can clearly be seen from the directions of the arrowheads. Note in Fig. 22.2 that the object and image are both pointing upward. (If the arrow and its image were both pointing downward, the image would still be described as upright to indicate that it has the same orientation as the object.)

With other types of mirrors, such as spherical mirrors (which we will consider shortly), it is possible to have inverted images. In this case, the image arrow points in the direction opposite to that of the object arrow.

One characteristic of a plane mirror that you may have noticed is a seeming right-left reversal of the images. If you stand in front of a plane mirror and raise your right hand, your image will raise its left hand. Also, if you part your hair, the part will appear to be on the wrong side (unless you part it down the middle). This right-left reversal of images in plane mirrors is illustrated in Figs. 22.3 and 22.4. From Fig. 22.3, we can see that the thumbs of the hand

FIGURE 22.3 Right–left reversal in a plane mirror An image in a plane mirror is said to show right-left reversal relative to the object. However, the object and image thumbs are both at the bottom of the mirror and point in the $+x$ direction. There is a front-to-back reversal, in that the palms of the object and image hands face in opposite directions. This gives rise to the apparent right-left reversal orientation.

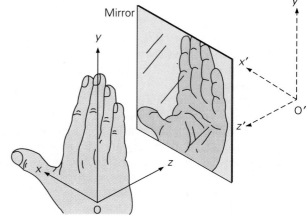

and its mirror image are both at the bottom of the mirror and both point in the $+x$ direction. However, there *is* a front-to-back reversal: The palm of the object hand faces the $+z$ direction, whereas the image hand faces the $-z$ direction. It is this reversal that gives rise to the apparent right-left reversal. This point would be easier to grasp if we were not symmetrical creatures with left hands that are mirror images of our right hands. Keep in mind that right and left are directional senses (like clockwise and counterclockwise) rather than fixed directions referenced to a coordinate system. If you face another person (a front-to-back reversal), his or her right is to your left, but you probably wouldn't think of this as a right-left reversal such as we ascribe to a mirror image.

Characteristics of the image formed by a plane mirror are summarized in Table 22.1.

TABLE 22.1 ■ Characteristics of Images Formed by Plane Mirrors

The image distance is equal to the object distance ($d_i = d_o$). That is, the image appears to be as far behind the mirror as the object is in front.

The image is virtual, upright, and unmagnified ($M = 1$).

The image shows right-left reversal.

EXAMPLE 22.1 ■ Seeing It All

What is the minimum vertical length of a plane mirror needed for a person to be able to see a complete (head-to-toe) image?

Solution. To determine this length, consider the situation shown in ● Fig. 22.5. With a mirror of minimum length, a ray from the top of the head would be reflected at the top of the mirror, and the ray from the feet would be reflected at the bottom of the mirror. The length L of the mirror is then the distance between the dashed lines perpendicular to the mirror at its top and bottom.

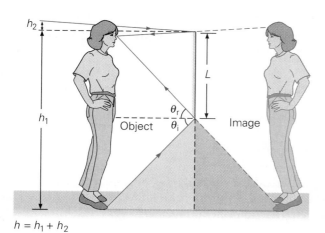

● **FIGURE 22.5 Seeing it all** The minimum height or vertical length of a plane mirror needed for a person to see his or her complete (head-to-toe) image turns out to be one-half the person's height.

However, these lines are also the normals for the ray reflections. By the law of reflection, they bisect the angles between incident and reflected rays, that is, $\theta_i = \theta_r$. Then, because the large shaded triangles are similar, the length of the mirror from its bottom to a point even with the person's eyes is $h_1/2$, where h_1 is the person's height from the feet to the eyes. Similarly, the small upper length of the mirror is $h_2/2$ (the vertical distance between the eyes and the top of mirror). Thus,

$$L = \frac{h_1}{2} + \frac{h_2}{2} = \frac{h_1 + h_2}{2} = \frac{h}{2}$$

where h is the person's total height.

Hence, for a person to see his or her complete image in a plane mirror, the minimum height or vertical length of the mirror must be one-half the height of the person.

It would appear so, but you know this is not possible. It's really a matter of reflection and an image.

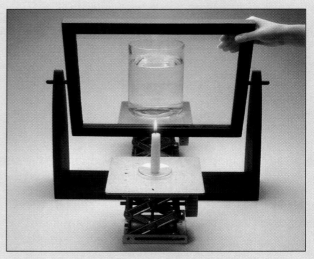

(a) The black frame holds a pane of glass, which acts as a plane mirror. The burning candle seen in the water is the image of the candle on the front stand. There is a container of water on a similar stand behind the glass, but no burning candle.

(b) The effect can be removed by tilting the glass—the image can no longer be seen from this viewing point. (Something to do with the law of reflection. What?)

22.2 ■ Spherical Mirrors

As the name implies, a **spherical mirror** is a reflecting surface with spherical geometry. As ● Fig. 22.6 shows, if a portion of a sphere of radius R is sliced off along a plane, the severed section has the shape of a spherical mirror. Either the inside or outside of such a section can be reflective. For inside reflections, the mirror is a **concave mirror** (think of looking into a cave to remember the indented surface). For outside reflections, the mirror is a **convex mirror.**

●**FIGURE 22.6 Spherical mirrors** A spherical mirror is a section of a sphere. Either the outside (convex) surface or the inside (concave) surface of the spherical section may be the reflecting surface.

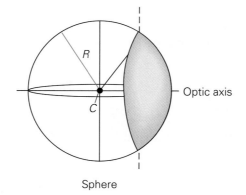

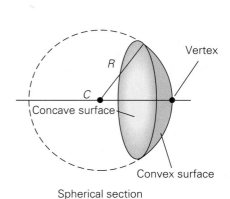

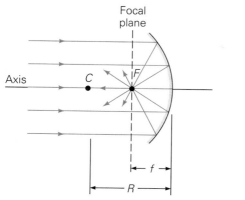

(a) Concave or converging mirror

(b) Convex or diverging mirror

● **FIGURE 22.7 Focal point**
(a) Rays parallel and close to the optic axis of a concave spherical converge at the focal point *F*. Off-axis parallel rays converge in the focal plane. **(b)** Rays parallel to the optic axis of a convex spherical mirror are reflected along paths as though they came from a focal point behind the mirror.

Concave mirror—converging mirror
Convex mirror—diverging mirror

The radial line through the center of the spherical mirror is called the optic axis, and it intersects the mirror surface at the vertex of the spherical section (Fig. 22.6). The point on the optic axis that corresponds to the center of the sphere of which the mirror forms a section is called the **center of curvature** (*C*). The distance between the vertex and the center of curvature is equal to the radius of the sphere and is called the **radius of curvature** (*R*).

When rays parallel to the optic axis are incident on a concave mirror, the reflected rays intersect at a common point called the **focal point** (*F*).* As a result, a concave mirror is called a **converging mirror** (● Fig. 22.7a). Similarly, a beam parallel to the optic axis of a convex mirror diverges on reflection, as though the reflected rays came from a focal point behind the mirror's surface (Fig. 22.7b). Thus, a convex mirror is called a **diverging mirror**. An example of a diverging mirror is shown in ● Fig. 22.8.

The distance from the vertex to the focal point of a spherical mirror is called the **focal length** (*f*). The focal length is related to the radius of curvature by this simple equation:

$$f = \frac{R}{2} \quad \begin{array}{l}\text{focal length for a}\\\text{spherical mirror}\end{array} \quad (22.2)$$

In some instances, spherical mirrors form virtual images that appear to be behind the surface just as images with plane mirrors do. However, in other instances, spherical mirrors form images that can be seen on a screen placed at certain positions. These are real images, formed by converging rays. The distinction between these two kinds of images is sometimes said to be that a real image can be formed on a screen, and a virtual image cannot. Stated another way, light actually emanates from a real image, but only appears to do so from a virtual image.

The characteristics of images formed by spherical mirrors can be determined using geometrical optics. The method involves drawing rays emanating from one or more points on an object. The law of reflection applies ($\theta_i = \theta_r$),

● **FIGURE 22.8 Diverging mirror**
Note by reverse-ray tracing in Fig. 22.7b that a diverging (convex) spherical mirror gives an expanded field of view, as can be seen in this store-monitoring mirror.

*This is the case for the approximation where the width of the mirror or the illuminated area is small compared to the radius of curvature, that is, for small angles of reflection.

Rays for spherical mirror diagrams

and certain rays are defined with respect to the mirror's geometry as follows:

- A **parallel ray** is a ray that is incident along a path parallel to the optic axis and is reflected through the focal point (as are all rays near and parallel to the axis).
- A **chief ray**, or **radial ray**, is a ray that is incident through the center of curvature (C). Since it is incident normal to the mirror's surface, this ray is reflected back along its incident path, through C.
- A **focal ray** is a ray that passes through (or appears to go through) the focal point and is reflected parallel to the optic axis. (It is a mirror image, so to speak, of a parallel ray.)

These rays are illustrated in the ray diagrams in ● Fig. 22.9 for concave and convex mirrors. It is customary to use the tip of the object (for example, the head of an arrow or the flame of a candle) as the origin of the rays. This makes it easy to see whether the image is upright or inverted. The corresponding point of the image is at the point of intersection of the rays. Also, the candle is arbitrarily taken to be upright with its base on the optic axis. Keep in mind, however, that properly traced ray from *any* point on the object can be used to find the image.

Note in Fig. 22.9a that for a concave mirror with an object at a distance greater than the radius of curvature ($d_o > R$), a real image is formed. That is, the image may be seen on a screen (for example, a piece of white paper) positioned at a distance d_i from the mirror. The image has the characteristics of being real, inverted, and smaller than the object.

The rays reflected from a convex mirror diverge (Fig. 22.9b). Projecting the rays behind the mirror to find where they intersect indicates that the image is virtual. (The candle is drawn with its flame at the point of intersection.) This image is analogous to the virtual image formed by a plane mirror, and it cannot be projected on a screen. Since the reflected rays for an object at any distance from a convex mirror diverge, *a diverging mirror cannot form a real image.*

For a converging spherical mirror, the characteristics of the image change with the distance of the object from the mirror. There are two points at which dramatic changes take place: C and F (the center of curvature and the focal point). These points divide the optic axis into three regions, as shown in ● Fig. 22.10*: $d_o > R$, $R > d_o > f$, and $d_o < f$. Let's start with an object in the region farthest from the mirror ($d_o > R$) and move progressively toward the mirror.

- The case of $d_o > R$ has already been dealt with in Fig. 22.9a.
- The case of $d_o = R$ is shown in Demonstration 15. The image is real, inverted, and the same size as the object.
- When $R > d_o > f$, we see from the ray diagram in Fig. 22.10b that an enlarged, inverted, real image is formed. As can be seen from the sequence, the image is magnified when the object is inside the center of curvature C.
- When $d_o = f$, so that the object is at the focal point (Fig. 22.10c), the reflected rays are parallel and the image is said to be formed at infinity. This expresses the idea that parallel lines converge at infinity (like parallel railroad tracks that appear to converge at a great distance). The focal point F is a special "cross-over" point. When $d_o > f$, the image is real, and (as we shall see below) when $d_o < f$, the image is virtual. We can't actually see an image formed at infinity, but we say the image is formed

*Only two rays are needed to determine the image, and we will use only two, the parallel and chief rays, for illustration clarity. However, you are encouraged (or may be required) to draw the third ray in your diagrams as a check.

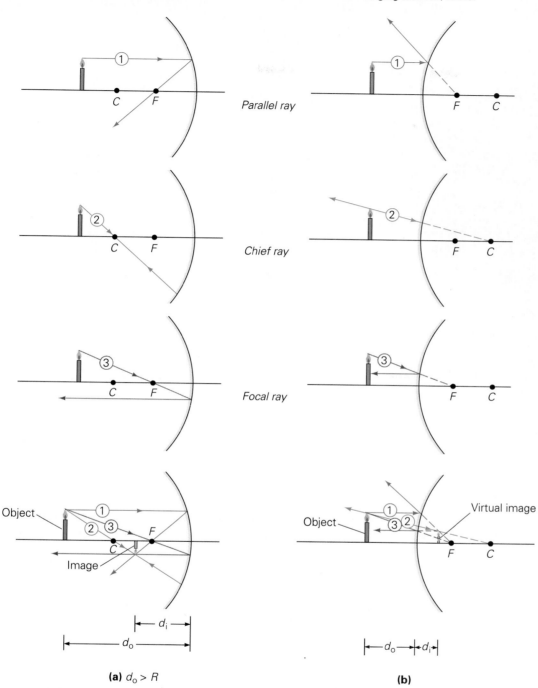

Converging (concave) mirror

Diverging (convex) mirror

Parallel ray

Chief ray

Focal ray

(a) $d_o > R$

(b)

• **FIGURE 22.9 Ray diagrams** Ray diagrams to find the images for spherical mirrors can be drawn using three rays: (1) A parallel ray is reflected through the focal point F for a converging mirror, and appears to come from the internal focal point for a diverging mirror. (2) A chief ray is incident through the center of curvature C and is reflected back along its path of incidence for a converging mirror, and appears to be reflected from the internal center of curvature for a diverging mirror. (3) A focal ray passing through the focal point F is reflected parallel to the optic axis for a converging mirror, and the incident ray appears to pass through the internal focal point for a diverging mirror. **(a)** Ray diagram for a converging mirror for $d_o > R$. With the rays coming from the tip of the object arrow, their intersection defines the location of the tip of the image arrow relative to the optic axis. **(b)** Ray diagram for a diverging mirror. Here the image is virtual and behind or inside the mirror.

Figure 22.9

Or is it? Notice that one flame is burning downward, which is rather strange. It's an illusion done with a spherical concave mirror.

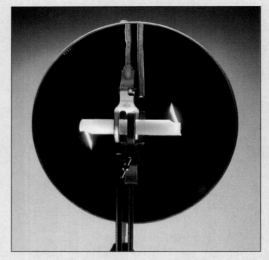

(a) When an object is at the center of curvature of a spherical concave mirror, a real image is formed that is inverted and the same size as the object, and the image distance is the same as the object distance. What is seen here is a horizontal candle burning at one end (flame up) and its overlapping image (flame down). Viewed at the same level, the inverted flame image appears to be at the opposite end of the candle.

(b) A side view showing the burning end of the horizontal candle in front of the spherical mirror.

there because of symmetry with the case represented in Fig. 22.7c—when an object is at "infinity," meaning that it is so far away that the rays emanating from it are essentially parallel, its image is formed at F. By reverse ray tracing, rays from an object at a great (taken as "infinite") distance from the mirror can be shown to be essentially parallel when they are near the mirror and to form an image on a screen aligned in the focal plane. This fact provides a method for determining the focal length of a concave mirror.

- When $d_o < f$, so that the object is inside the focal point (between the focal point and the mirror's surface), a virtual image is formed (Fig. 22.10d).

The position and size of the image may be determined graphically by ray diagrams drawn to scale. However, these can be determined more quickly by analytical methods. The distances and focal length can be shown to be related by what is known as the **spherical mirror equation**.

$$\frac{1}{d_o} + \frac{1}{d_i} = \frac{1}{f}$$ spherical mirror equation (22.3)

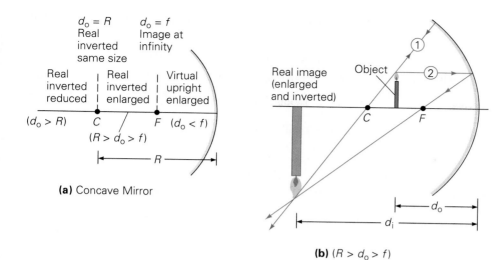

$d_o = R$
Real
inverted
same size

$d_o = f$
Image at
infinity

| Real inverted reduced | Real inverted enlarged | Virtual upright enlarged |

$(d_o > R)$ C F $(d_o < f)$

$(R > d_o > f)$

R

(a) Concave Mirror

Real image
(enlarged
and inverted)

Object

C F

d_o

d_i

(b) $(R > d_o > f)$

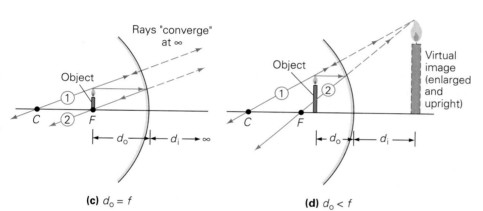

Rays "converge" at ∞

Object

C F

d_o $d_i \rightarrow \infty$

(c) $d_o = f$

Object

Virtual image
(enlarged
and upright)

C F

d_o d_i

(d) $d_o < f$

Figure 22.10

● **FIGURE 22.10 Concave mirrors** **(a)** For a concave or converging mirror, the location of the object may be within one of three regions defined by the center of curvature (C) and the focal point (F) or at these two points. For $d_o > R$, the image is real, inverted, and smaller, as shown by the ray diagram in Fig. 22.9a. **(b)** For $R > d_o > f$, the image will also be real and inverted, but enlarged or magnified. Only two rays are needed to locate the image, and the focal ray is omitted for clarity. **(c)** For an object at the focal point F or $d_o = f$, the image is said to be formed at infinity. **(d)** For $d_o < f$, the image will be virtual, upright, and enlarged. (See Demonstration 15 for $d_o = R$.)

Since the image distance d_i is often the quantity to be found, a convenient alternative form of this equation is

$$d_i = \frac{d_o f}{d_o - f} \qquad (22.4)$$

The **magnification factor** M can also be found analytically. This is expressed in terms of the image and object distances.

$$M = -\frac{d_i}{d_o} \qquad \begin{array}{l}\text{magnification equation} \\ \text{for a spherical mirror}\end{array} \qquad (22.5)$$

A helpful hint for remembering that the magnification is d_i over d_o is that the ratio is in alphabetical order (i over o).

The minus sign is added as a sign convention to indicate the orientation of the image as given below. Hence, if $|M| > 1$, the image is magnified, or larger than the object. If $|M| < 1$, the image is reduced, or smaller than the object.

The signs on the various quantities are very important in the application of these equations. This book uses these sign conventions for spherical mirrors:

$|M|$ is the absolute value of M, that is, its magnitude without regard to sign. For example, $|+2| = |-2| = 2$.

Sign convention for mirrors

TABLE 22.2 ■ Sign Convention for Spherical Mirrors

Concave mirror: f positive
Convex mirror: f negative
d_o always positive

d_i	Image	M	Image
+	Real	+	Upright
−	Virtual	−	Inverted

- The focal length f (or R) is positive for a concave mirror and negative for a convex mirror.
- The object distance d_o is always taken to be positive.
- The image distance d_i is positive for a real image (formed on the same side of the mirror as the object) and negative for a virtual image (formed behind the mirror).
- The magnification M is positive for an upright image and negative for an inverted image.

This sign convention is summarized in Table 22.2.

The use of the spherical mirror equations and sign convention can be illustrated using a plane mirror, which can be thought of as being a section of a sphere having an infinite radius of curvature ($R = \infty$). Then, since $f = R/2$,

$$\frac{1}{d_o} + \frac{1}{d_i} = \frac{1}{f} = \frac{2}{R} = \frac{2}{\infty} = 0$$

Thus,

$$\frac{1}{d_o} = -\frac{1}{d_i} \quad \text{and} \quad d_i = -d_o$$

This result indicates that the image distance is equal to the object distance and that the image is virtual (appears to be behind the mirror) because there is a minus sign. The magnification is

$$M = -\frac{d_i}{d_o} = -\frac{(-d_o)}{d_o} = +1$$

The image is the same size as the object. Since the magnification factor is positive, the image is upright. From experience, you know that these are the correct characteristics for an image that is formed by a plane mirror.

The following examples show how the spherical mirror equations are used.

EXAMPLE 22.2 ■ Images with a Concave Mirror

A concave mirror has a radius of curvature of 30 cm. If an object is placed (a) 45 cm, (b) 30 cm, (c) 20 cm, and (d) 10 cm from the mirror, where is the image formed and what are its characteristics? (Specify real or virtual, upright or inverted, and larger or smaller for each image.)

Solution.

Given: $R = 30$ cm
so $f = R/2 = 15$ cm
(a) $d_o = 45$ cm
(b) $d_o = 30$ cm
(c) $d_o = 20$ cm
(d) $d_o = 10$ cm

Find: d_i and image characteristics for all parts.

Note that these object distances correspond to the regions shown in Fig. 22.10a.

(a) In this case, the object distance is greater than the radius of curvature ($d_o > R$), and

$$\frac{1}{d_o} + \frac{1}{d_i} = \frac{1}{f}$$

or

$$\frac{1}{45} + \frac{1}{d_i} = \frac{1}{15}$$

where the units (cm) have been omitted for simplicity. Then

$$\frac{1}{d_i} = \frac{2}{45}$$

or

$$d_i = \frac{45}{2} = +22.5 \text{ cm}$$

and

$$M = -\frac{d_i}{d_o} = -\frac{22.5 \text{ cm}}{45 \text{ cm}} = -\frac{1}{2}$$

Thus, the image is real (positive d_i), inverted (negative M), and $\frac{1}{2}$ as large as the object ($|M| = \frac{1}{2}$).

(b) The object distance is equal to the radius of curvature in this case ($d_o = R$). Since $f = R/2$, for convenience, the spherical mirror equation can be put in terms of R: that is, in this case,

$$\frac{1}{d_o} + \frac{1}{d_i} = \frac{1}{f}$$

becomes

$$\frac{1}{R} + \frac{1}{d_i} = \frac{2}{R}$$

Thus,

$$\frac{1}{d_i} = \frac{1}{R}$$

or

$$d_i = R = +30 \text{ cm}$$

Then

$$M = -\frac{d_i}{d_o} = -\frac{R}{R} = -1$$

The image is therefore real, inverted, and the same size as the object. (Draw a ray diagram to prove to yourself that the object and image arrows are both at the center of curvature pointing in opposite directions.)

(c) Here $R > d_o > f$. Using Eq. 22.4,

$$d_i = \frac{d_o f}{d_o - f} = \frac{(20 \text{ cm})(15 \text{ cm})}{20 \text{ cm} - 15 \text{ cm}} = 60 \text{ cm}$$

and

$$M = -\frac{d_i}{d_o} = -\frac{60 \text{ cm}}{20 \text{ cm}} = -3$$

In this case, the image is real, inverted, and three times the size of the object. (d) For this case, $d_o < f$, and

$$d_i = \frac{d_o f}{d_o - f} = \frac{(10 \text{ cm})(15 \text{ cm})}{10 \text{ cm} - 15 \text{ cm}} = -30 \text{ cm}$$

Then

$$M = -\frac{d_i}{d_o} = -\frac{(-30 \text{ cm})}{10 \text{ cm}} = +3$$

In this case, the image is virtual (negative d_i), upright (positive M), and three times the size of the object.

From the denominator of the right-hand side of the equation for d_i (Eq. 22.4), you can see that d_i will always be negative when d_o is less than f. Therefore, a virtual image is always formed for an object inside the focal point of a converging mirror.

(Draw representative ray diagrams for each of these cases to see that the image characteristics are correct.)

■ Problem-Solving Hint

When using the spherical mirror equations to find image characteristics, it is helpful to first make a quick sketch of the ray diagram for the situation. Doing this shows you the image characteristics and helps you avoid making mistakes when applying the sign convention. *The ray diagram and the mathematical solution must agree.*

EXAMPLE 22.3 ■ Image with a Diverging Mirror

An object is 30 cm in front of a diverging mirror that has a focal length of 10 cm. Where is the image and what are its characteristics?

Solution.
   Given: $d_o = 30$ cm   *Find:* d_i and image characteristics
   $f = -10$ cm

Note that the focal length is negative for a convex mirror (see Table 22.2). Using Eq. 22.4,

Emphasize that the focal length is negative for a diverging mirror.

$$d_i = \frac{d_o f}{d_o - f} = \frac{(30 \text{ cm})(-10 \text{ cm})}{30 \text{ cm} - (-10 \text{ cm})} = -7.5 \text{ cm}$$

Then

$$M = -\frac{d_i}{d_o} = -\frac{(-7.5 \text{ cm})}{30 \text{ cm}} = +0.25$$

Thus, the image is virtual (negative d_o) and upright (positive M) and is 0.25 times (¼) the size (height) of the object.

EXAMPLE 22.4 ■ Finding the Focal Length

Where is the image formed for a concave mirror if the object is at infinity? (An object at a great distance from a mirror, relative to the mirror's dimensions, may be considered to be at infinity.)

Solution. With $d_o = \infty$, we have

$$\frac{1}{d_o} + \frac{1}{d_i} = \frac{1}{\infty} + \frac{1}{d_i} = \frac{1}{f} \qquad \text{or} \qquad d_i = f$$

Thus, the image is real ($+d_i$) and is formed at the focal point (or in the focal plane for an extended object).

This result provides an experimental means of determining the focal length of such a mirror. A screen (a piece of white paper) is adjusted until an image is formed on it. The screen is then in the focal plane of the mirror. (Does the magnification equation hold for this special case?)

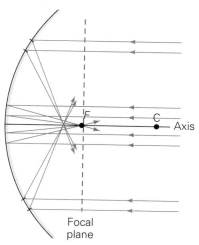

● **FIGURE 22.11 Spherical aberration for a mirror** According to the small-angle approximation, rays parallel to and near the mirror's axis converge at the focal point. However, when parallel rays not near the axis are reflected, they converge in front of the focal point. This effect is called spherical aberration and gives rise to blurred images.

The descriptions of image characteristics for spherical mirrors are true only for objects near the optic axis, that is, only for small angles of reflection. If these conditions do not hold, the images will be blurred (out of focus) or distorted, because not all of the rays will converge in the same plane. As illustrated in ● Fig. 22.11, incident parallel rays far from the optic axis do not converge at the focal point. The farther the incident ray is from the axis, the more distant is its reflected ray from the focal point. This effect is called **spherical aberration**.

Spherical aberration does not occur with a parabolic mirror. (As the name implies, a section through the center of a parabolic mirror has the form of a parabola.) All of the incident rays parallel to the optic axis of a parabolic mirror have a common focal point. However, parabolic mirrors are more difficult to make than spherical mirrors (and therefore more expensive).

22.3 ■ Lenses

The word "lens" comes from the Latin word for lentil, a seed whose shape is similar to that of a common lens. An optical **lens** is made from some transparent material (most commonly glass but sometimes plastics or crystals). One or both surfaces usually have a spherical contour. Biconvex spherical lenses (both surfaces convex) and biconcave spherical lenses (both surfaces concave) are illustrated in ● Fig. 22.12.

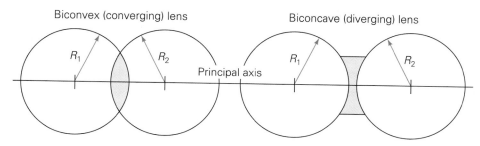

Biconvex (converging) lens

Biconcave (diverging) lens

Principal axis

● **FIGURE 22.12 Spherical lenses** Spherical lenses have surfaces defined by two spheres, and the surfaces may be either convex or concave. Biconvex and biconcave lenses are shown here. If $R_1 = R_2$, a lens is spherically symmetric.

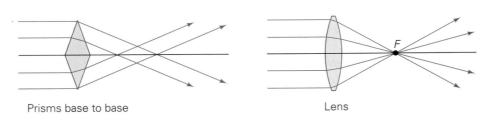

(a) Biconvex (converging) lens

Prisms base to base   Lens

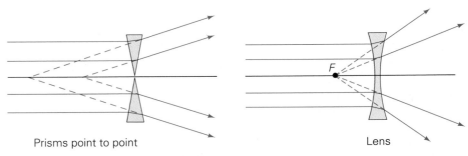

Prisms point to point   Lens

(b) Biconcave (diverging) lens

● **FIGURE 22.13 Converging and diverging lenses** **(a)** A biconvex, or converging, lens may be approximated by two prisms placed base to base. For a thin biconvex lens, rays parallel to the axis converge at the focal point *F*. **(b)** A biconcave, or diverging, lens may be approximated by two prisms placed tip to tip. Rays parallel to the axis of a biconcave lens appear to diverge from a focal point on the incident side of the lens.

Refraction was discussed in Section 21.3; see also Figure 21.20.

● **FIGURE 22.14 Burning glass** A magnifying glass can be used to focus the sun's rays to a point—with incendiary results.

686

The properties of lenses are due to the refraction of light passing through them. When light rays pass through a lens, they are bent, or deviated from their original paths, according to the law of refraction. You studied refraction by prisms in Chapter 21. To analyze lens refraction, a biconvex lens may be approximated by two prisms placed base to base, as shown in ● Fig. 22.13a. A biconvex lens is a **converging lens**: incident light rays parallel to the lens axis converge at a focal point on the opposite side of the lens. You may have focused the Sun's rays with a magnifying glass (a biconvex, or converging, lens) and have witnessed the concentration of radiant energy that results (● Fig. 22.14). The parallel rays coming from the Sun or some other distant object (at infinity) converge at the focal point. This fact provides a way for experimentally determining the focal length of a converging lens.

Conversely, a biconcave lens can be approximated by two prisms placed point to point (Fig. 22.13b). A biconcave lens is a **diverging lens**: incident parallel rays emerge from the lens as though they emanated from a focal point on the incident side of the lens.

There are several types of converging and diverging lenses (● Fig. 22.15). Meniscus lenses are the type most commonly used for corrective eyeglasses. In general, a converging lens is thicker at its center than at its periphery, and a diverging lens is thinner at its center than at its periphery. This discussion will be limited to spherically symmetric biconvex and biconcave lenses, that is, ones for which both surfaces have the same radius of curvature.

When light passes through a lens, it is refracted and displaced laterally, as shown in Example 21.2 (Fig. 21.9). If a lens is thick, this displacement may be fairly large and can complicate the analysis of the lens's characteristics. This problem does not arise with thin lenses, for which the refractive displacement of transmitted light is negligible. This discussion will be limited to thin lenses.

Like a spherical mirror, a lens with spherical geometry has *for each lens surface* a center of curvature, a radius of curvature, a focal point, and a focal length. If each surface has the same radius of curvature, the focal points are at equal distances on either side of the lens. However, for a spherical lens

$f \neq R/2$, as it is for a spherical mirror (see Section 22.5). Usually, only the focal length of a lens is specified rather than the radius of curvature.

The general rules for drawing ray diagrams for lenses are similar to those for spherical mirrors, but obviously some modifications are necessary since light passes through a lens in either direction. Opposite sides of a lens are generally distinguished as the object and image sides. The object side is, of course, the side on which an object is positioned, and the image side is the opposite side of the lens (where a real image would be formed). The rays from a point on an object are drawn as follows:

- A **parallel ray** is a ray that is parallel to the lens axis on incidence and that after refraction either passes through the focal point on the image side of a converging lens, *or* appears to diverge from the focal point on the object side of a diverging lens.

- A **chief ray** is a ray that passes through the center of the lens and is undeviated.*

- A **focal ray** is a ray that passes through the focal point of the object side of a converging lens, *or* appears to pass through the focal point of the image side of a diverging lens, and after refraction is parallel to the lens axis.

These rays are shown ● Fig. 22.16 for converging and diverging lenses. Focal rays are drawn only in the initial diagrams. As with spherical mirrors, only two rays are needed to determine the image, and we will use the parallel and chief rays. (As in the case of mirrors, however, it is generally a good idea to include a third ray in your diagrams as a check.)

For a lens, the image is real when formed on the side of the lens opposite the object (on the image side, see ● Fig. 22.17) and virtual when formed on the same side of the lens as the object (on the object side). Another way of looking at this is that rays converge at a real image and appear to diverge from a virtual image. Note that for a diverging lens (Fig. 22.16b), the rays appear to diverge from the virtual image on the object side of the lens after being refracted. Like diverging mirrors, diverging lenses can form only virtual images. Also, as with spherical mirrors, a real image from a lens can be formed on a screen but a virtual image cannot.

Regions for the object distance with a converging lens could be similarly defined, as was done for a converging mirror in Fig. 22.10a. Here, an object distance of $d_o = 2f$ for a converging lens has a significance similar to that of $d_o = R$ for a converging mirror. However, the center of curvature for a converging lens does not have the same distinction as it does for a mirror, since $R \neq 2f$ for a lens. (See Demonstration 16, p. 688.)

The image distances and characteristics for a lens can also be found analytically. The equations for thin symmetrical biconvex and biconcave lenses are identical to those for spherical mirrors (see Exercise 57). The **thin lens equation** is

$$\frac{1}{d_o} + \frac{1}{d_i} = \frac{1}{f} \qquad \text{or} \qquad d_i = \frac{d_o f}{d_o - f} \qquad (22.6)$$

thin lens equation

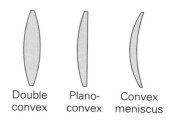

Double convex   Plano-convex   Convex meniscus

Converging lenses

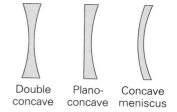

Double concave   Plano-concave   Concave meniscus

Diverging lenses

● **FIGURE 22.15 Lens shapes** The wide variety of lens shapes are generally categorized as converging or diverging. In general, a converging lens is thicker at its center than at the periphery, and a diverging lens is thinner at its center than at the periphery.

Demonstration/activity: Pass laser light through diverging and converging smoked lenses to show the path the light takes in a lens. Also use a lighted candle or other light source to show how the image changes in size, location and orientation with the position of the object.

Try to show what happens when the lens is placed under water. Use an aquarium filled with water. Does this make the focal length longer or shorter? Why?

---

*This incorporates the thin lens approximation (negligible lateral deviation).

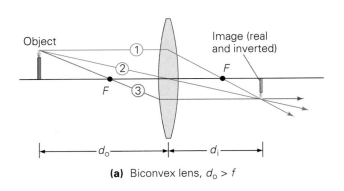

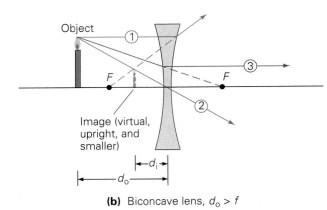

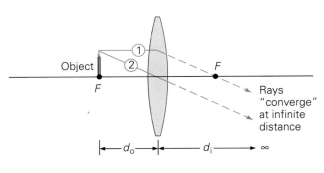

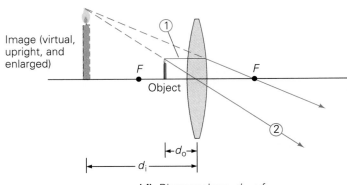

(a) Biconvex lens, $d_o > f$

(b) Biconcave lens, $d_o > f$

(c) Biconvex lens, $d_o = f$

(d) Biconvex lens, $d_o < f$

●**FIGURE 22.16 Ray diagrams for lenses** As in ray diagrams for mirrors, parallel rays (1), chief rays (2), and focal rays (3) may be drawn to analyze lenses. These rays are shown for a converging biconvex lens **(a)** and for a diverging biconcave lens **(b)** for $d_o > f$. Diagrams **(c)** and **(d)** for a converging lens show the ray diagrams for the cases of $d_o = f$ and $d_o < f$, respectively. The focal rays have been omitted in these diagrams for clarity. For $d_o = f$, the image is formed at infinity, and for $d_o < f$, the image is virtual.

Figure 22.16

DEMONSTRATION 16 ■ Inverted Image

A demonstration of how a cylindrical convex lens produces an inverted image.

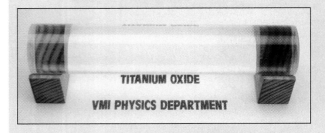

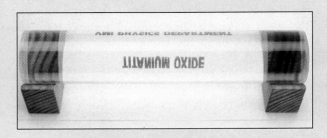

(a) Observe the words TITANIUM OXIDE as placed before a cylindrical plastic rod.
Note: a similar inversion occurs for spherical convex lenses.

(b) Viewed through the rod, the letters in TITANIUM are inverted, but the letters in OXIDE appear not to be. What is wrong, or is there something wrong?

The **magnification factor** is given by

$$M = -\frac{d_i}{d_o}$$ (22.7)

magnification equation for a thin lens

The sign conventions for these thin lens equations are similar to those for spherical mirrors.

A summary of sign conventions for the lens equations appears at the end of the chapter.

Sign conventions for lenses

- The focal length (f) is positive for a converging lens (sometimes called a positive lens) and negative for a diverging lens (sometimes called a negative lens).
- The object distance (d_o) is always taken as positive for a single lens.
- The image distance (d_i) is positive for a real image (which is located on the image side of the lens) and negative for a virtual image (which is located on the object side of the lens).
- The magnification (M) is positive for an upright image and negative for an inverted image.

Just as when working with mirrors, it is helpful to sketch a ray diagram before working a lens problem analytically.

FIGURE 22.17 Real image A converging lens forms a real image of a candle flame on a posterboard screen. Note that the image is inverted.

EXAMPLE 22.5 ■ Images with a Converging Lens

A biconvex lens has a focal length of 12 cm. Where is the image formed and what are its characteristics for an object (a) 18 cm from the lens and (b) 4 cm from the lens?

Solution.

Given: $f = 12$ cm   *Find:* d_i and image characteristics
   (a) $d_o = 18$ cm   for both parts
   (b) $d_o = 4$ cm

(a) Using fractional form of Eq. 22.6,

$$\frac{1}{d_o} + \frac{1}{d_i} = \frac{1}{f}$$

or

$$\frac{1}{18} + \frac{1}{d_i} = \frac{1}{12}$$

With a common denominator,

$$\frac{2}{36} + \frac{1}{d_i} = \frac{3}{36}$$

Thus,

$$\frac{1}{d_i} = \frac{1}{36}$$

or

$$d_i = 36 \text{ cm}$$

Then

$$M = -\frac{d_i}{d_o} = -\frac{36}{18} = -2$$

The image is real (positive d_i), inverted ($-M$), and twice as tall as the object ($|M| = 2$).

(b) Using the alternate form of Eq. 22.6 for d_i gives

$$d_i = \frac{d_o f}{d_o - f} = \frac{(4\ \text{cm})(12\ \text{cm})}{4\ \text{cm} - 12\ \text{cm}} = -6\ \text{cm}$$

Then

$$M = -\frac{d_i}{d_o} = -\frac{(-6\ \text{cm})}{4\ \text{cm}} = +1.5$$

In this case, the image is virtual (on the object side of the lens), upright, and magnified by a factor of 1.5.

You should draw ray diagrams for these two cases.

EXAMPLE 22.6 ■ Image with a Diverging Lens

A biconcave lens has a focal length of 15 cm. If an object is 30 cm from the lens, where is the image and what are its characteristics?

Solution.

Given: $f = -15\ \text{cm}$ *Find:* d_i and image characteristics
$d_o = 30\ \text{cm}$

Note that the focal length is negative since the lens is diverging (biconcave). Then

$$d_i = \frac{d_o f}{d_o - f} = \frac{(30\ \text{cm})(-15\ \text{cm})}{30\ \text{cm} - (-15\ \text{cm})} = -10\ \text{cm}$$

and

$$M = -\frac{d_i}{d_o} = -\frac{(-10\ \text{cm})}{30\ \text{cm}} = \frac{1}{3}$$

Thus, the image is virtual, upright, and $\frac{1}{3}$ of the size of the object.

EXAMPLE 22.7 ■ Magnified and Reduced Images

As the object distance of a biconvex lens is varied, at what point does the real image go from being reduced to being magnified?

Solution. Recall that for a converging mirror, the real image of an object at $d_o = R$ is the same size as the object ($M = 1$). Also, for $d_o > R$, the image is smaller, and for $R > d_o > f$, the image is larger. For a biconvex lens, R does

Again stress the importance of ray diagrams for locating and finding characteristics of images for lenses. Students often confuse diagrams for mirrors with those for lenses. This is why it is so important to show the direction of the light path with arrows.

not have the same significance, since $R \neq 2f$. However, an object distance $d_o = 2f$, or a distance $2f$ from the lens, does have this significance for the size of the real image. Substituting $2f$ for d_o in Eq. 22.6 gives

$$\frac{1}{d_o} + \frac{1}{d_i} = \frac{1}{2f} + \frac{1}{d_i} = \frac{1}{f}$$

which can be solved for d_i:

$$d_i = 2f$$

Thus,

$$M = -\frac{d_i}{d_o} = -\frac{2f}{2f} = -1$$

The real, inverted image formed equidistant from the lens is the same size as the object when the object is twice the focal length from the lens.

It can be easily shown that for $d_o > 2f$, the image is smaller than the object, and for $2f > d_o > f$, the image is larger. (You can do this analytically or by quickly sketching two ray diagrams.)

Combinations of Lenses

Many optical instruments such as microscopes and telescopes (Chapter 24) use a combination of lenses, or a compound lens system. When two or more lenses are used in combination, the overall image produced may be determined by considering the lenses individually in sequence. That is, the image formed by the first lens is the object for the second lens, and so on.

If the first lens produces an image in front of the second lens, that image is treated as a real object for the second lens (●Fig. 22.18a). If, however, the

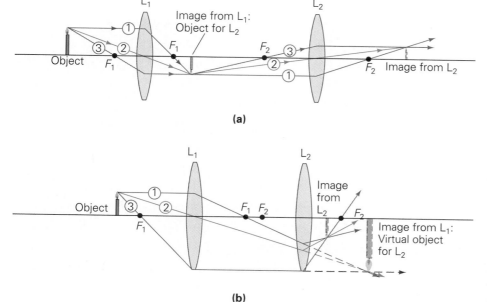

Figure 22.18

(a)

(b)

●**FIGURE 22.18 Lens combinations** The final image produced by a compound lens system may be found by treating the image of one lens as the object for the adjacent lens. **(a)** If the image of the first lens (L_1) is formed in front of the second lens (L_2), the object for the second lens is said to be real. **(b)** If the rays pass through the second lens before the image is formed, the object for the second lens is said to be virtual, and the object distance for the second lens is taken to be negative.

lenses are close enough together that the image from the first lens is not formed before the rays pass through the second lens (Fig. 22.18b), then a modification must be made in the sign convention. In this case, the image from the first lens is treated as a *virtual* object for the second lens, and the object distance for it is taken to be *negative* in the lens equation.

The total magnification (M_t) of a compound lens system is the product of the magnification factors (absolute values) of all the component lenses. For example, for a two-lens combination, as in Fig. 22.18,

$$M_t = |M_1| \times |M_2| \qquad (22.8)$$

You should be aware that not all lenses are spherical. A different type of lens is discussed in the Insight feature.

INSIGHT ■ Fresnel Lenses

To focus or to produce a large beam of parallel light rays, a sizable converging lens is necessary. The large mass of glass necessary to form such a lens is bulky and heavy; moreover, the thick lens absorbs some of the light, and is likely to show aberrations. A French physicist named Augustin Fresnel (Fre-nel'; 1788–1827) developed a solution to this problem for lenses used in lighthouses. Fresnel recognized that the refraction of light takes place at the surfaces of a lens. Hence, a lens could be made thinner—even flat—by removing glass from the interior, as long as this was done without changing the refracting properties of the surfaces.

This can be accomplished by cutting a series of concentric grooves in the surface of the lens (Fig. 1a). Note that the surface of each remaining curved segment is nearly parallel to the corresponding surface of the original lens. Together, the concentric segments refract light like the original biconvex lens (Fig. 1b). In effect, the lens has simply been slimmed down by the removal of unnecessary glass between the refracting surfaces.

A lens with such a series of concentric curved surfaces is called a Fresnel lens. Such lenses are widely used in overhead projectors and in beacons (Fig. 1c). A Fresnel lens is very thin and therefore much lighter than a conventional biconvex lens with the same optical properties. Also, Fresnel lenses are easily molded from plastic—often with one flat side (planoconvex) so that the lens can be attached to a glass surface.

One disadvantage of Fresnel lenses is that concentric circles are visible when an observer is looking through such a lens or when an image produced by one is projected on a screen as when using an overhead projector.

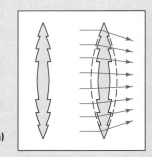

(a)

(b)

(c)

FIGURE 1 Fresnel lens (a) The focusing action of a lens comes from refraction at its surfaces. It is therefore possible to reduce the thickness of a lens by cutting away glass in concentric grooves, leaving a set of curved surfaces with the same refractive properties as the lens from which they were derived. (b) A flat Fresnel lens with concentric curved surfaces magnifies like a biconvex converging lens. (c) An array of Fresnel lenses is used to produce focused beams in this Boston harbor light. (Fresnel lenses were in fact originally developed for use in lighthouses.)

22.4 ■ Lens Aberrations

Lenses, like mirrors, also can have aberrations. Here are some common ones.

Spherical Aberration

The discussion of lenses thus far has focused on thin lenses, for which the lateral deviation of light due to refraction is negligible. In general, with spherical mirrors and lenses, light rays parallel to and near the axis will be reflected or refracted to converge at the focal point. Converging lenses may, however, like spherical mirrors, show **spherical aberration**, the effect that occurs when parallel rays passing through different regions of a lens do not come together on a common focal plane. In general, rays close to the axis of a converging lens are refracted less and come together at a point farther away from the lens than do rays passing through the periphery (●Fig. 22.19a). The place where the transmitted light beam has the smallest cross section is called the *circle of least confusion*, and the best (least distorted) image is formed at this location.

Spherical aberration can be minimized by using an aperture to reduce the effective area of the lens, so that only light rays near the axis are transmitted. Also, combinations of converging and diverging lenses may be used. The aberration of one lens can be compensated for (nullified) by the optical properties of another lens.

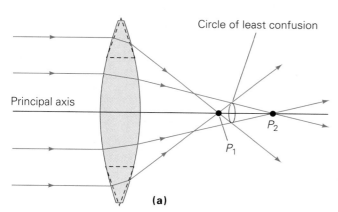

(a)

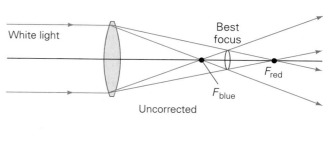

Uncorrected

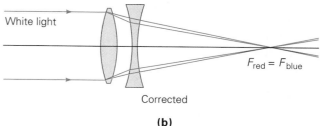

Corrected

(b)

●**FIGURE 22.19 Lens aberrations** **(a)** Spherical aberration. In general, rays closer to the axis of a lens are refracted less and come together at a point farther from the lens than do rays passing through the periphery of the lens. The smallest cross section of the transmitted beam is called the circle of least confusion, and the best (least distorted) image will be formed in its plane. **(b)** Chromatic aberration. Because of dispersion, different wavelengths (colors) of light are focused in different planes, which results in distortion of the overall image (top). This aberration may be corrected by using a converging lens and a diverging lens of different materials that compensate for each other's dispersion. The lens pair is called an achromatic doublet (bottom).

Chromatic Aberration

Chromatic aberration is an effect that occurs because the index of refraction of the material making up a lens is not the same for all wavelengths (that is, the material is dispersive). When white light is incident on a lens, the transmitted rays of different wavelengths (colors) do not have a common focal point, and images of different colors are produced at different locations (Fig. 22.19b). This dispersive aberration can be eliminated by using a compound lens system consisting of lenses of different materials, such as crown glass and flint glass. The lenses are chosen so that the dispersion produced by one is compensated for by opposite dispersion produced by the other. With a properly constructed two-component lens system, called an *achromatic doublet* (achromatic means without color), the images of selected colors will coincide.

Astigmatism

A circular beam of light along the lens axis forms a circular illuminated area on the lens; and when incident on a converging lens, the parallel beam converges at the focal point. However, when a circular cone of light from an off-axis source falls on the convex spherical surface of a lens some distance away, the light forms an elliptical illuminated area on the lens. The rays along the major and minor axes of the ellipse then focus at different points after passing through the lens. This condition is called **astigmatism**.

With different focal points in different planes, the images in both planes are blurred. For example, the image of a point is no longer a point, but two separated short line images (blurred points). As with spherical aberration, the best image is formed somewhere between the images at the location of the circle of least confusion. Astigmatism can also be reduced by reducing the effective area of the lens with an aperature.

22.5 ■ The Lens Maker's Equation ◇

The biconvex and biconcave lenses considered so far in this chapter have been symmetric, with the same focal length for each side because of equal radii of curvature. However, there are converging and diverging lenses that have surfaces with different radii of curvature (see Fig. 22.15 and ●Fig. 22.20). The focal lengths of such lenses are also important in optical analyses.

In general, the focal length of a thin lens in air ($n_{air} = 1$) is given by the so-called **lens maker's equation**:

$$\frac{1}{f} = (n - 1)\left(\frac{1}{R_1} - \frac{1}{R_2}\right) \tag{22.9}$$

More generally, *n* is the ratio of the refractive index of the lens to that of the surrounding medium. This explains why some converging lenses become diverging when submerged in water.

For this equation, n is the index of refraction of the material, and R_1 and R_2 are the radii of curvature of the first (front side) and second (back side) lens surfaces, respectively (that is, the first surface is the one on which light from the object is first incident). The equation locates the focal point at which light passing through a lens will converge for a converging lens or the focal point from which transmitted light appears to diverge for a diverging lens.

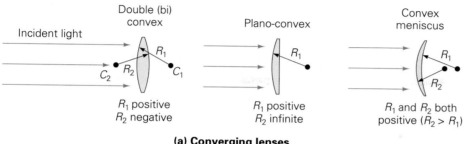

(a) Converging lenses

Double (bi) convex R_1 positive R_2 negative

Plano-convex R_1 positive R_2 infinite

Convex meniscus R_1 and R_2 both positive ($R_2 > R_1$)

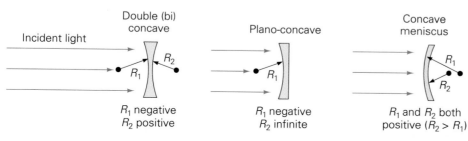

(b) Diverging lenses

Double (bi) concave R_1 negative R_2 positive

Plano-concave R_1 negative R_2 infinite

Concave meniscus R_1 and R_2 both positive ($R_2 > R_1$)

FIGURE 22.20 Sign convention for the lens maker's equation The radius of curvature is taken to be positive if it is on the emergent side of the lens and negative if it is on the incident side of the lens. A plane surface has an infinite radius of curvature (with no sign).

The sign conventions for the radii of curvature in the lens maker's equation are included in Fig. 22.20 and summarized in Table 22.3. Note that the signs depend on identifying the location of the center of curvature (C) of a surface relative to either the side of the lens on which light is incident or the side from which it emerges.

The lens maker's equation can be used to show why the focal length of a biconvex glass lens is not equal to half the radius of curvature ($f \neq R/2$), as it is for a spherical mirror. For a glass lens to have a focal length that was half of its radius of curvature would require an index of refraction of 2, which is greater than the values for known glasses (see Table 21.1).

TABLE 22.3 ■ Sign Convention for Lens Maker's Equation

$+R$	when C is on side of lens from which light emerges or on back side of lens.
$-R$	when C is on side of lens on which light is incident or on front side of lens.
$R = \infty$	for a plane (flat) surface
$+f$	converging (positive) lens
$-f$	diverging (negative) lens

EXAMPLE 22.8 ■ Focal Length of a Plano-convex Lens

A thin plano-convex lens made of crown glass has an index of refraction of 1.5. If the radius of curvature for the convex surface is 20 cm, what is the focal length of the lens when light is incident on (a) the convex surface and (b) the plane surface?

Solution.

Given: $n = 1.5$
(a) $R_1 = +20$ cm and $R_2 = \infty$
(b) $R_1 = \infty$ and $R_2 = -20$ cm

Find: (a) f (focal lengths)
(b) f

(a) The first surface in this case is the convex surface, so its radius of curvature is R_1. The center of curvature of R_1 is on the emergent side of the lens, so R_1 is positive by the sign convention (see Fig. 22.20a). The plane surface is the second surface, and $R_2 = \infty$ (with no positive or negative designation). Then

$$\frac{1}{f} = (n - 1)\left(\frac{1}{R_1} - \frac{1}{R_2}\right) = (1.5 - 1)\left(\frac{1}{20 \text{ cm}} - \frac{1}{\infty}\right) = \frac{0.5}{20 \text{ cm}}$$

and
$$f = \frac{20 \text{ cm}}{0.5} = 40 \text{ cm}$$

Thus, the lens is converging (f is positive), and a beam parallel to the axis and incident on the convex surface will converge at a point 40 cm beyond the plane surface of the lens.

(b) In this case, the first surface is the plane surface, and the radius of curvature is negative for the convex surface since its center of curvature is on the incident side of the lens. Using the lens maker's equation with these conditions gives

$$\frac{1}{f} = (n-1)\left(\frac{1}{R_1} - \frac{1}{R_2}\right) = (1.5 - 1)\left[\frac{1}{\infty} - \frac{1}{(-20 \text{ cm})}\right] = \frac{0.5}{20 \text{ cm}}$$

and
$$f = \frac{20 \text{ cm}}{0.5} = 40 \text{ cm}$$

The lens is converging, which you may find unexpected. Draw an off-axis parallel ray and show that convergence is the case when the ray is refracted away from the normal in emerging from the glass lens.

The focal length must be expressed in meters.

Optometrists express the power (P) of a lens in **diopters**, which is simply the reciprocal of the focal length of the lens expressed in meters.

$$P \text{ (diopters)} = \frac{1}{f} \text{ (in meters)} \qquad (22.10)$$

Note that the lens maker's equation (Eq. 22.9) gives a value in diopters ($1/f$) if the radii of curvature are expressed in meters.

When examining eyes for glasses, eye doctors sometimes talk about adding or subtracting so many "clicks." Each click changes the power by 1/4 diopter or the focal length by 25 cm.

Converging and diverging lenses are referred to as positive (+) and negative (−) lenses, respectively. Thus, if an optometrist prescribes a corrective lens with a power of +2 diopters, this means a converging lens with a focal length of

$$f = 1/P = 1/2 = 0.50 \text{ m} = 50 \text{ cm}$$

The greater the power of a lens in diopters, the shorter its focal length and the more converging or diverging it is.

EXAMPLE 22.9 ■ Diopters and Lens Grinding

A lens maker must fill a prescription for a corrective convexo-concave lens with a power of −2.1 diopters. The lens is to be ground from a glass blank ($n = 1.7$) having a convex front surface whose radius of curvature is 50 cm. To what radius of curvature should the second (back) surface of the lens be ground?

Solution.

Given: $P = 1/f = -2.1 \text{ diopters (m}^{-1})$ *Find:* R_2 (radius of curvature)
$n = 1.7$
$R_1 = 50 \text{ cm} = 0.50 \text{ m}$

Since the center of curvature of the convex surface is on the emergent side of the lens, R_1 is positive (see Fig. 22.20).

$$P = \frac{1}{f} = (n - 1)\left(\frac{1}{R_1} - \frac{1}{R_2}\right)$$

or

$$-2.1 = (1.7 - 1)\left(\frac{1}{0.50 \text{ m}} - \frac{1}{R_2}\right)$$

Solving for R_2 gives

$$\frac{1}{R_2} = \frac{1}{0.50 \text{ m}} + \frac{2.1}{0.7 \text{ m}} = 5.0 \text{ m}^{-1}$$

and

$$R_2 = \frac{1}{5.0 \text{ m}^{-1}} = 0.20 \text{ m} = 20 \text{ cm}$$

A positive value for R_2 indicates that the center of curvature for the second surface is on the emergent (or back) side of the lens, as is that for R_1. With $R_1 > R_2$, the lens is diverging.

Important Concepts

You should be able to define and/or explain these chapter concepts clearly.

plane mirror
virtual image
real image
lateral magnification
spherical mirror
concave (converging) mirror
convex (diverging) mirror
center of curvature
radius of curvature
focal point

focal length
parallel ray (mirror)
chief (radial) ray (mirror)
focal ray (mirror)
spherical mirror equation
magnification factor (spherical mirror)
spherical aberration (mirror)
lens
converging (biconvex) lens
diverging (biconcave) lens

parallel ray (lens)
chief ray (lens)
focal ray (lens)
thin lens equation
magnification factor (lens)
spherical aberration (lens)
astigmatism
chromatic aberration
lens maker's equation
diopters

Important Relationships for Review

These relationships are mathematical statements of the concepts and principles presented in the chapter. You should be able to identify the symbols and to explain the relationships before proceeding to the Exercises. In-text equation reference numbers are given for convenience.

Focal Length for Spherical Mirror:

$$f = \frac{R}{2} \tag{22.2}$$

Spherical Mirror Equation:

$$\frac{1}{d_o} + \frac{1}{d_i} = \frac{1}{f} \quad \text{or} \quad d_i = \frac{d_o f}{d_o - f} \tag{22.3-4}$$

Thin Lens Equation (where $f \neq R/2$):

$$\frac{1}{d_o} + \frac{1}{d_i} = \frac{1}{f} \quad \text{or} \quad d_i = \frac{d_o f}{d_o - f} \tag{22.6}$$

Magnification Factor (spherical mirror and lens):

$$M = -\frac{d_i}{d_o} \qquad \text{(22.5 and 7)}$$

Total Magnification With a Two-Lens System:

$$M_t = |M_1| \times |M_2| \qquad \text{(22.8)}$$

Sign Conventions for Spherical Mirrors and Lenses:

Concave mirror $\Big\}$ f positive   Convex mirror $\Big\}$ f negative
Biconvex lens $\Big\}$ Biconcave lens $\Big\}$

d_o is always positive.

When d_i is positive, image is real; when d_i is negative, image is virtual.

When M is positive, image is upright; when M is negative, image is inverted.

◊ *Lens Maker's Equation:*

$$\frac{1}{f} = (n - 1)\left(\frac{1}{R_1} - \frac{1}{R_2}\right) \qquad \text{(22.9)}$$

◊ *Lens Power in Diopters* (where f is in meters):

$$P = \frac{1}{f} \qquad \text{(22.10)}$$

◊ *Sign Convention for Lens Maker's Equation:*

$+ R$ when C is on side of lens from which light emerges
$- R$ when C is on side of lens on which light is incident
$R = \infty$ for a plane (flat) surface
$+ f$ converging (positive) lens
$- f$ diverging (negative) lens

Exercises

22.1 ■ Plane Mirrors

1 A plane mirror (a) can be used to magnify, (b) produces both real and virtual images, (c) always produces a virtual image, (d) forms images by diffuse reflection. (c)

● **2** A plane mirror (a) has a greater image distance than object distance, (b) produces a virtual, upright, unmagnified image, (c) changes the vertical orientation of an object, (d) reverses top and bottom. (b)

3 (a) A transparent window pane can serve as a mirror, but this is usually observed only when it is dusk or dark outside. Why? (b) When looking at a reflecting window pane at night, you may see two similar images. Why? (c) One-way mirrors reflect on one side and can be seen through on the other. (This effect is sometimes used on sunglasses.) What is the principle used in making a one-way mirror? [*Hint:* At night a windowpane may be a one-way mirror.] see ISM

4 In Example 22.1, the minimum length of a plane mirror for a person to see a complete (head-to-toe) image was derived. What effect does the person's distance from the mirror have on this result? no effect

5 ■ An object 7.5 cm tall is placed 50 cm from a plane mirror. Find (a) the distance from the object to the image, (b) the height of the image, and (c) its magnification.
(a) 1.0×10^2 cm   (b) 7.5 cm   (c) 1.0

6 ■■ A boy runs toward a plane mirror with a speed of 3.5 km/h. What is the speed of the boy's image in m/s relative to him?   1.9 m/s

● **7** ■■ A small dog sits in front of a plane mirror at a distance of 2.0 m. (a) Where is the dog's image? (b) If the dog jumps at the mirror with a speed of 0.5 m/s, how fast does it approach its image?   (a) 2.0 m behind the mirror   (b) 1.0 m/s

8 ■■ A man 1.8 m tall stands in front of a plane mirror. (a) What is the minimum height the mirror must be to allow the man to view his complete image from head to foot? Assume that his eyes are 10 cm below the top of his head.
(a) 0.90 m   (b) no

9 ■■ Show that the spherical mirror equation and magnification equation give the correct image characteristics for a plane mirror.   see ISM

10 ■■ A dance studio has plane mirrors on opposite walls, and multiple images are observed when a person stands between them. If a dancer stands 3.0 m from the mirror on the north wall and 5.0 m from the mirror on the south wall, what are the image distances for the first two images in both mirrors?   3m and 13m; 5m and 11m

11 ■■ Draw ray diagrams showing how three images of an object are formed in two plane mirrors at right angles as shown in ● Fig. 22.21a. [*Hint:* Consider rays from each end of the object arrow in the drawing for each image.] Fig. 22.21b shows a similar situation from a different point of view that gives four images. Explain the extra image in this case?
see ISM

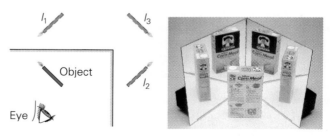

● **FIGURE 22.21 Two mirrors—multiple images**   See Exercise 11.

12 ■■■ A student who is 1.5 m tall stands in front of a plane mirror that is tilted at an angle of 20° away from her. If the student's eyes are 12 cm below the top of her head, find the minimum height the mirror must be to allow her to see her complete image (head to foot) in it. 0.70 m

13 ■■■ Find the minimum height of the mirror in Exercise 12 if it is tilted at an angle of 20° toward the student. 0.70 m

22.2 ■ Spherical Mirrors

14 A ray used in a ray diagram of a converging spherical mirror that is reflected back along the incident ray is (a) the parallel ray, (b) the chief ray, (c) the focal ray. (b)

15 A spherical concave mirror with an object at its center of curvature will produce (a) a virtual image behind the mirror at $d_i = 2f$, (b) a virtual image in front of the mirror at $d_i = 2R$, (c) a real, upright, magnified image at $d_i = f$, (d) a real, inverted, unmagnified image at $d_i = R$. (d)

16 (a) What is the purpose of using a dual mirror on a car or truck, such as the one shown in ● Fig. 22.22? (b) Some rear-view mirrors on the passenger side of automobiles have the written warning "OBJECTS IN MIRROR ARE CLOSER THAN THEY APPEAR." Explain why this is so. (c) Could a TV satellite dish as shown in Fig. 19.19 be considered a converging mirror? Explain. see ISM

● **FIGURE 22.22 Mirror applications**   See Exercise 16.

17 A popular novelty item consists of a concave mirror with a ball suspended at or slightly inside the center of curvature (see ● Fig. 22.23). When the ball swings toward the mirror, its image grows larger and suddenly fills the whole mirror. The effect is that the image appears to be jumping out of the mirror. Explain what is happening. see ISM

18 ■■ A candle with a flame 2.5 cm tall is placed 5.0 cm from the front of a concave mirror. A virtual image is produced that is 10 cm from the vertex of the mirror. (a) Find the focal length and radius of curvature of the mirror. (b) How tall is the image? (a) $f = 10$ cm, $R = 20$ cm (b) 5.0 cm

● **FIGURE 22.23 Spherical mirror toy**   See Exercise 17.

19 ■■ Show that the magnification for objects near the optic axis of a convex mirror is given by $M = d_i/d_o$. [Hint: Use a ray diagram with rays reflected at the mirror's vertex.] see ISM

● **20** ■■ An object 3.0 cm tall is placed 20 cm from the front of a concave mirror with a radius of curvature of 30 cm. Where is the image formed and how tall is it? $d_i = 60$ cm; $h_i = 9.0$ cm; image is real and inverted

21 ■■ If the object in Exercise 20 is moved to a position 10 cm from the front of the mirror, what will the characteristics of the image be? $d_i = -30$ cm; $h_i = 9.0$ cm; image is virtual, upright, and magnified.

22 A concave mirror with a radius of curvature of 40 cm reflects sunlight. Find (a) the location and (b) the diameter of the Sun's image. (a) 20 cm (b) 0.19 cm

23 The image of an object 18 cm from a convex mirror is half the size of the object. What is the focal length of the mirror? −18 cm

● **24** An object 3.0 cm tall is placed at different locations in front of a concave mirror whose radius of curvature is 30 cm. Determine the location of the image and its characteristics when the object distance is (a) 40 cm, (b) 30 cm, (c) 20 cm, (d) 15 cm, and (e) 5.0 cm. see ISM

25 Draw a ray diagram for each of the cases in Exercise 24. see ISM

26 ■■ A virtual image at a magnification of 0.50 is produced when an object is placed in front of a spherical mirror. (a) What type of mirror is it? (b) Find the radius of curvature of the mirror if the object is 7.0 cm from it. (a) convex (b) 14 cm

27 ■■ A concave shaving mirror is constructed so that a man at a distance of 20 cm from the mirror sees his image magnified 1.5 times. What is the radius of curvature of the mirror? 120 cm

28 ■■ If an object is 15 cm in front of a convex mirror that has a focal length of 30 cm, how far behind the mirror will the image appear to an observer, and how tall will it be? $d_i = -10$ cm; $M = (2/3)h_o$

29 ■■ A concave mirror has a magnification of 3.0 for an

object placed 50 cm in front of it. (a) What type of image is produced? (b) Find the radius of curvature of the mirror.
(a) virtual   (b) 150 cm

30 ■■ A dentist holds a concave mirror whose radius of curvature is 40 mm at a distance of 15 mm from a tooth to view a small cavity. How large is the image of the cavity?
$h_i = 4(h_o)$

31 ■■ A child looks at a reflecting Christmas tree ball that has a diameter of 9.0 cm and sees an image of her face that is half the real size. How far is the child's face from the ornament?   2.3 cm

● **32** ■■ A pill bottle 4.5 cm tall is placed 12 cm from the front of a mirror. An upright image 9.0 cm tall is formed. What kind of mirror is it, and what is its radius of curvature?
concave, 48 cm

33 ■■ A mirror at an amusement park shows anyone who stands 2.5 m from it an upright image three times the person's height. What is the mirror's radius of curvature?   7.5 m

34 ■■ A concave cosmetic mirror produces a virtual image 1.5 times the size of a person when his face is 20 cm from the mirror. (a) Draw a ray diagram of this situation. (b) What is the focal length of the mirror?   (b) 60 cm

35 ■■ A wooden cube 5.0 cm on a side is placed a distance of 30 cm in front of a converging mirror having focal length of 20 cm. Where is the image of the front surface of the block located and what are its characteristics? $d_i = 60$ cm; $h_i = 10$ cm; image is real, inverted, and larger

● **36** ■■ A spherical section is mirrored on both sides. If the magnification of an object is $+1.8$ when the device is used as a concave mirror, what is the magnification of an object at the same distance in front of the convex side?   0.69

37 ■■ A concave mirror has a radius of curvature of 20 cm. For what *two* object distances will the image have twice the height of the object?   5.0 cm, 15 cm

38 ■■■ (a) Sketch graphs of (1) d_i versus d_o and (2) M versus d_o for a concave mirror, for values of d_o from 0 to ∞. (b) Sketch similar graphs for a convex mirror.   see ISM

39 ■■■ A convex mirror is used on the exterior of the passenger side of many cars. If the focal length of such a mirror is -40.0 cm, what will the location and height of the image of a car that is 2.0 m tall be if the car is (a) 100 m behind and (b) 10.0 m behind? (See Exercise 16b.)   (a) $d_i = 39.8$ cm, $h_i = 0.80$ cm; $d_i = 38.5$ cm; $h_i = 7.7$ cm

40 ■■■ Two students in a physics laboratory each have a concave mirror with the same radius of curvature, 40 cm. Each student places an object in front of a mirror. The image in both mirrors is three times the size of the object. However, when the students compare notes, they find that the object distances are not the same. Is this possible? If so, what are the object distances?   yes: $d_o = 13.3$ cm; $d_o = 26.7$ cm

22.3 ■ Lenses

22.4 ■ Lens Aberrations

41 A diverging lens (a) must have at least one concave surface, (b) always produces a virtual image, (c) is thinner at the center than at the periphery, (d) all of these.   (d)

● **42** A virtual image will be formed by (a) a biconvex spherical lens with an object inside the focal point, (b) a biconvex spherical lens with the object at $d_o = 2f$, (c) a biconcave spherical lens with an object at $d_o = 2f$, (d) both (a) and (c).   (d)

43 A lens aberration that is caused by dispersion is called (a) spherical aberration, (b) chromatic aberration, (c) refractive aberration, (d) none of these.   (b)

44 How can the focal length be determined experimentally for (a) a concave mirror and (b) a biconvex lens?   see ISM

45 ■■ Find the location and magnification of the Sun's image with a convex lens whose focal length is 15 cm.
15 cm, 10^{-12}

● **46** ■■ An object 4.0 cm tall is placed in front of a converging lens whose focal length is 22 cm. Where is the image formed and what are its characteristics if the object distance is (a) 15 cm and (b) 36 cm? Sketch ray diagrams for each.
(a) $d_i = -47$ cm; $h_i = 12$ cm; image is virtual and upright
(b) $d_i = 57$ cm; $h_i = 6.3$ cm; image is real and inverted

47 ■■ An object is placed in front of a biconcave lens whose focal length is 18 cm. Where is the image located and what are its characteristics if the object distance is (a) 10 cm and (b) 25 cm? Sketch ray diagrams for each.   (a) $d_i = -6.4$ cm; $M = 0.64$ (b) $d_i = -10.5$ cm; $M = 0.42$

48 ■■ Show mathematically why application of the thin lens equation to a diverging lens always gives a virtual image.   see ISM

49 ■■ A light source and a screen are separated by a distance d, and a converging lens with a focal length f can be placed anywhere between them. Determine whether sharp images will be formed on the screen (a) when $d < 4f$ and (b) when $d > 4f$.   (a) no   (b) yes

50 ■■ A biconvex lens has a focal length of 0.12 m. Where on the lens axis should an object be placed in order to get (a) a real, enlarged image with a magnification of 2.5 and (b) a virtual, enlarged image with a magnification of 2.5?
(a) 16.8 cm   (b) 7.2 cm

51 ■■ An object 5.0 cm tall is 10 cm from a concave lens. The resulting image is $\frac{1}{5}$ as large as the object. What are the focal length and power of the lens?   (a) -2.5 cm   (b) -40 D

● **52** ■■ (a) Design a single-lens projector that will form a sharp image on a screen 4.0 m away with the transparent slides 6.0 cm from the lens. (b) If the object on a slide is 1.0 cm tall, how tall will the image on the screen be, and how should the slide be placed in the projector?   (a) 5.9 cm   (b) 67 cm, inverted

53 ■■ A single-lens camera (biconvex lens) is used to photograph a man 1.7 m tall who is standing 4.0 m from the camera. If the man's image fills the length of a frame of 35-mm film, what is the focal length of the lens? 8.2×10^{-2} m

54 ■■ A photographer uses a single-lens camera having a focal length of 60 mm to photograph a full moon. What will be the size of the moon's image on the film? [*Note:* Data on the moon may be found inside the back cover.]   0.56 mm

55 ■■ An object is placed 40 cm from a screen. At what point between the object and the screen should a converging lens with a focal length of 10 cm be placed so it will produce a sharp image on the screen and what is its magnification?
20 cm, $M = -1$

56 ■■ (a) Show that a ray parallel to the axis of a biconvex lens is refracted toward the axis at the incident surface and again at the exit surface. (b) Show that the refraction effect also holds for a biconcave lens, but with deflections away from the axis. *see ISM*

57 ■■ Using ● Fig. 22.24, derive (a) the thin lens equation and (b) the lens magnification equation. [*Hint:* Use similar triangles.] *see ISM*

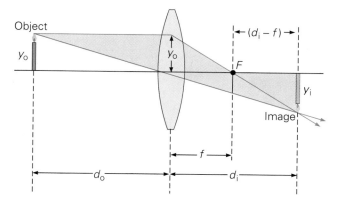

● **FIGURE 22.24 The thin lens equation** The diagram shows the geometry for deriving the thin-lens equation (and magnification factor). Note that the two triangles with vertical markings are similar, as are the two triangles with horizontal markings. See Exercise 57.

58 ■■ A stamp collector views a stamp with a magnifying glass (a convex lens). If she sees the stamp magnified by a factor of 2.5 when the glass is held 3.0 cm from it, what is the power of the lens? *20 D*

59 ■■ (a) If a book is held 30 cm from an eyeglass lens with a power of −1.75 diopters, where is the image of the print formed? (b) If an eyeglass lens with a power of +1.75 diopters is used, where is the image formed? *(a) −20 cm (b) −63 cm*

60 ■■ For the arrangement shown in ● Fig. 22.25, an object is placed 0.40 m in front of the converging lens, which has a focal length of 0.15 m. If the concave mirror has a focal length of 0.13 m, where is the final image formed and what are its characteristics? *0.26 m in front of the concave mirror, M_t = 0.60, real and upright*

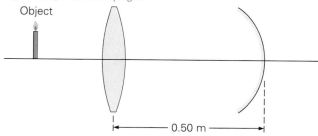

● **FIGURE 22.25 Lens-mirror combination** See Exercise 60.

● **61** ■■ A simple camera uses a single converging lens with a focal length of 0.045 m. A picture is taken of a 26-cm tall

physics book standing on a table at a distance of 1.5 m from the camera. If a sharp image is formed on the film, how far is the film from the lens? (b) What is the height of the image of the book on the film? *(a) 4.6 cm, (b) 0.80 cm*

62 ■■ Two positive lenses, each having a power of 10 diopters, are placed 20 cm apart along the same axis. If an object is 60 cm from the first lens on the side opposite the second lens, where is the final image relative to the first lens, and what are its characteristics? *20 cm on object side of L_1, virtual, inverted, M_t = 1.0*

63 ■■ The geometry of a compound microscope, which consists of two converging lens, is shown in ● Fig. 22.26. (More detail on microscopes is given in Chapter 24.) The objective lens and the eyepiece lens have focal lengths of 2.8 mm and 3.3 cm, respectively. If an object is located 3.0 mm from the objective lens, where is the final image located? *virtual image 18 cm from eyepiece*

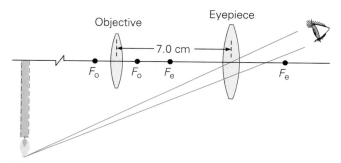

● **FIGURE 22.26 Compound microscope** See Exercise 63.

64 ■■■ Two converging lenses, L_1 and L_2, have focal lengths of 30 cm and 20 cm, respectively. The lenses are placed 60 cm apart along the same axis, and an object is placed 50 cm from L_1 on the side opposite to L_2. Where is the image formed relative to L_2, and what are its characteristics? *8.6 cm, real, inverted, M = 0.86*

65 ■■■ Show that for thin lenses with focal lengths f_1 and f_2 and in contact the effective focal length (f) is given by *see ISM*

$$\frac{1}{f} = \frac{1}{f_1} + \frac{1}{f_2}$$

22.5 ■ The Lens Maker's Equation ◇

66 The lens maker's equation (a) expresses spherical aberration, (b) applies only to lenses with equal radii of curvature, (c) shows that $f \neq R/2$ for glass lenses, (d) is the same as the thin-lens equation. *(c)*

67 The power of a lens is expressed in units of (a) watts, (b) diopters, (c) m^{-1}, (d) both (b) and (c). *(d)*

68 ■■ A symmetric biconvex lens has a focal length of 50 cm and is made of glass whose index of refraction is 1.5. What is the radius of curvature of both lens surfaces? *50 cm*

69 ■■ A biconvex lens made of glass whose index of refraction is 1.5 has radii of curvature of 30 cm for one surface and 40 cm for the other. What is the focal length of the lens for light incident on each side?   34 cm; same

● **70** ■■ An optometrist prescribes a corrective lens with a power of +1.5 diopters. The lens maker will start with glass blank with an index of refraction of 1.6 and a convex front surface whose radius of curvature is 20 cm. To what radius of curvature should the other surface be ground?   40 cm

71 ■■ A plastic plano-concave lens has a radius of curvature of 50 cm for its concave surface. If the index of refraction of the plastic is 1.35, what is the power of the lens?
−0.70 D

72 ■■■ A converging glass lens with an index of refraction of 1.62 has a focal length of 30 cm in air. What is the focal length when it is submerged in water?   64 cm

■ **Additional Exercises**

73 A person stands 3.0 m away from the reflecting surface of a plane mirror. (a) What is the apparent distance between the person and his or her image? (b) What are the image characteristics?   (a) 6.0 m   (b) upright, virtual, and same size

74 (a) For a biconvex lens, what is the minimum distance between an object and its image if the image is real? (b) What is the distance if the image is virtual? [*Hint:* Use ray diagrams and the thin lens equation.]   (a) $d_i = 4f$ (b) approaches 0

75 A bottle 6.0 cm tall is located 75 cm from the concave surface of a mirror with a radius of curvature of 50 cm. Where is the image located and what are its characteristics?
(a) 37.5 cm   (b) $h_i = 3.0$ cm; real and inverted

76 A 15-cm pencil is placed with its eraser on the optic axis of a concave mirror and its point directed upward at a distance of 20 cm in front of the mirror. The radius of curvature of the mirror is 30 cm. (a) Where is the image of the pencil formed, and what are its characteristics? (b) Draw a ray diagram for the situation where the pencil point is directed downward from the optic axis.   (a) $d_i = 60$ cm; $M = 3.0$; image is real and inverted

77 A method of determining the focal length of a diverging lens is called autocollimation. As ● Fig. 22.27 shows, first, a sharp image of a light source is projected on a screen by a converging lens. Second, the screen is replaced with a plane mirror. Third, a diverging lens is placed between the converging lens and the mirror. Light will then be reflected by the mirror back through the compound lens system, and an image will be formed on a screen near the light source. This image is made sharp by adjusting the distance between the diverging lens and the mirror. The distance at which the image is clearest is equal to the focal length of the lens. Explain why this is true.   see ISM

78 A biconvex lens produces a real, inverted image of an object that is magnified 2.5 times when the object is 20 cm from the lens. What is the focal length of the lens?   14 cm

79 A converging mirror with a radius of curvature of 90 cm produces an upright, virtual image 45 cm tall and located

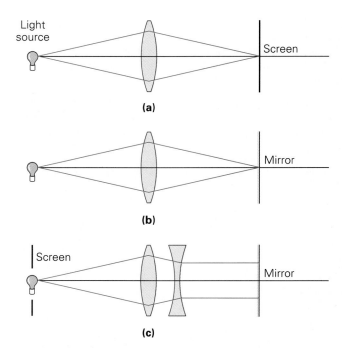

(a)

(b)

(c)

● **FIGURE 22.27 Autocollimation**   See Exercise 77.

30 cm from the vertex of the mirror. What are the height and location of the object? $h_o = 26$ cm; $d_o = 18$ cm

80 Show that $f = R/2$ for a converging mirror. [*Hint:* Draw a ray diagram for the situation $R > d_o > f$ using all three types of rays, as in Fig. 22.9a. Look for similar triangles.]   see ISM

81 (a) Sketch graphs of (1) d_i versus d_o and (2) M versus d_o for a converging lens, for values of d_o from 0 to ∞. (b) Sketch similar graphs for a diverging lens.   see ISM

82 A dentist uses a spherical mirror that produces an upright image that is magnified four times. What is the focal length of the mirror in terms of the object distance?
$f = (4/3)d_o$

83 An object is 15 cm from a converging lens whose focal length is 10 cm. On the opposite side of that lens at a distance of 60 cm from the object is another converging lens with a focal length of 20 cm. Where is the final image formed, and what are its characteristics?   60 cm on image side of L_2, real, upright, $M_t = 4.0$

84 An object 3.0 cm tall is placed 15 cm from the front of a convex mirror whose focal length is 25 cm. Find the location of the image and its height. $h_i = 1.9$ cm; $d_i = -9.4$ cm

85 A lens with a power of +5.0 diopters is used to produce an image on a screen that is 2.0 m from the lens. How many times is the object magnified?   9.0

86 The image of an object located 30 cm from a concave mirror is formed on a screen located 20 cm from the mirror. What is the mirror's radius of curvature?   24 cm

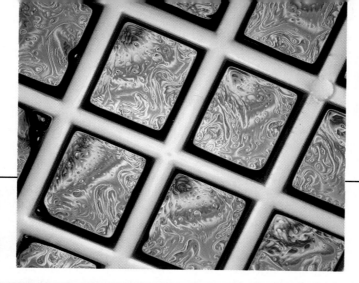

23

Physical Optics: The Wave Nature of Light

INSIGHTS

- **Nonreflecting Lenses**
- **LCD's and Polarized Light**

The phenomena of reflection and refraction are conveniently analyzed using geometrical optics. Ray diagrams show what happens when light is reflected from a mirror or passed through a lens. However, some other phenomena involving light, such as the interference patterns in the photo above, cannot be adequately explained or described using rays, since this technique ignores the wave nature of light. Other such phenomena include diffraction and polarization.

Topics of study in this chapter include Young's double-slit experiment, interference of thin films, diffraction, polarization, and atmospheric scattering of light.

   Physical optics, *or* **wave optics***, takes into account wave motion and characteristics. The wave theory of light leads to satisfactory explanations of those phenomena that cannot be analyzed with rays. Thus, this chapter again considers waveforms.*

23.1 ■ Young's Double-Slit Experiment

It has been stated in this book that light is a wave, but no proof of this has been given. How would you go about demonstrating the wave nature of light? One method that involves the use of interference was first carried out in 1801 by the English scientist Thomas Young (1773–1829). **Young's double-slit experiment** not only demonstrated the wave nature of light but also allowed him to measure its wavelengths.

   Recall from the discussion of wave interference in Chapters 13 and 14 that superimposed waves may interfere constructively or destructively. Total constructive interference occurs when two crests are superimposed, and total destructive interference occurs when a crest and a trough of two identical waves are superimposed. Interference can be observed with water waves

Interference is explained in Sections 13.4 and 14.4

Demonstrate interference of waves with a ripple tank.

FIGURE 23.1 Water wave interference The constructive and destructive interference of water waves from two coherent sources in a ripple tank produce interference patterns.

FIGURE 23.2 Double-slit interference (a) The coherent waves from two slits spread out and interfere, producing alternate maxima and minima, or bright and dark fringes, on the screen. (b) An actual interference pattern. Note the symmetry of the pattern about the central maxima ($n = 0$).

(● Fig. 23.1; compare with Fig. 14.7a), for which constructive and destructive interference produce obvious interference patterns.

The interference of light waves is not as easily observed because of their relatively short wavelengths ($\cong 10^{-7}$ m). Also, stationary interference patterns are produced with *coherent sources*, that is, sources that produce light waves having a constant phase relationship to one another. For example, for constructive interference to occur at some point, the waves meeting at that point must be in phase. As the waves meet, a crest must always overlap a crest and a trough must always overlap a trough. If a phase difference develops between the waves with time, the interference pattern changes, and a stable or stationary pattern will not be set up.

In an ordinary light source, the atoms are excited randomly, and the emitted light waves fluctuate in amplitude and frequency. Thus, light from two such sources is incoherent and does not produce a stationary interference pattern. Interference does occur, but the phase difference between the interfering waves changes so fast that the interference effects are not discernible.

To solve this problem, Young used light from a single source (sunlight) to illuminate two narrow, closely spaced slits (● Fig. 23.2a). The slits act as two sources, but light emerging from the slits is coherent because the slits merely separate the original beam into two parts. Any random changes in the light from the source will thus occur for the light passing through both slits, and the phase difference will be constant.

Demonstration/activity: Construct two identical transparencies of concentric circles representing waves radiating from a point source. Overlaying the transparencies on an overhead projector shows how waves from two point sources interfere with each other.

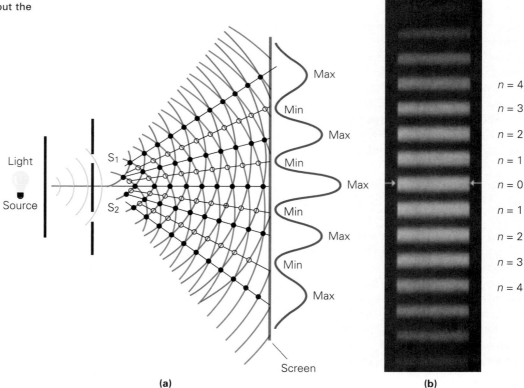

(a)

(b)

A popular theory, dating back to the ancient Greeks, considered light to be composed of "corpuscles" or particles. This theory survived relatively unquestioned for many centuries. Over the years, however, evidence to support a wave nature of light began to accumulate. Leonardo da Vinci (circa 1500), noting the similarity between sound echoes and the reflection of light, speculated that light might have a wave nature. By the 17th century a definite controversy existed over the nature of light. Isaac Newton favored a theory that treated light as a stream of particles.

If light consisted of a stream of tiny particles, then in Young's experiment one would expect to see two bright lines on a screen placed behind the slits. Instead, Young observed a *series* of bright lines (Fig. 23.2b), and he was able to explain this pattern in terms of *wave interference*.

To help analyze what Young observed, let's imagine that light with a single wavelength (monochromatic) is used. Because of diffraction, or the bending of the light around the corners of the slits, the waves spread out and interfere as illustrated in the figure. Coming from two coherent "sources," the interfering waves produce a stable interference pattern on the screen. The pattern consists of a bright central maximum (●Fig. 23.3a) and a series of symmetrical dark (Fig. 23.3b) and bright (Fig. 23.3c) side fringes, which mark the positions at which destructive and constructive interferences occur. The bright fringes decrease in intensity on either side of this central maximum. Hence, the interference pattern demonstrates the wave nature of light.

Measuring the wavelength of light requires looking at the geometry of Young's experiment, as shown in ●Fig. 23.4. Let the screen be a distance L from the slits, and P be a point at the center of an arbitrary maximum, or bright side fringe. P is located a distance y from the center of the central maximum and at an angle θ relative to a normal line between the slits. The slits S_1 and S_2 are separated by a distance d. Note that the light path from one slit to P is longer than the path from the other slit to P. As the figure shows, the path difference (Δ) is

$$\Delta = d \sin \theta$$

where the angle in the small shaded triangle is θ by similar triangles.

The relationship of the phase difference of two waves to their path difference was discussed in Chapter 14 for the interference of sound waves. (There the path difference was abbreviated PD.) The conditions for interference given for sound waves hold for any sinusoidal waves, including light waves. Recall that constructive interference occurs at any point where the path difference between two coherent, in-phase waves is an integral number of wavelengths.

$$\Delta = n\lambda \qquad \text{for } n = 0, 1, 2, 3, \dots \tag{23.1}$$

condition for constructive interference

Similarly, for destructive interference, the path difference is an odd number of half-wavelengths.

$$\Delta = \frac{m\lambda}{2} \qquad \text{for } m = 1, 3, 5, \dots \tag{23.2}$$

condition for destructive interference

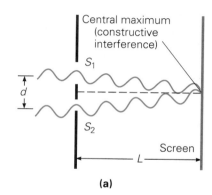

(a)

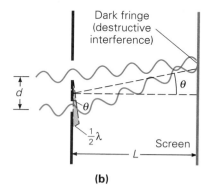

(b)

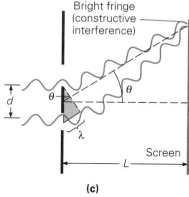

(c)

●**FIGURE 23.3 Interference** The interference that produces bright or dark fringes depends on the difference in the path lengths of the light from the two slits. **(a)** The path difference at the position of the central maximum is zero, so the waves arrive in phase and interfere constructively. **(b)** At the position of the first dark fringe, the path difference is $\lambda/2$, and the waves interfere destructively. **(c)** At the position of the first bright fringe, the path difference is λ, and the interference is constructive.

Figure 23.3

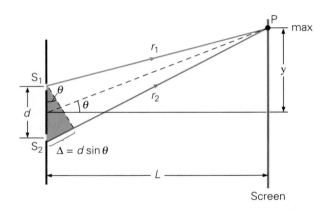

• **FIGURE 23.4 Geometry of Young's double-slit experiment** The difference in the path lengths for light from the two slits traveling to a position P of an arbitrary maximum, or bright fringe, is $r_2 - r_1 = d \sin \theta$. (The angle θ in the small shaded triangle can be shown geometrically to be the same as that formed by the dashed line to P with the horizontal.) By the condition for constructive interference, $d \sin \theta = n\lambda$, where $n = 0, 1, 2, 3, \ldots$ is the order of interference. (Drawing not to scale, $y \ll L$.)

Figure 23.4

Thus, in Fig. 23.4, point P will be the location of a bright fringe (constructive interference) if

$$d \sin \theta = n\lambda \qquad \text{for } n = 0, 1, 2, 3, \ldots \qquad (23.3)$$

where n is called the order number. The zeroth-order fringe ($n = 0$) corresponds to the central maximum, the first-order fringe ($n = 1$) is the first bright fringe on either side of the central maximum, and so on. Basically, varying path differences give rise to different phase differences at different points.

The wavelength can therefore be determined by measuring d and θ for a particular order bright fringe (other than the central maximum). The angle θ is related to the displacement of a fringe (y), which is conveniently measured from a photograph of the interference pattern. If P is a point at the center of a bright fringe, the distance from the center of the central maximum to that nth fringe is (see Fig. 23.4)

$$y = L \tan \theta = L\left(\frac{\sin \theta}{\cos \theta}\right) \qquad (23.4)$$

If θ is small ($y \ll L$), $\cos \theta \cong 1$ and $\tan \theta$ may be approximated by $\sin \theta$. Therefore,

$$y \cong L \sin \theta \qquad \text{or} \qquad \sin \theta \cong \frac{y}{L} \qquad (23.5)$$

Substituting this expression for $\sin \theta$ in Eq. 23.3 and then solving for y gives a good approximation of the distance of the nth bright fringe (y_n) from the central maximum on either side.

$$y_n \cong \frac{nL\lambda}{d} \qquad \text{(for small } \theta \text{ only)} \qquad (23.6)$$

lateral distance to bright fringe

Thus, the wavelength of the light is

Wavelength of light in Young's experiment from geometrical analyses

$$\lambda \cong \frac{y_n d}{nL} \qquad \text{for } n = 1, 2, 3, \ldots \qquad (23.7)$$

A similar analysis gives the distances to the dark fringes (see Exercise 13).

From Eq. 23.3, we see that, except for the zeroth order fringe, $n = 0$ (the central maximum), the positions of the fringes depend on wavelength— different values of λ give different values of $\sin \theta$, and therefore of θ. Hence, when the experimenter uses white light, as Young did, the central fringe is white, but the other orders contain a spectrum of colors similar to that formed by a prism. By measuring the positions of the color fringes within a particular order, Young was able to determine the wavelengths of the colors of visible light and thus show that physiological perception of color is related to a physical quantity: the wavelength of light.

You may wish to discuss the effect of the intensity of light on the order number.

EXAMPLE 23.1 ■ Measuring the Wavelength of Light

Monochromatic light passes through two narrow slits that are 0.050 mm apart. A picture of the interference pattern is taken with a camera 1.0 m from the slits, and the second-order bright fringe is measured as being 2.4 cm from the center of the central maximum on the photograph. (a) What is the wavelength of the light? (b) What is the distance between the second-order and third-order bright fringes?

Solution. From the problem we have

Given: $d = 0.050\,\text{mm}$
$L = 1.0\,\text{m} = 10^3\,\text{mm}$
$y_2 = 2.4\,\text{cm} = 24\,\text{mm}$
$n = 2$

Find: (a) λ (wavelength)
(b) $y_3 - y_2$ (distance between $n = 2$ and $n = 3$)

(a) Using Eq. 23.7 with distances in millimeters for convenience gives

$$\lambda = \frac{y_n d}{nL} = \frac{(24\text{ mm})(0.050\text{ mm})}{2(10^3\text{ mm})} = 6.0 \times 10^{-4}\text{ mm} = 6.0 \times 10^{-7}\text{ m}$$

This is 600 nm, which is the wavelength of yellow-orange light (see Fig. 19.16). (b) With the wavelength, y_3 could be computed, and then the distance between the second-order and third-order fringes ($y_3 - y_2$) could be found. However, the bright fringes for a given wavelength of light are evenly spaced; in general, the distance between adjacent bright fringes is constant:

$$y_{n+1} - y_n = \frac{(n + 1)L\lambda}{d} - \frac{nL\lambda}{d} = \frac{L\lambda}{d}$$

In this case,

$$\frac{L\lambda}{d} = \frac{(1.0\text{ m})(6.0 \times 10^{-7}\text{ m})}{0.050 \times 10^{-3}\text{ m}} = 1.2 \times 10^{-2}\text{ m} = 1.2\text{ cm}$$

23.2 ■ Thin-Film Interference

Have you ever wondered what causes the rainbowlike colors when light is reflected from a thin film of oil or a soap bubble? This effect is due to interference

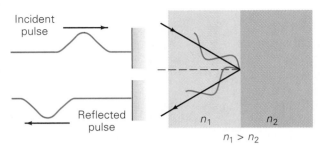

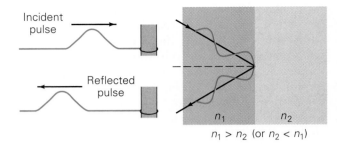

(a) Fixed end: 180° phase shift

(b) Free end: zero phase shift

● **FIGURE 23.5 Reflection and phase shifts** The phase changes that light waves undergo on reflection are analogous to those for pulses in strings. **(a)** The phase of a pulse in a string is shifted by 180° on reflection from a fixed end, and so is the phase of a light wave when it is reflected from a more optically dense medium. **(b)** A pulse in a string has a phase shift of zero (is not shifted) when reflected from a free end. Analogously, a light wave is not phase-shifted when reflected from a less optically dense medium.

● **FIGURE 23.6 Thin-film interference** For an oil film on water, there is a 180° phase shift for light reflected from the air-oil interface and a zero phase shift at the oil-water interface. **(a)** Destructive interference occurs if the oil film has a thickness of $\lambda'/2$. **(b)** Constructive interference occurs with a minimum film thickness of $\lambda'/4$. **(c)** Thin-film interference in an oil slick. Different film thicknesses give rise to the reflections of different colors.

of light reflected from opposite surfaces of the film and may be readily understood in terms of wave interference.

First, however, you need to see how the phase of a light wave is affected by reflection. Recall from Chapter 13 that a wave pulse undergoes a 180° phase shift (change) when reflected from a rigid support and no phase shift when reflected from a free support (● Fig. 23.5). Similarly, as the figure shows, the phase change for the reflection of light waves at a boundary depends on the optical densities, or the indices of refraction, of the two materials.

- A light wave traveling in one medium and reflected from the boundary of a second medium whose index of refraction is greater than that of the first medium ($n_2 > n_1$) undergoes a 180° phase change.

- If the reflecting medium has the smaller index of refraction ($n_2 < n_1$), there is no phase change.

Figure 23.6

To understand why you see colors in an oil film (on water or on a wet road), consider the reflection of monochromatic light from a thin film, as illustrated in ● Fig. 23.6. The path length of the wave in the film depends on the angle of incidence (why?), but for simplicity we will assume normal incidence for the light, even though the rays are drawn at an angle in the figure for greater clarity of illustration.

The oil film has a greater index of refraction than air, and the light reflected from the air-oil interface undergoes a 180° phase shift. The transmitted waves

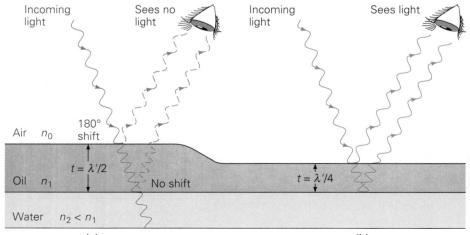

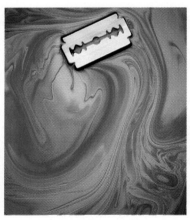

(a)

(b)

(c)

pass through the oil film and are reflected at the oil-water interface. In general, the index of refraction of oil is greater than that of water (see Table 21.1); that is, $n_2 < n_1$, so a reflected wave in this instance is *not* phase shifted.

You might think that if the path length of the wave in the oil film ($2t$, twice the thickness, down and back) were an integral number of wavelengths [$2t = 2(\lambda'/2) = \lambda'$ in figure], then the waves reflected from the two surfaces would interfere constructively. But keep in mind that the wave reflected from the top surface undergoes a 180° phase shift. The reflected waves from the two surfaces are therefore out of phase for this film thickness, and they interfere destructively. This means that the light of this wavelength is not reflected, but transmitted completely.

Similarly, if the path length in the film were an odd number of half-wavelengths [$2t = 2(\lambda'/4) = \lambda'/2$ in figure], the reflected waves would be in phase (as a result of a 180° phase shift of the incident wave), and they would interfere constructively. Light of this wavelength would be reflected from the oil film.

Keep in mind that the path length is expressed in terms of the wavelength of light *in the film*, λ'. Recall from Chapter 21 that the wavelength is different in different media. Here, $\lambda' = \lambda/n$, where λ' is the wavelength in the oil, λ the wavelength in air, and n the index of refraction of the oil. This may be seen from $n = c/v = f\lambda/f\lambda' = \lambda/\lambda'$.

Because oil and soap films generally have different thicknesses in different regions, particular wavelengths (colors) of white light interfere constructively in different regions and are reflected (see Demonstration 17). As a result, a vivid display of various colors appears, which may change if the film thickness changes with time. (A similar display of colors is seen if two glass slides are stuck together with an air film between them.)

Demonstration/activity: Blow soap bubbles and display them in light from an incandescent source. Set up a lens and light source to project reflected light from a soap film onto a large screen. A 1-in diameter ring will hold a soap film long enough to observe the thickness change as the liquid is pulled downward. Notice that just before the film breaks, the reflected light disappears. Why?

The wavelength of light in different media is discussed in Section 21.3. See also Table 21.1.

Demonstration/activity: View thin-film interference caused by oil on water. Also view the interference in both monochromatic and polychromatic light.

DEMONSTRATION 17 ■ Thin-Film Interference

A vivid display of colors from soap bubbles illuminated with white light from below.

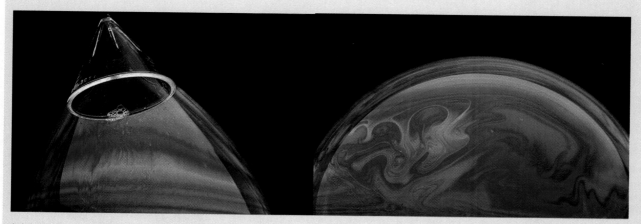

(a) Interference of light reflected from the inner and outer soap-bubble surfaces produces the colors. As the film thickness increases from top to bottom, the spectrum of colors is repeated several times.

(b) Swirling colors may be produced by blowing gently tangent to the film.

From this analysis of thin-film interference, you can see that the word "destructive" in this context does not imply that energy is destroyed. Destructive interference is simply a description of a physical fact—that a light wave is not present at a particular location, but is somewhere else. When destructive interference occurs for light incident on the surfaces of a thin film, this tells you that the light is not reflected, but that the incident beam (energy) is transmitted. Similarly, the mathematical description of Young's double-slit experiment in the preceding section tells you that you *should not expect* to find light at the locations of the dark fringes. The light is at the bright fringes, and the interference analysis shows that the light is merely redistributed.

A practical application of thin-film interference is described in the Insight feature dealing with nonreflective coatings for lenses. The situation discussed

INSIGHT ■ Nonreflecting Lenses

You have probably noticed the blue-purple tint of the coated optical lenses used in cameras and binoculars. The coating makes the lenses "nonreflecting." If a lens is nonreflecting, the incident light is totally transmitted. Complete transmission of light is desirable for the exposing of photographic film and for viewing objects with binoculars.

A lens is made nonreflecting by coating it with a thin film of a material that has an index of refraction between those of air and glass (Fig. 1). If the coating is a quarter-wavelength thick, the difference in path length between the reflected rays is $\lambda'/2$, where λ' is the wavelength of light in the coating. In this case, both reflected waves undergo a 180° phase shift, and they are out of phase for a path difference of $\lambda'/2$ and interfere destructively. That is, the incident light is transmitted, and the coated lens is nonreflecting.

FIGURE 2 **Coated lenses** The nonreflective coating on binocular and camera lenses generally produces a characteristic bluish-purple hue.

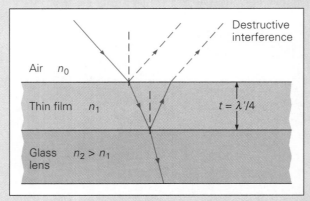

FIGURE 1 **Nonreflective coating** For a thin film on a glass lens, there is a 180° phase shift at each interface when the index of refraction of the film is less than that of the glass. As a result, destructive interference occurs for a minimum film thickness of $\lambda'/4$, and the waves are transmitted rather than reflected, making the lens surface nonreflecting.

A quarter-wavelength thickness of film is, of course, specific for a particular wavelength of light. The thickness is usually chosen to be a quarter-wavelength of yellow-green light ($\lambda \cong 550$ nm), to which the human eye is most sensitive. The wavelengths at the red and blue ends of the visible region are reflected, giving the coated lens its bluish-purple hue (Fig. 2). Sometimes other quarter-wavelength thicknesses are chosen, giving rise to other observed hues, such as amber or reddish purple.

The coating on a nonreflecting lens actually serves a double purpose. It promotes nonreflection from the front of the lens and cuts down on back reflection. Some of the light transmitted through a lens is reflected from the back surface. This could be reflected again from the front surface of an uncoated lens, giving rise to poor images. However, the proper film thickness allows back reflections to be transmitted through the coating.

Nonreflective coatings are also applied to the surfaces of solar cells, which convert light into electrical energy (Chapter 26). Since the thickness of such a coating is wavelength-dependent, like that on a nonreflecting lens, not all of the light is transmitted. However, the reflective losses may be decreased from around 30% to 10%, making the cell more efficient.

there differs from that of the previous oil-water example, because glass has a greater index of refraction than the nonreflecting film. Consequently, phase shifts of incident light take place at both the film and glass surfaces. In such a case, the condition for destructive interference is that the path difference for normally incident light is $\lambda'/2$, where λ' is the wavelength of the light in the film. Thus, $2t = \lambda'/2$, and the film thickness that will produce the least reflection is

$$t = \frac{\lambda'}{4}$$

The wavelength of the light in the film (λ') is related to that in air (λ) by

$$\lambda' = \frac{\lambda}{n}$$

Substituting for λ' in the equation for t gives

$$\boxed{t = \frac{\lambda}{4n}} \quad \text{minimum film thickness} \quad (23.8) \quad \textbf{Nonreflecting film thickness}$$

Demonstration/activity: Open a new box of microscope slides. Take out two slides at a time and pass them to students. Have the students press the two slides together and observe the reflection from the room light. If the slides are very clean, the air film is thin enough to show the phase reversal.

EXAMPLE 23.2 ■ Nonreflecting Coatings

A glass lens ($n = 1.6$) is coated with a thin transparent film of magnesium fluoride (MgF_2, $n = 1.38$) to make it nonreflecting. What is the minimum thickness of the film for the lens to be nonreflecting for incident light whose wavelength is 550 nm?

Solution.

Given: $n = 1.38$ (for film)   **Find:** t (film thickness)
       $\lambda = 550$ nm

Using Eq. 23.8,

$$t = \frac{\lambda}{4n} = \frac{(550 \text{ nm})}{4(1.38)} = 99.6 \text{ nm}$$

which is quite thin ($\approx 10^{-5}$ cm).

Optical Flats and Newton's Rings

The phenomenon of thin-film interference is used to check the smoothness and uniformity of optical components such as mirrors and lenses. **Optical flats** are made by grinding and polishing glass plates until they are as flat and smooth as possible. The degree of flatness may be checked by putting two such plates together so there is a thin air wedge between them (● Fig. 23.7). If the plates are smooth and flat, a regular interference pattern of bright and dark fringes, or bands, appears. This pattern is a result of the uniformly varying differences in path lengths between the plates. Any irregularity in the pattern indicates

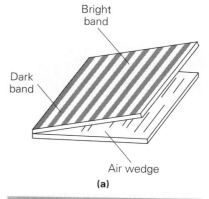

(a)

(b)

● **FIGURE 23.7 Optical flatness**
(a) An optical flat is used to check the smoothness of a reflecting surface. The flat is placed so that there is an air wedge between it and the surface. **(b)** If the surface is smooth, a regular or symmetric interference pattern is seen.

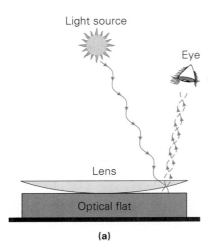

(a)

(b)

●**FIGURE 23.8 Newton's rings** **(a)** A lens placed on an optical flat forms a ring-shaped air wedge, which gives rise to interference. The interference pattern is a set of concentric rings called Newton's rings. Lens irregularities produce a distorted pattern. **(b)** The air film between two microscope slides in contact produces a Newton's ring pattern.

an irregularity in at least one of the plates. Once a good optical flat is verified, it can be used to check the flatness of a reflecting surface, such as that of a precision mirror.

A similar technique is used to check the smoothness and symmetry of lenses (●Fig. 23.8). When a curved lens is placed on an optical flat, the air wedge is a ring (below the periphery of the circular lens). The regular interference pattern in this case is a set of concentric bright and dark circular fringes. They are called **Newton's rings**, after Isaac Newton, who first described this interference effect. Lens irregularities give rise to a distorted fringe pattern.

23.3 ■ Diffraction

In geometrical optics, light is represented by rays and pictured as traveling in straight lines. If this model represented the real nature of light, however, there would be no interference effects in Young's double-slit experiment. Instead, there would be only two bright slit images on the screen with a well-defined shadow area (●Fig. 23.9a). In fact, there are interference patterns, which means that the light must deviate from a straight-line path and enter the regions that would otherwise be in shadow. Huygens' principle requires that the waves spread out from the slits (Fig. 23.9b), and this deviation, or bending, of light is called **diffraction**.

Diffraction generally occurs when waves pass through small openings or around sharp edges or corners. The diffraction of sound (Chapter 14) is quite evident. Someone can talk to you from another room or around the corner of a building, and even in the absence of reflections, you can easily hear them from your position in the acoustical shadow, so to speak. Diffraction phenomena for light waves often go unnoticed. However, careful observation will reveal that

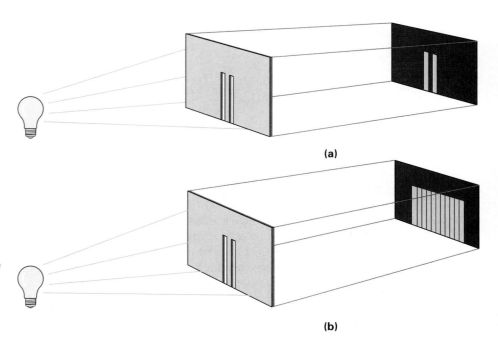

(a)

(b)

●**FIGURE 23.9 Slit diffraction** **(a)** If there were no diffraction of light, two bright slit images would be observed on a screen in Young's double-slit experiment. **(b)** Because an interference pattern does occur, the light must be diffracted, or bent, on passing through the slits.

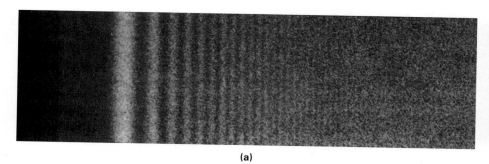

(a)

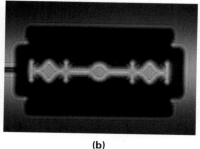

(b)

FIGURE 23.10 Diffraction in action (a) Fringes at a shadowy boundary. (b) Diffraction patterns produced by a razor blade.

a shadow boundary is blurred or fuzzy. On close inspection, you can see that there is a pattern of bright and dark fringes (Fig. 23.10). These interference patterns are evidence of the diffraction of the light around the edge of the object.

An illustrative example of diffraction is that with a single slit (Fig. 23.11). Suppose that a single slit of width w is illuminated using monochromatic light whose wavelength (λ) is much smaller than the slit width. An interference pattern consisting of a bright central maximum and a symmetrical array of bright fringes (regions of constructive interference) on both sides is observed on a screen at a distance L from the slit (where $L \ggg w$).

Because the width of the slit is very much greater than the wavelength of the light, the slit with light passing through cannot be treated as a point source of Huygens' wavelets. However, various points on the wave front passing through the slit may be considered to be such point sources. Then, the interference of those wavelets can be analyzed much as double-slit interference was analyzed earlier. The analysis will not be done here, but you will notice that the stated result is very similar in form to that for Young's double slit. However,

Demonstration/activity: Place a pin or some other small object in the path of laser light to produce a diffraction pattern. Compare this with the diffraction pattern produced by shining laser light through a small slit.

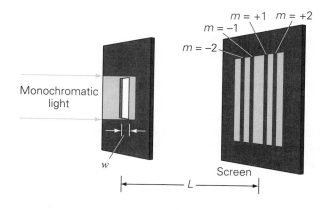

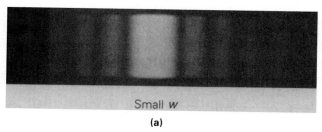

Small w

(a)

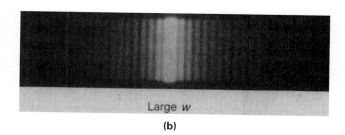

Large w

(b)

FIGURE 23.11 Single-slit diffraction (a) The diffraction of light by a single slit gives rise to an interference pattern consisting of a large bright central maximum and a symmetric array of side fringes. (b) The widths of the fringes depend on the width of the slit—the larger the slit width, the smaller the fringe widths.

for the single slit, dark fringes (regions of destructive interference) are analyzed, rather than bright fringes. From the geometry, the condition for *destructive interference*, or interference minima (dark fringes) can be shown to be

$$w \sin \theta = m\lambda \qquad \text{for } m = 1, 2, 3, \ldots \qquad (23.9)$$

where θ is the angle of a particular minimum designated by $m = 1, 2, 3, \ldots$ on either side of the central bright fringe. (There is no $m = 0$. Why?)

Also, a small-angle approximation, $\sin \theta \cong y/L$, can be made in some instances in obtaining a formula for the lateral displacement of a particular interference minimum on the screen. This gives a good approximation of the distances of the dark fringes on either side of the center of the central maximum.

> Compare the expression for a single-slit diffraction pattern with that for a double-slit interference pattern. Students often confuse these two patterns.

$$y_m = \frac{mL\lambda}{w} \qquad \text{for } m = 1, 2, 3, \ldots \qquad (23.10)$$

Note that Eq. 23.10 has the same form as Eq. 23.6, the equation for the displacement of bright fringes for double-slit interference. Here w is the width of the slit, and there d is the distance between the slits.

The qualitative predictions from Eq. 23.10 are quite interesting and instructive.

- For a given slit width (w), the greater the wavelength (λ), the wider the diffraction pattern.
- For a given wavelength (λ), the narrower the slit width (w), the wider the diffraction pattern.
- The width of the central maximum is twice the width of the side maxima.

Let's consider each of these.

As the slit is made narrower, the central maximum and the side fringes spread out and become larger (Fig. 23.11b). Eq. 23.10 is not applicable to very small slit widths (because of the small-angle approximation). If the slit is decreased until it is the same order of magnitude as the wavelength of the light, the central maximum spreads out over the whole screen. That is, diffraction effects are greater when the slit width (or object) is about the same size as the wavelength.

Thus, diffraction effects are easily observed when $\lambda/w \geq 1$, or $\lambda \geq w$. For example, FM radio waves (88–108 MHz, or $\lambda = 2.8$–3.4 m) tend to travel in straight lines and may be blocked by large objects such as hills or buildings. However, AM radio waves (525–1610 kHz, or $\lambda = 186$–570 m) may be diffracted around objects and received in the shadow areas.

Conversely, if the slit is made wider when using a particular wavelength of light, the diffraction pattern becomes narrower. The fringes move closer together and eventually become difficult to distinguish when λ is much smaller than w ($\lambda \ll w$). The pattern then appears as a fuzzy shadow around the central maximum, which is the illuminated image of the slit. This type of pattern is observed for the image produced by sunlight entering a dark room through a hole in a curtain. Such an observation led early experimenters to investigate the wave nature of light, and the acceptance of this theory was, in large part, due to the explanation of diffraction offered by physical optics.

Notice from Figs. 23.2 and 23.10 that the central maximum differs in width from the side maxima. The central maximum turns out to be twice as wide, which can be shown as follows. Taking the width of the central maximum to be the distance between the bounding minima on each side ($m = 1$), or a width of $2y_1$, we have from Eq. 23.10 with $y_1 = L\lambda/w$:

$$2y_1 = \frac{2L\lambda}{w} \qquad (23.11)$$

width of central maximum

Similarly, the width of the bright side fringes is given by

$$y_2 - y_1 = y_3 - y_2 = \frac{L\lambda}{w}$$

Or, in general,

$$y_{m+1} - y_m = \frac{L\lambda}{w} \qquad (23.12)$$

width of central maximum

Thus, the width of the central maximum is twice that of the side fringes.

EXAMPLE 23.3 ■ Width of the Central Maximum

Monochromatic blue light ($\lambda = 425$ nm) passes through a slit whose width is 0.50 mm. What is the width of the central maximum on a screen located 1.0 m from the slit?

Solution.

Given: $\lambda = 425$ nm $= 4.25 \times 10^{-5}$ cm Find: $2y_1$ (width of
  $w = 0.50$ mm $= 0.050$ cm   central maximum)
  $L = 1.0$ m $= 10^2$ cm

Eq. 23.10 can be used directly.

$$2y_1 = \frac{2L\lambda}{w} = \frac{2(4.25 \times 10^{-5}\text{ cm})(10^2\text{ cm})}{0.050\text{ cm}} = 0.17\text{ cm}$$

Note that diffraction effects would not be readily observed in this case since $\lambda \ll w$.

Diffraction Gratings

Bright and dark fringes result from diffraction and interference when light passes through a single slit or double slits. Fringe patterns increase as the number of slits is increased, and it is observed that the bright lines become sharper (narrower) and the dark lines broader. The intensity of the lines is less when the light has to pass through an increased number of slits, but even so, the sharp lines are useful in optical analysis of light sources and other applica-

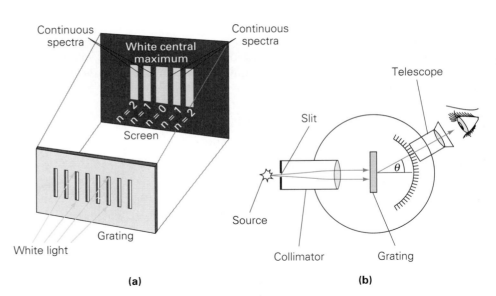

FIGURE 23.12 Diffraction grating (a) A diffraction grating produces a sharply defined interference pattern. In each of the side fringes, components of different wavelengths are separated, since the deviation depends on the wavelength: $\theta = \sin^{-1}(n\lambda/d)$. (b) As a result, gratings are used in spectrometers to determine the wavelengths present in a beam of light by measuring their angles of diffraction and to separate the various wavelengths for further analysis.

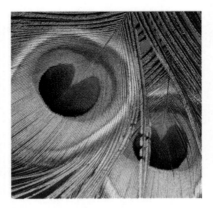

FIGURE 23.13 Feather diffraction Feathers act as diffraction gratings, producing colorful displays. Shown here are peacock feathers.

tions. As a result, arrangements of large numbers of parallel, closely spaced slits are fashioned in the form of **diffraction gratings** (Fig. 23.12).

Diffraction gratings were first made of fine strands of wire. Their effects were similar to what may be seen by viewing a candle flame through a feather held close to the eye. In fact, the brilliant, iridescent colors of many birds are the product of diffraction by their feathers (Fig. 23.13). Better gratings have a large number of fine lines or grooves on glass or metal surfaces. If light is transmitted through a grating, it is called a *transmission grating*.

However, *reflection gratings* are more common. These are made by depositing a thin film of aluminum on an optically flat surface and then removing some of the reflecting metal by cutting regularly spaced, parallel lines. Precision diffraction gratings are made using two coherent laser beams, intersecting at an angle. The beams expose a layer of photosensitive material, which is then etched. The spacing of the grating lines is determined by the intersection angle of the beams. Precision gratings may have 30,000 lines per centimeter or more and are therefore expensive and difficult to fabricate. Most gratings used in laboratory instruments are replica gratings, which are plastic castings of high-precision master gratings.

Diffraction gratings are widely used in spectroscopy (the study of spectra), where they have almost replaced prisms. The forming of a spectrum and the measurement of wavelengths by means of a grating depend only on geometrical measurements such as lengths and/or angles. Wavelength determination using a prism, on the other hand, also depends on the dispersive characteristics of the glass or other material of which the prism is made.

It can be shown that the condition for interference maxima for a grating illuminated with monochromatic light is identical to that for a double slit.

$$d \sin \theta = n\lambda \qquad \text{for } n = 0, 1, 2, 3, \ldots \tag{23.13}$$

interference maxima for a grating

Here n is the order of interference and θ is the angle of deviation of a particular wavelength. The zeroth-order maximum corresponds to the central maximum of the diffraction pattern. The spacing between adjacent slits (d) is obtained from the number of lines per unit length of the grating, $d = 1/N$. For example, if $N = 5000$ lines/cm,

$$d = \frac{1}{N} = \frac{1}{5000} = 2.0 \times 10^{-4} \, cm$$

In contrast to a prism, which deviates red light least and violet light most, a diffraction grating produces the least angle of deviation for violet light (short λ), and the greatest for red light (long λ). Also, a prism disperses white light into a single spectrum. A diffraction grating, however, produces a number of spectra, one for each order other than $n = 0$. There is no deviation of the components of the light for the zeroth order (sin $\theta = 0$ for all wavelengths), so the central maximum is a white image. However, a spectrum is produced for each higher order.

The sharp spectra produced by diffraction gratings are used in instruments called spectrometers. In a spectrometer, materials are illuminated with light of various wavelengths to find which ones are strongly absorbed. The pattern of absorption helps to identify the material. The grating is rotated so that the sample is illuminated with a succession of different wavelengths. The wavelengths may also be in the infrared and ultraviolet regions.

The number of spectral orders produced by a grating depends on the wavelength of the light (which may be infrared, visible, or ultraviolet) and on the grating's spacing (d). From Eq. 23.13, since sin θ cannot exceed 1 (sin $\theta \leq 1$),

$$\sin \theta = \frac{n\lambda}{d} \leq 1$$

The order number is therefore limited as follows:

$$n \leq \frac{d}{\lambda} \qquad\qquad (23.14) \quad \text{Limit of spectral orders}$$

EXAMPLE 23.4 ■ Grating Spacings and Spectral Orders

A particular diffraction grating produces an $n = 2$ spectral order at a deviation angle of 30° for light with a wavelength of 500 nm. (a) How many lines per centimeter does the grating have? (b) If the grating were illuminated with white light, how many orders of the *complete* visible spectrum would be produced?

Solution.

Given: $n = 2$ Find: (a) N (lines/cm)
 $\lambda = 500$ nm $= 5.0 \times 10^{-7}$ m   (b) Number of orders
 $\theta = 30°$

(a) The grating spacing may be found using Eq. 23.13:

$$d = \frac{n\lambda}{\sin \theta} = \frac{2(5.0 \times 10^{-7} \, m)}{\sin 30°}$$

$$= 20 \times 10^{-7} \, m = 2.0 \times 10^{-4} \, cm$$

Then

$$N = \frac{1}{d} = \frac{1}{2.0 \times 10^{-4}\,\text{cm}}$$

$$= 5000\,\text{lines/cm}$$

(b) The greatest angle of deviation for any order is for the longest wavelength, which is that of red light in the visible spectrum ($\lambda = 700$ nm $= 7.0 \times 10^{-7}$ m). With a spacing of $d = 2.0 \times 10^{-4}$ cm $= 2.0 \times 10^{-6}$ m, by Eq. 23.14, the order numbers for the visible spectrum for this grating are limited to

$$n \leq \frac{d}{\lambda} = \frac{2.0 \times 10^{-6}\,\text{m}}{7.0 \times 10^{-7}\,\text{m}} = 2.9$$

Thus, for a grating with this spacing, only two orders are observed, or four complete visible spectra—two on each side of the central maximum. A large portion of the spectrum is seen in the third order, but this is limited to wavelengths of

$$\lambda \leq \frac{d}{n} = \frac{2.0 \times 10^{-6}\,\text{m}}{3} = 6.7 \times 10^{-7}\,\text{m} = 670\,\text{nm}$$

Thus, the red end of the visible spectrum is not seen in the third spectral order.

Note that it is possible for spectra produced by diffraction gratings to overlap at higher orders. That is, the angles of deviation for different orders may be the same for two different wavelengths. The spacing must be greater than or equal to the wavelength ($d \geq \lambda$) to have interference for the first spectral order. However, if the spacing is much greater than the wavelength ($d \gg \lambda$), the difference in the deviation angles for nearby wavelengths is quite small. This difference increases for higher orders, but then the overlapping of spectra becomes a problem. Thus, there is an optimal spacing of the lines on a diffraction grating for each spectral region of interest—infrared, visible, or ultraviolet. The spacing is typically chosen to be between 3λ and 6λ, where λ is the median (middle) wavelength of the spectral region.

X-ray Diffraction

The wavelength of any electromagnetic wave may be determined if a diffraction grating with the appropriate spacing is available. Diffraction was used to determine the wavelengths of X-rays early in this century. Experimental evidence indicated that the wavelengths of X-rays were probably around 10^{-8} cm, but it is impossible to construct a diffraction grating with this spacing. Around 1913, Max von Laue, a German physicist, suggested that the regular spacing of the atoms in a crystalline solid might make it act as a diffraction grating for X-rays, since the atomic spacing in crystals is on the order of 10^{-8} cm. X-rays were directed at crystals, and diffraction patterns were observed.

● Figure 23.14a illustrates diffraction by the planes of atoms in a crystal such as sodium chloride. You can see that the path difference is $2d \sin \theta$, where d is the distance between the crystal's internal planes. Thus, the condition for

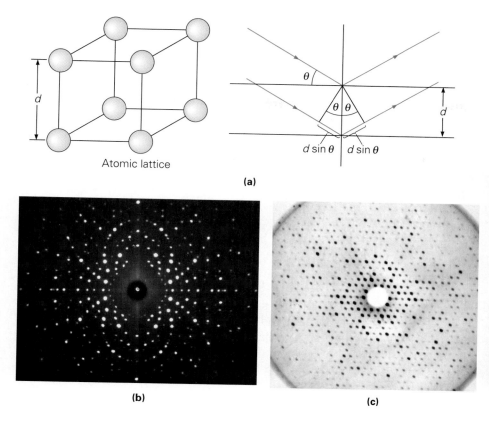

Atomic lattice

$d \sin \theta$ | $d \sin \theta$

(a)

(b)

(c)

(d)

constructive interference is

$$2d \sin \theta = n\lambda \qquad \text{for } n = 1, 2, 3, \ldots \qquad (23.15)$$

This relationship is known as **Bragg's law**, after W. L. Bragg, the British physicist who first derived it.

The wavelengths of X-rays were experimentally determined by this means, and X-ray diffraction is now used to investigate the internal structure, not only of simple crystals, but of large, complex biological molecules such as proteins and DNA. Because of their short wavelengths, X-rays provide a diffraction "probe" for investigating the interatomic spacings of molecules.

23.4 ■ Polarization

When you think of polarized light, you may visualize polarizing (or Polaroid) sunglasses, since this is one of the more common applications of polarization. When something is polarized, this means that it has a preferential direction, or orientation (think of a polarized electrical plug, as described in Chapter 17). In terms of light waves, **polarization** refers to the orientation of the transverse oscillations.

Recall that light is an electromagnetic wave with oscillating electric and magnetic field vectors (**E** and **B**, respectively) perpendicular to the direction of propagation (Chapter 19). Light from most sources consists of a large number

FIGURE 23.14 Crystal diffraction (a) The array of atoms in a crystal lattice structure acts like a diffraction grating, and X-rays are diffracted from the planes of atoms. With a lattice spacing of d, the path difference for the X-rays diffracted from adjacent planes is $2d \sin \theta$. (b) X-ray diffraction pattern of a crystal of potassium sulfate. By analyzing the geometry of such patterns, investigators can deduce the structure of the crystal and the position of its various atoms. (c) X-ray diffraction pattern of the protein hemoglobin that carries oxygen in the blood. (d) Equipment like this makes it possible to orient a crystal at any angle in the incident X-ray beam and to measure precisely the intensity of the radiation diffracted in any direction (the "spots" on the film in parts b and c).

Polarization proves that light is a transverse wave.

(a) Unpolarized

(b) Partially polarized

$\uparrow E$

(c) Plane (linearly) polarized

FIGURE 23.15 Polarization
Polarization is represented by the orientation of the plane of vibration of the electric field vectors. **(a)** When the vectors are randomly oriented (as viewed along the direction of propagation), the light is unpolarized. **(b)** With preferential orientation of the vectors, the light is partially polarized. **(c)** When the vectors are in a plane, the light is plane polarized, or linearly polarized.

of electromagnetic waves emitted by the atoms of the source. Each atom produces a wave with a particular orientation, corresponding to the direction of the atomic vibration. However, since electromagnetic waves are produced by numerous atoms, all orientations of the **E** and **B** fields are possible in the composite light emitted. When the field vectors are randomly oriented, the light is said to be *unpolarized*. This is commonly represented schematically in terms of the electric field vector as shown in Fig. 23.15.

If there is some preferential orientation of the field vectors, the light is said to be *partially polarized*. Also, if the field vectors oscillate in only one plane, the light is *plane polarized*, or *linearly polarized*. Note that polarization is evidence that light is a transverse wave. Longitudinal waves, such as sound waves, cannot be polarized because there are no two-dimensional vibrations.

Light can be polarized in several ways. Polarization by reflection and double refraction will be discussed here. Polarization by scattering will be considered in Section 23.5.

Polarization by Reflection

When a beam of unpolarized light strikes a smooth transparent medium such as glass, it is partially reflected and partially transmitted. The reflected light may be completely polarized, partially polarized, or unpolarized, depending on the angle of incidence. The unpolarized case occurs for 0° or normal incidence. As the angle of incidence is varied, it is found that the reflected light is partially polarized. The electric field components parallel to the surface are reflected more strongly, giving this partial polarization. However, at one particular angle of incidence, the reflected beam is completely polarized. Also, with polarization of the reflected beam, the refracted beam is partially polarized (Fig. 23.16).

David Brewster (1781–1868), a Scottish physicist, found that the complete polarization of the reflected beam occurs when the reflected and refracted beams are 90° apart. The incident angle for this to occur is called the **polarizing angle** (θ_p) or the **Brewster angle**, and it is specific for a given material. As

Figure 23.16

FIGURE 23.16 Polarization by reflection When the reflected and refracted components of a beam of light are 90° apart, the reflected component is linearly polarized, and the refracted component is partially polarized. This occurs when $\theta_1 = \theta_p = \tan^{-1} n$. With a stack of glass plates (or layers of thin film), there are multiple reflections which increase the intensity of the reflected polarized beam, and the refracted beam becomes more polarized.

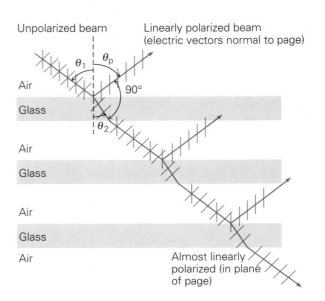

shown in Fig. 23.16, when $\theta_1 = \theta_p$, the reflected and refracted beams are 90° apart, and

$$\theta_1 + 90° + \theta_2 = 180°$$

Then

$$\theta_1 + \theta_2 = 90°$$

or

$$\theta_2 = 90° - \theta_1$$

By Snell's law (Chapter 21), for incidence in air,

$$\frac{\sin \theta_1}{\sin \theta_2} = n$$

In this case, $\sin \theta_2 = \sin (90° - \theta_1) = \cos \theta_1$. Thus,

$$\frac{\sin \theta_1}{\sin \theta_2} = \frac{\sin \theta_1}{\cos \theta_1} = \tan \theta_1 = n$$

With $\theta_1 = \theta_p$,

$$\boxed{\tan \theta_p = n} \qquad (23.16)$$

θ_p, **polarizing or Brewster angle:** incident angle for complete polarization of reflected light

(Since glass is dispersive, the Brewster angle also depends to some degree on the wavelength of the incident light.)

EXAMPLE 23.5 ■ Polarizing (Brewster) Angle

Sunlight is reflected from the smooth surface of a pond of water. What is the Sun's altitude when the polarization of the reflected light is the greatest?

Solution. Here, we obtain the index of refraction from Table 21.1, and so we can write:

Given: $n = 1.33$ (Table 21.1) *Find:* θ (altitude angle for complete polarization)

The altitude is the angle between the Sun and the horizon, so the angle of incidence is the complementary angle (draw yourself a sketch), and

$$\theta = 90° - \theta_p$$

where θ_p is the polarizing angle for maximum or complete polarization. Then using Eq. 23.16,

$$\theta_p = \tan^{-1} n = \tan^{-1} (1.33) = 53°$$

So

$$\theta = 90° - \theta_p = 90° - 53° = 37°$$

As shown in Fig. 23.16, in a stack of glass plates, the reflections from successive surfaces increase the intensity of the reflected polarized beam. Note that the refracted beam becomes more linearly polarized with successive refractions. This effect is sometimes referred to as polarization by refraction. In practical applications, several thin films of a transparent material are used instead of glass plates.

Polarization by Double Refraction (Birefringence and Dichroism)

When monochromatic light travels in glass, its speed is the same in all directions and is characterized by a single index of refraction. Any material like this is said to be *isotropic*, meaning that it has the same characteristics in all directions. Some crystalline materials, such as quartz, calcite, and ice, are *anisotropic* with respect to the speed of light; that is, the speed of light is different in different directions within the material. Anisotropy gives rise to some unique optical properties, one of which is that the index of refraction is dependent on the direction of propagation. Such materials are said to be doubly refracting, or to exhibit **birefringence**.

For example, a beam of unpolarized light incident on a birefringent crystal of calcite ($CaCO_3$, calcium carbonate) is illustrated in ● Fig. 23.17. When the beam propagates at an angle to a particular crystal axis, it is doubly refracted and separated into two components, or rays. Also, the two rays are linearly polarized in mutually perpendicular directions. One ray, called the **ordinary (o) ray**, passes through the crystal in an undeflected path and is characterized by an index of refraction (n_o) that is the same in all directions. The second ray, called the **extraordinary (e) ray**, is refracted and is characterized by an index of refraction (n_e) that varies with direction. The particular axis direction indicated by dashed lines in Fig. 23.17 is called the optic axis. Along this direction, $n_o = n_e$, and nothing extraordinary is noted about the transmitted light.

Some birefringent crystals, such as tourmaline, exhibit the interesting property of absorbing one of the polarized components more than the other. This property is called **dichroism**. If a dichroic crystal is sufficiently thick, the more strongly absorbed component may be completely absorbed. In that case, the emerging beam is plane polarized (● Fig. 23.18).

Another dichroic crystal is quinine sulfide periodide (commonly called herapathite, after W. Herapath, an English physician who discovered its polarizing properties in 1852). This crystal was of great practical importance in the development of modern polarizers. Around 1930, Edwin H. Land, an American

Demonstration/activity: Take a calcite crystal and place it over a transparency which has very small writing. View the transmitted light through a polaroid sheet to demonstrate birefringence.

● **FIGURE 23.17 Double refraction, or birefringence**
(a) Unpolarized light incident normal to the surface of birefringent crystal and at an angle to a particular crystal direction is separated into two components. The ordinary (o) ray and the extraordinary (e) ray are plane polarized in mutually perpendicular directions.
(b) Birefringence seen through a calcite crystal.

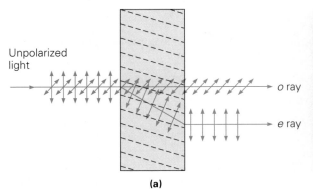

(a)   (b)

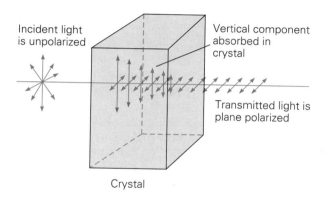

scientist, found a way to align tiny, needle-shaped dichroic crystals in sheets of transparent celluloid. The result was a thin sheet of polarizing material that was given the commercial name Polaroid.

Better polarizing films have been developed using synthetic polymer materials instead of celluloid. During the manufacturing process, this kind of film is stretched in order to align the long molecular chains of the polymer. With proper treatment, the outer (valence) electrons of the molecules can move along the oriented chains. As a result, light with E vectors parallel to the oriented chains is readily absorbed, but light with E vectors perpendicular to the chains is transmitted. The direction perpendicular to the orientation of the molecular chains is called the **transmission axis**, or the **polarization direction**. Thus, when unpolarized light falls on a polarizing sheet, the sheet acts as a polarizer and polarized light is transmitted (● Fig. 23.19).

The human eye cannot distinguish between polarized and unpolarized light. To tell whether or not light is polarized, we must use an analyzer, which may be a sheet of polarizing film. As shown in Fig. 23.19b, if the transmission axis of an analyzer is perpendicular to the plane of polarization of polarized light, little light (ideally none) will be transmitted.

Now you can understand the principle of polarizing sunglasses. Light

A common analogy compares a polarizer to a picket fence. A randomly oriented wave in a rope will only pass through the fence parallel to the pickets and will be polarized in that direction. However, keep in mind that in a polarizing sheet the polarization direction is perpendicular to the oriented molecules.

Demonstration/activity: Take two polaroid sheets which have their polarizing planes designated on the sheets and place them on an overhead projector to show polarization. Replace one of the sheets with a pair of polaroid sun glasses.

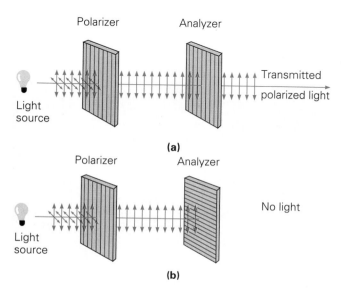

● **FIGURE 23.19 Polarizing sheets** (a) When polarizing sheets are oriented so that their polarization directions are the same, the emerging light is polarized. The first sheet acts as a polarizer and the second as an analyzer. (b) When one of the sheets is rotated 90° and the polarization directions are perpendicular (crossed Polaroids), little light (ideally, none) is transmitted.

reflected from a smooth surface is partially polarized, as you learned earlier in this section. The direction of polarization is chiefly in the plane of the surface (see Fig. 23.16). Light reflected from water or snow may be so intense that it gives rise to visual glare (Fig. 23.20). To reduce this, the polarizing lenses of glasses are oriented with their transmission axes vertical so that some of the partially polarized light from reflective surfaces is blocked or absorbed.

Polarizing glasses whose lenses show different polarization directions are used to view some 3-D movies. The pictures are projected on the screen by two projectors that transmit slightly different images. The projected light is linearly polarized, but in directions that are mutually perpendicular. The lenses of the 3-D glasses also have polarization directions that are perpendicular, and one eye sees the image from one projector and the other eye sees the image from the other projector. The brain interprets the image difference as depth, or a third dimension.

Some transparent materials have the ability to rotate the direction of polarization of linearly polarized light. This property is called **optical activity** and is due to the molecular structure of the material (Fig. 23.21). The rotation may be clockwise or counterclockwise, depending on the molecular orientation. Optically active molecules include those of certain proteins, amino acids, and sugars.

Glasses and plastics become optically active when under stress. The greatest rotation of the direction of polarization of the transmitted light occurs in the

FIGURE 23.20 Glare reduction (a) Light reflected from a horizontal surface is partially polarized in the horizontal plane. When sunglasses are oriented so that their polarizing direction is vertical, the horizontally polarized component of such light is not transmitted. The intensity of the reflected light reaching the eye is diminished, so glare is reduced. (b) Polarizing filters for cameras exploit the same principle. The photo at right was taken with such a filter. Note the reduction in reflections from the store window.

(a)

(b)

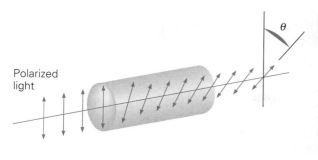

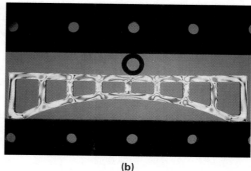

(a)

(b)

● **FIGURE 23.21 Optical activity** (a) Some substances have the property of rotating the polarization direction of linearly polarized light. This ability depends on the molecular structure and is called optical activity. (b) Glasses and plastics become optically active under stress, and the points of greatest stress are apparent when the material is viewed through crossed Polaroids. Engineers can thus test plastic models of structural elements to see where the greatest stresses will occur when they are "loaded" (subjected to the forces that they will experience in actual use). Here a model of a suspension bridge strut is being analyzed.

Demonstration/activity: Place pieces of plastic wrapping etc. between two polarizing sheets to show how different materials can rotate the plane of polarization.

Most students have a calculator which displays numbers with a liquid crystal. Have students rotate the plane of polarization of a Polaroid filter to check for the polarized light.

regions where the stress is the greatest. Viewing the stressed piece of material through crossed Polaroids allows the points of greatest stress to be identified. This determination is called optical stress analysis.

Another use of polarizing films is described in the Insight feature.

INSIGHT ■ LCD's and Polarized Light

LCD's (Liquid Crystal Displays) are commonplace on watches, calculators, gas pumps, and even some television screens. The name "liquid crystal" may seem self-contradictory. In general, when a crystalline solid melts, the resulting liquid no longer has an orderly atomic or molecular arrangement. Some organic compounds, however, pass through an intermediate state in which the molecules may rearrange somewhat but still maintain the overall order that is characteristic of a crystal.

A liquid crystal flows like a liquid, but its optical properties may depend on the order of its molecules. For example, certain liquid crystals with orderly arrangements are transparent. But the crystalline order can easily be disturbed by applied electrical forces. The disordered liquid then scatters light and is opaque.

A common type of LCD, called a twisted-nematic display, makes use of the effect of a liquid crystal on polarized light (Fig. 1). A display like those commonly used on wristwatches and calculators is made by sandwiching a layer of liquid crystal material between two glass plates that have fine parallel grooves, or channels, in their surfaces. One of the plates is then rotated or twisted 90°. The molecules in contact with the plates remain parallel to the grooves, and the result is a molecular orientation that rotates through 90° between the plates. In this configuration, the liquid crystal is optically active and will rotate the direction of polarization of linearly polarized light. The plates are then placed between crossed polariz-

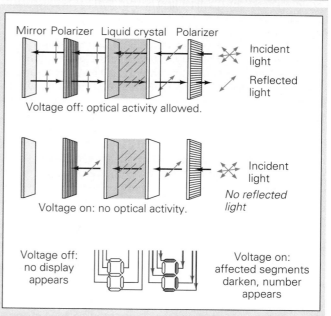

FIGURE 1 Liquid crystal display (LCD) A twisted-nematic display is an application involving the optical activity of a liquid crystal and crossed Polaroids. When the crystalline order is disoriented by an electric field from an applied voltage, the liquid crystal loses its optical activity in that region, and light is not transmitted. Numerals and letters are formed by applying voltages to segments of a block display.

Figure 1

FIGURE 2 Polarized light The light from an LCD is polarized, as can be shown by using polarizing sunglasses as an analyzer.

layer. The molecules of the liquid crystal are polar, so an electric field can disorient them. Transparent, electrically conductive film coatings arranged in a seven-block pattern are put onto the glass plates. Each block, or display segment, has a separate electrical connection. When a voltage is applied across one or more of the segments, the electric field disorients the molecules of the liquid crystal in that area, and their optical activity is lost. (Note that the numerals 0 through 9 can all be formed by the segmented display.) The incident polarized light passing through the disoriented regions of the liquid crystal is absorbed by the second polarizer. Thus, these regions are opaque and appear dark. To produce images on small TV screens, the display segments of the LCD are coupled so that many of them may be energized with a single lead.

ing sheets and backed with a mirrored surface. Light entering and passing through the LCD is polarized, rotated 90°, reflected, and again rotated 90° by its components. After the return trip through the liquid crystal, the polarization direction of the light is the same as that of the initial polarizer. Thus, the light is transmitted and leaves the display unit. Because of the reflection and transmission the display appears to be a light color (usually light gray) when illuminated with white, unpolarized light.

The light from the bright regions of an LCD can readily be shown to be polarized using an analyzer. The whole display appears dark when the analyzer is properly oriented. You may have noticed this effect if you have evertried to see the time on the LCD of a wristwatch while wearing polarizing sunglasses (Fig. 2).

The dark numbers or letters on an LCD are formed by applying an electric field to parts of the liquid crystal

One of the major advantages of LCD's is their low power consumption. Other similar displays, such as those using red light-emitting diodes (LED's), produce light themselves, using relatively large amounts of power. LCD's, on the other hand, produce no light, but instead use reflected light.

Some liquid crystals respond to temperature changes, and the orientation of the molecules affects the light-scattering properties of the crystal. Different wavelengths, or colors, of light are selectively scattered. The color of a liquid crystal can thus be an indication of the surrounding temperature, and liquid crystals are used to make thermometers.

Finally, there is a new and emerging display technology called the polymer dispersed liquid crystal (PDLC) display. This type of display consists of a liquid crystal and polymer mixture, which has the property of electrically controlled light scattering. That is, the transparency of the display can be varied from opaque to clear by changing the applied voltage. A major advantage of PDLC displays is that they do not require polarizers, making them more cost effective.

23.5 ■ Atmospheric Scattering of Light

When light is incident on a suspension of particles, such as the molecules of air, some of the light may be absorbed and reradiated. This process is called **scattering**. The scattering of sunlight in the atmosphere produces some interesting effects. Among these are the polarization of sky light (that is, sunlight that has been scattered by the atmosphere), the blueness of the sky, and the redness of sunsets and sunrises.

Atmospheric scattering causes the sky light to be polarized. When unpolarized sunlight is incident on air molecules, the electric field of the light wave sets electrons of the molecules into vibration. The vibrations are complex, but the accelerating charges emit radiation, like the vibrating charges in the antenna of a radio broadcast station (see Section 19.4). The intensity of this emitted radiation is strongest along a line perpendicular to the oscillation. And, as illustrated in ● Fig. 23.22, an observer viewing from an angle of 90° with respect to the direction of the sunlight will receive linearly polarized light because of the horizontal charge oscillations. The light also has a vertically polarized component. At other viewing angles, both components are present, and sky light seen through a polarizing filter appears partially polarized because of the stronger component.

Since the scattering of light with the greatest degree of polarization is at a right angle to the direction of the Sun, at sunrise and sunset the scattered light from directly overhead has the greatest degree of polarization. The polarization of sky light may be observed by viewing the sky through a polarizing filter (or a polarizing sunglass lens) and rotating the filter. Light from different regions of the sky will be transmitted in different degrees, depending on its degree of polarization. It is believed that some insects, such as bees, use polarized sky light to determine navigational directions relative to the Sun.

The scattering of sunlight by air molecules causes the sky to look blue. This is not a polarization effect but is caused by the selective absorption of light. As oscillators, air molecules have resonant frequencies in the ultraviolet region. Consequently, light in the nearby blue end of the visible region is scattered more than that in the red end. That is, the sunlight is preferentially scattered, with the light of the blue end of the visible spectrum being scattered almost ten times more than the light of the red end.

For particles such as air molecules, which are much smaller than the wavelength of light, the scattering is found to be inversely proportional to the

Why the sky is blue

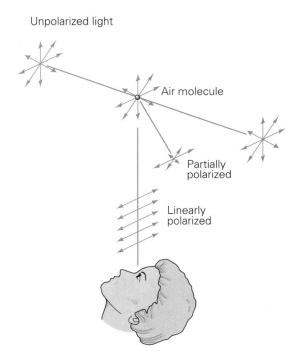

Figure 23.22

● **FIGURE 23.22 Polarization by scattering** When incident unpolarized sunlight is scattered by a molecule of gas of the air, the light in the plane perpendicular to the direction of the incident ray is linearly polarized. Light scattered at some arbitrary angle is partially polarized. An observer at a right angle to the direction of the incident sunlight receives plane-polarized light.

wavelength to the fourth power (that is, $1/\lambda^4$). This is called **Rayleigh scattering** after Lord Rayleigh (1842–1919), a British physicist who derived the wavelength relationship. This inverse relationship predicts that light of the shorter-wavelength, or blue, end of the spectrum will be scattered more than light of the longer-wavelength, or red, end. The scattered blue light is rescattered in the atmosphere and eventually is directed toward the ground. This is the sky light we see, so the sky appears blue.

EXAMPLE 23.6 ■ Rayleigh Scattering

How much more is light at the blue end of the visible spectrum scattered by air molecules than light at the red end?

Solution. As stated earlier, light at the blue end of the visible spectrum is scattered almost ten times more than light at the red end. This can be shown as follows. The Rayleigh scattering relationship is $S \propto 1/\lambda^4$, where S is the degree of scattering for a particular wavelength. Thus, you can form a ratio

$$\frac{S_b}{S_r} = \left(\frac{\lambda_r}{\lambda_b}\right)^4$$

where the subscripts stand for blue and red light. The blue end of the spectrum (violet light) has a wavelength of about $\lambda_b = 400$ nm and red light has a wavelength of about $\lambda_r = 700$ nm. Putting these values into the ratio gives

$$\frac{S_b}{S_r} = \left(\frac{\lambda_r}{\lambda_b}\right)^4 = \left(\frac{700 \text{ nm}}{400 \text{ nm}}\right)^4 = 9.4 \qquad \text{or} \qquad S_b = 9.4 \, S_r$$

Thus, blue light is scattered almost ten times as much as red light is.

You should keep in mind that all colors are present in sky light, but the dominant color is blue. (The sky doesn't appear violet because our eyes are more sensitive to blue and there is more blue than violet light in sunlight.) You may have noticed that the sky is more blue directly overhead than toward the horizon and appears white just above the horizon (look for this effect the next time you are outside on a clear day). This is because there are relatively fewer air molecules or scatterers directly overhead than toward the horizon. Multiple scatterings occur in the denser air near the horizon, and the recombination of light gives rise to the white appearance. Analogously, if a drop or two of milk is added to a glass of water and the suspension is illuminated with intense white light, the scattered light has a bluish hue. But a glass of undiluted milk is white, because of multiple scatterings. Similarly, atmospheric pollution may give rise to a milky white appearance of most or all of the sky.

When the Sun is near the horizon, the sunlight travels a greater distance through the denser air near the Earth's surface. Since the light therefore undergoes a great deal of scattering, you might think that only the least scattered light, the red light, would reach observers on the Earth's surface. This would explain red sunsets. However, it has been shown that the dominant color of white light after only molecular scattering is orange. Thus, there must be other scattering that shifts the light from the setting (or rising) Sun toward the red (Fig. 23.23).

See "Colors of the Sky," by C. F. Bohren and A. B. Fraser, in The Physics Teacher, *vol. 23, no. 5, p. 267.*

Red sunsets have been found to result from the scattering of sunlight by atmospheric gases *and* small foreign particles. These particles are not necessary for the blueness of the sky, but are necessary for deep red sunsets and sunrises. (This is why spectacular red sunsets are observed in the months after a large volcanic eruption that puts tons of particulate matter into the atmosphere.) Red sunsets occur most often when there is a high-pressure air mass to the west, since the concentration of particles is generally greater in high-pressure air masses than in low-pressure air masses. Similarly, red sunrises occur most often when there is a high-pressure air mass to the east.

Why sunsets and sunrises are red

Now you can understand the meaning behind the old weather saying, "Red sky at night, sailors' delight. Red sky in the morning, sailors take warning." (See also the Biblical quote Matthew 16:1-4.) Fair weather generally accompanies high-pressure air masses because they are associated with reduced cloud formation. Most of the United States lies in the Westerlies wind zone, in which air masses generally move from west to east. A red sky at night is thus likely to indicate a fair-weather, high-pressure air mass to the west that will be coming your way. A red sky in the morning means the high-pressure air mass has passed, and poor weather may set in.

Red sunrises and sunsets are often made more spectacular by clouds. The clouds are colored pink by reflected red light. In general, clouds are white (when not shadowed) because the water droplets of which they are composed are not preferential scatterers and scatter visible light of all wavelengths (white light). Clouds do not affect the color of the incident light, but simply diffusely reflect the light from the Sun.

Important Concepts

You should be able to define and/or explain these chapter concepts clearly.

physical (wave) optics	diffraction grating	transmission axis (polarization direction)
Young's double-slit experiment	Bragg's law	optical activity
thin-film interference	polarization	LCD (liquid crystal display)
optical flats	polarizing (Brewster's) angle	scattering
Newton's rings	birefringence	Rayleigh scattering
diffraction	dichroism	

Important Relationships for Review

These relationships are mathematical statements of the concepts and principles presented in the chapter. You should be able to identify the symbols and to explain the relationships before proceeding to the Exercises. In-text equation reference numbers are given for convenience.

Path Difference for Constructive Interference:

$$\Delta = n\lambda \qquad \text{for } n = 0, 1, 2, 3, \dots \qquad (23.1)$$

Path Difference for Destructive Interference:

$$\Delta = \frac{m\lambda}{2} \qquad \text{for } m = 1, 3, 5, \dots \qquad (23.2)$$

Bright Fringe Condition (double-slit interference):

$$d \sin \theta = n\lambda \qquad \text{for } n = 0, 1, 2, 3, \dots \qquad (23.3)$$

Wavelength Measurement (double-slit interference):

$$\lambda = \frac{y_n d}{nL} \qquad \text{for } n = 1, 2, 3, \dots \qquad (23.7)$$

Width of Bright Fringes (double-slit interference):

$$y_{n+1} - y_n = \frac{L\lambda}{d} \qquad \text{(Example 23.1)}$$

Nonreflecting Film Thickness:

$$t = \frac{\lambda}{4n} \qquad (23.8)$$

Dark Fringe Condition (single-slit diffraction):

$$w \sin \theta = m\lambda \qquad m = 1, 2, 3, \dots \qquad (23.9)$$

Lateral Displacement of Dark Fringes (single-slit diffraction):

$$y_m \cong \frac{mL\lambda}{w} \qquad \text{for } m = 1, 2, 3, \dots \qquad (23.10)$$

Width of Bright Fringes (single-slit diffraction):

$$\Delta = y_{m+1} - y_m = \frac{L\lambda}{w} \qquad (23.12)$$

Interference Maxima for a Diffraction Grating:

$$d \sin \theta = n\lambda \qquad \text{for } n = 0, 1, 2, 3, \dots \qquad (23.13)$$

Limit of the Order Number:

$$n \le \frac{d}{\lambda} \qquad (23.14)$$

Bragg's Law:

$$2d \sin \theta = n\lambda \qquad \text{for } n = 1, 2, 3, \dots \qquad (23.15)$$

Brewster (polarizing) Angle:

$$\tan \theta_p = n \qquad (23.16)$$

Exercises

23.1 ■ Young's Double-Slit Experiment

1 The formation of a stationary fringe pattern in Young's experiment depends on (a) diffraction, (b) interference, (c) coherent light, (d) all of these. (d)

2 If in a Young's experiment using monochromatic light the screen distance is doubled, then which of the following would change: (a) d, (b) λ, (c) y_n, (d) none of these? (c)

3 When white light is used in Young's experiment, bright fringes with spectrums of colors are seen. In a fringe spectrum, is the red end or the blue end closer to the central maximum? blue

● **4** What would be observed with the set-up for Young's experiment if the light passing through the slit were not coherent? no interference fringes

5 Television pictures are often seen to flutter when an airplane passes by (● Fig. 23.24). Explain the cause of this fluttering. see ISM

6 In the developments of Young's experiment and single-slit diffraction, a small-angle approximation was used to find the lateral displacements of the bright and dark fringes, respectively. How good is this approximation? That is, for what angles does it no longer apply? $\theta > 10°$

7 ■■ What is the shortest path difference for two waves of

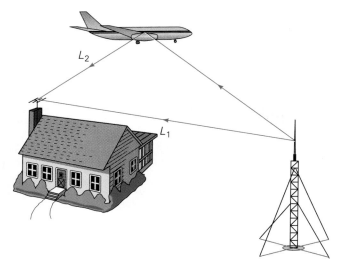

FIGURE 23.24 Interference See Exercise 5.

equal frequency that arrive at a point with a phase difference of (a) $\pi/3$, (b) $\pi/2$, and (c) π? [*Hint:* See Eq. 14.5.]
(a) $\lambda/6$ (b) $\lambda/4$ (c) $\lambda/2$

8 ■■ Two point sources of coherent sound are located at (2 m, 5 m) and (4 m, 3 m) in the xy plane. What is the wavelength of the sound waves at the point (15 m, 20 m) if (a) they interfere destructively to the first order and (b) they have a phase difference of 45°? (a) 0.80 m (b) 3.2 m

9 ■■ An interference pattern is formed when light whose wavelength is 550 nm is incident on two parallel slits 50 μm apart. The second-order bright fringe is 2.0 cm from the center of the central maximum. How far from the slits is the screen on which the pattern is formed? 0.91 m

● **10** ■■ Monochromatic light illuminates two parallel slits that are 0.20 mm apart. The adjacent bright lines of the interference pattern on a screen 1.5 m away from the slits are 0.45 cm apart. What is the color of the light? 600 nm (orange-yellow)

11 ■■ Blue light with a wavelength of 440 nm is used in a double-slit experiment where the slits are 0.75 mm apart. If the screen is 1.5 m from the slits, what is the angular separation of the first-order and third-order bright fringes? 1.2 × 10⁻³ rad

12 ■■ Two parallel slits 1.0 mm apart are illuminated with monochromatic light whose wavelength is 640 nm, and an interference pattern is observed on a screen 1.50 m away. (a) What is the separation between adjacent interference maxima? (b) If the distance between the slits were increased to 1.5 mm, how would this affect the separation of the maxima? (a) 9.6 × 10⁻⁴ m (b) 6.4 × 10⁻⁴ m

13 ■■ Derive a relationship giving the locations of the dark fringes in Young's double-slit experiment. What is the distance between the dark fringes? $y_m = m\lambda L/2d$, (m = 1, 3, 5, . . .); $\Delta = L\lambda/d$

14 ■■ Monochromatic light whose wavelength is 500 nm falls on two slits separated by a distance of 40 μm. What is the distance between the first-order and third-order bright fringes formed on a screen 1.2 m away from the slits? 3.0 × 10⁻² m

15 ■■ Monochromatic light passes through two narrow slits 0.25 mm apart and forms an interference pattern on a screen 1.5 m away. If light with a wavelength of 680 nm is used, what is the distance between the center of the central maximum and the center of the third-order bright fringe? 1.2 × 10⁻² m

16 ■■ In a double-slit experiment using monochromatic light, a screen is placed 1.25 m from the slits, which have a separation distance of 0.0250 mm. The third-order bright fringe (n = 3) is measured to be 6.60 cm from the center of the central maximum. Find (a) the wavelength of the light, and (b) the lateral displacement of the second-order bright fringe (n = 2). (a) 4.4 × 10⁻⁷ m (b) 4.4 × 10⁻² m

17 ■■ Yellow-orange light (λ = 550 nm) is used in a double-slit experiment with a slit separation distance of 1.75 × 10⁻⁴ m. With the screen located 2.00 m from the slits, determine (a) the angular separation distance between the central maximum and the second-order bright fringe (n = 2) and (b) the lateral displacement of the fringe. (a) 6.3 × 10⁻³ rad (b) 1.3 × 10⁻² m

● **18** ■■ In a double-slit experiment with a screen at a distance of 1.50 m and using monochromatic light, the angle between the second-order bright fringe and the central maximum is measured to be 0.0230 rad. If the separation distance of the slits is 0.0350 mm, (a) what is the color of the light? (b) What is the lateral displacement of the fringe? (a) n = 4.0 × 10⁻⁷ m, violet (b) 3.4 × 10⁻² m

19 ■■■ What would be the effect on the interference fringes if the apparatus for Young's double-slit experiment were completely immersed in still water? (b) What would the displacement in Exercise 15 be if the entire system were immersed in still water? (a) y'_n = (0.75)y_n (b) 9.0 × 10⁻³ m

20 ■■■ Light of two different wavelengths is used in a double-slit experiment. The location of the third-order bright fringe for yellow light (λ = 600 nm in air) coincides with the location of the fourth-order bright fringe for the other light. What is the wavelength of the other light? 450 nm

23.2 ■ Thin-Film Interference

21 For a thin film with $n_1 > n_0$ and $n_1 > n_2$, where n_1 is the index of refraction of the film, a film thickness for constructive interference of the reflected light is (a) $\lambda'/4$, (b) $\lambda'/2$, (c) λ', (d) both (a) and (b). (a)

● **22** For a thin film with $n_0 < n_1 < n_2$, where n_1 is the index of refraction of the film, the minimum film thickness for destructive interference of the reflected light is (a) $\lambda'/4$, (b) $\lambda'/2$, (c) λ', (d) both (a) and (b). (a)

23 What would be the effect of nondirect incidence on thin-film interference? see ISM

24 Suppose you wanted to make a reflecting surface using thin-film interference. How would you go about doing this? see ISM

25 ■■ Light with a wavelength of 550 nm in air is normally incident on a glass plate ($n = 1.5$) whose thickness is 10^{-5} cm. (a) What is the total path length of the light in wavelengths in glass? (b) What is the phase difference in the beams reflected from each surface (that is, will they interfere constructively or destructively)? (a) 27 λ (b) 180°, destructively

26 ■■ A camera lens is coated with a thin layer of a material that has an index of refraction of 1.35. This makes the lens nonreflecting for light with a wavelength of 450 nm (in air) that is normally incident on the lens. What is the thickness of the thinnest film that will make the lens nonreflecting? 8.33 × 10⁻⁶ cm

● **27** ■■ Magnesium fluoride ($n = 1.38$) is frequently used as a lens coating to make nonreflecting lenses. What is the difference in the minimum film thicknesses required for maximum transmission of blue light ($\lambda = 400$ nm) and red light ($\lambda = 700$ nm)? 5.5 × 10⁻⁸ m

28 ■■ A solar cell is to have a nonreflecting coating of a transparent material whose index of refraction is 1.25. (a) What is the minimum thickness of the film for light with a wavelength of 550 nm? (b) Does the index of refraction of the adjacent material in the solar cell make a difference? (a) 1.10 × 10⁻⁷ m (b) yes, $n_3 > n_2 > n_1$

29 ■■ A thin layer of oil ($n = 1.5$) floats on water. Destructive interference is observed for light with wavelengths of 480 nm and 600 nm. Find the two thicknesses of the oil film for normal incidence. $t_1 = 160$ nm; $t_2 = 200$ nm

30 ■■ It is desired to make a nonreflecting lens for an infrared radiation detector. If the coating material has an index of refraction of 1.20, what should the film thickness be for radiation with a frequency of 3.75 × 10¹⁴ Hz? 1.67 × 10⁻⁷ m

31 ■■ Two parallel plates are separated by a small distance as illustrated in Fig. 23.25. If the top plate is illuminated with red light ($\lambda = 680$ nm), for what minimum separation distances will the light be (a) reflected back to the observer,

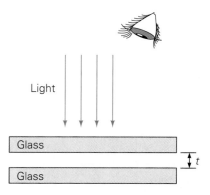

● **FIGURE 23.25 Reflection or transmission?** See Exercise 31.

and (b) transmitted through the plates? [$t = 0$ is not an answer for (b).] (a) 170 nm (b) 340 nm

32 ■■ An air wedge is shown in ● Fig. 23.26. If the top glass plate is illuminated with monochromatic light, describe the observed interference pattern. (a) Express the locations of the bright interference fringes in terms of wedge thicknesses measured from the apex of the wedge. (b) Show that the number of dark fringes is given by $m = 2t/\lambda_{air}$, where t is the maximum thickness of the wedge. (a) $2t = [m + (1/2)]\lambda$, where $m = 0, 1, 2, \ldots$ (b) see ISM

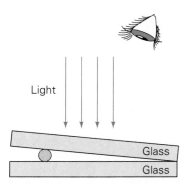

● **FIGURE 23.26 Air wedge** See Exercise 32.

33 ■■ The glass plates in Fig. 23.26 are separated by a thin, round filament. When the top plate is illuminated with light having a wavelength of 550 nm, it is observed that the filament lies directly below the sixth order bright fringe. What is the diameter of the filament? 1.8 × 10⁻⁶ m

23.3 ■ Diffraction

● **34** For single-slit diffraction, (a) all fringes are equal in width, (b) dark fringes occur for only odd integers, $m = 1, 3, 5, \ldots$, (c) for a given wavelength, the narrower the slit, the wider the fringes, (d) none of these. (c)

35 As the number of lines per unit length of a diffraction grating increases, the number of spectral orders for a given wavelength (a) increases, (b) decreases, (c) remains unchanged. (b)

36 (a) What does the equation for the width of the central maximum for single-slit diffraction predict when $w = \lambda$? Is the prediction correct? Explain. (b) Why is there no zeroth order ($n = 0$) in Bragg's law for crystal diffraction? see ISM

37 ■■ A slit of width 0.20 mm is illuminated with monochromatic light having a wavelength of 480 nm, and a diffraction pattern is formed on a screen 1.0 m away from the slit. (a) What is the width of the central maximum? (b) What is the distance between the second-order and third-order bright fringes? (a) 4.8 × 10⁻³ m (b) 2.4 × 10⁻³ m

38 ■■ Red light (λ = 700 nm) passes through a slit with a width of 1.0 mm. (a) Find the angular width and the linear width of the central maximum of the diffraction pattern on a screen 2.0 m from the slit. (b) Find the angular and linear widths if the slit is illuminated with blue light (λ = 440 nm). (a) 1.4×10^{-3} rad, 0.28 cm (b) 9.0×10^{-4} rad, 0.18 cm

39 ■■ When illuminated with monochromatic light, a diffraction grating with 10,000 lines/cm produces a first-order maximum 4.5 cm from the central maximum on a screen 1.0 m away. What is the wavelength of the light? 450 nm

40 ■■ A diffraction grating is designed to have the red end of the visible spectrum an angular distance of 7.5° from the center of the central maximum for the second order. How many lines per centimeter does the grating have? 9.3×10^2 lines/cm

41 ■■ In a particular diffraction pattern, the red component (λ = 700 nm) in the second-order spectrum is deviated at an angle of 20°. (a) How many lines per centimeter does the grating have? (b) If the grating is illuminated with white light, how many orders of the complete visible spectrum are produced? (a) 2.3 lines/cm (b) 5

42 ■■ Find the locations of the blue (λ = 420 nm) and red (λ = 680 nm) components of the first-order and second-order spectra in a pattern formed on a screen 1.5 m from a diffraction grating with 7500 lines/cm. blue: x_1 = 0.49 m, 18° x_2 = 1.2 m, 39° m; red: x_1 = 0.90 m, 31°; x_2 not possible

43 ■■ White light whose components have wavelengths from 400 nm to 700 nm illuminates a diffraction grating with 4000 lines/cm. Do the first and second orders overlap? Justify your answer. do not overlap, see ISM

44 ■■ Visible white light traveling in air contains components with wavelengths from about 400 nm to 700 nm. If the white light illuminates a diffraction grating with 8000 lines/cm, what is the angular width of the first-order spectrum produced? 15°

45 ■■ What is the angular width of the first-order spectrum in Exercise 44 if the whole system is immersed in water? 11°

46 ■■ A certain crystal gives a deflection angle of 25° for the first-order diffraction of monochromatic X-rays with a frequency of 5.0×10^{17} Hz. What is the lattice spacing of the crystal? 7.1×10^{-10} m

47 ■■ A diffraction grating with 8000 lines per centimeter is illuminated with a beam of monochromatic red light from a He-Ne laser (λ = 632.8 nm). How many side maxima would be formed in the diffraction pattern, and at what angles would they be observed? two at $\pm 30°$

48 ■■ A single slit with a width of 0.25 mm is illuminated with monochromatic light (λ = 680 nm). A screen is placed 1.80 m from the slit to observe the fringe pattern. (a) What is the angle between the second dark fringe (m = 2) and the central maximum? (b) What is the lateral displacement of this dark fringe? (a) 5.4×10^{-3} rad = 0.31° (b) 9.8×10^{-3} m

49 ■■ What is the width of the central maximum for the single slit arrangement in Exercise 48? 9.8×10^{-3} m

● **50** ■■■ Show that for a diffraction grating the violet (λ = 400 nm) portion of the third-order spectrum overlaps the yellow-orange (λ = 600 nm) portion of the second-order spectrum regardless of the grating's spacing. see ISM

51 ■■■ A teacher standing inside a doorway 1.0 m wide blows a whistle with a frequency of 1000 Hz to summon children from the playground. All the children come, except two boys playing near the swings, about 100 m away from the school building. Later the teacher begins to reprimand them, but the boys maintain rather convincingly that they did not hear the whistle. The swings are at an angle of 19.6° from a line normal to the doorway, and the speed of sound is 335 m/s. Could they be telling the truth? Justify your answer. yes, see ISM

23.4 ■ Polarization

● **52** Light may be polarized by (a) reflection, (b) refraction, (c) scattering, (d) all of these. (d)

53 Brewster's angle depends on (a) the index of refraction of a material, (b) Bragg's law, (c) internal reflection, (d) interference. (d)

54 Given two pairs of sunglasses, could you tell if one or both were polarizing? see ISM

55 Suppose that you held two polarizing sheets in front of you and looked through both of them. How many times would you see the sheets lighten and darken (a) if one of them were rotated through one complete rotation, (b) if both of them were rotated through one complete rotation at the same rate in opposite directions, (c) if both of them were rotated through one complete rotation at the same rate in the same direction, and (d) if both were rotated through one complete rotation in opposite directions, but one twice as fast as the other? (a) twice (b) four times (c) none (d) six times

● **56** ■ Some types of glass have a range of indices of refraction of about 1.4 to 1.7. What is the range of the polarizing angle for these glasses? 54° to 60°

57 ■ If the Brewster angle for a glass plate is 1.0 rad, what is the index of refraction of the glass? 1.6

58 ■■ The critical angle for internal reflection in a certain medium is 55°. What is the Brewster angle for light externally incident on the medium? 51°

59 ■■ The angle of incidence is adjusted so there is maximum linear polarization for the reflection of light from a transparent piece of plastic with n = 1.22. But some of the light is transmitted. What is the angle of refraction? 39°

60 ■■ What angle of incidence is necessary for reflected light to have the maximum linear polarization at a water–crown glass interface?   48.8°

61 ■■ Sunlight is reflected off a glass plate window ($n = 1.55$). What would be the Sun's altitude for the *reflected* light to be completely polarized?   57.2°

● **62** Brewster's angle in Eq. 23.16 was derived for the incident light in air. Show that in general, $\theta_p = \tan^{-1} n$, where n is the relative index of refraction ($n = n_2/n_1$).   see ISM

63 ■■ Find the Brewster angle for a piece of glass ($n = 1.60$) that is submerged in water.   50.3°

64 ■■■ A plate of crown glass is covered with a layer of water. A beam of light traveling in air is incident on the water and partially transmitted. Is there any angle of incidence for which the light reflected from the water–glass interface will have maximum linear polarization? Justify your answer mathematically.   no

23.5 ■ Atmospheric Scattering of Light

65 Which of the following colors is scattered the most in the atmosphere: (a) green, (b) yellow, (c) orange, (d) color makes no difference?   (a)

66 Red sunsets result from (a) polarization, (b) molecular scattering (c) scattering by foreign particles, (d) both (b) and (c).   (d)

67 (a) Why does the sky not have a uniform blueness? (b) What would an astronaut on the moon see when looking at the sky or into space?   (a) different air molecule density
(b) black

■ Additional Exercises

68 A slit 0.025 mm wide is illuminated with red light ($\lambda = 680$ nm). How wide are (a) the central maximum and (b) the side maxima of the diffraction pattern formed on a screen 1.0 m from the slit?   (a) 5.4×10^{-2} m   (b) 2.7×10^{-2} m

69 Sketch a graph of the Brewster angle versus the index of refraction.   see ISM

70 In a double-slit experiment with slits that are 0.25 mm apart, the interference pattern is formed on a screen 1.0 m away from the slits. The third-order bright fringe is 0.60 cm from the center of the central maximum. (a) What is the frequency of the monochromatic light? (b) What is the distance between the second-order and third-order bright fringes?   (a) 6.0×10^{14} Hz   (b) 2.0×10^{-3} m

71 Show that the film thicknesses for a "nontransmissive" lens are given by

$$t = \frac{(m+1)\lambda}{2n}$$

where $m = 0, 1, 2, 3, \ldots$, and n is the index of refraction of the film.   see ISM

72 If the slit width in a single-slit experiment were doubled, the distance to the screen reduced by a third, and the wavelength of the light changed from 600 nm to 450 nm, how would the width of the bright fringes be affected?
$\Delta y = (0.25)\Delta y_0$

73 A glass plate has an index of refraction of 1.5. At what angle of incidence will light be reflected from the plate with the maximum linear polarization?   56°

74 A thin air wedge between two flat glass plates forms bright and dark interference bands when illuminated with normally incident monochromatic light, as illustrated in Fig. 23.7. (a) Show that the thickness of the air wedge changes by $\lambda/2$ from one bright fringe to the next, where λ is the wavelength of the light. (b) What would the change in the wedge thickness between bright fringes be if the space were filled with a liquid with an index of refraction n?
see ISM

75 A light beam is incident on a glass plate ($n = 1.58$), and the reflected ray is completely polarized. What is the angle of refraction for the beam?   32.3°

76 A lens with an index of refraction of 1.6 is to be coated with a material ($n = 1.4$) that will make it nonreflecting for red light ($\lambda = 700$ nm). What is the minimum required thickness of the coating? 1.3×10^{-7} m

77 Two parallel slits 0.75 mm apart are illuminated with monochromatic light whose wavelength is 480 nm. Find the angle between the center of the central maximum and the center of an adjacent bright fringe formed on a screen 1.2 m away from the slits. 6.4×10^{-4} rad

78 A film on a lens is 1.0×10^{-7} m thick and is illuminated with white light. The index of refraction of the film is 1.4. For what wavelength of light will the lens be nonreflecting?
5.6×10^{-7} m

79 What is the spectral order limit for a diffraction grating with 9000 lines/cm when illuminated by white light?   one red; two violet

80 In a double-slit experiment using monochromatic light, the angular separation between the central maximum and the second-order bright fringe is measured to be 0.16°. What is the wavelength of the light if the slit separation distance is 0.35 mm? 4.9×10^{-7} m

81 How many orders of interference maxima occur when monochromatic light with a wavelength of 560 nm illuminates a diffraction grating with 10,000 lines/cm?   1

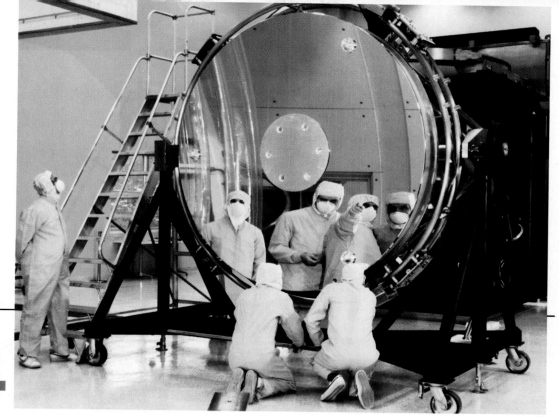

24

Optical Instruments

Optical instruments are used for a variety of purposes, but their basic function is to improve and extend our powers of visual observation. Indeed, vision is our chief means of acquiring information about the world. Without microscopes and telescopes to view objects that are not visible or undiscernible to the unaided eye, our knowledge of the world would be enormously impoverished. For example, biologists could not observe the cells of the human body or pathogenic bacteria. The science of astronomy too would be little advanced beyond the level attained by the ancient Greeks.

Although some optical instruments are extremely complex, they can generally be understood in terms of their basic components, usually mirrors and lenses. Without these reflective and refractive elements, our visual investigations would be severely limited. Nearly all large modern telescopes, for example, depend on mirrors like the one shown above, made for the Hubble Space Telescope and now in orbit around the Earth.

Mirrors and lenses were discussed in Chapter 22 and other optical phenomena in Chapter 23. You will find that the principles developed in those chapters can be easily applied to the study of optical instruments. This chapter first considers the optical properties of the human eye itself and then describes how optical instruments improve our view of the world.

INSIGHTS

- **A New Generation of Telescopes**
- **Telescopes Using Nonvisible Radiation**

Topics of study in this chapter include the human eye, microscopes, telescopes, diffraction and resolution, and color and optical illusions.

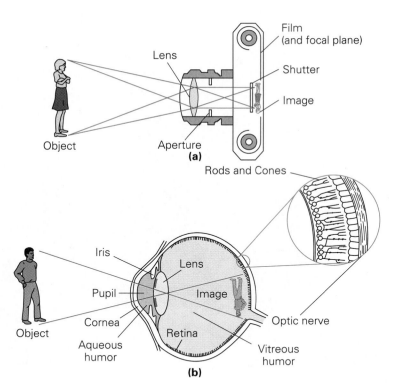

FIGURE 24.1 Camera and eye analogy In some aspects, a simple camera is similar to the human eye. An image is formed on the film in a camera and on the retina in the eye. (The complex refractive properties of the eye are not shown because multiple refractive media are involved. See text for description.)

Image formation by a converging lens is discussed in Section 22.3; see Figure 22.15a

Remove the back from a simple camera and place a piece of Scotch (Magic Plus) tape in the focal plane. Then fix the shutter open and pass the camera around to the students to view a large lighted object in the front of the room. An image of the object appears on the tape.

24.1 ■ The Human Eye

The human eye is the most fundamental optical instrument, since without it the field of optics would not exist. The eye is analogous to a simple camera in several respects (●Fig. 24.1). A simple camera consists of a converging lens, which is used to focus images on light-sensitive film at the back of the camera's interior chamber. (Recall from Chapter 22 that for relatively distant objects, a converging lens produces a smaller, inverted, real image.) An adjustable diaphragm, or opening, and a shutter control the amount of light entering the camera.

The eye too has a converging lens that focuses images on the light-sensitive lining on the rear surface inside the eyeball. (Other transparent media in the eye also help to create the image by refracting incoming light.) The iris is a circular diaphragm that opens and closes to adjust the amount of light entering the eye. The eyelid might be thought of as a shutter; however, the shutter of the camera, which controls the exposure time, is generally opened for only a fraction of a second, and the eyelid is normally open for continuous exposure. The human nervous system performs a function analogous to that of a shutter in analyzing image signals from the eye at a rate of about 30 times per second. The eye might therefore be likened to a movie or videocamera, which exposes a similar number of frames (images) per second.

Although the optical functions of the eye are relatively simple, its physiological functions are quite complex. As Fig. 24.1b shows, the eyeball is a nearly spherical chamber (with an internal diameter of about 1.5 cm) filled with a jellylike substance called the *vitreous humor*. The eyeball has a white outer covering called the sclera, part of which is visible as the white of the eye. Light enters the eye through a curved, transparent tissue called the cornea and passes into a clear fluid known as the *aqueous humor*. Behind the cornea is a circular diaphragm, the iris, whose central hole is called the pupil. The iris contains the pigment that determines eye color. Through muscle action, the iris can change the area of the pupil (2 to 8 mm in diameter), thereby controlling the amount of light entering the eye.

Behind the iris is a **crystalline lens**, a converging lens composed of microscopic glassy fibers. When tension is exerted on the lens by attached muscles, the glassy fibers slide over each other, causing the shape and curvature (focal length) of the lens to change, thus focusing the image properly. (Light passing through the eyeball travels through five different media with different indices of refraction: air [$n = 1.00$], the cornea [$n = 1.38$], the aqueous humor [$n = 1.33$], the crystalline lens [average $n = 1.40$], and the vitreous humor [$n = 1.34$]. The various refractions at boundaries all participate in the formation of an image.)

On the back interior wall of the eyeball is a light-sensitive surface called the **retina**. From the retina, the optic nerve relays signals to the brain. The retina is composed of nerves and two types of light receptors, or photosensitive cells, called **rods** and **cones** because of their shapes. The rods are more sensitive to light and distinguish light from dark in low light intensities (twilight vision). The cones, on the other hand, are less sensitive to light but resolve sufficiently intense light into its component wavelengths, thereby enabling us to see colors. The cells of the retina are present in large numbers (estimated at 125,000/mm²). Most of the cones are clustered together around a center (called the fovea centralis). The more numerous rods are outside this region and distributed nonuniformly over the retina.

The optical adjustments of the eye are truly amazing. Most of the refraction of light occurs at the front surface of the cornea. The crystalline lens makes fine adjustments for focusing the images. Nerve signals cause the attached muscles to change the radius of curvature and thus the focal length of the lens. Thus, images of objects at different distances are focused sharply on the retina. When the eye is focused on distant objects, the muscles are relaxed, and the crystalline lens has its thinnest shape and a power of about 20 D (diopters). Recall that the power of a lens in diopters is the reciprocal of its focal length in meters.

When the eye is focused on closer objects, the lens becomes thicker, and the radius of curvature and focal length are decreased. The lens power may increase to 30 D, or even more in young children. The adjustment of the focal length of the crystalline lens is called **accommodation**. (Look at a nearby object and then one in the distance and notice how fast accommodation takes place. It's practically instantaneous.)

The focusing adjustment of the eye is unlike that of a camera. A camera lens has a constant focal length, and the image distance is varied by moving the lens to produce sharp images on the film for different object distances. In the eye, the image distance is constant, and the focal length of the lens (or its radius of curvature) is varied to produce sharp images on the retina for different object distances.

Use a model of the human eye if available. Otherwise mount a single lens and screen on an optical bench.

Rods: dim light vision
Cones: color vision

This relationship is presented in Eq. 22.10, Section 22.5

Age (years)	Near Point (centimeters)
10	10
20	12
30	15
40	25
50	40
60	100

The extremes of the range over which distinct vision (sharp focus) is possible are known as the far point and the near point. The **far point** is the greatest distance at which the normal eye can see objects clearly and is taken to be infinity. The **near point** is the position closest to the eye at which objects can be seen clearly and depends on the extent the lens can be deformed (thickened) by accommodation. The range of accommodation gradually diminishes with age as the crystalline lens loses its elasticity. That is, the near point gradually recedes with age. The approximate positions of the near point at various ages are listed in Table 24.1.

Children can see sharp images of objects that are within 10 cm (4 in.) of their eyes, and the crystalline lens of a normal young adult eye can be deformed to produce sharp images of objects as close as 12–15 cm (5–6 in.). However, at about the age of 40, the near point normally moves beyond 25 cm (10 in.). You may have noticed people over the age of 40 holding reading material at some distance from their eyes to bring it within the range of accommodation. When the print gets too small or the arm too short, corrective reading glasses are the solution. The recession of the near point with age is not considered an abnormal defect of vision, since it proceeds at about the same rate in all normal eyes.

Vision Defects

Speaking of the "normal" eye implies the existence of eyes with defective vision. That this is the case is apparent from the number of people who wear glasses or contact lenses. The eyes of many people cannot accommodate within the normal range of 25 cm to infinity. These people have one of the two most common visual defects: nearsightedness (myopia) and farsightedness (hyperopia). Both of these conditions can usually be corrected (● Fig. 24.2).

Nearsightedness is the condition of being able to see nearby objects clearly but not distant objects. That is, the far point is not infinity but some nearer point. When an object beyond the far point is viewed, the rays come to focus in front of the retina (Fig. 24.2b). As a result, the image on the retina is blurred, or out of focus. As the object is moved closer to the eye, its image moves back toward the retina. If the object is moved within the far point, a sharp image is seen.

Nearsightedness arises because the eyeball is too long or perhaps because the curvature of the cornea is too great. Whatever the reason, the images of distant objects are focused in front of the retina. Appropriate diverging lenses will correct this. Such a lens causes the rays to diverge, and the eye focuses the image farther back so that it falls on the retina.

Farsightedness is the condition of being able to see distant objects clearly but not nearby objects. That is, the near point is not at the normal position but

● **FIGURE 24.2 Nearsightedness and farsightedness** **(a)** The normal eye produces sharp images on the retina for objects between the near point and the far point. **(b)** In a nearsighted eye, the image of a distant object is focused in front of the retina. This defect is corrected by using a diverging lens. **(c)** In a farsighted eye, the image of a nearby object is focused behind the retina. This defect is corrected by using a converging lens.

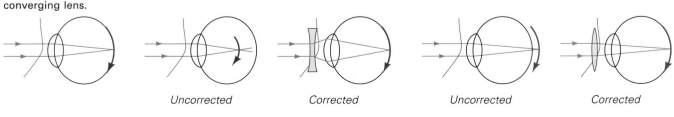

Uncorrected *Corrected* *Uncorrected* *Corrected*

(a) Normal **(b) Nearsightedness (myopia)** **(c) Farsightedness (hyperopia)**

at some point farther from the eye. The image of an object that is closer to the eye than the near point is formed behind the retina (Fig. 24.2c). Farsightedness arises because the eyeball is too short or perhaps because of insufficient curvature of the cornea. Also, there is a normal receding of the near point as discussed previously.

Farsightedness is usually corrected by appropriate converging lenses. Such a lens causes the rays to converge, and the eye focuses the image on the retina. Converging lenses are also used to correct the farsightedness associated with the natural recession of the near point with age. Older people often must wear reading glasses, which have converging lenses.

Have students find the near point by moving this book closer to the eye and then measuring the distance.

Explain the effects of aging upon the eye.

EXAMPLE 24.1 ■ Correcting Nearsightedness

A certain nearsighted person cannot see objects clearly when they are more than 78 cm from either eye. (a) What power must corrective lenses have if this person is to see distant objects clearly? (b) The person's near point is 23 cm. How is this affected when viewing through the corrective lenses? Assume that the lenses are in eyeglasses and are 3.0 cm in front of the eye. (See ● Figure 24.3.)

Solution. Listing the given distances, we have

Given: $d_f = 78$ cm   **Find:** (a) P (in diopters)
        $d_n = 23$ cm   (b) Near point (with glasses)
         $d = 3.0$ cm

(a) Here we have a lens problem similar to those in Chapter 22. Recall that the power of a lens is $P = 1/f$ (Eq. 22.10). We can use the thin-lens equation (Eq. 22.6) to find f or $1/f$ if we can determine the object and image distances, d_o and d_i.

The lens must effectively put the image of a distant object ($d_o = \infty$) at the far point d_f, which is 78 cm from the eye. The image, which acts as an object for the eye, is then within the range of accommodation. The image distance is *measured from the lens*, which is 3.0 cm from the eye, so $d_i = -75$ cm. ($d_i = d_f - d = 78$ cm $- 3.0$ cm $= 75$ cm; see the figure, which is not to scale.) A minus sign is used because the image is virtual, being on the object side of the lens. (You may recall from Chapter 22 that diverging lenses can form only virtual images.)

Review Examples 22.5 and 22.6

Image formation by a diverging lens is discussed in Section 22.3; see Figure 22.15d

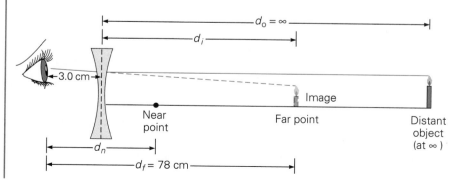

Figure 24.3

● **FIGURE 24.3 Correcting nearsightedness** A diverging lens is used. Example 24.1.

Then, using the thin lens equation,

$$\frac{1}{f} = \frac{1}{d_o} + \frac{1}{d_i} = \frac{1}{\infty} - \frac{1}{75 \text{ cm}} = -\frac{1}{75 \text{ cm}} \qquad \text{or} \qquad f = -75 \text{ cm}$$

The minus sign indicates a diverging lens. Recall that the lens power (P) in diopters (D) is the reciprocal of the focal length *in meters* (Eq. 22.10). Thus,

$$P = \frac{1}{f} = -\frac{1}{0.75 \text{ m}} = -1.33 \text{ D}$$

A negative, or diverging, lens with a power of 1.33 D is needed.

(b) To determine the effect of wearing glasses on near point vision, we must find the shortest distance at which an object can still be seen clearly—i.e., the minimum d_o with glasses. Thus we need to calculate where the *object* must be so that its *image* formed by the corrective lens is at the near point (d_n), 23 cm from the eye. The image is then 23 cm − 3.0 cm = 20 cm *from the lens*, so $d_i = -20$ cm (a minus sign is used since the image is virtual). Then, with $f = -75$ cm,

$$d_o = \frac{d_i f}{d_i - f} = \frac{(-20 \text{ cm})(-75 \text{ cm})}{-20 \text{ cm} - (-75 \text{ cm})} = 27 \text{ cm}$$

This means that the near point for this person is 27 cm in front of the lenses when the glasses are being worn. If an object is closer than this, its image (which acts as an object for the eye) will be inside the near point and so will not be seen clearly.

If the near point is changed by corrective lenses as in Example 24.1 and this causes a problem, bifocal lenses can be used. Bifocals were invented by Ben Franklin, who glued two lenses together. They are now made by grinding lenses with different curvatures in two different regions. Both nearsightedness and farsightedness can be treated at the same time with bifocals. (There are also trifocals, with lenses having three different curvatures.)

EXAMPLE 24.2 ■ Correcting Farsightedness

A farsighted person has a near point of 75 cm for one eye and a near point of 100 cm for the other. What powers should contact lenses have to allow the person to see an object clearly at a distance of 25 cm?

Solution.

Given: $d_{i_1} = -75 \text{ cm} = -0.75 \text{ m}$ *Find:* P_1 and P_2 (Lens power
$ d_{i_2} = -100 \text{ cm} = -1.0 \text{ m}$ for each eye)
$ d_o = 25 \text{ cm} = 0.25 \text{ m}$

The eyes are distinguished as 1 and 2. Keep in mind that the optics of a person's eyes are usually different, and a different lens may be needed for each eye. In this case, each lens is to form an image at its eye's near point of an object that is at a distance (d_o) of 25 cm. This image will then act as an

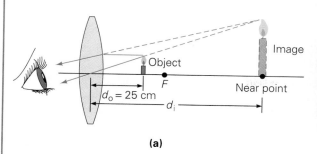

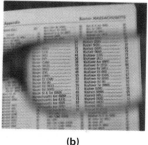

(a)

(b)

FIGURE 24.4 Reading glasses and correcting nearsightedness (a) When an object at the normal near point (25 cm) is viewed through reading glasses with converging lenses, the image is formed farther away but within the eye's range of accommodation (beyond the receded near point). (b) Small print as viewed through the lens of reading glasses. Example 24.2.

object within the eye's range of accommodation. This situation is illustrated in ●Fig. 24.4 and corresponds to a person wearing reading glasses. (For the sake of clarity, the lens in the figure is not in contact with the eye.)

 Figure 24.4

The image distances are negative since the images are virtual. With contact lenses, the distance from the eye and the distance from the lens are taken to be the same. Then

$$P_1 = \frac{1}{f_1} = \frac{1}{d_{o_1}} + \frac{1}{d_{i_1}} = \frac{1}{0.25 \text{ m}} - \frac{1}{0.75 \text{ m}}$$

$$= \frac{2}{0.75 \text{ m}} = +2.7 \text{ D}$$

and

$$P_2 = \frac{1}{f_2} = \frac{1}{d_{o_2}} + \frac{1}{d_{i_2}} = \frac{1}{0.25 \text{ m}} - \frac{1}{1.0 \text{ m}}$$

$$= \frac{3}{1.0 \text{ m}} = +3.0 \text{ D}$$

Note that these are positive, or converging, lenses.

Another common defect of vision is **astigmatism**, which is usually due to a refractive surface, most usually the cornea or crystalline lens, being out of round (nonspherical). As a result, the eye has different focal lengths in different planes (●Fig. 24.5a). Points may appear as lines, and the image of a line may be distinct in one direction and blurred in another or blurred in both directions in the circle of least confusion (the location of the least distorted image). A test for astigmatism is given in Fig. 24.5b.

Astigmatism may be corrected with lenses that have greater curvature in the plane in which the cornea or crystalline lens has deficient curvature (Fig.

FIGURE 24.5 Astigmatism When one of the eye's refracting components is not spherical, the eye has different focal lengths in different planes. (a) The effect occurs because rays along the transverse and radial axes are focused at different points, F_t and F_r, respectively. (b) If the eye is astigmatic, some or all of the lines in this diagram will appear blurred. (c) Nonspherical lenses, such as plano-convex cylindrical lenses, are used to correct astigmatism.

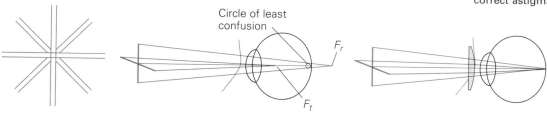

Circle of least confusion

F_r

F_t

(a) Test for astigmatism   **(b) Uncorrected astigmatism**   **(c) Corrected by lens**

24.5c). Astigmatism is lessened in bright light because the pupil of the eye becomes smaller and the circle of least confusion is reduced.

QUESTION: What is meant by 20/20 vision?

ANSWER: Visual acuity is a measure of how vision is affected by object distance. This is commonly measured using a chart of letters placed at a given distance from the eyes. The result is usually expressed as a fraction: the numerator is the distance at which the tested eye sees a standard symbol, such as the letter E, clearly, and the denominator is the distance at which the letter is seen clearly by a normal eye. A 20/20 rating, which is sometimes called perfect vision, means that at a distance of 20 ft, the eye being tested can see standard-sized letters as clearly as can a normal eye.

A person's eyes may have differing visual acuity; for example, one eye may have 20/20 vision and the other 20/30. The latter means that the eye can read standard-sized letters from a chart at a distance of 20 ft and someone with normal vision could do so at 30 ft. That is, the eye in question has to be closer to the chart than a normal eye would to see the letters clearly.

If time permits, use an eye chart to explain 20/20 vision, 20/30 vision, etc.

24.2 ▪ Microscopes

Microscopes are used to magnify objects so that we can see more detail or see features that are normally indiscernible. Two basic types of microscope will be considered here.

The Magnifying Glass (Simple Microscope)

When we look at an object in the distance, it appears to be very small. As it is brought or comes closer to our eyes, it appears larger. How large an object appears depends on the size of the image on the retina. This may be related to the angle subtended by the object (● Fig. 24.6); the greater the angle, the larger the image.

When we want to examine detail, or look at something closely, we bring it close to our eyes so that it subtends a greater angle. For example, you may examine the detail of a figure in this book by bringing it closer to your eyes. The greatest amount of detail will be seen when the book is at your near point (assuming you are not wearing glasses). If your eyes were able to accommodate to shorter distances, an object brought very close to them would be further magnified. However, as you can easily prove by bringing this book very close

●**FIGURE 24.6 Magnification and angle** (a) How large an object appears may be related to the angle subtended by the object. (b) The angle and the size of an object may be increased by using a converging lens.

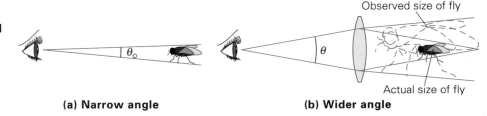

(a) Narrow angle   **(b) Wider angle**

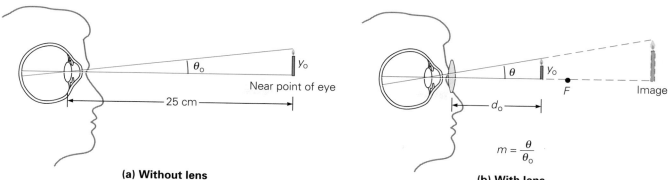

(a) Without lens

(b) With lens

$$m = \frac{\theta}{\theta_{\mathrm{o}}}$$

● **FIGURE 24.7 Angular magnification** The angular magnification (m) of a lens is defined as the ratio of the angular size of an object viewed through the lens and the angular size of the object viewed without the lens, $m = \theta/\theta_{\mathrm{o}}$.

to your eyes, images are blurred when objects are at distances inside the near point.

A **magnifying glass**, which is simply a single convex lens (sometimes called a simple microscope), allows a clear image to be formed of an object that is closer than the near point. In such a position, an object subtends a greater angle and therefore appears larger, or magnified. The lens produces a virtual image beyond the near point on which the eye focuses (● Fig. 24.7). If a hand-held magnifying glass is used, its position is usually adjusted until this image is seen clearly.

As illustrated in Fig. 24.7, the angle subtended by an object is much greater when a magnifying glass is used. The magnification of an object *viewed through a magnifying glass* is expressed in terms of this angle. This **angular magnification**, or magnifying power, is symbolized by m. The angular magnification is defined as the ratio of the angular size of the object viewed through the magnifying glass (θ) to the angular size of the object viewed without the magnifying glass (θ_{o}):

Angular magnification—different from lateral magnification (Chapter 22).

$$m = \frac{\theta}{\theta_{\mathrm{o}}} \qquad (24.1)$$

angular magnification

(Note that this is not the same as M, the lateral magnification, which is a ratio of heights: $M = h_{\mathrm{i}}/h_{\mathrm{o}}$.)

The maximum angular magnification occurs when the image seen through the glass is at the eye's near point, that is, $d_{\mathrm{i}} = -25$ cm, since this is as close as it can be clearly seen. (A value of 25 cm will be assumed to be typical for near point in this discussion. The minus sign is used because the image is virtual.) The corresponding object distance may be calculated from the thin lens equation (Chapter 22):

Lateral magnification is discussed in Section 22.3; see Eq. 22.7.

$$d_{\mathrm{o}} = \frac{d_{\mathrm{i}}f}{d_{\mathrm{i}} - f} = \frac{(-25 \text{ cm})f}{-25 \text{ cm} - f}$$

or

$$d_{\mathrm{o}} = \frac{25 f}{25 + f} \qquad (24.2)$$

where f is in centimeters.

The angular sizes of the object are related to its height by (see Fig. 24.7)

$$\tan \theta_{\mathrm{o}} = \frac{y_{\mathrm{o}}}{25} \qquad \text{and} \qquad \tan \theta = \frac{y_{\mathrm{o}}}{d_{\mathrm{o}}}$$

Assuming that a small-angle approximation ($\tan \theta \cong \theta$) is valid gives

$$\theta_o \cong \frac{y_o}{25} \quad \text{and} \quad \theta \cong \frac{y_o}{d_o}$$

Then the maximum angular magnification may be expressed as

$$m = \frac{\theta}{\theta_o} = \frac{y_o/d_o}{y_o/25} = \frac{25}{d_o}$$

Substituting for d_o from Eq. 24.2 gives

$$m = \frac{25}{25f/(25+f)}$$

which simplifies to

Magnifying glass magnification

$$\boxed{m = 1 + \frac{25 \text{ cm}}{f}} \qquad (24.3)$$

angular magnification
for image at near point (25 cm)

Thus, lenses with shorter focal lengths give greater angular magnifications.

In the derivation of Eq. 24.3, the object for the unaided eye was taken to be at the near point, as was the image viewed through the lens. Actually, the normal eye can focus on an image located anywhere between the near point and infinity. At the extreme where the image is at infinity, the eye is more relaxed (the muscles attached to the crystalline lens are relaxed, and the lens is thin). For the image to be at infinity, the object must be at the focal point of the lens. In this case,

$$\theta \cong \frac{y_o}{f}$$

and the angular magnification is simply

$$\boxed{m = \frac{25 \text{ cm}}{f}} \qquad (24.4)$$

angular magnification
for image at infinity

Mathematically, it seems that the magnifying power may be increased to any desired value by using lenses with short enough focal lengths. Physically, however, lens aberrations limit the practical range of a single magnifying glass to about $3\times$ or $4\times$ (read as "three ex" and "four ex"), or a sharp image magnification of three or four times the size of the object when used normally.

EXAMPLE 24.3 ■ Magnifying Glass Magnification

A person uses a converging lens with a focal length of 12 cm to examine the fine detail on a painting. (a) What is the maximum magnification given by the lens? (b) What is the magnification for relaxed eye viewing?

Solution.

Given: $f = 12$ cm

Find: (a) m (d_i = near point)
(b) m ($d_i = \infty$)

(a) The maximum magnification occurs when the image formed by the lens is at the near point of the eye, which for Eq. 24.3 was taken to be 25 cm. Using that equation,

$$m = 1 + \frac{25 \text{ cm}}{f} = 1 + \frac{25 \text{ cm}}{12 \text{ cm}} = 3.1$$

(b) The eye is most relaxed when viewing distant objects. Eq. 24.4 gives the magnification for the image formed by the lens at infinity:

$$m = \frac{25 \text{ cm}}{f} = \frac{25 \text{ cm}}{12 \text{ cm}} = 2.1$$

The Compound Microscope

A compound microscope provides greater magnification than is attained with a single lens, or simple microscope. A basic **compound microscope** consists of a pair of converging lenses, each of which contributes to the magnification (Fig. 24.8a). A converging lens having a relatively short focal length ($f_o < 1$ cm) is known as the **objective**. It produces a real, inverted, and enlarged image of an object positioned slightly beyond its focal point. The other lens, called the **eyepiece**, or **ocular**, has a longer focal length (f_e is a few centimeters) and is positioned so that the image formed by the objective falls just *inside* its focal point. This lens forms a magnified virtual image that is viewed by the observer. In essence, the objective acts as a projector, and the eyepiece is a simple magnifying glass used to view the projected image.

The total magnification (m_t) of a lens combination is the product of the magnifications produced by the two lenses. The image formed by the objective is greater in size than its object by a factor M_o equal to the lateral magnification ($M_o = d_i/d_o$, with the minus sign omitted). In Fig. 24.8a, note that the image distance for the objective lens is approximately equal to L, the distance between the lenses. That is, $d_i \cong L$. (The image I_o is formed by the objective just inside the focal point of the eyepiece, which has a very short focal length.) Also, since the object is very close to the focal point of the objective, $d_o \cong f_o$. With

Review Section 22.3 and Fig. 22.16.

Figure 24.8a

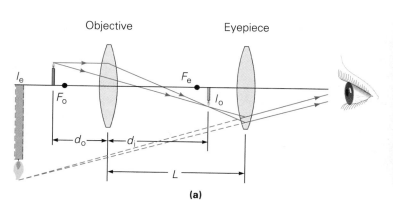

(a)

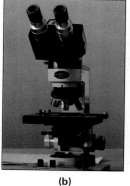

(b)

FIGURE 24.8 The compound microscope (a) In the optical system of a compound microscope, the real image formed by the objective falls just within the focal point of the eyepiece (F_e) and acts as an object for this lens. An observer looking through the eyepiece sees an enlarged image. **(b)** A compound microscope.

these approximations,

$$M_o \cong \frac{L}{f_o}$$

The angular magnification of the eyepiece for an object at the focal point is given by Eq. 24.4:

$$m_e = \frac{25 \text{ cm}}{f_e}$$

Since the object for the eyepiece (the image formed by the objective) is very near the focal point of the eyepiece, a good approximation is

$$m_t \simeq M_o m_e \simeq \left(\frac{L}{f_o}\right)\left(\frac{25 \text{ cm}}{f_e}\right)$$

Compound microscope magnification

or

$$\boxed{m_t \simeq \frac{(25 \text{ cm}) L}{f_o f_e}} \qquad (24.5)$$

angular magnification of
compound microscope

Remind students that all units are cm in this equation.

where f_o, f_e, and L are in centimeters.

EXAMPLE 24.4 ■ Magnification with a Compound Microscope

A microscope has an objective with focal length of 10 mm and an eyepiece with a focal length of 4.0 cm. The lenses are fixed at 20 cm apart in the barrel. Determine the approximate total magnification of the microscope.

Solution.

Given: $f_o = 10 \text{ mm} = 1.0 \text{ cm}$ *Find:* m_t (total magnification)
$f_e = 4.0 \text{ cm}$
$L = 20 \text{ cm}$

Using Eq. 24.5,

$$m_t \simeq \frac{(25 \text{ cm}) L}{f_o f_e} = \frac{(25 \text{ cm})(20 \text{ cm})}{(1.0 \text{ cm})(4.0 \text{ cm})} = 125 \times$$

(Note the relatively short focal length of the objective.)

A modern compound microscope is shown in Fig. 24.8b. Interchangeable eyepieces with magnifications from about $5\times$ to over $100\times$ are available. For standard microscopic work in biology or medical laboratories, $5\times$ and $10\times$ eyepieces are normally used. Microscopes are often equipped with rotating turrets, which usually contain three objectives for different magnifications, for example, $10\times$, $43\times$, and $97\times$. These objectives and the $5\times$ and $10\times$ eyepieces can be used in various combinations to provide magnifying powers from $50\times$ to $970\times$. The maximum magnification with a compound microscope is about $2000\times$.

Opaque objects are usually illuminated with a light source placed above them. In many instances, the specimens to be viewed are transparent, such as glass slides containing cells or thin sections of tissues. These are illuminated with a light source beneath the microscope stage so that light passes through the specimen. A modern microscope is usually equipped with a light condenser (converging lens) and diaphragm below the stage, which are used to concentrate the light and control its intensity. A microscope may have an internal light source. The light is reflected into the condenser from a mirror. Older microscopes have two mirrors with reflecting surfaces: one a plane mirror for reflecting light from a high-intensity external source, and the other a concave mirror for converging low-intensity light such as sky light.

Use an old microscope to point out features of the compound microscope.

24.3 ■ Telescopes

Telescopes apply the optical principles of mirrors and lenses to improve our ability to see distant objects. Used for both terrestrial and astronomical observations, telescopes allow some objects to be viewed in greater detail and other fainter or more distant objects simply to be seen. Basically, there are two types of telescopes: refracting and reflecting. These depend on the gathering and converging of light by lenses and mirrors, respectively.

Refracting Telescope

The principle behind one type of **refracting telescope** is similar to that behind a compound microscope. The major components of this type of telescope are objective and eyepiece lenses, as illustrated in ● Fig. 24.9. The objective is a large converging lens with a long focal length, and the movable eyepiece has a relatively short focal length. Rays from a distant object are essentially parallel and form an image (I_o) at the focal point (F_o) of the objective. This image acts as an object for the eyepiece, which is moved until the image lies just inside its focal point (F_e). A large, inverted image (I_e) is seen by an observer.

For relaxed viewing, the eyepiece is adjusted so that its image (I_e) is at

Demonstration/activity: Pass out two lenses to each student (or group), one lens with a focal length of about 10 cm and another with a focal length of about 20 cm. Show how to make a microscope by holding the long lens close to the eye and viewing a nearby object through the short lens. Then show how to make a telescope by holding the short lens close to the eye and viewing a distant object through the long lens.

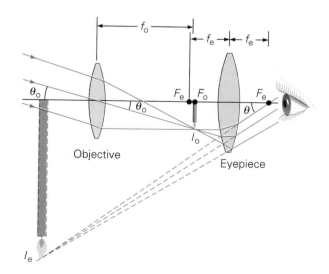

Objective

Eyepiece

Figure 24.9

● **FIGURE 24.9 The refracting astronomical telescope** In an astronomical telescope, rays from a distant object form an intermediate image at the focal point of the objective (F_o). The eyepiece is moved so that the image is at or slightly inside its focal point (F_e). An observer sees an enlarged image at infinity (shown at a finite distance here for illustration).

infinity, which means that the objective image (I_o) is at the focal point of the eyepiece (f_e). As you can see in the figure, the distance between the lenses is then the sum of the focal lengths ($f_o + f_e$), which is the length of the telescope tube. The magnifying power of a telescope focused for the final image at infinity can be shown (see Exercise 49) to be

Refracting telescope magnification

$$m = -\frac{f_o}{f_e} \qquad (24.6)$$

angular magnification
of refracting telescope

where the minus is inserted to indicate the image is inverted as in our lens sign convention in Chapter 22. Thus, to achieve the greatest magnification, the focal length of the objective should be made as great as possible and the focal length of the eyepiece as short as possible.

Astronomical telescope—inverted image

The telescope illustrated in Fig. 24.9 is called an **astronomical telescope**. Its final image is inverted, but this poses little problem to astronomers. (Why?) However, someone viewing an object on Earth through a telescope finds it more convenient to have an upright image. A telescope in which the final image is upright is called a **terrestrial telescope**. An upright final image can be obtained in several ways—two are illustrated in Fig. 24.10.

Galileo did not invent the telescope, but he built one after hearing about a Dutch instrument that could be used for distant viewing.

In the telescope diagrammed in ● Fig. 24.10a, a diverging lens is used as an eyepiece. This type of terrestrial telescope is referred to as a Galilean telescope, because Galileo built one in 1609. The image of the objective is formed at the

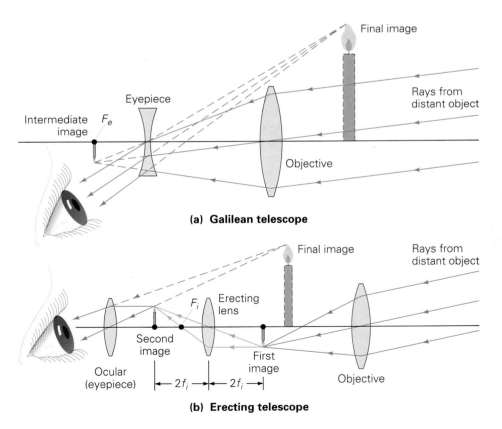

●**FIGURE 24.10 Terrestrial telescopes** (a) A Galilean telescope uses a diverging lens as an eyepiece, producing upright images. (b) Another way to produce upright images is to use a converging "erecting" lens between the objective and eyepiece in an astronomical-type telescope. This elongates the telescope, but the length may be shortened by using internally reflecting prisms.

(a) **Galilean telescope**

(b) **Erecting telescope**

focal point of the eyepiece (F_e) on the side opposite the objective. Thus, the light rays pass through the eyepiece before the image is formed, and the image acts as a "virtual" object for the eyepiece (see Section 22.3). An observer sees a magnified, upright, virtual image. (Note that with a diverging lens and negative focal length ($-f_e$), Eq. 24.6 gives a $+m$, indicating an upright image.)

Galilean telescopes have several disadvantages, most notably very narrow fields of view and limited magnification. A better type of terrestrial telescope, illustrated in Fig. 24.10b, uses a third lens, called the erecting lens, between converging objective and eyepiece lenses. If the image is formed by the objective at a distance that is twice the focal length of the intermediate erecting lens ($2f_i$), then the lens merely inverts the image without magnification, and the telescope magnification is still given by Eq. 24.6.

However, to achieve the upright image in this way requires a greater telescope length. Using the intermediate erecting lens to invert the image increases the length of the telescope by four times the focal length of the erecting lens ($2f_i$ on each side). An erecting lens is used in a spyglass, which is telescoped to achieve the necessary length. The inconvenient length can be avoided by using internally reflecting prisms. This is the principle of prism binoculars, which are double telescopes—one for each eye (● Fig. 24.11).

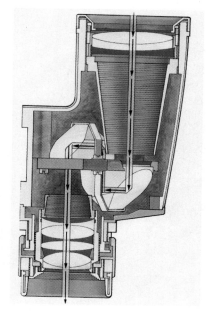

● **FIGURE 24.11 Prism binoculars**
A schematic cutaway view of one ocular showing the internal reflections in the prisms, which reduce the overall length.

EXAMPLE 24.5 ■ Astronomical and Terrestrial Telescopes

An astronomical telescope has an objective lens with a focal length of 30 cm and an eyepiece with a focal length of 9.0 cm. (a) What is the magnification of the telescope? (b) If an erecting lens with a focal length of 7.5 cm is used to convert the telescope to a terrestrial type, what is the overall length of the telescope tube?

Solution. Listing the data, we have

Given: $f_o = 30$ cm   *Find:* (a) m (magnification)
$f_e = 9.0$ cm   (b) L (length of
$f_i = 7.5$ cm (intermediate   telescope tube)
erecting lens)

(a) The magnification is given by Eq. 24.6:

$$m = -\frac{f_o}{f_e} = -\frac{30 \text{ cm}}{9.0 \text{ cm}} = -3.3\times$$

where the minus sign indicates that the final image is inverted.
(b) Taking the length of the astronomical tube to be the distance between the lenses, this is the sum of their focal lengths.

$$L_1 = f_o + f_e = 30 \text{ cm} + 9.0 \text{ cm} = 39 \text{ cm}$$

Adding the intermediate erecting lens adds four times the focal length of this lens ($4f_i$) to the length of the scope (Fig. 24.10b). The overall length is then

$$L = L_1 + L_2 = 39 \text{ cm} + 4f_i = 39 \text{ cm} + 4(7.5 \text{ cm}) = 69 \text{ cm}$$

Hence the telescope length is over two-thirds of a meter, with an upright image, but the same magnification. (Why?)

EXAMPLE 24.6 ■ Optical Instruments

A student is given two converging spherical lenses, one with a focal length of 5.0 cm and the other with a focal length of 20 cm. To construct a telescope to best view distant objects with these lenses, the student should hold the lenses (a) more than 25 cm apart, (b) less than 25 cm but more than 20 cm apart, (c) less than 20 cm but more than 5.0 cm apart, or (d) less than 5.0 cm apart. *Clearly establish the reasoning and physical principle(s) used in determining your answer before checking it below. That is, **why** did you select your answer?*

Reasoning and Answer. The only type of telescope that can be constructed with two converging lenses is an astronomical telescope. In this type of telescope, the lens with the longer focal length is used as an objective lens to produce a real image of a distant object. That image is then viewed with the lens with the shorter focal length (the eyepiece), used as a simple magnifier. If the object is at a great distance, a real image is formed by the objective lens in the focal plane of the lens. This image acts as the object for the eyepiece, which is positioned so that the image/object lies just inside its focal point so as to produce a large, inverted second image (Fig. 24.9).

This means that the two lenses must be *slightly* less than 25 cm apart. Thus, answer (a) is not correct. Answers (c) and (d) are also not correct because the eyepiece would be too close to the objective to get the large secondary image needed for optimal viewing of a distant object. (In these cases, the rays would pass through the second lens before the image was formed, and a *reduced* image might be produced (see Section 22.3 and Fig. 22.17b). Thus, (b) is the correct answer with the objective image just inside the eyepiece focal point.

Follow-up Exercise. A third converging lens with a focal length of 4.0 cm is used with the above two lenses to produce a terrestrial telescope in which the third lens does nothing more than invert the image. How should the lenses be positioned, and what distances should there be between them, for the final image to be of maximum size and upright? (*Reasoning and answer may be found in the Answers to Exercises section at the back of the book.*)

Reflecting Telescope

For viewing the Sun, moon, and nearby planets, magnification is necessary to see details. However, even with the highest feasible magnification, stars appear only as faint points of light. For distant stars and galaxies, it is more important to gather enough light so that the object can be seen at all. The intensity of light from a distant source is very weak. In many instances, such a source may be detected only when the light is gathered and focused on a photographic plate over a long period of time.

Recall that intensity is energy per *area* per unit time. Thus, more light can be gathered if the size of the objective is increased. This increases the distance at which the telescope can detect faint objects such as distant galaxies. (Recall that light intensity of a point source is inversely proportional to the *square* of the distance between the source and the observer.) However, producing a large lens involves difficulties associated with glass quality, grinding, and polishing. Compound lens systems are required to reduce aberrations, and a very large lens may sag under its own weight, producing further aberrations. The largest objective lens in use has a diameter of 40 in. (102 cm, or about 1 m) and is part of the refracting telescope of the Yerkes Observatory at Williams Bay, Wisconsin (●Fig. 24.12).

These problems are reduced with a **reflecting telescope** utilizing a large, concave, front-surface parabolic mirror (●Fig. 24.13). A parabolic mirror does not exhibit spherical aberration and a mirror has no inherent chromatic aberration. High-quality glass is not needed, since the light is reflected by a mirrored front surface. And only one surface has to be ground, polished, and silvered.

As shown in Fig. 24.13a, light from a distant source entering the telescope tube is focused by the concave mirror. A small plane mirror is sometimes used to direct the converging rays to a more convenient viewing location (Fig. 24.13b). Such a mirror blocks only a small percentage of the incoming light. A third type of focus is shown in Fig. 24.13c.

The largest reflecting telescope in the United States, located at Hale Observatories on Palomar Mountain in California, has a mirror with a 200-in. (5.1-m) diameter (●Fig. 24.14a). In the Hale design, the observer can actually sit in a cage at the prime focus of the telescope (Fig. 24.14b). The world's largest single mirror reflecting telescope has a reflector with a 6-m diameter and is located in the former Soviet Union (Fig. 24.14c).

Even though reflecting telescopes have advantages over refracting telescopes, they also have their own problems. Like a large lens, a large mirror may sag under its own weight, and the weight necessarily increases with the size of the mirror. The weight factor is also a cost factor for construction of a telescope, since the supporting elements for a heavier mirror must be more massive.

These problems are addressed by two new technologies described in the first of the chapter Insight features. The second Insight feature points out that all telescopes are not optical in nature.

●**FIGURE 24.12 Yerkes refracting telescope** The largest refracting telescope, at the Yerkes Observatory at Williams Bay, Wisconsin, has an objective lens with a diameter of 40 in. or 102 cm.

●**FIGURE 24.13 Reflecting telescopes** A concave mirror may be used in a telescope to converge light to form an image of a distant object. The image may be at the prime focus as shown in **(a)**; or a small mirror and lens may be used to focus the image outside the telescope, as in **(b)**, called a Newtonian focus. **(c)** Another arrangement, called a Cassegrain focus, uses a mirror to form the image below the mirror.

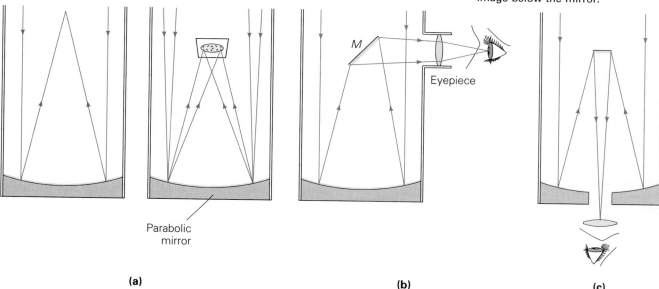

Parabolic mirror

(a)

Eyepiece

(b)

(c)

(a)

(b)

(c)

● **FIGURE 24.14 Reflecting telescopes** **(a)** The Hale telescope on Palomar Mountain in California, with a mirror 200 in. (5.1 m) in diameter, is the largest single-mirror reflecting telescope in the United States. The mirror is at the bottom center, just above the heads of the people. **(b)** An astronomer in the observing capsule near the top of the telescope tube, at the prime focus. (Compare Figure 24.13a.) **(c)** The world's largest single-mirror reflecting telescope, located in the Caucasus Mountains of the former Soviet Union, has a mirror with a diameter of 6 m. (Note the size of the person standing on the observatory floor for comparison.) The dark flaps that surround the mirror fold down to protect it when not in use.

INSIGHT ■ A New Generation of Telescopes

As mentioned in the text, large mirrors are more easily constructed than large lenses. This fact is crucial for the production of astronomical telescopes, since the larger the mirror, the more light-gathering capability it has, allowing faint objects to be seen more clearly and distant objects to be discovered. Even mirrors, however, cannot be made ever larger without encountering technical problems. To overcome these obstacles, several new technologies are being employed to build reflecting telescopes with greater light-gathering power. The trick is to build large mirrors that are still light enough so that they don't sag under their own weight, or to find ways of simulating the performance of such mirrors. Cost is of course also a factor.

One approach is the use of an array of small mirrors, coordinated so as to function as a large, single mirror. An example is the Multiple Mirror Telescope (MMT) on Mount Hopkins, Arizona (Fig. 1a). In operation since 1979, the MMT utilizes six 1.8-m mirrors to simulate the light-

gathering and resolving power of a single mirror of 4.5 m. A more sophisticated recent example is the Keck telescope at Mauna Kea in Hawaii (Fig. 1b). Its mirror consists of 36 hexagonal segments that are computer positioned to give the effect of a 10-m mirror.

A similar method has been proposed in Europe by the eight-nation European Southern Observatory (ESO). The plan calls for the construction of the Very Large Telescope, which would be the world's largest optical telescope. It will consist of four 8-m mirrors, the combined images of which will be equivalent to that produced by a single mirror 16 m in diameter. The four mirrors will stand in a row, each housed in a separate structure.

A mirror 8 m in diameter made from a single glass slab of normal thickness would be too heavy to maintain its shape with the precision that such a telescope requires. Instead, the ESO telescope mirrors will be thin and mounted on a support system controlled by a computer to maintain their overall shape. A technique for the pro-

(a)

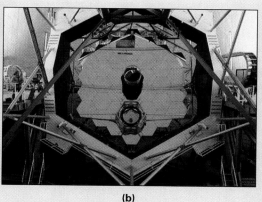

(b)

FIGURE 1 Multiple-mirror telescopes **(a)** The Multiple Mirror Telescope (MMT) on Mount Hopkins in Arizona employs six separate mirrors, each 1.8 m in diameter. Together, they perform like a single mirror 4.5 m in diameter. It is much cheaper and easier to construct a number of distortion-free small mirrors than a single equivalent large mirror. **(b)** The recently-completed Keck telescope on Mauna Kea, Hawaii, has a mirror consisting of 36 separate segments, positioned by computer to function like a single 10-m mirror.

duction of large mirrors is being worked on in this country too. The new design calls for a single piece of ultrathin glass that is kept rigid by a honeycomb structure behind its polished surface. Roger Angel at the University of Arizona makes these large, lightweight mirrors by "spin-casting" them in a rotating furnace (Fig. 2). The melted glass takes on a parabolic shape because of the rotation.

Another way of extending our view into space is to put telescopes into orbit about the Earth. Above the atmosphere, the view is unaffected by the twinkling effect of atmospheric turbulence and refraction, and there is no background problem from city lights. An infrared telescope was first put into orbit in the late 1980s. It surveyed the "infrared sky" and sent back a great deal of data before it stopped transmitting.

In 1990, the optical Hubble space telescope (HST) was launched into orbit (Fig. 3). Even with a mirror of only 2.4 m, its privileged position was expected to allow the HST to produce images seven times clearer than those formed by Earthbound telescopes.

The HST's mirror was constructed in 1980, and apparently the optical system used to test it was faulty. The unfortunate result was spherical aberration (Section 22.2), which causes the images to be somewhat blurred. Even so, the HST is still able to make many observations that are superior to those made from Earth's surface. Scientists can also use computers to process the images and reduce the blur in some cases.

A mission is planned for astronauts to carry up a replacement for the main camera and correcting lenses for instruments so as to bring the HST into full operation. Such missions to make repairs and to install updated equipment should make the HST a useful observatory well into the 21st century.

FIGURE 2 Spin-cast mirror Roger Angel checks the curvature of a 1.8-m spun-cast disk. Such lightweight mirrors are made in a rotating furnace, visible at the top of the photo.

FIGURE 3 Hubble Space Telescope (HST) A photo taken by the crew of the Space Shuttle Discovery showing the HST being deployed on April 25, 1990.

The word "telescope" usually brings to mind visual observations. However, the visible region is a very small part of the electromagnetic spectrum, and celestial objects emit radiation of many other types, including radio waves. This fact was discovered accidentally in 1931 by an electrical engineer named Carl Jansky while he was working on the problem of static interference with intercontinental radio communications. Jansky found an annoying static hiss that came from a fixed direction in space, apparently from a celestial source. It was soon apparent that radio waves are another source of astronomical information, and radio telescopes were built.

A radio telescope operates similarly to a reflecting light telescope. A reflector with a large area collects and focuses the radio waves at a point where a detector picks up the signal (Fig. 1). The parabolic collector, called a disk, is covered with metal wire mesh. Since the wavelengths of radio waves range from a centimeter or so to several meters, the wire mesh is a good reflecting surface for such waves.

Radio telescopes supplement optical telescopes and provide some definite advantages. Radio waves pass freely through the huge clouds of dust in our galaxy that hide a large part of it from visual observation. Also, radio

FIGURE 2 Several of the dish antennae that make up the Very Large Array (VLA) radio telescope near Socorro, New Mexico. There are 27 movable dishes, each 25 m in diameter, forming the array along a Y-shaped railway network. The data from all the antennae is combined to produce a single radio image. In this way it is possible to attain a resolution equivalent to that of one giant radio dish.

waves easily penetrate the Earth's atmosphere, which reflects and scatters a large percentage of the incoming visible light (Fig. 2).

The Earth's atmosphere absorbs many wavelengths of electromagnetic radiation and completely blocks some from reaching the ground. Water vapor, a strong absorber of infrared radiation, is concentrated in the lower part of the atmosphere near the surface of the Earth. Thus, observations with infrared telescopes are made from high-flying aircraft or from orbiting spacecraft.

The first orbiting infrared observatory was launched in 1983. Not only are atmospheric interferences eliminated in space, but a telescope may be cooled to a very low temperature without becoming coated with condensed water vapor. Cooling the telescope helps eliminate interfering infrared radiation generated by the telescope itself. The orbiting telescope launched in 1983 was cooled with liquid helium to about 10 K; it carried out an infrared survey of the entire sky.

The atmosphere is virtually opaque to ultraviolet radiation, X-rays, and gamma rays from distant sources. Orbiting satellites with telescopes sensitive to these types of radiation have mapped out portions of the sky, and other surveys are planned.

At the altitudes reached by orbiting satellites, even observations in the visible region are not affected by air motion and temperature refraction. Perhaps in the not too distant future, a permanently staffed orbiting observatory carrying a variety of telescopes will help expand our knowledge of the universe.

FIGURE 1 The 300-m diameter radio telescope at Arecibo, Puerto Rico, is the largest single radio dish in the world. The wire mesh of its reflecting surface (inset) hugs the concave interior of a natural bowl in the jungle. The receiver is seen suspended above the dish. Because it is fixed and unsteerable, the Arecibo telescope relies on the Earth's rotation to move its field of view across the sky.

24.4 ■ Diffraction and Resolution

The diffraction of light places a limitation on our ability to distinguish objects that are close together when using microscopes or telescopes. This effect can be understood by considering two point sources located far from a narrow slit of width d (● Fig. 24.15). The sources could be distant stars, for example. In the absence of diffraction, two bright spots, or images, would be observed on a screen. As you know from Section 23.3, however, the slit diffracts the light, and each image consists of a central maximum with a pattern of weaker bright and dark fringes on either side. If the sources are close together, the two central maxima may overlap. Then the images cannot be distinguished, or are unresolved. For the images to be resolved, the central maxima must not overlap appreciably.

In general, images of two sources can be resolved if the central maximum of one falls at or beyond the first minimum (dark fringes) of the other. This generally accepted limiting condition for the **resolution** of two diffracted images was first proposed by Lord Rayleigh (1842–1919), a British physicist, and is known as the **Rayleigh criterion**:

> Two images are said to be just resolved when the central maximum of one image falls on the first minimum of the diffraction pattern of the other image.

The Rayleigh criterion may be expressed in terms of the angular separation (θ) of the sources (see Fig. 24.15). The first minimum ($m = 1$) for a single-slit diffraction pattern satisfies this relationship:

$$d \sin \theta = \lambda$$

or

$$\sin \theta = \frac{\lambda}{d}$$

This gives the minimum angular separation for two images to be just resolved according to the Rayleigh criterion. In general, for visible light, the wavelength is much smaller than the slit width ($\lambda \ll d$), so a good approximation is $\sin \theta \cong \theta$. The limiting, or minimum, angle of resolution (θ_{min}) for a slit of width d is then

$$\boxed{\theta_{min} = \frac{\lambda}{d}} \quad (24.7)$$

minimum angle of
resolution for a slit

(Note that θ_{min} is a pure number and is therefore expressed in radians.) Thus, the images of two sources will be *distinctly* resolved if the angular separation of the sources is greater than λ / d.

The apertures (openings) of microscopes and telescopes are generally circular. Thus, there is a circular diffraction pattern around the central maximum, which is a bright circular disk (● Fig. 24.16). Detailed analysis for a circular aperture shows that the minimum angular separation for two objects for their images to be just resolved is

$$\boxed{\theta_{min} = \frac{1.22\lambda}{D}} \quad (24.8)$$

minimum angle of resolution
for a circular aperture

where D is the diameter of the aperture, and θ_{min} is in radians.

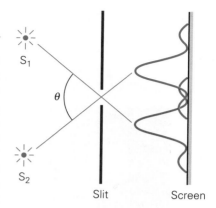

Slit   Screen

(a) Resolved

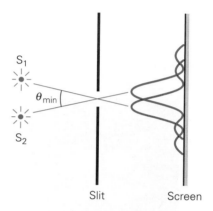

Slit   Screen

(b) Just resolved

● **FIGURE 24.15 Resolution** Two light sources in front of a slit produce diffraction patterns. **(a)** When the angle subtended by the sources at the slit is large enough for the diffraction patterns to be distinguishable, the images are said to be resolved. **(b)** At smaller angles, the central maxima are closer together. At θ_{min} the central maximum of one pattern falls on the first dark fringe of the other, and the images are said to be just resolved. For smaller angles, the patterns are unresolved.

 Figure 24.15

> *Estimate the diameter of the human iris and determine what features an astronaut in a 100-mile-high orbit can identify in her home town.*

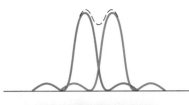

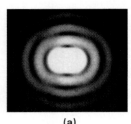

(a)

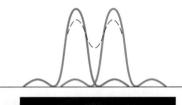

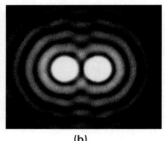

(b)

● **FIGURE 24.16 Circular aperture resolution** When the angular separation of two objects is very small, the images are not resolved. **(a)** The minimum angular separation for two images to be just resolved is given by the Rayleigh criterion: the central maximum of the diffraction pattern of one image falls on the first minimum of the diffraction pattern of the other image. **(b)** When the objects are farther apart, the images are well resolved.

Eq. 24.8 applies to the objective lens of a microscope or telescope, which may be considered to be a circular aperture for light. According to the equation, to make θ_{min} small so objects close together can be resolved, it is advantageous to make the aperture as large as possible. This is another reason for using large lenses (and mirrors) in telescopes, in addition to their greater light-gathering power.

EXAMPLE 24.7 ■ Resolution

Determine the minimum angle of resolution by the Rayleigh criterion for (a) the pupil of the eye (diameter of about 4.0 mm), (b) the Yerkes Observatory 40-in. (102-cm) refracting telescope for visible light with a wavelength of 660 nm, and (c) a radio telescope of 25-m diameter for radiation with a wavelength of 21 cm.

Solution.

Given: (a) $D = 4.0$ mm $= 4.0 \times 10^{-3}$ m   *Find:* (a) θ_{min} (minimum
(b) $D = 102$ cm $= 1.02$ m   angles of
$\lambda = 660$ nm $= 6.60 \times 10^{-7}$ m   resolution)
(c) $D = 25$ m   (b) θ_{min}
$\lambda = 21$ cm $= 0.21$ m   (c) θ_{min}

Eq. 24.8 applies to all three cases.
(a) For the eye,

$$\theta_{min} = \frac{1.22\,\lambda}{D} = \frac{(1.22)(6.60 \times 10^{-7}\,\text{m})}{4.0 \times 10^{-3}\,\text{m}} = 2.0 \times 10^{-4}\,\text{rad}$$

(b) For the light telescope,

$$\theta_{min} = \frac{(1.22)(6.60 \times 10^{-7}\,\text{m})}{1.02\,\text{m}} = 7.9 \times 10^{-7}\,\text{rad}$$

(c) For the radio telescope,

$$\theta_{min} = \frac{(1.22)(0.21\,\text{m})}{25\,\text{m}} = 0.010\,\text{rad}$$

The smaller the angular separation, the better the resolution. What do the results tell you?

For a microscope, it is convenient to specify the actual separation (s) between two point sources. Since the objects are usually near the focal point of the objective, to a good approximation

$$\theta_{min} = \frac{s}{f} \quad \text{or} \quad s = f\theta_{min}$$

where f is the focal length of the lens. (Here s is taken as the arc length subtended by θ_{min}, and $s = r\theta_{min} = f\theta_{min}$.) Then, using Eq. 24.8,

$$\boxed{s = f\theta_{min} = \frac{1.22\,\lambda f}{D}}$$

(24.9)

resolving power of microscope

This minimum distance between two points whose images can be just resolved is called the **resolving power** of the lens. Practically, the resolving power of a microscope indicates the ability of the objective to distinguish fine detail in specimens' structures.

As with a telescope, the *useful* magnification of a microscope is not limited by the power (focal lengths) of its lenses but by the aperture, that is, the size of the objective lens. However, making the objective lens larger leads to another limiting factor on resolution by a lens—aberrations. As you learned in Chapter 22, spherical and other aberrations give rise to blurred images. Compound lens systems can be used to reduce these effects significantly, so the limit of resolution is determined mainly by diffraction.

Note from Eq. 24.8 that greater resolution may be gained by using radiation of a shorter wavelength. Thus, a telescope with a particular objective will show greater resolution with violet light than with red light. For microscopes, an overall reduction in wavelength may be accomplished. One of the objectives has an *oil immersion lens*. A film of transparent oil is used between the objective and the specimen. Recall that the wavelength of light in the oil is $\lambda_m = \lambda/n$, where n is the index of refraction of the oil and λ the wavelength of the light in air. With values of n of about 1.50 or higher, the wavelength is significantly reduced and the resolving power increased.

The wavelength can be further reduced by using ultraviolet light and photographic film. However, ordinary glass absorbs ultraviolet light, so the lenses of ultraviolet microscopes must be made of other materials, such as quartz.

The fact that the resolving power of a microscope can be increased by decreasing the wavelength of the incident light opened up new possibilities for seeing fine details after it was discovered that a beam of electrons has wave properties (Chapter 27). These beams have wavelengths that can be adjusted to be very much smaller than those of light in or near the visible region. Using a short-wavelength electron beam, an electron microscope can magnify up to 1000 times more than a visible-light microscope. X-rays have even shorter wavelengths than electron beams and are used to attain even greater magnification and resolution.

Useful magnification refers to the magnification of resolvable details. Magnification beyond the limit of usefulness offers no further disclosure of the details of an object and is called empty magnification. For example, enlarging a newspaper picture is empty magnification.

The relationship between wavelength and index of refraction is given in Section 21.3.

24.5 ■ Color and Optical Illusions ◇

In general, physical properties are fixed or absolute. For example, a particular radiation has a certain frequency or wavelength. However, visual perception of this radiation may vary from person to person. How we see radiation gives rise to the following interesting topics.

Color Vision

Color is perceived because of a physiological response to light excitation of cone receptors in the retina of the human eye. (Many animals have no cone cells and live in a black-and-white world.) The cones are sensitive to light with frequencies between 7.5×10^{14} Hz and 4.3×10^{14} Hz (wavelengths between 400 nm and 700 nm). Different frequencies of light are perceived by the brain as different colors. The association of a color with a particular frequency is subjective and may vary from person to person. As pitch is to sound and hearing, color is to light and vision.

The relationship between pitch and frequency for sound is discussed in Sections 14.1 and 14.5.

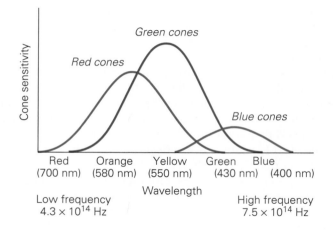

• FIGURE 24.17 Sensitivity of cones Different types of cones in the retina of the human eye respond to different frequencies of light to give three general color responses, red, green, and blue.

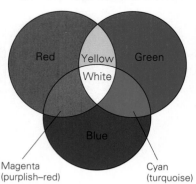

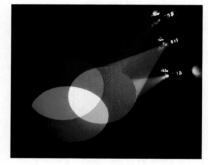

• FIGURE 24.18 Additive method of color production When light beams of the primary colors (red, blue, and green) are projected onto a white screen, mixtures of them produce other colors. Varying the intensities of the beams allow most colors, or hues, to be produced.

Color vision is not well understood. One of the most widely accepted theories is that there are three different types of cones in the retina, responding to different parts of the visible spectrum, particularly in the red, green, and blue regions (• Fig. 24.17). Presumably, the types of cones absorb light of specific wavelengths and functionally overlap one another to form combinations that are interpreted by the brain as various colors of the spectrum. For example, when red and green cones are stimulated equally by light of a particular frequency, the brain interprets the color as yellow. But when the red cones are stimulated more strongly than the green cones, the brain senses orange.

Color-blindness results when one type of cone is missing. This condition is genetically inheritable, and most color-blind people are males (about 4% of the male population). For example, if a man lacks red cones, he can see green through orange-red by means of his green cones. However, he is unable to distinguish among these colors satisfactorily because there are no red cones to provide contrast. A similar condition occurs if the green cones are missing. In either case, this is termed red-green color-blindness because the person cannot effectively distinguish those colors.

The theory of color vision described above is based on the experimental fact that beams of varying intensities of red, green, and blue light will produce most colors, or hues. The red, blue, and green from which we interpret a full spectrum of colors are called the **additive primary colors** (or the additive primaries). When light beams of the additive primaries are projected on a white screen so they overlap, other colors will be produced, as illustrated in • Fig. 24.18. This is called the **additive method of color production** or mixing.

Triad dots consisting of three phosphors that emit the additive primary colors when excited are used in television picture tubes to produce colored images.

Note in Fig. 24.18 that the proper combination of the primary colors appears white to the eye. Also, many *pairs* of colors appear white to the eye when combined. The colors of such pairs are said to be **complementary colors**. For example, the complement of blue is yellow, that of red is cyan, and that of green is magenta. As you can see from the figure, the complementary color of a particular primary is the combination or sum of the other two primaries. Hence, the primary and its complement together appear white.

Edwin H. Land (the developer of Polaroid film) has shown that when the proper mixtures of only two wavelengths (colors) of light are passed through black and white transparencies (no color), they produce images of various colors. Land wrote, *"In this experiment we are forced to the astonishing conclusion that the rays are not in themselves color-making. Rather they are bearers of information that the eye uses to assign appropriate colors to various objects in an image."** From Land's experiments, it seems that information about colors, other than for the two wavelengths used, is developed in the brain via the cone cells. However, knowledge about color vision is far from complete.

Objects have color when they are illuminated with white light because they reflect (scatter) or transmit the light of the wavelength of the color they appear to be. The other wavelengths of the white light are absorbed. For example, when white light strikes a ripe red apple, only the waves in the red portion of the spectrum are reflected—all others (and thus all other colors) are absorbed. Similarly, when white light passes through a piece of transparent red glass, or a filter, only the red rays are transmitted. This occurs because the color pigments in the glass are selective absorbers.

Pigments are mixed to form various colors, such as in the production of paints and dyes. You are probably aware that mixing yellow and blue paints produces green. This is because the yellow pigment absorbs most of the wavelengths except those in the yellow and nearby regions of the visible spectrum, the blue pigment absorbs most of the wavelengths except those in the blue and nearby regions. The wavelengths in the intermediate green region, adjacent to both the yellow and blue wavelengths, are not strongly absorbed by either pigment, and therefore the mixture appears green. The same effect may be accomplished by passing white light through stacked yellow and blue filters. The light coming through both filters appears green.

Mixing pigments results in the subtraction of colors, and the color that is perceived is the one not absorbed, or subtracted. This is an example of what is called the **subtractive method of color production** or mixing. Three particular pigments—cyan, magenta, and yellow—are called the **subtractive primary pigments** (or subtractive primaries). Various combinations of two of the three subtractive primaries produce the three additive primary colors (red, blue, and green), as illustrated in ● Fig. 24.19. When the subtractive primaries are mixed in the proper proportions, the mixture appears black (all wavelengths are absorbed). Painters often refer to the subtractive primaries as red, yellow, and blue. They are loosely referring to magenta (purplish-red), yellow, and cyan "true" blue). Mixing these paints in the proper proportions produces a broad spectrum of colors.

Note in Fig. 24.19 that the magenta pigment essentially subtracts the color green where it overlaps with cyan and yellow. As a result, magenta is sometimes referred to as "minus green." If a magenta filter were placed in front of a green light, no light would be transmitted. Similarly, cyan is called "minus red," and yellow is called "minus blue."

Visual Images and Illusions

Everything you see and therefore most of what you know depends on the information conveyed through the tiny pupils of your eyes. Thus, the better you understand the function of your eyes, the better you can interpret your

*From "Experiments in Color Vision," by Edwin H. Land, in *Scientific American*, May 1959, pp. 84–99.

Recall that the colors of the visible spectrum may be remembered by ROY G. BIV (Chapter 21).

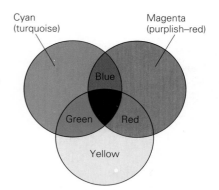

● **FIGURE 24.19 Subtractive method of color production** When the primary pigments (cyan, magenta, and yellow) are mixed, different colors are produced by subtractive absorption; for example, the mixing of yellow and magenta produces red. When all three pigments are mixed and all the wavelengths of visible light are absorbed, the mixture appears black.

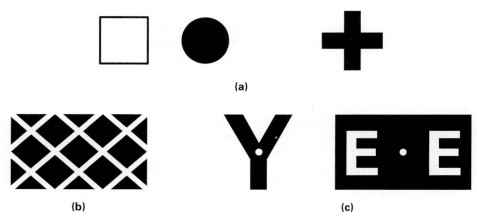

(a)

(b)   (c)

FIGURE 24.20 Visual images **(a)** To experience the blind spot, hold the book at arm's length and with the right eye closed, look intently at the black cross. Then bring the book slowly toward your face. The square and dot will alternately disappear and reappear. **(b)** Induced stimulation causes fleeting patches of gray to appear on the white areas between the points of the black diamonds. **(c)** Similarly, an image of a Y can be induced between the E's by looking intently at the white dot in the black Y, and then shifting your gaze to the white dot between the E's.

visual observations. Even though you often hear people say "seeing is believing," you can't take this statement too literally. For example, from the discussion on converging lenses, you know that the images of objects formed on the retinas of your eyes are inverted. But you don't perceive the world as upside down—by some means, the brain interprets the inverted image of the world as being right side up.

Knowledge of the eye will help you understand some of the things you see and do not see. For example, look intensely at the black cross in Fig. 24.20a with this book held at arm's length and your right eye closed. Then slowly bring the book toward you. At certain points, you will see the square and dot alternately disappear and reappear. Is there something wrong with your vision? No, the experiment demonstrates the eye's blind spot. This is the region where the optic nerve attaches to the retina and where there are no rods or cones. Consequently, there is no optical response. You don't ordinarily notice the blind spot because of movements of the eye and objects and because you have binocular vision.

The stimulation of one area of the retina may affect the sensations in an adjacent area, producing an illusion. For example, if you look at the pattern in Fig. 24.20b, you will probably see fleeting patches of gray where the white lines intersect between the black diamonds. Also, stimulation of a particular area of the retina may be "remembered" after your gaze has shifted to another object. Look intensely at the white dot in the black Y in Fig. 24.20c for about 30 seconds. Then transfer your gaze to the white dot between the E's. You will see a ghostly white Y between the E's. This is an example of an afterimage. Afterimages are commonly seen when you shift your eyes from a bright object to a dark background, for example, looking at a light in a darkened room and then looking away from it.

Keep in mind that there are many ways people can be visually fooled. For example, your depth perception is tricked in 3-D movies when objects give the illusion of jumping out of the screen. Other optical illusions are caused by the geometry of the situation in which they occur. Several common illusions are shown in Demonstration 18. How does your eye (or brain) interpret them?

What do you see?

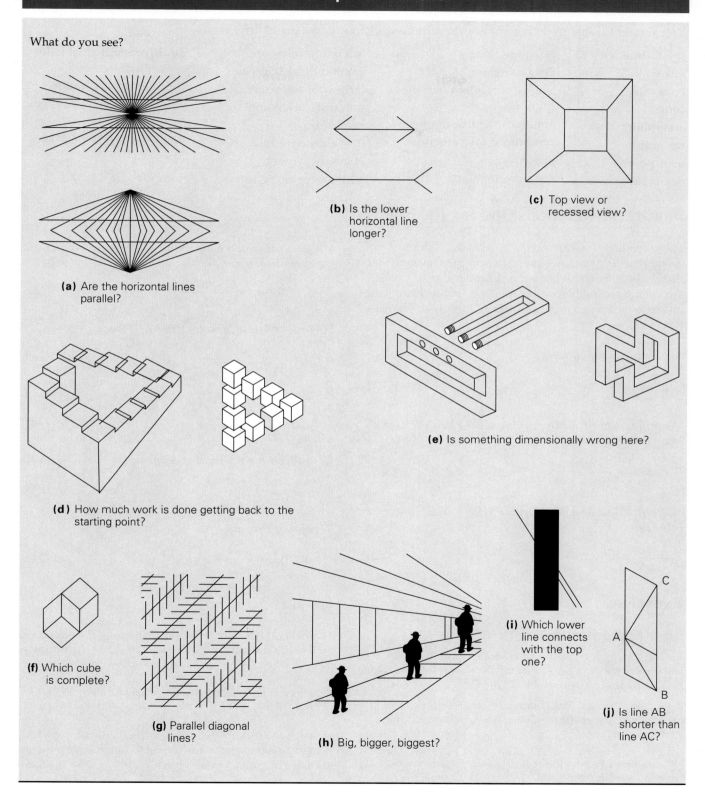

(a) Are the horizontal lines parallel?

(b) Is the lower horizontal line longer?

(c) Top view or recessed view?

(d) How much work is done getting back to the starting point?

(e) Is something dimensionally wrong here?

(f) Which cube is complete?

(g) Parallel diagonal lines?

(h) Big, bigger, biggest?

(i) Which lower line connects with the top one?

(j) Is line AB shorter than line AC?

Important Concepts

You should be able to define and/or explain these chapter concepts clearly.

crystalline lens	farsightedness	refracting telescope	additive method of color
retina	astigmatism	astronomical telescope	production
rods	magnifying glass (simple	terrestrial telescope	complementary colors
cones	microscope)	reflecting telescope	subtractive method of color
accommodation	angular magnification	resolution	production
far point	compound microscope	Rayleigh criterion	subtractive primaries
near point	objective	resolving power	
nearsightedness	eyepiece (ocular)	additive primary colors	

Important Relationships for Review

These relationships are mathematical statements of the concepts and principles presented in the chapter. You should be able to identify the symbols and to explain the relationships before proceeding to the Exercises. In-text equation reference numbers are given for convenience.

Angular Magnification:

$$m = \frac{\theta}{\theta_o} \tag{24.1}$$

Magnification of a Magnifying Glass With Image at Near Point (25 cm):

$$m = 1 + \frac{25 \text{ cm}}{f} \tag{24.3}$$

Magnification of a Magnifying Glass With Image at Infinity:

$$m = \frac{25 \text{ cm}}{f} \tag{24.4}$$

Magnification of a Compound Microscope (with L and f's in centimeters):

$$M_t \simeq M_o m_e \simeq \frac{(25 \text{ cm}) L}{f_o f_e} \tag{24.5}$$

Magnification of a Refracting Telescope:

$$m = -\frac{f_o}{f_e} \tag{24.6}$$

Resolution for a Slit:

$$\theta_{min} = \frac{\lambda}{d} \tag{24.7}$$

Resolution for a Circular Aperture:

$$\theta_{min} = \frac{1.22\lambda}{D} \tag{24.8}$$

Resolving Power:

$$s = f\theta_{min} = \frac{1.22\lambda f}{D} \tag{24.9}$$

Exercises

24.1 ■ The Human Eye*

1 The cones of the retina (a) conically focus light, (b) are responsible for twilight vision, (c) are responsible for color vision, (d) are responsible for 20/20 vision. (c)

● **2** A vision defect that commonly occurs with age is (a) astigmatism, (b) nearsightedness, (c) farsightedness, (d) accommodation. (c)

*Assume corrective lenses are in contact with the eye (contact lenses) unless otherwise stated.

3 The focal length of the crystalline lens of the human eye varies with muscle action. Discuss the shape (curvature) of the lens when looking at distant and close objects. see ISM

4 Which ratio indicates better vision, 20/15 or 15/20? Explain. see ISM

5 ■■ A certain farsighted person has a near point of 45 cm. What type and power of lens should an optometrist prescribe to enable the person to see objects clearly as close as 25 cm? +1.8 D

6 ■■ The far point of a certain nearsighted person is 250 cm. What type of lens with what focal length will allow this person to see distant objects clearly? diverging, −250 cm

● **7** ■■ If the normal near point of the crystalline lens is taken to be 25 cm, what is the power of the lens? (Assume all refraction is done by the lens and that it is thin and spherical. Need more data? See text.) +71 D

8 ■■ A woman cannot see objects clearly when they are farther than 7.5 m away. (a) Is she nearsighted or farsighted? (b) What type of lens of what power (in diopters) will allow her to see distant objects clearly?
(a) nearsighted (b) −0.13 D

9 ■■ A farsighted professor can just see the print in a book clearly when holding the book at arm's length (0.80 m from the eyes). What type of lens with what focal length will allow the professor to read the text at the normal near point?
+2.8 D

10 ■■ To correct a case of hyperopia, an optometrist prescribes positive contact lenses that effectively move the patient's near point from 100 cm to 25 cm. (a) What is the power of the lenses? (b) Will the person be able to see distant objects clearly when wearing the lenses or will she have to take them out? (a) +3.0 D (b) can leave them on

11 ■■ A farsighted person with a near point of 1.15 m gets contact lenses and can then read a newspaper held at a distance of 25 cm. What is the power of the lenses? (Assume that the lenses are the same for both eyes.) +3.1 D

● **12** ■■ A farsighted person wears glasses whose lenses have a power of +1.25 D and is then able to see objects clearly as close as 25 cm. What is the person's near point?
0.36 m

13 ■■ A nearsighted person wears glasses whose lenses have a power of −0.15 D. What is the person's far point?
6.7 m

14 ■■ A myopic person has glasses with a lens that corrects for a far point of 130 cm. What is the power of the lens?
−0.77 D

15 ■■ A nearsighted person has a far point located 752 cm from one eye. (a) If a corrective lens is worn 2.0 cm from the eye, what would be the power of the lens necessary for the person to see distant objects? (b) What would be the power if a contact lens were used? (a) −0.133 D (b) −0.133 D

16 ■■ On being examined by an optometrist, it is found that a person's near point has receded from 40 cm to 50 cm. What is the difference in the powers of the lenses prescribed so that the person would be able to read a newspaper at a distance of 25 cm? (Assume the same near point for both eyes.) ΔP = +0.50 D

17 ■■■ Bifocal glasses are used to correct a person's nearsightedness and farsightedness (●Fig. 24.21). If the person's near points in the right and left eyes are 35.0 cm and 45.0 cm, respectively, and the far point is 220 cm for both eyes, what are the powers of the lenses prescribed for the glasses? (Assume the glasses are worn 3.00 cm from the eyes.)
right: +1.14 D, −0.46 D; left: +1.78 D, −0.46 D

●**FIGURE 24.21 Bifocals** See Exercise 17.

18 ■■■ The amount of light reaching the film in a camera depends on the lens aperture (the effective area) as controlled by the diaphragm. The f number is the ratio of the focal length of the lens to its effective diameter. For example, an f/8 setting means that the diameter of the aperture is $\frac{1}{8}$ of the focal length of the lens. The lens setting is commonly referred to as the f-stop. (a) Determine how much light each of the following lens settings will admit to the camera compared to f/8: (1) f/3.2, and (2) f/16. (b) The exposure time of a camera is controlled by the shutter speed. If a photographer correctly uses a lens setting of f/8 with a film exposure time of $\frac{1}{60}$ s, what exposure time should he use to get the same amount of light *exposure* if he sets the f-stop at f/5.6?
see ISM

24.2 ■ Microscopes*

19 A magnifying glass (a) is a concave lens, (b) allows a clear image to be formed of an object that is closer than the near point, (c) magnifies by effectively increasing the angle the object subtends, (d) both (b) and (c). (d)

20 A compound microscope has (a) unlimited magnification, (b) two lenses of the same focal length, (c) a diverging objective lens, (d) an eyepiece of relatively long focal length. (d)

21 With an object at the focal point of a magnifying glass, the magnification is given by $m = 25 \text{ cm}/f$ (Eq. 24.4). According to this, the magnification could be increased indefinitely by using lenses with shorter focal lengths. Why, then, have compound microscopes? see ISM

● **22** ■■ A diamond cutter uses a jeweler's lens to examine a diamond. If the focal length of the lens is 16 cm, what is the maximum angular magnification of the diamond? 2.6

23 ■■ Using the small-angle approximation, compare the angular sizes of a car 2.0 m in height when at distances of 500 m and 1025 m. 0.0040 rad and 0.0020 rad

24 ■■ An object is placed 10 cm in front of a converging lens with a focal length of 18 cm. What are (a) the lateral magnification and (b) the angular magnification?
(a) 2.3× (b) 2.5×

*The normal near point should be taken as 25 cm unless otherwise specified.

25 ■■ A certain adult has a near point of 50 cm. What is the maximum magnification this person can obtain using a magnifying glass with a focal length of 10 cm? 6.0×

● **26** ■■ A physics student uses a converging lens with a focal length of 15 cm to read a small measurement scale. (a) What is the maximum magnification that can be obtained? (b) What is the magnification for viewing with a relaxed eye? (a) 2.7× (b) 1.7×

27 ■■ A student uses a magnifying glass to examine the details of a microcircuit in the lab. If the lens has a focal length of 8.0 cm and a virtual image is formed at the student's near point (25 cm), (a) how far from the circuit is the lens held, and (b) what is the magnification? (a) 6.1 cm (b) 4.1×

28 ■■ A detective looks at a fingerprint with a magnifying glass whose power is +2.5 D. What is the maximum magnification of the print? 1.6×

29 ■■ What is the maximum magnification of a magnifying glass with a power of +3.0 D for (a) a person with a near point of 25 cm and (b) a person with a near point of 10 cm? (a) 1.8× (b) 1.3×

30 ■■ The focal length of the objective lens of a compound microscope is 4.5 mm. The eyepiece has a focal length of 3.0 cm. If the distance between the lenses is 18 cm, what is the magnification of a viewed image? $3.3 \times 10^2 \times$

31 ■■ A compound microscope has a 15-cm barrel and an ocular with a focal length of 8.0 mm. What power should the objective have to give a total magnification of 360×? +77 D

● **32** ■■ The tube of a compound microscope is 16 cm long. The focal length of the eyepiece is 3.0 cm, and the focal length of the objective is 0.45 cm. Find the total magnification of the microscope. $3.0 \times 10^2 \times$

33 ■■ A compound microscope has an objective lens with a focal length of 0.50 cm and an ocular with a focal length of 3.25 cm. The separation distance between the lenses is 22 cm. For a person with a normal near point using the microscope, (a) what is the total magnification? (b) Compare this (as a percentage) with the magnification of the eyepiece alone as a simple magnifying glass. (a) 340× (b) 3900%

34 ■■ A 1500× microscope has an objective whose focal length is 0.75 cm. If the tube of the microscope is 20 cm long, find the focal length of the eyepiece. 0.44 cm

35 ■■■ A specimen is 5.0 mm from the objective of a compound microscope, which has a power of +250 D. What must the magnifying power of the eyepiece be if the total magnification of the specimen is 100×? 25×

36 ■■■ Referring to ● Fig. 24.22, show that the magnifying power for a magnifying glass held at a distance d from the eye is given by

$$ m = \left(\frac{25}{f}\right)\left(\frac{1-d}{D}\right) + \frac{25}{D} $$

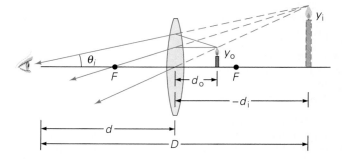

● **FIGURE 24.22 Power of amagnifying glass** See Exercise 36.

when the actual object is located at the near point (25 cm). [*Hint:* Use a small-angle approximation and note that $y_i/y_o = d_i/d_o$ by similar triangles.] see ISM

37 ■■■ A magnifying glass with a focal length of 10 cm is held 4.0 cm from the eyes to view the small print of a book. What is the magnification if the magnifying glass is 5.0 cm from the book? [*Hint:* see Exercise 36.] 3.6×

38 ■■■ The objective of a compound microscope has a focal length of 4.0 mm and is located 4.5 mm from a specimen. The specimen is viewed through an eyepiece ($f_e = 3.0$ cm) adjusted for minimum eye strain. (a) What is the magnification of the specimen? (b) How far apart are the lenses? (a) 200×; (b) 6.6 cm

39 ■■■ A modern microscope is equipped with a turret having three objectives with focal lengths of 16 mm, 4.0 mm, and 1.6 mm and interchangeable eyepieces of 5× and 10×. A specimen is positioned so each objective produces an image 150 mm from the objective. What are the least and greatest magnifications possible? $M_{(max)} = 930 \times$; $M_{min} = 43 \times$

24.3 ■ Telescopes

● **40** An inverted image is produced by (a) a terrestrial telescope, (b) an astronomical telescope, (c) a Galilean telescope, (d) all of these. (b)

41 Compared to large refracting telescopes, large reflecting telescopes have the advantage of (a) greater light-gathering capability, (b) freedom from chromatic aberration, (c) lower cost, (d) all of these. (d)

42 ■■ Find the magnification of a telescope whose objective has a focal length of 50 cm and whose eyepiece has a focal length of 2.7 cm. 19×

43 ■■ An astronomical telescope has an objective and an eyepiece whose focal lengths are 60 cm and 15 cm, respectively. What are (a) the magnifying power and (b) the length of the telescope? (a) 4.0× (b) 75 cm

44 ■■ A telescope has an ocular with a focal length of 10 mm. If the length of the tube is 1.5 m, what is the angular magnification of the telescope when it is focused for an object at infinity? $1.5 \times 10^2 \times$

45 ■■ A refracting telescope has an objective with a focal length of 50 cm and an eyepiece with a focal length of 15 mm. The telescope is used to view an object that is 10 cm high and located 50 m away. What is the apparent angular height of the object as viewed through the telescope? $3.8°$

● **46** ■■ An astronomical telescope has objective and eyepiece lenses with focal lengths of 87.5 cm and 7.50 mm, respectively. (a) What is the magnification of the telescope, and (b) what is its approximate length?   (a) $117 \times$ (b) 88.3 cm

47 ■■ The three lenses of a terrestrial telescope have focal lengths of 40 cm, 20 cm, and 15 cm for the objective, erecting lens, and eyepiece, respectively. (a) What is the magnification of the telescope for an object at infinity? (b) What is the length of the telescope?   (a) $2.7\times$ (b) 135 cm

48 ■■ Two astronomical telescopes have the following characteristics:

Telescope	Objective focal length (cm)	Eyepiece focal length (cm)	Objective diameter
1	90.0	0.84	75
2	85.0	0.77	60

Which telescope would you choose (a) for magnification? (b) For resolution?   (a) T_2 (b) T_1

49 ■■■ Referring to ● Fig. 24.23, show that the angular magnification of a refracting telescope focused for the final image at infinity is $m = f_o/f_e$. (Since telescopes are designed for viewing distant objects, the angular size of an object viewed with the unaided eye is the angular size of the object at its actual location rather than at the near point, as is true for a microscope.)   see ISM

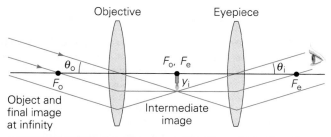

● **FIGURE 24.23 Angular magnification of a refracting telescope**   See Exercise 49.

24.4 ■ Diffraction and Resolution

50 The images of two sources are said to be resolved when (a) the central maxima of the diffraction patterns fall on each other, (b) the first bright fringes of the diffraction patterns

fall on each other, (c) the central maximum of one diffraction pattern falls on the first dark fringe of the other, (d) none of these.   (d)

51 For a telescope with a circular aperture, the minimum angle of resolution (a) is greater for red light than for blue light, (b) is independent of the frequency of the light, (c) is directly proportional to the radius of the aperture, (d) is independent of the area of the aperture.   (a)

● **52** The purpose of using oil-immersion lenses on microscopes is to reduce (a) the spherical aberration, (b) the useful magnification, (c) the chromatic aberration, (d) the wavelength of light so as to increase the resolving power.   (d)

53 ■■ According to the Rayleigh criterion, what is the minimum angle of resolution for two point sources of red light ($\lambda = 680$ nm) in the diffraction pattern produced by a single slit with a width of 0.35 mm? 1.9×10^{-3} rad

54 ■■ Light from two monochromatic 550-nm point sources is incident on a slit that is 0.050 mm wide. What is the minimum angle of resolution in the diffraction pattern formed by the slit? 1.1×10^{-2} rad

55 ■■ The minimum angular separation of the images of two monochromatic point sources in a single-slit diffraction pattern is 0.0055 rad. If a slit width of 0.10 mm is used, what is the wavelength of the sources? 5.5×10^{-7} m

● **56** ■■ What is the resolution limit due to diffraction for the 40-in. (102-cm) Yerkes Observatory refracting telescope for light with a wavelength of 550 nm? 6.58×10^{-7} rad

57 ■■ What is the resolution due to diffraction for the Hale telescope at Mount Palomar with its 200-in. diameter mirror for light with a wavelength of 550 nm? Compare this to the resolution limit for the Yerkes Observatory telescope. 1.32×10^{-7} rad

● **58** ■■ The objective of a microscope is 2.5 cm in diameter and has a focal length of 30 mm. (a) If yellow light with a wavelength of 570 nm is used to illuminate a specimen, what is the minimum angular separation of two fine details of the specimen for them to be just resolved? (b) What is the resolving power of the lens?   (a) 2.8×10^{-5} rad (b) 8.4×10^{-4} mm

59 ■■ For the microscope in Exercise 58, how great a difference is there in (a) the minimum angular separation and (b) the resolving power for blue light ($\lambda = 400$ nm) and red light ($\lambda = 700$ nm)?   (a) $\Delta\theta = 1.47 \times 10^{-5}$ rad (b) $\Delta s = 4.40 \times 10^{-5}$ cm

60 ■■ A refracting telescope with a lens whose diameter is 30 cm is used to view a binary star that emits light in the visible region. (a) What is the minimum angular separation of the pair of stars for them to be barely resolved? (b) If the binary star is a distance of 6.0×10^{20} km from the Earth, what is the lateral separation between the stars of the pair?   (a) 1.6×10^{-6} rad   (b) 9.6×10^{17} m

61 ■■■ A microscope with an objective 1.20 cm in diameter is used to view a specimen using light from a mercury source with a wavelength of 546.1 nm. (a) What is the limiting angle of resolution? (b) What color of light in the visible spectrum would give the maximum limit of resolution? (c) If an oil-immersion lens were used (n_{oil} = 1.40), what would be the change (expressed as a percentage) in the resolving power?   (a) 5.55×10^{-5} rad   (b) red   (c) 70.2%

◊ 24.5 ■ Color and Optical Illusions

62 An additive primary color is (a) blue, (b) green, (c) red, (d) all of these.   (d)

63 A subtractive primary pigment is (a) cyan, (b) yellow, (c) magenta, (d) all of these.   (d)

● **64** Describe how the American flag would appear if it were illuminated with light of each of the primary colors.
see ISM

65 Can white be obtained by the subtractive method of color production? Explain. It is sometimes said that black is the absence of all colors, or that a black object absorbs all incident light. If so, why do we see black objects?   see ISM

66 White light is incident on two filters as shown in ● Fig. 24.24. Complete the color rays to show what color light emerges from the yellow filter.   see ISM

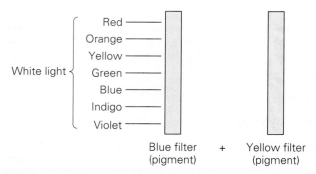

● **FIGURE 24.24 Color absorption**   See Exercise 66.

■ Additional Exercises

67 A lens with a power of +10 D is used as a simple microscope. (a) How close can an object be brought to the eye to be examined through the lens? (b) What is the angular magnification at this point?   (a) 7.1 cm   (b) 3.5 ×

68 A man cannot see objects clearly when they are closer than 250 cm. (a) Is he nearsighted or farsighted? (b) What type of lens of what power (in diopters) will allow him to see objects clearly at 25 cm?   (a) farsighted   (b) +3.6 D

69 A middle-aged man starts to wear glasses with lenses of +2.0 D that allow him to read a book held as closely as 25 cm. Several years later, he finds that he must hold a book no closer than 33 cm to read it clearly using the same glasses, so he gets new glasses. What is the power of the new lenses? (Assume both lenses are the same.)   +3.0 D

70 In a terrestrial telescope, an objective and eyepiece with focal lengths of 45 cm and 15 cm, respectively, are used. What should the focal length of the erecting lens be if the overall length of the telescope is to be 0.75 m?   3.8 cm

71 When viewing an object with a magnifying glass whose focal length is 20 cm, a person positions the lens so there is minimum eyestrain. What is the observed magnification?   1.3×

72 A certain myopic person has a far point of 150 cm. (a) What power must a lens have to allow him to see distant objects clearly? (b) If he is able to read print at 25 cm while wearing his glasses, is his near point less than 25 cm? (c) Give an indication of his age with normal recession of near point.   (a) −0.67 D   (b) 21 cm; 35–40 years old

73 A compound microscope has an objective with a focal length of 4.0 mm and an eyepiece with a magnification of 10×. If the objective and eyepiece are 15 cm apart, what is the total magnification of the microscope?   375×

74 A person using a magnifying glass with a focal length of 12 cm views an object at the focal point of the lens. What is the magnification?   2.1×

75 An eyeglass lens with a power of +2.8 D allows a far-sighted person to read a book held at a distance of 25 cm from her eyes. At what distance must the person hold the book to read it without glasses?   82 cm

76 The minimum angular separation of two stars emitting red light (λ = 680 nm) for a particular telescope is 5.6×10^{-6} rad. What is the diameter of the telescope's lens?   15 cm

77 A student views the details of a dollar bill with a magnifying glass, achieving its maximum magnification of 3×. What is the focal length of the magnifying glass?   13 cm

78 A microscope whose tube is 15 cm long has objective and eyepiece lenses with focal lengths of 7.5 mm and 10 mm, respectively. What is the total magnification of the microscope? 5.0×10^3×

79 A college professor can see objects clearly only if they are between 70 and 500 cm from her eyes. Her optometrist prescribes bifocals that enable her to see distant objects through the top half of the lenses and read students' papers at a distance of 25 cm through the lower half. What are the powers of the bifocal lenses? (Assume both lenses are the same.)   top: −0.20 D   bottom: +2.6 D

25

Relativity

There was a bustle of scientific activity in so many areas at the beginning of the twentieth century that the study of modern physics could begin with any one of a variety of topics. We begin here with the popular and intriguing subject of relativity, which was an outgrowth of nonclassical observations and thought.

Relativity originated from the analysis of physical phenomena when speeds approach that of light. This was not a subject of study in classical Newtonian kinematics and dynamics because objects with speeds of these magnitudes cannot ordinarily be observed. However, some particles in nature do travel at speeds comparable to the speed of light and they can now be accelerated to such speeds in particle accelerators.

When such speeds are obtained, measurements show changes in fundamental properties such as length, time, and mass. It is difficult to imagine observing metersticks becoming shorter, clocks running slower, or the masses of objects increasing—all because of traveling at very fast speeds—but this is what we observe. Indeed, relativity created a revolution in physics by causing us to rethink our basic understanding of space, time, and gravitation. It successfully challenged Newtonian concepts that had dominated scientific thinking for 250 years.

The impact of relativity has been especially great in those branches of science concerned with the extremes of physical reality—the subatomic realm (discussed in Chapters 28 and 29), where time intervals and distances are almost inconceivably small, and the cosmic realm, where time intervals and distances are almost unimaginably large. (The dark "horsehead" shape in the photograph above is part of a cloud of light-absorbing interstellar dust. Although it is about 5×10^{18} km from Earth—a distance that light takes 5000 years to travel—it is virtually in our own back yard by astronomical standards.) All modern theories about the birth, evolution, and ultimate fate of our universe are inextricably linked to our understanding of relativity.

In this chapter, you will learn how Einstein's theory of relativity explains the changes in length and time that we observe in rapidly-moving objects, the equivalence of energy and mass, and the bending of light rays by gravitational fields—phenomena that appear very strange from the classical viewpoint. We will begin by exploring some of the deficiencies in classical concepts, and then describe how new ideas were developed.

Studies in Chapter 25 include classical relativity, the Michelson-Morley experiment, the special theory of relativity, relativistic mechanics, and the general theory of relativity.

25.1 ■ Classical Relativity

Physics is concerned with the description of the world around us, and this endeavor depends on observations and measurements. We expect some aspects of nature to be consistent and unvarying. That is, the ground rules by which nature plays are consistent, and our descriptions of nature or physical principles do not change from observation to observation. We emphasize this fact by referring to such principles as "laws"—for example, the laws of motion. Not only have physical laws proved valid over time, but they are the same for all observers.

This means that a physical principle or law is the same regardless of the observer's frame of reference. When we make a measurement or perform an experiment, we do so with reference to a particular frame or coordinate system, usually the laboratory, which is considered to be at rest. The experiment may also be observed by a person passing by (in motion). On comparing notes, the experimenter and the observer should find the results of the experiment or the physical principles involved to be the same in both frames of reference or coordinate systems. Basically, you can't change the laws of nature by the way you observe them. Descriptions may be different, but they describe the same thing.

For example, suppose that you observe two cars traveling in the same direction on a straight road at speeds of 60 km/h and 90 km/h. These speeds are relative to your reference frame. However, a person in the car traveling at 60 km/h observes the other car in front of it traveling at a speed of 30 km/h, *relative to her reference frame*—the car in which she is riding. (What does someone in the car traveling at 90 km/h observe?) In general, what one observes is *relative velocity*—that is, the velocity relative to the observer's frame of reference.

In measuring relative velocity, there seems to be no "true" rest frame. Any reference frame may be considered to be at rest *relative* to another. We can, however, make a distinction between what are called inertial and noninertial reference frames. An **inertial reference frame** is one in which Newton's first law of motion holds. That is, in an inertial system, an *isolated* object (one on which there is no net force) is stationary or moves with a constant velocity. If Newton's first law holds, then his second law in its customary $F = ma$ form applies to the system.

A noninertial reference frame, on the other hand, is one in which an *isolated* object is accelerating, so we speak of an accelerated object or system. If an isolated object is accelerating, then Newton's second law, $F = ma$, cannot be applied directly.

Any reference frame moving with a constant velocity relative to an inertial reference frame is also an inertial frame. Given a constant relative velocity, no acceleration effects are introduced in comparing one frame to another. In such cases, $F = ma$ can be used by observers in either frame (with their own coordinate values) to analyze a dynamic situation, and both observers will come up with similar results. Thus, there is no one inertial frame of reference that is preferred over another. This is sometimes called the **principle of classical, or Newtonian, relativity**.

The laws of mechanics are the same in all inertial reference frames.

Review the discussion of relative velocity in Section 3.3.

Review the discussion of Newton's laws in Sections 4.2 and 4.3.

Discuss the importance of Newton's first law of motion. Can the classroom in which we are sitting be considered an inertial reference frame? Why?

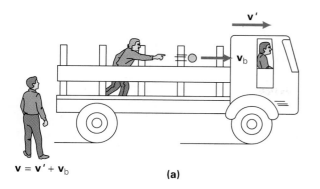

$\mathbf{v} = \mathbf{v}' + \mathbf{v}_b$

(a)

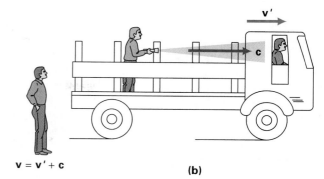

$\mathbf{v} = \mathbf{v}' + \mathbf{c}$

(b)

● **FIGURE 25.1 Relative velocity** (a) According to a stationary observer on the ground, the velocity of the ball is $\mathbf{v} = \mathbf{v}' + \mathbf{v}_b$. (b) Similarly, the velocity of light would classically be measured to be $\mathbf{v} = \mathbf{v}' + \mathbf{c}$, with a magnitude greater than c.

Around the turn of the century, with the development of the theories of electricity and magnetism, some serious questions arose. Maxwell's equations (Section 19.4) predicted light to be an electromagnetic wave that traveled with a speed of 3.0×10^8 m/s in vacuum. But relative to *what reference frame* does light have this definite speed? Classically, the speed of light in different frames of reference would be expected to differ—possibly to be even greater than 3.0×10^8 m/s— by simple vector addition.

Consider the situation illustrated in ● Fig. 25.1. If a person in a reference frame moving in relation to another with a constant velocity $\mathbf{v}'$ threw a ball with a velocity $\mathbf{v}_b$, then the "stationary" observer would say that the ball had a velocity of $\mathbf{v} = \mathbf{v}' + \mathbf{v}_b$ relative to the stationary reference frame. For example, suppose that a truck were moving with a speed of 20 m/s relative to the ground and a ball were thrown with a speed of 10 m/s relative to the truck in the direction of the truck's motion. Then the ball would have a speed of 20 m/s + 10 m/s = 30 m/s relative to the ground.

Now suppose that a person on the truck turned on a flashlight, projecting a beam of light. In this case we have $\mathbf{v} = \mathbf{v}' + \mathbf{c}$, and the speed of light measured by an observer on the ground would be greater than 3.0×10^8 m/s. Classically then, the measured speed of light could have almost any value, depending on the observer's frame of reference.

A particular speed of $c = 3.0 \times 10^8$ m/s must then be referenced to a particular frame. As a wave, light was thought to require some medium of transport. This was quite natural. All experience showed that physical disturbances must be transmitted through the stress or distortion of some medium, as when sound waves travel in air. Since the Earth receives light from the Sun and from distant stars and galaxies, it was thought that this special light-transporting medium in which $c = 3.0 \times 10^8$ m/s must permeate all space. The medium was given the name *luminiferous ether* and referred to simply as **ether**.

The idea of an undetected ether became popular in the latter part of the nineteenth century. Maxwell, whose work laid the foundations for our understanding of electromagnetic waves (Chapter 19), believed in the necessity of an etherlike substance:

> Whatever difficulties we may have in forming a consistent idea of the constitution of the ether, there can be no doubt that the interplanetary and interstellar spaces are not empty, but are occupied by a material substance or body which is certainly the largest, and probably the most uniform body of which we have any knowledge.

Don't confuse this ether with the anesthetic of the same name.

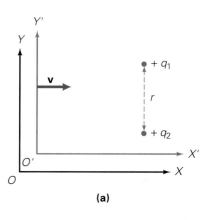

(a)

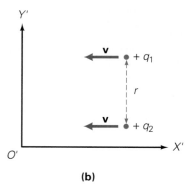

(b)

FIGURE 25.2 Nonrelativity
(a) The repulsive forces between the stationary charges in the O reference frame are given by Coulomb's law: $F = kq_1q_2/r^2$. (b) For an observer in the O' reference frame, moving toward the charges with a velocity **v**, the charges would be seen to be moving. In addition to the repulsive Coulomb forces, there would be attractive forces due to the production of and interaction with magnetic fields (similar to the forces between parallel current-carrying wires).

Discuss the swimmer analogy of the Michelson-Morley experiment. Have one group of students calculate the time to swim across the river and back, and another group calculate the time to swim down the river and return. (See Exercise 15.)

It would seem then that Maxwell's equations, which describe the propagation of light, did not satisfy the Newtonian relativity principle as did other physical laws. On the basis of the preceding discussion, a preferential or special inertial reference frame would seem to exist, one that could be considered absolutely at rest.

As another example of the apparent violation of the relativity principle, consider two stationary charges separated by a distance r as illustrated in ● Fig. 25.2—stationary relative to the O reference frame. In this frame, an observer would measure a force of repulsion between the charges as given by Coulomb's law ($F = kq_1q_2/r^2$).

Now suppose the charges were investigated by an observer in the O' frame moving with a velocity **v** relative to the O frame and perpendicularly to a line separating the charges. This observer would see the two charges approaching with a velocity $-\mathbf{v}$. The repulsive Coulomb force would be evident as it was to the observer in the O frame. But, since a moving charge corresponds to a current, the observer in the O' system would also note attractive forces between the charges (similar to the forces for parallel current-carrying wires). In other words, observers in different inertial frames would measure *different* effects for the same situation, and the physical laws would not be the same in all inertial reference frames. Here, the laws of electricity and magnetism seem not to satisfy the classical relativity principle.

This was the state of affairs toward the end of the nineteenth century when scientists set out to investigate whether they had come upon a new dimension of physics or perhaps a flaw in what were considered to be established principles. One of the first attempts was to test the ether theory. If it could be proved that the ether existed, then presumably a true rest frame would be identified. This was the purpose of the famous Michelson-Morley experiment.

25.2 ■ The Michelson-Morley Experiment

If the ether permeated all space, the Earth itself would be surrounded by the medium. In its orbit around the Sun, the Earth presumably moved through the ether, like an airplane moving through still air. Relative to the airplane, there is air motion or wind, and similarly the orbiting Earth would experience an "ether wind."

If the ether wind existed, this would provide a means to detect the ether experimentally. Observers could carefully measure the speed of light on Earth in several directions, for example, parallel and perpendicular to the orbital velocity of the Earth, and compare the results. Because of the relative motion, the speeds should be different in different directions.

As an analogy, consider two airplanes flying at the same air speed and the same distances back and forth as illustrated in ● Fig. 25.3a. The plane flying perpendicular to the wind must be directed slightly into the wind on both legs to stay on a straight-line course, an action that slows the plane somewhat. The other plane flies into the wind on the first leg and with the wind on the return leg. The velocities at which the planes fly relative to the ground are given by vector addition. It is a simple matter to show that the plane flying perpendicular to the wind takes less time to complete the two-leg trip than does the other plane.

A similar effect should be seen if the airplanes are replaced by light beams and the atmospheric wind by the ether wind (Fig. 25.3b). However, this is more easily said than done. The average speed of the Earth in its orbit is about 30 km/s, compared to 300,000 km/s for the speed of light. In the late nineteenth century, the American physicist Albert A. Michelson devised a technique capable of detecting the required differences using the interference of light and designed an extremely sensitive **interferometer**.

The basic principle of Michelson's interferometer is shown in ● Fig. 25.4. A monochromatic light beam is incident on a partially silvered glass plate (P), which acts as a beam splitter, dividing the incident beam into transmitted and reflected components. The beams advance to the plane mirrors (M_1 and M_2), and are reflected back toward P. There, part of each beam is reflected and transmitted to the detector screen (D). Another glass plate (P') is inserted in one beam, so both beams pass through an equal amount of glass.

Since the beams are from the same initial beam, they are coherent and, on combining, interfere according to their phase relationship. This is determined by the difference in the path lengths of the beams.

Consider now the interferometer (and the Earth) traveling through the stationary ether with a velocity **v** with the interferometer arranged so one of its perpendicular arms is parallel to **v** (●Fig. 25.5). The speed of the beam traveling to M_2 is $c - v$. On the return trip the relative velocity has a magnitude of $c + v$. The total time for the round trip is

$$t_2 = \frac{L_2}{c - v} + \frac{L_2}{c + v} = \frac{2L_2 c}{c^2 - v^2} \qquad (25.1)$$

For the light beam traveling between P and M_1, the speed of light in the direction perpendicular to **v** is $v_\perp = \sqrt{c^2 - v^2}$ (as shown in the vector diagram in the figure). The speed is the same for the return path, so the time for the round trip on this arm is $t_1 = 2d/v_\perp$, or

$$t_1 = \frac{2L_1}{\sqrt{c^2 - v^2}} \qquad (25.2)$$

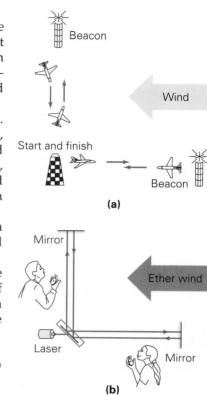

(a)

(b)

● **FIGURE 25.3 Ether wind detection** **(a)** When two airplanes fly with equal air speeds over equal distances, the plane flying perpendicular to the wind direction takes less time to fly a round trip. **(b)** Classically, the ether wind should give rise to a similar difference in the times for light to be reflected back and forth.

Figure 25.4a

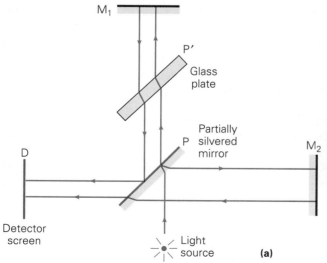

(a)

(b)

● **FIGURE 25.4 Michelson interferometer** **(a)** If one of the arms of the interferometer were in the direction of the ether wind, the beams should arrive at the detector out of phase, and an interference pattern should be produced. See text for details. **(b)** A laboratory model of the Michelson interferometer.

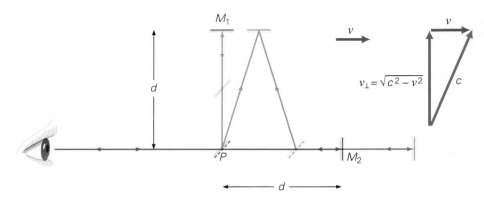

Figure 25.5

The mirrors are arranged so that
they are not quite perpendicular to
the beams. As a result, the path
difference of the two beams
depends upon where the light
strikes the mirrors, which gives rise
to a pattern of fringes.

Then, if $L_1 = L_2 = L$, we have with some rearranging of terms

$$\Delta t = t_2 - t_1 = \frac{2L}{c}\left[\frac{1}{1 - v^2/c^2} - \frac{1}{\sqrt{1 - v^2/c^2}}\right] \qquad (25.3)$$

If there were a time difference, the beams would arrive at the detector out of
phase, and an interference pattern would be observed. From the spacing of
the fringes, one could determine the velocity of the light source relative to the
ether.

However, it is difficult to ensure experimentally that $L_1 = L_2$ to the required
degree of accuracy. This problem may be resolved by rotating the apparatus
by 90°, which interchanges the arms. (With equal arms, the time difference,
Δt, when the arms are interchanged becomes the negative of the value obtained
using Eq. 25.3.) What would be observed then is a *shift* in the interference
pattern or fringes.

Michelson performed this experiment with his colleague E. W. Morley in
1887. The interferometer was more than sensitive enough to detect the predicted
fringe shift. But much to their surprise, Michelson and Morley found *no fringe
shift at all*. Perhaps the experiment had been done just at a time when the
Earth was nearly at rest relative to the ether. To check this possibility, many
measurements were made at different times (day and night and at different
seasons), with always the same null result.

Where then was the ether? Several hypotheses were suggested to explain
the null result of the Michelson-Morley experiment. One suggested that the
ether in the vicinity of the Earth was dragged along with the Earth, so the
interferometer was at rest with respect to the ether (that is, $v = 0$). But because
of the Earth's orbital motion, the viewing of a star perpendicular to the plane
of the orbital plane throughout the year requires the telescope to be consistently
tilted in the direction of the Earth's motion. This effect is called the *aberration
of starlight* and was explained in 1725. The aberration requires that the Earth
move with respect to the ether and *not* be stationary relative to it (as would
be the case if the ether were dragged along).

An Irish physicist, George F. FitzGerald, proposed that as a result of the
motion through the ether, perhaps the linear dimensions of objects in the
direction of motion contracted or became smaller. If this occurred by a fractional
amount of $\sqrt{1 - v^2/c^2}$, then the null result would be explained. Of course, this
hypothesis cannot be experimentally investigated, since the measurement instrument would also contract by the same fractional amount.

Another scientist, H. A. Lorentz, who also suggested this possibility, consid-

*Students should be able to
explain the anticipated goal of the
Michelson-Morley experiment, the
outcome, and its importance to
modern physics.*

ered it in terms of possible changes in the electromagnetic forces between the atoms of a material due to the motion. This suggested length contraction, which is known as the Lorentz-FitzGerald contraction, might be termed the right idea, but for the wrong reason. Oddly enough, such a length contraction was predicted by a theory proposed by Albert Einstein in 1905. His special theory of relativity is described in the next section.

25.3 ■ The Special Theory of Relativity

The failure of the Michelson-Morley experiment to detect the ether left the scientific community in a quandary. The inconsistencies between Newtonian mechanics and electromagnetic theory remained unexplained. These problems were resolved, however, by a theory introduced in 1905 by Albert Einstein. Apparently, the Michelson-Morley experiment was not a motivation for the development of his theory of relativity. Einstein later could not recall whether he had even known about the experiment at the time when he was working as a clerk in the Swiss Patent Office (see ●Fig. 25.6).

Einstein's approach was that the inconsistencies in electromagnetic theory were due to the assumption that an absolute rest frame (the ether) existed. His theory, based on two postulates, did away with the need for an ether. The first postulate is an extension of the Newtonian principle of relativity that applies to the laws of mechanics. In Einstein's theory, the **principle of relativity** applies to *all* the laws of physics, including those of electricity and magnetism:

●**FIGURE 25.6 Einstein and Michelson** A 1931 photo shows Einstein (center) with Michelson (front left) during a meeting in Pasadena, California. At right is Robert A. Millikan, the recipient of the Nobel Prize in 1923 for his experimental determination of the value of the electronic charge.

Postulate I (principle of relativity): The laws of physics are the same in all inertial reference frames.

The first postulate implies that all inertial reference frames are equivalent, with physical laws being the same in all of them. This means that there is no *absolute* reference frame with some unique physical property to distinguish it from other inertial reference frames. The first postulate seems quite reasonable, since one would expect the basic laws of nature to be the same for all inertial observers. One would not expect nature to play favorites.

Einstein's second postulate involves the speed of light. As he saw it, the problem came from the vector addition of velocities. With the appropriate relative velocity between inertial systems, the observed speed of light could have any value. With $(c + v)$, it could be greater than c; with $(c - v)$, it could be less than c; and with $(c - c)$, it could even reduce to zero! The last possibility was the source of Einstein's question "What would I see if I rode a beam of light?" In the same frame as the electromagnetic wave, electric and magnetic field vectors would be seen that varied in space but *not* with time. Wave motion is fundamental to waves, and such static fields were not consistent with electromagnetic theory.

Such variations in the speed of light were unacceptable to Einstein, so he made the speed constant in his second postulate:

Postulate II (constancy of the speed of light): The speed of light in a vacuum has the same value in all inertial systems.

These two postulates form the basis of Einstein's **special theory of relativity**. The "special" designation is to indicate that it deals only with the particular case of inertial reference frames. The *general* theory of relativity, which will be discussed later, deals with noninertial or accelerating reference frames.

The second postulate is perhaps more difficult to accept than the first. It implies that two observers in different inertial reference frames would measure the speed of light to be c independent of the speed of the source and/or the observer. For example, if a person moving toward you with a constant velocity turned on a light, both you and the other person would measure the speed of light to be c, irrespective of the magnitude of the relative velocity.

Relativistic addition of velocities is dealt with in optional Section 25.6.

The second postulate is essential to the validity of the first, and the null result of the Michelson-Morley experiment is consistent with the constancy of the speed of light. By doing away with the ideas of the ether and an absolute reference frame, Einstein was able to reconcile the differences between mechanics and Maxwell's equations. Even so, the second postulate seemed to go against common sense. But then we have very little everyday experience in dealing with velocities at or near the speed of light. A theory may beautifully postulate a solution to a problem, but it must stand the test of the scientific method. What does Einstein's theory predict, and has it been experimentally verified?

The predictions of the special theory of relativity can be understood by imagining simple situations to see what the theory predicts. Einstein did this himself in what he called *Gedanken*, or thought, experiments. We also look at some experimental evidence that supports the nonclassical predictions of the theory.

Simultaneity and Time Dilation

The special theory of relativity, particularly the second postulate, offers new views of the concepts of space and time. To illustrate the nonclassical nature of the second postulate, consider the situation shown in ●Fig. 25.7. In this thought experiment, two observers, O and O', each with lasers, simultaneously turn the lasers on when the observers are next to each other. Now let O be stationary and O' be in motion to the right in the same direction as the light beams are traveling. Assume that O' is traveling with a constant speed of $\frac{1}{3}$ of the speed of light ($c/3$), or 100,000 km/s.

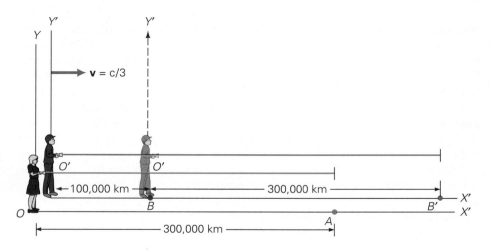

●**FIGURE 25.7 Constancy of the speed of light** Two inertial observers send out light beams in the same direction when at the same location. In 1 second, the light beam from O travels to point A. During the 1 second, O' travels with a speed of $c/3$ to point B, and by the constancy of the speed of light, his beam has traveled to point B'.

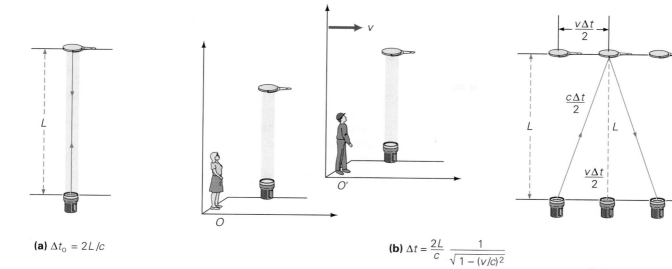

(a) $\Delta t_o = 2L/c$

(b) $\Delta t = \dfrac{2L}{c} \dfrac{1}{\sqrt{1 - (v/c)^2}}$

●**FIGURE 25.8 Time dilation**
(a) A light clock that measures time in units of round-trip reflections of light. The time for light to travel up and back is $\Delta t_o = 2L/c$. **(b)** An observer O measures a time interval of $\Delta t = (2L/c)\,[1/\sqrt{1 - (v/c)^2}\,]$ on the clock in the relatively moving O' frame, so it appears to be running more slowly. See text for further description.

Figure 25.8

At the end of 1 second, O would say that the light from his laser had traveled 300,000 km to point A. According to the second postulate, the speed of light is independent of the speed of the source, so O' would say that in 1 second the light from her laser also traveled this distance relative to her. But at the end of 1 second, O' has traveled 100,000 km to point B, so relative to O', her light beam has reached point B'.

But how about O looking at the O' laser light? He too measures the speed of the light to be c (as required by the second postulate), and in 1 s relative to him the light from the other laser travels 300,000 km to point A. How can the light beam be *simultaneously* at two locations?

This paradox can be resolved when we realize that two events that are simultaneous to one observer are not necessarily simultaneous to another observer in a different reference frame (see Exercise 80). Such radical results appear to be contradictory, but they lead to the conclusion that time and space are not absolute. The reason that the light beam appears to be at different points to the two observers in the preceding example is because the 1-second time interval is not absolute, or the same for both observers. What Einstein's theory predicts is that time is measured differently by different inertial observers.

To see this, try another thought experiment. For the comparison of time intervals in moving inertial reference frames, let's use a special "relativistic" light clock as illustrated in ●Fig. 25.8a. A tick or unit time interval on the clock corresponds to the time for a light pulse to make a complete round trip between two mirrors. It is assumed that two observers, O and O' again, have light clocks and that the clocks have been synchronized and run at the same rate when both systems are at rest. In this case, the time interval for a light round trip is

$$\Delta t_o = \frac{2L}{c} \qquad (25.4)$$

Now consider O' to be in motion with a constant velocity of magnitude v relative to O. In the moving system, O' still measures the unit time interval as Δt_o. But, as seen by O, the clock in the O' system is moving, and the path of the light pulse is along the sides of the triangle, as shown in Fig. 25.8b. Thus, O sees a longer light path for the clock in the moving system. From O's point of view, the geometry of the path of the light pulse is, by the triangle in the figure,

$$\frac{c^2 \, \Delta t^2}{2} = \frac{v^2 \, \Delta t^2}{2} + L^2$$

It is now clear why O and O' will measure time differently. If the light path in the moving clock, viewed from O's frame of reference, is longer, and the speed of light is constant (as it must be for *all* observers), then the light in the moving clock must take a longer time to cover the path. Thus from O's point of view, the moving clock seems to be running slower. We can calculate how much slower by solving the preceding equation for Δt:

$$\Delta t = \frac{2L}{c} \left[\frac{1}{\sqrt{1 - (v/c)^2}} \right] \tag{25.5}$$

But the same time interval as measured by either observer in his or her *own* frame is given by Eq. 25.4. So the measured time intervals are different, and combining the equations, we have

$$\boxed{\Delta t = \frac{\Delta t_o}{\sqrt{1 - (v/c)^2}}} \tag{25.6}$$

Since $\sqrt{1 - (v/c)^2}$ is less than 1 (why?), we have $\Delta t > \Delta t_o$, and O measures a longer time interval on the O' clock than does observer O'. This effect is called **time dilation**. With longer ticks, the O' clock appears to O to run more slowly than his own. Of course, the situation is relative, and O' would say that the clock in O's system ran slow. Thus, because of the relativistic time dilation, *moving* clocks are measured to run more slowly than clocks in the observer's system.

To distinguish between the two time intervals, we refer to **proper time**. It is usually "proper" or normal to be at rest with respect to a clock when making a measurement. In our development the proper time can be seen to be Δt_o, or the time interval in the reference frame (O') in which the clock was at rest. Stated another way, the proper time interval between two events is that interval measured by an observer who is at rest relative to the events and sees the events occur *at the same location or point in space*. In Fig. 25.8, observer O sees the events by which the time interval is measured at different locations (because of the moving clock), so Δt is not the proper time.

Many of the equations of special relativity can be written more simply if we represent the expression $1/\sqrt{1 - (v/c)^2}$ as a factor, gamma (γ):

$$\gamma \equiv \frac{1}{\sqrt{1 - (v/c)^2}} \tag{25.7}$$

Gamma (γ) is always greater than or equal to 1 ($c > v$ and $v = 0$, respectively). The values of γ for several values of v expressed as fractions of c are given in Table 25.1. The values illustrate how the speed of an object must be an appreciable fraction of the speed of light for relativistic effects to be observed. The time dilation equation (Eq. 25.6) may then be written

$$\boxed{t = \gamma t_o = \frac{t_o}{\sqrt{1 - (v/c)^2}}} \tag{25.8}$$

time dilation

The proper time t_o is always less than the dilated time t.

where the deltas have been eliminated and it is understood that the t's are intervals. In words, the time interval measured on a moving clock is γ times the proper time.

For example, suppose you were observing a clock that was in a system moving at a constant velocity magnitude of $0.6c$ relative to you. The gamma factor would then be 1.25 (Table 25.1). Because of the time dilation effect, when 20 minutes had elapsed on the moving clock, you would observe a time of $t = \gamma t_o = (1.25)(20 \text{ min}) = 25 \text{ min}$ to have elapsed on your clock. The 20 minutes is the proper time, since the events defining the 20-minute interval took place at the same location (that of the clock, for example for readings at 8:00 and 8:20). As was pointed out earlier, the moving clock runs more slowly (20 minutes elapsed as opposed to 25 minutes on your clock).

TABLE 25.1 ■ Some Values of $\gamma = \dfrac{1}{\sqrt{1 - (v/c)^2}}$

v	γ
0	1.00
$0.100c$	1.01
$0.200c$	1.02
$0.300c$	1.05
$0.400c$	1.09
$0.500c$	1.15
$0.600c$	1.25
$0.700c$	1.40
$0.800c$	1.67
$0.900c$	2.29
$0.950c$	3.20
$0.990c$	7.09
$0.995c$	10.0
$0.999c$	22.4
c	∞

EXAMPLE 25.1 ■ Time Dilation Verified by Experiment

There are many subatomic elementary particles. One of these, called a muon, has the same charge as an electron but is about 200 times more massive. Muons are created in the Earth's atmosphere when cosmic rays (mostly protons) from outer space collide with the nuclei of the gas molecules of the air. The muons then approach the Earth with a speed near that of light (say about $0.998\,c$).

However, muons are unstable and quickly decay into other particles. The average lifetime of a muon has been measured in the laboratory as 2.20×10^{-6} s. In this time, a muon would travel a distance of

$$d = v_o t = (0.998c)(2.20 \times 10^{-6} \text{ s})$$
$$= [(0.998)(3.00 \times 10^8 \text{ m/s})](2.20 \times 10^{-6} \text{ s}) = 659 \text{ m}$$

This is only 0.659 km, or less than half a mile. Since muons are created at altitudes of about 15–20 km, none of them should reach the Earth's surface.

However, an appreciable number of muons do reach the Earth. The paradox arises because the preceding calculation does not take time dilation into account. That is, a muon decays by its own clock, or in a time measured by a clock in its own reference frame (proper time, since in the rest frame of the muon, birth and death events take place at the same location).

To an observer on Earth, a clock in the moving muon reference frame would appear to run more slowly than a clock on Earth (Fig. 25.9). So in a time $t_o = 2.20 \times 10^{-6}$ s on the muon clock, the Earth clock reads a time interval that is greater by a factor of

$$\gamma = \frac{1}{\sqrt{1 - (v/c)^2}} = \frac{1}{\sqrt{1 - (0.998c/c)^2}} = 15.8$$

and with $t = \gamma t_o$, we have the muon traveling a distance of

$$d = vt = \gamma v t_o = \gamma(0.660 \text{ km}) = (15.8)(0.659 \text{ km}) = 10.4 \text{ km}$$

Therefore, the theory of relativity predicts that the muons would travel a much greater distance in our frame than is predicted by classical physics. Since t_o is an *average* value, some muons do travel greater distances and are observed at the Earth's surface.

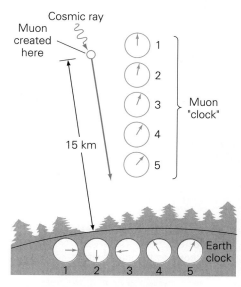

FIGURE 25.9 Experimental evidence of time dilation Muons are observed at the surface of the Earth as predicted by the theory of relativity. Example 25.1.

Length Contraction

As was mentioned earlier, the null result of the Michelson-Morley experiment could be explained if it were assumed that the length of the interferometer arm in the direction of motion contracted by a factor of $\sqrt{1 - (v/c)^2}$, the Lorentz-FitzGerald contraction. The same length contraction also turned out to be a consequence of the special theory of relativity. Its origin is not some ether interaction with interatomic forces but is explained by the two postulates of the relativity theory.

Since time is not uniform as observed in different reference frames, one might expect that lengths would also differ as observed in different frames. To see how this arises, consider a measuring rod of length L_o, as measured by an observer in his reference frame (●Fig. 25.10). L_o is called the **proper length**, being the length or distance between two points as measured by an observer *at rest relative to them*. In the case of a measuring rod, the proper length is measured by the observer at rest relative to the rod (observer O in Fig. 25.10).

A measuring rod of length L_o is also used as a length standard by observer O' in the figure, who is traveling with a constant velocity of magnitude v in the direction parallel to the rod. Observer O', wishing to compare her length standard with that of observer O, takes a measurement of the rod in his reference frame. To do this, O' measures the time interval between her passing the ends of the rod. The length of the rod is then given by $L = vt$, where t is the measured time interval.

Observer O could also measure the length of the rod by the same means. If he noted the times on his clock when O' passed by the ends of the rod, he

Figure 25.10

●**FIGURE 25.10 Length contraction** Observer O' measures the length of the rod by measuring the time interval between her passing the ends of the rod. Observer O similarly measures the length of the rod by using the same events. It is found that the measured lengths are not the same. See text for description.

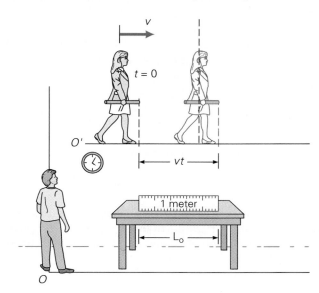

would measure a time interval t_o, and for him the length of the rod would be given by $L_o = vt_o$. Then, dividing one length by the other, we have

$$\frac{L}{L_o} = \frac{vt}{vt_o} = \frac{t}{t_o} \qquad (25.9)$$

But in this case, the *proper time* is in the O' reference frame (t, the time measurement done at the same point as the rod in O's frame moved by). That is, the time t_o is greater than t or there is a time dilation, since moving clocks run more slowly. This changes the notation in Eq. 25.8, and $t_o = \gamma t$, or $t/t_o = 1/\gamma$. Eq. 25.9 then becomes

$$L = \frac{L_o}{\gamma} = L_o \sqrt{1 - (v/c)^2} \qquad (25.10)$$

length contraction

Since γ is always greater than 1 with relative motion, we have $L < L_o$ or a **length contraction** by a factor of $1/\gamma$ or $\sqrt{1 - (v/c)^2}$, which is precisely the Lorentz-FitzGerald contraction.

Thus, not only time intervals but also space intervals, that is, lengths and distances, are affected by relative motion. It should be noted that the length contraction occurs only along the direction of motion. As a result, objects would appear to be contracted in the direction of their motion in a relativistic world.

Sketch graphs of L and t as functions of the speed.

EXAMPLE 25.2 ■ Length Contraction and Time Dilation

An observer sees a spaceship that is measured to be 100 m long when at rest (proper length) pass by in uniform motion with a speed of $0.500c$ (● Fig. 25.11). While the observer is watching the spaceship, a time of 2.00 s elapses on a clock on board the ship (proper time). (a) What would the observer measure the length of the moving spaceship to be? (b) What time interval elapses on the observer's clock for the 2.0 s interval measured on the ship's clock?

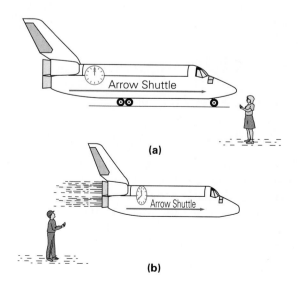

(a)

(b)

● **FIGURE 25.11 Length contraction and time dilation** As a result of length contraction, moving objects appear to be shorter, or contracted in the direction of motion, and moving clocks appear to run more slowly because of time dilation. Example 25.2.

Solution.

  Given: $L_o = 100$ m (proper length)   *Find:* (a) L (contracted length)
        $t_o = 2.00$ s (proper time)   (b) t (dilated time)
        $v = 0.500c$

(a) By calculation or from Table 25.1, $\gamma = 1.15$ for $v = 0.500c$, and the length contraction is given by Eq. 25.10:

$$L = \frac{L_o}{\gamma} = \frac{100 \text{ m}}{1.15} = 87.0 \text{ m}$$

Thus the observer measures the length of the spaceship to be considerably shorter than its proper length.

(b) The time dilation is given by Eq. 25.8:

$$t = \gamma t_o = (1.15)(2.00 \text{ s}) = 2.30 \text{ s}$$

EXAMPLE 25.3 ■ Muon Decay Revisited

In Example 25.1, we discussed a way of verifying time dilation using muon decay. A muon traveling at $v = 0.998c$ decays by its own clock (proper time) in $t_o = 2.20 \times 10^{-6}$ s. As shown in Example 25.1, a muon would travel a distance of only 659 m in this time. Yet muons are observed on Earth, and so must travel more than 10 km, on the average, before decaying. This paradox was explained by a time dilation effect for an Earth observer, who sees the muon's clock as running slow.

  But let us consider for a moment how this situation looks to a hypothetical observer riding with the muon (a thought experiment). For such an observer, the muon clock will appear to be correct—and it registers a time that is not sufficient for the muon to reach the Earth's surface. How can this be reconciled with the observation of the Earth-bound observer?

  When length contraction is also taken into account, however, both observations can be seen to be consistent with the experimental results. For our imaginary observer riding on the muon, the muon clock reads correctly ($t_o = 2.20 \times 10^{-6}$ s), but the observed travel distance—a length—is shorter because of length contraction. With $\gamma = 15.8$ for $v = 0.998c$, a travel length of 10.0 km in the Earth frame (proper length L_o) is measured by the observer on the muon to be L,

$$L = \frac{L_o}{\gamma} = \frac{10.0 \text{ km}}{15.8} = 0.633 \text{ km} = 633 \text{ m}$$

To travel this distance would take a time of

$$t = \frac{L}{v} = \frac{L}{(0.998c)} = \frac{633 \text{ m}}{(0.998)(3.00 \times 10^8 \text{ m/s})} = 2.11 \times 10^{-6} \text{ s}$$

Thus, through relativistic considerations both observers agree that the muon lives long enough to reach the Earth (the experimental result). The Earth observer explains this result by saying that the muon clock is running slowly (time dilation). The observer on the muon says, "No, my clock is fine—it's your meterstick that is too short" (length contraction).

The Twin Paradox

Time dilation gave rise to another popular relativistic topic—the so-called **twin, or clock, paradox**. According to the special theory, a clock in a moving system runs more slowly than, or not as fast as, one viewed by an inertial observer in another reference frame. For example, with $\gamma = 4$, for a proper time interval of 15 minutes, an hour would elapse on the observer's clock. Similarly, 1 year of proper time in the moving system takes 4 years in the observer's time frame. Since our heartbeat and age are measured by proper time, the question arises: Does an observer in one system age more quickly than a person in a system that is in motion relative to the first system?

One way to explore this question is to consider identical twins, one of whom goes on a high-speed space journey. The question then is: Will the space traveler come back younger than his Earth-bound twin? You might say that everything is relative and that the space twin sees the Earth clock run more slowly. He would claim that the Earth twin would age more slowly. They can't both be right.

The solution to the paradox lies in the fact that in leaving and returning to Earth, the space twin must experience accelerations and so is not in an inertial reference frame. However, if it is assumed that the acceleration periods occupy only a small part of the total trip time, the special theory can be applied to get an indication of what is likely to happen. (Einstein's general theory considers accelerating systems and confirms the result.) The prediction is that the space-traveling twin does return younger than the Earth-bound twin.

While traveling at a constant velocity on the outward and return trips, which occupy most of the total trip, both twins measure the same relative speed. However, the *proper length* of the trip is in the Earth twin's reference frame with fixed beginning and end points. The traveling twin thus measures a length contraction or shorter trip distance. Traveling at the same relative speed, the space twin travels a shorter time, and returns home younger than his Earth-bound twin.

In a sense, the twin paradox has been experimentally tested. In 1971, atomic clocks were flown around the world in jet planes. It was necessary to use these extremely accurate clocks to detect any effects because of the small γ factors, or low relative velocities. They were accurate enough to measure the predicted effect (with corrections made for accelerations, etc.), and a time dilation was experimentally verified. As was stated in the reporting article, "These results provide an unambiguous empirical resolution of the famous clock 'paradox' with macroscopic clocks."*

25.4 ■ Relativistic Mass and Energy

The ramifications of the theory of relativity are particularly important in particle physics, in which the speeds of the particles may be appreciable fractions of the speed of light. Such speeds are possible in particle accelerators, in which

---

*Hafele and Keating, "Around the World Atomic Clocks: Relativistic Time Gains Observed," *Science*, vol. 117, no. 4044 (July 14, 1972), pp. 166–170.

charged particles are accelerated through electrical potential differences. Of the three fundamental properties—length, time, and mass—the preceding discussion has shown that the first two are relativistic quantities. That is, their values depend on the measurement frame of reference. What about mass? Is it too a relativistic quantity?

Einstein showed that it is, and the mass of a particle or object increases with speed according to the relationship

$$m = \gamma m_0 = \frac{m_0}{\sqrt{1 - (v/c)^2}} \qquad (25.11)$$

Here m_0 is the mass of the particle at rest, or its **rest mass** (proper mass), and m is the **relativistic mass**, or the measured particle mass in an inertial reference frame moving with a speed v.

There is experimental evidence for this relativistic mass increase. Particle accelerators accelerate beams of charged particles to very high speeds using electric fields. Linear beams are deflected, or turned, by deflecting magnets. It is found that greater magnetic fields are needed to deflect high-speed particles than would be needed as predicted by using the particle rest mass. This is explained by a mass increase. Also, in circular accelerators called cyclotrons, accelerating particles are found to arrive at particular points in the orbital paths increasingly late. The lateness is explained by a relativistic increase in the masses of the particles.

According to Eq. 25.11, as v approaches c, the mass (m) of an object approaches infinity (Fig. 25.12). A basic result of the special theory of relativity is that no object can travel as fast as or faster than the speed of light. This may be seen from Eq. 25.11. As a particle approaches the speed of light, its mass approaches infinity. To accelerate an object until $v = c$ would require an infinite amount of energy, which makes c the maximum and unattainable speed limit.

Time dilation and length contraction (Eqs. 25.5 and 25.10) also imply this. If a reference frame could travel with a speed c, then time would be infinite, or would stand still, and the lengths of objects would contract to nothing, or the objects would vanish.

In the theory of relativity, the variation of mass with speed is a condition for the conservation of momentum. The basic conservation principle is valid if the **relativistic momentum** is defined as

$$p = mv = \gamma m_0 v \qquad (25.12)$$

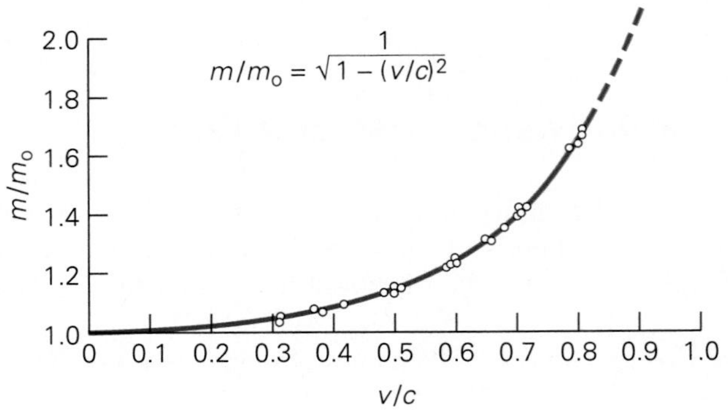

FIGURE 25.12 Relativistic mass The data points on the curve are measured masses of accelerated particles. There is good agreement between the theoretical prediction for relativistic mass and the experimental data.

Einstein used the conservation of momentum to derive the relativistic mass equation.

Since the relativistic momentum has a different form from its classical counterpart, we might expect the expression for kinetic energy also to be different, and this is true. In the relativistic case, the kinetic energy K of a particle is no longer equal to $\frac{1}{2}mv^2$ (or $p^2/2m$), even if the relativistic mass and momentum are used. The mathematics is beyond the scope of this text, but Einstein showed, using the work-energy theory, that the **relativistic kinetic energy** is given by

$$K = mc^2 - m_o c^2 \qquad (25.13)$$

This expression may be written in other forms, for example,

$$\boxed{K = \gamma m_o c^2 - m_o c^2 = m_o c^2 (\gamma - 1)} \qquad (25.14)$$

The relativistic expression is more complicated than the classical $K = \frac{1}{2}mv^2$, but it must be used if the speed of a particle is comparable to c. Of course, if $v \ll c$, the relativistic expression reduces to the classical form.*

In classical mechanics, the total mechanical energy is $E = K + U$. When there is no potential energy, this reduces to $E = K$; that is, the total energy is simply the kinetic energy. Equation 25.13, however, suggests a different way of looking at energy. We can see from this equation that an increase in the kinetic energy would be manifested as an increase in the relativistic mass m. (The other factors in the expression, m_o and c, do not vary.) We know that an increase in the kinetic energy would increase the total energy by a corresponding amount. Now, let us rearrange Eq. 25.13 to read

$$mc^2 = K + m_o c^2.$$

The right hand side of this equation consists of a variable quantity (the kinetic energy K) plus a constant quantity ($m_o c^2$) that is a function of the body's rest mass. Mathematical techniques beyond the scope of this book can be used to show that mc^2 is the **relativistic total energy**. This means that even when a body is at rest, and thus has no kinetic energy ($K = 0$), it still has an energy of $m_o c^2$. This minimum energy that a body always possesses is called its **rest energy**, E_o:

$$\boxed{E_o = m_o c^2} \qquad (25.15)$$

In the absence of potential energy, therefore, we can express the total energy as

or
$$\boxed{\begin{array}{l} E = K + E_o \\ E = K + m_o c^2 = mc^2 \end{array}} \qquad (25.16)$$

> Sketch graphs of kinetic energy and momentum as functions of the speed.

> When a particle has potential energy U, the total energy is $E = mc^2 = K + U + m_o c^2$.

*This can be shown by expanding the $\gamma - 1$ term in the preceding equation in a series expansion:

$$K = m_o c^2 (\gamma - 1) = m_o c^2 (-1 + \gamma) = m_o c^2 \left[-1 + \left(1 - \frac{v^2}{c^2} \right)^{-\frac{1}{2}} \right]$$

$$= m_o c^2 \left[-1 + 1 + \frac{1}{2} \left(\frac{v^2}{c^2} \right) + \frac{3}{8} \left(\frac{v^4}{c^4} \right) + \cdots \right] \quad (v < c)$$

and

$$K \cong m_o c^2 \left[\frac{1}{2} \left(\frac{v^2}{c^2} \right) \right] = \frac{1}{2} m_o v^2 \quad \text{(to a first-order approximation, where } v \ll c)$$

A direct relationship between the total energy and the rest energy may be obtained by noting that the relativistic mass is given by $m = \gamma m_o$ (Eq. 25.11). Multiplying both sides of the equation by c^2, we have

$$mc^2 = \gamma m_o c^2$$

or
$$E = \gamma E_o \tag{25.17}$$

and we have another gamma relationship. For a case with $v = 0$ (or $K = 0$) and $\gamma = 1$, we have $E = E_o$, as noted above.

Equation 25.15 shows that the relativistic mass is a direct measure of the total energy of the particle. This is Einstein's famous **mass-energy equivalence** formula, which points out that *mass is a form of energy*. Classically, when work is done on an object, its speed and energy increase. But, relativistically, the mass of an object increases with increasing speed. Thus, the work goes not only into increasing the speed but also to increasing the mass, and we are led to the concept of a mass-energy equivalence: that mass is a form of energy.

The mass-energy equivalence does not mean that we can convert rest mass into useful energy at will. If we could, our energy problems would be solved, since we have a lot of mass. Mass-energy conversion does take place on a limited scale in nuclear reactions (discussed in Chapter 29). For now, the important point is that any variation (increase or decrease) in the kinetic energy of a particle will give a variation in its relativistic mass.

The energy equivalent of a particle at rest depends on its rest mass, $E_o = m_o c^2$. For example, the rest mass of an electron is 9.109×10^{-31} kg, and the rest energy is (using four significant figures)

$$E_o = m_o c^2 = (9.109 \times 10^{-31}\text{ kg})(2.998 \times 10^8\text{ m/s})^2 = 8.187 \times 10^{-14}\text{ J}$$

or $E_o = (8.187 \times 10^{-14}\text{ J})\left(\dfrac{1\text{ eV}}{1.602 \times 10^{-19}\text{ J}}\right) = 5.110 \times 10^5\text{ eV} = 0.511\text{ MeV}$

In modern physics, particle energies are commonly expressed in units of electron volts because charged particles are often given energy by accelerating them through potential differences.

EXAMPLE 25.4 ■ Accelerating Energy

How much energy is required to give an electron initially at rest a speed of $0.9c$?

Solution. The energy needed is equal to the kinetic energy of the electron traveling at $0.9c$. However, we cannot use the classical expression $K = \frac{1}{2}mv^2$, since the increase in the energy of an electron traveling this fast significantly increases its mass, and the relativistic expression must be used. Since the rest mass of an electron has been shown to be $E_o = m_o c^2 = 0.511$ MeV, a convenient form of the relativistic kinetic energy is given by Eq. 25.14:

$$K = m_o c^2(\gamma - 1) = E_o(\gamma - 1)$$

With $v = 0.9c$, by calculation or from Table 25.1, $\gamma = 2.29$, and

$$K = E_o(\gamma - 1) = (0.511\text{ MeV})(2.29 - 1) = 0.659\text{ MeV}$$

Thus, 0.659 MeV would be required to accelerate an electron to a speed of $0.9c$.

EXAMPLE 25.5 ■ Total Energy and Speed

An electron is accelerated from rest through a voltage difference of 500 kV. (a) What is the total energy of the electron? (b) What is its speed?

Solution.

Given: $V = 500$ kV   *Find:* (a) E (total energy)
$E_o = 0.511$ MeV   (b) v (speed)

(a) The kinetic energy gained by the electron is

$$K = eV = (1e)(500 \text{ kV}) = 500 \text{ keV}$$

Notice why it is convenient to work in electron volts.
The rest energy of an electron is $E_o = 0.511$ MeV $= 511$ keV. Thus,

$$E = K + E_o = 500 \text{ keV} + 511 \text{ keV} = 1011 \text{ keV} (= 1.011 \text{ MeV})$$

(b) Finding the speed v may appear to be difficult with the known data. But remember that the speed is contained in the gamma factor, which appears in so many special relativity equations. One of these involving energies is Eq. 25.17, $E = \gamma E_o$, so we can easily find γ:

$$\gamma = \frac{E}{E_o} = \frac{1011 \text{ keV}}{511 \text{ keV}} = 1.98$$

Then, with $\gamma = 1/\sqrt{1 - (v/c)^2}$,

$$\frac{v}{c} = \left(1 - \frac{1}{\gamma^2}\right)^{\frac{1}{2}} = \left[1 - \frac{1}{(1.98)^2}\right]^{\frac{1}{2}} = 0.863$$

and $v = 0.863c = (0.863)(3.00 \times 10^8 \text{ m/s}) = 2.59 \times 10^8 \text{ m/s}$

QUESTION: Given the kinetic energy of a particle, how do we know whether to use the relativistic formula or the classical formula to find its speed?

ANSWER: The relativistic formula always applies, but if $v \ll c$, the simpler classical formula may be used. As an example, let us compare the kinetic energy and the rest energy, expressing the kinetic energy as in Eq. 25.14:

$$K = m_oc^2(\gamma - 1) = m_oc^2\left[\frac{1}{\sqrt{1 - (v/c)^2}} - 1\right]$$

Solving for $(v/c)^2$, we have

$$\left(\frac{v}{c}\right)^2 = 1 - \frac{1}{[1 + K/m_oc^2]^2} = 1 - \frac{1}{[1 + K/E_o]^2}$$

Thus, if $K \ll E_o$, or the kinetic energy is much less than the particle's rest energy, then $v \ll c$, and the classical formula $K = \frac{1}{2}m_ov^2$ is adequate. The relativistic formula could be used, but more calculation would be required to give essentially the same result. (If $v \leq 0.14c$, the error is less than 1 percent.)
The relationship also shows the reverse result. That is, if $v \ll c$, then $K \ll E_o$, or the kinetic energy is much less than the rest energy.

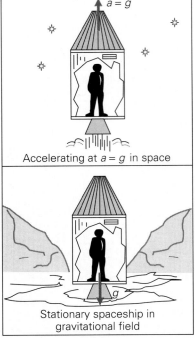

Accelerating at $a = g$ in space

Stationary spaceship in gravitational field

(a)

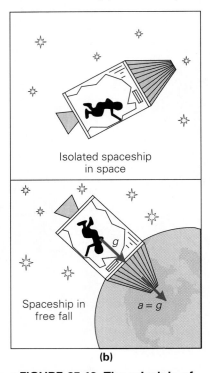

Isolated spaceship in space

Spaceship in free fall

(b)

● **FIGURE 25.13 The principle of equivalence** **(a)** The astronaut can perform no experiment in a closed spaceship that would determine whether he was in a gravitational field or an accelerating system. **(b)** Similarly, an inertial frame without gravity cannot be distinguished from free fall in a gravitational field.

Figure 25.13

25.5 ■ The General Theory of Relativity

Einstein's special theory of relativity applies to inertial systems, but not to accelerating systems. Accelerating systems require an extremely complex theory, which was described by Einstein in several papers published around 1915. Called the **general theory of relativity**, this surprisingly is essentially a gravitational theory.

The Equivalence Principle

An important aspect of the general theory is the **principle of equivalence**:

> An inertial reference frame in a uniform gravitational field is equivalent to a reference frame in the absence of a gravitational field that has a constant acceleration with respect to the inertial frame.

What this basically means is that:

> No experiment performed in a closed system can distinguish between the effects of a gravitational field and the effects of an acceleration.

According to the principle of equivalence, an observer in an accelerating system would find the effects of a gravitational field and those of an acceleration to be equivalent or indistinguishable. Rotating frames are excluded, since rotational acceleration can be distinguished from gravitational acceleration by releasing an object on a smooth, almost frictionless table. (Why?)

To better understand the equivalence principle, consider the situations illustrated in ● Fig. 25.13. Imagine yourself as an astronaut in a closed spaceship. You drop a pencil and it falls to the floor. What does this mean? According to the equivalence principle, it could mean (a) that you are in a gravitational field, or (b) that you are in an accelerating system (Fig. 25.13a). You can't determine which. If the spaceship were in free space and accelerating with an acceleration a, or if it were stationary in a gravitational field with $g = -a$, the pencil would fall to the floor in either case. In your closed system, there is nothing you could do to determine whether this was a gravitational or an acceleration effect.

What if the pencil did *not* fall and remained suspended in midair? This could mean that (a) you are in an inertial frame with no gravity, or (b) you are in free fall in a gravitational field. In the closed spaceship (Fig. 25.13b), you would have no way of knowing whether you were in free space (negligible gravity) or accelerating toward the Earth in free fall. (Think of how a released pencil would appear to float to an observer in a closed, freely falling elevator.) In this case, the equivalent effects of gravitation and acceleration cancel each other out, and there is no way to tell whether both are present, or neither.

Light and Gravitation

The principle of equivalence of the general theory of relativity leads to an important prediction—that light is bent in a gravitational field. To see how this prediction arises, let's use our imaginations in the thought experiment illustrated in ● Fig. 25.14. Suppose a beam of light traverses a spaceship that is accelerating at a very fast rate. If the spaceship were stationary, light emitted at point A would arrive at point B. However, because the spaceship is accelerating, the ship moves upward during the finite time it takes for the light to traverse the ship, and the light arrives at point C.

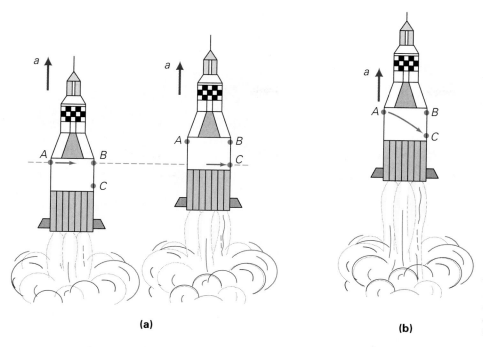

●FIGURE 25.14 Light bending
FIGURE 25.14 Light bending
(a) Light traversing an accelerating, closed rocket from point *A* arrives at point *C*. (b) In the moving system, the light would appear to be bent. Since an acceleration produces this effect, by the principle of equivalence, light should also be bent in a gravitational field.

Now consider the situation as observed by a person on board the spaceship. From his point of view, he would observe the same effect of light emitted at *A* and arriving at *C*—*as though the light were bent.* The acceleration of the rocket produces the effect, so by the equivalence principle we conclude that light should also be bent in a gravitational field.

Of course, this effect must be very small, since we have observed no evidence of gravitational light bending in the Earth's gravitational field. However, this prediction of Einstein's theory was experimentally verified in 1919 during a solar eclipse (see ●Fig. 25.15). Distant stars appear to be motionless and are measured to have a constant angular distance separating them at night when they can be clearly seen. Light from one of the stars may pass near the Sun, which has a relatively strong gravitational field. However, any bending would not be observed on Earth because most of the starlight would be masked by the glare of the Sun itself and by the sunlight scattered in the atmosphere (which is why we can't see stars during the day).

But during a total solar eclipse, when the moon comes between the Earth and the Sun, an observer in the moon's shadow (the umbra) can see stars. If the light from a star passing near the Sun were bent, then the star would have an apparent location different from its actual position. As a result, the angular distance between stars would be measured to be slightly larger. Einstein's theory predicted the angular difference in the apparent and actual positions of the stars to be 1.75 seconds of arc, or an angle of about 0.00005°. The experimental angular distance was found to be 1.61 ± 0.30 seconds of arc.

General relativity views a gravitational field as a "warping" of space and time, as illustrated in Fig. 25.15c. A light beam follows the curvature of space like a ball rolling on a surface. The bending of light was verified during subsequent solar eclipses and also by signals from space probes. For example, signals from a Viking lander on the surface of Mars passed through the gravitational field of the Sun when Mars was on the far side of the Sun. The signals were observed to be delayed by about 100 μs. The delay was caused by the signals passing through the space-time warp, or "sink," of the Sun.

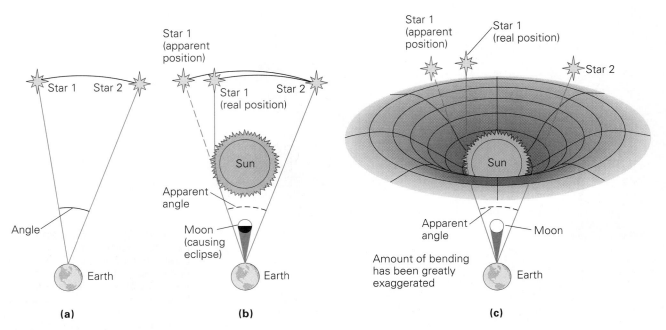

FIGURE 25.15 Gravitational attraction of light **(a)** Normally, two distant stars are observed to be a certain angular distance apart. **(b)** During a solar eclipse, the star behind the Sun can still be seen because of the effect of solar gravity on the starlight, which is evidenced by a larger measured angular separation. **(c)** General relativity views a gravitational field as a warping of space and time. A simplified analogy is the surface of a warped rubber diaphragm or sheet.

Gravitational Lensing. Another effect of the gravitational bending of light is known as *gravitational lensing.* In the late 1970s, a double quasar was discovered. (A quasar is an astronomical radio source.) The fact that it was a double quasar was not unusual, but everything about the two quasars seemed to be exactly the same, except that one was fainter than the other. It was suggested that perhaps there was only one quasar and that somewhere between it and the Earth, a massive object had deflected its electromagnetic radiation, producing multiple images. The subsequent detection of a faint galaxy between the two quasar images confirmed this hypothesis, and other examples have since been discovered (● Fig. 25.16).

Gravitational lenses can be considered to be a test of general relativity, but this concept has also given general relativity a new role in astronomy. By examining multiple images of a distant galaxy or quasar, their relative intensities and so on, astronomers can gain information about an intervening galaxy or cluster of galaxies whose gravitational field causes the bending of light.

●**FIGURE 25.16 Gravitational lensing** **(a)** The bending of light by a massive object such as a galaxy or cluster of galaxies can give rise to multiple images of a more distant object. **(b)** The discovery of what appeared to be four images of the same quasar (the "Einstein cross") suggested the possibility of gravitational lensing. On investigation, a faint intervening galaxy was in fact found.

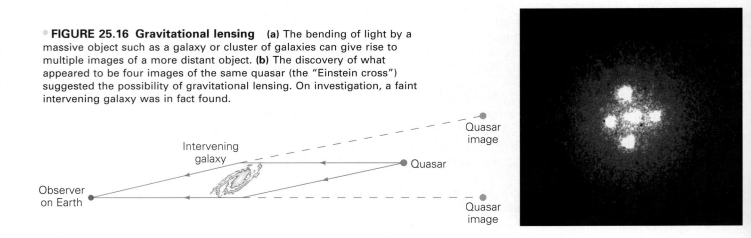

Such an effect has been observed by the Hubble Space Telescope. On a photo of a remote galaxy, two mirror images of the structure were observed on opposite sides of the picture (Fig. 25.17). These were believed to be caused by the gravitational lensing of an intervening cluster of galaxies containing much *dark matter*: matter that does not emit electromagnetic radiation, and so cannot be detected by regular observations. How much matter the universe contains is an important question for scientists. Is there enough matter for gravitational attraction to slow and stop the Big Bang expansion of the universe? It is estimated that there might be 10 times more matter in the universe than is actually observed, but we do not know what this dark matter consists of, although several theories have been proposed. Gravitational lensing may provide a method to detect such dark matter.

Black Holes. The idea that gravity can affect light finds its most extreme application in the concept of a black hole. A **black hole** is generally considered to form from the gravitationally collapsed remnant of a star.* Such an object has a density so great and a gravitational field so intense that nothing can escape it. In terms of the space-time warp analogy presented above, a black hole would be a bottomless pit in the fabric of space time. Even light can't escape the intense gravitational field of a black hole—hence the blackness.

An idea of the size of a black hole may be obtained by using the escape speed. Recall from Section 7.6 that the escape speed from a spherical body of mass M and radius R is given by

$$v_e = \sqrt{\frac{2GM}{R}} \qquad (25.18)$$

If light does not escape from a black hole, this means that the hole's escape speed must exceed the speed of light, 3.0×10^8 m/s. The critical radius of a sphere around a black hole from which light will not be able to escape may

*It is speculated that black holes may originate in other ways, such as the collapse of entire star clusters in the center of a galaxy. We will limit our discussion to stellar collapse.

 FIGURE 25.17 Gravitational lensing and dark matter (a) Two similar images of a remote galaxy on either side of an intervening cluster of galaxies were observed by the Hubble Space Telescope. (b) The effect was thought to result from gravitational lensing by dark matter—matter that emits no electromagnetic radiation and so cannot be detected by ordinary observations. Gravitational lensing may offer a method to detect dark matter.

HUBBLE SPACE TELESCOPE VIEW

(a)

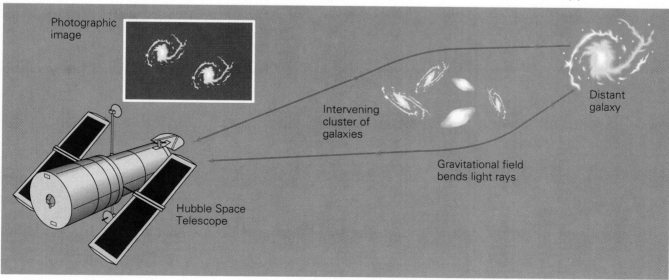

(b)

be obtained by substituting $v_e = c$ in Eq. 25.18. Solving for R,

$$R = \frac{2GM}{c^2} \tag{25.19}$$

A collapsing star becomes a black hole if it reaches the Schwarzschild radius, and collapses beyond this event horizon.

This is called the **Schwarzschild radius** after Karl Schwarzschild (1873–1916), a German astronomer who developed the concept. The boundary of a sphere of radius R defines what is called the **event horizon**. Any event occurring within this horizon is invisible to an observer outside, since light cannot escape.

It should be noted that the event horizon is not necessarily the radius of a black hole. It gives only the limiting distance within which light is unable to escape from a black hole. The mass inside would continue to collapse, making the black hole smaller than its event horizon.

EXAMPLE 25.6 ■ Blacking Out

Assuming our Sun collapsed to a black hole, what would be its radius when it reached black hole status? ($M_s = 2.0 \times 10^{30}$ kg)

Solution.

> **Given:** $M_s = 2.0 \times 10^{30}$ kg
> $G = 6.67 \times 10^{-11}$ N-m^2/kg^2 (known)
> $c = 3.0 \times 10^8$ m/s (known)
>
> **Find:** R (Schwarzschild radius)

This is a straightforward calculation using Eq. 25.19. However, it should be mentioned that our Sun will not become a black hole. This is a fate that befalls only stars that are considerably more massive than the Sun.

$$R = \frac{2GM}{c^2} = \frac{2(6.67 \times 10^{-11} \text{ N-m}^2/\text{kg}^2)(2.0 \times 10^{30} \text{ kg})}{(3.0 \times 10^8 \text{ m/s})^2}$$
$$= 3.0 \times 10^3 \text{ m} = 3.0 \text{ km}$$

(The Sun currently has a radius of about 7.0×10^5 km.)

If nothing, including radiation, escapes a black hole from inside its event horizon, how then might we observe or locate one? This is addressed in the Insight feature on black holes.

Another result of the general theory is that gravity affects time by causing it to slow down—the greater the gravitational field, the greater the slowing of time. There is evidence of this effect in what is called the *gravitational red shift* of light from the Sun. The gravitational field of the Sun is quite large, and if the slowing of time in a gravitational field is correct, the electronic vibrations in the Sun's atoms should be slower. As a result, light from these atoms would have a lower frequency and be shifted toward the red end of the visible spectrum.

The measurement of a solar gravitational red shift was unsuccessful for many years. This was because of a lack of understanding of the conditions at the solar surface. For example, violent motions, with convection columns of rising hot gases and descending cooler gases, give rise to Doppler shifts. If wasn't until 20 to 30 years ago, after a variety of effects were taken into account,

The idea of black holes has caught the public's fancy. Imagine—something so dense that neither light nor anything else can escape from it. So dense that a tablespoon full of its matter would probably weigh more than Mt. Everest. Black holes are becoming increasingly common in science fiction, but in actuality they are still largely theoretical. In the theory of stellar evolution, black holes result from the collapse of stars having much greater mass than the Sun. But experimental proof of the existence of a black hole is another matter.

The boundary of the "blackness" of the hole is the surface of the event horizon at its Schwarzschild radius (see discussion in text). The matter within can continue to contract, but the force of gravity at the event horizon would still be the same. (Why?) Thus, the boundary of a black hole is not a sphere of matter, but the radius at which the gravitational force is sufficiently strong to keep light from escaping. The actual radius of the matter of a black hole may be much less. What form matter takes inside a black hole is not known, and may never be known. If we ever determine the location of a black hole in space, any probe we might send to check this could never report back when it arrived. (Can you see why this is so?)

How then do you observe a black hole? Light passing nearby would be bent, but it is unlikely that we would ever observe this, considering the vastness of space. The most likely possibility comes from binary star systems, which are quite common. Cygnus X-1, the first X-ray source discovered in the constellation Cygnus, provides

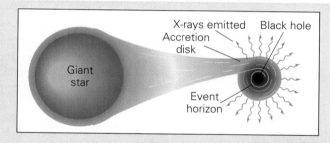

FIGURE 2 X-rays and black holes   Matter drawn from the other member of a binary star system forms a spiraling accretion disk around the hole. Matter falling into the disk is accelerated, and collisions give rise to the emission of X-rays.

the best evidence so far for a black hole in our galaxy. (Approximately 100 or more X-ray sources have been discovered through observations from rockets and satellites. Recall that X-rays do not penetrate the atmosphere.) If we look at the location of the Cygnus X-1 X-ray source in visible light (Fig. 1), we observe a giant star whose spectrum shows Doppler shifts, indicating an orbital motion. (Binary stars orbit about each other—more precisely, about their common center of mass.)

It is speculated that in binary star systems such as Cygnus X-1, one member has become a black hole. Matter drawn from the other member would create a so-called *accretion disk* of spiraling matter around the hole (Fig. 2). The matter falling into the disk would be accelerated and heated. Collisions and deceleration would produce X-rays, and the black hole would appear as an X-ray source.

So, do black holes exist? Time (and further investigation) will presumably tell. In the meantime, there will be a lot of speculation. Perhaps the situation is best summed up in the words of the scientist J. B. S. Haldane:

"My suspicion is that the universe is not only queerer than we suppose, but queerer than we *can* suppose."

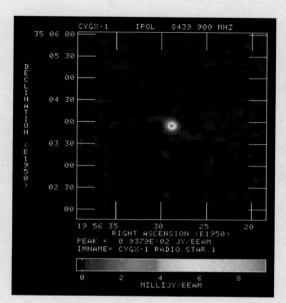

FIGURE 1 Cygnus X-1   The overexposed large dark spot at the center is a giant star believed to be a companion of the X-ray source and possibly a black hole, Cygnus X-1.

that the gravitational red shifts of solar lines were measured. The gravitational red shift was also verified in the laboratory in the 1960s by using complicated nuclear techniques.

Relativistic effects apply to all systems, no matter how complicated. Although these effects are most readily observed for atomic particles and light, they apply as well to real clocks, human beings, and so on. Although we don't commonly observe them, relativistic effects form the bases of some good science fiction and even some (not so good?) poetry, such as the following variation on a limerick that appeared in the British magazine *Punch* in 1923:

A precocious student named Bright
Could travel much faster than light;
He departed one day
In relative way
And arrived home the previous night.

25.6 ■ Relativistic Velocity Addition ◇

As we have seen, the postulates of special relativity affect our classical concepts of distance (and displacement) and time. Since velocity involves displacement and time, we might expect that measured relative velocities are also affected, and this is the case.

Recall from Section 25.1 that there was a problem with vector addition when light was involved. Ordinarily, if someone on a train moving at 20 m/s threw a ball at 10 m/s in the direction of motion, an observer on the ground would say that the velocity of the ball was simply 20 m/s + 10 m/s = 30 m/s in the direction of the motion. (Review Fig. 25.1 if necessary.) However, if the ball is replaced by a beam of light, such vector addition gives a speed greater than c, which violates the second postulate of relativity.

Review Section 6.6 and Fig. 6.22

The same problem occurs with objects moving at appreciable fractions of the speed of light. Consider the rocket separation shown in ● Fig. 25.18. After separation, the jettisoned stage has a velocity **v** with respect to the Earth and the rocket has a velocity **u′** with respect to the jettisoned stage. Then, assuming straight line motion, by ordinary vector addition we would say that the velocity **u** of the rocket payload with respect to Earth was given by

$$\mathbf{u} = \mathbf{v} + \mathbf{u}'$$

However, assuming relativistic speeds, suppose $v = 0.50c$ and $u' = 0.60c$. Then we would have $u = v + u' = 0.50c + 0.60c = 1.1c$. Thus, classical velocity addition says that an Earth observer would see the rocket traveling at a speed greater than the speed of light in vacuum. However, as we saw in Section 25.4, no object can travel faster than the speed of light.

Einstein recognized that since length and time are different with respect to different reference frames, classical velocity vector addition is not applicable in relativistic cases. Instead, he showed the correct formula for motion in a straight line to be

$$u = \frac{v + u'}{1 + \dfrac{vu'}{c^2}} \tag{25.20}$$

where the velocities have the same meanings as above and sign notation is

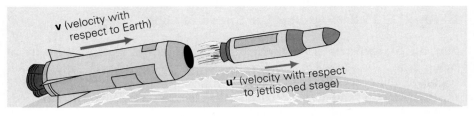

FIGURE 25.18 Relativistic velocity addition After jettisoning, the rocket has a velocity u′ with respect to the jettisoned stage. The jettisoned stage has a velocity **v** with respect to the Earth. For relativistic velocities, the velocity u of the rocket with respect to the Earth is not given by ordinary vector velocity addition and must be obtained using a special formula. See text for description.

used to indicate velocity directions. Notice that the observed velocity u is reduced by a factor of $1/(1 + vu'/c^2)$ from that of the classical formula.

It is important to clearly identify the velocities in working problems, so we list them again generally:

v = velocity of object 1 with respect to an observer (e.g., on Earth)

u' = velocity of object 2 with respect to object 1

u = velocity of object 2 with respect to observer (e.g., on Earth)

Signs are used to indicate velocity directions. For example, if object 2 were approaching object 1, then we would have a negative u' and the plus signs in Eq. 25.20 would change to minus signs. Let's test Eq. 25.20 with some examples.

EXAMPLE 25.7 ■ Faster Than The Speed of Light?

For the rocket separation in Fig. 25.8, let the velocities be $v = 0.50c$ and $u' = 0.60c$, as given in the text example. According to relativistic velocity addition, what is the velocity of the rocket nose cone as seen by an observer on Earth?

Solution. Listing the data,

Given: $v = 0.50c$ (velocity of jettisoned stage with respect to Earth)

Find: u (velocity of rocket with respect to Earth)

$u' = 0.60c$ (velocity of rocket with respect to jettisoned stage)

The motions were taken to be in the positive direction, and using Eq. 25.20,

$$u = \frac{v + u'}{1 + \dfrac{vu'}{c^2}} = \frac{0.50c + 0.60c}{1 + \dfrac{(0.50c)(0.60c)}{c^2}}$$

$$= \frac{1.1c}{1.3} = 0.85c$$

Thus we have a speed less than c. (Note how it is convenient to carry c along in symbol form, which includes units.)

EXAMPLE 25.8 ■ Adding The Speed of Light

Suppose a spacecraft is moving in free space with a constant velocity of magnitude 0.25c relative to and directly away from an observer. An astronaut on board the spacecraft sends out two light beams, one in the direction of the craft's motion and the other back toward the observer. What are the speeds of the light beams relative to the observer?

Solution. Here, if you apply the second postulate of special relativity, you already know the answers. But let's see if Eq. 25.20 gives us the correct answers.

Given: $v = 0.25c$
      (a) $u' = c$ (in direction of motion)
      (b) $u' = -c$ (opposite direction of motion)

Find: u (velocity of the light for both cases)

(a) Using Eq. 25.20, we have

$$u = \frac{v + u'}{1 + \dfrac{vu'}{c^2}} = \frac{0.25c + c}{1 + \dfrac{(0.25c)(c)}{c^2}}$$

$$= \frac{1.25c}{1.25} = c$$

(b) Similarly, with $u' = -c$,

$$u = \frac{v + u'}{1 + \dfrac{vu'}{c^2}} = \frac{0.25c - c}{1 + \dfrac{(0.25c)(-c)}{c^2}}$$

$$= \frac{-0.75c}{0.75} = -c$$

Hence our relativistic velocity addition formula gives answers consistent with the special theory's second postulate—the constancy of the speed of light.

Important Concepts

You should be able to define and/or explain these chapter concepts clearly.

relativity
inertial reference frame
principle of classical (Newtonian) relativity
ether
Michelson-Morley experiment
interferometer
principle of relativity
special theory of relativity

constancy of the speed of light
time dilation
proper time
proper length
length contraction
twin (clock) paradox
rest mass
relativistic mass
relativistic momentum

relativistic kinetic energy
relativistic total energy
rest energy
mass-energy equivalence
general theory of relativity
principle of equivalence
black hole
Schwarzschild radius
event horizon

Important Relationships for Review

These relationships are mathematical statements of the concepts and principles presented in the chapter. You should be able to identify the symbols and to explain the relationships before proceeding to the Exercises. In-text equation reference numbers are given for convenience.

Time Dilation:

$$t = \gamma t_o = \frac{t_o}{\sqrt{1 - (v/c)^2}} \tag{25.8}$$

Length Contraction:

$$L = \frac{L_o}{\gamma} = L_o \sqrt{1 - (v/c)^2} \tag{25.10}$$

Relativistic Mass:

$$m = \gamma m_o = \frac{m_o}{\sqrt{1 - (v/c)^2}} \tag{25.11}$$

Relativistic Momentum:

$$p = mv = \gamma m_o v \tag{25.12}$$

Relativistic Kinetic Energy:

$$K = \gamma m_o c^2 - m_o c^2 = E_o(\gamma - 1) \tag{25.14}$$

Rest Energy:

$$E_o = m_o c^2 \tag{25.15}$$

Relativistic Total Energy:

$$E = K + E_o \tag{25.16}$$
$$E = K + m_o c^2 = mc^2$$

Total Energy and Rest Energy:

$$E = \gamma E_o \tag{25.17}$$

Schwarzschild Radius:

$$R = \frac{2GM}{c^2} \tag{25.19}$$

◇ **Relativistic Velocity Addition:**

$$u = \frac{v + u'}{1 + \dfrac{vu'}{c^2}} \tag{25.20}$$

Exercises

25.1 ■ Classical Relativity

25.2 ■ The Michelson-Morley Experiment

1 If an isolated object in a system exhibits an increasing velocity, then (a) it is an inertial reference frame, (b) $F = ma$ applies to the system, (c) the laws of mechanics are the same in this reference frame as in all inertial frames, (d) none of these. (d)

● **2** The principle of classical relativity (a) did not seem to apply to electricity and magnetism, (b) required a constant speed of light different from that predicted by Maxwell's equations, (a) implied that nothing could travel as fast as light, (d) provided an absolute frame of reference for the description of motion. (a)

3 The existence of the ether (a) would provide a special or absolute reference frame, (b) is necessary for light propagation, (c) made Maxwell's equations relativistically correct, (d) both (a) and (b). (a)

4 The Michelson interferometer (a) was based on refraction, (b) depended on coherent light beams, (c) showed the expected fringe shift, (d) was based on the Lorentz-FitzGerald contraction. (b)

5 The Michelson-Morley experiment (a) worked only at night, (b) showed the existence of the ether, (c) gave a null result, (d) none of these. (c)

6 We live on a rotating Earth and therefore are in an accelerating, noninertial system. How then can we apply Newton's laws of motion on Earth? see ISM

7 How does the constancy of the speed of light explain the null result of the Michelson-Morley experiment? see ISM

8 ■■ Car A is traveling eastward at 90 km/h. Car B is traveling with a speed of 65 km/h. Find the relative velocity of car B with respect to car A if car B is traveling (a) westward and (b) eastward. (a) 155 km/h eastward (b) −25 km/h westward

9 ■■ A person 1.5 km from your location fires a gun; you observe the muzzle flash. If a 10 m/s wind is blowing, how long will it take the sound to reach you if the wind is (a)

toward you and (b) toward the other person? (The speed of sound is 345 m/s.) (a) 4.22 s (b) 4.48 s

10 ■■ A small airplane has an air speed of 200 km/h (speed with respect to air). Find the airplane's ground speed if there is (a) a head wind of 35 km/h and (b) a tail wind of 25 km/h. (a) 165 km/h (b) 225 km/h

11 ■■ A speedboat can travel 50 m/s in still water. If the boat is in a river that has a flow speed of 5 m/s, find the maximum and minimum values of the boat's speed relative to an observer on the river bank. 55 m/s and 45 m/s

12 ■■ A small boat can travel 0.45 m/s in still water. The boat heads directly across a 50-m-wide river, which has a flow speed of 0.15 m/s. (a) How long will it take the boat to reach the opposite shore? (b) How far downstream will the boat travel? (c) How would the boat have to be steered if it were to arrive at a point across the river directly opposite the departure point? (a) 1.1×10^2 s (b) 17 m (c) 18° upstream

13 ■■■ The apparatus used for measuring the speed of light by the French scientist Fizeau in 1849 is illustrated in ● Fig. 25.19. Teeth on a rotating disk periodically interrupt a beam of light. The light flashes travel to a plane mirror and are reflected back to an observer. Show that if the disk is rotated so that light passing through one tooth gap travels to the mirror and returns to be seen by the observer through the adjacent gap, the speed of light is given by $c = 2fNL$, where N is the number of gaps in the disk, f is the frequency of, or number of revolutions per second made by, the rotating disk, and L is the distance between the disk and the mirror. see ISM

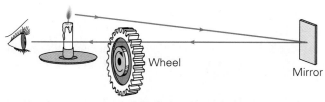

● FIGURE 25.19 Fizeau's apparatus See Exercise 13.

14 ■■■ One proposed explanation of the null result of the Michelson-Morley experiment suggested that the interferometer arm in the direction of motion is shortened by a factor of $[1 - (v/c)^2]^{\frac{1}{2}}$ (the Lorentz-FitzGerald contraction). Show how this contraction would give a null result. (This hypothesis was disproved by making the lengths of the arms of the interferometer unequal.) see ISM

15 ■■■ The swimmer analogy of the Michelson-Morley experiment is illustrated in ● Fig. 25.20. Two swimmers start out from point A at the same time and swim with a constant speed c relative to the water. One swimmer swims across the river to point C and back to A, while the other swimmer swims to point B and back to A. Do the swimmers arrive back at A at the same time? Justify your answer mathematically. see ISM

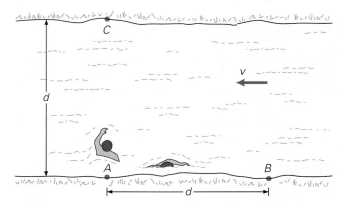

● FIGURE 25.20 The swimmer analogy See Exercise 15.

25.3 ■ The Special Theory of Relativity

16 The special theory of relativity (a) applies to inertial systems, (b) applies to noninertial systems, (c) applies to both inertial and noninertial systems, (d) disproves the constancy of the speed of light. (a)

17 Proper time is the interval (a) measured by an observer on a clock in another reference frame, (b) determined by events that occur at the same location, (c) measured by an observer on a clock in his or her reference frame, (d) both (b) and (c). (d)

18 In a thought experiment, assume that you rode a beam of light coming from the face of a clock. How would time measured on the clock appear to you? see ISM

19 An observer sees a friend passing by her in a rocketship with a uniform velocity that has a magnitude of an appreciable fraction of the speed of light. The observer knows her friend to be 5 ft 8 in. tall. Does he still appear the same to the observer? Explain. see ISM

20 ■■ A person observes 5.0 min to elapse on her clock. How much time does she observe to elapse during this interval on a clock in another reference frame moving with a relative velocity of $0.9c$? 2.2 min

21 ■■ A person has a pulse rate of 60 beats/min. What would the person's pulse rate be according to an observer moving with a velocity of $0.85c$ relative to the person? 32 beats/min

22 ■■ A time of 15 min elapses on a clock, but an observer in another inertial reference frame measures 20 min to elapse. What is the relative velocity of the reference frames of the clocks? 0.66 c

23 ■■ A spaceship travels with a speed of $c/3$. What time change does a clock *on board* the spaceship show in 1 hour according to an observer in another inertial frame? 1.06 h

24 ■■ Alpha Centauri, the binary star closest to this solar system, is about 4.30 light years away. If a spaceship traveled a similar distance with a constant speed of $0.60c$ relative to Earth, how much time would elapse on a clock on the spaceship?   5.7 y

25 ■■ One 25-year-old twin decides to take a space trip while the other remains on Earth. The traveling twin travels at a speed of $0.95c$ for 39 years according to Earth time. Assuming that special relativity applies, what are their ages when the traveling twin returns to Earth?   64 y; 37 y

26 ■■ The proper lifetime of a muon is 2.20 μs. If a muon approaches the Earth with a speed of $0.998c$, how long is its lifetime according to an observer on Earth?   34.8 μs

27 ■■ An astronaut in a spaceship sets up a 25 g mass on a spring ($k = 0.28$ N/m) and starts it oscillating on a frictionless surface. If the spaceship travels with a constant speed of $0.60c$ relative to Earth, what would be the period of the mass as measured by an observer on Earth?   2.4 s

● **28** ■■ A cylindrical-shaped spaceship has a length of 35.0 m and a diameter of 8.25 m. If it passes by the Earth with a relative speed of 2.44×10^8 m/s, what are the dimensions of the ship measured by an Earth observer?
20.3 m and 8.25 m

29 ■■ A "flying wedge" spaceship has a side view of a right triangle. At rest, the base and altitude (which form the 90° angle) measure 40.0 m and 15.0 m, respectively. As the ship moves by an observer on Earth with a speed of $0.90c$, (a) what does she measure the area of the side of the ship to be? (b) What does she calculate the angles of the triangular form to be?   (a) 131 m² (b) 40.6°

30 ■■ How fast must a meterstick be moving relative to an observer so that he measures its length to be 50 cm?
0.87 c

31 ■■ How fast must a meterstick travel for its rest length to contract by (a) 1%, (b) 10%, and (c) 99%?   (a) 0.14 c
(b) 0.44 c (c) 0.9999 c

32 ■■ According to a navigator on a spaceship, the ship has traveled one light year. How long does it take for the spaceship to travel this distance according to an inertial observer in a frame to which the relative speed of the spaceship is $0.8c$?   1.67 y

33 ■■ A pole vaulter at the Relativistic Olympics sprints past you down the runway for a vault with a speed of $0.65c$. When he is at rest, his pole is 7.0 m long. What does the length of the pole appear to be to you as he passes you?
5.3 m

34 ■■■ What is the length contraction (ΔL) of an automobile 5.00 m long when it is traveling at 100 km/h? [Hint: $\sqrt{1 - x^2} \cong 1 - (x^2/2)$ for $x \ll 1$.] 2.15×10^{-14} m

35 ■■■ An observer in a certain reference frame cannot understand why there is a problem in converting to the metric system because she notes the length of a meterstick in another moving reference frame to be the same length as her yardstick. Explain this situation quantitatively.   0.40 c

36 ■■■ Two reference frames move in relation to each other with a constant relative speed v in the x direction. Considering the O' frame to move in the $+x$ direction relative to the O frame, (a) show that the coordinate systems are classically related by the transformation equations: $x = x' + vt$, $y = y'$, and $z = z'$. These are the so-called Galilean transformations, together with $t = t'$, since time classically is a universal quantity. (b) Taking into account the relativistic length contraction, show that the transformation equations are given by: $x' = \gamma(x - vt)$, $y' = y$, and $z' = z$. These relativistic transformations are called the Lorentz transformations. (Time is also affected, but the derivation of this transformation is beyond the scope of this book.)   see ISM

25.4 ■ Relativistic Mass and Energy

37 The special theory of relativity shows relativistic effect(s) for the fundamental properties of (a) length, (b) mass, (c) time, (d) all of these.   (d)

38 The total energy for a relativistic-moving particle is equal to (a) $E = \frac{1}{2}mv^2$, (b) $E = \gamma E_o$, (c) $E = K - m_o c^2$, (d) $E = m_o c^2$.   (b)

39 If an electron has a kinetic energy of 2 keV, could the classical expression for kinetic energy be used to compute its speed? What if the kinetic energy of the electron were 2 MeV?   yes; no

40 ■■ An electron travels at a speed of $c/2$. What is its total energy?   0.588 MeV

41 ■■ An electron is accelerated from rest through a potential difference of 2.5 MV. Find the (a) speed, (b) kinetic energy, and (c) momentum of the electron.   (a) 0.985 c (b) 2.5 MeV   (c) 1.59×10^{-21} kg-m/s

42 ■■ A home uses approximately 15,000 kWh of electricity per year. How much matter would have to be converted directly to energy to supply this amount of energy?
6.0×10^{-7} kg

43 ■■ How fast must an object travel for its mass to have an increase of (a) 1.0% and (b) 99% of its rest mass?   (a) 0.14 c
(b) 0.87 c

44 ■■ The United States uses over 3.0 trillion kWh of electricity annually. If 3.0 trillion kWh of electrical energy were supplied by nuclear generating plants, how much nuclear mass would have to be converted to energy, assuming a production efficiency of 35%?   340 kg

45 ■■ A person whose mass is 48 kg wishes to gain 12 kg relativistically with respect to another reference frame. (a) How fast would the person have to travel? (b) How would her height and width change? (c) How would her overall density change?   (a) 0.60 c (b) same height; smaller width (c) increase

● **46** ■■ The kinetic energy of an electron is 60% of its total energy. Find the speed and momentum of the electron.
(a) 0.917 c (b) 6.26×10^{-22} kg-m/s

47 ■■ An electron has a total relativistic energy of 2.8 MeV. What is its relativistic momentum? 1.5×10^{-21} kg-m/s

48 ■■ How much energy is required to accelerate an electron from rest to half the speed of light? 7.7×10^{-2} MeV

49 ■■ Sketch graphs of the (a) momentum, (b) energy, and (c) mass of an object as a function of speed.   see ISM

50 ■■ A proton moves with a speed of $0.35c$. What are its (a) total energy, (b) kinetic energy, and (c) relativistic momentum?   (a) 1.00×10^3 MeV   (b) 69 MeV
(c) 1.88×10^{-19} kg-m/s

51 ■■ A proton moving with a constant velocity has a total energy 2.5 times its rest energy. (a) What is the proton's speed? (b) What is its kinetic energy?   (a) $0.92\ c$
(b) 1.4×10^3 MeV

● **52** ■■ The Sun has a mass of 1.989×10^{30} kg and radiates at a rate of 3.827×10^{23} kW. (a) How much mass does the Sun lose in one hour? (b) How long will it take the Sun to lose one percent of its mass?   see ISM

53 ■■ A beam of electrons is accelerated from rest to a speed of $0.950c$ in a particle accelerator. What are (a) the kinetic energy and (b) the total energy in MeV of an electron in the beam?   (a) 1.1 MeV   (b) 1.6 MeV

54 ■■ Phase changes require latent heat (Chapter 11). If one kilogram of water at 100°C is converted to steam, how much more mass would the steam have as a result of absorbing energy? 2.5×10^{-11} kg

55 ■■ A small package of artificial sweetener (the kind you see in restaurants) contains one gram. If this mass could be completely converted to energy, how long would it keep a 1600 W hair dryer running? 1.79×10^3 y

56 ■■■ Show that the *classical* kinetic energy and momentum of a particle can be expressed in terms of the rest energy of the particle as $K = \frac{1}{2}E_o\left(\dfrac{v^2}{c^2}\right)$ and $p = \dfrac{E_o v}{c^2}$.
see ISM

57 ■■■ Using the relation $m^2 c^4 - m_o^2 c^4$, show that the total energy E and relativistic momentum p are related by $E^2 = p^2 c^2 + (m_o c^2)^2$.   see ISM

58 ■■■ The National Accelerator Laboratory at Batavia, Illinois, can accelerate particles to energies of giga-electron volts (1 GeV = 10^9 eV). (a) What is the speed of a proton accelerated from rest through a potential of 1 GeV? ($m_o = 1.67 \times 10^{-27}$ kg.) (b) The planned Superconducting Super Collider (SSC) accelerator would produce particles with energies of 40 GeV. What would be the speed of a proton in this case?   (a) $0.351\ c$ (b) $0.9997\ c$

25.5 ■ The General Theory of Relativity

59 The general theory of relativity (a) is essentially a gravitational theory, (b) applies to rotating systems, (c) applies only to inertial systems, (d) refutes the principle of equivalence.   (a)

60 One of the predictions of the general theory is (a) black holes, (b) the Schwarzschild radius, (c) the event horizon, (d) the bending of light in a gravitational field.   (d)

61 Suppose a meterstick were dropped through the event horizon of a black hole. Describe what the effects might be on the meterstick.   see ISM

62 If the Sun gravitationally collapsed to become a black hole, what would be its density at the time it became a black hole? (See Example 25.6.) 1.8×10^{19} kg/m³

63 Compare the radii of Earth's and Jupiter's event horizons if they contracted to black holes. ($M_E = 6.0 \times 10^{24}$ kg, $M_J = 318\ M_E$)   8.9 mm and 2.8 m

● **64** A black hole has an event horizon of 5.00×10^3 m. (a) What is the mass of the "hole"? (b) Set a lower limit on the density of the black hole.   (a) 3.37×10^{27} kg
(b) 6.44×10^{15} kg/m³

65 ■■ Assume that a collapsing star has a mass of 5 solar masses. (a) What would the radius of the collapsed star have to be for it to qualify as a black hole? (b) What would be the density of the black hole?   (a) 1.5×10^4 m
(b) 7.1×10^{17} kg/m³

66 ■■■ An apparatus like the one in ● Fig. 25.21 was given to Albert Einstein on his 76th birthday by Eric M. Rogers, a physics professor at Princeton University. The goal is to get the ball into the cup without touching the ball. (Jiggling the pole up and down will not do it.) Einstein solved the puzzle immediately and then confirmed his answer with an experiment. How did Einstein get the ball into the cup? [*Hint:* He used a fundamental concept of general relativity.]*   see ISM

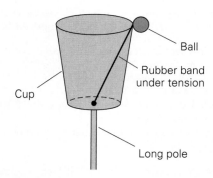

● **FIGURE 25.21 How to get the ball in the cup?**   See Exercise 66.

25.6 ■ Relativistic Velocity Addition ◇

67 ■■ After jettisoning a stage, a rocket has a velocity of $+0.20c$ relative to the jettisoned stage. An observer on Earth sees the jettisoned stage moving with a velocity of

---

*This problem is adapted from R. T. Weidner, *Physics*. Allyn and Bacon, 1985, p. 333.

$+7.5 \times 10^7$ m/s relative to her. What is the velocity of the rocket relative to the Earth observer?   0.43 c

68 ■■ In moving away from a planet, a spacecraft fires back a probe toward the planet with a speed of 0.15c relative to the spacecraft. If the speed of the spacecraft is 0.40c relative to the planet, what is the velocity of the probe as seen by an observer on the planet?   0.27 c

69 ■■ A rocket launched outward from Earth has a speed of 0.100c relative to the Earth. It is directed toward an incoming meteor that may hit the planet. If the meteor moves directly toward the rocket with a speed of 0.250c relative to it, what is the velocity of the meteor as observed from Earth?
−0.154 c

70 ■■■ In a linear particle accelerator, a proton with a speed of 0.80c moves in one direction and an electron moves with a speed of 0.90c in the opposite direction. Both speeds are relative to the laboratory frame of reference. (a) What is the velocity of the proton relative to the electron? (b) What is the velocity of the electron relative to the proton?   (a) −0.988 c (b) +0.988 c

■ **Additional Exercises**

71 What is the rest mass of a proton in mega-electron volts?   939 MeV

72 If the kinetic energy of a particle is increased by 20 keV, what is the corresponding increase in its relativistic mass?
3.6×10^{-32} kg

73 A spaceship travels with a speed of 1.5×10^8 m/s. How much time does a clock on board the spaceship appear to lose in one day according to an observer in another inertial frame?   3.1 h

74 An electron travels at a speed of 9.5×10^7 m/s. What is its total energy?   0.54 MeV

75 A relativistic rocket is measured to be 50 m long, 2.5 m high, and 2.0 m wide. What are its dimensions to an inertial observer when the rocket travels at 0.75c?   length 33 m; height 2.5 m; width 2.0 m

76 If the mass of 1.0 kg of coal (or any substance) could be completely converted into energy, how many kilowatt-hours of energy would be produced? 2.5×10^{10} kWh

77 Imagine that you are moving with a velocity of 0.80c past a person who is reading this chapter in the textbook. If he takes 30 min to read the chapter by his clock, how much time would be observed to elapse on your clock?   50 min

78 Each second the Sun gives off about 4.0×10^{26} J of radiant energy produced by converting mass to energy in nuclear reactions. (a) How many kilograms of mass are lost by the Sun each second? (b) If the Sun's mass is 2.000×10^{30} kg, at the rate calculated in part (a), how long in years will it be until the Sun's mass is reduced to 1.999×10^{30} kg?
7.3×10^9 y

79 A spaceship has a length of 150 m in its rest frame. An observer in another inertial frame sees the length of the spaceship as 110 m. What is the relative velocity of the reference frames?   0.67 c

80 Events that are simultaneous to an observer in one inertial frame (O) are not necessarily simultaneous when viewed by an observer in another inertial frame (O'). If the frames are at rest, as illustrated in ● Fig. 25.22, both observers would say that the lightning flash events were simultaneous, or occurred at the same time, because the signals are received at the same instant. However, let the O' frame be moving to the right with a constant velocity and the simultaneous events occur just as O' passes the O position. Show that the observers would disagree about the simultaneity of the events in this case.   see ISM

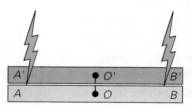

● **FIGURE 25.22 Simultaneity or not?**   See Exercise 80.

81 In the Michelson-Morley experiment, a difference in phase for the beams could be detected if the apparatus were rotated by 90°, since the interference pattern should change. Also, this would take care of the problem of assuming that the interferometer arm lengths are exactly equal, which should not be done arbitrarily. Show that the difference in the time differences for beam travel for the nonrotated and rotated conditions is

$$\Delta t - \Delta t' = \frac{2}{c}(l_1 + l_2)\left[\frac{1}{1 - v^2/c^2} - \frac{1}{\sqrt{1 - v^2/c^2}}\right]$$

where $\Delta t'$ is the time difference for the rotated condition and l_1 and l_2 are the lengths of the interferometer arms.
see ISM

26

Quantum Physics

Einstein's theory of special relativity (Chapter 25) helped to resolve the problems that classical physics had in describing particles moving at speeds comparable to that of light. However, there were other troublesome areas in which classical theory did not agree with experimental results. As scientists explored the nature of the atom in the early years of the twentieth century, attempts to apply classical physics to explain phenomena on the atomic level were unsuccessful in several major areas of study.

The period from about 1900 to 1930 was perhaps one of the most productive in the history of physics. Scientists looked at the problems of the day and came up with new hypotheses that explained them. These hypotheses were new in the respect that they required nonclassical approaches. It was during this period that the principles of modern physics were developed, and a so-called revolution in physics took place.

Chief among these new theories was the idea that light is quantized, that it is made up of discrete amounts of energy. This concept led to the formulation of a new set of principles and a new branch of physics, quantum mechanics. Quantum mechanics introduced many seemingly strange, paradoxical concepts that nevertheless proved to be extremely powerful and far-reaching. Quantum theory forced physicists to reexamine the relationship between particles and waves. More specifically, it demonstrated that particles often exhibit wave properties and waves frequently behave like particles. It also showed that in the realm of the atom, explanations had to deal in probabilities rather than the precisely determined values of classical theory.

This does not mean that quantum theory does not predict accurate results. On the contrary, it has proven highly successful in analyzing the behavior of atoms and explaining many atomic phenomena where classical theory fails. For example, quantum theory is needed to account for the photoelectric effect and the discrete spectrum of hydrogen. Also, it was central in the development of the laser, and thus to such practical devices as compact discs, bar-code scanners, and holograms (an example of which is shown in the photo above), along with a host of other applications in communications, medicine, and industry.

> Topics in this chapter include Planck's hypothesis, the photoelectric effect, the Compton effect, the Bohr theory of the hydrogen atom, and a study of the laser.

A detailed treatment of quantum mechanics is quite complex mathematically and beyond the scope of this text. However, a general overview of the important results is essential in the understanding of physics as we know it today. Toward this end, the important developments of "quantum" physics are presented in this chapter, and an introduction to quantum mechanics will be given in Chapter 27.

26.1 ■ Quantization: Planck's Hypothesis

One of the problems scientists were having at the end of the nineteenth century was how to explain the spectra of radiation emitted by hot objects. This is sometimes called **thermal radiation**. You learned in Chapter 11 that the total intensity of the emitted radiation is proportional to the fourth power of the absolute Kelvin temperature (T^4). Since the temperatures of all objects are above absolute zero, all objects emit thermal radiation.

See Eq. 11.8, Section 11.4

At normal temperatures, this radiation is in the infrared region and so is not detected visibly. At temperatures of about 1000 K, an object emits radiation in the long-wavelength end of the visible spectrum and has a reddish glow. A hot electric stove burner is a good example. Increased temperature causes the radiation and color to shift to a shorter wavelength, such as yellow-orange. Above a temperature of about 2000 K, an object glows with a yellowish-white color, like the glowing filament of a light bulb.

Analysis shows that, although we see only a particular color, there is a general continuous spectrum at all these temperatures as illustrated in the graph in Fig. 26.1a. The curves shown here do not correspond to a real radiating body; they are for an idealized blackbody. A **blackbody** is an ideal system that absorbs (and emits) all radiation that is incident on it.

Blackbody—an ideal absorber and emitter

An absorbing blackbody may be approximated by a small hole leading to an internal cavity inside a block of material (see Fig. 26.1b). Radiation falling

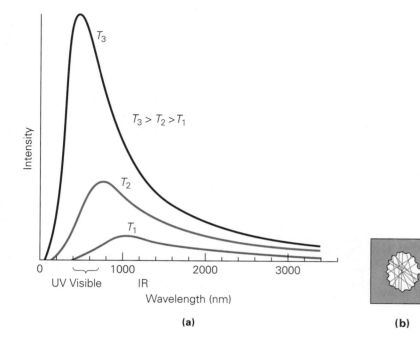

Figure 26.1

FIGURE 26.1 Thermal radiation **(a)** Intensity versus wavelength curves for the thermal radiation from an idealized blackbody at different temperatures. Notice how the location of maximum intensity shifts to a shorter wavelength with increasing temperature. **(b)** A blackbody may be approximated by a small hole leading to an interior cavity in a block of material.

The concept of a blackbody was introduced in Section 11.4

on the hole enters the cavity and is reflected back and forth by the cavity walls. If the hole is very small in comparison to the surface area of the cavity, only a small amount of radiation will be reflected through the hole. Since nearly all radiation incident on the hole is absorbed and very little comes back out, the hole, viewed from the outside, is an excellent approximation to the surface of a blackbody.

Notice in the figure that as the temperature increases, more radiation is emitted at every wavelength and that the wavelength of the maximum-intensity component becomes shorter. This shift is found to obey a relationship, named after the scientist who developed it in 1893, known as **Wien's displacement law**:

Wien's displacement law— displacement of λ_{max} with temperature

$$\boxed{\lambda_{max}\, T = \text{a constant} = 2.90 \times 10^{-3}\ \text{m-K}} \tag{26.1}$$

where λ_{max} is the wavelength of the radiation (in meters) at which maximum intensity occurs and T is the temperature of the body (in kelvins).

Wien's law can be used to determine the wavelength of the maximum spectral component if the temperature is known, or the temperature of the emitter if the wavelength of the strongest emission is known. Thus, it can be used to estimate the temperatures of stars from their radiations.

Point out the relationship between the frequency and the absolute temperature.

EXAMPLE 26.1 ■ Wien's Law: Temperature and Wavelength

When we look at the Sun with the unaided eye, what we see is the gaseous photosphere from which radiation escapes over a range of depths and temperatures. At the top of the photosphere, the temperature is 4500 K and all the radiation escapes. At a depth of about 260 km, the temperature is 6800 K, and only 13 percent of the radiation escapes. What are the wavelengths of the radiations of maximum intensity for these temperatures?

Solution.

Given: $T_1 = 4500$ K   *Find:* λ_{max} (for different
$\qquad\quad T_2 = 6800$ K   temperatures)

The wavelengths are given by Wien's law (Eq. 26.1), so we have

$$\text{At top:}\quad \lambda_{max} = \frac{2.90 \times 10^{-3}\ \text{m-K}}{4500\ \text{K}} = 6.44 \times 10^{-7}\ \text{m} = 644\ \text{nm}$$

$$\text{At depth of 260 km:}\quad \lambda_{max} = \frac{2.90 \times 10^{-3}\ \text{m-K}}{6800\ \text{K}} = 4.26 \times 10^{-7}\ \text{m} = 426\ \text{nm}$$

Hence, at the Sun's surface, the emitted radiation of maximum intensity is in the orange region of the visible spectrum (Section 19.4). As the temperature increases with photosphere depth, the wavelength of the radiation of maximum intensity shifts towards the blue end of the spectrum; at 260 km, λ_{max} is near the violet end of the spectrum. Some of the emitted radiation will be in the ultraviolet region, most of which will hopefully be absorbed by the ozone layer in the atmosphere (see the Insight feature in Chapter 19).

Classically, thermal radiation is pictured as resulting from the thermal agitation and acceleration of electric charges near the surface of an object. Electric charges in thermal agitation would undergo many different accelerations, so a continuous spectrum of emitted radiation would be expected. However, classical theoretical calculations based on the number of standing wave modes in a blackbody cavity predicted the intensity to be proportional to $1/\lambda^4$, that is, $I \propto 1/\lambda^4$.

At long wavelengths, the classical prediction agrees fairly well with experimental data. However, at shorter wavelengths, there is little correlation. Contrary to experimental observations (including Wien's law), classical theory predicts that the radiation intensity should increase as the wavelength goes to zero (Fig. 26.2). This classical prediction is sometimes called the *ultraviolet catastrophe*—"ultraviolet" because the difficulty occurs for short wavelengths beyond the violet end of the visible spectrum and "catastrophe" because it predicts that the emitted intensity or energy will become infinitely large.

The failure of classical physics to explain the characteristics of thermal radiation led Max Planck, a German physicist, to reexamine the phenomenon. In 1900, Planck formulated a theory that correctly predicted the observed distribution of the blackbody radiation spectrum, but only by introducing a radical new idea. Planck showed that if it were assumed that the thermal oscillators could have only *discrete*, or particular, amounts of energy, rather than a continuous distribution of energies, or any arbitrary energy, then theory would agree with experiment.

These discrete amounts of energy of thermal oscillators were related to the frequency f of oscillation by

$$E_n = nhf \qquad \text{for } n = 1, 2, 3, \dots \qquad (26.2)$$

that is, the energy is in integral multiples of hf. The symbol h represents a constant known (quite naturally) as **Planck's constant**, and has a value of

$$h = 6.63 \times 10^{-34} \text{ J-s}$$

This relationship is commonly called **Planck's hypothesis**. Rather than considering oscillator energy to be a continuous quantity as it is classically, by Planck's hypothesis the energy was *quantized*; that is, an oscillator could have only discrete amounts of energy. The smallest possible amount of oscillator energy is

$$E = hf \qquad (26.3)$$

which is called a **quantum** of energy (from the Latin word *quantus*, meaning "how much"). Thus, each oscillator could absorb or emit energy in quantum amounts.

Although the theoretical predictions agreed with experiment, Planck himself was not convinced of the validity of his quantum hypothesis. However, the concept of quantization was quickly used to explain other phenomena that could not be explained classically. His quantum hypothesis earned Planck the Nobel Prize in 1918.

26.2 ■ Quanta of Light: Photons and the Photoelectric Effect

The concept of the quantization of light was introduced in 1905 by Albert Einstein in a paper on light absorption and emission. Einstein reasoned that if the energy of the molecular oscillators in a hot substance is quantized, then,

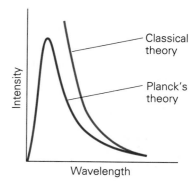

● FIGURE 26.2 The ultraviolet catastrophe Classical theory predicts the intensity of thermal radiation to be proportional to $1/\lambda^4$, and thus for an infinitely large amount of radiation energy as the wavelength approaches zero. Planck's quantum theory, on the other hand, predicts the observed radiation distribution.

Planck's hypothesis—the birth of quantum physics

Energy of a photon of light with frequency *f*.

to conserve energy, the emitted radiation should also be quantized. He therefore proposed that

> ... the radiant energy from a point source is not distributed continuously throughout an increasingly larger region, but, instead, this energy consists of a finite number of spatially localized energy quanta which, moving without subdividing, can only be absorbed and created in whole units.

A quantum or packet of light is referred to as a **photon**, and each photon has an energy

$$E = hf \tag{26.4}$$

This idea suggests that light is transmitted as discrete quanta (plural of quantum), or "particles" of energy, rather than as waves.

Einstein used this quantum concept to explain what is called the **photoelectric effect**, another area in which classical physical description was inadequate. Certain metallic materials are *photosensitive*. That is, when light strikes the surface of the material, electrons are emitted. The radiant energy supplies the work necessary to free the electrons from the material surface. Such materials are used to make photocells, as illustrated in ● Fig. 26.3a. A voltage is maintained between the anode and the cathode. When light strikes the cathode which is the photosensitive material, electrons are emitted. These emitted photoelectrons complete the circuit, and a current is registered on the ammeter.

When a photocell is illuminated with monochromatic (single-frequency) light of different intensities, characteristic curves are obtained as shown in Fig. 26.3b. The photocurrent increases as the voltage across the cell increases until a saturation current is reached. At this point, all of the emitted electrons are reaching the anode, and any further increase in voltage has no effect on the magnitude of the current.

As would be expected classically, the current is proportional to the intensity of the incident light—the greater the intensity ($I_2 > I_1$ in the figure), the more energy there is to free electrons.

An idea of the kinetic energy of the electrons being emitted from the photoelectric material can be gained by reversing the voltage across the elec-

Figure 26.3

●**FIGURE 26.3 The photoelectric effect and characteristic curves** (a) Incident light on the photomaterial in a phototube (or cell) causes the emission of electrons, and a current flows in the circuit. The voltage applied to the tube is varied by means of the variable resistor. (b) As the plots of current versus voltage for two intensities of monochromatic light show, the current becomes saturated, or constant, as the voltage is increased. For negative voltages (by a reversal of the battery polarity), the current goes to zero at a particular stopping potential $-V_o$.

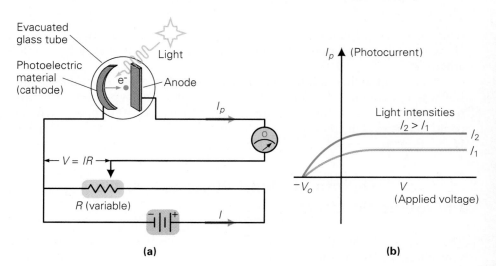

(a)

(b)

trodes. As the voltage is decreased and then made negative, the photocurrent decreases. Because of the retarding voltage, only the electrons with kinetic energy greater than $e|V|$ can make it to the negative plate and contribute to the current. (The kinetic energy goes into the work of overcoming the negative potential difference, $W = U = -eV$.)

At some voltage V_o, called the **stopping potential**, no current will flow between the electrodes. The maximum kinetic energy (K_{max}) of the electrons emitted from the photomaterial is then related to the stopping potential by

Point out the significant features on an E versus f graph.

$$\boxed{K_{max} = eV_o} \tag{26.5}$$

since eV_o is the work needed to stop the most energetic electrons.

When the frequency of the incident light is varied, the maximum kinetic energy of the electrons is found to depend linearly on the frequency (Fig. 26.4). No photoemission at all is observed to occur below a certain *cutoff frequency* f_o. Moreover, the electron emission begins instantaneously, with no observable time delay, when a photomaterial is illuminated by light with $f > f_o$, even if the light intensity is very low.

These four key characteristics of photoemission are summarized in Table 26.1 along with their correspondence with classical theoretical predictions. Note that only the first observation is predicted correctly by classical wave theory. Classically, the electric field of a light wave interacts with the electrons at the surface of the material and sets them into oscillation. The intensity (energy) of an oscillating electric field is proportional to the amplitude of the field vector, so the average kinetic energy of the electrons is proportional to the intensity of the light. Therefore, it is difficult to understand why the maximum kinetic energy of the photoelectrons is independent of the intensity, as is observed. Also, according to wave theory, photoemission should occur for any frequency of light, provided that the intensity is sufficient.

The immediate emission of electrons raises an even more serious problem for classical theory. Classically, the time required for an electron to acquire enough energy to be freed and have a typically observed value of maximum kinetic energy is on the order of minutes for low light intensities.

An explanation of the photoelectric effect was put forth by Einstein in 1905 using the quantum idea suggested by Planck in his theory of thermal oscillators. As was stated previously, he considered the energy of light to be in quantum bundles or photons, each with an energy $E = hf$. Thus, light of frequency f

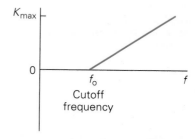

FIGURE 26.4 Maximum kinetic energy and frequency The maximum kinetic energy (K_{max}) of the photoelectrons is a linear function of the frequency of the incident light. Below a certain cutoff frequency f_o, no photoemission occurs.

Explain why classical theories cannot account for the photoelectric effect.

TABLE 26.1 ■ **Photoelectric Effect Characteristics**

Characteristic	Predicted by classical theory?
1. The photocurrent is proportional to the intensity of the light.	Yes
2. The maximum kinetic energy of the emitted electrons is dependent on frequency but not on intensity.	No
3. No photoemission occurs for light with a frequency below a certain cutoff frequency f_o regardless of intensity.	No
4. A photocurrent is observed immediately when the light frequency is greater than f_o even if the intensity is low.	No

could give energy to the electrons in a photomaterial only in discrete amounts equal to hf. An electron absorbs either the whole photon or nothing at all.

The electrons of a photomaterial are bound by attractive forces. Therefore, work must be done to free an electron, and some of the absorbed photon energy goes into doing this. Then, when a photon of energy hf is absorbed, by the conservation of energy we have

$$hf = K + \phi \tag{26.6}$$

where ϕ is the work or energy needed to free a surface electron from the material. Thus, *part of the energy of the absorbed photon goes into the work of freeing the electron, and the rest is carried off by the emitted electron as kinetic energy.*

The least tightly bound electron will have the maximum kinetic energy (K_{max}), and in this case, the energy needed to free the electron is called the **work function** ϕ_0 of the material. The energy conservation equation is then

Work function—the minimum energy needed to dislodge an electron.

$$\boxed{hf = K_{max} + \phi_0} \tag{26.7}$$

incident photon energy　maximum kinetic energy of dislodged electron　minimum work needed to dislodge electron

Other electrons will require more energy than the ϕ_0 minimum amount, and the kinetic energy of these electrons will be less than K_{max}.

EXAMPLE 26.2 ■ The Photoelectric Effect

The work function of a particular metal is 2.0 eV. (a) If the metal is illuminated with monochromatic light having a wavelength of 550 nm, what will be the maximum speed of the emitted electrons? (b) What is the stopping potential?

Solution.

Given: $\phi_0 = 2.0$ eV $\times$ $(1.6 \times 10^{-19}$ J/eV$)$　　*Find:* (a) v_{max} (maximum
　　　　$= 3.2 \times 10^{-19}$ J　　　　　　　　　　　speed)
　　　　$\lambda = 550$ nm $= 5.50 \times 10^{-7}$ m　　　(b) V_0 (stopping
　　　　　　　　　　　　　　　　　　　　　　　potential)

(a) The maximum speed is obtained from the maximum kinetic energy, which is given in Eq. 26.7. First computing the energy of a photon of light with the given wavelength, with $\lambda f = c$, we have

$$hf = \frac{hc}{\lambda} = \frac{(6.63 \times 10^{-34} \text{ J-s})(3.0 \times 10^8 \text{ m/s})}{5.50 \times 10^{-7} \text{ m}}$$

$$= 3.6 \times 10^{-19} \text{ J}$$

Then

$$K_{max} = hf - \phi_0 = 3.6 \times 10^{-19} \text{ J} - 3.2 \times 10^{-19} \text{ J}$$

$$= 0.4 \times 10^{-19} \text{ J} = \tfrac{1}{2}mv_{max}^2$$

where m is the mass of an electron and is a known quantity (m =

Chapter 26 Quantum Physics

9.11 × 10^{-31} kg). Solving for v_{max}, we get

$$v_{max} = \sqrt{\frac{2K_{max}}{m}} = \left[\frac{2(0.4 \times 10^{-19})}{9.11 \times 10^{-31}\,\text{kg}}\right]^{\frac{1}{2}}$$

$$= 3.0 \times 10^5\,\text{m/s}$$

(b) The stopping potential is related to K_{max}, that is, $K_{max} = eV_o$, and

$$V_o = \frac{K_{max}}{e} = \frac{0.4 \times 10^{-19}}{1.6 \times 10^{-19}\,\text{C}} = 0.25\,\text{V}$$

Einstein's quantum theory was consistent with experimental observation. Increasing the light intensity increases the number of photons and thus increases the number of electrons dislodged (the photocurrent). An increase in intensity does not change the energy of *individual* photons, however, which depends solely on the frequency of the light: $E = hf$. Therefore K_{max} is independent of intensity but linearly dependent on the frequency of the incident light as had been observed experimentally.

Einstein's theory also explained the existence of a limiting frequency. In the Einstein interpretation, the photon energy depends on frequency. Below a certain (cutoff) frequency, the photons do not have enough energy to dislodge electrons, so no current is observed. When $K_{max} = 0$, we can find the minimum cutoff frequency f_o from Eq. 26.7:

$$hf_o = K_{max} + \phi_o = 0 + \phi_o$$

$$\boxed{f_o = \frac{\phi_o}{h}} \qquad (26.8) \qquad \text{Cutoff frequency and work function}$$

In this case, the photon has just enough energy to free an electron from the material, but no extra energy to give it kinetic energy. (Notice that the graph in Fig. 26.4 provides a means to determine the value of h, which is the slope of the line.)

A photon with energy hf_o will just dislodge an electron, but the electron will have no kinetic energy

The frequency f_o is sometimes called the **threshold frequency**. Only light with a frequency above the threshold set by f_o will dislodge electrons. If the incident light has a frequency less than f_o, then no matter how many photons are available (that is, no matter how intense the light), the photons will not have enough energy to cause photoemission. No time delay in the emission of electrons would be expected in the quantum theory, since the energy is "dumped" on an electron in a concentrated bundle instead of being continuously supplied and distributed over an area as in wave theory.

EXAMPLE 26.3 ■ Threshold Frequency

What is the threshold frequency of the metal described in Example 26.2?

Solution.

Given: $\phi_o = 2.0$ eV   *Find:* f_o (threshold frequency)
$\qquad\quad = 3.2 \times 10^{-19}$ J
(from Example 26.1)

The threshold (or cutoff) frequency is related directly to the work function, $\phi_o = h f_o$ (Eq. 26.8), so finding f_o is a straightforward calculation:

$$f_o = \frac{\phi_o}{h} = \frac{3.2 \times 10^{-19}\,\text{J}}{6.63 \times 10^{-34}\,\text{J-s}} = 4.8 \times 10^{14}\,\text{Hz}$$

Notice that this lies in the visible region of the electromagnetic spectrum. (Which end of that region is it near?)

There are many applications of the photoelectric effect. The fact that the current produced by photocells is proportional to the light intensity makes them ideal for use in photographers' light meters. Photocells are also used in solar energy applications to convert sunlight to electricity. They are not yet efficient enough (about 20–25%) to produce electricity commercially, but this pollution-free method has great potential.

Another common application of the photocell is in the so-called "electric eye." ● Figure 26.5a illustrates the principle of the electric eye. As long as light strikes the photocell, current flows in the circuit. Interrupting or blocking the light opens the circuit in the relay (magnetic switch), which controls some device. A common application of the electric eye is to turn on street lights automatically at night. A safety application is shown in Fig. 26.5b. The threshold frequencies of some metals lie outside the visible region. Hence, nonvisible light (infrared or ultraviolet) may be used in such applications as burglar alarms or home protection systems.

Einstein's theory of the photoelectric effect represented a great success for the idea of quantization. It is interesting to note that Einstein won the Nobel Prize in physics in 1921 for his theory of the photoelectric effect rather than for his more famous theory of relativity (Chapter 25). Two years later, the quantum theories scored another triumph, which we will look at in the following section.

●**FIGURE 26.5 Photoelectric applications** (a) A diagram of an electric eye circuit. When light strikes a phototube or cell, there is a current in the circuit. Interrupting the light beam opens the circuit in the relay or magnetic switch, which controls some device. (b) An electric eye may be used in an automatic garage door application. When the door starts downward, any interruption of the electric eye beam will cause the door to stop, protecting children or pets that may be passing under the descending door.

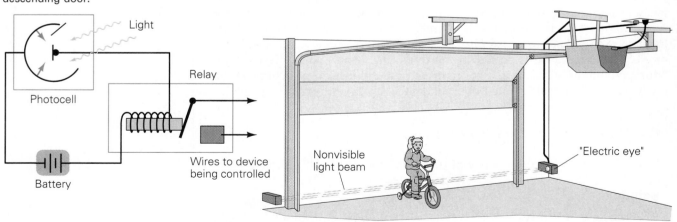

(a) Electric eye diagram

(b) Electric eye application

26.3 ■ Quantum "Particles": The Compton Effect

In 1923, the American physicist Arthur H. Compton (1892–1962) explained a phenomenon he observed in the scattering of X-rays from a graphite (carbon) block by considering the radiation to be composed of quanta. This explanation of the observed effect provided convincing evidence that light, and electromagnetic radiation in general, is composed of quanta, or "particles" of energy.

Compton had observed that when a beam of monochromatic X-rays was scattered by a material, the scattered radiation had a wavelength slightly longer than the wavelength of the incident beam. Also, the change in the wavelength was dependent on the angle through which the X-rays were scattered but not on the scattering material (see ● Fig. 26.6). This phenomenon came to be known as the **Compton effect**.

Classically, scattered radiation should have the same frequency or wavelength as the incident radiation. The electrons in the atoms of the scattering material absorb the radiation and oscillate at the same frequency as the incident wave. On emission, the scattered radiation should have the same frequency in all directions.

According to Einstein's photon theory, the frequency of a quantum is directly proportional to its energy ($E = hf$), so a change in frequency or a shift in wavelength would indicate a change in energy. The effect of the scattering angle on energy reminded Compton of scattering in the elastic collision of particles. Could the same principles apply in the scattering of quantum "particles"?

Pursuing this idea, Compton was able to explain the observed effect. He assumed that a quantum behaves like a particle or billiard ball in collision with other particles. If an incident quantum collides with an electron initially at rest, the quantum loses some energy and momentum to the electron. Hence, the energy and frequency of the scattered quantum is less ($E = hf$) and its wavelength is greater ($\lambda = c/f$). The collision is elastic, so both energy and momentum must be conserved. However, in this case it is the *relativistic* energy and momentum which are conserved. Applying these principles, Compton showed that the shift in the wavelength of the scattered quantum is given by

$$\Delta\lambda = \lambda - \lambda_o = \lambda_C(1 - \cos\theta) \qquad (26.9)$$

Figure 26.6a

Elastic collisions are discussed in Section 6.4

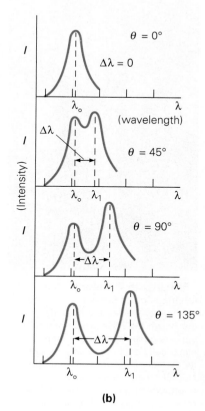

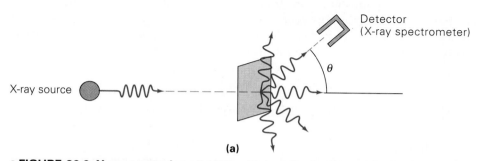

(a)

● **FIGURE 26.6 X-ray scattering** **(a)** When X-rays of a single wavelength are scattered by a material, the wavelengths of the scattered X-rays are longer than the wavelength of the incident beam. **(b)** The wavelength shift depends on the scattering angle.

(b)

where λ_o is the wavelength of the incident quantum and λ_1 that of the scattered quantum (see Exercise 67). The constant $\lambda_C = h/m_o c^2 = 2.43 \times 10^{-12}$ m is called the *Compton wavelength*. The equation correctly predicted the observed wavelength shift, and Compton was awarded a Nobel Prize in 1927 for his work.

EXAMPLE 26.4 ■ The Compton Effect

A monochromatic beam of X-rays with a wavelength of 1.35×10^{-10} m is scattered by a metal foil. By what percentage is the wavelength shifted for the scattered component observed at an angle of 90°?

Solution.

Given: $\lambda_o = 1.35 \times 10^{-10}$ m   *Find:* % shift
$\quad\quad\quad \theta = 90°$

The change or shift in the wavelength, $\Delta\lambda$, is given by Eq. 26.9, and the fractional change is $\Delta\lambda/\lambda_o$. Then

$$\frac{\Delta\lambda}{\lambda_o} = \frac{\lambda_C}{\lambda_o}(1 - \cos\theta)$$

$$= \frac{2.43 \times 10^{-12} \text{ m}}{1.35 \times 10^{-10} \text{ m}}(1 - \cos 90°) = 1.80 \times 10^{-2}$$

and

$$\frac{\Delta\lambda}{\lambda_o} \times 100\% = 1.80\%$$

Einstein's and Compton's successes in explaining electromagnetic phenomena in terms of quanta left scientists with two theories of electromagnetic radiation. Classically, the radiation is pictured as a continuous wave, and classical theory explains satisfactorily such things as interference and diffraction. On the other hand, quantum theory was necessary to satisfactorily explain the photoelectric and Compton effects.

The two theories gave rise to a description that is called the **dual nature of light**. That is, light apparently behaves sometimes as a wave and at other times as photons or "particles." The relationship between the wave and particle nature of light will be considered in the next chapter.

●**FIGURE 26.7 Colors** The luminous glass tubes that we see everywhere today are actually gas discharge tubes, in which atoms of various gases emit light when electrically excited. Each gas radiates its own characteristic wavelengths. Not all "neon lights" actually contain neon, which glows with a red hue; other gases are used to produce other colors.

26.4 ■ The Bohr Theory of the Hydrogen Atom

In the 1800s, a great deal of experimental work was done with gas discharge tubes—for example, those containing hydrogen, neon, and mercury vapor. (The common neon "lights," such as those in ● Fig. 26.7, are gas discharge tubes.) Normally, light from an incandescent source exhibits a *continuous spectrum*. However, when light emissions from these tubes were analyzed, discrete, or

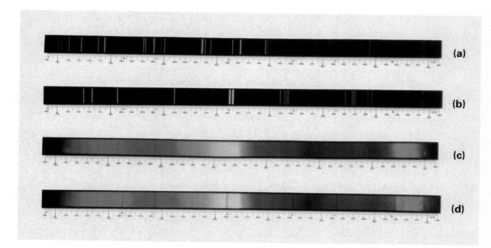

(a)

(b)

(c)

(d)

• FIGURE 26.8 Spectra When a gas is excited by heat or electricity and the light that it emits separated into its component wavelengths with a prism or diffraction grating, the result is a bright-line, or emission, spectrum. Each atom or molecule emits a characteristic pattern of discrete wavelengths. These spectra are produced by (a) barium and (b) neon. When a continuous spectrum consisting of all wavelengths (c) is passed through a cool gas, a series of dark lines is observed. Each line represents a "missing" wavelength—a particular wavelength that has been absorbed by the gas. The wavelengths absorbed by any substance are the same ones it emits when excited. The spectrum in (d) is that of the Sun. The absorption lines are produced by the gases of the solar atmosphere.

line, spectra were observed (•Fig. 26.8). Light coming directly from a tube gives a *bright-line* or **emission spectrum**, indicating that only certain specific wavelengths are present. In general, a discrete line spectrum is characteristic of the atoms or molecules of a particular material and is used to identify that material spectroscopically (Fig. 26.8a).

But atoms absorb light as well as emitting it. If white light is passed through a relatively cool gas, certain frequencies or wavelengths are found to be missing. The result is an **absorption spectrum**—a series of dark lines superimposed on a continuous spectrum (Fig. 26.8b). When the absorption and emission lines of a particular gas were compared, they were found to occur at the same frequencies. The reason for line spectra was not understood at the time, but they provided a clue to the electronic structure of the atoms.

Hydrogen, with a relatively simple visible spectrum, received much attention. Hydrogen is also the simplest atom, with only one electron and one proton. In the late nineteenth century, a Swiss physicist, J. J. Balmer, found an empirical formula that gives the wavelengths of the spectral lines of hydrogen in the visible region:

$$\frac{1}{\lambda} = R\left(\frac{1}{2^2} - \frac{1}{n^2}\right) \qquad \text{for } n = 3, 4, 5, \ldots \qquad (26.10)$$

where R is called the *Rydberg constant* and has a value of $1.097 \times 10^{-2} \text{ nm}^{-1}$. The spectral lines of hydrogen in the visible region, called the **Balmer series**, were found to fit the formula, but it was not understood why. Similar formulas were found to fit the spectral lines in the ultraviolet and infrared regions.

An explanation of the spectral lines was given in a theory of the hydrogen atom put forth in 1913 by the Danish physicist Niels Bohr. Bohr assumed that the hydrogen electron orbited the nuclear proton in a circular orbit (analogous to a planet orbiting the Sun). The electrical Coulomb force (Chapter 15) supplies the necessary centripetal force for the circular motion, and we may write

$$F = \frac{mv^2}{r} = \frac{ke^2}{r^2} \qquad (26.11) \qquad \text{Review Eq. 7.11 and Eq. 15.2}$$

where e is the charge of the proton and the electron, v the electron's orbital speed, and r the radius of the orbit (•Fig. 26.9).

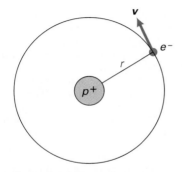

FIGURE 26.9 The Bohr model of the hydrogen atom The electron of the hydrogen atom is pictured as revolving around the nuclear proton in a circular orbit, similar to the way a planet orbits the Sun.

The total energy of the electron is the sum of its kinetic and potential energies:

$$E = K + U = \tfrac{1}{2}mv^2 - \frac{ke^2}{r}$$

From Eq. 26.11, the kinetic energy may be written $\tfrac{1}{2}mv^2 = ke^2/2r$. Using this, the total energy becomes

$$E = \frac{ke^2}{2r} - \frac{ke^2}{r} = -\frac{ke^2}{2r} \tag{26.12}$$

Notice that as the radius approaches infinity, E approaches zero. With $E = 0$, the electron would no longer be bound to the proton, and the atom, having lost its electron, would be ionized.

Thus far, only classical principles have been applied. But at this point in the theory, Bohr made a radical assumption—radical in the sense that he introduced a quantum concept. His assumption was that the angular momentum of the electron was quantized and could have only discrete values that were integral multiples of $h/2\pi$, where h is Planck's constant. In equation form,

$$mvr = \frac{nh}{2\pi} \qquad \text{for } n = 1, 2, 3, 4, \ldots \tag{26.13}$$

The integer n is called a *quantum number*, specifically the **principal quantum number**. According to this assumption, the orbital speed of the electron (v) may have only certain values given by

$$v = \frac{nh}{2\pi m r}$$

Putting this expression for v into Eq. 26.11 and solving for r, we find that

$$\frac{ke^2}{r^2} = \frac{mv^2}{r} = \frac{m(nh/2\pi m r)^2}{r}$$

and

$$r = \left(\frac{h^2}{4\pi^2 ke^2 m}\right)n^2 \tag{26.14}$$

Thus, electron orbits of only certain radii are possible as determined by the principle quantum number n. The energy for a particular orbit can be found by substituting this expression for r into Eq. 26.12:

$$E = -\frac{ke^2}{2r} = -\frac{ke^2}{2}\left(\frac{4\pi^2 ke^2 m}{h^2 n^2}\right)$$

and

$$E_n = -\left(\frac{2\pi^2 k^2 e^4 m}{h^2}\right)\frac{1}{n^2} \tag{26.15}$$

where the energy is written as E_n to show its dependence on n. The quantities in the parentheses on the right-hand sides of Eqs. 26.14 and 26.15 are constants and are easily evaluated. The important results are

$$r_n = 0.53\, n^2 \text{ Å} = 0.053\, n^2 \text{ nm} \tag{26.16}$$

$$E_n = \frac{-13.6}{n^2} \text{ eV} \qquad \text{for } n = 1, 2, 3, 4, \ldots \tag{26.17}$$

where the radius is commonly expressed in angstroms or nanometers and the energy in electron volts. The n subscripts indicate that r and E are different for different values of n.

EXAMPLE 26.5 ■ Radius and Energy of a Bohr Orbit

Find the orbital radius and energy of an electron in a hydrogen atom characterized by the principal quantum number $n = 2$.

Solution. For $n = 2$,

$$r_2 = 0.53\, n^2\, \text{Å} = 0.53\,(2)^2\, \text{Å} = 2.12\, \text{Å} = 0.212\, \text{nm}$$

and

$$E_2 = \frac{-13.6}{n^2}\, \text{eV} = \frac{-13.6}{(2)^2}\, \text{eV} = -3.40\, \text{eV}$$

However, there is still a classical problem with Bohr's theory. Classically, an accelerating electron radiates electromagnetic energy, and even in discrete circular orbits the electron would be centripetally accelerating. Thus, an orbiting electron would lose energy and spiral into the nucleus, as an Earth satellite might go into a decaying orbit because of frictional losses. This doesn't happen in the hydrogen atom, so Bohr had to make another nonclassical assumption. He postulated that the hydrogen electron does not radiate energy when it is in a bound, discrete orbit but does so only when it makes a transition to another orbit.

Energy Levels

The possible allowed orbits of the hydrogen electron are commonly expressed in terms of energy levels as illustrated in ● Fig. 26.10. In this context, we simply refer to the electron as being in a particular energy level or state. The electron, being bound to the nuclear proton, is in a potential energy well, much like a gravitational potential well. The principal quantum number designates the energy level. The lowest energy level ($n = 1$) is called the **ground state**. The energy levels above the ground state are called **excited states**. For example, $n = 2$ is the first excited state and so on.

Review the concept of potential energy wells presented in Sections 5.4 and 7.5 (see Figs. 5.14 and 7.18)

Ground state $n = 1$
Excited states $n \geqslant 2$

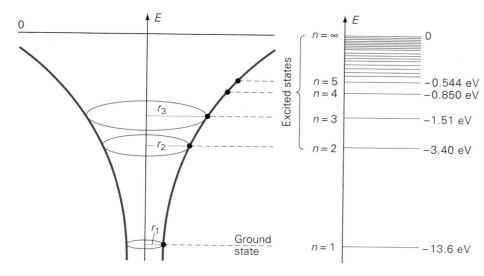

Figure 26.10

● **FIGURE 26.10 Orbits and energy levels of the hydrogen electron** The Bohr theory predicts that the hydrogen electron can be in only certain orbits having discreet radii (drawn here in a 1/r potential well for illustration). Each allowed orbit has a corresponding energy. These are conveniently displayed as energy levels. The lowest energy level ($n = 1$) is called the ground state, and those above ($n > 1$) are called excited states.

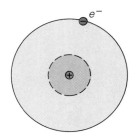

Excited atom

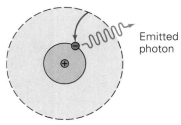

Emitted photon

De-excitation

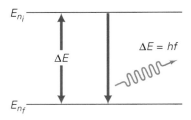

FIGURE 26.11 Electron transitions and photon emission
When the electron in an excited atom decays or makes a transition to a lower orbit or energy level, a photon is emitted. The energy of the photon is equal to the energy difference in the energy levels.

Wavelength of emitted photon

The electron is normally in the ground state and must be given energy to excite it, or to raise it up in the well to an excited state. The hydrogen electron can be excited only by discrete amounts. The energy levels are somewhat analogous to the rungs or steps of ladder. Analogous to a person going up and down a real ladder, an electron goes up and down the energy ladder in discrete steps. Notice, however, that the energy rungs of the hydrogen atom are not evenly spaced.

If enough energy is absorbed to raise the electron to the top of the energy well so that the electron is no longer bound, the atom is ionized. For example, to raise a hydrogen electron from the ground state to $n = \infty$ requires 13.6 eV of energy. However, if the electron is already part way up the ladder (or higher up in the well)—that is, in an excited state—less energy is needed to get it to the top of the well. The energy of the electron in any state is E_n, and the energy necessary needed to raise it to the top of the well is $-E_n$, which is called the **binding energy** of the electron. (Since E_n is negative—the electron is in a negative potential well—then $-E_n$ is positive, or the amount of energy needed to be *given* to the electron to raise it to the top of the well.)

An electron generally does not remain in an excited state for long and decays, or makes a transition to a lower energy level, in a short time. The time an electron spends in an excited state is called the **lifetime** of the excited state. For many states, the lifetime is about 10^{-8} s. In making a transition to a lower state, energy is emitted as a photon of light (● Fig. 26.11). The energy of the photon is equal to the energy difference of the levels, that is,

$$\Delta E = E_{n_i} - E_{n_f} = \frac{-13.6}{n_i^2} \text{ eV} - \left(\frac{-13.6}{n_f^2} \text{ eV} \right)$$

or

$$\boxed{\Delta E = 13.6 \left(\frac{1}{n_f^2} - \frac{1}{n_i^2} \right) \text{ eV}} \qquad (26.18)$$

where the subscripts of n_i and n_f refer to the initial and final states, respectively. Since $\Delta E = hf = hc/\lambda$, only photons of particular frequencies and wavelengths may be emitted. These correspond to the various discrete transitions. Solving for the wavelength λ, we get

$$\boxed{\lambda = \frac{hc}{\Delta E} = \frac{12,400}{\Delta E} \text{ Å} = \frac{1,240}{\Delta E} \text{ nm}} \qquad (26.19)$$

where ΔE is expressed in electron volts and the constant hc has been evaluated to give the wavelength in either angstroms or nanometers.

EXAMPLE 26.6 ■ Discrete Spectral Lines

What is the wavelength of the emitted light when excited electrons in hydrogen atoms make transitions from the $n = 3$ energy level to the $n = 2$ energy level?

Solution.

Given: $n_i = 3$ *Find:* λ (wavelength of
        $n_f = 2$ emitted light)

The energy of the emitted photon is

$$\Delta E = 13.6\left(\frac{1}{n_i^2} - \frac{1}{n_f^2}\right) \text{eV} = 13.6\left(\frac{1}{4} - \frac{1}{9}\right) \text{eV} = 1.89 \text{ eV}$$

and, using Eq. 26.19,

$$\lambda = \frac{12{,}400}{\Delta E} \text{ Å} = \frac{12{,}400}{1.89} \text{ Å} = 6561 \text{ Å} \ \ (= 656.1 \text{ nm})$$

which is light in the red portion of the visible spectrum.

Therefore, since the hydrogen electron can make transitions only between discrete energy levels, discrete wavelengths, producing spectral lines, are emitted (● Fig. 26.12). The wavelengths of the various transitions were computed and they were found to correspond to the experimentally observed spectral lines.

Distinguish between the Balmer, Lyman, and Paschen series.

Although Bohr's results gave exact agreement with the experiment for hydrogen as well as singly ionized helium and doubly ionized lithium (Why?), it could not successfully handle multielectron atoms. Bohr's theory was incomplete in the sense that it patched new quantum ideas into a basically classical framework. The theory contains many correct ideas, but a complete description of the atom did not come until quantum mechanics was developed (Chapter 27).

Fluorescence. Note in Fig. 26.12 that an electron excited to a higher energy level does not necessarily have to return to the ground state in a single jump. Often, several routes involving different sequences of transitions may be possible. For example, an electron in the $n = 4$ excited level could theoretically drop to either of the intermediate levels ($n = 3$ or $n = 2$)—or even to both in succession—before falling back to the ground state. This fact underlies the phenomenon of *fluorescence*, exhibited by a number of natural and synthetic substances.

In **fluorescence**, an electron that has been excited by absorbing a photon returns to the ground state in two or more steps, like a ball bouncing down a flight of stairs. At each step a photon is emitted. Since each drop represents a smaller energy transition than the original upward boost, the emitted photons

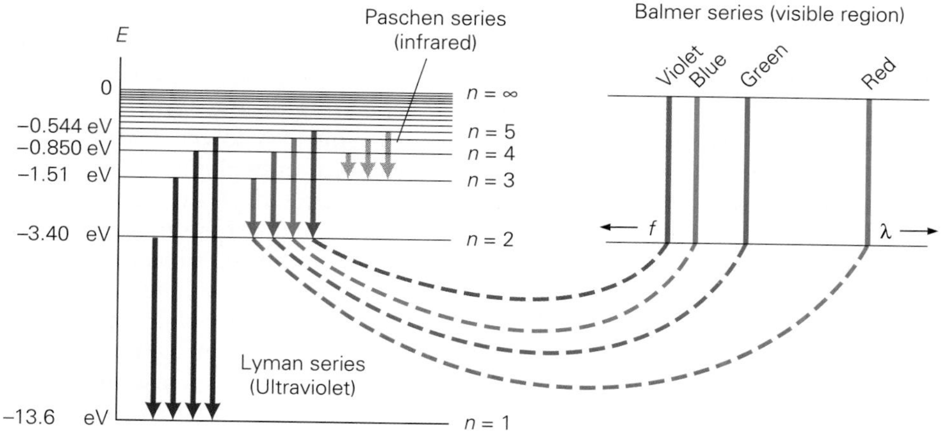

Figure 26.12

● **FIGURE 26.12 Hydrogen spectrum** Transitions may occur between two or more energy levels as the excited electron returns to the ground state. Transitions to the $n = 2$ state give spectral lines that lie in the visible region of the spectrum and are called the Balmer series. Transitions to other levels give other series as shown.

● FIGURE 26.13 Fluorescence
Many minerals will glow at visible wavelengths when illuminated by invisible ultraviolet light ("black light"). The visible light is produced when electrons excited by the uv fall back to lower energy levels in two or more smaller steps.

must have a lower energy, and thus a longer wavelength, than the exciting photon. For example, many minerals are excited by ultraviolet (uv) light and fluoresce, or glow, in the visible region (● Fig. 26.13). A variety of living organisms, from corals to butterflies, manufacture fluorescent pigments that emit visible light.

Our familiar fluorescent lights make use of the same phenomenon. Fast-moving electrons, accelerated by the voltage across the tube, collide with atoms of gaseous mercury in the tube and excite them to higher energy levels. As the excited mercury atoms return to their normal energy levels they emit uv radiation, which in turn excites atoms in the fluorescent coating that lines the tube. It is this mineral that fluoresces, emitting visible light as described previously.

26.5 ■ A Quantum Success: The Laser

The development of the laser was a major technological success derived from theoretical atomic physics. Many scientific discoveries have been made accidentally, and many applications or inventions have come about by trial and error. Roentgen's discovery of X-rays and Edison's electric lamp are examples. However, the laser was developed on purpose. Using quantum physics, the principle of the **laser** was first predicted theoretically, then practically applied. (The word "laser" is an acronym and stands for *l*ight *a*mplification by *s*timulated *e*mission of *r*adiation. Stimulated emission will be discussed shortly.)

This relatively new invention has found widespread usefulness. Laser light beams are used in medicine to control bleeding, to weld torn retinas, and to treat skin cancer. In industry, lasers are used to drill holes in materials, to weld, and to cut cloth. They are used in surveying, and laser light is used to carry telephone and television signals over fiber optic cables. There are laser printers, laser pickups in video and compact disc players, and lasers in supermarket checkouts.

The concept of atomic energy levels allows us to understand this important technological instrument. Bohr's model of the atom explains spectral absorption and emission lines in terms of quantum jumps between energy levels. However, other questions arise. Since there are usually several lower energy levels, what determines the energy level to which an excited electron will decay? Also, how long does an electron stay in an excited state before making a transition? These questions were answered by the use of quantum mechanics, a way of dealing with waves that is described in Chapter 27. The general results are that different transitions have different probabilities and, crucial to laser application, the time an electron stays in an excited state varies, depending on the atom.

Usually, an excited electron makes a transition to a lower energy level almost immediately, remaining in an excited state for only about 10^{-8} s. However, the lifetimes of an excited electron in some energy levels are appreciable. An energy level in which an excited electron remains for some time is called a **metastable state**. *Phosphorescent* materials are examples of substances with metastable states. These materials are used on luminous watch dials, toys, and items that "glow in the dark." When a phosphorescent material is exposed to light, atomic electrons are excited to higher energy levels. Many of the excited electrons return to their normal state fairly soon, but there are metastable states in which electrons remain for seconds, minutes, even more than an hour.

Consequently, the material, being made of billions of atoms, emits light or glows for some time (Fig. 26.14).

A major consideration in laser operation is the emission process. As Fig. 26.15 shows, spontaneous absorption and emission of radiation occur between two energy levels. That is, a photon is absorbed and a photon is emitted. However, there is another possible emission process called **stimulated emission**. Einstein proposed this process in 1919. If a photon with an energy equal to an allowed transition strikes an atom in an excited state, it may stimulate an electron to make a transition to a lower energy level, with the emission of another photon. Two photons with the same frequency and phase then go off in the same direction.

Notice that stimulated emission is an amplification process—one photon in, two out. Of course, this is not a case of getting something for nothing, since the atom must be initially excited, and energy is needed to boost an electron to a higher state. This excitation process is somewhat analogous to pumping water to a roof-top reservoir for later use. However, stimulated emission does provide a way to amplify light.

Ordinarily, when light passes through a material, photons are more likely to be absorbed than to give rise to stimulated emission. This is because there are normally many more atoms in the ground state than in excited states.* However, under special circumstances, it is possible to have more atoms in an excited state than in the ground state. This condition is known as a **population inversion**. In this case, there may be more stimulated emission than absorption, and we have amplification. *Population inversion and stimulated emission are two basic conditions necessary for laser action.*

There are several types of lasers, but the helium-neon (He-Ne) gas laser is most commonly used for classroom demonstrations and laboratory experiments. The characteristic reddish-pink light produced by the He-Ne laser ($\lambda = 632.8$ nm) can be observed in an optical scanning system at supermarket checkouts. The gas mixture is about 85% helium and 15% neon. Essentially, the helium is used for energizing and the neon for amplification. The gas mixture is subjected to a high dc voltage discharge rectified from a radio-frequency power supply, and helium atoms are excited (Fig. 26.16a). This process is referred to as *pumping*. Energy is pumped into the system, and the helium atoms are pumped into an excited state (20.61 eV).

The excited state in the He atom has a relatively long lifetime (about 10^{-4} s) and has almost the same energy as an energy level in the Ne atom (20.66 eV). There is a good chance that before an excited He atom spontaneously emits a photon, it will collide with a Ne atom. When a collision occurs, energy is transferred to the Ne atom. Light is produced when the Ne electron drops

*A multielectron atom is referred to as being in the ground state when its electrons occupy the lowest energy levels.

FIGURE 26.14 Phosphorescence
When electrons in a phosphorescent material are excited, some of them do not immediately return to the ground state but remain in metastable states for periods of time. In this exhibit at the San Francisco Exploratorium, phosphorescent walls and floor continue to glow for about 30 seconds after being illuminated, retaining the shadows of children present when the phosphors were exposed to light.

Population inversion—more excited atoms than unexcited atoms

FIGURE 26.15 Photo absorption and emission (a) In absorption, a photon is absorbed, and the electron is excited to a higher energy level. (b) After a short time, the electron spontaneously decays to a lower energy level with the emission of a photon. (c) If another photon with an energy equal to an allowed transition strikes an excited atom, stimulated emission occurs, and two photons with the same frequency and phase go off in the same direction as that of the incident photon.

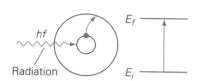

(a) Absorption

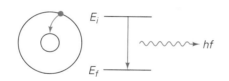

(b) Spontaneous emission

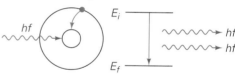

(c) Stimulated emission

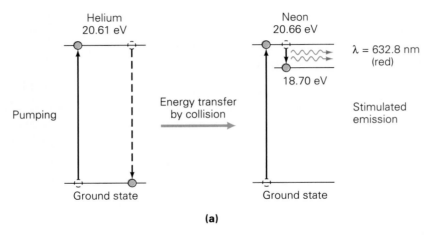

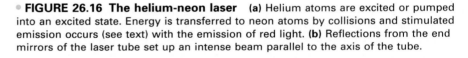

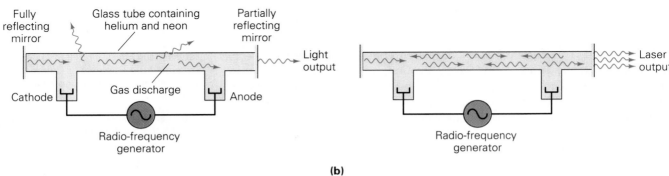

Helium
20.61 eV

Neon
20.66 eV

$\lambda = 632.8$ nm
(red)

18.70 eV

Pumping

Energy transfer
by collision

Stimulated
emission

Ground state

Ground state

(a)

Fully
reflecting
mirror

Glass tube containing
helium and neon

Partially
reflecting
mirror

Light
output

Laser
output

Cathode

Gas discharge

Anode

Radio-frequency
generator

Radio-frequency
generator

(b)

Figure 26.16

● **FIGURE 26.16 The helium-neon laser** **(a)** Helium atoms are excited or pumped into an excited state. Energy is transferred to neon atoms by collisions and stimulated emission occurs (see text) with the emission of red light. **(b)** Reflections from the end mirrors of the laser tube set up an intense beam parallel to the axis of the tube.

(a) Coherent

(b) Incoherent

● **FIGURE 26.17 Coherent light**
(a) Laser light is monochromatic (single frequency or color) and coherent, or in phase. **(b)** Light waves from sources in which atoms emit randomly are incoherent, or out of phase.

down to a specific lower level. However, the lifetime of the 20.66-eV neon state is relatively long, and the delay causes a piling up of neon atoms in this state, leading to a population inversion.

The stimulated emission and amplification of the light emitted by the neon atoms are enhanced by reflections from mirrors placed at the end of the laser tube (Fig. 26.16b). Some excited Ne atoms spontaneously emit photons in all directions, and these photons induce stimulated emissions. In stimulated emission, the two photons leave the atom in the same direction as that of the incident photon. Photons traveling in the direction of the tube axis are reflected back through the tube by the end mirrors. The photons, in reflecting back and forth, cause more stimulated emissions, and an intense, highly directional, coherent, monochromatic beam develops along the tube axis. Part of the beam emerges through one of the end mirrors, which is only partially silvered.

The monochromatic (single-frequency), coherent (same-phase) and directional properties of laser light give it unique properties (● Fig. 26.17). Light from sources like incandescent lamps is incoherent. The atoms emit randomly and at different frequencies (many different transitions). As a result, the light is out of phase or incoherent. Such beams spread out and become less intense. The properties of laser light allow the formation of a very tight beam, which with amplification can be very intense.

We hear about the development of "Star Wars" laser weapons that would shoot down missiles, and lasers powerful enough to induce nuclear fusion

(Chapter 29). Such lasers must have large power outputs. Carbon dioxide gas lasers can have a continuous 25,000-W output, which is capable of cutting steel. Pulsed ruby lasers may deliver as much as 50 MW, but only for a short fraction of a second. The He-Ne gas lasers used in college classrooms and laboratories have typical outputs of a few milliwatts or less.

Some laser applications are shown in ● Fig. 26.18. It is important to remember that, although laser beams are used to weld detached retinas in the eye, viewing a laser beam directly can be quite hazardous. The beam is focused on the retina in a very small spot. If the beam intensity and the viewing time are sufficient, the photosensitive cells of the retina may be burned and destroyed.

You might own a laser yourself—see the Insight feature.

Another interesting application of laser light is the production of three-dimensional images in a process called **holography** (from the Greek word *holos*, meaning "whole"). The process does not use lenses as ordinary image-forming processes do, yet it re-creates the original scene in three dimensions. The key to holography is the coherent property of laser light, which gives the light waves a definite spatial relationship to each other.

In the photographic process of making a *hologram*, an arrangement such as that illustrated in ● Fig. 26.19 is used. Part of the light from the laser passes through a partial mirror to the object. The other part, or reference beam, is reflected to the film. The light incident on the object is reflected to the film, and it interferes with the reference beam. The film records the interference pattern of the two light beams, which essentially imprints on the film the information carried by the light wavefronts from the object.

When the film is developed, the interference pattern bears no resemblance to the object and appears as a meaningless pattern of light and dark areas. However, when the wavefront information is reconstructed by passing light through the film, a three-dimensional image is perceived (● Fig. 26.20). If part of the three-dimensional image is hidden from view, you can see it by moving your head to one side, just as you would to see a hidden part of a real object.

The technique of holography has many potential applications. Using coherent ultrasonic waves to form a hologram could provide a three-dimensional view of internal structures of the human body. Also, holograms made with X-ray lasers (yet to be developed) could provide better resolution for an in-depth view of microscopic structures. Holographic three-dimensional television may some day be commonplace.

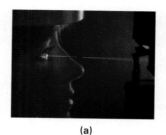

(a)

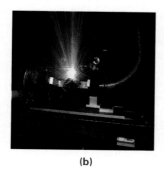

(b)

● **FIGURE 26.18 Laser applications**
(a) Laser eye surgery. Such procedures, which minimize bleeding and scarring, are used to repair detached retinas, treat glaucoma, correct retinal abnormalities produced by diabetes, and alleviate various other problems. **(b)** A laser is used to drill holes with great precision in the manufacture of jet engine turbine blades.

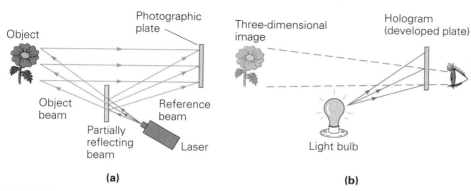

(a)   **(b)**

● **FIGURE 26.19 Holography**   **(a)** The coherent light from a laser is split into reference and object beams. The interference patterns of these beams are recorded on a photographic plate. **(b)** When the developed plate or hologram is illuminated, the viewer sees a three-dimensional image.

● **FIGURE 26.20 Holographic images**   A three-dimensional hologram can be viewed from any angle, just as if it were a real object.

The compact disc (CD) is a relatively new method of data storage, and most of us are familiar with it in terms of musical recordings. The CD system was introduced in 1980. There are now millions of CD players, and musical CD's outsell tapes and have nearly surplanted long-playing records. The disc itself is only 12 cm (5 in.) in diameter, but it can store more than 6 billion bits of information. This is the equivalent of more than 1000 floppy disks or over 275,000 pages of text, which could be stored on a CD for display on a television monitor. For audio use, a CD can store 74 minutes of music, and the sound reproduction is virtually unaffected by dust, scratches, or fingerprints on the disc.

The information on the disc is in the form of raised areas called "pits," which are separated by flat areas called "land." The surface is coated with a thin layer of aluminum to reflect the laser beam that "reads" the information (Fig. 1). The pits are arranged in a spiral track like the grooves in a phonograph record, but the track is much narrower. The tracks on a CD are about 60 times closer together than the grooves on a long-playing phonograph record.

In the read-out system, a laser beam, which comes from a small semiconductor (solid state) laser, is applied from below the disc and focused on the aluminum coating of the track. The disc rotates at about 3.5 to 8 revolutions per second as the laser beam follows the spiral track. When the beam strikes a land area between two pits, it is reflected. When the beam spot overlaps a land area and a pit, the light reflected from the different areas interferes, causing fluctuations in the reflected beam. To make the fluctuations more distinct, the raised pit thickness is chosen to be one-quarter wavelength of the laser beam. Then, reflected light from land areas travels an additional path length of one-half wavelength, and destructive interference occurs when the two parts of the reflected beam combine (Section 23.1). As a result, there is less reflected intensity when the beam passes over the edge of a pit than when passing over a land area alone.

The reflected beam of varying intensity strikes a photodiode (solid state photocell), which reads the information. The fluctuations of the reflected light convey the coded information as a series of binary numbers (zeros and ones). A pit represents a binary 1 and a land area is read as a binary 0. The signals are electronically converted back into sound.

Research is underway to produce an erasable disc so that CD's can be used to record sound as well as play it back.

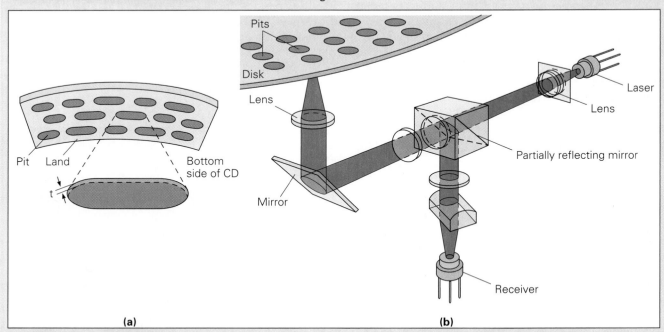

FIGURE 1 The compact disc (a) The information on the disc is recorded in the form of raised areas called "pits," which are separated by flat areas called "lands." The pits or imprints are actually on the bottom of the disc. (b) The surface is coated with a thin layer of aluminum to reflect the laser beam which "reads" the information.

Important Concepts

You should be able to define and/or explain these chapter concepts clearly.

thermal radiation	stopping potential	Bohr theory of the hydrogen	fluorescence
blackbody	work function	atom	metastable state
Wien's displacement law	threshold frequency	principal quantum number	stimulated emission
Planck's constant	Compton effect	ground state	population inversion
Planck's hypothesis	dual nature of light	excited states	holography
quantum	emission spectrum	binding energy	
photon	absorption spectrum	lifetime	
photoelectric effect	Balmer series	laser	

Important Relationships for Review

These relationships are mathematical statements of the concepts and principles presented in the chapter. You should be able to identify the symbols and to explain the relationships before proceeding to the Exercises. In-text equation reference numbers are given for convenience.

Wien's Displacement Law:

$$\lambda_{max}T = 2.90 \times 10^{-3} \text{ m-K} \tag{26.1}$$

Photon Energy:

$$E = hf \tag{26.4}$$

(where Planck's constant is $h = 6.63 \times 10^{-34}$ J-s)

K_{max} and Stopping Potential in the Photoelectric Effect:

$$K_{max} = eV_o \tag{26.5}$$

Energy Conservation in the Photoelectric Effect:

$$hf = K + \phi \tag{26.6}$$

Energy Conservation (with work function) in the Photoelectric Effect:

$$hf = K_{max} + \phi_o \tag{26.7}$$

Work Function and Threshold Frequency in the Photoelectric Effect:

$$\phi_o = hf_o \tag{26.8}$$

Compton Equation:
$$\Delta\lambda = \lambda - \lambda_o = \lambda_c(1 - \cos\theta) \tag{26.9}$$
(where $\lambda_c = h/m_oc = 2.4 \times 10^{-12}$ m)

Bohr Theory Orbit Radius:

$$r_n = 0.53\, n^2 \text{ Å} = 0.053\, n^2 \text{ nm} \tag{26.16}$$

Bohr Theory Electron Energy:

$$E_n = \frac{-13.6}{n^2} \text{ eV} \qquad n = 1, 2, 3, \dots \tag{26.17}$$

Bohr Theory Transition Energy (in eV):

$$\Delta E = 13.6\left(\frac{1}{n_f^2} - \frac{1}{n_i^2}\right) \tag{26.18}$$

Bohr Theory Photon Wavelength:

$$\lambda = \frac{12,400}{\Delta E} \text{ Å} = \frac{1,240}{\Delta E} \text{ nm} \tag{26.19}$$

Exercises

26.1 ■ Quantization: Planck's Hypothesis

1 A blackbody (a) is an ideal system, (b) re-emits all radiation incident on it, (c) has a wavelength of maximum radiation intensity inversely proportional to its temperature, (d) all of these. (d)

2 Planck's hypothesis (a) was the basis of Wien's law, (b) called for the intensity of blackbody radiation to be proportional to $1/\lambda^4$, (c) quantized the energy of thermal oscillators, (d) predicted the ultraviolet catastrophe. (c)

3 Does a blackbody at 200 K emit twice as much total radiation as when its temperature is 100 K? Explain.   no

4 ■■ What is the wavelength and frequency of the most intense radiation component from a blackbody with a temperature of 0°C? 1.06×10^{-5} m, 2.83×10^{13} Hz

5 ■■ Sketch a graph of temperature versus (a) frequency and (b) wavelength of the most intense radiation component of blackbody radiation.   see ISM

● **6** ■■ Find the approximate temperature of a blue star that emits light with a wavelength of maximum emission of 450 nm. 6.4×10^3 K

7 ■■ Assuming the skin to be the same temperature as normal body temperature (37°C), what is the wavelength of the radiation component of maximum intensity emitted by our bodies? What region of the EM spectrum is this in?
9.4×10^{-6} m; infrared

8 ■■ The walls of a blackbody cavity are at a temperature of 327°C. What is the frequency of the radiation of maximum intensity? 6.2×10^{13} Hz

9 ■■ What would be the approximate temperature of a blackbody if it appeared to be (a) red and (b) blue?
(a) 4.1×10^3 K   (b) 7.3×10^3 K

10 ■■ What would be the change in the frequency of the most intense spectral component of a blackbody that was initially at 200°C and was heated to 400°C?
$\Delta f = 2.1 \times 10^{13}$ Hz

● **11** ■■ What is the energy of a thermal oscillator in a blackbody with a temperature of 212°F? 2.56×10^{-20} J

12 ■■■ The temperature of a blackbody is 1000 K. If the total intensity of the emitted radiation of 1.0 W were due to the most intense frequency component, how many quanta would be emitted per second per square meter?
1.4×10^{19} quanta/s-m²

26.2 ■ Quanta of Light: Photons and the Photoelectric Effect

13 In the photoelectric effect, classical theory predicted that (a) no photoemission can occur below a threshold frequency, (b) the photocurrent is proportional to the light intensity, (c) the kinetic energy of the emitted electron is dependent on frequency, but not on intensity, (d) a photocurrent is observed immediately.   (b)

● **14** A photocurrent is observed when (a) the light frequency is above the threshold frequency, (b) the energy of the photons is greater than the work function, (c) the light frequency is below the threshold frequency, (d) both (a) and (b).   (d)

15 If electromagnetic radiation is made up of quanta, why don't we detect the discrete packets of energy, for example, when listening to a radio?   see ISM

16 ■ What is the energy of a quantum of light with a frequency of 5.5×10^{14} Hz? 3.6×10^{-19} J

17 ■ A photon has an energy of 3.3×10^{-15} J. Of what type of light is this photon? 6.0×10^{-11} m; X-ray

18 ■■ Which has more energy and how many times more: a quantum of violet light or a quantum of red light from the ends of the visible spectrum? $E_v = (1.75) E_r$

19 ■■ A photon of ultraviolet light has a wavelength of 300 nm. How much energy does the photon have in (a) joules and (b) electron volts?   (a) 6.63×10^{-19} J   (b) 4.1 eV

20 ■■ Light with an intensity I produces a photocurrent I_p. If the intensity of the light is doubled, how is the current affected?   doubled

21 ■■ Light with an intensity I produces a photoelectric emission, and the photoelectrons have a kinetic energy of 5.0 eV. If the intensity of the light is tripled, what is the kinetic energy of photoelectrons?   5.0 eV

● **22** ■■ A 100-W light bulb gives off 5.0% of its energy as visible light. How many photons of visible light are given off in 1 minute? (Use an average wavelength of 550 nm.)
8.3×10^{20} quanta

23 ■■ A solar cell is 20% efficient and the intensity of sunlight received at the Earth's surface is 2.5×10^{12} J/km²-h. Assuming an average wavelength of 550 nm and a cell area of 1.0 cm², what is the maximum number of photons converted to electrical energy each second? 3.8×10^{28} photons

24 ■■ ● Fig. 26.21 shows a graph of stopping potential versus frequency for a photoelectric material. Determine (a) Planck's constant and (b) the work function of the material from the graph data.   (a) 6.4×10^{-34} J-s   (b) 2.9×10^{-19} J

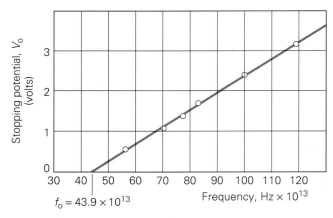

● **FIGURE 26.21 Stopping potential versus frequency**   See Exercise 24.

25 ■■ A metal with a work function of 2.6 eV is illuminated with a beam of monochromatic light. It is found that the stopping potential for the emitted electrons is 2.8 V. What is the wavelength of the light? 2.3×10^{-7} m

26 ■■ The threshold wavelength for emission from a metallic surface is 500 nm. Calculate the maximum speed of emitted photoelectrons for light having a wavelength of (a) 400 nm and (b) 600 nm.   (a) 4.7×10^5 m/s   (b) no emission

27 ■■ The work function of a photoelectric material is 3.5 eV. If the material is illuminated with monochromatic light ($\lambda = 300$ nm), (a) find the stopping potential of emitted photoelectrons. (b) What is the cutoff frequency?   (a) 0.63 V (b) 8.4×10^{14} Hz

28 ■■ What is the longest wavelength of light that can cause the release of electrons from a metal that has a work function of 2.48 eV? 5.01×10^{-7} m

29 ■■ Blue light with a wavelength of 420 nm is observed to cause the emission of photoelectrons with a maximum kinetic energy of 1.0×10^{-19} J. Will red light ($\lambda = 700$ nm) cause the emission of photoelectrons from the material? Justify your answer mathematically.   no, see ISM

● **30** ■■ Gold has a work function of 4.82 eV. If a gold surface is illuminated with ultraviolet light ($\lambda = 160$ nm), what is (a) the maximum kinetic energy of the emitted photoelectrons and (b) the threshold frequency for gold?
(a) 4.70×10^{-19} J; (b) 1.16×10^{15} Hz

31 ■■ Sodium and silver have work functions of 2.46 eV and 4.73 eV, respectively. (a) If the surfaces of the metals are both illuminated with the same monochromatic light, the photoelectrons from which metal will have the greater speed on emission? (b) What wavelengths are required for photoelectrons to be emitted from each metal surface?
(a) sodium (b) $\lambda_{silver} < 2.63 \times 10^{-7}$ m; $\lambda_{sodium} < 5.0 \times 10^{-7}$ m

32 ■■■ When a certain photoelectric material is illuminated with red light ($\lambda = 680$ nm) and then blue light ($\lambda = 420$ nm), it is found that the photoelectrons resulting from the blue light have a maximum kinetic energy 1.8 times greater than those from red light. What is the work function of the material?   0.42 eV

33 ■■■ When the surface of a particular material is illuminated with monochromatic light of various frequencies, the stopping potentials for the photoelectrons are determined to be:

Frequency (Hz)			
7.0×10^{15}	5.5×10^{15}	4.0×10^{15}	3.5×10^{15}
Stopping potential (V)			
12.6	10.3	4.10	1.97

Plot these data and determine the values of Planck's constant and the work function of the metal.   see ISM

26.3 ■ Quantum "Particles": The Compton Effect

34 The Compton effect was observed for (a) visible light, (b) infrared radiation, (c) ultraviolet light, (d) X-rays.   (d)

35 The wavelength shift for Compton scattering is maximum when (a) the photon scattering angle is 90°, (b) the electron scattering angle is 90°, (c) the shift is equal to the Compton wavelength, (d) the incident photon is back scattered.   (d)

36 ■ Calculate the maximum wavelength shift for Compton scattering. 4.8×10^{-12} m

37 ■ What is the change in wavelength when monochromatic X-rays are scattered through an angle of 20°? 1.4×10^{-13} m

38 ■■ A photon with an energy of 10 keV is scattered by a free electron. If the electron recoils with a kinetic energy of 5.0 keV, what is the wavelength of the scattered photon? 2.5×10^{-10} m

39 ■■ Show that the Compton wavelength has units of length and a value of 2.43×10^{-12} m.   see ISM

● **40** ■■ X-rays scattered from a carbon target show a wavelength shift of 0.000711 nm. What is the scattering angle?   45°

41 ■■ A monochromatic beam of X-rays with a wavelength of 2.80×10^{-10} m is scattered through a metal foil. What is the wavelength of the scattered X-rays observed at an angle of 45° from the direction of the incident beam? 2.81×10^{-10} m

42 ■■ X-rays with a wavelength of 0.0045 nm are used in a scattering experiment. If the X-rays are scattered through an angle of 53°, what is the wavelength of the scattered radiation? 5.46×10^{-12} m

43 ■■ The Compton effect also occurs for protons. (a) What is the value of the Compton wavelength for a proton? (b) Compare the wavelength shifts for maximum electron and proton scatterings by forming a ratio.   (a) 1.32×10^{-15} m (b) 1.8×10^{3}

26.4 ■ The Bohr Theory of the Hydrogen Atom

● **44** In his theory of the hydrogen atom, Bohr postulated the quantization of (a) energy, (b) centripetal acceleration, (c) light, (d) angular momentum.   (d)

45 An excited hydrogen atom emits light when its electron (a) makes a transition to a lower energy level, (b) is excited to a higher energy level, (c) makes a transition between any two excited states, (d) is in the ground state.   (d)

46 Explain why the Bohr theory is applicable only to the hydrogen atom and hydrogen-like atoms, such as singly ionized helium, doubly ionized lithium, and other one-electron systems.   see ISM

47 ■ Find the energy required to excite a hydrogen electron from (a) the ground state to $n = 3$ and (b) $n = 2$ to $n = 3$.
(a) 12.1 eV   (b) 1.89 eV

48 ■ Find the energy needed to ionize a hydrogen atom whose electron is in the (a) $n = 4$ state and (b) $n = 5$ state.
(a) 0.850 eV   (b) 0.544 eV

49 ■ What is the frequency of a photon that would excite the electron of a hydrogen atom from (a) $n = 2$ to $n = 5$ and (b) $n = 2$ to $n = \infty$?   (a) 6.9×10^{14} Hz   (b) 8.2×10^{14} Hz

50 ■■ Use the Bohr theory to find the value of the Rydberg constant for the empirical formula (Eq. 26.10).
1.097×10^{-2} nm^{-1}

51 ■■ Give the radius of the electron orbit of the hydrogen atom for each of the following states: (a) $n = 3$, (b) $n = 6$, (c) $n = 10$.   (a) 0.477 nm   (b) 1.91 nm   (c) 5.30 nm

● **52** ■■ Find the energy of the hydrogen electron for each of the following states: (a) $n = 3$, (b) $n = 6$, (c) $n = 10$.
(a) -1.51 eV   (b) -0.378 eV   (c) -0.136 eV

53 ■■ Find the binding energy of the hydrogen electron for each of the following states: (a) $n = 3$, (b) $n = 5$, (c) $n = 10$.

54 ■■ The ionization energy of a hydrogen atom is 13.6 eV. Would the absorption of a photon having a frequency of 7.00×10^{15} Hz cause a hydrogen atom to be ionized? If so, what would be the kinetic energy of the emitted electron?

55 ■■ The hydrogen spectrum has a series of lines called the Brackett series, which results from transitions to the $n = 4$ energy level. What is the longest wavelength in this series and in what region of the EM spectrum does it lie?

● **56** ■ A hydrogen electron in the ground state is excited to the $n = 5$ energy level. The electron makes a transition to the $n = 2$ level before returning to the ground state. (a) What are the wavelengths of the emitted photons? (b) Would the light emitted from a group of such hydrogen atoms be visible?

57 ■■ For which one of the following transitions of a hydrogen electron is the photon of greatest energy emitted: (a) $n = 5$ to $n = 3$, (b) $n = 6$ to $n = 2$, (c) $n = 2$ to $n = 1$? Justify your answer mathematically.

58 ■■ What is the binding energy for the electron in the ground state in the following hydrogen-like ions: (a) He^+; (b) Li^{2+}?

59 ■■ A hydrogen atom absorbs a photon of wavelength of 434.1 nm. (a) How much energy did the atom absorb? (b) What were the initial and final states of the hydrogen electron?

● **60** ■■ Show that the total energy of an electron in a hydrogen atom in an orbit with a radius of 0.0529 nm is -13.6 eV.

61 ■■ Show that the speeds of an electron in the Bohr orbits are given by $v_n = (2.2 \times 10^6 \text{ m/s})/n$.

62 ■■■ A muon is a particle that has the same charge as an electron but a mass 207 times that of an electron. Using a muon to replace the electron in a hydrogen atom, muonium atom is possible. For such an atom, (a) what would be the muon's energies and orbital radii for the $n = 1$ and $n = 2$ states? (b) What is the wavelength of the photon emitted for a muon transition from $n = 2$ to $n = 1$?

26.5 ■ A Quantum Success: The Laser

63 Which of the following is *not* an essential for laser action? (a) population inversion, (b) phosphorescence, (c) pumping, (d) stimulated emission.

64 What do the letters *se* in the laser acronym stand for? Explain the process to which this term refers.

■ **Additional Exercises**

65 An FM radio station broadcasts at a frequency of 98.9 MHz and radiates 750 kW of power. How many quanta are radiated from the station's antenna each second?

● **66** A photon is a "massless" particle, i.e., it has no rest mass. Show that the momentum of a photon is given by $p = h/\lambda$. [*Hint:* Use the relativistic relationship $E^2 = p^2 c^2 + (m_o c^2)^2$ which relates the total energy and momentum (Exercise 57, Chapter 25)].

67 Referring to ● Fig. 26.22, derive the equation for the wavelength shift in Compton scattering by applying the conservation of momentum and energy in an elastic collision. [*Hint:* Consider total relativistic energy. Also, analyze momentum classically with components as in Chapter 6, but remember that in the end $p = h/\lambda$ for a photon.]

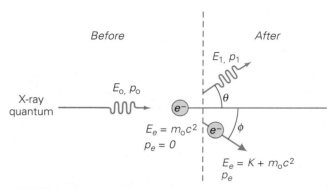

● **FIGURE 26.22 Compton scattering** See Exercise 67.

68 What are the energies of the photons required to excite a hydrogen electron in the ground state to states (a) $n = 4$ and (b) $n = 8$?

69 How many quanta of red light ($\lambda = 700$ nm) would it take to have 1.0 J of energy?

70 What is the incident energy of a beam of monochromatic X-rays that are observed to have a wavelength of 4.5×10^{-10} m at a Compton scattering angle of 37°?

71 What is the frequency of the most intense spectral component from a blackbody at room temperature (20°C)?

72 The work function of a particular metal is 5.0 eV. (a) What is the frequency of light that causes electrons to be emitted with a maximum kinetic energy of 2.0×10^{-19} J? (b) What is the stopping potential of the electrons?

73 For what scattering angle would the wavelength shift for Compton scattering be 25 percent of the Compton wavelength?

74 How many transitions of the electron in a hydrogen atom will result in the emission of light in the visible region of the spectrum (400–700 nm)?

27

Quantum Mechanics

The initial successes of the quantum theory were impressive, but they also created a perplexing situation. They seemed to show that light (that is, electromagnetic radiation) has a dual nature. On one hand, in many experiments, light exhibits classical wave behavior. On the other hand, explanations of certain phenomena require light to have a quantum or particle nature. It was as though each theory worked well for its own experiments. Moreover, as you learned in Chapter 26, the classical and quantum theories of light are mutually contradictory in some instances. For example, classical wave theory predicts a time delay in the emission of photoelectrons, while the quantum theory does not.

Around 1925, a new kind of physics based on the synthesis of wave and quantum ideas was introduced. This new theory, called **quantum mechanics***, attempts to combine the wave-particle duality into a single consistent description. It revolutionized scientific thought and provided the basis of our present understanding of microscopic (and submicroscopic) phenomena.*

Quantum mechanics deals mainly with the minute aspects of the world of atoms. It has replaced the mechanistic view of the universe, in which all things moved according to exact natural laws, with a new concept of probability. All physical measurements are now accepted as being to some degree uncertain. Even so, when quantum mechanics is applied to macroscopic phenomena, it must reproduce the results of classical physics in order to be a consistent theory. This means that there is no need to abandon the laws and principles of classical physics, since they provide an accurate description of everyday phenomena.

In a sense, the two theories complement each other. A more complete understanding of phenomena is obtained if the two theories are used in conjunction. It is more convenient to describe macroscopic phenomena by classical physics. But when dealing with phenomena in the atomic and subatomic worlds, we use quantum mechanics. One could make an analogy here. When objects are moving with speeds that are appreciable fractions of the speed of light, we must use the theory of relativity. But when things slow down, classical principles can be applied.

The application of quantum mechanics to specific phenomena is mathematically detailed and beyond the scope of the text. In this chapter we will generally discuss some of the basic ideas of quantum mechanics to see how they describe waves and particles. Also, some of the results of this new approach to describing phenomena will be presented—for example, the electron microscope (shown in use on page 825).

27.1 ■ Matter Waves: The de Broglie Hypothesis

These relationships are discussed in
Section 25.4

Since a photon travels at the speed of light, we treat it relativistically as a "massless" particle—that is, a particle having no rest mass m_0. Otherwise, it would have infinite mass ($m = m_0 / \sqrt{1 - (v/c)^2}$), and therefore infinite energy ($E = mc^2$). It can be shown that the total energy and relativistic momentum are related to each other by the equation $E^2 = p^2c^2 + (m_0c^2)^2$. (See Exercise 57, Chapter 25). A photon without rest mass then has a total energy of $E = pc$, or a relativistic momentum of $p = E/c$. The energy may also be written $E = hf = hc/\lambda$, so the momentum of a photon is related to its wavelength by

See Eq. 26.4 in Section 26.2

$$ p = \frac{E}{c} = \frac{hf}{c} = \frac{h}{\lambda} \tag{27.1} $$

Thus, the energy in electromagnetic waves of wavelength λ is carried by photon "particles" having momenta of h/λ.

Since nature exhibits a great deal of symmetry, the French physicist Louis de Broglie thought that there might be a wave-particle symmetry. That is, if light sometimes behaves like a particle, perhaps material particles, such as electrons, also had wave properties. In 1924, de Broglie put forth a hypothesis that a moving particle has a wave associated with it. He proposed that the wavelength of a material particle was related to the particle's momentum by an equation similar to that for a photon (Eq. 27.1). More specifically, the **de Broglie hypothesis** is

Whenever a particle has momentum p, a wave is associated with its motion with a wavelength of

$$ \lambda = \frac{h}{p} = \frac{h}{mv} \tag{27.2} $$

These waves associated with moving particles were called **matter waves** or, more commonly, **de Broglie waves**, and were thought to somehow influence or "guide" particle motion. One might think of an electromagnetic wave as the de Broglie wave for a photon, but the de Broglie waves associated with particles such as electrons and protons are *not* electromagnetic waves. The de Broglie equation for the wavelength gives no clue as to what kind of wave is associated with a particle of matter.

De Broglie's hypothesis met with a great deal of skepticism. The idea that

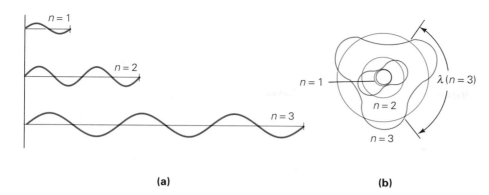

The Bohr model is outlined in Section 26.4

Compare the discussion of standing waves in Section 13.5

FIGURE 27.1 De Broglie waves and Bohr orbits (a) Similar to standing waves in a stretched string, de Broglie waves form circular standing waves on the circumferences of the Bohr orbits (b). Notice that the number of wavelengths in a particular orbit is equal to the principle quantum number of that orbit.

the motions of photons were somehow governed by the electromagnetic wave properties of light did not seem unreasonable. But to extend this idea and say that the motion of a mass particle is governed by the properties of an associated wave was difficult to accept. Moreover, there was no evidence that particles had wave properties.

In support of his hypothesis, de Broglie showed how it could give an interpretation of the quantization of the angular momentum in Bohr's theory of the hydrogen atom. For a free particle, the associated wave would have to be a traveling wave. However, the bound electron of a hydrogen atom travels repeatedly in discrete circular orbits according to the Bohr theory. The associated matter wave would then be expected to be a standing wave. That is, only an integral number of wavelengths could fit into a given orbit. The wave must meet itself, so to speak, with the same phase. This is a boundary condition, similar to those in the linear case for a standing wave in a stretched string fixed at both ends (see ● Fig. 27.1).

The circumference of a Bohr orbit of radius r_n is $2\pi r_n$, so this length would be equal to an integral number of wavelengths

$$2\pi r_n = n\lambda \qquad \text{for } n = 1, 2, 3, \ldots$$

Using the de Broglie equation, $\lambda = h/mv$, for the wavelength, we have

$$2\pi r_n = \frac{nh}{mv} \qquad \text{or} \qquad mvr_n = \frac{nh}{2\pi}$$

Thus, the angular momentum is quantized just as Bohr proposed. For orbits other than those allowed by the Bohr theory, the de Broglie wave for the orbiting electron would not close on itself. That is, the circumference of a disallowed orbit would not be equal to an integral number of wavelengths. This would give rise to destructive interference, and the wave would die out or have zero amplitude. The amplitude of the de Broglie wave must then be related in some way to the location of the electron.

EXAMPLE 27.1 ■ Short de Broglie Waves

A pitcher throws a fast ball with a speed of 40 m/s. If the mass of the ball is 0.15 kg, what is the wavelength of the de Broglie wave associated with the moving ball?

Solution.

> *Given:* $m = 0.15$ kg   *Find:* λ (de Broglie wavelength)
> $\qquad\quad v = 40$ m/s

This is a straightforward calculation. Using the de Broglie equation (Eq. 27.2), we have

$$\lambda = \frac{h}{mv} = \frac{6.63 \times 10^{-34}\,\text{J-s}}{(0.15\,\text{kg})(40\,\text{m/s})} = 1.1 \times 10^{-34}\,\text{m}$$

Notice the small wavelength for the baseball in the example. It is little wonder that we don't observe any effects of such matter waves. Recall that wave effects such as diffraction interference take place only when the size of the objects or slits is comparable to the wavelength, and $\lambda \cong 10^{-34}$ m is just too small. We cannot detect the wave properties of ordinary objects. For a 1000-kg jet plane flying near the speed of sound, say 350 m/s, the wavelength of the de Broglie wave is on the order of 10^{-39} m.

However, particles with very small masses traveling at relatively low speeds are another matter, as the following example shows.

EXAMPLE 27.2 ■ Not-So-Short de Broglie Waves

What is the de Broglie wavelength of the wave associated with an electron that has been accelerated through a potential of 50.0 V?

Solution.

> *Given:* $V = 50.0$ V   *Find:* λ (de Broglie wavelength)

The work done in accelerating the electron is $W = eV$ and is equal to its kinetic energy ($\frac{1}{2}mv^2$). We could therefore use this relationship to calculate the magnitude of the electron's velocity, and then use $\lambda = h/mv$ to find the wavelength. However, the calculation is simpler if we begin by expressing the kinetic energy in terms of momentum. Since $p = mv$, the kinetic energy, $\frac{1}{2}mv^2$, can be written as $p^2/2m$. Then,

$$\frac{p^2}{2m} = eV \qquad \text{or} \qquad p = (2meV)^{\frac{1}{2}}$$

The de Broglie wavelength is then

$$\lambda = \frac{h}{p} = \frac{h}{(2meV)^{\frac{1}{2}}}$$

Putting in the values of h, e, and m leads to a convenient general formula for the de Broglie wavelength of an electron accelerated through a potential difference.

$$\lambda = \left(\frac{150}{V}\right)^{\frac{1}{2}} \times 10^{-10}\,\text{m} = \left(\frac{150}{V}\right)^{\frac{1}{2}}\,\text{Å} \qquad (27.3)$$

where V is in volts. Then, for $V = 50.0$ V,

$$\lambda = \left(\frac{150}{50.0}\right)^{\frac{1}{2}} \times 10^{-10} \text{ m} = 1.73 \times 10^{-10} \text{ m} = 1.73 \text{ Å}$$

Although wavelengths on the order of 10^{-10} m are extremely small, such waves can at least be detected. Slit widths on this order cannot be fabricated. However, recall from Chapter 23 that nature provides such slits in the form of crystal lattices.

Diffraction from crystal lattices is discussed in Section 23.3

In 1927, two physicists in the United States, C. J. Davisson and L. H. Germer, used a crystal to diffract a beam of electrons, thereby demonstrating a wavelike property of particles. A single crystal of nickel was used in the experiment. The crystal was cut to expose a spacing of $d = 2.15$ Å between the lattice planes. When a beam of electrons was directed normally on the crystal face, a maximum in the intensity of the scattered electrons was observed at an angle of $50°$ relative to the surface normal (Fig. 27.2). The scattering was most intense for an accelerating potential of 54 V.

According to wave theory, the first-order maximum should be observed at an angle given by

$$d \sin \theta = \lambda$$

This condition requires a wavelength of

$$\lambda = d \sin \theta = (2.15 \text{ Å}) \sin 50° = 1.65 \text{ Å}$$

Using the general formula from the previous example (Eq. 27.3) to determine the de Broglie wavelength of the electrons, we find

$$\lambda = \left(\frac{150}{V}\right)^{\frac{1}{2}} = \left(\frac{150}{54}\right)^{\frac{1}{2}} = 1.67 \text{ Å}$$

The agreement of the wavelengths was excellent within the limits of experimental accuracy, and the Davisson-Germer experiment gave convincing proof of the validity of de Broglie's hypothesis.

Another experiment carried out by G. P. Thomson in Great Britain in the same year added further proof. Thomson passed a beam of energetic electrons through a thin metal foil. The diffraction pattern of the electrons was the same as that of X-rays. A comparison of such patterns, as shown in Fig. 27.3, leaves little doubt that particles exhibit wavelike properties. An example of a practical application of this fact is given in the Insight feature.

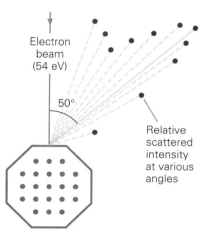

FIGURE 27.2 The Davisson-Germer experiment When a beam of 54-eV electrons is incident on the face of a nickel crystal, a maximum in the scattering intensity is observed at an angle of $50°$.

Experimental proof of the wave nature of particles

Figure 27.2

(a) **(b)**

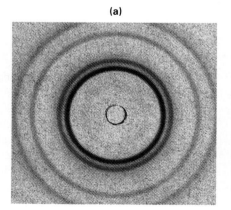 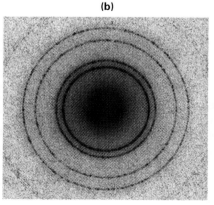

FIGURE 27.3 Diffraction patterns Comparison of an X-ray diffraction pattern (a) and an electron diffraction pattern (b) leaves little doubt that electrons exhibit wavelike properties.

The de Broglie hypothesis helped to set the stage for the development of an important practical application of electron beams—the *electron microscope*. As we have seen from the discussion of the Davisson-Germer and Thomson experiments, the wave properties of the electron are not just conceptual or mathematical abstractions. Electrons can undergo diffraction, as do light waves. You might wonder whether this implies that electrons can also be focused like light waves. The answer is yes; and in fact, we can focus electron "waves" to form images, though the means employed are somewhat different from those used with light waves in a light microscope. More-

over, as you can see from Example 27.2, accelerated electrons have very short wavelengths. With such short wavelengths, it is possible to obtain greater magnification and finer resolution than any light microscope can provide. (Recall the discussion of the relationship of resolving power to wavelength in Chapter 24.) The resolving power

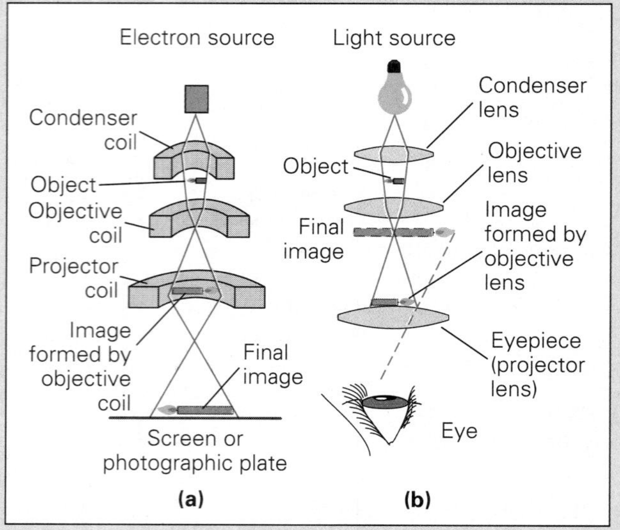

FIGURE 1 Electron and light microscopes A comparison of the elements of **(a)** an electron microscope and **(b)** a compound light microscope. The light microscope is drawn upside down for a better comparison.

FIGURE 2 An electron microscope The microscope is housed in the cylindrical vacuum chamber.

27.2 ■ The Schrödinger Wave Equation

De Broglie's hypothesis predicts that moving particles have associated waves that somehow govern or describe their behavior. However, it does not tell us the form of these waves, only their wavelengths. To have a useful theory, we need an equation that will give us the form of the matter waves, in particular the wave form of a particle moving under the influence of a force. Also, we must know how these waves govern particle motion. In 1926, Erwin Schrödinger, an Austrian physicist, presented a general equation that describes the de Broglie matter waves.

As a wave, the de Broglie wave form is described by some mathematical function. (Recall how wave motion was described using equations in Chapter 13.) In general, this function is commonly denoted by the symbol ψ (the Greek

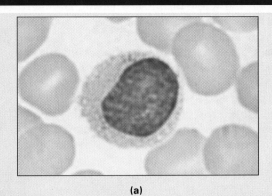

(a)

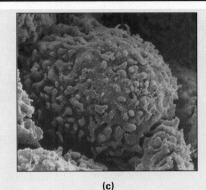

(b)   (c)

FIGURE 3 Lymphocytes (white blood cells) Images produced by **(a)** a light microscope, **(b)** a transmission electron microscope (TEM), and **(c)** a scanning electron microscope (SEM).

of standard electron microscopes is on the order of a few nanometers.

Technological developments made in the 1920s, in particular the focusing of electron beams by magnetic coils, permitted the construction of the first electron microscope in Germany in 1931. In a *transmission electron microscope*, an electron beam is directed onto a very thin specimen. Different numbers of electrons pass through different parts of the specimen depending on its structure. The transmitted beam is then brought to focus by a magnetic objective lens as illustrated in Fig. 1. The general components of electron and light microscopes are analogous, but an electron microscope is housed in a high-vacuum chamber so that the electrons are not deflected by air molecules. As a result, an electron microscope looks nothing like a light microscope (Fig. 2). Magnifications up to 100,000X can be achieved with an electron microscope (Fig. 3b), whereas a light microscope image (Fig. 3a) is limited to a magnification of about 2000X.

Another difference is that the final lens in an electron microscope, called the projector lens, has to project a real image onto a fluorescent screen or photographic film, since the eye cannot perceive an electron image directly. Specimens for transmission electron microscopy must be very thin. Special techniques allow the preparation of specimen sections as thin as 100 Å to 200 Å (only about 100 atoms thick).

The surfaces of thicker objects may be examined by the reflection of the electron beam from the surface. This is done with the more recently developed *scanning electron microscope*. A beam spot is scanned across the specimen by means of deflecting coils, much as is done in a television tube. Surface irregularities cause directional variations in the intensity of the reflected electrons, which gives contrast to the image. The specimens have to be coated with a thin layer of metal (such as gold or aluminum) to make them conducting. Otherwise, they would charge up nonuniformly from the electron beam and distort the image. Through such techniques, an electron microscope gives pictures with a remarkable three-dimensional quality such as those shown in Fig. 3c.

letter psi, pronounced "sigh") and is called the **wave function**. It describes the wave as a function of time and space. For example, the wave function for a (classical) traveling wave is $\psi = A \sin(kx - \omega t)$.

Since the de Broglie wave governs a particle's motion, it would seem reasonable that the wave function was associated with the particle's energy, both kinetic and potential. Recall from the discussion of conservation of energy in Chapter 5 that for a conservative mechanical system, we had

$$K + U = E$$

Equation 5.9, Section 5.5

where the sum of the kinetic and potential energies equals the total energy. Schrödinger proposed a similar equation for the de Broglie matter waves involving the wave function ψ. This is known as **Schrödinger's wave equation**, and has the general form

$$(K + U)\psi = E\psi \qquad (27.4)$$

Explain when the de Broglie wave equation is used and when the Schrödinger wave equation is used.

That is, the wave function is associated with the energy of a system. When applied to various situations with known values of K and U, complex mathematical equations result which are beyond the scope of this text. Using advanced mathematical procedures, the idea is to solve an equation for ψ and thus determine its form. But what is the physical significance of ψ?

In the early development of quantum mechanics (sometimes called wave mechanics), it was not clear how ψ should be interpreted. After much thought and investigation, it was concluded that ψ^2 (the wave function squared)* represents the probability of finding the particle at a certain position and time.

This interpretation involves the amplitude of ψ. Recall from Chapter 13 that the energy or intensity of a classical wave (function) is proportional to the square of its amplitude. Similarly, the intensity of a light wave is proportional to E^2, where E is the electric field amplitude. Looking at this in terms of "particle" photons, the *intensity* of a light beam is proportional to the number (n) of photons in the beam, so $n \propto E^2$. That is, the number of photons is proportional to the square of the electric field amplitude of the wave.

Review Eq. 13.5 in Section 13.1

The wave function ψ is interpreted in an analogous manner. The wave function generally varies in magnitude in space and time. If ψ describes an electron beam, then ψ^2 will be proportional to the number of electrons that may be expected to be found in a small volume around a point at some time. However, when there are only a few electrons or a single particle, ψ^2 represents the probability of finding an electron at the point. Thus, in quantum mechanics the square of the wave function is proportional to the probability of finding a particle in space and time, or the **probability density**.

The square of the absolute value of the wave function solution to Schrödinger's equation, ψ^2, gives the probability of finding a particle at a location.

The interpretation of ψ^2 as the probability of finding a particle at a particular place altered the idea that the hydrogen electron could be found only in orbits at discrete distances from the nucleus, as described in the Bohr theory. When the Schrödinger equation was solved for the hydrogen atom, the radial wave functions for each energy level were found to have values at any radial distance from the nucleus. Thus, quantum mechanically, there is some finite probability (ψ^2) of finding the electron in a given energy level at *any* distance from the nucleus. For example, the relative probability that an electron in the ground state ($n = 1$) is at a given radius is shown in ●Fig. 27.4a. The *maximum* probability coincides with the Bohr radius of 0.053 nm, but it is possible (that is, there is a finite probability) that the electron could be at various distances from the nucleus. The wave function exists for distances beyond 0.20 nm, but there is little chance (probability) of finding an electron in the ground state beyond this distance. The probability density distribution gives rise to the idea of an *electron cloud* around the nucleus (Fig. 27.4b). The cloud density reflects the probability density that the electron is in a particular region.

The Bohr model of the hydrogen atom is discussed in Section 26.4

Thus, quantum mechanics uses the wave functions of particles to predict the probability of particle phenomena. This also applies to light and photon "particles." For example, the square of a classical electromagnetic wave function, which is the de Broglie wave for its photons, gives the probability of finding the photons at a point in space. In a single-slit diffraction experiment, we know that many photons will strike the region of the central maximum on a screen, but fewer photons will arrive in the regions of the first bright fringes, even fewer in the regions of the second bright fringes, and so on. The quantum mechanical probability distribution or density on the screen has the same relative distribution as the classical intensity.

Single-slit diffraction is discussed in Section 23.3.

*More precisely, the absolute value of the wave function squared, $|\psi|^2$.

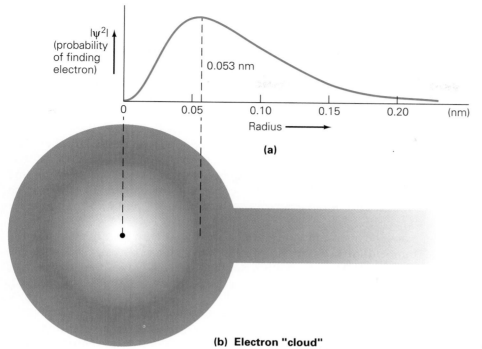

$|\psi^2|$ (probability of finding electron)

0.053 nm

0   0.05   0.10   0.15   0.20   (nm)

Radius ⟶

(a)

(b) Electron "cloud"

• FIGURE 27.4 Probability
(a) The square of the wavefunction or the probability of finding the hydrogen electron in the ground state ($n = 1$) as a function of radial distance. The electron has the greatest probability of being found at distance of 0.053 nm from the nuclear proton, which is the radius of the first Bohr orbit. **(b)** The probability distribution gives rise to the idea of an electron cloud around the nucleus. The cloud density reflects the probability density that the electron is in a particular region.

However, if the experiment could be done with only one photon, we could not predict exactly to which region it would go. We could only give the probability that it would be found in a particular region.

Another somewhat strange result of the fact that quantum mechanics deals exclusively with the *probability* of finding a particle in a particular region is called *tunneling*. Classically, there are regions where particles are forbidden because of energy considerations. For example, an electron in a conductor cannot jump across an empty space to another conductor unless it has enough energy to do so. Without adequate energy, we say that there is a potential barrier that the electron cannot cross. However, in certain instances, quantum mechanics predicts a small, but finite, probability of a particle being found in a classically forbidden region. This means that there is a probability of particles penetrating the barrier, or "tunneling" through. Such tunneling does occur and forms the basis of the scanning electron microscope (STM), which creates images with resolution on the order of the size of a single atom (Fig. 27.5). (Such barrier penetration is also used to explain a nuclear process in Chapter 28.)

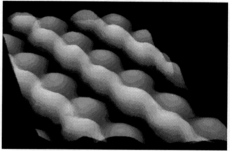

• FIGURE 27.5 Scanning Tunneling Microscopy The scanning tunneling microscope, developed in the 1970s and 1980s, makes use of the quantum phenomenon called *vacuum tunneling.* The tip of a metal needle probe is moved across the contours of a metalized specimen surface. Applying a small voltage between the probe and the surface causes electrons to tunnel through the vacuum. The tunneling current is extremely sensitive to the separation of the needle tip and the surface. As a result, when the probe is scanned across the sample, surface features as small as atoms show up as variations in the tunneling current that can be processed to produce 3-dimensional images. This photograph shows atoms of the semiconductor gallium arsenide.

27.3 ■ Atomic Quantum Numbers and the Periodic Table

The Hydrogen Atom

When the Schrödinger equation was solved for the hydrogen atom, the results predicted the allowed energy levels to be the same as those in the Bohr theory

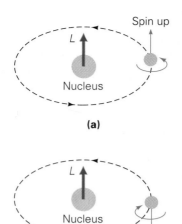

• FIGURE 27.6 Electron spin
Electron spin is sometimes described by analogy with the classical angular momentum of a spinning object. The spin can be either **(a)** up or **(b)** down, depending on the rotational sense of the electron. However, electron spin is strictly a quantum mechanical property and cannot be physically identified with the spin of a macroscopic body. (L is the angular momentum vector of the electron's revolution about the nucleus.)

(Section 26.4). Also, the energy values depended only on the principal quantum number n. However, in addition, the solution gave two other quantum numbers, which are designated as ℓ and m_ℓ.

The quantum number ℓ is called the **orbital quantum number**. It is associated with the angular momentum of the electron due to its orbital motion. The ℓ quantum number has integer values for each n quantum number from zero up to $\ell = (n - 1)$. For example, if $n = 3$, the values of ℓ are $\ell = 0, 1$, and 2. The number of different ℓ values is then equal to n for a given principal quantum number. As just shown, for $n = 3$, there are 3 ℓ values.

The quantum number m_ℓ is called the **magnetic quantum number**. The term "magnetic" has an experimental origin. It was found that a strong magnetic field influences the energy of the hydrogen electron, as observed in its line spectrum. The quantum number m_ℓ describes this "magnetic" effect, hence the name. In the absence of an external magnetic field, m_ℓ plays no role in determining the energy.

The magnetic quantum number m_ℓ is associated with the orientation of the angular momentum (a vector) in space. It is therefore associated with the ℓ quantum number. The m_ℓ quantum number has positive *and* negative integer values for each ℓ quantum number from zero up to $\pm\ell$. That is $m_\ell = 0, \pm 1, \pm 2, \pm 3, \ldots, \pm\ell$. For example, for $\ell = 3$, we have $m_\ell = -3, -2, -1, 0, 1, 2, 3$; or a total of $(2\ell + 1) = (2 \times 3 + 1) = 7$ values of m_ℓ.

But that's not all. There's one more quantum number for the hydrogen atom. Using high resolution spectrometers, it was found that each spectral line of hydrogen is in fact two very closely spaced lines. This is taken into account by a **spin quantum number**, m_s. It is associated with the intrinsic angular momentum of the electron. This property is called *electron spin*, and is sometimes described by analogy with the angular momentum of a spinning object in classical mechanics (• Fig. 27.6).

If an electron were a spinning particle, it would have either angular momentum "spin up" or "spin down," or two possible values. These are used to describe the so-called two-line spectral *fine structure*. However, electron spin is strictly a quantum mechanical property and cannot be accurately described by the classical analogy. Spin is an internal property of an electron that is characterized by the spin quantum number. The m_s quantum number has only two values, which are taken to be $m_s = \pm\frac{1}{2}$, for each value of m_ℓ. For example, for $\ell = 1$, there are three values of m_ℓ, and each of these has two m_s values: for $m_\ell = -1$, $m_s = \pm\frac{1}{2}$; for $m_\ell = 0$, $m_s = \pm\frac{1}{2}$, and for $m_\ell = 1$, $m_s = \pm\frac{1}{2}$.

The four quantum numbers for the hydrogen atom are summarized in Table 27.1. It should be noted that except in the case of the spin quantum number, the restrictions on their values came directly from mathematical theory. The value of the spin quantum number was dictated by experimental observations.

TABLE 27.1 ■ Hydrogen Atom Quantum Numbers

Quantum Number		Allowed Values	Number of Allowed Values
Principal	n	$1, 2, 3, \ldots, \infty$	No limit
Orbital angular momentum	ℓ	$0, 1, 2, 3, \ldots, (n-1)$	n (for each n)
Orbital magnetic	m_ℓ	$0, \pm 1, \pm 2, \pm 3, \ldots, \pm\ell$	$2\ell + 1$ (for each ℓ)
Spin magnetic	m_s	$\pm\frac{1}{2}$	2 (for each m_ℓ)

Multielectron Atoms

The Schrödinger equation for atoms with more than one electron (multielectron atoms) cannot be solved exactly. However, a solution can be found, to a workable approximation, in which each electron occupies a state characterized by a set of hydrogen-type quantum numbers. As might be expected with repulsive forces between electrons, the description of electron energy in a multielectron atom is more complicated. For one thing, the energy depends not only on the principal quantum number n, but also on the orbital quantum number ℓ. This gives rise to a subdivision of energy levels.

It is common to refer to the energy level of a given n as a **shell**, and ℓ levels of that shell as **subshells**. That is, atomic electrons with the same n value are said to be in the same electron shell. Electrons with the same n and ℓ values are said to be in the same electron subshell.

The ℓ subshells can be designated by numbers, as discussed previously. However, it is common to use letters instead. The letters s, p, d, f, g, ... correspond to the values of $\ell = 0, 1, 2, 3, 4, \ldots$, respectively. (This designation derives from historical spectroscopic notation.) After f, the letters go alphabetically (Table 27.2). This letter designation helps avoid confusion with n numbers, as we shall see.

Since an electron's energy depends on both n and ℓ in multielectron atoms, both quantum numbers are used to label atomic energy levels (a shell and subshell). For distinction, we write n as a number, followed by the letter that stands for the value of ℓ. For example, $1s$ denotes an energy level with $n = 1$ and $\ell = 0$; $2p$ is for $n = 2$ and $\ell = 1$; $3d$ for $n = 3$ and $\ell = 2$, and so on.

Also, it is common to refer to the m_ℓ values as representing *orbitals*. For example, a $2p$ energy level has three orbitals corresponding to the m_ℓ values of -1, 0, and 1 (for the p subshell or $\ell = 1$). The spin quantum number also applies to electrons in multielectron atoms, but no special name is given to it.

As we saw in the Bohr theory, the energy levels for the hydrogen atoms are not evenly spaced, but run sequentially upward. However, in multielectron atoms, the numerical sequence of the energy levels has numbers out of order. The shell-subshell (n-ℓ) energy level sequence for a multielectron atom is shown in Fig. 27.7a. Notice, for example, how the $4s$ level is below the $3p$ level. Such

TABLE 27.2 ■ Subshell Designations

ℓ Value	Letter Designation
$\ell = 0$	s
$\ell = 1$	p
$\ell = 2$	d
$\ell = 3$	f
$\ell = 4$	g
$\ell = 5$	h
.	.
.	.
.	.

n quantum numbers designate energy shells.
ℓ quantum numbers designate energy subshells.

Figure 27.7

FIGURE 27.7 Energy levels for a multielectron atom **(a)** The shell-subshell (n-ℓ) sequence shows that the energy levels are not evenly spaced and that the sequence of energy levels has numbers out of order. For example, note that the $4s$ level lies below the $3d$ level. The maximum number of electrons for a subshell, $2(2\ell + 1)$, is shown in parentheses on representative levels. **(b)** A convenient way to remember the energy level order for a multielectron atom is to list the n versus ℓ values as shown here. The diagonal lines then give the energy levels in ascending order.

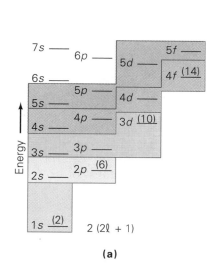

(a)

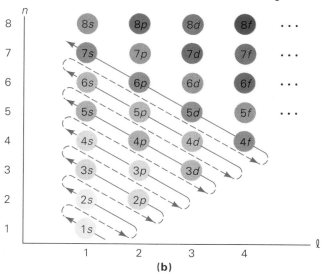

(b)

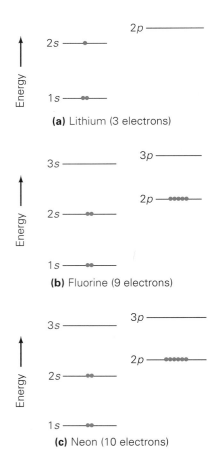

(a) Lithium (3 electrons)

(b) Fluorine (9 electrons)

(c) Neon (10 electrons)

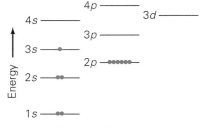

(d) Sodium (11 electrons)

FIGURE 27.8 Filling subshells The electron subshell distributions for several unexcited atoms according to the Pauli exclusion principle. The *s* subshell can have a maximum of two electrons and the *p* subshell a maximum of six electrons.

variations result in part from electrical forces between electrons in multielectron atoms. Also, the electrons in the outer orbits are shielded from the attractive force of the nucleus by the electrons in the inner orbits. A convenient way to remember the level order is given in Fig. 27.7b.

The ground state for a multielectron atom is similar to that of the hydrogen atom, with the electron in the 1s or lowest energy level. For a multielectron atom, the ground state is the combination of energy levels with the lowest total energy. That is, the electrons are in the lowest possible energy levels. But to identify those levels, we must know how many electrons can occupy a particular energy level. For example, the lithium (Li) atom has three electrons. Can they all be in the 1s level? As we shall see, the answer is no.

The Exclusion Principle. How the electrons of a multielectron atom distribute themselves in the ground-state energy levels is governed by a principle set forth in 1928 by the Austrian physicist Wolfgang Pauli. The **Pauli exclusion principle** states

No two electrons in a multielectron atom can have the same set of quantum numbers (n, ℓ, m_ℓ, m_s). That is, no two electrons can be in the same quantum state.

What this means is that each different set of quantum numbers (n, ℓ, m_ℓ, m_s) corresponds to a different energy state that can be occupied by only one electron.

The limits on the quantum numbers set the limits on the number of states for a given energy level. For example, the 1s ($n = 1$, $\ell = 0$) can have only one m_ℓ value, $m_\ell = 0$, along with only two m_s values, $m_s = \pm\frac{1}{2}$. Thus, there are only two unique sets of quantum numbers (n, ℓ, m_ℓ, m_s) for the 1s level: (1, 0, 0, $+\frac{1}{2}$) and (1, 0, 0, $-\frac{1}{2}$), so only two electrons can be in the 1s level. When this is the case, the shell is full; all other electrons are excluded from it by Pauli's principle. Thus, for a Li atom, with three electrons, the third electron must occupy the next higher level (2s) when the atom is in the ground state. This is illustrated in ● Fig. 27.8, along with the ground-state energy levels for some other atoms.

EXAMPLE 27.3 ■ **Number of Electron States**

How many possible sets of quantum numbers or electron states are there in (a) the 3p subshell and (b) the 4d subshell?

Solution.
  Given: (a) 3p level ($n = 3$, $\ell = 1$)   ***Find:*** sets of quantum
     (b) 4d level ($n = 4$, $\ell = 2$)   numbers or electron states

(a) For a particular subshell, it is the ℓ value that determines the number of states. Recall that there are ($2\ell + 1$) possible m_ℓ values for a given ℓ. Thus, for $\ell = 1$, there are [(2 × 1) + 1] = 3 values for m_ℓ. Each of these can have two m_s values ($\pm\frac{1}{2}$), making 6 different combinations of (n, ℓ, m_ℓ, m_s), or 6 states.
Notice that the number of possible states for a given value is then $2(2\ell + 1)$.

> Number of possible electron
> states for a given ℓ value = $2(2\ell + 1)$

TABLE 27.3 ■ **Possible Sets of Quantum Numbers and States**

Electron Shell n	Subshell ℓ	Subshell Notation	Orbitals m_ℓ	Number of Orbitals (m_ℓ) in Subshell ($2\ell + 1$)	Number of States (m_s) in Subshell $2(2\ell + 1)$	Total Electron States for n Shell $2n^2$
1	0	1s	0	1	2	2
2	0	2s	0	1	2	8
	1	2p	1, 0, −1	3	6	
3	0	3s	0	1	2	18
	1	3p	1, 0, −1	3	6	
	2	3d	2, 1, 0, −1, −2	5	10	
4	0	4s	0	1	2	32
	1	4p	1, 0, −1	3	6	
	2	4d	2, 1, 0, −1, −2	5	10	
	3	4f	3, 2, 1, 0, −1, −2, −3	7	14	

Table 27.3

(b) This part is now easy. The 4*d* level with $\ell = 2$ has

$$\text{Number of states} = 2(2\ell + 1) = 2[(2 \times 2) + 1] = 10$$

This development is summarized in Table 27.3. Notice that the number of states in all of the subshells in a given shell (given n) is $2n^2$. For example, for the $n = 2$ shell, the total number of states for its *s* and *p* subshells ($\ell = 0, 1$) is $2n^2 = 2(2)^2 = 8$. This means that up to 8 electrons can be accommodated in the $n = 2$ shell.

Electron Configurations. We can build up atoms, so to speak, by putting an increasing number of electrons in the lower energy levels [(hydrogen (H), 1 electron; helium (He), 2 electrons; lithium (Li), 3 electrons, etc.)], as was done for four particular electrons in Fig. 27.8. However, rather than draw diagrams each time we want to represent the electron arrangement in a ground-state atom, we can use a shorthand notation called the **electron configuration**.

In this notation, we write the energy levels in increasing order, and designate the number of electrons in each level with a superscript. For example, $3p^5$ means that a 3*p* subshell has 5 electrons:

Electron configuration—a shorthand notation for the electron quantum states occupied in a ground-state atom.

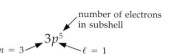

The electron configurations for the atoms shown in Fig. 27.8 are

Li	(3 electrons)	$1s^2 2s^1$
F	(9 electrons)	$1s^2 2s^2 2p^5$
Ne	(10 electrons)	$1s^2 2s^2 2p^6$
Na	(11 electrons)	$1s^2 2s^2 2p^6 3s^1$

In writing an electron configuration, when one subshell is filled, you go on to the next higher one. Notice that the superscripts in a configuration add up to the total number of electrons in the atom.

PERIODS	s		d										p					
	GROUP I	GROUP II	Transition elements										GROUP III	GROUP IV	GROUP V	GROUP VI	GROUP VII	GROUP VIII
1	1 H 1.01 $1s^1$																	2 He 4.00 $1s^2$
2	3 Li 6.94 $2s^1$	4 Be 9.01 $2s^2$											5 B 10.81 $2p^1$	6 C 12.01 $2p^2$	7 N 14.01 $2p^3$	8 O 16.00 $2p^4$	9 F 19.00 $2p^5$	10 Ne 20.18 $2p^6$
3	11 Na 22.99 $3s^1$	12 Mg 24.31 $3s^2$											13 Al 26.98 $3p^1$	14 Si 28.09 $3p^2$	15 P 30.97 $3p^3$	16 S 32.07 $3p^4$	17 Cl 35.45 $3p^5$	18 Ar 39.95 $3p^6$
4	19 K 39.10 $4s^1$	20 Ca 40.08 $4s^2$	21 Sc 44.96 $3d^14s^2$	22 Ti 47.88 $3d^24s^2$	23 V 50.94 $3d^34s^2$	24 Cr 52.00 $3d^54s^1$	25 Mn 54.94 $3d^54s^2$	26 Fe 55.85 $3d^64s^2$	27 Co 58.93 $3d^74s^2$	28 Ni 58.69 $3d^84s^2$	29 Cu 63.55 $3d^{10}4s^1$	30 Zn 65.39 $3d^{10}4s^2$	31 Ga 69.72 $4p^1$	32 Ge 72.61 $4p^2$	33 As 74.92 $4p^3$	34 Se 78.96 $4p^4$	35 Br 79.90 $4p^5$	36 Kr 83.80 $4p^6$
5	37 Rb 85.47 $5s^1$	38 Sr 87.62 $5s^2$	39 Y 88.96 $4d^15s^2$	40 Zr 91.22 $4d^25s^2$	41 Nb 92.91 $4d^45s^1$	42 Mo 95.94 $4d^55s^1$	43 Tc (98) $4d^55s^2$	44 Ru 101.07 $4d^75s^1$	45 Rh 102.91 $4d^85s^1$	46 Pd 106.42 $4d^{10}5s^6$	47 Ag 107.87 $4d^{10}5s^1$	48 Cd 112.41 $4d^{10}5s^2$	49 In 114.82 $5p^1$	50 Sn 118.71 $5p^2$	51 Sb 121.76 $5p^3$	52 Te 127.60 $5p^4$	53 I 126.90 $5p^5$	54 Xe 131.29 $5p^6$
6	55 Cs 132.91 $6s^1$	56 Ba 137.33 $6s^2$	57 La 138.91 $5d^16s^2$ *	72 Hf 178.49 $5d^26s^2$	73 Ta 180.95 $5d^36s^2$	74 W 183.85 $5d^46s^2$	75 Re 186.21 $5d^56s^2$	76 Os 190.2 $5d^66s^2$	77 Ir 192.22 $5d^76s^2$	78 Pt 195.08 $5d^96s^1$	79 Au 196.97 $5d^{10}6s^1$	80 Hg 200.59 $5d^{10}6s^2$	81 Tl 204.36 $6p^1$	82 Pb 207.2 $6p^2$	83 Bi 208.98 $6p^3$	84 Po (209) $6p^4$	85 At (210) $6p^5$	86 Rn (222) $6p^6$
7	87 Fr (223) $7s^1$	88 Ra 226.03 $7s^2$	89 Ac 227.03 $6d^17s^2$ †	104 Rf (261) $6d^27s^2$	105 Ha (262) $6d^37s^2$	106 Unh (263) $6d^47s^2$	107 Uns (262)	108 Uno (265)	109 Une (266)									

Atomic number — 26 Fe — Symbol
58.85 — Atomic mass
Outer electron configuration — $3d^64s^2$

	f															
*	58 Ce 140.12 $5d^14f^16s^2$	59 Pr 140.91 $4f^36s^2$	60 Nd 144.24 $4f^46s^2$	61 Pm (145) $4f^56s^2$	62 Sm 150.36 $4f^66s^2$	63 Eu 151.96 $4f^76s^2$	64 Gd 157.25 $5d^14f^76s^2$	65 Tb 158.93 $5d^14f^86s$	66 Dy 162.50 $4f^{10}6s^2$	67 Ho 164.93 $4f^{11}6s^2$	68 Er 167.26 $4f^{12}6s^2$	69 Tm 168.93 $4f^{13}6s^2$	70 Yb 173.04 $4f^{14}6s^2$	71 Lu 174.97 $5d^14f^{14}6s^2$	(Lanthanides)
†	90 Th 232.04 $6d^27s^2$	91 Pa 231.04 $5f^26d^17s^2$	92 U 238.03 $5f^36d^17s^2$	93 Np 237.05 $5f^46d^17s^2$	94 Pu (244) $5f^66d^07s^2$	95 Am (243) $5f^76d^07s^2$	96 Cm (247) $5f^76d^17s^2$	97 Bk (247) $5f^86d^17s^2$	98 Cf (251) $5f^{10}6d^07s^2$	99 Es (252) $5s^{11}6d^07s^2$	100 Fm (257) $5f^{12}6d^07s^2$	101 Md (258) $5f^{13}6d^07s^2$	102 No (259) $5f^{14}6d^07s^2$	103 Lw (260) $5f^{14}6d^17s^2$	(Actinides)

Figure 27.9

FIGURE 27.9 The periodic table The elements are arranged in order of increasing atomic or proton number. Horizontal rows are called periods, and vertical columns are called groups. The elements in a group have similar chemical properties. Each atomic mass represents an average of that element's isotopes, weighted to reflect their relative abundance in our immediate environment. The masses have been rounded to two decimal places; more precise values can be found in Appendices IV and V. (A value in parentheses represents the mass number of the best-known or longest-lived isotope of an unstable element.)

The spacings between the energy levels are not equal, as can be seen from Figs. 27.7a or 27.8. In general, there are relatively large energy gaps between the s levels and the levels below them. The lower levels are p levels, with the exception of the lowest case—the $1s$ level below the $2s$ level (and there are no levels below the $1s$). The gaps between other levels—for example, between an s level and the p level above, or between the d and p levels—are smaller and do not differ greatly in energy.

The periodic large energy gaps in the levels may be represented by

$$1s^2 \mid 2s^2 2p^6 \mid 3s^2 3p^6 \mid 4s^2 3d^{10} 4p^6 \mid 5s^2 4d^{10} 5p^6 \mid 6s^2 4f^{14} 5d^{10} 6p^6 \mid \text{etc.}$$

(number of states)   (2)   (8)   (8)   (18)   (18)   (32)

where the vertical lines indicate the energy gaps. The states between the lines have similar energies. We refer to such sets of energy levels that have about the same energy as an **electron period**.

These electron periods are the basis of the periodic table of elements. With your present knowledge of electron configurations, you are now in a position to understand the periodic table better than the person who originally developed it.

> Electron periods—sets of energy levels with about the same energy, separated by energy gaps.

The Periodic Table of Elements

By 1860, over 60 chemical elements had been discovered. Several attempts had been made to classify the elements or put them into some orderly arrangement, but none proved to be very satisfactory. It had been noted in the early 1800s that the elements could be listed in such a way that similar chemical properties recurred periodically throughout the list. Following this idea, in 1869 a Russian chemist, Dmitri Mendeleev (pronounced men-duh-*lay*-eff), formulated an arrangement of the elements based on this periodic property. The modern version of his periodic table of elements is used today and can be seen on the walls of just about every science building, as well as in ● Fig. 27.9.

Mendeleev arranged the known elements in rows, which are called **periods**, in order of increasing atomic masses. When he came to an element that had chemical properties similar to those of one of the previous elements, he went back and put this element below the previous similar one. In this manner, he formed both horizontal rows of elements, and vertical columns called **groups**, or families of elements with similar properties. The table was later rearranged in order of increasing atomic or proton number (the number of protons in an atom and the numbers at the top of the element boxes in Fig. 27.9) to resolve some inconsistencies. (Notice that if atomic masses were used, cobalt and nickel, atomic numbers 26 and 27, would fall in different groups.

> Periods—horizontal rows

> Groups—vertical columns

With only 65 elements, there were vacant spaces in Mendeleev's table. The elements for these spaces were yet to be discovered. Since the missing elements were part of a sequence and had properties similar to those of other elements in a group, Mendeleev was able to predict their masses and chemical properties. Less than 20 years after Mendeleev devised his table, showing chemists what to look for to find the undiscovered elements, three of the missing elements were discovered.

Notice that the periodic table puts the elements into seven horizontal rows or periods, in order of increasing atomic or proton number. The first period has only two elements. Periods 2 and 3 have eight elements, while periods 4 and 5 have 18 elements. Recall that the s, p, d, and f, subshells can contain a maximum of 2, 6, 10, and 14 electrons [$2(2\ell + 1)$], respectively. You should begin to see a correlation between these numbers and the arrangements of elements in the periodic table.

The periodicity of the periodic table can be understood in terms of the electron configurations of the atoms. For $n = 1$, the electrons are in one of two s states ($1s$); for $n = 2$, we have the last electrons going into the $2s$ and $2p$ states, which gives 10 electrons; and so on. Thus, for a given element, its period number is equal to the highest n shell containing electrons in the atom. Notice the electron configurations for the elements in Fig. 27.9. Also, compare the electron periods given earlier, as defined by energy gaps, and the periods in the periodic table (Fig. 27.10). There is a one-to-one correlation, so the periodicity comes from energy level considerations in atoms.

Chemists refer to *representative elements*, which we see from Fig. 27.9 are those in which the last electron enters an s or p subshell. In *transition elements*, the last electron enters a d subshell; and in *inner transition elements*, the last electron enters an f subshell. So that the periodic table is not unmanageably wide, the f subshells are usually placed in two rows at the bottom of the table. Each row is given a name—the *lanthanide series* and the *actinide series*—based on where it is positioned within the period.

Finally, you can also understand why elements in vertical columns or groups have similar chemical properties. As you probably know from chemistry classes, the chemical properties of an atom, such as its ability to react and form compounds, depends on the outermost electrons in the atom—that is, the number of electrons in the outermost *unfilled* shell. It is these electrons, called *valence electrons*, that form chemical bonds. Because of the way the elements

Shell (last to be filled)	Subshells	Number of electrons in subshell, $2(2\ell + 1)$	Corresponding period in periodic table
n = 7	7p	6	Period 7 (32 elements)
	6d	10	
	5f	14	
	7s	2	
n = 6	6p	6	Period 6 (32 elements)
	5d	10	
	4f	14	
	6s	2	
n = 5	5p	6	Period 5 (18 elements)
	4d	10	
	5s	2	
n = 4	4p	6	Period 4 (18 elements)
	3d	10	
	4s	2	
n = 3	3p	6	Period 3 (8 elements)
	3s	2	
n = 2	2p	6	Period 2 (8 elements)
	2s	2	
n = 1	1s	2	Period 1 (2 elements)

Energy ↑

FIGURE 27.10 Electron periods The periods of the periodic table are related to electron configurations. The last n shell to be filled is equal to the period number. The electron periods and the corresponding periods of the table are defined by energy gaps in the energy levels of the atoms.

are arranged in the table, the outermost electron configuration of all the atoms in any one group are the same or very similar. The atoms would therefore be expected to have similar chemical properties, as indeed they do. For example, notice the first two groups at the left of the table. They have one and two outermost electrons in an *s* subshell, respectively. These elements are all highly reactive metals that form compounds having many similarities. The group at the far right, the noble gases, have completely filled subshells *and* come at the ends of electron periods, or just before a large energy gap. These gases are all chemically very nonreactive and form compounds only under very special conditions.

27.4 ■ The Heisenberg Uncertainty Principle

Another important aspect of quantum mechanics has to do with measurement and accuracy. In classical mechanics, there is no limit to the accuracy of a measurement. Theoretically, by continual refinement of a measurement instrument and procedure, the accuracy could be improved to any degree so as to give *exact* values. This resulted in a *deterministic* view of nature. For example, if you know position and velocity of an object *exactly* at a particular time, you can determine where it will be in the future and where it was in the past (assuming no future or past unknown forces).

However, quantum theory predicts otherwise and sets limits on the possible accuracy of measurements. This idea was introduced by the German physicist Werner Heisenberg in 1927, who had developed another approach to quantum mechanics that complemented Schrödinger's wave theory. The **Heisenberg uncertainty principle** as applied to position and momentum (or velocity) may be stated as follows:

> It is impossible to know simultaneously an object's exact position and momentum.

This concept is often illustrated with a simple thought experiment. Suppose that you wanted to measure the position and momentum (actually the velocity) of an electron. In order for you to "see," or locate, the electron, at least one photon must bounce off the electron and come to your eye, as illustrated in ● Fig. 27.11. However, in the collision process, some of the photon's energy and momentum are transferred to the electron (similar to the Compton effect in Section 26.3).

After the collision, the electron recoils. Thus, in the very process of trying to locate the position very accurately (trying to make the uncertainty of position Δx very small), you induce more uncertainty into your knowledge of the electron's velocity or momentum ($\Delta p = m \Delta v$), since the process of determining its position sent the electron flying off. In the macroscopic world, the uncertainty due to viewing an object would be negligible, because light does not appreciably alter the motion or position of an ordinary-sized object.

For our subatomic case, the position of an electron could be measured at best to an accuracy of about the wavelength λ of the incident light, that is, $\Delta x \cong \lambda$. The photon "particle" has a momentum of $p = h/\lambda$. Since we cannot tell how much of this momentum might be transferred during collision, the final momentum of the electron would have an uncertainty of $\Delta p \cong h/\lambda$.

The total uncertainty is given by the product of the individual uncertainties,

Incident photon

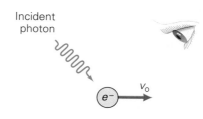

v_o

e^-

(a) Before collision

Scattered photon

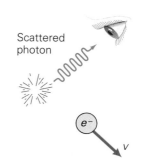

e^-

v

(b) After collision

● **FIGURE 27.11 Measurement-induced uncertainty** (a) To measure the position and momentum (or velocity) of an electron, at least one photon must collide with the electron and be scattered toward the eye. (b) In the collision process, energy and momentum are transferred to the electron, which induces uncertainty in the velocity.

and is at least,

$$(\Delta p)(\Delta x) \cong \left(\frac{h}{\lambda}\right)(\lambda) = h$$

This equation gives an estimate of the minimum uncertainties or maximum accuracies of simultaneous measurements of the momentum and position. In actuality, the uncertainties could be much worse depending on the amount of light (number of photons) used, apparatus, and technique. Through theoretical calculations, Heisenberg found that, at very best,

$$\boxed{(\Delta p)(\Delta x) \geq \frac{h}{2\pi}} \qquad (27.5)$$

Thus, Heisenberg's uncertainty principle states that the product of the *minimum* uncertainties of position and momentum is on the order of Planck's constant ($\cong 10^{-34}$). These are the minimum uncertainties, or the *best degree of accuracies* we can ever hope to achieve for *simultaneous* measurements. In the process of trying to locate the position of the particle accurately (that is, to make Δx small), the uncertainty in the momentum is made larger ($\Delta p \cong h/2\pi \, \Delta x$) and vice versa. Thus, the measurement procedure itself limits the accuracy to which we can simultaneously measure position and momentum. If we could measure the exact location of a particle ($\Delta x \to 0$), we would have no idea about its momentum ($\Delta p \to \infty$).

According to Heisenberg, "Since the measuring device has been constructed by the observer . . . we have to remember that what we observe is not nature in itself but nature exposed to our method of questioning."

EXAMPLE 27.4 ■ Uncertainty Principle

An electron and a 20-g bullet, both moving linearly, are measured to have equal speeds of 300 m/s to an accuracy of ±0.010%. What is the minimum uncertainty in the position of each?

Solution.

Given: $m_b = 20$ g $= 0.020$ kg   *Find:* Δx's (minimum uncertainties
($m_e = 9.11 \times 10^{-31}$ kg)   in position)
$v = 300$ m/s $\pm 0.010\%$

An uncertainty of 0.010% in velocity is

$$(300 \text{ m/s})(0.00010) = 0.030 \text{ m/s} \quad (= 3.0 \text{ cm/s})$$

and $v = 300$ m/s $\pm 0.010\% = 300$ m/s ± 0.030 m/s.

The total uncertainty in velocity is twice this amount, since the measurements can be off by 0.010% above or below the actual values and

$$\Delta v = 0.060 \text{ m/s} \quad (= 6.0 \text{ cm/s})$$

Then, for the electron,

$$\Delta x = \frac{h}{2\pi \Delta p} = \frac{h}{2\pi m_e \Delta v}$$
$$= \frac{6.63 \times 10^{-34} \text{ J-s}}{2\pi (9.11 \times 10^{-31} \text{ kg})(0.060 \text{ m/s})} = 0.0019 \text{ m} \quad (= 0.19 \text{ cm})$$

Similarly for the bullet,

$$\Delta x = \frac{h}{2\pi m_b \Delta v} = \frac{6.63 \times 10^{-34}\ \text{J-s}}{2\pi(0.020\ \text{kg})(0.060\ \text{m/s})} = 8.8 \times 10^{-32}\ \text{m}$$

Notice that the uncertainty in the position of the bullet is considerably less than that of the electron. The uncertainty for relatively massive objects traveling at ordinary speeds is practically negligible.

Another form of the uncertainty principle relates energy and time. To understand this, consider the position of the electron in the previous thought experiment to be known with an uncertainty of $\Delta x \cong \lambda$. The photon used to detect the particle travels with a speed c, and it takes a time of $\Delta x/c \cong \lambda/c$ for this photon to traverse a distance equal to the uncertainty in the particle's position. Thus, the time when the particle is at the measured position is uncertain by about

$$\Delta t \cong \lambda/c$$

Since we can't tell whether the photon transfers some or all of its energy ($E = hf = hc/\lambda$) to the particle, the uncertainty in the energy is

$$\Delta E = \frac{hc}{\lambda}$$

Then the total uncertainty is at least

$$(\Delta E)(\Delta t) \cong \left(\frac{hc}{\lambda}\right)\left(\frac{\lambda}{c}\right) = h$$

Similar to the position-momentum relationship, at very best we have

$$\boxed{(\Delta E)(\Delta t) \geq \frac{h}{2\pi}} \qquad (27.6)$$

This form of the uncertainty principle indicates that the energy of an object may be uncertain by an amount ΔE for a time $\Delta t \cong h/2\pi\Delta E$. During this time, the energy is uncertain and might not even be conserved. This is an important consideration in particle interactions, as you will learn in a later chapter.

Also, we cannot measure the energy of a particle exactly unless we take an infinite amount of time to do so. If a measurement of energy is carried out in a time Δt, then it must be uncertain by an amount ΔE. For example, the measurement of the frequency of a photon emitted by an atomic electron is a measurement of the energy associated with the transition from an excited state to the ground state. The measurement must be carried out in a time comparable to the time the electron is in the excited state—that is, the lifetime of the excited state. As a result, the observed emission line in the frequency spectrum has a finite width, since $\Delta E = h\Delta f$ (● Fig. 27.12).

This so-called *natural broadening* was ignored in Chapter 26, where spectral lines were considered to have widths of single frequencies (that is, no width at all). This is the same as assuming that the excited states of the Bohr atom have infinite lifetimes.

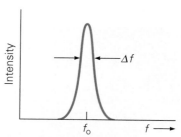

● **FIGURE 27.12 Natural line broadening**   Because a measurement must be carried out in a time comparable to the lifetime (Δt) of an electron in an excited state, the energy is uncertain by an amount $\Delta E = h\Delta f$. The observed emission line has a width of Δf, rather than being a line of single frequency f_o.

Minimum uncertainties in energy and time

Spectral lines are discussed in Section 26.4

EXAMPLE 27.5 ■ Natural Line Broadening

An electron in an excited state has a lifetime of 1.00×10^{-8} s. (a) What is the minimum uncertainty in the energy of the photons emitted on de-excitation? (b) What is the magnitude of the natural broadening of the spectral line?

Solution.

Given: $\Delta t = 1.00 \times 10^{-8}$ s   *Find:* (a) ΔE (minimum energy uncertainty)
(b) Δf (frequency broadening)

(a) The minimum uncertainty in the energy is given by Eq. 27.6:

$$\Delta E = \frac{h}{2\pi \Delta t} = \frac{6.63 \times 10^{-34} \text{ J-s}}{2\pi (1.00 \times 10^{-8} \text{ s})} = 1.06 \times 10^{-26} \text{ J}$$

(b) The uncertainty, or broadening, of the frequency of the observed spectral line is then

$$\Delta f = \frac{\Delta E}{h} = \frac{1.06 \times 10^{-26} \text{ J}}{6.63 \times 10^{-34} \text{ J-s}} = 1.60 \times 10^{7} \text{ Hz}$$

The uncertainty in the frequency of a spectral line is called the *natural line width*. The lifetimes of atomic electrons in their excited states are on the order of 10^{-8} s, and it might appear from the magnitude of the value of Δf that the spectral lines would be quite wide.

However, recall that the frequency of visible light is on the order of 10^{14} Hz. Expressed as a percentage, the uncertainty is about $10^{7}/10^{14}$ ($\times$ 100%) = 0.00001%. Thus, the spectral line has a frequency width on the order of $10^{14} \pm 0.00001\%$, which is quite narrow.

● **FIGURE 27.13 Cloud-chamber photograph of pair production** In this false-color bubble-chamber photograph, a gamma-ray photon (not visible) interacts with an atomic nucleus to produce an electron and a positron (green and red spiral tracks at top). In the process, it also dislodges an orbital electron (the vertical green track). An external magnetic field causes the electron and positron to be deflected in paths of opposite curvature. A similar event is recorded in the bottom half of the photo. (Why do you suppose the paths of the particles created in this case show less deflection?)

27.5 ■ Particles and Antiparticles

The quantum mechanics of Schrödinger and Heisenberg were successful in explaining observations and in predicting new atomic phenomena. When the quantum theory was extended to include relativistic considerations by the British physicist Paul A. M. Dirac in 1928, something new and very different was predicted—a particle called the **positron**. The positron should have the same mass as the electron but should carry a *positive* charge. The oppositely charged positron is said to be the **antiparticle** of the electron.

The positron was first observed experimentally in 1932 by the American physicist C. D. Anderson in cloud chamber experiments with cosmic rays. The curvature of the particle tracks in a magnetic field showed two types of particles (● Fig. 27.13). The tracks indicated that both particles had the same mass, but their spiral curvatures were opposite. From the magnetic force relationship, $F = qvB$, this observation requires the particles to be oppositely charged. Thus, Anderson discovered the positron, a particle with a mass equal to that of the electron and with the same magnitude of electrical charge, but opposite in sign.

By the conservation of charge, a positron can be created only with the simultaneous creation of an electron in a process called **pair production**. In Anderson's experiment, positrons were observed to be emitted from a thin lead plate exposed to cosmic rays from outer space, which contain highly energetic X-rays. Pair production occurs when an X-ray "photon" collides with a nucleus—the nuclei of the lead atoms of the plate in the Anderson experiment. In the collision process, the photon goes out of existence, and an "electron pair" (an electron and a positron) is created, as illustrated in ● Fig. 27.14, in a conversion of energy into mass. By the conservation of energy,

$$hf = 2m_ec^2 + K_{e^-} + K_{e^+} + K_{\text{nuc}}$$

where hf is energy of the photon, $2m_e$ is the rest mass of the electron pair, and the K's are the kinetic energies of the particles and the recoil nucleus. Because of its relatively large mass, the kinetic energy of the recoil nucleus can usually be considered negligible, and

$$hf \cong 2m_ec^2 + K_{e^-} + K_{e^+}$$

From the energy equation, we can see that a photon cannot create an electron pair unless

$$hf \geq 2m_ec^2 = 1.022 \text{ MeV} \qquad (27.7)$$

(Recall that the rest mass of an electron is $m_ec^2 = 0.511$ MeV.) This minimum energy is called the *threshold energy for pair production*.

But if positrons are created by cosmic rays, why are they not commonly found in nature? For example, why are they not evident in ordinary chemical processes? The answer is because positrons are taken out of existence by a process called **pair annihilation**. When an energetic positron appears in pair production, it loses kinetic energy in collisions as it passes through matter. Finally, moving at a low speed, it combines with an electron of the material and forms a hydrogenlike atom, called a *positronium atom*, in which a positron substitutes for a proton. The positronium atom is unstable and quickly decays ($\cong 10^{-10}$ s) into two 0.511-MeV photons (● Fig. 27.15). Pair annihilation is then a direct conversion of (rest) mass into electromagnetic energy, the inverse of pair production so to speak. These processes are striking examples of the mass-energy equivalence predicted by Einstein ($E = mc^2$).

All subatomic particles have been found to have antiparticles, which are observed in cosmic rays from outer space and/or are produced in nuclear processes. For example, there is an antiproton with same mass as a proton but with a negative charge. There are also antineutrons. In our environment, there is a preponderance of electrons, protons, and neutrons. When antiparticles are created, they quickly combine with their respective particles in annihilation processes.

It is conceivable that antiparticles predominate in some parts of the universe. If so, the atoms of the **antimatter** in this region would consist of negatively charged nuclei composed of antiprotons and antineutrons, surrounded by positively charged positrons (antielectrons). It would be difficult to distinguish a region of antimatter visibly, since it would appear the same as ordinary matter. The physical behavior of antimatter atoms would presumably be the same as those of ordinary matter. (Recall that the assignment of plus and minus signs to electric charges is an arbitrary convention.) However, if antimatter and ordinary matter came into contact, they would annihilate each other with an explosive release of energy.

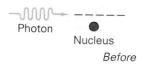

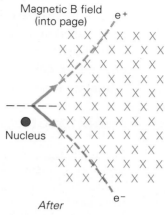

After

● **FIGURE 27.14 Pair production** An electron and a positron are created when an energetic photon interacts with a nucleus.

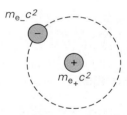

Before annihilation
(positronium atom)

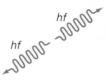

After annihilation
(photons)

● **FIGURE 27.15 Pair annihilation** The disappearance of a positronium atom is signaled by the appearance of two 0.511-MeV photons.

Important Concepts

You should be able to define and/or explain these chapter concepts clearly.

quantum mechanics
de Broglie hypothesis
de Broglie (matter) waves
wave function
Schrödinger's wave equation
probability density (ψ^2)
orbital quantum number (ℓ)

magnetic quantum number
(m_ℓ)
spin quantum number (m_s)
shell
subshell
Pauli exclusion principle
electron configuration

electron period
periodic table of
elements
period
group
Heisenberg uncertainty
principle

positron
antiparticle
pair production
pair annihilation
antimatter

Important Relationships for Review

These relationships are mathematical statements of the concepts and principles presented in the chapter. You should be able to identify the symbols and to explain the relationships before proceeding to the Exercises. In-text equation reference numbers are given for convenience.

Momentum of a Photon:

$$p = \frac{E}{c} = \frac{hf}{c} = \frac{h}{\lambda} \tag{27.1}$$

de Broglie Wavelength:

$$\lambda = \frac{h}{p} = \frac{h}{mv} \tag{27.2}$$

Electron Wavelength When Accelerated Through Potential V:

$$\lambda = \left(\frac{150}{V}\right)^{\frac{1}{2}} \text{Å} = \left(\frac{150}{V}\right)^{\frac{1}{2}} \times 10^{-10}\,\text{m} \tag{27.3}$$

Heisenberg Uncertainty Principle:

$$(\Delta p)(\Delta x) \geq \frac{h}{2\pi} \tag{27.5}$$

$$(\Delta E)(\Delta t) \geq \frac{h}{2\pi} \tag{27.6}$$

Condition for Electron Pair Production:

$$hf \geq 2m_e c^2 = 1.022\,\text{MeV} \tag{27.7}$$

Condition for Electron Pair Production:

$$hf \geq 2m_e c^2 = 1.022\,\text{MeV} \tag{27.7}$$

Exercises

27.1 ■ Matter Waves: The de Broglie Hypothesis

1 The momentum of a photon is (a) zero, (b) equal to c, (c) inversely proportional to its wavelength, (d) given by the de Broglie hypothesis. (c)

2 The Davisson-Germer experiment (a) was concerned with X-ray spectra, (b) verified the Heisenberg principle, (c) supported the Pauli exclusion principle, (d) demonstrated the wavelike properties of particles. (d)

3 ■ What is the de Broglie wavelength associated with photons of red light ($\lambda = 680$ nm)? 680 nm

● **4** ■■ What is the de Broglie wavelength of (a) an electron and (b) a proton, both moving with a speed of 300 m/s? (a) 2.43×10^{-6} m (b) 1.32×10^{-9} m

5 ■■ Calculate the de Broglie wavelength of a 70-kg person running with a speed of 20 m/s. 4.7×10^{-37} m

6 ■■ A proton and an electron are accelerated from rest through a potential difference V. What is the ratio of the de Broglie wavelength of an electron to that of a proton? $\lambda_e/\lambda_p = 43$

7 ■■ An electron is accelerated from rest through a potential difference that gives it a matter wave with a wavelength of 0.10 nm. What is the potential difference? 150 V

8 ■■ Electrons are accelerated from rest through a potential of 250 kV. If the potential is increased to 600 kV, how is the de Broglie wavelength of the electrons affected? $\lambda_2/\lambda_1 = 0.65$

9 ■■ Charged particles are accelerated through a potential difference V. By what factor would the de Broglie wavelength of the particles change if the voltage were tripled? 0.58

10 ■■ A proton traveling with a speed of 4.5×10^4 m/s is accelerated through a potential difference of 37 V. By what percentage does the de Broglie wavelength of the proton change? -66%

11 ■■ What is the energy of a beam of electrons that exhibits a first-order maximum at an angle of 30° when diffracted by a crystal grating with a spacing between the lattice planes of 0.15 nm? 2.7×10^2 eV

12 ■■ What is the de Broglie wavelength of the Earth in its orbit about the Sun? (Assume a circular orbit.) 3.7×10^{-63} m

13 ■■ The resolution of a light microscope is directly proportional to the radiation it uses (Section 24.4). Similarly for an electron microscope, the waves associated with the electrons set a limit on the resolution. (a) If the electrons in a microscope are accelerated through a potential difference of 10^5 V and the microscope has a circular aperture with a diameter of 0.10 mm, what is the minimum angular separation of two viewed objects? (b) Through what voltage difference would the electrons have to be accelerated to improve the angular resolution by a factor of two? (a) 5.1×10^{-16} rad (b) 2×10^5 V

14 ■■■ According to the Bohr theory of the hydrogen atom, the speed of the electron in the first Bohr orbit is 2.19×10^8 cm/s. (a) What is the wavelength of the matter wave associated with the electron? (b) How does this compare with the circumference of the first Bohr orbit? (a) 3.32×10^{-10} m (b) $\lambda = 2\pi r_1$

● **15** ■■■ It is desired to observe details whose size is of the order of 1.0 Å with an electron microscope. Through what potential must the electrons be accelerated so that they have a de Broglie wavelength of this order? 150 V

27.2 ■ The Schrödinger Wave Equation

16 The wave function solution to the Schrödinger equation (a) can never be found, (b) is the probability of finding a particle, (c) functionally describes the de Broglie wave of a particle, (d) none of these. (c)

17 The square of a particle's wave function is interpreted as being (a) the energy of the particle, (b) the probability of locating the particle, (c) the quantum number of a state, (d) the basis of the Pauli exclusion principle. (b)

● **18** ■■■ A particle in a box is constrained to move in one dimension, like the bead on a wire illustrated in ● Fig. 27.16. Assuming no forces act on the particle in the interval $0 < x < L$ and that it hits a perfectly rigid wall, it may be thought of as a particle at the bottom of an infinite well with $U = 0$. (a) Show that the spatial wave function for the particle is $\psi_n = A \sin n\pi x/L$ for $n = 1, 2, 3, \ldots$ (b) Show that the kinetic energy of the particle is given by $K = n^2 h^2/8mL^2$, where m is its mass. [*Hint:* (a) Consider boundary conditions like those of a standing wave. (b) Recall that $K = p^2/2m$ and that the wave function involves the de Broglie wavelength.] see ISM

19 ■■ (a) Sketch the wave functions of the particle for the first three states in Exercise 18. (b) Where is the particle most likely to be found, and what is the probability of finding it there? Sketch the probability densities for these states. see ISM

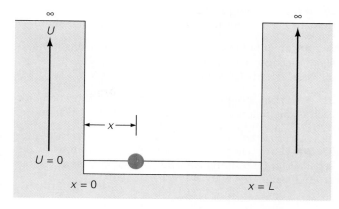

● **FIGURE 27.16 Particle in a box** See Exercise 18.

20 ■■ Sketch the form you would expect for the probability of the position of the hydrogen electron in the first three excited states as a function of the radial distance from the nucleus. see ISM

27.3 ■ Atomic Quantum Numbers and the Periodic Table

21 The ℓ quantum number of the hydrogen atom (a) determines the total energy of a state, (b) is associated with the angular momentum of the electron, (c) is associated with the orientation of the angular momentum, (d) is associated with the electron spin. (b)

● **22** The quantum number m_s (a) is purely a quantum mechanical concept, (b) arises from the orbital motion of an electron, (c) is due to actual electron spinning, (d) all of these. (a)

23 Niels Bohr set forth a *correspondence principle*, which states that quantum mechanics and classical physics are in general agreement when the quantum numbers are very large. Discuss this principle in terms of the hydrogen atom. see ISM

24 What is the basis of the periodic table of elements in terms of quantum theory, and what do the elements in a particular group have in common? see ISM

25 ■ (a) How many possible sets of quantum numbers are there for $n = 2$ and $n = 3$? (b) Explicitly write the (n, ℓ, m_ℓ, m_s) sets for these levels. (a) 8; 18 (b) see ISM

26 ■ How many possible sets of quantum numbers are there for the given subshells (a) $\ell = 0$ and (b) $\ell = 3$? (a) 2 (b) 14

27 ■ Which has more possible sets of quantum numbers, $n = 2$ or $\ell = 3$? 8 for $n = 2$; 14 for $\ell = 3$

28 ■■ An electron in a multielectron atom has a magnetic quantum number of $m_\ell = 3$. What are the minimum values of (a) ℓ and (b) n the electron could have? (a) $\ell = 3$ (b) $n = 4$

29 ■■ Draw the ground-state energy level diagrams like those in Fig. 27.8 for (a) nitrogen, N, and (b) potassium, K.
see ISM

● **30** ■■ Identify the atoms of each of the following ground-state electron configurations: (a) $1s^2 2s^2$, (b) $1s^2 2s^2 2p^3$, (c) $1s^2 2s^2 2p^6$, (d) $1s^2 2s^2 2p^6 3s^2 3p^4$. (a) Be (b) N (c) Ne (d) S

31 ■■ Write the ground-state electron configurations for each of the following atoms: (a) boron, B, (b) calcium, Ca, (c) zinc, Zn, (d) tin, Sn. (a) $1s^2 2s^2 2p^1$ (b) $1s^2 2s^2 2p^6 3s^2 3p^6 4s^2$ (c) $1s^2 2s^2 2p^6 3s^2 3p^6 3d^{10} 4s^2$ (d) $1s^2 2s^2 2p^6 3s^2 3d^{10} 4s^2 4p^6 4d^{10} 5s^2 5p^2$

32 ■■ Draw schematic diagrams for the electrons in the subshells of (a) sodium, Na, and (b) argon, Ar atoms in the ground state. see ISM

27.4 ■ The Heisenberg Uncertainty Principle

33 If the uncertainty in the position of a moving particle increases, (a) it may be located more exactly, (b) the uncertainty in its momentum decreases, (c) the uncertainty in its velocity increases, (d) none of these. (b)

34 According to the uncertainty principle, to measure the exact energy of a particle requires (a) special equipment, (b) an infinite time, (c) uncertainty in the momentum, (d) none of these. (b)

35 ■■ What is the minimum uncertainty in the velocity of an electron that is known to be somewhere between 0.50 Å and 1.0 Å from a proton? 2.3×10^6 m/s

● **36** ■■ What is the minimum uncertainty in the velocity of a 0.50-kg ball that is known to be at 1.0000 ± 0.0005 cm from the edge of a table? 2.1×10^{-29} m/s

37 ■■ An electron travels with a speed of 3.60560 ± 0.00021 m/s. To what minimum uncertainty can its position be measured? 0.28 m

38 ■■ The energy of a 2.00-keV electron is known to $\pm 3.00\%$. How accurately can its position be measured? 1.5×10^{-10} m

39 ■■ Using a photoelectric gate, it is possible to determine the speed of a 0.15-kg ball rolling on a table within a minimum uncertainty of 5.0×10^{-4} m/s. What are the corresponding uncertainties in (a) the momentum and (b) the position of the ball? (a) 7.5×10^{-5} kg-m/s (b) 1.4×10^{-30} m

40 ■■ If an excited state of an atom is known to have a lifetime of 10^{-7} s, what is the minimum error within which the energy of the state can be measured? 1.1×10^{-27} J

41 ■■ The energy of the first excited state of a hydrogen atom is measured to be -0.34 ± 0.0003 eV. What is the order of the average lifetime for this state? 1.1×10^{-12} s

42 ■■ How much greater is the width of a spectral line due to natural broadening for a transition from an excited state with a lifetime of 10^{-12} s than for one from a state with a lifetime of 10^{-8} s? 10^4

27.5 ■ Particles and Antiparticles

43 Pair production involves (a) the production of two electrons, (b) the production of two positrons, (c) a positronium atom, (d) a certain threshold energy. (d)

● **44** Pair annihilation involves (a) an electron, (b) a proton, (c) a positronium atom, (d) both (a) and (c). (d)

45 It has been suggested in science fiction that matter and antimatter could be combined as a source of energy. (The starship Enterprise on *Star Trek* had antimatter engines with the antimatter being stored in antimatter "pods.") Speculate how or in what antimatter might be stored for such use. [*Hint:* See confinement methods for fusion in Chapter 29.] see ISM

46 ■■ A photon with a frequency of 1.25×10^{18} Hz collides with a nucleus. Will pair production occur? Justify your answer mathematically. no

● **47** ■■ What is the frequency of the photons produced in electron pair annihilation? 1.23×10^{20} Hz

48 ■■ What would be the required energy for a photon that could cause the production of a proton-antiproton pair? 1.9 GeV

49 ■■ A muon, or μ meson, has a negative charge like that of an electron, but its mass is 207 times greater. What would be the required energy for a photon that could cause the pair production of a meson and an antimeson? 212 MeV

■ Additional Exercises

50 If the spacing between the lattice planes of a particular crystal is 1.90 Å, what voltage is needed to accelerate electrons (from rest) into a beam that exhibits a first-order diffraction maximum at an angle of 50°? 70.4 V

51 With what accuracy would you have to measure the position of a moving electron so that its velocity is uncertain by 10^{-3} cm/s? 11.5 m

52 A photon has an energy of 7.5 MeV. What are (a) its momentum and (b) the wavelength of the associated de Broglie wave? (a) 4.0×10^{-21} N-s (b) 1.7×10^{-13} m

53 An electron is accelerated from rest through a potential of 5.00 kV $\pm 3.0\%$. What is the minimum uncertainty in the position of the electron after the acceleration? 9.59×10^{-11} m

● **54** Show that for a particle moving in one dimension between the fixed boundaries $-L < x < L$ the wave functions are given by $\psi_n = A \cos n\pi x/2L$, where $n = 1, 3, 5, \ldots$. see ISM

55 Where is the particle of Exercise 54 most likely to be found? What is the probability of finding it there? $\psi = A^2$; $x = 0$

56 What is the de Broglie wavelength for the matter wave associated with (a) a 250-g ball thrown at 30 m/s and (b) a 1800-kg automobile traveling at 80 km/h? (a) 8.8×10^{-35} m (b) 1.7×10^{-38} m

57 Is it possible for the number of quantum states for a given n quantum number to be equal to the number of states for a given ℓ quantum number? Justify your answer. see ISM

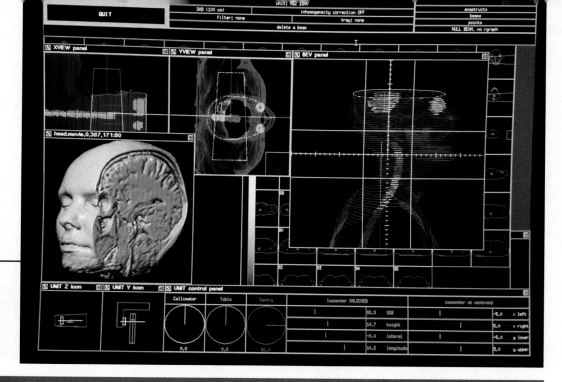

28

The Nucleus

28.1 **Nuclear Structure and the Nuclear Force**

28.2 **Radioactivity**

28.3 **Decay Rate and Half-Life**

28.4 **Nuclear Stability and Binding Energy**

28.5 **Radiation Detection and Applications**

Great advances have been made in nuclear physics since the introduction of the nuclear model of the atom by Lord Rutherford in 1911. However, not everything is known about the nucleus. This small integral part of the atom presents a great challenge to the ingenuity of scientists, who would like to know its secrets and to apply its potential.

How does one study such a minute entity? Because of its unimaginably small size, all our knowledge of the nucleus hinges on indirect observations that are not fully explained by classical theory. Quantum mechanics, which was so helpful to our understanding of atomic physics, can be applied to the nucleus, but with limitations. The potential energy of a system must be known before we can attempt to solve the Schrödinger equation. Knowing the energy of the system requires an understanding of the force that holds the nucleus together. The exact form of this nuclear force is not known.

Even so, a great deal can be learned about the nucleus by looking at the properties of various nuclei. One of the most revealing phenomena of the nucleus is radioactivity. Some nuclei are unstable and decay into nuclei of other elements with the telltale emission of detectable particles. These energetic particles, though they can be highly dangerous, also have many beneficial uses—for example, in treating cancer. (The photograph above shows a display of computerized tomography information used for the precise targeting of radiation therapy.) In addition to such practical applications, the study of radioactivity and nuclear stability gives us some general ideas about the nature of the nucleus, the energy it possesses, and how this energy can be released. The release of nuclear energy has become one of our major energy sources. This will be considered in the next chapter. First, let's take a look at the nucleus itself.

> *Topics studied in this chapter include nuclear structure and force, radioactivity, decay and half-life, nuclear stability and binding energy, and radioactivity detection and applications.*

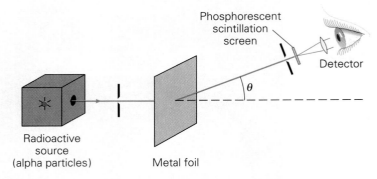

● **FIGURE 28.1 Rutherford's scattering experiment** A beam of alpha particles from a radioactive source is scattered by a thin foil, and the scattering is observed to be a function of the scattering angle θ. The detector was a scintillating phosphorescent screen.

Phosphorescent scintillation screen

Detector

θ

Radioactive source (alpha particles)

Metal foil

28.1 ■ Nuclear Structure and the Nuclear Force

It is evident from the emission of electrons from heated filaments (thermionic emission) and the photoelectric effect that the atoms of certain materials, probably all materials, contain electrons. Since atoms are normally electrically neutral, they must also contain positive charge equal in magnituide to that of the electrons in an atom. Also, since the mass of an electron is small in comparison to the mass of even the lightest atoms, most of the mass appears to be associated with the positive charge.

Based on these observations, J. J. Thomson, a British physicist who had experimentally proven the existence of the electron in 1897, proposed a model of the atom. In the Thomson model, the negatively charged electrons were pictured as being uniformly distributed within a continuous sphere of positive charge. It was called the "plum pudding" model because the electrons in the positive charge were thought of as analogous to the raisins in a plum pudding. The region of positive charge was assumed to have a radius on the order of 10^{-8} cm, based on calculations from the bulk properties of matter.

Our modern model of atomic structure is quite different. This model pictures all of the atom's positive charges, and practically all of its mass, as being concentrated in a central "nucleus," which is surrounded by the negatively charged electrons. The concept of an atomic nucleus was proposed by the British physicist Ernest Rutherford (1871–1937). Combined with Bohr's theory of orbiting electrons (Section 26.4), this idea led to the simplistic "solar system" or **Rutherford-Bohr model** of the atom.

Rutherford's insight came from the results of alpha particle scattering experiments performed in his laboratory around 1911. An alpha particle (α particle) is a doubly positively charged particle that is naturally emitted from some radioactive materials (to be discussed in more detail in Section 28.2). A beam of alpha particles from a radioactive source was directed at a thin gold foil "target" and the scattering angles of the particles were observed (● Fig. 28.1). Such experiments were designed to investigate the distribution of mass and electric charge in atoms.

An alpha particle is over 7000 times more massive than an electron. Thus, the Thomson model would predict only small deflections, the result of collisions with the light electrons as an alpha particle passed through a large gold atom (● Fig. 28.2). Surprisingly, however, Rutherford and his colleagues observed appreciable scattering angles. In some instances (about 1 in 8000), the alpha particles were actually found to be *back-scattered*, that is, scattered through angles greater than 90° (● Fig. 28.3).

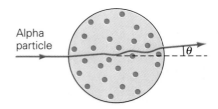

Alpha particle

θ

● **FIGURE 28.2 The plum pudding model** With Thomson's plum pudding model of the atom, the alpha particles would be expected to be only slightly deflected by collisions with the electrons in the atom. The experimental results were different.

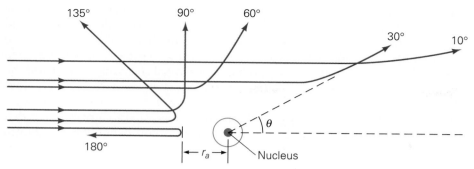

FIGURE 28.3 Rutherford scattering A compact, dense atomic nucleus with a positive charge accounts for the observed scattering. An alpha particle in a head-on collision with the nucleus would be scattered directly backwards ($\theta = 180°$) after coming within a distance r_a of the nucleus.

Calculations showed that the probability of this taking place with a Thomson model of the atom and such a thin foil (10^{-4} cm) was minuscule—certainly much less than 1 in 8000. As Rutherford described the back-scattering, "It was almost as incredible as if you had fired a 15-inch shell at a piece of tissue paper and it came back and hit you."

The experimental results led Rutherford to the concept of an atomic nucleus:

> *On consideration, I realized that this scattering backward must be the result of a single collision, and when I made calculations I saw that it was impossible to get anything of that order of magnitude unless you took a system in which the greater part of the mass of the atom was concentrated in a minute nucleus. It was then that I had the idea of an atom with a minute massive center carrying a charge.*

If all of the positive charge of a target atom were concentrated in a very small region in the atom, then an alpha particle coming very close to this region would experience a large deflecting force. The mass of this nucleus of charge would be larger than that of the alpha particle, since most of the atomic mass is associated with the positive charge. Thus, back-scattering would be possible.

A simple calculation can give an idea of the size of the atomic nucleus. For a head-on collision, an alpha particle would attain its distance of closest approach to the nucleus (r_a in Fig. 28.3). That is, the alpha particle approaching the nucleus would be stopped at a distance r_a by the repulsive Coulomb force and would be accelerated back along its original path. Assuming a spherical or point charge distribution, the electric potential is $V = kZe/r$, where Z is the *atomic number* or the number of protons in the nucleus and Ze is the total charge. The work done by the Coulomb force in stopping the alpha particle is $W = qV = 2eV$, since the alpha particle has a double positive electronic charge. This is equal to the initial kinetic energy of the alpha particle, that is,

A review of electrostatics may be necessary.

$$\tfrac{1}{2}mv^2 = \frac{k(2e)Ze}{r_a}$$

and

$$r_a = \frac{4\,k\,Ze^2}{mv^2} \qquad (28.1)$$

By using the known value of the energy of the alpha particles from the particular source and other known values, r_a is found to be about 10^{-12} cm. This value is an *upper limit* for the nuclear radius, since the alpha particle does not reach the nucleus itself.

Although the nuclear model of the atom is useful, the nucleus is much more than simply a region of positive charge. For one thing, we now know that the atomic nucleus is composed of two types of particles—protons and neutrons—which are collectively referred to as **nucleons**. The nucleus of the common hydrogen atom is a proton. Rutherford suggested that the hydrogen nucleus be give the name *proton* (from a Greek word meaning "first") after he became convinced that there was no positively charged particle in an atom lighter than the hydrogen nucleus. As you know from an earlier chapter, a neutron is an electrically neutral particle with a mass slightly greater than that of a proton. The existence of the neutron was not experimentally verified until 1932.

The Nuclear Force

Review Example 15.4, Section 15.3

Of the forces in the nucleus, we know there is an attractive gravitational force between nucleons (protons and neutrons). But in Chapter 15, we saw that the magnitude of the gravitational force is negligible in comparison with the mutually repulsive electrical force between positively charged protons. Taking only these repulsive forces into account, one might expect the nucleus to fly apart. Yet the nuclei of most atoms are stable, so there must be an additional force that holds the nucleus together. This strongly attractive force is called the **strong nuclear force**, usually referred to simply as the *nuclear force*.

The exact mathematical expression for the nuclear force is not known, and approximations indicate that it is extremely complicated. However, some general features of this force are as follows:

There is also a weak nuclear force, which is much weaker than the strong nuclear force and is involved in certain types of radioactive decay (more on this force in Chapter 29).

- The nuclear force is strongly attractive and much larger in relative magnitude than the electrostatic and gravitational forces.

- The nuclear force is very short-ranged; that is, a nucleon interacts only with its nearest neighbors.

- The nuclear force acts between any two nucleons within a short range, that is, between two protons, a proton and a neutron, or two neutrons.

Thus, nuclear protons in close proximity repel each other by the electric force but attract each other (and nearby neutrons) by the strong nuclear force. Having no electric charge, neutrons are attracted to nearby protons and neutrons only by the nuclear force (gravitational force being negligible).

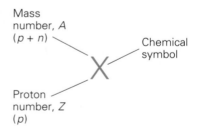

Nuclear Notation

To describe the nuclei of different atoms, it is convenient to use the notation illustrated in Fig. 28.4. The chemical symbol of the element is used with subscripts and a superscript. The left subscript is the **atomic number** (Z), which indicates the number of protons in the nucleus. For an atom, the atomic number is the number of electrons in an atom. An alternative (and more descriptive) term, which we will generally use in this book, is **proton number**. For electrically neutral atoms, this is also the number of orbital electrons.

Recall that the number of protons in the nucleus of an atom defines the species of the atom—that is, the element to which the atom belongs. In the example in Fig. 28.4, the proton number $Z = 6$ indicates that this is a carbon

FIGURE 28.4 Nuclear notation A nucleus is represented by the chemical symbol of the particular element with the mass number A (number of protons and neutrons) as a left superscript and the proton (atomic) number Z as a left subscript. This nuclear notation is shown for a nucleus of the most common isotope of carbon.

nucleus. The proton number defines which chemical symbol is used. Electrons can be removed (or added) to an atom to form an ion, but this does not change its species. For example, a nitrogen atom with an electron removed, N^+, is still nitrogen—a nitrogen ion. The number of electrons can vary by ionization, but the proton number cannot vary without producing an atom of a different element.

The left superscript on the chemical symbol is called the **mass number** (A) and the total number of protons and neutrons in the nucleus. Since protons and neutrons have approximately equal masses, the mass numbers of nuclei give a relative comparison of nuclear masses. For the example in Fig. 28.4, the carbon nucleus, the mass number is $A = 12$, since there are 6 protons and 6 neutrons. The number of neutrons, called the **neutron number** (N), is sometimes indicated by a subscript on the right side, but this is usually omitted, since it can easily be found from the other numbers: $N = A - Z$.

The atoms of an element, all of which have the same number of protons in their nuclei, may have different numbers of neutrons. For example, nuclei of different carbon atoms ($Z = 6$) may contain either 6, 7, or 8 neutrons and in nuclear notation would be written

$$^{12}_{6}C \qquad ^{13}_{6}C \qquad ^{14}_{6}C$$

Atoms whose nuclei have the same number of protons but different numbers of neutrons are called **isotopes**, in this case isotopes of carbon.

Isotopes are like members of a family; they all have the same Z number and the same surname, but the members of the family are distinct on the basis of the number of neutrons in their nuclei. However, it should be noted that different isotopes of the same family have the same electronic structure and thus the same chemical properties. Isotopes are referred to by their mass numbers, for example, the previous isotopes of carbon are called carbon-12, carbon-13, and carbon-14. There are also other isotopes of carbon, ^{11}C, ^{15}C, and ^{16}C. A particular nuclear species, or isotope of any element is called a **nuclide**. Thus, we have six nuclides, or isotopes, of carbon.

Another family of isotopes is that of hydrogen, which has three isotopes or nuclides: ^1H, ^2H, ^3H. These isotopes are not generally referred to as hydrogen-1, and so on, but are given special names. ^1H is called ordinary hydrogen or, simply hydrogen. ^2H is called *deuterium*. Deuterium, sometimes known as heavy hydrogen, can combine with oxygen to form heavy water (commonly written D_2O). The third isotope of hydrogen, ^3H, is called *tritium*.

Mass number—the total number of protons and neutrons in a nucleus

Isotopes of an element—same number of protons, different numbers of neutrons

Nuclide—a particular nuclear species or isotope

28.2 ▪ Radioactivity

Most elements have stable isotopes. It is the atoms of these stable nuclides with which we are most familiar in the environment. However, the nuclei of some isotopes are unstable and disintegrate spontaneously (decay) of their own accord, emitting energetic particles. Such isotopes are said to be *radioactive* or to exhibit **radioactivity**. For example, the tritium isotope of hydrogen just mentioned is radioactive. The prefix "radio" refers to the emitted nuclear radiation, which in the modern context may be a particle or a wave. A radioactive isotope is "active" when it is emitting radiation. Of the nearly 1200 known

FIGURE 28.5 The Curies Marie Sklodowska Curie (1867–1934) was born in Poland and studied in France. There she met and married Pierre Curie (1859–1906), who was a physicist well known for his work on crystals and magnetism. In 1903, Madame Curie (as she is commonly known) and Pierre shared the Nobel Prize in physics with Henri Becquerel for their work on radioactivity. Mme. Curie was also awarded the Nobel prize in chemistry in 1911 for the discovery of radium and the study of its properties. The Curies are shown here on the cover of a 1904 magazine.

Compare and contrast the three types of radioactive decay.

Alpha particle—a helium nucleus

unstable nuclides, only a small number occur naturally. The others are produced artificially (Chapter 29).

We know that radioactivity is completely unaffected by normal physical or chemical processes, such as heat and pressure or chemical reactions. Since chemical processes involve the outer electrons of atoms, the source of radioactivity must lie deeper in the atom, that is, in the nucleus. This instability cannot be explained directly by a simple imbalance of attractive and repulsive forces in the nucleus because the nuclear disintegrations of a given isotope occur as a function of time at a fixed rate. Classically, one would expect identical nuclei to do the same thing at the same time. Therefore, radioactive decay processes suggest quantum mechanical probability effects.

The discovery of radioactivity is credited to the French scientist Henri Becquerel. In 1896, while studying the fluorescence of a uranium compound, Becquerel discovered that a photographic plate in the vicinity of a sample had been darkened when the compound had not been activated by exposure to light and was not fluorescing. Apparently, the darkening was caused by some new type of radiation being emitted from the compound. In 1893, Pierre and Marie Curie announced the discovery of two radioactive elements, radium and polonium, which they had isolated from uranium pitchblende ore (Fig. 28.5).

Experiment easily shows the radiation emitted by radioactive isotopes to be of three different kinds. When a radioisotope is placed in a chamber so that the emitted radiation passes through a magnetic field to a photographic plate (Fig. 28.6), the radiations expose the plate, producing identifying spots. The positions of the spots show that some isotopes emit radiation that is deflected to the left; some, radiation that is deflected to the right; and some, radiation that is undeflected.

These spots are characteristic of what is known as alpha, beta, and gamma radiations. From the deflections of two of the types of radiation in the magnetic field, it is evident that positively charged particles are emitted from nuclei undergoing alpha decay and that negatively charged particles are emitted in beta decay. Also, the degree of deflection shows that alpha particles must be more massive than beta particles. The undeflected gamma radiation (gamma rays) is electrically neutral.

Investigations of the different radiations reveal that

- **Alpha particles** are doubly charged (2 +) particles containing two protons and two neutrons. They are identical to the nucleus of the helium atom ($^{4}_{2}He$).
- **Beta particles** are electrons.*
- **Gamma rays** are particles, or quanta, of electromagnetic energy.

For a few radioactive elements, two spots are found on the film, indicating that some elements decay by two different modes.

Let's now look at what happens to a decaying nucleus in each of these three decay modes of radioactivity.

Alpha Decay

When an alpha particle is ejected from a radioactive nucleus, the nucleus loses two protons and two neutrons, so the mass number (A) is decreased by four,

---

*In the most common type of beta decay, there are different types, as will be discussed shortly.

that is, $\Delta A = -4$, and the proton number (Z) is decreased by two, $\Delta Z = -2$. Since the parent nucleus loses two protons, the resulting daughter nucleus must then be the nucleus of another element as defined by the proton number. (The original and resulting nuclei are commonly referred to as the parent and daughter nuclei, respectively.) Thus, the decay process is one of nuclear *transmutation* in which the nuclei of one element change into the nuclei of a lighter element.

An example of an isotope or nuclide that undergoes **alpha decay** is polonium-214. The decay process is represented in the form of a nuclear equation, similar to a chemical equation.

$$\underset{\text{polonium}}{^{214}_{84}\text{Po}} \rightarrow \underset{\text{lead}}{^{210}_{82}\text{Pb}} + \underset{\substack{\text{alpha particle}\\ \text{(helium nucleus)}}}{^{4}_{2}\text{He}}$$

Note that the totals of the mass numbers and the proton numbers are equal on each side of the equation: ($214 = 210 + 4$) and ($84 = 82 + 2$), respectively. This reflects the fact that *two conservation laws apply to all nuclear processes.* The first is the **conservation of nucleons**. That is, the total number of nucleons (A) remains constant in any process. The second is the familiar **conservation of charge**, when applied to nuclear reactions.

EXAMPLE 28.1 ■ Alpha Decay

A $^{238}_{92}\text{U}$ nucleus undergoes alpha decay. What is the resulting daughter nucleus?

Solution. Since $\Delta Z = -2$ for alpha decay, the uranium-238 (^{238}U) nucleus loses two protons, and the daughter nucleus has a proton number of $Z = 92 - 2 = 90$, which is the proton number of thorium (see the periodic table, Fig. 27.9). The equation for the process is

$$^{238}_{92}\text{U} \rightarrow {}^{234}_{90}\text{Th} + {}^{4}_{2}\text{He}$$

The energies of the alpha particles from radioactive sources are on the order of a few mega-electron volts (MeV). For example, the energy of the alpha particle emitted from the decay of ^{214}Po is about 7.7 MeV, and that from ^{238}U decay is about 4.2 MeV. Alpha particles from such radioactive sources were used in the scattering experiments that led to the Rutherford nuclear model (Section 28.1). Scattering experiments give some idea about the size of the nucleus, as you learned in the preceding section. They also shed light on its structure and the forces within it.

Outside the nucleus, the repulsive electric force increases as an alpha particle gets close to the nucleus. Inside the nucleus, however, the strongly attractive nuclear force dominates. These conditions are depicted in ● Fig. 28.7 in a graph of the electrical potential energy, U, as a function of r, the distance from the center of the nucleus. Outside the nucleus, U is proportional to, or "falls off," as $1/r$. Within the nucleus, U is opposite in sign and has the form of a negative potential well because of the strongly attractive nuclear force. This could be the representation of a uranium-238 nucleus.

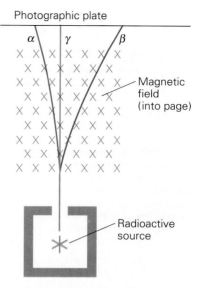

● **FIGURE 28.6 Nuclear radiation** The radiations from radioactive sources can be distinguished by passing them through a magnetic field. Alpha and beta particles are deflected. Applying the right-hand rule for the equation $F = qvB$, we find that alpha particles are positively charged and beta particles are negatively charged. The radii of curvature (not to scale) allow the particles to be distinguished by mass. Gamma rays are not deflected, and so are uncharged, being in fact quanta of electromagnetic energy.

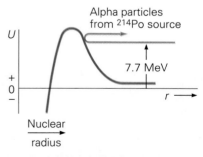

● **FIGURE 28.7 Nuclear potential barrier** Alpha particles from radioactive polonium with energies of 7.7 MeV do not have enough energy to overcome the nuclear potential barrier of ^{238}U and are scattered.

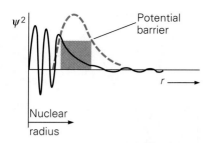

FIGURE 28.8 Tunneling effect, or barrier penetration The probability (ψ^2) of finding an alpha particle tails off through the nuclear potential barrier, and there is a finite probability of finding the alpha particle outside the nucleus.

Recall the discussion of the uncertainty principle in Section 27.4

Beta particle—an electron

Consider alpha particles from a ^{214}Po source incident on a thin foil of ^{238}U. As illustrated in the figure, Rutherford scattering takes place. We say that this results from a potential barrier that the incident ^{214}Po alpha particles do not have enough energy to overcome. Instead, they are scattered away. If an incident particle did have enough energy to overcome, or cross, this Coulomb barrier, it would enter the nucleus. In this case, a nuclear reaction would result (to be discussed in Chapter 29).

However, the ^{238}U nucleus itself undergoes alpha decay, emitting an alpha particle with an energy of 4.4 MeV. This fact appears to contradict the scattering experiment. How can these lower-energy alpha particles cross a potential barrier that higher-energy incident alpha particles cannot? Classically, this is impossible, since it violates the conservation of energy, and the inside of the potential barrier is referred to as a classically forbidden region. However, quantum mechanics offers an explanation.

Quantum mechanics predicts a finite probability of finding a particle in a classically forbidden region as a result of a wave function existing for this region. If the alpha particle exists as an entity in the nucleus, its wave function may tail off in the barrier region and make it through to the outside (Fig. 28.8). There would then be a finite probability (ψ^2, Section 27.2) of finding the alpha particle outside the nucleus. This is called **tunneling**, or **barrier penetration**, since the alpha particle seems to tunnel through the barrier.

You might point out that there is still a violation of the conservation of energy. However, the uncertainty principle tells us that energy conservation can be violated by an amount ΔE for a time Δt. Quantum mechanics thus allows the conservation of energy to be violated for brief periods, which may be long enough for the alpha particle to tunnel through the barrier. The probability of finding an alpha particle outside the ^{238}U nucleus is extremely small, but this is reflected in the extremely slow decay rate of ^{238}U. The height and width of the potential barrier determine the time or probability an alpha particle has to escape from the nucleus. Thus, the barrier would also control the decay rate (which will be considered in the next section).

Beta Decay

The emission of an electron (a beta particle) in a nuclear decay process might seem contradictory to the proton-neutron model of the nucleus. It should be noted that the electron emitted in beta decay is not an orbital electron, but comes from the nucleus. In fact, the process of **beta decay** indicates that the electron is created in the nucleus itself. For example, the equation for carbon-14 beta decay is

$$^{14}_{6}\text{C} \rightarrow ^{14}_{7}\text{N} + ^{0}_{-1}\text{e}$$
$$\text{carbon} \quad \text{nitrogen} \quad \text{beta particle}$$
$$\text{(electron)}$$

The parent carbon nucleus has six protons and eight neutrons, whereas the daughter nitrogen nucleus has seven protons and seven neutrons. Notice that the notation for the electron specifies a mass number of zero and a charge number of -1. This is in keeping with the conservation of nucleons and charge.

In this and similar beta decay processes, the neutron number of the parent nucleus decreases by one ($\Delta N = -1$), and the proton number of the daughter nucleus increases by one ($\Delta Z = +1$). The mass number is unchanged. This indicates that a neutron within the nucleus decays into a proton and an electron.

$$\,_{0}^{1}n \rightarrow \,_{1}^{1}p + \,_{-1}^{0}e$$

neutron   proton   electron

In nuclear notation, the neutron has a mass number of 1 and a proton number of zero. (Why?)

There is another elementary particle, called a neutrino, emitted in beta decay, but for the sake of simplicity, it is not shown in the nuclear equation. Its place in beta decay will be discussed in Chapter 29.

There are actually three modes of beta decay: β^-, β^+, and electron capture. The β^- decay involves the emission of an electron as above. Isotopes that decay by this means have more neutrons than protons. However, unstable isotopes that undergo **β^+ decay**, which involves the emission of a positron ($\,_{+1}^{0}e$), have more protons than neutrons. An example of β^+ decay is

$$\,_{8}^{15}O_{7} \rightarrow \,_{7}^{15}N_{8} + \,_{+1}^{0}e$$

oxygen   nitrogen   positron

As in β^- decay, the mass numbers of the parent and daughter nuclei do not change, but the proton number of the daughter nucleus in this case is one less than that of the parent nucleus. This implies that a proton disintegrates in the process into a neutron and a positron.

$$\,_{1}^{1}p \rightarrow \,_{0}^{1}n + \,_{+1}^{0}e$$

The third related process, **electron capture**, involves the absorption of one of the orbital electrons by a nucleus. An example of electron capture is

$$\,_{-1}^{0}e + \,_{4}^{7}Be \rightarrow \,_{3}^{7}Li$$

electron   beryllium   lithium

As in β^+ decay, a proton changes into a neutron, but no beta particle is emitted in the electron capture process. An electron in the innermost shell, called the K shell from spectroscopic notation, is usually captured, an event referred to as K-capture. Since there is no charged particle emission, the process is detected by observing the emission of characteristic X-rays produced when an electron from an outer shell moves into a K-shell vacancy. Because X-ray emission must take place *after* the K-capture, the X-rays are characteristic of the daughter nucleus, not the parent.

Electron shells are discussed in Section 27.3

Gamma Decay

In **gamma decay**, the nucleus emits a gamma (γ) ray, or a "particle" of electromagnetic energy. The emission of a gamma ray by a nucleus in an excited state is analogous to the emission of a photon by an excited atom. This may come about because of an energetic collision with another particle or, more commonly, because the nucleus is in an excited state after a previous radioactive decay.

Gamma ray—a quantum of electromagnetic energy

Nuclei are thought of as having energy levels like atomic electrons. However, the nuclear energy levels are much farther apart than those of an atom. The nuclear energy levels are on the order of kilo-electron volts (keV) and mega-electron volts (MeV) apart, compared to a few electron volts of difference between atomic energy levels. As a result, gamma rays are very energetic, having frequencies greater than those of X-rays.

$$\,_{28}^{61}Ni^* \rightarrow \,_{28}^{61}Ni + \gamma$$

nickel   nickel   gamma ray

The asterisk indicates that the ^{61}Ni nucleus is in an excited state. The de-excitation process results in the emission of a gamma ray. Note that *in the gamma decay process, the mass and proton numbers do not change.* The daughter nucleus in this case is simply the parent nucleus with less energy.

Radiation Penetration

The absorption or degree of penetration of nuclear radiations is an important consideration in applications such as radioisotope treatment of cancer and nuclear shielding (for example, around a nuclear reactor). Also, as we will see in a later section, the absorption of nuclear radiation is used to monitor and automatically control the thickness of metal and plastic sheets and films in fabrication processes.

The three types of radiation (alpha, beta, and gamma) are absorbed quite differently. The electrically charged alpha and beta particles interact with the atoms of a material and may produce ionizations along their paths. The greater the charge and the slower the particle, the greater is the energy transfer and ionization along the path, which determines the degree of penetration. Also, the penetration depends on the density of the material.

Alpha particles are doubly charged, have a larger mass, and generally move more slowly. A few centimeters of air or a sheet of paper will usually completely stop or absorb them.

Beta particles are singly charged and can travel a few meters in air or a few millimeters in aluminum before being stopped.

Gamma rays are uncharged and are therefore more penetrating. A significant portion of a beam of high-energy gamma rays can penetrate a centimeter or more of a dense material such as lead. (Lead is commonly used as shielding for harmful X-rays, which are slightly less energetic than gamma rays.)

Radiation passing through matter can do considerable damage. In biological tissue, the radiation damage is chiefly due to ionizations in living cells (Section 28.5). We are exposed to normal background radiation from radioisotopes in the environment and cosmic radiation (radiation from outer space). The intensity of most such radiation is too low to be harmful. Another recent radiation concern involves airplanes. In the discussion of telescopes using nonvisible radiation, it was mentioned that X-rays and gamma rays from distant sources do not penetrate the Earth's atmosphere. However, concern has been expressed that flight crews who are frequently on board high-flying jet aircraft may receive excessive exposure to such radiations.

Metals and other structural materials may become brittle and lose their strength when exposed to strong radiation, such as that in nuclear reactors used for electrical generation (Chapter 29) and the intense cosmic radiation that space vehicles are exposed to.

As pointed out earlier, of the nearly 1200 known unstable nuclides, only a small number occur naturally. Most of the radioactive nuclides found in nature occur as products of the decay series of heavy nuclei, that is, a series of continual radioactive decay into lighter elements. The ^{238}U decay series is given in ● Fig. 28.9. It stops when the stable isotope ^{206}Pb is reached. Note that some nuclides in the series decay by two modes. As can be seen, radon (^{222}Rn) is part of this decay series. This radioactive gas has received a great deal of attention because it can accumulate in dangerous amounts in poorly ventilated buildings.

Recall the Insight feature in Chapter 24

See also Figs. 28.23 (neptunium-237) and 28.24 (plutonium-239).

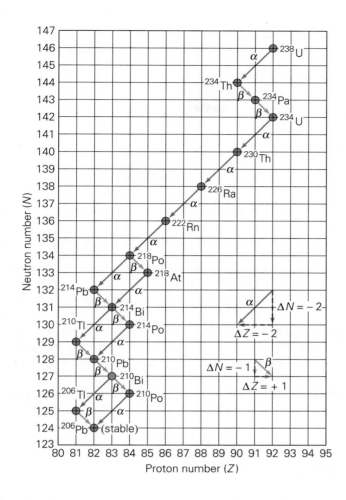

Figure 28.9

FIGURE 28.9 Decay series for uranium-238 On this plot of *N* versus *Z*, a diagonal transition from right to left is an alpha decay process and a diagonal transition from left to right is a beta decay process. The decay series continues until a stable nucleus is reached.

28.3 ■ Decay Rate and Half-Life

The nuclei of a sample of a radioactive isotope do not decay all at once, but do so randomly at a characteristic rate that is unaffected by any external stimulus. No one can tell exactly when particular nuclei will decay. All that can be determined is how many nuclei in a sample will decay over a period of time.

The **activity** of a sample of radioactive isotope is expressed in terms of the number (ΔN) of disintegrations, or decays, per time, that is, $\Delta N/\Delta t$. For a given amount of material, the activity decreases with time, since fewer and fewer radioactive nuclei remain. Each isotope has its own characteristic rate of decrease. Since the activity is proportional to the number of nuclei (N) present,

$$\frac{\Delta N}{\Delta t} \propto N$$

This can be written in equation form with a constant of proportionality.

$$\frac{\Delta N}{\Delta t} = -\lambda N \qquad (28.2)$$

859

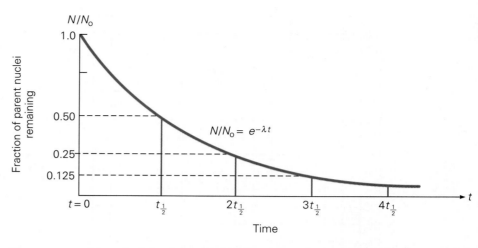

Figure 28.10

● **FIGURE 28.10 Radioactive decay versus time** When the fraction of the remaining parent nuclei (N/N_o) in a radioactive sample is plotted as a function of time, an exponential decay curve is obtained, the shape or steepness of which depends on the decay constant λ.

The constant λ is called the **decay constant** and is different for different isotopes. The greater the decay constant, the greater the rate of decay and the more radioactive the isotope. The minus sign in the equation indicates that N is decreasing.

It turns out that the number of parent nuclei N decreases with time as illustrated in ● Fig. 28.10. The curve and N are said to decay *exponentially* since the graph follows an exponential function $e^{-\lambda t}$ with time, where λ is the decay constant.*

The number N of the undecayed parent nuclei at a particular time is given by

$$N = N_o e^{-\lambda t} \qquad (28.3)$$

where N_o is the initial number of nuclei ($t = 0$). The activity decreases proportionally.

Point out how to determine the half-life for Figure 28.10.

The decay rate of an isotope is commonly expressed in terms of its half-life rather than the decay constant. The **half-life** ($t_{\frac{1}{2}}$) is defined as the time it takes for half of the radioactive nuclei in a sample to decay. This is the time corresponding to $N_o/2$ parent nuclei in Fig. 28.10, that is, $N/N_o = \frac{1}{2}$. The activity also decreases to the fractional amount N/N_o in a time t, since it is proportional to the number of nuclei present. That is, the number of decays per second decreases by half. This is what is usually measured to determine the half-life, rather than counting nuclei.

For example, as may be seen from the graph in ● Fig. 28.11, the half-life of strontium-90 (^{90}Sr) is 28 years. Another way of looking at what occurs in a half-life is to consider the amount, or mass, of the parent material. As illustrated in Fig. 28.11, if one had 100 micrograms (μg) of ^{90}Sr initially, only 50 μg would remain at the end of 28 years. The other 50 μg would have beta decayed by the reaction

$$^{90}_{38}\text{Sr} \rightarrow {}^{90}_{39}\text{Y} + {}^{0}_{-1}\text{e}$$
$$\text{strontium} \quad \text{yttrium} \quad \text{electron}$$

and the sample would contain both strontium and yttrium nuclei. After another 28 years, another half of the strontium nuclei would decay, leaving only 25 μg, and so on.

*The value $e = 2.718$ is the base of natural logarithms.

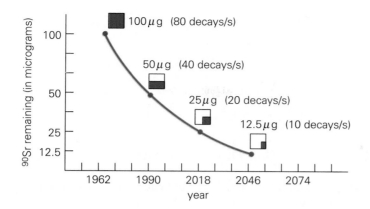

● **FIGURE 28.11 Radioactive decay and amount of isotope** As illustrated here for strontium-90, after each half-life ($t_\frac{1}{2}$ = 28 years), only half of the amount of ^{90}Sr present at the start of that time period remains, and the activity (decays/s) has decreased by one half. The other half of the sample has decayed into ^{90}Y via beta decay.

Figure 28.11

The half-lives of radioactive isotopes vary greatly, as may be seen from Table 28.1. Isotopes with very short half-lives are generally created in nuclear reactions, which will be considered in the next chapter. If these isotopes did exist when the Earth was formed, they would have long since decayed away. In fact, the elements technetium (Tc) and promethium (Pm) are not found on Earth (but can be produced in laboratories). On the other hand, the half-life of the naturally occurring ^{238}U isotope is 4.5 billion years, which is close to the estimated age of the Earth. Thus, we would expect about half of the original ^{238}U present when the Earth was formed to exist today.

The longer the half-life of an isotope, the more slowly it decays and the smaller the decay constant. The half-life and the decay constant have an inverse relationship ($t_\frac{1}{2} \propto 1/\lambda$). This can be seen from Eq. 28.3. The half-life is the time it takes for the number of nuclei present or the activity to decrease by half, that is, $N = N_0/2$. Thus,

$$\frac{N}{N_0} = e^{-\lambda t_\frac{1}{2}} = \frac{1}{2}$$

TABLE 28.1 ■ The Half-Lives of Some Radioactive Isotopes

Nuclide	Decay Mode	Half-Life
Beryllium-8 ($^{8}_{4}$Be)	α	1×10^{-16} s
Polonium-213 ($^{213}_{84}$Po)	α	4×10^{-6} s
Oxygen-19 ($^{19}_{8}$O)	β^-	27 s
Fluorine-17 ($^{17}_{9}$F)	β^+	66 s
Polonium-218 ($^{218}_{84}$Po)	α, β	3.05 min
Technetium-104 ($^{104}_{43}$Tc)	β^-	18 min
Krypton-76 ($^{76}_{36}$Kr)	Electron capture	14.8 h
Magnesium-28 ($^{28}_{12}$Mg)	β^-	21 h
Iodine-123 ($^{123}_{53}$I)	α	13.3 h
Radon-222 ($^{222}_{86}$Rn)	γ	3.82 days
Cobalt-60 ($^{60}_{27}$Co)	β^-	5.3 years
Strontium-90 ($^{90}_{38}$Sr)	β^-	28 years
Radium-226 ($^{226}_{88}$Ra)	α	1600 years
Carbon-14 ($^{14}_{6}$C)	β^-	5730 years
Plutonium-239 ($^{239}_{94}$Pu)	α	2.4×10^4 years
Uranium-238 ($^{238}_{92}$U)	α	4.5×10^9 years
Rubidium-87 ($^{87}_{37}$Rb)	β^-	4.7×10^{10} years

But

$$e^{-0.693} = \frac{1}{2}$$

so by comparison,

$$t_{\frac{1}{2}} = \frac{0.693}{\lambda} \qquad (28.4)$$

EXAMPLE 28.2 ■ Half-Life and Activity

The half-life of iodine-131 used in medical thyroid treatments is 8.0 days. After a certain time, an amount of ^{131}I containing 4.0×10^{22} nuclei accumulates in a patient's thyroid gland. (a) What will be the observed activity in 24 hours? (b) How many of the ^{131}I nuclei remain at this time?

Solution.

Given: $t_{\frac{1}{2}} = 8.0$ days (86,400 s/day)   *Find:* (a) $\dfrac{\Delta N}{\Delta t}$ (activity, decays/s)
$\qquad\qquad = 6.9 \times 10^5$ s
$\qquad N_o = 4.0 \times 10^{22}$ nuclei   (b) N (number of
$\qquad \Delta t = 24$ h $= 8.64 \times 10^4$ s $\qquad\qquad$ nuclei)

(a) First, Eq. 28.4 is used to find the decay constant with $t_{\frac{1}{2}}$ in seconds:

$$\lambda = \frac{0.693}{t_{\frac{1}{2}}} = \frac{0.693}{6.9 \times 10^5 \text{ s}} = 1.0 \times 10^{-6} \text{ s}^{-1}$$

Then, using Eq. 28.2, we have

$$\Delta N / \Delta t = -\lambda N_o = -(1.0 \times 10^{-6} \text{ s}^{-1})(4.0 \times 10^{22})$$
$$= -4.0 \times 10^{16} \text{ decays/s}$$

where the minus indicates that the activity is decreasing.
(b) The actual number of parent nuclei present can be found from Eq. 28.3. With $t = 1$ day and $\lambda = 0.693/t_{\frac{1}{2}} = 0.693/8.0$ day $= 0.087$ day^{-1}, we have

$$N = N_o e^{-\lambda t} = (4.0 \times 10^{22} \text{ nuclei})e^{-(0.087 \text{ day}^{-1})(1 \text{ day})}$$
$$= (4.0 \times 10^{22} \text{ nuclei})(0.917) = 3.7 \times 10^{22} \text{ nuclei}$$

Units of radioactivity

The strength of a radioactive sample or source may be specified at a given time by its activity. A common unit of radioactivity is named in honor of Pierre and Marie Curie. One **curie (Ci)** is defined as

$$1 \text{ Ci} \equiv 3.70 \times 10^{10} \text{ decays/s}$$

This is based on the activity of one gram of radium.

The curie is the traditional unit for expressing radioactivity. However, the proper SI unit is the **becquerel (Bq)**, which is simply defined as

$$1 \text{ Bq} \equiv 1 \text{ decays/s}$$

Therefore,

$$1 \text{ Ci} = 3.70 \times 10^{10} \text{ Bq}$$

Even with the emphasis on the SI, radioactive sources are commonly rated in curies. The curie is a relatively large unit, however, so the millicurie (mCi), the microcurie (μCi), and sometimes even smaller multiples are used.

EXAMPLE 28.3 ■ Source Strength

A ^{90}Sr beta source has a strength of 10.0 mCi. How many decays/s will be observed to take place at the end of 84 years, or 3 half-lives?

Solution.

Given: strength = 10.0 mCi   *Find:* $\dfrac{\Delta N}{\Delta t}$ (activity, decays/s)
          Δt = 84 years
          $t_{\frac{1}{2}}$ = 28 years (from Table 28.1)

After 3 half-lives, the activity will be only $\frac{1}{8}$ as great as ($\frac{1}{2} \times \frac{1}{2} \times \frac{1}{2} = \frac{1}{8}$), and the strength of the source will then be

$$\frac{\Delta N}{\Delta t} = 10.0 \text{ mCi} \times \frac{1}{8} = 1.25 \text{ mCi} = 1.25 \times 10^{-3} \text{ Ci}$$

and

$$\frac{\Delta N}{\Delta t} = (1.25 \times 10^{-3} \text{ Ci}) \left(3.70 \times 10^{10} \frac{\text{decays/s}}{\text{Ci}} \right)$$

$$= 4.63 \times 10^7 \text{ decays/s} \ (= 4.63 \times 10^7 \text{ Bq})$$

Radioactive Dating

Because of their constant decay rates, radioactive isotopes can be used as nuclear clocks. As we have seen, the half-life of a radioactive isotope may be used to project how much of a given amount of material will exist in the future. Similarly, by using the half-life to project backward in time, scientists can determine the ages of objects containing radioactive isotopes. As you might surmise, some idea of the initial compositon or intial amount of an isotope must be known.

To illustrate the principle of radioactive dating, let's take a look at how it is done with ^{14}C. **Carbon-14 dating** is used on materials that were once part of living things, or the remnants of objects made from or containing such materials (such as wood, bone, leather, or parchment). The process depends on the fact that living things—plants and animals (including yourself)—contain a known amount of radioactive ^{14}C. The concentration of carbon-14 is very small, about one ^{14}C atom for 7.2×10^{11} atoms of ordinary ^{12}C. Even so, this concentration cannot be due to an abundance of carbon-14 present when the Earth was formed, since a half-life of $t_{\frac{1}{2}}$ = 5730 years for ^{14}C is brief in comparison to the estimated age of the Earth (over 4 billion years).

The observed concentration of ^{14}C is accounted for by its continuous production in the upper atmosphere. Cosmic rays from outer space cause reactions that produce neutrons (Fig. 28.12). The neutrons react with the nuclei of the nitrogen atoms of the air to produce ^{14}C by the reaction

$$^{14}_{7}\text{N} + ^{1}_{0}\text{n} \rightarrow ^{14}_{6}\text{C} + ^{1}_{1}\text{H}$$

Recall that the carbon-14 then decays by beta decay ($^{14}_{6}\text{C} \rightarrow ^{14}_{7}\text{N} + ^{0}_{-1}\text{e}$). Although the intensity of incident cosmic rays may not be constant, the concentration of ^{14}C in the atmosphere is relatively constant because of atmospheric mixing and the fixed decay rate.

FIGURE 28.12 Carbon-14 radioactive dating The diagram illustrates the formation of carbon-14 in the atmosphere and its entry into the biosphere.

Figure 28.12

The ^{14}C is oxidized into carbon dioxide (CO_2), so a small fraction of the CO_2 molecules of the air are radioactive. Plants take in this radioactive CO_2 by photosynthesis, and animals ingest the material produced. As a result, the concentration of carbon-14 in living organic matter is the same as the concentration in the atmosphere, 1 part in 7.2×10^{11}. However, once an organism dies, the ^{14}C is not replenished, and the concentration decreases with radioactive decay (with $t_{\frac{1}{2}} = 5730$ years). Measurement of the concentration of carbon-14 in dead matter relative to that in living things can then be used to establish when the organism died.

Since radioactivity is generally measured in terms of activity, we must first know the ^{14}C activity in a living organism. This is derived in the following example.

EXAMPLE 28.4 ■ **Natural Carbon-14 Activity**

Determine the average ^{14}C activity, in decays per minute per gram of natural carbon, found in living organisms if the concentration of carbon-14 is the same as that in the atmosphere.

Solution. The relative concentration of carbon-14 in living organisms is one part in 7.2×10^{11} (given in text) or

$$\frac{^{14}C}{^{12}C} = \frac{1}{7.2 \times 10^{11}} = 1.4 \times 10^{-12}$$

Carbon has an atomic mass of 12.0, so the number of nuclei (atoms) in 1 g of carbon may be found by using Avogadro's number N_A (6.02×10^{23}) and the number of moles ($n = N/N_A$, Chapter 10). With $n = 1\text{g}/(12 \text{ g/mole}) = 1/12$ mole,

$$N = nN_A = \frac{1}{12} \text{ mole } (6.02 \times 10^{23} \text{ nuclei/mole})$$

$$= 5.0 \times 10^{22} \text{ nuclei (per gram)}$$

The number of ^{14}C nuclei per gram is

$$N\left(\frac{^{14}C}{^{12}C}\right) = (5.0 \times 10^{22} \text{ nuclei/g})(1.4 \times 10^{-12})$$

$$= 7.0 \times 10^{10} \ ^{14}C \text{ nuclei/g}$$

With a half-life of $t_{\frac{1}{2}} = (5730 \text{ y})(5.26 \times 10^5 \text{ min/y}) = 3.01 \times 10^9$ min, the decay constant is

$$\lambda = \frac{0.693}{t_{\frac{1}{2}}} = \frac{0.693}{3.01 \times 10^9 \text{ min}} = 2.30 \times 10^{-10} \text{ min}^{-1}$$

Then the activity, or the number of decays per minute per gram of carbon, is given by

$$\Delta N/\Delta t = \lambda N = (2.30 \times 10^{-10} \text{ min}^{-1})(7.0 \times 10^{10}) = 16 \text{ decays/g-min}$$

Thus, if we determined that an artifact such as a bone or a piece of cloth had an activity of 8.0 counts/min per gram of carbon, the original living organism would have died one half-life or about 5700 years ago. This would put the date of the artifact near 3700 B.C.

EXAMPLE 28.5 ■ **Carbon-14 Dating**

An old bone is unearthed in an archeological dig. On analysis in the laboratory, it is found that there are on the average 2 beta emissions per minute per gram of carbon in the bone. What is the approximate age of the bone?

Solution.

> *Given:* Activity = 2 decays/g-min   *Find:* Age of bone

Assuming that the organism had the normal concentration of carbon-14 when it died, at the time of death the carbon-14 activity would be 16 decays/g/min (Example 28.4). Afterward, the decay rate would decrease by one half for each half-life:

$$16 \xrightarrow{t_{\frac{1}{2}}} 8 \xrightarrow{t_{\frac{1}{2}}} 4 \xrightarrow{t_{\frac{1}{2}}} 2 \text{ decays}$$

So with the observed activity, the carbon-14 in the bone would have gone through three half-lives, or the bone would be three half-lives old. Thus with $t_{\frac{1}{2}} = 5730$ y (Table 28.1),

$$\text{Age} = 3t_{\frac{1}{2}} = 3(5730 \text{ y}) = 17,190 \text{ y}$$

The limit of radioactive carbon dating depends on the ability to measure the very low activity after an appreciable number of half-lives. Current techniques give an age dating limit of about 40,000–50,000 years, depending on the size of the sample. After about 9 half-lives, the radioactivity of a carbon sample has decreased so much that it is barely measurable (about 2 counts per gram per *hour*).

Another radioactive dating process uses lead-206 (^{206}Pb) and uranium-238 (^{238}U). This dating method is used extensively in geology because of the long half-life of ^{238}U. Lead-206 is the stable end isotope of the ^{238}U decay series (see Fig. 28.9). If a rock sample contains both these uranium and lead isotopes, the lead is assumed to be a result of the decay of the uranium that was there when the rock was first formed. Thus, the ratio of ^{206}Pb/^{238}U is a measure of the geologic age of the rock.

28.4 ■ Nuclear Stability and Binding Energy

Now that we have considered some of the properties of unstable isotopes, we turn out attention to the more common stable isotopes. Stable isotopes exist naturally for all elements having proton numbers from 1 to 83, except for those with $Z = 43$ (technetium) and $Z = 61$ (promethium), as was pointed out earlier. The nuclear interactions that give rise to nuclear stability are extremely complicated. However, by looking at some of the general properties of stable nuclei, it is possible to obtain some qualitative and quantitative criteria for nuclear stability. This will give us some general rules for determining whether or not a particular nuclide would be expected to be stable or unstable.

Nucleon Populations

One of the easiest things to consider is the relative number of protons and neutrons in stable nuclei. Nuclear stability must be related in some way to the dominance of either the repulsive Coulomb force between protons or the attractive nuclear force between nucleons, and this force dominance depends on the relative numbers, or ratio, of protons and neutrons.

For stable nuclei of low mass numbers for about $A < 40$, the nucleon ratio is approximately 1. That is, the number of protons and the number of neutrons are equal or nearly equal. For example, $^{4}_{2}\text{He}$, $^{12}_{6}\text{C}$, $^{23}_{11}\text{Na}$, and $^{27}_{13}\text{Al}$ have the same or about the same number of protons as neutrons. For stable nuclei of higher mass numbers ($A > 40$), the number of neutrons exceeds the number of protons. The heavier the nuclei, the more the neutrons outnumber the protons.

This trend is illustrated in ● Fig. 28.13, which shows a plot of the neutron number (N) versus proton number (Z) for stable nuclei. Notice how the heavier stable nuclei lie above the 45° or $N = Z$ line. Examples of heavy stable nuclei include $^{62}_{28}\text{Ni}$, $^{114}_{50}\text{Sn}$, $^{208}_{82}\text{Pb}$, and $^{209}_{83}\text{Bi}$. (Bismuth is the heaviest element that has a stable isotope.)* The extra neutrons of such heavy stable nuclei presumably allow the attractive forces among the nucleons to be greater than the repulsive forces between the numerous protons.

Radioactive decay adjusts the proton and neutron numbers of an unstable isotope until a stable isotope on the neutron-proton stability curve in Fig. 28.13 is reached. Since alpha decay decreases the numbers of protons and neutrons by equal amounts, alpha decay alone would give nuclei with neutron populations that are larger than that of one of the stable isotopes on the curve. However, beta decay following alpha decay could lead to a stable combination, since the effect of beta decay is the loss of a neutron and the gain of a proton. Very heavy nuclei undergo a chain, or sequence, of alpha and beta decays until a stable nucleus is reached, as was illustrated in Fig. 28.9 for ^{238}U.

Pairing Effect

A fact that is not too evident from Fig. 28.13 is that many stable nuclei have even numbers of both protons and neutrons, and very few have odd numbers of both protons and neutrons. A survey of the stable isotopes (Table 28.2) shows that 168 stable nuclei have this even-even combination, 107 are even-odd or odd-even, and only four contain odd numbers of both protons and neutrons. These four are isotopes of the elements with the four lowest odd proton numbers: $^{2}_{1}\text{H}$, $^{6}_{3}\text{Li}$, $^{10}_{5}\text{B}$, and $^{14}_{7}\text{N}$.

The even and odd combinations seem to indicate that the protons and neutrons in stable nuclei above the very lightest elements tend to pair up (even-even). That is, two protons pair up and two neutrons pair up, but a proton and neutron are less likely to do so. Except for the very lightest elements, when there are odd number of nucleons (odd-odd), there is general instability. The

TABLE 28.2 ■ Pairing Effect of Stable Nuclei

Proton Number	Neutron Number	Number of Stable Nuclei
Even	Even	168
Even	Odd ⎫	
Odd	Even ⎬	107
Odd	Odd	4

*Bismuth-209 does undergo alpha decay, but with a half-life of 2×10^{18} y; so for all practical purposes, it is stable.

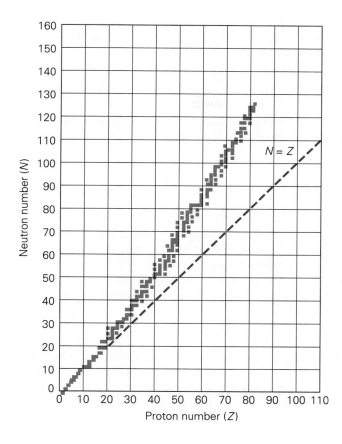

Figure 28.13

•**FIGURE 28.13 A plot of N versus Z for stable nuclei** For nuclei with mass numbers $A < 40$ ($Z < 20$ and $N < 20$), the number of protons and the number of neutrons are equal or nearly equal. For nuclei with $A > 40$, the number of neutron exceeds the number of protons and the nuclei lie above the $N = Z$ line.

instability is less if there is only one odd nucleon (odd-even or even-odd combination).

This so-called **pairing effect** gives a qualitative criterion for stability. For example, you would expect the sodium isotope $^{23}_{11}Na$ to probably be stable, but not $^{22}_{11}Na$. This is actually the case.

The **general criteria for nuclear stability** can be summarized as follows:

1. All isotopes with proton number greater than 83 ($Z > 83$) are unstable.

2. (a) Most even-even nuclei are stable.
   (b) Many odd-even or even-odd nuclei are stable.
   (c) Only four odd-odd nuclei are stable (^2_1H, ^6_3Li, $^{10}_5B$, and $^{14}_7N$).

3. (a) Stable nuclei with mass numbers less than 40 ($A < 40$) have approximately the same number of protons and neutrons.
   (b) Stable nuclei with mass numbers greater than 40 ($A > 40$) have more neutrons than protons.

State the general criteria for stability and give some examples.

EXAMPLE 28.6 ■ Nuclear Stability

Is the sulfur isotope $^{38}_{16}S$ likely to be stable?

Solution. Applying the general criteria:

1. *Satisfied.* Isotopes with $Z > 83$ are immediately known to be unstable. With $Z = 16$, this criterion is satisfied.

2. *Satisfied.* The isotope $^{38}_{16}S_{22}$ has an even-even nucleus and so has a good probability of being stable.

3. *Not satisfied.* $A < 40$, and $Z(= 16)$ and $N(= 22)$ are not approximately equal.

Therefore, the ^{38}S isotope is likely to be unstable. (The nucleus actually is unstable to beta decay.)

Binding Energy

An important quantitative aspect of nuclear stability is the binding energy of the nucleons. This can be calculated by considering the sums of nuclear and electron masses and of their corresponding energies.

Since the masses of nuclei are so small in relation to the standard kilogram mass unit, another standard, the *unified* **atomic mass unit (u)** is used to measure nuclear masses. A *neutral atom* of carbon-12 is defined as having an exact value of 12.000000 u. Thus,

$$1 \text{ u} = 1.6606 \times 10^{-27} \text{ kg}$$

The various atomic particles then have masses in atomic mass units as shown in Table 28.3. The listed energy equivalents reflect Einstein's $E = m_o c^2$ relationship.

Note that the proton and the hydrogen atom (^1_1H) are listed separately, having different masses. The difference is the mass of the atomic electron. We deal almost exclusively with the masses of neutral atoms (nucleons plus Z electrons) rather than strictly with nuclei, since the atomic masses are what can be measured. Keep this in mind. We will soon be interested in very small mass differences, and the electronic mass is significant. We must balance out the masses of the electrons when we compare masses in nuclear processes.

Given this information, we can look at the effect of mass-energy relationships on nuclear stability. For example, if you compare the mass of a helium nucleus to the total mass of nucleons that make it up, a significant inequity emerges: a neutral helium atom has a mass of 4.002603 u. (Atomic masses of various isotopes are given in Appendix V.) The total mass of two protons (with two electrons, or actually two ^1H atoms) and two neutrons is by addition,

$$2m(^1H) \, 1 = 2.015650 \text{ u}$$

$$2m_n = \underline{2.017330 \text{ u}}$$

$$4.032980 \text{ u}$$

TABLE 28.3 ■ Particle Masses and Energy Equivalents

| Particle | Mass | | Energy (MeV) |
	u	kg	
	1	1.6606×10^{-27}	931.50
Electron	0.000548	9.1095×10^{-31}	0.511
Proton	1.007276	1.67265×10^{-27}	938.28
^1_1H atom	1.007825	1.67356×10^{-27}	938.79
Neutron	1.008665	1.67500×10^{-27}	939.57

Nucleus
(a certain mass)

$+$

28.30 MeV

$\longrightarrow$

Separated nucleons
(greater mass)

FIGURE 28.14 Binding
energy   As illustrated here, 28.30
MeV or work is required to separate
a helium nucleus into free protons
and neutrons. Conversely, if two
protons and two neutrons could be
combined to form a helium nucleus,
28.30 MeV of energy would be given
up. This is the binding energy of the
nucleus.

Notice that this is greater than the mass of the helium atom. The helium nucleus
is then less massive than the sum of its parts by an amount

$$\Delta m = [2m(^1\text{H}) + 2m_n] - m(^4\text{He})$$

$$= 4.032980 \text{ u} - 4.002603 \text{ u} = 0.030377 \text{ u}$$

where the electronic mass cancels out, since the mass of a hydrogen atom was
used for the proton. This mass difference, called the *mass defect*, has an energy
equivalent of

$$(0.030377 \text{ u})(931.5 \text{ MeV/u}) = 28.30 \text{ MeV}$$

which proves to be the same as the **total binding energy** (E_b) of the nucleus,
or the amount of energy/mass given up when the separate nucleons combined
to form a nucleus. Then, in general, for any nucleus,

$$E_b = (\Delta m)c^2 \qquad (28.5)$$

where Δm is the mass defect.

Another way of looking at the binding energy is that it is the energy that
would be required to separate the constituent nucleons into free particles. This
is illustrated in ● Fig. 28.14 for the helium nucleus. Exactly 28.30 MeV of energy
would be necessary to do the work of separating the nucleons. If the mass of
a nucleus and the mass of its nucleons were the same, then the nucleons would
not be bound together, and the nucleus would come apart without energy
input.

We can gain some insight into the nature of the nuclear force by considering
the **average binding energy per nucleon** for stable nuclei. This is simply the
total binding energy of a particular nucleus divided by its number of nucleons,
of E_b/A, where A is the mass number (number of protons plus number of
neutrons). For example, for the helium nucleus (^{4}He), the average binding
energy per nucleon is

$$\frac{E_b}{A} = \frac{28.30 \text{ MeV}}{4} = 7.075 \text{ MeV}$$

Compared to the binding energy of atomic electrons, for example, 13.6 eV for
a hydrogen electron in the ground state, the nuclear binding energy is about
10^6 times larger, which indicates a very strong binding force.

EXAMPLE 28.7 ■ Binding Energy per Nucleon

Compute the average binding energy per nucleon for the iron-56 nucleus
($^{56}_{26}$Fe).

Solution. We can find the atomic mass of the iron isotope in Appendix V,
and we have the other needed masses from Table 28.3.

Given: $^{56}_{26}$F mass $= 55.934939$ u   *Find:* E_b/A
$^1_1$H mass $= 1.007825$ u   (average binding
$^1_0$n mass $= 1.008665$ u   energy per nucleon)

(Notice that we use the mass of the hydrogen atom rather than the proton mass. Why?)

We then compute the mass defect or difference between the iron atom and its parts for $^{56}_{26}$Fe$_{30}$. The particles have a total mass of

$$26m(^1\text{H}) = 26(1.007825 \text{ u}) = 26.203450 \text{ u}$$

$$30m_n = 30(1.008665 \text{ u}) = \underline{30.259950 \text{ u}}$$

$$56.463400 \text{ u}$$

and

$$\Delta m = [26m(^1\text{H}) + 30m_n] - m(^{56}\text{Fe})$$
$$= 56.463400 \text{ u} - 55.934939 \text{ u} = 0.528461 \text{ u}$$

The total binding energy is then

$$E_b = (\Delta m)c^2$$

or

$$= (0.528461 \text{ u})(931.5 \text{ MeV/u}) = 492.3 \text{ MeV}$$

where the conversion factor for MeV/u was used instead of making the c^2 calculation. (Why?)

Our iron nuclide has 56 nucleons, so the average binding energy per nucleon is

$$\frac{E_b}{A} = \frac{492.3 \text{ MeV}}{56} = 8.791 \text{ MeV}$$

If E_b/A is calculated for various nuclei and plotted versus mass number, the values generally lie along a curve as shown in ● Fig. 28.15. The value of E_b/A rises rapidly with increasing A for light nuclei and starts to level off (around $A = 15$) at about 8.0 MeV, with a maximum value of about 8.8 MeV. The maximum of the curve occurs in the vicinity of iron, which has a very stable nucleus. For $A > 60$, the E_b/A values decrease slowly for heavier nuclei, a finding indicating that the nucleons are less tightly bound.

The maximum in the curve gives an important indication that will be considered in the next chapter. If a large nucleus could be split or fissioned into two lighter nuclei, the nucleons would be more tightly bound, and energy would be given up in the process—more than would be needed to induce the fission. (Recall how an atom gives up energy when one of its electrons becomes more tightly bound or makes a transition to a lower energy level.) Similarly, on the other side of the maximum, if we could fuse together two very light nuclei into a heavier nucleus, this would raise it up on the curve, and energy would be released. Such fission and fusion processes are the sources of nuclear energy (Chapter 29).

In general, the E_b/A curve shows that, except for very light nuclei, the binding energy per nucleon does not change a great deal and is on the order of

$$E_b/A \cong 8 \text{ MeV}$$

Since E_b/A is relatively constant for most nuclei, an approximation is that

$$E_b \propto A$$

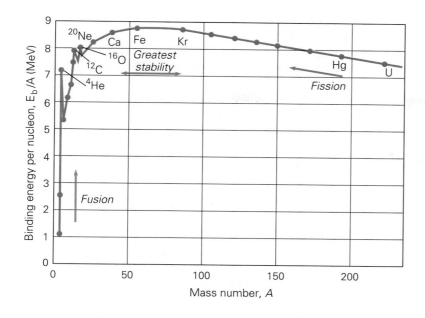

Figure 28.15

FIGURE 28.15 A plot of binding energy per nucleon versus mass number If the binding energy per nucleon (E_b/A) is plotted versus mass number (A), the curve has a maximum, which indicates that the nuclei in this region are on the average the most tightly bound together and have the greatest stability.

In other words, the total binding energy is proportional to the mass number.

This proportionality indicates a characteristic of the nuclear force that was mentioned earlier, one that makes it quite different from the electrical Coulomb force. Imagine that the attractive nuclear force acts among *all* nucleons (protons and neutrons) in a nucleus. Each pair of nucleons would then contribute to the total binding energy. In a nucleus containing A nucleons, there are $A(A - 1)/2$ pairs, so there would be $A(A - 1)/2$ contributions to the total binding energy. For nuclei for which $A \gg 1$ (heavier nuclei), then, $A(A - 1) \cong A^2$, and we would expect that

$$E_b \propto A^2$$

But in fact, $E_b \propto A$ for most nuclei, which indicates that a given nucleon is not bound to all the other nucleons. This phenomenon, called *saturation*, implies that the nuclear forces act over a very short range.

28.5 ■ Radiation Detection and Applications

Detecting Radiation

Since the products of radioactive decay cannot be detected directly by our senses, detection must be done by some indirect means. We have already mentioned the detection of nuclear radiation by photographic film. People who work with radioative materials and X-rays usually wear film badges, which indicate cumulative exposure to radiation by the degree of darkening of the film when it is developed. However, more immediate and quantitative methods of detection are desirable, and a variety of instruments have been developed for this purpose.

These **radiation detectors** are based on the ionization or excitation of atoms by the passage of energetic particles through matter. Alpha and beta particles are electrically charged particles, and they transfer energy to atoms along their

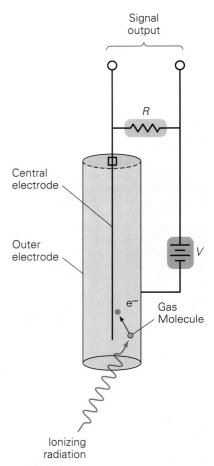

FIGURE 28.16 The Geiger counter A diagram showing the basic elements of a Geiger counter. See text for description.

paths by Coulomb electrical interactions. Gamma rays, on the other hand, have no electrical charge and cannot produce direct ionization. In this case, the gamma ray photons lose energy by Compton scattering or pair production (Chapter 27). The energetic particles produced by these interactions cause the ionization in a detector.

One of the most common radiation detectors is the *Geiger counter*, which was developed chiefly by Hans Geiger, a student and then colleague of Lord Rutherford. The principle of the Geiger counter is illustrated in ● Fig. 28.16. A voltage of 800–1000 V is applied across the wire electrode and metal tube of the Geiger tube. The tube contains a gas (such as argon) at low pressure. When an ionizing particle enters the tube through a thin window at one end, it ionizes a few atoms of the gas. The freed electrons are attracted and accelerated toward the positive wire anode. On their way, they strike and ionize other gas atoms. This process multiplies, and an "avalanche" discharge results that produces a current pulse. The pulse is amplified and sent to an electronic counter that counts the pulses or the number of particles detected. The pulses may also be used to drive a loudspeaker so that particle detection is heard as an audible click.

A major disadvantage of the Geiger counter is its relatively long "dead time." This is the recovery time required between successive electrical discharges in the tube. The dead time of a Geiger tube is about 200 μs, which limits the counting rate to a few hundred counts per second. If particles are incident on the tube at a faster rate, not all of them will be counted.

Another method of detection, one of the oldest, has a much shorter dead time. This is used in the *scintillation counter* (● Fig. 28.17). In the counter, atoms of a phosphor material (such as sodium iodide, NaI) are excited by an incident particle, and visible light is emitted when the atoms return to their ground state. The light pulse is converted to an electrical pulse by a photoelectric material. This is amplified in a *photomultiplier tube*, which consists of a series of successively higher potential electrodes. The photoelectrons are accelerated toward the first electrode and acquire sufficient energy to cause several secondary electrons to be emitted when they strike the electrode. This process continues, and relatively weak scintillations are converted into sizable electrical pulses, which are counted electronically.

The dead time of a scintillation counter is a few microseconds. Also, the magnitude of the photomultiplier current pulse is proportional to the number of photons generated by the incident particle, which in turn is proportional to its energy. Thus, the magnitude of the counted pulses gives a measure of the energy of the incident particle. Counters with this feature are called *proportional counters.*

FIGURE 28.17 The scintillation counter A photon emitted by a phosphor atom excited by an incoming particle causes the emission of a photoelectron. Accelerated through a potential difference in a photomultiplier tube, the photoelectrons free secondary electrons when they collide with successive electrodes at higher potentials. After several steps, a relatively weak scintillation is converted into a measurable electric pulse.

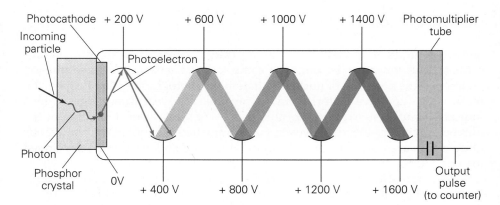

Semiconductor materials provide a relatively new method of radiation detection. In a *solid state* or *semiconductor detector*, charged particles passing through semiconductor material produce electron-hole pairs (solid state ionization, so to speak). With a voltage across the material, the electron-hole pairs give rise to electric signals, which may be amplified and counted. Solid state detectors are very fast—that is, they are capable of very high counting rates.

Counting methods are used to determine the number of particles produced by decaying nuclei of a radioactive isotope. Other methods of detection allow the tracks of charged particles to be seen or recorded visually. Those most commonly used are the cloud chamber, the bubble chamber, and the spark chamber. In the first two, vapors and liquids are supercooled and superheated, respectively, by suddenly varying the volume and pressure. The *cloud chamber* was developed early in this century by C.T.R. Wilson, a British atmospheric physicist. In the chamber, supercooled vapor condenses into droplets on ionized molecules along the path of an energetic particle. When the chamber is illuminated, the droplets scatter the light, making the path visible (Fig. 28.18).

The *bubble chamber*, which was invented by the American physicist D.A. Glazer in 1952, uses a similar principle. A reduction in pressure causes a liquid to be superheated and able to boil. Ions produced along the path of an energetic particle become sites for bubble formation and a trail of bubbles is created. When a magnetic field is applied across the chamber, charged particles are deflected, and the energy of a particle can be calculated from the radius of curvature of the particle's path measured from a photograph. Since the bubble chamber uses a liquid, commonly liquid hydrogen, the density of atoms in it is much greater than in the vapor of a cloud chamber. Thus tracks are more readily observable, and bubble chambers have displaced cloud chambers.

Gamma rays do not leave visible tracks in a bubble chamber. However, their presence can be detected indirectly, for example, by electrons that have been Compton scattered and by the creation of electron-positron pairs.

The path of a charged particle is registered by a series of sparks in a *spark chamber*. Basically, a charged particle passes between a pair of electrodes with a high potential difference that are immersed in an inert (noble) gas. The charged particle is accelerated and causes the ionization of gas molecules, giving rise to a visible spark or flash between the electrodes. A spark chamber is merely an array of such electrode pairs in the form of parallel plates or wires. A series of sparks, which can be photographed, marks the particle's path.

FIGURE 28.18 Cloud chamber tracks The circular track in this cloud chamber photograph was made by a positron in a strong magnetic field. (Can you explain the path of the particle and deduce the orientation of the field?)

Early detection was done with a scintillating phosphorescent screen that produced tiny flashes of light bright enough to be seen in the dark with the aid of a magnifying glass. Such a screen was used in the Rutherford scattering experiments with slow and tiring visual counting. It has been said that whenever Rutherford himself did the actual counting, he would "damn" vigorously for a few minutes and then have an assistant take over. More than a million scintillations or flashes were counted in the overall experiment.

Biological Effects and Medical Applications of Radiation

In medicine, nuclear radiation has both advantages and disadvantages. It can be used beneficially in the diagnosis and treatment of some diseases but can also be potentially harmful if it is not properly handled and administered. Nuclear radiation and X-rays can penetrate human tissue without pain or any other sensation. However, early investigators quickly learned that large doses or repeated small doses led to red skin, lesions, and other conditions.

The chief hazard of radiation is damage to living cells, primarily due to ionization. Ions, particularly complex ions or radicals produced by ionizing radiation, may be highly reactive (for example, a hydroxyl ion OH^- from water). These interfere with the normal chemical operations of the cell. If enough cells are damaged or killed by radiation, cell reproduction might not be fast enough, and the irradiated tissue could eventually die.

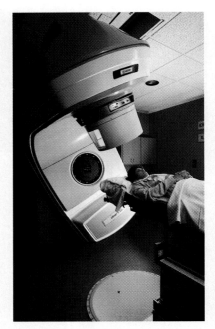

FIGURE 28.19 Nuclear medicine A particle accelerator used in cancer treatment.

The gray unit was named in honor of Louis Harold Gray, a British radiobiologist whose studies laid the foundation for measuring absorbed dose.

TABLE 28.4 ▪ Typical Relative Biological Effectiveness (RBE) of Radiations

Type	RBE (or QF)
X-rays and gamma rays	1
Beta particles	1.2
Slow neutrons	4
Fast neutrons and protons	10
Alpha particles	20

In other instances, there may be damage to a chromosome in the cell nucleus (genetic damage or mutation). If the affected cells are sperm or egg cells (or their precursors), any children that they produce may suffer from various birth defects. If the damaged cells are ordinary body cells, they may become cancerous, losing their normal form and reproducing in a rapid and uncontrolled manner. Such transformed cells grow at the expense of the surrounding tissues, producing a malignant tumor. Skin cancer and leukemia (cancer of the white blood cells, or leukocytes,) often result from excessive radiation exposure. The human cells most susceptible to radiation damage are those of the reproductive organs, bone marrow, and lymph nodes.

Radiation can cause cancer, but it can also be used to treat cancer. Localized doses of radiation are used to destroy cancer cells. The radiation may be gamma rays from a radioactive ^{60}Co source or X-rays from a machine. Other particles produced in particle accelerators are also used to treat cancer (● Fig. 28.19). The electrical charge and energy of the particles determine the penetrating power of the radiation. X-rays and gamma rays are deeply penetrating, but generally, beta particles penetrate only a few millimeters into biological tissue, and alpha particles penetrate only a fraction of a millimeter. Localization is achieved by using tightly focused beams. Also, radioactive compounds may be administered that are preferentially absorbed by certain tissues or organs.

Radiation Dosage. An important consideration in radiation therapy and radiation safety is the amount, or *dose*, of radiation. Several quantities are used to describe this in terms of *exposure*, *absorbed dose*, and *equivalent dose*. The earliest unit of dosage, the **roentgen** (R), was based on exposure and was defined in terms of the ionization produced in air. (One roentgen is the quantity of X-rays or gamma rays required to produce an ionization charge of 2.58×10^{-4} C/kg of air.)

The roentgen has been largely replaced by the **rad** (*rad*iation *a*bsorbed *dose*), which is an absorbed dose unit. One **rad** is an absorbed dose of radiation of 10^{-2} J of energy per kilogram in any absorbing material. The SI unit for absorbed dose is the **gray** (Gy).

$$1 \text{ Gy} \equiv 1 \text{ J/kg} \equiv 100 \text{ rad}$$

These *physical* units give the energy absorbed per mass, but it is helpful to have some means for measuring the biological damage produced by radiation, since equal doses of different types of radiation produce different effects. For example, a relatively massive alpha particle with a charge of 2+ moves through tissue rather slowly with a great deal of Coulomb interaction. The ionizing collisions thus occur close together along a short penetration path, and more localized damage is done than by a fast-moving electron or gamma ray.

This effect, or effective dose, is measured in terms of the **rem** unit (*r*oentgen or *r*ad *e*quivalent *m*an). The different degrees of effectiveness of different particles are characterized by the **relative biological effectiveness (RBE)** or *quality fact* (QF), which has been tabulated for the various particles in Table 28.4. The RBE is defined in terms of the number of rads of X-rays or gamma rays that produces the same biological damage as one rad of a given radiation. (Note in Table 28.4 that gamma rays have an RBE of 1.)

The effective dose is then given by the product of the dose in rads and the appropriate RBE.

$$\text{effective dose (in rem)} = \text{dose (in rad)} \times \text{RBE} \qquad (28.6)$$

Thus, one rem of any type of radiation does approximately the same amount of biological damage. For example, a 20-rem effective dose of alpha particles

TABLE 28.5 ■ Radiation Units

Unit	Basis
roentgen (R)	1 R—the quantity of X-rays or gamma rays that produces an ionization charge of 2.58×10^{-4} C/kg of air
rad (*r*adiation *a*bsorbed *d*ose)	1 rad—an absorbed dose of radiation of 10^{-2} J/kg
gray (Gy)	SI absorbed dose unit $1 \text{ Gy} = 1 \text{ J/kg} = 100 \text{ rad}$
rem (*r*ad *e*quivalent *m*an) effective dose (in rem) = dose (in rad) $\times$ RBE	Effective dose. Relative effectiveness depends on type of radiation and is characterized by RBE (relative biological effectiveness). See Table 28.4.
sievert (Sv) effective dose (in Sv) = dose (in Gy) $\times$ RBE	SI unit of effective dose. $1 \text{ Sv} = 100 \text{ rem}$

does the same damage as a 20-rem dose of X-rays, but 20 rad of X-rays are needed compared to 1 rad of alpha particles.

The SI unit of absorbed dose is the gray, and the effective dose with this unit is called the **sievert** (Sv):

$$\text{effective dose (in Sv)} = \text{dose (in Gy)} \times \text{RBE} \qquad (28.7)$$

Since $1 \text{ Gy} \equiv 100 \text{ rad}$, it follows that $1 \text{ Sv} \equiv 100 \text{ rem}$.

A summary of the radiation units is given in Table 28.5.

It is difficult to set a maximum permissible radiation dosage, but the general standard for humans is an average dose of 5 rem/year after the age of 18 with no more than 3 rem in any 3-month period. In the United States, the normal average annual dose per capita is about 200 mrem (millirem). About 125 mrem comes from the natural background of cosmic rays and naturally occurring radioactive isotopes in the soil, building materials, and so on. The remainder is chiefly from diagnostic medical applications, mostly X-rays (●Fig. 28.20), and from miscellaneous sources such as television tubes.

Diagnostic Applications. Radioactive isotopes offer an important technique for diagnostic procedures. For example, a radioactive isotope behaves chemically like a stable isotope of the element, and can participate in its normal chemical reactions. This makes it possible to label or tag molecules with radioisotopes, which can then be used as tracers.

By using tracers, many body functions can be studied simply by monitoring the location and activity of the labeled molecules as they are absorbed during body processes. For example, the activity of the thyroid gland in hormone production can be determined by monitoring its iodine uptake using radioactive iodine-123. This isotope emits gamma rays and has a half-life of 13.3 hours. Iodine is required by the body, and when ingested it collects in the thyroid gland near the throat. The uptake of radioactive iodine can be monitored by a detector to see if there is an abnormality.

Similarly, radioactive solutions of iodine and gold are quickly absorbed by the liver. This can be done safely because the isotopes have short half-lives. One of the most commonly used diagnostic tracers is technetium-99 (^{99}Tc). It has a convenient half-life of 6 hours and combines with a large variety of compounds. Detectors outside the body can scan over a region and record the activity so that a complete activity image may be reconstructed.

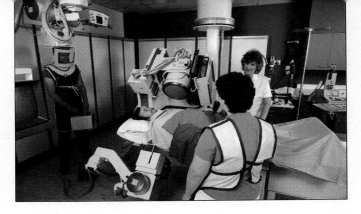

•**FIGURE 28.20 X-ray precautions** Though the radiation dosage represented by an individual diagnostic X-ray is generally relatively small, there is no threshold level for radiation exposure. That is, no exposure, no matter how low, can be guaranteed to be totally safe. Moreover, radiation effects are cumulative over time. For this reason, lead-lined aprons are used to protect parts of the patient's body that are not being photographed, as well as the technicians operating the X-ray equipment. It is especially important to shield the reproductive organs, since damage to reproductive cells could result in the production of children with birth defects, or even cause sterility.

(a)

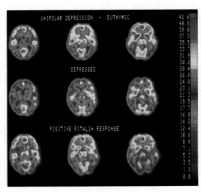

(b)

•**FIGURE 28.21 PET** (a) A PET scanner monitors brain activity after administration of glucose containing radioactive isotopes. (b) A series of PET scans of the brain showing normal metabolic activity (top), the effects of depression (middle), and the response to a drug that stimulates the central nervous system (bottom).

It is possible to image gamma ray activity in a single plane, or "slice," through the body. A gamma detector is moved around the patient to measure the emission intensity from many angles. A complete image can then be constructed by using computer-assisted tomography as in X-ray CT. This is referred to as single-photon emission tomography (SPET). Another technique, positron emission tomography (PET), uses tracers that are positron emitters such as ^{11}C and ^{15}O. When a positron is emitted, it is quickly annihilated, and two gamma rays are produced. Recall that the photons have equal energies and fly off in nearly opposite directions so as to conserve momentum (Section 27.4) The detection of the gamma rays is by a ring of detectors surrounding the patient (•Fig. 28.21)

Domestic and Industrial Applications of Radiation

The major application of radioactivity in the home is the smoke detector. In the smoke detector, a weak radioactive source ionizes the air molecules, setting up a small current in the detector circuit. If smoke enters the detector, the ions there become attached to the smoke particles. This causes a reduction in the current because the heavier, charged smoke particles move more slowly. The current difference is electronically sensed, and an alarm is triggered (•Fig. 28.22).

Industry also makes good use of radioactive isotopes. Radioactive tracers are used to determine flow rates in pipes, to detect leaks, and to study corrosion and wear. Also, it is possible to radioactivate certain compounds at a particular stage in a process by irradiating them with particles, generally neutrons. This technique is called **neutron activation analysis**, and is also an important method of identifying elements in a sample. Until its development, the chief methods of identification were chemical and spectral analyses. In both of these methods, a fairly large amount of a sample has to be destroyed in the analysis procedure. As a result, a sample may not be large enough for analysis, or small traces of elements in a sample may go undetected. Neutron activation analysis has the advantage over these methods on both scores: Only minute samples are needed, and the method can detect even tiny amounts of an element.

A neutron activation process is as follows:

$$(\text{source}) \quad ^{252}_{98}\text{Cf} \rightarrow ^{251}_{98}\text{Cf} + ^{1}_{0}\text{n}$$

$$^{1}_{0}\text{n} + ^{14}_{7}\text{N} \rightarrow ^{15}_{7}\text{N}^* \rightarrow ^{15}_{7}\text{N} + \gamma$$
$$\quad\quad\quad\quad\quad \underset{\text{(excited}}{} \quad\quad \underset{\text{(gamma}}{}$$
$$\quad\quad\quad\quad\quad \underset{\text{nucleus)}}{} \quad\quad \underset{\text{ray)}}{}$$

Here, neutrons from a californium-252 source bombard nitrogen-14, which absorbs a neutron and becomes an excited nitrogen-15 nucleus, which decays with the emission of a gamma ray with distinctive energy. Other neutron-activated nuclei may decay by other modes, such as beta decay.

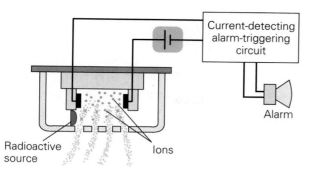

FIGURE 28.22 Smoke detector In the common smoke detector, a weak radioactive source ionizes the air and sets up a small current. Smoke particles in the detector reduce the current, causing an alarm to sound.

The nitrogen activation process is used to screen for bombs in airport luggage. Virtually all explosives contain nitrogen. By using neutron activation and analyzing the energy and pattern of any gamma ray emission coming from a suitcase, the possibility of an explosive device being in the suitcase can be checked. Of course, other materials in the suitcase may contain nitrogen too, so manual checks are made to confirm any suspicious findings.

Finally, the federal goverment recently gave permission for the use of gamma radiation in the processing of poultry. The radiation kills bacteria and helps preserve the food.

Important Concepts

You should be able to define and/or explain these chapter concepts clearly.

Rutherford-Bohr model	beta particle	gamma decay	average binding energy per nucleon
strong nuclear force	gamma ray	activity	
nucleons	alpha decay	decay constant	radiation detectors
proton (atomic) number	conservation of nucleons	half-life	roentgen (R)
mass number	conservation of charge	curie (Ci)	rad
neutron number	tunneling	becquerel (Bq)	gray (Gy)
isotopes	barrier penetration	carbon-14 dating	rem
nuclide	beta (β^-) decay	pairing effect	relative biological effectiveness (RBE)
radioactivity	β^+ decay	atomic mass unit (u)	
alpha particle	electron capture	total binding energy	sievert (Sv)
			neutron activation analysis

Important Relationships for Review

These relationships are mathematical statements of the concepts and principles presented in the chapter. You should be able to identify the symbols and to explain the relationships before proceeding to the Exercises. In-text equation reference numbers are given for convenience.

Radius of Nucleus (upper limit):

$$r_a = \frac{4kZe^2}{mv^2} \tag{28.1}$$

Activity of a Radioisotope:

$$\frac{\Delta N}{\Delta t} = -\lambda N \qquad (28.2)$$

$$(1 \text{ Ci} = 3.70 \times 10^{10} \text{ decays/s} = 3.70 \times 10^{10} \text{ Bq})$$

Number of Undecayed Nuclei:

$$N = N_o e^{-\lambda t} \qquad (28.3)$$

Half-life and Decay Constant:

$$t\tfrac{1}{2} = \frac{0.693}{\lambda} \qquad (28.4)$$

Total Binding Energy:

$$E_b = (\Delta m)c^2 \qquad (28.5)$$

Effective Dose:

$$\text{Dose (in rem)} = \text{dose (in rad)} \times \text{RBE} \qquad (28.6)$$

$$\text{Dose (in Sv)} = \text{dose (in Gy)} \times \text{RBE} \qquad (28.7)$$

Exercises

28.1 ▪ Nuclear Structure and the Nuclear Force

1 The concept of the atomic nucleus resulted from the work of (a) Bohr, (b) Thompson, (c) Rutherford, (d) Geiger. (d)

● 2 Carbon-12 (a) has the same number of protons and neutrons, (b) is a nuclide, (c) is an isotope, (d) all of these. (d)

3 The strong nuclear force (a) binds the orbital electrons to the atomic nucleus, (b) has a longer range than the electrostatic force, (c) acts only between identical particles, (d) overcomes the force of repulsion between protons in the nucleus. (d)

4 Calcium (atomic number 20) has 14 isotopes. Write the nuclear notations for the isotopes of calcium with mass numbers from 40 to 46. (The complete family of isotopes has mass numbers from 37 to 50.) $^{40}_{20}$Ca; $^{41}_{20}$Ca; $^{42}_{20}$Ca; $^{43}_{20}$Ca; $^{44}_{20}$Ca; $^{45}_{20}$Ca; $^{46}_{20}$Ca

5 ■■ Two isotopes of cobalt are ^{60}Co and ^{59}Co. What is the number of protons, neutrons, and electrons in each if (a) the atom is electrically neutral, (b) the atom has a -2 charge, and (c) the atom has a $+1$ charge? (a) 27 p, 33 n, 27 e; 27 p, 32 n, 27 e (b) 27 p, 33 n, 29 e; 27 p, 32 n, 29 e (c) 27 p, 33 n, 26 e; 27 p, 32 n, 26 e

6 ■■ (a) Write the nuclear notations for the isotopes of hydrogen using the symbols H, D, and T. (b) Write the chemical symbols for six different forms of water using these symbols. Identify the radioactive forms. (a) ^{1}H, ^{2}D, ^{3}T (b) H_2O, D_2O, T_2O (c) HDO, HTO, DTO

7 ■■ For the isotopes of uranium that has mass numbers from 232 through 239, what are the numbers of protons, neutrons, and electrons in a neutral atom of each isotope? 92 p and 92 e for each with n from 140 to 147

● 8 ■■ Oxygen has three stable isotopes with 8, 9, and 10 neutrons, respectively. Write the nuclear notations for these isotopes. $^{16}_{8}$O, $^{17}_{8}$O, $^{18}_{8}$O

9 ■■ An isotope of potassium has the same number of neutrons as the nuclide argon-40. Write the nuclear notation for the potassium isotope. $^{41}_{19}$K

10 ■■ Tin ($_{50}$Sn) has 23 isotopes that have mass numbers from 108 to 130. Write the complete nuclear notations for the isotopes of tin with mass numbers from 115 to 120. $^{115}_{50}$Sn; $^{116}_{50}$Sn; $^{117}_{50}$Sn; $^{118}_{50}$Sn; $^{119}_{50}$Sn; $^{120}_{50}$Sn

11 ■■ A theoretical result for the approximate nuclear radius (R) is $R = R_o A^{1/3}$, where $R_o = 1.2 \times 10^{-15}$ m and A is the mass number of the nucleus. Find the nuclear radii of atoms of the noble gases: He, Ne, Ar, Kr, Xe, and Rn. He: 1.9×10^{-15} m Ne: 3.3×10^{-15} m Ar: 4.1×10^{-15} m Kr: 5.3×10^{-15} m Xe: 6.1×10^{-15} m Rn: 7.2×10^{-15} m

28.2 ▪ Radioactivity

12 The conservation of nucleons and the conservation of charge apply to (a) only alpha decay, (b) only beta decay, (c) only gamma decay, (d) all nuclear processes. (d)

13 The neutron number is not conserved in (a) alpha decay, (b) beta decay, (c) gamma decay, (d) none of these. (b)

14 ■■ Write the nuclear equations expressing (a) the beta decay of $^{60}_{27}$Co, and (b) the alpha decay of $^{226}_{88}$Ra. (a) $^{60}_{27}$Co → $^{60}_{28}$Ni + $^{0}_{-1}$e (b) $^{226}_{88}$Ra → $^{218}_{86}$Rn + $^{4}_{2}$He

15 ■■ Write the nuclear equations for (a) the alpha decay of neptunium-237, (b) the β^- decay of phosphorus, (c) the β^+ decay of cobalt-56, (d) electron capture in cobalt-56, and (e) the gamma decay of potassium-42. (a) $^{237}_{93}$Np → $^{233}_{91}$Pa + $^{4}_{2}$He (b) $^{32}_{15}$P → $^{32}_{16}$S + $^{0}_{-1}$e (c) $^{56}_{27}$Co → $^{56}_{26}$Fe + $^{0}_{1}$e (d) $^{56}_{27}$Co + $^{0}_{-1}$e → $^{56}_{26}$Fe (e) $^{42}_{19}$K* → $^{42}_{19}$K + γ

16 ■■ Tritium is radioactive. Would you expect it to β^+ or β^- decay? [*Hint:* Write the nuclear equation for each.] see ISM

17 ■■ Bismuth-210 undergoes both alpha and beta decay, and its daughter nuclei decay into lead-206. Write the nuclear equations for these decay processes. $^{210}_{83}$Bi → $^{206}_{81}$Tl + $^{4}_{2}$He; $^{206}_{81}$Tl → $^{206}_{82}$Pb + $^{0}_{-1}$e

● 18 ■■ Radon-222 ($^{222}_{86}$Rn), a radioactive gas, undergoes alpha decay. (a) What is the daughter nucleus? Write the nuclear equation for the decay process. (b) The daughter nucleus decays by either alpha decay or beta decay. What is the granddaughter nucleus for each of these processes? (a) $^{222}_{86}$Rn → $^{218}_{84}$Po + $^{4}_{2}$He (b) $^{214}_{82}$Pb; $^{218}_{85}$At

19 ■■ A lead-209 nucleus results from both alpha-beta sequential decays and beta-alpha sequential decays. What was the grandparent nucleus? (Show this for both decay routes by writing the nuclear equations for these decay processes.) $^{213}_{83}$Bi

20 ■■ Complete the following nuclear decay equations:

(a) $^{8}_{4}$Be → $^{4}_{2}$He + $\underline{\text{(a) } ^{4}_{2}\text{He}}$

(b) $^{240}_{94}$Po → $^{97}_{38}$Sr + $^{139}_{56}$Ba + $\underline{\text{(b) } 4(^{1}_{0}\text{n})}$

(c) $^{47}_{21}$Sc* → $^{47}_{21}$Sc + $\underline{\text{(c) } \gamma}$

(d) $^{29}_{11}$Na → $^{0}_{-1}$e + $\underline{\text{(d) } ^{29}_{12}\text{Mg}}$

21 ■■ Complete the following nuclear equations:

(a) $^{238}_{92}$U → $^{234}_{90}$Th + $\underline{\text{(a) } ^{4}_{2}\text{He}}$

(b) $^{40}_{19}$K → $^{40}_{20}$Ca + $\underline{\text{(b) } ^{0}_{-1}\text{e}}$

(c) $^{236}_{92}U \rightarrow ^{131}_{53}I + 3^1_0n + $ (c) $^{102}_{39}Y$ (e) $^{22}_{11}Na + $ (e) $^{0}_{-1}e \rightarrow ^{22}_{10}Ne$

(d) $^{23}_{11}Na + \gamma \rightarrow$ (d) $^{23}_{11}Na*$

22 ■■■ Actinium-227 ($^{227}_{89}Ac$) decays by alpha decay or by beta decay and is part of a decay sequence like the one in Fig. 28.9. Each daughter nucleus then decays to radium-223 ($^{223}_{88}Ra$), which subsequently decays to polonium-215 ($^{215}_{84}Po$). Write the nuclear equations for the decay process in the decay series from ^{227}Ac to ^{215}Po.   (a) alpha to ^{224}Fr; beta to ^{227}Th (b) beta to ^{223}Ra; alpha to ^{223}Ra (c) both alpha to ^{219}Rn (d) alpha to ^{215}Po

23 ■■■ The decay series for neptunium-237 is shown in Fig. 28.23. (a) Identify the decay modes and each of the nuclei in the sequence. (b) Tell why each nucleus is likely to decay.   see ISM

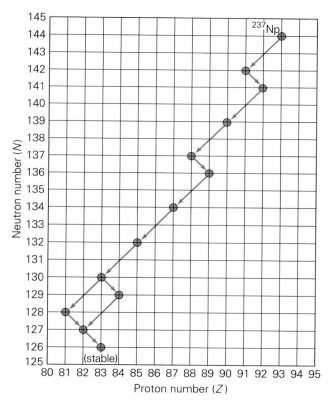

FIGURE 28.23 Neptunium-237 decay series   See Exercise 23.

Figure 28.23

28.3 ■ Decay Rate and Half-Life

24 After one half-life, a sample of a particular radioactive material (a) is half as massive, (b) has its half-life reduced by one half, (c) is no longer radioactive, (d) has its activity reduced by one half.   (d)

25 In three half-lives, the activity of a radioactive sample will have *decreased* by what percent? (a) 33.3%, (b) 50.0%, (c) 66.7%, (d) 87.5%.   (d)

26 What physical or chemical properties affect the decay rate or half-life of a radioactive isotope?   see ISM

27 If one radioactive nuclide has a decay constant that is half that of another, will twice as many of its nuclei decay in a given time? Explain.   no, decay is exponential

28 ■ A particular radioactive sample is observed to undergo 2.50×10^5 decays per second. What is the activity of the sample in (a) curies and (b) becquerels?   (a) 6.76 μCi   (b) 2.50×10^5 Bq

29 ■ A radioactive beta source has an activity of 20 mCi. How many beta particles per minute are emitted from the source? 4.4×10^{10} decays/min

● **30** ■ The half-life of a radioactive isotope is 1 h. What fraction of a sample of this isotope would be left after (a) 2 h and (b) 1 day?   (a) 1/4   (b) $1/2^{24}$

31 ■■ A sample of technetium-104 has an activity of 10 mCi. What will be the activity of the sample in one hour? ($t_{\frac{1}{2}} = 18$ min.)   0.993 mCi

32 ■■ What period of time is rquired for a sample of radioactive tritium (^3_1H) to lose $\frac{4}{5}$ of its activity? Tritium has a half-life of 12.3 years.   28.6 y

33 ■■ Which has the greater decay constant, $^{28}_{12}Mg$ or $^{104}_{43}Tc$? By a factor of how much is it greater? T_c smaller; $\lambda_{Tc}/\lambda_{Mg} = 70$

34 ■■ If an amount of ^{123}I is introduced into a patient in a medical diagnostic procedure, what percentage of the sample remains in 2 hours if all of the ^{123}I is retained in the patient's thyroid gland?   90.1%

35 ■■ A sample of old bone is found to have 4 beta decays/min for each gram of carbon. Approximately how old is the bone?   11,460 y

● **36** ■■ Show that the number N of radioactive nuclei remaining in a sample after n half-lives is given by

$$N = \frac{N_o}{2^n}$$

where N_o is the initial number of nuclei.   see ISM

37 ■■ Some ancient writings on parchment are found sealed in a jar in a cave. If carbon-14 dating shows the parchment to be 28,650 years old, what percentage of the carbon-14 atoms still remain in the sample compared to the number when the parchment was made?   3.1%

38 ■■ For a sample of cobalt-60, how long would it take for its activity to reduce to 10% of the original activity?   18 y

39 ■■ The activity of an unknown beta source is observed to decrease by 87.5% in 54 min. What is the radioactive isotope? ^{104}Tc

40 ■■ A soil sample contains 40 μg of ^{90}Sr. Approximately how much ^{90}Sr will be in the sample in the year 2098?   3.0 μg

41 ■■ (a) What is the decay constant of fluorine-17? (b) How long will it take for the activity of a sample of ^{17}F to decrease to 12.5 percent of its value at $t = 0$? ($t_{\frac{1}{2}} = 66$ s.)
(a) 1.1×10^{-2} s^{-1} (b) 189 s

● **42** ■■ Francium-223 ($^{223}_{87}$Fr) has a half-life of 21.8 min. (a) If there is initially a 25.0-mg sample of this isotope, how many nuclei are present? (b) How many nuclei will be present 1 h and 49 min later? (a) 6.74×10^{19} nuclei
(b) 2.11×10^{18} nuclei

43 ■■ A basement room containing radon gas is sealed air-tight. If there are 7.50×10^{10} radon atoms trapped in the room at the time, (a) using half-life periods ($t_{\frac{1}{2}} = 3.82$ days) estimate how many radon atoms would remain in the room at the end of 30 days. (b) Compute the number of atoms remaining. (c) Radon undergoes alpha decay. Is there the same number of daughter nuclei as decayed parent nuclei at the end of 30 days? Explain. (a) $7.50 \times 10^{10}/2^8$ atoms
(b) 3.25×10^8 atoms (c) no

44 ■■ An old artifact is found to contain 500 g of carbon and has an activity of 950 decays per minute. What is the approximate age of the artifact? 17,000 to 18,000 y

45 ■■ The recoverable reserves of high-grade uranium-238 ore (high-grade ore contains about 10 kg of ^{238}U$_3$O$_8$ per ton) in the United States are estimated to be about 500,000 tons. Neglecting any geological changes, how much ^{238}U existed in this high-grade ore when the Earth was formed about 4.6 billion years ago? [Hint: See Appendix V.] 1.9×10^6 kg

46 ■■ In 1898, Pierre and Marie Curie isolated about 10 mg of radium-226 from 8 tons of uranium ore. If this sample had been placed in a museum how much of the radium would remain in the year 2000? 9.6 mg

47 ■■■ Nitrogen-13, with a half-life of 10 minutes, decays with the emission of positrons. (a) Write the nuclear equation for this decay process. (b) If a pure sample of ^{13}N has a mass of 0.0015 kg, what is the activity in 35 minutes? (c) What percentage of the sample is ^{13}N at this time?
(a) $^{13}_{7}$N $\rightarrow$ $^{13}_{6}$C $+$ $^{0}_{+1}$e (b) 64.26×10^{20} decays/min (c) 8.84%

28.4 ■ Nuclear Stability and Binding Energy

● **48** For nuclei with mass number greater than 40, which of the following answers is correct? (a) The number of protons is approximately equal to the number of neutrons, (b) the number of protons exceeds the number of neutrons, (c) all are stable up to $A = 83$, (d) none of these. (d)

49 The average binding energy per nucleon of the daughter nucleus in a decay process is (a) greater than, (b) less than, (c) equal to that of the parent nucleus. (a)

50 ■■ Determine whether each of the three isotopes of hydrogen is likely to be stable or unstable. ^{1}H and ^{2}H: stable;
^{3}H: unstable

51 ■■ Only two isotopes of Sb (antimony, $Z = 51$) are stable. Pick the two stable isotopes from the following: (a) ^{120}Sb, (b) ^{121}Sb, (c) ^{122}Sb, (d) ^{123}Sb, (e) ^{124}Sb. (b) and (d) ^{123}Sb;
others are odd-odd

52 ■■ Determine whether the following isotopes are likely to be unstable: (a) ^{23}Na, (b) ^{50}V, (c) ^{209}Bi, (d) ^{209}Po, (e) ^{44}Ca.
(b) and (d)

53 ■■ The total binding energy of $^2_1$H is 2.224 MeV. What is the mass of a ^{2}H nucleus? 2.013553 u

54 ■■ Use Avogadro's number $N_A = 6.02 \times 10^{23}$ atoms/mole (see Section 10.3) to show that 1 u $= 1.66 \times 10^{-27}$ kg. (Recall a ^{12}C atom has a mass of exactly 12 u.) see ISM

55 ■■ The mass of $^3_2$He is 3.016029 u. (a) What is the total binding energy of this nucleus? (b) What is the average binding energy per nucleon? (a) 7.72 MeV (b) 2.57 MeV/nucleon

● **56** ■■ The mass of $^9_4$Be is 9.012183 u. What is E_b/A for this nucleus? 6.46 MeV/nucleon

57 ■■ Which isotope of hydrogen has the greatest average binding energy per nucleon? (Justify your answer mathematically.) tritium

58 ■■ How much energy would be required to separate the nucleons of an oxygen-16 nucleus, the atom of which has a mass of 15.994915 u? 127.6 MeV

59 ■■ Determine the average binding energy per nucleon of sulfur-32 ($m = 31.972072$ u). 8.49 MeV/nucleon

● **60** ■■ If an alpha particle could be removed intact from an aluminum-27 nucleus ($m = 26.981541$ u), a sodium-23 nucleus ($m = 22.989770$ u) would remain. How much energy would be required to do this? 10.1 MeV

61 ■■ On the average, are nucleons more tightly bound in an ^{27}Al nucleus or in a ^{23}Na nucleus? (See Exercise 60.)
Al: 8.33 MeV/nucleon; Na: 8.57 MeV/nucleon

62 ■■ The total binding energy of $^6_3$Li is 32.0 MeV. What is the mass of ^{6}Li nucleus? 6.013470 u

63 ■■ The atomic mass of $^{238}_{92}$U is 238.050786 u. Find the average binding energy per nucleon for this isotope.
7.57 MeV/nucleon

64 ■■ The mass of $^8_4$Be is 8.005305 u. (a) Which is less, the total mass of two alpha particles or the mass of the ^{8}Be nucleus? (b) Which is greater, the total binding energy of ^{8}Be nucleus or the total binding energy of an alpha particle? (c) Would you expect the ^{8}Be nucleus to decay spontaneously into two alpha particles? (a) two alphas
(b) 28.3 MeV (c) no

28.5 ■ Radiation Detection and Applications

65 The detector with the greatest dead time is the (a) Geiger counter, (b) scintillation counter, (c) solid state detector, (d) all have the same dead time. (a)

● **66** A unit of radiation dosage is the (a) rad, (b) gray, (c) sievert, (d) all of these. (d)

67 A basic assumption of radiocarbon dating is that the cosmic ray intensity has been generally constant for the last 40,000 years or so. Suppose it were found that the intensity was much less 10,000 years ago than it is today. How would this affect carbon-14 dating? see ISM

68 ■■ A patient receives a 2.0-rad dose of radiation from X-rays and a 2.0 rad dose from an alpha source. (a) Which has the greater biological effectiveness? (b) What are the doses in rems?   (a) alpha particles   (b) 2.0 rem; 40 rem

69 ■■ A technician working at a nuclear reactor facility is exposed to a slow neutron radiation and receives a dose of 1.25 rad. (a) How much energy is absorbed by 200 g of the worker's tissue? (b) Was the maximum permissible radiation dosage exceeded?   (a) 0.25 J   (b) yes

● **70** ■■ A person working with nuclear isotopes for a 2-month period receives a 0.5-rad dose from a gamma source, a 0.3-rad dose from a slow-neutron source, and a 0.1-rad dose from an alpha source. Was the maximum permissible radiation dosage exceeded?   yes

71 ■■ In a diagnostic procedure, a patient in a hospital ingests 80 mCi of gold-198 ($t_\frac{1}{2}$ = 2.7 days). What is the approximate activity at the end of two weeks if none is eliminated from the body by biological functions?   ≈ 2.3 mCi

72 Neutron activation analysis was performed on small pieces of hair that had been taken from the exiled Napoleon after he died on the island of St. Helena in 1821. The samples were found to contain abnormally high levels of arsenic, which supported a theory that his death was not due to natural causes. If this evidence was derived through beta emissions coming from germanium-76 nuclei in the sample, what was the arsenic nuclide in the hair? ^{76}As

■ **Additional Exercises**

73 A fossil specimen is believed to be about 18,000 years old. Explain how this could be confirmed. ^{14}C dating

74 The binding energy per nucleon for $^4_2$He is 7.075 MeV. (a) What is the total binding energy? (b) What is the mass of a ^{4}He nucleus?   (a) 28.3 MeV   (b) 4.001501 u

75 Starting with uranium-234 ($^{234}_{92}$U), there is a decay sequence of four alpha decays and two beta decays. What is the resulting nucleus of the sequence? $^{222}_{86}$Rn

76 A sample of ^{215}Bi, which beta decays ($t_\frac{1}{2}$ = 2.4 min), contains Avogadro's number of nuclei. How many bismuth nuclei are present after (a) 10 min and (b) 1 h? (c) What are the activities in curies and becquerels at these times, and is this a realistic radioactive source?   (a) 3.3 × 10^{22} nuclei   (b) 1.7 × 10^{16} nuclei   (c) see ISM

77 An old bone yields, on the average, one beta emission per minute per gram of carbon. About how old is the bone?   22,900 y

78 Complete the following nuclear decay equations:

(a) $^{47}_{21}$Sc → $^{47}_{22}$Ti + <u>(a) $^{0}_{-1}$e</u>

(b) $^{226}_{88}$Ra → $^4_2$He + <u>(b) $^{222}_{86}$Rn</u>

(c) $^{237}_{93}$Np → $^{0}_{-1}$e + <u>(c) $^{237}_{94}$Pu</u>

(d) $^{210}_{84}$Po* → $^{210}_{84}$Po + <u>(d) γ</u>

(e) $^{11}_6$C → $^{11}_5$B + <u>(e) $^{0}_{+1}$e</u>

79 ● Figure 28.24 shows the decay series for plutonium-239. (a) Identify each of the nuclei in this decay process. (b)

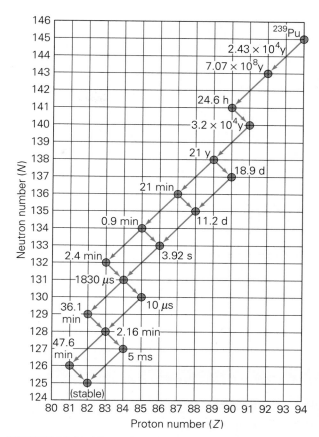

146
145
144
143
142
141
140
139
138
137
136
135
134
133
132
131
130
129
128
127
126
125
124

Neutron number (N)

^{239}Pu
2.43 × 10^{4}y
7.07 × 10^{8}y
24.6 h
3.2 × 10^{4}y
21 y
18.9 d
21 min
11.2 d
0.9 min
2.4 min
3.92 s
1830 μs
36.1 min
10 μs
2.16 min
47.6 min
5 ms
(stable)

80 81 82 83 84 85 86 87 88 89 90 91 92 93 94
Proton number (Z)

● **FIGURE 28.24 Plutonium-239 decay series** See Exercise 79.

Compare the totals of the half-lives before and after the ^{227}Ac nuclei.   see ISM

80 The value of E_b/A for $^{238}_{92}$U is 7.58 MeV. (a) What is the total binding energy of this nucleus? (b) What is its mass?
(a) 1804 MeV   (b) 237.997821 u

81 What are the equivalent rest energies of (a) a carbon-12 atom and (b) an alpha particle?   (a) 1.118 × 10^4 MeV
(b) 3.728 × 10^3 MeV

82 Determine whether the following isotopes are likely to be stable (a) ^{20}C, (b) ^{100}Sn, (c) ^{6}Li, (d) ^{9}Be, (e) ^{22}Na.
(c) and (d)

83 Thorium-229 ($^{229}_{90}$Th) undergoes alpha decay. (a) What is the daughter nucleus of this process? Write the decay equation. (b) This daughter nucleus undergoes beta decay. What is the resulting granddaughter nucleus? Write the decay equation.   (a) $^{225}_{88}$Ra   (b) $^{225}_{89}$Ac

84 Bismuth-214 ($^{214}_{83}$Bi) decays by alpha decay or beta decay. Identify the daughter nucleus in each case by writing the decay equations.   alpha: $^{210}_{81}$Tl; beta: $^{214}_{84}$Po

85 Gold-198 beta decays with a half-life of 2.7 days. (a) What is the activity of a sample of 1.0 g of pure $^{198}_{79}$Au? (b) What is the activity of the sample of the end of 1 week?
(a) −5.4 × 10^{17} decays/min   (b) −8.9 × 10^{16} decays/min

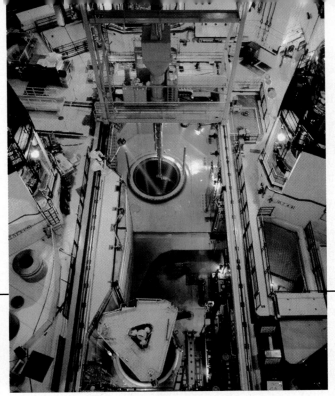

29

Nuclear Reactions and Elementary Particles

Thus far, we have considered radioactive processes in which the nucleus of one element decays into the nucleus of another with the emission of energetic particles. As might be expected, scientists wondered whether the reverse was possible. That is, could the nucleus of one element be bombarded with energetic particles and thus be converted into the nucleus of another element with the absorption of a particle (reverse radioactivity, so to speak)?

Indeed, such processes are possible, and physicists have learned how to manipulate nuclear particles to create one isotope from another by nuclear reactions. Such reactions have made available the many artificial radioactive isotopes that do not occur naturally.

Also, the knowledge of nuclear reactions has made it possible to release energy from the nucleus. The dropping of the so-called atomic bomb on the Japanese city of Hiroshima on August 6, 1945, was stark demonstration of the devastating power that could be released from the nucleus. On the more peaceful side, nuclear reactors such as the one shown in the photograph above are now an important source of energy for generating electricity.

In this chapter, we will look at these topics and at elementary particles. Investigations have shown that a variety of particles other than the proton and neutron are associated with the nucleus. The discovery of the existence of these particles has given some insight into the still puzzling world of the nucleus and what holds it together.

Topics in Chapter 29 include nuclear reactions, fission and fusion, beta decay and the neutrino, force and exchange particles, and elementary particles.

29.1 ■ Nuclear Reactions

In chemical reactions, substances react with each other to form compounds. Since an ordinary chemical reaction involves only the atomic electrons, the atoms of the reactants do not lose their nuclear identities in the process. On the other hand, in **nuclear reactions**, nuclei are converted into the nuclei of other isotopes, which in general are nuclides of completely different elements. From the study of the nucleus, scientists have learned to initiate nuclear reactions by bombarding nuclei with energetic particles.

The first induced nuclear reaction was produced by Lord Rutherford in 1919. Nitrogen was bombarded with alpha particles from a natural source (bismuth-214). The occasional particles that came from the reactions were identified as protons. Rutherford reasoned that an alpha particle colliding with a nitrogen nucleus must sometimes induce a reaction that produces a proton. As a result, the nitrogen nucleus is *artificially* transmuted into an oxygen nucleus:

$$\underset{\substack{\text{nitrogen} \\ (14.003074 \text{ u})}}{^{14}_{7}\text{N}} \quad + \quad \underset{\substack{\text{alpha particle} \\ (4.002603 \text{ u})}}{^{4}_{2}\text{He}} \quad \rightarrow \quad \underset{\substack{\text{oxygen} \\ (16.999133 \text{ u})}}{^{17}_{8}\text{O}} \quad + \quad \underset{\substack{\text{proton} \\ (1.007825 \text{ u})}}{^{1}_{1}\text{H}}$$

where the atomic masses are given for a later purpose.

Nuclear reactions are sometimes written with intermediate compound nuclei. For example, the preceding reaction can be written,

$$^{14}_{7}\text{N} + {}^{4}_{2}\text{He} \rightarrow ({}^{18}_{9}\text{F}^{*}) \rightarrow {}^{17}_{8}\text{O} + {}^{1}_{1}\text{H}$$

The fluorine nucleus, $^{18}_{9}\text{F}$, is formed in an excited state, as is indicated by the asterisk. A compound nucleus rids itself of excess energy by ejecting a particle. This occurs in a very short time, so the compound nucleus is commonly omitted from the equation for the nuclear reaction.

Think of the implication of Rutherford's discovery. One element had been converted into another. This was the age-old dream of the alchemists, although their main concern was changing common metals, such as mercury and lead, into gold. Even this transmutation can be done today through nuclear reactions. Large machines called **particle accelerators** use electric and magnetic fields to accelerate charged particles to very high energies (● Fig. 29.1). When the particles strike target nuclei, they initiate nuclear reactions. Different reactions require different particle energies. One nuclear reaction that is initiated when protons strike nuclei of mercury is

$$\underset{\substack{\text{mercury} \\ (199.968321 \text{ u})}}{^{200}_{80}\text{Hg}} \quad + \quad \underset{\substack{\text{proton} \\ (1.007825 \text{ u})}}{^{1}_{1}\text{H}} \quad \rightarrow \quad \underset{\substack{\text{gold} \\ (196.96656 \text{ u})}}{^{197}_{79}\text{Au}} \quad + \quad \underset{\substack{\text{alpha particle} \\ (4.002603 \text{ u})}}{^{4}_{2}\text{He}}$$

In this reaction, mercury is converted into gold, so it would seem that modern physics has fulfilled the alchemists' dream. However, making such small amounts of gold in an accelerator costs far more than the gold is worth.

Reactions like those written above have the general form

$$A + a \rightarrow B + b$$

where the uppercase letters represent the nuclei and the lowercase letters represent the particles. Such reactions are often written in a shorthand notation like this:

$$A(a, b)B$$

For example, in this form, the two previous reactions are written as

$$^{14}\text{N}(\alpha, p)^{17}\text{O} \quad \text{and} \quad ^{200}\text{Hg}(p, \alpha)^{197}\text{Au}$$

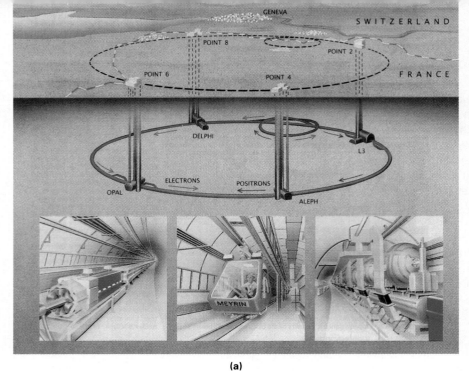

(a)

(b)

●FIGURE 29.1 Particle accelerators (a) The Large Electron-Positron (LEP) Collider on the border between France and Switzerland, near Geneva, has a circumference of 26.7 km. The electrons and positrons that it accelerates in opposite directions to speeds approaching that of light collide with an energy of over 100 billion electron volts (100 GeV). (b) A map showing the location of the planned Superconducting Super Collider (SSC). Several billion dollars had already been spent on research and development, and construction of the 87-km long tunnel for the facility had already begun, when Congress discontinued funding in the fall of 1993. The accelerator was to have used some 10,000 superconducting magnets to guide and focus beams of protons accelerated to energies of 20 trillion electron volts (20 TeV).

Q value—energy released or absorbed in a reaction

In general, the periodic table lists 105 elements, but only 90 elements occur naturally on Earth. Elements with proton numbers greater than uranium ($Z = 92$), as well as technetium ($Z = 43$) and promethium ($Z = 61$), are created artificially by nuclear reactions. The name technetium comes from the Greek word *technetos*, meaning "artificial," and technetium was the first unknown element to be created by artificial means. Elements up to $Z = 109$ have been confirmed.

Conservation of Mass-Energy and the Q Value

In every nuclear reaction, the total relativistic energy is conserved ($E = K + m_o c^2$, as shown in Chapter 25). Consider the reaction by which nitrogen is converted into oxygen, $^{14}N(\alpha, p)^{17}O$. By the conservation of total relativistic energy,

$$(K_N + m_N c^2) + (K_\alpha + m_\alpha c^2) = (K_O + m_O c^2) + (K_p + m_p c^2)$$

where the subscripts refer to the kinetic energy and mass of a particular particle and the respective masses are rest masses. Rearranging the equation, we have

$$K_O + K_p - K_N - K_\alpha = (m_N + m_\alpha - m_O - m_p)c^2 \qquad (29.1)$$

The **Q value** of the reaction is defined as

$$Q = (K_O + K_p) - (K_N + K_\alpha) \qquad (29.2)$$

The Q value is a measure of the total energy released or absorbed in a reaction, that is, the kinetic energy of the products of the reaction minus the kinetic energy of the reactants of the reaction. In terms of final and initial kinetic energies, $Q = K_f - K_i$. The Q value is an important quantity in nuclear reactions. However, the kinetic energies of the reactants and products of the reaction do not have to be measured, since by Eqs. 29.1 and 29.2,

$$Q = (m_N + m_\alpha - m_O - m_p)c^2 \qquad (29.3)$$

Similarly, in terms of a general reaction of the form $A + a \rightarrow B + b$, or $A(a, b) B$,

$$Q = (m_A + m_a - m_B - m_b)c^2 = (\Delta m)c^2 \qquad (29.4)$$

Thus, the Q value is given by the difference in the rest energies of the reactants and the products of a reaction. Note that this means that mass is generally converted into energy or vice versa. The amount of mass involved in these transformations, Δm, is sometimes referred to as the *mass defect* of the reaction.

Using the masses given in the parentheses under the reactants and products in the previous equation for $^{14}\text{N}(\alpha,p)^{17}\text{O}$ reaction, we obtain

$$
\begin{aligned}
Q &= (m_N + m_\alpha - m_O - m_p)c^2 \\
&= [(14.003074 \text{ u} + 4.002603 \text{ u}) - (16.999133 \text{ u} + 1.007825 \text{ u})]c^2 \\
&= (-0.001281 \text{ u})c^2
\end{aligned}
$$

or $\quad Q = (-0.001281 \text{ u})c^2[931.5 \text{ MeV}/(1 \text{ u})c^2] = -1.193 \text{ MeV}$

The negative Q value indicates that energy was absorbed in the reaction. When Q is less than zero (negative), the reaction is said to be **endoergic** (or endothermic). That is, energy (*ergic*) must be put into (*endo*) the reaction for it to proceed. In endoergic reactions, the kinetic energy of the reacting particles is at least partially converted into mass.

Endoergic—energy in

When the Q value of a reaction is positive ($Q > 0$), energy is released, and the reaction is said to be **exoergic** (or exothermic). That is, energy is produced by (*exo*) the reaction. In this case, mass is converted into energy and is carried away as kinetic energy of the products of the reaction.

Exoergic—energy out

EXAMPLE 29.1 ■ Endoergic or Exoergic Reaction?

Is the following reaction endoergic or exoergic?

$$
\begin{array}{ccccccc}
{}^{2}_{1}\text{H} & + & {}^{2}_{1}\text{H} & \rightarrow & {}^{3}_{2}\text{He} & + & {}^{1}_{0}\text{n} \\
\text{deuteron} & & \text{deuteron} & & \text{helium} & & \text{neutron} \\
(2.014102 \text{ u}) & & (2.014102 \text{ u}) & & (3.016029 \text{ u}) & & (1.008665 \text{ u})
\end{array}
$$

Point out examples of endoergic and exoergic reactions.

Solution. The nature of the reaction is determined by the Q value: endoergic, $Q < 0$; exoergic, $Q > 0$. The Q value is given by Eq. 29.4, so we first calculate the mass defect (Δm) of the reaction using the given masses:

$$
\begin{aligned}
\Delta m &= 2m_H - m_{He} - m_n \\
&= 2(2.014102 \text{ u}) - 3.016029 \text{ u} - 1.008665 \text{ u} = 0.00351 \text{ u}
\end{aligned}
$$

Then

$$
\begin{aligned}
Q &= (\Delta m)c^2 \\
&= (0.00351 \text{ u})(931.5 \text{ MeV}/\text{u}) = 3.27 \text{ MeV}
\end{aligned}
$$

A positive mass defect gives a positive Q, so the reaction is exoergic ($Q > 0$) —is converted to energy. (Notice that the Q value was computed directly from the mass difference using the mass-energy conversion factor, which eliminates using c^2 numerically and gives the energy in MeV.)

The Q value of radioactive decay is always positive ($Q > 0$), since energetic particles are emitted spontaneously. This is sometimes called the *disintegration*

TABLE 29.1 ■ Interpretation of Q Values

Q value	Effect
Positive ($Q > 0$)	Exoergic, mass converted into energy (mass of reactants greater than mass of products)
Negative ($Q < 0$)	Endoergic, energy converted into mass (mass of products greater than mass of reactants)

energy. The meanings of the positive and negative Q values are summarized in Table 29.1.

When the Q value of a reaction is negative, or the mass of the products is greater than the mass of the reactants, then a certain amount of kinetic energy is converted into rest mass. Such reactions will not occur unless enough kinetic energy is initially available. One might think that if a particle incident on a stationary nucleus had kinetic energy equal in value to the Q of the reaction ($K = |-Q|$)*, then the reaction would occur, since this energy is equivalent to the mass gain. However, if all the kinetic energy were converted to mass, there would be none left over, so the particles would be at rest after the reaction (all $v = 0$). But then momentum would not be conserved, as it must: the incoming particle had momentum, but the products wouldn't.

Hence, for the products of the reaction to have the same total momentum as the incident particle, the kinetic energy of the incident particle must be greater than $K = |-Q|$.

Threshold energy for an endoergic reaction: $K_{min} > |-Q|$

The minimum kinetic energy (K_{min}) that an incident particle needs to have to initiate an endoergic reaction is called the **threshold energy**. For classical (nonrelativistic) collisions, the threshold energy can be shown to be

$$K_{min} = \left(1 + \frac{m_a}{M_A}\right)|-Q| \tag{29.5}$$

where m_a and M_A are the masses of the incident particle and the *stationary* target nucleus, respectively.

EXAMPLE 29.2 ■ Threshold Energy

What is the threshold energy for the reaction $^{14}N(\alpha, p)^{17}O$?

Solution. The masses and the Q value for this reaction were given previously; thus we have

Given: $m_a = m_\alpha = 4.002603$ u   *Find:* K_{min} (threshold energy)
$M_A = m_N = 14.003074$ u
$|-Q| = 1.193$ MeV

Using Eq. 29.5, we can write

$$K_{min} = \left(1 + \frac{m_a}{M_N}\right)|-Q| = \left(1 + \frac{4.002603 \text{ u}}{14.003074 \text{ u}}\right)(1.193 \text{ MeV}) = 1.534 \text{ MeV}$$

Note that K_{min} is significantly larger than $K = |-Q|$.

---

*The kinetic energy is written as the absolute value of $-Q$, that is, $|-Q|$, because the kinetic energy cannot be negative. The sign of Q arises from the mass defect and merely indicates the gain or loss of energy to mass.

886

Reaction Cross Sections

More than one reaction is possible when a particle collides with a nucleus. As we have seen for an endoergic reaction, the incident particle must have a minimum kinetic energy to initiate a particular reaction. When a particle has more kinetic energy than the threshold energies of the several possible reactions, any of the reactions may occur. A measure of the probability that a particular reaction will occur is called the **cross section** of the reaction. If we knew the exact expression for the nuclear force and the form of the nuclear structure, we might be able to calculate the cross section for each possible reaction. However, at present we must determine the probabilities, or cross sections, of the possible reactions experimentally.

The cross section of a particular reaction is observed to vary with the kinetic energy of the incident particle. For positively charged incident particles, such as protons and alpha particles, the Coulomb barrier of the nucleus influences the reaction cross section. Thus the probability of a given reaction increases as the kinetic energy of the incident particle increases. Since reactions do not occur for relatively low energy particles, there must be barrier penetration (from the outside in). The increase in the cross section with particle energy may be thought of as being due to more energetic particles having a greater probability of tunneling through the Coulomb barrier.

Being electrically neutral, neutrons are unaffected by the Coulomb barrier. As a result, the cross section for a given reaction may be quite large for low-energy neutrons, in particular for reactions such as ^{27}Al(n, γ)^{28}Al. For low energies, the neutron reaction cross sections are often found to be proportional to $1/v$, where v is the speed of the neutron. That is, the probability of a reaction appears to be proportional to the time a neutron spends in the vicinity of a nucleus ($t \propto 1/v$). As the neutron energy increases, the cross section varies a great deal, as ● Fig. 29.2 shows. The peaks in the curve are referred to as resonances. The resonances are believed to be associated with nuclear energy levels.

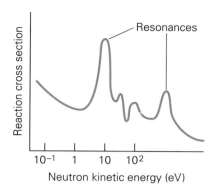

FIGURE 29.2 Reaction cross section A typical graph of neutron reaction cross section versus energy. The peaks where the probabilities of reactions are greatest are called resonances.

Review Figs. 28.6 and 28.7

29.2 ■ Nuclear Fission

In early attempts to make heavier elements artificially, uranium, the heaviest element known at the time, was bombarded with particles, mainly neutrons. An unexpected result was that the uranium nuclei sometimes broke apart or "split" into fragments. In the later 1930s, these fragments were identified as the nuclei of lighter elements. The process was dubbed "fission" after biological fission (the dividing of living cells).

Thus, in a **fission reaction**, a heavy nucleus divides into two lighter nuclei with the emission of two or more neutrons. Energy is emitted in the process, being carried off primarily by the neutrons and fission fragments. Some heavy nuclei undergo spontaneous fission, but at very slow rates. However, fission may be *induced*, and this is the important process in energy production. For example, when a ^{235}U nucleus absorbs an incident neutron, it may fission according to the reaction

Nuclear fission—"splitting" of the nucleus.

$$^{235}_{92}\text{U} + {}^{1}_{0}\text{n} \rightarrow ({}^{236}_{92}\text{U*}) \rightarrow {}^{140}_{54}\text{Xe} + {}^{94}_{38}\text{Sr} + 2({}^{1}_{0}\text{n})$$

The ^{235}U nucleus is fissionable.

The capture of a neutron results in the formation of an excited uranium-236 nucleus (● Fig. 29.3).

887

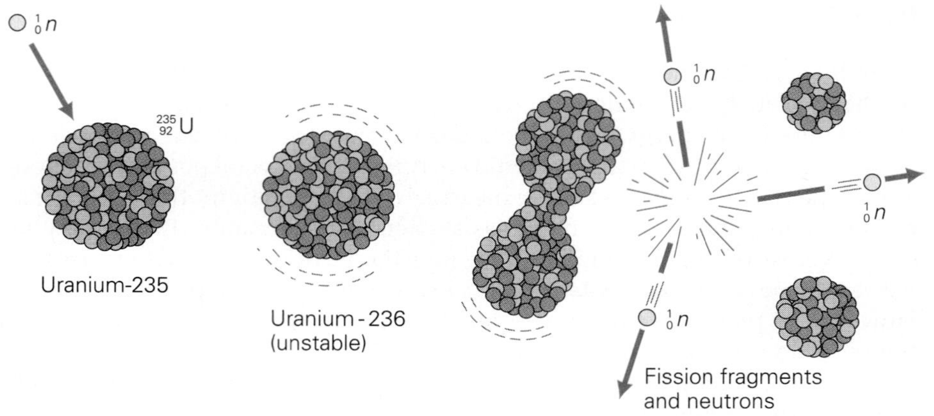

FIGURE 29.3 Fission When an incident neutron is absorbed by a fissionable nucleus, such as ^{235}U, the unstable nucleus (^{236}U) undergoes violent oscillations and breaks apart like a liquid drop, emitting two or more neutrons.

Uranium-235

Uranium-236
(unstable)

Fission fragments
and neutrons

According to the so-called liquid drop model, the ^{236}U nucleus undergoes violent oscillations and becomes distorted like a liquid drop. The separation of the nucleons into different parts of the "drop" weakens the nuclear force, and the repulsive electrical force of the protons between the parts causes the nucleus to split or fission. Some isotopes, however, capture neutrons without fissioning.

The preceding reaction is only one of a variety of ways in which uranium-235, the fissionable isotope of uranium, can fission. Others include (compound nuclei omitted)

$$ {}^{1}_{0}\text{n} + {}^{235}_{92}\text{U} \rightarrow {}^{141}_{56}\text{Ba} + {}^{92}_{36}\text{Kr} + 3({}^{1}_{0}\text{n}) $$

$$ {}^{1}_{0}\text{n} + {}^{235}_{92}\text{U} \rightarrow {}^{150}_{60}\text{Nd} + {}^{81}_{32}\text{Ge} + 5({}^{1}_{0}\text{n}) $$

Only certain nuclei undergo fission, and the probability of a fission reaction for a fissionable isotope depends on the energy of the incident neutrons. For example, the greatest cross sections for fission reactions for ^{235}U and ^{239}Pu are for "slow" neutrons with energies less than 1 eV. However, for ^{232}Th, the greatest cross section is for "fast" neutrons with energies of 1 MeV or more.

Rather than calculating the energy released in an exoergic fission reaction, we can obtain an estimate of its magnitude from the E_b/A curve for stable nuclei (Fig. 28.14). When a nucleus with a large mass number A, such as uranium, splits into two nuclei, it is in effect moving up along the downward-sloping tail of the E_b/A curve to a more stable state. As a result, the average binding energy per nucleon increases from a value of about 7.8 MeV to a value of approximately 8.8 MeV. Thus, the energy liberated is about 1 MeV per nucleon. In the first of the preceding fission reactions, there are 140 + 94 = 234 nucleons in the fission products, so the accompanying energy release is approximately

$$ \frac{1\text{ MeV}}{\text{nucleon}} \times 234 \text{ nucleons} \cong 234 \text{ MeV} $$

On a relative basis, this might seem to be a lot of energy. For example, the energy released *per atom* in a chemical exothermic process is on the order of a few hundred electron volts. On the other hand, this might not seem like much energy, since 200 MeV is only about 3×10^{-11} J. When you pick up your textbook from the desk, you do about 5 J of work or expend 5 J of energy. But keep in mind that there are billions and billions of nuclei in a small sample of

fissionable material. We simply need to have enough nuclei fissioning to produce practical amounts of energy.

This is accomplished by means of a **chain reaction.** For example, suppose
a ^{235}U nucleus fissions with the release of two neutrons (● Fig. 29.4). Ideally,
the neutrons may then initiate two more fission reactions, a process that results
in the availability of four neutrons. These neutrons may initiate more reactions,
and the process multiplies, the number of neutrons doubling with each generation. When this occurs uniformly with time, we have exponential growth.

To have a sustained chain reaction, there must be an adequate quantity of
fissionable material. The minimum mass required to produce a chain reaction
is called the **critical mass.** In effect, this means that there is enough fissionable
material that one nuetron from each fission event, on the average, goes on to
fission another nucleus. With more neutrons, the chain and energy output
would grow.

Several factors determine the critical mass. Most evident is the amount of
fissionable material. If the quantity of material is small, many neutrons will
escape from the sample before inducing a fission event, and the reaction will
die out. Also, the nuclei of other isotopes in the sample may absorb neutrons,
thereby limiting the chain reaction. As a result, the purity of the fissionable
isotope affects the critical mass.

In general, natural uranium is made up of the isotopes ^{238}U and ^{235}U. The
concentration of the fissionable ^{235}U isotope is only 0.7%. The remaining 99.3%
is ^{238}U, which may absorb neutrons for a nuclear reaction other than fission,
thereby preventing those neutrons from contributing to a chain reaction. To
have more fissionable ^{235}U nuclei in a sample and reduce the critical mass, the
^{235}U is concentrated. This enrichment varies from 3–5% ^{235}U for reactor-grade
material used in electrical generation to over 99% for weapons-grade material.
This is an important difference, since it is desirable that a nuclear reactor used
for electrical generation not explode like an atomic bomb.

Chain reactions take place almost instantaneously. If such a reaction proceeds uncontrolled in a fissionable sample of critical mass, the quick and enor-

A chain reaction requires a critical
mass.

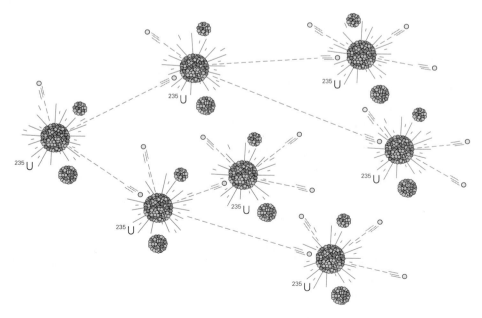

Figure 29.4

● **FIGURE 29.4 Fission chain
reaction** The neutrons that result
from one fission event may initiate
other fission reactions, and so on.
When enough fissionable material is
present, the sequence of reactions is
self-sustaining (a chain reaction).

A simulation of how neutrons from reactions induce reactions in other nuclei so that the process grows in a chain reaction.

  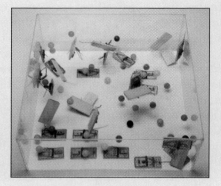

(a) Here, the "nuclei" are mousetraps loaded with a single Super Ball (neutron) that will be emitted when a reaction occurs. A Plexiglas enclosure ensures a "critical mass," or that the balls (neutrons) are reflected instead of "escaping."

(b) The reaction begins.

(c) The reaction grows, with an increasing number of fissions and release of energy.

mous release of energy from billions and billions of fissioning nuclei causes an explosion. (See Demonstration 19.) This is the principle of the atomic bomb. For the practical production of nuclear energy, the chain reaction process must be controlled. This is done in nuclear reactors.

Nuclear Reactors

Currently, the only type of practical nuclear reactor is based on the fission chain reaction. Research reactors designed primarily to produce artifical isotopes exist, but we will focus on the type of reactor that is commonly used to generate electricity. The technology for this is over 30 years old, the first nuclear reactor for the purpose of electrical generation having gone into operation in 1957 in Shippingport, Pennsylvania. A typical design for the components of a nuclear reactor is shown in ● Fig. 29.5. There are four key elements to a reactor: fuel rods, coolant, control rods, and moderator.

Tubes packed with pellets of uranium oxide form the **fuel rods**, which are located in the reactor core. A typical commercial reactor contains fuel rods bundled in fuel assemblies of around 200 rods each. A fuel rod assembly is constructed so that a coolant can flow around the rods to remove the energy emitted from the fission chain reaction. The reactors used in the United States are light water reactors, which means that ordinary water is used as a **coolant**. However, the hydrogen nuclei of ordinary water have a tendency to capture neutrons. The reaction is

$$\,^1_0n + \,^1_1H \to \,^2_1H + \gamma$$

This removes neutrons from the chain reaction, so enriched uranium with 3 to 5% ^{235}U is used to achieve a critical mass. A nuclear reactor can run for 3 or 4 years before having to be refueled.

The chain reaction and energy output of a reactor are controlled by means of boron or cadmium **control rods**, which can be inserted into and withdrawn from the reactor core. Cadmium and boron have a high cross section for absorbing neutrons. By inserting the control rods between the fuel rod assemblies, neutrons are absorbed and removed from the chain reaction. The control rods can be adjusted so that energy is released at a relatively steady rate. If more energy is needed, the rods are further withdrawn from the core. When the control rods are fully inserted, enough neutrons are removed to curtail the chain reaction and shut down the reactor.

Control rods—absorb neutrons and control chain reaction

The water flowing through the fuel rod assemblies acts not only as a coolant, but also as a **moderator**. The fission cross section of ^{235}U is largest for slow neutrons (kinetic energies less than 1 eV). The neutrons emitted from a fission reaction are fast neutrons with average energies of 2 MeV. These are slowed down, or their speed is moderated, by collisions. The hydrogen atoms in water are very effective in slowing down neutrons because their masses are nearly equal to the neutrons' masses. Recall from Section 6.4 that there is maximum energy transfer in a head-on collision of particles of equal mass. Not all collisions are head-on, but it takes only about 20 collisions to moderate fast neutrons to energies of less than 1 eV.

Moderator slows neutrons to increase reaction cross-section for uranium-235

This result was obtained in Section 6.4

Heavy water (D_2O, where D stands for deuterium) can also be used as a moderator, eliminating the need for enriched uranium. Unlike ordinary hydrogen, deuterium does not readily absorb neutrons, so natural uranium (0.7% ^{235}U) can be used as fuel. Most of the reactors in Canada are heavy water reactors.

Other materials, such as graphite (carbon), may be used as moderators. Because of their relatively heavy nuclei, not as much energy is transferred per collision. On the average, about 120 collisions with carbon atoms are needed

● **FIGURE 29.5 Nuclear reactor**
(a) A schematic diagram of a reactor showing a fuel rod assembly and a fuel rod. (b) The core of a reactor seen through shielding water. The characteristic blue glow is called Cerenkov radiation, which is produced by particles traveling at speeds greater than the speed of light in a transparent medium. [Nothing travels faster than c, the speed of light in vacuum, but in a transparent medium, the speed of light is less than c (Chapter 21) and so can be exceeded.]

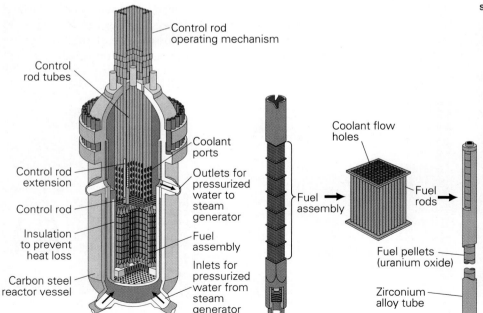

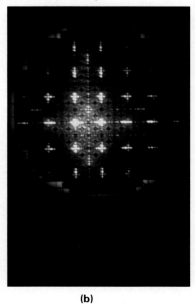

(b)

to produce slow, or thermal, neutrons. (A thermal neutron is one that has acquired the average thermal speed of the atoms of the moderator.)

Although not the best moderator, carbon in the form of graphite also permits a chain reaction to occur for natural (unenriched) uranium. The first experimental proof that a chain reaction was feasible was accomplished in such a reactor in 1942 by a team of scientists working with Enrico Fermi on the World War II Manhattan Project. The reactor was called a "pile" because it essentially consisted of a pile of graphite blocks. (A carbon reactor was involved in the 1986 accident at Chernobyl in the former Soviet Union, discussed below.)

The Breeder Reactor. In general, the ^{238}U in a reactor goes along for the ride, so to speak, without reacting. However, ^{238}U has an appreciable cross section for fast neutrons. Not all the neutrons are moderated, so there is some ^{238}U reaction, and there is conversion into ^{239}Pu via radioactive decay:

$$^1_0n + {}^{238}U \rightarrow {}^{239}U \rightarrow {}^{239}Np \rightarrow {}^{239}Pu$$

Plutonium-239, with a half-life of 24,000 years, is fissionable and serves as additional fuel that prolongs the time to reactor refueling. It is possible to promote the reaction of ^{238}U in a reactor by reduced moderation. When, on the average, one or more neutrons from ^{235}U fission are absorbed by ^{238}U nuclei to produce ^{239}Pu, then the same amount or more of fissionable fuel is produced (^{239}Pu) as is consumed (^{235}U). This is the principle of the **breeder reactor**, which produces more fissionable fuel than it consumes.

Work on the breeder reactor in the United States was stopped in the 1970s. France went on to develop an operational breeder reactor which provides nuclear fuel for its reactors. France and other nations are highly dependent on nuclear energy, as we will see below.

Electrical Generation. The components of a typical pressurized water reactor used in the United States are illustrated in ● Fig. 29.6. The heat generated by the controlled chain reaction is carried away by the water moderator surrounding the rods in the fuel assembly. The water in the reactor is pressurized to between 100 and 200 atmospheres. This is done so that it can be heated to temperatures over 300°C for more efficient heat removal. (Recall that the boiling point of water increases with pressure.) The hot water is pumped to a heat

The relationship between pressure and boiling point is discussed in Section 11.5

Figure 29.6

● **FIGURE 29.6 Pressurized water reactor** The diagram shows the components of a pressurized water reactor. The heat energy from the reactor core is carried away by the circulating water moderator. The water in the reactor is pressurized so that it can be heated to high temperatures for more efficient heat removal. The energy is used to generate steam, which drives the turbine that turns the generator.

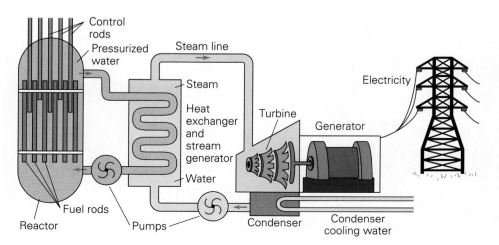

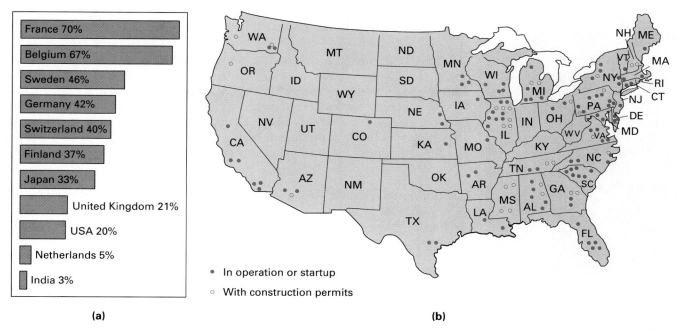

France 70%

Belgium 67%

Sweden 46%

Germany 42%

Switzerland 40%

Finland 37%

Japan 33%

United Kingdom 21%

USA 20%

Netherlands 5%

India 3%

• In operation or startup

○ With construction permits

(a)   **(b)**

exchanger, where the energy is transferred to the water of a steam generator. Notice that the reactor coolant and exchanger water are in separate, closed systems. (Why do you think this is?)

High-pressure steam then turns a turbine that operates a generator, as would be the case for a coal- or oil-fired plant. The steam is cooled and condensed after turning the turbine.

Nuclear energy is used to generate a substantial amount of the electricity in the world (Fig. 29.7a). Some twenty-five countries now produce nuclear-generated electricity, and twelve more plan to do so in the 1990s. More than 430 nuclear reactor units are in operation throughout the world, with about 110 operating units in the United States (Fig. 29.7b). With the increasing number of nuclear-generating facilities comes the fear of a possible nuclear accident and the release of radioactive materials into the environment. We now hear such terms as **LOCA** (loss-of-coolant accident) and **meltdown**. Should the reactor coolant be lost or stop flowing through the core, the reactor would shut down because of the loss of the moderator. However, residual reactions might cause the core to melt into a hot fissioning mass that could go through the floor of the containment building into the environment.

A partial meltdown occurred at the Three Mile Island (TMI) generating plant near Middletown, Pennyslvania. This was a LOCA in which fortunately only a relatively small amount of radioactive steam was vented into the atmosphere. The nuclear accident at Chernobyl was a meltdown caused by human error. The reactor involved was a graphite reactor, with some 1660 fuel assemblies encased in 1700 tons of graphite blocks (Fig. 29.8). Chemical explosions blew the top off the building, and when the fuel rods melted down, the graphite blocks burned like a massive coal pile. The radioactive smoke was carried over Europe by the wind. In the region near the plant, more than 30 people died. It is estimated that thousands will die of cancer related to the Chernobyl radiation.

Another problem with nuclear reactors is their leftovers, or radioactive waste. The by-products of the fission reactions are radioactive and have long half-lives. As a result, nuclear waste will be a continuing problem for years.

FIGURE 29.7 Nuclear electrical generation (a) A comparison of the percentages of nuclear electrical generation in various countries. Keep in mind that 20% of the electrical generation in a large country such as the United States represents more nuclear reactors than 67% of the electrical generation in a relatively small country such as Belgium. (b) Nuclear plant locations in the United States.

Severe reactor accident—LOCA and meltdown

29.3 ▪ Nuclear Fusion

● **FIGURE 29.8 Nuclear accident** The photo shows the damage to the reactor at Chernobyl in the former USSR. This accident released large amounts of radioactive materials into the environment with very dire consequences.

The other important type of reaction for the production of nuclear energy is fusion. In a **fusion reaction**, light nuclei fuse together to form a heavier nucleus, releasing energy in the process ($Q > 0$). A simple fusion reaction—the fusion of two deuterium nuclei ($^2_1$H), sometimes called a D-D reaction—was given in Example 29.1. This as we saw in the Example, releases 3.27 MeV of energy.

Another example is the fusion of deuterium and tritium (a D-T reaction):

$$\underset{(2.014102\ u)}{^2_1\text{H}} + \underset{(3.016049\ u)}{^3_1\text{H}} \rightarrow \underset{(4.002603\ u)}{^4_2\text{He}} + \underset{(1.008665\ u)}{^1_0\text{n}}$$

Computing the Q value for this reaction, it is found that there is a release of $Q = 17.6$ MeV.

Notice that the energy releases from these fusion reactions are very small in comparison to the more than 200 MeV energy release from a fission reaction. However, think about samples of hydrogen and uranium with equal masses. A given mass of hydrogen isotopes has many, many more nuclei than an equivalent mass of a heavy fissionable isotope. The complete fusion of a quantity of hydrogen gives almost three times the energy released from the complete fissioning of an equivalent amount of uranium.

We might say that our lives are dependent on nuclear fusion. Fusion is the source of energy for stars, including the Sun. In the initial stages of a star's life, there is hydrogen burning, or the fusion of hydrogen into helium. One such sequence of fusion reactions that is believed to be possible for solar energy output is as follows:

$$^1_1\text{H} + {}^1_1\text{H} \rightarrow {}^2_1\text{H} + {}^0_{+1}\text{e} + \nu$$

$$[{}^1_1\text{H} + {}^2_1\text{H} \rightarrow {}^3_2\text{He} + \gamma]$$

$$^3_2\text{He} + {}^3_2\text{He} \rightarrow {}^4_2\text{He} + {}^1_1\text{H} + {}^1_1\text{H}$$

The net effect of this sequence, called the *proton-proton cycle*, is the combining of four protons ($^1_1$H) to form one helium ($^4_2$He) nucleus plus two positrons ($^0_{+1}$e), two gamma rays (γ), and two neutrinos (ν), a particle to be discussed in the next section) and the release of energy:

$$4({}^1_1\text{H}) \rightarrow {}^4_2\text{He} + 2({}^0_{+1}\text{e}) + 2\gamma + 2\nu + \text{energy}$$

Fusion as a Source of Energy

In several ways, fusion appears to be the ideal energy source of the future. Enough deuterium exists in the oceans, in the form of heavy water, to supply our needs for centuries. Controlled fusion does not depend on a chain reaction, as we shall see, so there is less danger of loss of control and the release of radioactive material through a meltdown. Also, the light nuclei of fusion products have relatively short half-lives. For example, tritium has a half-life of 12.3 years, as compared to hundreds or thousands of years for fission by-products.

However, there are many unresolved technical problems in the production of fusion energy for practical use.

Very high temperatures are needed to initiate fusion reactions. This reflects the energy needed to overcome the Coulomb repulsion of the fusing nuclei. Temperatures on the order of millions of degrees are needed to initiate **thermonuclear reactions**, as they are called. Because of this, practical fusion reactors have not been achieved. The problem is in confining sufficient energy in a reaction region to maintain the necessary high temperatures. Uncontrolled fusion has been demonstrated in the form of the hydrogen (H) bomb. In this case, the fusion reaction was initiated by a small atomic or fission bomb.

At such high temperatures, electrons are stripped from their nuclei, and a gas of positively charged ions and free, negatively charged electrons is obtained. Such a gas of charged particles is called a **plasma**. Plasmas have a number of special physical properties. As a result, they are sometimes referred to as the fourth phase of matter, a term used in 1879 by William Crookes, an English chemist, who generated plasmas in gas-discharge or Crookes tubes.

Plasma—a fourth phase of matter

The problem of plasma confinement is being approached in two ways: magnetic confinement and inertial confinement. Since a plasma is a gas of charged particles, it can be controlled and manipulated by using electric and magnetic fields. In **magnetic confinement**, magnetic fields are used to hold the plasma in a confined space, a so-called magnetic bottle (see Fig. 18.22).

Electric fields are used to produce electric currents in the plasma that raise its temperature. Temperatures of 100 million kelvins have been achieved in a design that uses a donut-shape called a *tokamak* (Fig. 29.9).

However, to initiate fusion, it is not enough to have a very high temperature; there are also requirements on the plasma density and confinement time. The trick is to put all these together in a fusion reactor. The generation of 1.7 MW of power for less than a second in a magnetically confined plasma, achieved in 1991, represents the best result obtained so far.

Inertial confinement depends on an implosion technique similar to that used in the ^{239}Pu bomb mentioned earlier. Hydrogen fuel pellets would be either dropped or positioned in a reactor chamber (Fig. 29.10). Pulses of laser, electron, or ion beams would then be used to implode the pellet, producing compression and high temperatures. Fusion would occur if the pellet stayed together for a sufficient time, which depends on its inertia (hence the name, inertial confinement). At this time, lasers and particle beams are not powerful enough to sustain fusion by this means.

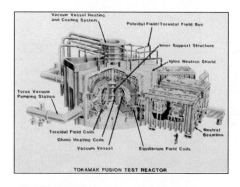

TOKAMAK FUSION TEST REACTOR

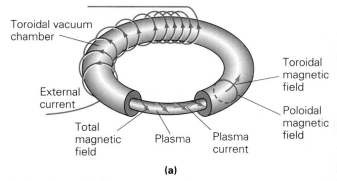

(a)

(b)

 FIGURE 29.9 Magnetic confinement Magnetic confinement is one method by which controlled nuclear fusion might be achieved. **(a)** Tokamak configuration showing the B field generated by external currents. The magnetic field confines the plasma in the ring. **(b)** Photo of the Princeton Tokamak Fusion Test Reactor (TFTR) and an artists's cutaway drawing.

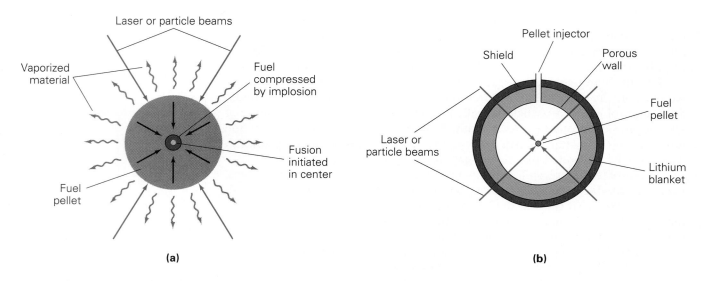

(a)

(b)

(c)

●**FIGURE 29.10 Inertial confinement** Another method by which controlled nuclear fusion might be achieved is inertial confinement. **(a)** An illustration of fuel pellet implosion. The compression is enhanced by the vaporization of the outer shell material of the pellet. **(b)** A schematic diagram of a possible fusion reactor cavity. The lithium blanket captures neutrons from the fusion reactions, producing tritium, which may be used as a fuel. **(c)** The world's most powerful laser, Nova, created a brief burst of fusion in a tiny fuel capsule. In the 50 trillionths of a second it lasted, the reaction gave off ten trillion neutrons.

However, research continues. The technological problems of controlled thermonuclear fusion are enormous, but so are the benefits. Practical energy production from fusion is not expected until well into the next century.

29.4 ■ Beta Decay and the Neutrino

As described in Section 28.2, beta decay appears to be a straightforward process. That is, certain radioactive nuclei naturally decay and emit an electron or a positron in the process. Examples of β^- and β^+ reactions are

$$\underset{(14.003242 \text{ u})}{^{14}_{6}\text{C}} \rightarrow \underset{(14.003074 \text{ u})}{^{14}_{7}\text{N}} + {}^{0}_{-1}\text{e}$$

$$\underset{(13.005739 \text{ u})}{^{13}_{7}\text{N}} \rightarrow \underset{(13.003355 \text{ u})}{^{13}_{6}\text{C}} + {}^{0}_{+1}\text{e}$$

However, more than this is going on in these beta decay processes. When they are analyzed in detail, there seem to be some violations of physical principles. One of the difficulties arises with energy. The energy released in the above β^- process may be calculated from the mass difference in the nuclei*:

$$\Delta m = 14.003242 \text{ u} - 14.003074 \text{ u} = 0.000168 \text{ u}$$

and

$$E = (0.000168 \text{ u})(931.5 \text{ MeV/u}) = 0.156 \text{ MeV}$$

Thus, the decay reaction has a Q value or disintegration energy of 0.156 MeV. The electron, being a light particle, would be expected to carry away virtually all of this energy or to have a $K_{max} = Q$. However, this is not what happens.

When the kinetic energies of the electrons from a given beta decay process are determined, a continuous spectrum of energies is observed up to a cutoff point of $K_{max} \cong Q$ (●Fig. 29.11). That is, electrons are emitted that have less

*Using the atomic mass of ^{14}N takes into account the emitted electron, since the daughter atom in beta decay would have only six electrons like the parent ^{14}C atom.

kinetic energy than expected; in fact, most of the beta particles have $K < Q$. This would appear to be a violation of the conservation of energy, since not all the energy is accounted for.

Nor is this the only difficulty. Observations of individual beta decay processes show that the emitted electron and the daughter nucleus from a stationary parent nucleus do not always leave the disintegration site in opposite directions. Thus, we have an apparent violation of the conservation of linear momentum.

To top it off, there is also an apparent violation of the conservation of angular momentum. Nucleons and electrons have intrinsic (spin) quantum numbers of $\frac{1}{2}$. For a nucleus with an even number of nucleons (such as ^{14}C), the total angular momentum quantum number will be an integer, since an even number of spin $\frac{1}{2}$ particles is present. When an electron is created in the decay process, an odd number of spin $\frac{1}{2}$ particles is present, and the total angular momentum quantum number will be a half-integer. Therefore, the sums of the total angular momentum quantum numbers are not equal before and after spontaneous decay.

What, then, is the problem? We would hope that our conservation laws are not invalid. The other alternative is that these apparent violations are telling us something about nature that we do not know or do not yet recognize. The difficulties would be resolved if we assumed that, in addition to the electron, another unobserved particle of intrinsic spin $\frac{1}{2}$ were created and emitted in beta decay.

This explanation and the existence of such a particle were first suggested in 1930 by Wolfgang Pauli. The particle was christened the **neutrino** (meaning "little neutral one"). The name suggests that, by the conservation of charge for beta decay, an additional particle must necessarily be elecrtrically neutral. A charged particle would be easily detectable, and since the neutrino had not been observed, it must interact very weakly with matter, that is, by a *weak interaction* or a *weak nuclear force*. (This force is discussed in the next section.)

The observations of beta decay suggest that the neutrino has zero rest mass and therefore travels with the speed of light like a photon. It also has linear momentum p with a total relativistic energy $E = pc$ and has an intrinsic spin quantum number of $\frac{1}{2}$. In 1956, a particle with these properties was detected experimentally, and the existence of the neutrino was established. Thus, we add the neutrino to the list of subatomic particles. The nuclear equations for the previous beta decay reactions become

$$^{14}_{6}\text{C} \rightarrow {}^{14}_{7}\text{N} + {}_{-1}^{0}\text{e} + \bar{\nu}_e$$

and

$$^{13}_{7}\text{N} \rightarrow {}^{13}_{6}\text{C} + {}_{+1}^{0}\text{e} + \nu_e$$

where the Greek letter ν (nu) is the symbol used for the neutrino. The symbol with a bar over it represents an antineutrino. This bar method is a common way of indicating an antiparticle. In general, a neutrino is emitted in β^+ decay, and an antineutrino is emitted in β^- decay. The e subscript identifies the neutrinos as associated with beta decay. As you will learn in a later section, there is another type of neutrino.

Note that the antineutrino is associated with the decay of a neutron:

$$\text{n} \rightarrow \text{p} + \text{e}^- + \bar{\nu}_e$$

The neutrino ν_e is associated with the decay of a proton into a neutron and a positron. A free proton cannot decay by positron emission because of the

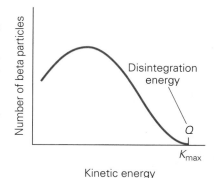

FIGURE 29.11 **Beta ray spectrum** For a typical beta decay process, most beta particles are emitted with $K < Q$, leaving energy unaccounted for.

You may wish to discuss with students the problem of the "missing mass" and the theory that neutrinos may possess a very small rest mass. If this is true, the ocean of neutrinos that pervades the universe could account for the missing mass. They could provide enough gravitational attraction to ultimately halt the Big Bang expansion and perhaps reverse it, producing a closed universe that collapses back on itself.

conservation of energy (rest masses). But because of binding energy effects, this does occur in the nucleus in β^+ decay with the emission of a neutrino. We will discuss neutrinos and their relationship to the weak nuclear force in a later section.

29.5 ■ Fundamental Forces and Exchange Particles

All interactions take place by means of forces. The forces involved in everyday activities are complicated because of the large numbers of atoms or molecules that make up ordinary objects. Frictional forces, the forces holding this book together, and so on, are actually due to the electromagnetic forces between atoms. Looking at the fundamental interactions between particles makes things simpler. On this level, there are only four known **fundamental forces**: the *gravitational force*, the *electromagnetic force*, the *strong nuclear force*, and the *weak nuclear force*.

The most familiar of these four are the gravitational and electromagnetic forces. Gravity acts between all particles, while the electromagnetic force is restricted to charged particles. Both forces have an inverse square relationship for the separation distance between interacting particles, and an infinite range.

Classically, the action-at-a-distance of a force is described by using the concept of a field. For example, a charged particle is considered to be surrounded by an electric field that *interacts* with other charged particles. Quantum mechanics, on the other hand, provides an alternative description of how forces are transmitted. This process is pictured as an exchange of particles, analogous to you and another person interacting by tossing a ball back and forth.

The creation of such particles would classically violate the conservation of energy. However, within the limitations of the uncertainty principle, a particle can be created for a *short time* with no extra energy. For ordinary time intervals, the conservation of energy is obeyed. But for *extremely* short time intervals, the uncertainty principle *permits* a large uncertainty in energy ($\Delta E \Delta t \geq h/2\pi$), so energy conservation may be briefly violated. A particle created in such a manner and time is called a **virtual particle**.

The fundamental forces are considered to be carried by virtual **exchange particles**. The exchange particle for each force is different in mass. The greater the mass of the particle, the greater the amount of energy required to create it and the shorter time it exists. Since a massive particle can exist for only a short time, the range of the interaction for the associated force would be small. That is, the range of an exchange particle is inversely proportional to its mass.

The exchange particle for the electromagnetic force is a (virtual) **photon**. As a "massless" particle, it has an infinite range, as does the electromagnetic force. The idea of a particle exchange is represented by the Feynman diagram of the collision of two electrons in ●Fig. 29.12. Such space-time diagrams are named after American scientist and Nobel Prize winner Richard Feynman (1918–1988), who used them to analyze electrodynamic interactions. The important points are the vertices of the diagram. One electron may be considered to create a virtual photon at point A and the other electron to absorb it at point B. Each of the two interacting particles undergoes a change in energy and momentum by virtue of the exchange photon and force interaction.

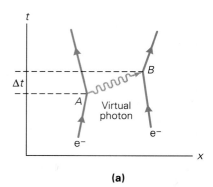

●**FIGURE 29.12 Feynman diagram for an electron-electron interaction** The interacting electrons undergo a change in energy and momentum due to the exchange of a virtual photon, which is created at A and absorbed at B in a time (Δt) that is consistent with the uncertainty principle.

Using the idea of exchange particles, Japanese physicist H. Yukawa in 1935 proposed that the short-range strong nuclear force between two nucleons is associated with an exchange particle called the **meson**. An estimate of the mass of this particle may be gained from the uncertainty principle. If a nucleon were to create a meson, the conservation of energy would have to be violated by an amount of energy at least as great as the meson rest mass or

$$\Delta E = (\Delta m)c^2 = m_m c^2$$

where m_m is the rest mass of the meson.

By the uncertainty principle, the meson would be absorbed in the exchange process in a time on the order of

$$\Delta t = \frac{h}{2\pi \Delta E} = \frac{h}{2\pi m_m c^2}$$

In this time, the meson could travel a distance not greater than

$$R = c\Delta t = \frac{h}{2\pi m_m c} \tag{29.6}$$

Taking this distance to be the range of the nuclear force ($R \cong 1.4 \times 10^{-15}$ m) and solving for m_m gives

$$m_m \cong 274 m_e$$

where m_e is the electron rest mass. Thus, if the meson existed, it would be expected to have a rest mass about 270 times that of an electron.

Of course, virtual mesons of an exchange process cannot be observed. But if sufficient energy were supplied to colliding nucleons, real mesons might be created through the energy made available in the collision process. The real mesons could then leave the nucleus and be detected. At the time of Yukawa's prediction, there were no known particles with masses between that of the electron (m_e) and that of the proton ($m_p = 1836 m_e$).

In 1936, Yukawa's prediction seemed to come true when a new particle with a mass of about 200 m_e was discovered in cosmic rays. Originally called the μ meson, and now just **muon**, the particle was shown to have two charged varieties, positive and negative of electronic magnitude, with a rest mass of

$$m_{\mu^\pm} = 207 m_e$$

However, further investigations showed that the muon did not behave like the particle of Yukawa's theory. In particular, the interaction of muons with matter (nuclei) was very weak. The muons from cosmic radiation could penetrate a large mass of material as evidenced by their detection in deep mines.

This situation was a source of controversy and confusion for several years. But in 1947, two more charged particles of the appropriate mass range were discovered in cosmic radiation. These particles were called π mesons (primary mesons), and now more commonly **pions**, together with a less massive electrically neutral π meson. Measurement showed the masses of the pions to be

$$m_{\pi^\pm} = 274 m_e \quad \text{and} \quad m_{\pi^0} = 264 m_e$$

Moreover, the pion interacted strongly with matter. The pions fulfilled the requirements of Yukawa's theory, and it has been generally accepted that this meson is the particle that transmits or mediates the strong nuclear force. The Feynman diagrams of nucleon-nucleon interactions are illustrated in ● Fig. 29.13.

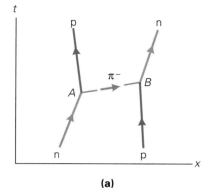

(a)

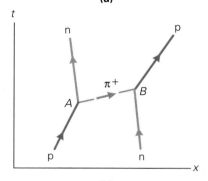

(b)

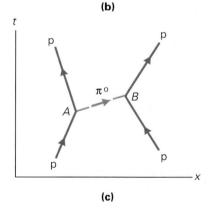

(c)

● **FIGURE 29.13 Feynman diagrams for nucleon-nucleon interactions** Diagrams for nucleon-nucleon interactions through the exchange of virtual pions. **(a)** n-p reaction, **(b)** p-n reaction, **(c)** p-p reaction.

Free mesons are unstable. For example, the π^+ particle quickly decays (in about 10^{-8} s) into a muon:

$$\pi^+ \rightarrow \mu^+ + \nu_\mu$$

Here ν_μ is a μ neutrino which differs from the electron neutrino produced in beta decay. The muons also decay into positrons and electrons with the emission of both types of neutrinos, for example,

$$\mu^+ \rightarrow e^+ + \nu_e + \overline{\nu}_\mu$$

The discrepancies in beta decay discussed in the preceding section led to another discovery. Electrons and neutrinos are emitted from unstable nuclei, but there was evidence that they did not exist *inside* the nucleus. Enrico Fermi proposed that these particles did not exist before being emitted but were instantaneously created in the decaying radioactive nucleus. For β^- decay, this would mean that a single neutron was in some way transmuted into three particles:

$$n \rightarrow p^+ + e^- + \overline{\nu}_e$$

That a neutron could do this was confirmed by the observation of free neutrons. Free neutrons disintegrate after a few minutes into a proton, an electron, and a neutrino. But the question arose as to what force could cause a neutron to disintegrate in this manner. None of the known forces was applicable, so some other force must be acting in beta decay. Decay rate measurements indicated that the force was extraordinarily weak—much weaker than the electromagnetic force, but still much stronger than the gravitational force. Thus, the **weak nuclear force** was discovered.

It was originally thought that the weak interaction was localized, without any range at all. However, we now know that the weak force has a range of about 10^{-17} m. In terms of an exchange particle, this means that the virtual carriers must be much more massive than the pions of the strong nuclear force. The virtual exchange particles for the weak nuclear force were called **W particles**. W (*weak*) particles have masses about 100 times that of a proton, which explains the extremely short range and weakness of the force. The existence of the W particle was confirmed in the 1980s when accelerators were built with enough energy to create the first real W particles.

The weak force is the only force that acts between neutrinos, which explains why they are so difficult to detect. Research has shown that the weak force is involved in the transmutation of other subatomic particles. In general, the weak force is limited to transmuting the identities of particles within the nucleus. The only way it manifests its existence in the outside world is through the emitted neutrinos. One highly noticeable but infrequent announcement of the weak force at work is during the explosion of a supernova. The collapse of the core of an aging star gives rise to a huge release of neutrinos. The weak force of the neutrinos blasts the outer layers of the star into space in a cataclysmic explosion of a supernova.

Finally, let's not forget the gravitational force. The exchange particles of the gravitational force are called **gravitons**. There is still no firm evidence of the existence of this massless particle. Several searches have been made to detect the graviton or gravity waves, but the relative weakness of its interaction makes it a very elusive particle. A comparison of the relative strengths of the fundamental forces is given in Table 29.2

W particle—the exchange particle for the weak nuclear force

Graviton—the exchange particle for the gravitational force

TABLE 29.2 ■ Fundamental Forces

Force	Relative Strength	Action Distance	Exchange Particle	Particles with Interaction
Strong nuclear	1	Short range ($\cong 10^{-15}$ m)	Pion (π meson)	Hadrons*
Electromagnetic	10^{-3}	Inverse square (infinite)	Photon	Electrically charged
Weak nuclear	10^{-8}	Extremely short range ($\cong 10^{-17}$ m)	W particle†	All
Gravitational	10^{-45}	Inverse square (infinite)	Graviton	All

*Hadrons are elementary particles (see Section 30.5).
†Actually, three particles are involved: W^+, W^-, and Z. These are described in Sections 29.5 and 29.6.

29.6 ■ Elementary Particles

The fundamental particles, the building blocks of atoms, are referred to as **elementary particles**. Simplicity reigned when it was thought that an atom was an indivisible particle and thus *the* elementary particle. However, scientists now know of a variety of subatomic particles, many of which have not been mentioned here. Indeed, scientists are working to simplify and reduce a long list of elementary particles, since some may be variations of one type of particle.

There are several systems for classifying elementary particles on the basis of their various properties. One classification uses the distinction of nuclear force interactions. Particles that experience or interact by the strong nuclear force are called **hadrons**. These include the proton, neutron, and pion. Other particles, which interact via the weak nuclear force but not the strong force, are called **leptons** ("light ones"). The lepton family includes the electron, muon, and their neutrinos.

Hadrons—strong nuclear force interactions

Leptons—weak nuclear force interactions

Note in Table 29.2 that gravity and the weak force are the only forces acting between *all* particles. But the force of gravity is so weak in elementary particle interactions that it can, in general, be neglected. This leaves the weak force as the only measurable force that interacts with all particles.

Let's take a brief look at the lepton and hadron families.

Leptons

The most familiar lepton is the electron. It is apparently the only member of the lepton family that exists naturally in atoms. There is no evidence of any internal structure, and, measured to be smaller than 10^{-17} m in size, the electron is considered to be a point particle.

Muons were first observed in cosmic rays. They are electrically charged (μ^-) and 200 times heavier than an electron. Since they appear not to have any internal structure, they are sometimes called heavy electrons. Muons are unstable and decay in about 10^{-6} s. This is the decay reaction:

$$\mu^- \rightarrow e^- + \bar{\nu}_e + \nu_\mu$$

Recall that muon decay was used as evidence for relativistic time dilation in Chapter 25.

A third charged lepton has been discovered. Known as a tau (τ^-) particle or **tauon**, it has a mass twice that of a proton. The electron, muon, and tauon are all negatively charged and appear to have no internal structure. Their antiparticles are positively charged.

The only other known leptons are neutrinos, which are present in cosmic rays and emitted in some radioactive decays. Neutrinos appear to have no mass and to travel at the speed of light. They feel neither the electromagnetic force nor the strong forces and so pass through matter as if it weren't there. Neutrinos are so weakly interacting that most neutrinos striking the Earth pass right through it.

Neutrinos come in several varieties. The electron neutrino (ν_e) and muon neutrino (ν_μ) are well documented, and it is believed that a tau neutrino (ν_τ) also exists. There is an antineutrino for each of these types.

This completes the list of leptons. With a total of six leptons *plus antiparticles*, there are twelve different leptons in all. Current theories predict that there should be no others.

Hadrons

Another category of elementary particles is called hadrons. All hadrons interact by the strong force, the weak force, and gravity. Also, some are electrically charged. The best known hadrons are nucleons: the proton and the neutron. All others are short-lived and decay via the weak force or more rapidly under the influence of the strong force.

The number of hadrons suggests that they are perhaps composites of other elementary particles. Some help came to sorting out the hadron "zoo" in 1963 when Murray Gell-Mann and George Zweig of Caltech put forth the quark theory. It suggested that **quarks** were elementary charged particles that made up hadrons. They could combine only in two possible ways, either in trios or in quark-antiquark pairs. Combinations of three quarks produce relatively heavy hadrons called **baryons** ("heavy ones"), which include the proton and neutron. Quark-antiquark pairs form lighter particles called *mesons*.

To account for the hadrons that were known at that time, the theory proposed three types or "flavors" of quarks. These were given the names "up" (u), "down" (d), and "strange" (s). In addition, the quarks carried fractional electronic charges. The u, d, and s quarks had charges of $+\frac{2}{3}e$, $-\frac{1}{3}e$, and $-\frac{1}{3}e$, respectively, antiquarks having charges of opposite signs (for example $\bar{u}$ with a charge of $-\frac{2}{3}e$). Thus, the quark combinations for the proton and neutron are uud and udd, respectively. A positive pion (π^+) is a ud combination.

Despite much experimental effort, no isolated quark has been observed, and they are not believed to exist freely outside of the nucleus. This would explain why we do not observe fractional electronic charges in nature. Quarks can also exist in excited states, similar to the excited states of an atom. It is thought that many of the observed hadrons might be excited states of certain combinations of quarks.

Quarks interact by the strong force, but they are also subject to the weak force. A weak force acting on a quark changes its flavor and gives rise to the decay of hadrons.

The discovery of new elementary particles in the 1970s led to the addition of more quark flavors, which were called "charm" (c), "top" (or "truth," t), and "bottom" (or "beauty," b). A summary of the quark flavors is given in Table 29.3. There is an oppositely charge antiquark for each quark.

Since quarks interact, it is postulated that they too have interacting exchange particles. The exchange particle for quarks has been dubbed the **gluon**. Gluons

TABLE 29.3 ■ Types of Quarks

Name	Symbol	Charge
Up	u	$+\frac{2}{3}e$
Down	d	$-\frac{1}{3}e$
Strange	s	$-\frac{1}{3}e$
Charm	c	$+\frac{2}{3}e$
Top (truth)	t	$+\frac{2}{3}e$
Bottom (beauty)	b	$-\frac{1}{3}e$

bind or "glue" hadrons together, thus replacing the pion as the hadron exchange particle.

The quark theory was further extended in terms of a force field. To give the strong force a field representation, each quark is said to possess the analog of electric charge that is the source of the gluon field. Instead of calling this property "charge," it was called "color," with no relationship to ordinary color. Each quark can come in one of three colors: red, green, and blue. There are corresponding anticolors for antiquarks.

When a quark emits or absorbs a gluon, it changes color. The effect is to change the identity of a quark—for example, a blue quark to a red quark. In this respect, the strong force resembles the weak force (for which there is a change of one particle into another with the exchange of a W particle).

Actually, the quark color scheme was developed after a major discovery was made concerning the weak force. Like Maxwell's combining the electric force and the magnetic force into a single electromagnetic force, the weak force and the electromagnetic force were combined, or shown to be two parts of a single **electroweak force**. This came about as the result of a theory put forth by Sheldon Glashow, Abdus Salam, and Steven Weinberg, for which they received a Nobel Prize in 1979. In this theory, weakly interacting particles such as electrons and neutrinos carry a weak charge that gives rise to a weak force field. The exchange particles for the weak force interaction are the heavy, electrically charged W^+ and W^- particles, along with a neutral Z particle for reactions in which there is no change in or transfer of charge. Two weak interactions are illustrated in ● Fig. 29.14.

In 1983, the existence of the Z was confirmed, and the four fundamental forces were reduced to three. Of course, scientists would like to reduce the list even further. A theory that would merge the strong nuclear force and the electroweak force into a single unified force is the so-called *grand unified theory* (GUT). Some two dozen exchange particles are required, including the 12 already known particles.

Should the three fundamental forces be reduced to two by the GUT or a similar theory, there is the hope of a further reduction with the idea that all forces are part of a single *superforce*. The combining of a grand unified force with gravity is a real challenge. While the three components of a grand force may be represented as force fields in space and time, our current view is that gravity *is* space and time.

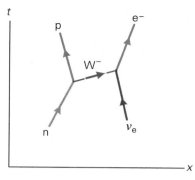

(a)

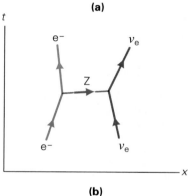

(b)

● **FIGURE 29.14 Weak force interactions** The Feynman diagrams for **(a)** the decay of a neutron into a proton and an electron through the exchange of a W^- particle and **(b)** the scattering of a neutrino by an electron through the exchange of a neutral Z particle.

Important Concepts

You should be able to define and/or explain these chapter concepts clearly.

nuclear reactions	fuel rods	inertial confinement	elementary particles
particle accelerators	coolant	neutrino	hadrons
Q value	control rods	fundamental forces	leptons
endoergic	moderator	virtual particle	tauon
exoergic	breeder reactor	exchange particle	quarks
threshold energy	LOCA	photon	baryons
cross section	meltdown	muon	gluon
fission reaction	fusion (thermonuclear)	pion	electroweak force
chain reaction	reaction	weak nuclear force	
critical mass	plasma	W particle	
nuclear reactor	magnetic confinement	graviton	

Important Relationships for Review

These relationships are mathematical statements of the concepts and principles presented in the chapter. You should be able to identify the symbols and to explain the relationships before proceeding to the Exercises. In-text equation reference numbers are given for convenience.

Q Value:

$$Q = (m_A + m_a - m_B - m_b)c^2 = (\Delta m)c^2 \qquad (29.4)$$

Threshold Energy:

$$K_{\min} = \left(1 + \frac{m_a}{M_A}\right)|Q| \qquad (29.5)$$

Range of Exchange Particle:

$$R = c\Delta t = \frac{h}{2\pi m_m c} \qquad (29.6)$$

Exercises

29.1 ■ Nuclear Reactions

1 The Q value of a reaction (a) is given by the difference in the kinetic energies of the products and reactants, (b) is equal to the mass-energy defect, (c) may be positive or negative, (d) all of these. (d)

● **2** To initiate an endoergic reaction, a particle incident on a stationary nucleus must have (a) a minimum disintegration energy, (b) a kinetic energy equal to the Q value, (c) a kinetic energy less than a certain threshold energy, (d) a kinetic energy equal to the Q value plus a factor of m_a/M_A of the Q value. (d)

3 How does the threshold energy of an incident particle in a nuclear reaction vary with the mass of the target nuclei? (Sketch a graph if you can.) What is the interpretation of $Q = 0$? see ISM

4 Complete the following nuclear reactions:

(a) $^{1}_{0}n + {}^{40}_{18}Ar \rightarrow \underline{{}^{(a)}\,{}^{41}_{19}K} + {}^{0}_{-1}e$

(b) $^{1}_{0}n + {}^{235}_{92}U \rightarrow {}^{98}_{40}Zr + \underline{{}^{(b)}\,{}^{135}_{52}Te} + 3({}^{1}_{0}n)$

(c) $^{1}_{0}n + {}^{235}_{92}U \rightarrow {}^{133}_{51}Sb + {}^{99}_{41}Nb + \underline{{}^{(c)}\,4({}^{1}_{0}n)}$

(d) $\underline{{}^{(d)}\,{}^{14}_{7}N}.\ (\alpha, p)\ {}^{17}_{8}O$

(e) $^{10}_{5}B(\underline{{}^{(e)}\,n}, \alpha){}^{7}_{3}Li$

5 Write the compound nuclei for the reactions in Exercise 4. (a) $^{41}_{18}Ar^*$ (b) $^{236}_{92}U^*$ (c) $^{236}_{92}U^*$ (d) $^{18}_{9}F^*$ (e) $^{11}_{5}B^*$

6 ■■ Complete the following nuclear reactions:

(a) $^{13}_{6}C + {}^{1}_{1}H \rightarrow \gamma + \underline{{}^{(a)}\,{}^{14}_{7}N}$

(b) $^{10}_{5}B + {}^{4}_{2}He \rightarrow {}^{12}_{6}C + \underline{{}^{(b)}\,{}^{2}_{1}H}$

(c) $^{27}_{13}Al(\alpha, n)\underline{{}^{(c)}\,{}^{30}_{15}P}$

(d) $^{14}_{7}N(\alpha, p)\underline{{}^{(d)}\,{}^{17}_{8}O}$

(e) $^{13}_{6}C(p, \alpha)\underline{{}^{(e)}\,{}^{10}_{5}B}$

7 ■■ Give the compound nuclei for the reactions in Exercise 6. (a) $^{14}_{7}N^*$ (b) $^{14}_{7}N^*$ (c) $^{31}_{15}P^*$ (d) $^{18}_{9}F^*$ (e) $^{14}_{7}N^*$

● **8** ■■ Lead-211 undergoes beta decay. Find the energy released in the decay reaction. 0.868 MeV

9 Determine what the daughter nuclei in the following decay equations might be and whether the reactions occur spontaneously:

(a) $^{27}_{13}Al \rightarrow \underset{(26.981541\ u)}{\underline{{}^{(a)}\,{}^{27}_{14}Si}} + {}^{0}_{-1}e$

(b) $^{226}_{88}Ra \rightarrow \underset{(226.025406\ u)}{\underline{{}^{(b)}\,{}^{222}_{86}Rn}} + {}^{4}_{2}He$

(c) $^{197}_{79}Au \rightarrow \underset{(196.96656\ u)}{\underline{{}^{(c)}\,{}^{193}_{77}Ir}} + {}^{4}_{2}He$

10 Find the Q value for the alpha decay of uranium-238:

$$\underset{(238.050786\ u)}{{}^{238}_{92}U} \rightarrow \underset{(234.043583\ u)}{{}^{234}_{90}Th} + \underset{(4.002603\ u)}{{}^{4}_{2}He}$$

Would you expect the Q to be positive or negative? 4.28 MeV

11 Show that the Q value for the D-T reaction given in Section 29.3 is 17.6 MeV. see ISM

12 ■■ Find the threshold energy for the following reaction:

$$\underset{(15.994915\ u)}{{}^{16}_{8}O} + \underset{(1.008665\ u)}{{}^{1}_{0}n} \rightarrow \underset{(13.003355\ u)}{{}^{13}_{6}C} + \underset{(4.002603\ u)}{{}^{4}_{2}He}$$

2.36 MeV

13 Determine the minimum kinetic energy of an incident alpha particle that will initiate the following reaction:

$$\underset{(14.003074\ u)}{{}^{14}_{7}N} + \underset{(4.002603\ u)}{{}^{4}_{2}He} \rightarrow \underset{(16.999131\ u)}{{}^{17}_{8}O} + \underset{(1.007825\ u)}{{}^{1}_{1}H}$$

1.53 MeV

14 Is the following reaction endoergic or exoergic?

$$\underset{(7.016005\ u)}{{}^{7}_{3}Li} + \underset{(1.007825\ u)}{{}^{1}_{1}H} \rightarrow \underset{(4.002603\ u)}{{}^{4}_{2}He} + \underset{(4.002603\ u)}{{}^{4}_{2}He}$$

exoergic

15 ■■ Determine the Q value of the following reaction:

$$\underset{(9.012183\ u)}{{}^{9}_{4}Be} + \underset{(4.002603\ u)}{{}^{4}_{2}He} \rightarrow \underset{(12.000000\ u)}{{}^{12}_{6}C} + \underset{(1.008665\ u)}{{}^{1}_{0}n}$$

5.70 MeV

16 ■■ Is the reaction ^{200}Hg (p, α)^{197}Au endoergic or exoergic? (See the equation in chapter for mass values.) exoergic

17 ■■ Find the threshold energy of the following reaction:

$$^{13}_{6}\text{C} \quad + \quad ^{1}_{1}\text{H} \quad \rightarrow \quad ^{1}_{0}\text{n} \quad + \quad ^{13}_{7}\text{N}$$

(13.003355 u) (1.007825 u) (1.008665 u) (13.005739 u)
3.23 MeV

18 ■■ Find the threshold energy for the following reaction:

$$^{3}_{2}\text{He} \quad + \quad ^{1}_{0}\text{n} \quad \rightarrow \quad ^{2}_{1}\text{H} \quad + \quad ^{2}_{1}\text{H}$$

(3.016029 u) (1.008665 u) (2.014102 u) (2.014102 u)
4.36 MeV

●19 ■■ What is the minimum kinetic energy of a proton that will initiate the reaction $^{3}_{1}$H(p,d)$^{2}_{1}$H? (d stands for a deuterium nucleus.) 5.37 MeV

20 ■■ ^{226}Ra decays and emits a 4.706-MeV alpha particle. Find the velocity of the recoiling daughter nucleus from the decay of a stationary radium-226 nucleus. 2.7×10^5 m/s

21 ■■ The same type of incident particle is used for two endoergic reactions. In one reaction, the mass of the target nucleus is 25 times greater, and in the other reaction 40 times greater. If the Q value of the first reaction is twice that of the second, which has the greater threshold energy and how many times greater? $(K_1/K_2)_{min} = 2.03$

22 ■■ Consider n ceramic pie plates of radius r randomly fixed on a rectangular wall with dimensions L and W. If you threw a baseball at the wall, what would be the percent probability of hitting a pie plate (or your reaction cross-section) in terms of these parameters? [*Hint:* Think in terms of area.] see ISM

23 ■■ Assume that the average kinetic energy of ions in a plasma is given by the equation for the kinetic energy of the atoms in an ideal gas ($\frac{1}{2}mv^2 = 3/2kT$) and that fusion occurs when the ions approach each other within a distance of the upper limit for the nuclear diameter ($R = 10^{-12}$ cm). Calculate the temperature required for fusion of two deuterium ions. (Boltzmann's constant is $k = 1.38 \times 10^{-23}$ J/K.)
5.6×10^8 K

29.2 and 29.3 ■ Nuclear Fission and Fusion

●24 Nuclear fission (a) is endoergic, (b) occurs only for uranium-235, (c) releases about 500 MeV of energy per fission, (d) requires a critical mass for a sustained reaction. (d)

25 A nuclear reactor (a) can operate on natural (unenriched) uranium, (b) has its chain reaction controlled by neutron-absorbing materials, (c) can be partially controlled by the amount of moderator, (d) all of these. (d)

26 A nuclear fusion reaction (a) has a negative Q value, (b) may occur spontaneously, (c) is an example of "splitting" the atom, (d) releases less than 50 MeV of energy per fusion process. (d)

27 Controlled fusion requires (a) no critical mass, (b) confinement, (c) formation of a plasma, (d) all of these. (d)

28 The energy produced in fission reactions is carried off as kinetic energies of the products. How is this converted to heat in a nuclear reactor? see ISM

●29 ■■ Find the approximate energy released in the following fission reactions:

(a) $^{235}_{92}$U $+ ^{1}_{0}$n $\rightarrow$ with the release of 5 neutrons (a) 231 MeV

(b) $^{235}_{94}$Pu $+ ^{1}_{0}$n $\rightarrow$ with the release of 3 neutrons (b) 237 MeV

30 ■■ Calculate the amounts of energy released in the following fusion reactions:

(a) $^{2}_{1}$H $+ ^{2}_{1}$H $\rightarrow ^{3}_{2}$He $+ ^{1}_{0}$n (a) 3.27 MeV

(b) $^{3}_{2}$He $+ ^{3}_{2}$He $\rightarrow ^{4}_{2}$He $+ 2^{1}_{1}$H (b) 12.9 MeV

31 Find the Q values for the (a) H-D reaction and (b) He-He reaction in the proton-proton cycle in Section 29.3.
(a) 5.494 MeV (b) 12.86 MeV

29.4 ■ Beta Decay and the Neutrino

●32 In the absence of a neutrino, what is not conserved in beta decay: (a) energy, (b) linear momentum, (c) angular momentum, (d) all of these? (d)

33 A neutrino interacts with matter by (a) an electrical interaction, (b) a strong interaction, (c) a weak interaction, (d) both (b) and (c). (c)

34 Why is it so difficult to detect neutrinos experimentally? see ISM

35 ■ A neutrino created in a beta decay process has an energy of 26.5 MeV. What is the de Broglie wavelength of the neutrino? 4.69×10^{-14} m

●36 ■■ Show that the disintegration energy for β^- decay is

$$Q = (m_p - m_d - m_e)c^2 = (M_p - M_d)c^2$$

where the m's represent the masses of the parent and daughter nuclei and the M's represent the masses of the neutral atoms. see ISM

37 ■■ What is the maximum kinetic energy of the electron emitted when a ^{12}B nucleus beta decays into a ^{12}C nucleus? (See Exercise 36.) 13.34 MeV

38 ■■ The kinetic energy of an electron emitted from a ^{32}P nucleus that beta decays into a ^{32}S nucleus is observed to be 1 MeV. What is the energy of the accompanying neutrino of the decay process? (See Exercise 36.) 0.71 MeV

39 ■■ Show that the disintegration energy for β^+ decay is

$$Q = (m_p - m_d - m_e)c^2 = (M_p - M_d - 2m_e)c^2$$

where the m's represent the masses of the parent and daughter nuclei and the M's represent the masses of the neutral atoms. see ISM

40 ■■ The kinetic energy of a positron emitted from the β^+ decay of a ^{13}N nucleus into a ^{13}C nucleus is measured to be 1.190 MeV. What is the energy of the accompanying neutrino in the process? (Neglect the recoil energy of the nucleus and see Exercise 39.)   0.008 MeV

41 ■■ On the basis of the Q values given in Exercises 36 and 39, what are the mass requirements of the parent atoms for β^- and β^+ processes? $\beta^-: M_p > M_d; \beta^+: M_p > M_d + 2m_e$

29.5 ■ Fundamental Forces and Exchange Particles

42 Virtual particles (a) form virtual images, (b) exist only in a time for the violation of the conservation of energy permitted by the uncertainty principle, (c) make up positrons, (d) can be observed in exchange processes.   (b)

43 The exchange particle for the strong nuclear force is the (a) pion, (b) W particle, (c) muon, (d) positron.   (a)

●**44** If virtual exchange particles are unobservable, how is their existence verified?   see ISM

45 ■■ Draw the Feynman diagrams for (a) the Compton effect and (b) electron pair annihilation.   see ISM

46 ■■ Assuming the range of the nuclear force to be on the order of the Bohr radius (about 0.050 nm), predict the mass of the exchange particle.   see ISM

47 ■■ In a reaction process, a high-speed proton collides with a nucleus and travels a distance of 5.0×10^{-16} m on the average before the reaction takes place. What type of interaction is this and during what time period does the interaction take place? 1.7×10^{-24} s

●**48** ■■ By how much energy is the conservation of energy violated in a meson exchange process?   138 MeV

●**49** ■■ How long is the conservation of energy violated in a meson exchange process? 4.71×10^{-24} s

50 ■■ A W particle in a weak interaction is found to have an energy of 100 MeV. What is the approximate range for the weak interaction with this particle? 1.98×10^{-15} m

29.6 ■ Elementary Particles

51 Particles that interact by the strong nuclear force are called (a) muons, (b) hadrons, (c) W particles, (d) leptons.   (b)

●**52** Quarks make up which of the following particles? (a) baryons, (b) muons, (c) Z particles, (d) all of these.   (a)

53 What is meant by quark flavor and color? Can these be changed? Explain.   see ISM

54 With so many types of hadrons, why aren't fractional electronic charges observed?   see ISM

■ **Additional Exercises**

55 Complete the following nuclear reactions:

(a) $^6_3\text{Li} + ^1_1\text{H} \rightarrow ^3_2\text{He} + \underline{\quad (a)\ ^4_2\text{He} \quad}$

(b) $^{58}_{28}\text{Ni} + ^2_1\text{H} \rightarrow ^{59}_{28}\text{Ni} + \underline{\quad (b)\ ^1_1\text{H} \quad}$

(c) $^{235}_{92}\text{U} + ^1_0\text{n} \rightarrow ^{138}_{54}\text{Xe} + 5^1_0\text{n} + \underline{\quad (c)\ ^{93}_{38}\text{Sr} \quad}$

(d) $^9_4\text{Be}(\alpha, n)\ \underline{\quad (d)\ ^{12}_6\text{C} \quad}$

(e) $^{16}_8\text{O}(n, p)\ \underline{\quad (e)\ ^{16}_7\text{N} \quad}$

56 Give the compound nuclei for the reactions in Exercise 56.   (a) $^7_4\text{Be}^*$ (b) $^{60}_{29}\text{Cu}^*$ (c) $^{236}_{92}\text{U}^*$ (d) $^{13}_6\text{C}^*$ (e) $^{17}_8\text{O}^*$

57 Compute the Q values of the following fusion reactions:

(a) $^2_1\text{H} + ^3_1\text{H} \rightarrow ^4_2\text{He} + ^1_0\text{n}$ (a) 17.6 MeV

(b) $^2_1\text{H} + ^2_1\text{H} \rightarrow ^3_1\text{H} + ^1_1\text{H}$ (b) 4.03 MeV

58 Determine the threshold energy of the following reaction: 2.36 MeV

$$\underset{(15.994915\ u)}{^{16}_8\text{O}} + \underset{(1.008665\ u)}{^1_0\text{n}} \rightarrow \underset{(13.003355\ u)}{^{13}_6\text{C}} + \underset{(4.002603\ u)}{^4_2\text{He}}$$

59 Polonium-210 undergoes alpha decay. Find the energy released in the decay reaction.   5.40 MeV

●**60** Show that the Q value for electron capture is given by

$$Q = (m_p + m_e + m_d)c^2 = (M_p - M_d)c^2$$

where the m's represent the masses of the parent and daughter nuclei and the M's represent the masses of the neutral atoms.   see ISM

61 (a) In the electron capture process of ^{7}Be nucleus converting into a ^{7}Li nucleus, what is the energy of the emitted neutrino? (See Exercise 62.) (b) Is it energetically possible for ^{7}Be to β^+ decay into ^{7}Li?   (a) 0.86 MeV   (b) not possible

62 Assuming the ratio of the ranges of the virtual W particle and the virtual pion to be equal to the ratio of the average lifetimes of the decay processes involving these interactions, estimate the mass of the W particle.   see ISM

63 Complete the following nuclear reactions:

(a) $^{27}_{13}\text{Al} + ^1_0\text{n} \rightarrow ^1_1\text{H} + \underline{\quad (a)\ ^{27}_{12}\text{Mg} \quad}$

(b) $^{13}_7\text{N} + ^1_0\text{n} \rightarrow ^4_2\text{He} + \underline{\quad (b)\ ^{10}_5\text{B} \quad}$

(c) $^{92}_{40}\text{Zr}(p, \alpha)\ \underline{\quad (c)\ ^{89}_{39}\text{Y} \quad}$

(d) $^{25}_{12}\text{Mg}(\gamma, p)\ \underline{\quad (d)\ ^{24}_{11}\text{Na} \quad}$

APPENDICES

APPENDIX I

Mathematical Relationships

ALGEBRAIC RELATIONSHIPS

$(a + b)^2 = a^2 + 2ab + b^2$
$(a - b)^2 = a^2 - 2ab + b^2$
$(a^2 - b^2) = (a + b)(a - b)$

Quadratic Formula

if $ax^2 + bx + c = 0$,

$$x = \frac{-b \pm \sqrt{b^2 - 4ac}}{2a}$$

Powers and Exponents

$x^0 = 1$

$x^1 = x$ $x^{-1} = \dfrac{1}{x}$

$x^2 = x \cdot x = x^2$ $x^{-2} = \dfrac{1}{x^2}$ $x^{\frac{1}{2}} = \sqrt{x}$

$x^3 = x \cdot x \cdot x = x^3$ $x^{-3} = \dfrac{1}{x^3}$ $x^{\frac{1}{3}} = \sqrt[3]{x}$

etc.   etc.   etc.

$x^a \cdot x^b = x^{(a+b)}$
$x^a / x^b = x^{(a-b)}$
$(x^a)^b = x^{ab}$

Logarithms

if $x = a^n$, then $n = \log_a x$

$\log xy = \log x + \log y$
$\log x/y = \log x - \log y$
$\log x^y = y \log x$

common logarithms: base 10
(assumed when abbreviation "log" is used unless another base is specified)

$\log 10^x = x$

natural logarithms: base $e = 2.71828 \ldots$ (abbreviated "ln")

$\ln e^x = x$
$\log x = 0.43429 \ln x$
$\ln x = 2.3026 \log x$

GEOMETRIC AND TRIGONOMETRIC RELATIONSHIPS

Areas and Volumes of Some Common Shapes

Circle: $A = \pi r^2 = \dfrac{\pi d^2}{4}$ (area)

$c = 2\pi r = \pi d$ (circumference)

Triangle: $A = \frac{1}{2}ab$

Sphere: $A = 4\pi r^2$

$V = \frac{4}{3}\pi r^3$

Cylinder: $A = \pi r^2$ (end)

$A = 2\pi rh$ (body)

$A = 2(\pi r^2) + 2\pi rh$ (total)

$V = \pi r^2 h$

Definitions of Trigonometric Functions

$$\sin \theta = \frac{y}{r} \qquad \cos \theta = \frac{x}{r} \qquad \tan \theta = \frac{\sin \theta}{\cos \theta} = \frac{y}{x}$$

$\theta°$ (rad)	$\sin \theta$	$\cos \theta$	$\tan \theta$
0° (0)	0	1	0
30° ($\pi/6$)	0.500	0.866	0.577
45° ($\pi/4$)	0.707	0.707	1.00
60° ($\pi/3$)	0.866	0.500	1.73
90° ($\pi/2$)	1	0	$\to \infty$

See trigonometric tables for other angles.

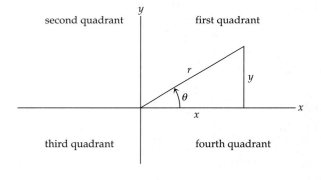

For very small angles:

$\cos \theta \cong 1$ $\sin \theta \cong \theta$ (radians)

$\tan \theta = \dfrac{\sin \theta}{\cos \theta} \cong \theta$ (radians)

The sign of a trigonometric function depends on the quadrant, or the signs of x and y; for example, in the second quadrant $(-x, y)$, $-x/r = \cos \theta$ and $y/r = \sin \theta$. The sign can also be assigned using the reduction formulas.

Reduction Formulas

	(θ in second quadrant)	(θ in third quadrant)	(θ in fourth quadrant)
$\sin \theta =$	$\cos(\theta - 90°) =$	$-\sin(\theta - 180°) =$	$-\cos(\theta - 270°)$
$\cos \theta =$	$-\sin(\theta - 90°) =$	$-\cos(\theta - 180°) =$	$\sin(\theta - 270°)$

Fundamental Identities

$\sin^2 \theta + \cos^2 \theta = 1$

$\sin 2\theta = 2 \sin \theta \cos \theta$

$\cos 2\theta = \cos^2 \theta - \sin^2 \theta = 2 \cos^2 \theta - 1 = 1 - 2 \sin^2 \theta$

$\sin^2 \theta = \frac{1}{2}(1 - \cos 2\theta)$

$\cos^2 \theta = \frac{1}{2}(1 + \cos 2\theta)$

For half-angle ($\theta/2$) identities, replace θ with $\theta/2$, for example:

$\sin^2 \theta/2 = \frac{1}{2}(1 - \cos \theta)$ $\cos^2 \theta/2 = \frac{1}{2}(1 + \cos \theta)$

$\sin(\alpha \pm \beta) = \sin \alpha \cos \beta \pm \cos \alpha \sin \beta$

$\cos(\alpha \pm \beta) = \cos \alpha \cos \beta \mp \sin \alpha \sin \beta$

$\tan(\alpha \pm \beta) = \dfrac{\tan \alpha \pm \tan \beta}{1 \mp \tan \alpha \tan \beta}$

Law of Cosines

For a triangle with angles X, Y, and Z and opposite sides x, y, and z, respectively:

$$x^2 = y^2 + z^2 - 2yz \cos X$$

For $X = 90°$, this reduces to the Pythagorean theorem:

$$x^2 = y^2 + z^2$$

Trigonometric Tables

Angle					Angle				
Degrees	Radians	Sine	Cosine	Tangent	Degrees	Radians	Sine	Cosine	Tangent
0°	.000	0.000	1.000	0.000					
1°	.017	.018	1.000	.018	46°	0.803	0.719	0.695	1.036
2°	.035	.035	0.999	.035	47°	.820	.731	.682	1.072
3°	.052	.052	.999	.052	48°	.838	.743	.669	1.111
4°	.070	.070	.998	.070	49°	.855	.755	.656	1.150
5°	.087	.087	.996	.088	50°	.873	.766	.643	1.192
6°	.105	.105	.995	.105	51°	.890	.777	.629	1.235
7°	.122	.122	.993	.123	52°	.908	.788	.616	1.280
8°	.140	.139	.990	.141	53°	.925	.799	.602	1.327
9°	.157	.156	.988	.158	54°	.942	.809	.588	1.376
10°	.175	.174	.985	.176	55°	.960	.819	.574	1.428
11°	.192	.191	.982	.194	56°	.977	.829	.559	1.483
12°	.209	.208	.978	.213	57°	.995	.839	.545	1.540
13°	.227	.225	.974	.231	58°	1.012	.848	.530	1.600
14°	.244	.242	.970	.249	59°	1.030	.857	.515	1.664
15°	.262	.259	.966	.268	60°	1.047	.866	.500	1.732
16°	.279	.276	.961	.287	61°	1.065	.875	.485	1.804
17°	.297	.292	.956	.306	62°	1.082	.883	.470	1.881
18°	.314	.309	.951	.325	63°	1.100	.891	.454	1.963
19°	.332	.326	.946	.344	64°	1.117	.899	.438	2.050
20°	.349	.342	.940	.364	65°	1.134	.906	.423	2.145
21°	.367	.358	.934	.384	66°	1.152	.914	.407	2.246
22°	.384	.375	.927	.404	67°	1.169	.921	.391	2.356
23°	.401	.391	.921	.425	68°	1.187	.927	.375	2.475
24°	.419	.407	.914	.445	69°	1.204	.934	.358	2.605
25°	.436	.423	.906	.466	70°	1.222	.940	.342	2.747
26°	.454	.438	.899	.488	71°	1.239	.946	.326	2.904
27°	.471	.454	.891	.510	72°	1.257	.951	.309	3.078
28°	.489	.470	.883	.532	73°	1.274	.956	.292	3.271
29°	.506	.485	.875	.554	74°	1.292	.961	.276	3.487
30°	.524	.500	.866	.577	75°	1.309	.966	.259	3.732
31°	.541	.515	.857	.601	76°	1.326	.970	.242	4.011
32°	.559	.530	.848	.625	77°	1.344	.974	.225	4.331
33°	.576	.545	.839	.649	78°	1.361	.978	.208	4.705
34°	.593	.559	.829	.675	79°	1.379	.982	.191	5.145
35°	.611	.574	.819	.700	80°	1.396	.985	.174	5.671
36°	.628	.588	.809	.727	81°	1.414	.988	.156	6.314
37°	.646	.602	.799	.754	82°	1.431	.990	.139	7.115
38°	.663	.616	.788	.781	83°	1.449	.993	.122	8.144
39°	.681	.629	.777	.810	84°	1.466	.995	.105	9.514
40°	.698	.643	.766	.839	85°	1.484	.996	.087	11.43
41°	.716	.656	.755	.869	86°	1.501	.998	.070	14.30
42°	.733	.669	.743	.900	87°	1.518	.999	.052	19.08
43°	.751	.682	.731	.933	88°	1.536	.999	.035	28.64
44°	.768	.695	.719	.966	89°	1.553	1.000	.017	57.29
45°	.785	.707	.707	1.000	90°	1.571	1.000	.000	∞

Kinetic Theory of Gases

The basic assumptions are as follows:

1. The molecules of a pure gas all have the same mass (m) and are in continuous and completely random motion. (The mass of each molecule is so small that the effect of gravity on it is negligible.)

2. The gas molecules are separated by large distances and occupy a volume that is negligible compared to these distances; that is, they are point particles.

3. The molecules exert no forces on each other except when they collide.

4. The collisions of the molecules with one another and the walls of the container are perfectly elastic.

The magnitude of the force exerted on the wall of the container by a gas molecule colliding with it is $F = \Delta p / \Delta t$. Assuming that the direction of the velocity ($\mathbf{v}_x$) is normal to the wall, the magnitude of the average force is

$$F = \frac{\Delta(mv)}{\Delta t} = \frac{mv_x - (-mv_x)}{\Delta t} = \frac{2mv_x}{\Delta t} \qquad (1)$$

After striking one wall of the container, which for convenience is assumed to be cubical with sides of dimensions L, the molecule recoils in a straight line. Suppose that the molecule reaches the opposite wall without colliding with any other molecules along the way. The molecule then travels the distance L in a time equal to L/v_x. After the collision with that wall, again assuming no collisions on the return trip, the round trip will take $\Delta t = 2L/v_x$. Thus, the number of collisions per unit time a molecule makes with a particular wall is $v_x/2L$, and the average force of the wall from successive collisions is

$$F = \frac{2mv_x}{\Delta t} = \frac{2mv_x}{2L/v_x} = \frac{mv_x^2}{L} \qquad (2)$$

The random motions of the many molecules produce a relatively constant force on the walls, and the pressure (p) is the total force on a wall divided by its area:

$$p = \frac{\sum F_i}{L^2} = \frac{m(v_{x_1}^2 + v_{x_2}^2 + v_{x_3}^2 + \cdots)}{L^3} \qquad (3)$$

The subscripts refer to individual molecules.

The average of the squares of the speeds is given by

$$\overline{v_x^2} = \frac{v_{x_1}^2 + v_{x_2}^2 + v_{x_3}^2 + \cdots}{N}$$

where N is the number of molecules in the container. In terms of this average, Eq. 3 may be written

$$p = \frac{Nm\overline{v_x^2}}{L^3} \qquad (4)$$

However, the molecules' motions occur with equal frequency along any one of the three axes, so $\overline{v_x^2} = \overline{v_y^2} = \overline{v_z^2}$, and $\overline{v^2} = \overline{v_x^2} + \overline{v_y^2} + \overline{v_z^2} = 3\overline{v_x^2}$. Then,

$$\sqrt{\overline{v^2}} = \overline{v}$$

where $\overline{v}$ is called the root-mean-square (rms) speed, and we write $(\overline{v})^2 = \overline{v^2}$. Substituting this result into Eq. 4 along with V for L^3 (since L^3 is the volume of the cubical container) gives

$$p = \frac{Nm\overline{v}^2}{3V} \qquad \text{or} \qquad pV = \tfrac{1}{3}Nm\overline{v}^2 \qquad (5)$$

This result is true even though collisions between molecules were ignored. Statistically, these collisions average out, so the number of collisions with each wall is as described. This result is also independent of the shape of the container. A cube merely simplifies the derivation.

Combining this result with the empirical perfect gas law gives

$$pV = NkT = \tfrac{1}{3}Nm\overline{v}^2$$

The average kinetic energy per gas molecule is thus proportional to the absolute temperature of the gas:

$$\tfrac{1}{2}m\overline{v}^2 = \tfrac{3}{2}kT \qquad (6)$$

The collision time is negligible compared with the time between collisions. Some kinetic energy will be momentarily converted to potential energy during a collision; however, this potential energy can be ignored because each molecule spends a negligible amount of time in collisions. Therefore, by this approximation, the total kinetic energy is the internal energy of the gas, and the internal energy of a perfect gas is directly proportional to its absolute temperature.

APPENDIX III

Planetary Data

Name	Equatorial Radius (km)	Mass (Compared to Earth's)*	Mean Density ($\times 10^3$ kg/m³)	Surface Gravity (Compared to Earth's)	Semimajor Axis $\times 10^6$ km	Semimajor Axis AU	Period Years	Period Days	Eccentricity	Inclination to Ecliptic
Mercury	2,439	0.0553	5.43	0.378	57.9	0.3871	0.24084	87.96	0.2056	7°00'26"
Venus	6,052	0.8150	5.24	0.894	108.2	0.7233	0.61515	224.68	0.0068	3°23'40"
Earth	6,378.140	1	5.515	1	149.6	1	1.00004	365.25	0.0167	0°00'14"
Mars	3,397.2	0.1074	3.93	0.379	227.9	1.5237	1.8808	686.95	0.0934	1°51'09"
Jupiter	71,398	317.89	1.36	2.54	778.3	5.2028	11.862	4,337	0.0483	1°18'29"
Saturn	60,000	95.17	0.71	1.07	1427.0	9.5388	29.456	10,760	0.0560	2°29'17"
Uranus	26,145	14.56	1.30	0.8	2871.0	19.1914	84.07	30,700	0.0461	0°48'26"
Neptune	24,300	17.24	1.8	1.2	4497.1	30.0611	164.81	60,200	0.0100	1°46'27"
Pluto	1,500–1,800	0.02	0.5–0.8	~0.03	5913.5	39.5294	248.53	90,780	0.2484	17°09'03"

* Planet's mass/Earth's mass, where $M_E = 6.0 \times 10^{24}$ kg.

APPENDIX IV

Alphabetical Listing of the Chemical Elements

Element	Symbol	Atomic Number (Proton Number)	Atomic Mass	Element	Symbol	Atomic Number (Proton Number)	Atomic Mass	Element	Symbol	Atomic Number (Proton Number)	Atomic Mass
Actinium	Ac	89	227.0278	Curium	Cm	96	(247)	Lawrencium	Lr	103	(260)
Aluminum	Al	13	26.98154	Dysprosium	Dy	66	162.50	Lead	Pb	82	207.2
Americium	Am	95	(243)[a]	Einsteinium	Es	99	(252)	Lithium	Li	3	6.941
Antimony	Sb	51	121.757	Erbium	Er	68	167.26	Lutetium	Lu	71	174.967
Argon	Ar	18	39.948	Europium	Eu	63	151.96	Magnesium	Mg	12	24.305
Arsenic	As	33	74.9216	Fermium	Fm	100	(257)	Manganese	Mn	25	54.9380
Astatine	At	85	(210)	Fluorine	F	9	18.998403	Mendelevium	Md	101	(258)
Barium	Ba	56	137.33	Francium	Fr	87	(223)	Mercury	Hg	80	200.59
Berkelium	Bk	97	(247)	Gadolinium	Gd	64	157.25	Molybdenum	Mo	42	95.94
Beryllium	Be	4	9.01218	Gallium	Ga	31	69.72	Neodymium	Nd	60	144.24
Bismuth	Bi	83	208.9804	Germanium	Ge	32	72.561	Neon	Ne	10	20.1797
Boron	B	5	10.81	Gold	Au	79	196.9665	Neptunium	Np	93	237.048
Bromine	Br	35	79.904	Hafnium	Hf	72	178.49	Nickel	Ni	28	58.69
Cadmium	Cd	48	112.41	Hahnium	Ha	105	(262)	Niobium	Nb	41	92.9064
Calcium	Ca	20	40.078	Helium	He	2	4.00260	Nitrogen	N	7	14.0067
Californium	Cf	98	(251)	Holmium	Ho	67	164.9304	Nobelium	No	102	(259)
Carbon	C	6	12.011	Hydrogen	H	1	1.00794	Osmium	Os	76	190.2
Cerium	Ce	58	140.12	Indium	In	49	114.82	Oxygen	O	8	15.9994
Cesium	Cs	55	132.9054	Iodine	I	53	126.9045	Palladium	Pd	46	106.42
Chlorine	Cl	17	35.453	Iridium	Ir	77	192.22	Phosphorus	P	15	30.97376
Chromium	Cr	24	51.996	Iron	Fe	26	55.847	Platinum	Pt	78	195.08
Cobalt	Co	27	58.9332	Krypton	Kr	36	83.80	Plutonium	Pu	94	(244)
Copper	Cu	29	63.546	Lanthanum	La	57	138.9055	Polonium	Po	84	(209)

Element	Symbol	Atomic Number (Proton Number)	Atomic Mass	Element	Symbol	Atomic Number (Proton Number)	Atomic Mass	Element	Symbol	Atomic Number (Proton Number)	Atomic Mass
Potassium	K	19	39.0983	Silicon	Si	14	28.0855	Tungsten	W	74	183.85
Praseodymium	Pr	59	140.9077	Silver	Ag	47	107.8682	Uranium	U	92	238.0289
Promethium	Pm	61	(145)	Sodium	Na	11	22.98977	Vanadium	V	23	50.9415
Protactinium	Pa	91	231.0359	Strontium	Sr	38	87.62	Xenon	Xe	54	131.29
Radium	Ra	88	226.0254	Sulfur	S	16	32.066	Ytterbium	Yb	70	173.04
Radon	Rn	86	(222)	Tantalum	Ta	73	180.9479	Yttrium	Y	39	88.9059
Rhenium	Re	75	186.207	Technetium	Tc	43	(98)	Zinc	Zn	30	65.39
Rhodium	Rh	45	102.9055	Tellurium	Te	52	127.60	Zirconium	Zr	40	91.22
Rubidium	Rb	37	85.4678	Terbium	Tb	65	158.9254			106	(263)
Ruthenium	Ru	44	101.07	Thallium	Tl	81	204.383			107	(262)
Rutherfordium	Rf	104	(261)	Thorium	Th	90	232.0381			108	(265)
Samarium	Sm	62	150.36	Thulium	Tm	69	168.9342			109	(266)
Scandium	Sc	21	44.9559	Tin	Sn	50	118.710				
Selenium	Se	34	78.96	Titanium	Ti	22	47.88				

APPENDIX V

Properties of Selected Isotopes

Atomic Number (Z)	Element	Symbol	Mass Number (A)	Atomic Mass*	Abundance (%) or Decay Mode† (if radioactive)	Half-Life (if radioactive)
0	(Neutron)	n	1	1.008665	β^-	10.6 min
1	Hydrogen	H	1	1.007825	99.985	
	Deuterium	D	2	2.014102	0.015	
	Tritium	T	3	3.016049	β^-	12.33 years
2	Helium	He	3	3.016029	0.00014	
			4	4.002603	~100	
3	Lithium	Li	6	6.015123	7.5	
			7	7.016005	92.5	
4	Beryllium	Be	7	7.016930	EC, γ	53.3 days
			8	8.005305	2α	6.7×10^{-17} s
			9	9.012183	100	
5	Boron	B	10	10.012938	19.8	
			11	11.009305	80.2	
			12	12.014353	β^-	20.4 ms
6	Carbon	C	11	11.011433	β^+, EC	20.4 ms
			12	12.000000	98.89	
			13	13.003355	1.11	
			14	14.003242	β^-	5730 years
7	Nitrogen	N	13	13.005739	β^-	9.96 min
			14	14.003074	99.63	
			15	15.000109	0.37	
8	Oxygen	O	15	15.003065	β^+, EC	122 s
			16	15.994915	99.76	
			18	17.999159	0.204	
9	Fluorine	F	19	18.998403	100	

Atomic Number (Z)	Element	Symbol	Mass Number (A)	Atomic Mass*	Abundance (%) or Decay Mode† (if radioactive)	Half-Life (if radioactive)
10	Neon	Ne	20	19.992439	90.51	
			22	21.991384	9.22	
11	Sodium	Na	22	21.994435	β^+, EC, γ	2.602 years
			23	22.989770	100	
			24	23.990964	β^-, γ	15.0 h
12	Magnesium	Mg	24	23.985045	78.99	
13	Aluminum	Al	27	26.981541	100	
14	Silicon	Si	28	27.976928	92.23	
			31	30.975364	β^-, γ	2.62 h
15	Phosphorus	P	31	30.973763	100	
			32	31.973908	β^-	14.28 days
16	Sulfur	S	32	31.972072	95.0	
			35	34.969033	β^-	87.4 days
17	Chlorine	Cl	35	34.968853	75.77	
			37	36.965903	24.23	
18	Argon	Ar	40	39.962383	99.60	
19	Potassium	K	39	38.963708	93.26	
			40	39.964000	β^-, EC, γ, β^+	1.28×10^9 years
20	Calcium	Ca	30	39.962591	96.94	
24	Chromium	Cr	52	51.940510	83.79	
25	Manganese	Mn	55	54.938046	100	
26	Iron	Fe	56	55.934939	91.8	
27	Cobalt	Co	59	58.933198	100	
			60	59.933820	β^-, γ	5.271 years
28	Nickel	Ni	58	57.935347	68.3	
			60	59.930789	26.1	
			64	63.927968	0.91	
29	Copper	Cu	63	62.929599	69.2	
			64	63.929766	β^-, β^+	12.7 h
			65	64.927792	30.8	
30	Zinc	Zn	64	63.929145	48.6	
			66	65.926035	27.9	
33	Arsenic	As	75	74.921596	100	
35	Bromine	Br	79	78.918336	50.69	
36	Krypton	Kr	84	83.911506	57.0	
			89	88.917563	β^-	3.2 min
38	Strontium	Sr	86	85.909273	9.8	
			88	87.905625	82.6	
			90	89.907746	β^-	28.8 years
39	Yttrium	Y	89	89.905856	100	
43	Technetium	Tc	98	97.907210	β^-, γ	4.2×10^6 years
47	Silver	Ag	107	106.905095	51.83	
			109	108.904754	48.17	
48	Cadmium	Cd	114	113.903361	28.7	
49	Indium	In	115	114.90388	95.7; β^-	5.1×10^{14} years
50	Tin	Sn	120	119.902199	32.4	
53	Iodine	I	127	126.904477	100	
			131	130.906118	β^-, γ	8.04 days
54	Xenon	Xe	132	131.90415	26.9	
			136	135.90722	8.9	
55	Cesium	Cs	133	132.90543	100	

Atomic Number (Z)	Element	Symbol	Mass Number (A)	Atomic Mass*	Abundance (%) or Decay Mode† (if radioactive)	Half-Life (if radioactive)
56	Barium	Ba	137	136.90582	11.2	
			138	137.90524	71.7	
			144	143.922673	β^-	11.9 s
61	Promethium	Pm	145	144.91275	EC, α, γ	17.7 years
74	Tungsten (wolfram)	W	184	183.95095	30.7	
76	Osmium	Os	191	190.96094	β^-, γ	15.4 days
			192	191.96149	41.0	
78	Platinum	Pt	195	194.96479	33.8	
79	Gold	Au	197	196.96656	100	
80	Mercury	Hg	202	201.97063	29.8	
81	Thallium	Tl	205	204.97441	70.5	
			210	209.990069	β^-	1.3 min
82	Lead	Pb	204	203.973044	β^-, 1.48	1.4×10^{17} years
			206	205.97446	24.1	
			207	206.97589	22.1	
			208	207.97664	52.3	
			210	209.98418	α, β^-, γ	22.3 years
			211	210.98874	β^-, γ	36.1 min
			212	211.99188	β^-, γ	10.64 h
			214	213.99980	β^-, γ	26.8 min
83	Bismuth	Bi	209	208.98039	100	
			211	210.98726	α, β^-, γ	2.15 min
84	Polonium	Po	210	209.98286	α, γ	138.38 days
			214	213.99519	α, γ	164 μs
86	Radon	Rn	222	222.017574	α, β	3.8235 days
87	Francium	Fr	223	223.019734	α, β^-, γ	21.8 min
88	Radium	Ra	226	226.025406	α, γ	1.60×10^3 years
			228	228.031069	β^-	5.76 years
89	Actinium	Ac	227	227.027751	α, β^-, γ	21.773 years
90	Thorium	Th	228	228.02873	α, γ	1.9131 years
			232	232.038054	100; α, γ	1.41×10^{10} years
92	Uranium	U	232	232.03714	α, γ	72 years
			233	233.039629	α, γ	1.592×10^5 years
			235	235.043925	0.72; α, γ	7.038×10^8 years
			236	236.045563	α, γ	2.342×10^7 years
			238	238.050786	99.275; α, γ	4.468×10^9 years
			239	239.054291	β^-, γ	23.5 min
93	Neptunium	Np	239	239.052932	β^-, γ	2.35 days
94	Plutonium	Pu	239	239.052158	α, γ	2.41×10^4 years
95	Americium	Am	243	243.061374	α, γ	7.37×10^3 years
96	Curium	Cm	245	245.065487	α, γ	8.5×10^3 years
97	Berkelium	Bk	247	247.07003	α, γ	1.4×10^3 years
98	Californium	Cf	249	249.074849	α, γ	351 years
99	Einsteinium	Es	254	254.08802	α, γ, β^-	276 days
100	Fermium	Fm	253	253.08518	EC, α, γ	3.0 days
101	Mendelevium	Md	255	255.0911	EC, α	27 min
102	Nobelium	No	255	255.0933	EC, α	3.1 min
103	Lawrencium	Lr	257	257.0998	α	$\approx$35 s
104	Rutherfordium	Rf	261	261.1087	α	1.1 min
105	Hahnium	Ha	262	262.1138	α	0.7 min

* The masses given throughout this table are those for the neutral atom, including the Z electrons.
† EC stands for electron capture.

Answers to Follow-up Exercises

• **1.4** **(a)** To get the smallest numerical value, we must make the numerator number as small as possible and the denominator as large as possible. It is clear from Example 1.4 that, since 1 mi = 1.61 km, the length unit in the numerator should be miles. Similarly, since there are about four liters to one gallon, the unit used in the denominator should be liters. Expressing fuel economy in mi/L will give the smallest numerical value. **(b)** To demonstrate the comparative numerical values, we convert one set of units to the other three. Arbitrarily selecting 1 mi/gal, we have:

$$1 \text{ mi/gal } (1.61 \text{ km/mi}) = 1.61 \text{ km/gal}$$
$$1 \text{ mi/gal } (0.264 \text{ gal/L}) = 0.264 \text{ mi/L}$$
$$1 \text{ mi/gal } (1.61 \text{ km/mi})(0.264 \text{ gal/L}) = 0.425 \text{ km/L}$$

Thus 1 mi/gal = 1.61 km/gal = 0.264 mi/L = 0.425 km/L.

• **2.6:** The speeds of the cars as a function of time are given by Eq. 2.7, $v = v_o + at$. With $v_o = 0$, this reduces to $v = at$. The speed is proportional to t (i.e., $v \propto t$), so if car B accelerates twice as long, it will have twice the speed. (No t^2 dependence here.)

• **2.11:** As long as the ball is in flight, its position is given by Eq. 2.9′, $y = v_o t - \frac{1}{2}gt^2$. We know that at $t = 2.28$ s, the ball has returned to the starting point ($y = 0$), so the negative t^2 term in the equation is equal in magnitude to the positive t term. For $t > 2.28$ s, the negative t^2 term dominates and negative y values are obtained, which means that the ball's position is below the starting or zero reference point.

• **4.9:** There is the apple's downward weight force, of course, and the tree must exert an equal and opposite force on the apple's stem since there is no net force on the apple. (If there were, it would not remain at rest.) Can you identify the other forces in the third law force pairs?

• **4.12:** Taking F_1 and F_2 as the pulling and pushing forces, respectively, we have $N = mg - F_1 \sin \theta_1 + F_2 \sin \theta_2$, where the θ's are the direction angles of the respective forces. Thus, the magnitude of the normal force depends on both the magnitudes of the forces and the angles at which they act. If $F_1 = F_2$ and $\theta_1 = \theta_2$, the vertical components of the applied forces cancel each other. Otherwise, the normal force could be affected. (It is possible for both the forces and the angles to be different *without* affecting the normal force? How?)

• **4.14:** Air resistance depends not only on speed, but also on size and shape. If the heavier ball were larger, it would have more exposed area to collide with air molecules, and the retarding force of air resistance would build up faster. Depending on the size difference, the heavier ball might reach terminal velocity first and the lighter ball strike the ground first. Also, the balls might reach terminal velocity together. How would they strike the ground in this case?

• **5.6:** Here we have $m_s = m_g/2$ as before, but $v_s/v_g = (6.0 \text{ m/s})/(4.0 \text{ m/s}) = \frac{3}{2}$. Using a ratio as in the example,

$$K_s/K_g = (m_s/m_g)(v_s/v_g)^2 = \left(\frac{1}{2}\right)\left(\frac{3}{2}\right)^2 = \frac{9}{8}$$

or

$$K_s = \frac{9}{8}K_g$$

Thus the safety still has more kinetic energy than the guard, $\frac{9}{8}$ times as much. The answer could also be obtained from direct calculations of the kinetic energies, but for a relative comparison, a ratio is usually quicker.

• **5.9:** To determine whether the speed depends on the mass of a ball *in principle*, one needs to look at the physical relationships that apply to the situation. Here, energy is the major consideration, and as always, one should keep in mind the versatility of the conservation of energy in analyzing phenomena. The mechanical energy is conserved while the balls are in flight, so let's consider the initial energy (E_o) of a ball and its energy (E) just before striking the ground. By the conservation of energy,

$$E_o = E$$

or

$$\tfrac{1}{2}mv_o^2 + mgh = \tfrac{1}{2}mv^2$$

and

$$\tfrac{1}{2}v_o^2 + gh = \tfrac{1}{2}v^2 \quad \text{(cancellation of } m\text{'s)}$$

Then,

$$v = \sqrt{v_o^2 - 2gh}$$

As you can see, the mass doesn't appear in the equation. Thus, the speed is independent of mass. (Recall that objects or projectiles in free fall all fall with the same vertical acceleration g, Section 2.5.)

• **6.9:** No. In an inelastic collision, kinetic energy is not conserved, but momentum is. Recall that a collision impulse is equal to the *change in momentum* ($\overline{\mathbf{F}}\Delta t = \Delta\mathbf{p}$). Thus if the total momentum is conserved ($\Delta\mathbf{p} = 0$), so is the impulse, $\overline{F}_2\Delta t_2 - \overline{F}_1\Delta t_1 = \Delta p = 0$, and $\overline{F}_2\Delta t_2 = \overline{F}_1\Delta t_1$.

• **7.6:** The string cannot be exactly horizontal, but must make some small downward angle to the horizontal so that there will be an upward component of the tension force to balance the ball's weight. Stated another way, if the string were exactly horizontal, the total tension force would supply the centripetal force on the ball, and the ball would accelerate downward because of its unbalance weight force, which doesn't happen. Sketch yourself a diagram to help see this.

• **7.16:** A backward or reverse thrust would do negative work and so cause the spacecraft to go lower in its potential energy well, decreasing the potential energy. With a decrease in r, the kinetic energy increases, and the total energy decreases by becoming more negative. (See Eqs. 7.28 and 7.29.)

• **8.7:** The long pole carried by a tightrope walker (or your extended arms) increases the moment of inertia I by placing more mass farther from the axis of rotation (the tightrope or rail). When the walker leans to the side, a gravitational torque tends to cause a rotation about the rope—that is, he or she begins to fall. However, with greater rotational inertia (greater I), the walker has time to shift his or her body so that its center of gravity is again over the rope or wire, and thus again in (unstable) equilibrium.

• **9.7:** The object would sink, so the buoyant force is less than the object's weight. Hence the scale would have a reading greater than 40 N. Note that with a greater density, the object would not be as large, and less water would be displaced. (For additional related follow-up exercises, see Exercises 51 and 52 at the end of the chapter.)

• **10.7:** Again focusing on the sphere of metal removed from the block, a decrease in temperature would cause it to contract or decrease in volume. The block of metal would react as though the sphere were still a part of it, so the spherical cavity would become smaller.

• **10.9:** Forming a ratio from Eq. 10.14 with T constant and $m_O = 16m_{He}$,

$$\overline{v}_O/\overline{v}_{He} = \sqrt{m_{He}/m_O} = \sqrt{1/16} = \tfrac{1}{4}$$

Thus the more massive oxygen molecule moves with $\tfrac{1}{4}$ the rms speed of the helium atom ($\overline{v}_O = \overline{v}_{He}/4$) under the same conditions.

• **10.10:** Here we can use the first part of Eq. 10.15, leaving the temperatures in the expression. (Since they are not equal, they do not cancel out in this case.) With $R_N = R_O$, the equation becomes $R_N/R_O = \sqrt{(T_N/T_O)(m_O/m_N)} = 1$. Rather than solving explicitly, note that $m_O/m_N = 1.14$, so for the relationship to hold, $T_N/T_O = 1/1.14$, or $T_O = (1.14)T_N = (1.14)(293\text{ K}) = 334\text{ K}$ (or 61°C).

• **11.9:** No; the air spaces are essential, because air is a poor conductor. The many small pockets of air between a person's body and the outer shell of the garment or quilt cover forms an insulating layer that minimizes conduction and so retards the loss of body heat to the environment. (There is little convection in the small spaces.) Similarly, for a diver in a wet suit, a thin film of water acts as an insulating layer.

• **12.5:** We know that the entropy of the universe increases in *every* natural process, including the death of your amoeba. We know, too, that *local* decreases in entropy are possible, provided they are "paid for" with greater increases in entropy elsewhere. Living things "purchase" such decreases by capturing and using energy to grow and increase the complexity of their organization (see the Insight on Life, Order, and the Second Law). However, when an organism dies, it loses the ability to utilize energy and can no longer maintain its highly ordered structure. It decays, and eventually its elements return to the environment ("dust to dust"). Thus the death of an amoeba represents a local as well as a global increase in entropy.

• **13.7:** You could tune the string to 264 Hz by tightening it, increasing the tension. Assuming any change in the cross-

sectional area of the string to be negligible, we have $f_2 = f_1\sqrt{F_2/F_1}$, and $F_2 = (f_2/f_1)^2 F_1 = (264\text{ Hz}/220\text{ Hz})^2 F_1 = (1.44)F_1$. Hence, the tension force would have to be increased by a factor of 1.44.

• **14.7:** With the source and the observer traveling in the same direction at the same speed, their relative velocity would be zero. That is, the observer would consider the source to be stationary relative to him or her. Since their speed is subsonic, the sound from the source would simply overtake the observer without a shift in frequency. Keep in mind that generally for the motions involved in a Doppler shift, the word *toward* is associated with an *increase* in frequency, and *away* with a *decrease* in frequency. Here, the source and observer remain a constant distance apart. (What would be the case if the speeds were supersonic?)

• **15.6:** In effect, positive charge would be transferred from the rod to the container and would neutralize the induced negative charge on its inside surface. (In reality, it's the electrons that actually move, and they neutralize the positive charge on the rod.) This leaves a net positive charge on the outside of the container. Thus, the deflection of the outside-connected electroscope is unchanged, while the inside-connected electroscope's leaves collapse a neutral state. Basically, the *excess* charge acquired by the metal container appears on its outside surface.

• **16.7:** The resistance of a wire is directly proportional to its length ($R \propto L$, Eq. 16.3). So, with $L_2 = (0.95)L_1$, then $R_2 = (0.95)R_1$. Hence, since $P = V^2/R$, the R_2 coil, with less resistance, would have the greater power output. A factor comparison may be obtained by forming the ratio, $P_2 = (R_1/R_2)P_1 = (1/0.95)P_1 = (1.1)P_1$, about 10% more power output.

• **17.2:** **(a)** As more bulbs burn out, the total resistance decreases and the current in the circuit increases. If the bulbs are not replaced, the circuit (and particularly the bulbs) will heat up because of joule heat of I^2R losses, creating a potential fire hazard. This is particularly true if the lights are on a dry tree or in contact with combustible decorations. **(b)** If the shunt were wired in parallel with the filament, current would normally be divided between the two elements. If a bulb blew out, all of the current would then flow through the shunt. We know that a combination of elements wired in parallel must always have *less* resistance than any one of its individual components. Therefore the resistance of the bulb would be greater with only the shunt in the circuit. (In fact, it would most likely be considerably greater, because the shunt would probably be designed to have a relatively high resistance. If it were low, most of the current would ordinarily flow through the shunt rather than through the bulb, wasting electricity and producing dangerous heat.)

Since the bulb is in series with the rest of the circuit, the total resistance of the string of bulbs would be slightly increased when a bulb blew out. There would thus be slightly less current in the entire string, so all the bulbs would glow a little more dimly.

• **18.1:** A beam of electrons traveling in the same direction as the beam of protons (north) would experience an initial force in the opposite direction to that on the protons, or to the west. Reversing the direction of the electron beam would reverse the direction of the force, so the electron beam should be directed south to experience the same initial force in an easterly direction.

• **20.2:** If the voltage were reduced to 120 V by the proper converter, the operation of a hair dryer would be generally unaffected. However, electric clocks and some other appliances depend on the line frequency as a timing mechanism or standard. With a lower frequency, a clock designed for 60 Hz would run slower and not keep the correct time.

• **21.3:** **(a)** The frequency of the light is unchanged in the different media, so it has the same frequency as that of the light source. **(b)** The wavelength in air is independent of the water and glass media, as can be shown by adding another step (medium) to the Example solution. By reverse analysis, $\lambda_{air} = n_{water}\lambda_{water} = (c/v_{water})\lambda_{water} = c/f$. Thus, the wavelength in air is simply $\lambda = c/f$, where c is the speed of light (in vacuum) and f is the frequency of the light source.

• **24.6:** The erecting lens should be positioned between the objective and the eyepiece. The image formed by the objective should be at an object distance of $2f_e$ from the erecting lens, where f_e is the focal length of this lens. The erecting lens then produces an inverted image of the same size at an image distance of $2f_e$ on the opposite side of the lens. This image is used as an object for the eyepiece. The use of the erecting lens lengthens the telescope by $4f_e$.

Answers to Odd-Numbered Exercises

Note to the student: The answers given here were reached by solving the problems step-by-step and rounding the result at each step. If you work the problems fully and then round only your result, your correct answer may differ slightly from the answer you find here. Variations may be due either to rounding or to calculator differences.

Chapter 1

1. **(c)**
3. simplest entities, and remain the same regardless of where measured
5. 12¢ per dime, 12 dimes per dollar, 144¢ per dollar
7. **(d)**
9. yes; unit analysis is the stronger condition and tells both
11. $[L] = [L] + [L]$
13. no, $[L]^3 \neq [L]^2$; $V \propto d^3$
15. $[L] = [L] + [L] + [L]$
17. yes; m/s = m/s − m/s
19. yes; $[L]^2 = [L]^2 + [L]^2$
21. **(a)** dimensionless **(b)** 1/m; **(c)** m²
23. m/s²
25. **(a)** kg-m/s²; **(b)** kg-m/s
27. **(a)**
29. cm
31. **(a)** 9.1 m **(b)** 1.61×10^3 m
33. **(a)** 70 kg
35. 50 mi by 2.1×10^4 m
37. 27 mL more in 500 mL
39. 25.0 mi/gal; 10.6 km/L; 10.6 m/mL
41. 6.2 cm/y
43. **(a)** 54 m²; **(b)** 65 yd²
45. 9.4×10^2 m²
47. (137 m)(22.9 m)(13.7 m) = 4.30×10^4 m³
49. **(a)** 62 lb/ft³; **(b)** 8.3 lb
51. **(d)**

53. **(a)** 4; **(b)** 3; **(c)** 5; **(d)** 2;
55. **(b)** and **(d)**; **(a)** has four and **(c)** has six
57. 5.06 cm; 5.06×10^{-1} dm; 5.06×10^{-2} m
59. 46 cm²
61. 1.29 m³
63. 1.12×10^{21} m³
65. **(c)**
67. compare to typical, familiar value
69. 1.36×10^4 g
71. 3.1×10^2 km
73. 9.5×10^{12} km
75. larger is the better buy
77. **(a)** 21 mi/gal and 8.9 km/L; **(b)** 5.5 gal/h and 21 L/h; **(c)** 1.1×10^2 mi/h and 1.8×10^2 km/h
79. **(a)** 1/m **(b)** dimensionless **(c)** m
81. metric ton, by 200 lb or 91 kg
83. **(a)** 1.8×10^3 cm²; **(b)** 0.18 m²; **(c)** 5.6×10^2 kg/m³
85. **(a)** cm; **(b)** 0.946 m²
87. not reasonable (56 mi/h)
89. 7.3×10^{22} kg
91. 4.48×10^3 cm³

Chapter 2

1. **(c)**
3. yes, for a round trip; no, the distance

must always be greater than or equal to the magnitude of the displacement
5. **(d)**
7. zero velocity and speed; constant velocity
9. 1.0 m/s
11. 23 km
13. $s_{AC} = 1.90$ m/s, $\bar{s} = (s_{AB} + s_{BC})/2 = (2.00$ m/s + 1.67 m/s)/2 = 1.84 m/s; not equal
15. **(a)** 60 mi/h **(b)** 27 m/s
17. 4.4×10^2 s
19. **(a)** $s_{0-2} = 1.0$ m/s, $s_{2-3} = 0$, $s_{3-4.5} = 1.3$ m/s, $s_{4.5-6.5} = 2.8$ m/s, $s_{6.5-7.5} = 0$, $s_{7.5-9} = 1.0$ m/s; **(b)** $v_{0-2} = 1.0$ m/s, $v_{2-3} = 0$, $v_{3-4.5} = 1.3$ m/s, $v_{4.5-6.5} = -2.8$ m/s, $v_{6.5-7.5} = 0$, $v_{7.5-9} = 1.0$ m/s; **(c)** $v_{1.0} = 1.0$ m/s, $v_{2.5} = 0$, $v_{4.5} = 0$, $v_{6.0} = -2.8$ m/s; **(d)** −0.89 m/s
21. **(a)** $v = 74.6$ mi/h; $v = 134.4$ mi/h; **(b)** +80%
23. **(a)** 58.0 km/h; **(b)** 53.0 km/h
25. 12.5 s; 56.3 m (relative to first runner)
27. **(c)**
29. v_0
31. 1.39 m/s²
33. −2.2 m/s each s
35. $a_{0-4} = 2.0$ m/s²; $a_{4.0-10} = 0$; $a_{10-17} = -1.1$ m/s²
37. no; −5.0 m/s²
39. **(a)**

41. no; would accelerate object having $-v$ or $v_o = 0$

43. 7.7 s

45. (a) 129 m; (b) 36.8 m/s

47. -6.2×10^{-3} m/s²

49. 1.43×10^{-3} s

51. 23 s

53. (a) 8.8 m (29 ft); 30 m (98 ft); 45.3 m (149 ft); (b) 17.5 m (57 ft); 55 m (180 ft); 84.3 m (277 ft)

55. $x = (v_o^2/a) - (1/2)at_s^2$

57. 96 m

59. (a) 27.5 m/s; (b) 10.0 s

61. (a) 8.45 s; (b) $x_m = 157$ m; $x_c = 132$ m; (c) motorcycle is 13 m ahead

63. (c)

65. yes; y up and down

67. (a) inverted V with tip on horizontal axis; (b) parabola

69. 2.8 s

71. (a) 2.02 s; (b) 19.8 m/s downward

73. $y_1 = 0.20$ cm; $y_2 = 0.78$ cm; $y_3 = 1.8$ cm; $y_4 = 3.1$ cm

75. slightly less than 8.0 m/s

77. (a) -21.3 m; (b) 20.6 m/s downward

79. (a) 36 m; (b) -3.8 m/s (c) $v = -27$ m/s; $t = 3.4$ s

81. (a) 0.84 s; (b) 4.1 m/s

83. (a) 10.5 m below the initial point; (b) 3.46 s; -22.7 m/s

85. (a) -20.9 m/s; (b) 2.87 s

87. (a) 3.38 s; (b) 31.4 m/s

89. hits 14 cm in front

91. -3.2 m/s

93. (a) 18.5 m/s; (b) 15.9 m

95. (a) -4.2 m/s²; (b) 365 m

97. (a) 0.20 s and 1.4 s; (b) v_o is not enough to reach the height; (c) 0.20 s; the negative root has no physical significance

99. 1.2×10^2 m

101. 52.4 m

103. 3.09 s and 13.8 s

Chapter 3

1. (a)

3. (a) velocity increases or decreases; (b) moves in a parabolic path; (c) moves in a circle

5. 1.3×10^2 m, 27° N of E

7. 60 m; 21°

9. (a) 8.4 m/s; (b) 7.5 m/s

11. $x = 3.34$ m; $y = -2.52$ m

13. (a) 65.0 m; (b) 86.7 m

15. (a) $v_x = 19$ m/s, $v_y = 16$ m/s; (b) $x = 48$ m; $y = 40$ m

17. (b)

19. could obtain negative angles and require use of double angle formulas.

21. 4.9 m 59° below the $+x$ axis

23. (a) 35 m x; (b) 5 m x (c) -5 m x

25. -2.0 m x

27. (a) 19 N x; (b) -19 N x; (c) yes

29. -6.0 m y

31. six possibilities

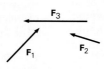

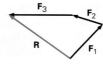

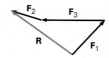

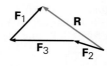

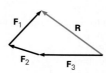

33. $\mathbf{d} = 170$ m x + 23 m y

35. $\mathbf{d} = 3.5$ m x $- 1.5$ m y or 3.8 m, 23° below the $+x$ axis

37. $\mathbf{d} = 1.3$ m x + 1.2 m y

39. (a) 40 km/h (b) -50 km/h

41. (a) 18.7° up the river (b) 238 s = 3.97 min

43. $v_{br}(\sin \theta)$ is not greater than v_{rs}

45. (a) $x = 50.0$ m, $y = 6.25$ m; (b) 25.0 s

47. $(x, y) = $ (25.0 km, W; 100 km, N)

49. (a) 58.3 km/h, 59° S of E; (b) 58.3 km/h, 59° W of N

51. (d)

53. see Fig. 3.16

55. 1.6 m

57. yes; 0.11 m

59. 6.02 m/s

61. (a) 6.75 m (b) 38.5 m

63. $(x, y) = (1.02 \times 10^3$ m, 227 m)

65.

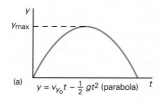

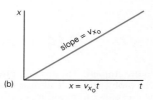

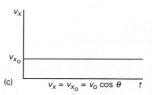

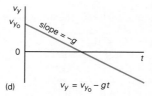

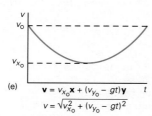

67. $R_{35°} = (1.1) R_{60°}$

69. No

71. 8.7 m/s

73. (a) 161 m; (b) 159 m

75. $(x, z) = $ (13.2 m, 39.3 m)

A-12

79.

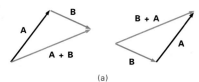

(a)

(b)

81. (a) 68 m;   **(b)** 79°
83. (a) $v = 1.8$ m/s x + 2.5 m/s y
(b) $x = 3.6$ m; $y = 11$ m
85. (a) $v_x = 21.7$ m/s   **(b)** $v = 33.3$ m/s;
$\theta = -49.2°$
87. (a) 4.1 m   **(b)** 22 m
89. 10.9 m/s, 1.6° in the 1st quadrant
91. 3.6 m x; -2.7 y
93. (a) 53°;   **(b)** 8.0 m/s
95. 306 km

Chapter 4

1. (d)
3. no: same mass, same inertia
5. (a) forward, backward;   **(b)** liquid inertia
7. $w_{large} = (2.553)\ w_{small}$
9. $F_3 = -1.0$ N x + 2.5 N y
11. $F_3 = -7.4$ N x
13. (a)
15. measure y and t and use $y = (1/2)at^2$ to find a; then find g using the masses of the bodies, a, and the equation: $(m_1 + m_2)\ a = (m_1 - m_2)\ g$
17. 6.75 N
19. (a) $w = 690$ N, $m = 70.4$ kg
21. 20 m/s² in direction of force
23. (a) 762 kg;   **(b)** 1.68×10^3 lb
25. (a) 0.200 m/s²;   **(b)** 0.200 m/s²
27. $F_{net} = 0$
29. 3.7×10^6 N
31. (a) -8.0×10^3 N or 8.0×10^3 N in opposite direction of v_o;   **(b)** -1.0×10^4 N or 1.0×10^4 N in opposite direction of v_o
33. 4.45×10^4 N
35. 90 N
37. 5.90 m/s²
39. (a) 3.7×10^2 N;   **(b)** $T = 5.3 \times 10^2$ N;
(c) $T = (75)(9.8)/2(\sin \theta)$; T becomes larger as θ becomes smaller
41. 1.4 m/s²
43. (a) 5.4 kg;   **(b)** 5.2 kg
45. (a) 2.5 m/s² to the right;   **(b)** 2.0 m/s² to the left

47. 2.1 m/s²
49. (a) 2.0 m/s²;   **(b)** 23 N
51. (a) 686 N;   **(b)** 723 N;   **(c)** 649 N
53. (a) m_1;   **(b)** $m_1 = (1.2)\ m_2$
55. (c)
57. flying bird's weight is supported by reaction force of air due to wing motion; it would weigh less, since most air would pass through floor and little force would register on scale.
59. masses and magnitudes change
61. 1.50×10^3 N
63. (d)
65. $\tan \theta$
67. (a) opposes slipping;   **(b)** can increase or decrease depending on the direction of the wind
69. 1.7°
71. (a) 30°   **(b)** 22°
73. 0.586
75. (a) net force is zero if mass is originally at rest;   **(b)** net force is 92 N
77. 0.720
79. 1.0
81. no
83. (a) 26 N;   **(b)** 21 N
85. 0.34
87. 0.25
89. 2.3 m/s²
91. 10 m/s²
93. (a) 31 N   **(b)** $(0.16)f_s$
95. 0.31 N
97. (a) $F = 2.8$ N x $- 2.4$ N y;   **(b)** object would move with constant velocity if originally in motion; if initially at rest, would remain at rest.
99. $y = 0.66$ m for both, since acceleration is independent of mass
101. 28 m

Chapter 5

1. (d)
3. No motion, no displacement, therefore no work
5. 9.80×10^4 J
7. 8.0 m
9. 15 J
11. 1.47×10^5 J
13. (a) 2.2×10^5 J;   **(b)** -2.2×10^5 J
15. 2.5×10^3 J
17. 3.0×10^2 J
19. (d)
21. 1.1×10^{-2} J
23. 0.21 J
25. 0.50 J more
27. 0.38 m
29. (a) 16 J;   **(b)** 20 J;   **(c)** use more segments
31. (c)
33. (a) 45 J;   **(b)** 21 m/s
35. 1.3×10^7 m/s

37. (a) 240 m;   **(b)** 58 km/h
39. -6.8 J
41. (d)
43. 0.21 m
45. 2.9 J
47. (a) -1.8×10^2 J;   **(b)** 8.5×10^2 J;
(c) 8.5×10^2 J
49. 3.4×10^3 J; swings horizontally
51. 21 J
53. (d)
55. (a) 282 J;   **(b)** $K = 184$ J, $U = 98$ J;
(c) $E = 282$ J, $v = 47.5$ m/s;   **(d)** 0, 184 J,
47.5 m/s, 0
57. (a) $K = 9.8$ J, $U = 4.2$ J;   **(b)** $K = 10$ J
59. 17 m/s
61. (a) 32 m;   **(b)** 22 m
63. (a) $h = L (1 - \cos 25°)$;
(b) 9.0×10^{-2} J;   **(c)** 1.4 m/s
65. no since $K_{bottom} < U_{top}$
67. (a) 0.166 m;   **(b)** 4.82 m/s
69. (b)
71. same work; shorter time means more power expended
73. (a) no, does not involve time;   **(b)** no loss of energy; creation of energy
75. (a) 9.90×10^5 J;   **(b)** \$0.67
77. (a) 8.19×10^5 J;   **(b)** 3.89×10^3 W
79. 7.7×10^3 W
81. 6.7×10^2 J/s
83. (a) 3.2 W;   **(b)** -13 J;   **(c)** 0.21
85. (a) 2.4 hp;   **(b)** 10 hp
87. \$0.36
89. 13 J
91. 42.5 J
93. 2.7×10^5 J
95. 41 %
97. (a) 3.0 J;   **(b)** 0.33
99. (a) 5.8 ft-lb;   **(b)** 7.6 J
101. 29 W

Chapter 6

1. (d)
3. air and boat move in opposite directions—conservation of momentum
5. 1.5×10^2 kg-m/s
7. 6.9×10^2 kg-m/s
9. 7.2 kg-m/s in opposite direction of v_o
11. (a) 0.35 kg-m/s x;
(b) 9.2×10^{-2} kg-m/s y
13. 8.1×10^3 kg-m/s
15. (a) 2.1 kg-m/s, 45° above the $+x$ axis, 1st quadrant;   **(b)** a collision does not have to occur; momentum would be the same
17. $\Delta p = -2.8$ kg-m/s x $- 1.1$ kg-m/s y
19. (a) -3.1×10^2 kg-m/s;   **(b)** yes;
-3.5×10^2 kg-m/s
21. (a)
23. no, because net force is present
25. 1.7 m/s in opposite direction
27. 7.7×10^{-23} m/s

29. $p = 1.3 \times 10^2$ kg-m/s x + 1.6×10^2 kg-m/s y

31. moves at 0.25 m/s in opposite direction

33. (a) 5.0 km/h in direction of moving car (b) 2.5 km/h in direction of faster car (c) 18 km/h in same direction as initial motion

35. (a) 0.95 m/s in sled's direction (b) 0.95 m/s;

39. (c)

41. in each case, difference in result depends partly on difference in Δt

43. momentum is a vector; kinetic energy is a scalar

45. uranium would have far less speed due to much larger mass

47. 50 m/s

49. 28 m/s

51. 11 N-s

53. $v_1 = 1.3$ m/s; $v_2 = 5.3$ m/s

55. $v_1 = -48$ cm/s; $v_2 = 2.0$ cm/s

57. (a) 77 N-s; (b) 5.5×10^2 N; (c) 33 m/s

59. 3.0×10^2 N

61. $K_{lost} = 1.1 \times 10^2$ J

63. 19.6%

65. (a) 28%; (b) 1.3×10^7 m/s

67. Since $K_o \neq K$, not elastic; primarily heat.

71. (a) $v = v_o/90$; (b) 2.5×10^2 m/s; (c) 98.9%

73. (b) $e_{(elastic)} = 1.0$; $e_{(inelastic)} = 0$

75. (d)

77. yes; use strings to hang from three different locations

79. -0.35 m

81. (a) 4.6×10^6 m from center of Earth (b) 8×10^6 m below surface of Earth

83. (8.0 m, 0)

85. (8.7 cm, 15 cm)

89. 97.6 m from the shore

91. $v_1 = -4.0$ m/s, $v_2 = 2.0$ m/s

93. $v_1 = v_2 = 2.0$ m/s

95. (a) 6.3×10^5 kg-m/s; (b) 1.6×10^5 N

97. 1.1×10^{12} m; none

99. (a) (2.0 m, 2.0 m); (b) (2.0 m, 2.0 m); (c) (2.8 m, 2.0 m)

101. 3.3 N

103. 2.0 m/s; 53°

Chapter 7

1. (c)

3. $r = 2.0$ m; $\theta = 45°$

5. (a) 0.13 rad; (b) 4.71 rad; (c) 0.79 rad; (d) 9.42 rad

7. 4.2 cm

9. 2.4×10^8 km

11. 120 s

13. (a) 2.0×10^2 rad; (b) 60 m

15. (c)

17. into the turntable

19. 0.084 rad/s

21. A is faster

23. (a) $T_h = 4.3 \times 10^4$ s, $T_s = 60$ s; (b) $f_s = 1.7 \times 10^{-2}$ s^{-1}, $f_h = 2.3 \times 10^{-5}$ s^{-1}; (c-d) into clock face

25. (a) 11.1 rad/s; (b) 102 ft

27. (a) 1.16×10^{-5} s^{-1}; (b) 7.29×10^{-5} rad/s, in direction of N pole; (c) $v_{45}/v_{eq} = 70.7\%$

29. (d)

31. only if $a_t = 0$

33. the outward force does not exist in an inertial system.

35. 1.1 m/s^2

37. (a) 2.2×10^{-4} m/s^2; (b) moon 10 times greater

39. (a) 5.9 m/s; (b) 23 m/s^2

41. (a) 3.0×10^4 m/s; (b) no; $a_c = 3.4 \times 10^{-2}$ m/s^2

43. (a) $a_t = 2.5$ m/s^2, $a_c = 9.7$ m/s^2; (b) at lowest point of swing; $a_t = 0$

45. (d)

47. if angular acceleration is constant, angular velocity decreases until it stops, then both are zero.

49. (a) 0.38 m/s^2; (b) 2.7 m/s

51. (a) 4.26 rad; (b) 4.26 ft

53. (a) 8.40×10^{-3} rad/s^2; (b) 0

55. (a) 1.7×10^3 N; (b) gravitational force

57. 10 rad

59. (a) friction not enough to provide centripetal force for circular motion; (b) 9.8×10^{-2}

61. (a) 4.9 m/s^2; (b) 8.7 m/s^2; (c) 5.0 m

63. (d)

65. no water would run out of cup as it fell

67. 2.0×10^{30} kg

69. (a) 9.74 m/s^2; (b) 3.4×10^5 m

71. 3.4×10^8 m from Earth; if other gravitational forces are negligible

73. 3.7 m/s^2

75. 3 g's

77. (a) -2.5×10^{-10} J; (b) 0

79. 1.63 m/s^2

81. (c)

83. none

85. (a) fire rockets for forward thrusts; (b) no; would have same speed

87. 1.4×10^9 J

89. $4.4 - 10^{11}$ m

91. (a) 0.16 rad/s; (b) speed decrease by a factor of 0.71

93. 1.9×10^{27} kg

95. when moon is on near side, it produces a bulge for a high tide; attraction is greater on the Earth than on the water on the far side, producing a high tide there also.

99. 1.02 m/s^2

103. (a) 3.4×10^{-2} m/s^2; (b) 2.6×10^{-2} m/s^2; (c) 0

105. (a) 4.6×10^{-11} N toward m_2; (b) 1.8×10^{-10} m/s^2; (c) 0.27 m from m_1; no other point

107. 115° or 2 rad

109. (a) 4.42 m/s; (b) a = 4.42 m/s^2 r +

0.766 m/s^2 t

111. 3.8 m

113. (a) $v = (Rg)^{\frac{1}{2}}$; (b) $h = (5/2) R$

Chapter 8

1. (d)

3. yes, e.g., rolling

5. (a) 0.45 m/s; (b) 0.90 m/s

7. 1.5 m

9. 1.7 rad/s

11. no

13. (b)

15. (a) closer to ground; (b) yes

17. (a) 6 stable; 20 unstable (b) 5 always stable (4 sides, and on 2 inside edges simultaneously); 22 always unstable (12 edges and 10 corners); 2 sometimes stable (ends may be both stable, both unstable, or 1 stable and 1 unstable, depending on lengths of arms and location of CM)

19. 81 N

21. 1.1×10^2 N-m

23. 98 cm-N = 98 cm-N

25. 22 N, for each case

27. (a) 88.2 N; (b) 10.5 kg

29. $F_L = 4.26 \times 10^7$ N; $F_R = 4.26 \times 10^7$ N

31. (a) $\tan \theta = f_s/N$; (b) 30°

33. $T_1 = 15$ N; $T_2 = 21$ N

35. $x_{CM} = 7.9$ cm; $y_{CM} = 16$ cm; $z_{CM} = 1.0$ cm

37. 27°

39. $T_1 = 261$ N; $T_2 = 572$ N

41. (d)

43. (a) yes; (b) no

45. (a) in direction of force; (b) torque decreases, and reverses at critical angle

47. lengthen day

49. 3.4×10^4 rad/s^2

53. 2.5×10^{38} kg-m^2 or twice as much

55. 1.1 rad

57. (a) 0.89 m/s^2; (b) 0.44 m/s^2

59. 3.9×10^2 rad/s

61. $\theta = \tan^{-1}(7\mu/2)$

63. (b)

65. 7.5×10^2 J

67. 5.9 m/s

69. (a) 1.39×10^8 J; (b) 1.54×10^6 W

71. (a) 29% (b) 40% (c) 50%

73. (a) 2.8×10^2 J; (b) 0.25 s

75. 3.4×10^{-3} J; the work is supplied by the motor

77. (a) $v = [Rg]^{1/2}$; (b) $h = (2.7)R$ (c) weightlessness

79. (c)

81. to decrease I and increase angular velocity

83. gravity exerts a force to produce a torque

85. 3.4 rad/s

87. (a) $v_p > v_a$; (b) 1.03; (c) 3.1×10^6 mi

89. $L_{rev} = 2.6 \times 10^{34}$ m-N-s; $L_{rot} = 2.2 \times 10^{30}$ m-N-s

91. $I = (0.60) I_o$

93. (a) 4.3 rad/s; **(b)** $K = (1.1) K_o$

95. $d = b (v_o/v)$

97. 0.13 J

99. 0.833 rad/s

101. 5.4 rad/s

103. (a) 3.8 rad/s; **(b)** 1.2 m-N-s

105. net torque is zero

107. (a) 2.5 cm; **(b)** 36 cm

109. (a) 1.3 m/s^2; **(b)** 1.1×10^{-2} m-N

111. (a) 35°; **(b)** 0.18 m from the end

Chapter 9

1. (d)

3. less strain for given stress

5. no

7. (a) 2.5×10^5 N/m^2; **(b)** 4.3×10^5 N/m^2

9. 2.1×10^3 N/m^2

11. 9.5×10^{10} N/m^2

13. bends toward copper

15. 6.7×10^3 N/m^2

17. 1.3×10^4 N

19. (a) ethyl alcohol; **(b)** $\rho_w/\rho_{Al} = 2.2$

21. 7.0×10^4 N

23. 0.13 cm

25. (b)

27. (a) force on the bottom of the container is pressure times area; areas are equal and pressures are equal due to equal depths **(b)** weights would differ (different volumes) but pressure increase same for all.

29. (a) When the liquid is poured from an unvented can, a partial vacuum develops and opposing atmospheric pressure makes pouring difficult. By opening the vent, you are allowing the pressure to equalize, and the liquid can easily be poured. **(b)** When you squeeze a medicine dropper before inserting it into a liquid, you are forcing the air out and reducing the pressure inside the dropper. When you release the top with the dropper in a liquid, the liquid rises in the dropper due to atmospheric pressure. **(c)** When we inhale, the diaphragm relaxes and the rib cage expands. The volume of the lungs increases, so the pressure decreases and air rushes into the lungs. When we exhale, the diaphragm tenses and the rib cage contracts. The volume of the lungs decreases, so the pressure increases, and air rushes out.

31. 26 atm or 3.8×10^2 lb/in^2

33. 10 m high

35. 2.1×10^3 Pa

37. 2.0×10^5 Pa

39. 1.07×10^5 Pa

41. (a) 1.1×10^8 Pa; **(b)** 2.0×10^6 N

43. 2.4 g/cm^3

45. 3.7×10^{-2} N

47. 1.6 cm

49. (a)

51. no; no

53. no

55. 10,000 metric tons

57. no

59. 2.6×10^3 kg

61. 1.5×10^{-5} m^3; 6.0×10^3 kg/m^3

63. 2.0 m

65. no, density less than solid gold

67. (d)

69. There are many more capillaries than arteries, with total cross-sectional area greater than that of the arteries. If Av is constant, then if A increases, v must decrease.

71. (a) 4.7×10^{-3} N; **(b)** 8.4×10^{-2} N

73. $h = 0$

75. (a) water; **(b)** $h_w = (1.3) h_b$

77. $h = -0.010$ m

79. 0.65 cm

81. (c)

83. (a) if velocity increases over the top of the wing, pressure is reduced and lift force is produced. **(b)** air is forced downward and a third law reaction force acts on blades.

85. increase in air speed over top produces a decrease in pressure, so net force is upward.

87. $\Delta p = \rho g (y_2 - y_1)$, with $v = 0$

89. $\rho_b = 3\rho_a$

91. (a) 3.5 cm^3/s **(b)** 0.031%

93. 5.7×10^3 Pa

95. 2.2 Pa

97. (c)

101. 2.0×10^2 Pa

103. 7.9×10^{-5} m^3/s

105. (b) 4.7×10^{-5} m^3/s

107. 5.9×10^4 Pa

109. through capillary action, water is taken up by wooden pegs, which swell and split the rock.

111. 8.2×10^{-5} m^3/s

113. 99.97 cm

115. yes

117. 8.0×10^3 m

Chapter 10

1. (a)

3. no

5. (a) 18 F°; **(b)** 5.6 C°

7. (a) 90°F; **(b)** 257°F; **(c)** 5°F; **(d)** $-460°$

9. 57°C and $-62°$C

11. 102°F

13. (c)

15. (a) decreases; **(b)** remains constant

17. pressure kept constant by changing the volume

19. (a) advances; **(b)** $+0.11$

21. (a) $-273°$C; **(b)** $-23°$C; **(c)** $-185°$C; **(d)** 0°C

23. 27×10^6 °F; 15×10^6 °C

25. (a) 1.3×10^{24} molecules; **(b)** 1.5×10^{24} molecules; **(c)** 1.8×10^{24} molecules; **(d)** 1.5×10^{24} molecules

27. 40°C

29. $p_2 = 4p_1$

31. 3.1×10^{23} molecules

33. (a) 5.0 L; **(b)** 2.4×10^2 kPa; **(c)** -132°C

35. 574.61 K

37. (c)

39. (a) when ball is heated, it becomes too large to fit through ring; **(b)** when ring is heated it expands outward, the hole gets larger, and the ball passes through.

41. 6.8×10^{-4} m

43. 4.8 C°

45. 4.3×10^{-3} m

49. (a) 1.02×10^3 °C **(b)** no

51. $\beta = [(p_1/p_2)(T_2/T_1) - 1]/[(T_2 - T_1) + 273]$

53. 76°C

55. (d)

57. If red blood cells were placed in hypertonic solutions, the cells would collapse and shrink up. The solution has a lower concentration of water than cells, so water would leave the cells via osmosis. In a hypotonic solution, the cells would swell up with water and undergo lysis (blow up). This is because the solution has a much higher concentration of water than the cells, so water would move into the cells via osmosis.

59. 12% greater

61. 4.8×10^2 m/s

63. 7% increase

65. $R_{oxygen} = (1.2) R_{ozone}$

67. 273°C

71. -45°C by 4 F°

73. 33 C°

75. (a) 6.7°C, -49°C, $\Delta T = -56$ C°; -20°C, 7.2°C, $\Delta T = 27$ C° **(b)** -0.07 F°/min, -0.039 C°/min; 25 F°/min, 14 C°/min

77. 126 cm^3

79. yes; low coefficient of linear expansion

81. 136°F; -128°F

83. 3.1 cm

85. $+2.4$%

87. (a) 1.9×10^5 Pa; **(b)** -130°C

Chapter 11

1. (b)

3. intrinsic energy value of substance (chemical potential energy)

5. 4.0×10^5 J

7. 4.4×10^3 W

9. (a) < 4.0 Btu; **(b)** 8.3×10^2 Btu

11. 210 kg

13. 4% difference

15. (a)

17. water has a high specific heat; water does not lose heat quickly

19. 0.20 cal/g-C°

21. 32.6°C

23. 0.33 cal/g-C°

25. 77 C°

27. $m_{Pb} = 6\,(m_{Cu})$
29. 0.074 kcal/kg-C°, final T higher
31. (a) 7.2×10^2 W; (b) 9.4×10^2 W
33. 7.6×10^{-2} cm
35. (c)
37. no, no change in temperature
39. Under pressure, the boiling point of the coolant is increased and the engine may be operated at a higher temperature for increased efficiency. If the cap is removed and the pressure suddenly reduced, the coolant flash boils and can spurt out, causing burns.
41. 6.2×10^2 kcal greater
43. 12 kcal
45. water, 92 s; alcohol, 28 s
47. 1.7×10^{-2} kg
49. 0.98 kcal
51. 52.4 kcal
53. 0.17 L
55. (d)
57. Bridge is exposed to cold air all around, while the road is in contact with ground, which does not cool below freezing point until weather has been very cold for a long time.
59. Heat loss from an object is greater when the temperature difference is greater; adding cream right away will reduce the temperature difference with less heat loss.
61. $t_o = (4.5)\,t_t$
63. $k_{silver}/k_{iron} = 9.1$
65. $d_{Cu} = (1.6)\,d_{Al}$
67. (a) 5.1 in.; (b) 5.7 in.
69. 1.4
71. 2.3 cm
73. 1.4×10^4 J
75. 2.0×10^4 kcal
77. 23°C
79. (c)
81. Fogs due to cooler temperatures found in valley during radiation losses at night; sun increases temperature above the dew point which dissipates the fog.
83. (a) 100 g (b) 1.7 kg
85. 11°C
87. 62 Cal
89. 27°C
91. (a) 34 kcal (b) 0.43 kg (c) No since ice is less dense than water
93. 0.18 C°
95. 110°C
97. 0.12 C°
99. 38 kcal or 1.6×10^5 J

Chapter 12

1. (d)
3. (a) On the liquid-vapor surface, water exists in two phases, liquid and vapor, in equilibrium. On the solid-vapor surface, the solid phase and the vapor phase coexist in equilibrium. (b) At certain combinations of p, V, and T, all three phases of water coexist in equilibrium. These coordinates lie along the triple line.

5.

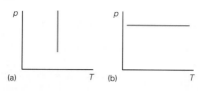

7.

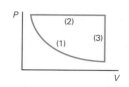

9. (c)
11. (a) As you pump, external force compresses gas and does work on system; heat leaves system, causing pump to heat up (b) The rapid expansion is nearly adiabatic, and work is done chiefly at the expense of the internal energy of the gas, lowering its temperature and cooling the valve.
13. -2.5×10^3 J; temperature decreases
15. 3.0×10^5 J
17. 8.4×10^3 J
19. (a) 1.13×10^3 J; (b) heat in
21. 2.09×10^3 J
23. 3.6×10^4 J
25. (a) 1.0×10^5 J, 0; (b) -7.5×10^4 J; (c) -7.5×10^4 J
27. (a)
29. energy created
31. 1.2×10^3 J/K
33. $S_i = -9.2 \times 10^2$ J/K; $S_s = 1.5 \times 10^3$ J/K; ice has more order
35. 11 J/K
37. 1.5×10^3 J/K
39. -2.2×10^2 J/K
41. -17 J/K
43. 0; -5.0×10^3 J
45. 0.2 J/K
47. (b)
49. complete conversion possible for single processes; work needed for cyclic process.
51. no
53. (a) 33%; (b) 2.1×10^5 J or 50 kcal
55. 49%
57. (a) 6.6×10^8 J; (b) 27%
59. (a) 145 kcal; (b) 1.89×10^6 J

61. (a) 4.7×10^6 J; (b) 1.3×10^7 J
65. (b)
67. when outside temperature is low, heat pump is less efficient
69. 39%
71. 273 K
73. 2.00×10^3 J; 265°C
75. no, not possible
77. 71°C
79. if $e_{th} = 1$, then $Q_c/Q_h = 0 = T_c/T_h$, and $e_c = 1$
81. (a) 67%; (b) $T_h = 3\,T_c$
83. 40 kW / cycle
87. 8.0 L
89. (a) isobaric; (b) 145 J
93. 3.1×10^3 J
95. (a) 3.4×10^3 J; (b) $\Delta U = 0$ since the process is isothermal; (c) 12 J/K

Chapter 13

1. (b)
3. (a) four times greater; (b) twice as large
5. 20 Hz
7. 2.0 s
11. (a) 10^{-12} s; (b) 63 m/s
13. 9.0×10^{-2} J
15. (a) 2.4 m/s; (b) 2.4 m/s; (c) 2.6 m/s; equilibrium
17. Fig. 13.22 by a factor of 4
19. (a) 0.38 m; (b) 8.4×10^{-3} m
21. (d)
23. no; tangent goes to infinity
25. (a) $x = A \sin \omega t$; (b) $x = \pm A \cos \omega t$
27. (a) 0.50 %; (b) 2%; (c) 3%
29. (a) 4.8 cm; (b) 4.4 cm/s; (c) -1.2 cm/s^2
31. (a) 14 cm; (b) 89 cm/s; (c) 0.17 J
33. (a) $v_{max} = A\omega$ at $y = 0$ and $t = nT/4$, where $n = 1, 3, 5 \ldots$; (b) $a_{max} = A\omega^2$ at $y = \pm A$ and $t = nT/2$, where $n = 0, 1, 2, \ldots$
35. (a) $y = 0$; (b) $v = 94$ cm/s; (c) $a = 0$
37. $T_1 = (1.4)\,T_2$
39. (a) $y = -(0.10 \text{ m}) \sin 10\pi t/3$; (b) $k = 38$ N/m
41. (d)
43. (a) transverse and longitudinal; (b) longitudinal; (c) transverse
45. 45 m/s
47. 7.9×10^{-7} m
49. 3.2 m/s
51. (a) 0.20 m; (b) 13 cm
53. (a) 0.20 s; (b) 2.0 s
55. $y = A \sin (kx - \omega t)$; $v = A \sin 2\pi (x/\lambda - t/T)$; $v = A \sin (2\pi/\lambda)(x - \lambda t/T)$ $y = \sin (2\pi/\lambda)(x - vt)$; Consider motion from another reference frame traveling with the wave. $y = A \sin (2\pi/\lambda)x'$ since it is not moving, and $x' = x - vt$ or $x = x' + vt$, so wave and traveling reference frame moving to the right or in the positive x direction (Galilean transformation). (b) This process is similar

for $x' = x + vt$ and motion to the left or in the negative x direction.
57. (b)
59. (a)
61. For example, for $f_o/2$, a push is given every other time; for $2f_o$, a push is given when the swing is not there.
63. (a) 660 Hz; **(b)** 880 Hz
65. 4.0 s
67. 38 Hz; 76 Hz; 114 Hz; 152 Hz
69. (a) 12 cm; **(b)** 3 cm, 9 cm, 15 cm
71. 480 Hz
73. $f = (2n + 1)(8.8 \times 10^3 \text{ Hz})$, where $n = 0, 1, 2, \ldots$
77. 1.7 s; 0.60 Hz
79. 2.8×10^2 Hz; 1.2 m
81. 210 Hz
83. 1.1 cm/s
85. 1.1 m/s
87. 3.0 s
89. (a) 4×10^2 s; **(b)** 1.9×10^3 km; **(c)** 1.6×10^3 s; shear waves not transmitted through outer liquid core

Chapter 14

1. (d)
3. (a)
5. sound is not all in the audible range
7. depends on the density
9. (a) 337 m/s; **(b)** 346 m/s
11. 15°C
13. 1.4×10^2 m/s
15. 10 s
17. (a) 0.81 s; **(b)** 0.78 s
19. 4.5 s
21. 12°C
23. $v = 331 + 0.6\,T_c$ (m/s)
25. 2.8%
27. (b)
29. yes, an intensity below the threshold intensity
31. (a) 100 dB; **(b)** 80 dB; **(c)** -10 dB
33. (a) 5.1 **(b)** 5.7 m
35. 2.0×10^5 times greater
37. (a) $I_s/I_o = 10^5$; **(b)** 1.0×10^{-3} W/m²; 1.0×10^{-8} W/m²
39. 10 bands
41. 15 dB
43. (a) 3.2×10^{-3} W/m² **(b)** 16
47. (a) 2.5×10^2 m; **(b)** 2.5×10^5 m
49. (d)
51. no; beat of the music refers to rhythm and tempo
53. a "jet fish" would have to exceed the speed of sound in water
55. 0.286 m
57. 259 Hz and 253 Hz
59. 16 m/s
61. 28 m/s
63. 42°

65. 2.027 m
67. combine individual expressions by eliminating v
69. (d)
71. (a) The snow absorbs sound, so there is little reflection. **(b)** In an empty room, reflections die out slowly (little absorption), giving a hollow, echoing sound. **(c)** Standing wave resonances are set up, giving rise to more harmonics and richer sound quality.
73. not possible
75. 1.1 m
77. (a) f_2 does not exist—only odd harmonics; **(b)** 322 Hz
79. 716 Hz
81. 1.92×10^3 Hz
83. 30°
85. (a) 3.72×10^{-4} W/m², 1.00×10^{-1} W/m²; **(b)** 9.55×10^{-3} W/m², 6.03×10^{-2} W/m²
87. (a) 0.39 m; **(b)** 4 Hz
89. $l_2 = (1/9)\,l_1$
91. 7.5×10^{-3} J
93. 38 vibrations
95. $R_{60} = 10(R_{80})$
97. 1.8%

Chapter 15

1. (d)
3. (a) We know there are two types of electrical charge because attractive and repulsive forces can be produced by different combinations. **(b)** none
5. 1.6×10^{-15} C
7. (a) $+2.4 \times 10^{-9}$ C **(b)** 1.4×10^{-20} kg
9. (a)
11. You would have to ground the electroscope and then bring a positively charged rod close to it, causing electrons to move from the ground to the electroscope. You then remove the ground. You could prove it was negatively charged by bringing a rod of either charge close to the bulb. A negatively charged rod would cause the leaves to separate further, a positively charged rod would cause them to come closer together.
13. (c)
15. (a) 4 times larger; **(b)** 1/9 as large
17. (a) 2.3×10^{-26} N **(b)** zero
19. 4.0 N
21. (a) no place **(b)** no place
23. 5.6×10^{-11} N/C downward
25. $F_e = (2.3 \times 10^{39})F_g$
27. (a) 96 N 39° below the $-x$ axis; **(b)** 61 N 84° above $-x$ axis
29. (b)
31. by the relative density of the field lines
33. The field vectors point away from the positive charge. Because of the mutually repelling forces between them, the like charges will be distributed evenly over the sphere. The electric field outside the sphere behaves as though all the excess charge on the sphere

were concentrated at its center. The electric field is zero inside the sphere. If the charge were negative, the field vectors would all point toward the charge. The field inside would still be zero, and the electric field outside would still behave as though all excess charge were concentrated in the center, even though the excess charge is located on the surface.
35.

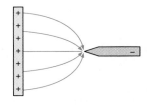

37. 5.8×10^9 N/C
39. (a) (0.080 m, 0); **(b)** any point above x axis where $E_x = 0$
41. 12 cm from the 4.0 μC charge (between the charges)
43. 1.2×10^6 N/C toward the -4.0 μC charge
45. 0.30 m
47. $\mathbf{E} = (-4.2 \times 10^6 \text{ N/C})\,\mathbf{x} + (7.3 \times 10^7 \text{ N/C})\,\mathbf{y}$
49. (c)
51. when work done on charge in moving to other position
53. Person accumulates charge; hair acts like eletroscope leaves. Must be insulated to accumulate charge.
55. 6.0×10^{-19} J; work is done against the field
57. -0.72 J
59. 3.1×10^5 V
61. 5.9×10^6 m/s
63. 0.52 m
65. (a) 27 V; **(b)** 4.3×10^{-18} J; **(c)** no; also has kinetic energy
67. -1.3×10^6 V
69. (a) 1.0×10^2 V; **(b)** 1.0×10^{-4} J in both situations
71. -7.4×10^5 V
73. (b)
75. the charged rubber rod deflects polarized water molecules
77. 1.2×10^{-5} C
79. $K = 3.5$
81. $q = 1.1 \times 10^{-8}$ C; $U = 6.5 \times 10^{-8}$ J
83. 2.4
85. (a) 0.34 μF; **(b)** 1.40 μF
87. 0.19 μF
89. 2.0 μF
91. 6.5 μF; 0.67 μF
93. (a) 4.0×10^{-16} N in the $-y$ direction; **(b)** 4.0×10^{-17} J
95. 0.20 J
97. (a) 0.17 m; **(b)** 0.17 m

Chapter 16

1. (a)
3. no; energy per unit charge
5. (a) 3.0 V (b) 1.5 V
7. 0.25 A
9. one: 1.5 V, 6.0 V, 12 V; two: 7.5 V, 13.5 V, 18 V; three: 4.5 V, 10.5 V, 19.5 V
11. (a) 0.18 C; (b) 0.54 J
13. 2nd wire by 0.060 A
15. 7.5×10^{20} electrons
17. 6.6×10^{-6} m/s
19. (a)
21. (a) same; (b) 1/4 the original current
23. 1.0 V
25. 21 V
27. 0.18 A
29. (a) 2.5×10^{-4} Ω-m; (b) 4.0×10^{3} Ω-m
31. $R_{thick}/R_{thin} = 1/4$
33. 54%
35. $\Delta R/R_o = 0.43$
37. 105 m
39. 4.4 A less
41. 0.12 A
43. 0.77 A
45. decrease of 0.27 A
47. (d)
49. 3.0×10^{2} W
51. (a) 10 A; (b) 12 Ω
53. 1.2 Ω
55. (a) 4.3×10^{3} J/s; (b) 13 Ω
57. 85 Ω
59. (a) 1.2 kWh; (b) $0.14
61. (a) 0.15 A; (b) 1.4×10^{-4} Ω-m; (c) 2.3 W
63. 1.8×10^{3} W
65. 96 Ω
67. $P_2 = 9P_1$
69. 53 s
71. (a) 3.6×10^{2} m; (b) 3.8×10^{2} W
73. 64°C
75. $0.36
77. 5.3×10^{3} A
79. 4.1×10^{-3} Ω
81. (a) 1.0×10^{-7} Ω-m; (b) iron or platinum
83. 9.4×10^{17} electrons
85. 1.8 Ω
87. (a) $R_{Cu}/R_{Al} = 0.61$; (b) $R_{Cu}/R_{Al} = 4.9$
89. 2.4×10^{-2} V

Chapter 17

1. (a)
3. no, only if all resistors are equal
5. (a) 90 Ω; (b) 9.2 Ω
7. parallel: 1.3 Ω; series: 12 Ω; series-parallel: 2.7 Ω; parallel-series: 6.0 Ω
9. series: 15 Ω, 20 Ω, 25 Ω, 30 Ω; parallel: 3.3 Ω, 3.8 Ω, 6.0 Ω, 2.7 Ω; series-parallel: 11 Ω, 13.8 Ω, 18.3 Ω, 4.2 Ω, 6.7 Ω, 7.5 Ω
11. 6.0 Ω

13. No
15. (a) 11 A; (b) 2.0 Ω: 6.0 A, 4.0 Ω: 3.0 A, 6.0 Ω: 2.0 A; (c) 2.0 Ω: 72 W, 4.0 Ω: 36 W, 6.0 Ω: 24 W; (d) $P_{sum} = P_{total} = 132$ W
17. 0.80 Ω
19. 6.9 Ω
21. 30 bulbs
23. 2.6 V; 1.5 A
25. two in parallel, in series with other resistors
27. 5 resistors
29. (a) 4.3×10^{2} s $= 7.2$ min
(b) 2.5×10^{3} s $= 42$ min
31. (a) 8.5×10^{-2} A (b) $P_{15} = 6.9$ W; $P_{40} = 2.6$ W; $P_{60} = 0.25$ W; $P_{100} = 0.41$ W
33. 12 different ways
35. 8.1 Ω
39. (d)
41. $I_1 = I_2 = 0.33$ A
43. $I_1 = 1.0$ A; $I_2 = 0.50$ A; $I_3 = 0.50$ A
45. $I_1 = 3.75$ A; $I_2 = -1.25$ A; $I_3 = 5.00$ A
47. $I_1 = 0.664$ A; $I_2 = 0.786$ A; $I_3 = 1.450$ A; $I_4 = 0.770$ A; $I_5 = 0.016$ A
49. (d)
51. circuit is still closed and when tube ceases to conduct, voltage rises again
53. 1.5 s
55. 1.5 μF; 1.5 MΩ
57. (a) 7.5×10^{5} Ω; (b) 11.4 V
59. (a) 4.7×10^{-4} C; (b) $V_C = 12$ V; $V_R = 0$
61. (a) 2.0 μA; (b) 1.7×10^{-6} C
63. (c)
65. (a) If a low-resistance ammeter were connected in parallel and the circuit current was significantly high, the galvanometer could burn out. (b) If a voltmeter was connected in series, it would not be able to measure voltage drop across the circuit element, but as a high-resistance component, it would affect the voltage drops and current in the circuit.
67. 6.5×10^{2} Ω
69. 49.98 kΩ
71. 0.30 mA
75. (c)
77. the high voltage is necessary to produce a harmful current
79. 23.7 V
81. 9.5 J/s
83. (a) $I_1 = 1.0$ A; $I_2 = I_4 = 0.40$ A; $I_3 = 0.20$ A; $P_1 = 100$ W (b) $P_3 = 2.0$ W; $P_2 = P_4 = 4.0$ W
85. (a) $I = 0.34$ A (for each), $P = 36$ W; (b) $I = 0.69$ A (for each), $P = 150$ W
87. (a) 4.47 V (b) 0.447 A
89. 0.30 A; 0.03334 Ω; 3.0 A: 0.0033334 Ω; 30 A: 0.000333334 Ω

Chapter 18

1. (c)
3. The magnet would attract the unmagnetized iron bar when a pole end is placed at the center of its long side. If the end of the

unmagnetized bar were placed at the center of the long side of the magnet, it would not be attracted (field lines perpendicular).
5. (a)
7. If the charge carriers are electrons (negative), an excess negative charge accumulates on the left side of the strip; if positive, on the right side of the strip. The sign of the voltage drop across the strip would then indicate the type of charge.
9. 6.7×10^{-7} T
11. 3.1×10^{2} T in $+z$ direction
13. (a) 0; (b) down (c) west
15. 2.0×10^{-14} T, left, looking in direction of beam
17. 5.0 A
19. (a) bottom half—directed into page; top half—out of page; (b) same
21. (a) 0; (b) 1.28×10^{-5} T
23. 2.9×10^{-6} T
25. (a) 1.4×10^{-2} m; (b) 1.2×10^{7} N/C
27. (a) 3.3×10^{-5} T; (b) no place
29. 1.6 A
31. 10 cm
33. 1.4×10^{-5} m
35. 4.2×10^{-5} T, 45° from page
37. 4.4×10^{-2} T
39. (b)
41. no; two magnets with N and S poles
43. 9.3×10^{-24} A-m²
45. 13 T
47. (a)
49. no force, no work
51. 3.0 N, $-y$ direction (taking $+z$ as upward)
53. (a) to right; (b) toward top of page; (c) into page; (d) to left; (e) into or out of page
55. 5.6×10^{-2} N
57. 6.7×10^{-2} A; $-z$ direction
59. 3.0×10^{-4} N/m, repulsive
61. 1.8×10^{-5} N, repulsive
63. 2.7×10^{-5} N/m toward the other wire
65. short: 25 N; long: 75 N
67. (a) 9.0×10^{-2} A-m²; (b) with plane of coil parallel to B
69. 0.39 m-N
71. (d)
73. Pushing the button in both cases completes the circuit. The current through the wires activates the electromagnet, causing the clapper to be attracted and ring the bell. However, this breaks the armature contact and opens the circuit. Holding the switch causes this to repeat and the bell rings continuously. For the chimes, when the circuit is completed, the electromagnet attracts the core and compresses the spring. Inertia causes it to hit one tone bar and the spring force then sends the core in the opposite direction to strike the other bar.
75. 1.0×10^{5} m/s
77. (a) 1.8×10^{3} V; (b) same speed, independent of charge

79. (a) 1.6×10^4 V/m; (b) 1.6×10^3 V;
(c) top is negative
81. (a) 4.8×10^{-26} kg; (b) 2.4×10^{-18} J;
(c) no, work equals zero
83. (d)
85. 3.3×10^{-3} T
87. 0.44 T
89. 1.0×10^{-14} J
91. (a) 1.0 m; (b) 1.5×10^{-5} s
93. 6.3×10^{-8} T north
95. 1.3×10^{-3} m
97. 3.0×10^{-2} m-N

Chapter 19

1. (d)
3. (a)

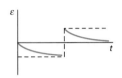

(b) no, Lenz's law
5. (a) 0; (b) 2.3×10^{-3} T-m²; (c) 3.8×10^{-3} T-m²
7. (a) 0; (b) 1.9×10^{-2} T-m²; (c) 1.4×10^{-2} T-m²
9. (a) 0.20 m²; (b) 0.17 m³
11. 40 V
13. 1.8×10^3 A
15. 0.35 T
17. (a) -50 V; (b) 13 V; (c) 0; (d) 12 V
19. -0.96 V
21. (a) change in direction; (b) 4.0×10^4 V
23. -4.0 V
25. lower: 0.12 T-m²; upper 0.11 T-m²; back 0.22 T-m²; others zero
27. 0.13 V
29. (c)
31. The magnet moving through the coils produces a current (Faraday's law). As it moves up and down it will induce a current in the coil which will light the bulb. However, the current is produced at the expense of the magnet's kinetic energy and potential energy. The magnet's motion will damp out rather quickly (Lenz's law).
33. (a) 3.4×10^{-2} V; (b) 0.51 V
35. ± 104 V
37. (a) 0; (b) 87%
39. 16 Hz
41. 108 V
43. (a) 216 V; (b) 160 A; (c) 8.1 Ω
45. (d)
47. (a) step-down; (b) 90 V, 5.3 A
49. (a) 16; (b) 5.0×10^2 A
51. $N_p/N_s = 2.0 \times 10^{-2}$
53. (a) 11 A; (b) 17 V

55. 1.6×10^2 turns
57. (a) $N_s/N_p = 1/12$; (b) 4.2×10^{-2} A
59. 128 kWh
61. (a) 1/2.3, 1/13.9, 1/30; (b) 2.3, 13.9, 30
63. (a) 53 W; (b) $N_p/N_s = 200$
65. (c)
67. (a) 3.0×10^8 Hz; (b) 1.2×10^7 Hz;
(c) 2.0×10^6 Hz
69. 3.6×10^4 m
71. (a) 800 nm > 700 nm: infrared (b) 300 nm < 400 nm: ultraviolet
73. AM: 74 m; FM: 0.77 m
75. primary, 13
77. 100%; ideal conditions
79. (a) 30 A; (b) 4 Ω
81. (a) 840 V; (b) 0.57 A
83. (a) 0.67 A; (b) 54 V
85. 50 Hz

Chapter 20

1. (c)
3. The magnitude of V_o is greater than I_o because the expression for peak current is $I_o = V_o/R$. Since V and I are in phase and reach peak values at the same time, V is always greater than I, since $I = V/R$.
5. 170 V; 340 V
7. 110 V
9. (a) $V_{rms} = 3.8$ V; $V_o = 5.3$ V; (b) 2.8 W
11. (a) 85 V; (b) 19 Hz
13. (a) 60 Hz; (b) 1.4 A; (c) 1.2×10^2 W;
(d) $V = 120 \sin(380t)$ V
15. $V = 170 \sin(340 \pi t)$ V
17. 10 Hz
19. (a) 3.0×10^2 A; (b) 3.4×10^2 V
21. 1.5 rad/s
23. (d)
25. (d)
27. In an ac circuit, a capacitor can oppose current flow. This is because, as the capacitor charges, the voltage across its plates increases, opposing the current. Also, an inductor can oppose current flow because the induced emf opposes the change in flux and thus opposes the current flow in the circuit.
29. 5.9×10^{-8} F
31. 6.6×10^2 Ω
33. 1.7 A
35. 0.24 H
37. (a) 91 V; (b) 90°
39. 0.70 H
41. (d)
43. (c)
45. The current would simply increase with frequency, approaching the vertical asymptote at the resonant frequency (f_o). Above f_o, the current recedes from the vertical asymptote and approaches a horizontal asymptote as f gets very large. Physically, this implies the current would become infinitely large, or a short-circuit condition, at f_o.

47. 68 Hz
49. $-65°$
51. 1.0×10^2 Hz
53. 2.0 H for any resistor
55. ab: 1.3 A; ac: 1.2 A; bc: 4.0 A; cd: 1.8 A; bd: 1.7 A; ad: 2.9 A
57. 3.6×10^{-11} F to 3.4×10^{-12} F
59. (a) 4.4×10^2 Ω; (b) 1.1×10^2 Hz
61. $V_R = 12$ V; $V_L = 2.7 \times 10^2$ V; $V_C = 2.7 \times 10^2$ V
63. (a) 0.895; (b) 13 μF
65. (a) 3.8×10^2 W; (b) 24 Ω; (c) -18 Ω
67. -91 W
69. 0.35 A
71. 1.7×10^{-17} F
75. 37°
77. The magnitude of the capacitance and inductance are chosen so that the capacitor offers a large impedance to low frequencies but passes high frequencies to the load, and the inductor offers a low impedance to low frequencies so as to provide a low path frequency path. Thus the low frequencies are filtered out of the signal and the circuit passes high frequencies. The circuit is both hi-pass and low-pass filter, so only a certain band of frequencies on either side of the common resonant frequency are passes to the load. Frequencies above and beyond this and are filtered out.

Chapter 21

1. (c)
3. (d)
5. diffuse reflection by dust particles
7. 50°
9. 34°
11. 23°
13. 90°
15. (c)
17. (b)
19. (a) Different angles of refraction at air-glass and water-glass interfaces. (b) The sun appears flattened because light from the top and bottom portions is refracted differently as it passes through different atmospheric densities. Light from sides is not refracted.
21. 1.40
23. 0°; not refracted
25. 47°
27. 57°
29. 1.6
31. $f = 6.5 \times 10^{14}$ Hz; $\lambda = 3.0 \times 10^{-7}$ m
33. 4.74×10^{14} Hz; 465.3 nm
35. 32°
37. (a) (0.96 m, -6.0 m) (b) 9.0°
39. (a) 1.41 (b) 1.88
41. (a) 33°; it is reflected for 45° (b) No
43. 75%
45. 1.6
47. >48.8°

49. 1.6 cm

51. (b)

53. no, dispersion by second prism also

55. red end of the spectrum is internally reflected, yet blue end of the spectrum is transmitted

57. 1.4980

59. 0.45°

63. (a) 25° **(b)** 1.97×10^8 m/s **(c)** 362 nm

65. seen for 40° but not for 50°

Chapter 22

1. (c)

3. (a) Reflections in a mirror are seen more clearly against a dark background. During the day, light passes both ways through the pane, and although some is reflected, it is difficult to see the reflection due to the light coming through. At night, there is little light coming through the pane, so the reflections are much clearer. **(b)** The two images are due to reflection on both sides of the plane glass, producing two similar images.

5. (a) 1.0×10^2 cm; **(b)** 7.5 cm; **(c)** 1.0

7. (a) 2.0 m behind the mirror; **(b)** 1.0 m/s

11. (a)

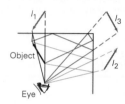

(b) I_4 is counterpart of I_3 with rays reflected from the mirrors in opposite sequence.

13. 0.70 m

15. (d)

17. as the ball swings toward the mirror and approaches the focal point, the image enlarges and moves toward the observer; this produces the effect of appearing to jump out of the mirror

21. $d_i = -30$ cm; $h_i = 9.0$ cm; image is virtual and upright.

23. -18 cm

25. see pp. 677–681. **(a)** $d_o > R$: Fig. 22.9a; **(b)** $d_o = R$: Demonstration 15; **(c)** $R > d_o > f$: Fig. 22.10b; **(d)** $d_o = f$: Fig. 22.10c; **(e)** $d_o < f$: Fig. 22.10d

27. 120 cm

29. (a) virtual; **(b)** 150 cm

31. 2.3 cm

33. 7.5 m

35. $d_i = 60$ cm, $h_i = 10$ cm; image is real and inverted

37. 15 cm and 5.0 cm

39. (a) $d_i = 39.8$ cm, $h_i = 0.80$ cm; **(b)** $d_i = 38.5$ cm, $h_i = 7.7$ cm

41. (d)

43. (b)

45. 15 cm; 10^{-12}

47. (a) $d_i = 6.4$ cm, $M = 0.64$ **(b)** $d_i = 10.5$ cm, $M = 0.42$

49. (a) no; **(b)** yes

51. (a) -2.5 cm; **(b)** -40 D

53. 8.2×10^{-2} m

55. 20 cm, $M = -1$

59. (a) -20 cm; **(b)** -63 cm

61. (a) 4.6 cm; **(b)** 0.80 cm

63. virtual image 18 cm from eyepiece

67. (d)

69. 34 cm; same

71. -0.70 D

73. (a) 6.0 m; **(b)** upright, virtual, and same size

75. (a) 37.5 cm; **(b)** $h_i = 3.0$ cm, real and inverted

77. A sharp image on the screen means the diverging lens is positioned so that the transmitted light rays are parallel to its principal axis. The rays are reflected directly back through the lens system to the position of the source. Rays incident on a diverging lens are transmitted parallel to its principal axis if they apparently converge at the lens's focal point on the image side of the lens. Since the rays are originally converged at the mirror (screen), the mirror is at the focal point of the diverging lens.

79. $h_o = 26$ cm; $d_o = 18$ cm

81. (a) Converging lens

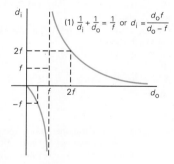

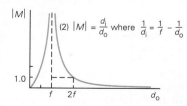

(b) Diverging lens

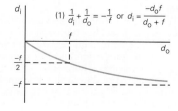

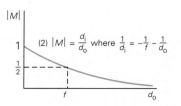

83. 60 cm image sides of L_2, real, upright, $M_t = 4.0$

85. 9.0

Chapter 23

1. (d)

3. blue

5. because of changing path differences there is constructive and destructive interference, giving rise to variations in reception intensity.

7. (a) $\lambda/6$; **(b)** $\lambda/4$; **(c)** $\lambda/2$

9. 0.91 m

11. 1.2×10^{-3} rad

13. $y_m = m\lambda L/2d$, $m = 1,3,5$; $\Delta = L\lambda/d$

15. 1.2×10^{-2} m

17. (a) 6.3×10^{-3} rad; **(b)** 1.3×10^{-3} m

19. (a) $y_n' = (0.75) y_n$; **(b)** 1.2×10^{-2} m

21. (a)

23. The path length through the film would be different (rather than directly down and back in terms of t alone), and the angular dependence and Snell's law would have to be used to determine the conditions for interference.

25. (a) 27 λ; **(b)** 180° destructively

27. 5.5×10^{-8} m

29. $t_1 = 160$ nm, $t_2 = 200$ nm

31. (a) 170 nm; **(b)** 340 nm

33. 1.8×10^{-6} m

35. (b)

37. (a) 4.8×10^{-3} m; **(b)** 2.4×10^{-3} m

39. 450 nm

41. (a) 2.4×10^3 lines/cm **(b)** 5

43. do not overlap

45. 11°

47. 2 at $\pm 30°$

49. 9.8×10^{-3} m

51. yes

53. (d)

55. (a) twice; **(b)** four times; **(c)** none; **(d)** six times

57. 1.6

59. 39°

61. 57.2°

63. 50.3°

65. (a)

67. (a) different air molecular density; **(b)** sky would be black

69.

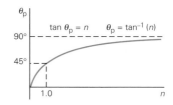

73. 56°
75. 32.3°
77. 6.4×10^{-4} rad
79. one red; two violet
81. 1

Chapter 24

1. (c)
3. The eye focuses by changing the shape of its lens to change the focal length. The focal length is adjusted to produce a sharp image.
5. +1.8 D
7. +71 D
9. +2.8 D
11. +3.1 D
13. 6.7 m
15. (a) −0.133 D; (b) −0.133 D
17. right: +1.14 D, −0.46 D; left: +1.78 D, −0.46 D
19. (d)
21. no; lens aberrations limit practical range of single magnifying glass to about 3 X to 4 X
23. 0.0040 rad; 0.0020 rad
25. 6.0 X
27. (a) 6.1 cm (b) 4.1 X
29. (a) 1.8 X (b) 1.3 X
31. +77 D
33. (a) 340 X (b) 3900%
35. 25 X
37. 3.6 X
39. $M_{max} = 930$ X, $M_{min} = 43$ X
41. (d)
43. (a) 4.0 X; (b) 75 cm
45. 3.8°
47. (a) 2.7 X; (b) 135 cm
51. (a)
53. 1.9×10^{-3} rad
55. 5.5×10^{-7} m
57. 1.32×10^{-7} rad
59. (a) $\Delta\theta = 1.47 \times 10^{-5}$ rad; (b) $\Delta s = 4.40 \times 10^{-5}$ cm
61. (a) 5.55×10^{-5} rad; (b) red; (c) 70.2%
63. (d)
65. (a) Since white is obtained by adding colors, it cannot be obtained by the subtractive method. That method subtracts colors, and the one we see is the one that is not absorbed. (b) Black objects absorb most wavelengths of light. We see the objects because we perceive the extremely faint reflected light as black.

67. (a) 7.1 cm; (b) 3.5 X
69. +3.0 D
71. 1.3 X
73. 375 X
75. 82 cm
77. 13 cm
79. top: −0.20 D; bottom: +2.6 D

Chapter 25

1. (d)
3. (a)
5. (c)
7. Since there was no difference in the interference pattern, the speed of light was not different for the two directions. The speed of light is constant in all directions through all frames of reference. No ether (wind) would be present.
9. (a) 4.22 s; (b) 4.48 s
11. 55 m/s and 45 m/s
13. The time for light to travel to and return from the mirror is $t = 2L/c$. The disk advances one tooth gap in $t = 1/nN$. Equating these times, we obtain $c = 2nN\,L$
15. no; swimmer takes dv^2/c^2 less time ($v \ll c$)
17. (d)
19. Her friend would appear same height if standing and perpendicular to the direction of motion. However, he would be thinner or contracted in the direction of motion.
21. 32 beats/min
23. 1.06 h
25. 64 y; 37 y
27. 2.4 s
29. (a) 131 m²; (b) 40.6°
31. (a) 0.14c; (b) 0.44c; (c) 0.9999c
33. 5.3 m
35. 0.40c
37. (d)
39. yes; no
41. (a) 0.985 c; (b) 2.5 MeV; (c) 1.59×10^{-21} kg-m/s
43. (a) 0.14c; (b) 0.87c
45. (a) 0.60c; (b) same height, smaller width; (c) increase
47. 1.5×10^{-21} kg-m/s
51. (a) 0.92 c (b) 1.4×10^3 MeV
53. (a) 1.1 MeV; (b) 1.6 MeV
55. 1.79×10^3 y
59. (a)
61. The stick would be elongated or "stretched" by the gravity gradient.
63. 8.9 mm and 2.8 m
65. (a) 1.5×10^4 m/s; (b) 7.1×10^{17} kg/m³
67. 0.43 c
69. −0.154 c
71. 939 MeV
73. 3.1 h
75. length 33 m, height 2.5 m, width 2.0 m
77. 50 min
79. 0.67 c

Chapter 26

1. (d)
3. no
5.

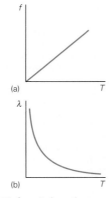

7. 9.4×10^{-6} m; infrared
9. (a) 4.1×10^3 K (b) 7.3×10^3 K
11. 2.56×10^{-20} J
13. (b)
15. The packets of radio signal arrive so quickly and in such quantity that our ear cannot distinguish between discrete arrivals.
17. 6.0×10^{-11} m; X-ray
19. (a) 6.63×10^{-19} J; (b) 4.1 eV
21. 5.0 eV
23. 3.8×10^{28} photons
25. 2.3×10^{-7} m
27. (a) 0.63 V; (b) 8.4×10^{14} Hz
29. no
31. (a) sodium (b) $\lambda_{silver} < 2.63 \times 10^{-7}$ m; $\lambda_{sodium} < 5.01 \times 10^{-7}$ m
33. slope = 5.6×10^{-34} J-s; y-intercept = 9.0 eV
35. (d)
37. 1.4×10^{-13} m
41. 2.81×10^{-10} m
43. (a) 1.32×10^{-15} m; (b) 1.8×10^3
45. (d)
47. (a) 12.1 eV; (b) 1.89 eV
49. (a) 6.9×10^{14} Hz; (b) 8.2×10^{14} Hz
51. (a) 0.477 nm; (b) 1.91 nm; (c) 5.30 nm
53. (a) 1.51 eV; (b) 0.544 eV; (c) 0.136 eV
55. 4.05×10^3 nm; infrared
57. (c)
59. (a) 2.86 eV; (b) $n_i = 2$, $n_f = 5$
63. (d)
65. 1.14×10^{31} quanta
69. 3.5×10^{18} quanta
71. 3.0×10^{13} Hz
73. 41°

Chapter 27

1. (c)
3. 680 nm
5. 4.7×10^{-37} m
7. 150 V
9. 0.58
11. 2.7×10^2 eV

13. (a) 5.1×10^{-16} rad; **(b)** 2×10^5 V

15. 150 V

17. (b)

19. curve with maxima at 0.212 nm, 0.477 nm, and 0.848 nm

21. (b)

23. when quantum numbers are large (>1000), the transitions between energy levels are effectively continuous, and in agreement with classical predictions.

25. (a) 8; 18

(b)

$(2,1,1,+1/2)$	$(3,2,2,+1/2)$
$(2,1,1,-1/2)$	$(3,2,2,-1/2)$
$(2,1,0,+1/2)$	$(3,2,1,+1/2)$
$(2,1,0,-1/2)$	$(3,2,1,-1/2)$
$(2,1,-1,+1/2)$	$(3,2,1,+1/2)$
$(2,1,-1,-1/2)$	$(3,2,0,+1/2)$
$(2,0,0,+1/2)$	$(3,2,0,-1/2)$
$(2,0,0,-1/2)$	$(3,2,-1,+1/2)$
	$(3,2,-1,-1/2)$

$(3,2,-2,+1/2)$
$(3,2,-2,-1/2)$
$(3,1,1,+1/2)$
$(3,1,1,-1/2)$
$(3,1,0,+1/2)$
$(3,1,0,-1/2)$
$(3,1,-1,+1/2)$
$(3,1,-1,-1/2)$
$(3,0,0,+1/2)$
$(3,0,0,-1/2)$

27. 8 for $n = 2$; 14 for $\ell = 3$

29. N = 7 electrons; K = 19 electrons

31. (a) $1s^2 2s^2 2p^1$
(b) $1s^2 2s^2 2p^6 3s^2 3p^6 4s^2$
(c) $1s^2 2s^2 2p^6 3s^2 3p^6 3d^{10} 4s^2$
(d) $1s^2 2s^2 2p^6 3s^2 3d^{10} 4s^2 4p^6 4d^{10} 5s^2 5p^2$

33. (b)

35. 2.3×10^6 m/s

37. 0.28 m

39. (a) 7.5×10^{-5} kg-m/s; **(b)** 1.4×10^{-30} m

41. 1.1×10^{-12} s

43. (d)

45. ionize the antimatter and use magnetic confinement.

47. 1.23×10^{20} Hz

49. 212 MeV

51. 11.5 m

53. 9.59×10^{-11} m

55. $\Psi = A^2$; $x = 0$

57. yes, for $n = \ell = 1$ only

Chapter 28

1. (d)

3. (d)

5. (a) 27 p, 33 n, 27 e; 27 p, 32 n, 27 e
(b) 27 p, 33 n, 29 e; 27 p, 32 n, 29 e
(c) 27 p, 33 n, 26 e; 27 p, 32 n, 26 e

7. 92 p and 92 e for each with n from 140 to 147

9. $^{41}_{19}$K

11. He: 1.9×10^{-15} m; Ne: 3.3×10^{-15} m; Ar: 4.1×10^{-15} m; Kr: 5.3×10^{-15} m; Xe: 6.1×10^{-15} m; Rn: 7.2×10^{-15} m

13. (b)

15. (a) $^{237}_{93}$Np $\rightarrow$ $^{233}_{91}$Pa $+$ $^4_2$He
(b) $^{32}_{15}$P $\rightarrow$ $^{32}_{16}$S $+$ $^{0}_{-1}$e
(c) $^{56}_{27}$Co $\rightarrow$ $^{56}_{26}$Fe $+$ $^{0}_{1}$e
(d) $^{56}_{27}$Co $+$ $^{0}_{-1}$e $\rightarrow$ $^{56}_{26}$Fe
(e) $^{42}_{19}$K* $\rightarrow$ $^{42}_{19}$K $+$ γ

17. $^{210}_{83}$Bi $\rightarrow$ $^{206}_{81}$Tl $+$ $^4_2$He; $^{206}_{81}$Tl $\rightarrow$ $^{206}_{82}$Pb $+$ $^{0}_{-1}$e

19. $^{213}_{83}$Bi

21. (a) $^4_2$He; **(b)** $^{0}_{-1}$e; **(c)** $^{102}_{39}$Y; **(d)** $^{23}_{11}$Na*; **(e)** $^{0}_{-1}$e

23. (a) alpha,^{233}Pa; beta,^{233}U; alpha,^{229}Th; alpha,^{225}Ra; alpha^{221}Fr; alpha,^{217}At; alpha,^{213}Bi; (alpha,^{209}Tl; beta,^{209}Pb) or (beta,^{213}Po; alpha,^{209}Pb); beta,^{209}B; **(b)** essentially all these decays occur because the nuclei contain too many neutrons

25. (d)

27. no; decay is exponential, not linear

29. 4.4×10^{10} decays/min

31. 0.993 mCi

33. Tc; $\lambda_{Tc}/\lambda_{Mg} = 70$

35. 11,460 y

37. 3.1%

39. ^{104}Tc

41. (a) 1.1×10^{-2} s^{-1}; **(b)** 189 s

43. (a) $7.50 \times 10^{10}/2^8$ atoms; **(b)** 3.25×10^8 atoms; **(c)** no

45. 1.9×10^6 kg

47. (a) $^{13}_{7}$N $\rightarrow$ $^{13}_{6}$C $+$ $^{0}_{+1}$e; **(b)** 64.26×10^{20} decays/min; **(c)** 8.84%

49. (a)

51. (b) ^{121}Sb, **(d)** ^{123}Sb; others are odd-odd

53. 2.013553 u

55. (a) 7.72 meV; **(b)** 2.57 MeV/nucleon

57. tritium

59. 8.49 MeV/nucleon

61. Al: 8.33 MeV/nucleon; Na: 8.57 MeV/nucleon

63. 7.57 MeV/nucleon

65. (a)

67. less ^{14}C would have been produced, so a sample measured now would have the age overestimated.

69. (a) 0.25 J; **(b)** yes

71. ≈ 2.3 mCi

73. ^{14}C dating

75. $^{222}_{86}$Rn

77. 22,900 y

79. (a) see Fig. 28.25; **(b)** before: over 700 million years; after: 30 days (one route) and 67 min (other route)

81. (a) 1.118×10^4 MeV; **(b)** 3.728×10^3 MeV

83. (a) $^{225}_{88}$Ra **(b)** $^{225}_{89}$Ac

85. (a) -5.4×10^{17} decays/min; **(b)** -8.9×10^{16} decays/min

Chapter 29

1. (d)

3. As target mass is increased, $K_{min} \rightarrow |Q|$; if $Q = 0$, the incident particle would not need any kinetic energy and reaction would occur spontaneously.

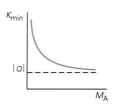

5. (a) $^{41}_{18}$Ar*; **(b)** $^{236}_{92}$U*; **(c)** $^{236}_{92}$U*; **(d)** $^{18}_{9}$F*; **(e)** $^{11}_{5}$B*

7. (a) $^{14}_{7}$N*; **(b)** $^{14}_{7}$N*; **(c)** $^{31}_{15}$P*; **(d)** $^{18}_{9}$F*; **(e)** $^{14}_{7}$N*

9. (a) $^{27}_{14}$Si; **(b)** $^{222}_{86}$Rn; **(c)** $^{193}_{77}$Ir

11. 17.6 MeV

13. 1.53 MeV

15. 5.70 MeV

17. 3.23 MeV

19. 5.37 MeV

21. $(K_1/K_2)_{min} = 2.03$

23. 5.6×10^8 K

25. (d)

27. (d)

29. (a) 231 MeV; **(b)** 237 MeV

31. (a) 5.494 MeV; **(b)** 12.86 MeV

33. (c)

35. 4.69×10^{-14} m

37. 13.34 MeV

41. β^-: $M_p > M_d$; β^+: $M_p > M_d + 2m_e$

43. (a)

45.

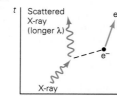

(a) Compton effect

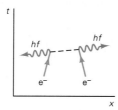

(b) Electron-pair annihilation

47. 1.7×10^{-24} s

49. 4.71×10^{-24} s

51. (b)

53. Quark flavor has one of six values. It can be changed by a weak interaction. Quark color has one of three values. It can be changed by gluons

55. (a) $^4_2$He; **(b)** $^1_1$H; **(c)** $^{93}_{38}$Sr; **(d)** $^{12}_6$C; **(e)** $^{16}_9$N

57. (a) 17.6 MeV; **(b)** 4.03 MeV

59. 5.40 MeV

61. (a) 0.86 MeV; **(b)** not possible

63. (a) $^{27}_{12}$Mg; **(b)** $^{10}_5$B; **(c)** $^{89}_{39}$Y; **(d)** $^{24}_{11}$Na

INDEX

I-2

I-6

Photo Credits

Chapter 1 **CO.1** George Whiteley/Photo Researchers **Fig. 1.2** Rieger Communications **Fig. 1.3c** National Institute of Standards and Technology **Fig. 1.5** Jerry Wilson **Fig. 1.7** John Smith **Fig. 1.14** Jerry Wilson **Fig. 1.17a** Jerry Wilson **Fig. 1.17b** Jerry Wilson

Chapter 2 **CO.2** Marc Romanelli/The Image Bank **IN, p. 52 Fig. 1** North Wind Picture Archives; p. 52, **Fig. 2** J.M. Charles-Rapho/Photo Researchers **Fig. 2.4** John Smith **Fig. 2.6** David Madison/Duomo Photography **Fig. 2.14b** James Sugar/Black Star **Fig. 2.15** John Smith

Chapter 3 **CO.3** Globus Brothers/Stock Market **Fig. 3.15b** Richard Megna/Education Development Center Fundamental Photographs **Fig. 3.17a** John Gillmoure/Stock Market **Fig. 3.17b** William Waterfall/Stock Market **Fig. 3.18b** The Rare Book Room, New York Public Library

Chapter 4 **CO.4** Globus Brothers/Stock Market **Fig. 4.1** Bill Sanderson/Science Photo Library/Photo Researchers **Fig. 4.4** John Smith **Fig. 4.19a** Charles Krebs/Stock Market **Fig. 4.19b** Gabe Palmer/Mug Shots/Stock Market **Fig. 4.25** Photri/Stock Market **Fig. 4.26** John Smith **Fig. 4.34** David Lissy/Focus on Sports

Chapter 5 **CO.5** John Gillmoure/Stock Market **Fig. 5.9a** John Gillmoure/Stock Market **Fig. 5.9b** Francois Gohier/Photo Researchers **5.11a** Ken Straiton/Stock Market **Fig. 5.11b** Joan Menschenfreund/Stock Market **Fig. 5.11c** Focus on Sports **Fig. 5.19** DiMaggio/Kalish/Stock Market **Fig. 5.21** Greg Vaughn/Tom Stack & Associates **Fig. 5.27** David Woods/Stock Market

Chapter 6 **CO.6** Focus on Sports **Fig. 6.1a** Gary S. Settles/Photo Researcher **Fig. 6.1b** Harvey Lloyd/Stock Market **Fig. 6.7a** Omikron/Science Source/Photo Researchers **Fig. 6.8a** M. Hans/Vandystadt/Photo Researchers **Fig. 6.8b** Globus Brothers/Stock Market **Fig. 6.14a** Richard Megna/Fundamental Photographs **Fig. 6.14b** Jeffry W. Myers/Stock Market **Fig. 6.17c** John Smith **Fig. 6.19** Focus on Sports **Fig. 6.22a** NASA **Fig. 6.22b** NASA **Fig. 6.23a** Runk/Schoenberger/Grant Heilman Photography **IN, p. 186 Fig. 1** Ford Motor Company

Chapter 7 **CO.7** Dr. Fred Espenak/Science Photo Library/Photo Researchers **IN, p. 230** NASA **Fig. 7.6b** Tom Tracy/Stock Market **Fig. 7.11c** Michael Livenston/Stock Market **Fig. 7.12** Lewis Portnoy/Stock Market **Fig. 7.16b (cartoon)** © 1991 Dan Birtcher/David Adams **Fig. 7.22** Biophoto Associates/Photo Researchers **Fig. 7.29** NASA **Fig. 7.30** Michael Freeman **Fig. 7.32** Erich Lessing/Art Resource **IN, p. 240 Fig. 1** Photo Researchers

Chapter 8 **CO.8** Saloutos/Stock Market **IN, p. 264** California Institute of Technology Archives, Pasadena **Fig. 8.11a** Richard Hutchings/Photo Researchers **Fig. 8.11b** Dr. E.R. Degginger, EPSA **Fig. 8.11c** Jean-Marc Loubat/Agence Vandystadt/Photo Researchers **Fig. 8.12b** Richard Megna/Fundamental Photographs **Fig. 8.23a&b** Jerry Wilson **Fig. 8.23c** NASA **Fig. 8.26a** Jerry Wilson **Fig. 8.26b** Jerry Wilson **Fig. 8.26d** Jerry Wilson **Fig. 8.35b** Tony Savino/The Image Works **Fig. 8.37b** NASA

Chapter 9 **CO.9** E.R. Degginger/Photo Researchers **Fig. 9.11** NASA **Fig. 9.12** Balloon Excelsior, Inc. **Fig. 9.14** Bill Curtsinger/Photo Researchers **Fig. 9.15** John Smith **Fig. 9.16** Hermann Eisenbeiss/Photo Researchers **Fig. 9.17d** David Spears/Science Photo Library/Photo Researchers **Fig. 9.21b** Diane Schiumo/Fundamental Photographs **Fig. 9.28** John Smith **Fig. 9.31a** John Smith **Fig. 9.31b** John Smith **Fig. 9.34** John Smith **IN, p. 297 Fig. 1** Jonathan Watts/Science Photo Library/Photo Researchers **IN, p. 324 Fig. 2** Blair Seitz/Photo Researchers

Chapter 10 **CO.10** Richard Megna/Fundamental Photographs **Fig. 10.2b** John Smith **Fig. 10.3a** John Smith **Fig. 10.3b** John Smith **Fig. 10.5b** John Smith **Fig. 10.6a** Sinclair Stammers/Science Photo Library/Photo Researchers **Fig. 10.6b** Sinclair Stammers/Science Photo Library/Photo Researchers **Fig. 10.11** Richard Choy/Peter Arnold **Fig. 10.15a** John Smith **Fig. 10.15b** John Smith **Fig. 10.20a** John Smith **Fig. 10.20b** John Smith

Chapter 11 **CO.11** A. De Menil/Science Source/Photo Researchers **Fig. 11.2** Jerry D. Wilson **Fig. 11.3a** North Wind Picture Archives **Fig. 11.3b** North Wind Picture Archives **Fig. 11.4** John Smith **Fig. 11.5** John Smith **Fig. 11.12c** Richard Lowenberg/Science Source/Photo Researchers **Fig. 11.16a** Adam Hart-Davis/Science Photo Library/Photo Researchers **Fig. 11.16b** Dr. Ray Clark & Mervyn Goff/Science Photo Library/Photo Researchers **Fig. 11.17** John Smith **Fig. 11.18b** Martin Dohrn/Science Photo Library/Photo Researchers **Fig. 11.20** John Smith **Fig. 11.21** John Smith **Fig. 11.24** Randy Montoya/Sandia National Laboratories

Chapter 12 **CO.12** David Weintraub/Photo Researchers **Fig. 12.8** Pacific Gas and Electric Company **Fig. 12.9a** Photo Researchers

Chapter 13 **CO.13** Richard Megna/Fundamental Photographs **IN, p. 447 Fig. 1** Special Collection Division, University of Washington Libraries **Fig. 13.9** John Smith **Fig. 13.12a** Richard Megna/Fundamental Photographs **Fig. 13.12b** Richard Megna/Fundamental Photographs **Fig. 13.17a** Fundamental Photographs **Fig. 13.17b** Fundamental Photographs **Fig. 13.18a** Richard Megna/Fundamental Photographs **Fig. 13.20** Dr. E. R. Degginger, EPSA **Fig. 13.24** CENCO

Chapter 14 **CO.14** Scala/Art Resource **Fig. 14.1b** John Smith **Fig. 14.3a** NASA **Fig. 14.3b** Merlin D. Tuttle/Photo Researchers **Fig. 14.11a** Peter Arnold **Fig. 14.11b** Philippe Plailly/Science Photo Library/Photo Researchers **Fig. 14.11c** Martin Rogers/Tony Stone Images **Fig. 14.14b** John Smith **Fig. 14.15** Scala/Art Resource **Fig. 14.17c** Royce Blair/Stock Market **Fig. 14.17** Erich Lessing/Art Resource

Chapter 15 **CO.15** Gordon Gore **Fig. 15.5** Alberto Culver **Fig. 15.8b** Jerry D. Wilson **Fig. 15.23** Spencer Grant/Photo Researchers **IN, p. 499 Fig. 1b** Kent Wood/ Photo Researchers; **Fig. 1c** Grant Heilman Photography **IN, p. 504 Fig. 1c** Jerry Wilson; **Fig. 1b** Rainbow

Chapter 16 **Fig. 16.8** Richard Megna/Fundamental Photographs **Fig. 16.10a** John Smith **Fig. 16.10b.1** Jerry D. Wilson **Fig. 16.10b.2** Jerry D. Wilson **Fig. 16.11** NASA/Mark Marten/Photo Researchers **Fig. 16.12** John Smith **IN, p. 528 Fig. 1** IBM Research **IN, p. 530 Fig.1, Fig. 2** Jerry D. Wilson

Chapter 17 **CO.17** Richard Megna/Fundamental Photographs **IN, p. 554 Fig. 2** Zircon **Fig. 17.12** CENCO **Fig. 17.15c** CENCO **Fig. 17.17b** Jerry Wilson **Fig. 17.17c** Paul Silverman/Fundamental Photographs **Fig. 17.18b** Paul Silverman/Fundamental Photographs **Fig. 17.18c** U.S. Dept. of Energy/Science Photo Library/Photo Researchers **Fig. 17.21a** Jerry Wilson **Fig. 17.21b** Jerry Wilson **Fig. 17.40** M. Antman/The Image Works

Chapter 18 **CO.18** CENCO **Fig. 18.1** CENCO **Fig. 18.3a** Richard Megna/Fundamental Photographs **Fig. 18.3b** Richard Megna/Fundamental Photographs **Fig. 18.3c** Richard Megna/Fundamental Photographs **Fig. 18.4b** Richard Megna/Fundamental Photographs **Fig. 18.8a** Richard Megna/Fundamental Photographs **Fig. 18.9a** Richard Megna/Fundamental Photographs **Fig. 18.10a** Richard Megna/Fundamental Photographs **Fig. 18.13** Grant Heilman Photography **Fig. 18.20b** Jerry Wilson **Fig. 18.24a** CENCO **Fig. 18.28** Pekka Parviainen/Science Photo Library/Photo Researchers

Chapter 19 **CO.19** U.S. Dept. of Energy/Science Photo Library/Photo Researchers **Fig. 19.1** Richard Megna/Fundamental Photographs **Fig. 19.9** Thomas Dimock/Stock Market **Fig. 19.11d** Westinghouse Electric Corp. **Fig. 19.12b** Jerry Wilson **Fig. 19.14a** Dan Guravich/Photo Researchers **Fig. 19.14b** U.S. Dept. of Energy/Mark Marten/Photo Researchers **Fig. 19.18** Peter Ryan/Science Photo Library/Photo Researchers **Fig. 19.19** Willie L. Hill, Jr./Stock Boston **IN, p. 625 Fig. 1** NASA/IBM **Fig. 19.21** Simon Fraser/Science Photo Library/Photo Researchers

Chapter 20 **CO.20** Doug Handel/Stock Market **Fig. 20.4** Jerry Wilson **Fig. 20.12** The Image Works

Chapter 21   **CO.21** Alese/Mort Pechter/Stock Market   **Fig. 21.5a** David Barnes/Stock Market   **Fig. 21.6** Richard Megna/Fundamental Photographs   **Fig. 21.8a** Richard Megna/Fundamental Photographs   **Fig. 21.10b** Runk/Schoenberger/Grant Heilman Photograph   **Fig. 21.13** Ken Kay/Fundamental Photographs   **Fig. 21.14** W. Eastep/Stock Market   **Fig. 21.15** Gemological Institute of America   **Fig. 21.17** Richard Megna/Fundamental Photographs   **Fig. 21.18a** C. Falco/Photo Researchers   **Fig. 21.18b** Hank Morgan/Photo Researchers   **Fig. 21.19** P. Gontier/The Image Works   **Fig. 21.20** Alexander Tsiaras/Science Source/Photo Researchers   **Fig. 21.21b** David Parker/Science Photo Library/Science Source/Photo Researchers   **Fig. 21.22** Martin Bond/Science Photo Library/Photo Researchers   **Fig. 21.24a** SIU/Photo Researchers   **Fig. 21.24b** J. Barry O'Rourke/Stock Market   **IN, p. 666 Fig. 1** Doug Johnson/Science Photo Library/Photo Researchers

Chapter 22   **CO.22** Peticolas/Fundamental Photographs   **Fig. 22.4a** John Smith   **Fig. 22.4b** John Smith   **Fig. 22.8** John Smith   **Fig. 22.14** Photo Researchers   **Fig. 22.17** John Smith   **Fig. 22.21** Michael Freeman   **Fig. 22.22** John Smith   **IN, p. 692** Bohdan Hrynewych/Stock Boston; John Smith

Chapter 23   **CO.23** Adrienne Hart-Daves/Science Photo Library/Photo Researchers   **Fig. 23.1** Richard Megna/Fundamental Photographs   **Fig. 23.2b** From the "Atlas of Optical Phenomena," Michel Cagnet, Maurice Francon, Jean Claude Thrierr. © by Springer-Verlag OHG, Berlin, 1962. Published by Prentice-Hall, Inc., Englewood Cliffs, NJ.   **Fig. 23.8b** David Parker/Science Photo Library/Photo Researchers   **Fig. 23.10b** Ken Kay/Fundamental Photographs   **Fig. 23.13** Gregory G. Dimijian, M.D./Photo Researchers   **Fig. 23.14b** Prof. A.J. Stosick, University of Southern California   **Fig. 23.14c** Medical Research Council Centre, Cambridge   **Fig. 23.14d** Courtesy Merck & Co.   **Fig. 23.17b** Dr. E.R. Degginger/Dr. E.R. Degginger   **Fig. 23.19b** Diane Schiumo/Fundamental Photographs   **Fig. 23.20b** Nina Barnett   **Fig. 23.21b** Peter Aprahamian/Sharples Stress/Engineers Ltd./SPL/Photo Researchers   **Fig. 23.23** Roger Ressmeyer/Starlight   **IN, p. 726 Fig. 2** John Smith   **IN, p. 710 Fig. 2** Kristen Brochmann/Fundamental Photographs

Chapter 24   **CO.24** Perkin-Elmer   **Fig. 24.4b** John Smith   **Fig. 24.8b** Larry Mulvehill/Photo Researchers   **Fig. 24.11** Courtesy of Bausch & Lomb Sports Optics Div.   **Fig. 24.12** Yerkes Observatory Photographic Services   **Fig. 24.14a** Hale Observatory   **Fig. 24.14b** Hale Observatory   **24.14c** Science Photo Library/Photo Researchers   **Fig. 24.18** Fritz

Goro/Life Magazine, © Time, Inc.   **IN, p. 753 Fig. 1a** Gary Ladd.Science Source/Photo Researchers; **Fig. 1b** Roger Ressmeyer/Starlight; **Fig. 2** Roger Ressmeyer/Starlight; **Fig. 3** NASA   **IN, p. 754 Fig. 1a** Hencoup Enterprises/Science Photo Library/Photo Researchers; **Fig. 1b** Hencoup Enterprises/Science Photo Library/Photo Researchers; **Fig. 2d** Tony Craddock/Science Photo Library/Photo Researchers

Chapter 25   **CO.25** David Malin/Royal Observatory, Edinburgh   **Fig. 25.4b** PASCO Scientific   **Fig. 25.6** Science Photo Library/Photo Researchers   **Fig. 25.17b** NASA   **Fig. 25.18a** NASA   **IN, p. 791 Fig. 1** David Hardy/Science Photo Library/Photo Researchers

Chapter 26   **CO.26** Philippe Plailly/Science Photo Library/Photo Researchers   **Fig. 26.7** Ann Purcell/Photo Researchers   **Fig. 26.8a** Bausch and Lomb   **Fig. 26.8b** Bausch and Lomb   **Fig. 26.13a** Gary Retherford/Photo Researchers   **Fig. 26.13b** Gary Retherford/Photo Researchers   **Fig. 26.14** Dan McCoy/Rainbow   **Fig. 26.18a** Will & Deni McIntyre/Photo Researchers   **Fig. 26.18b** Ray Ellis/Photo Researchers   **Fig. 26.20a** Philippe Plailly/Science Source/Photo Researchers   **Fig. 26.20b** Chuck O'Rear/WESTLIGHT

Chapter 27   **CO.27** Sinclair Stammers/Science Photo Library/Photo Researchers   **Fig. 27.5** IBM Research   **Fig. 27.13** Lawrence Berkeley/Science Photo Library/Photo Researchers   **Fig. 27.13a** Courtesy of Lawrence Berkeley Laboratory   **IN, p. 830 Fig. 2** Hank Morgan/Science Source/Photo Researchers   **IN, p. 831 Fig. 3a** Manfred Kage/Peter Arnold; **Fig. 3b** Dr. R. Kessel/Peter Arnold; **Fig. 3c** David Scharf/Peter Arnold

Chapter 28   **CO.28** Will & Deni McIntyre/Photo Researchers   **Fig. 28.5** Mary Evans Picture Library/Photo Researchers   **Fig. 28.18** Lawrence Berkeley/Photo Researchers   **Fig. 28.19** Larry Mulvehill/Photo Researchers   **Fig. 28.20** Larry Mulvehill/Photo Researchers   **Fig. 28.21a** Dan McCoy/Rainbow   **Fig. 28.21b** Drs. Mazziotta, Phelps, et al/UCLA School of Medicine

Chapter 29   **CO.29** Y. Arthus Bertrand/Explorer/Photo Researchers   **Fig. 29.1a** George Retseck/George Retseck from *Scientific American*, July, 1990   **Fig. 29.1b** Universities Research Association, Washington, DC   **Fig. 29.6b** EPRI/PH.A.I./Science Source/Photo Researchers   **Fig. 29.8** Igor Kostin/Imago/Sygma   **Fig. 29.9a** Princeton University Plasma Physics Laboratory   **Fig. 29.9b** Princeton University Plasma Physics Laboratory   **Fig. 29.10** Gary Stone/Lawrence Livermore National Laboratory

Interactive Physics II™

A Complete Motion Lab on the Computer!

Draw It.

Any motion experiment you can imagine. Create this air drop experiment, for example, to investigate projectile motion.

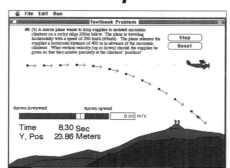

Click RUN.

Your experiment moves acording to real-world physical laws. The plane moves across the screen and releases a survival packet in smooth movie-like animation.

The 'Word Processor' of Motion Simulation.

Interactive Physics II is a complete motion lab on the computer. Choose an experiment — **any experiment you can think of** — at any level of sophistication. Select it from the library of ready-to-run experiments or build it yourself using the simple Macintosh® palette and pull-down menus. Roll a ball down an inclined plane, launch a rocket between orbiting planets, send a stream of electrons through a varying magnetic field, build a working model of a combustion engine. Model virtually any experiment, visualize any principle, all in smooth animation.

Model with Powerful Tools.

Mass objects include circles, squares, rectangles, and polygons. Simulation elements include ropes, motors, actuators, pulleys, pin joints, springs, dampers, and forces. Change object mass, elasticity, friction, and charge. Vary and modify gravity and electrostatics, or create your own custom forces. Use **pulleys** and **pin joints** to connect one object to another. Attach **motors** and **actuators** to create advanced models such as dump trucks and motorcycles. Paste in your own graphics and watch them work in your simulations. Modify and control your simulations while they are running.

What You Get:

- Three 3.5-inch disks (one program disk and two experiment disks).
- A complete curriculum guide on disk with experiments.
- Interactive Physics II User's Guide with step-by-step tutorials.

System Requirements:

- Apple Macintosh
- System 6.0.5 or greater
- System 7.0 Savvy
- 2 MB RAM (2.5 MB RAM for System 7)

Knowledge Revolution

15 Brush Place
San Francisco, CA 94103 USA

800-766-6615

What Can You Create with Interactive Physics II?

Build simple machines. Then track their motion as they run.

Model your own solar system with as many planets as you like.

Visualize complex phenomena that are difficult to understand.

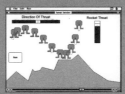

Design your own simulations and control them as they run.

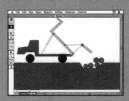

Make QuickTime™ movies of any simulation. Export your data.

Build Working Textbook Problems

Get an extra edge by creating your own visual experiments!

Interactive Physics II is an advanced motion simulation program that allows you to build and simulate virtually any mechanics, statics, dynamics, kinematics, electrostatics, electrodynamics, gravitational, and planetary gravitational problem from your textbook. Get an edge by building experiments and <u>seeing your problems in motion</u>.

Special Student Offer!

We know that students don't have a lot of money, that's why we've made a special discount version of Interactive Physics II available to students at a special reduced price. Remember though, the student version will not entitle you to technical support or upgrades to our future products. See the offer below for more information.

Cut here, then fold and tape.

---

fold here

I Want to Get an Edge Today!   (800) 766-6615

❏ **Yes,** Please send me my copy of Interactive Physics II. I am an educator and qualify for the educational discount of $319 which is $80 off the list price of $399.

Educational Discount $319

❏ **Yes,** I am a registered student and have enclosed a photocopy of my student ID. I understand that the student version of Interactive Physics II **does not** entitle me to upgrades or technical support.

Student Discount $99

Subtotal	$_____
CA Res. Add 8.25% Sales Tax	$_____
Shipping and Handling	$_____
TOTAL	$_____

Name _____

Organization _____

Address _____

City _____ State_____ Postal Code_____

Country _____ Phone Number (___)_____

Payment by: ❏ Check ❏ Visa ❏ MC

Card Number_____Exp.Date_____

BEFORE MAILING: **1.** Be sure to sign and enclose your check drawn on a U.S. bank and in U.S. dollars, payable to Knowledge Revolution; or, complete the credit card information section. **2. Do not send cash. 3.** Add $6.00 for shipping and handling. **4.** California residents add 8.25 percent for sales tax. **5.** Canadian customers, add $10.00 for shipping and handling; all other countries, add $30.00.

Mathematical Symbols

$=$	is equal to
$\neq$	is not equal to
$\approx$	is approximately equal to
$\propto$	is proportional to
$>$	is greater than
$\geq$	is greater than or equal to
$\gg$	is much greater than
$<$	is less than
$\leq$	is less than or equal to
$\ll$	is much less than
$\bar{x}$	average value
Δx	change in x
$|x|$	absolute value of x
Σ	sum

The Greek Alphabet

Alpha	A	α	Nu	N	ν
Beta	B	β	Xi	Ξ	ξ
Gamma	Γ	γ	Omicron	O	o
Delta	Δ	δ	Pi	Π	π
Epsilon	E	ϵ	Rho	P	ρ
Zeta	Z	ζ	Sigma	Σ	σ
Eta	H	η	Tau	T	τ
Theta	Θ	θ	Upsilon	Y	υ
Iota	I	ι	Phi	Φ	ϕ, φ
Kappa	K	κ	Chi	X	χ
Lambda	Λ	λ	Psi	Ψ	ψ
Mu	M	μ	Omega	Ω	ω

Quadratic Formula

$$x = \frac{-b \pm \sqrt{b^2 - 4ac}}{2a}$$

where $ax^2 + bx + c = 0$

Values of Some Useful Numbers

$\pi = 3.14159\ldots$ $\qquad \sqrt{2} = 1.41421\ldots$

$e = 2.71828\ldots$ $\qquad \sqrt{3} = 1.73205\ldots$

Trigonometric Relationships

Definitions of Trigonometric Functions

$$\sin\theta = \frac{y}{r} \qquad \cos\theta = \frac{x}{r} \qquad \tan\theta = \frac{\sin\theta}{\cos\theta} = \frac{y}{x}$$

$\theta°$ (rad)	$\sin\theta$	$\cos\theta$	$\tan\theta$
0° (0)	0	1	0
30° ($\pi/6$)	0.500	0.866	0.577
45° ($\pi/4$)	0.707	0.707	1.00
60° ($\pi/3$)	0.866	0.500	1.73
90° ($\pi/2$)	1	0	$\to \infty$

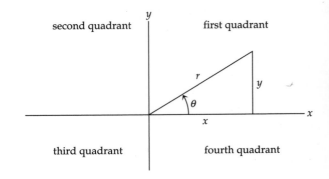